Phenotypic Plasticity of Insects

Mechanisms and Consequences

Phenotypic Plasticity of Insects

Mechanisms and Consequences

Editors

Douglas W. Whitman
Department of Biological Sciences
Illinois State University
Normal, IL, USA

T.N. Ananthakrishnan
Formerly Director of
Entomology Research Institute
Chennai, India

Science Publishers

Enfield (NH) Jersey Plymouth

Science Publishers www.scipub.net
234 May Street
Post Office Box 699
Enfield, New Hampshire 03748
United States of America

General enquiries : *info@scipub.net*
Editorial enquiries : *editor@scipub.net*
Sales enquiries : *sales@scipub.net*

Published by Science Publishers, Enfield, NH, USA
An imprint of Edenbridge Ltd., British Channel Islands
Printed in India

ISBN: 978-1-57808-423-4

Library of Congress Cataloging-in-Publication Data

Phenotypic plasticity of insects: mechanisms and consequences/
editors, Douglas Whitman, T.N. Ananthakrishnan.
p. cm.
Includes bibliographical references.
ISBN 978-1-57808-423-4 (hardcover)
1. Insects. 2. Phenotypic plasticity. I. Whitman, Douglas
(Douglas W.) II. Ananthakrishnan, T.N., 1925-

QL463.P48 2008
595.7135--dc22

2008029198

Preface

Our impression of how life works is changing. Old paradigms give way to new, and this is the case with phenotypic plasticity and evolutionary theory. The Modern Synthesis and Neo-Darwinism of the 1930's – 1940's culminated in a genocentric view of evolution, whereby genetic variability produced by genetic mutation keyed evolution. As emphasized by Hamilton, Maynard Smith, Willims, and especially Dawkins, the selfish gene was king and the true unit of selection. For geneticists, focused on mutations and population genetics, phenotypic variation was an inconvenience – non-heritable developmental noise, to be minimized or ignored. Today, this genocentric paradigm is being challenged by the concept of phenotypic plasticity.

A convergence of events has spurred this change. First is that phenotypic plasticity was staring us in the face all along – castes in social insects, locust swarms, butterfly and caterpillar polyphenisms, alteration of generations in aphids, beetle horns and other diet-based allometries, acclimation, induced defenses, and alternative reproductive strategies – all of these examples of phenotypic plasticity require explanations, and all explanations need to be brought under a single conceptual framework joined with evolutionary theory.

Second, the recent surge in interest in evolutionary developmental biology (evo-devo) elevated the importance of development in evolution and highlighted the idea that a single genotype gives rise to not a single phenotype, but to a range of possible phenotypes, determined by the environment (see Chapters 3, 10-13, this volume). As Fred Nijhout and Mary Jane West-Eberhard suggest, all gene expression is ultimately environmentally dependent: phenotypes are always products of genes by environment interactions. Coupled to this was a growing appreciation that natural selection selects not among genotypes, but among phenotypes (see Chpt.1). Evolution requires heritable phenotypic variation, and that variation can arise from underlying genetic variation *or* from phenotypic plasticity (which is itself inherited). A phenotype that never occurs cannot be selected for. Hence, the environment comes to play a greater role in evolution, both creating phenotypic variation and then selecting among that variation. Plasticity then becomes both a cause and a consequence of evolution, as reaction norms evolve, just like any other trait (Chpts 12 & 21). Indeed, as suggested by Carl Schlichting and Massimo Pigliucci, the reaction norm may be a central object of natural selection.

The adaptive value[1] of plasticity also requires consideration. Many cases of phenotypic plasticity are irrefutably adaptive, and this represents adaptation by individuals within a single generation. This contrasts with traditional genocentric adaptation, which concerns adaptations of populations between generations. Environments are dynamic, and individuals that can adjust their phenotypes accordingly can achieve high fitness. Indeed, the only way an *individual* can adapt to a changing environment is by altering its phenotype. Plasticity allows a single genotype to assume multiple adaptive phenotypes. A comprehensive theory of adaptive variation must address both within- (individual) and between-generation adaptation.

Individual flexibility can have many benefits. For both individuals and species, plasticity can broaden environmental tolerance, niche, and, presumably, geographic range. The first increases survival of individuals and the last reduces the possibility of extinction (Chpts. 5, 15-17). Beyond the individual, plasticity expands evolutionary potential: it allows genotypes and populations to explore new designs by trial and error, with the advantage that specific new phenotypes are matched to specific environments. Furthermore, as Waddington pointed out 65 years ago, and Schlichting and Harry Smith have recently reemphasized, phenotypes originally initiated via a plastic response, can be fixed through genetic assimilation as alternative regulatory pathways are shut off (see Chpts. 1, 13, & 20). Hence, individual change may be a pathway to evolutionary change.

There is also greater realization that canalization, homeostasis, and plasticity are closely related (see Chpt. 15). Jason Wolf suggests that plasticity and canalization can be viewed as alternate sides to the same basic process, lying at opposite ends of a spectrum of sensitivity to context. In fact, homeostasis or canalization in one process usually requires phenotypic plasticity in another (see Chpt. 1).

---

[1] Here "adaptive" refers to beneficial, and "adaptation" refers to a beneficial change that increases fitness, as opposed to adaptive evolution via natural selection. Of course, plasticity often includes both types of adaptation.

Phenotypic plasticity has numerous important consequences that also must be considered. As this book documents, plasticity influences survival and fitness, population dynamics, ecological relationships, and community structure. It creates phenotypic diversity and novelty, and may stimulate evolutionary adaptation and speciation. For researchers, phenotypic plasticity impacts virtually every biological sub-discipline. In developmental biology, ontogeny is no longer fixed, but subject to the environment (Chpts. 3-5, 9-14). In physiology, plasticity counters stress and underlies most physiological regulation (Chpts. 1, 15 & 16). Rapid and persistent acclimation warns physiologists and behaviourists that they must consider both present and past conditions of their experimental animals, including parental conditions (Chpts 16 & 19). In taxonomy, species morphology is not static (Chpts. 4 & 5). In ecology, the interactions of individuals and populations with the environment change in space and time (Chpts. 5-12, 16-19), and such changes can have profound ecosystem consequences (Chpts. 1, 5-8). Behavior also can be analyzed within a plasticity framework (Chpts. 8, 11, & 18). For evolution, West-Eberhard states that in regard to phenotypic plasticity, genes are useful followers, not leaders in evolutionary change. Phenotypic plasticity theory also has practical value: it aids our understanding of pest control, medicine, conservation, invasive species, climate change, etc. Lastly, powerful new technologies have greatly increased our ability to understand the molecular underpinnings of plasticity. For the first time, high-throughput screens can tell us the transcriptional and translational events and signaling pathways triggered by environmental stimuli, as well as identifying events that regulate or accompany development and plasticity. These are exciting times for biologists. The new molecular and bioinformatics tools have sparked a knowledge explosion across the entire discipline. These technological breakthroughs come at just the right time for phenotypic plasticity research, which requires an unparalleled integration across virtually all biological hierarchies and disciplines, from genes to ecology and evolution, and back to genes. Equally exciting are the ideas that two individuals may interact via reciprocal phenotypic plasticity, leading to "plasticity coevolution" (Chpt. 1), or that plasticity allows organisms to self-redesign.

As this book documents, phenotypic plasticity is an inherent and ubiquitous consequence of life, and deserves greater attention in biology. As always with paradigm changes, the new ideas and approaches derived from plasticity theory will generate new questions and hypotheses, leading to a better understanding of life.

This book could not have been accomplished without the help of many people. We especially wish to thank Anurag Agrawal, Rachel Bowden, Kenneth Bowler, David Borst, Jean David, Douglas Emlen, Paul Garris, Erick Greene, Cheryl Hickey, James Hunt, Shajahan Johny, Steven Juliano, Sally Little, Laurence Mound, Toru Miura, Ebony Murrell, Wade Nichols, Frederik Nijhout, Massimo Pigliucci, Barbara Reinagle, Kenneth Ross, Scott Sakaluk, Oswald Schmitz, John Sedbrook, Melissa Stauffer, Tim Stauffer, David Stern, Charles Thompson, Joseph Tomkins, David Tuss, Laura Vogel, John Weitholder, Diana Wheeler, Jessie Whitman, and Leigh Whitman. Armin Moczek kindly prepared the front cover figure and Michelle Elekonich the back cover figure. Leigh Whitman composed the book jacket. We thank Chad Buckley and the Milner Library staff for bibliographic help. Supported by NSF CRUI grants BIR 9510979A000 and DBI-0442412. This book dedicated to Kathy, Leigh, and Jessica.

Douglas W. Whitman
Normal, IL USA
dwwhitm@ilstu.edu

T.N. Ananthakrishnan
Chennai, India
ananthrips@yahoo.com

Contents

1

What is Phenotypic Plasticity and Why is it Important?

Douglas W. Whitman[1] and Anurag A. Agrawal[2]

[1]*Department of Biological Sciences, Illinois State University, Normal, IL, USA 61790-4120, E-mail: dwwhitm@ilstu.edu*

[2]*Department of Ecology and Evolutionary Biology, Cornell University, Ithaca, New York, USA 14853, E-mail: agrawal@cornell.edu*

Abstract

Phenotypic plasticity, the capacity of a single genotype to exhibit variable phenotypes in different environments, is common in insects and is often highly adaptive. Here we review terminology, conceptual issues, and insect plasticity research, including variance partitioning, reaction norms, physiological mechanisms, adaptive value, and evolution. All plasticity is physiological, but can manifest as changes in biochemistry, physiology, morphology, behavior, or life history. Phenotypic plasticity can be passive, anticipatory, instantaneous, delayed, continuous, discrete, permanent, reversible, beneficial, harmful, adaptive or non-adaptive, and generational. Virtually any abiotic or biotic factor can serve to induce plasticity, and resulting changes vary from harmful susceptibilities to highly integrated and adaptive alternative phenotypes. Numerous physiological mechanisms accomplish plasticity, including transcription, translation, enzyme, and hormonal regulation, producing local or systemic responses. The timing, specificity, and speed of plastic responses are critical to their adaptive value. Understanding plasticity requires knowing the environment, physiological mechanisms, and fitness outcomes. Plasticity is thought to be evolutionarily favored under specific conditions, yet many theoretical predictions about benefits, costs, and selection on plasticity remain untested. The ecological consequences of plasticity range from simple environmental susceptibilities to mediating interspecific interactions, and extend to structuring of ecological communities, often through indirect effects. Phenotypic plasticity, through its ecological effects, can facilitate evolutionary change and speciation. Plasticity is important because it is an encompassing model to understand life on earth, it can increase fitness, generate novelity, and facilitate evolution, it structures ecological communities, and it has numerous practical applications. As such, all biologists should understand phenotypic plasticity.

Introduction

A young caterpillar feeds on oak flowers and develops into a stunning mimic of an oak catkin (Fig 1b.). A second caterpillar from the same egg batch feeds on leaves and becomes a twig mimic (see Chapter 4, this volume). In response to low-quality, fibrous food, a grasshopper develops larger mandibles and mandibular muscles (Thompson 1992), and another develops a larger gut (Yang and Joern 1994). A different grasshopper alters the number of chemosensilla on its antennae in response to the number of plant chemicals it encounters (Chapman and Lee 1991, Rogers and Simpson 1997). In a nearby aphid colony, females are busy adjusting the future morphology and behavior of their offspring in response to predator threats. When ant bodyguards are absent, females rapidly produce soldier offspring (Shingleton and Foster 2000), and produce winged offspring when predators invade the colony (Weisser et al. 1999). Close by, a gravid fly, unable to locate her normal host plant, deposits her eggs on a novel host. Surprisingly, the larvae survive on the new host, and chemically imprint on it before dispersing as adults. The flies subsequently orient to the novel plant to mate and oviposit, instead of their ancestral plant (Feder et al. 1994, see Chapter 18, this book). In the same tree, a caterpillar bites into a leaf. A plant sensory mechanism detects the caterpillar saliva and signals the entire plant to begin synthesis of anti-herbivore toxins and the release of volatile pheromones. The latter dissipate to neighboring plants, alerting them to the presence of herbivores, and stimulating them to synthesize their own chemical defenses. But, the plant's clever counter-ploys do not go unchallenged; in response to increasing plant toxins, the caterpillar synthesizes detoxifying gut enzymes, effectively negating the plant's chemical escalation (see Chapter 7). On the ground below, a *Drosophila*

Fig. 1 Morphological phenotypic plasticity in insects. (a, b) Discrete seasonal polyphenisms in *Nemoria arizonaria* caterpillars (fam. Geometridae). Summer brood feeds on oak leaves and resembles an oak twig. Spring brood feeds on and resembles oak catkins. Photos: E. Greene (Greene 1989). (c) Wet-season (left) and dry-season (right) *Precis octavia* (fam. Nymphalidae) butterflies, from Africa (McLeod 2007). Photos courtesy of F. Nijhout. (d) Many insects alter body color in response to rearing temperature: *Romalea microptera* grasshoppers (fam. Romaleidae) from south Florida reared at 35°C (top) and at 25°C (bottom). (e) Harlequin bugs, *Murgantia histrionica* (fam. Pentatomidae). Black and yellow individuals were reared at 22 and 30°C, respectively. (f) Nutrition strongly influences insect body size. *Taeniopoda eques* grasshoppers (fam. Romaleidae), from the Chihuahuan Desert in SE Arizona, showing plasticity in body size to nutrition. Males on left; females on right. Bottom four individuals from site that received ample rains and had lush vegetation; top four individuals from a site 15-km distant

Fig. 1 Contd. ...

Fig. 1 Contd. ...

that received poor rains and had poor vegetation. In previous years, rain, vegetation, and grasshopper size patterns were reversed at these two sites (d-f: Whitman, unpubl.).

maggot, feeding inside a sun-exposed fruit, responds to near-lethal temperatures by mounting a full-fledged biochemical counter-response. Rapid transcription and translation floods the cells with protective heat-shock proteins that stabilize thermal-labile proteins, preventing death. In a nearby shaded sibling, no heat shock proteins are produced (Chapter 17). Across the meadow, a different insect is trapped on a poor-quality host. This inadequate diet profoundly alters her life history and fecundity by reducing her development and growth rates, body size (Figs. 1f, 4i), number of ovarioles, and clutch size and egg size, which, in turn, alters the life history and fecundity of her offspring (Chapter 11). A beetle larva, sensing its fungal competitor, accelerates its development (Roder et al. 2008). As fall turns to winter, the adult, exposed to short day lengths, radically switches its behavior and physiology. It stops feeding, burrows into the soil, changes color, dramatically lowers its metabolism, and fortifies its tissues with cyroprotectants, enabling survival at frigid temperatures. Its sibling, kept in long-day conditions, exhibits none of these changes and is killed by mildly cold temperatures (Chapter 16).

The above insects share a singular commonality: in each case, an individual has changed its morphology, physiology, behavior, or life history in response to changing environmental conditions. Such phenotypic plasticity is universal among living things and derived from the fact that environments vary. These environmental changes, be they temporal, spatial, abiotic, or biotic, are challenging because they can destabilize homeostasis and development, and disrupt the match between an organism's phenotype and the environment, thereby lowering fitness. Organisms counter environmental variation with their own adaptive variation of two types: between- and within-generation variation (Meyers and Bull 2002, DeWitt and Langerhans 2004). The former is mostly genetic and can result in adaptive change within a population. Between-generation variation has been the primary focus of evolutionary biologists and is based on natural selection acting on heritable variation caused by mutation, recombination, genetic drift, etc. In contrast, within-generation variation is almost always non-genetic, occurs in individuals, and is frequently adaptive, because it allows individuals to adjust to environmental variation in real time.

Interest in phenotypic plasticity has grown exponentially in the last 20 years, igniting an explosion of literature. Most of the ideas expressed in this chapter are derived from the following excellent reviews, and readers should consult these sources for a more comprehensive understanding of plasticity: Bradshaw 1965, Scheiner 1993, Nylin and Gotthard 1998, Schlichting and Pigliucci 1998, Tollrian and Harvell 1999, Agrawal 2001,

2005, Pigliucci 2001, Zera and Harshman 2001, Schlichting and Smith 2002, Nijhout 2003a, West Eberhard 2003, Benard 2004, Dewitt and Scheiner 2004, Ohgushi 2005, Emlen et al. 2007, Shingleton et al. 2007, Sultan 2007.

What is Phenotypic Plasticity?

The concept of phenotypic plasticity is deceptively simple. Numerous authors have defined phenotypic plasticity (Box 1), and, at face value, these definitions seem fairly similar. However, the devil is in the details, and we consider these details, below. For our purposes, we define phenotypic plasticity as the capacity of a single genotype to exhibit a range of phenotypes in response to variation in the environment (Fordyce 2006).

Box 1 Definitions of Phenotypic Plasticity

- Plasticity is shown by a genotype when its expression is able to be altered by environmental influence . . . it does not have any implications concerning the adaptation value of the change occurring . . . (Bradshaw 1965).
- A change in the expressed phenotype of a genotype as a function of the environment *or* when an individual's phenotype is influenced by its environment (Scheiner 1993).
- The capacity of an organism to develop any of several phenotypic states, depending on the environment; usually this capacity is supposed to be adaptive (Futuyma 1998).
- The ability of an organism to express different phenotypes depending on the environment (Agrawal 2001).
- The property of a given genotype to produce different phenotypes in response to distinct environmental conditions (Pigliucci 2001).
- Any change in an organism's characteristics in response to an environmental signal (Schlichting & Smith 2002).
- Condition-sensitive development *or* the ability of an organism to react to an environmental input with a change in form, state, movement, or rate of activity (West-Eberhard 2003).
- Environment-dependent phenotype expression *or* the environmentally sensitive production of alternative phenotypes by given genotypes (Dewitt & Scheiner 2004).
- The expression of different phenotypes in a single genotype when subjected to different environments (Ananthakrishnan & Whitman 2005).
- Variation, under environmental influence, in the phenotype associated with a genotype (Freeman & Herron 2007).
- Environmental sensitivity for a trait (Various authors).

Variance Partitioning

Phenotypic plasticity represents measureable variation, and as such can often be expressed and analyzed by Analysis of Variance (ANOVA) (Pigliucci 2001). A statistical measure of variation is variance, which quantifies the deviation of values around a mean. The variance of a phenotypic trait can be partitioned as follows:

$$V_P = V_G + V_E + V_{G\times E} + V_{\text{error}}$$

Where:

V_P = Total phenotypic variance for a trait
V_G = Genetic variance (proportion of phenotypic variation attributable to genes)
V_E = Environmental variance (proportion of variation caused by the environment)
$V_{G\times E}$ = Genotype × environment interaction (Genetic variation for phenotypic plasticity)
V_{error} = Unexplained variance, including developmental noise, measurement error, etc.

ANOVA can partition phenotypic variation into the above components. However, these terms, especially the expression of genetic variance, are often further divided into component parts (Debat and David 2001, Piersma and Drent 2003). Thus, experimental designs with some form of genetic structure (i.e., using clones, half-sibling families, multiple populations, etc.) and environmental treatments are extremely powerful for studying phenotypic plasticity. Nonetheless, genetic structure is not required for the study of plasticity. A simple design of several individuals of a species, randomly assigned to different environments, can often yield a robust estimate of plasticity. Here, V_G and $V_{G\times E}$ are unknown, but V_P can still be partitioned into what is explained by V_E (i.e., phenotypic plasticity) and all other sources of phenotypic variation. $V_{G\times E}$ is an important term because it shows that different genotypes express different plastic responses. Such genetic variance in plasticity allows plasticity to evolve.

Graphic Representation: Reaction Norms

Phenotypic plasticity can be visualized by the use of reaction norms, which plot values for a specific phenotypic trait across two or more environments or treatments (Schlichting and Pigliucci 1998, Sarkar 2004). Figure 2 shows hypothetical reaction norms, for a specific trait (in this case, let's say horn length), for five genotypes in a population. Each genotype expresses a different mean value for horn length in Environment 1 (V_G). However, when subjected to a new environment, most genotypes alter their horn length. In this case, when comparing the grand means (the triangles) in each environment, we see that horn length generally increases in Environment 2 (V_E). However, each genotype exhibits a different reaction norm (i.e., a different response to environment, or different slopes in Figure 2). Genotype 4 shows no plasticity for this particular trait: mean horn length remains the same in both environments. In contrast, Genotype 3 shows extreme phenotypic plasticity for mean horn length, growing long horns in Environment 2. Alternatively, for Genotype 1, mean horn length decreases in Environment 2. The fact that each Genotype shows a different response (non-parallel reaction norms) represents genotype × environment interaction ($V_{G \times E}$), indicating genetic variation in plasticity itself, upon which natural selection can act to alter the shape and variance of the species' reaction norm. Figure 3 shows real reaction norms from real animals; additional examples can be found throughout this book. Note that when multiple environments or continuous environmental gradients are included, reaction norms may be highly curvilinear or discontinuous (Roff 1996, Emlen and Nijhout 2000, David et al. 2004). One problem with both variance partitioning and reaction norms is that they do not explain the evolution, underlying mechanisms, or consequence of phenotypic plasticity

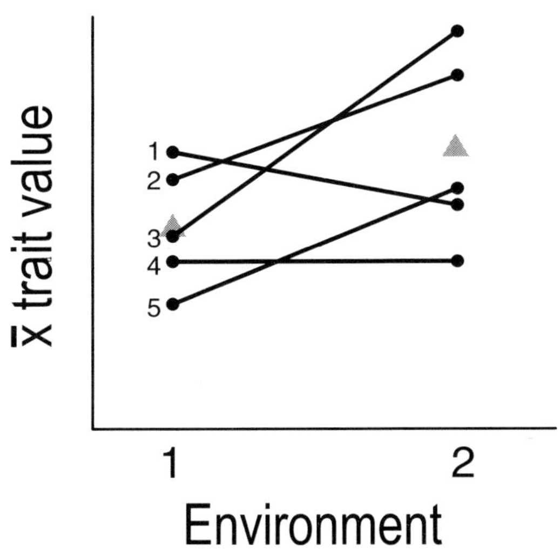

Fig. 2 Hypothetical reaction norms for five genotypes in one population. Triangles show mean population trait value at two different environments. See text for explanation.

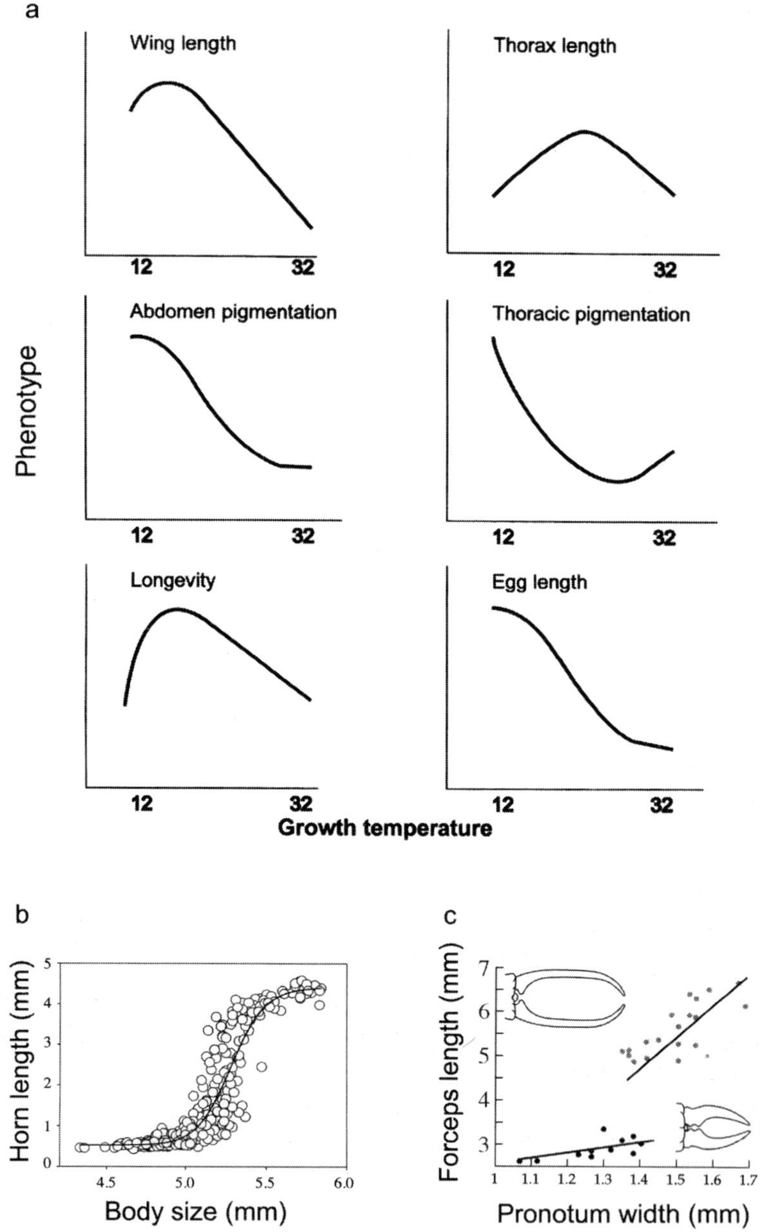

Fig. 3 Reaction norms from insects, showing the great diversity in phenotypic plasticity response. (a) Reaction norms for various traits in *Drosophila* in response to growth temperature (David et al. 2004; by permission of Oxford University Press, Inc.). (b) Sigmoid allometry for

Fig. 3 Contd. ...

— they simply depict their variable and heritable natures (Nijhout 2003a, Frankino and Raff 2004).

Characterizing Phenotypic Plasticity

Scientists agree that phenotypic plasticity concerns environmentally induced changes to phenotypes (see Box 1). Also, most consider discrete morphological polyphenisms (Figs. 1a-d, 4a,b,k,l, 5) as good examples of this concept. However, environments can influence phenotypes in diverse and complicated ways, and it is among these varied effects that opinions about plasticity begin to diverge. Below, we discuss some of the complexities and controversies surrounding phenotypic plasticity.

What Kinds of Traits Exhibit Phenotypic Plasticity?

Virtually any trait can show phenotypic plasticity. The concept was first applied to morphological traits (Woltereck 1909, Schlichting and Pigliucci 1998), and some authors still link phenotypic plasticity to morphology. However, it is clear that organisms can also alter biochemistry, physiology, behavior, and life history in response to the environment (B. Agarwala 2007, see Chapters 5, 11, 12, this book), and such changes are now generally accepted as phenotypic plasticity. Hence, such diverse phenomena as heat shock reaction (Chapter 17), acclimatizations (Chapter 16), diapause, immunology, learning and imprinting (Chapter 18), host-plant switching (Chapter 18), enzyme induction, predator-induced defense (Chapters 7, 8), maternal effects (Chapter 19), homeostasis (Chapter 15), mate choice and hybridization (Pfennig 2007), dispersal (Chapter 14), environmentally induced transcription and translation, and general stress responses are now often analyzed under the rubric of phenotypic plasticity. However, because virtually all phenotypic traits result from underlying biochemical-physiological processes, virtually all phenotypic plasticity represents (or results from) altered physiology.

Many authors view plasticity as a developmental process (Cronk 2005), and even ontogeny can be considered a continuous reaction norm of the

Fig. 3 Contd. ...

horn length in male *Onthophagus taurus* beetles in response to body size, which is largely determined by larval nutrition (after Moczek et al. 2004, see Chpt.3). (c) Allometry for nutrition-influenced forceps length in *Eluanon bipartitus* male earwigs, showing two discrete morphs with no intermediaries (Tompkins & Simmons 1996, Schlichting & Piglucci 1998, Tomkins 1999).

entire genotype (Schlichting and Smith 2002), with time as the "environmental" variable. Others link phenotypic plasticity to environmental induction of gene or allele expression (e.g., Czesak et al. 2006). However, these all represent biochemical-physiological processes. Development is particularly susceptible to deviating perturbations, with manifold downstream consequences, and this is why plasticity theory is closely tied to development (see Chapters 3, 4, 12, 13, 14).

Organisms are complex networks of interacting systems. As such, altered environments induce not single, but manifold changes, altering suites of independent and interconnected traits that range across multiple levels of biological organization (Relyea 2004a, Gorur et al. 2005, de Kroon et al. 2005, Chapters 2, 14). An example is locust polyphenism, in which solitary and gregarious phenotypes differ in behavior, morphology, food selection, body color, gene expression, neuro-, endocrine, and nutritional physiology, metabolism, immune responses, pheromone production, reproduction, and longevity (Simpson et al. 2005, Song 2005, see Chapters 5, 6, this book). A developmental evolutionary challenge is the integration of numerous plasticities into a functioning individual of high fitness (Pigliucci and Preston 2004, Shingleton et al. 2007).

Fig. 4 Morphological phenotypic plasticity in insects. (a) Soldier vs. worker in *Reticulitermes flavipes* termites (Klausnitzer 1987; Courtesy: Edition Leipzig). (b) Non-estivating (left) and estivating (right) nymphs of *Periphyllus granulatus* aphids (Hille Ris Lambers 1966. Reprinted, with permission, from *Annual Review of Entomology*, Vol. 11 (1966) by Annual Reviews, www.anualreviews.org. (c) Soldier (top) and non-soldier *Pseudoregma alexanderi* aphids. Reprinted from: Minks & Harrewijn 1987, courtesy of Elseiver Ltd. See Stern & Foster 1996, Shibau et al. 2003, 2004). (d) Oedymerous (left) and gynaecoid (right) male *Tiarothrips subramanii* thrips (Ananthakrishnan 2005). (e) Small and large male *Phoxothrips pugilator* thrips (Haga & Okajima 1975, Mound 2005). (f) Polyphenism in *Forficula auricularia* male earwig cerci (Carpenter 1899, Tomkins & Simmons 1996, Tomkins 1999). (g) Horned and hornless *Onthophagus taurus* dung beetles. (h) Polyphenism in male *Cladognathus giraffe* stag beetles (Otte & Stayman 1979). (i) Male *Brentus anchorago* weevils from Costa Rica exhibit enormous plasticity in body length (7 to 49 mm) (Johnson 1982; Courtesy of John Wiley & Sons Ltd). (j) Heads of small and large male *Mecynothrips kraussi* thrips (Palmer & Mound 1978). (k) Phenotypic plasticity to host in a trichogrammatid egg parasitoid, *Trichogramma semblidis*: small winged male (left) from moth eggs. Large wingless male (right) from alder fly eggs (Salt 1937). (l) Dispersing and non-dispersing forms of male *Pseudidarnes minerva* fig wasps (Cook et al. 1997; Courtesy of Royal Society of London. See also Pienaar & Greef 2003a,b).

Fig. 4 Contd. ...

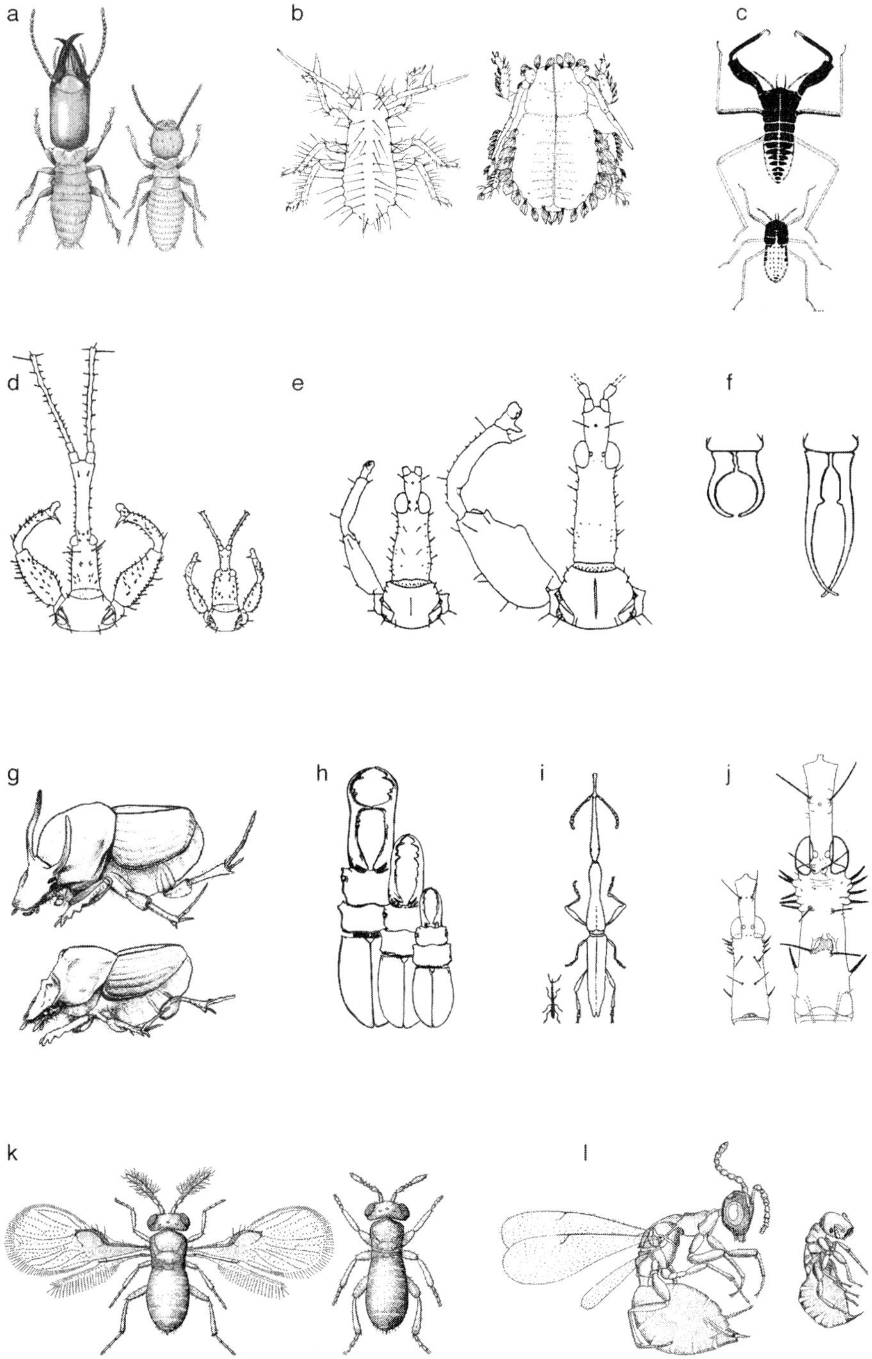

Fig. 4

Phenotypic Modulation vs. Developmental Conversions

Phenotypic plasticities range from graded, continuous responses (phenotypic modulation), to discrete switches in phenotype with no intermediate forms (developmental conversion or threshold traits) (Roff 1996, Windig et al. 2004) (see Glossary). The former are sometimes assumed to be non-adaptive, reversible "susceptibilities," and produce continuous linear or curvilinear reaction norms. Examples include nutrition and temperature effects on growth rate and body size (Figs. 1f, 3a). Developmental conversions are sometimes assumed to be beneficial, permanent adaptations, and produce discontinuous or sigmoid reaction norms. Examples include discrete polyphenisms (Figs. 1a-c, 3c, 4a, b, k, l, 5). Most plasticities fall somewhere between these extremes. Importantly, a plastic trait may be erroneously designated a developmental conversion due to improper sampling or failure to expose experimental organisms to intermediate environments (Fig. 7e) (Nijhout 2003a). In addition, a continuous, graded process may underlie a discontinuous plasticity, such as when a trait responds to a gradual change in an underlying hormone concentration, via a threshold mechanism (Roff 1996, Nijhout, 2003a). Note that developmental conversions can be alternatives (such as in social castes) or sequential, as in sequential sex change (Munday et al. 2006).

Cues

Phenotypic plasticity can be initiated by either environmental stimuli or cues. The former are often environmental factors such as temperature or oxygen level that directly disrupt homeostasis or development in non-adaptive ways. In contrast, organisms can evolve mechanisms to sense and adaptively respond to certain cues that predict environmental change (Nijhout 2003a). Hence, cues are generally considered to be specific environmental signals that predict environmental change, and induce adaptive plasticities. Cues tend to be non-harmful stimuli (i.e., photoperiod or a predator-released chemical) that do not harm the individual directly, whereas stimuli, themselves, are often harmful selective agents (toxin, high temperature). However, the division between these two is blurred, and the same environmental factor, such as temperature, can simultaneously initiate a highly adaptive plastic response and harmful physiological disruption. In general, organisms should evolve mechanisms to detect and respond to environmental stimuli or signals that accurately predict future environmental conditions. Hence, stress factors and correlated predictive signals should evolve into cues.

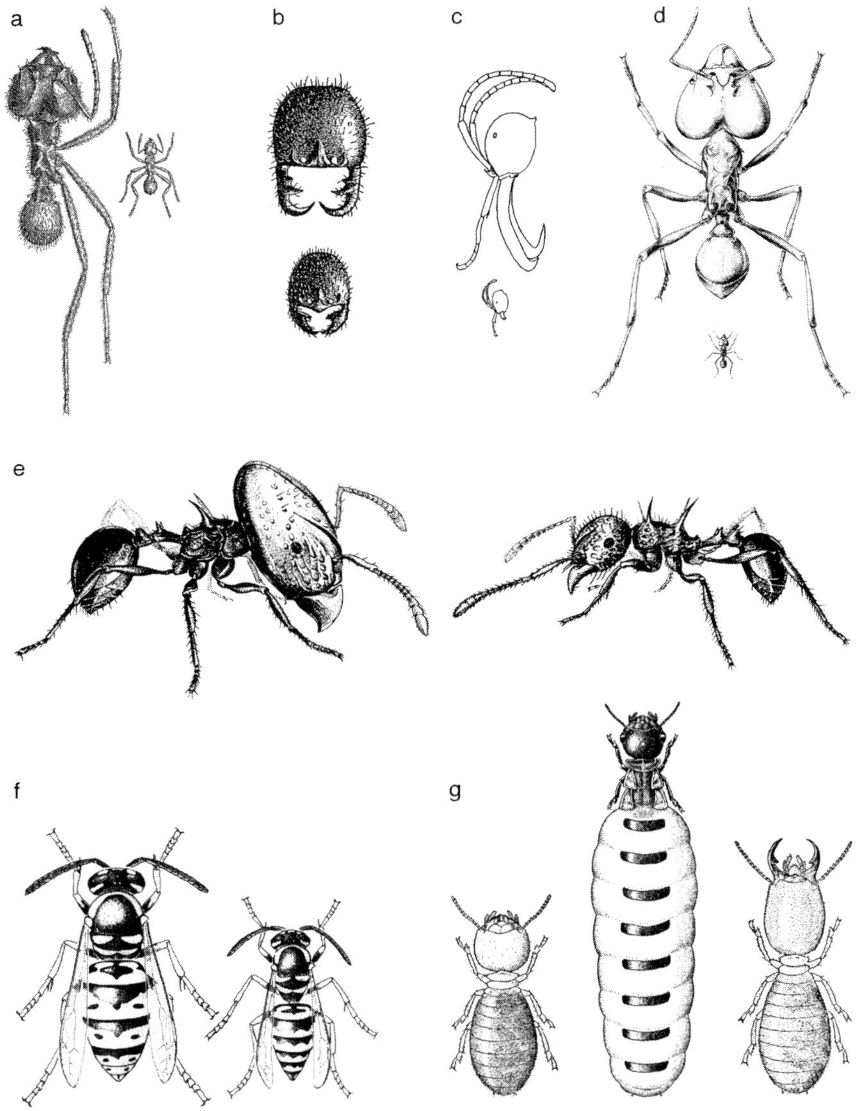

Fig. 5 (a) Soldier (left) and small worker caste (right) in *Atta texana* leafcutter ants (Wheeler 1910). (b) Heads of soldier vs. small worker castes in *Cheliomyrmex nortoni* driver ants (Wheeler 1910). a and b courtesy Columbia University Press. (c) Heads of large soldier (top) and small worker (bottom) of *Eciton burchelli* army ants (after Schneirla & Topoff 1971). (d) Caste polyphenism in *Atta laevigata* leafcutter ants. (e) Minor and major workers of an *Acanthomyrmex* species from the Celebes (d & e drawn by Turid Hölldobler; Oster & Wilson 1978); courtesy of Princeton University Press. (f) *Vespula maculifrons* queen and worker illustrated by S. Landry (Evans & West-Eberhard 1970). Courtesy University Michigan Press. (g) Worker (left) queen (middle) and soldier (right) of *Amitermes hastatus* (after Skaife 1954).

Both stimuli and cues can originate internally or externally. For example, initial hatchling size, growth rates, nutrient titers, or pathogen presence may serve as internal cues that determine alternative developmental outcomes (Nijhout 2003a,b, Mirth and Riddiford 2007, Shingleton et al. 2007). Virtually any factor can serve as a stimulus or cue to initiate a plastic response, and can be received via any sensory modality [chemical, visual, thermal, mechanical (tactile, acoustic), electrical, etc.] (see other chapters, this book). In *Diacamma* ants, queens induce young adults to become workers by chewing off their vestigial wings (Peeters and Higashi 1986, Baratte et al. 2006). Water force cues development rate in stonefly nymphs (Franken et al. 2008) and penis length in barnacles (Neufeld and Palmer 2008).

Specific vs. General Plasticities

Some plastic responses are highly specific in either requisite stimuli or response. For example, some plants possess receptor proteins that detect only their most common natural enemy (Zhao et al. 2005). Such specificity is seen in corn plants that increase defense in response to saliva from young, but not old armyworm caterpillars, perhaps because the plastic defense is only effective against young caterpillars (Takabayahshi et al. 1995). Elm trees produce volatiles attractive to egg parasitoids, in response to oviposition by its primary beetle herbivore, but not to beetle feeding (Meiners and Hilker 2000). Following fires, some grasshoppers will respond to altered light quality by adaptively changing their body color to black (see Uvarov 1966). Other grasshopper species fail to respond to light, but change color specifically in response to temperature, humidity, food, or crowding, or to some combination of these cues (Rowell 1971, Tanaka 2004, Chapters 5, 6).

Other elicitors and responses are more general, such as temperature (Chapters 12, 16, 17) and nutrition (Chapters 3, 10, 11, 19) which can influence nearly every aspect of an animal's phenotype and ecology. In some aphids, alate production is induced by any combination of photoperiod, crowding, nutrition, or presence of natural enemies (B. Agrawal 2008). Learning is a general form of plasticity that can respond to manifold environmental stimuli, and produce a great variety of plastic responses (Kukas 2004, Chapter 18). Likewise, growth and development rates are plastic to innumerable environmental factors (Chapter 10). Complex plasticities, such as locust polyphenism and life history plasticities, represent composites of numerous underlying plastic traits (Song 2005). The species characteristics and environmental factors that favor the evolution of general vs. specific cues and responses, and the underlying physiological and ecological constraints that shape these responses are currently unknown.

Adaptive Plasticity

Many examples of phenotypic plasticity are clearly adaptive (i.e., beneficial as the result of past selection), such as some immune responses, antipredator defenses, acclimatizations, diapause, life-history shifts, dispersals, etc. (West-Eberhard 2003, Lyytinen et al. 2004, Schmid-Hempel 2005). Other plasticities are non-adaptive. These include many susceptibilities to abiotic factors, and manipulations of hosts by parasites and pathogens (Hurd and Lane 1998, Roy et al. 2006, Kenyon and Hunter 2007, Poinar and Yanoviak 2008). For example, some leaf miners and gall makers induce maladaptive resource allocations and leaf retention in host plants (Prichard and James 1984, Oishi and Sato 2007). However, the environment can influence phenotypes in complex ways, and it is often difficult to determine whether or not altered phenotypes are beneficial or adaptive (van Kleunen and Fischer 2005, Pigliucci 2005, see Chapters 7, 10). Plasticities are under conflicting selective pressures (Sih 2004) and carry numerous costs and tradeoffs (DeWitt et al. 1998, Fordyce 2001, Chapters 3, 7, 10, 11, 12, 14), and some have argued that it is nearly impossible to ever know their total cost/benefit ratios. First is that a great many traits may be altered by a single environmental factor, and not all of these changes may be recognized or studied, including their numerous and complex physiological and environmental interactions and consequences (Relyea 2004a, Agrawal 2005). A specific altered trait may be highly beneficial in one context, but overwhelmingly detrimental in another. For example, plastic production of large spines or heavy armor in a prey (Fig. 6b) in response to the presence of predators may aid antipredator defense, but reduce feeding, migration, mating, fecundity, etc. (Roff 1996). Hence, the benefit of any phenotype is relative to a specific time and place and presence or absence of interacting individuals (Nykänen and Koricheva 2004, Thompson 2005). To understand adaptive plasticity, one must consider benefits and costs of plastic phenotypes in several environments.

Genetic and environmental correlations are themselves plastic to the environment (Piggliucci 2005). A particular plastic response may be highly advantageous in one season and detrimental in the the next. Indeed, a specific plastic response might be evolutionarily favored, and thus maintained in a population even if its expression produces great fitness benefits only once every 10 years; i.e., uncommon, periodic events may drive some evolution. In most years, a researcher would have little chance of observing an uncommon, but powerful selective event (e.g., Stephen 2005). Also, cost/benefit analysis should continue into the next generation, because of parental effects (Agrawal 2001, Mondor et al. 2005). Most cost/

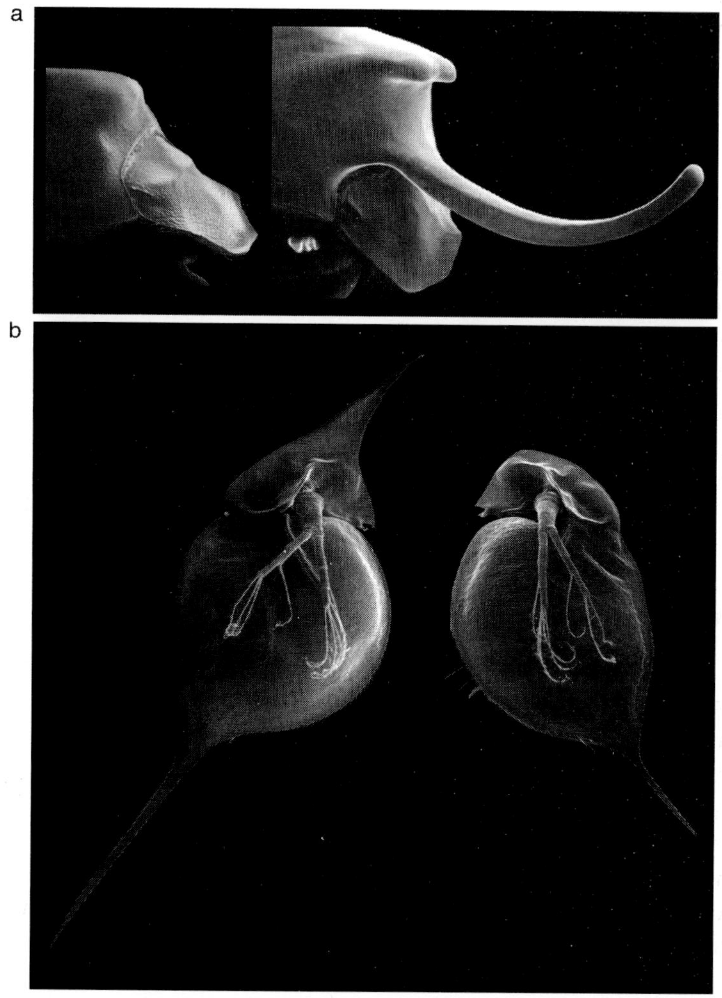

Fig. 6 (a) Horn polyphenism in *Onthophagus nigriventris*. Courtesy of D. Emlen. See Emlen et al. 2006, 2007. (b) Predator-induced plasticity in *Daphnia lumholtzi*. Left individual was exposed to fish-predator chemicals, right individual was not. The long spines reduce predation (Agrawal 2001).

benefit studies are conducted in the laboratory, greenhouse, or outdoor cages, and may not accurately reflect the realities of nature. Finally, "adaptive" implies past selection, but a population's history is often opaque (Doughty and Reznick 2004). Hence, it is difficult to know the adaptive value of phenotypic plasticity, and, for that reason, "adaptiveness" cannot be the criterion for judging if an environmentally altered trait represents plasticity (van Klunen and Fischer 2005).

There is another reason for not restricting phenotypic plasticity to only "adaptive" traits. This is because, whether adaptive or not, all environment-induced changes to phenotypes are similar in that they place those individuals into a different selective regime, with potential fitness and evolutionary consequences. Indeed, the evolution of many adaptive plasticities, such as diapause, alternative morphologies, mating and life history strategies in small individuals, and even sociality, may have been stimulated by detrimental plastic responses to harmful factors such as low temperature or poor nutrition (West-Eberhard 2003, Emlen et al. 2006, Chapter 3).

Of course, we should continue to test the Beneficial Plasticity Hypothesis, and to evaluate both the specific and overall value of plasticity (Wilson and Franklin 2002, Doughty and Reznick 2004, van Kleunen and Fischer 2005, Chapter 10). Two features that often imply adaptation are anticipatory and active plasticities (see below).

Anticipatory vs. Responsive Phenotypic Plasticity

Some plastic responses are anticipatory, in that individuals initiate phenotypic change before the appearance of a harmful (or beneficial) environmental factor. Examples include diapause induction before the onset of winter, and detoxification induction in caterpillars. Some plants exhibit defense plasticity whereby caterpillar feeding induces the production of the plant hormones jasmonate and salicylate, which, in turn, triggers synthesis of anti-herbivore toxins (Chapter 7). Amazingly, *Helicoverpa zea* caterpillars have deciphered the plant's chemical signaling, and apparently monitor the plant's hormone concentration, which allows them to preempt poisoning. Consumption of the plant hormones activates four genes in the caterpillar that code for cytochrome P450 detoxifying enzymes, preparing it for the oncoming plant defensive onslaught (Li et al. 2002). Other plastic responses are non-anticipatory and are only triggered after the appearance of the new environment. Anticipatory and responsive plasticities are sometimes termed cued plasticity and direct plasticity, respectively (West-Eberhard 2003).

Because non-anticipatory plasticity may allow damage before the individual has a chance to change, we would expect direct plasticity to evolve into cued (anticipatory) plasticity, when possible. Likewise, if a particular trait's differential expression is strongly associated with fitness, we would expect organisms to evolve to respond to multiple predictive cues, as in the case of aphids, which produce winged, dispersing phenotypes in response to photoperiod, crowding, nutrition, and densities of both natural enemies and ant body guards (B. Agarwala 2007). A requirement for anticipatory plasticity is that the cue must reliably predict the environmental change (Karban et al. 1999). Consistent abiotic cues of seasonal change (i.e., photoperiod) are perhaps the most reliable cues favoring the evolution of anticipatory plasticity (Bradshaw and Holzapfel 2007).

Active vs. Passive Phenotypic Plasticity and Susceptibilities

Many environment-induced phenotypic changes are active in that the response involves multiple regulatory genes and processes acting at different hierarchies to produce a complex, coordinated change. Good examples are discussed throughout this book, and include locust polyphenisms (Chapter 5) and environmentally induced diapause, which are plastic, highly integrated responses, involving behavior, physiology, morphology, and life history, regulated by specific genes and feedback mechanisms, and complex coordinated physiological-endocrine processes. As previously noted, diapause plasticity is also often anticipatory, in that the insect responds to environmental cues that predict future stressful environmental changes. Active, anticipatory phenotypic plasticities provide strong circumstantial support for adaptive plasticity.

In contrast to active plasticity, other environmentally induced phenotypic alterations appear to be simple susceptibilities to physical or chemical environmental stresses. Toxins, poor nutrition, and extreme temperatures, pH, O_2 levels, and osmolarities can directly alter chemical, enzymatic, cellular, and developmental process, producing passive (not regulated by the organism) changes to the phenotype. Small size resulting from poor nutrition is perhaps the classic example of passive plasticity. Nonetheless, most forms of plasticity likely contain active and passive components, and distinguishing them can be difficult. Even for poor nutrition-induced small size, one could make the argument that smaller individuals are more fit than larger individuals, given the current environment. Active and passive plasticity can act simultaneously on the same trait in the same individual, and can be in similar or different directions (van Kleunen and Fischer 2005). Both types represent phenotypic plasticity.

Period of Responsiveness

Species vary greatly as to when in their development they can respond to environmental change. Some species remain responsive throughout much of their lives. In others, developmental processes create specific windows when plasticity is possible. This is particularly true in arthropods because of their discrete life stages (i.e., metamorphosis) and their external skeleton, which is not amenable to change after scleritization (Frankino and Raff 2004). Hence, for insects, phenotypic plasticity in external morphology must be initiated before molting (Shingleton et al. 2007). Many species have evolved precise temporal windows of responsiveness, and if they do not receive the appropriate environmental stimuli during that critical period, plasticity does not occur. Examples are some butterfly polyphenisms (Nijhout 1991, 2003a) and insects that imprint on their host or habitat immediately following adult eclosion (Davis and Stamps 2004, Chapter 18).

Speed of Induction

If one includes behavior (Sih 2004) and transgenerational plasticity (Mousseau and Dingle 1991, Mousseau and Fox 1998), then the speed of plastic responses ranges from instantaneous to generational. Phenotypic change should match environmental change, and too long of a lag time can be maladaptive. Thus, a key is again the match between the response, no matter what the time scale, to the environment experienced: rapidly changing environments should select for rapid plastic response, and slowly changing environments for graded or slowed responses. The former would require behavioral or physiological phenotypic plasticity, and the latter could be met with slower acting developmental plasticity, including altered morphologies or life histories.

Reversibility

Plastic traits vary in their permanency. In general, behavioral and physiological traits are rapidly reversible within individuals, whereas morphology and life history can be permanent. A great many traits fall somewhere in-between. It is difficult to shed morphologies, but even more so in insects because of their hardened exoskeleton, and because they are relatively short lived, and thus have little time or need to reverse phenotypes. However, some insect morphological plasticities are reversible. Adult Thysanura and krill (Crustaceae) molt to a smaller size in response to poor nutrition (Marinovic and Mangel 1999, Piersma and Drent 2003). *Kosciuscola tristis* grasshoppers can repeatedly alter their body color in as little as 1 hr (Key and Day 1954a,b). Conversely, induced plasticities in plants and humans can last for years (Tollrian and Harvell 1999), or across generations (Mousseau and Fox 1998, Agrawal et al. 1999, Bateson et al. 2004).

Physiological Mechanisms underlying Phenotypic Plasticity

Adaptive phenotypic plasticity is accomplished via a vast diversity of mechanisms, involving virtually all physiological levels and systems. Detailed understanding of biochemical pathways and mechanisms exists for pathogen-induced plasticity in vertebrates and plants (Frost 1999, Defranco 2007, Chapter 7), beetle horns (Emlen et al. 2007, Chapter 3), butterfly polyphenisms (Chapter 9), body size and allometry (Emlen and Allen 2004, Shingleton et al. 2007, Chapters 10, 13), wing polyphenisms (Chapter 14), some acclimations (Chapter 16), stress proteins (Chapter 17), and social castes (Page and Amdam 2007). For adaptive, coordinated phenotypic plasticity, the process involves cue recognition, stimulus transduction, and complex effector systems (Schlichting and Smith 2002, Windig et al. 2004). Cues can originate from in- or outside the individual. In some cases environmental cues are specific and are detected by specialized sense organs or mechanisms designed primarily for that purpose. In other cases, eliciting cues are more generic and are received by the general sense organs, such as eyes or mechanoreceptors. For example, tactile stimulation of sensory hairs on the legs of locusts triggers behavioral phase change in response to high density (Chapter 5). Translated signals may be sent to specific tissues and used immediately, or stored for later induction. Phenotype alteration can be accomplished from a single, unchanging genome via any combination of transcription, translation, enzyme, hormone, and morphogen regulation, morphogenesis, apoptosis, and neural control, with appropriate regulation and feedbacks between subcomponents of the overall process (Miura 2005, Amdam 2007, Emlen et al. 2007, Wolschin and Amdam 2007, Zhou et al. 2007, Chapter 7). Between cue reception and production of the ultimate phenotype, may lie dozens of steps, influenced by hundreds of genes and untold environmental/ physiological factors. This, in part, is what makes understanding of both genetic and physiological control of plasticity so difficult. In insects, environmental factors can directly turn on or off genes (Ellers et al. 2008, Chapters 7, 16, 17) or hormones (Emlen and Nijhout 2000), and hormones induce differential gene expression and development (Nijhout 1994, Evans and Wheeler 1999). Hormones lie at the base of virtually all insect developmental conversions (discrete polyphenisms) (Nijhout 1994, 1999, 2003a, Chapters 9, 13, 14), and small evolutionary changes in thresholds or timing of hormone release or sensitivity periods of specific tissues produce different reaction norms in different taxa (Emlen et al. 2007, Shingleton et al. 2007, Chapters 13, 14). In insects, parts of the endocrine and nervous systems are one-and-the-same, which probably aids transduction of environmental cues into physiological responses (Nijhout 2003a).

Selection may act anywhere along this chain. However, reaction norm evolution is often accomplished by altering timing of physiological mechanisms that control developmental switches (Moczek and Nijhout 2003, Chapter 13). Brief and disparate sensitivity periods for different tissues, such as various imaginal disks, allow independent, compartmentalized regulation and evolution of traits, fostering great diversity of plastic responses (Emlen et al. 2007, Chapter 13).

Inability to elucidate the physiological mechanisms underlying phenotypic plasticity greatly hampered past plasticity research, and a complete understanding of plasticity will require knowing its physiology (Ricklefs and Wiekelski 2002, Frankino and Raff 2004, Lessells 2008, Chapters 11, 13, 14). However, multiple approaches for physiological understanding are now available, and modern molecular tools are stimulating rapid progress (Frankino and Raff 2004, Cossins et al. 2006, Emlen et al. 2006, Shiu and Borevitz 2008). For example, microarrays provide for the simultaneous monitoring of the expression of thousands of genes during induction and expression of phenotypic plasticity, and, when coupled with knockouts and other technologies, will allow identification of the specific genes and pathways responsible for adaptive responses.

In social insects, caste determination in different species ranges from environmental to genetic (O'Donnell 1998, Miura 2004, 2005, Hayashi et al. 2007, Hunt 2007, Whitfield 2007). Caste ratios are often determined by positive and negative feedback mechanisms, whereby increasing numbers of one caste feed back to reduce production of that caste, often accomplished via pheromones or nutrition (Shibao et al. 2004).

Physiological Homeostasis is Phenotypic Plasticity

Rapid, short-term physiological homeostasis such as regulation of blood pH and osmolarity represents phenotypic plasticity. Somewhat counter intuitively, homeostasis is derived from monitoring internal and external conditions, and manipulating physiology, i.e, keeping some aspect of the phenotype constant by altering enzyme activity or other physiological or behavioral parameters, in response to a varying environment (Chapters 13, 15). Some traditional homeostatic mechanisms and phenotypic plasticities share similar physiological mechanisms (Chapter 16). Physiological changes, be they rapid and short-term, or delayed and long-term, represent altered phenotypes to altered environments, and, as such, have the potential to produce the same evolutionary effect – increased fitness for those genotypes that can show the beneficial plasticity. Hence, phenotypic plasticity can not be defined by velocity and reversibility of responses.

Phenotypic Plasticity vs. Canalization

Phenotypic plasticity is often considered the opposite of canalization. However, reaction norms can be canalized (Scheiner 1993). In addition, to hold one trait constant in the face of a changing environment often requires change (plasticity) in another trait. For example, some insects exhibit canalized egg size, and when confronted with poor nutrition or end of season, such insects maintain egg size, but express plasticity in clutch size or oocyte development rates (Chapter 11). In other species, clutch size or oocyte development may be canalized (Stearns 1992, Nylin and Gotthard 1998, Fox and Czesak 2000). Given trade-offs, and that particular traits can evolve to be plastic or canalized, the evolutionary outcome is presumably based on the relative advantages of different strategies in different habitats. Furthermore, what at first may appear to be a non-adaptive passive response (for example, lowered clutch size under poor nutrition), may in fact be an evolved plastic response to maintain egg size, oocyte development rate, or female survival. As mentioned above, physiological homeostasis also requires an underlying plasticity. As such, canalizations in physiology, life-history, and development are often accomplished via phenotypic plasticity.

Why is Phenotypic Plasticity Important?

Phenotypic plasticity is important because it expands the existing "genocentric" evolutionary theory, producing an encompassing paradigm to explain life on earth. Plasticity was once considered "noise" but is now widely recognized as potentially adaptive under a wide array of circumstances. As with any major shift in scientific thinking, phenotypic plasticity engenders new ideas, causing us to ask new questions and test hypotheses that would not otherwise be examined, leading us to productive new scientific insights.

Phenotypic plasticity is a counterbalance to mutation-driven evolution — It is not surprising that during the first half of the 20th Century, scientists, flushed with excitement about Mendelian genetics, viewed evolution primarily as a mutational process. However, this bias largely ignored an important reality of evolution – that natural selection selects not among genotypes, but among phenotypes. Thus, the phenotype, and variation among phenotypes, plays a major role in evolution. And, because the environment in which an individual develops determines its phenotype, the environment also assumes a greater role in evolution, and may, in fact, produce more viable phenotypic variation than do mutations (West-Eberhard 2003, 2005). This is because mutations are not only rare, but usually deleterious. In contrast, a single environmental factor may alter the phenotypes of an entire population, providing natural selection with access to perhaps thousands of environmentally altered individuals, as opposed to a single mutant individual. In addition, mutations generally arise randomly with no correlation to specific environments, whereas new environmentally induced phenotypes are both directional and highly correlated with the specific new environment, allowing new environments to immediately produce and select among new phenotypes (Badyaev 2005). Altered environments may influence a diversity of traits that are not genetically linked, and hence may rearrange phenotypes in novel ways unavailable to single mutations. Unlike most mutations, a developmental rearrangement is likely to include both the altered trait and its background regulation (West-Eberhard 2005). And, because the inducing environmental factor may recur year after year, the new phenotype will recur often. Recurrence of a novel phenotype among large numbers of individuals that differ in numerous genetic, phenotypic, and environmental characteristics provides a fertile substrate for selection to act. Indeed, selection cannot act on a trait, if that trait is not exposed (i.e., the trait must be expressed in the phenotype). By producing novel combinations of phenotypic traits, the environment creates new raw products for selection. This process is believed to lead to adaptive phenotypic plasticity that we see today, and even to the generation of new species (West-Eberhard 2003, Schlichting 2004, Fordyce 2006, but see deJong 2005).

Under traditional evolutionary theory, the environment acts after phenotypic variation is produced, and plays a *single* role: selecting among genetically produced variation. With phenotypic plasticity, the environment plays a *dual* role in evolution: it both creates phenotypic variation and selects among that variation. This is a major change in how we view evolution. As such, environmentally induced phenotypic variation comes to assume a more important (perhaps dominant) position in evolutionary theory (West-Eberhard 2003). Similarly, theories of how organisms adapt to environmental heterogeneity previously emphasized between-generation adaptation by populations. In contrast, phenotypic plasticity emphasizes how individuals adapt within their lifetimes. Merging within- and between-generation, individual and population adaptation produces a more comprehensive theoretical framework of adaptive variation to environmental heterogeneity (Pigliucci 2001, Meyers and Bull 2002, West-Eberhard 2003), and may contribute to a new grand unifying theory of biology (Pigliucci 2007).

Including phenotypic plasticity produces a better model — As suggested above, the inclusion of phenotypic plasticity can result in a better model than mutation-allelic substitution alone in explaining the production of organismal diversity. For example, the initial evolution of warning color (aposematism), starting as a rare mutation is problematic because conspicuous prey should be quickly found and removed by predators (Lindström et al. 2001). In contrast, evolution of aposematism is easily explained by phenotypic plasticity (Sword 2002). Likewise, for development, phenotypic plasticity explains the evolution of allometry and exaggerated morphologies (Emlen and Nijhout 2000, Shingleton et al. 2007). For physiology, phenotypic plasticity explains adaptive, beneficial plasticities such as acclimation (Chapter 16), and response to exercise (Swallow et al. 2005), quite well. In ecology, it aids our understanding of life-history variation (Beckerman et al. 2002), population dynamics (Haukioja 1990, Gardner and Agrawal 2002), community structure (Werner and Peacor 2003, Agrawal 2005), and modeling of ecological and evolutionary processes (DeAngelis and Mooij 2005). Phenotypic plasticity also helps explain some sexual selection (Qvarnström and Price 2001), alternative mating strategies (Pfennig 2007, Chapter 3), and evolution of sociality (West-Eberhard 2003, Page and Amdam 2007).

Phenotypic plasticity elevates the importance of stress — Viewing evolution through environment-induced phenotypic plasticity elevates stress as a major ecological and evolutionary concept (Badyaev 2005, Roelofs et al. 2007). Environmentally induced stress is a constant reality for most individuals. How do organisms respond to stress and what are the physiological, ecological, and evolutionary consequences of stress? Are there commonalities among the responses to osmotic, thermal, temporal, nutritional, social, predator, and competitive stresses? The response of individuals to environmental stress may have stimulated the evolution of stress proteins, homeostasis, acclimation, canalization, immune response, learning, and the numerous phenotypic plasticities noted throughout this book (Gabriel 2005, Emlen et al. 2006). Furthermore, there are remarkable consequences of understanding stress plasticity in the context of the ecology and environmental impacts of interactions between humans and the environment (e.g., Relyea 2003a, Relyea and Hoverman 2006).

Phenotypic plasticity merges genes and environment — Phenotypic plasticity also addresses the nature vs. nurture controversy, because it merges these two polar concepts to reduce this exaggerated dichotomy. Under phenotypic plasticity, nature cannot be separated from nurture. Even at the lowest developmental level (transcription), gene activity is influenced by the surrounding internal and external environment. Environment influence on phenotype only increases down the epigenetic cascade of development. Even gamete genes are encased in a cytoplasmic environment that was presumably influenced by the parental environment, and continues to be influenced by current environmental conditions. Epigenetic inheritance further blurs genes and environment (Jablonka and Lamb 2005). Hence, genes and gene activities can never be separated from direct environmental influence, and most traits represent a gene-by-environment interaction. This realization elevates the role of environment in gene expression and development, and the role of development in evolution, and is partially responsible for the recent surge in evolutionary developmental biology (evo-devo) (Brakefield and French 2006, Sultan 2007).

Phenotypic plasticity alters environments and structures communities — Recent studies support the importance of phenotypic plasticity in shaping communities. An example is spittlebug-induced plasticity in growth form in willows, which subsequently alters the abundance of more than 30 willow-associated insects (Nakamura and Ohgushi 2003, Ohgushi 2005). An important realization is that impacts in communities need not be regulated by competition and predation in the classic sense of these factors shaping resource availability or the densities of organisms (Inbar et al. 2004, Schmitz et al. 2004, Van Zandt and Agrawal 2004, Agrawal 2005, Miner et al. 2005, Fordyce 2006, Schmitz 2006, Ohgushi et al. 2007, Chapter 7).

Phenotypic plasticity has practical importance — It aids systematics and taxonomy by helping to correct erroneous synonyms. Indeed, highly plastic genotypes have often been considered different species (Schlichting and Pigliucci 1998). This is especially problematic in entomology, where environmentally induced phenotypes are confused as distinct species (Uvarov 1966, Greene 1989, Mound 2005). For example, more than 20 divergent phenotyes of the thrips *Ecacanthothrip tibialis* were previously assigned species status (Ananthakrishnan 1969). Inaccurate species identification or failure to recognize phenotypic plasticity can hamper basic research, disease diagnosis and medical and agricultural pest control.

Phenotypic plasticity may help us forecast establishment and spread of invasive species (Peacor et al. 2006, Richards et al. 2006, Muth and Pigliucci 2007, Slabber et al. 2007), aid conservation (Beckerman et al. 2002, Davis and Stamps 2004), and help us understand the consequence of environmental disruption (Bradshaw and Holzapfel 2006, 2007, Hendry et al. 2008). Differential plastic responses among interacting species may alter ecosystem interactions (Visser et al. 2006). Plasticity may aid environmental monitoring (Ellison and Gotelli 2002, DeCoen et al. 2006, Lee et al. 2006). For example, herbivore attack often induces a plastic defense response in plants, including the release of novel volatile compounds (Chapter 7). Different plant taxa generally release different volatile blends. As such, scientists could monitor community stress levels by analyzing the local atmosphere (DeMoraes et al. 2004). In industry, phenotypic plasticity is already aiding bioprospecting, as companies expose species to extreme environments or specific elicitors, to induce synthesis of novel bioactive substances (Poulev et al. 2003, Li and Barz 2005). In the future, induction of plastic biochemical pathways in plants or tissue cultures will be used to produce any number of commercially useful substances (Al-Tawaha et al. 2005, de Jeong and Park 2005, Zhao et al. 2005).

In agriculture, phenotypic plasticity helps us to understand variation in crop performance vis-à-vis herbivores, pathogens, anthropogenic inputs, and seasonal and spatial variability (Karban et al. 2004, Agrawal 2005). Phenotypic plasticity in pests, crop plants, or natural enemies can influence pest control (Bean et al. 2007, Luczynski et al. 2007, Pereira et al. 2007, Chapters 6, 7, 8), and the evolution of resistance (Adler and Karban 1994), and might be employed to our benefit (Davis and Stamps 2004). Once we understand the biochemical pathways and regulatory genes controlling induced defenses in crop plants, we can manipulate them, use genetic engineering to increase those beneficial responses, or transfer the capability to produce beneficial plasticities to other species (Kliebenstein et al. 2002, Edreva 2004, Agrawal 2005, Kappers 2005, Von Rad et al. 2005, Dana et al. 2006). Fisheries and animal husbandry also benefit from understanding phenotypic plasticity (de Jong and Bijma 2002, Collier et al. 2006).

Knowledge of how humans respond to stress, disease, carcinogens and drugs will continue to aid Medicine (Bateson et al. 2004, Nadeau and Topol 2006, Swynghedauw 2006, Calderwood et al. 2007). Likewise, phenotypic plasticity in human performance in response to exercise, altitude, nutrition, temperature, space travel, etc, are of great interest (Flück 2006, Geurts et al. 2006, Asea and DeMaio 2007, Radakovic et al. 2007). Phenotypic plasticity influences racial disparity in IQ (Flynn 1987, Hernstein and Murray 1994, Pigliucci 2001), and confounds anthropology (Collard and Lycett 2008).

Benefits of Phenotypic Plasticity

Phenotypic plasticity is beneficial when it allows an individual to alter its phenotype to adaptively match a changing environment (DeWitt and Langerhans 2004). Plasticity can also be a beneficial self-reinforcing process (a self-induced adaptive plasticity) such as when voluntary exercise increases heart and muscle mass, which increases exercise ability (Swallow et al. 2005), or when sampling a new, but toxic food induces detoxifying enzymes which then allows the individual to switch to that new food. Because it can increase fitness in multiple environments, phenotypic plasticity widens niche breadth and geographic range, and may aid dispersal and colonization (Price et al. 2003, Schlichting 2004, Pigliucci et al. 2006, Chapters 16, 19), and evolutionary transitions (Aubret et al. 2007). Plastic species should be able to survive ecological catastrophes and avoid extinctions, not only because of their presumed broader geographic ranges, but because they already live in and have adapted to different habitats and express different phenotypes (Schlichting 2004). For interacting species, possession of phenotypic plasticity may retard coevolution in an antagonistic species, because of a "moving target effect" (Alder and Karban 1994, Agrawal 2001), and it may prevent competitive exclusion (Pijanowska et al. 2007).

The greatest benefit of phenotypic plasticity may be that it generates adaptive genetic change (see below), an essential long-term strategy for evolutionary persistence. Plasticity may foster adaptive evolution by allowing genotypes to jump maladaptive valleys to reach fitness peaks in adaptive landscapes (Price 2006). It may also protect hidden genetic diversity from elimination, allowing that stored diversity to be exposed under specific conditions (Schlichting 2004, Suzuki and Nijhout 2006). Indeed, maintenance of genetic variation is so essential to life that costly mechanisms to achieve it (recombination and sexual reproduction) are nearly universal. Phenotypic plasticity may serve a similar role by both shielding genetic diversity, and by producing organic novelty that can then be incorporated into the genome via genetic assimilation (see Box 2). By maintaining a capacity for plasticity, heredity may provide for modification of its own machinery (Baldwin 1896).

Box 2 Genetic Assimilation

Genetic assimilation (**GA**) is a process by which an environmentally induced trait comes, after selection, to be constitutively expressed. Conrad H. Waddington proposed the idea in 1942, and then went on to demonstrate it experimentally, twice, using *D. melanogaster*. In the first case, he applied heat shock to fly pupae to induce a new adult phenotype with a reduced cross vein. After 14 generations of artificial selection under heat shock for expression of the plastic trait, some flies produced the veinless condition without heat shock (Waddington 1952, 1953a,b). In the second case, Waddington exposed fly eggs to ether to induce a novel phenotyopic abnormality, "bithorax," in the adult. After 29 generations of selection, the flies produced the bithorax phenotype in the absence of ether, and this new phenotype was heritable (Waddington 1956, 1961). In a third case, Waddington induced large anal papillae by exposing fly larvae to high salt levels. After 21 generations, the maggots expressed both large papillae and greater plasticity in low salt media (Waddington 1959).

Waddington (1953a) proposed that selection had altered the regulation of trait expression, such that the thresholds for expressing these traits were lowered to the point that the traits were expressed in all environments (Fig. 8). Examples from nature might include fixation of extrafloral nectar production in *Acacia* (Heil et al. 2004), and fixation of aposematism (Sword 2002). Suzuki & Nijhout (2006) showed GA of body color in the lab.

GA is an important idea because it suggests that acquired, phenotypic-plastic traits can become genetically fixed (Schmalhausen 1949). Hence, environmental induction can initiate evolutionary change (Pigliucci & Murren 2003). Furthermore, because the bithorax condition (above) created a second pair of wings, it mimics macroevolution, and thus suggests that macroevolutionary jumps might occur via genetic assimilation. GA in one trait might favor plasticity evolution in other traits, because as one trait becomes invariable to environmental conditions, it may increase conditional expression or selection pressure for plasticity in another (Jablonka & Lamb 2005). GA, its occurrence in nature, and its role in evolution are controversial subjects (de Jong 2005, Pigliucci et al. 2006, Crispo 2007), in part because of its similarity to Lamarckian evolution, the inheritance of environmentally acquired traits. However, GA is assumed to proceed via traditional Mendelian and Darwinian processes (see main text).

Plasticity Evolution

Phenotypic plasticity is probably ancestral, in the sense that environments have always changed and all living things are susceptible to abiotic and biotic factors (Nijhout 2003a). Although plasticity is not required to be beneficial or to have undergone adaptive evolution, it often has. One hypothetical pathway for the evolution of adaptive plasticity is through susceptibility. In this scenario, an environmental variable disrupts physiological homeostasis and development, creating new traits and new trait values, and rearranging phenotypes to produce novel trait combinations (Eshel and Matessi 1998). Most organisms contain large amounts of "non-functional" genetic material in their genomes. These genes are normally repressed, via genetic canalization. However, extreme environmental or biochemical conditions may disrupt such inhibition, allowing expression of these hidden genes (releasing hidden phenotypes), while reducing expression of others, leading to novel phenotypes. Although some such changes are beneficial, most are probably not. Recurrence and selection would then presumably adjust the regulation of gene expression and select for gene combinations that produced either increased canalization or adaptive plasticities (Nijhout 2003a). Plasticities to diet, disease, and abiotic factors may have evolved this way.

There are many other hypothetical routes for the evolution of adaptive phenotypic plasticity. For example, plasticity could evolve as an exaptation, when a previously existing plasticity comes to serve a new function, is induced by a different cue, or is shifted in its expression (e.g., when a biochemical plasticity evolves to produce a morphological plasticity or when plasticity in overall body size is co-opted for a single structure) (Emlen et al. 2006). For example, eusociality in wasps may have evolved from diapause or nutritional plasticity (Page and Amdam 2007, Hunt et al. 2007). Likewise, plasticity could evolve after hybridization of two populations, each of which has evolved different fixed phenotypes, if expression of the different phenotypes in the new hybrid population now becomes environmentally controlled. In this case, the hybrid population already possessed the capacity to produce both phenotypes; all that is required is to link differential production to environment. Epigenetic processes (McCaffery et al.1998, Brodie and Agrawal 2001, Kirschner and Gerhart 2005), and extraneous sources of hormones may have influenced plasticity evolution (Chapter 20). A non-adaptive plasticity could evolve via genetic correlation with other traits under strong selection (Scheiner 1993).

A key to understanding how phenotypic plasticity can evolve is the concept of interchangeability (Fig. 8). Most traits are both genetic and environmentally influenced (Roundtree and Nijhout 1994, Bradshaw and Holzapfel 2001). An example is the pigment melanin, which is the end product of well-known enzyme chains (Baraboy 1999, Sugumaran 2002, Ito 2003). The sequence instructions for these enzymes are coded by DNA and are heritable (True 2003). But in many animals, melanin production and deposition are also environmentally influenced, whereby colder

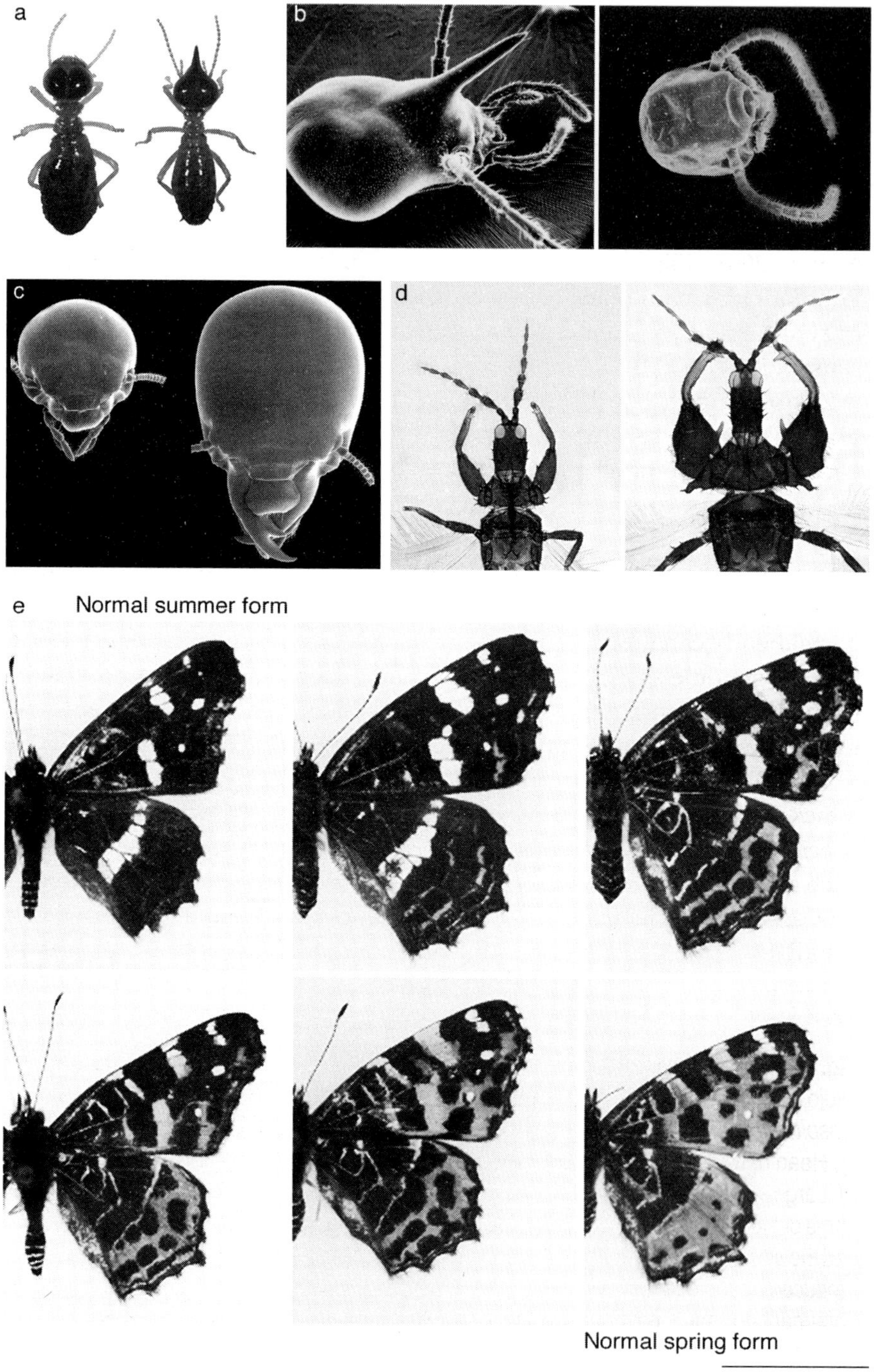

Fig. 7 Contd. ...

temperatures increase melanin and thus darken the body (Figs. 1d,e). This benefits individuals via solar heating, which counters the negative effects of cold temperatures (May 1984, Heinrich 1993). Hence, melanin production is both genetic and environmentally controlled, and this control is evolutionarily interchangeable: when there is genetic variation for degree of environmental influence, natural selection can select for either increased or decreased environmental sensitivity (West-Eberhard 2003, Suzuki and Nijhout 2006). Elimination of all flexibility produces a genetically fixed trait. Regulation of many traits is easily altered by adjusting response thresholds, enzyme saturation kinetics, timing of endocrine or development events or sensitivity periods of target tissues to hormones and morphogens (Meiklojohn and Hartl 2002, Moczek and Nijhout 2003a,b, Chapter 13). Hence, selection can easily slide trait regulation anywhere between total genetic and seemingly total environmental control (Fig. 8). The evolution of new phenotypes does not require the evolution of new gene complexes, but only the repatterning of existing genetic architecture and epigenetc interaction (Schlichting and Pigliucci 1998, Suzuki and Nijhout 2006, Emlen et al. 2007). Subsequent evolutionary loss of flexibility can permanently canalize the trait. Hence, a plastic trait can become subsumed into the genome as a canalized trait. An example may be extrafloral nectar (**EFN**) production in *Acacia* plants (Heil et al. 2004). In this genus, herbivore leaf damage induces the plant hormone jasmonic acid (**JA**) which induces EFN production, which attracts carnivorous plant bodyguards, which attack the herbivores. EFN inducibility is ancestral. But, in some *Acacia* species that have obligate ant bodyguards, the response to JA has evolved to such a low threshold, that individuals always produce EFN, in response to low, endogenous levels of JA. Hence, a plastic trait has been converted to a canalized trait via adjustments to the regulation of trait expression.

Fig. 7 Contd. ...

Fig. 7 (a) Worker and nasute-soldier of *Nasutitermes takasagoensis* termites from Japan (Hojo et al. 2004). Photo by Masaru Hojo. (b) Head of soldier (left) and minor worker (right) of *Hospitalitermes medioflavus* termites (Miura & Matsumoto 1995, 2000, Miura et al. 1998). (c) Head of worker (left) and soldier (right) of *Hodotermopsis sjostedti* termite (Miura 2005). (d) Large and small males of *Ecacanthothrips tibialis* thrips (Mound 2005). (e) Only two discrete forms of the nymphalid butterfly *Araschnia levana* are found in nature: the summer form (top left) and the spring form (bottom right). However, in the laboratory, intermediate phenotypes can be produced by subjecting individuals to intermediate environments or timed ecdysone injections, documenting that a continuous reaction norm lies at the base of this seasonal, diphenic polyphenism (Nijhout 2003a).

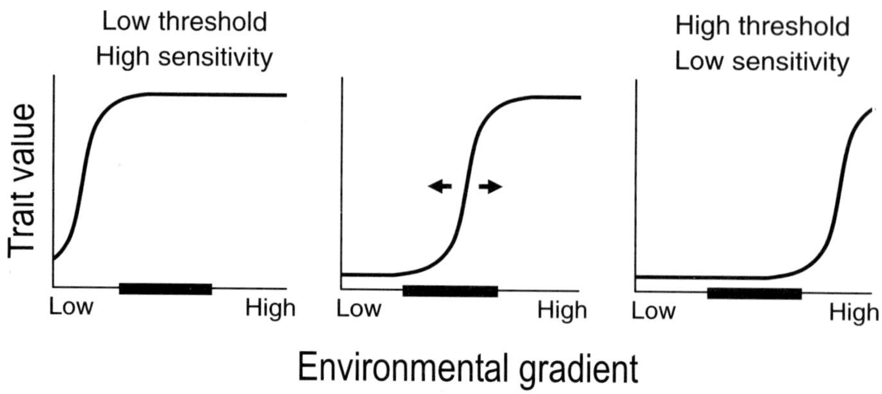

Fig. 8 Interchangeability between genetic and environmental control of a trait over evolutionary time. For each graph, the center (bold) of the horizontal axis denotes the normal range of values for the given environmental factor. Seldom encountered extreme conditions are shown by thin lines. In the right graph, the trait has a low value in virtually all normal environments and is thus considered a genetic trait. In the left graph, the trait has a high value in nearly all environments and is also considered genetically fixed. In the middle graph, trait value flips between high and low, depending on the environment, and hence is considered an environmentally determined trait.

Interchangability explains the phenomenon of phenocopies, which are environmentally induced phenotypes that resemble genetically determined ones (Figs. 1d,e) (Goldschmidt 1935, West-Eberhard 2003). Exposing a species to extreme conditions can elicit hidden phenocopies (Suzuki and Nijhout 2006, Otaki 2007, 2008).

When Should Phenotypic Plasticity Evolve and What Form Will it Take?

If environments were unchanging, then fixed phenotypes would be favored. But, because environments are constantly changing, plasticity is often favored. Indeed, the only way for an *individual* to adapt to a changing environment is by changing its phenotype. A plastic individual can achieve high fitness in two or more environments, whereas a fixed-phenotype specialist that is highly adapted to only one environment would be less fit in a different environment. Likewise, a fixed-phenotype generalist would presumably have only moderate fitness in all environments.

In general, phenotypic plasticity should be favored when it produces higher fitness than a fixed strategy across all environments (Berrigan and Scheiner 2004). A reaction norm (RN) is a trait of the genotype, and like all traits, a RN should evolve given directional selection on heritable additive genetic variance for plasticity (Doughty and Reznick 2004). Reaction norms respond to both artificial (Scheiner 2002, Suzuki and Nijhout 2006, Chapter 21) and natural selection (Hairston et al. 2001, Scheiner 2002, Bradshaw and Holzapfel 2006, Parsons and Robinson 2006), showing that plasticity can evolve. Note that plasticity evolution can be reversed (Chapter 21), and a flat RN (canalization or no plasticity) might evolve if it produced the highest fitness.

Theoreticians, modelers, and empiricist have proposed and examined numerous factors that favor or restrict plasticity evolution, alter reaction norms, or select for one type of plasticity over another (de Jong and Behera 2002, Scheiner 2002, Sultan and Spencer 2002, Berrigan and Scheiner 2004, David et al. 2004, de Jong 2005, Van Kleunen and Fischer 2005, Gabriel 2005, Gabriel et al. 2005, Garland and Kelly 2006, Chapters 15, 21), and these factors divide roughly into environmental factors, genetic (population and species) factors, and gene x environment interaction factors.

Environmental characteristics — Researchers interested in plasticity have examined a wide range of environmental factors including temporal vs. spatial heterogeneity, fine- vs. coarse-grain environments, predictability, speed, pattern, and permanency of change, number of selective factors, intensity of selection, frequency and strength of selection in alternative habitats, and reliability of cues that predict or signal environmental change (Chapter 7). By definition, phenotypic plasticity is a response to temporal or spatial environmental variation, and high variation should favor its expression and evolution Scheiner 1993, Sultan and Spencer 2004). Models and empirical studies suggest that plasticity should be more likely to evolve in temporal vs. spatial heterogeneity (Moran 1992), when cues are reliable (Karban et al.1999, DeWitt and Langerhans 2004), and in response to selective agents that slowly harm individuals, such as disease, cold, desiccation, etc., vs. those that act instantly with no warning, such as a tornado (Järemo et al. 1999, Sultan and Spencer 2002, Garland and Kelly 2006). Different environmental factors should select for different plasticities (Relyea 2003b, 2004b Boege and Marquis 2005). Speed of induction should correspond with speed of environmental change, and this may determine the type of plasticity that evolves (Van Buskirk 2002). When environmental changes are permanent, plastic change should be permanent (Relyea 2003c). Transgenerational plasticity should evolve when parent's environment predicts that of the offspring (Galloway 2005). When environmental variation is great and random, but cues are unreliable, then plasticity will not be favored and individuals should employ bet hedging strategies (Seger and Brockmann 1987, DeWitt and Langerhans 2004).

Genetic (population & taxon) characteristics — see de Jong and Behera 2002, DeWitt and Scheiner 2004. Phylogenetic constraints clearly prohibit certain plasticities in certain taxa. For example, plants are limited in behavioral plasticity, and in arthropods, molting, metamorphosis, and exoskeletons may preclude or favor certain plasticities. Plastic responses should change with ontogeny (Relyea 2003c), and decline with age, because of diminished developmental capability or because of impending senescence (Frechette et al. 2004). Some suggest that restricted gene flow favors plasticity evolution (Karban and Nagasaka 2004, Van Buskirk and Arioli 2005), and others opine that migration and panmixis favor plasticity (Tufto 2000, Sultan and Spencer 2002, Zhang 2006). *K*-strategists should be plastic, because they are long-lived, and thus encounter more temporal variability, and, with low fecundity, cannot afford to lose a single offspring. In contrast, large size and ample reserves in *K*-strategists may buffer environmental variation, obviating the need for plasticity. Short-lived *r*-strategists should have little need for plasticity, and can afford to lose some of their many offspring through bet-hedging. In contrast, plasticity may be favored in *r*-strategists because of their high rates of dispersal into new habitats. Polygenic quantitative traits should be more plastic than single locus traits (Roff 1996). Some models suggest that heterozygosity inhibits plasticity, because heterozygosity buffers environmental effects (see Scheiner 1993), but others disagree (Pigliucci 2005). If individuals cannot exhibit a reaction norm but groups can, then that reaction norm could feasibly evolve via group selection (Via et al. 1995, Sih 2004).

Gene x environment interaction factors — Plasticity evolution is presumably influenced by how individuals interact with the environment, including relative fitness benefits of plastic change in different environments, ecological tradeoffs, inclusive fitness, relative lengths of lifetime vs. stress period, dispersal range vs patch size, ecological feedbacks (i.e., when altered phenotypes alter the environment, which then alters selection on plasticity), including reciprocal plasticity interactions between genomes.

The type of plasticity that evolves should hinge on the ratio between generation time and environment fluctuation time (Gabriel and Lynch 1992, Schlichting and Pigliucci 1998, Meyers et al. 2005). Rapid, reversible behavioral and physiological plasticity should evolve when the environment rapidly switches back and forth and when life-span is much longer than environmental change (Gabriel 2005, Gabriel et al. 2005). With slower environmental change, morphological and life history plasticities should evolve, including once per lifetime developmental conversions. Longer cycle environmental fluctuations might select for transgenerational plasticities. Plasticity itself may or may not have high costs (Padilla and Adolph 1996, Van Tienderen 1997, Gabriel 2005), and therefore a plastic trait should be more likely to evolve when there are weak genetic or ecological correlations with other traits that are under selection in a different direction (de Jong 2005, Garland and Kelly 2006). However, plasticity evolution should increase when it produces large benefits, and when there is a positive correlation with other favorable traits, such as in plants when increased plastic antiherbivore defense aids pollination, allelopathy or disease resistance. Phylogenetic constraints on performance tradeoffs may prohibit certain plasticities in certain taxa. A correlation between habitat selection and trait plasticity should favor evolution of plasticity (Scheiner 1993). Selection can act directly on the shape of the reaction norm (Harshman et al. 1991, Scheiner and Lyman 1991, Scheiner 2002). However, directional selection on the mean constitutive value of a trait can also increase plasticity of that trait (alter the reaction norm), in the direction of selection (Swallow et al. 2005, Garland and Kelly 2006). Some suggest that traits highly correlated with fitness should have low plasticity (Schlichting and Smith 2002), and a flat reaction norm may be highly adaptive. However, some traits strongly linked to fitness, such as antipredator defenses and seasonal adaptations are often highly plastic (Chapters 4-9 & 16). In species with wide geographic ranges, different populations exhibit different, adaptive plasticities (Winterhalter and Mousseau 2007).

In sum, plasticity evolution is favored by environmental variation, strong differential selection in alternative environments, cues that accurately signal environmental change, high fitness benefits and low costs to plasticity, and heritable genetic variance for plasticity (Berrigan and Scheiner 2004).

A seldom discussed concept is that mutations lie at the base of phenotypic plasticity. This is because all trait expression is embedded in a particular genetic background. Different genotypes produce different reaction norms – one consequence of underlying genetic variation. The ultimate origin of such genetic variation is mutation. Hence, phenotypic plasticity is a consequence of mutational evolution. However, there is a fundamental difference between traditional mutational evolution and evolution via phenotypic plasticity. The mutations that produce specific plastic responses may remain hidden from the phenotype (or at least to that specific trait state) for millions of years.

It is only when the environment exposes that trait that selection on that trait state can begin. In general, mutations have little evolutionary impact until they are exposed in the phenotype.

There has been confusion regarding the relation between genetic variability for a specific plastic trait vs. overall genetic variability, and the occurrence vs. the evolvability of phenotypic plasticity. Phenotypic plasticity does not require genetic variability. For example, (except for developmental noise) all individuals of a genetically identical clone would exhibit the same phenotypic plasticity to the same altered environment. In this case, lack of genetic variability in all traits would preclude selection on both the reaction norm and associated traits (genetic accommodation). Thus, excepting for new mutations, this genotype could not evolve. In the case of background genetic variation, but no genetic variation for the plastic trait, the reaction norm could not evolve, but the new phenotype could evolve to be more fit via genetic accommodation. Hence, an invariant and initially detrimental plastic response could, over evolutionary time come to be imbedded in a highly fit phenotype. If genetic variation existed for both the plastic trait and most other traits, then the reaction norm, background traits, and fitness could evolve, to produce a highly intetgrated and adaptive plasticity.

In insects, evolution of plasticity is aided by their modularity and metamorphosis. For holometabolous insects in particular, future adult structures such as wings and legs derive from small clumps of cells (imaginal disks) that persist through immature development and are only activated via hormones during the pupal stage. Differences in timing of induction and in response of different imaginal disks allow independent expression and evolution of the resulting organs (Nijhout 2003a, Emlen et al. 2007).

Plasticity as a Factor in Evolution

Scientists debate whether or not phenotypic plasticity speeds or retards evolution (Chapter 21). Some suggest that plasticity shields traits from evolution because selection chooses among phenotypes (Huey et al. 2003, Price et al. 2003, de Jong 2005). Individual adaptation may preclude genetic selection. An example might be when a plastic behavior such as solar basking, microhabitat shift, or seasonal migration moderates body temperature, preempting selection for fur, melanin or thermal-adapted enzymes. Others suggest that plasticity stimulates evolutionary diversification by generating novelty (West-Eberhard 2003, Schlichting

2004), and/or via genetic accommodation. An example is dung beetles, where plasticity in body size may have subsequently stimulated evolution of testis size and alternative mating tactics (Simmons et al. 2007). Phenotypic plasticity may act as an evolutionary capacitor to shield genetic variation from elimination, only to release it under extreme environmental conditions (Masel 2005, Feder 2007). Price (2006) argues that phenotypic plasticity can either retard or accelerate rates of evolution, based on relative fitness of the new phenotype. If an environmentally induced plastic change has high fitness, then there should be little subsequent selection on either the plastic trait or associated traits (no genetic change), as long as the population is exposed to both environments. If the plastic change is highly detrimental, then selection should act on genes to reduce the plastic response or compensate in other ways (Nijhout 2003a, Grether 2005). If the plasticity is slightly or moderately favorable, then subsequent selection should produce genetic change that alters the reaction norm and associated traits to bring the genome to an adaptive peak.

How Phenotypic Plasticity Contributes to Speciation

A remarkable claim is that phenotypic plasticity stimulates evolution and contributes to speciation. But how can environment-induced changes to the phenotypes of individuals influence evolution? Isn't this Lamarckian, the assimilation of environmentally acquired traits into the genome? Surprisingly, phenotypic plasticity theory suggests that environment-induced changes to individuals can become absorbed into the genome (Jablonka et al. 1998, ten Cate 2000), but via traditional Medelian processes (West-Eberhard 2003, Schlichting 2004, Pigliucci et al. 2006). There are different hypothetical pathways for this to occur (e.g., Grether 2005, Rodríguez et al. 2007), but one possible pathway would be the following:

(1) *Trait origin via phenotypic plasticity* - the production of an environment-induced alteration of the phenotype. This could be passive, and be detrimental, neutral, or beneficial with regard to fitness.

(2) *Phenotypic accommodation*, whereby the individual accommodates the changed phenotype by adaptively altering additional phenotypic traits, such as physiology, behavior, or morphology (West-Eberhard 2003). Such accommodation increases survival of the new phenotype in the new environment, allowing reproduction. A hypothetical example is when the environment (such as a new food or infection by a non-lethal microorganism) induces a darker body color, which increases diurnal heating, thus limiting diurnal foraging for a

normally diurnal insect. An individual might accommodate this new phenotype by altering its foraging pattern to forage during the cooler night. Note that in this hypothetical example, an environmental factor has exposed a new trait (dark body color) that was not previously present. Expression of this novel trait (a different phenotype) places the population in a new selective regime (greater susceptibility to sunlight, nocturnal foraging, different predators, etc.).

(3) *Genetic accommodation* — Assuming population genetic variation in most traits, the recurrence of this particular environmental induction (dark color) in numerous individuals and generations would allow this novel phenotype to be tested repeatedly in the new environment and among a vast assortment of genetic variants. Over time, this would allow natural selection to select for alleles and gene combinations that improve regulation, form and side effects of the novel trait and its genetic background, increasing survival and fitness of individuals expressing the new environmentally induced traits (dark color and nocturnal foraging). Examples of genetic accommodation in this hypothetical case might be better nocturnal eyesight or longer antennae for non-visual sensing. Genetic accommodation can shift the overall fitness value of the environmentally induced phenotype, moving it from detrimental to beneficial.

(4) *Adjustment of the capacity and shape of the reaction norm via the Baldwin Effect* (see Box 3) — Here, natural selection alters the frequency of genes and gene combinations that influence the expression of the plasticity – genes that do not produce the optimal plastic response are eliminated.

(5) *Genetic assimilation* — An evolutionary mechanism by which environmentally induced (facultative) traits become genetically fixed (obligatory) (Box 2; Fig. 8). In our hypothetical case, the population would evolve to always express dark body color, under virtually all conditions and in all environments.

(6) *Speciation* — Phenotypic plasticity and some combination of steps 2-5 (above) produces differences that increase assortative mating or otherwise restrict gene flow. Continual natural selection, genetic drift, and mutation of the population increase habitat, mating, and genetic divergence from adjacent populations, leading to eventual speciation.

In the above scenario, the environment exposes a new trait, which eventually becomes genetically fixed in a population. Moreover, an environmentally induced change to the phenotype sends a population down a different evolutionary pathway, leading to speciation. Note that evolutionary divergence begins not with genetic change, but with environmentally induced change to the phenotype. Reproductive isolation does not initiate speciation. The process is Mendelian and Darwinian, because it relies on preexisting genetic variation and traditional natural selection, and may be as important as mutation for producing Earth's diversity (Schlichting and Pigliucci 1998, West-Eberhard 2003).

A real life example of how phenotypic plasticity might lead to evolutionary divergence is the famous case of sympatric diversification in *Rhagoletis pomonella* fruit flies. Adults of this New World species oviposit

Box 3 The Baldwin Effect

A proposed process by which traits acquired during an individual's lifetime influence subsequent evolution of those acquired traits.

James Baldwin was an American psychologist and evolutionary theorist who was interested in the development and evolution of cognition. He proposed that a beneficial learned behavior acquired during an individual's lifetime contributed to the fitness of that individual, and thus favored the evolution of increased capacity to acquire or perform that behavior in the population (Baldwin 1896): *"The most plastic individuals will be preserved to do the advantageous things for which their variations show them to be the most fit, and the next generation will show an emphasis of just this direction in its variations."* Here, phenotypic plasticity (a switch in behavior) of an individual during its life is a factor in evolution. Although Baldwin was primarily concerned with learning, his ideas could be applied to any number of plastic traits that are acquired by individuals in response to variable environments, such as increased muscle mass, melanization, calluses, and O_2 capacity in individuals exposed to exercise, sunlight, dermal irritation, or high altitudes, respectively. Natural selection should select for greater or lesser capacity for such plastic responses, depending on their contribution to fitness.

Baldwin's ideas are controversial (see Nortman 2003, Webber & Depew 2003, Crispo 2007), but important for a number of reasons. He was one of the first to recognize plasticity in individuals and to link such plasticity to evolution. He suggested that the expression of environmentally induced traits (the shape of the reaction norm) can evolve. His theory also approached Lamarckism in that acquired beneficial traits, induced by the environment, could (in some cases) become more genetically determined. Baldwin (1896) suggested that his process was an example of, *". . . heredity providing for the modification of its own machinery. Heredity not only leaves the future free for modifications, it also provides a method of life in operation of which modifications are bound to come."* The Baldwin Effect might explain sympatric speciation, instinct, fixation of learned songs in birds, host plant preferences in insects, and numerous other phenomena (see Webber & Depew 2003), and was recently demonstrated by Suzuki & Nijhout (2006).

into fruits, which the resulting maggots consume. After the colonization of North America by Europeans, some populations of *R. pomonella* jumped from native hawthorn fruits to introduced apple, and then to introduced cherry (Bush 1975, Prokopy et al. 1988, Feder et al. 1994), and this may have occurred via phenotypic plasticity (Chapter 18). Hawthorn fruit flies prefer hawthorn fruits. However, naïve adults that experience apple alter their phenotype (via behavioral phenotypic plasticity) to prefer apple (Prokopy et al. 1988), leading to habitat-specific mating and ovipostion (Feder et al. 1994). Recurrence in the new environment (apple or cherry), allowed natural selection to alter allele frequencies to fit the new habitat. For example, each fruit species ripens at a different time of year, and each host-population has evolved to emerge at the appropriate time (Bush 1975). The different populations also evolved different allele frequencies for thermally adapted enzymes that match the respective temperatures of their fruits (Feder et al. 1997, Filchak et al. 2000). In this case, phenotypic plasticity has apparently initiated a rapid evolutionary divergence in *R. pomonella*, which now exists as distinct races (Feder et al. 1994, Dambrowski et al. 2005). See also Bolnick and Fitzpatrick (2007).

Note that a well-established plasticity is a bridge to speciation. Indeed, a highly integrated and adapted developmental conversion is already ideally suited to stand on its own as an independent species.

Can Stress Induce Adaptive Mutations in Individuals? Is Genetic Variability a Plastic Trait?

Stress (poor nutrition, xenobiotics, radiation, extreme temperatures, etc.) can alter genes in individuals (Badyaev 2005). The best examples are the non-heritable, but adaptive mutations forming the mammalian immune response (Frost 1999). Stress also increases mutation rates in bacteria (Kidwell and Lisch 2001, Bjedov et al. 2003, Saint-Ruf and Matic 2006), often by induction of DNA mutases (Radman 1999). Although these mutations are non-directional, and mostly harmful, some are beneficial (Bjedov et al. 2003). *Escherichia coli* can switch phenotypes from high- to low-mutagenesis forms depending on environmental stress levels. This ability (hypermutation) varies among strains and is thus under genetic control (Bjedov et al. 2003). This is not restricted to bacteria. Individual flax plants mutate when nutritionally stressed, altering both their and their progeny's morphologies (Marx 1984, Cullis 1987, 1988). Hence, organisms may have evolved mechanisms to increase both genetic and phenotypic variation when exposed to harmful environments, and thus the probability of generating adaptive variants. However, even if the new mutations were not immediately adaptive, environmentally induced mutagenesis creates more raw product upon which selection can act. Moreover, the mutations occur at exactly the time when increased variation is needed – i.e., during profound environmental change. Because stress alters the phenotype, these new mutations are already imbedded in a plastic response. Presumably, both mutagenic capability/susceptibility and phenotypic sensitivity to mutation evolve and may be more common in populations or species that have evolved in highly variable environments (Bjedov et al. 2003, Meyers et al. 2005, Jones et al. 2007, Landry et al. 2007).

An example of this phenomenon in insects might be locusts, which display phenotypic plasticity in response to population density (Chapters 5, 6). The low-density phenotype is sedentary, whereas the high-density, gregarious phenotype typically migrates hundreds of kilometers to new environments. It is interesting that in the gregarious morph, recombination increases during meiosis, perhaps as an adaptation to increase genetic variability prior to dispersal into new, unknown environments (Nolte 1974).

Any number of mechanisms could increase genetic variation in individuals following environmental change (Mckenzie and Rosenberg 2001, Badyaev 2005), including stress-induced transposable elements (Kidwell and Lisch 2001), or lowered immunity, which fosters mutation-inducing viruses. Finally, genomes often contain large amounts of presumably neutral, non-functional genes. This largely unstudied and unknown genetic "dark matter" may represent a vast storehouse of potential genetic alternatives, waiting to be employed when changing environmental conditions necessitate altered phenotypes (Eshel and Matessi 1998). In general, genetic (V_G) and mutational (V_M) variability varies among environments and genotypes (Schlichting 2004, Landry et al. 2007).

Phenotypic Plasticity in Outcomes, Including Niche Construction

Although ecological and behavioral outcomes have not traditionally been considered phenotypic plastic traits (Thompson 1988), they are a direct extension of the phenotype, vary with genotype, and are greatly influenced by environmental conditions, and, hence, should be considered as plastic traits. Thus, competitive outcomes such as dominance and territory size, performance outcomes, such as mating success, fecundity, distance dispersed, and resources harvested, and survival under alternative environmental conditions can be plastic (Gorur et al. 2005, Engqvist 2008).

One outcome of the phenotype is niche construction, which refers to an organism's ability to alter the environment so as to influence or produce a niche (Day et al. 2003, Odling-Smee et al. 2003). Examples include structures produced by habitat architects, such as gallers, beavers, and nest-building social insects (Inbar et al. 2004). Habitat architects and ecosystem engineers may create new habitats, which feedback to dramatically induce phenotypic changes in those same individuals. For example, nest construction in social insects may influence any number of phenotypic traits of nest builders, such as body size, development rate, caste, fecundity, time spent in defense vs. foraging, etc. (Hölldobler and Wilson 1990). Likewise, by constructing reefs, corals alter local wave force and turbulence, temperature, oxygen and light levels, and associated biota, including predators, pathogens, and prey, and conspecific densities. Hence, there can be a continuous, reciprocal interplay between phenotype and environment. Altered environments induce phenotypic changes in individuals, and altered individuals may alter their environment in a continuous phenotypic plasticity-environment feedback loop.

Reciprocal Phenotypic Plasticity among Interacting Individuals

Individuals respond to changing environments by phenotypic plasticity, but what happens when a changing environment includes other individuals that are themselves changing? Phenotypic change in one individual may induce a change in a second individual, which induces further change in the first, in a continuous, reciprocal, phenotypic plasticity loop (Agrawal 2001, Fordyce 2006, Mooney and Agrawal 2007). Such interactions occur both within and between species, and can be mutualistic, antagonistic, or commensal. Conspecific examples are seen in certain social wasps and ants in which physical aggression between individuals determines queen vs. non-queen developmental trajectories, including differences in morphology, pheromone release, fecundity, and life history (Premnath et al. 1996, Heinze 2004). Here, an individual's phenotype changes over time in response to phenotype changes in an antagonist. Heterospecific cases include mutualistic reciprocal interactions between reward-producing herbivores or plants and their insect bodyguards (Huxley and Cutler 1991, Whitman 1994, Chapter 7). For example, *Piper cenocladum* plants produce more food bodies when bodyguard ants are present, which induces increased residency, feeding, and guarding by attendant *Pheidole bicornis* ants, which, in turn, presumably induces more food body production (Risch and Rickson 1981). Antgonistic reciprocal phenotypic plasticity can occur between predators and prey (Chapter 7). For example, some mussels increase shell thickness and muscle strength in response to crab presence, and some mussel-eating crabs produce larger claws in response to thick muscle shells (Smith and Palmer 1994, Reimer and Tedengren 1996, Leonard et al. 1999).

Reciprocal phenotypic plasticity requires the addition of a new term to the equation for phenotype variance, because now the environment includes the genes and plastic responses of one or more interacting organisms (Strauss and Irwin 2004, Wolf et al. 2004). Plotting the phenotype of one genotype vs. either density or phenotype of the interacting genotype provides a traditional reaction norm. Plotting the changing phenotypes of two interacting genotypes over time produces a plastic interaction norm (Thompson 1988, Agrawal 2001). Evolution of such plastic interaction norms could lead to runaway phenotypes, whereby each population continues to evolve ever greater capacity for phenotypic plasticity (more sensitive or extreme reaction norms) (Adler and Grünbaum 1999, Agrawal 2001). Such "plasticity coevolution" might have resulted in the phenomenal ploy-counter ploy interactions that we see among some antagonists (Chapter 7). Plasticity coevolution should be common among symbiotic species.

Transgenerational Effects on Offspring Phenotype

Parents can influence offspring phenotypes through a variety of non-genetic processes (Bernardo 1996, Mousseau and Fox 1998). Parental effects can derive from the nutritional status of the parent, or from more elaborate environmental effects generated by the habitat or the parent itself (Stelgenga and Fischer 2007). Females may adaptively vary size, quality, and diapause state of eggs (Fox and Czesak 2000, Chapters 11, 19), and allocate resources to offspring based on mate quality (Sheldon 2000). Egg size may determine plastic capacity of larvae (McAlister 2007). Parents also produce galls, nests, and habitats in which populations densities of interacting species have been radically altered. Galls, nests (Hölldobler and Wilson 1990) and altered habitats can persist for generations, serving as an ecological inheritance (Day et al. 2003). Lemon ants produce devil's gardens that can contain monocultures of 350 trees and last for 800 years (Frederickson et al. 2005). Parents can also pass culture to offspring (Sherry and Galef 1990, Grant and Grant 1996), and cultural divergence can lead to genetic divergence (ten Cate 2000, Slabbekoorn and Smith 2002).

In locusts, phase state is passed to the next generation by a water-soluble factor placed in the foam of the egg pod (Miller et al. 2008, Chapter 5). Mothers also adaptively control offspring sex and morphology (Pienaar and

Greef 2003b). In *Orthophagus* dung beetles, fathers influence son's mating strategy (Hunt and Summons 2000, Chapter 3). Large males help females to produce larger dung balls, which produce larger male offspring with horns. Only horned males fight for females; small males sneak copulations. In other insects, food- or habitat-imprinting causes epigenetic inheritance of habit preference, with numerous phenotypic plasticity consequences for offspring (Bernardo 1996, Davis and Stamps 2004). Repeated cycles of habitat-induction or imprinting in successive generations allow habitat-specific genetic adaptation (Davis and Stamps 2004).

Females also bequeath offspring with specific detrimental (O'Neill et al. 1997, Boucias and Pendland 1998) or beneficial (Baumann 2005) microorganisms, such as mutualistic symbionts, which alter host phenotype in numerous ways, such as increasing fitness under cold conditions (Dunbar et al. 2007). Mothers may choose to pass or not pass endosymbionts, depending on local conditions (e.g., Stern and Foster 1996). Changes induced by symbiotic microorganisms may drive genetic divergence (Wade 2001, Flor et al. 2007, Riegler and O'Neill 2007).

Conclusion

Phenotypic plasticity theory suggests that existing paradigms may be incomplete. It shifts our thinking, because it requires that virtually all biological fields consider individual flexibility. In developmental biology, ontogeny is no longer fixed, but is a flexible process subject to environmental input. We can not assume that differences in organisms are genetic, because the genotype does not determine a set phenotype, but a range of possible phenotypes. In systematics, species morphology is not static, but can vary in time and space. In physiology, canalization is often accomplished via underlying plasticity, and performance is conditional on present, past, and parent's environments. In ecology, phenotypic plasticity tells us that environmental interactions are much more complex and dynamic than previously imagined, and that plasticity can lead to altered community structure. Individuals can rapidly and adaptively alter their (and their offspring's) relationships with the environment, with profound fitness and ecosystem consequences. This process is reciprocal when altered environments, including interacting individuals, feedback to induce additional plasticity in individuals. In evolution, plasticity theory suggests that phenotypic plasticity generates novelty, and hence stimulates adaptation and speciation. As West-Eberhard (2003) notes, individual adaptation can lead to population adaptation, and genes are useful followers, not leaders in evolutionary change. Indeed, the developmental reaction norm may be the main object of selection on phenotypes (Dobzhansky 1951, Schlichting and Pigliucci 1998).

Phenotypic plasticity makes us realize that genes and environment are forever intimately linked. Biological existence is an iterative reciprocal process between genes, individuals, and environment. Genes provide a menu of developmental possibilities and phenotypes, but the environment determines the phenotypic outcome. The environment subsequently selects among altered individuals to alter population gene frequencies, which determine how future individuals will respond to environmental variability.

Glossary of Terms Used in Phenotypic Plasticity

(see Chapter 2, this Volume for additional discussion)

Baldwin effect – Stabilizing selection on the shape of a reaction norm.

Canalization – The operation of internal factors during development, physiology, or behavior, that reduces the influence of environmental stimuli to produce one outcome (Waddington 1940, 1942). Environmental canalization—The production of a single phenotype despite environmental variability (see Debat & David 2001).

Coevolution – Reciprocal genetic change in species engaged in an interaction.

Developmental conversion – An adaptive, discrete, and (normally) permanent phenotypic plasticity, usually with no intermediate forms. Thought to be produced via a developmental switch (Smith-Gill 1983).

Developmental plasticity – Irreversible phenotypic plasticity resulting from environmental influence on development of an individual (Piersma & Drent 2003).

Developmental switch – A threshold mechanism that produces a discrete phenotype (polyphenism), e.g., worker vs. queen determination in honeybees (Levins 1968).

Epigenetic – Development and interactions of products and processes downstream of primary gene products.

Epistasis – The effects of two or more genes on a single trait.

Genetic accommodation – A change in gene frequencies due to selection on the regulation, form, or associated effects of a novel trait (West-Eberhard 2003).

Genetic assimilation – When environmentally induced phenotypic variation becomes constitutively produced (no longer requires the environmental stimulus to be induced).

Genotype – The genes possessed by an individual, including all the genes or a specific defined subset of the total genome.

Homeostasis – Maintenance of an equilibrium state by some self-regulating capacity of an individual (see Debat & David 2001).

Lamarckian evolution – The theory that traits induced in individuals by the environment during their lifetime could be inherited by their offspring, thus causing evolution. This idea was discredited because it could not be demonstrated and lacked a mechanism by which an environmentally altered phenotypic trait could alter gametes. Nonetheless Lamarck had two very important contributions that often get lost in the discredit of "inheritance of acquired characteristics": 1) the recognition of phenotypic plasticity as an important fitness enhancing strategy within a generation, and 2) the many non-genetic parental environment effects that do influence the phenotype of offspring.

Life-cycle staging – Cyclic, reversible phenotypic plasticity in a long-lived individual in response to predictable seasonal changes, such as winter color change in fur of sub-Arctic birds and mammals.

Phenotype – The manifestation of an organism including its morphological, physiological, behavioral, and life history traits, exclusive of genetic composition, but inclusive of genetic expression.

Phenotypic accommodation – The immediate adaptive adjustments of an individual to the appearance of a new trait, without genetic change (see West-Eberhard 2003).

Phenotypic flexibility – Phenotypic plasticity that is reversible within individuals (Piersma & Drent 2003), e.g., gain or loss of muscle mass with exercise.

Phenotypic integration – The coordination in the expression of individual traits in response to environmental variation.

Phenotypic modulation – A continuous, often passive and reversible, non-adaptive phenotypic plasticity (Smith-Gill 1983).

Pleiotropy – When one gene influences multiple traits.

Polymorphism (genetic) – The existence of two or more genotypes in a population. Genetic polymorphism can lead to divergent phenotypes (see West-Eberhard 2003).

Polymorphism (phenotypic) – The existence of two or more (often discrete) morphological forms in a population, caused by either genetic polymorphism or phenotypic plasticity, but generally not ontogeny. Some authors include any variable phenotypic trait including allozymes, physiology, behavior, life history, etc. This definition of polymorphism is no longer widely used.

Polyphenism – Phenotypic variation a single genotype, due to phenotypic plasticity or development. Includes morphological, physiological, behavioral, and life history variation (see Mayr 1963, West-Eberhard 2003). Some authors (Nijhout 2003a, Piersma & Drent 2003) restrict polyphenism to discrete, adaptive, or irreversible alternative morphologies (see Chapters 2 & 13, this Volume).

Reaction norm (= norm of reaction) – The set of phenotypes expressed by a genotype when maintained under different environments. Usually illustrated as a line graph plotting phenotypes vs. environment for different genotypes. Some authors restrict reaction norms to continuous plasticities (e.g., Nijhout 2003a).

Threshold trait – A trait that exists in two or more discrete states (phenotypes), determined by a threshold level of an underlying continuously variable morphogen, such as hormone titer (Roff 1996).

References

Adler, F. R. and D. Grünbaum. 1999. Evolution of forager responses to inducible defenses. pp. 257–285 *in* R. Tollrian and C. Harvell (eds.). *The Ecology and Evolution of Inducible Defenses*. Princeton Univ. Press, Princeton, NJ.

Adler, F. R. and R. Karban. 1994. Defended fortresses or moving targets? Another model of inducible defenses inspired by military metaphors. Am. Nat. 144: 813–832.

Agrawal, A. A. 2001. Phenotypic plasticity in the interactions and evolution of species. Science 294: 321–325.

Agrawal, A. A. 2001. Transgenerational consequences of plant responses to herbivory: an adaptive maternal effect? Am. Nat. 157: 555–569.

Agrawal, A. A. 2005. Future directions in the study of induced plant responses to herbivory. Entomol. Exp. Appl. 115: 97–105.

Agrawal, A A., C. Laforsch and R. Tollrian. 1999. Transgenerational induction of defenses in animals and plants. Nature 401: 60–63.

Agrawal, B. K. 2007. Phenotypic plasticity in aphids (Homoptera: Insecta): components of variation and causative factors. Curr. Sci. 93: 308–313.

Al-Tawaha, A. M., P. Seguin, D. L. Smith and C. Beaulieu. 2005. Biotic elicitors as a means of increasing isoflavone concentration of soybean seeds. An. Appl. Biol. 146: 303–310.

Amdam, G. V., K. A. Nilsen, K. Norberg, M. K. Fondrk and K. Hartfelder. 2007. Variation in endocrine signaling underlies variation in social life history. Am. Nat. 170: 37–46.

Ananthakrishnan, T. N. 1969. Indian Thysanoptera. CSIR Zoological Monograph 1: 102–117.

Ananthakrishnan, T. N. 2005. Perspectives and dimensions of phenotypic plasticity in insects. pp. 1-23 *in* T. N. Ananthakrishnan and D. W. Whitman (eds.). *Insect Phenotypic Plasticity: Diversity of Responses*. Science Publishers, Enfield, NH.

Ananthakrishnan, T. N. and D. W. Whitman (eds.). 2005. *Insect Phenotypic Plasticity: Diversity of Responses*. Science Publishers, Enfield, NH.

Asea, A. A. and A. DeMaio (eds.). 2007. *Heat Shock Proteins: Potent Mediators of Inflammation and Immunity*. Springer, New York.

Aubret, F., X. Bonnet, R. Shine. 2007. The role of adaptive plasticity in a major evolutionary transition: early aquatic experience affects locomotor performance of terrestrial snakes. Functional ecology 21: 1154–1161.

Badyaev, A. V. 2005. Stress-induced variation in evolution: from behavioural plasticity to genetic assimilation. Proc. R. Soc. B 272: 877–886.

Baldwin, J. M. 1896. A New Factor in Evolution. Am. Nat. 30: 441–451, 536–553.

Baraboy, V. A. 1999. Melanin: the structure, biosynthesis, biological functions. Ukrainskii Biokhimicheskii Zhurnal 71: 5–14.

Baratte, S., M. Cobb, C. Peeters. 2006. Reproductive conflicts and mutilations in queenless *Diacamma* ants. Anim. Behav. 72: 305–311.

Bateson, P., D. Barker, T. Clutton-Brock, D. Deb, B. D'Udlna, R. A. Foley, P. Gluckman, K. Godfrey, T. Kirkwood, M. M. Lahr, J. McNamara, N. B. Metcalfe, P. Monaghan, H. G. Spencer and S. E. Sultan. 2004. Developmental plasticity and human health. Nature 430: 419–421.

Baumann, P. 2005. Biology of bacteriocyte-associated endosymbionts of plant sap-sucking insects. Annu. Rev. Microbiol. 59: 155–189.

Bean, D. W., T. L. Dudley and J. C. Keller. 2007. Seasonal timing of diapause induction limits the effective range of *Diorhabda elongata deserticola* (Coleoptera: Chrysomelidae) as a biological control agent for tamarisk (*Tamarisk* spp.). Environ. Entomol. 36: 15–25.

Beckerman, A., T. G. Benton, E. Ranta, V. Kaitala and P. Lundberg. 2002. Population dynamic consequences of delayed life-history effects. Trends Ecol. Evol. 17: 263–269.

Benard, M. F. 2004. Predator-induced phenotypic plasticity in organisms with complex life histories. Annu. Rev. Ecol. Evol. Syst. 35: 651–673.

Bernardo, J. 1996. Maternal effects in animal ecology. Am. Zool. 36: 83–105.

Berrigan, D. and S. M. Scheiner. 2004. Modeling the evolution of phenotypic plasticity. pp. 82-97 *in* T. J. DeWitt and S. M. Scheiner (eds.). *Phenotypic Plasticity*. Oxford Univ. Press, Oxford.

Bjedov, I., O. Tenaillon, B. Gérard, V. Souza, E. Denamur, M. Radman, F. Taddel and I. Matic. 2003. Stress-induced mutagenesis in bacteria. Science 300: 1404–1409.

Boege, K. and R. J. Marquis. 2005. Facing herbivory as you grow up: the ontogeny of resistance in plants. Trends Ecol. Evol. 20: 441–448.

Bolnick, D. I. and B. M. Fitzpatrick. 2007. Sympatric speciation: models and empirical evidence. Annu. Rev. Ecol. Evol. Syst. 38: 459–487.

Boucias, D. G. and J. C. Pendland. 1998. *Principles of Insect Pathology*. Klewer, Boston.

Bradshaw, A. D. 1965. Evolutionary significance of phenotypic plasticity in plants. Advances in Genetics. 13: 115–155.

Bradshaw, W. E. and C. M. Holzapfel. 2001. Genetic shift in photoperiodic response correlated with global warming. Proc. Nat. Acad. Sci. 98: 14509–14511.

Bradshaw, W. E. and C. M. Holzapfel. 2006. Evolutionary response to rapid climate change. Science 312: 1477–1478.

Bradshaw, W. E. and C. M. Holzapfel. 2007. Evolution of animal photoperiodism. Annu. Rev. Ecol. Evol. Syst. 38: 1–25.

Brakefield, P. M. and V. French. 2006. Evo-devo focus issue. Heredity 97: 137–138.

Brodie III, E. D. and A. F. Agawal. 2001. Maternal effects and the evolution of aposematic signals. Proc. Natl. Acad. Sci. USA. 98: 7884–7887.

Bush, G. L. 1975. Sympatric speciation in phytophagous parasitic insects. pp. 187-206 *in* P. Price (ed.). *Evolutionary Strategies of Parasitic Insects and Mites*. Plenum, New York.

Calderwood, S. K., M. Y. Sherman and D. R. Ciocca (eds.). 2007. *Heat Shock Proteins in Cancer*. Springer, New York.

Carpenter, G. H. 1899. *Insects their Structure & Life*. J. M. Dent & Co., London.

ten Cate, C. 2000. How learning mechanisms might affect evolutionary processes. Trends Evol. Ecol. 15: 179–181.

Chapman, R. F. and J. C. Lee. 1991. Environmental effects on numbers of peripheral chemoreceptors on the antennae of a grasshopper. Chem. Senses 16: 607–616.

Collard, M. and S. J. Lycett. 2008. Does phenotypic plasticity confound attempts to identify hominin fossil species? Folia Primatologia 79: 111-122.

Collier, R. J., G. E. Dahl and M. J. VanBaale. 2006. Major advances associated with environmental effects on dairy cattle. J. Dairy Sci. 89: 1244–1253.

Cook, J. M., S. G. Compton, E. A. Herre and S. A. West. 1997. Alternative mating tactics and extreme male dimorphism in fig wasps. Proc. R. Soc. Lond. B 264: 747–754.

Cossins, A., J. Fraser, M. Hughes and A. Gracey. 2006. Post-genomic approaches to understanding the mechanisms of environmentally induced phenotypic plasticity. J. Exp. Biol. 209: 2328–2336.

Crispo, E. 2007. The Baldwin effect and genetic assimilation: revisiting two mechanisms of evolutionary change mediated by phenotypic plasticity. Evolution 61: 2469–2479.

Cronk, Q. C. B. 2005. Plant eco-devo: the potential of popular as a model organism. New Phytol. 166: 39–48.

Cullis, C. A. 1987. The generation of somatic and heritable variation in response to stress. Am. Nat. 130 (supplement): S62–S73.

Cullis, C. A. 1988. Control of variation in higher plants. pp. 49-62 *in* M-W. Ho and S. Fox (eds.). *Evolutionary Processes and Metaphors*. Wiley, Chichester, NY.

Czesak, M. E., C. W. Fox and J. B. Wolf. 2006. Experimental evolution of phenotypic plasticity: how predictive are cross-environment genetic correlations? Am. Nat. 168: 323–335.

Dambroski, H. R., C. Linn, S. H. Berlocher, A. A. Forbes, W. Roelofs and J. L. Feder. 2005. The genetic basis for fruit odor discrimination in *Rhagoletis* flies and its significance for sympatric host shifts. Evolution 59: 1953–1964.

Dana, M. D., J. A. Pintor-Toro and B. Cubero. 2006. Transgenic tobacco plants overexpressing chitinases of fungal origin show enhanced resistance to biotic and abiotic stress agents. Plant Physiol. 142: 722–730.

David, J. R., P. Gibert and B. Moreteau. 2004. Evolution of reaction norms. pp. 50-63 *in* T. J. DeWitt and S. M. Scheiner (eds.). *Phenotypic Plasticity: Functional and Conceptual Approaches*. Oxford University Press, Oxford.

Davis, J. M. and J. A. Stamps. 2004. The effect of natal experience on habitat preferences. Trends Ecol. Evol. 19: 411–416.

Day, R. L., K. N. Laland and J. Odling-Smee. 2003. Rethinking adaptation: the niche-construction perspective. Perspectives in Biology and Medicine 46: 80–95.

DeAngelis, D. L and W. M. Mooij. 2005. Individual-based modeling of ecological and evolutionary processes. Annu. Rev. Ecol. Evol. Syst. 36: 147–168.

Debat, V. and P. David. 2001. Mapping phenotypes: canalization, plasticity and developmental stability. Trends Ecol. Evol. 16: 555–561.

De Coen, W., J. Robbens and C. Janssen. 2006. Ecological impact assessment of metallurgic effluents using in situ biomarker assays. Environmental Pollution. 141: 283–294.

Defranco, A. L., R. M. Locksley and M. Robertson. 2007. *Immunity*. New Science Press, London.

De Moraes, C. M., J. C. Schultz, M. C. Mescher and J. H. Tumlinson. 2004. Induced plant signaling and its implications for environmental sensing. J. Toxicol. Environ. Health 67: 819–834.

DeWitt, T. J. and R. B. Langerhans. 2004. Integrated solutions to environmental heterogeneity. pp. 98-111 *in* T. J. DeWitt and S. M. Scheiner (eds.). *Phenotypic Plasticity: Functional and Conceptual Approaches.* Oxford University Press, Oxford.

DeWitt, T. J. and S. M. Scheiner (eds.). 2004. *Phenotypic Plasticity: Functional and Conceptual Approaches*. Oxford Univ. Press, Oxford.

DeWitt, T. J. and S. M. Scheiner. 2004. Phenotypic variation from single genotypes: a primer. pp. 1-9 *in* T. J. DeWitt and S. M. Scheiner (eds.). *Phenotypic plasticity: Functional and Conceptual Approaches*. Oxford University Press. Oxford.

DeWitt, T. J., A. Sih and D. S. Wilson. 1998. Costs and limits of phenotypic plasticity. Trends Ecol. Evol. 13: 77–81.

Dobzansky, T. 1951. *Genetics and the Origin of Species.* 3rd edition. Columbia Univ. Press, New York.

Doughty, P. and D. N. Reznick. 2004. Patterns and analysis of adaptive phenotypic plasticity in animals. pp. 126-150 *in* T. J. DeWitt and S. M. Scheiner (eds.). *Phenotypic Plasticity: Functional and Conceptual Approaches.* Oxford Univ. Press, Oxford.

Dukas, R. 2004. Evolutionary biology of animal cognition. Annu. Rev. Ecol. Syst. 35: 347–374.

Dunbar, H. E., A. C. C. Wilson, N. R. Ferguson and N. A. Moran. 2007. Aphid thermal tolerance is governed by a point mutation in bacterial symbionts. PLoS Biol. 5: 1006–1015.

Edreva, A. 2004. A novel strategy for plant protection: induced resistance. Cell and Molecular Biology 3: 61–69

Ellers, J., J. Marien, G. Driessen and N. M. Van Straalen. 2008. Temperature-induced gene expression associated with different thermal norms for growth rate. J. Exp. Zool. B. 310: 137-147.

Ellison, A. M. and N. J. Gotelli. 2002. Nitrogen availability alters the expression of carnivory in the northern pitcher plant, *Sarracenia purpurea*. Proc. Natl. Acad. Sci. USA. 99: 4409–4412.

Emlen, D. J. and C. E. Allen. 2004. Genotype to phenotype: Physiological control of trait size and scaling in insects. Integr. Comp. Biol. 43: 617–634.

Emlen, D. J. and H. F. Nijhout. 2000. The development and evolution of exaggerated morphologies in insects. Annu. Rev. Entomol. 45: 661–708.

Emlen, D. J., Q. Szafran, L. S. Corley and I. Dworkin. 2006. Insulin signaling and limb-patterning: candidate pathways for the origin and evolutionary diversification of beetle 'horns.' Heredity 97: 179–191.

Emlen, D. J., L. C. Lavine and B. Ewen-Campen. 2007. On the origin and evolutionary diversification of beetle horns. Proc. Natl. Acad. Sci. USA 104: 8661–8668.

Engqvist, L. 2008. Genetic variance and genotype reaction norms in response to larval food manipulation for a trait important in scorpionfly sperm competition. Functional Ecology 22: 127–133.

Eshel, I. and C. Matessi. 1998. Canalization, genetic assimilation and preadaptation: a quantitative genetic model. Genetics 149: 2119–2133.

Evans, H. E. and M. J. West-Eberhard. 1970. *The Wasps*. University of Michigan Press, Ann Arbor.

Evans, J. D. and D. E. Wheeler. 1999. Differential gene expression between developing queens and workers in the honey bee, *Apis mellifera*. Proc. Natl. Acad. Sci. USA. 96: 5575–5580.

Feder, J. L., S. B. Opp, B. Wlazlo, K. Reynolds, W. Go and S. Spisak. 1994. Host fidelity is an effective premating barrier between sympatric races of the apple maggot fly, *Rhagoletis pomonella*. Proc. Nat. Acad. Sci. USA. 91: 7990–7994.

Feder, J. L., J. B. Roethele, B. Wlazlo and S. H. Berlocher. 1997. Selective maintenance of allozyme differences among sympatric host races of the apple maggot fly. Proc. Nat. Acad. Sci. USA. 94: 11417–11421.

Feder, M. E. 2007. Evolvability of physiological and biochemical traits: evolutionary mechanisms including and beyond single-nucleotide mutation. J. Exp. Biol. 210: 1653–1660.

Filchak, K. E., J. B. Roethele and J. L. Feder. 2000. Natural selection and sympatric divergence in the apple maggot *Rhagoletis pomonella*. Nature 407: 739–742.

Flor, M., P. Hammerstein and A. Telschow. 2007. *Wolbachia*-induced unidirectional cytoplasmic incompatibility and the stability of infection polymorphism in parapatric host populations. J. Evol. Biol. 20: 696–706.

Flück, M. 2006. Functional, structural and molecular plasticity of mammalian skeletal muscle in response to exercise stimuli. J. Exp. Biol. 209: 2239–2248.

Flynn, J. R. 1987. Massive IQ gains in 14 nations: what IQ tests really measure. Psychol. Bull. 101: 171–1191.

Fordyce, J. A. 2001. The lethal plant defense paradox remains: inducible host-plant aristolochic acids and the growth and defense of the pipevine swallowtail. Entomol. Exp. Appl. 100: 339–346.

Fordyce, J. A. 2006. The evolutionary consequences of ecological interactions mediated through phenotypic plasticity. J. Exp. Biol. 209: 2377–2383.

Fox, C. W. and M. E. Czesak. 2000. Evolutionary ecology of progeny size in arthropods. Annu. Rev. Entomol. 45: 341–369.

Franken, R. J., J. J. Gardeniers, J. A. Beijer and E. T. Peeters. 2008. Variation in stonefly (*Nemoura cinerea* Retzius) growth and development in response to hydraulic and substrate conditions. J. N. Am. Benthol. Soc. 27: 176-185.

Frankino, W. A. and R. A. Raff. 2004. Evolutionary importance and pattern of phenotypic plasticity. pp. 64-97 *in* T. J. DeWitt and S. M. Scheiner (eds.). *Phenotypic plasticity: Functional and Conceptual Approaches*. Oxford University Press. Oxford.

Fréchette, B., A. F. G. Dixon, C. Alauzet, and J-L. Hemptinne. 2004. Age and experience influence patch assessment for oviposition by an insect predator. Ecol. Entomol. 29: 578–583.

Frederickson, M. E., M. J. Greene and D. Gordon. 2005. 'Devil's gardens' bedevilled by ants. Nature 437: 495–496.

Freeman, S. and J. C. Herron. 2007. *Evolutionary Analysis*. Prentice Hall, Upper Saddle River, NJ.

Frost, S. D. W. 1999. The immune system as an inducible defense. pp. 104-126 *in* R. Tollrian and C. D. Harvell (eds.). *The Ecology and Evolution of Inducible Defenses*. Princeton University Press, Princeton, NJ.

Futuyma, D. J. 1998. *Evolutionary Biology*. Sinauer, Sunderland, MA.

Gabriel, W. 2005. How stress selects for reversible phenotypic plasticity. J. Evol. Biol 18: 873–883.

Gabriel, W. and M. Lynch. 1992. The selective advantage of reaction norms for environment tolerance. J. Evol. Biol. 5: 41–59.

Gabriel, W., B. Luttbeg, A. Sih and R. Tollrian. 2005. Environmental tolerance, heterogeneity, and the evolution of reversible plastic responses. Am. Nat. 166: 339–353.

Galloway, L. F. 2005. Maternal effects provide phenotypic adaption to local environmental conditions. New Phytol. 166: 93–100.

Gardner, S. N. and A. A. Agrawal. 2002. Induced plant defence and the evolution of counter-defences in herbivores. Evol. Ecol. Res. 4: 1131–1151.

Garland, T. Jr. and S. A. Kelly. 2006. Phenotypic plasticity and experimental evolution. J. Exp. Biol. 209: 2344–2361.

Geurts C. L. M., G. G. Sleivert and S. S. Cheung. 2006. Local cold acclimation during exercise and its effect on neuromuscular function of the hand. Physiologie Appliquee, Nutrition et Metabolisme. 31: 717–725.

Goldschmidt, R. 1935. Gen und Ausseneigenschaft, Z. Induktive Abstammungs Verbunglehre 69: 38–131.

Gorur, G., C. Lomonaco and A. Mackenzie. 2005. Phenotypic plasticity in host-plant specialization in *Aphis fabae*. Ecol. Entomol. 30: 657–664.

Grant, B. R. and P. R. Grant. 1996. Cultural inheritance of song and its role in the evolution of Darwin's finches. Evolution 50: 2471–2487.

Greene, E. 1989. A diet-induced developmental polymorphism in a caterpillar. Science 243: 643–646.

Grether, G. F. 2005. Environmental change, phenotypic plasticity, and genetic compensation. Am. Nat. 166: E115–E123.

Haga, K. and S. Okajima. 1975. Redescription and status of the genus *Phoxothrips* Karny (Thysanoptera). Kontyu. Tokyo. 42: 375–384.

Hairston, N. G. Jr., C. L. Holtmeier, W. Lampert, L. J. Weider, D. M. Post, J. M. Fischer, C. E. Cáceres, J. A. Fox, and U. Gaedke. 2001. Natural selection for grazer resistance to toxic cyanobacteria. Evolution 55: 2203–2214.

Haukioja, E. 1990. On the role of plant defences in the fluctuations of herbivore populations. Oikos 35: 202–213.

Hayashi, Y., N. Lo, H. Miyata and O. Kitade. 2007. Sex-linked genetic influence on caste determination in a termite. Science 318: 985–987.

Heil, M., S. Greiner, H. Meimberg, R. Krüger, J-L. Noyer, G. Heubl, K. E. Linsenmair and W. Boland. 2004. Evolutionary change from induced to constitutive expression of an indirect plant resistance. Nature 430: 205–208.

Heinrich, B. 1993. The Hot-Blooded Insects. Harvard University Press, Cambridge.

Heinze, J. 2004. Reproductive conflict in insect societies. Advances in the Study of Behavior. 34: 1–57.

Hendry, A. P., T. J. Farrugia, M. T. Kinnison. 2008. Human influences on rates of phenotypic change in wild animal populations. Molecular Ecology 17: 20–29.

Herrnstein, R. J. and C. Murray. 1994. The Bell Curve. Free Press Paperbacks, New York.

Hille Ris Lambers, D. 1966. Polymorphism in Aphididae. Annu. Rev. Entomol. 11: 47–78.

Hojo, M., S. Koshikawa, T. Matsumoto and T. Miura. 2004. Developmental pathways and plasticity of neuter castes in *Nasutitermes takasagoensis* (Isoptera: Termitidae). Sociobiology 44: 433–441.

Hölldobler, B. and E. O. Wilson. 1990. *The Ants*. Harvard University Press, Cambridge, MA.

Huey, R. B., P. E. Hertz and B. Sinervo. 2003. Behavioral drive versus behavioral inertia in evolution: a null model approach. Am. Nat. 161: 357–366.

Hunt, J. H. 2007. *The Evolution of Social Wasps*. Oxford University Press, New York.

Hunt, J. and L. W. Simmons. 2000. Maternal and paternal effects on offspring phenotype in the dung beetle *Onthophagus taurus*. Evolution 54: 936–941.

Hunt, J. H., B. J. Kensinger, J. A. Kossuth, M. T. Henshaw, K. Norberg, F. Wolschin and G. V. Amdam. 2007. A diapause pathway underlies the gyne phenotype in *Polistes* wasps, revealing an evolutionary route to caste-containing insect societies. Proc. Nat. Acad. Sci. 104: 14020–14025.

Hurd, H. and R. Lane (eds.). 1998. Parasite-insect interactions: reciprocal manipulation. Parasitology 116 (Supplement): 1–128.

Huxley, C. R. and D. F. Cutler. 1991. *Ant-Plant Interactions*. Oxford University Press, Offord.

Inbar, M., M. Wink and D. Wool. 2004. The evolution of host plant manipulation by insects: molecular and ecological evidence from gall-forming aphids on *Pistacia*. Mol. Phylogenet. Evol. 32: 504–511.

Ito, S. 2003. A chemist's view of melanogenesis. Pigment Cell Research 16: 230–236.

Jablonka, E., M. Lamb and E. Avital. 1998.'Lamarckian' mechanisms in Darwinian evolution. Trends Ecol. Evol. 13: 206–210.

Jablonka, E. and M. J. Lamb. 2005. Evolution in Four Dimensions: Genetic, Epigenetic, Behavioral, and Symbolic Variation in the History of Life. MIT Press, Cambridge, MA.

Järemo J., J. Tuomi and P. Nilsson. 1999. Adaptive status of localized and systemic defense response in plants. pp. 33-44 in Tollrian R and Harvell CD (eds.). The Ecology and Evolution of Inducible Defenses. Princeton University Press, Princeton, NJ.

Johnson, L. K. 1982. Sexual selection in a brentid weevil. Evolution 36: 251–262.

Jones, A. G., S. J. Arnold and R. Bürger. 2007. The mutation matrix and the evolution of evolvability. Evolution 61: 727–745.

de Jong, G. 2005. Evolution of phenotypic plasticity: patterns of plasticity and the emergence of ecotypes. New Phytol. 166: 101–118.

de Jong, G. and N. Behera. 2002. The influence of life-history differences on the evolution of reaction norms. Evol. Ecol. Res. 4: 1–25.

de Jong G. and P. Bijma. 2002. Selection and phenotypic plasticity in evolutionary biology and animal breeding. Livest. Prod. Sci. 78: 195–214.

de Jeong, G. T. and D. H. Park. 2005. Enhancement of growth and secondary metabolite biosynthesis: effect of elicitors derived from plants and insects. Biotechnol. Bioprocess Eng. 10: 73–77.

Kappers, I. F., A. Aharoni, T. W. J. M. van Herpen, L. L. P. Luckerhoff, M. Dicke and H. J. Bouwmeester. 2005. Genetic engineering of terpenoid metabolism attracts bodyguards to *Arabidopsis*. Science 309: 2070–2072, 2067.

Karban, R. and K. Nagasaka. 2004. Are defenses of wild radish populations well matched with variability and predictability of herbivory? Evol. Ecol. 18: 283–301.

Karban, R., A. A. Agrawal, J. S. Thaler and L. S. Adler. 1999. Induced plant responses and information content about risk of herbivory. Trends Ecol. Evol. 14: 443–447.

Karban R., M. Huntzinger and A. C. McCall. 2004. The specificity of eavesdropping on sagebrush by other plants. Ecology 85: 1846–1852.

Kenyon, S. G. and M. S. Hunter. 2007. Manipulation of oviposition choice of the parasitoid wasp, *Encarsia pergandiella*, by the endosymbiotic bacterium *Cardinium*. J. Evol. Biol. 20: 707–716.

Key, K. H. L. and M. F. Day. 1954a. A temperature-controlled physiological colour response in the grasshopper, *Kosciuscola tristis* Sjöst. (Orthoptera, Acrididae). Australian J. Zool. 2: 309–339.

Key, K. H. L. and M. F. Day.1954b. The physiological mechanism of colour change in the grasshopper, *Kosciuscola tristis* Sjöst. (Orthoptera, Acrididae). Australian J. Zool. 2: 340–363.

Kidwell, M. G. and D. R. Lisch. 2001. Perspective: transposable elements, parasitic DNA, and genome evolution. Evolution 55: 1–24.

Kirschner, M. W. and J. C. Gerhart. 2005. *The Plausibility of Life: Resolving Darwin's Dilemma*. Yale University Press, New Haven.

Klausnitzer, B. 1987. *Insects their Biology and Cultural History*. Universe Books/Edition Leipzig, NY.

van Kleunen, M. and M. Fischer. 2005. Constraints on the evolution of adaptive phenotypic plasticity in plants. New Phytol. 166: 49–60.

Kliebenstein, D. J., A. Figuth and T. Mitchell-Olds. 2002. Genetic architecture of plastic methyl jasmonate responses in *Arabidopsis thaliana*. Genetics 161: 1685–1696.

de Kroon, H., H. Huber, J. Stuefer and J. M. van Groenendael. 2005. A modular concept of phenotypic plasticity in plants. New Phytol. 166: 73–82.

Landry, C. R., B. Lemos, S. A. Rifkin, W. J. Dickinson and D. L. Hartl. 2007. Genetic properties influencing the evolvability of gene expression. Science 317: 118–121.

Lee, S. M., S. B. Lee, C. H. Park and J. Choi. 2006. Expression of heat shock protein and hemoglobin genes in *Chrionomus tentans* (Diptera, Chironomidae) larvae exposed to various environmental pollutants: A potential biomarker of freshwater monitoring. Chemosphere. 65: 1074–1081.

Leonard, G. H., M. D. Bertness and P. O. Yund. 1999. Crab predation, waterborne cues, and inducible defenses in the blue mussel, *Mytilus edulis*. Ecology 80: 1–14.

Lessells, C. M. 2008. Neuroendocrine control of life histories: what do we need to know to understand the evolution of phenotypic plasticity? Phil. Trans. R. Soc. Lond. B 363: 1589-1598.

Levins, R. 1968. Evolution in Changing Environments. Princeton University Press, Princeton.

Li, W. W. and W. Barz. 2005. Biotechnological production of two new 8,4′-oxynorneolignans by elicitation of *Echinacea purpurea* cell cultures. Tetrahedron Lett. 46: 2973–2977.

Li, X., M. A. Schuler and M. R. Berenbaum. 2002. Jasmonate and salicylate induce expression of herbivore cytochrome P450 genes. Nature 419: 712–715.

Lindström, L., R. V. Alatalo, A. Lyytinen and J. Mappes. 2001. Strong antiapostatic selection against novel rare aposematic prey. Proc. Natl. Acad. Sci. USA 98: 9181–9184.

Luczynski A, J. P. Nyrop and A. Shi. 2007. Influence of cold storage on pupal development and mortality during storage and on post-storage performance of *Encarsia formosa* and *Eretmocerus eremicus* (Hymenoptera : Aphelinidae). Biological Control. 40: 107–117.

Lyytinen, A., P. M. Brakefield, L. Lindström and J. Mappes. 2004. Does predation maintain eyespot plasticity in *Bicyclus anynana*? Proc. R. Soc. Lond. B 271: 279–283.

Marinovic, B. and M. Mangel. 1999. Krill can shrink as an ecological adaptation to temporarily unfavourable environments. Ecol. Lett. 2: 338–343.

Marx, J. L. 1984. Instability in plants and the ghost of Lamarck. Science 224: 1415–1416.

Masel, J. 2005. Evolutionary capacitance may be favored by natural selection. Genetics 170: 1359–1371.

May, M. L. 1984. Thermoregulation. pp. 507-552 *in* G. Kerkut and L. Gilbert (eds.). *Comprehensive Insect Physiology Biochemistry and Pharmacology*, Vo. 4. Pergamon Press, Oxford.

Mayr, E. 1963. *Animal Species and Evolution*. Belknap Press, Cambridge, MA.

McAlister, J. S. 2007. Egg size and the evolution of phenotypic plasticity in larvae of the echinoid genus *Strongylocentrotus*. Exp. Marine Biol. Ecol. 352: 306–316.

McCaffery, A. R., S. J. Simpson, M. S. Islam, P. Roessingh. 1998. A gregarizing factor present in the egg pod foam of the desert locust *Schistocerca gregaria*. J. Exp. Biol. 201: 347–363.

McKenzie, G. J. and S. M. Rosenberg. 2001. Adaptive mutations, mutator DNA polymerases and genetic change strategies of pathogens. Current Opinion in Microbiology 4: 586–594.

McLeod, L. 2007. Further investigations of the effect of low temperatures on the phenotype of the adults of *Precis octavia* (Cramer) (Lepidoptera: Nymphalidae). Metamorphosis 18: 48–55.

Meiklejohn, C. D. and D. L. Hartl. 2002. A single mode of canalization. Trends Ecol. Evol. 17: 468–473.

Meiners, T. and M. Hilker. 2000. Induction of plant synomones by oviposition of a phytophagous insect. J. Chem. Ecol. 26: 221–232.

Meyers, L. A. and J. J. Bull. 2002. Fighting change with change: adaptive variation in an uncertain world. Trends Ecol. Evo. 17: 551–557.

Meyers, L. A., F. D. Ancel and M. Lachmann. 2005. Evolution of genetic potential. PLoS Comput. Biol. 1: 236–243.

Miller, G. A., M. S. Islam, T. D. Claridge, T. Dodgson and S. J. Simpson. 2008. Swarm formation in the desert locust *Schistocerca gregaria*: isolation and NMR analysis of the primary maternal gregarizing agent. J. Exp. Biol. 211: 370-376.

Miner, B. G., S. E. Sultan, S. G. Morgan, D. K. Padilla and R. A. Relyea. 2005. Ecological consequences of phenotypic plasticity. Trends Ecol. Evol. 20: 685–692.

Minks, A. K. and P. Harrewijn. 1987. *Aphids their Biology, Natural Enemies and Control*. Elsevier, Amsterdam.

Mirth, C. K. and L. M. Riddiford. 2007. Size assessment and growth control: how adult size is determined in insects. BioEssays 29: 344–355.

Miura, T. 2004. Proximate mechanisms and the evolution of caste polyphenism in social insects: From sociality to genes. Ecol. Res. 19: 141–148.

Miura, T. 2005. Developmental regulation of caste-specific characters in social-insect polyphenism. Evolution & development 7: 122–129.

Miura, T. and T. Matsumoto. 1995. Worker polymorphism and division of labor in the foraging behavior of the black marching termite *Hospitalitermes medioflavus*, on Borneo Island. Naturwissenschaften 82: 564–567.

Miura, T. and T. Matsumoto. 2000. Soldier morphogenesis in a nasute termite: discovery of a disk-like structure forming a soldier nasus. Proc. R. Soc. Lond. B 267: 1185–1189.

Miura, T., Y. Roisin and T. Matsumoto. 1998. Developmental pathways and polyethism of neuter castes in the processional nasute termite *Hospitalitermes medioflavus* (Isoptera: Termitidae). Zoo. Sci. 15: 843–848.

Moczek, A. P. and F. H. Nijhout. 2003. Rapid evolution of a polyphenic threshold. Evol. Dev. 5: 259–268.

Mondor, E. B., J. A. Rosenheim and J. F. Addicott. 2005. Predator-induced transgenerational phenotypic plasticity in the cotton aphid. Oecologia 142: 104–108.

Mooney, K. A. and A. A. Agrawal. 2007. Phenotypic plasticity. pp. 43-57 *in* K. J. Tilmon (ed.). *The evolutionary biology of herbivorous insects: Specialization, speciation, and radiation*. University of California Press, Berkeley, CA.

Moran, N. A. 1992. The evolutionary maintenance of alternative phenotypes. Am. Nat. 139: 971–989.

Mound, L. A. 2005. Fighting, flight and fecundity: Behavioural determinants of Thysanoptera structural diversity. pp. 81-105 *in* T. N. Ananthakrishnan and D. W. Whitman (eds.). *Insect Phenotypic Plasticity*. Science Publishers, Enfield, NH, USA.

Mousseau, T. A. and H. Dingle. 1991. Maternal effects in insect life histories. Annu. Rev. Entomol. 36: 511–534.

Mousseau, T. A. and C. W. Fox (eds.). 1998. *Maternal Effects as Adaptation*. Oxford University Press, Oxford.

Munday, P. L., P. M. Buston and R. R. Warner. 2006. Diversity and flexibility of sex-change strategies in animals. Trends Ecol. Evol. 21: 89–95.

Muth, N. Z. and M. Pigliucci. 2007. Implementation of a novel framework for assessing species plasticity in biological invasions: responses of *Centaurea* and *Crepis* to phosphorus and water availability. Ecology 95: 1001–1013.

Nadeau, J. H. and E. J. Topol. 2006. The genetics of health. Nat. Genet. 38: 1095–1098.

Nakamura, M. and T. Ohgushi. 2003. Positive and negative effects of leaf shelters on herbivorous insects: linking multiple herbivore species on a willow. Oecologia 136: 445–449.

Neufeld, C. J. and A. R. Palmer. 2008. Precisely proportioned: intertidal barnacles alter penis form to suit costal wave action. Phil. Trans. R. Soc. Lond. B. 275: 1081-1087.

Nijhout, H. F. 1991. *The Development and Evolution of Butterfly Wing Patterns*. Smithsonian Institution Press, Washington, D.C.

Nijhout, H. F. 1994. *Insect Hormones*. Princeton Univ. Press, Princeton.

Nijhout, H. F. 1999. Control mechanisms of polyphenic development in insects. BioScience 49: 181–192.

Nijhout, H. F. 2003a. The development and evolution of adaptive polyphenisms. Evolution & Development 5: 9–18.

Nijhout, H. F. 2003b. The control of body size in insects. Developmental Biology 261: 1–9.

Nolte, D. J. 1974. The gregarization of locusts. Biol. Rev. Cambridge Philos. Soc. 49: 1–14.

Nortman, D. 2003. The evolution of phenotypic plasticity through the Baldwin Effect. Noesis 6: 1–13.

Nunney, L. and W. Cheung. 1997. The effect of temperature on body size and fecundity in female *Drosophila melanogaster*: evidence for adaptive plasticity. Evolution 51: 1529–1535.

Nykänen, H. and J. Koricheva. 2004. Damage-induced changes in woody plants and their effects on insect herbivore performance: a meta-analysis. Oikos 104: 247–268.

Nylin, S. and K. Gotthard. 1998. Plasticity in life-history traits. Annu. Rev. Entomol. 43: 63–93.

Odling-Smee, F. J., K. N. Laland and M. W. Feldman. 2003. Niche construction: the neglected process in evolution. Monographs in Population Biology 37. Princeton University Press, Princeton.

O'Donnell, S. 1998. Reproductive caste determination in eusocial wasps (Hymenoptera: Vespidae). Annu. Rev. Entomol. 43: 323–346.

Ohgushi T. 2005. Indirect interaction webs: herbivore-induced effects through trait change in plants. Annu. Rev. Ecol. Evol. Syst. 36: 81–105.

Ohgushi, T., T. P. Craig and P. Price (eds.). 2007. *Ecological Communities: Plant Mediation in Indirect Interaction Webs*. Cambridge University Press, Cambridge.

Oishi, M. and H. Sato. 2007. Inhibition of premature leaf abscission by a leafminer and its adaptive significance. Environ. Entomol. 36: 1504–1511.

O'Neill, S. L., A. A. Hoffmann and J. H. Werren (eds.). 1997. Influential Passengers: Inherited Microorganisms and Arthropod Reproduction. Oxford University Press, New York.

Oster, G. F. and E. O. Wilson. 1978. *Caste and Ecology in the Social Insects*. Princeton University Press, New Jersey.

Otaki, J. M. 2007. Stress-induced color-pattern modifications and the evolution of the painted lady butterflies *Vanessa cardui* and *Vanessa kershawi*. Zoological Science 24: 811–819.

Otaki, J. M. 2008. Phenotypic plasticity of wing color patterns revealed by temperature and chemical applications in a nymphalid butterfly *Vanessa indica*. 2008. J. Thermal Biol. 33: 128-139.

Otte, D. and K. Stayman. 1979. Beetle horns: some patterns in functional morphology. pp. 159-292 *in* M. Blum and N. Blum (eds.). *Sexual Selection and Reproductive Competition in Insects*. Academic Press, New York.

Padilla, D. K. and S. C. Adolph. 1996. Plastic inducible morphologies are not always adaptive: the importance of time delays in a stochastic environment. Evol. Ecol. 10: 105–117.

Page, R. E. Jr. and G. V. Amdam. 2007. The making of a social insect: developmental architectures of social design. BioEssays 29: 334–343.

Palmer, J. M. and L. A. Mound. 1978. Nine genera of fungus-feeding *Phlaeothripidae* (Thysanoptera) from the Oriental Region. Bull. British Mus. (Nat. History). Ent. 37: 153–215.

Parsons, K. J. and B. W. Robinson. 2006. Replicated evolution of integrated plastic responses during early adaptive divergence. Evolution 60: 801–813.

Peacor, S. D., S. Allesina, R. L. Riolo and M. Pascual. 2006. Phenotypic plasticity opposes species invasions by altering fitness surface. PLoS Biology 4(11) e372: 1–21.

Peeters, C. and S. Higashi. 1989. Reproductive dominance controlled by mutualition in the queenless ant *Diacamma australe*. Naturwissenschaften 76: 177–180.

Pereira R, N. Silva, C. Quintal, R. Abreu, J. Andrade and L. Dantas. 2007. Effect of acclimation to outdoor conditions on the sexual performance of mass-produced medflies (Diptera : Tephritidae). Florida Entomologist. 90: 171–174.

Pfenning, K. S. 2007. Facultative mate choice drives adaptive hybridization. Science 318: 965–967.

Pienaar, J. and J. M. Greeff. 2003a. Different male morphs of *Otitesella pseudoserrata* fig wasps have equal fitness but are not determined by different alleles. Ecology Letters 6: 286–289.

Pienaar, J. and J. M. Greeff. 2003b. Maternal control of offspring sex and male morphology in the *Otitesella* fig wasps. J. Evol. Biol. 16: 244–253.

Piersma, T. and J. Drent. 2003. Phenotypic flexibility and the evolution of organismal design. Trends in Ecology and Evolution 18: 228–233.

Pigliucci, M. 2001. Phenotypic Plasticity: Beyond Nature and Nurture. Johns Hopkins University Press, Baltimore, MD.

Pigliucci, M. 2005. Evolution of phenotypic plasticity: where are we going now? Trends Ecol. Evol. 20: 481–486.

Pigliucci, M. 2007. Do we need an extended evolutionary synthesis? Evolution 61: 2743–2749.

Pigliucci, M. and C. J. Murren. 2003. Genetic assimilation and a possible evolutionary paradox: can macroevolution sometimes be so fast as to pass us by? Evolution 57: 1455–1464.

Pigliucci, M. and K. Preston. 2004. Phenotypic Integration. Oxford University Press, Oxford.

Pigliucci, M., C. J. Murren and C. D. Schlichting. 2006. Phenotypic plasticity and evolution by genetic assimilation. J. Exp. Biol. 209: 2362–2367.

Pijanowska, J., P. Bernatowicz and J. Fronk. 2007. Phenotypic plasticity within *Daphnia longispina* complex: differences between parental and hybrid clones. Polish J. Ecol. 55: 761–769.

Poinar, G. and S. P. Yanoviak. 2008. *Myrmeconema neotropicum* n. g. sp., a new tetradonematid nematode parasitizing South American populations of *Cephalotes atratus* (Hymenoptera: Formicidae), with the discovery of an apparent parasite-induced host morhp. Systematic Parasitolgy 69: 145–153.

Poulev, A., J. M. O'Neal, S. Logendra, R. B. Pouleva, V. Timeva, A. S. Garvey, D. Gleba, I. S. Jenkins, B. T. Halpern, R. Kneer, G. M. Cragg and I. Raskin. 2003. Elicitation, a new window into plant chemodiversity and phytochemical drug discovery. J. Med. Chem. 46: 2542–2547.

Premnath, S. A. Sinha and R. Gadagkar. 1996. Dominance relationship in the establishment of reproductive division of labour in a primitively eusocial wasp (*Ropalidia marginata*). Behav. Ecol. Sociobiol. 39: 125–132.

Price, T. D. 2006. Phenotypic plasticity, sexual selection and the evolution of color patterns. J. Exp. Biol. 209: 2368–2376.

Price, T. D., A. Qvarnstöm and D. E. Irwin. 2003. The role of phenotypic plasticity in driving genetic evolution. Proc. R. Soc. B 270: 1433–1440.

Prichard, I. M. and R. James. 1984. Leaf mines: their effect on leaf longevity. Oecologia 64: 132–139.

Prokopy, R. J., S. R. Diehl and S. S. Cooley. 1988. Behavioral evidence for host races in *Rhagoletis pomonella* flies. Oecologia 76: 138–147.

Qvarnström, A. and T. D. Price. 2001. Maternal effects, paternal effects and sexual selection. Trends Ecol. Evol. 16: 95–100.

Radakovic SS, Maric J, Surbatovic M, Radjen S, Stefanova E, Stankovic N and Filipovic N 2007. Effects of acclimation on cognitive performance in soldiers during exertional heat stress. Military Medicine. 172: 133–136.

Radman, M. 1999. Enzymes of evolutionary change. Nature 401: 866–869.

Reimer, O. and M. Tedengren. 1996. Phenotypical improvement of morphological defences in the mussel *Mytilus edulis* induced by exposure to the predator *Asterias rubens*. Oikos 75: 383–390.

Relyea, R. A. 2003a. Predator cues and pesticides: a double dose of danger for amphibians. Ecol. Appl. 13: 1515–1521.

Relyea, R. A. 2003b. How prey respond to combined predators: a review and an empirical test. Ecology 84: 1827–1839.

Relyea, R. A. 2003c. Predators come and predators go: the reversibility of predator-induced traits. Ecology 84: 1840–1848.

Relyea, R. A. 2004a. Integrating phenotypic plasticity when death is on the line. pp. 176-190 *in* M. Pigliucci and K. Preston (eds.). *Phenotypic Integration*. Oxford University Press, Oxford.

Relyea, R. A. 2004b. Fine-tuned phenotypes: tadpole plasticity under 16 combinations of predators and competitors. Ecology 85: 172–179.

Releya, R. A. and J. Hoverman. 2006. Assesing the ecology in exotoxicology: a review and synthesis in freshwater systems. Ecol. Let. 9: 1157–1171.

Richards, C. L., O. Bossdorf, N. Z. Muth, J. Gurevitch and M. Pigliucci. 2006. Jack of all trades, master of some? On the role of phenotypic plasticity in plant invasions. Ecol. Lett. 9: 981–993.

Ricklefs, R. E. and M. Wikelski. 2002. The physiology/life-history nexus. Trends Ecol. Evol. 17: 462–468.

Riegler, M. and S. L. O'Neill. 2007. Evolutionary dynamics of insect symbiont associations. Trends Ecol. Evol. 22: 625–627.

Risch, S. J. and F. R. Rickson. 1981. Mutualism in which ants must be present before plants produce food bodies. Nature 291: 149–150.

Roder, G., M. Rahier and R. E. Naisbit. 2008. Counter-intuitive developmental plasticity induced by host quality. Phil. Trans. R. Soc. Lond. B. 275: 879-885.

Rodríguez, R. L., L. M. Sullivan, R. L. Snyder and R. B. Cocroft. 2007. Host shifts and the beginning of signal divergence. Evolution 62: 12–20.

Roelofs, D., M. G. M. Aarts, H. Schat and N. M. Van Straalen. 2007. Functional ecological genomics to demonstrate general and specific responses to abiotic stress. Funct. Ecol. 21: 1–12.

Roff, D. A. 1996. The evolution of threshold traits in animals. Quarterly Rev. Biol. 71: 3–35.

Rogers, S. M. and S. J. Simpson. 1997. Experience-dependent changes in the number of chemosensory sensilla on the mouthparts and antennae of *Locusta migratoria*. J. Exp. Biol. 200: 2313–2321.

Rountree, D. B. and H. F. Nijhout. 1994. Tenetic control of a seasonal morph in *Precis coenia* (Lepidoptera: Nymphalidae). J. Insect Physiol. 41: 1141–1145.

Rowell, C. H. F. 1971. The variable coloration of the Acridoid grasshoppers. Adv. Insect Physiol. 8: 145–198.

Roy, H. E., D. C. Steinkraus, J. Eilenberg, A. E. Hajek and J. K. Pell. 2006. Bizarre interactions and endgames: entomopathogenic fungi and their arthropod hosts. Annu. Rev. Entomol. 51: 331–357.

Saint-Ruf, C. and I. Matic. 2006. Environmental tuning of mutation rates. Environ. Microbiol. 8: 193–199.

Salt, G. 1937. The egg-parasite of *Sialis lutaria*: a study of the influence of the host upon a dimorphic parasite. Parasitology 29: 539–553.

Sarkar, S. 2004. From the *Reaktionsnorm* to the evolution of adaptive plasticity a historical sketch, 1909-1999. pp. 10-30 *in* T. J. DeWitt and S. M. Scheiner (eds.). *Phenotypic plasticity: Functional and Conceptual Approaches*. Oxford University Press. Oxford.

Scheiner, S. M. 1993. Genetics and evolution of phenotypic plasticity. Annu. Rev. Ecol. Syst. 24: 35–68.

Scheiner, S. M. 2002. Selection experiments and the study of phenotypic plasticity. J. Evol. Biol. 15: 889–898.

Scheiner, S. M. and R. F. Lyman. 1991. The genetics of phenotypic plasticity. II. responses to selection. J. Evol. Biol. 4: 23–50.

Schlichting, C. D. 2004. The role of phenotypic plasticity in diversification. pp. 191-200 *in* T. DeWitt and S. Scheiner (eds.). Phenotypic plasticity: functional and conceptual approaches. Oxford Univ. Press, Oxford.

Schlichting, C. D. and H. Smith. 2002. Phenotypic plasticity: linking molecular mechanisms with evolutionary outcomes. Evol. Ecol. 16: 189–211.

Schlichting, C. D. and M. Pigliucci. 1998. Phenotypic Evolution a Reaction Norm Perspective. Sinauer Associates, Sunderland, MA.

Schmalhausen, I. I. 1949. Factors of Evolution. Blakiston, Philadelphia.

Schmid-Hempel, P. 2005. Evolutionary ecology of insect immune defenses. 2005. Annu. Rev. Entomol. 50: 529–551.

Schmitz, O. J. 2006. Predators have large effects on ecosystem properties by changing plant diversity, not plant biomass. Ecology 87: 1432–1437.

Schmitz, O. J., V. Krivan and O. Ovadia. 2004. Trophic cascades: the primacy of trait-mediated indirect interactions. Ecol. Lett. 7: 153–163.

Schneirla, T. C. and H. R. Topoff. 1971. *Army Ants A Study in Social Organization*. W. H. Freeman and Company, San Francisco.

Seger, J. and H. J. Brockmann. 1987. What is bet-hedging? pp. 182-211 *in* P. Harvey and L. Partridge (eds.). Oxford Surveys in Evolutionary Biology 4. Oxford University Press, Oxford.

Sheldon, B. C. 2000. Differential allocation: tests, mechanisms and implications. Trends Ecol. Evol. 15: 397–402.

Sherry, D. F. and B. G. Galef. 1990. Social learning without imitation: more about milk bottle opening by birds. Anim. Behav. 40: 987–989.

Shibao, H., M. Kutsukake and T. Fukatsu. 2003. Density triggers soldier production in a social aphid. Proc. R. Soc. Lond. B (Suppl.) (Biology Letters). 03BL0217: S1-S4.

Shibao, H., M. Kutsukake and T. Fukatsu. 2004. Density-dependent induction and suppression of soldier differentiation in an aphid social system. J. Insect Physiol. 50: 995–1000.

Shingleton, A. W. and W. A. Foster. 2000. Ant tending influences soldier production in a social aphid. Proc. R. Soc. Lond. B 267: 1863–1868.

Shingleton, A. W., W. A. Frankino, T. Flatt, H. F. Nijhout and D. J. Emlen. 2007. Size and shape: the developmental regulation of static allometry in insects. BioEssays 29: 536–548.

Shiu, S. H. and J. O. Borevitz. 2008. The next generation of microarray research: applications in evolutionary and ecological genomics. Heredity 100: 141–149.

Sih, A. 2004. A behavioral ecological view of phenotypic plasticity. pp. 112-125 *in* T. DeWitt and S. Scheiner (eds.). *Phenotypic Plasticity*. Oxford University Press, Oxford.

Simmons, L. W., D. J. Emlen and J. L. Tomkins. 2007. Sperm competition games between sneaks and guards: a comparative analysis using dimorphic male beetles. Evolution 61: 2684–2692.

Simpson, S. J., G. A. Sword and A. De Loof. 2005. Advances, controversies and consensus in locust phase polyphenism research. J. Orthopt. Res. 14: 213–222.

Skaife, S. H. 1954. The black-mound termite of the Cape, *Amitermes atlanticus* Fuller. Trans. R. Soc. S. Afr. 34: 251–272.

Slabbekoorn, H. and T. B. Smith. 2002. Bird song, ecology and speciation. Philos. Trans. R. Soc. Lond. B 357: 493–503.

Slabber S, M. R. Worland, H. P. Leinaas and S. L. Chown. 2007. Acclimation effects on thermal tolerances of springtails from sub-Antarctic Marion Island: Indigenous and invasive species. J. Insect Physiol. 53: 113–125.

Smith, L. D. and A. R. Palmer. 1994. Effects of manipulated diet on size and performance of brachyuran crab claws. Science 264: 710–712.

Smith-Gill, S. J. 1983. Developmental plasticity: developmental conversion versus phenotypic modulation. Am Zool. 23: 47–55.

Song, H. 2005. Phylogenetic perspectives on the evolution of locust phase polyphenism. J. Orthopt. Res. 14: 235–245.

Stearns, S. C. 1992. *The Evolution of Life Histories*. Oxford Univ. Press, Oxford.

Stelgenga, M. J. and K. Fischer. 2007. Within- and between-generation effects of temperature on life-history traits in a butterfly. J. Thermal Biol. 32: 396–405.

Stephen, W. P. 2005. A serendipitous observation of a massive kill of the soil nesting bee, *Nomia melanderi* Cockerell (Hymenoptera: Apoidea: Halictidae). Pan-Pac. Entomol. 81: 181–183.

Stern, D. L. and W. A. Foster. 1996. The evolution of soldiers in aphids. Biol. Rev. 71: 27–76.

Strauss, S. Y. and R. E. Irwin. 2004. Ecological and evolutionary consequences of multispecies plant-animal interactions. Annu. Rev. Ecol. Evol. Syst. 35: 435–466.

Sugumaran, M. 2002. Comparative biochemistry of eumelanogenesis and the protective roles of phenoloxidase and melanin in insects. Pigm. Cell Res. 15: 2–9.

Sultan S. E. 2005. An emerging focus on plant ecological development. New Phytol. 166: 1–5.

Sultan S. E. 2007. Development in context: the timely emergence of eco-devo. Trends Eco. Evol. 22: 575–582.

Sultan S. E. and H. G. Spencer. 2002. Metapopulation structure favors plasticity over local adaptation. Am. Nat. 160: 271–283.

Suzuki, Y. and F. Nijhout. 2006. Evolution of a polyphenism by genetic accommodation. Science 311: 650–652.

Swallow, J. G., J. S. Rhodes and T. Garland, Jr. 2005. Phenotypic and evolutionary plasticity of organ masses in response to voluntary exercise in house mice. Integr. Comp. Biol. 45: 426–437.

Sword, G. A. 2002. A role for phenotypic plasticity in the evolution of aposematism. Proc. R. Soc. Lond. B. 269: 1639–1644.

Swynghedauw, B. 2006. Phenotypic plasticity of adult myocardium. Molecular mechanisms. J. Exp. Biol. 209: 2320–2327.

Takabayashi, J., S. Takahashi, M. Dicke and M. A. Posthumus. 1995. Developmental stage of herbivore *Pseudaletia separate* affects production of herbivore-induced synomone by corn plants. J. Chem. Ecol. 21: 273–287.

Tanaka, S. 2004. Environmental control of body-color polyphenism in the American Grasshopper, *Schistocerca americana*. Ann. Entomol. Soc. Am. 97: 293–301.

Thompson, D. B. 1992. Consumption rates and the evolution of diet-induced plasticity in the head morphology of *Melanoplus femurrubrum* (Orthoptera: Acrididae). Oecologia 89: 204–213.

Thompson, J. N. 1988. Variation in interspecific interactions. Annu. Rev. Ecol. Syst. 19: 65–87.

Thompson, J. N. 2005. The Geographic Mosaic of Coevolution. University of Chicago Press, Chicago.

Tollrian, R. and C. D. Harvell (Eds). 1999. The Ecology and Evolution of Inducible Defenses. Princeton, NJ: Princeton University Press.

Tomkins, J. L. 1999. Environmental and genetic determinants of the male forceps length dimorphism in the European earwig *Forficula auricularia* L. Behav. Ecol. Sociobiol. 47: 1–8.

Tomkins, J. L. and L. W. Simmons. 1996. Dimorphisms and fluctuating asymmetry in the forceps of male earwigs. J. Evol. Biol. 9: 753–770.

True, J. R. 2003. Insect melanism: the molecules matter. Trends Ecol. Evol. 18: 640–647.

Tufto, J. 2000. The evolution of plasticity and nonplastic spatial and temporal adaptations in the presence of imperfect environmental cues. Am. Nat. 156: 121–130.

Uvarov, B. 1966. Grasshoppers and Locusts. Vol. 1. Cambridge University Press, Cambridge.

Van Buskirk, J. 2002. A comparative test of the adaptive plasticity hypothesis: relationships between habitat and phenotype in anuran larvae. Am. Nat. 160: 87–102.

Van Buskirk, J. and M. Arioli. 2005. Habitat specialization and adaptive phenotypic divergence of anuran populations. J. Evol. Biol. 18: 596–608.

van Kleunen, M. and M. Fischer. 2005. Constraints on the evolution of adaptive phenotypic plasticity in plants. New Phytol. 166: 49–60.

Van Tienderen, P. H. 1997. Generalists, specialists, and the evolution of phenotypic plasticity in sympatric populations of distinct species. Evolution 51: 1372–1380.

Van Zandt, P. A. and A. A. Agrawal. 2004. Community-wide impacts of herbivore induced plant responses in milkweed (*Asclepias syriaca*). Ecology 85: 2616–2629.

Via, S., R. Gomulkiewicz, G. de Jong, S. M. Scheiner, C. D. Schlichting and P. H. Van Tienderen. 1995. Adaptive phenotypic plasticity: consensus and controversy. Trends Ecol. Evol. 10: 212–217.

Visser, M. E., L. J. M. Holleman and P. Gienapp. 2006. Shifts in caterpillar biomass phenology due to climate change and its impact on the breeding biology of an insectivorous bird. Oecologia 147: 164–172.

Von Rad, U., M. J. Mueller and J. Durner. 2005. Evaluation of natural and synthetic stimulants of plant immunity by microarray technology. New Phytol. 165: 191–202.

Waddington, C. H. 1940. *Organisers and Genes.* Cambridge University Press, Cambridge.

Waddington, C. H. 1942. Canalisation of development and the inheritance of acquired characters. Nature 150: 563–565.

Waddington, C. H. 1952. Selection for the genetic basis of an acquired character. Nature 169: 278.

Waddington, C. H. 1953a Genetic assimilation of an acquired character. Evolution 7: 118–126.

Waddington, C. H. 1953b. The "Baldwin Effect," "genetic assimilation" and "homeostasis." Evolution 7: 386–387.

Waddington, C. H. 1956. Genetic assimilation of the *bithorax* phenotype. Evolution 7: 1–13.

Waddington, C. H. 1959. Canalisation of development and genetic assimilation of acquired characters. Nature 183: 1654–1655.

Waddington, C. H. 1961. Genetic assimilation. Adv. Genet. 10: 257–290.

Wade, M. J. 2001. Infectious speciation. Nature 409: 675.

Webber, B. H. and D. J. Depew. 2003. Evolution and Learning: the Baldwin Effect Reconsidered. MIT Press, Cambridge.

Werner, E. E. and S. D. Peacor. 2003. A review of trait-mediated indirect interactions in ecological communities. Ecology 84: 1083–1100.

West-Eberhard, M. J. 1969. The social biology of polistine wasps. Misc. Publ. Mus. Zool., Univ. Michigan No. 140: 1–101.

West-Eberhard, M. J. 2003. Developmental Plasticity and Evolution. Oxford University Press, Oxford.

West-Eberhard, M. J. 2005. Developmental plasticity and the origin of species differences. Proc. Nat. Acad. Sci. USA. 102: 6543–6549.

Wheeler, W. M. 1910. *Ants*. Columbia University Press, New York.

Whitfield, J. 2007. Who's the queen? Ask the genes. Science318: 910–911.

Whitman, D. W. 1994. Plant bodyguards: mutualistic interactions between plants and the third trophic level. Pp. 207-248 *in* T. N. Ananthakrishnan (ed.). *Functional Dynamics of Phytophagous Insects*. Oxford & IBH Publishing, New Delhi.

Wilson, R. S. and C. E. Franklin. 2002. Testing the beneficial acclimation hypothesis. Trends Ecol. Evol. 17: 66–70.

Windig, J. J., C. G. F. De Kovel and G. De Jong. 2004. Genetics and mechanics of plasticity. Pp. 31-97 *in* T. J. DeWitt and S. M. Scheiner (eds.). *Phenotypic plasticity: Functional and Conceptual Approaches*. Oxford University Press, Oxford.

Winterhalter, W. E. and T. A. Mousseau. 2007. Patterns of phenotypic and genetic variation for the plasticity of diapause incidence. Evolution 61: 1520–1531.

Wolf, J. B., E. D. Brondie and M. J. Wade. 2004. The genotype-environment interaction and evolution when the environment contains genes. pp. 173-190 *in* T. DeWitt and S. Scheiner (eds.). Phenotypic Plasticity. Oxford University Press, Oxford.

Wolschin, F. and G. V. Amdam. 2007. Plasticity and robustness of protein patterns during reversible development in the honey bee (*Apis mellifera*). Analytical & Bioanalytical Chemistry 389: 1095–1100.

Woltereck, R. 1909. Weitere experimentelle Untersuchungen über Artveränderung, speziell über das Wesen quantitativer Artunterschiede bei Daphniden. Verh. D. Tsch. Zool. Ges. 1909: 110–172.

Yang, Y. and A. Joern. 1994. Gut size changes in relation to variable food quality and body size in grasshoppers. Funct. Ecol. 8: 36–45.

Zera, A. J. and L. G. Harshman. 2001. The physiology of life-history trade-offs in animals. Annu. Rev. Ecol. Syst. 32: 95–126.

Zhang, X. S. 2006. The phenotypic variance within plastic traits under migration-mutation-selection balance. Evolution 60: 1125–1136.

Zhao, J., L. C. Davis and R. Verpoorte. 2005. Elicitor signal transduction leading to production of plant secondary metabolites. Biotechnology Advances 23: 283–333.

Zhou X, M. R Tarver and M. E. Scharf. 2007. Hexamerin-based regulation of juvenile hormone-dependent gene expression underlies phenotypic plasticity in a social insect. Development 134: 601–610.

2

Phenotypic Plasticity and the Semantics of Polyphenism: A Hisorical Review and Current Perspectives

Michael Canfield* and Erick Greene**

**Department of Organismic and Evolutionary Biology, Harvard University, 26 Oxford Street, Cambridge, MA USA 02138, (617) 495-4012, E-mail: mcanfield@oeb.harvard.edu*

***Division of Biological Sciences, The University of Montana, Missoula, MT USA 59801-1002, (406) 243-2179, E-mail: erick.greene@mso.umt.edu*

Abstract

A diverse vocabulary now exists to describe aspects of phenotypic plasticity. One term in particular, polyphenism, has experienced significant drift since it was coined by Ernst Mayr in 1963. Current uses of the term polyphenism do not adhere to a common definition, and this hinders communication about the complex processes and systems that exhibit environmentally-induced phenotypic variation. We suggest that this term should be returned to its original meaning, and other more precise terms should be employed to describe specific types of phenotypic plasticity.

"Several distinguished naturalists maintain with much confidence that organic beings tend to vary . . . independently of the conditions to which they and their progenitors have been exposed; whilst others maintain that all variation is due to such exposure, though the manner in which the environment acts is as yet quite unknown. At the present time there is hardly any question in biology of more importance than this of the nature and cause of variability..."

- Charles Darwin, Prefatory notice to August Weismann's *Studies in the Theory of Descent* (1882)

Darwin clearly understood that the causes and consequences of phenotypic variation lie at the heart of evolutionary biology. The role of environmental factors in this variation, now termed phenotypic plasticity, integrates the fields of developmental biology, genetics, ecology and behavior. Given its wide ranging applicability, the fast-moving study of plasticity draws on a wide diversity of approaches and perspectives, and accommodates questions at many levels of analysis. In talking with colleagues, we have been surprised by the rapid divergence of perspectives and the use of terms associated with the study of phenotypic plasticity. In particular, there has been a significant drift in how the term ***polyphenism*** is used since it was first defined by Ernst Mayr:

"In order to make the term 'polymorphism' more useful and precise, there is now a tendency to restrict it to genetic polymorphism. Since this would leave nongenetic variation of the phenotype without a designation, the term 'polyphenism' is here proposed for it. Polyphenism is discontinuous when definite castes are present (certain social insects) or definite stages in the life cycle (larvae vs. adults; sexual vs. parthenogenetic) or definite seasonal forms (dry vs. wet; spring vs. summer). Polyphenism may be continuous, as in the cyclomorphosis of fresh-water organisms and some other seasonal variation" (Mayr 1963, p. 150).

In the 40 years that have elapsed since then, the term has changed from being an overarching category encompassing many loosely related phenomena to a narrow and restricted term that has been used in a variety of ways. A plethora of terms have been proposed to describe environmentally-induced variation, and the purpose of this chapter is to:

- provide a brief overview of the history and terminology associated with phenotypic plasticity;
- specifically point out the semantic drift in the word polyphenism, and some of the discrepancies in its current use; and
- argue that polyphenism's current usage is not helpful in understanding phenotypic plasticity, and that it should be once again be used in accordance with its original definition.

Historical Overview

The literature on evolution and phenotypic plasticity is extensive, and for broad overviews we steer the interested reader to the growing number of insightful reviews (e.g. the collection of papers in Issue 39 of BioScience (1989) devoted to evolution of the phenotype; Schlichting and Pigliucci 1998, West-Eberhard 2003, DeWitt and Scheiner 2004, Pigliucci 2005, and other chapters in this book).

Even though the term "phenotypic plasticity" was formalized in the mid-20th century (Bradshaw 1965, see discussion in Debat and David 2001), research concerning the flexibility of the phenotype began much earlier. Although best remembered for his role in distinguishing the characteristics of the germ line from that of the soma, August Weismann conducted pioneering research on how the phenotypes of butterflies and moths are affected by environmental cues (Weismann 1882). Weismann's research inspired others, such as Edward Poulton, who performed classic experiments that investigated the effects of rearing environment on the phenotypes of caterpillars and pupae (Poulton 1885, 1887, 1892).

By the beginning of the 1900s, it was becoming clear that a variety of proximate genetic and developmental mechanisms can give rise to phenotypic variation. During the last century, biologists focused primarily on two ways that morphological variation is generated (Greene 1999): genetic polymorphisms, in which individuals have fixed phenotypes that depend only upon their genetic makeup; and reaction norms (described by Woltereck 1909), in which an individual's phenotype depends upon both its genotype and environmental conditions.

In contrast to the long interest in genetic polymorphisms and heredity, the study of environmentally-induced plasticity received relatively little attention after the initial work of Weismann and Poulton. The historical reasons for this bias are many (Wu and Morris 2001), but perhaps most importantly, biologists were uncomfortable with the Lamarckian overtones suggested by phenotypic plasticity. Nowhere is the negative impact of Lamarckism on the study of environmentally-induced variation more evident than in the tragic fall from grace of Paul Kammerer in 1926 (see Koestler 1971). Kammerer conducted several sets of experiments on amphibians that purportedly demonstrated the inheritance of acquired characteristics, the most famous of which he performed on the midwife toad, *Alytes obstetricans*. Kammerer's experiments were later exposed as fraudulent, and this high-profile scandal tainted the study of environmental effects and further marginalized plasticity studies.

Subsequently, workers such as Goldschmidt (1940) and Schmalhausen (1949) made substantial contributions to our understanding of how the environment influences phenotypes (see discussion in Schlichting and Pigliucci 1998), but the importance of their work was not fully recognized at the time. Environmentally-induced phenotypic variation was sidelined partly due to the backlash against Lamarckism, and also because of the degree to which genetic determinism had risen to the dominant explanation for the cause of phenotypic variation in the early part of the 20th century.

Many influential scientists accepted that environmentally-induced variation represented nothing more than rather embarrassing developmental defects, as illustrated by R.A. Fisher's quip to V. Wigglesworth, "It is not surprising that such elaborate machinery should sometimes go wrong" (Wigglesworth 1961, quoted in West-Eberhard 2003). Indeed, the range of environmentally-induced variation was not explicitly defined until Ernst Mayr coined the term polyphenism.

Note that Mayr's original definition of polyphenism was very broad, and encompasses many types of variation, including reversible and non-reversible developmental phenomena, seasonal and aseasonal traits, continuous and discrete variation, adaptive or non-adaptive traits, and ontogenetic variation and metamorphosis. The diversity of phenomena included under this original definition can be illustrated with schematic phenotype-time plots that illustrate how the phenotype can change over time (Fig. 1). The phenotype represented on the y-axis can represent a single character, or can represent a complex phenotype composed of a number of individual characters.

The development of an individual may be influenced by environmental cues or triggers (Fig. 1a). Bifurcation points represent moments in time in which a developing organism can be shunted along different pathways. If developmental bifurcations result in different phenotypes (for example, the two upper pathways shown in red), this can produce discretely different alternative morphs. Examples of such discrete alternative forms are temperature dependent sex determination in some turtles (Bull and Vogt 1979) and fish (Conover and Kynard 1981). In some cases, however, bifurcations may result in phenotypes that are more similar (for example, the lower cluster of pathways shown in green). Such an arrangement of developmental pathways produces more continuous morphological variation. Examples of more graded variation are the intermediate forms of locusts (see discussion below).

Phenotypes may also vary over time, not as a result of changes during early development, but as the organism responds to environmental fluctuations throughout life (Fig. 1b). Examples involving longer time scales are the seasonal white and brown forms of snowshoe hares, ermine, and ptarmigan, and the conspicuous breeding plumage versus the duller, more cryptic winter plumage of many birds. Examples involving shorter time scales include organisms such as squid, octopus, flounder, and chameleons that can change their colors extremely rapidly (e.g. Hanlon 1999).

Finally, many developing organisms undergo substantial morphological changes during development (Fig. 1c). Spectacular examples are the

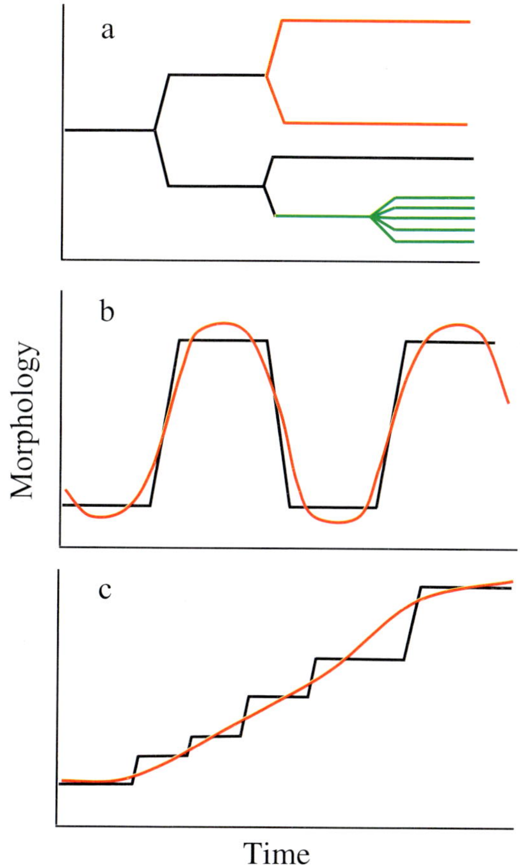

Fig. 1 Schematic diagrams of morphology versus time that illustrate different forms of phenotypic variation that were included in Mayr's original definition of polyphenism. ***Panel a*** represents possible developmental trajectories during the life of an individual. ***Panel b*** shows cyclical variation in morphology over time. ***Panel c*** shows the progression of an individual through different life history stages. See text for further description and explanation.

metamorphic transitions during the life of many insects, such as the egg, various instars of caterpillars, pupae, and the final adult moth or butterfly of Lepidoptera. Such variation in morphology during development can be fairly discrete (black line with steps), or more continuous (smoother red line). Ontogenetic changes, such as metamorphosis in insects, are part of a genetically-driven developmental program. However, the phenotypic variation that results in the population (e.g. larval forms and adult forms) is not the result of genetic differences among organisms, but is simply variation due to the stage of the life cycle of the individual organisms. In defining polyphenism as the set of phenotypic variation in a population that is not the result of genetic polymorphisms, Mayr clearly intended to include this developmental variation as part of the environmentally-induced differences among organisms. The timing and degree of expression of these developmental characters can also be influenced by nutrition, temperature, photoperiod and social factors.

Thus, Mayr's original conception of polyphenism encompassed all types of non-genetic phenotypic variation. This "grab bag" certainly includes many different types of environmental induction cues, proximate developmental and physiological mechanisms, outcomes, and adaptive benefits. However, the use of the term polyphenism has drifted significantly since its original broad definition, and the term is currently used in a confusing variety of ways and contexts. Although a few authors use polyphenism as a synonym for non-genetic phenotypic plasticity that is close to Mayr's original definition (e.g., Hanlon 1999, Kelmanson and Matz 2003), polyphenism is most commonly used to refer only to adaptive, irreversible, discrete alternative phenotypes (with no intermediate forms). For example:

"Polyphenism is the situation in which one genotype produces two or more discrete phenotypes in response to an environmental signal. The term polyphenism is the analog among phenotypes of the term polymorphisms among genotypes" (Stearns 1989).

"Adaptive phenotypic plasticity is carried to extremes in polyphenic organisms. Here, individual genotypes are able to express two or more discrete phenotypes in response to differences in the internal or external environment experienced by the developing organism (Nijhout 1999)" (Moczek and Nijhout 2002).

"...among many others, plasticity is limited to a period of development and is nonreversible (polyphenism)" (Wente and Phillips 2003).

In addition to the confusing semantic shift in the meaning of polyphenism and the proliferation of other terms, there is a striking taxonomic bias in the current use of the term polyphenism. Even though environmentally-induced plasticity is ubiquitous (West-Eberhard 2003), polyphenism is now almost exclusively used by those who study butterflies, moths, aphids, locusts, beetles, or frogs, toads and salamanders. We performed a literature search of papers published between 1995-2004 with polyphenism in the title or abstract. In a sample of 159 papers, 87% of the papers were on invertebrates (and virtually all of these were on Lepidoptera, aphids, orthopterans or beetles); the remaining papers represented only two groups of vertebrates (12% anurans—frogs, toads and salamanders, and 1%

salmonid fishes). Polyphenism is rarely used to describe environmentally-cued plasticity in other organisms, and it is virtually absent from the plant literature.

Mayr coined the term polyphenism to refer broadly to the large set of phenotypes that one genome may produce, either sequentially through the course of development (e.g. the stages of holometabolous insects), or by producing differential outcomes through alternative developmental pathways (e.g. castes of ants). There has been progress during the last century, and particularly in the last two decades, in understanding a large variety of developmental and genetic mechanisms that create these phenotypes. In the process of studying how these mechanisms cause phenotypic variation, a multitude of new terms were created: developmental plasticity, developmental stability, developmental conversion, phenotypic flexibility, life-cycle staging, acclimation, acclimatization, threshold traits, and phenotypic modulation to name a few (Smith-Gill 1983, Roff 1996, Debat and David 2001, Loescheke and Hoffman 2002, Wilson and Franklin 2002, Piersma and Drent 2003). These and other terms provide definitions relevant to the variation that polyphenism originally described, and some of their precise meanings are still being honed. We are not arguing that these terms are unnecessary: a highly varied and complex phenomenon such as phenotypic plasticity needs its own vocabulary. However, Mayr defined polypenism to include a large set of phenotypic outcomes, but the term is now often used to refer to only irreversible and adaptive environmentally-induced variation of the phenotype. These uses of polyphenism can be visualized in a Venn diagram, as subsets of Mayr's original definition (Fig. 2). These newer uses of polyphenism are often more specific, containing a range of concepts such as seasonality, discrete forms, and irreversibility. Some of these uses, such as Hanlon's use of polyphenism (1999) and Nijhout's use of sequential polyphenism are contained within Mayr's original meaning, but are outside what is normally considered phenotypic plasticity. New definitions of polyphenism have also caused confusion because they have been applied differently by biologists considering variation from various levels of analysis.

Traits versus Forms

Consider the green and brown seasonal forms of a caterpillar species. Although these forms may be easily differentiated by their apparent differences in basic morphology, simple labels (e.g. "green" versus "brown") may mask the degree of independence of the traits that together

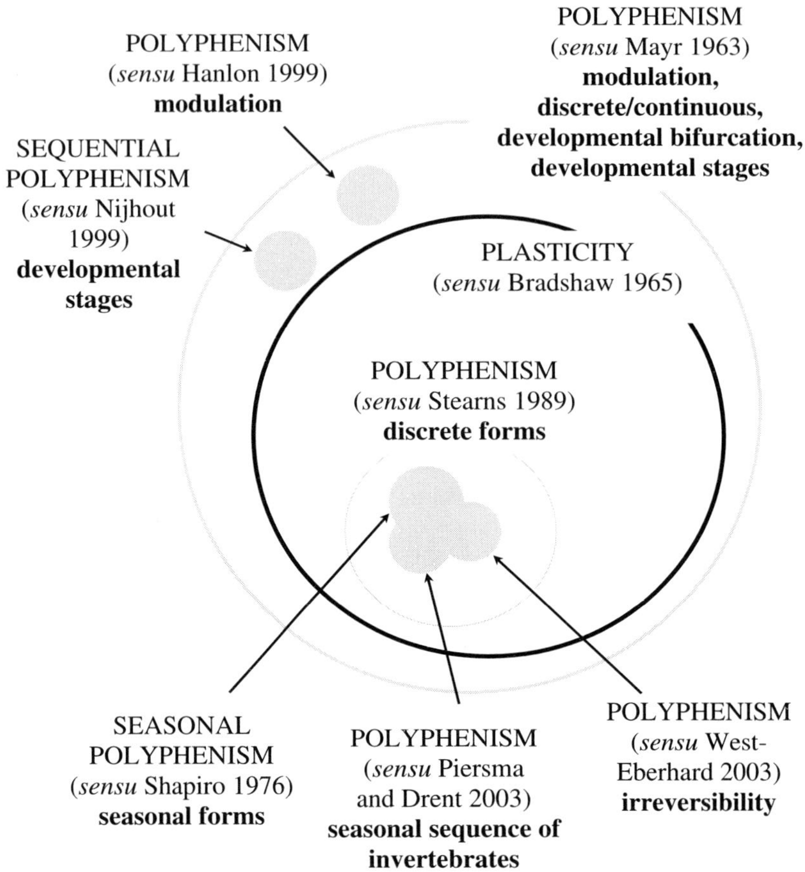

Fig. 2 A Venn diagram exhibits overlapping uses of the term *polyphenism* relative to the concept of phenotypic plasticity. A variety of elements are included in different definitions such as irreversibility, discrete forms and seasonality.

create the different forms, including, differences in skin morphology, bristle structure and development, the size, location and shape of epidermal protuberances, and the location, size and color of patches on the skin, often involving many different pigment pathways. In this way a "form" of an organism is not a unitary character trait, but rather depends upon the summation of the expression of many contributing traits.

This distinction between the "form" or "phenotype" of an individual, and the component traits that together create the form can be illustrated using an analogy to the equalizer on a sound system (Fig. 3). Each toggle or lever can be thought of as the expression of a separate developmental trait or character (e.g. skin texture, bristle structure, pigment pathway 1, pigment

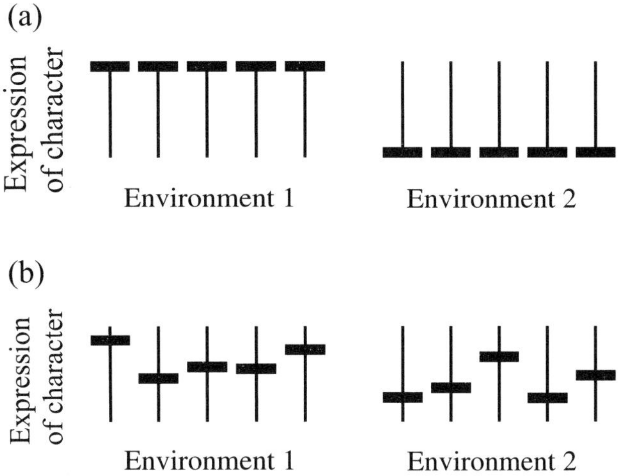

Fig. 3 Schematic analogy showing differing degrees of phenotypic integration. In (a), all traits have a high degree of integration and respond uniformly to different environments. A lower degree of integration is shown in (b), where traits are more independent in how they respond to changes in environment.

pathway 2, etc.); the overall form or appearance of the individual is a summation of all the contributing traits. Discretely different forms may thus be the result of highly coordinated developmental responses of individual toggles to environmental variation, perhaps with sensitive threshold responses to environmental variation (Fig. 3a). Alternatively, each developmental trait may respond fairly independently of the others, so that the developmental toggles have a low degree of covariation in response to environmental variation (Fig. 3b). In such a case, we would expect to see more intermediate forms with more independence among individual traits. The degree of coordination in the expression of individual traits in response to environmental variation is called phenotypic integration (Schlichting 1989, Pigliucci 2003, Pigliucci and Preston 2004). Figure 3a represents a case with high phenotypic integration. In many cases, the term polyphenism has recently been used to describe the production of discrete phenotypes. However, using polyphenism to describe two different "forms" may obscure differences in how the individual character traits that create those forms respond to environmental variables.

Discrete vs. Continuous Forms

Perhaps the largest consequence of the semantic shift of polyphenism has been the growing distinction between discrete alternative morphs and more continuous plasticity. We suggest that this is an artificial distinction in many cases, and may impede a more comprehensive understanding of the evolution of plasticity and phenotypes.

Nijhout points out that phenotypic plasticity is most likely the ancestral state for many taxa, and in many cases is "obtained gratis," since chemical, physical, metabolic and developmental processes are sensitive to environmental variation, such as temperature, pH and nutrients (Nijhout 2003). Non-plastic traits, therefore, are likely to result from genetic assimilation. Discrete alternative morphs (with no intermediate forms found in nature) can be produced from continuously varying reaction norms if the organisms typically develop in different environments. In contrast, discrete morphs may be the result of sparsely sampled reaction norms (Nijhout 2003). Examples of this include the butterfly *Araschnia levana,* which occurs in nature in discretely different spring and summer forms. However, a continuous gradient of intermediate forms can be produced in the lab by rearing caterpillars under intermediate environmental conditions that they do not usually experience in nature (Nijhout 2003). In this case, substantial phenotypic plasticity is "hidden," but might be expressed in nature if the environment changed. Alternatively, discretely different morphs, with no intermediate forms, can be produced by bifurcations in developmental pathways that produce discontinuous reaction norms (e.g. step functions or sharply sigmoidal responses). Thus, using polyphenism to describe only the latter creates a false dichotomy.

An example of continuous phenotypic plasticity is seen in phase polymorphism (polyphenism) in locusts (see Simpson and Sword, this volume). Worldwide, there are approximately 13,000 species of grasshoppers, virtually all of which are solitary. However, ~20 species, termed locusts, have the ability to switch from solitary to gregarious phenotypes. This "phase transformation" includes dramatic changes in physiology, morphology, behavior, ecology, and life history (Uvarov 1966, 1977, Loher 1990, Sword 2002). The changes are complex, and vary among locust species and environments, but generally, gregarious phase locusts tend to be smaller, more pigmented, undergo fewer instars, have fewer ovarioles and egg pods, lay smaller clutches of larger eggs, be more active, gregarious, and polyphagous, develop faster, have longer wings with more flight muscles than solitary phase locusts, have a tendency to gather in swarms of millions of individuals that fly great distances, and devastate crops wherever they settle. Interestingly, the solitary and gregarious types were initially assigned to different species, and it was not until 1921 when Uvarov discovered that the different forms are actually a result of plasticity.

With this understanding, our ability to control plague locusts began. We now know that this phase shift results from changes in population density and environmental quality, and appears to be adaptive. The two phases were first thought to be discrete forms; however, subsequent research has demonstrated that the plasticity is, in fact, continuous, and that virtually any intermediate phenotype is possible, given the correct environmental stimuli (Uvarov 1966, 1977).

Adaptation

Possibly the most common usage of polyphenism refers to the seasonal forms of butterflies and moths, which was termed seasonal polyphenism by Shapiro (1976). A wide range of lepidopterans exhibit phenotypic plasticity in traits such as the degree of wing melanization in pierids and the striking presence or absence of "eyespots" in satyrids (Shapiro 1976, Brakefield and Larsen 1984, see Brakefield and Frankino, this volume). In several of these cases, the adaptive value of the plasticity of the wing patterns has been explored, such as the importance of wing melanin to thermoregulatory ability in pierids (Kingsolver 1995, 1996) and the value of the "eyed" form of *Bicylclus anynana* under wet season conditions when adults are active, versus the cryptic form in the dry season (Lyytinen et al. 2004, Brakefield and Frankino, this volume). In many current uses of polyphenism, authors have implicitly or explicitly included adaptation as a criterion for a phenomenon to be an example of polyphenism. However, many examples of seasonal differences in pigmentation in butterflies are more subtle than those of pierids such as *Pieris occidentalis* (Kingsolver 1996) and they may or may not be adaptive. Such differences in pigmentation could arise from seasonal difference in larval diet or temperature, but may not translate into any specific adaptive benefit for the organism. In using polyphenism only to refer to adaptive examples of plasticity, we would drive an artificial wedge between closely related phenomena, and possibly obscure the fact that one (non-adaptive variation) may be simply a precursor for the other (adaptive variation). Moreover, adaptive value has only been established in a small fraction of systems that exhibit such phenotypic plasticity, so a definition of polyphenism that depended upon demonstrating the adaptive value of the trait would only be applicable to a few cases.

Irreversibility

The concept of irreversibility has also been attached to polyphenism, and this is consistent with a distinction between physiological and developmental changes in the phenotype that has been discussed in its relationship to the definition of phenotypic plasticity (see Pigliucci 2001). In the case of phenotypic plasticity, a distinction is made between physiological changes, such as the how the pigmentation of squid or plants may be altered in response to an environmental cue [often referred to as physiological acclimation or phenotypic modulation (Smith-Gill 1983)], in contrast to other irreversible changes in morphology that result from an underlying shift in gene expression (known as developmental conversion). However, some physiological responses, such as those to temperature, are based on changes in gene expression (e.g. heat shock responses; see related discussion and references in Pigliucci 2001, Williams et al., this volume). The line between physiological plasticity (acclimation or modulation) and morphological plasticity seems to be arbitrary and is biased by the biology of the organism. For example, pigments such as melanin produced in insects can be permanently deposited in the integument. In other organisms, such as mammals, an analogous process deposits pigments in hair and epidermal cells, but these may be shed over time and replaced with more or less pigmented alternatives. Despite the obvious similarities in these processes (environmental cues, gene regulation, pigment deposition), one is considered physiological plasticity and the other morphological (phenotypic) plasticity because the insect does not modify pigmentation over its short lifespan. This is partly because the insect's life cycle and structure does not allow for the process to be reversed, as it is in mammals. There are certainly some good arguments to separate fields of inquiry into physiological responses and morphological plasticity. However, discrete forms can be produced by a variety of processes, and a rich discussion of how irreversibility fits in to phenotypic plasticity already exists. Returning to Mayr's original definition allows one to refer to the whole set of environmentally-induced outcomes, while applying more restrictive conditions may artificially separate related phenomena.

Tempo of Plastic Response

Some definitions of polyphenism have variously incorporated the tempo of the plastic response, and the phenomena described generally take place over an extended course of development. Seasonal fluctuations such as plumage and coat modification in birds and mammals have not been termed polyphenism, although the term has been used to discuss the rapid response of cephalopods to environmental changes (Hanlon 1999). Given the vast differences in life history, ecology and development that living organisms display, it is hard to imagine a clean division between rapid and slow responses to the environment. We suggest that this again is an argument for maintaining an overarching concept of polyphenism that includes all environmental effects on the phenotype and different time scales of response.

We believe that returning to the original definition of polyphenism will make it easier for biologists who approach plasticity from different viewpoints to converge on some of the more interesting questions about how the evolution of the phenotype relates to different forms of plasticity. For example, how do organisms with discretely different morphs evolve from those with more continuous variation in response to environmental differences (or vice versa)? How labile or conservative are the developmental programs underlying the expression of traits? How labile or conservative are patterns of phenotypic integration? How does the natural history and ecology of species influence plasticity? Do patterns of plasticity influence speciation? We suggest that the distinction between the current narrow use of polyphenism including features such as irreversibly, alternate discrete morphs, and other types of phenotypic plasticity erects an artificial barrier between related phenomena, and makes it difficult to address these questions. For example, if one uses the term polyphenism to refer to discretely different phenotypes seen in nature, such as the different forms of caterpillars we study, one immediately begins to find ways to fit that phenotypic outcome with underlying mechanisms that would accommodate this concept. If the polyphenism label is used for discontinuous variation viewed from an organismic perspective, one begins by assuming that the system is composed of switch-controlled, irreversible, discrete characters, when in fact the system may be much more complicated.

Polyphenism has had an interesting journey since its birth in 1963. From a term that originally encompassed all types of environmentally-cued plasticity, it is currently used in a confusing variety of ways that are typically much more restrictive in their meaning. Instead of facilitating understanding, its current use can lead to artificial distinctions that may obscure relationships among related developmental, genetic and evolutionary phenomena. We suggest that this term be returned to its original broad definition, referring to the entire set of phenotypic variation that is the result of environmental effects, and that discrete, alternative morphologies should instead be described using the wealth of terminology that now exists to describe such complex and diverse processes.

Acknowledgements

We would like to thank Hans Hofman and Naomi Pierce for their thoughtful reviews of this essay.

References

Bradshaw, A. D. 1965. Evolutionary significance of phenotypic plasticity in plants. Advances in Genetics 13: 115–155.

Brakefield, P. M. and T. B. Larsen. 1984. The evolutionary significance of dry and wet season forms in some tropical butterflies. Biological Journal of the Linnean Society 22: 1–12.

Bull, J. J. and R. C. Vogt. 1979. Temperature-dependent sex determination in turtles. Science 206: 1186–1188.

Conover, D. O. and B. E. Kynard. 1981. Environmental sex determination: Interaction of temperature and genotype in a fish. Science 213: 577–579.

Debat, V. and P. David. 2001. Mapping phenotypes: canalization, plasticity and developmental stability. Trends in Ecology and Evolution 16: 555–561.

DeWitt, T. J. and S. M. Scheiner. 2004. *Phenotypic Plasticity: Functional and Conceptual Approaches*. Oxford University Press, Oxford.

Greene, E. 1999. Phenotypic variation in larval development and evolution: Polymorphism, polyphenism, and developmental reaction norms. In: *The Origin and Evolution of Larval Forms*. B. K. Hall and M. H. Wake eds., pp. 379–406. Academic Press, San Diego.

Goldschmidt, R. B. 1940. *The Material Basis of Evolution*. Yale University Press, New Haven.

Hanlon, R. T. 1999. Crypsis, conspicuousness, mimicry and polyphenism as antipredator defences of foraging octopuses on Indo-Pacific coral reefs, with a method of quantifying crypsis from video tapes." Biological Journal of the Linnean Society 66: 1–22.

Kelmanson, I. V. and M. V. Matz. 2003. Molecular basis and evolutionary origins of color diversity in great star coral *Montastraea cavernosa* (Scleractinia: Faviida). Molecular Biology and Evolution 20(7): 1125–1133.

Kingsolver, J. G. 1995. Fitness consequences of seasonal polyphenism in western white butterflies. Evolution 49: 942–954.

Kingsolver, J. G. 1996. Experimental manipulation of wing pigment pattern and survival in western white butterflies. American Naturalist 147: 296–306.

Koestler, A. 1971. *The Case of the Midwife Toad*. Vintage Books, New York.

Levins, R. 1968. *Evolution in Changing Environments*. Princeton University Press, Princeton, NJ.

Loeschcke, V. and A.A. Hoffmann. 2002. The detrimental acclimation hypothesis. Trends in Ecology and Evolution 17: 407–408.

Loher, W. and A. M. Giannakakis. 1990. Allatectomy in the penultimate larval instar of *Teleogryllus commodus*—morphological, physiological and behavioral consequences. Zoologische Jahrbucher-Abteilung fur Allgemeine Zoologie und Physiologie der Tiere 94(2): 167–179.

Lyytinen, A., P. M. Brakefield, L. Lindstrom and J. Mappes. 2004. Does predation maintain eyespot plasticity in *Bicyclus anynana*? Proceedings of the Royal Society B: Biological Sciences 271: 279–283.

Mayr, E. 1963. *Animal Species and Evolution*. Belknap Press of Harvard University, Cambridge, MA.

Moczek, A. P. and H. F. Nijhout. 2002. Developmental mechanisms of threshold evolution in a polyphenic beetle. Evolution and Development 4(4): 252–264.

Nijhout, H. F. 1999. Hormonal control in larval development and evolution. In: *The Origin and Evolution of Larval Forms* (B.K. Hall and M. Wake, eds.), pp. 217–254. Academic Press, New York.

Nijhout, H. F. 2003. Development and evolution of adaptive polyphenisms. Evolution and Development 5(1): 9–18.

Piersma, T. and J. Drent. 2003. Phenotypic flexibility and the evolution of organismal design. Trends in Ecology and Evolution 18: 228–233.

Pigliucci, M. 2001. *Phenotypic Plasticity: Beyond Nature and Nurture*. The Johns Hopkins University Press, Baltimore.

Pigliucci, M. 2003. Phenotypic integration: studying the ecology and evolution of complex phenotypes. Ecology Letters 6: 265–272.

Pigliucci, M. and K. Preston. 2004. *Phenotypic Integration: Studying the Ecology and Evolution of Complex Phenotypes*. Oxford University Press, Oxford.

Pigliucci, M. 2005. Evolution of phenotypic plasticity: where are we going now? Trends in Ecology and Evolution 20(9): 481–486.

Poulton, E. B. 1885. Further notes upon the markings and attitudes of lepidopterous larvae, together with a complete account of the life-history of *Sphinx ligustri* and *Selenia illunaria*. Transactions of the Entomological Society of London (2): 281–329.

Poulton, E. B. 1887. An enquiry into the cause and extent of a special colour-relation between certain exposed lepidopterous pupae and the surfaces that immediately surround them. Philosophical Transactions of the Royal Society of London B 178: 311–444.

Poulton, E. B. 1892. Further experiments upon the colour-relation between certain lepidopterous larvae, pupae, cocoons, and imagines and their surroundings. Transactions of the Entomological Society of London (4): 293–487.

Roff, D. A. 1996. The evolution of threshold traits in animals. The Quarterly Review of Biology 71(1): 3–35.

Schlichting, C. D. 1989. Phenotypic integration and environmental change. BioScience 39(7): 460–464.

Schlichting, C. D. and M. Pigliucci. 1998. *Phenotypic Evolution: A Reaction Norm Perspective*. Sinauer, Sunderland, MA.

Schmalhausen, I. I. 1949. *Factors of Evolution*. Blakiston, Philadelphia.

Shapiro, A. M. 1976. Seasonal polyphenism. Evolutionary Biology 9: 259–333.

Smith-Gill, S. J. 1983. Developmental plasticity: Developmental conversion *versus* phenotypic modulation. American Zoologist 23: 47–55.

Stearns, S. C. 1989. The evolutionary significance of phenotypic plasticity. BioScience 39(7): 436–445.

Sword, G. A. 2002. A role for phenotypic plasticity in the evolution of aposematism. Proceedings of the Royal Society B: Biological Sciences 269: 1639–1644.

Uvarov, B. V. 1966. *Grasshoppers and Locusts*. Vol. I. Cambridge University Press, London.

Uvarov, B. V. 1977. *Grasshoppers and Locusts*. Vol. II. Centre for Overseas Pest Research, London.

Weismann, A. 1882. *Studies in the Theory of Descent, Vols. I and II*. Sampson Low, Searle and Rivington, London, reprinted by AMS Press, New York. 1975.

Wente, W. H. and J. B. Phillips. 2003. Fixed green and brown color morphs and a novel color-changing morph of the Pacific tree frog *Hyla regilla*. American Naturalist 162(4): 461–473.

West-Eberhard, M. J. 2003. *Developmental Plasticity and Evolution*. Oxford University Press, Oxford.

Wigglesworth, V. B. 1961. Insect polymorphism—a tentative synthesis. In: *Insect Polymorphism*. J. S. Kennedy (ed.), Symposia of the Royal Entomological Society of London 1: 103–113.

Wilson, R. S. and C. E. Franklin. 2002. Testing the beneficial acclimation hypothesis. Trends in Ecology & Evolution 17: 66–70.

Woltereck, H. R. 1909. Weitere experimentelle Untersuchungen über Artveränderung, Speziell über das Wesen quantitativer Artunterschiede bei Daphniden. Deutsche Zoologische Gessellschaft Verhandlungen (Leipzig) 19: 100–172.

Wu, C.-T. and J. R. Morris. 2001. Genes, genetics and epigenetics: a correspondence. Science 293: 1103–1105.

3

Phenotypic Plasticity and the Origins of Diversity: A Case Study on Horned Beetles

Armin P. Moczek

Department of Biology, Indiana University, 915 E. 3rd Street, Myers Hall 150 Bloomington, IN 47405-7107, phone: (812) 856 1468; Fax: (812) 856 6705 E-mail: armin@indiana.edu

Abstract

Phenotypic plasticity has long been proposed to play a major role in the origin and subsequent diversification of morphological and behavioral novelties. Polyphenism, as an extreme yet common case of phenotypic plasticity, is thought to be a particularly important facilitator of rapid evolution and diversification. In the first part of this chapter I review how phenotypic plasticity and polyphenic development are thought to contribute to and shape patterns of morphological and behavioral innovation and diversification in insects. I then apply these insights to a highly diverse and speciose group of insects: horned beetles. In horned beetles polyphenic development is involved in the production of alternative horned and hornless male morphologies which are used in the context of alternative reproductive tactics. It is in these species that phenotypic plasticity produces some of the most exaggerated and diverse secondary sexual traits known in insects. In the second part of this chapter I explore the roles of developmental and behavioral plasticity in the origins of diversity among horned beetles. I review the physiological and developmental mechanisms that regulate the expression of male phenotype and examine the ecological and behavioral context within which horn polyphenic beetles function. I then explore how ecology and behavior have shaped aspects of phenotypic plasticity in natural populations, and how plasticity in turn has contributed to and directed the evolutionary diversification of horn polyphenic beetles. I end by discussing how such insights, combined with recent novel approaches, can help in uncovering the evolutionary origins of phenotypic novelties and the causes and mechanisms of their subsequent diversification.

Introduction

Phenotypic plasticity is a universal property of all forms of life, from phages to higher multicellular organisms (e.g. Ptashne 1992, Sultan 1987, Nijhout 1999). It transcends taxonomic boundaries as well as organizational levels within individuals. Transcription, translation, cell proliferation, organ function, endocrine and neuronal regulation, mating behavior and so on—all are phenotypically plastic in one way or another, and being able to respond flexibly to changes in external conditions is an integral part of their proper functioning (West-Eberhard 2003). Even though its evolutionary implications were recognized early on, phenotypic plasticity has only recently regained attention from evolutionary biologists (Baldwin 1902, Schmalhausen 1949, Waddington 1953, West-Eberhard 1989, 1992, Stearns 1989, Sultan 1992, Nijhout 1999, Schlichting and Pigliucci 1998, Pigliucci 2001). Opinions on its contributions to evolutionary processes, however, vary widely (reviewed in West-Eberhard 2003). In this chapter, I will review the role of phenotypic plasticity in insect evolution. In particular, I will examine the importance of phenotypic plasticity for one of the most significant, and arguably most poorly understood, phenomena in evolutionary biology: the origin and diversification of phenotypic novelties. I will then apply these insights to a group of insects that has been among the focal taxa for studies on the evolution and development of phenotypic plasticity: horned beetles. Because much of the debate about the importance of phenotypic plasticity in evolution has at least in part been due to an inconsistent use of terminology I will begin this chapter with a few definitions.

Definitions

For the most part my definitions of key terms such as *phenotype* and *plasticity* follow those of West-Eberhard (2003). Accordingly, I use the term *phenotypic plasticity* in its broadest definition, that is, I consider the phenotype as including all traits of an organism, whether they are physiological, morphological, or behavioral. Furthermore, while many authors use separate terms in reference to adaptive and non-adaptive, active and passive, reversible and irreversible phenotypic plasticity, or plasticity that generates a continuous or discontinuous range of phenotypes, I consider all of the above different manifestations of the same fundamental property of an organism, namely its ability to respond to environmental stimuli via changing the expression of a phenotype. I will, however, use *polyphenism* and *polyphenic development* to refer to organisms in which individuals are able to express two or more discretely different morphologies in response to external conditions. This is not meant to qualitatively separate polyphenism from plasticity, but rather to emphasize its extreme nature in many taxa. Furthermore, I will use terms such as *developmental* or *behavioral* plasticity to emphasize the context within which a particular plastic response is generated.

Phenotypic Plasticity and the Origins of Diversity

The main purpose of this section is to outline the mechanisms by which phenotypic plasticity can mediate the origin, exaggeration, and diversification of novel phenotypic traits in insects. However, to evaluate whether these mechanisms are likely to be of general importance in evolution or only relevant for special cases, and whether the phenotypic diversity generated by them is sufficient to mediate the evolution of complex traits, we must first understand the phenotypic and taxonomic range of phenotypic plasticity.

The Pervasiveness of Plasticity

Examples of phenotypic plasticity abound in the insects. They include extreme cases such as social castes in the Hymenoptera, termites and aphids (Wheeler and Nijhout 1983; Lüscher 1960, Stern and Foster 1996), seasonal polyphenisms in butterflies (Shapiro 1976), dispersal poyphenisms in a wide range of insects (Zera and Denno 1997), alternative asexual and sexual reproductive phases in aphids (Moran 1991), and alternative male morphologies in thrips (Crespi 1988) and beetles (Emlen 1994). In all these cases individuals have the ability to develop into one of two or more very different phenotypes, and decide based on genetic and environmental inputs which one to express. Many of these cases also involve the facultative expression of what in other contexts are considered important evolutionary transitions, such as the absence or presence of wings in ant castes (Abouheif and Wray 2002), the absence or presence of pattern elements on butterfly wings (Nijhout 1991), or the alternation between asexual and sexual modes of reproduction in aphids (Moran 1992). Distinct life stages are another form of extreme phenotypic plasticity. Here, each individual has the ability to consecutively express two or more discrete phenotypes. The holometabolous insects are thought to owe their evolutionary success in part to the extreme division of larval and adult stages (Yang 2001). The drastic differences in morphology, physiology, and behavior between the larval and adult stages of groups such as the butterflies, bees or beetles attest to the remarkable phenotypic range that alternative phenotypes can accommodate (Gullan and Cranston 2000). By focusing on extreme cases of phenotypic plasticity it is, however, easy to overlook that phenotypic plasticity and alternative phenotypes are in reality far more widespread, if not ubiquitous. For example, many aspects of insect growth and development are influenced profoundly by external conditions such as temperature and nutrient availability (Allegret 1964, Beck 1971a, Blakley and Goodner 1978), and many insects can respond to changes in environmental conditions flexibly and adaptively by altering their behavior, physiology, and development (Beck 1971b, Shafiei et al. 2001). Alternative reproductive tactics, in which individuals switch facultatively between different behaviors to acquire mates, were once thought of as special cases that evolve only under rather unique conditions (e.g. Gadgil 1972), but have now been described in many insect orders and appear commonplace (Thornhill and Alcock 1983, Shuster and Wade 2003). Other types of behavior such as foraging, feeding or provisioning behavior, are similarly plastic in many insects (Mitchel 1975, Tanaka 1985, Field 1992, Sowig 1996, Moczek 1998, 1999). Phenotypic plasticity and alternative phenotypes are thus not only taxonomically widespread, they also occur at every part of the phenotype, whether behavioral, physiological, or developmental.

The processes that mediate such widespread plasticity themselves operate on a wide range of levels. For example, the expression of different castes in social insects involves differential gene expression (Evans and Wheeler 1999, 2001), differences in endocrine physiology (Wheeler and Nijhout 1983, 1984), differences in the regulation of appendage growth and development (Abouheif and Wray 2002), differential development of reproductive organs (Passera and Suzzoni 1979, Otto 1962), differences in behavioral repertoires (Wilson 1976) etc. The expression of seasonal morphs in butterflies involve differences in behavior (Brakefield and Reitsma 1991), endocrine physiology (Koch and Bückmann 1987, Roundtree and Nijhout 1995), pigment synthesis (Koch 1995), and so on. At each of these levels genetic and environmental inputs onto the already existing phenotype determine subsequent patterns of phenotype expression. As we will see, the range of levels at which plastic responses can be mediated, and the joint contributions of genetic and environmental factors in guiding phenotype expression, have important consequences for understanding the contributions of plasticity to patterns of phenotypic diversification.

Diversity without Specialization

Phenotypic plasticity allows individual genotypes to express multiple phenotypes as a function of environmental inputs (Nijhout 1999). How often a given phenotype is expressed then becomes a function of the frequency with which certain environmental conditions recur. Whenever a particular phenotype is expressed, it can be subject to selection and modification through subsequent generations. In the absence of the inducing environment, however, a given phenotype may disappear from a population and be replaced by an alternate phenotype, which itself becomes the subject of selection. Phenotypic plasticity thus provides the opportunity for the independent evolution and adaptation of different phenotypes to different sets of environmental circumstances (West-Eberhard 1989, 2003). Exactly how independent different evolutionary trajectories can be is likely to depend, among others, on the extent to which genetic and developmental regulatory mechanisms are shared among alternatives. That alternative phenotypes can, at least in some cases, evolve rather independently of one another is suggested by the often extreme physiological and morphological differences that exist between alternative morphs, e.g. between queens, workers and soldier in ants (Hölldobler and Wilson 1990). In such cases the evolution and elaboration of developmental switch mechanisms are thought to play a central role in mediating the diversification of alternative phenotypes (Nijhout 1999). For example, hormonally mediated threshold responses are an extremely widespread component of many developmental switch mechanisms in insects, e.g. in determining the timing of pupation in butterflies (Nijhout 1976), the expression of winged and wingless morphs in crickets (Cisper et al. 2000), the determination of castes in social insects (Wheeler and Nijhout 1983, 1984) or the expression of alternative male phenotypes in horned beetles (Moczek and Nijhout 2002a). In these cases, developmental switches permit the coordinated and integrated expression of a large number of phenotypic traits in response to changes in environmental conditions, and as such largely decouple two or more suites of phenotypic characters from each other (Nijhout 1999). This has several important implications for the evolution of alternative phenotypes. Once uncoupled by a developmental switch, alternatives can follow evolutionary trajectories that are less dependent on each other, which in turn can facilitate the further specialization and divergence between alternatives. At the same time, the switch mechanism itself can become a target of selection, opening up a previously unavailable avenue for phenotype evolution.

Switches as Targets of Selection

Switch mechanisms, whether behavioral, developmental or physiological, allow individuals to adjust the expression of a large suite of phenotypic traits in response to changes in external conditions. As such switches are potent modifiers of phenotype expression. Not surprisingly, even subtle evolutionary changes in exactly how a switch operates can have a profound impact on patterns of phenotype expression. Particularly illuminating examples come from studies of scaling relationships in polyphenic organisms (Emlen and Nijhout 2000). For example, body parts of castes in social insects, or body size and length of secondary sexual traits in males in numerous insects, often exhibit species-specific scaling relationships, or allometries (Emlen and Nijhout 2000). The exact shape of a given allometry is in part a product of the developmental switch mechanisms involved in the production of body parts during larval development (Wheeler 1991, Nijhout 1994, Nijhout and Wheeler 1996). In ants, caste determination occurs as early as during embryonic development in case of queens (Passera and Suzzoni 1979) or relatively early in larval development in case of worker and soldier castes (Wheeler and Nijhout 1983, 1984, Wheeler 1991). As a consequence, the developmental trajectories of different castes are decoupled while individuals have yet to undergo a significant portion of their growth. This relatively early onset of independent growth trajectories in different incipient castes allows the resulting adult phenotypes to be at times extremely discontinuous and scaling relationships of body parts to be non-overlapping (Fig. 1a; Wilson 1978, 1985, Moffett 1987, Wheeler 1991). In contrast, whether or not a male beetle develops into a horned, major morph or a hornless, minor morph is determined relatively late in larval development. Here the developmental switch involved in determining the subsequent fate of a male larva may occur as little as 72 h before pupation and thus after larvae have already completed almost all of their growth (Moczek and Nijhout 2002a). As a consequence, while major and minor morphs may differ dramatically in the degree of horn development, the remaining morphology is, for the most part, unaffected. Similarly, the resulting scaling relationships between horn length and body size may be highly non-linear, but typically remain continuous and with broad overlap between alternative morphs (Fig. 1b; Moczek 1998). The ontogenetic timing of a developmental switch thus can have important consequences for patterns of phenotype expression.

Changes in ontogenetic timing are not the only means by which evolutionary changes in developmental switches can contribute of phenotypic

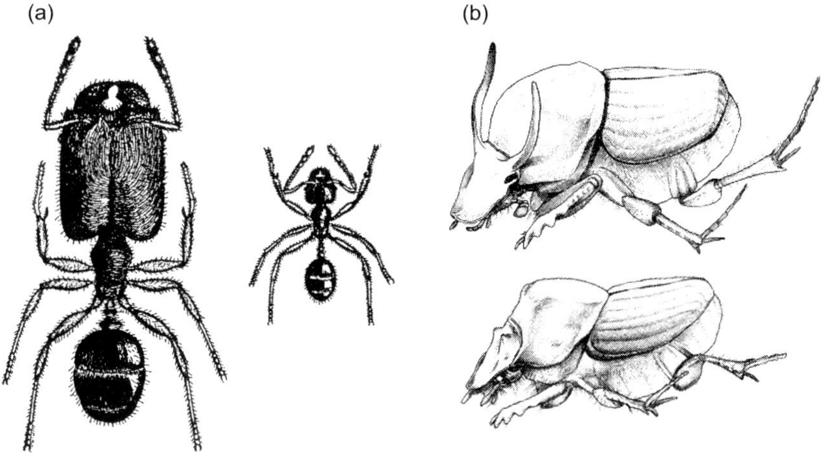

Fig. 1 Alternative phenotypes produced by developmental switches. **(a)** Different worker forms in the ant *Pheidole tepicana*. Developmental switches operating relatively early during larval development allow ants to produce highly divergent alternative phenotypes. Note relative sizes of head and alitrunk (fused thorax and first abdominal segment; after Wheeler 1910). **(b)** Alternative horned and hornless males in the beetle *Onthophagus taurus*. A developmental switch operating late during larval development causes adult males to differ dramatically in horn length while the remaining morphology is largely unaffected (drawings by Shane Richards).

diversity. Switches between alternative phenotypes typically employ a particular response threshold, e.g. to crowding (Denno et al. 1986), food quality (Moczek 1998) or photoperiod (Tauber and Tauber 1970), and usually generate a phenotypic transition of a certain well-defined magnitude and steepness. Threshold values, steepness and magnitude are all at least in part properties of the developmental switch itself (Nijhout and Wheeler 1996). Evolutionary changes in the developmental switch mechanism thus can modify one or more of these aspects of phenotype expression (Fig. 2). For example, size-dependent expression of alternative morphs in horned beetles is regulated at least in part via juvenile hormone titers present during certain sensitive periods (Emlen and Nijhout 1999). Experimental changes in JH titers (Emlen and Nijhout 1999), or evolutionary modifications in the sensitivity to JH (Moczek and Nijhout 2002, 2003), drastically change the body size at which adults switch between alternative morphs, which in turn results in substantial alterations of the average scaling relationship between body size and horn length (Moczek 2003, see also below). Developmental switches thus provide evolutionary mechanisms with an additional powerful set of targets by which phenotype expression can potentially be modified.

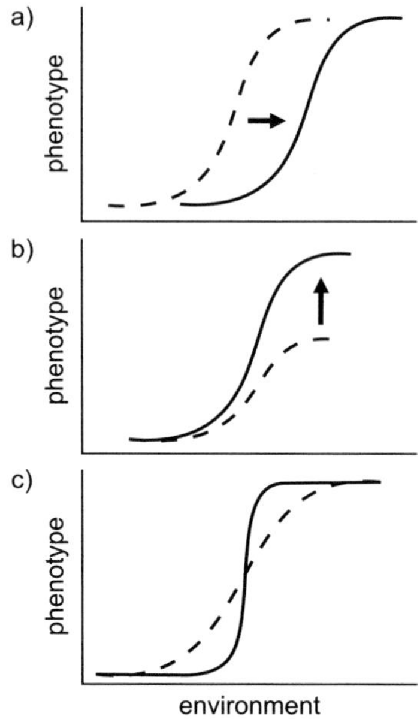

Fig. 2 Different aspects of developmental switches can become modified by selection. **(a)** Changes in the critical environment in which the switch between phenotypes occurs. **(b)** Changes in the magnitude of the phenotypic response to changes in environmental conditions. **(c)** Changes in the steepness of phenotypic transitions as environments change.

The next section highlights several important consequences of evolutionary changes in developmental switches.

Developmental Divergence and Speciation

An important consequence of evolutionary changes in developmental plasticity involves its potential to initiate the evolution of reproductive isolation between developmentally divergent populations. Even though plasticity *per se* increases an organism's ability to respond to a range of environmental conditions, different sets of environments are likely to favor different kinds of plastic responses. For example, geographic comparisons and breeding experiments on a wide range on insect taxa have illustrated that response thresholds can vary heritably and adaptively among populations (e.g. Tauber and Tauber 1972, 1982, 1987, Harrison 1979, Hazel and West 1982, Denno et al. 1986, Semlitsch and Wilbur 1989, Semlitsch et al. 1990, Emlen 1996, Ahlroth et al. 1999). Such between-population divergence in plastic phenotype expression may have important consequences once divergent populations establish contact. Divergence in particular traits, such as timing of pupation, eclosion and mating (Shafiei et al. 2001), or choice of host organism (Moran 1991) may alone be sufficient to reproductively isolate divergent populations spatially or temporally, even if they co-occur geographically. Even if premating isolation is incomplete or absent at first, between-population divergence in plastic phenotype expression may still facilitate speciation by reducing hybrid fitness. Hybrids may exhibit reduced fitness due to the expression of a sub-optimal phenotype (in this case a sub-optimal plastic response), which in turn may favor the spread of alleles that facilitate assortative mating within each parental population (Porter and Johnson 2002). While such a scenario should apply to all traits with important fitness consequences, there may be reasons to believe that phenotypically plastic trait expression, and especially developmental-switch mediated alternative phenotypes, might be particularly prone to initiate reproductive isolation through developmental divergence. Because developmental switches regulate the simultaneous expression of whole suites of phenotypic traits, between-population divergence in developmental switches instantly causes populations to diverge in a large number of phenotypic traits. Increasing the number of phenotypic traits involved in a divergence increases the probability that hybrids will express some type of sub-optimal phenotype for at least some traits under at least some conditions, which should intensify selection for premating or prezygotic isolation. While theoretical models (e.g. Porter and Johnson 2002) lend support to such scenarios, more comparative and experimental work is clearly needed to examine if and how plasticity contributes to the origins of reproductive isolation.

Loss and Recurrence of Alternatives

One important avenue through which phenotypic plasticity and polyphenic development can influence phenotypic diversity is through the loss of alternatives. For example, the loss of alternative phenotypes is thought to have played a major role in life cycle evolution of aphids (Moran and Whitham 1988, Moran 1991) and social evolution in halticid bees (Richards 1994). Loss of alternatives, though at first sight a reduction in phenotypic diversity, can nonetheless facilitate phenotypic diversification through a variety of very interesting mechanisms. First, the ability to produce alternative phenotypes may limit the kinds of genetic modifications that a polyphenic population can accommodate. Deletion of one or several alternative phenotypes would remove such constraints. The remaining phenotypes would then be free to incorporate genetic modifications previously selected against, and as a consequence should become increasingly specialized to their particular environment (Moran 1988, West-Eberhard 1989, 2003). Similar to the argument presented in the last section, *differential* deletion of alternatives in different populations of the same species has the potential to instantly generate reproductive barriers and facilitate speciation (West-Eberhard 1989, 2003).

Loss of alternative phenotypes does, however, not have to be permanent. Instead, lost alternatives may reappear in temporarily monomorphic lineages, possibly leading to the subsequent fixation of the recurrent phenotype. Well-known examples of recurrent gain and loss of phenotypes include the paedomorphic and metamorphic populations of salamanders (Shaffer 1984), sword-bearing and sword-less species of swordfish (Schluter et al. 1997) and directly and indirectly developing sea urchins (Raff 1996). Among the insects, a remarkable example of recurrence is the repeated reappearance of wings among secondarily wingless stick insects (Whiting et al. 2003). Recent phylogenetic analyses of the Phasmatodea provide strong evidence that wings, and thus flight, have been re-gained as many as four times independently, reversing the earlier loss of wings in these insects prior to their diversification (Whiting et al. 2003). Interestingly, in two clades recurrence of wings appears to have then been followed again by loss of wings in a subset of taxa. While estimates of the exact number of gains and losses depend on the weighing of their respective probabilities, these data nonetheless provide striking support for the idea that loss of complex traits is reversible and need not be an evolutionary dead end. What determines whether the loss of a phenotype is permanent or potentially reversible? The answer to this question most likely lies in the developmental genetic basis of a given trait. The recurrence of wings among wingless stick insects would appear, at first, highly unlikely. Wings are complex traits whose development and function require the coordinated integration of wing tissue growth, muscle growth and attachment, innervation and so on (Dudley 2000a,b). Once wings are lost we would expect the requisite genetic regulators to accumulate mutations that should increasingly disrupt their function. However, if we look more carefully at how wing development is regulated, we may understand why in this case, maintenance of at least large portions of the machinery necessary for wing production might be feasible even when the final phenotype is not expressed. First, wing development in insects relies in large part on the same regulatory genes involved in patterning legs and other appendages (Campbell et al. 1993, Campbell and Tomlinson 1998, but see Kubota et al. 2003). Since the protein products of these genes function in other important developmental contexts this should prevent the accumulation of mutations in at least the protein coding regions of these genes. Furthermore, in at least some insects, wing and leg primordia are derived from the same pool of embryonic cells (Cohen et al. 1993, Kim et al. 1996). In the early *Drosophila* embryo, interactions between two diffusible morphogens, *wingless* and *decapentaplegic*, determine which cells will develop into imaginal disks. This pool of cells subsequently undergoes a separation into wing and leg imaginal disk precursor cells. Leg and wing imaginal disks thus have a common origin in at least some insects (Cohen et al. 1993, Kim et al. 1996). One implication of these observations is that even though a complex phenotype may be absent in adults, large portions of the embryonic and larval developmental machinery required for its expression will remain intact because they are shared with, and integrated into, other developmental processes. If this perspective turns out to be correct, we would predict that phenotype recurrence should be more likely the greater the extent to which underlying developmental mechanisms are shared with and integrated into other developmental processes.

Phenotypic Plasticity and Diversity in Horned Beetles

The preceding sections outlined some of the major mechanisms through which phenotypic plasticity is thought to influence and contribute to phenotypic evolution. We are now in a position to apply these insights to horned beetles, a highly diverse and speciose group of insects that recently has been among the focal taxa for studies on the evolution and development of morphological diversity (Emlen 2000). The following sections introduce several important aspects of the biology of horned beetles. I begin by describing the most important patterns of morphological variation in horned beetles. I then summarize what is known about the proximate genetic and physiological mechanisms that generate this variation and review the behavioral and ecological context within which different morphologies function. In the following sections I draw heavily from work on the scarabaeid genus *Onthophagus*, where recent studies have accumulated the most extensive knowledge of the evolutionary ecology and physiology of any group of horned beetles. I include, also, studies on other beetle taxa and attempt to determine the extent to which patterns found in onthophagine beetles are likely to be true for horned beetles in general.

Morphological Diversity in Horned Beetles: within Species

Horns and horn-like structures in beetles have attracted the attention of biologists since Darwin's time (Wallace 1869, Darwin 1871, Arrow 1899). Early naturalists recognized not only the absolute size but also extraordinary variability in the expression of horns between and within species (Figs. 3 and 4;Arrow 1951). However, it was not until the 1990's that experimental evidence demonstrated the nature and source of some of this extreme variability (Emlen 1994; see below). Within species of horned beetles scientists noted two general patterns of morphological diversity: the relative absence of horns in females, and the often extraordinary variability of horns in males (Fig. 3; Paulian 1935). In the vast majority of species females show no or greatly reduced expression of horns compared to their male counterparts (von Reichenau 1881). If females do develop horns it is typically a similar horn *type* as in males, e.g. a paired head horn, or a single pronotal horn (Balthasar 1963). Both patterns extend across all groups of horned beetles, and exceptions are rare (Arrow 1951). One such exception is *Onthophagus sagittarius*, in which females not only develop relatively larger horns than males, but also a completely different type of horn (Fig. 3c). But in the vast majority of species horn expression is either restricted to, or much more pronounced, in males. Furthermore, it is also exquisitely variable. So much, in fact, that males with low and high levels of horn development were sometimes classified as belonging to separate species (Paulian 1935). Because of this extreme variability in trait expression, one common way of describing the morphology of a given species of horned beetle is by use of a static allometry, or scaling relationship, in which the horn length of individuals of different body sizes is graphed as a function of body size (Eberhard and Gutiérrez 1991, Emlen and Nijhout 2000, Moczek and Nijhout 2003). The shape of such scaling relationships can range from linear to broken and sigmoidal in different species of horned beetles (Fig. 5; Rasmussen 1994, Hunt and Simmons 1997, Moczek 2002). Because linear scaling relationship typically exhibit a slope >1, all three types of scaling relationships cause large males to not just be scaled-up, enlarged versions of their smaller conspecifics, but to develop a fundamentally different morphology (Emlen and Nijhout 2000). Interestingly, the exact shape of a given scaling relationship can vary subtly to dramatically between populations of the same species, suggesting that conspecific populations can differ in switch mechanisms and resulting developmental trajectories (Moczek and Nijhout 2003, Moczek 2003). The following sections therefore explore the relationship between intra- and interspecific patterns of variation among horned beetles.

Fig. 3 Examples of intraspecific variation in horn development in *Onthophagus* beetles. **(a)** *Onthophagus nigriventris:* males above a certain body size develop a large, pronotal horn (left), whereas smaller males develop only a small, rudimentary horn (top right) and females remain entirely hornless regardless of body size (bottom right). **(b)** *O. watanabei:* Large (left) and small (center) males develop a pair of head horns, though horn development is relatively greater in large males. Large, but not small, males also express a central pronotal horn. Female *O. watanabei* develop a relatively small paired head horn and no pronotal horn. **(c)** *O. sagittarius*: This species is very unusual in that large (left) and small (right) males develop only a pair of minor head horns, while large females develop a single head and pronotal horn much larger in size than horns of males of similar body sizes.

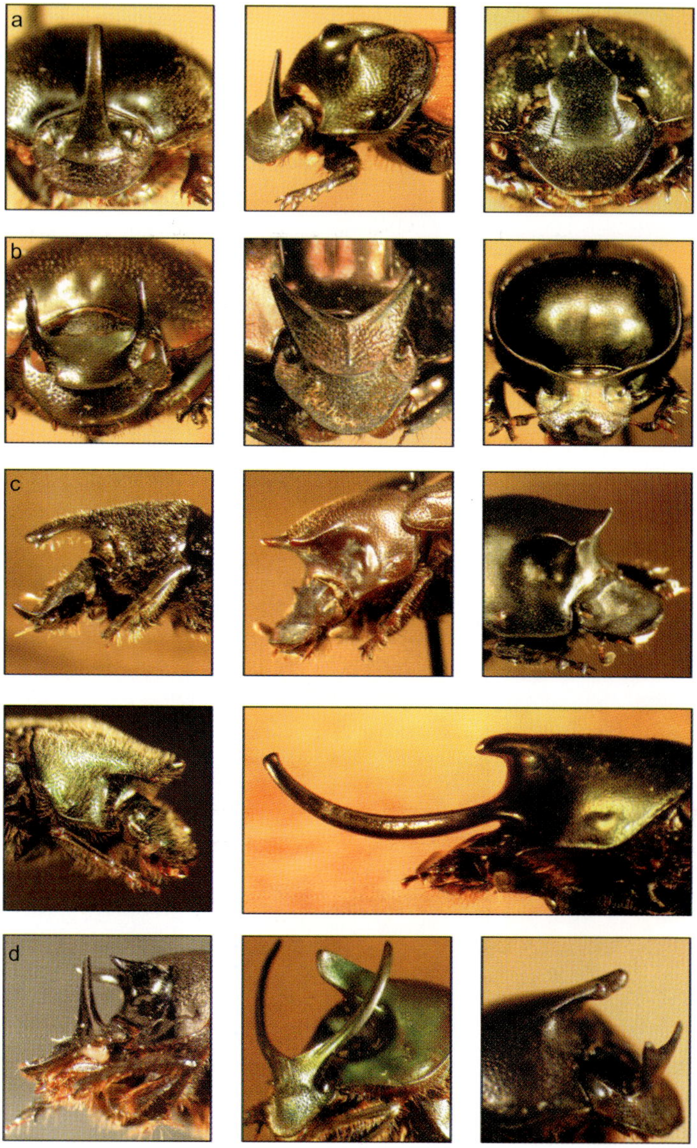

Fig. 4 Interspecific variation in size, shape, location, and number of horns in *Onthophagus* beetles. **(a)** Single head horns in (left to right): *O. spec* (unknown species; Vietnam), *O. insignis* (Malawi), *O. vacca* (India). **(b)** Paired head horns in *O. gazella* (S-Africa), *O. taurus* (U.S.A.), *O. watanabei* (Borneo). **(c)** Single pronotal horns in (top): *O. hecate* (U.S.A.), *O. turbatus* (U.S.A.), *O. binodis* (S-Africa); (bottom) *O. medorensis* (U.S.A.), *O. nigriventris* (Kenya). **(d)** Various combinations of head horns and pronotal horns in *O. ferox* (Australia), *O. atripennis* (Thailand), *O. lunatus* (Vietnam).

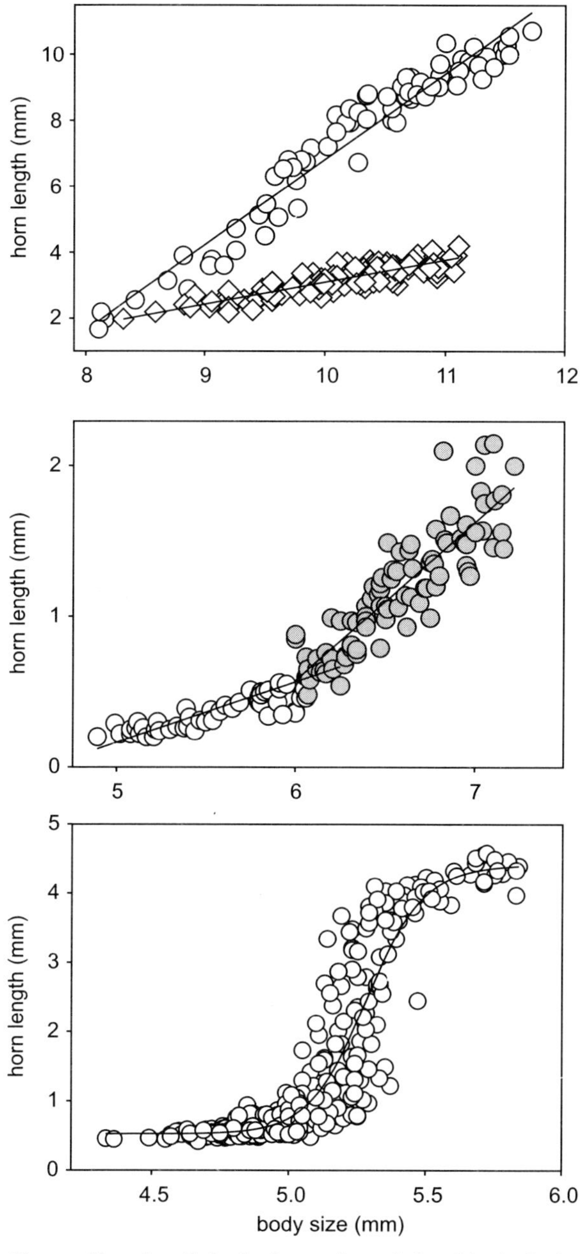

Fig. 5 Different types of horn length-body size scaling relationships in *Onthophagus* beetles. Top: linear allometries of paired head horns in male (open circles) and female (open diamonds) *O. watanabei*. Center: broken allometry of pronotal horn length in male *O. binodis*. Bottom: sigmoidal allometry of paired head horns in male *O. taurus* (after Moczek et al. 2004).

Morphological Diversity in Horned Beetles: between Species

Horns or horn-like structures have evolved independently in a number of beetle families, such as the Tenebrionidae (e.g. Pace 1967, Conner 1989), Staphilinidae (Darwin 1871), Passalidae (MacGown and MacGown 1996), Curculionidae (Eberhard and Garcia-C. 2000, Eberhard et al. 2000) or Chrysomelidae (Eberhard 1981, Windsor 1987). Beetle horns are, however, most extreme and most diverse in the chafers or scarab beetles (Scarabaeidae, e.g. Arrow 1951, Balthasar 1963, Matthews 1972). Horns or horn-like structures can develop from the clypeus (mouth plate), head, or thorax, horns may appear singly or paired, and different species may exhibit different combinations of single or paired horns produced by different regions of the body. In some cases such extreme variation in horn types may exist in a single genus, such as *Onthophagus* (Fig. 4; see also Balthasar 1963, Matthews 1972, Howden and Young 1981). Typically, however, closely related species exhibit similarities in horn development, allowing us to make inferences about the processes that have mediated this spectacular diversification. For example, closely related species usually express the same type of horn. What then distinguishes these species are differences in the exact *scaling* between horn length and body size. For example, species in the *Onthophagus incensus* group (Fig. 6; Emlen 1996) all develop paired head horns, yet differ widely in the range of horn lengths (*amplitude*), the body size at which the scaling relationship transitions rapidly from hornless to horned morphologies (*switch point* or *threshold*), and the steepness of the slope that characterizes this transition (*slope*). Differences in body size thresholds and slope have also been documented between species that otherwise cannot be distinguished by external characters alone. For example, *O. taurus* and *O. illyricus* are sympatric through large portions of the Mediterranean, and taxonomists have debated whether they should be classified as variants, subspecies, or true species (Balthasar 1963, Baraud 1992, Lohse and Lucht 1992). Currently, both are considered separate species, largely because of consistent differences in male genital morphology (Lohse and Lucht 1992). Apart from genital characters, however, both species are extremely difficult to distinguish. Males of either species develop very similar hornless and horned morphologies and transition from one to the other over a very narrow range of body sizes. Allometric analyses, however, revealed species-specific differences in slope and the exact location of body size thresholds (Fig. 7; Moczek and Nijhout 2003). Combined, these data suggest that changes in certain aspects of the scaling relationships between body size and horn length may constitute important avenues for phenotype diversification. However, to understand

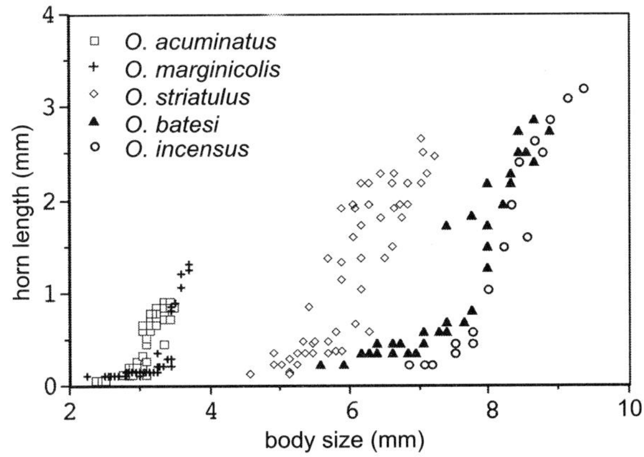

Fig. 6 Horn length-body size scaling relationships in the *Onthophagus acuminatus* group (after Emlen 1996, with permission).

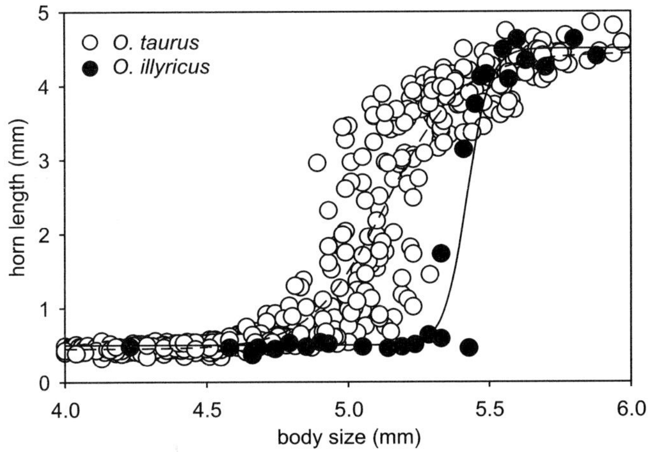

Fig. 7 Horn length-body size scaling relationships in *Onthophagus* sister species, *O. taurus* and *O. illyricus* (after Moczek and Nijhout 2002).

how and why scaling relationships might evolve on their own we first have to understand how and why beetles develop horns in the first place.

The Developmental Basis of Horns and Horn Dimorphisms

Whether or not a male beetle develops horns as an adult depends in large part on the nutritional conditions experienced during larval development (Emlen 1994, Hunt and Simmons 1997, Moczek 1998). This has been particularly well demonstrated in dung beetles where parents provision each egg with a discrete amount of dung, called a brood ball, in underground tunnels. Brood balls can be weighed and manipulated and thus offer an excellent means of quantifying the effects of nutritional environment on adult beetle phenotype (Moczek 1998). Using experimental manipulation of brood ball mass in combination with a controlled breeding design Emlen (1994) showed in the horn polyphenic beetle *Onthophagus acuminatus* that male body size and horn length *per se* exhibit no significant heritable variation. However, experimental manipulation of brood ball mass had a profound effect on offspring horn phenotypes, and male larvae with access to large brood balls developed into large, horned males with few exceptions (Emlen 1994). A subsequent study (Moczek and Emlen 1999) on *O. taurus* explored the effects of natural variation in brood ball weights in combination with a controlled breeding design, with similar results (Fig. 8). While horn length and body size again exhibited no significant heritability, natural variation in brood ball weights explained 39% and 36% of variation in body size and horn length, respectively (Moczek and Emlen 1999). While brood ball weight affected body size in a continuous fashion, length of horns

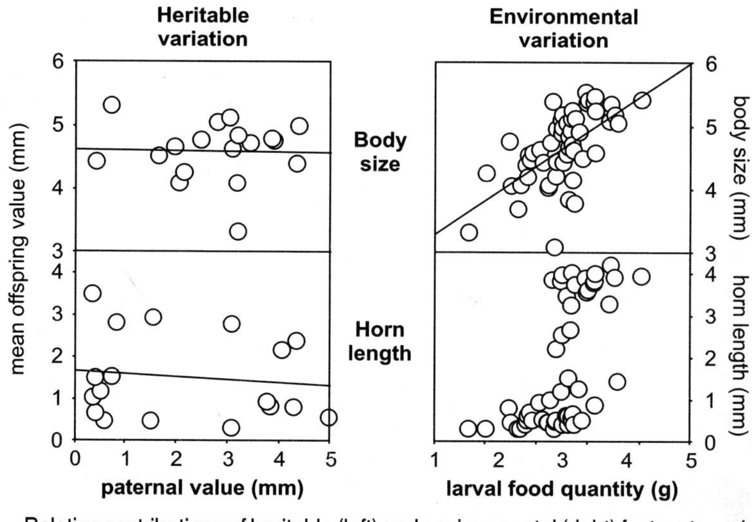

Fig. 8 Relative contributions of heritable (left) and environmental (right) factors to variation in male offspring phenotype in the horn polyphenic beetle *Onthophagus taurus*. Panels on the left show parent-offspring regressions of paternal (x-axis) body size (top) and horn length (bottom) against corresponding mean male offspring values (y-axis). Neither regression is significantly different from 0. Panels on the right show regressions of larval food quantity approximated as brood ball mass (x-axis) against individual male body size (top) and male horn length (bottom). Variation in brood ball mass explained 39% of the variation in male body size and 36% of the variation in male horn length. Note sigmoid distribution of data points in lower right panel. Modified after Moczek and Emlen (1999).

developed via a threshold response. Only males that exceeded a certain critical body size due to good nutrition developed horns, whereas males that developed to smaller sizes remained hornless (Moczek and Emlen 1999). Alternative horned and hornless morphologies are therefore not the manifestations of different genotypes. Instead, each individual male larva has the ability to develop into either morph. Horn dimorphisms are thus examples of polyphenic development, and as such are similar to caste determination in social Hymenoptera or seasonal polyphenism in butterflies. As we will see later on, individuals can, however, differ heritably in other aspects of horn development, with important consequences for patterns of phenotype diversification.

Horns themselves develop from imaginal disk-like tissues that undergo rapid and massive cell proliferation during the prepupal stage of late larval development (Emlen and Nijhout 1999). Because this growth occurs underneath the larval cuticle the resulting tissue cannot expand and instead undergoes massive folding underneath the larval cuticle (Fig. 9). Once the animal is ready to molt into a pupa and sheds its larval skin the folded-up horn tissue then becomes free to telescope outwards and to form the future adult horn. The timing and speed of horn development therefore resembles that of more conventional appendages such as legs, mouthparts and wings in holometabolous insects (Kim 1959, Schubiger 1971, Fristrom and Fristrom

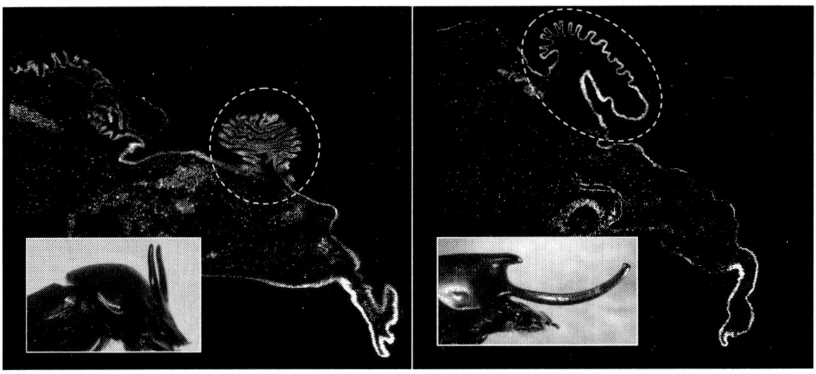

Fig. 9 Developmental basis of horn development. Horns develop during the prepupal stage at the end of larval development. Certain regions of the larval epidermis undergo rapid cell proliferation, which causes the resulting tissue to fold up underneath the larval cuticle. Once the animals molts into the pupa the horn tissue is free to expand into the pupal and subsequent adult horn. Shown are sagittal sections (DAPI stained to highlight nuclei) through head and thorax of incipient horned male *O. taurus* (left) and *O. nigriventris* (right). Future horns are highlighted by dashed line (after Moczek and Nagy 2005).

1993), which will become important later when we will explore the developmental origins of horns.

Recent research on *Onthophagus* beetles has identified some important components of the regulation of facultative, size-dependent expression of horns. *Onthophagus* larvae develop in underground brood balls, which are discrete and finite amounts of food provisioned for them by their parents (Halffter and Edmonds 1982). Larvae appear to use food availability as a cue to determine when to prepare for pupation. If larvae raised in artificial brood balls are removed from their food source at any time during the third and final instar they will initiate a stereotyped sequence of developmental transitions ultimately leading to pupation (Shafiei et al. 2001). Unlike numerous holometabolous insects, which require the attainment of a critical weight in order to pupate (e.g. Nijhout 1975), *Onthophagus* larvae can pupate at a wide range of body sizes and metamorphose into a wide range of adult sizes (Shafiei et al. 2001). This behavior appears adaptive since *Onthophagus* larvae do not have the option to locate additional food sources once their own brood ball is exhausted.

During development larvae are also somehow able to assess their own body size, predict their future adult body size, and adjust the subsequent development of horns accordingly (Emlen and Nijhout 2000). An important component of the regulatory mechanisms behind this appears to be juvenile hormone (JH), which is know to regulate a wide array of developmental processes in insects (Nijhout 1994, 1999). Several studies recently implicated juvenile hormone (JH) as an important endocrine regulator of horn development in beetles and changes in JH action as an important avenue for evolutionary diversification in phenotype expression (Emlen and Nijhout 1999, 2001; Moczek and Nijhout 2002, see below). In particular, earlier studies identified two brief sensitive periods in the last larval instar during which JH appears to determine the fate of developing larvae. During the first sensitive period, which occurs toward the end of the active feeding stage, application of the JH analogue methoprene causes larvae fated to develop into medium sized, horned males to suppress horn development and, instead, to develop into hornless males (Emlen and Nijhout 2001). During the second sensitive period around the gut purge and the onset of the prepupal stage, methoprene application has the opposite effect. Here, methoprene application to larvae fated to develop into small, hornless males causes them to develop into horned individuals instead (Moczek and Nijhout 2002). The presence or absence of sufficient JH titers during these two sensitive periods is therefore thought to determine which morph male larvae will develop into. Furthermore, the restriction of tissue sensitivity to very brief periods, combined with the late, explosive growth of the presumptive horn tissue in the prepupal stages, is thought to allow horned beetles to generate the highly non-linear, broken or S-shaped allometries discussed earlier (Nijhout and Wheeler 1996, Emlen and Nijhout 2000, Moczek and Nijhout 2002). The notion that JH and the relative timing of sensitive and growth periods influence scaling relationships is further supported by the observation that populations that differ in patterns of morph expression also differ in the degree and timing of sensitivity to JH during the second sensitive period (Moczek and Nijhout 2002, 2003). We will return to this point when we explore the mechanisms of morphological diversification in horned beetles. But first we have to understand what, if anything, horns might be good for.

The Behavioral Ecology of Horned Beetles

Several hypotheses have been proposed to explain the evolution and potential adaptive significance of beetle horns (reviewed in Arrow 1951). Horns have been suggested to serve as indicators of male quality to choosy females (Darwin 1871), a hypothesis that recently has been re-examined without generating supporting evidence (Cook 1990, Kotiaho 2002). Alternatively, horns were thought to protect against predators (Wallace 1869, in Arrow 1951), serve as digging implements (Lameere 1904), or allow beetles to perforate and lacerate plants to feed on their sap (Doane 1913). Arrow (1951) himself suggested that beetle horns might be functionless, selectively neutral, and possibly the incidental outcome of selection towards larger body size. Eberhard was among the first to present substantial evidence that beetles in a range of families use their horns primarily in male-male competition (Eberhard 1978, 1979, 1981, 1982, 1987, Eberhard and Garcia-C. 2000, Eberhard et al. 2000). Many subsequent studies have since confirmed this conclusion (Rasmussen 1994, Otronen 1988, Siva-Jothy 1987, Windsor 1987, Cook 1990, Emlen 1997a, Moczek and Emlen 2000, Hunt and Simmons 2002). Regardless of the variation in sizes, location, and number of horns in different species of horned beetles, horns are used largely, if not entirely, as weapons in male-male combat over access to females. In species that fight inside tunnels fights typically occur head to head and are largely shoving contests (Palmer 1978, Emlen 1997a, Moczek and Emlen 2000). In these species horns appear to serve mainly as positioning devices which allow fighting males to deliver powerful blows with their heads and thorax, but also as means to prevent intruders from passing in tunnels (Moczek and Emlen 2000). In species that fight above ground, males often use their horns to grab, lift and throw opponents (e.g. Beebe 1944, Siva-Jothy 1987), sometimes inflicting serious and occasionally fatal injuries. Using their horns male *Allomyrina dichotoma* may puncture the exoskeleton of their rivals and tear off elytra and hind wings. Throwing an opponent off a tree and onto the ground can result in the loss of appendages or massive cracks to the exoskeleton (Siva-Jothy 1987). A particularly interesting type of fight involving a particularly remarkable type of horn occurs in the weevil *Parisoschoenus expositus* (Eberhard and Garcia 2000, Eberhard et al. 2000). Here large males not only possess a pair of large, forward projecting, prothoracic horns, but also a forked tube or sheath that invaginates deep into the males prothoracic cuticle. During fights males interlock by inserting one of their horns into the sheath of his opponent. Males cannot use their own horns in fights unless they receive the other male's horn in their own sheath. Interlocked in this fashion males try to twist each other and lift each other from the substrate (Eberhard and Garcia 2000, Eberhard et al. 2000).

Horns are not only used in the context of fights but also measurably improve a male's chances of winning a fight. For example, in *Onthophagus taurus* males fight in subterranean tunnels underneath dung pads (Fig. 10).

Fig. 10 Mating system and alternative reproductive tactics in males and females of the horn polyphenic beetle *Onthophagus taurus* (drawings by Barrett Klein). Big picture: Adults beetles colonize dung pads and dig tunnels into the soil underneath, creating a complex, interconnected tunnel system. Females pack dung into the blind ends of tunnels to provision food for their offspring in the form of brood balls. Each brood ball contains one egg only and constitutes the sole amount of food available for a developing larva. Males compete with each other for access to females during tunneling and brood ball production (see below). Once females stop producing brood balls males desert and females fill the remaining tunnel space with the previously excavated soil. **(a-d)** Alternative male reproductive tactics: Large, horned males try to monopolize access to breeding tunnels and females through aggressive fighting behavior. Males guard tunnel entrances and engage in head-to-head combat with other males that try to enter the tunnel using their horns as weapons. Small, hornless males employ alternative sneaking behaviors to gain access to females when confronted with a physically superior opponent. Sneaking behaviors include **(a)** passing guarding males engaged in fights and **(b)** waiting near tunnel exits for females that collect dung for brood balls and mating above ground with these females while guarding males remain inside tunnels. Hornless males are also able to access breeding tunnels and females underneath guarding males via **(c)** the use of tunnel interceptions created by the digging activity of breeding females and **(d)** actively digging horizontal side tunnels to intercept breeding tunnels. **(e)** Alternative female reproductive tactics: Females typically reproduce by provisioning dung for larvae in the form of brood balls at the end of tunnels, but will switch opportunistically to intraspecific kleoptoparasitic behavior when encountering a brood ball produced by another female (after Moczek 1996, 1998, 1999; Moczek and Emlen 2000; Moczek and Cochrane, 2006).

Fig. 10 Contd. ...

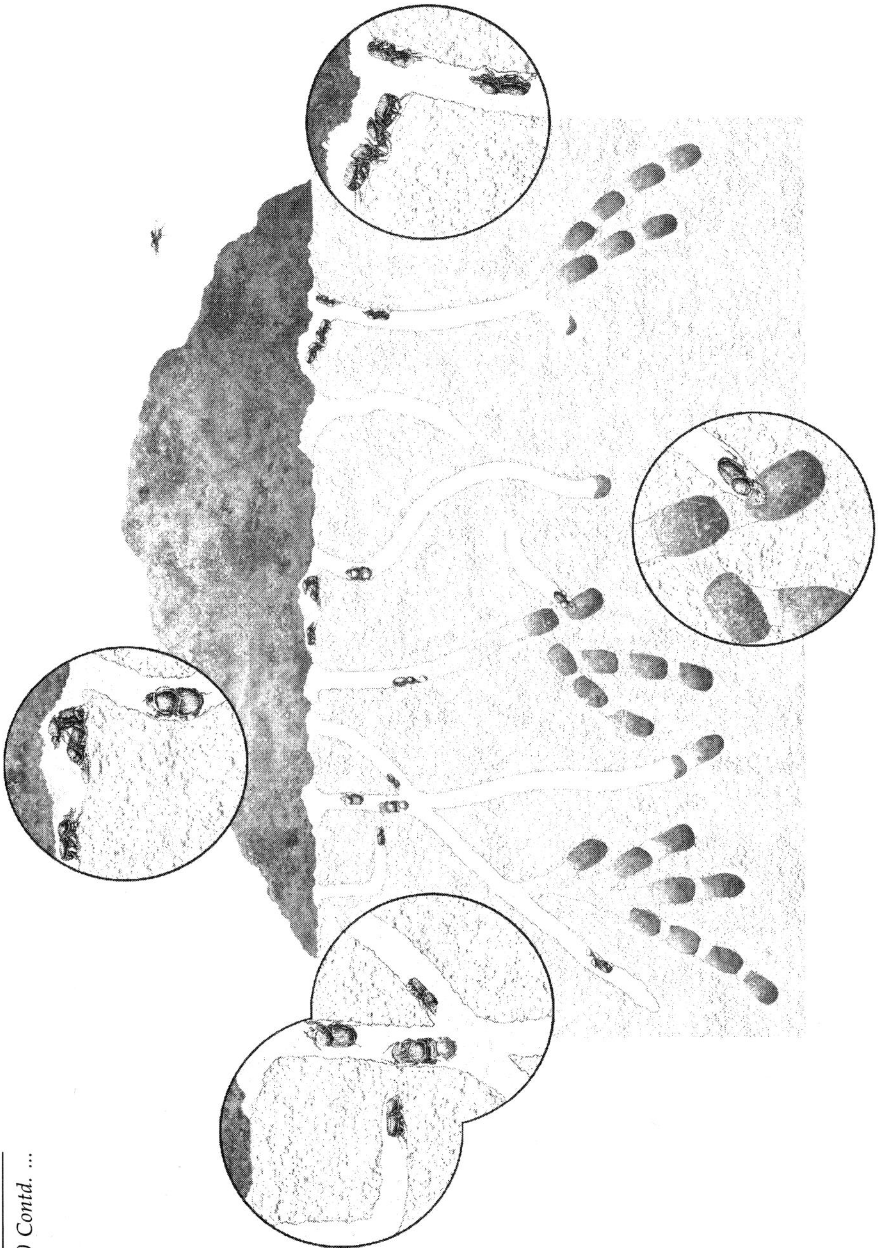

Fig. 10 Contd. ...

Large males almost always win fights against smaller males. However, in fights between similar sizes males horn length is an excellent predictor of outcome and large-horned males almost always defeat their small-horned but otherwise equally-sized contestants (Moczek and Emlen 2000). However, not all males engage in fights to access females. In horn-dimorphic species, small hornless males commonly withdraw from fights with physically superior males and engage in non-aggressive sneaking behaviors to gain access to females. In dung-breeding, tunneling species such sneaking behaviors may include the use of naturally occurring tunnel intersections or the digging of a shallow, horizontal tunnel to intercept a breeding tunnel underneath a guarding male (Cook 1990, Emlen 1997a, Moczek 1999, Moczek and Emlen 2000). Sneaker males may also wait next to tunnel entrances for females searching for dung and mate with them above ground, or wait for the guarding male to emerge to help in brood provisioning, in which case sneaker males will quickly enter the breeding tunnel and mate with the female underground (Moczek 1999, Moczek and Emlen 2000). In one species, *Onthophagus taurus*, the absence of horns in small, sneaking males has been shown to significantly improve their agility and maneuverability inside tunnels, which in turn is thought to increase their chances of locating and fertilizing breeding females despite the presence of a guarding male (Moczek and Emlen 2000). While individual sneaker males may not be successful at circumventing large, guarding males, a group of sneaker males will eventually overwhelm a guarding male and at least one sneaker male will mate with the female (Hunt and Simmons 2002). It is important to note, however, that the presence of horns is not a prerequisite for fighting. In *O. taurus*, hornless males will fight for hours over access to females, provided the opponent is himself a hornless male. Such fights are indistinguishable from those of their horned counterparts, except of course, for the use of horns. Nonetheless, combined the available evidence to date suggests that beetle horns are adaptive in the context of male-male combat. It also suggests that fighting and sneaking behaviors and corresponding horned and hornless male morphologies reflect alternative solutions to the same problem: securing breeding opportunities in the presence of competing males.

In summary, three important conclusions can be drawn from the above: 1) Horns are expressed in response to environmental conditions, in particular larval feeding conditions. In all horn polyphenic species studied so far individual male larvae have the potential to develop into a horned and hornless adult, and "decide" during late larval development which morph to develop into. 2) The decision which morph to develop into is mediated in part by JH, which acts during brief sensitive periods late in larval development. The dynamic interplay between JH mediated morph determination and the explosive, imaginal-disk like growth of future horns during prepupal development is thought to give rise to the species-specific, often highly non-linear scaling relationships observed in many species of horned beetles. 3) Alternative horned and hornless male morphs function in the context of alternative reproductive tactics. The success of a horned, fighting male has been shown to depend on his own body size relative to that of his opponents, the size of his horns, and the number of males he has to compete against at the same time. Hornless males in turn may benefit from the absence of horns through increased agility in locating females. We are now in a position to integrate the preceding chapters and to explore the consequences of polyphenic development for the origins of diversity among horned beetles.

The Consequences of Polyphenic Development in the Evolution of Horned Beetle Diversity

Scaling Relationships as Targets of Selection

Comparisons of horn length-body size scaling relationships between closely related species presented the first evidence that suggested that allometric parameters such as switch points or slopes might evolve independent of horns length *per se* (Kawano 1995, 1997, Emlen 1996). We now know from several studies that some of these components of scaling relationships indeed exhibit heritable variation and present important avenues for phenotypic diversification in horned beetles (Fig. 11). For example, artificial selection experiments (Emlen 1996), common garden breeding (Moczek et al. 2002) and large-scale geographic comparisons between isolated populations (Moczek 2003) have shown that switch points, or the critical body size at which males transition from hornless to horned morphologies, vary heritably between individuals and can evolve in different directions in different populations (Moczek and Nijhout 2003, Moczek 2003). The latter studies also indicated that such evolution can be extraordinarily rapid. Geographically isolated *Onthophagus taurus* populations have, since introduction to a new habitat less than 40 years ago, diverged in switch points to a degree typically only observed between species (Figs. 6, 7 and 11). Slight geographic differences were also detected in the steepness of the slope at the switch, though whether these differences are heritable remains to be investigated (Moczek and Nijhout 2003).

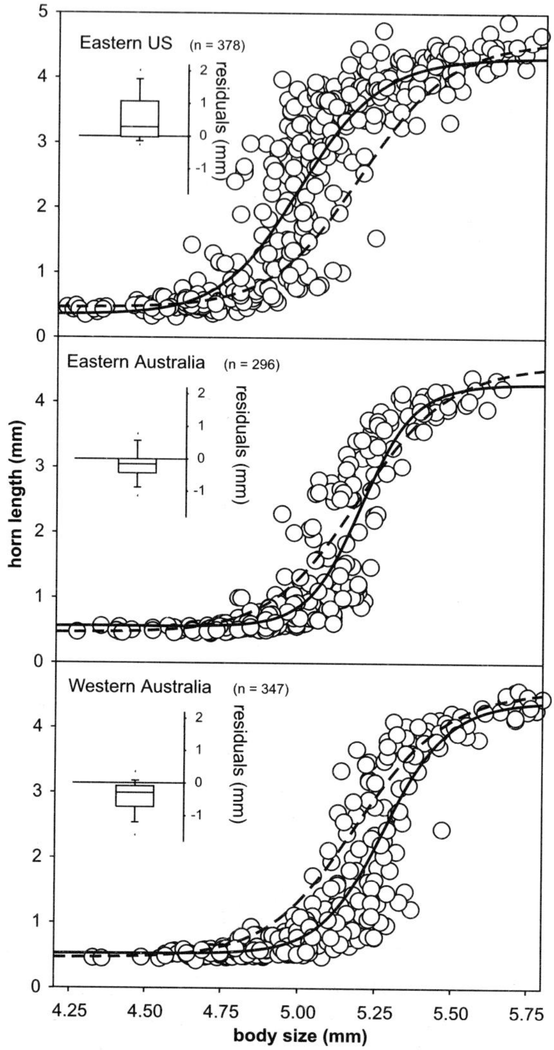

Fig. 11 Allometric divergence in exotic and geographically isolated populations of *O. taurus*. Plots show scaling relationship between horn length and body size for male *O. taurus* in the Eastern US (top), Eastern Australia (center) and Western Australia (bottom). For ease of interpretation also plotted is a common reference curve (dashed; identical in all three panels). This curve indicates a best-fit non-linear regression for all three ranges combined. This regression was used to calculate residual horn lengths (shown in inserts, see below). Solid curves indicate best-fit non-linear regressions calculated separately for each exotic range. Inserts: Box plot of horn length residuals (with 90/10% confidence intervals) for each exotic range based on best-fit non-linear regression for all three exotic ranges combined (dashed curve; after Moczek 2003).

The reason why body size thresholds may be particularly prone to rapid evolution is most likely due to the behavioral ecological context in which alternative male morphs function, and the factors that determine their relative success. Three alternative, but not mutually exclusive, hypotheses have been proposed and tested so far (Fig. 12). The *differential body size hypothesis* is derived from the observation that male fighting success is in part determined by male body size (Emlen 1996, Moczek and Emlen 2000). A male's overall competitive ability therefore can be considered a function of its own body size relative to the mean body size of males in the population within which he competes for mates. If the mean male body size in a population changes, e.g. via seasonal or geographic changes in larval feeding conditions, so should the competitive status of a given male, even if his absolute body size remains the same. In a population with a relatively small mean male body size genotypes would be favored that switch from the hornless to the horned morph at a relatively small threshold body size. Individuals in populations with relatively large mean male body size are instead predicted to delay the switch to a relatively larger body size. This hypothesis therefore predicts a positive correlation between male threshold body sizes and mean male body size in a population. This hypothesis received initial support in a study by Emlen (1996) on *Onthophagus acuminatus*, which showed a partial correlation between seasonal changes in mean male body size and body size thresholds. A more recent study on three exotic and highly threshold-divergent populations of *O. taurus* found no differences in average male body sizes in one comparison, and differences in the direction opposite to what was predicted by this hypothesis in two comparisons (Moczek 2003).

Alternatively, the *intraspecific competition hypothesis* argues that differences in the intensity of intraspecific competition for breeding opportunities has the potential to select for different threshold body sizes in different populations, via two different mechanisms. The first mechanism is derived from the observation that a horned male's ability to gain and maintain access to females through fighting decreases with an increase in the number of males with which he has to compete (Hunt and Simmons 2002). Under low density conditions, male-male encounter frequencies are likely to be low, and even medium-sized horned males may be able to deter rival males effectively enough to gain relatively higher fitness through fighting and the development of horns rather than through sneaking. Under such conditions, selection may favor a relatively low threshold body size. As male density increases, however, the likelihood that a guarding male will be challenged by one or more intruders at a given time increases as well. Under

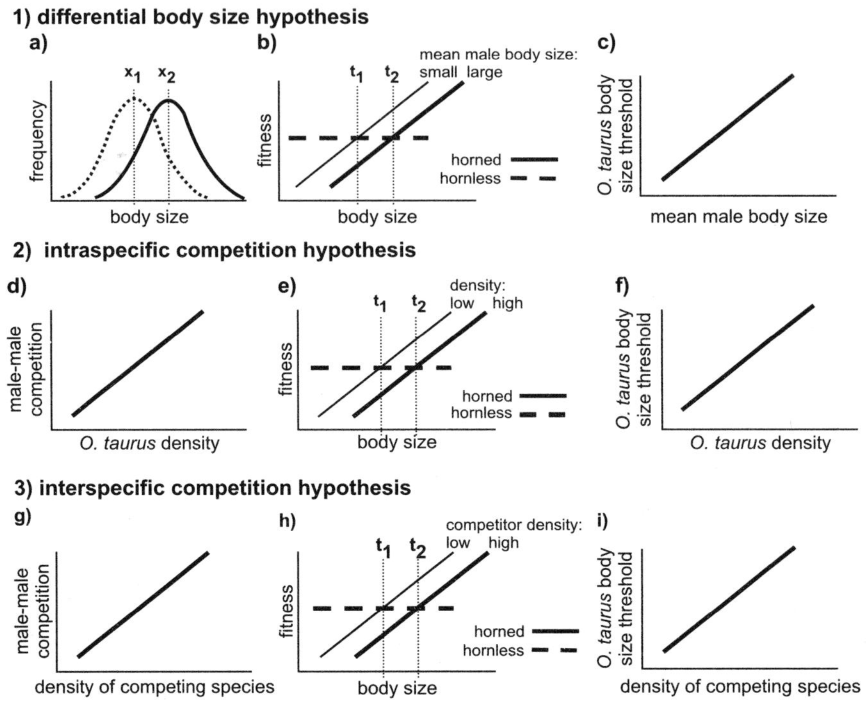

Fig. 12 Ecological mechanisms of threshold evolution in onthophagine beetles. **I:** Differential body size hypothesis: **(a)** Changes in the mean body size of competing males (**x_1** to **x_2**) alter the average body size of males with which a given male has to compete for access to females. If the competitive status of a male is determined by its own body size relative to that of other males in the same population, then changes in mean male body size should alter the competitive status of a given male, even if his own body size remains the same. **(b)** In a population with a relatively small mean male body size, males of an intermediate body size—on an absolute scale—may maximize their fitness through fighting and the development of horns, whereas in a population with a relatively large mean male body size these same intermediate-sized males may fare better by remaining hornless and engaging in sneaking behaviors. Increases in the mean male body size in a population are therefore predicted to cause subsequent increases in the threshold body size that separates horned and hornless male morphs. **(c)** The differential body size hypothesis therefore predicts a positive correlation between male threshold body sizes and mean male body size in a population. **II:** Intraspecific competition hypothesis: **(d)** Increasing local densities intensify male-male competition via increasing male encounter rates inside dung pads. Increased local densities also result in an increase in the relative proportion of females that fail to secure breeding opportunities due to resource limitation. This in turn causes the ratio of competing males to breeding females to become more male biased and levels of male-male competition to intensify even further. **(e)** As local densities increase from low to high and male-male competition intensifies, sneaking behavior becomes more profitable than fighting behavior over a wider range of male body sizes, selecting for a shift of the critical threshold body size **t_1**

Fig. 12 Contd. ...

high-density conditions, medium-sized males may no longer be able to maintain access to females through fights. Instead, such males may maximize fitness by remaining hornless and engaging in sneaking behaviors. Under such conditions, selection may therefore favor a relatively high threshold body size. While this first mechanisms emphasizes changes in the nature of male-male interactions as a cause of threshold evolution, the second mechanism emphasizes that female-female interactions may be just as important, especially in species where females depend on patchy and ephemeral resources such as dung. It argues that under low-density conditions, most females will be able to secure enough resource to reproduce, resulting in a relatively even ratio of competing males to breeding females and relatively low levels of male-male competition. As the densities of competing females per resource patch increases, females will eventually become resource-limited, and a growing proportion of females will fail to secure enough resources to reproduce. Under such conditions a relatively large number of males will compete for access to a relatively small number of breeding females. As a consequence, the overall intensity of male-male competition will increase. As before, increased levels of male-male competition should in turn limit the profitability of fighting behavior to only but the largest males and favor a corresponding shift to a relatively high threshold body size. Combined, the intraspecific competition hypothesis predicts a positive correlation between male threshold body sizes and local population densities. Comparing three threshold divergent exotic *O. taurus* populations, Moczek (2003) found strong support for this hypothesis (Fig. 13).

Fig. 12 Contd. ...

to a larger body size $\mathbf{t_2}$. **(f)** The intraspecific competition hypothesis therefore predicts a positive correlation between male threshold body sizes and *O. taurus* densities. **III:** Interspecific competition hypothesis: **(g)** Increasing competition from other species that compete for the same ephemeral resource crucial for reproduction (dung) indirectly intensifies male-male competition by increasing the proportion of females that are unable to breed due to resource limitation. This in turn causes males to compete for relatively fewer breeding females and levels of male-male competition to increase. Consequently, as interspecific competitor densities increase from low to high, male-male competition intensifies. **(h)** This in turn limits the profitability of fighting behavior to only but the largest males, causes sneaking behaviors to become more profitable over a wider range of body sizes, which in turn selects for a shift of the critical threshold body size to larger body sizes. **(i)** The interspecific competition hypothesis therefore predicts a positive correlation between threshold body sizes of male *O. taurus* and the densities of competing dung beetle species (modified after Moczek 2003).

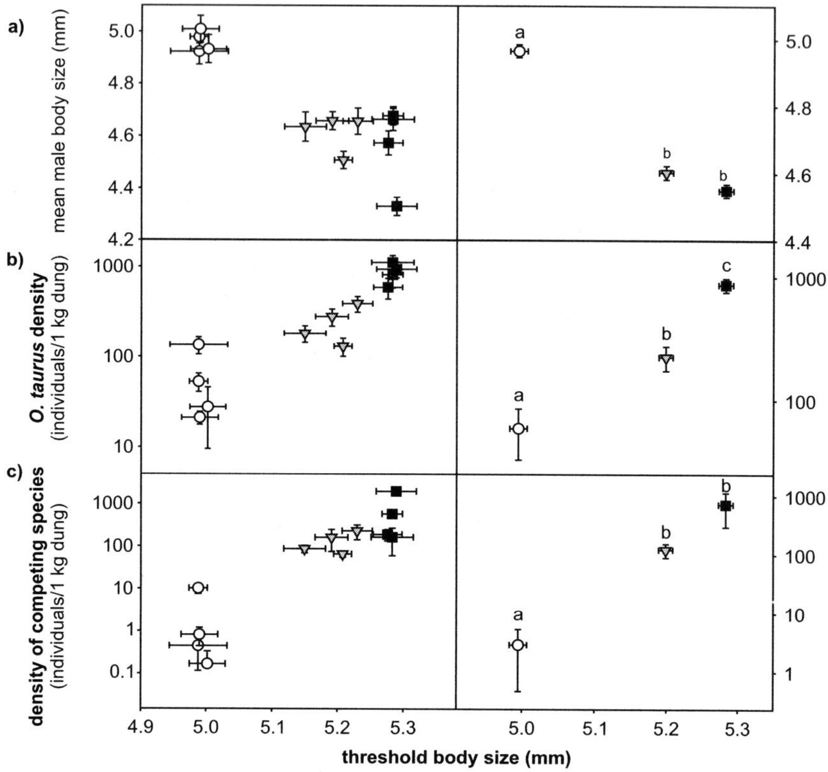

Fig. 13 Relationship between threshold body size (x-axis) and **(a)** mean male body size, **(b)** *O. taurus* density, and **(c)** competitor density within and between exotic ranges of the horn polyphenic beetle *Onthophagus taurus*. Left: sites within each exotic range. Right: means for each exotic range (open circles: Eastern US, gray triangle: Eastern Australia, solid squares: Western Australia). All three exotic ranges expressed highly significantly different threshold body sizes. Samples collected from different sites within each exotic range did not differ significantly in threshold body sizes, even though some sites differed considerably in densities or male body sizes. Different letters in the exponent denote significant differences in range-wide mean male body sizes, *O. taurus* densities, or competitor densities, respectively. Note that densities are plotted on a logarithmic scale (modified after Moczek 2003).

Lastly, the *interspecific competition hypothesis* rests on the observation that many species compete with other species over access to patchy and ephemeral breeding opportunities (e.g. Ridsdill-Smith 1993, Giller and Doube 1989). Low levels of interspecific competition for breeding opportunities should allow a relatively large portion of females to secure the resources necessary to breed. In such a population males will compete for access to a relatively large number of breeding females, resulting in relatively low levels of male-male competition. As before, under such conditions selection is predicted to favor a relatively low threshold body size. As levels of interspecific competition increase, however, an increasing proportion of females will be denied the opportunity to breed. Under such conditions males will compete for access to a relatively small number of breeding females, causing levels of male-male competition to increase. As before, increased levels of male-male competition should in turn limit the profitability of fighting behavior to only the largest males and favor a shift of the threshold body size to relatively larger body sizes. The interspecific competition hypothesis argues that increased competition from other species that compete for breeding resources can intensify male-male competition indirectly by increasing the proportion of females that are unable to breed due to resource limitation. The interspecific competition hypothesis thus predicts a positive correlation between male threshold body sizes and the densities of competing species. So far this hypothesis has received partial support through the comparison of exotic, threshold divergent populations of *O. taurus*. While some populations differed highly significantly in competitor densities as predicted by this hypothesis, others exhibited at least a tendency in the expected direction (Moczek 2003).

Combined, the available evidence supports the hypothesis that differences in the degree of intra- and interspecific competition may indeed be able to drive threshold divergences between geographically isolated populations. However, so far the available evidence is entirely correlational and experimental results are strongly needed. In particular, quantification of fitness ratios of horned and hornless males of identical body sizes under a range of external conditions are necessary to allow for a more rigorous examination of these and other hypotheses. If the intra- and interspecific competition hypotheses receive further support by future studies, this would have important implications for our understanding the relative ease with which divergences in scaling relationships could be initiated. For example, the degree of intra- and interspecific competition present at a given site is in likely to depend on a variety of ecological factors, such as resource availability, predation pressure, or parasite density, which are themselves likely to differ in intensity from one local to another (Giller and Doube 1989, Ridsdill-Smith 1991, 1993). As a consequence, different populations of horn polyphenic beetles are bound to differ in at least some of these factors, and thus between-population divergences in threshold body sizes may be far more widespread than currently appreciated. As we will see next, threshold divergences driven by different ecologies may also have additional more subtle, but possibly more far-reaching consequences than just a change in threshold, due to nature of the developmental mechanisms that regulate threshold responses in horned beetles.

The Physiological Basis of Threshold Evolution and the Integration of Development

The preceding sections discussed the role of juvenile hormone as an important regulator of threshold responses in polyphenic beetles. The implication of JH in the regulation of beetle horn development raises the possibility that changes in JH metabolism and physiology may provide an important avenue for the evolution of allometries. Results of a recent study support this notion. Investigating the same populations of *O. taurus* that were used to study the behavioral ecology of threshold evolution, Moczek and Nijhout (2002) explored whether evolved changes in how JH regulates morph expression during the second sensitive period in late larval development may have contributed to the divergence in body size thresholds present between these populations. As mentioned before, during this second sensitive period artificial JH application induces horn expression in males otherwise fated to develop into the hornless morph (Emlen and Nijhout 1999). Based on these findings, Emlen and Nijhout (1999) developed a model of the endocrine control of horn expression during this period, which suggests that male larvae differ in their JH titers depending on their body mass (Fig. 14a). According to the model, small male larvae exhibit JH titers below a certain threshold concentration during a well-defined sensitive period, and consequently develop into the hornless morph. Larger male larvae express JH titers above this threshold and develop into the horned morph (Emlen and Nijhout 1999). This model suggests at least two major developmental avenues for threshold evolution. First, changes in the *sensitivity* to JH could alter the location of the body size threshold (Fig. 14b). Reduced sensitivity, for example, would cause males that would have expressed JH titers just above the threshold to now develop into the hornless, instead of the horned, morph. Second, changes in the *timing of sensitivity* to JH relative to the temporal pattern of JH secretion could also result in a modification of the body size threshold (Fig. 14c). For example, if the JH sensitive period normally occurs during a high but falling phase of JH titers, then a delay in the sensitive period could now cause it to coincide with JH titers that fall below the threshold required to induce horn growth. As a consequence, males who previously expressed JH titers just above the threshold now fall below the threshold and consequently, will express the hornless male morph. At the level of a population, both mechanisms would be manifest as a shift of the body size threshold to larger

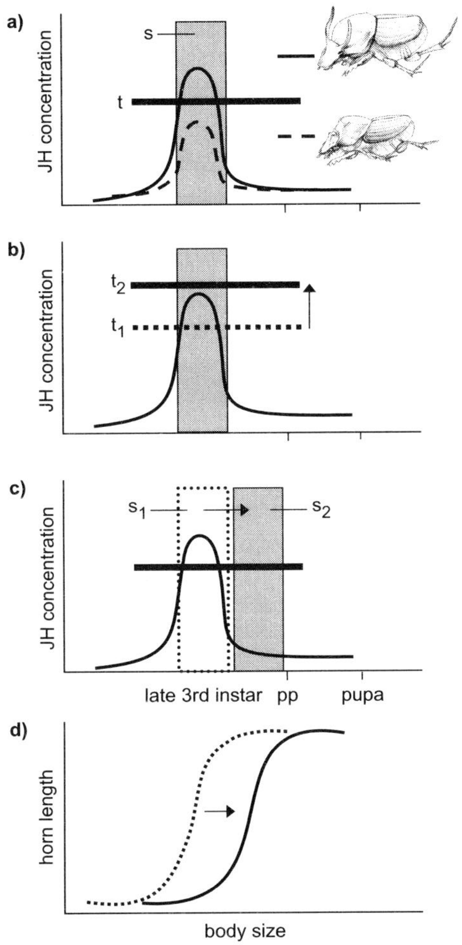

Fig. 14 Potential developmental mechanisms that mediate threshold evolution in exotic populations of *O. taurus.* **(a)** Endocrine control of male horn dimorphism (modified after Emlen and Nijhout 1999). Males are assumed to differ in juvenile hormone (JH) titers depending on their body size. Only large males express JH-titers above a threshold (**t**) during a certain sensitive period (**s**), and will develop horns as adults, whereas smaller males with JH-titers below the threshold will remain hornless. **(b)** Elevation in the JH-threshold (**t_1** to **t_2**) causes a medium-sized male larva to express JH-titers below the threshold necessary for horn development and to express the hornless instead of horned morph as adults. **(c)** A delay in the JH sensitive period (**s_1** to **s_2**) relative to JH-secretion results in JH titers of medium-sized male larvae to fall below the JH-threshold necessary for horn development before the horn-developing tissue acquires JH-sensitivity, causing these males to express the hornless instead of horned morph as adults (pp = prepupal stage). **(d)** On the level of a population both developmental modifications would be manifest in a shift of the critical threshold body size to larger body sizes. Modified after Moczek and Nijhout (2002).

body sizes (Fig. 14d). Contrasting population specific responses to the JH analogue methoprene, Moczek and Nijhout (2002) found support for both hypotheses. Males derived from a population with a high body size threshold, i.e. in which only very large males develop horns, were less sensitive to JH and exhibited their sensitive period later in larval development, compared to males derived from a population with a low body-size threshold. Strain-specific differences in the sensitivity to JH have previously been suggested to be responsible for differences in morph expression patterns in hemipterans (Dingle and Winchell 1997), suggesting that evolutionary modification of JH sensitivity may be a common mechanism that mediates the evolution of novel response thresholds in insects. However, strain specific differences in timing of tissue sensitivity have so far not been reported in any insect, but may possibly play an equally important role in the diversification of response thresholds. Combined, these results also suggest that relatively simple and subtle changes in the regulation of a threshold response can have profound consequence for patterns of phenotype expression.

Interestingly, threshold evolution via changes in JH regulation may bring about a number of correlated responses in other developmental and life history events. High-threshold populations in *O. taurus* did not only exhibit reduced and delayed sensitivity to JH, they also required more time to complete larval development and exhibited delayed pupation, metamorphosis, and eclosion. This may not be surprising as JH is involved in the regulation of numerous larval developmental events, and plays a central role in the coordination of molting, pupation, and metamorphosis (Nijhout 1994, 1999). For example, pupation generally requires the *absence* of JH during a particular sensitive period during late larval development (Nijhout 1994, 1999). In *O. taurus* this latter period is preceded by the sensitive period for horn induction. Here, JH has to be *present* above a certain concentration to induce horn expression (Emlen and Nijhout 1999). A delay in the sensitive period for JH-mediated horn expression, as is the case in high-threshold *O. taurus* populations, may cause a correlated delay in subsequent JH-sensitive periods, such as the one involved in regulating pupation, which in turn would result in an extension of the larval stage. If this hypothesis is correct, delayed pupation and an extended larval stage would reflect correlated responses to an evolutionary modification of the threshold response that mediates horn expression. Interestingly, a delay in pupation was also seen in hornless males and females of high threshold populations, even though neither express horns. This indicates that, while the evolutionary alteration of the developmental threshold for horns has changed the morphology of only large males, the underlying developmental modifications required to achieve this alteration may have had consequences for all members of the population. A close genetic or developmental correlation between morph expression and other developmental and life-history events has also been implicated in earlier studies on wing-polyphenic crickets and hemipterans (Zera and Zhang 1995, Dingle and Winchell 1997, Roff et al. 1997, 1999).

The amount of developmental differentiation that accumulates between horn polyphenic populations may become important once populations re-establish contact and hybridize. Hybrids may express intermediate thresholds suboptimal for competition within either parental population. Furthermore, hybrids may have to contend with reduced developmental integration as their ontogeny is now controlled by developmental mechanisms derived from two developmentally divergent parental strains. Consequently, hybrids may suffer reduced fitness, which may favor the spread of alleles that facilitate assortative mating among members of both parental populations. This, in turn, may then lead to the subsequent evolution of reproductive isolation between these populations, an outcome also observed in recent theoretical models (Porter and Johnston 2002). While this scenario is, at this point, largely speculative, it provides a plausible and experimentally testable avenue for how divergent social regimes can cause geographically isolated populations to diverge rapidly in certain developmental and morphological properties, which in turn can foster the evolution of reproductive isolation once these populations come into secondary contact.

Body Size Thresholds are Themselves Phenotypically Plastic

An interesting aspect of body size thresholds in horn polyphenic beetles is that they themselves exhibit a certain degree of phenotypic plasticity. Emlen (1997b) showed that *Onthophagus acuminatus* reared on artificially low diet switched to the horned morph at slightly but significantly smaller body sizes compared to animals reared on regular diet. Mean body size of males in the low quality food treatment was also smaller, and Emlen interpreted this diet-induced threshold plasticity as an adaptive mechanism by which developing larvae adjust the optimal body size threshold to the range of body sizes likely to be present in the adult generation within which they have to compete, analogous to the differential body-size hypothesis outlined above. A more recent study used natural variation in feeding conditions in *O. taurus* and, too, observed that threshold body sizes covaried slightly but significantly with feeding conditions in laboratory reared as well as natural populations (Moczek 2002). Some, but not all of the populations tested also exhibited corresponding differences in mean male body size. This study, however, emphasized an alternative explanation for this phenomenon. A certain degree of diet induced plasticity in body size thresholds should be expected simply because of how adult horn size and body size are determined during larval development, independent of whether such plasticity may be adaptive or not. In particular, diet induced plasticity may emerge simply because whether or not a male larva will develop horns as an adult is determined *before* the animal ceases to accumulate body mass. As a consequence male larvae will still increase in body mass even after their future horn phenotype has been specified. Exactly how much larvae change body mass after morph determination should depend on larval feeding conditions. A larvae with access to good feeding conditions will gain relatively more weight than a larvae restricted to poorer conditions. However, if both of these larvae exceed the critical larval weight required for horn expression at the time of morph determination, they will both develop horns as adults. Because their post-morph determination mass gain is different, however, they will differ in the final weight with which they will pupate, the body size at which they will eclose as adults, and therefore the lengths of horns relative to their body size. At the level of a population this alone may be sufficient to bring about a change in the critical body size threshold that separates alternative male morphologies, causing populations with access to relatively poor conditions to initiate horn expression at relatively small body sizes (Moczek 2002). The main implication of this alternative explanation is, however, not about the adaptiveness of threshold-plasticity. True, it is plausible that plasticity in body size thresholds could be adaptive under certain conditions, e.g. in the context of population-wide changes in larval food quality or availability (Edwards 1991, Emlen 1997b), in which case this alternative explanation may illustrate the physiological means by which such an adaptive response could be achieved. Instead, the main implication of this alternative explanation is to illustrate another example for the increasingly common observation that plasticity may emerge initially simply as a by-product of development, without requiring any initial changes in genotypes and gene frequencies (West-Eberhard 2003). Even though non-heritable at first, if the change in phenotype expression happens to be in a direction favored by selection, subsequent genetic changes would then have the opportunity to assimilate and stabilize this new response. It is intriguing to speculate whether the dramatic divergences in body size threshold observed in different exotic *O. taurus* populations discussed above might have been initiated though population-wide changes in larval feeding conditions, e.g. through changes in food quality (Edwards 1991) or via changes in the intensity of resource competition between provisioning females (Hirschberger 1999).

Developmental Trade Offs as a Source of Phenotypic Diversity

An additional and possibly very important mechanism contributing to phenotypic diversification in horned beetles may arise from resource allocation trade offs during the growth of horns. Allocation trade offs during development may arise when two or more structures compete for a shared and limited pool of resources necessary to sustain their growth. A shared limiting resource could be nutrient that is used up during the physical growth of a tissue, or a hormone, growth factor, or morphogen that is sequestered by competing binding sites in different tissues (Kawamura et al. 1999, Gibson and Schubiger 2000, Oldham et al. 2000, Brogiolo et al. 2001, Nijhout and Grunert 2002). Such limiting resources may thus constrain the absolute sizes to which a structure can grow, and the presence of absence of a growing structure may therefore, theoretically, influence the size to which other structures are able to develop (Nijhout and Wheeler 1996). Recent work on butterfly wings and beetle horns lend support to this hypothesis (Kawano 1995, 1997; Klingenberg and Nijhout 1998, Nijhout and Emlen 1998). Studying giant rhinoceros beetles (*Chalcosoma* and *Dynastes* species) Kawano (1995) was the first to describe a negative correlation between relative horn size and wing area in males, i.e. males that developed disproportionally large horns also expressed relatively smaller wings. Kawano (1997) found similar results when studying a large number of stag beetle species in which males with disproportinally large mandibles developed relatively smaller wings, and suggested that resource allocation trade offs during development between mandibles and wings and horns and wings might account for these results (Kawano 1997). In *Onthophagus* beetles Nijhout and Emlen (1998) and Emlen (2002) showed that males that develop relatively long horns also develop relatively smaller antennae, eyes, or wings compared to their hornless female counterparts. Interestingly, exactly which structures participated in this interaction appeared to depend on exactly where horns developed. For example, individuals with large head horns developed relative smaller eyes without antennae or wings being affected (Emlen 2001). In one species, *Onthophagus acuminatus*, artificial selection for relatively long horns resulted in a correlated response in the expression of relatively small eyes, demonstrating that evolutionary changes in one trait, horns, can bring about evolutionary changes in another, eye size in this case (Nijhout and Emlen 1998). Since eyes, antennae, and wings are functionally very important traits, reduction in their size might carry certain costs with it, which in turn may limit the extent to which horn growth can be enlarged at certain locations (Emlen 2001). Similarly, this might bias which types of horns evolve in a given taxon depending on its ecology. For example, species that rely heavily on flight might be biased to evolve head rather than thoracic horns as the former are more likely not to negatively affect wing size (Emlen 2001). Trade offs between competing structures could also influence selection on other systems. In the above case, evolution of head horns and, consequently, reduced eye size might cause such a population to rely increasingly on pheromones rather than vision in mate location. Trade offs between competing structures thus have the potential to cause populations with different phenotypes to follow very different evolutionary trajectories.

Evaluating the significance of developmental trade offs in horned beetles faces several challenges. Emlen (2001) used females as controls to establish whether a given negative correlation between the relative sizes of horns and other structures is indeed due to the presence of horns in males. In most species females do not express horns but develop the same morphological landmarks, which can be used to obtain morphological measurements. While this is a logical approach in theory, in practice it poses a challenge due to the extremely short relative horn length measurements in females and the correspondingly large effects of measurement errors. This in turn may obscure horn growth-independent phenotypic correlations in females, especially given the moderate sample sizes used in this study (Emlen 2001). Furthermore, not all species examined so far showed the predicted trade offs. For example, large *O. nigriventris* males develop one of the relatively largest thoracic horns of any species in the genus (see 4c), yet without any corresponding reduction in relative wing size (Emlen, pers. communication). Most importantly, however, evaluating the evolutionary significance of resource allocation trade offs requires a solid understanding of the fitness consequences, if any, that reductions in the sizes of eyes, antennae, and wings might bring with them. To date no such data are available, and it may therefore be premature to label the relatively moderate reductions in the relative sizes of antennae, eyes and wings that accompany the development of horns as functional costs of horn expression (Emlen 2001).

Interestingly, at this point no mechanism has been identified that could account for why primarily neighboring structures should trade off during development. So far, nutrients, hormones, and growth factors have been shown to limit the growth of imaginal-disk derived structures, yet all these factors can circulate freely in the haemolymph and it is unclear why competition for them should be restricted to adjacent structures (Kawamura et al. 1999, Oldham et al. 2000, Brogiolo et al. 2001, Nijhout and Grunert 2002). This is not to suggest that such tradeoffs do not exist, but that proximity may not be the most important determinant of trade off intensity. To test whether distant structures can, in fact, engage in resource allocation tradeoffs, Moczek and Nijhout (2004) examined interactions between head horns and genitalia in *O. taurus*. Both structures develop on opposite ends of the animal. This study not only uncovered evidence in support of significant resource allocation tradeoffs between these distant structures, but also showed that trade off intensity depended on exactly when one of the structures was removed from the competition. The more the growth periods of genitalia and head horns overlapped, the stronger was the tradeoff. This suggests that timing of growth and resource consumption might be significant in determining whether or not two structures will exhibit resource allocation tradeoffs. An important implication of these findings is that changes in the relative timing of growth periods may provide an important avenue for morphological evolution to escape potentially constraining developmental tradeoffs.

Fixation and Recurrence of Alternative Phenotypes

An important avenue by which phenotypic plasticity is thought to contribute to phenotypic diversity is through the temporary fixation of one of several alternative phenotypes, followed by rapid specialization and, under certain circumstances, the subsequent recurrence of lost alternatives. Whether these mechanisms have contributed to phenotypic diversity and speciation in horned beetles is an open question. Recent phylogenetic analyses of horned beetle taxa, however, are beginning to lend tentative support to the notion that the ability to express horns may have been gained and lost repeatedly and independently in certain groups of horned beetles. For example, if patterns of horn expression are mapped onto a recent molecular phylogeny of Iberian dung beetles, including members of the tribe Onthophagini (Villalba et al. 2002), single head horns appear to either have evolved four times (Fig. 15a) or twice independently followed by 3 independent losses (Fig. 15b). The latter scenario would require one additional independent event, which, depending on how losses and gains are weighed, presents a credible possibility. Interestingly, the same phylogeny provides relatively strong evidence that paired head horns evolved independently at least twice in this clade (Fig. 15c). A single origin of paired head horns would require at least four independent, subsequent losses to explain present differentiation patterns, which appears considerably less probable

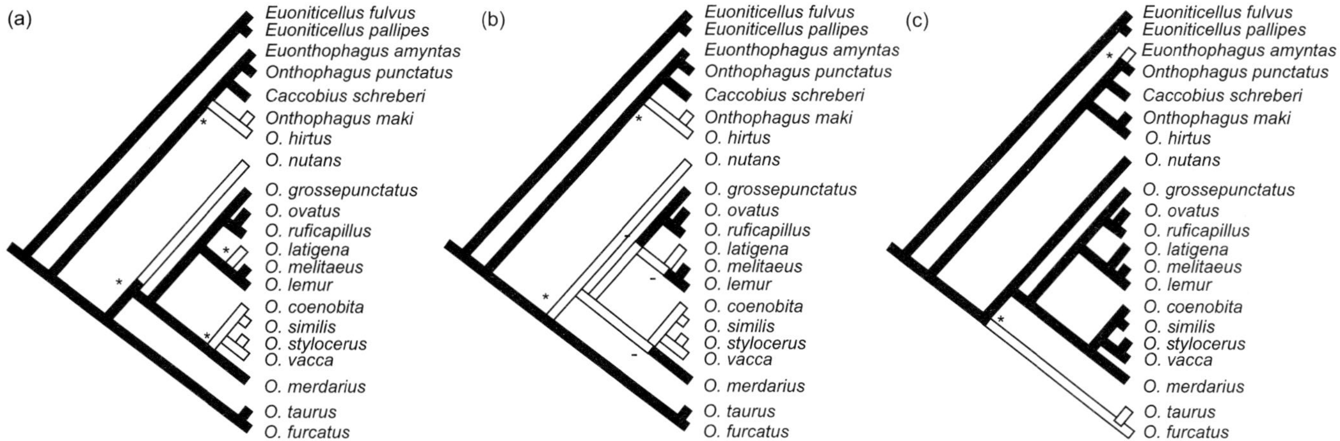

Fig. 15 Alternative phylogenetic scenarios for the evolution of single and paired head horns in *Onthophagus* beetles. Stars indicate gains of horns, whereas (-) symbols indicate possible losses. **(a)** Independent gains of a single head horns in four lineages; **(b)** two independent gains of single head horns followed by three independent losses. **(c)** Independent gains of paired head horns in two lineages. Species descriptions were obtained from Balthasar (1963) and Zardoya (pers. communication). Phylogeny after Villalba et al. 2002, with permission from R. Zardoya.

(Fig. 15c). Clearly, more detailed phylogenetic analyses of horned beetles are strongly needed to get a better insight into patterns of horn evolution and the rates of gains, losses, and possibly recurrence of horn morphologies. Ideally, such studies should be accompanied by behavioral and biogeographic studies to learn more about the ecological factors that might facilitate such transitions. Given the environmental determination of male horn dimorphism, population-wide losses of especially the horned morph appear feasible, even without any initial genetic changes. For example, changes in population-wide larval feeding conditions (Edwards 1991), or changes in climatic conditions (hotter, dryer conditions result in larger portions of brood balls drying out and cause adults to emerge at smaller body sizes; Moczek unpublished) could result in temporary loss of the horned morph, without requiring any changes in gene frequencies. Subsequent genetic changes could then stabilize and assimilate this initially environmental monomorphism. Ideally, such studies should also be accompanied by developmental genetic approaches which would be ideal for detecting developmental remnants of lost alternative phenotypes (Abouheif and Wray 2002, and see below).

The Origin(s) of Horns

The preceding sections have dealt primarily with mechanisms of diversification, rather than origin, of horns and horn-like structures in beetles. In this last sections I will explore how horns and horn polyphenisms originated, highlight some of the most interesting, open questions, and point out promising new approaches that have the potential to address them.

A Horn is a Horn is a Horn?

Horns are not a prerequisite for fights. Head to head shoving contests are a common form of male-male combat in many non-horned beetles. In horn polyphenic species, hornless males fight just as intensely against other males as their horned counterparts, provided their opponent is themselves hornless (Moczek 1999, Moczek and Emlen 2000). Beetle horns are therefore likely to be an example of a trait where the evolution of a behavior prepared the way for the evolution of a corresponding, adaptive morphology.

In many species, horns develop in places where hornless females and males develop ridges or bumps (Balthasar 1963). In fact, in some species large males differ from small males just by slight elevations of the corners of ridges that run across the head (Balthasar 1963). Developmentally, ridges and bumps originate through simple folds in the epidermis during the prepupal stage prior to the secretion of the pupal and subsequent adult cuticle (Moczek and Nagy 2005). As such, ridges are qualitatively not too different from horns, which originate as massive folds of the epidermis during the prepupal stage. Horns might therefore have originated from ridges, via the localized addition of epidermal folds during prepupal development. It is conceivable that even an initially minor increase in ridge hight might already have been sufficient to improve performance in aggressive encounters. Such a scenario is supported by results from staged fights between males of identical body size but different horn lengths, which have shown that even horn length differences as small as 0.5–1 mm significantly improve a male's chances of winning a fight (Moczek and Emlen 2000). Interestingly, for at least two species there is anecdotal evidence that pointy outgrowths can appear spontaneously in some individuals even though members of the species normally do not express horns in that location (Carpaneto and Piatella 1988, Ziani 1994).

Other aspects of horn development, its timing and speed, are reminiscent of the development of regular appendages such as legs, antennae and mouthparts in holometabolous insects (Kim 1959, Schubiger 1971, Fristrom and Fristrom 1993). In fact, incipient horns first become discernible around the same time as mouthparts and antennae during the larvae-prepupal transition. Preliminary results suggest that some of the same regulatory genes involved in the development of traditional appendages are also associated with the development of horns. For example, the transcription factor *Distal-less* has been shown to play a central role in patterning the distal portion of arthropod appendages, and *Distal-less* protein is also expressed in the distal portion of future beetle horns (Moczek and Nagy 2005). Beetle horns thus offer an exciting possibility to explore how regulatory genes used in a traditional developmental process such as appendage formation became redeployed and reorganized in a novel developmental and evolutionary context. Modern developmental genetic approaches provide all the tools necessary for such an exploration.

Not all horns, however, need to be created equally. Different types of horns may have evolved independently, and the same type of horn may have evolved more than once in a given clade (Fig. 15b,c). Thus, it is conceivable that horns develop by different means in different species. Preliminary evidence in favor of this hypothesis comes from comparative studies of pupal and adult morphologies in species with different horn types (Fig. 16). Many *Onthophagus* species develop a thoracic horn as pupae. In *O. taurus*, the epidermis that produced this horn recedes during the pupal stage before

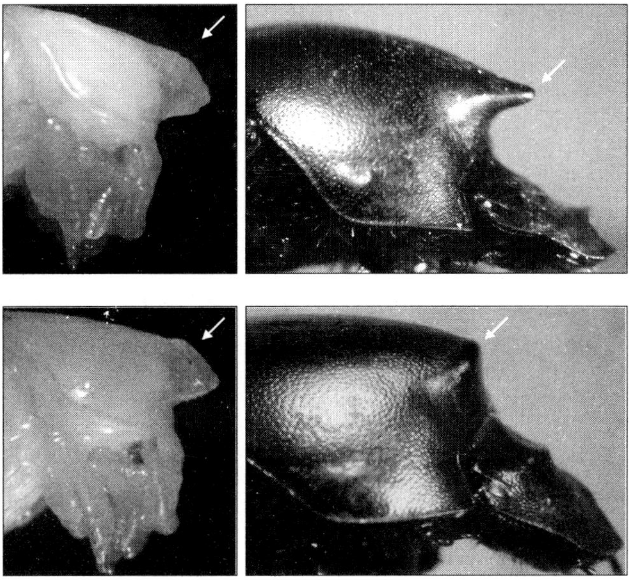

Fig. 16 Regulation of horn development during the pupal stage in *O. nigriventris.* Pupae in minor males (top left) and females (bottom left) both express a single pronotal outgrowth (arrow). In the males, the pupal epidermis underneath this outgrowth stays in place prior to depositing the adult cuticle, and these males also express a corresponding pronotal horn as adults (top right). In females, the pupal epidermis underneath the outgrowth retracts before the adult cuticle is deposited and adult females do not develop a corresponding pronotal horn (bottom right).

it produces the adult cuticle. As a consequence, neither male nor female *O. taurus* develop a thoracic horn as an adult. In *O. nigriventris,* pupae of both sexes also develop a thoracic horn. In fact, large male pupae develop an extra-large, down curved and coiled thoracic horn, which later gives rise to the large thoracic horn present in the adult. Smaller males and females develop regular-sized thoracic horns as pupae much like *O. taurus,* however, only in females *O. nigriventris* does the thoracic epidermis recede and give rise to a completely hornless adult (Fig. 16). In small to medium-sized male *O. nigriventris* this epidermis instead stays in place and gives rise to a significant, pointy thoracic horn in the adult. This modulation of horn growth is quite different from the regulation of head horns in *O. taurus.* Here, future horns grow through the prepupal stage only. Once the animal pupates no further major modulation of horn size takes place. Combined, this suggests that different processes, operating at different times during development, may regulate when and where a horn develops (Moczek 2005).

The Origins of Horn Polyphenisms

Two alternative hypotheses have been proposed to explain the evolutionary origin of horn polyphenism (Emlen and Nijhout 2000). Polyphenic development in horned beetles may have evolved directly from uniformly hornless ancestors, that is, genotypes acquired the ability to "turn on" horn development once a certain body size could be attained (Fig. 17). In this scenario hornlessness would be the ancestral state for all males, and facultative expression of horns in large males would constitute the more derived state. Alternatively, the facultative, polyphenic expression of horns may have evolved from obligately horned ancestors, in which all males expressed horns proportional to their body size. Genotypes then acquired the ability to suppress horn development below a certain body size threshold (Fig. 17). In this case, facultative expression of horns could have evolved much later than horns themselves. In this scenario, the secondarily hornless, small males would constitute the more derived state. The presence of rudimentary and possibly functionless horn remnants in small males of many horn polyphenic species could be interpreted as evidence in favor of this second scenario. Both hypotheses thus differ with respect to which morphology they predict to be the more derived state, which could be used to distinguish between them through comparative developmental genetic studies. For example, if hornlessness in small males of horn polyphenic species constitutes a secondary, derived state, certain components of the ancestral gene network involved in horn expression should still be detectable in small males during larval and pupal development, even through they do not express horns as adults. Recent elegant studies on the evolution of wing loss in ants illustrate the power of such an approach (Abouheif and Wray 2002).

Conclusions

Over the past decade, horned beetles have been the focus of a variety of research approaches designed to explore the behavioral ecology, endocrine physiology or evolutionary biology of beetle horns and horn polyphenisms. The recurring theme that emerges from all these studies is one of dramatic developmental plasticity and phenotypic flexibility in all aspects of the beetles' life. At the same time horned beetles have been known for a long time to be among the most speciose taxa in the insects, producing some of the most exaggerated and diverse secondary sexual traits in existence. Here I have attempted to explore how ecology and behavior have shaped aspects of

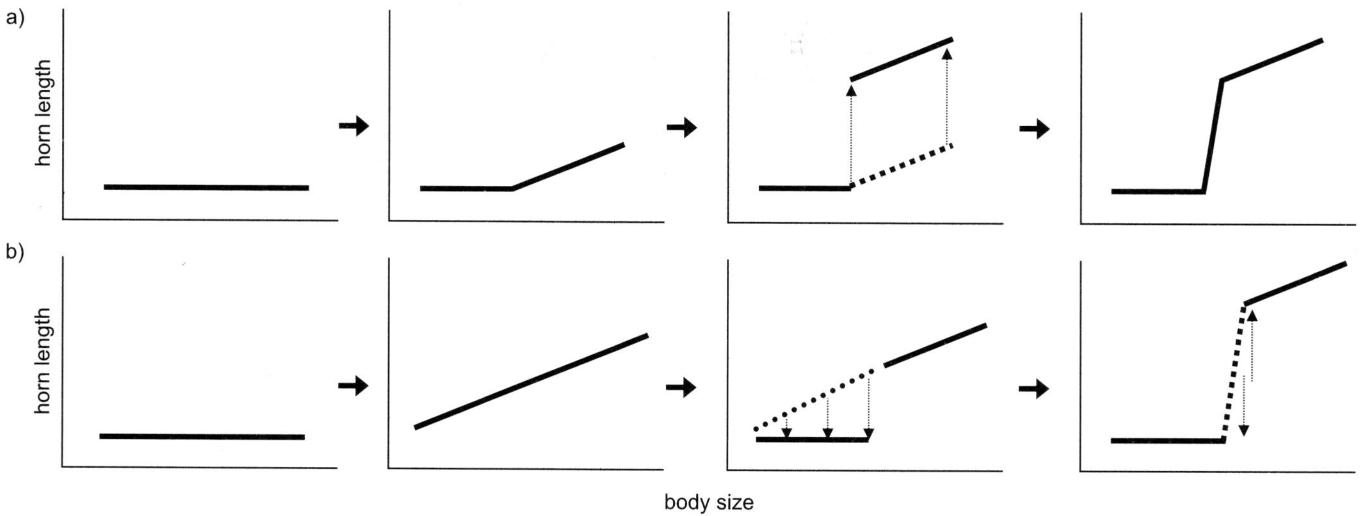

Fig. 17 Alternative origins of horn polyphenic development. **(a)** Horn polyphenic development originated from uniformly hornless ancestors. Genotypes initially expressed no horns regardless of body size, but subsequently evolved the ability to express horns in phenotypes above a certain body size. The ability to induce horns above a certain size threshold then becomes accentuated over time. **(b)** As before, horn polyphenic development originated from uniformly hornless ancestors, however, genotypes first evolved the ability to express horns as a linear function of body size, i.e. small males ancestrally represented a small version of larger males. Genotypes subsequently evolved the ability to repress horn development in males below a certain body size threshold. The ability to repress horns below and induce horns above a certain threshold body size then becomes accentuated over time.

developmental and behavioral plasticity in horned beetles, such as the evolution of flexible timing of pupation to accommodate unpredictable larval feeding conditions. At the same time I have tried to highlight how developmental and behavioral plasticity in turn have contributed to and directed the evolutionary diversification of horned beetles, for example by enabling simple ecological factors to shape patterns of morph expression through evolutionary modifications of response thresholds. Horned beetles emerge as an outstanding opportunity for integration, not only of genetics, physiology, ecology and behavior, but also of external conditions and their role in shaping phenotypes and the environment in which they function. Many of the interactions between phenotypic plasticity and evolution presented here are likely not to be unique to horned beetles, but may be relevant to the numerous taxa in which alternative morphologies rely on alternative reproductive tactics, and whose performances themselves depend on external, social, and ecological conditions. As horned beetles demonstrate, integrating the role of phenotypic plasticity and environment in the evolution of phenotypes makes our understanding of the origins of diversity not only more complete, but also by far more interesting.

Acknowledgements

Laura Mojonnier, Emilie Snell-Rood, Diana Wheeler and Douglas Whitman provided helpful comments on earlier versions of this manuscript, and R. Reed and M. Terry helped with phylogenetic questions. I would like to thank Barrett Klein for the outstanding drawing of *Onthophagus* reproductive behavior shown in Figure 10. Funding was provided in part through a National Institutes of Health Postdoctoral Excellence in Research and Teaching Fellowship (NIH Training Grant # 1K12GM00708) and National Science Foundation Grant IOB 0445661 to the author. I also would like to thank the Department of Biology, Duke University, and the Center for Insect Science at the University of Arizona for financial and logistic support.

References

Abouheif, E., and G. A. Wray. 2002. Evolution of the gene network underlying wing polyphenism in ants. Science, 297: 249–252.

Abu-Shaar, M. and R. S. Mann. 1998. Generation of multiple antagonistic domains along the proximodistal axis during *Drosophila* leg development. Development, 125: 3821–3830.

Allegret, P. 1964. Interrelationship of larval development, metamorphosis and age in a pyralid lepidopteran, *Galleria melloneella* (L), under the influence of dietetic factors. Experimental Gerontology, 1: 49–66.

Ahlroth, P., Alatalo, R. V., Hyvärinen, E., and J. Suhonen. 1999. Geographical variation in wing polymorphism in the water strider *Aquarius najas* (Heteroptera: Gerridae). Journal of Evolutionary Biology, 12: 156–160.

Arrow, G. H. 1899. On sexual dimorphism in beetles of the family Rutelidae. Transaction of the Entomological Society of London, 1899: 255–269.

Arrow, G. H. 1951. Horned Beetles. Junk Publishers, The Hague.

Baldwin, J. M. 1902. Development and Evolution. Macmillan, New York.

Balthasar, V. 1963. Monographie der Scarabaeidae und Aphodiidae der palaearktischen und orientalischen Region (Coleoptera: Lamellicornia). Band 2, Coprinae. Verlag der tschechoslowakischen Akademie der Wissenschaften, Prag.

Baraud, J. 1992. Coléoptères scarabaeiodea d'Europe. Fédéracion Française des Sociétiés des Sciences Naturelles, Paris.

Beck, S. D. 1971a. Growth and retrogression in larvae of *Trogoderma glabrum* (Coleoptera: Dermestidae). 1. Characteristics under feeding and starvation. Annals of the Entomological Society of America, 64: 149–155.

Beck, S. D. 1971b. Growth and retrogression in larvae of *Trogoderma glabrum* (Coleoptera: Dermestidae). 2. Factors influencing pupation. Annals of the Entomological Society of America, 64: 946–949.

Blakley, N. and S. R. Goodner. 1978. Size-dependent timing of metamorphosis in milkweed bugs (*Oncopeltus*) and its life history implications. Biological Bulletin, 155: 499–510.

Brakefiel, P. M. and Reitsma, N. 1991. Phenotypic plasticity, seasonal climate and the population biology of *Bycyclus* butterflies (Satyridae) in Malawi. Ecological Entomology, 16: 291–303.

Brogiolo, W., Stocker, H., Ikeya, T., Rintelen, F., Fernandez, R., and E. Hafen. 2001. An evolutionarily conserved function for the *Drosophila* insulin receptor and insulin-like peptides in growth control. Current Biology, 11: 213–221.

Campbell, G. and A. Tomlinson. 1998. The roles of the homeobox genes *aristaless* and *Distal-less* in patterning the legs and wings of *Drosophila*. Development, 125: 4483–4493.

Campbell, G., Weaver, T., and A. Tomlinson. 1993. Axis specification in the developing *Drosophila* appendage: the role of *wingless*, *decapentaplegic*, and the homeobox gene *aristaless*. Cell, 74: 1113–1123.

Carpaneto, G. M. and E. P. Piatella. 1988. Due casi di teratologia bilaterale simmetrica in coleotteri scabeoidei. Bollettino dell'Associazione Romana di Entomologia 43: 29–32.

Cisper, G., Zera, A. J., and D. W. Borst. 2000. Juvenile hormone titer and morph-specific reproduction in the wing-polymorphic cricket, *Gryllus firmus*. Journal of Insect Physiology, 46: 585–596.

Cohen, B., Simcox, A. A., and S. M. Cohen. 1993. Allocation of the thoracic imaginal primordia in the *Drosophila* embryo. Development, 117: 597–608.

Conner, J. 1989. Density dependent sexual selection in the fungus beetle, *Bolitotherus cornutus*. Evolution, 43: 1378–1386.

Cook, D. 1990. Differences in courtship, mating and postcopulatory behavior between male morphs of the dung beetle *Onthophagus binodis* Thunberg (Coleoptera: Scarabaeidae). Animal Behavior, 40: 428–436.

Crespi, B. J. 1988. Adaptation, compromise and constraint: the development, morphometrics, and behavioral basis of a fighter-flier polymorphism in male *Hoplothrips karni* (Insecta: Thysanoptera). Behavioral Ecology and Sociobiology, 23: 93–104.

Darwin, C. 1871. The descent of man and selection in relation to sex. The Modern Library, Random House 1959. Toronto, Canada.

Denno, R. F., Douglass, L. W., and D. Jacobs. 1986. Effects of crowding and host plant nutrition on a wing-dimorphic planthopper. Ecology, 67: 116–123.

Dingle, H. and R. Winchell. 1997. Juvenile hormone as a mediator of plasticity in insect life histories. Archives of Insect Biochemistry and Physiology, 35: 359–373.

Doane, R. W. 1913. The rhinoceros beetle *Oryctes rhinoceros* in Samoa. Journal of Economic Entomology, 6: 437–442.

Dudley, R. 2000a. The evolutionary physiology of animal flight: paleobiological and present perspectives. Annual Review of Physiology, 62: 135–155.

Dudley, R. 2000b. The Biomechanics of Insect Flight. Princeton University Press. Princeton, NJ.

Eberhard, W. G. 1978. Fighting behavior of male *Golofa porteri* beetles (Scarabaeidae: Dynastinae). Psyche, 83: 292–298.

Eberhard, W. G. 1979. The functions of horns in *Podischnus agenor* Dynastinae and other beetles, pp. 231–258. *In* M. S. Blum and N. A. Blum [eds] 1979. Sexual Selection and Reproductive Competition in Insects. Academic Press, New York.

Eberhard, W. G. 1981. The natural history of *Doryphora* sp. (Coleoptera: Chrysomelidae) and the function of its sternal horn. Annals of the Entomological Society of America, 74: 445–448

Eberhard, W. G. 1982. Beetle horn dimorphism: making the best of a bad lot. American Naturalist, 119: 420–26.

Eberhard, W. G. 1987. Use of horns in fights by the dimorphic males of *Ageopsis nigricollis* (Coleoptera, Scarabaeidae, Dynastinae). Journal of the Kansas Entomological Society, 60: 504–509.

Eberhard, W. G. and E. Gutierrez. 1991. Male dimorphisms in beetles and earwigs and the question of developmental constraints. Evolution, 45: 18–28.

Eberhard, W. G. and J. M. Garcia-C. 2000. Ritual jousting by horned *Parisoschoenus expositus* weevils (Coleoptera, Curculionidae, Baridinae). Psyche, 103: 55–84.

Eberhard W. G., Garcia, J. M., and J. Lobo 2000. Size-specific defensive structures in a horned weevil confirm a classic battle plan: avoid fights with larger opponents. Proceedings of the Royal Society of London Series B, 267: 1129–1134.

Edwards, P. 1991. Seasonal variation in the dung of African grazing mammals, and its consequences for coprophagous insects. Functional Ecology, 5: 617–628.

Emlen D. J. 1994. Environmental control of horn length dimorphism in the beetle *Onthophagus acuminatus* (Coleoptera: Scarabaeidae). Proceedings of the Royal Society of London Series B, 256: 131–136.

Emlen, D. J. 1996. Artificial selection on horn length-body size allometry in the horned beetle *Onthophagus acuminatus* (Coleoptera: Scarabaeidae). Evolution, 50: 1219–1230.

Emlen, D. J. 1997a. Alternative reproductive tactics and male-dimorphism in the horned beetle *Onthophagus acuminatus* (Coleoptera: Scarabaeidae). Behavioral Ecology and Sociobiology, 141: 335–341.

Emlen, D. J. 1997b. Diet alters male horn allometry in the beetle *Onthophagus acuminatus* (Coleoptera: Scarabaeidae). Proceedings of the Royal Society of London Series B, 264: 567–574.

Emlen, D. J. 2000. Integrating development with evolution: a case study with beetle horns. Bioscience, 50: 403–418.

Emlen, D. J. 2001. Costs and the diversification of exaggerated animal structures. Science, 291: 1534–1536.

Emlen, D. J., and H. F. Nijhout. 1999. Hormonal control of male horn length dimorphism in the dung beetle *Onthophagus taurus* (Coleoptera: Scarabaeidae). Journal of Insect Physiology, 45: 45–53.

Emlen, D. J., and H. F. Nijhout. 2000. The development and evolution of exaggerated morphologies in insects. Annual Review of Entomology, 45: 661–708.

Emlen, D. J., and H. F. Nijhout. 2001. Hormonal control of male horn length dimophism in the dung beetle *Onthophagus taurus* (Coleoptera: Scarabaeidae): a second critical period of sensitivity to juvenile hormone. Journal of Insect Physiology, 47: 1045–1054.

Evans, J. D. and D. E. Wheeler. 1999. Differential gene expression between developing queens and workers in the honey bee, *Apis mellifera*. Proceedings of the National Academy of Science, 96: 5575–5580.

Evans, J. D. and D. E. Wheeler. 2001. Expression profiles during honeybee caste determination. Genome Biology 2: Research 0001.

Field, J. 1992. Intraspecific parasitism as an alternative reproductive tactic in nestbuilding wasps and bees. Biological Reviews, 67: 79–126.

Fristrom, D. and J. W. Fristrom. 1993. The metamorphic development of the adult epidermis, pp. 843–897. *In* Bate, M. and A. M. Arias. [eds] 1993. The Development of *Drosophila melanogaster*. Cold Spring Harbor Laboratory Press, New York.

Gadgil, M. 1972. Male dimorphism as a consequence of natural selection. American Naturalist, 106: 574–580.

Gibson, M. C. and G. Schubiger. 2000. Peripodial cells regulate proliferation and patterning of *Drosophila* imaginal discs. Cell, 103: 343–350.

Giller, P. S. and B. M. Doube. 1989. Experimental analysis of inter- and intraspecific competition in dung beetle communities. Journal of Animal Ecology, 58: 129–144.

Gullan, P. J. and P. S. Cranston. 2000. The Insects. An Outline of Entomology. Blackwell Science, Oxford, UK.

Halffter, G. and W. G. Edmonds, 1982. The nesting behavior of dung beetles (Scarabaeidae): an ecological and evolutive approach. Instituto de Ecologica, Mexico, FD, Mexico.

Harrison, R. G. 1979. Flight polymorphism in the field cricket *Gryllus pennsylvanicus*. Oecologia, 40: 125–132.

Hazel, W. N. and D. A. West. 1982. Pupal colour dimorphism in swallowtail butterflies as a threshold trait: selection in *Eurytides marcellus* (Cramer). Heredity, 49: 295–301.

Hirschberger, P. 1999. Larval population density affects female weight and fecundity in the dung beetle *Aphodius ater*. Ecological Entomology, 24: 316–322.

Howden, H. F. and O. P. Young. 1981. Panamanian Scarabaeinae: taxonomy, distribution, and habits (Coleoptera: Scarabaeidae). Contributions of the American Entomological Institute, 18: 1–204.

Hunt, J. and L. W. Simmons. 1997. Patterns of fluctuating asymmetry in beetle horns: an experimental examination of the honest signalling hypothesis. Behavioral Ecology and Sociobiology, 41: 109–114.

Hunt, J. and L. W. Simmons. 2002. Confidence of paternity and paternal care: covariation revealed through experimental manipulation of a mating system in the beetle *Onthophagus taurus*. Journal of Evolutionary Biology, 15: 784–795.

Hölldobler, B. and E. O. Wilson. 1990. The Ants. Springer Verlag, Berlin.

Kawamura, K. Shibata, T., Saget, O., Peel, D., and P. J. Bryant. 1999. A new family of growth factors produced by the fat body and active on *Drosophila* imaginal disk cells. Development, 126: 211–219.

Kawano, K. 1995. Horn and wing allometry and male dimorphism in giant rhinozeros beetles (Coleoptera: Scarabaeidae) of tropical Asia and America. Annals of the Entomological Society of America, 88: 92–99.

Kawano, K. 1997. Cost of evolving exaggerated mandibles in stag beetles (Coleoptera: Lucanidae). Annals of the Entomological Society of America, 90: 453–461.

Kubota, K., Goto, S., and Hayashi, S. 2003. The role of Wg signaling in the patterning of embryonic leg primordium in *Drosophila*. Developmental Biology, 257: 117–126.

Kim, C. W. 1959. The differentiation centre inducing the development from larval to adult leg in *Pieris brassicae* (Lepidoptera). Journal of Embryology and Experimental Morphology, 7: 172–582.

Kim, J., Sebring, A., Esch, J. J., Kraus, M. E., Vorwerk, K., Magee, J., and S. B. Carroll. 1996. Integration of positional signals and regulation of wing formation and identity by *Drosophila* vestigial gene. Nature, 382: 133–138.

Klingenberg, C. P. and H. F. Nijhout. 1998. Competition among growing organs and the developmental control of morphological asymmetry. Proceedings of the Royal Society of London, Series B, 265: 1135–1139.

Koch, P. B. 1995. Color pattern specific melanin synthesis is controlled by ecdysteroids via dopa dicarboxylase activity in a butterfly wing of *Precis coenia* Hubner. Insect Biochemistry and Molecular Biology, 25: 73–82.

Koch, P. B. and D. Bückmann. 1987. Hormonal control of seasonal morphs by the timing of ecdysteroid release in *Araschnia levana* (Nymphalidae: Lepidoptera). Journal of Insect Physiology, 33: 823–829.

Kotiaho, J. S. 2002. Sexual selection and condition dependence of courtship display in three species of horned dung beetles. Behavioral Ecology, 13: 791–799.

Lameere, A. 1904. L'Evolution des ornaments sexuels. Bulletin Academe Belgique, 1904: 1327–1364.

Lohse, G. A. and W. H. Lucht. 1992. Die Käfer Mitteleuropas. Volume 13. Goecke and Evers, Krefeld, Germany.

Lüscher, M. 1960. Hormonal control of caste differentiation in termites. Annals of the New York Academy of Science, 89: 549–563.

MacGown, J. and M. MacGown. 1996. Observation of a nuptial flight of the horned *Passalus* beetle, *Odontotaenius disjunctus* (Illiger) (Coleoptera: Passalidae). Coleopterist Bulletin, 50: 201–203.

Matthews, E. G. 1972. A revision of the Scarabaeinae dung beetles of Australia. I. Tribe Onthophagini. Australian Journal of Zoology Supplemental Series, 9: 1–330.

Mitchel, R. 1975. The evolution of oviposition tactics in the bean weevil, *Callosobruchus maculatus* (F.). Ecology, 56: 696–702.

Moczek, A. P. 1996. Male dimorphism in the scarab beetle *Onthophagus taurus* Schreber, 1759 (Scarabaeidae, Onthophagini): Evolution and plasticity in a variable environment. Ms-Thesis, Julius-Maximilians-University Würzburg, Germany.

Moczek, A. P. 1998. Horn polyphenism in the beetle *Onthophagus taurus*: larval diet quality and plasticity in parental investment determine adult body size and male horn morphology. Behavioral Ecology, 9: 636–641.

Moczek, A. P. 1999. Facultative paternal investment in the polyphenic beetle *Onthophagus taurus*: the role of male morphology and social context. Behavioral Ecology, 10: 641–647.

Moczek, A. P. 2002. Allometric plasticity in a polyphenic beetle. Ecological Entomology, 27: 58–67.

Moczek, A. P. 2003. The behavioral ecology of threshold evolution in a polyphenic beetle. Behavioral Ecology, 14: 831–854.

Moczek, A. P. 2005. The evolution and development of novel traits, or how beetles got their horns. BioScience, 11: 935–951.

Moczek, A. P. and L. M. Nagy. 2005. Diverse developmental mechanisms contribute to different levels of diversity in horned beetles. Evolution and Development, 7: 175–185

Moczek, A. P. and D. J. Emlen. 1999. Proximate determination of male horn dimorphism in the beetle *Onthophagus taurus* (Coleoptera: Scarabaeidae). Journal of Evolutionary Biology, 12: 27–37.

Moczek, A. P. and D. J. Emlen. 2000. Male horn dimorphism in the scarab beetle *Onthophagus taurus*: do alternative reproductive tactics favor alternative phenotypes? Animal Behavior, 59: 459–466.

Moczek, A. P. and H. F. Nijhout. 2002a. Developmental mechanisms of threshold evolution in a polyphenic beetle. Evolution and Development, 4: 252–264.

Moczek, A. P. and H. F. Nijhout. 2002b. A method for sexing third instar larvae of the genus *Onthophagus* Latreille (Coleoptera: Scarabaeidae). Coleopterist Bulletin, 56: 279–284.

Moczek, A. P. and H. F. Nijhout. 2003. Rapid evolution of a polyphenic threshold. Evolution and Development, 5: 259–268.

Moczek, A. P. and H. F. Nijhout. 2004. Trade offs during the development of primary and secondary sexual traits in a horn dimorphic beetle. American Naturalist, 163: 184–191.

Moczek, A. P., Krell, F. T., and Bruehl, C. 2004. Gradual and threshold-dependent expression of secondary sexual traits in the same individual: insights from a horned beetle. Biological Journal of the Linnean Society, 83: 473–480.

Moczek, A. P. and J. Cochrane. Intraspecific female brood parasitism in the dung beetle *Onthophagus taurus*. Ecological Entomology 31: 1–6.

Moczek, A. P., Hunt, J., Emlen, D. J. and L. W. Simmons. 2002. Evolution of a developmental threshold in exotic populations of a polyphenic beetle. Evolutionary Ecology Research, 4: 587–601.

Moffett, M. W. 1987. Division of labor and diet in the extremely polymorphic ant *Pheidologeton diversus*. National Geographic Research, 3: 282–304.

Moran, N. A. 1991. Phenotype fixation and genotypic diversity in the complex life cycle of the aphid *Pemphigus betae*. Evolution, 45: 957–970.

Moran, N. A. 1992. The evolutionary maintenance of alternative phenotypes. American Naturalist, 139: 971–989.

Moran, N. A. and T. G. Whitham. 1988. Evolutionary reduction of complex life cycles: loss of host alternation in *Pemphigus* (Homoptera: Aphididae). Evolution, 42: 717–728.

Nijhout, H. F. 1975. A threshold size for metamorphosis in the tabacco hornworm, *Manduca sexta*. Biological Bullettin, 149: 214–225.

Nijhout H. F. 1976. The role of Ecdysone in pupaption of *Manduca sexta*. Journal of Insect Physiology, 22: 453–463.

Nijhout, H. F. 1991. The Development and Evolution of Butterfly Wing Patterns. Smithonian Institution Press, Washington, USA.

Nijhout, H. F. 1994. Insect Hormones. Princeton: Princeton University Press, USA.

Nijhout, H. F. 1999. Control mechanisms of polyphenic development in insects. Bioscience, 49: 181–192.

Nijhout, H. F. and D. E. Wheeler. 1996. Growth models of complex allometries in holometabolous insects. American Naturalist, 148: 40–56.

Nijhout, H. F. and D. J. Emlen. 1998. Competition among body parts in the development and evolution of insect morphology. Proceedings of the National Academy of Science, 95: 3685–3689.

Nijhout, H. F. and L. W. Grunert. 2002. Bombyxin is a growth factor for wing imaginal disks in Lepidoptera. Proceedings of the National Academy of Science, 99: 15446–15450.

Oldham, S., Böhni, R., Stocker, H., Broglio, W., and E. Hafen. 2000. Genetic control of size in *Drosophila*. Philosophical Transaction of the Royal Society, London, Series B, 355: 945–952.

Otronen, M. 1988. Intra- and intersexual interactions at breeding burrows in the horned beetle, *Coprophaneus ensifer*. Animal Behavior, 36: 741–748.

Otto, D. 1962. Die roten Waldameisen. A. Ziemsen Verlag, Wittenberg, Germany.

Pace, A. 1967. Life history and behavior of a fungus beetle, *Bolitotherus cornutus* (Tenebrionidae). Occasional papers of the Museum of Zoology at the University of Michigan, 653: 1–15.

Palmer, T. J. 1978. A horned beetle which fights. Nature, 274: 583–584.

Panganiban, G., Nagy, L., and S. B. Carroll. 1994. The role of the Distal-less gene in the development and evolution of insect limbs. Current Biology, 4: 671–675.

Passera, M. P. and J. P. Suzzoni. 1979. Le rôle de la reine de *Pheidole pallidula* (Nyl.) (Hymenoptera: Formicidae) dans la sexualization du couvain après traitement par l'hormone juvénile. Insectes Sociaux, 26: 343–353.

Paulian, R. 1935. Le polymorphisme des males de coléopteres. In: Exposés de biométrie et statistique biologique IV, pp. 1–33. *In* G. Tessier [ed.] 1935.Actualités scientifiques et industrielles 255. Hermann and Cie., Paris, France.

Pigliucci, M. 2001. Phenotypic Plasticity. John Hopkins University Press, Maryland, USA.

Porter, A. H. and N. A. Johnson. 2002. Speciation despite gene flow when developmental pasthways diverge. Evolution, 56: 2103–2111.

Ptashne, M. 1992. Genetic Switch: Phage Lambda and Higher Organisms (2nd ed.). Blackwell Sciences, San Diego, CA, USA.

Raff, R. 1996. The Shape of Life: Genes Development, and the Evolution of Animal Form. University of Chicago Press, Chicago, USA.

Rasmussen, J. L. 1994. The influence of horn and body size on the reproductive behavior of the horned rainbow scarab beetle *Phanaeus difformis* (Coleoptera: Scarabaeidae). Journal of Insect Behavior, 7: 67–82.

Reichenau, W. von. 1881. Üeber den Ursprung der secundären männlichen Geschlechtscharakteren, insbesondere bei den Blatthornkäfern. Kosmos, 10: 172–194.

Richards, M. H. 1994. Social evolution in the genus *Halictus*: a phylogenetic approach. Insectes Sociaux, 41: 315–325.

Ridsdill-Smith, T. J. 1991. Competition in dung-breeding insects, pp. 264–292. *In* W. J. Baily and T. J. Ridsdill-Smith [eds] 1991. Reproductive Behaviour of Insects. Chapman & Hall, London, UK.

Ridsdill-Smith, T. J. 1993. Asymmetric competition in cattle dung between two species of *Onthophagus* dung beetle and the bush fly, *Musca vetustissima*. Ecological Entomology, 18: 241–246.

Roff, D. E., Stirling, G., and D. J. Fairbairn. 1997. The evolution of threshold traits: a quantitative genetic analysis of the physiological and life-history correlates of wing dimorphism in the sand cricket. Evolution, 51: 1910–1919.

Roff, D. E., Tucker, J., Stirling, G., and D. J. Fairbairn. 1999. The evolution of threshold traits: effects of selection on fecundity and correlated response in wing dimorphism in the sand cricket. Journal of Evolutionary Biology, 12: 535–546.

Roundtree, D. B. and H. F. Nijhout. 1995. Hormonal control of a seasonal polyphenism in *Precis coenia* (Lepidoptera: Nymphalidae). Journal of Insect Physiology, 41: 987–992.

Shaffer, H. B. 1984. Evolution in a paedomorphic lineage I. An electrophoretic analysis of the Mexican ambystomatid salamanders. Evolution, 38: 1194–1206.

Schlichting, C. and M. Pigliucci. 1998. Phenotypic Evolution: A Reaction Norm Perspective. Sinauer Associates, Sunderland, MA.

Schluter, D., Price, T. D., Mooers, A. Ø., and D. Ludwig. 1997. Likelihood of ancestor states in adaptive radiations. Evolution, 51: 1699–1711.

Schmalhausen II. 1949. *Factors of Evolution*. University of Chicago Press, Chicago.

Schubiger, G. 1971. Regeneration, duplication and transdetermination in fragments of the leg disk of *Drosophila melanogaster*. Developmental Biology, 26: 277–295.

Semlitsch, R. D., Harris, R. N., and H. M. Wilbur. 1990. Paedomorphosis in *Ambystoma talpoideum*: maintenance of population variation and alternative life-history pathways. Evolution, 44: 1604–1613.

Semlitsch, R. D. and H. M. Wilbur. 1989. Artificial selection for paedomorphosis in the salamander *Ambystoma talypoideum*. Evolution, 43: 105–112.

Shafiei, M., Moczek, A. P., and H. F. Nijhout. 2001. Food availability controls the onset of metamorphosis in the dung beetle *Onthophagus taurus* (Coleoptera: Scarabaeidae). Physiological Entomology, 26: 173–180.

Shapiro, A. M. 1976. Seasonal polyphenism. Evolutionary Biology, 9: 259–333.

Shuster, S. M. and M. J. Wade. 2003. Mating Systems and Strategies. Princeton University Press. Princeton, NJ, USA.

Siva-Jothy, M. T. 1987. Mate securing tactics and the cost of fighting in the Japanese horned beetle, *Allomyrina dichotoma* L. (Scarabaeidae). JEthol, 5: 165–172.

Sowig, P. 1996. Duration and benefits of biparental brood care in the dung beetle *Onthophagus vacca* (Coleoptera: Scarabaeidae). EcolEntomol, 21: 81–86.

Stearns, S. C. 1989. The evolutionary significance of phenotypic plasticity. Bioscience, 39: 436–445.

Stern, D. L. and W. A. Foster. 1996. The evolution of soldiers in aphids. Biological Reviews, 71: 27–79.

Sultan, S. E. 1987. Evolutionary implications of phenotypic plasticity in plants. Evolutionary Biology, 20: 127–178.

Sultan, S. E. 1992. Phenotypic plasticity and the neo-Darwinian legacy. Evolutionary Trends in Plants, 6: 61–71.

Tanaka, Y. 1985. Alternative manners of prey carrying in the fossorial wasp, *Oxybelus strandi* Yasumatsu (Hymenoptera: Sphecoidae). Kontyû, 53: 277–283.

Tauber, M. J. and C. A. Tauber. 1970. Photoperiodic induction and termination of diapause in an insect: Response to changing day lengths. Science, 167: 170.

Tauber, M. J. and C. A. Tauber. 1972. Geographic variation in critical photoperiod and in diapause intensity of *Chrysopa carnea* (Neuroptera). Journal of Insect Physiology, 18: 25–29.

Tauber, C. A. and M. J. Tauber. 1982. Evolution of seasonal adaptations and life history traits in Chrysopa: response to diverse selective pressures, pp. 51–72. *In* H. Dingle and J. P. Hegmann [eds] 1982. Evolution and Genetics of Life Histories. Springer, New York, USA.

Tauber, C. A. and M. J. Tauber. 1987. Inheritance of seasonal cycles in *Chrysoperla* (Insecta: Neuroptera). Genetical Research Cambridge, 49: 215–223.

Thornhill, R. and J. Alcock. 1983. The Evolution of Insect Mating Systems. Harvard University Press, Cambridge, Mass., USA.

Villalba, S., Lobo, J. M., Martin-Piera, F., and R. Zardoya. 2002. Phylogenetic relationships of Iberian dung beetles (Coleoptera: Scarabaeinae): insights on the evolution of nesting behavior. Journal of Molecular Evolution, 55: 116–126.

Waddington, C. H. 1953. Genetic assimilation of an acquired character. Evolution, 7: 118–126.

Wallace, A. R. 1869. The Malay Archipelago. Dover Publications 1962, New York, USA.

West-Eberhard, M. J. 1989. Phenotypic plasticity and the origins of diversity. Annual Review of Ecology and Systematics, 20: 249–278.

West-Eberhard, M. J. 1992. Behavior and evolution, pp. 57–75. *In* P. R. Grant and H. S. Grant [eds] 1992. Molds, Molecules and Metazoa: Growing Points in Evolutionary Biology. Princeton Univ Press, Princeton, NJ, USA.

West-Eberhard, M. J. 2003. Developmental Plasticity and Evolution. Oxford University Press, New York.

Wheeler, W. M. 1910. Ants: their structure, development and behavior. Columbia University Press, NY, USA.

Wheeler, D. E. 1991. Developmental basis of worker caste polymorphism in ants. American Naturalist, 138: 1218–1238.

Wheeler, D. E. and H. F. Nijhout. 1983. Soldier determination in *Pheidole bicarinata*: effect of methophrene on caste and size within castes. Journal of Insect Physiology, 29: 847–854.

Wheeler, D. E. and H. F. Nijhout. 1984. Soldier determination in the ant *Pheidole bicarinata*: inhibition by adult soldiers. Journal of Insect Physiology, 30: 127–135.

Whiting, M. F., Bradler, S., and T. Maxwell. 2003. Loss and recovery of wings in stick insects. Nature, 421: 264–267.

Wilson, E. O. 1976. Behavioral discretization and the number of castes in an ant species. Behavioral Ecology and Sociobiology, 1: 141–154.

Wilson, E. O. 1978. Division of labor in fire ants based on physical castes (Hymenoptera: Fomicidae: *Solenopsis*). Journal of the Kansas Entomological Society, 51: 615–636.

Wilson, E. O. 1985. The sociogenesis of insect colonies. Science, 228: 1489–1495.

Windsor, D. M. 1987. Natural history of a subsocial tortoise beetle, *Acromis sparsa* Boheman (Chrysomelidae, Cassidinae) in Panama. Psyche, 94: 127–150.

Yang, A. S. 2001. Modularity, evolvability, and adaptive radiations: a comparison of the hemi- and holometabolous insects. Evolution and Development, 3: 59–72.

Zera, A. J. and R. F. Denno, 1997. Physiology and ecology of dispersal polymorphism in insects. Annual Review of Entomology, 42: 207–231.

Zera, A. J. and C. Zhang. 1995. Evolutionary endocrinology of juvenile hormone esterase in *Gyllus assimilis*: direct and correlated responses to selection. Genetics, 141: 1125–1134.

Ziani, S. 1994. Un interessante caso di teraologia simmetrica in *Onthophagus (Paleonthophagus) fracticornis* (Coleoptera, Scarabaeidae). Bollettino dell'Associazione Romana di Entomologia, 49: 165–167.

4

Developmental Flexibility, Phenotypic Plasticity, and Host Plants: A Case Study with *Nemoria* Caterpillars

Erick Greene*, Michael Canfield and Adam Ehmer*****

**Division of Biological Sciences, The University of Montana, Missoula, MT USA 59801-1002, E-mail: erick.greene@mso.umt.edu*

***Department of Organismic and Evolutionary Biology, Harvard University, Cambridge, MA USA 02138, E-mail: mcanfield@oeb.harvard.edu*

****Department of Ecology and Evolution State University of New York at Stony Brook, Stony Brook, NY USA 11794-5245, E-mail: aehmer@life.bio.sunysb.edu*

Abstract

Ernst Mayr originally defined "polyphenism" to include all nongenetic variation of the phenotype (1963). Since his definition, the term polyphenism has undergone substantial semantic drift, so that today it is generally used in a much more restricted way to refer to discrete, irreversible seasonal forms rather than more general forms of plasticity. In a companion chapter we argue that this change in the use of "polyphenism" has created an artificial dichotomy between discrete alternate morphs and more gradual plasticity with intermediate forms. This has obscured some of the important developmental and ecological links between different forms of phenotypic plasticity. Using examples from inchworm caterpillars (Geometridae) in the genus *Nemoria*, we illustrate some of these points. Closely-related *Nemoria* species show a wide range of diet breadth, geographic range, and developmental plasticity. Host specialists tend to show discrete alternative forms that are triggered by a single developmental induction cue, while host generalists tend to show more continuous plasticity with intergrading forms, and respond to more different kinds of induction triggers. Developmental plasticity in *Nemoria* is evolutionarily labile, and illustrates how underlying developmental and genetic mechanisms may respond easily to selection to produce a wide range of plastic responses.

Introduction

Understanding the ecological influences on the evolution of phenotypic plasticity remains a synthetic challenge to current biology. It is obvious at some level that the environment experienced by a population shapes the form of phenotypic plasticity in that population. We know little, however, about how specific environmental factors relate to or predict the evolution of specific plastic outcomes, or how the mechanisms underlying the expression of plasticity relate to specific environments or life histories. For example, how might degree of polyphagy, voltinism, or geographic range influence the type of phenotypic plasticity that evolves? In this chapter, we address such questions, using caterpillars in the genus *Nemoria*.

Mayr (1963) originally coined the term polyphenism as "nongenetic variation of the phenotype." Mayr's definition was extremely broad, and included all non-genetic forms of plasticity. In contrast, current usage of polyphenism tends to refer only to discretely different, irreversible phenotypes generated in response to environmental cues. This semantic drift has lead to confusion, and has created an artificial dichotomy between threshold developmental traits and more gradual developmental responses that generate intermediate forms (see Canfield and Greene, this volume). In turn, this has obscured some potentially important relationships between different patterns of phenotypic plasticity. In this chapter, we retain the broad definition of polyphenism.

Caterpillars of *Nemoria* (Geometridae) moths are widespread throughout the New World, and exhibit a great diversity in geographic range, host plant range, voltinism, and life history. The genus shows interesting geographic patterns of radiation. Most relevant for these discussions is that caterpillars of different *Nemoria* species display different patterns of developmental plasticity—larvae of some species can develop into discrete alternative morphs with no intermediate forms, whereas closely-related species develop along a more gradual phenotypic continuum, and still other species exhibit little phenotypic plasticity. As such, the genus *Nemoria* is an ideal system to investigate the following questions:

- Are there relationships between diet breadth (host generalist versus host specialist) and developmental plasticity?
- Are different types of environmental induction cues co-opted to guide development in different species, and is this a function of diet?
- Do patterns of developmental plasticity relate to patterns and modes of speciation?

The Genus *Nemoria*

The genus *Nemoria* is a fascinating group of geometrid moths. There are currently 134 described species restricted to the New World (Ferguson 1969, Ferguson 1985, Piktin 1993), ranging from northern South America, Central America, and North America into Canada. At least 36 species of *Nemoria* occur north of Mexico. *Nemoria* adults tend to have green wings, which gives rise to the common name of the emerald moths.

As a group, *Nemoria* feed on many distantly-related plant taxa across the Angiosperms and some Gymnosperms. Individual species range from extremely broad host generalists to host specialists that feed on only one species of plant. Information on the geographic ranges and host associations of some North American *Nemoria* is summarized in Figure 1. This figure is not meant to be exhaustive, but rather to illustrate some common patterns of distribution and host associations. Notice that there are two broad constellations of species in North America—a group of western species and a group of eastern species. There are broad host generalists (represented by species with many host arrows) in both the west (e.g. *N. darwiniata*) and the east (*N. mimosaria*), whose caterpillars readily feed on plants in many distantly-related orders. At the other extreme, there are also many host specialists (represented by a single host arrow). These specialists are scattered widely across the plant phylogeny, from persimmon (*Diospyros*, Ebenaceae: *N. zygotaria*), scrub rosemary (*Ceratiola ericoides*, Ericaceae: *N. outina*), oaks (*Quercus*, Fagaceae: *N. arizonaria* and *N. pulcherrima;*), and gooseberry (*Ribes*, Saxifragaceae: *N. unitaria*).

There have been few systematic studies of the patterns of larval plasticity in relation to host breadth. What sort of phenotypic plasticity is shown by closely-related dietary specialists versus generalists? Individual *Nemoria* caterpillars usually feed only on the one plant where their egg was laid, rather than moving between different plants before they pupate. However, a single female of a generalist species may lay her eggs on a range of host plants, depending on the specific location and season in which she emerges. In such a case, her offspring would develop in many different types of backgrounds of twigs, foliage and flowers. A schematic hypothesis that relates plasticity to diet breadth is shown in Figure 2: specialist caterpillars, occurring in only a few microsites on one species of plant, are hypothesized to have threshold, step-functions in their developmental responses that generate only a few discrete forms. In this case, selection may cause developmental alternatives to evolve that are tightly linked to a specific triggering cue. Generalist species that occur on many different types of

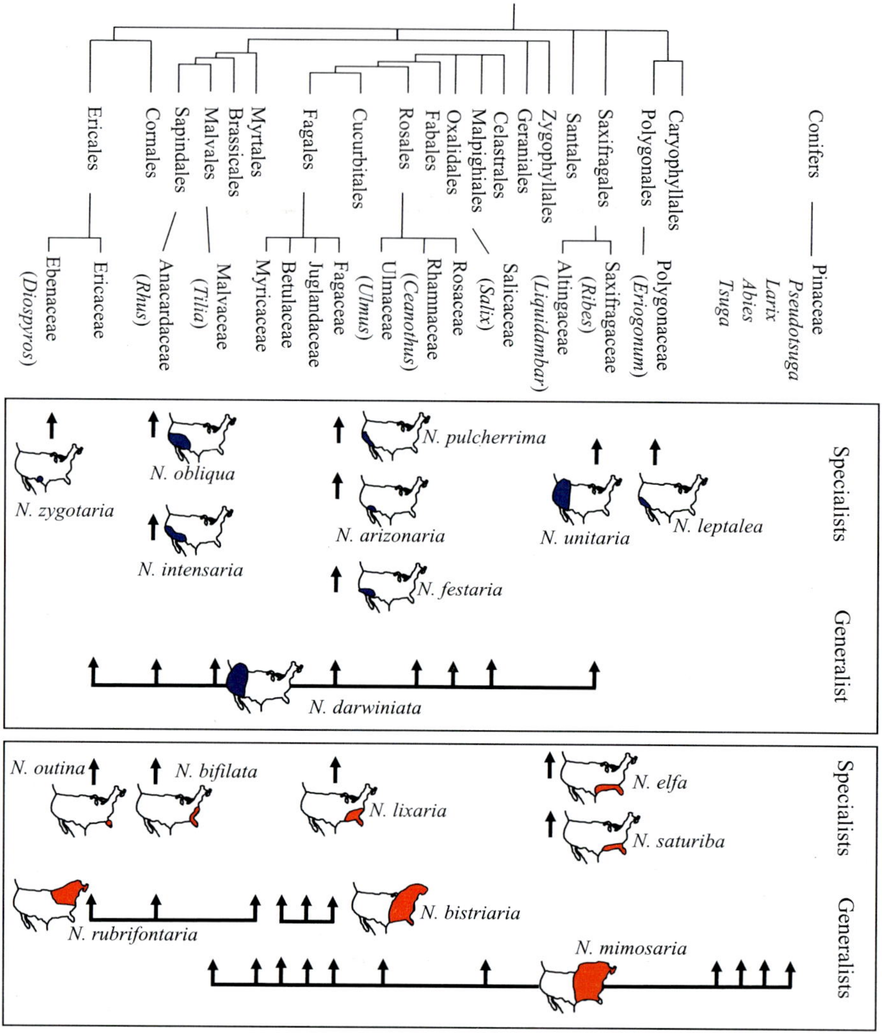

Fig. 1 Summary of geographic distributions and host plant relationships for some species of North American *Nemoria* caterpillars. Geographic distributions are separated into a group of western species (upper panel—blue ranges) and eastern species (lower panel—red ranges). Plant phylogeny is indicated along the top, with major groups of Angiosperms on the left, and some conifers within the gymnosperms on the right (compiled from Angiosperm Phylogeny Group 1998). Arrows (next to species distribution maps) pointing to plant groups indicate host associations; data compiled from Ferguson (1969, 1985).

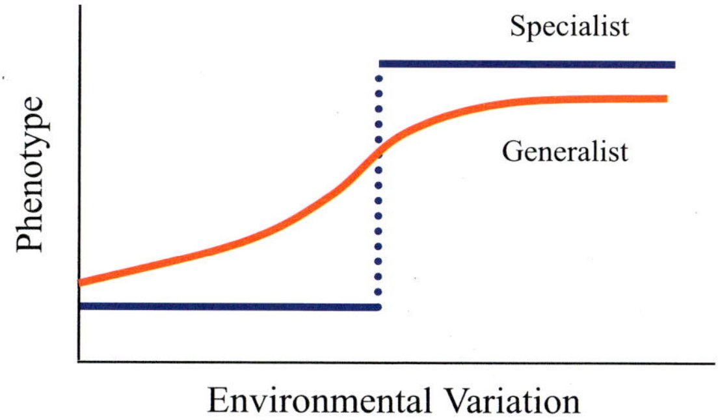

Fig. 2 Schematic hypothesis showing two possible patterns of developmental plasticity. The step function shown in blue would represent a situation in which individuals developed into two different forms with no intermediates. A developmental threshold is cued by a critical value in an environmental variable. The more gradual reaction norm, shown in red, would generate more continuous variation with many intermediate forms. Selection may act on dietary specialists to be well matched with a few species of host plants or microsites, and exhibit developmental thresholds. Dietary generalists that feed on many host plants would be predicted to have more gradual developmental reaction norms.

plants are hypothesized to have smoother reaction norms that generate a continuous gradation of phenotypes with intermediate forms. The developmental plasticity in these generalist species may be less tightly linked to a specific background, and may be more sensitive to a wider range of developmental induction cues.

Many *Nemoria* caterpillars are extremely cryptic. In addition, some species exhibit remarkable adaptive seasonal polyphenisms, which are some of the best examples of phenotypic plasticity known. For example, the oak specialist *N. arizonaria* is bivoltine, with a spring brood of caterpillars that emerges when the oaks are in flower, followed by a summer brood several months later. In the spring, the caterpillars develop a remarkable resemblance to the oak catkins (staminate flowers), complete with false stamens running down their backs (Figure 3a). In contrast, caterpillars of the summer brood develop into superb mimics of oak twigs (Figure 3b). This polyphenism appears to be adaptive because if the summer brood developed into the catkin form they would likely be very conspicuous (McFarland 1988, Greene 1989).

Nemoria outina is another dietary specialist, feeding only on scrub rosemary (*Ceratiola ericoides*, Ericaceae). Similar to *N. arizonaria*, *N. outina* caterpillars develop into two discretely different morphs—a green foliage mimicking form (Figure 3c), and a brown twig mimicking form (Figure 3d)

Fig. 3 Two *Nemoria* dietary specialists develop into discrete forms, with no intermediate forms. Caterpillars of the oak specialist *Nemoria arizonaria* develop into catkin morphs (a) or twig morphs (b) depending on which plant tissue they consume. These two caterpillars were siblings, reared on diets of catkins and oak leaves, respectively. Photos from Greene (1989). Lower panels: Discrete morphs in caterpillars of the *Nemoria outina*, which feeds only on scrub rosemary (*Ceratiola*) in Florida. A foliage mimic form (c) is very different from the twig morph (d), with no intermediate forms (photos and unpublished data, M. Canfield).

(Deyrup and Eisner 1993). This species also has a step developmental threshold, since intermediate forms are not observed (Canfield 2006).

What environmental cues are used by dietary specialists to trigger the development into the different discrete morphs? Split-brood rearing experiments with the oak specialist *N. arizonaria* showed that these very different developmental trajectories are induced by differences in larval diet: all caterpillars that were reared on catkins developed into catkin morphs, whereas all those reared on oak leaves developed into twig morphs (Greene 1989). Furthermore, the discrete larval forms in this dietary specialist are produced by a developmental threshold: caterpillars reared on experimental mixed diets (containing varying proportions of catkins and leaves) do not develop into intermediate forms; rather, there is threshold in the dietary ratio of leaves to catkins that determines which morph is produced.

What is the mechanism by which environmental differences are translated into morphological differences? In a set of elegant experiments, Edward Poulton demonstrated that some geometrid caterpillars develop different colors depending on the background colors they perceive (Poulton 1885, 1887, 1892). Could the developmental trajectory of specialists, such as *N. arizonaria*, be influenced by the color of their larval diet (i.e. predominantly yellow for catkins, predominantly green for leaves) rather than the chemistry of the larval diet? To test this hypothesis, *N. arizonaria* caterpillars were reared on catkin and leaf diets, and either in the dark, under green light, or under yellow light. The spectra of reflected light they perceived had no effect on their phenotype: all caterpillars, independent of light treatment, developed into the catkin form if they ate catkins, or developed into the twig form if they ate oak leaves (Figure 4; Greene 1996). Other potential environmental induction cues, such as temperature, humidity, and photoperiod likewise did not influence the developmental "decision" between the two morphs (Greene 1989, 1996, 1999). Current evidence suggests that, in this plant specialist, the food consumed determines the morph produced.

What sort of developmental plasticity is shown by dietary generalists? To address this question, we studied a host generalist. In western Montana *N. darwiniata* caterpillars feed on over 20 species of plants in at least 8 distantly related families (Greene and Ehmer, in prep.). The hiding sites found in these different host plants range from snowy white *Ceanothus* flowers, leafy green backgrounds with varying amounts of silvery-white (especially in willows (*Salix*)), reds (especially in sumacs (*Rhus*) and roses (*Rosa*)), and browns (for example in kinnikinick (*Arctostaphylos*), hawthorn (*Crataegus*) and serviceberry (*Amelanchier*)). In split-brood rearing experiments (to control for

Fig. 4 Spectral qualities of light does not influence the development of the oak specialist *Nemoria arizonaria*. Caterpillars in the top row were reared on oak catkins; those in the bottom row were reared on oak leaves. Caterpillars shown in the left column (a and d) were reared in total darkness; caterpillars in the middle column (b and e) were reared under yellow light; caterpillars in the right column (c and f) were reared under green light (photos from Greene, 1996).

genetic effects), sibling caterpillars developed into a wide range of forms, from almost pure white when reared on white *Ceanothus* flowers, olive green on *Ceanothus* leaves, mottled caterpillars with different amounts of red, green and brown, to very dark brown caterpillars when reared on *Crataegus* (Figure 5). The general color of the caterpillars' skin and the pattern of patches of color tend to match the feeding microsites within each particular host plant. Thus the dietary generalist *N. darwiniata* caterpillars exhibit considerable phenotypic plasticity in their larval development. However,

Fig. 5 Phenotypic plasticity in *Nemoria darwiniata*, a dietary generalist. Figure shows sibling caterpillars raised on a few of their different host plants. The caterpillars show gradual variation in color, ranging from creamy white to dark splotchy brown, with many intermediate forms. Individual caterpillars are shown on the host plants on which they were reared: a, *Ceanothus velutinus* flowers (Rhamnaceae); b, *Salix exigua* (Salicaceae); c, *Rhus glabra* (Anacardaceae); d, *Arctostaphylos uva-ursi* (Ericaceae); e, *Amelanchier alnifolia* (Rosaceae); f, *Crataegus columbiana* (Rosaceae).

unlike the specialists *N. arizonaria* and *N. outina*, the phenotypic variation of this host generalist is more continuous, with smooth intergrading of intermediate forms.

In addition, the generalist appears to use a wider range of induction cues than the specialists. For *N. darwiniata*, in addition to host plant effects, the spectra of reflected light they perceive influences their phenotype: caterpillars develop more red, green, or yellow if they perceive those wavelengths during development; they develop into very dark brown-blackish forms if reared in the dark (Figure 6).

Fig. 6 Spectral qualities of light does influence the development of the dietary generalist *Nemoria darwiniata*. These photos show siblings, reared either under ambient light conditions (top row), or reared in total darkness (bottom row). The diets were *Ceanothus velutinus* flowers (a and f), *Ceanothus velutinus* leaves (b and g), *Salix exigua* (c and h), *Rhus glabra* (d and i), and *Crataegus columbiana* (e and j).

These examples with just a few species of *Nemoria* caterpillars demonstrate that closely-related species can have very different patterns of developmental plasticity, and even use different sets of environmental induction cues to guide their development. These patterns may be related to host breadth and plant associations: dietary specialists seem to show developmental thresholds, and develop into just a few highly-integrated discrete phenotypes (sensu Schlichting and Pigliucci 1998, Pigliucci 2003); a dietary generalist shows more gradual developmental responses with many intermediate forms along a phenotypic continuum. Although these different patterns of developmental plasticity appear quite different, the fact that these are closely-related congeners suggests that both phenotypic plasticity and the environmental cues used to elicit plasticity evolve easily. A broad, comparative approach, such as presented here, will shed light on many important questions about the evolution of plasticity, and its influence on speciation and diversification (West-Eberhard 2003).

Acknowledgements

We thank David Haig, Dan Janzen, Naomi Pierce, and Dave Wagner for helpful discussions and help in the field. All the credit for the original discovery and description of seasonal variation in the morphology of the caterpillars and moths belongs entirely to Noel McFarland. Mr. McFarland shared this information with E.G. Some of the subsequent research focused on the environmental cues that trigger these developmental differences, but this research was possible because of Mr. McFarland's pioneering work. Readers are urged to read his book "Portraits of South Australian Geometrid Moths," which sets the gold standard for the careful documentation of the life histories, development, behavior, and ecology of Lepidoptera.

References

Angiosperm Phylogeny Group. 1998. An ordinal classification for the families of flowering plants. Annals of the Missouri Botanical Garden 85: 531–553.

Canfield, M. R. 2006. The ecology and evolution of phenotypic plasticity: case studies in Lepidoptera. Ph.D thesis, Harvard University, Cambridge, MA.

Deyrup, M. and T. Eisner. 1993. Last stand in the sand. Natural History 102(12): 42–47.

Ferguson, D. C. 1969. A revision of the moths of the subfamily Geometrinae of America north of Mexico (Insecta, Lepidoptera). Peabody Museum of Natural History (Yale) 29: 1–251.

Ferguson, D. C., in Dominick, R. B. et al. 1985. *The Moths of America North of Mexico*, Fasc. 18.1, Geometroidea: Geometridae (in part).

Greene, E. 1989. A diet-induced developmental polymorphism in a caterpillar. Science 243: 643–646.

Greene, E. 1996. Effect of light quality and larval diet on morph induction in the polymorphic caterpillar *Nemoria arizonaria* (Lepidoptera: Geometridae). Biological Journal of the Linnean Society 58: 277–2865.

Greene, E. 1999. Phenotypic variation in larval development and evolution: Polymorphism, polyphenism, and developmental reaction norms. In: *The Origin and Evolution of Larval Forms*. B. K. Hall and M. H. Wake eds., pp. 379–406. Academic Press, San Diego.

Mayr, E. 1963. *Animal Species and Evolution*. Belknap Press of Harvard University, Cambridge, MA.

McFarland, N. 1988. *Portraits of South Australian Geometrid Moths*. Allen Press, Lawrence, KS.

Pitkin, L. M. 1993. Neotropical Emerald moths of the genera *Nemoria, Lissochlora* and *Chavarriella,* with particular reference to the species of Costa Rica (Lepidoptera: Geometridae, Geometrinae). Natural History Society of London (Entomology) 62(2): 39–159.

Pigliucci, M. 2003. Phenotypic integration: studying the ecology and evolution of complex phenotypes. Ecology Letters 6: 265–272.

Poulton, E. B. 1885. Further notes upon the markings and attitudes of lepidopterous larvae, together with a complete account of the life-history of *Sphinx ligustri* and *Selenia illunaria.* Transactions of the Entomological Society of London (2): 281–329.

Poulton, E. B. 1887. An enquiry into the cause and extent of a special colour-relation between certain exposed lepidopterous pupae and the surfaces that immediately surround them. Philosophical Transactions of the Royal Society of London B 178: 311–444.

Poulton, E. B. 1892. Further experiments upon the colour-relation between certain lepidopterous larvae, pupae, cocoons, and imagines and their surroundings. Transactions of the Entomological Society of London (4): 293–487.

Schlichting, C. D. 1989. Phenotypic integration and environmental change. BioScience 39(7): 460–464.

Schlichting C. D. and M. Pigliucci. 1998. *Phenotypic Evolution.* Sinauer, Sunderland, MA.

West-Eberhard, M. J. 2003. *Developmental Plasticity and Evolution*. Oxford University Press, Oxford.

5

Phase Polyphenism in Locusts: Mechanisms, Population Consequences, Adaptive Significance and Evolution

Stephen J. Simpson* **and Gregory A. Sword****

**School of Biological Sciences, Heydon-Laurence Building, A08 University of Sydney, Sydney, NSW 2006 Australia, Phone: +61 2 9351 5633, Fax: +61 2 9351 4119, E-mail: stephen.simpson@bio.usyd.edu.au*

***School of Biological Sciences, Macleay Building, A12, University of Sydney, Sydney, NSW 2006, Australia, Phone +61293512386, Fax: +61293514119, E-mail: greg.sword@bio.usyd.edu.au*

Abstract

Locusts express a remarkable form of phenotypic plasticity called phase polyphenism in which local population density affects the expression of a variety of behavioural, physiological, and morphological traits. Behaviour is the most labile of the various density-dependent phenotypic changes and provides positive feedbacks that drive the process at a population level. Locusts avoid each other in the low-density 'solitarious' phase, but they actively aggregate in the high-density 'gregarious' phase. Individuals change behavioural phase rapidly, in a matter of hours. Behavioural phase change is mediated primarily by direct physical contact among locusts, with the major site of mechanosensory input being the hind legs—a finding that has opened the possibility to study phase change as a model system for neuronal plasticity. Behavioural phase state is also transmitted epigenetically across generations. Females can manipulate the phenotypes of their developing offspring to an extent that reflects both maternal and paternal experience of crowding. This maternal effect is mediated during oviposition by a chemical agent introduced into the foam surrounding the eggs. Laboratory and field experiments in conjunction with individual-based computer models have elucidated the relationship between individual behaviour, population responses, and the spatial distribution and chemical quality of resources within the local environment. Whether a local population of solitarious phase locusts will gregarize, and hence potentially seed larger scale outbreaks, depends critically on the fine-scale distribution and quality of resources. A consequence of fluctuating population densities is that locusts are intermittently exposed to periods of increased disease and predation risk. Recent insights into the expression of density-dependent anti-predator and pathogen resistance strategies indicate that a number of traits involved in locust phase change are specific adaptations to these risks.

Phase Polyphenism in Locusts

All locusts are grasshoppers, but not all grasshoppers are locusts. The distinction between the two lies in the expression of a form of phenotypic plasticity known as phase polyphenism. In locust species, local population density induces the expression of graded phenotypic changes in an array of traits that include colouration, morphometry, anatomy, egg mass, food selection, nutritional physiology, reproductive physiology, metabolism, neurophysiology, endocrine physiology, molecular biology, immune responses, longevity and pheromone production (reviewed by Pener 1991, Pener and Yerushalmi 1998, Simpson et al. 1999, 2005; De Loof et al. 2006, see also Simpson et al. 2002, Wilson et al. 2002, Ferenz and Seidelmann 2003, Kang et al. 2004, Hassanali et al. 2005, Simpson and Sword 2008, Wilson and Cotter, this Volume). At their extremes, the resulting phenotypic forms vary between the sedentary, cryptic "solitarious" phase that is produced under low population density conditions, and the swarming "gregarious" phase produced at high population densities (Fig. 1). This form of density-dependent phenotypic plasticity has independently evolved to different degrees in a number of unrelated insect lineages (Pener 1991, Applebaum and Heifetz 1999), but it is most commonly associated with locust species in the Orthoptera (family: Acrididae).

Phase change is central to the biology of locusts and presents itself as a model system with which to analyze the proximate mechanisms underlying the expression of phenotypic plasticity, as well as its ecological and evolutionary consequences. The study of locust phase polyphenism is also of particular economic importance in light of the fact that swarming locust outbreaks regularly affect the livelihoods of people on five different continents.

The theory of locust phases was first proposed by Uvarov (1921) in his taxonomic revision of the genus *Locusta*. He suggested that two distinct species, the swarming *L. migratoria* and solitary-living *L. danica*, were really

Fig. 1 Final-instar nymphs of the two extreme forms (phases) of the desert locust, *Schistocerca gregaria*. Locusts have the genetic potential to exist in either phase or in various transitional forms, depending on their experience of crowding during their own lifetime and also that of their parents and grandparents.

alternative phenotypic forms or "phases" of just one species, the African migratory locust, *L. migratoria*. The environmental cue mediating phenotypic change between the different phases was unknown to Uvarov at the time. In controlled rearing experiments, Faure (1923) soon thereafter showed that the degree of nymphal crowding during rearing was responsible. Nymphs of the brown locust, *Locustana pardalina,* reared in isolation or near-isolation developed phenotypes characteristic of solitarious phase locusts, whereas nymphs reared together in a crowd with conspecifics developed gregarious phase phenotypes. Faure (1932) went on to demonstrate a similar phenotypic effect of rearing density in *L. migratoria* as well as the desert locust, *Schistocerca gregaria*, and it has since been shown to occur in many different locust species (Uvarov 1966, 1977). Initially, the phenomenon was widely referred to as phase polymorphism, but has recently come to be more appropriately referred to as phase polyphenism—reflecting the fact that the expression of the different phases is environmentally as opposed to genetically determined (Dingle 1996).

Despite being considered a classic example of phenotypic plasticity, locust phase polyphenism has been vastly underused as a model study system since the renaissance of interest in phenotypic plasticity that began in the late 1980s. This is evidenced, for example, by the fact there are no references to locusts or related species in four major texts on phenotypic plasticity (Schlichting and Pigliucci 1998, Pigliucci 2001, Pigliucci and Preston 2004, DeWitt and Scheiner 2004). In addition, only a handful of studies on locusts or related species have adopted a reaction norm perspective when considering the evolution of phase polyphenism (Sword

2002, 2003). West-Eberhard (2003) mentions locusts as "among the most striking coordinated alternative phenotypes known" (pg. 132), but refers in the main to older literature. We suspect the paucity of coverage is due in large part to the applied theme of much phase polyphenism research, especially the earlier studies. One of our aims in writing this chapter is to place locust phase polyphenism in its rightful place as an exemplar system for studying phenotypic plasticity—not simply as a major economic threat, management of which will benefit from a better understanding of its biology.

Here we focus on the study of two of the primary components of locust phase polyphenism, density-dependent changes in behaviour and colouration, as a means to provide insight into the use of locusts as model systems to address a variety of physiological, ecological and evolutionary questions. We begin with the physiological mechanisms underlying the expression of density-dependent traits within the lifetime of an individual locust, as well as epigenetically across generations. We then examine the ecological and evolutionary consequences of locust phase change as it applies to the study of population dynamics, solitary versus group living, and the evolution of warning colouration among insects in general, Unless specifically stated, our discussion pertains to studies of the desert locust, *S. gregaria*, the focus of a majority of the experimental work on the subject. We do not mean to slight the vast body of work focusing on other density-dependent traits and we encourage interested readers to refer to the reviews cited above (see also chapter by Wilson and Cotter, this book). Of course, we do attempt to discuss plasticity in other traits as well as in other species with respect to the underlying physiological mechanisms and adaptive significance of locust phase change whenever relevant.

Behaviour: The Engine of Phase Change

The behavior of solitarious locusts is strikingly different to that of gregarious locusts. They are much less active, move away from rather than towards other locusts, groom less frequently, and generally behave in a manner that is consistent with their solitary, cryptic lifestyle. This is a striking contrast to the behaviour of gregarious phase locusts which readily form aggregations, are much more active, and often migrate *en masse* (Ellis 1953, 1963, Gillett 1973, Roessingh et al. 1993, Simpson et al. 1999). These multiple behavioural differences can be encapsulated into a single measure of behavioural phase state, referred to as '*p*(solitarious)'. *P*(solitarious) is the probability of an individual behaving as a fully solitarious phase locust and is calculated using multiple logistic regression analysis of the behavioural responses of individual locusts in a test arena (Roessingh et al. 1993, Simpson et al. 1999).

Behaviour is the first phase characteristic to change overtly in response to variation in population density, with a shift in behaviour becoming evident within an hour of crowding (Roessingh and Simpson 1994). Once locusts begin to become attracted rather than repelled by others, a positive feedback is established. Under appropriate environmental circumstances, this feedback loop can drive the change of an initially solitarious locust population to behavioural gregariousness (Fig. 2). Other phase changes occur after the transition to the gregarious behavioural phase despite the fact that, individually, these traits are under separate physiological control. However, due to the autocatalytic nature of behavioural phase change, the entire suite of phase characteristics can change in a coordinated fashion. In effect, behavioural phase change serves to couple a diverse suite of continuous traits into a coordinated threshold trait (a point that is developed further on pp. 110-111).

The Mechanisms of Behavioural Phase Change

The Time-course of Behavioural Phase Change

Understanding the mechanisms of behavioural phase change requires first describing in detail its time-course. When solitarious nymphs (Roessingh and Simpson 1994) or adults (Bouaïchi et al. 1995) are crowded, their behaviour rapidly changes. Within 4-8 h of crowding they cannot be distinguished behaviourally from gregarious phase locusts that have been crowded for many generations. Solitary-reared locusts that have been crowded for 48 h or less and undergone behavioural phase change will rapidly shift back to behaving solitariously when removed from the crowd. However, when locusts that have been reared for several generations under crowd-reared conditions are isolated they show an initially rapid behavioural solitarization, but this does not proceed to completion. Instead they remain at an intermediate p(solitarious) value for the rest of that developmental stage. Reaching higher levels of solitariousness (i.e. completely losing gregarious phase behaviours) requires rearing in isolation over several stadia or generations (Roessingh et al. 1993, Roessingh and Simpson 1994).

Thus, behavioural gregarization and solitarization proceed at different rates. Gregarization occurs rapidly, but behavioural solitarization is a two-stage process. It initially proceeds rapidly to a level that is a function of the

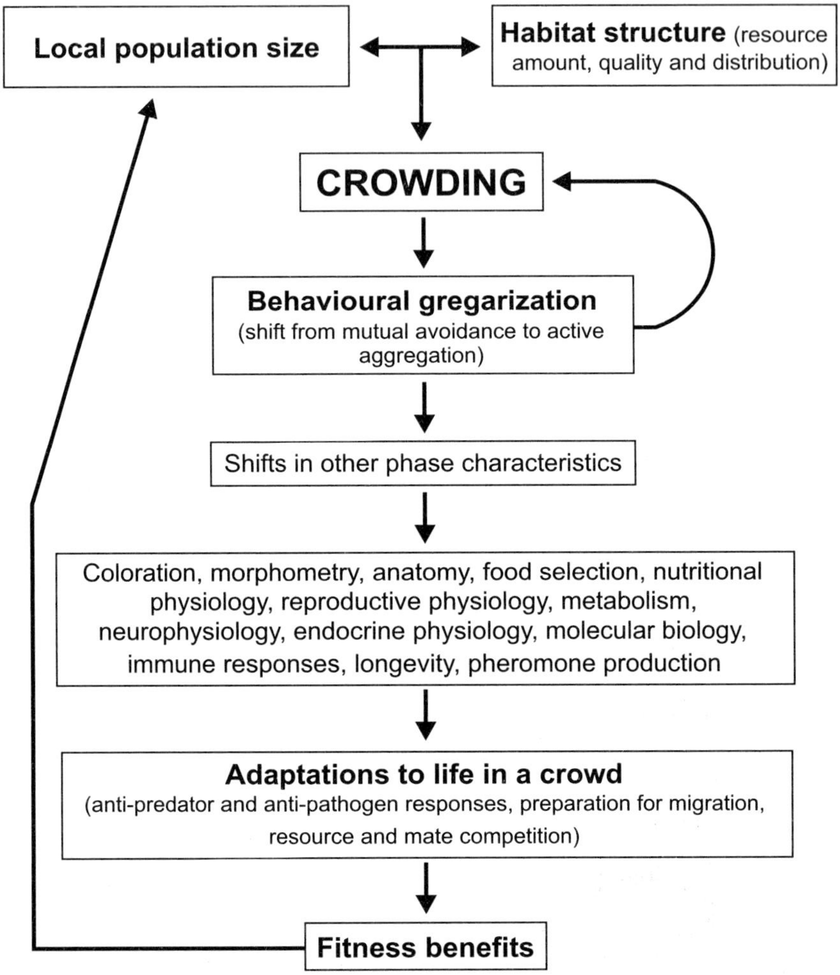

Fig. 2 A summary flowchart showing the effects of local ecology and crowding on the expression of phase polyphenism in behaviour and other traits in the desert locust. Behavioural gregarization (involving, *inter alia*, the transition from avoiding other locusts to actively aggregating and marching) occurs rapidly in response to an increase in local population density and serves to drive the coordinated expression of the full suite of phase characteristics. From Simpson et al. (2005).

previous time spent crowded, but it also requires several stadia or even generations to complete in insects that have been crowded for extended periods (Fig. 3). This hysteresis effect is likely to be adaptive. Longer periods of crowding should indicate the persistent presence of considerable number

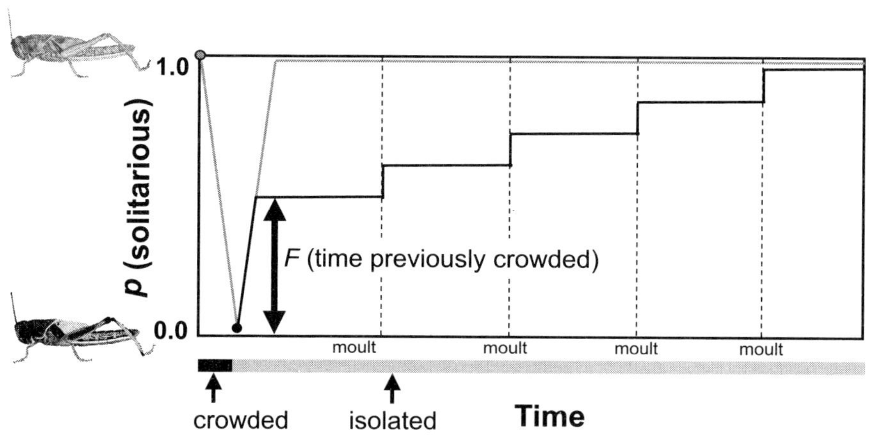

Fig. 3 A summary of the temporal dynamics of behavioural phase change in the desert locust. When solitarious nymphs (grey line) are crowded they undergo rapid behavioural gregarization, within a few hours. Solitarization is a two-phase process. There is an initial rapid phase, the extent of which is a function of the time spent previously crowded. If re-isolated after only a few days of crowding, locusts return rapidly to the fully solitarious behavioural state (shown by the grey line). In contrast, when fully gregarious locusts that have been reared under crowded conditions for several generations (black line) are isolated, they initially solitarize behaviourally to a degree, but not completely, remaining in a transitional state until the next moult (vertical dotted lines). From this point they solitarize further if continued to be reared in isolation. This step-wise return to the solitarious behavioural state continues from moult to moult and continues across generations via a maternal effect.

of nearby individuals. Given that the longer a locust has been crowded the more resistant it is to losing gregariousness, a gregarious locust that becomes temporarily separated from the group will continue to be active and attracted by others, enhancing its chances of rejoining the safety of the group. However, if isolation occurs after only a brief period of environmentally imposed crowding, a cryptic, solitary-reared locust will rapidly reassume cryptic behaviour and hence avoid attracting predators.

Stimuli Inducing Behavioural Phase Change and Their Mode of Action

Behavioural phase transition occurs rapidly, but what are the stimuli provided by other locusts that elicit the change? Experiments in which tactile, olfactory and visual stimuli were provided separately and in combination showed that contact with other locusts is the critical stimulus (Fig. 4). The combination of visual and olfactory stimuli is also

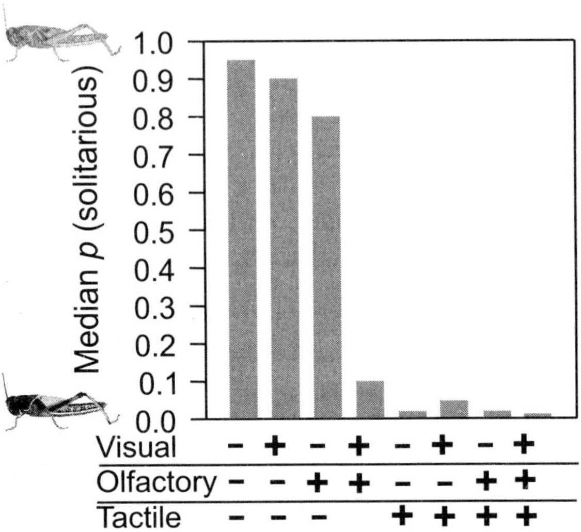

Fig. 4 The results of an experiment to tease apart the individual and interactive effects of possible behaviourally gregarizing stimuli provided by other desert locusts. The presence of a given stimulus is indicated by a cross, its absence by a minus sign. The combination of the sight and smell of other locusts induced behavioural gregarization in solitarious desert locust nymphs within a 4-h period, although neither alone was effective. The most powerful gregarizing stimulus was contact, simulated by repeated buffeting with paper spheres. After Roessingh et al. (1998).

behaviourally gregarizing (Roessingh et al. 1998, Lester et al. 2005), but neither is effective when presented singly. Subjecting solitarious locust nymphs to buffeting with small papier mâchè balls for 4 h induces complete behavioural gregarization (Roessingh et al. 1998). Repeated contact between the paper balls and the cuticle of the test insect would have provided both mechanical stimulation and contact chemical self-stimulation. Subsequent experiments that controlled for chemical self-stimulation indicated that the primary stimulus is mechanical, not chemical (Hägele and Simpson 2000) (Fig. 5).

The hind femur has recently been identified as the key site of tactile input triggering behavioural phase change (Simpson et al. 2001). Repeatedly touching as little as one quarter of the outer surface area of a hind femur (referred to by some as the gregarization- or G-spot) with a fine paintbrush produces full behavioural gregarization within 4 h (Fig. 6) (Simpson et al. 2001, Rogers et al. 2003). Interestingly, solitarious locusts have approximately 30% more mechanosensory hairs on the hind femora than do gregarious locusts, but have similar or fewer numbers of hairs elsewhere on

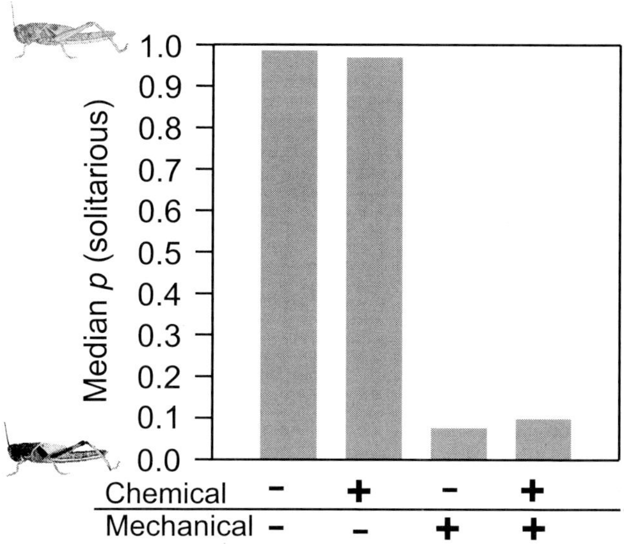

Fig. 5 The results of an experiment to establish whether contact stimulation exerts its behaviourally gregarizing effects via surface chemicals or mechanical stimulation. Desert locust nymphs were either kept isolated without chemical or mechanical stimulation, or showered with millet seeds and/or exposed to surface wax extracts of other locusts. The presence of a given stimulus is indicated by a cross, its absence by a minus sign. Mechanical stimulation alone was fully effective, with no evidence of a response to chemical stimulation. After Hägele and Simpson (2000).

the legs (Rogers et al. 2003). Patterned electrical stimulation of metathoracic nerve 5, which innervates the hind leg, produces full behavioural gregarization in restrained locusts (Rogers et al. 2003). Furthermore, a full response was observed only if the femur was allowed to move toward the body during stimulation, indicating a role for proprioception. Hence, the gregarizing input was shown to combine both exteroceptive and proprioceptive components, which travel in both nerves 5B1 and 5B2 (Rogers et al. 2003; Fig. 7). These findings have provided a powerful experimental method with which to elicit and study the neural bases of behavioural plasticity.

A recent study compared relative amounts of neurotransmitters and neuromodulators in the central nervous systems of solitarious and gregarious locusts as they underwent behavioural phase transition (Rogers et al. 2004). Of 13 substances that were measured, 11 differed between long-term solitarious and gregarious locusts, including important excitatory (glutamate, acetylcholine) and inhibitory (GABA) neurotransmitters as well

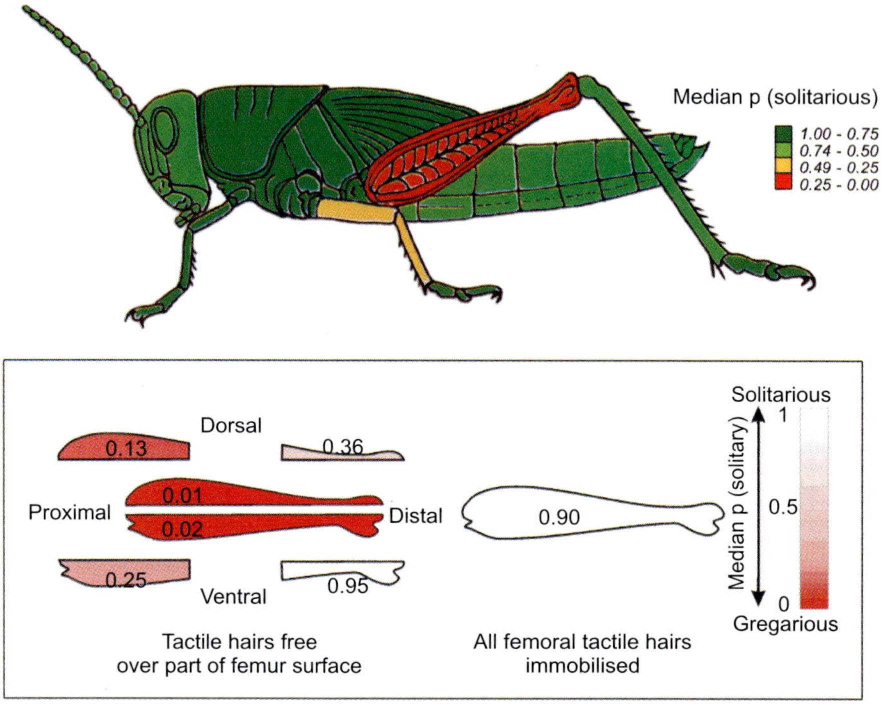

Fig. 6 Localisation of the sites of mechanical stimulation of behavioural gregarization in desert locust nymphs. Insects were tickled with a fine paintbrush on various body regions. The hind femur was shown to be the site of gregarizing input. Immobilizing all tactile hairs on the femur blocked induction of behavioural phase change, whereas localizing stimulation to quartiles of the outer surface of a single hind femur was sufficient to evoke gregarization, except for the lower distal quartile. Further experiments showed that in addition to input from leg mechanosensory hairs, it was necessary to allow the femur to move in toward the body during stimulation, indicating a role for proprioceptive inputs, probably from the coxal area. After Simpson et al. (2001) and Rogers et al. (2003).

as the neuromodulators/neurohormones dopamine and serotonin. Of all these substances, however, only serotonin underwent a substantial increase (8-fold) during the critical 4-h period during which behavioural gregarization is established (Fig. 8). This increase occurred in the thoracic ganglia (which receive inputs from leg mechanoreceptors), but not the brain.

Thus, of the 13 putative gregarizing substances measured by Rogers et al. (2004), only serotonin has the possibility of being an enabling/causal agent in a phase transition, and this is the focus of ongoing analysis. It is tempting to draw the analogy between the mechanisms of behavioural phase change

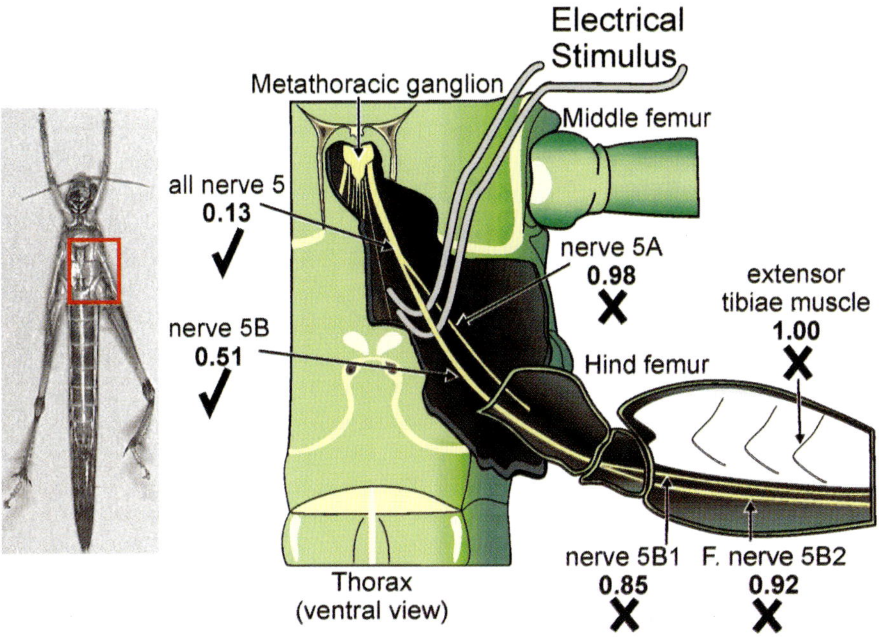

Fig. 7 Results from experiments to localize the neural pathways carrying behaviourally gregarizing stimuli from the hind femur of desert locust nymphs. Patterned electrical stimulation delivered directly to Nerve 5 in the thorax produced full behavioural gregarization in completely restrained locusts, whereas electrical stimulation applied to distal leg nerve branches or to the extensor tibiae muscle did not elicit gregarization. Values indicate median *p* (solitarious) after 4 h of electrical stimulation. After Rogers et al. (2003), with thanks to Steve Rogers for use of his figure.

and those underlying learning and memory. Both processes involve an initial, labile short-term phase, which with continued reinforcement becomes translated into a change that takes longer to reverse. Serotonin is known to be involved in the establishment of short-term memory via its modulatory effects on synaptic transmission, and leads via second messenger cascades (PKA, MAPK) and transcription factors (CREB) to synthesis of proteins that produce structural synaptic changes that result in long-term memory (Kandel and Pittenger 1999). Perhaps the same is true for behavioural phase change.

Mechanisms Controlling Changes in Other Phase Characters

Stimuli that evoke other phase characteristics are as yet unknown. Changes in morphometry and nymphal colour pattern take longer than behavioural

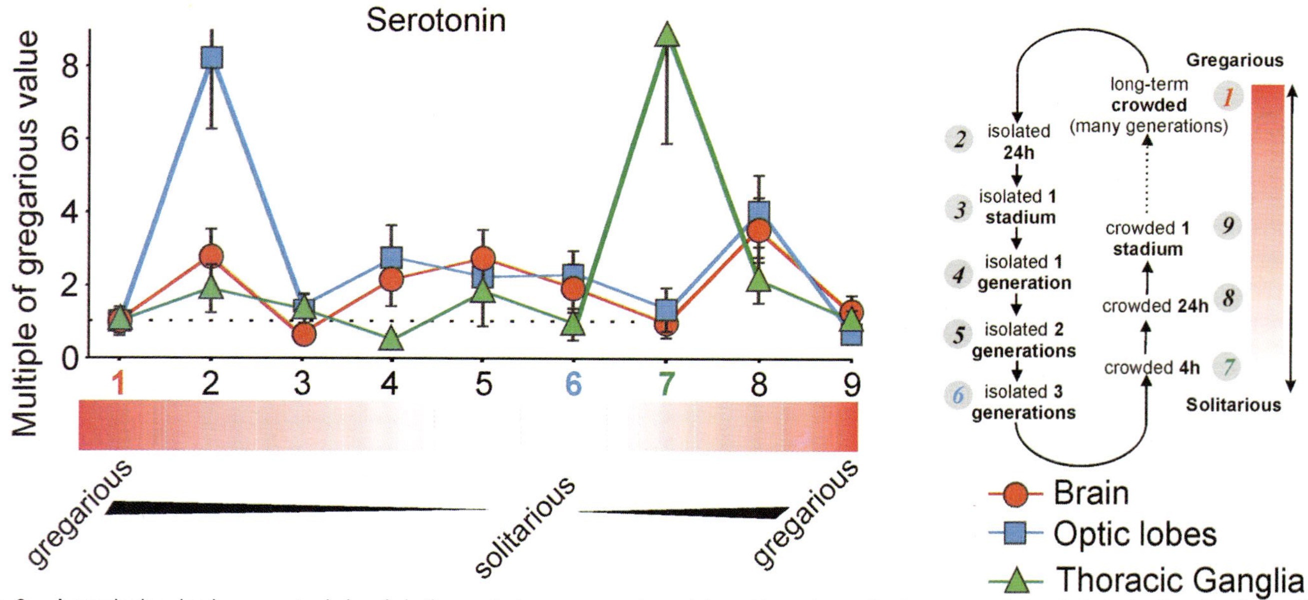

Fig. 8 A graph showing how serotonin levels in the central nervous system of desert locust nymphs increase substantially in the first 4 h of crowding, during which behavioural gregarization occurs. Values are presented as multiples of the serotonin titre in an average, fully gregarious insect.

The rise occurred only in the thoracic ganglia, the destination of mechanosensory inputs from the hind leg. Coupled with the hysteresis effect in the time-course of behavioural phase change, with rapid and long-term stages (Fig. 3), this suggests similarities between the mechanisms of behavioural phase change and those known to be involved in learning and memory. Interestingly, serotonin levels also rose sharply during the initial phase of solitarization—not in the thoracic ganglia but in the optic lobes. After Rogers et al. (2004), with thanks to Steve Rogers for use of his figure.

gregarization and need not be induced by the same stimuli or be mediated by the same physiological mechanisms as behavioural phase change. The smell alone of other locusts causes the development of black patterning in solitarious nymphs, but does not cause the production of yellow background colouration or behavioural phase change (Lester et al. 2005). The combination of the sight and smell of other locusts is behaviourally gregarizing but does not evoke yellowing, which requires rearing among a crowd of conspecific nymphs, presumably indicating the presence of a contact chemical cue (Lester et al. 2005).

The production of nymphal black patterning is controlled by the neuropeptide [His^7]-corazonin, also termed dark-colour-inducing neurohormone or dark pigmentotropin (Tawfik et al. 1999, Tanaka 2000, 2001). [His^7]-corazonin also changes morphometry towards the gregarious condition, but has no effect on behavioural phase state (Hoste et al. 2002, Tanaka et al. 2002).

Much attention has been paid to the possible roles of ecdysteroids and juvenile hormone in phase change, but it is clear that these are not controlling agents of gregarization, either within a generation (Pener et al. 1992, Applebaum et al. 1997, Pener and Yerushalmi 1998) or between generations (Hägele et al. 2004). Hartfelder and Emlen (2005) have restated the view that juvenile hormone is involved in phase change, but this view was founded on earlier published claims that did not take account of substantial contrary evidence (Dorn et al. 2000). There are several other hormonal differences between the phases, e.g. in brain-derived peptides (e.g. Ayali et al. 1996a,b, Wedekind-Hirschberger et al. 1999, Clynen et al. 2001, Rahman et al. 2002), but none has been shown to cause phase change. See De Loof et al. (2006) for a recent review of the molecular evidence.

The Maternal and Paternal Transmission of Phase State

Phase characteristics such as nymphal colour, hatchling mass, morphometry and reproductive variables not only change within the life of an individual, but also accumulate across generations through an epigenetic mechanism (Gunn and Hunter-Jones 1952, Hunter-Jones 1958, Injeyan and Tobe, 1981, Dale and Tobe, 1990). This was also found to be the case for behavioural phase state (Roessingh et al. 1993). By subjecting adults to various crowding treatments and then observing the effects on the behavioural phase-state of their newly hatched offspring, it was found that experience of crowding in either the mother or the father resulted in hatchlings that emerged from the eggs behaving gregariously (Islam et al.

1994a). Crowding as late as at the time of oviposition causes solitarious females to produce gregariously behaving offspring, while isolation during oviposition of gregarious females leads to partial behavioural solitarization of the offspring (Islam et al. 1994b). The extent to which the offspring show gregarious behavioural characteristics is a positive function of how recently the mother was last crowded (Bouaïchi et al. 1995, Fig. 9).

Hence, female locusts use their own experience of being crowded, as well as indirect evidence from the phase of their mate, to predict the population density that their larvae will experience upon emergence. As such, they predispose their offspring to behave appropriately. Such transgenerational transfer of behavioural state will result in continued accumulation of phase characteristics across generations.

The mechanisms of transgenerational phase change are partly understood for the maternal effect, but the nature of the paternal influence is

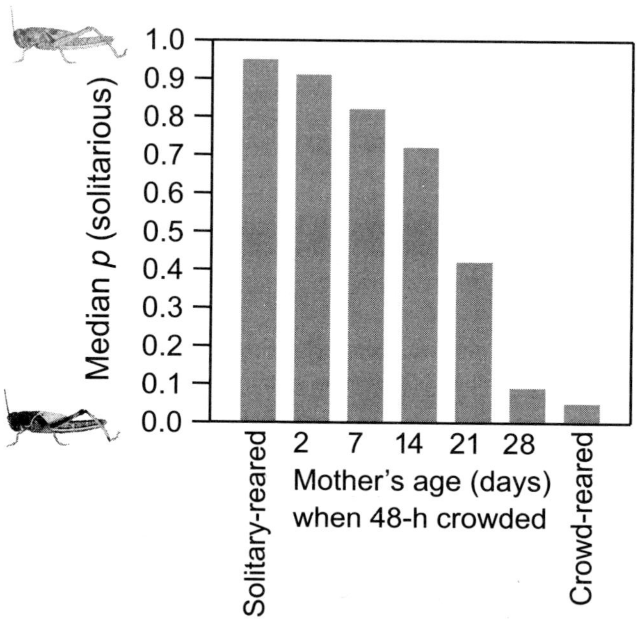

Fig. 9 Graph indicating the extent to which hatchlings emerged from the egg pod in the gregarious behavioural state, after their solitarious mother had been subjected to a 48-h period of crowding at one of four ages after the adult moult (2, 7, 14 or 28 days), prior to when she laid her 1st eggs, on day 30. Hatchlings from solitarious mothers that were not crowded, and from fully gregarious (crowd-reared) mothers are also included. The more recently the mother was crowded prior to egg laying, the more gregarious were her hatchlings. After Bouaïchi et al. (1995).

not yet known. The maternal gregarizing agent is found in the egg froth that is deposited with the pod of 50-100 eggs as they are laid in the soil and forms a plug filling the hole drilled by the female's abdomen. The female reproductive accessory glands are a source of compounds essential to the gregarizing agent (Hägele et al. 2000). Activity resides in aqueous extracts of fresh gregarious egg foam, with some activity in ethanol extracts but none in hexane extracts (indicating that the substance is not a C8 ketone, as suggested by Malual et al. (2001)). Treating the eggs of a solitarious female with aqueous extracts of egg foam from gregarious females induces behavioural gregarization of the offspring (Fig. 10). The gregarizing agent influences egg development for only a short period immediately following oviposition. Larvae emerge behaving solitariously when the eggs from a freshly ovulated pod laid by a gregarious female are washed with saline during the first few hours after oviposition. Application of saline extracts of egg washings and foam from crowd-reared females can reinstitute gregarious behaviour in larvae from washed gregarious egg pods, but only if the material is applied within the first day of laying (McCaffery et al. 1998). The gregarizing agent is heat sensitive, and less than 3 kDa in size, but its chemical identity is not yet established. See Simpson and Miller 2007 for a detailed review of the data on maternal gregarization.

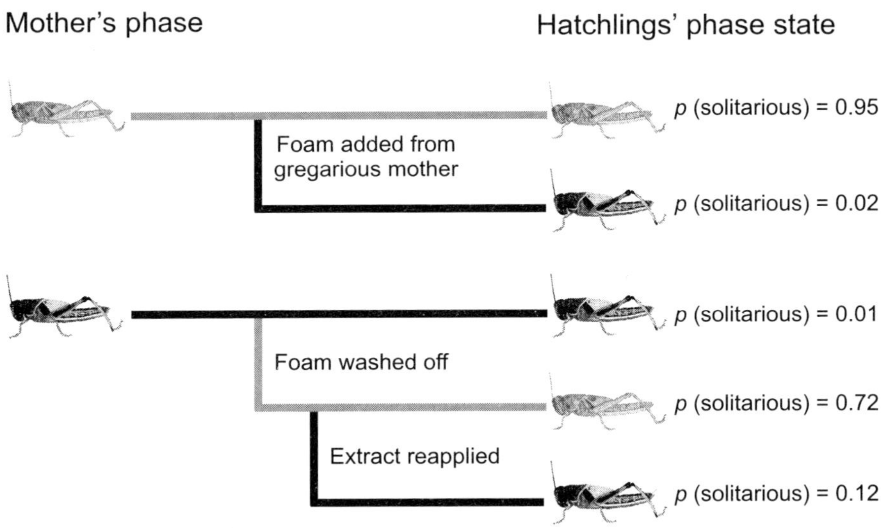

Fig. 10 Summary of experiments showing that the source of the maternal gregarizing agent is an aqueous substance found in the froth surrounding the egg pod of gregarious females. After McCaffery et al. (1998), Simpson et al. (1999).

Population Consequences of Behavioural Phase Change

Behavioural phase change is an individual-level phenomenon, but its expression has important implications for population-level responses. Hence, if the environment forces solitarious locusts to come together against their predisposition to avoid one other, congregation and close contact between individuals will rapidly induce active aggregation, which will in turn promote further gregarization. Given that gregarious phase locusts are migratory and move together in either marching bands (nymphs) or swarms (adults), their mobility can lead to coalescence with other local groups. Ultimately, this process may seed the production of massive swarms across an entire region. In contrast, when previously aggregated individuals become separated, they will begin to solitarize, hence reducing their tendency to aggregate and so promoting further solitarization. If the habitat tends to keep locusts apart by virtue of its structure or resource quality, then this will ultimately lead to resolitarization of a gregarious population.

The Importance of Resource Distribution at a Fine Spatial Scale

Initial experiments in the laboratory and the field showed that small-scale features of resource distribution are critical in determining the extent to which phase change occurs in a local population of desert locusts. Clumping of resources such as food plants, roosting sites, and areas of favourable microclimate encourages solitarious locusts to come together and as a consequence to behaviourally gregarize (Bouaïchi et al. 1996, Fig. 11). Such effects are also transmitted across generations, with the degree of clumping of food plants in the parental environment influencing the phase state of hatchlings (Despland and Simpson 2000a).

The relationships between resource distribution, resource abundance, and locust population size have been explored using individual-based computer simulations. These models were parameterised using experimental data on the time-course and mechanisms of behavioural phase change (Collett et al. 1998). The extent of gregarization within a simulated population increases with rising locust density and increasing clumping of food resources. Significantly, there were shown to be critical zones across particular combinations of resource abundance, resource distribution and population size where a solitarious population would rapidly gregarize (Fig. 12). Experimental verification of the model yielded results that were fully consistent with the predictions from the simulation (Despland et al. 2000).

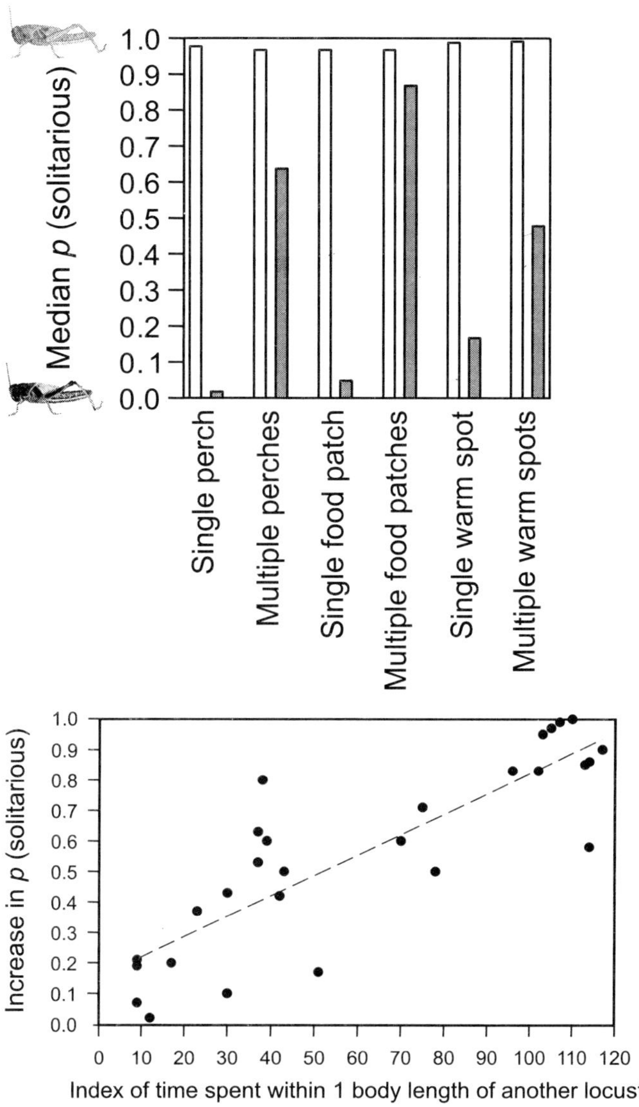

Fig. 11 Results from a laboratory experiment in which 10 solitarious desert locust nymphs were confined in an arena with either a single or multiple resource sites (food patches, perches or warm spots). In the upper graph, open bars indicate the behavioural phase state of insects prior to being placed in the arena, while dark bars show their phase state after 4 h. A single resource site encouraged insects to come together against their predisposition to avoid each other, and led to behavioural gregarization of the population. Multiple sites allowed locusts to avoid one another and hence remain more solitarious. The lower graph combines data across these treatments to show that the effects of resource distribution were explicable in terms of the amount of time individuals spent in close contact with other locusts. After Bouaïchi et al. (1996).

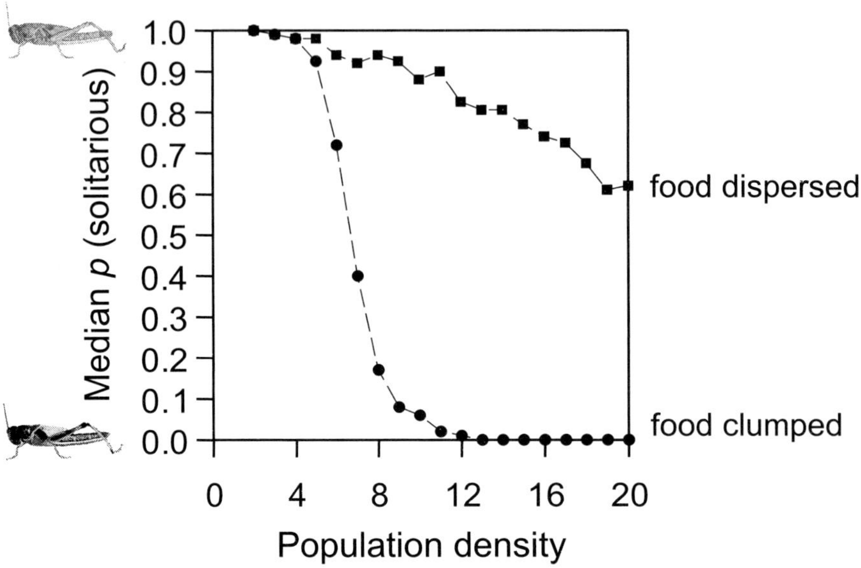

Fig. 12 Results from an individual-based simulation of the interactions between population size, resource abundance and resource distribution on the tendency for a local population of solitarious locusts to gregarize. The graph indicates the situation where a fixed amount of resource is distributed in either a widely dispersed or a clumped fashion and population size increases. Note that when resources are dispersed, the number of locusts in the environment can increase substantially without triggering full gregarization of the population. However, when those same resources are aggregated, gregarization spreads rapidly when locust population size crosses a threshold value. After Collett et al. (1998).

These models demonstrate how vegetation distribution and architecture at small spatial scales and can have profound effects on population dynamics over periods of just a few hours through their influences on the behaviour of individual locusts. They have provided a new tool for making predictions about the likelihood of local populations gregarizing, and thus potentially contributing to larger scale swarms.

The simulation models also showed how behavioural phase shift serves to make phase change a threshold trait at the population level. When the behavioural characteristics of phase change were removed from the model (i.e. the tendency to avoid or be attracted by other locusts), gregarization of individuals in the population still occurred as resource distribution became more clumped, but the population-level response was gradual. Including attraction and repulsion in the model shifted the response to a much more critical function. The resulting sigmoidal shape of the response curve effectively leads to the expression of behavioural change as a threshold trait at the population level (Collett et al. 1998; Fig. 13).

The Interaction between Resource Distribution and Resource Quality

It is not simply resource distribution that determines the spread of phase change at a local scale. So too does resource quality. Experiments using chemically defined artificial diets showed that nutritional quality interacts with spatial food distribution patterns (Despland and Simpson 2000b). When all food patches were of uniform, high nutritional quality (high in both protein and digestible carbohydrate), gregarization was inhibited even

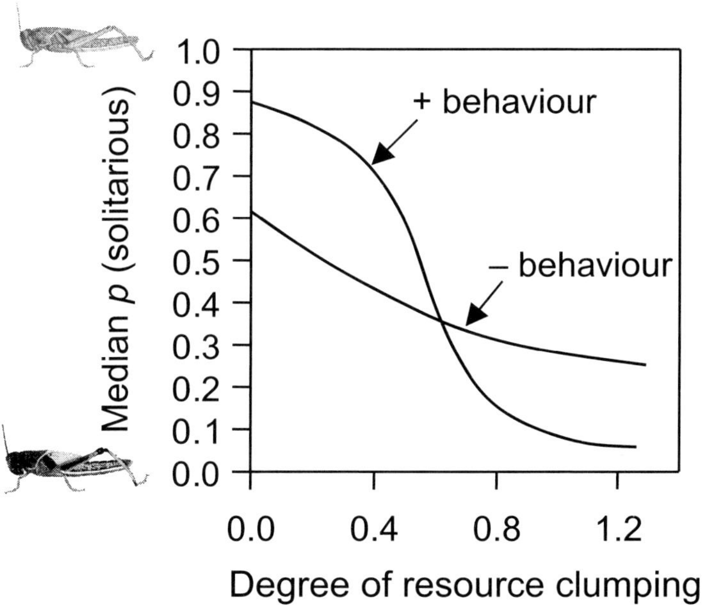

Fig. 13 The effect of behavioural phase change on the extent to which a population of locusts will undergo gregarization in response to clumping of resources. When behavioural phase change (avoidance of others in the solitarious state and active aggregation in the gregarious state) was removed from the simulation model, the response to resource concentration became a less critical function. When resources are widely dispersed, solitarious behavioural responses help keep locusts apart and hence remain solitarious, whereas when a critical level of inter-individual interactions occur as resources become more clumped, locusts switch into the gregarious behavioural state and begin to aggregate, which drives gregariousness in an autocatalytic fashion—both within individuals (in which the diverse suite of phase traits (see Fig. 2) become coupled) and across the population. After Collett et al. (1998).

when food patches were highly clumped in the environment. In contrast, when foods were uniformly of low quality (low in both protein and carbohydrate), substantial gregarization occurred even when foods were widely dispersed. When half of the foods contained high protein but low carbohydrate and the other half contained the reverse, gregarization was low in a dispersed habitat and high when food patches were clumped. These different outcomes were all mediated through the effect of food quality on locomotion. Those insects with the dilute food moved far more than those on other food treatments, while locusts with the two nutritionally complementary foods were forced to move between food patches to balance their intake. Locusts with nutritionally optimal food patches did not move far from a patch after feeding, resulting in limited interactions between locusts even when food patches were clumped. These results make sense in light of what is known about the physiological control of feeding and nutrient balancing in locusts (Simpson and Raubenheimer 2000). Locusts tightly regulate the nutritional composition of their diets and their nutritional state influences their activity levels. Thus, both their endogenous nutritional status as well as their tendency to switch between complementary food resources will influence their probability of contact with other locusts.

Another important food quality is the concentration of secondary plant metabolites. As is discussed below shifts, in diet selection occur with phase change and are significant in the context of protection against predators.

Scaling-up

Small-scale features of the habitat such as resource abundance, quality and distribution, are therefore critical in either promoting or deterring phase transition within local populations (Collett et al. 1998, Despland et al. 2000, Despland and Simpson 2000a,b). When scaled up from metres to a small number of kilometres, the same pattern seems to apply. Babah and Sword (2004) found differences in the distribution patterns of two dominant tussock grasses (locust resources upon which contact and gregarization are known to occur) between adjacent regions in Mauritania that differ in their history of supporting gregarious phase locust populations. As predicted, the resources were aggregated to a greater extent along 2-km^2 belt transects in the region with the higher historical frequency of locust gregarization (Fig. 14).

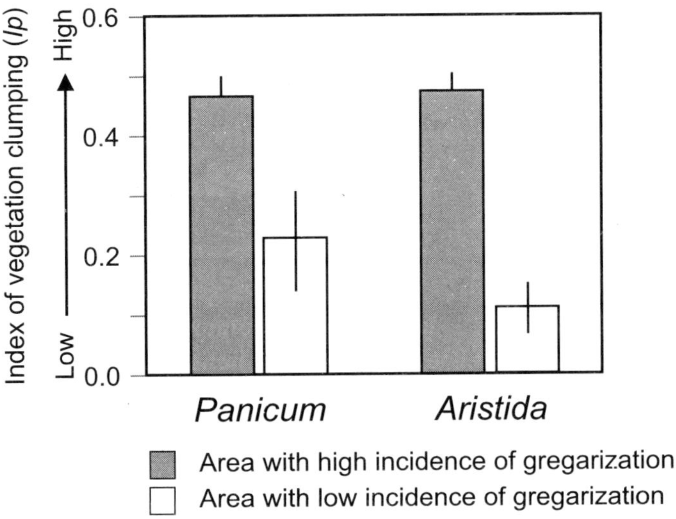

Fig. 14 Data indicating a relationship between resource distribution and gregarization in desert locust populations in Mauritania. As predicted from the simulation models shown in Figs 12 and 13, resources were clumped to a greater extent along 2-km^2 belt transects in a region with a high historical frequency of locust gregarization than in an adjacent region with a low frequency of gregarization. After Babah and Sword (2004). *Panicum* and *Aristida* are tussock grasses that are key locust resources.

At higher spatial scales the relationship between vegetation distribution and desert locust outbreaks changes as a result of the different ecological processes at work. Whereas at the scale of individual plants, a fragmented habitat with multiple dispersed patches encourages solitarization, at the landscape scale the pattern is reversed: habitat fragmentation brings migrating locusts together and encourages outbreaks (Despland et al. 2004). Understanding patterns of movement between resource patches at the landscape scale will require elucidating what causes bands of marching hoppers and flying adults to remain as cohesive groups and to move synchronously and collectively between patches. Recent advances in the use of techniques such as self-propelled particles (SPP) modelling (e.g. Couzin et al. 2005) have proven useful in this regard (See Buhl et al. 2006).

The Adaptive Significance of Locust Phase Polyphenism

As we have shown, considerable progress has been made in recent years in understanding how various phase traits in the desert locust are controlled by different, overlapping sets of stimuli and can involve different underlying physiological mechanisms. In addition, the recognition of the fundamental role of behavioural phase change and its cascading effects on the expression of other phase traits led to important insights into the population-level consequences of phase change and its implications for swarm formation. Interestingly, these major advances in our understanding of desert locust biology occurred in large part without an understanding of the adaptive significance of some of its most conspicuous phase traits.

Phase polyphenism in locusts is related to extreme fluctuations in population size and the probability of swarm formation. The observed correlation between phase polyphenism and propensity of locusts to swarm led to the idea that phase change is an adaptation for migration in the face of environmental heterogeneity (Kennedy 1956). Indeed, many density-dependent phase traits can be interpreted in terms of their value as migratory adaptations (Fescemyer 1993). However, populations of gregarious phase locusts do not necessarily swarm (Uvarov 1977) and solitarious phase locusts also migrate, albeit not *en masse* (Farrow 1990).

Not surprisingly given their economic impact, locusts have been most intensively studied during outbreak periods. Since locusts express gregarious phase traits at high population densities, it is no surprise that the functional significance of these traits is often considered in terms of their function under outbreak conditions. What this perspective overlooks is that density-dependent traits are expressed by locusts in small, isolated populations well before outbreaks ever occur. Adaptive reaction norms of phenotypically plastic traits should be shaped by natural selection to express phenotypes in environments in which they are favoured (Schlichting and Pigliucci 1998). This implies that the adaptive value and resultant effects of phase traits should be investigated when they are initially expressed (i.e. prior to outbreaks).

Phase Change as an Anti-predator Strategy

Recent field, lab, and simulation studies have shown that some density-dependent changes in desert locust nymphs, most notably changes in colour and behaviour, function in concert as an anti-predator strategy based on the expression of warning colouration (aposematism) (Sword 1999, 2001, 2002, Sword and Simpson 2000, Sword et al. 2000, Despland and Simpson 2005a,b). These insights have resulted in an even greater appreciation of the interaction between local ecology and the expression of phase change, as well as its consequences for locust population dynamics.

There are many examples of adaptive phenotypic plasticity functioning as an inducible predator defence in a variety of organisms, both invertebrates and vertebrates (see Tollrian and Harvell 1999). What appears to set density-dependent changes apart from these other examples of inducible predator defence is the nature of the environmental cue that mediates the phenotypic response. In most if not all other instances of induced predator defence, prey phenotypes change in response to the immediate presence of a predator, usually mediated by kairomones (Tollrian and Harvell 1999). However, density-dependent phenotypic changes that function as part of an anti-predator strategy are not induced by cues emanating directly from a predator. Instead, phenotypic changes are induced by the presence of other conspecifics. In this case, population density acts as a surrogate for increased risk of predation when prey become numerous in a given habitat.

Density-dependent Warning Colouration

The catalyst for understanding the adaptive significance of density-dependent colour change and behavioural gregarization in desert locust nymphs came not from the desert locust itself, but rather from a relatively innocuous and little-studied species from North America, *Schistocerca lineata* (taxonomy according to Song 2004a; previously referred to as *Schistocerca emarginata*). Importantly, one of the most critical insights proved to be a detailed understanding of local host plant use ecology. *S. lineata* grasshoppers exhibit both developmental and geographic variation in host plant use patterns (Sword and Dopman 1999). The adults tend to be dietary generalists, a feeding strategy that is quite common among grasshoppers (Chapman 1990, Chapman and Sword 1997). However, the nymphs were found to be dietary specialists associated with specific host plants in different geographically and genetically distinct populations (Sword and Dopman 1999, Dopman et al. 2002).

Using *S. lineata*, Sword (1999) provided the first ever demonstration of density-dependent warning colouration. *S. lineata* nymphs from populations feeding primarily on *Ptelea trifoliata* (Rutaceae) express density-dependent phenotypic plasticity in colouration similar to that known to occur in the in the desert locust. They are commonly green when reared at low population density, but can become a conspicuous yellow-and-black at high density (Fig. 15). Grasshoppers in these populations derive gut-content mediated toxicity to vertebrate predators simply by consuming their primary host plant, *P. trifoliata* (Sword 2001). Predators of *S. lineata*, in turn, learn to associate the toxic effects of the nymphs' gut contents with their conspicuous high-density phenotypes and subsequently avoid attacking yellow-and-black individuals.

Fig. 15 Examples of the differential expression of density-dependent colour polyphenism between different host plant-associated forms of *Schistocerca lineata*. Nymphs from *Ptelea trifoliata*-associated populations are unpalatable to vertebrate predators by virtue of feeding on their host plant and express density-dependent warning colouration (aposematism). Closely related, but genetically distinct nymphs from *Rubus trivialis*-associated populations do not derive unpalatability from their host plant. They should not benefit from the expression of density-dependent warning colouration, and accordingly their ability to express density-dependent colour change has been much reduced by natural selection. After Sword (2002).

By linking the effects of host plant use on palatability to the expression of density-dependent colour change, it became evident that the *S. lineata* grasshoppers exhibit a shift in anti-predator strategy from crypsis at low population density to warning colouration at high density (Sword 1999; Fig. 15). Although density-dependent warning colouration in *S. lineata* was a novel find, the ability to express warning colouration only at high population densities as a form of induced predator defence is predicted by warning colouration theory and has important implications for understanding the evolution of warning colouration among insects in general (Sword 1999, Sword 2002, Ruxton et al. 2004). In addition to its theoretical relevance, results on *S. lineata* suggested that the desert locust might express density-dependent warning colouration as well, with host plant use mediating its deterrence to predators.

Density-dependent Warning Colouration in the Desert Locust

The obvious conspicuousness of gregarious phase desert locust nymphs led many early investigators to suspect that their colouration functioned as warning colouration (Key 1957, Kennedy 1962). However, repeated laboratory and field observations of their palatability failed to support the warning colouration hypothesis (Gillett and Gonta 1978). Importantly, these studies did not consider the potential effects of host plant use on locust palatability (Sword et al. 2000, Sword and Simpson 2000). In the absence of support for the warning colouration hypothesis, it became widely assumed that density-dependent colour phenotypes functioned intraspecifically as visual signals involved in either nymphal aggregation (Ellis and Pearce 1962, Gillett 1973) or group cohesion while marching in migratory bands (Gillett and Gonta 1978). Visual stimuli are clearly important as cues mediating the process of behavioural phase change (Roessingh et al. 1998, Simpson et al. 1999), but empirical studies have failed to support the notion that colour change alone plays a role in either the behavioural gregarization process or nymphal aggregation and group cohesion (Sword and Simpson 2000).

As with *S. lineata*, understanding local patterns of host plant use also proved to be the key to understanding the adaptive significance of density-dependent colour change in the desert locust. Field experiments conducted in Mauritania demonstrated that desert locust nymphs in pre-outbreak populations derive gut-content mediated toxicity to vertebrate predators by feeding on native toxic host plants (Sword et al. 2000). The black and yellow gregarious phase colour pattern of desert locust nymphs (Fig. 1) was shown to function as warning colouration, signalling to potential predators that locusts had been feeding on toxicity-conferring plants. In addition, predators associated locust toxicity much more readily with conspicuous gregarious phase colouration than they did with cryptic solitarious phase phenotypes (Sword et al. 2000).

Density-dependent Gregariousness

Cryptic solitarious phase locusts are repelled by and try to move away from other individuals. In contrast, gregarious phase locusts orient toward other individuals and actively form aggregations (See above reviewed in Simpson et al. 1999). Understanding the anti-predator role of density-dependent colour change provided a simple adaptive explanation for this major change in behaviour. There is a strong evolutionary association between warning colouration and gregariousness in insects. Many aposematic organisms live in aggregations (Vulinec 1990, Ruxton et al. 2004) and gregariousness is thought to enhance the effectiveness of visual warning signals (Gagliardo and Guilford 1993, Gamberale and Tullberg 1998, Krause and Ruxton 2002). In addition, the available phylogenetic evidence to date suggests that the evolution of gregariousness often follows that of secondary defences and warning colouration (Sillén-Tullberg 1988; Sillén-Tullberg and Leimar 1988; Tullberg and Hunter 1996). This suggests that density-dependent gregariousness in the desert locust quite likely evolved as a means to further aggregate conspicuous gregarious phase phenotypes. Density-dependent plasticity in colouration and gregarious behaviour, therefore, can be readily interpreted as two related components of an overall anti-predator strategy based on warning colouration (Sword and Simpson 2000).

Food Choice as a Predator Defence

Density-dependent changes in food selection behaviour provide further support for a suite of locust phase traits functioning together as an inducible anti-predator strategy. Recent studies have shown that newly crowded solitarious phase nymphs shift from avoiding to preferentially ingesting foods containing plant secondary compounds that are powerful vertebrate toxins (Despland and Simpson 2005a,b). This rapidly induced change in feeding behaviour serves to protect cryptically coloured solitarious phase nymphs against predators during the vulnerable intermediate period when they are behaviourally gregarious (and thus more apparent to predators), but do not yet express warning colouration. This enhanced secondary defence during the transition to warning colouration is yet another aspect of locust phase change that is consistent with warning colouration theory (Despland and Simpson 2005a).

Desert locusts are generalist feeders. Their populations are ephemeral and can occur across a vast geographic range encompassing a number of different plant communities. In addition, both solitarious and gregarious phase individuals are migratory and can fly great distances as adults (Farrow 1990), presumably precluding local adaptation to specific host plants (but see Ibrahim et al. 2000 for some suggestion of local genetic population structure). In the absence of host plant specificity, phenotypic plasticity in food selection behaviour enables desert locust nymphs to behaviourally regulate their toxicity to predators. This subtle but important behavioural change is critical to the function of density-dependent warning colouration in a highly mobile species such as the desert locust that must utilize for defence a variety of host plant species whose chemistry is unpredictable in both space and time.

Population-level Consequences

Once desert locust nymphs become behaviourally gregarious, chemically protected, and conspicuously coloured, they possess the full suite of characteristics in which warning colouration acts as an anti-predator strategy. A reduction in per capita predation rate resulting from the combination of gut content-mediated toxicity and the expression of warning colouration will contribute to increases in local population density and subsequent migratory band formation (Sword et al. 2000). Predation rates will be reduced further as bands increase in size, both through the diluting effect of numbers and also the likely amplification of the warning signal in larger groups (Gamberale and Tullberg 1998, Krause and Ruxton 2002).

The protection from predators afforded to insects by migratory band formation was recently demonstrated in the migratory band-forming Mormon cricket, *Anabrus simplex* (Tettigoniidae). Using radiotelemetric methods, Sword et al. (2005) showed that individuals in migratory bands are much less likely to be killed by predators than insects removed from the band. Thus, in locusts the expression of a suite of density-dependent changes functioning together as an anti-predator strategy can influence local population dynamics and lead to migratory band formation. Migratory bands themselves then also function as an anti-predator strategy and likely play an important role in local population dynamics by further reducing predation on band members.

Ironically, a consequence of band formation is an elevated risk of another form of predation: cannibalism. Recent work on Mormon crickets, and locusts shows that mass movement of migratory bands is driven by cannibalism (Simpson et al. 2006, Bazazi et al. 2008). Migratory bands are a "Forced March" in which individuals must keep moving both to find new resources and avoid being attacked by canibalistic conspecifics approaching from behind.

Density-dependent Warning Colouration in other Locusts?

Does the warning colouration hypothesis apply to other locust species? *Locusta migratoria* nymphs, for example, express dark, conspicuously coloured gregarious phase phenotypes, yet there is no evidence to date suggesting that theses colour patterns function as a visual warning signals to predators. There are a number of potential explanations for this lack of evidence. One is simply that the requisite experiments to demonstrate aposematism have not yet been carried out (see Sword et al. 2000 for an explanation of why aposematism had been overlooked for so long in

S. gregaria). *Locusta migratoria* feeds primarily on grasses, a group not generally considered to be rich in noxious secondary compounds. However, some grasses such as *Cynodon dactylon* are cyanogenic and may be candidates for conferring toxicity to predators (Wilson 2000). In addition, Whitman (1990) suggests that *L. migratoria* possess an eversible pronotal gland capable of emitting an odorous secretion. The effect of host plant use on the secretion composition has not been addressed and its possible anti-predator function remains unknown (Schmidt et al. 1987). Conversely, it may be that *L. migratoria* nymphs are entirely palatable and that their high density phenotypes either perform some as yet undetermined function or are selectively neutral by-products of another adaptive density-dependent response such as density-dependent prophylaxis (Sword 2002, Wilson et al. 2002, Wilson and Cotter, this Volume).

The Adaptive Significance of other Phase Changes

Not all phase characteristics serve an anti-predator function, although it seems apparent that they are all adaptations to the demands of living either alone or in a crowd. Up-regulated immune responses in gregarious locusts protect against higher rates of pathogen transmission at high density, with obvious implications for population dynamics (Wilson et al. 2002, Wilson and Cotter, this Volume). Changes in nutrient balancing strategies, metabolism and lipid deposition reflect preparation for migratory behaviour and competition for limiting food resources (Applebaum and Heifetz 1999, Simpson et al. 2002). Changes in pheromonal chemistry suit a crowded lifestyle in which there is increased competition among males for access to females (Seidelmann et al. 2005), and differences between the phases in longevity, egg size, and number of eggs reflect life-history trade-offs between sedentary and migratory lifestyles (Uvarov 1966).

Evolution of Locust Phase Polyphenism

Very few studies have explicitly examined the evolution of phase polyphenism in locusts. The expression of phase polyphenism has convergently evolved multiple times in acridid grasshoppers. It is expressed to varying degrees, and in different combinations of traits, in species from several different subfamilies (Uvarov 1966, 1977, Pener 1991). This variation reflects the fact that density-dependent changes in different traits are often independently regulated and are, therefore, capable of differentially evolving in different taxa. However, given that controlled rearing experiments examining the phenotypic effects of rearing density are lacking in the overwhelming majority of grasshoppers species, it is in all likelihood inappropriate to say that the expression of phase polyphenism itself has evolved numerous times. Rather, it seems more reasonable to state that the extreme expression of phase polyphenism has evolved independently many times, but that genetic variation for the expression of density-dependent traits is present, to some degree, in a multitude of grasshopper species.

The expression of density-dependent phenotypic plasticity in locusts (or any other organism) should not necessarily be considered as a discrete, all or nothing trait. Although the ability to express plasticity can certainly be lost and gained, reaction norms can also be shaped by natural selection so that the phenotypic response to the environment is reduced to ecologically relevant levels, but not eliminated altogether. Additionally, individuals of some species might rarely or never experience population densities high enough to induce the expression of density-dependent traits. In such cases, we would expect reaction norms to evolve via genetic drift rather than selection, barring a cost to maintaining plasticity (Schlichting and Pigliucci 1998). Density-dependent phenotypic changes are known to be expressed in both swarming and non-swarming acridid species. However, for reasons that have yet to be fully explained, their expression seems to be particularly favoured in swarming locusts.

Phase Polyphenism and Swarming: Comparative Studies

The expression of phase polyphenism correlates with the development of locust outbreaks and subsequent mass migration, but the cause and effect relationship between these phenomena can be difficult to establish (Key 1950, Sword 2003). To further confuse matters, this correlation may appear to be weak or even absent in some species commonly referred to as locusts. For example, the Australian plague locust, *Chortoicetes terminifera*, was thought to express little or no obvious discernible phase changes (Uvarov 1966), despite the fact that it forms large and frequent swarms (Hunter 2004). However, field observations by Clark (1949) describe what clearly appeared to be behavioural phase change in nymphs and these observations have been supported in recent laboratory behavioural assays (L. Gray, G. Sword and S.J. Simpson, unpublished data). As mentioned above, this situation highlights the need for detailed experimental analyses to be conducted in order to establish the presence or absence of phase polyphenism in specific traits among different species.

Differences among taxa in phase characteristics may prove informative in understanding the phase polyphenism-swarming correlation. Comparative studies both within and between locust species hold considerable promise for elucidating the respective roles of genetic factors such as the expression of phase changes versus environmental factors such as weather in promoting locust swarm formation. Different strains or populations of locusts of the same species are known to vary in their expression of phase polyphenism (reviewed in Pener and Yerushalmi 1998, see also Yerushalmi *et al.* 2001). These differences highlight the fact that while the expression of phase traits is environmentally determined, the underlying mechanisms and degree of plasticity are under genetic control and can differentially evolve via natural selection or genetic drift in different populations or species (Schlichting and Pigliucci 1998).

Laboratory-based comparisons of the expression of phase polyphenism between swarming and non-swarming species, or strains within a species, provide correlative support for an association between phase change and swarming. Chapuis et al. (unpublished data) found that the degree of expression of behavioural phase change in *Locusta migratoria* nymphs from a historically non-swarming population in France was reduced relative to that of nymphs from a genetically distinct swarming Madagascan population. Similarly, Sword (2003) compared the expression of nymphal behavioural phase change between populations of the non-swarming grasshopper *Schistocerca americana*, as well as between *S. americana* and the swarming desert locust, *S. gregaria*. The former exhibited both developmental and geographic variation in its expression of behavioural phase change, but overall the degree of its expression was much reduced relative to that observed in *S. gregaria* (Fig. 16). In both of these cases, the patterns are consistent: the swarming species or strain expresses behavioural phase polyphenism to a greater extent than related, non-swarming taxa. Although informative and suggestive, these studies unfortunately fail to provide any direct evidence of a causal role for phase polyphenism in locust swarm formation.

Perhaps the most suggestive evidence of a causal role for locust phase change in swarm formation came about as a result of the comparative studies of density-dependent warning colouration between *S. lineata* and the desert locust described earlier. The expression of density-dependent changes in colour and behaviour acting as an anti-predator strategy clearly has the potential to influence local population dynamics and swarm formation in the desert locust. However, *S. lineata* does not swarm, suggesting that the evolution of density-dependent warning colouration

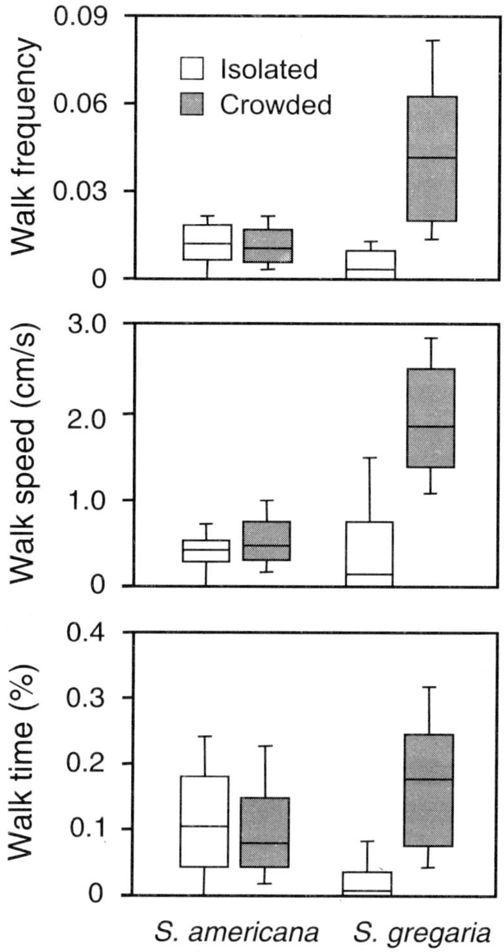

Fig. 16 Box plots depicting differences in the expression of behavioural phase polyphenism between final instar nymphs of non-swarming *Schistocerca americana* and the swarming desert locust, *S. gregaria*. Insects were reared under isolated and crowded conditions. The three locomotory behaviours shown are known to vary with rearing density in *S. gregaria* and were quantified in both species using the behavioural phase assay of Roessingh et al. (1993). After Sword (2003).

alone does not necessarily predispose a species to outbreaks and swarm formation.

Locust outbreaks are likely to result from a complex interaction between biotic and abiotic ecological factors (Joern and Gaines 1999). Thus, an important question arises about the generality of the relationship between locust phase traits and their role in locust swarm formation. What is the relative importance in swarm formation of genetic factors such as the expression of phase polyphenism in certain traits versus biotic (host plants, habitat structure, predators), and abiotic ecological factors (weather)? The absence of swarming in any species may simply be due the absence of suitable environmental or habitat conditions even though the insects retain the hidden genetic variation to mount the relevant density-dependent responses. Conversely, a species may lack the genetic capacity to express phenotypic plasticity in response to changes in its population density. In these cases, species may exist in habitats conducive to swarm production, but lack the ability to mount the density-dependent responses that can lead to swarm formation (Sword 2003). These questions can, at least in principle, be addressed using phylogenetic analyses.

Phase Polyphenism and Swarming: Phylogenetic Studies

The study of the evolution of locust phase polyphenism as well as the relationship between phase polyphenism and swarm formation will benefit immensely from comparative analyses framed in a phylogenetic context. Phylogenetic analyses of phenotypic plasticity are rare, due in large part to the difficulty of measuring plastic responses in numerous taxa under controlled conditions. However, such studies hold considerable promise for understanding the ecology and evolution of plastic traits (Pigliucci 2001). To this end, different researchers utilizing different techniques have been investigating the evolutionary origins of *Schistocerca* (Song 2004b, Lovejoy et al. 2006) as a means to provide a phylogenetic framework for testing hypotheses about the evolution of phase polyphenism and locust swarm formation. *Schistocerca* serves as an excellent model system for the purpose because it contains both swarming and non-swarming species (Dirsh 1974), and the expression of phase polyphenism varies among its members in various traits from extreme to absent (Duck 1939, Kevan 1943, Rowell and Cannis 1971, Dirsh 1974, Sword 2002, 2003). Unfortunately, the results of a morphological cladistic analysis conducted by Song (2004b) differ dramatically from the mitochondrial DNA-based phylogeny of Lovejoy et al. (2006), and more work is needed to establish a robust assessment of the phylogenetic relationships among *Schistocerca* species.

Nevertheless, both phylogenies do suggest that swarming has arisen independently multiple times within the genus (Song 2004b, Lovejoy et al. 2006). Future phylogenetically controlled analyses will enable us to answer distinct questions about the relationship between plasticity and swarming as well as the evolution of density-dependent traits in general. For example, does an inability to express phase polyphenism in traits such as colour and behaviour correlate with the absence of swarming in a given lineage? Also, has plasticity *per se* in colour and behaviour been lost and re-evolved multiple times in *Schistocerca* or is it ubiquitous among species and simply fine-tuned within them via reaction norm evolution?

Plasticity and the Evolution of Warning Colouration

The elucidation of density-dependent warning colouration in *Schistocerca* provided a novel perspective on the evolution of warning colouration, a controversial debate in evolutionary biology dating back to Darwin and Wallace (see reviews by Guilford 1990, Mallet and Joron 1999, Speed 2003, Ruxton et al. 2004). The difficulty in explaining the initial spread of alleles for conspicuous phenotypes in a population of initially cryptic but deterrent prey lies in the selective disadvantage suffered by the first rare conspicuous mutants. Rare conspicuous individuals are by definition more conspicuous than the cryptic majority, and therefore more likely to be attacked by predators. At the same time, their initial frequency in the population is too low to facilitate predator aversion learning through multiple experiences with conspicuous and deterrent prey. Hypotheses proposed to explain the evolution of aposematism have traditionally been framed as allelic substitution models with conspicuous prey phenotypes being genetically determined and constitutively expressed. Prior to the discovery of density-dependent warning colouration, the possibility that a genetically determined, but environmentally expressed phenotype could act as an adaptive intermediate stage in the evolution of warning colouration had not been considered. The density-dependent expression of conspicuousness acts as a best of both worlds anti-predator strategy in which deterrent prey can be cryptic when rare but express warning colouration at locally high population densities when it becomes advantageous (Sword 1999).

A likely role for adaptive reaction norm evolution in the evolution of warning colouration was demonstrated by Sword (2002). As discussed above, *S. lineata* grasshoppers were shown to express density-dependent warning colouration with deterrence to predators mediated by toxic gut contents obtained by feeding on their primary host plant, *Ptelea trifoliata* (Rutaceae) (Fig. 15) (Sword 1999, Sword and Dopman 1999, Sword 2001). By contrast, *S. lineata* juveniles from genetically distinct populations associated with a different host plant, *Rubus trivialis* (Rosaceae), do not derive host plant-mediated deterrence to vertebrate predators (Dopman et al. 2002, Sword 2002). Juveniles in these palatable *Rubus*-feeding populations should not benefit from the expression of warning colouration at high population density, and as expected, the degree to which they change colour with crowding is much less extensive than that of insects from unpalatable *Ptelea*-feeding populations (Fig. 15) (Sword 2002). Importantly, the cues mediating colour change are independent of host plant chemistry (Sword 2002) and differences in the ability to change colour reflect genetic differences between the populations (Dopman et al. 2002). Thus, the ability to express density-dependent changes in juvenile colouration has differentially evolved between palatable and unpalatable *S. lineata* populations. In this case, the expression patterns of density-dependent colour change were causally linked to local patterns of host plant use and predator defence (Fig. 15).

The differential reaction norm evolution in density-dependent colour change observed between palatable and unpalatable *S. lineata* populations suggests that plasticity in colouration may have played an important role in the evolution of warning colouration by providing an adaptive intermediate stage for the evolution of many extant constitutively expressed warning colouration phenotypes (Sword 1999, 2002; Fig. 17). Recent simulation studies suggest that alleles for density-dependent conspicuousness are much more likely to go to fixation in initially cryptic, but defended populations, than are those for constitutively expressed conspicuous phenotypes. In addition, the success of plastic versus constitutive alleles is robust to a variety of assumptions about population demography and predator learning models (Fielding, D.J. and G.A. Sword, unpublished). Once alleles for density-dependent colour phenotypes have gone to fixation in a population, constitutively expressed conspicuous phenotypes can then arise from the initially plastic ancestral state via reaction norm evolution and genetic assimilation (Fig. 17) (Sword 2002, Pigliucci and Murren 2003).

Some investigators have discounted a widespread role for phenotypic plasticity in the evolution of warning colouration. How can plasticity be important in the evolution of most animals' colour patterns when theses traits are genetically determined and constitutively expressed? (Mallet and Joron 1999). This perspective fails to consider the fact that current modes of gene expression do not necessarily reflect their history. Constitutive phenotypic expression may in many cases be a derived state (Pigliucci 2001). In fact, it has been argued that the genetic assimilation of initially plastic phenotypes may occur so rapidly within species that its role in many types of phenotypic evolution has simply gone unnoticed (Pigliucci and Murren 2003). Importantly, density-dependent colour change is not limited to *Schistocerca* nymphs. It has evolved in many unrelated insect lineages (reviewed in Fuzeau-Braesch 1995, Pener 1991, Applebaum and Heifetz

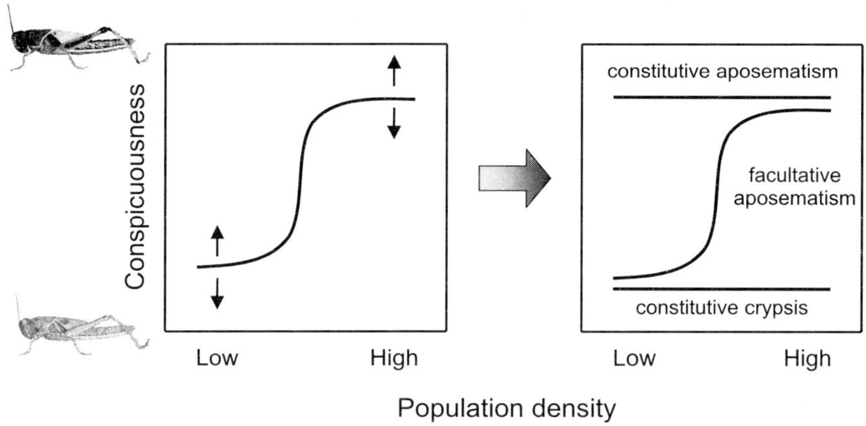

Fig. 17 A conceptual model of the role of density-dependent colour polyphenism in the evolution of warning colouration (aposematism). The sigmoidal curve represents a population-level reaction norm for the expression of density-dependent conspicuousness representing underlying variable and heritable individual reaction norms. The density-dependent expression of conspicuousness provides a 'best of both worlds' anti-predator strategy for insects in a chemically defended population. They can be cryptic when rare, but also benefit from the expression of warning coloration when it becomes advantageous at high local population densities (Sword 1999). One of the three possible outcomes of subsequent reaction norm evolution is the loss of plasticity by either drift or natural selection resulting in the constitutive expression of aposematic phenotypes (Sword 2002). After Sword (2002).

1999), and can affect both larval and adult colour patterns in hemimetabolous as well as holometabolous insects (Applebaum and Heifetz 1999; Barnes and Siva-Jothy 2000). In species that once expressed density-dependent warning colouration, genetic assimilation of the initially plastic phenotypes can result in the constitutive expression of warning colouration observed in extant organisms (Sword 2002, Pigliucci and Murren 2003). The elucidation of extant cases of density-dependent warning colouration may represent fortuitous observations of this process in action (Sword 1999). At the same time, the fact that relatively few contemporary cases are known may reflect the speed at which the genetic assimilation of initially plastic aposematic phenotypes can occur.

Studies investigating the evolution of warning colouration are often not just concerned with colour patterns, but also the evolution of secondary defences and gregariousness (see Ruxton et al. 2004 for many examples). Locust phase polyphenism can be exploited as a powerful model system for empirical research into the evolution of warning colouration. The transition between multiple phenotypic states can be examined within a single organism in a system unencumbered with phylogenetic effects. For example, due to their plastic response to population density in multiple traits, individual desert locusts can be manipulated to be any combination of cryptic or aposematic in colouration (Sword et al. 2000), solitary-living or gregarious in behaviour (Simpson et al. 1999), and toxic or non-toxic to predators (Sword et al. 2000, Despland and Simpson 2005a,b).

Conclusion

Locust phase change is indeed "among the most striking coordinated alternative phenotypes known" (West-Eberhard 2003). It is composed of a series of separately controlled, continuous traits that collectively equip locusts for either a solitary existence or life within a crowd. These traits are coupled and converted into a coordinated threshold response, both at the level of the individual and the population, by behavioural phase change in response to local population density. We now understand a considerable amount about the mechanisms controlling some key phase characteristics, notably behavioural phase transition. We also know how the fine-scale structure of the habitat influences the propensity of local populations to change phase, providing some of the most compelling examples of the power of individual-based approaches in ecology. Additionally, work on the adaptive significance of density-dependent phase characteristics has established the locust system as a model for studies of inducible anti-predator and pathogen strategies as well as the evolution of warning colouration.

References

Applebaum, S. W. and Y. Heifetz. 1999. Density-dependent physiological phase in insects. Annual Review of Entomology 44: 317–341.

Applebaum, S. W., E. Avisar and Y. Heifetz. 1997. Juvenile hormone and locust phase. Archives of Insect Biochemistry and Physiology 35: 375–391.

Ayali, A., M. P. Pener and J. Girardie. 1996a. A comparative study of neuropeptides from the corpora cardiaca of solitary and gregarious *Locusta*. Archives of Insect Biochemistry and Physiology 31: 439–450.

Ayali, A., M. P. Pener, S. M. Sowa and L. L. Keeley. 1996b. Adipokinetic hormone content of the corpora cardiaca in gregarious and solitary migratory locusts. Physiological Entomology 21: 167–172.

Babah, M. A. O. and G. A. Sword. 2004. Linking locust gregarization to local resource distribution patterns across a large spatial scale. Environmental Entomology 33: 1577–1583.

Barnes, A. I. and M. T. Siva-Jothy. 2000. Density-dependent prophylaxis in the mealworm beetle *Tenebrio molitor* L. (Coleoptera: Tenebrionidae): cuticular melanization is an indicator of investment in immunity. Proceedings of the Royal Society of London Series B-Biological Sciences 267: 177–182.

Bazazi, S., J. Buhl, J. J. Hale, M. L. Anstey, G. A. Sword, S. J. Simpson and I. D. Couzin. 2008. Collective motion and cannibalism in locust marching bands. Current Biology 18: 735–739.

Bouaïchi, A., P. Roessingh and S. J. Simpson. 1995. An analysis of the behavioural effects of crowding and re-isolation on solitary-reared adult desert locusts (*Schistocerca gregaria*) and their offspring. Physiological Entomology 20: 199–208.

Bouaïchi, A., S. J. Simpson and P. Roessingh. 1996. The influence of environmental microstructure on the behavioural phase state and distribution of the desert locust *Schistocerca gregaria*. Physiological Entomology 21: 247–256.

Buhl, J., D. J. T. Sumpter, I. D. Couzin, J. J. Hale, E. Despland, E. R. Miller, and S. J. Simpson. 2006. From disorder to order in marching locusts. Science 312: 1402–1406.

Chapman, R.F. 1990. Food selection. *In* Chapman R. F. and A. Joern (Eds.) *Biology of Grasshoppers*. John Wiley & Sons, New York. Pp. 39–72.

Chapman, R. F. and G. A. Sword. 1997. Polyphagy in the Acridomorpha. *In* S. K. Gangwere, M. C. Muralirangan and M. Muralirangan (Eds.) *The Bionomics of Grasshoppers, Katydids and Their Kin*. CABI, Wallingford. Pp. 183–195.

Clark, L. R. 1949. Behaviour of swarm hoppers of the Australian plague locust (*Chortoicetes terminifera* Walk.). CSIRO Bulletin 245: 1–27.

Clynen, E., G. Baggerman, D. Veelaert, A. Cerstiaens, D. Van der Horst, L. Harthoorn, R. Derua, E. Waelkens, A. De Loof and L. Schoofs. 2001. Peptidomics of the pars intercerebralis-corpus cardiacum complex of the migratory locust, *Locusta migratoria*. European Journal of Biochemistry 268: 1929–1939.

Collett, M., E. Despland, S. J. Simpson and D. C. Krakauer. 1998. Spatial scales of desert locust gregarization. Proceedings of the National Academy of Sciences of the United States of America 95: 13052–13055.

Couzin, I. D., J. Krause, N. R. Franks and S. A. Levin. 2005. Effective leadership and decision-making in animal groups on the move. Nature 433: 513–516.

Dale, J. F. and S. S. Tobe. 1990. The endocrine basis of locust phase polymorphism. *In* R. F. Chapman and A. Joern (Eds.) *Biology of Grasshoppers*. John Wiley & Sons, New York. Pp. 393–414.

De Loof, A., I. Claeys, G. Simonet, P. Verleyen, T. Vandermissen, F. Sas, and J. Huybrechts. 2006. Molecular markers of phase in locusts. Insect Science 13: 3–12.

Despland, E. and S. J. Simpson. 2000a. Small-scale vegetation patterns in the parental environment influence the phase state of hatchlings of the desert locust. Physiological Entomology 25: 74–81.

Despland, E. and S. J. Simpson. 2000b. The role of food distribution and nutritional quality in behavioural phase change in the desert locust. Animal Behaviour 59: 643–652.

Despland, E. and S. J. Simpson. 2005a. Surviving the change to warning colouration: density-dependent polyphenism suggests a route for the evolution of aposematism. Chemoecology 15: 69–75.

Despland, E. and S. J. Simpson. 2005b. Food choices of solitarious and gregarious locusts reflect cryptic and aposematic antipredator strategies. Animal Behaviour 69: 471–479.

Despland, E., J. Rosenberg, and S. J. Simpson. 2004. Landscape structure and locust swarming: a satellite's view. Ecography 27: 381–391.

Despland, E., M. Collett, and S. J. Simpson. 2000. Small-scale processes in desert locust swarm formation: how vegetation patterns influence gregarization. Oikos 88: 652–662.

DeWitt, T. J. and S. M. Scheiner. 2004. *Phenotypic Plasticity: Functional and Conceptual Approaches*. Oxford University Press, Oxford.

Dingle, H. 1996. *Migration: The Biology of Life on the Move*. Oxford University Press, Oxford.

Dirsh, V. M. 1974. Genus *Schistocerca* (Acridomorpha, Insecta). W. Junk, The Hague.

Dopman, E. B., G. A. Sword and D. M. Hillis. 2002. The importance of the ontogenetic niche in resource-associated divergence: Evidence from a generalist grasshopper. Evolution 56: 731–740.

Dorn, A., C. Ress, S. Sickold and S. Wedekind-Hirschberger. 2000. Arthropoda – Insecta: endocrine control of phase polymorphism. *In* Dorn, A. (Ed.) *Progress in Developmental Endocrinology*. John Wiley, Chichester Pp. 205–253.

Duck, L. D. 1944. The bionomics of *Schistocerca obscura* (Fabr.). Journal of the Kansas Entomological Society 17: 105–119.

Ellis, P. E. 1953. Social aggregation and gregarious behaviour in hoppers of *Locusta migratoria migratorioides* (R. & F.). Behaviour 5: 225–260.

Ellis, P. E. 1963. Changes in the social aggregation of locust hoppers with changes in rearing conditions. Animal Behaviour 11: 152–160.

Ellis, P. E. and A. Pearce. 1962. Innate and learned behavioural patterns that lead to group formation in locust hoppers. Animal Behavior 10: 305–318.

Farrow, R. A. 1990. Flight and migration in acridoids. Pp. 227–314. *In* R.F. Chapman and A. Joern (Eds.). *The Biology of Grasshoppers*. John Wiley & Sons, New York.

Faure , J. C. 1923 . The Life History of the Brown Locust (*Locustana pardalina* (Walker)). Bulletin of the Faculty of Agriculture, Transvaal University College, No. 4. 30 pp.

Faure, J. C. 1932. The phase of locusts in South Africa. Bulletin of Entomological Research 23: 293–424.

Ferenz, H.-J. and K. Seidelmann. 2003. Pheromones in relation to aggregation and reproduction in desert locusts. Physiological Entomology 28: 11–18.

Fescemyer, H. W. 1993. Influence of phase polymorphism on the morphometric and physiological processes in preparation for insect migration. Journal of Agricultural Entomology 10: 219–237.

Fuzeau-Braesch, S. 1985. Colour changes. Comprehensive Insect Physiology, Biochemistry and Pharmacology 9: 549–589.

Gagliardo, A. and T. Guilford. 1993. Why do warning-coloured prey live gregariously? Proceedings of the Royal Society of London Series B-Biological Sciences 251: 69–74.

Gamberale, G. and B. S. Tullberg. 1998. Aposematism and gregariousness: the combined effect of group size and coloration on signal repellence. Proceedings of the Royal Society of London Series B-Biological Sciences 265: 889–894.

Gillett, S. D. 1973. Social determinants of aggregation behavior in adults of the desert locust. Animal Behavior 21: 599–606.

Gillett, S. D. and E. Gonta. 1978. Locust as prey: factors affecting their vulnerability to predation. Animal Behavior 26: 282–289.

Guilford, T. 1990a. The evolution of aposematism. Pp. 23–61. *In* D. L. Evans and J. O. Schmidt (Eds.) *Insect Defenses: Adaptive Mechanisms and Strategies of Prey and Predators*. State University of New York Press, Albany.

Gunn, D. L. and P. Hunter-Jones. 1952. Laboratory experiments on phase differences in locusts. Anti-locust Bulletin 12: 1–29.

Hägele, B. F. and S. J. Simpson. 2000. The influence of mechanical, visual and contact chemical stimulation on the behavioural phase state of solitary desert locusts (*Schistocerca gregaria*). Journal of Insect Physiology 46: 1295–1301.

Hägele, B. F., V. Oag, A. Bouaïchi, A. R. McCaffery and S. J. Simpson. 2000. The role of the female accessory glands in maternal inheritance of phase in the desert locust, *Schistocerca gregaria*. Journal of Insect Physiology 46: 275–280.

Hägele, B. F., F-H. Wang, F. Sehnal and S. J. Simpson. 2004. Effects of crowding, isolation, and transfer from isolation to crowding on the total ecdysteroid content of eggs in *Schistocerca gregaria*. Journal of Insect Physiology 50: 621–628.

Hartfelder, K. and D. J. Emlen. 2005. Endocrine control of insect polyphenism. *In* L. I. Gilbert, K. Iatrou and S. S. Gill (Eds.) *Comprehensive Molecular Insect Science, Vol. 3, Endocrinology*. Elsevier Pergamon, Oxford Pp. 651–703.

Hassanali, A., P. G. N. Njagi and M. O. Bashir. 2005. Chemical ecology of locusts and related acridids. Annual Review of Entomology 50: 223–245.

Hoste, B., S. J. Simpson, S. Tanaka, D-H. Zhu, A. De Loof and M. Breuer. 2002. Effects of [His7]-corazonin on the phase state of isolated-reared (solitarious) *Schistocerca gregaria*. Journal of Insect Physiology 48: 981–990.

Hunter, D. M. 2004. Advances in the control of locusts (Orthoptera: Acrididae) in eastern Australia: from crop protection to preventative control. Australian Journal of Entomology 43: 293–303.

Hunter-Jones, P. 1958. Laboratory experiments on the inheritance of phase characters in locusts. Anti-locust Bulletin 29: 1–32.

Ibrahim K. M., P. Sourrouille and G. M. Hewitt. 2000. Are recession populations of the desert locust (*Schistocerca gregaria*) remnants of past swarms? Molecular Ecology 9: 783–791.

Injeyan, H. S. and S. S. Tobe. 1981. Phase polymorphism in *Schistocerca gregaria*: reproductive parameters. Journal of Insect Physiology 27: 97–102.

Islam, M. S., P. Roessingh, S. J. Simpson and A. R. McCaffery. 1994a. Effects of population density experienced by parents during mating and oviposition on the phase of hatchling desert locusts, *Schistocerca gregaria*. Proceedings of the Royal Society of London Series B-Biological Sciences 257: 93–98.

Islam, M. S., P. Roessingh, S. J. Simpson and A. R. McCaffery. 1994b. Parental effects on the behaviour and colouration of nymphs of the desert locust *Schistocerca gregaria*. Journal of Insect Physiology 40: 173–181.

Joern, A. and S. B. Gaines. 1990. Population dynamics and regulation in grasshoppers. *In* Chapman, R. F. and A. Joern (Eds.) *Biology of Grasshoppers*. John Wiley and Sons, New York Pp. 415–482.

Kandel, E. R. and C. Pittenger. 1999. The past, the future and the biology of memory storage. Philosophical Transactions of the Royal Society of London, Series B – Biological Sciences 354: 2027–2052.

Kang, L., X. Y. Chen, Y. Zhou, B. W. Liu, W. Zheng, R. Q. Li, J. Wang and J. Yu. 2004. The analysis of large-scale gene expression correlated to the phase changes of the migratory locust. Proceedings of the National Academy of Sciences of the United States of America 101: 17611–17615.

Kennedy, J. S. 1956. Phase transformation in locust biology. Biological Reviews of the Cambridge Philosophical Society 31: 349–370.

Kennedy, J. S. 1962. The division of labour between the phases of locusts. Colloques Internationaux de Centre National de la Recherche Scientifique 114: 269–297.

Kevan, D. K. McE. 1943. An account of *Schistocerca flavofasciata* (De Geer 1773) in Trinidad (Orthoptera: Acrididae). Bulletin of Entomological Research 34: 291–310.

Key, K. H. L. 1950. A critique on the phase theory of locusts. Quarterly Review of Biology 25: 363–407.

Key, K. H. L. 1957. Kentromorphic phases in three species of Phasmatodea. Australian Journal of Zoology 5: 247–285.

Krause, J. and G. D. Ruxton. 2002. *Living in Groups*. Oxford University Press, Oxford.

Lester, R. L., C. Grach, M. P. Pener and S. J. Simpson. 2005. Stimuli inducing gregarious colouration and behaviour in nymphs of *Schistocerca gregaria*. Journal of Insect Physiology 51: 737–747.

Lovejoy, N., S. P. Mullen, G. A. Sword, R. F. Chapman and R. Harrison. 2006. Ancient trans-Atlantic flight explains locust biogeography: Molecular phylogenetics of the locust genus *Schistocerca*. Proceedings of the Royal Society of London Series B-Biological Sciences 273: 767–774.

Mallet, J. and M. Joron. 1999. Evolution of diversity in warning color and mimicry: polymorphisms, shifting balance, and speciation. Annual Review of Ecology and Systematics 30: 201–233.

Malual, A. G., A. Hassanali, B. Torto, Y. O. H. Assad and P. G. N. Njagi. 2001. The nature of the gregarizing signal responsible for maternal transfer of phase to the offspring in the desert locust, *Schistocerca gregaria*. Journal of Chemical Ecology 27: 1423–1435.

Matheson, T., S. M. Rogers and H. G. Krapp. 2004. Plasticity in the visual system is correlated with a change in lifestyle of solitarious and gregarious locusts. Journal of Neurophysiology 91: 1–12.

McCaffery, A. R., S. J. Simpson, M. S. Islam and P. Roessingh. 1998. A gregarizing factor present in egg pod foam of the desert locust *Schistocerca gregaria*. Journal of Experimental Biology 201: 347–363.

Pener, M. P. 1991. Locust phase polymorphism and its endocrine relations. Advances in Insect Physiology 23: 1–79.

Pener, M. P. and Y. Yerushalmi. 1998. The physiology of locust phase polymorphism: an update. Journal of Insect Physiology 44: 365–377.

Pener, M. P., A. Ayali and E. Ben-Ami. 1992. Juvenile hormone is not a major factor in locust phase changes. *In* B. Mauchamp, F. Couillaud and J. C. Baehr (Eds.) *Insect Juvenile Hormone Research*. Institut National de la Recherché Agronomique, Paris Pp. 125–134.

Pigliucci, M. 2001. *Phenotypic Plasticity: Beyond Nature and Nurture*. Johns Hopkins University Press, Baltimore.

Pigliucci, M. and K. Preston. 2004. *Phenotypic Integration. Studying the Ecology and Evolution of Complex Phenotypes*. Oxford University Press, Oxford.

Rahman, M. M., L. Vanden Bosch, G. Baggerman, E. Clynen, K. Hens, B. Hoste, K. Meylaers, T. Vercammen, L. Schoofs, A. De Loof and M. Breuer. 2002. Search for peptidergic molecular markers in hemolymph of crowd- (gregarious) and isolated-reared (solitary) desert locusts, *Schistocerca gregaria*. Peptides 23: 1907–1914.

Roessingh, P., S.J. Simpson and S. James. 1993. Analysis of phase-related changes in behaviour of desert locust nymphs. Proceedings of the Royal Society of London Series B-Biological Sciences 252: 43–49.

Roessingh, P. and S. J. Simpson. 1994. The time-course of behavioural phase change in nymphs of the desert locust, *Schistocerca gregaria.* Physiological Entomology 19: 191–197.

Roessingh, P., A. Bouaïchi and S. J. Simpson. 1998. Effects of sensory stimuli on the behavioural phase state of the desert locust, *Schistocerca gregaria.* Journal of Insect Physiology 44: 883–893.

Rogers, S. M., T. Matheson, E. Despland, T. Dodgson, M. Burrows and S. J. Simpson. 2003. Mechanosensory-induced behavioural gregarization in the desert locust *Schistocerca gregaria.* Journal of Experimental Biology 206: 3991–4002.

Rogers, S. M., T. Matheson, K. Sasaki, K. Kendrick, S. J. Simpson and M. Burrows. 2004. Substantial changes in central nervous system neurotransmitters and neuromodulators accompany phase change in the locust. Journal of Experimental Biology 207: 3603–3617.

Rowell, C. H. F. and T. L. Cannis. 1971. Environmental factors affecting the green/brown polymorphism in the cyrtacanthacridine grasshopper *Schistocerca vaga* (Scudder). Acrida 1: 69–77.

Ruxton, G. D., T. N. Sherratt and M. P. Speed. 2004. *Avoiding Attack: The Evolutionary Ecology of Crypsis, Warning Signals and Mimicry.* Oxford University Press, Oxford.

Schlichting, C. D. and M. Pigliucci. 1998. *Phenotypic Evolution: A Reaction Norm Perspective.* Sinauer Associates, Inc., Sunderland.

Schmidt, G. H., W. Krempien and B. Johannes. 1987. Studies on the secretion of the prothoracic epidermal gland in *Acrotylus patruelis* (Insecta: Saltatoria: Acrididae). Zoologischer Anzeige 219: 357–368.

Seidelmann, K., K. Warnstorff and H.-J. Ferenz. 2005. Phenylacetonitrile is a male-specific repellent in gregarious desert locusts, *Schistocerca gregaria.* Chemoecology 15: 37–43.

Seidelmann, K., H. Weinert and H.-J. Ferenz. 2003. Wings and legs are production sites for the desert locust courtship inhibition pheromone, phenylacetonitrile. Journal of Insect Physiology 49: 1125–1133.

Sillén-Tullberg, B. 1988. Evolution of gregariousness in aposematic butterfly larvae: a phylogenetic analysis. Evolution 42: 293–305.

Sillén-Tullberg, B. and O. Leimar. 1988. The evolution of gregariousness in distasteful insects as a defense against predators. American Naturalist 132: 723–734.

Simpson, S. J. and D. Raubenheimer. 2000. The hungry locust. Advances in the Study of Behavior 29: 1–44.

Simpson, S. J. and G. A. Miller. 2007. Maternal effects on phase characteristics in the desert locust, *Schistocerca gregaria:* A review of current understanding. Journal of Insect Physiology 53: 869-876.

Simpson, S. J. and G.A. Sword. 2008. Locusts. Current Biology 18: R364–366.

Simpson, S. J., E. Despland, B. F. Hägele and T. Dodgson. 2001. Gregarious behavior in desert locusts is evoked by touching their back legs. Proceedings of the National Academy of Sciences of the United States of America 98: 3895–3897.

Simpson, S. J., A.R. McCaffery and B. F. Hägele. 1999. A behavioural analysis of phase change in the desert locust. Biological Reviews 74: 461–480.

Simpson, S. J. G. A. Sword and A. De Loof. 2005. Advances, controversies and consensus in locust phase polymorphism research. Journal of Orthoptera Research 14: 213–222.

Simpson, S. J., D. Raubenheimer, S. T. Behmer, A. Whitworth and G. A. Wright. 2002. A comparison of nutritional regulation in solitarious- and gregarious-phase nymphs of the desert locust, *Schistocerca gregaria.* Journal of Experimental Biology 205: 121–129.

Simpson, S. J., G. A. Sword, P. D. Lorch and I. D. Couzin. 2006. Cannibal crickets on a forced march for protein and salt. Proceedings of the National Academy of Science USA 103: 4152–4156.

Song, H. 2004a. Revision of the alutacea group of genus *Schistocerca* (Orthoptera : Acrididae : Cyrtacanthacridinae). Annals of the Entomological Society of America 97: 420–436.

Song, H. 2004b. On the origin of the desert locust *Schistocerca gregaria* (Forskål) (Orthoptera : Acrididae : Cyrtacanthacridinae). Proceedings of the Royal Society of London, Series B-Biological Sciences 271: 1641–1648.

Speed, M. P. 2003. Theoretical developments in the understanding of warning signals. Comments on Theoretical Biology 8: 207–224.

Sword, G. A. 1999. Density-dependent warning coloration. Nature 397: 217.

Sword, G. A. 2001. Tasty on the outside, but toxic in the middle: Grasshopper regurgitation and host plant-mediated toxicity to a vertebrate predator. Oecologia 120: 416–421.

Sword, G. A. 2002. A role for phenotypic plasticity in the evolution of aposematism. Proceedings of the Royal Society of London Series B-Biological Sciences 269: 1639–1644.

Sword, G. A. 2003. To be or not to be a locust? A comparative analysis of behavioral phase change in nymphs of *Schistocerca americana* and *S. gregaria*. Journal of Insect Physiology 74: 709–717.

Sword, G. A. and E. B. Dopman. 1999. Developmental specialization and geographic structure of host plant use in a polyphagous grasshopper, *Schistocerca emarginata* (=*lineata*) (Orthoptera: Acrididae). Oecologia 120: 437–445.

Sword, G. A. and S. J. Simpson. 2000. Is there an intraspecific role for density-dependent colour change in the desert locust? Animal Behaviour 59: 861–870.

Sword, G. A., P. D. Lorch and D. T. Gwynne. 2005. Migratory bands give crickets protection. Nature 433: 703.

Sword, G. A., S. J. Simpson, O. T. M. El Hadi, and H. Wilps. 2000. Density-dependent aposematism in the desert locust. Proceedings of the Royal Society of London Series B-Biological Sciences 267: 63–68.

Tanaka, S. 2000. The role of [His^7]-corazonin in the control of body-colour polymorphism in the migratory locust, *Locusta migratoria* (Orthoptera: Acrididae). Journal of Insect Physiology 46: 1169–1176.

Tanaka, S. 2001. Endocrine mechanisms controlling body-color polymorphism in locusts. Archives of Insect Biochemistry and Physiology 47: 139–149.

Tanaka, S., D-H. Zhu, B. Hoste and M. Breuer. 2002. The dark-color inducing neuropeptide [His^7]-corazonin, causes a shift in morphometric characteristics towards the gregarious phase in isolated-reared (solitarious) *Locusta migratoria*. Journal of Insect Physiology 48: 1065–1074.

Tawfik, A. I., S. Tanaka, A. De Loof, L. Schoofs, G. Baggerman, E. Waelkens, R. Derua, Y. Milner, Y. Yerushalmi, and M.P. Pener. 1999. Identification of the gregarization-associated dark-pigmentotropin in locusts through an albino mutant. Proceedings of the National Academy of Sciences of the United States of America 96: 7083–7087.

Tollrian, R. and C. D. Harvell. 1999. *The Ecology and Evolution of Inducible Defenses*. Princeton University Press, Princeton.

Tullberg, B. S. and A. F. Hunter. 1996. Evolution of larval gregariousness in relation to repellent defences and warning coloration in tree-feeding Macrolepidoptera: a phylogenetic analysis based on independent contrasts. Biological Journal of the Linnean Society 57: 253–276.

Uvarov, B. P. 1921. A revision of the genus *Locusta* (L.) (=*Patchytylus* Fieb.), with a new theory as to the periodicity and migrations of locusts. Bulletin of Entomological Research 12: 135–163.

Uvarov, B. 1966. *Grasshoppers and Locusts*, Vol. I. Cambridge University Press, Cambridge.

Uvarov, B. 1977. *Grasshoppers and Locusts*. Vol. II. Centre for Overseas Pest Research, London.

Vulinec, K. 1990. Collective security: aggregation by insects as a defense. *In* D. L. Evans and J.O. Schmidt (Eds.) *Insect Defenses*. State University of New York Press, Albany Pp. 251–288.

Wedekind-Hirschberger, S., S. Sickold and A. Dorn. 1999. Expression of phase-specific haemolymph polypeptides in a laboratory strain and field catches of *Schistocerca gregaria*. Journal of Insect Physiology 45: 1097–1103.

West-Eberhard, M. J. 2003. *Developmental Plasticity and Evolution*. Oxford University Press, Oxford.

Whitman, D. W. 1990. Grasshopper chemical communication. *In* Chapman R.F. and A. Joern (Eds.) *Biology of Grasshoppers*. John Wiley & Sons, New York Pp. 357–391.

Wilson, K. 2000. How the locust got its stripes: the evolution of density-dependent aposematism. Trends in Ecology and Evolution 15: 88–90.

Wilson, K., M. B. Thomas, S. Blanford, M. Doggett, S. J. Simpson and S. L. Moore. 2002. Coping with crowds: Density-dependent disease resistance in desert locusts. Proceedings of the National Academy of Sciences of the United States of America 99: 5471–5475.

Yerushalmi, Y., E. Tauber and M. P. Pener. 2001. Phase polymorphism in *Locusta migratoria*: the relative effects of geographical strain and albinism on morphometrics. Physiological Entomology 26: 95–105.

6

Density-Dependent Prophylaxis in Insects

Kenneth Wilson[1] and Sheena C. Cotter[2,3]

[1] *Department of Biological Sciences, Lancaster Environment Centre, Lancaster University, Lancaster LA1 4YQ, United Kingdom; E-mail: ken.wilson@lancaster.ac.uk*

[2] *CSIRO Entomology, Private Bag 5, PO Wembley, WA 6913, Australia*

[3]*Department of Zoology, University of Cambridge, Downing St, Cambridge, CB2 3EJ, United Kingdom*

Abstract

Parasites and pathogens are a ubiquitous threat facing all organisms. Life-history theory predicts that if investment in parasite resistance mechanisms is costly (as suggested by numerous studies), then organisms should tailor their investment in them to match their perceived risk of infection. Because most parasites are transmitted in a positively density-dependent manner, the threat from parasites tends to increase as population density increases. As a result, it is predicted that organisms should use population density as a cue to the risk of becoming infected and should increase investment in disease resistance mechanisms as the degree of crowding increases—this is known as *density-dependent prophylaxis* (DDP). This phenomenon has been experimentally tested in a number of insect species, and in most cases support for the DDP hypothesis has been forthcoming. DDP is likely to be particularly prevalent in species exhibiting density-dependent phase polyphenism (i.e. the phenotype adopted by the insect is plastic and dependent on the population density it experiences during its early development). We discuss the hormonal and genetic mechanisms underlying phase polyphenism and DDP, and speculate on the circumstances leading to their evolution. We end by discussing how future research into DDP might develop.

Corresponding Author: Dr Kenneth Wilson, Department of Biological Sciences, Lancaster Environment Centre, Lancaster University, Lancaster LA1 4YQ, United Kingdom, Tel: +44 (0)1524 593349, Fax: +44 (0)1524 593192, e-mail: ken.wilson@lancaster.ac.uk

Introduction

All animals and plants share their biotic environment with parasites and pathogens (generically, 'parasites'). As a consequence, parasites represent an important selective force on their hosts. Indeed, they have been implicated in the evolution of host secondary sexual traits (e.g. Hamilton and Zuk 1982); the manipulation of host behavior (e.g. Moore 1984); the maintenance of host genetic diversity (e.g. van Valen 1973; Hamilton 1980) and the evolution of host optimal life-history strategies (e.g. Sheldon and Verhulst 1996). In this chapter, we review the evidence for, and the mechanisms underpinning, a phenomenon termed "density-dependent prophylaxis" (DDP; see Table 1; Wilson and Reeson 1998).

The density-dependent prophylaxis hypothesis argues that because the potential threat posed to organisms by their parasites generally changes as a function of population density, if disease resistance is costly to maintain, then prophylactic investment in disease resistance mechanisms should be phenotypically plastic and individuals should tailor them to meet this predictable threat (Wilson and Reeson 1998). More specifically, because parasite transmission tends to be positively density-dependent, the *per capita* risk of infection generally increases with population density and individuals in crowds are generally under greatest threat from parasites (Anderson and May 1981). As a consequence, it is predicted that organisms will increase their investment in disease resistance mechanisms (immunological, behavioral, chemical and/or physical) as population density increases, and that this will result in a positive relationship between population density and *per capita* parasite resistance. This prediction is particularly interesting because it is counter-intuitive; it is generally assumed that crowding is "stressful" and will lead to an increase in susceptibility to parasites (Steinhaus 1958,1963).

Whilst it is true that parasitism and disease are both more likely to be prevalent under crowded conditions, the DDP hypothesis argues that this is largely a consequence of the nature of the density-dependent transmission process. Thus, as population density increases, and the risk of infection rises, so optimal investment in disease resistance mechanisms increases, dose-dependent *per capita* mortality rate falls, but *overall* mortality due to parasitism may continue to rise. Crucially, however, the rate at which parasite-induced mortality increases with population density will be slower than that observed in the absence of density-dependent prophylaxis (Fig. 1). The DDP hypothesis makes three specific predictions: (i) for a given challenge of parasites, the *per capita* mortality rate will decline as previous

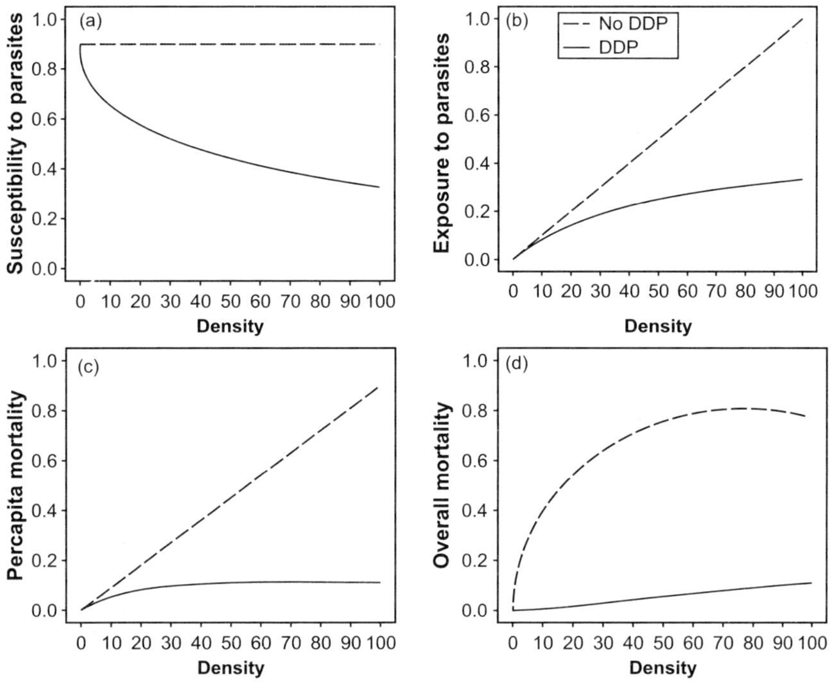

Fig. 1 A graphical representation of the possible effect of (a) density-dependent investment in parasite resistance on (b) exposure to parasites, (c) *per capita* mortality and (d) overall mortality. The dashed lines show the patterns expected in the absence of density-dependent prophylaxis, and the solid lines show those predicted by the DDP hypothesis. Note that the exact patterns depend on the precise functions used to describe DDP and parasite transmission. All scales are arbitrary.

experience of crowding increases; (ii) as the degree of crowding increases, so *per capita* investment in disease resistance mechanisms will increase (assuming that resistance increases in proportion to investment levels); and (iii) under field conditions, parasite-induced host mortality will be a saturating function of host density. Of these three predictions, the third is the weakest because it may also be generated by non-adaptive mechanisms (D'Amico et al. 1996; Dwyer et al. 1997). In all cases, it is assumed that there is no degree of crowding that constrains the optimal decision. However, there may be an upper threshold above which "stresses" limit the options set (e.g. see Goulson and Cory 1995, and below).

The DDP hypothesis emerges from the interaction of life-history theory (e.g., Stearns 1992; Roff 2002) and epidemiological theory (e.g., Anderson and May 1991), and forms part of a growing body of theory generated by the relatively new field of 'ecological immunology' (Sheldon and Verhulst 1996;

Wilson 2004). It relies on three important assumptions. First, that parasite transmission is generally positively density-dependent. Second, that potential hosts can alter their phenotype in response to cues associated with population density. And third, the DDP hypothesis implicitly assumes that parasite defense is costly.

Assumptions

Parasite Transmission is Positively Density-Dependent

Density-dependent transmission of parasites has been a fundamental assumption of most epidemiological models for nearly a century (see review by McCallum et al. 2001). The reasoning is fairly straightforward—as population density increases, so too does the *per capita* risk of an individual encountering an infectious conspecific/cadaver or parasite infective stage. This idea is based on the "mass action" concept; namely that infectious and susceptible individuals within a population behave like the molecules of two chemical reagents in a closed system, moving in space randomly in relation to each other, such that the encounter rate between the two types of molecules is directly proportional to their relative densities. Thus, if the density of susceptible hosts is represented by S, and the density of infected hosts (or other infectious agents) is I, then the number of new infected hosts per unit area per unit time will be βSI, where β is the so-called *transmission coefficient*—a constant representing the probability of a new infection arising *per contact* between a susceptible and infectious host.

Although most epidemiological models have assumed that the transmission rate is dependent on host density, not all parasites are expected to exhibit density-dependent transmission. For some parasites, the contact rate between susceptible and infected individuals is predicted to be independent of host density (i.e., 'frequency-dependent' or 'density-independent') or there may be an asymptotic relationship between contact rate and host density (McCallum et al. 2001). For example, it is often assumed in models of sexually-transmitted diseases that because the number sexual partners of an individual usually depends on its attractiveness and/or the mating system of the species, parasite transmission will be only weakly related to host density (and the number of new infections per unit time and area will approximate $\beta SI/N$, where $N = S + I$ (but see Ryder et al. 2005)).

Although 'mass action' is commonly assumed in epidemiological models, it has only rarely been empirically challenged. In insect-parasite studies, tests of the mass action assumption have generally involved manipulating the density of infectious and/or susceptible hosts and determining the relationship between I and/or S and the transmission coefficient, β (Dwyer 1991).

The constant β can be calculated using a protocol developed by Dwyer (1991), which is derived from the basic Anderson and May (1981) model.

$$\beta = \frac{-1}{I_0 t} \ln\left[1 - \left(\frac{I_t}{S_0}\right)\right] \tag{1}$$

where I_0 is the initial density of infected hosts/cadavers; t = number of days of exposure to the infected hosts/cadavers; I_t is the density of infected hosts at time t; and S_0 is the density of susceptible hosts at the start of the experiment. This calculation assumes that the change in density of infectious hosts over the course of the experiment is negligible, that all infections are lethal, and that there are a negligible number of deaths due to non-parasite causes; all of these assumptions can be tested. If the mass action assumption applies, then β will be independent of both I and S and the rate of acquisition of new infections will increase linearly with the density of infectious and susceptible hosts (e.g. Anderson and May 1981; Dwyer 1991).

Experimental tests of the mass action assumption (reviewed in McCallum et al. 2001) have shown that although parasite transmission usually exhibits some form of positive density-dependence (Fig. 2), in most cases the strict assumption of mass action is violated (and β tends to decline with I and may decline or increase with S; McCallum et al. 2001). Thus, although it is intuitively appealing to assume that the *per capita* risk of becoming

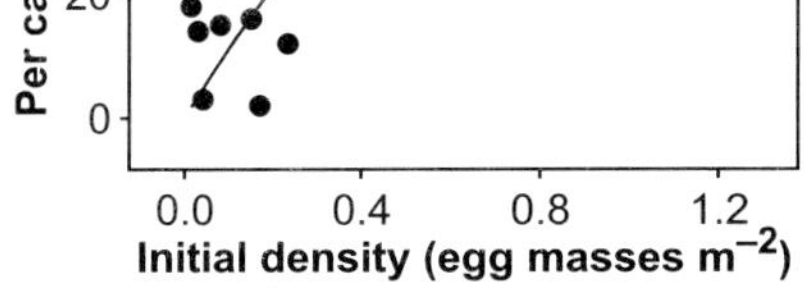

Fig. 2 Relationship between larval density and *per capita* infection rate in the gypsy moth, *Lymantria dispar* (data extracted from Woods and Elkinton 1987). Solid line is a second order polynomial constrained to pass through the origin. As host density increases, so too does the *per capita* risk of infection.

infected increases with local population density, and the available evidence is broadly in support of this idea, more studies explicitly testing this assumption are clearly required.

Host Phenotype is Phenotypically Plastic with Respect to Conspecific Population Density

A second assumption underlying the DDP hypothesis is that organisms can alter their phenotype in response to cues associated with population density (and hence the risk of infection). Perhaps more than any other taxonomic group, insects are renowned for their great phenotypic plasticity in response to population density. This is typified by nymphs of the desert locust, *Schistocerca gregaria*, which exist as a green, cryptic form under low density conditions and a conspicuous yellow-and-black form under high density conditions (e.g., Uvarov 1966; Pener 1991; Applebaum and Heifetz 1999) (see chapter 5). Density-dependent phenotypic changes are not restricted to color however; they also include alterations in morphology, physiology, behavior, development and feeding ecology (Applebaum and Heifetz 1999).

When an individual's phenotype is altered in response to perceived changes in the local density of conspecifics, it is usually known as "density-dependent phase polyphenism" (Table 1); or sometimes as "phase polymorphism". However, this latter expression is misleading, since polymorphisms usually refer to genetically-based changes in phenotype, as opposed to those that occur in response to changes in environmental conditions (Table 1) (Chapman 1998). Density-dependent phase polyphenism has been characterized in insects belonging to a wide range of insect taxa, including Lepidoptera, Phasmida, Orthoptera, Coleoptera, Hemiptera and Homoptera. In some phase polyphenic species, the phenotype adapted to low density conditions is referred to as the *solitaria* phase, and that adapted to high density conditions as the *gregaria* phase (e.g. Uvarov 1921, 1966; Iwao 1968). Of course, these are the extreme phenotypes that individuals may adopt under very low or high densities, but intermediate forms are usually common, and may reflect differences in exposure to density-dependent cues or genetic differences in the phenotypic response to density (i.e., genotype x environment interactions). Although density-dependent phase polyphenism has been characterized in many insect species, some of which are manifested in rather extreme color or morphological changes, it is likely that most insect species have the potential to alter their phenotype in response to density-dependent cues, though these may not always be apparent in their visible phenotype (i.e., morphology or behavior). Thus, whilst the

Table 1 Glossary of terms

Density dependent prophylaxis (DDP)	A form of phenotypic plasticity. A phenomenon in which individuals up-regulate their parasite resistance mechanisms under conditions of conspecific crowding to counteract the increased risk of parasitic infection at high densities.
Ecdysteroids	Ecdysone and related hormones. The molting hormone, ecdysone is a steroid produced by the prothoracic glands in juvenile insects. In adults, the gonads produce ecdysteroids, which are involved in gonad maturation (Nijhout 1994).
Encapsulation	The formation of an envelope of immune cells around an invading organism that is too large to be phagocytosed. The cells become melanized via the action of phenoloxidase (Götz 1986).
Juvenile hormone (JH)	A methyl-ester sesquiterpene hormone produced by the *corpora allata* and involved in many aspects of reproduction and development including metamorphosis, regulation of diapause, phase polyphenism and ovarian development (Nijhout 1994).
LPS	Lipopolysaccharide—a component of bacterial cell walls that can trigger the prophenoloxidase cascade, and is often used in assays of immune function.
Lysozyme	The first antibacterial factor purified from insect hemolymph; its structure is very similar to lysozymes found in chickens. Lysozyme is bactericidal to some gram-positive bacteria, but mostly works in concert with other antibacterial proteins (Boman and Hultmark 1987).
Phase polyphenism	A form of phenotypic plasticity by which a single genotype can produce *distinct* phenotypes depending on the environmental conditions experienced.
Phenoloxidase (PO) and the proPO cascade	Phenoloxidase, a key enzyme in the synthesis of melanin, is stored in the hemolymph as the inactive precursor, prophenoloxidase. The prophenoloxidase cascade is an enzyme cascade that can be triggered by microbial cell wall components, such as peptidoglycan and zymosan, via a serine protease cascade.
Polymorphism	General sense: variation in body form or color within a species (see also *Polyphenism*). Specific sense: genetically-based (as opposed to environmentally regulated) variation in body color or morphology within a species.
Polyphenism	A form of phenotypic plasticity. The occurrence of different phenotypes within a species, where the development of the phenotype is governed by environmental conditions experienced during development, including temperature, humidity and population density (see also *Polymorphism*).
Prophylaxis	Prevention of, or protection from, disease causing agents.

DDP hypothesis is most likely to be supported by insect species exhibiting overt density-dependent phase polyphenism, it would not be surprising if supporting evidence was forthcoming from non-phase polyphenic species.

Parasite Defense is Costly

A third assumption implicit in the DDP hypothesis is that the maintenance of parasite defense mechanisms is costly. This is because, if parasite defense was cost-free, then selection would favor organisms that maintained them at high levels at all times. If parasite defense is costly, then selection will favour individuals that alter their investment in resistance mechanisms to match the perceived risk that they will be required (such as when levels of crowding are high). There is now good evidence from a number of studies indicating that both the maintenance and deployment of parasite resistance mechanisms are costly (see reviews by Lochmiller and Deerenberg 2000; Kraaijeveld et al. 2002; Schmid-Hempel 2003, 2005; Wilson 2004).

One of the clearest examples illustrating the costs of maintaining an effective parasite resistance mechanism comes from the fruit fly, *Drosophila melanogaster* L., and two of its parasitoids. Kraaijeveld and Godfray (1997) set up four genetic lines that were selected for resistance to attack by the braconid parasitoid wasp *Asobara tabida* Nees and compared them with four control lines that were not exposed to the parasitoid. They found that selected lines rapidly increased their cellular encapsulation response to parasitoid eggs from 5% at the start of the experiment to greater than 60% after 5 generations of selection. After extensive investigation, they discovered that the main cost of resistance in this system was a decline in the competitive ability of larvae in the selected lines relative to controls when food was in short supply. Similar experiments with the eucoilid wasp, *Leptopilina boulardi* Forster, showed a similarly rapid response to selection, from less than 1% encapsulation at the start of selection to 45% after just 5 generations (Fellowes et al. 1998). Significantly, increased resistance to parasitism by *L. boulardi* was also achieved at the expense of competitive ability when food was limited. Similar experiments in other insect-parasite systems have also indicated that parasite defense is costly to maintain, though these costs may be revealed only under certain environmental conditions (e.g., when animals are food stressed, McKean et al. 2008; also see reviews by Schmid-Hempel 2003; Wilson 2004).

Empirical Examples

Density-dependent prophylaxis has been tested and/or observed in a range of insect species, including both phase polyphenic and non-polyphenic species (Table 2). Amongst the phase polyphenic species, most of the examples come from Lepidoptera, Coleoptera and Orthoptera. Amongst the non-polyphenic species, examples are from the Hymenoptera and Isoptera.

Table 2 Insect species tested for density-dependent resistance to pathogens and/or immune function

Order	*Species*	*Phase polyphenic*	*Density-dependent resistance to pathogens*	*Density-dependent immune response*
Lepidoptera	*Spodoptera littoralis*	Y	Fungus resistance increases with melanism and rearing density (Wilson et al. 2001)	Cuticular PO increases and antibacterial activity decreases with density (Cotter et al. 2004a) Hemolymph PO, cuticular PO and capsule melanization increase and antibacterial activity decreases with melanism (Cotter et al. 2004a)
Lepidoptera	*Spodoptera exempta*	Y	Virus resistance increases with density (Reeson et al. 1998) Fungus resistance increases with density at low doses but not high (Wilson et al. 2001) Parasitoid resistance increases with melanism (Wilson et al. 2001)	Hemolymph PO, cuticular PO and midgut PO increase with density (Reeson et al. 1998; Wilson et al. 2001)
Lepidoptera	*Mamestra brassicae*	Y	Virus resistance increases with density but falls at very high densities (Goulson and Cory 1995)	
Lepidoptera	*Mythimna separata*	Y	Resistance to fungus increases with melanism (Mitsui and Kunimi 1988) Resistance to virus increases with melanism and rearing density (Kunimi and Yamada 1990)	
Coleoptera	*Tenebrio molitor*	Y	Resistance to fungus increases with melanism (Barnes and Siva-Jothy 2000)	No effect of density on hemolymph PO (Barnes and Siva-Jothy 2000)
Orthoptera	*Locusta migratoria*	Y	Resistance to fungus increases with density (Wilson et al. 2002)	No effect of density on hemolymph PO, antibacterial activity increases with density (Wilson et al. 2002)
	Gryllus texensis	N	No effect of density on resistance to bacteria (Adamo and Parsons 2006)	No effect of density on PO or lysozyme activity (Adamo and Parsons 2006)
Hymenoptera	*Acromyrmex echinatior*	N	Resistance to fungus increases with density (Hughes et al. 2002)	
Isoptera	*Zootermopsis angusticollis*	N	Resistance to fungus increases with density (Rosengaus et al. 1998)	No effect of density on encapsulation (Traniello et al. 2002)

Phase Polyphenic Species

Lepidopteran Larvae Pathogen resistance has been examined in relation to phase and/or rearing density in a number of phase-polyphenic, lepidopteran species. In many cases, these two factors can be hard to separate as phenotype and density may be inextricably linked. However, in some species, the typical high-density phenotype can occur at low densities and vice-versa, thus allowing the effects of rearing density and phase to be disentangled.

Mythimna separata Using larvae of the true armyworm, *Mythimna (Pseudaletia) separata* Walker, Kunimi and Yamada (1990) examined the relationship between density, color phase and pathogen resistance. Larvae were reared in solitary (1 larva per container) or crowded conditions (2, 5 or 20 larvae per container) from the onset of the second instar until the onset of the fifth instar, at which time a nucleopolyhedrovirus (NPV) was administered orally. They found that susceptibility to NPV decreased with rearing density, such that larvae in the highest density treatment were 20 times more resistant to the virus than those reared solitarily (Fig. 3a). The experiment was then repeated with solitary and crowd-reared (20 larvae per container) treatments, but this time the density experienced by the larvae was switched *after* administration of the virus, giving four treatment groups: solitary-solitary (SS), solitary-crowded (SC), crowded-solitary (CS), and crowded-crowded (CC). LC_{50} values showed that whilst SS larvae were most susceptible and CC most resistant to the virus, the two groups for which the rearing density was switched after viral administration (SC and CS) showed similar, but intermediate, levels of resistance to the other two groups. This highlighted the fact that rearing density both prior to and during infection were important factors in determining an individual's resistance levels. When they injected non-occluded virus directly into the haemocoel (i.e., virus that was not enclosed in a proteinaceous occlusion body), they found that there was no difference between solitary and crowd-reared larvae in their susceptibility, suggesting that resistance mechanisms in the gut may be responsible for the increased resistance shown by crowd-reared larvae.

Density-dependent prophylaxis in this species was reflected not only in resistance to its NPV, but also in resistance to its granulovirus virus (GV) (Kunimi and Yamada 1990) and to the generalist entomopathogenic fungus *Nomuraea rileyi* (Farlow) Samson (Mitsui and Kunimi 1988). Relative to gregarious larvae, solitary *M. separata* larvae were about 5 times more susceptible to oral doses of GV (Kunimi and Yamada 1990; Fig. 3b), and twice as susceptible to percutaneous doses of *N. rileyi* conidia (Mitsui and Kunimi 1988; Fig. 3c).

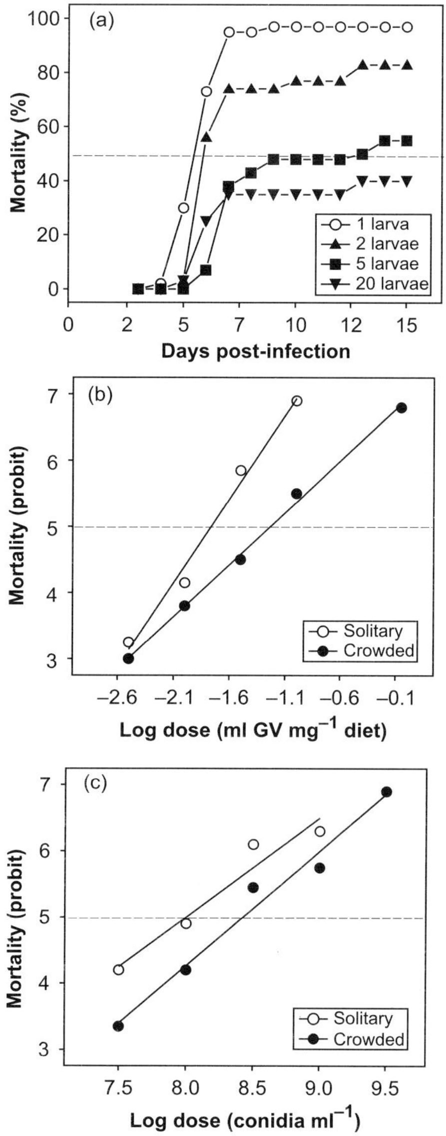

Fig. 3 Relationship between larval density and resistance to parasites in the true armyworm, *Mythimna* (*Pseudaletia*) *separata*. (a) Cumulative mortality for larvae reared at 1, 2, 5 or 20 larvae per container and orally infected with its nucleopolyhedrovirus, NPV (Kunimi and Yamada 1990); (b) Dose-mortality relationship for solitary and crowded larvae orally infected with its granulovirus, GV (Kunimi and Yamada 1990); (c) Dose-mortality relationship for solitary and crowded larvae percutaneously infected with the fungus, *Nomuraea rileyi* (Mitsui and Kunimi 1988). The dashed line in each graph indicates 50% mortality.

The role of color phase in pathogen resistance was also examined in this species, whilst controlling for rearing density. Although melanism generally increases with density, it can occur in low-density animals. Low-density melanic larvae were 5 times more resistant to the entomopathogenic fungus *Nomuraea rileyi* (Mitsui and Kunimi 1988), whilst melanic larvae that occurred at high densities were twice as resistant to an NPV than pale larvae (Kunimi and Yamada 1990).

Spodoptera spp. More recent studies have tried to quantify both pathogen resistance and investment in immune function in relation to phase and rearing density. In the African armyworm, *Spodoptera exempta* Walker (Fig. 4), resistance to its NPV was lowest in non-melanic, solitary-reared larvae (average mortality over a range of viral doses = 70%; LD50 = 1,325 OB/larva) and highest in melanic crowd-reared larvae (42%; 14,188 OB/larva), whilst melanic, solitary-reared larvae showed intermediate levels of resistance (59%; 3,082 OB/larva; Fig. 5a; Reeson et al. 1998). Thus, the typical solitary form of *S. exempta* was approximately 10 times more susceptible to NPV than the typical gregarious form. Furthermore, it has also been shown that vertical transmission of the virus is more likely from sub-lethally infected solitary than gregarious individuals (Vilaplana et al. 2008).

Fig. 4 Cuticular melanism in the African armyworm, *Spodoptera exempta* (Wilson et al. 2001). The two larvae are siblings; the one on the left was reared solitarily, the one on the right in crowded conditions. Photograph courtesy of Ken Wilson.

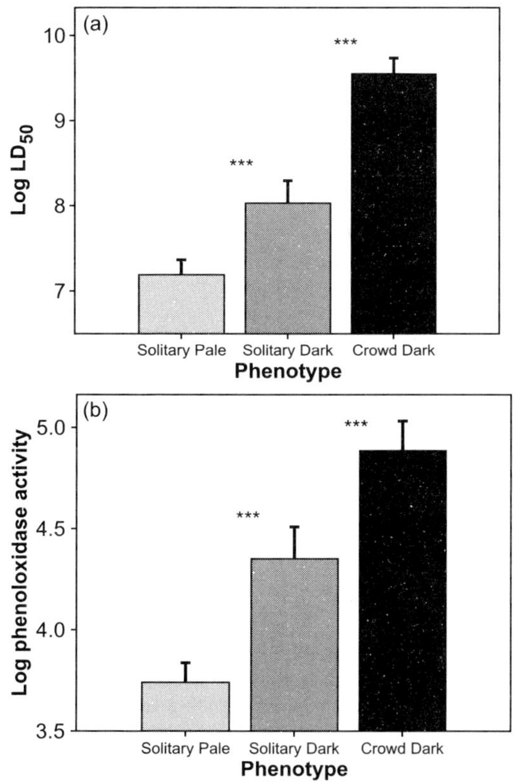

Fig. 5 The relationship between larval density/phase the African armyworm *Spodoptera exempta* and (a) resistance to a nucleopolyhedrovirus, measured as the dose of virus required to kill 50% of the larvae, LD_{50} (Reeson et al. 1998), (b) immune function measured as hemolymph phenoloxidase activity (Wilson et al. 2001). Bars show means ± standard errors.

Investment in hemolymph phenoloxidase, a key enzyme in the production of melanin and an important component of the insect immune system, followed the same patterns, increasing from solitary-reared, non-melanic larvae to crowd-reared, melanic larvae (Fig. 5b; Reeson et al. 1998). A further examination of parasite resistance and immune function in this species found that crowd-reared, melanic larvae were more resistant both to the entomopathogenic fungus, *Beauveria bassiana* (Balsamo) Vuillemin (at low doses, but not high), and to the ectoparasitoid, *Euplectrus laphygmae* Ferrière, than solitary-reared, non-melanic larvae (Wilson et al. 2001). Moreover, investment in phenoloxidase in the midgut and cuticle, important sites of entry for many parasites and pathogens, increased with increasing melanism (Fig. 6a), suggesting that it could be the phenoloxidase in these sites that is providing the resistance (Fig. 6b).

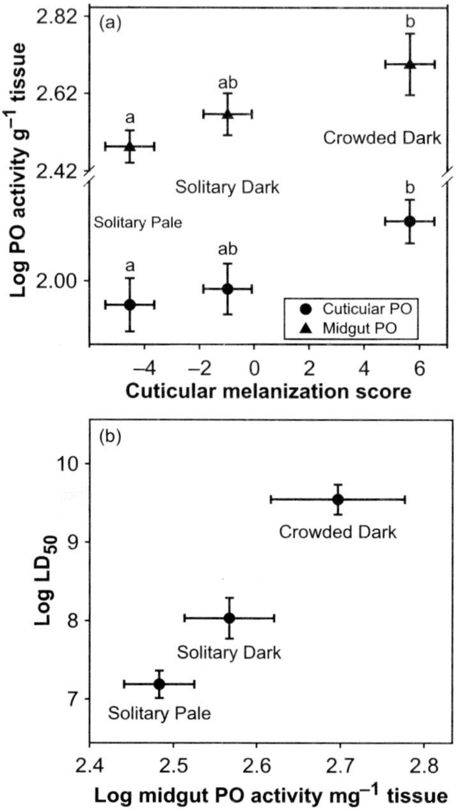

Fig. 6 Cuticular melanism, phenoloxidase activity and resistance to nucleopolyhedrovirus in the African armyworm, *Spodoptera exempta* (Wilson et al. 2001). (a) Cuticlar melanization is associated with increasing levels of phenoloxidase activity in the cuticle and midgut; (b) elevated levels of phenoloxidase activity in the midgut (and elsewhere) are associated with enhanced resistance to nucleopolyhedrovirus, as measured by LD_{50} (for details see legend to Fig. 5a). In each graph, means ± standard errors are shown.

The only study to examine the relationship between phase, rearing density and a suite of immune system components found surprising results. Using larvae of the Egyptian cotton leafworm, *Spodoptera littoralis* Boisduval, it was shown that, as predicted by the DDP hypothesis, levels of phenoloxidase activity in the hemolymph and cuticle were higher in melanic than non-melanic larvae, as was melanization of an artificial parasite (Figs. 7a-c). However, (lysozyme-like) antibacterial activity levels in the hemolymph were highest in non-melanic and solitary-reared larvae (Fig. 7d), suggesting a potential trade-off in the insect immune system (Cotter et al. 2004a).

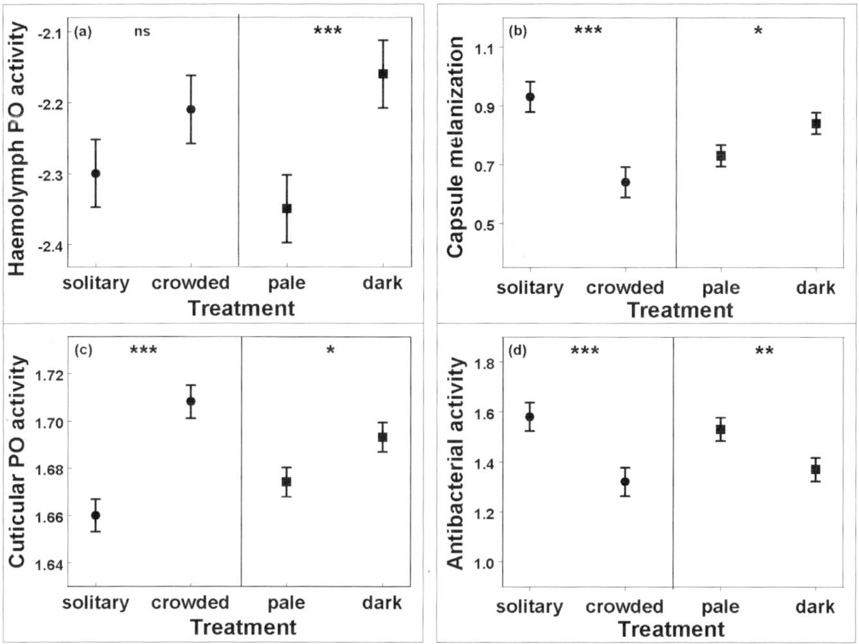

Fig. 7 Effect of color and larval density on immune function parameters in larvae of the Egyptian cotton leafworm, *Spodoptera littoralis*. Larvae were reared in solitary or crowded conditions and scored for color: pale or dark. The figures show mean (+/- SE) levels of investment in four different immune parameters: (a) hemolymph phenoloxidase activity, (b) cuticular phenoloxidase activity, (c) capsule melanization of a foreign implant, and (d) lysozyme-like antibacterial activity (Cotter et al. 2004a).

Other Lepidopteran species Goulson and Cory (1995) examined the relationship between pathogen resistance and rearing density in the cabbage moth, *Mamestra brassicae* L. and found that resistance to NPV (and larval melanism) increased with rearing density, although in this case larvae that were reared at exceptionally high densities experienced a decrease in resistance, suggesting that under very crowded conditions adaptive responses to density may break down.

Mealworm beetles Pathogen resistance was examined in the weakly phase-polyphenic coleopteran *Tenebrio molitor* L. Adult mealworm beetles range in color from tan to black, with melanic (black) beetles being more common at high densities. Melanic *T. molitor* were up to three times more resistant to the entomopathogenic fungus *Metarhizium anisopliae* than the paler beetles (Barnes and Siva-Jothy 2000). However, in this species, there was no association between hemolymph phenoloxidase activity and rearing density. Unfortunately, the correlation between phenoloxidase activity and melanism was not determined directly and phenoloxidase activity was not measured in the cuticle, so the basis for the increased resistance of melanic beetles is not clear (Barnes and Siva-Jothy 2000).

Desert locusts The desert locust (*Schistocerca gregaria* Forskal) is the archetypal phase polyphenic species. Therefore, if the DDP hypothesis is to be credible, then it should apply to this species. Wilson et al. (2002) tested the DDP hypothesis using the desert locust and found that locusts reared under crowded conditions were significantly more resistant than solitary-reared locusts to the entomopathogenic fungus, *Metarhizium anisopliae* (Metschnikov) Sorokin *var acridum*, a key natural disease of acridids and an important agent in locust and grasshopper biocontrol. After accounting for body mass differences between the phases, the relative daily mortality risk for the *solitaria* phase locusts was nearly twice that of *gregaria* phase conspecifics (mean = 1.76; 95% confidence interval = 1.35–2.31; Fig. 8). The enhanced pathogen resistance observed in the crowded locusts was associated with significantly elevated antimicrobial (lysozyme-like) activity, a marginally greater total hemocyte count, and non-significant differences in hemolymph phenoloxidase activity and cellular encapsulation response.

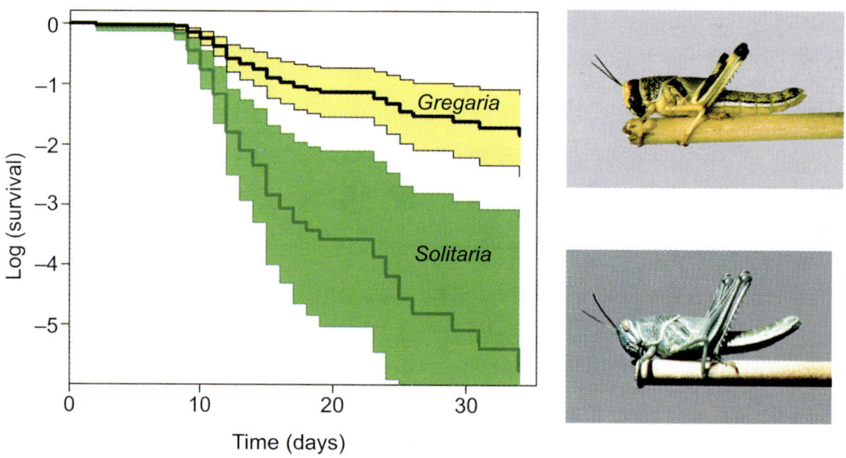

Fig. 8 Density-dependent changes in the desert locust, *Schistocerca gregaria*. Log-survival curves for *solitaria* and *gregaria* phase locusts infected with the *Metarhizium anisopliae* var *acridum* (Wilson et al. 2001). The two bold lines show the fitted values from the Cox's Proportional Hazard Model and the narrow lines and shading represent the 95% confidence intervals. Photographs courtesy of Stephen J. Simpson.

Thus, as predicted by the DDP hypothesis, the high-density form of the desert locust exhibits significantly greater pathogen resistance than the solitary form. However, the precise immunological and/or physical mechanisms underpinning this observation have yet to be determined.

Non-phase Polyphenic Species

Crickets. A study using adult *Gryllus texensis* crickets found no effect of crowding on lysozyme activity, PO activity or resistance to the bacterium *Serratia marcescens* (Adamo and Parsons 2006).

Leafcutting ants. Studies examining the effects of density on pathogen resistance in non-phase polyphenic species are rare. However, a recent study found that after inoculating *Acromyrmex echinatior* Forel workers with the fungus *Metarhizium anisopliae*, those that were kept with groups of nest-mates showed reduced mortality compared to those kept in isolation (Hughes et al. 2002). It was proposed that this was due to a combination of hygienic allogrooming and increased antibacterial secretions stimulated by the presence of fungal spores. Previous work has shown that the prevalence of allogrooming increases with colony size across ant species (Schmid-Hempel 1998).

Termites A similar study was conducted using nymphs of the dampwood termite, *Zootermopsis angusticollis* Hagen (Rosengaus et al. 1998). After exposure to spores of the entomopathogenic fungus, *Metarhizium anisopliae*, nymphs were maintained individually or in groups of 10 or 25. A strong effect of grouping was apparent with nymphs that were kept in groups showing a 90% reduction in their daily mortality risk compared to those kept in isolation. By monitoring the behavior of nymphs that had been exposed to spores, it was shown that the frequency of allogrooming was greatly increased in those exposed to the fungus compared to controls. However, the frequency of self-grooming did not increase, either in nymphs maintained in groups or in isolation, suggesting that grooming by nestmates is necessary to remove fungal spores and so reduce the risk of infection.

A more recent study using the same species examined further adaptations to reduce the spread of disease at high densities (Traniello et al. 2002). Termites respond to a non-lethal dose of a bacterial or fungal pathogen by improving their physiological response to a later infection, i.e. they can become immunized against particular pathogens (Rosengaus et al. 1999). By grouping naive and immunized termite nymphs prior to a challenge with a lethal dose of fungal spores, it was found that there was a "social transfer" of immunity from immunized to naïve nymphs. Nymphs that had been immunized by this social transfer displayed a 13-fold reduction in susceptibility compared to controls (Traniello et al. 2002). The mechanisms underlying this social transfer of immunity are unclear. Naïve nymphs may be exposed to a sub-lethal dose of spores during allogrooming of immunized individuals or there could be a transfer of immune factors during trophallactic exchanges between naïve and immunized individuals (Traniello et al. 2002). However, it appears that nestmate density *per se* does not affect susceptibility to fungal pathogens in this species. A study examining the effects of high versus low density on susceptibility to *M. anisopliae* found no significant difference in mortality between the treatments (Pie et al., 2005). Furthermore, an examination of the encapsulation response found no difference between solitary and grouped termites, suggesting that it is not an up-regulation of immunity in response to high density that causes the reduction in mortality in grouped versus solitary termites (Traniello et al. 2002).

Social insects naturally occur at high densities and so behavioral mechanisms such as those described above may reduce the risk of parasitism in groups and as such fall within the remit of the DDP hypothesis. However, the DDP hypothesis has yet to be tested in a species that undergoes strong fluctuations in population density and does not display density-dependent phase polyphenism.

Field Experiments

The non-linearity observed in the larval density-infection rate plots generated from field experiments is consistent with the DDP hypothesis (e.g. Woods and Elkinton 1987; D'Amico et al. 1996; Dwyer et al. 1997; Fig. 2). However, to our knowledge, there has so far been just one direct field-based test of the DDP hypothesis. Reeson et al. (2000) reared larvae of the African armyworm (*S. exempta*) in the laboratory under either solitary or crowded conditions, and when they reached the fourth instar they were introduced into 1 m^3 field cages containing 9 maize plants. The larvae were introduced to the plants at a density of one, three or nine larvae per plant. A few days prior to this, the maize plants were 'seeded' with two larvae that had been lethally-infected with nucleopolyhedrovirus (NPV). Thus, the introduced larvae were exposed to infectious cadavers for several days before being reclaimed and reared again in the laboratory under solitary conditions until pupation or death.

Using this experimental design, it was possible to estimate the effect of rearing density (solitary versus crowded) and exposure density (one, three or nine larvae per plant) on both the *per capita* mortality rate and the viral transmission parameter, β (see Dwyer 1991, and Eq. 1 above). As predicted by the DDP hypothesis, Reeson and colleagues found that both the mortality rate and the viral transmission parameter were significantly lower for larvae that were reared under crowded conditions prior to exposure to the virus than for larvae that were reared under solitary conditions (Fig. 9). Thus, the mass action assumption was not upheld for *S. exempta* and its NPV. Local density during the experiment (i.e., the number of larvae per plant) had little

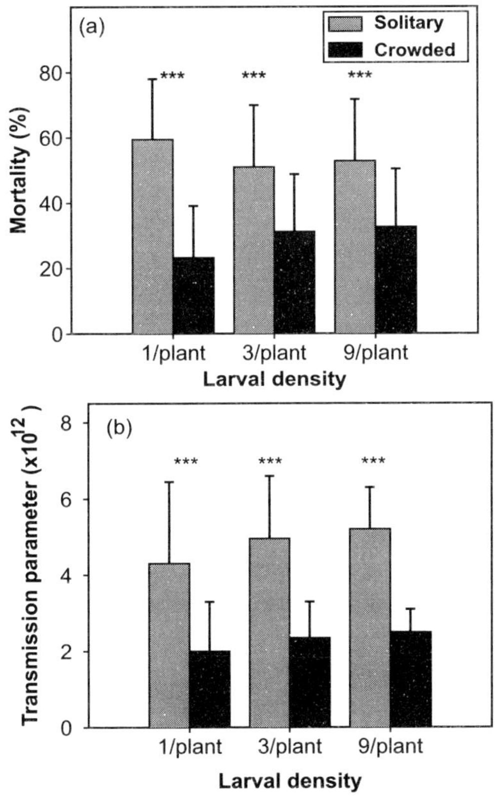

Fig. 9 Relationship between larval density and (a) average mortality and (b) the transmission parameter, for larvae of the African armyworm, *Spodoptera exempta,* exposed to its nucleopolydrovirus (NPV) under field conditions. Mortality and viral transmission were greatest for larvae reared under solitary, rather than crowded conditions until the start of the 4th instar, but did not vary significantly with the density of larvae during the few days that the larvae were exposed to the virus (Reeson et al. 2000). Means $\pm$ standard errors are shown.

effect on viral transmission (Reeson et al. 2000). It is important to realize here that the transmission parameter, β, comprised two components, namely the rate at which the host encounters parasites (i.e., the contact rate) and the rate at which contact between hosts and parasites results in infection (i.e., host susceptibility). The experiment described here indicates that differences in host susceptibility generated by DDP may be large enough to produce density-dependent changes in β.

Two further field studies have considered the effects of phase and population density on resistance in natural populations. The strength of the encapsulation response was found to be independent of current and past population density in peak-phase autumnal moth, *Epirrita autumnata* (Klemola et al. 2007), whilst, contrary to expectation, rates of parasitoid attack were positively correlated with cuticular melanism in field-caught larvae of the winter moth *Operophtera brumatta* (Hagen et al. 2006).

Mechanisms

Sensory Systems Underlying Plasticity: Perception of Density

A key question in the study of DDP, and of phase polyphenism in general, is the relative importance of different mechanisms for the perception of population density. This issue has been most thoroughly addressed in studies of the desert locust, *Schistocerca gregaria*. Experimental studies have shown that locust phase determination is governed by the combined effects of genetic, maternal and environmental effects. For example, having a crowd-reared mother, a crowd-reared father, parents which were crowded only for the period of mating and oviposition, or being reared from birth in a crowd, all had a marked gregarizing effect on the behavior and color of desert locust hatchlings (Islam et al. 1994). However, the most important factor determining locust phase was the individual's perception of local population density.

When desert locust nymphs are switched between low and high-density conditions, different aspects of the suite of phase characteristics change at different rates. Changes in metabolism and behavior occur within just 1–4 hours of the switch, whereas color and morphological changes can occur only following molting to subsequent instars (e.g. Roessingh and Simpson 1994; Applebaum and Heifetz 1999). Olfactory, visual and tactile stimuli all play a role in determining whether individuals develop into *solitaria* or *gregaria* phase locusts (Uvarov 1966). In an elegant series of experiments, Roessingh et al. (1998) examined the relative importance of these different stimuli in determining behavioral phase transition in fifth-instar nymphs of the desert locust. They found that whilst tactile stimulation (by rolling paper spheres) was highly gregarizing, and olfactory and visual stimuli together caused partial behavioral gregarization, visual and olfactory stimulation provided alone were only weakly gregarizing, if at all. A number of other studies on lepidopteran species also appear to support the notion that tactile stimulation is the key trigger for density-dependent phase changes (e.g. Drooz 1966; Sasakawa 1973; Kazimirova 1992; Gunn 1998).

In subsequent experiments on the desert locust, Simpson et al. (2001) observed that a significant switch from solitarious to gregarious behavior occurred when the outer face of a hind femur was stimulated (by a paint brush), but mechanical stimulation of 10 other body regions (including other parts of the legs, the abdomen, thorax, antennae and mouthparts) did not result in significant behavioral changes (see Chapter 5, this volume). Thus, it appears that stimulation of mechanoreceptors on the hind femora is the most important factor determining behavioral phase change in desert locusts, probably because this part of the body is least likely to be self-stimulated or stimulated during normal behavioral activities, such as feeding. It remains to be established whether this also serves as the trigger for the up-regulation of parasite resistance mechanisms in this same species (Wilson et al. 2002, see above).

Genetics of Phase and Density-Dependent Prophylaxis

Surprisingly little is known about the genetic basis of phase transition. However, in the desert locust *Schistocerca gregaria,* a recent study examined phase-specific gene expression in the brains of solitary and gregarious adults using differential display and reverse transcriptase PCR (Rahman et al. 2003b). This revealed one solitary-specific gene (SSG) and one gregarious-specific gene (GSG). Although the SSG could not be identified, the GSG showed homology with the SPARC gene (Secreted Protein Acidic, Rich in Cysteine) known to regulate growth factors and cell adhesion in vertebrates (Brekken and Sage 2000). An examination of the phase-dependent peptide profile in desert locusts revealed differentially-expressed serine protease inhibitors belonging to the pacifastin peptide family (Rahman et al. 2002). Quantification of the transcription levels of the precursors of these peptides (SGPP-1–3) in solitary and crowd-reared locusts revealed higher expression in crowd-reared than solitary-reared individuals (Simonet et al. 2004). Furthermore, an immunoregulatory role for these peptides has been posited; injection of locusts with fungal elicitors of the immune response resulted in changes in the abundances of the SGPP transcript (Simonet et al. 2004).

A more detailed study examined ESTs isolated from the head, gut and hind leg of solitary and gregarious *Locusta migratoria* (Kang et al. 2004). Kang et al. found a number of genes that were differentially expressed, both between the phases and between the different tissue types. A number of muscle-related genes from the hind-leg library were up-regulated in solitary hoppers, which is consistent with the observation that solitary hoppers tend to have longer hind legs and greater jumping ability than gregarious hoppers. Over 100 genes from the midgut library were uniquely expressed in solitary hoppers, the majority of which were metabolism-related. The ESTs isolated from the head showed striking differences between phases in the JHPH superfamily of genes, which includes juvenile hormone binding protein, hemocyanin and phenoloxidase amongst others. Three JHPH families were identified based on sequence similarity, all of which were highly expressed in the head of gregarious hoppers and in the hind leg of solitary hoppers. JHPH1 was up-regulated in gregarious hoppers whilst JHPH3 was up-regulated in solitary hoppers; a number of genes from the JHPH2 family were either expressed in the gregarious head or in the solitary head but not in both. Further identification of these genes and their possible function should provide a valuable insight into the genetic control of phase change in this species.

In Lepidoptera, phase-polyphenism is mainly determined by rearing density, but it also has a genetic component (Tojo 1991; Goulson 1994; Cotter et al. 2004a). However, to date there has been just one study examining the quantitative genetics of phase and immunity. Using larvae of the Egyptian armyworm, *Spodoptera littoralis*, Cotter et al. (2004b) calculated heritability estimates for several immune function and life-history traits and the genetic and phenotypic correlations between them (Table 3). Based on the correlations between traits, notably cuticular melanization, it was possible to predict theoretical life-history trajectories based on color-phase, i.e., pale or dark larvae (Table 4). Pale larvae are characterized by a slow larval development rate and short adult lifespan, both traits known to be common to solitary phase Lepidoptera. An important discovery was that whilst hemocyte density was positively genetically correlated with phenoloxidase activity, it was negatively genetically correlated with lysozyme-like antimicrobial activity (Table 3), providing further evidence for a potential trade-off in the insect immune system identified in a previous phenotypic study (Cotter et al. 2004a). In terms of the DDP hypothesis, this suggests that insects could not simultaneously up-regulate all components of the immune

Table 3 Heritabilities and genetic and phenotypic correlations between traits in solitary-reared *Spodoptera littoralis* larvae

	Larval development rate	*Pupal weight*	*Pupal development rate*	*Adult longevity*	*Cuticular melanization*	*Hemolymph PO activity*	*Antibacterial activity*	*Hemocyte density*
Larval development rate	**0.42** ***	0.21 ***	0.09 *	–0.11 *	0.04 ns	–0.10 *	0.08 ns	0.07 ns
Pupal weight	0.24 ***	**0.49** ***	–0.09 *	0.11 *	–0.02 ns	0.10 *	–0.08 ns	–0.07 ns
Pupal development rate	0.20 *	–0.34 ***	**0.20** **	0.02 ns	–0.05 ns	–0.05 ns	0.08 ns	–0.06 ns
Adult longevity	–0.08 ns	0.41 ***	–0.49 ***	**0.22** **	–0.10 *	0.06 ns	0.05 ns	0.02 ns
Cuticular melanization	0.18 ***	0.00 ns	–0.43 ***	–0.22 ns	**0.36** ***	0.04 ns	–0.01 ns	0.09 *
Hemolymph PO activity	–0.05 ns	0.41 ***	–0.38 ***	0.56 ***	–0.08 ns	**0.65** ***	0.10 *	0.23 ***
Antibacterial activity	–0.29 ***	0.01 ns	0.35 ***	–0.04 ns	–0.06 ns	0.01 ns	**0.63** ***	0.07 ns
Hemocyte density	–0.04 ns	–0.17 **	–0.66 ***	0.04 ns	0.55 ***	0.21 *	–0.23 *	**0.36** ***

Values on the leading diagonal indicate the narrow-sense heritability estimates (after accounting for any significant maternal effects); values above the diagonal show (Pearson's) phenotypic correlations between traits; and values below show genetic correlations. Significance levels were determined with t-tests: ns $p < 0.05$, * $p < 0.05$. ** $p < 0.01$, *** $p < 0.001$. Modified after Cotter et al. (2004b)

Table 4 Theoretical life-history trajectories for pale- or dark-phase *Spodoptera littoralis* larvae

Trait	*Pale*	*Dark*
Life-history traits		
Larval development rate	Slow larval development	Fast larval development
Pupal weight	*Low pupal weight*	*High pupal weight*
Pupal development rate	Fast pupal development	Slow pupal development
Adult longevity	*Short adult lifespan*	*Long adult lifespan*
Immune function traits		
Hemolymph PO activity	*Low PO*	*High PO*
Antibacterial activity	*High lysozyme*	*Low lysozyme*
Haemocyte density	Low HC	High HC

Theoretical life-history trajectories were predicted from the point of view of color-phase, i.e. pale or dark larvae, based on the genetic correlations between traits (Table 3). Traits in italics are not directly genetically correlated with cuticular melanization and so the correlations were inferred from relationships with other traits.

system in response to crowding, but may up-regulate hemocyte density and phenoloxidase activity at the expense of lytic activity (Table 4). This may represent an important cost of DDP to this insect.

Hormonal Regulation Underlying Plasticity of Phase and Immune Function

As DDP is predicted to occur most strongly in phase polyphenic species, an understanding of the hormonal regulation of phase transition and immune function may elucidate the mechanisms of increased parasite resistance observed in crowded phase individuals.

Phase transition Although the hormonal regulation of phase polyphenism is far from clear, there is compelling evidence, particularly from studies on locusts, that juvenile hormone (JH) plays a key role. The solitary phase of many phase polyphenic Lepidoptera and locusts have high JH titres compared to the gregarious phase (Yagi and Kuramochi 1976; Nijhout and Wheeler 1982). High JH titres are associated with a lack of melanization of the cuticle, with solitary phases tending towards pale or cryptic coloration (Ikemoto 1983; Fescemyer and Hammond 1988). Indeed, exogenous application of JH to gregarious-phase locusts results in a reversion to the green coloration typical of solitary-phase locusts (Pener 1991). Furthermore, as JH is involved in regulation of the molting cycle, high titres usually result in longer juvenile development times (Yagi and Kuramochi 1976; Fescemyer and Hammond 1988), a common characteristic of solitary phase Lepidoptera.

Low JH titres are also associated with migration (Nijhout 1994); gregarious adults of many species tend to have larger lipid reserves than solitaries and have a greater propensity to migrate (Iwao 1968; Gunn and Gatehouse 1993). Solitary and gregarious phase locusts display a number of phase-specific polypeptides of unknown function in their hemolymph (Wedekind-Hirschberger et al. 1999; Clynen et al. 2002; Rahman et al. 2002; Rahman et al. 2003a). Treatment of gregarious adult locusts with a single dose of the JH analogue, Fenoxycarb, resulted in suppression of nine of the 17 *gregaria*-specific polypeptides and expression of two of the three *solitaria*-specific polypeptides (Wedekind-Hirschberger et al. 1999). Studies such as these underline the importance of JH in phase transition; however it is clear that JH alone is not responsible for the myriad of morphological, physiological and behavioral changes brought on by high densities.

The role of ecdysteroids in phase-transition has also been investigated. The ecdysteroid content of the eggs produced by gregarious desert locust females is significantly higher than in eggs produced by solitaries (Tawfik et al. 1999). This trend is maintained in newly hatched larvae with gregarious larvae having five times the ecdysteroid content of solitaries (Tawfik et al. 1999). However, during larval development, *solitaria* larvae show a higher peak ecdysteroid titer than *gregaria* larvae, though the duration of the hormonal surge is longer in gregarious larvae (Tawfik and Sehnal 2003). Solitary adults also show higher hemolymph ecdysteroid titers than gregarious adults; it may be that low titers stimulate, whilst high titers suppress the production of aggregation pheromones (Tawfik and Sehnal 2003).

Biogenic amines may play a role in density-dependent phase change. Rogers et al (2004) examined the changes in potential neurotransmitters/ neuromodulators in the central nervous system of the desert locust, *Schistocerca gregaria* across several stages of gregarization. Crowding of solitarious locusts typically resulted in a rapid decrease in the neurochemicals measured with the exception of serotonin levels in the thoracic ganglia which increased nine-fold during the first four hours of crowding. Increased serotonin levels may be linked to the sensitivity of receptors on the hind leg to mechanical stimulation during this timeframe.

Certain neurosecretory peptide hormones have also been implicated in the regulation of density-dependent color change. Melanization and Reddish Coloration Hormone (MRCH) is responsible for cuticular melanization in a number of lepidopteran species (Ogura 1975; Matsumoto 1981; Matsumoto et al. 1984; Altstein et al. 1994). A neurohormone isolated from the locust *Locusta migratoria* known as Dark Color Inducing

Neurohormone (DCIH) or [His^7]-Corazonin, and the related neurohormone, Corazonin, were found to induce melanization in a number of orthopteran species (Tanaka 2000; Yerushalmi and Pener 2001). Interestingly, Corazonin, which appears to be highly conserved throughout the Insecta, has recently been shown to induce the release of ecdysis-triggering hormones in *Manduca sexta* and *Bombyx mori* (Zitnan D. 2002), suggesting a potential interaction with ecdysone. Furthermore, as neurohormones can stimulate or inhibit JH production, it is possible that there is an interaction between the neurohormones responsible for melanization in different insect species and JH, though this remains to be investigated.

Immune function To date, there have been few studies investigating the hormonal regulation of immune function in insects. However, recently, mediation of the encapsulation response by JH was investigated using the Egyptian cotton leafworm, *Spodoptera littoralis* and its parasitoid *Microplitis rufiventris* Kok. (Khafagi and Hegazi 2001). Application of JH reduced the encapsulation response of *S. littoralis* larvae to *M. rufiventris*, with higher doses showing a stronger effect. Application of anti-JH compounds to parasitised *S. littoralis* larvae resulted in an increased encapsulation rate and enhanced melanization of the capsules, suggesting a possible interaction with phenoloxidase (PO), a key enzyme in the synthesis of melanin.

The link between JH and PO activity is further strengthened by a study showing that JH could inhibit PO activity in *Tenebrio molitor* (Rolff and Siva-Jothy 2002). JH produced by mated adults caused a down-regulation of PO activity, though whether this was direct or indirect is unclear. Using the same species, Rantala et al. (2003) showed that females were more attracted to the pheromones produced by males that had had their JH levels experimentally increased. The same males showed reduced PO levels and encapsulation ability, whilst lysozyme levels were unaffected by the treatment.

There have been several reports of parasitoids depressing PO activity in their hosts (e.g. Kitano et al. 1990; Beck et al. 2000; Shelby et al. 2000; Asgari et al. 2003). Teratocytes, cells that originate from the membrane of parasitoid eggs, have been shown to reduce JH esterase and ecdysone titres in *H. virescens* (Zhang et al. 1992; Dong et al. 1996). The subsequent increase in circulating JH levels may result in a reduced immune response to the parasitoid. However, decreased ecdysone titres may also play a role in inhibiting the immune response.

In larvae of the flesh fly, *Neobellieria bullata,* application of 20-hydroxyecdysone increased nodulation activity in a dose-dependent fashion, whilst application of JH decreased activity (Franssens et al. 2006). In addition, there is evidence for a synergistic effect of juvenile hormone and 20-hydroxyecdysone on hemocytes, increasing their adhesive capacity and altering their morphology (Hoshino et al. 2004). In *Drosophila,* the application of ecdysone increased the phagocytic activity of hemocytes *in vitro,* and up-regulated the expression of inducible antimicrobial peptides such as diptericin and drosomysin (Dimarcq et al. 1997). Ecdysone also caused the induction of genes expressing membrane receptors, which are involved in the recognition of microorganisms. In the yellow fever mosquito, *Anopheles gambiae* Giles, an ecdysone receptor site has been identified in the pro-PO 1 gene (Ahmed et al. 1999). Ecdysone up-regulated prophenoloxidase two hours after application and levels increased up to 24 hours later before returning to normal. Furthermore, in ecdysone-deficient *Drosophila,* the encapsulation response is severely compromised, suggesting a key role for ecdysone in this important immune response (Sorrentino et al. 2002).

To date, the evidence seems to point to either ecdysone-mediated *up-regulation* of immune function, or JH-mediated *inhibition* of immune function. However, it is unlikely that these two mechanisms are mutually exclusive. It is commonly accepted that JH can inhibit the ecdysteroid-induced expression of certain genes, such as those involved in metamorphosis (Nijhout 1994) and oogenesis (Soller et al. 1999). It is therefore possible that JH could inhibit the ecdysteroid-mediated expression of certain genes relating to immune function. This would provide an explanation for JH-induced color change, as it is ecdysone that is responsible for cuticular melanization (Curtis et al. 1984; Nijhout 1994). Furthermore, in JH-deficient *Manduca sexta* larvae, levels of dopa decarboxylase (DDC), an enzyme that plays a role in cuticular melanization, were found to be twice those found in control larvae (Hiruma and Riddiford 1993). It could be JH-mediated inhibition of PO expression by ecdysone that is responsible for the reduced PO activity in mated *Tenebrio molitor*. Mating was also found to reduce antibacterial activity in *Drosophila* males, measured as the clearance of *E. coli* from the hemocoel (McKean and Nunney 2001). Again, if mated *Drosophila* males have increased JH titres and this inhibits the ecdysone-mediated regulation of phagocytosis and the expression of inducible antimicrobial peptides such as diptericin and drosomysin, we would expect these males to have reduced antibacterial activity.

It seems that there is a clear relationship between the hormonal regulation of phase transition and immune function in insects. The high-density phase of many orthopteran and lepidopteran species is characterized by high JH titres, which are linked, directly or indirectly, to a number of the physiological and behavioral traits typical of the phenotype. Similarly, many immune function traits are either inhibited by JH or up-regulated by ecdysteroids. In terms of the DDP hypothesis, it may be that increased parasite resistance is more likely to occur in phase-polyphenic species simply as a by-product of the hormonal regulation of phase. However, it is also possible that selection for increased parasite resistance in crowds preceded the other behavioral and physiological changes associated with phase polyphenism.

Population Level Responses

Inducible, Constitutive and Density-Dependent Defenses

A key question relating to density-dependent prophylaxis is under what circumstances is its adoption favored? Defense mechanisms may be broadly defined as either inducible (plastic) or constitutive. *Inducible defenses* are those that become fully active only after some time delay (for example, it may require time for the fat body to synthesize enough antibacterial peptides to be effective). Whilst inducing these defense mechanisms may be costly, the maintenance of the ability to respond is likely to be relatively cost-free. On the other hand, *constitutive defenses* (such as those associated with the prophenoloxidase enzyme cascade) are effective immediately, but the host must pay the cost of their maintenance even if there is no parasite attack (see Lochmiller and Deerenberg 2000; Kraaijeveld et al. 2002; Schmid-Hempel 2003; Wilson 2004, for reviews of the costs of immune defense). Thus, in the absence of parasites, inducible defense is cheaper than constitutive defense. Shudo and Iwasa (2001) constructed a theoretical model to examine the circumstances favoring the evolution of inducible defense versus constitutive defense versus a combined inducible and constitutive defense. They concluded that constitutive defense is favored over inducible defense when it is relatively effective and cheap; when the initial parasite abundance is large and its growth rate is high; when the difference in the time delay between inducible and constitutive defenses is large; and when the parasite is virulent and frequently attacks. Adopting both constitutive and inducible defense is optimal if the parasite is highly virulent (abundant, high growth rate, causes severe damage) and if the inducible defense is more cost-effective than constitutive defense.

Density-dependent prophylaxis is inducible in response to population density, but it is a constitutive defense in the sense that, once adopted, it is maintained even in the absence of parasites. However, these maintenance costs are modulated by the perceived risk of them needing to be deployed (i.e. costs are assumed to increase with increasing population density). Thus, DDP represents an alternative defense strategy to purely constitutive or inducible defenses. The circumstances favoring DDP over these alternatives have yet to be formally modeled. However, it seems likely that the adoption of DDP will be favored when: population density fluctuates markedly between generations and reliably predicts infection risk; when the error associated with the perception of population density is low; when the parasite is virulent; when the delay in inducible defenses is long; and when density-dependent investment in defenses is cheaper than density-independent constitutive defense.

Population Dynamic Consequences of Density-Dependent Prophylaxis

As well as affecting the life-histories of the individuals practicing it, density-dependent prophylaxis may also impact on the stability of the host-parasite interaction. White and Wilson (1999) constructed a series of models to examine the impact of DDP on the dynamics of a simple discrete-time host-parasite model in which individuals were born either resistant or susceptible to the parasite (modeled as a micro-parasite or pathogen), and the proportion of individuals born into each host type was potentially density-dependent (characterizing DDP). They found that, under some circumstances, inclusion of a density-dependent resistant class of hosts (i.e. those practicing DDP) might stabilize inherently unstable host-parasite interactions, though greatest regulation was achieved when parasite resistance was density-independent. The stabilizing effect of DDP is enhanced when there is a cost to parasite resistance. Interestingly, the inclusion of DDP in the model favors bi-stable dynamics, in which the final outcome is determined by the initial conditions for the model and either the parasite is driven to extinction or the density-dependent resistant host population grows exponentially. Insect outbreaks, like those common to many phase polyphenic species, are observed in this model only when density-dependent resistance carries no costs, suggesting that the costs of resistance may be trivial in these species (see Discussion).

Density-Dependent Prophylaxis and Biocontrol Strategies

There is substantial interest in reducing reliance on chemical insecticides for the control of insect pests and adopting more environmentally sustainable technologies (Lomer et al. 2001). As a result, dozens of biopesticides have been, or are being developed, including some to be used against insect species that exhibit DDP (e.g. Cherry et al. 1997; Lomer et al. 2001; Thomas et al. 2001). Ongoing studies are developing an isolate of a nucleopolyhedrovirus against the African armyworm (*S. exempta*) in eastern Africa (Cherry et al. 1997), and biopesticides based on isolates of *M. anisopliae* var *acridum* have been developed for the control of the desert locust (*S. gregaria*) and other acridids throughout Africa, Australia and parts of Europe and Central/ Southern America (Lomer et al. 2001; Thomas et al. 2001).

The implications of DDP for these new biopesticides remain to be seen. Certainly, increased transmission efficiency among low-density hosts may allow parasites to persist in low-density host populations better than expected, while high-density populations may be more resistant than expected to invasion by parasites. Thus, counter-intuitively, it is possible that the efficacy and economics of biocontrol strategies may be improved by targeting low-density populations, as opposed to high-density, outbreak populations.

Discussion

Related Prophylaxis Phenomena

Although the DDP hypothesis is framed around adaptive responses to population density, the same rationale can be applied to the adaptive allocation of resources to parasite defense in relation to any cue that reliably predicts risk of parasitic infection. For example, recently there has been interest in examining seasonal variation in immune defense (e.g. Nelson 2004). Møller et al. (2003) have argued that since many parasites time their reproduction to coincide with that of their hosts, there will be strong selection on hosts to exhibit an annual peak in their ability to mount an immune response during the breeding season. As predicted, they found that in a sample of temperate bird species, between the breeding and non-breeding seasons, spleen mass decreased by 18 percent (the spleen is the main organ for B-cell differentiation and proliferation) and T-cell mediated immunity dropped by 33 percent. Whether insects exhibit similar seasonal variation in immune function remains to be determined.

The risk of becoming parasitized may be most reliably predicted by previous infection, since infection may indicate that parasites are becoming more abundant. Thus, the adaptive immune system of vertebrates may be viewed as a prophylactic immune defense mechanism. Although acquired immunity does not exist in invertebrates, this logic has been applied to the study of the insect immune system by Moret and Siva-Jothy (2003). When an insect is subjected to immune challenge, it produces an immune response that appears to persist for longer than is strictly necessary to clear the infection. Moret and Siva-Jothy argue that this is because these long-lasting immune responses provide increased resistance to later infections. They tested this idea by experimentally mimicking a primary immune insult (pre-challenge) in larvae of the mealworm beetle, *Tenebrio molitor*, with lipopolysaccharides (LPS; see Table 1) prior to exposure to spores of the entomopathogenic fungus, *Metarhizium anisopliae*. They found that these pre-challenged larvae produced a long-lasting antimicrobial response, which provided a survival benefit when the larvae were subsequently exposed to fungal infection. This result suggests that the long-lasting immune response of insects protects them from secondary challenges, and hence may serve a prophylactic function.

A similar logic may also apply to anti-predator defenses (D. Whitman, pers. comm.). Thus, crypsis and aposematism may be viewed as adaptive phenotypic responses to density-dependent changes in the risk of predation. Extrapolating further, we may also observe parallels to the seasonal prophylactic responses, discussed above, in terms of adaptive responses to the seasonal risks of predation. For example, the spring brood of the caterpillar *Nemoria arizonaria* selects, feeds on, and resembles oak catkins, whereas the summer brood selects and feeds on oak leaves, and resembles oak leaf petioles. Thus, these divergent seasonal forms may be viewed as seasonal 'prophylactic' responses to predation risk (Greene 1989). Many age-, stage- or size-related polyphenisms (e.g., Reiskind 1970) may also be viewed as pre-emptive measures to reduce predictable changes in predation risk.

Group Living and Prophylactic Immune Defense

The DDP hypothesis concerns the allocation of resources to parasite resistance mechanisms in relation to the density-dependent increase in infection risk. It is tempting to extrapolate this hypothesis to make predictions regarding the relationship between group living and investment in parasite defense. A naïve prediction might be that, across species, we would expect group-living insects to invest more in prophylactic disease resistance than solitary-living insects. This is because group-living insects will typically experience higher local densities than solitary-living ones and hence, presumably, higher *per capita* infection risk. It has long been assumed that increased parasitism is a cost of group-living. However, the evidence for it is equivocal at best (Freeland 1979; Davies et al. 1991; Coté and Poulin 1995). Moreover, a recent experimental comparative study indicated that in larval Lepidoptera, at least, solitary-feeding species appear to invest *more* in immune defense than do closely-related gregariously-feeding species reared under similar conditions (Wilson et al. 2003).

To examine this counter-intuitive result further, Wilson et al. (2003) developed a dynamic, susceptible/infected spatially-explicit model in which different degrees of host-clustering were generated to simulate the effect of group-living on infection risk. Using this model, Wilson and colleagues showed that, for a significant area of parameter space, host clustering could reduce the *per capita* infection risk. Thus, it appears that aggregating in clusters might actually reduce the probability of becoming infected by parasites, rather than increase it. The reason for this is as follows: if parasite transmission requires close proximity between infectious and susceptible hosts, then any process that increases the distance between individuals will lead to reduced parasite transmission. By increasing the *variance* in nearest-neighbor distance, host clustering increases the probability that the parasite will fail to breach the gap between the host it is infecting and the nearest susceptible hosts. Thus, the model indicates that part of the advantage of group-living in these scenarios is attributable to the fact that any disease epidemics will tend to fade out faster within populations of group-living animals than within populations of solitary ones (Wilson et al. 2003, see also Watve and Jog 1997). However, the costs and benefits of group living will depend critically on both the mode of parasite transmission and on the spatial structure of the host population (Wilson et al. 2003, see also Pie et al. 2003).

Melanism and Immunity in Insects

An intriguing aspect of the DDP hypothesis is its link with melanism in insects. Many phase polyphenic species are characterized by increased melanization of the cuticle, the adaptive value of which has been debated for many years. Two key hypotheses are that there could be potential thermoregulatory benefits of a dark cuticle (Clusella Trullas et al. 2007 and references therein) or that the conspicuous black cuticle could play a role in aposematic signalling (Iwao 1968; Wilson 2000). However, the relationship between cuticular melanization, pathogen resistance and immune function (Kunimi and Yamada 1990; Reeson et al. 1998; Barnes and Siva-Jothy 2000; Wilson et al. 2001; Cotter et al. 2004a, b) suggests that the black cuticle common in high-density phenotypes may be a by-product of the up-regulation of phenoloxidase activity in the hemolymph and cuticle; and of the strengthening role of melanin in protection against parasitoids and parasites that enter via the cuticle (e.g. entomopathogenic fungi).

As such, cuticular melanization could be seen as an immune parameter in its own right. But does cuticular melanization carry the costs expected of other immune traits? A recent study found little evidence of costs associated with cuticular melanization in the Egyptian cotton leafworm, *S. littoralis*, at least under laboratory conditions (Cotter 2004b). However, it may be that the costs of cuticular melanization are apparent only under conditions of resource competition, as has been shown for other parasite resistance traits (e.g., Moret and Schmid-Hempel 2000; Kraaijeveld and Godfray 1997; Fellowes et al. 1998). Alternatively, the costs may be associated with the increased conspicuousness to predators that is bound to occur with black larvae feeding on green foliage. Further studies examining the effects of crowding on parasite resistance in non-phase polyphenic species are required in order to elucidate the role of melanism in density-dependent prophylaxis.

Future Directions

Although evidence supporting the DDP hypothesis is slowly accumulating, the most convincing examples come from studies of larval Lepidoptera, and this taxonomic bias needs to be addressed. DDP could also apply to vertebrates and indeed a recent study found that T-cell responses were higher in bird species from areas of high population density than the same species in areas of low population density (Moller et al. 2006). As well as broadening the taxonomic scope of the studies, the mechanisms underpinning DDP need to be examined at a much finer scale. Although there is some evidence relating to the immunological mechanisms that are regulated in response to changes in population density, these have so far been investigated at a fairly crude level, and there is a need for these to be refined to examine DDP responses at the cellular and molecular genetic level. There is also a need to determine precisely how gross changes in population density are translated into neurological and hormonal signals that trigger immunological and other resistance mechanisms.

Finally, although the maintenance costs of resistance have been well characterized in some species exhibiting genetically-regulated differences in levels of constitutive resistance (e.g., replicated lines of *D. melanogaster* selected for resistance to parasitoids; see above), few studies have attempted to measure the costs of resistance in species exhibiting DDP. In the Egyptian cotton leafworm, *S. littoralis*, cuticular melanism is associated with increased resistance to an entomogenous fungus (Wilson et al. 2001; see above). In this species, the melanic phenotype is smaller and has lower hemolymph protein levels (Cotter 2004a). Thus, there do appear to be small, but detectable, costs of maintaining high levels of investment in immune function in this phase polyphenic species. In contrast, in the related African armyworm, *S. exempta*, the costs of resistance are less obvious. In this species, resistance to a range of entomopathogens (including nucleopolyhedrovirus and an ectoparasitoid) is positively associated with larval density and cuticular melanism, and the melanic, high-density phenotype has significantly higher levels of phenoloxidase activity than the non-melanic, low-density phenotype (Reeson et al. 1998; Wilson and Reeson 1998; Reeson et al. 2000). Contrary to expectation, when adult females are denied access to a carbohydrate source, moths raised under high-density conditions as larvae produce around 25% *more* eggs than those reared under low-density conditions (there was no phase difference in fecundity when the adult moths were allowed access to sucrose solution; Mensah and Gatehouse 1998). It is clear, therefore, that under laboratory conditions at least, high-density, melanic females do not appear to incur a fecundity cost to investing in immune function. Thus, a priority for future studies is to identify the costs associated with density-dependent prophylaxis.

References

Adamo, S. A. and N. M. Parsons. 2006. The emergency life-history stage and immunity in the cricket, *Gryllus texensis*. Animal Behaviour, 72: 235-244.

Ahmed, A., D. Martin, A. G. O. Manetti, S. J. Han, W. J. Lee, K. D. Mathiopoulos, H. M. Muller, F. C. Kafatos, A. Raikhel and P. T. Brey. 1999. Genomic structure and ecdysone regulation of the prophenoloxidase 1 gene in the malaria vector *Anopheles gambiae*. Proceedings of the National Academy of Sciences of the United States of America, 96: 14795–14800.

Altstein, M., O. Benaziz and Y. Gazit. 1994. Pheromone Biosynthesis Activating Neuropeptide (PBAN) and color polymorphism—an immunochemical study in *Spodoptera littoralis*. Journal of Insect Physiology, 40: 303–309.

Anderson, R. M. and R. M. May 1981. The population dynamics of microparasites and their invertebrate hosts. Philosophical Transactions of the Royal Society of London Series B-Biological Sciences, 291: 451–524.

Anderson, R. M. and R. M. May 1991. Infectious Disease of Humans: Dynamics and Control. Oxford University Press. Oxford.

Applebaum, S. W. and Y. Heifetz 1999. Density-dependent physiological phase in insects. Annual Review of Entomology, 44: 317–341.

Asgari, S., G. M. Zhang, R. Zareie and O. Schmidt 2003. A serine proteinase homolog venom protein from an endoparasitoid wasp inhibits melanization of the host hemolymph. Insect Biochemistry and Molecular Biology, 33: 1017–1024.

Barnes, A. I. and M. T. Siva-Jothy 2000. Density-dependent prophylaxis in the mealworm beetle *Tenebrio molitor* L (Coleoptera: Tenebrionidae): cuticular melanization is an indicator of investment in immunity. Proceedings of the Royal Society of London Series B-Biological Sciences, 267: 177–182.

Beck, M., U. Theopold and O. Schmidt 2000. Evidence for serine protease inhibitor activity in the ovarian calyx fluid of the endoparasitoid *Venturia canescens*. Journal of Insect Physiology, 46: 1275–1283.

Boman, H. G. and D. Hultmark 1987. Cell-Free Immunity in Insects. Annual Review of Microbiology, 41: 103–126.

Brekken, R. A. and E. H. Sage 2000. SPARC, a matricellular protein: at the crossroads of cell-matrix. Matrix Biology, 19: 569–580.

Chapman, R. F. 1998. The Insects: Structure and Function, 4th Edition. Cambridge University Press.

Cherry, A. J., M. A. Parnell, D. Grzywacz and K. A. Jones 1997. The optimization of in vivo nuclear polyhedrosis virus production in *Spodoptera exempta* (Walker) and *Spodoptera exigua* (Hubner). Journal of Invertebrate Pathology, 70: 50–58.

Clusella Trullas, S., J.H. van Wyk and J.R. Spotila. 2007. Thermal melanism in ectotherms. Journal of Thermal Biology, 32: 235–245.

Clynen, E., D. Stubbe, A. De Loof and L. Schoofs 2002. Peptide differential display: a novel approach for phase transition in locusts. Comparative Biochemistry and Physiology B-Biochemistry & Molecular Biology, 132: 107–115.

Coté, I. M. and R. Poulin 1995. Parasitism and group-size in social animals—a meta-analysis. Behavioral Ecology, 6: 159–165.

Cotter, S. C., R. S. Hails, J. S. Cory and K. Wilson 2004a. Density-dependent prophylaxis and condition-dependent immune function in Lepidopteran larvae: a multivariate approach. Journal of Animal Ecology, 73: 283–293.

Cotter, S. C., L. E. B. Kruuk and K. Wilson 2004b. Costs of resistance: genetic correlations and potential trade-offs in an insect immune system. Journal of Evolutionary Biology, 17: 421- 429.

Curtis, A. T., M. Hori, J. M. Green, W. J. Wolfgang, K. Hiruma and L. M. Riddiford 1984. Ecdysteroid regulation of the onset of cuticular melanization in allatectomized and black mutant *Manduca sexta* larvae. Journal of Insect Physiology, 30: 597–606.

D'Amico, V., J. S. Elkinton, G. Dwyer, J. P. Burand and J. P. Buonaccorsi 1996. Virus transmission in gypsy moths is not a simple mass action process. Ecology, 77: 201–206.

Davies, C. R., J. M. Ayres, C. Dye and L. M. Deane 1991. Malaria infection rate of Amazonian primates increases with body weight and group size. Functional Ecology, 5: 655–662.

Dimarcq, J., J. Imler, R. Lanot, R. Alan, B. Ezekowitz, J. A. Hoffmann, C. A. Janeway and M. Lagueux 1997. Treatment of 1(2)mbn *Drosophila* tumorous blood cells with the steroid hormone ecdysone amplifies the inducibility of antimicrobial peptide gene expression. Insect Biochemistry and Molecular Biology, 27: 877–886.

Dong, K., D. Q. Zhang and D. L. Dahlman 1996. Down-regulation of juvenile hormone esterase and arylphorin production in *Heliothis virescens* larvae parasitized by *Microplitis croceipes*. Archives of Insect Biochemistry and Physiology, 32: 237–248.

Drooz, A. 1966. Color studies of reared elm spanworm larvae and pupae. Annals of the Entomological Society of America, 59: 568–573.

Dwyer, G. 1991. The roles of density, stage, and patchiness in the transmission of an insect virus. Ecology, 72: 559–574.

Dwyer, G., J. S. Elkinton and J. P. Buonaccorsi 1997. Host heterogeneity in susceptibility and disease dynamics: Tests of a mathematical model. American Naturalist, 150: 685–707.

Fellowes, M. D. E., A. R. Kraaijeveld and H. C. J. Godfray 1998. Trade-off associated with selection for increased ability to resist parasitoid attack in *Drosophila melanogaster*. Proceedings of the Royal Society of London Series B-Biological Sciences, 265: 1553–1558.

Fescemyer, H. W. and A. M. Hammond 1988. The relationship between population density, juvenile hormone, juvenile hormone esterase and phase variation in larvae of the migrant insect, *Anticarsia gemmatalis* Hubner. Journal of Insect Physiology, 34: 29–35.

Franssens, V., G. Smagghe, G. Simonet, I. Claeys, B. Breugelmans, A. De Loof, and J. Vanden Broeck. 2006. 20-Hydroxyecdysone and juvenile hormone regulate the laminarin-induced nodulation reaction in larvae of the flesh fly, *Neobellieria bullata*. Developmental and Comparative Immunology, 30: 735–740.

Freeland, W. J. 1979. Primate social groups as biological islands. Ecology, 60: 719–728.

Götz, P. 1986. Encapsulation in arthropods. *In* M. Brehelin [ed.]. Immunity in invertebrates—cells, molecules and defense reactions. Springer Verlag, Berlin Heidelberg, pp.

Goulson, D. and J. S. Cory 1995. Responses of *Mamestra brassicae* (Lepidoptera, Noctuidae) to crowding-interactions with disease resistance, color phase and growth. Oecologia, 104: 416–423.

Greene, E. 1989. A diet-induced developmental polymorphism in a caterpillar. Science, 243: 643–646.

Gunn, A. and A. G. Gatehouse 1993. The migration syndrome in the African armyworm moth, *Spodoptera exempta*—allocation of resources to flight and reproduction. Physiological Entomology, 18: 149–159.

Hagen, S. B., O. Sorlibraten, R. A. Ims and N. G. Yoccoz. 2006. Density-dependent melanism in winter moth larvae (Lepidoptera: Geometridae): A countermeasure against parasitoids? Environmental Entomology, 35: 1249–1253.

Hamilton, W. D. 1980. Sex versus non-sex versus parasite. Oikos, 35: 282–290.

Hamilton, W. D. and M. Zuk 1982. Heritable true fitness and bright birds: a role for parasites? Science, 218: 384–387.

Hiruma, K. and L. M. Riddiford 1993. Molecular mechanisms of cuticular melanization in the tobacco hornworm, *Manduca sexta* (L.) (Lepidoptera, Sphingidae). International Journal of Insect Morphology & Embryology, 22: 103–117.

Hoshino, K., T. Hiraoka, and K. Iwabuchi. 2004. Adhesion and cytoplasmic extension of *Xylotrechus pyrrhoderus* (Coleoptera; Cerambycidae) hemocytes induced by 20-hydroxyecdysone and juvenile hormone in vitro. Applied Entomological Zoology, 39: 209–216.

Hughes, W. O. H., J. Eilenberg and J. J. Boomsma 2002. Trade-offs in group living: transmission and disease resistance in leaf-cutting ants. Proceedings of the Royal Society of London Series B-Biological Sciences, 269: 1811–1819.

Ikemoto, H. 1983. The role of juvenile hormone in the density related color variation in larvae of *Cephonodes hylas* L. (Lepidoptera, Sphingidae). Applied Entomology and Zoology, 18: 57–61.

Islam, M. S., P. Roessingh, S. J. Simpson and A. R. McCaffery 1994. Parental effects on the behavior and coloration of nymphs of the desert locust *Schistocerca gregaria*. Journal of Insect Physiology, 40: 173–181.

Iwao, S. 1968. Some effects of grouping in Lepidopterous insects. Colloques internationaux du Centre national de la recherche scientifique, 173: 185–210.

Kang L., X. Y. Chen, Y. Zhou, B. W. Liu, W. Zheng, R. Q. Li, J. Wang and J. Yu 2004. The analysis of large-scale gene expression correlated to the phase changes of the migratory locust. Proceedings of the National Academy of Sciences of the United States of America, 101: 17611–17615.

Kazimirova, M. 1992. The role of physical contact in the induction of phase polymorphism of *Mamestra brassicae* (Lepidoptera, Noctuidae). Acta Entomologica Bohemoslovaca, 89: 87–95.

Khafagi, W. E. and E. M. Hegazi 2001. Effects of juvenile hormones and precocenes on the immune response of *Spodoptera littoralis* larvae to supernumerary larvae of the solitary parasitoid, *Microplitis rufiventris* Kok. Journal of Insect Physiology, 47: 1249–1259.

Kitano, H., H. Wago and T. Arakawa 1990. Possible role of teratocytes of the gregarious parasitoid, *Cotesia* (= *Apanteles*) *glomerata* in the suppression of phenoloxidase activity in the larval host, *Pieris rapae*. Archives of Insect Biochemistry and Physiology, 13: 177–185.

Klemola, T., N. Klemola and T. Andersson. 2007. Does immune function influence population fluctuations and level of parasitism in the cyclic geometrid moth? Population Ecology, 49: 165–178.

Kraaijeveld, A. R., J. Ferrari and H. C. J. Godfray 2002. Costs of resistance in insect-parasite and insect-parasitoid interactions. Parasitology, 125: S71–S82.

Kraaijeveld, A. R. and H. C. J. Godfray 1997. Trade-off between parasitoid resistance and larval competitive ability in *Drosophila melanogaster*. Nature, 389: 278–280.

Kunimi, Y. and E. Yamada 1990. Relationship of larval phase and susceptibility of the armyworm, *Pseudaletia separata* Walker (Lepidoptera, Noctuidae) to a Nuclear Polyhedrosis Virus and a Granulosis Virus. Applied Entomology and Zoology, 25: 289–297.

Lochmiller, R. L. and C. Deerenberg 2000. Trade-offs in evolutionary immunology: just what is the cost of immunity? Oikos, 88: 87–98.

Lomer, C. J., R. P. Bateman, D. L. Johnson, J. Langewald and M. Thomas 2001. Biological control of locusts and grasshoppers. Annual Review of Entomology, 46: 667–702.

Matsumoto, S. 1981. Purification and properties of the melanization and reddish coloration hormone (MRCH) in the armyworm, *Leucania separata* (Lepidoptera). Insect Biochemistry, 11: 725–733.

Matsumoto, S., A. Isogai and A. Suzuki 1984. Isolation of the melanization and reddish coloration hormone (MRCH) in the armyworm, *Leucania separata*, from the silkworm, *Bombyx mori*. Agricultural and Biological Chemistry, 48: 2401–2403.

McCallum, H., N. Barlow and J. Hone 2001. How should pathogen transmission be modeled? Trends in Ecology & Evolution, 16: 295–300.

McKean, K. A. and L. Nunney 2001. Increased sexual activity reduces male immune function in *Drosophila melanogaster*. Proceedings of the National Academy of Sciences of the United States of America, 98: 7904–7909.

McKean, K.A., C.P. Yourth, B.P. Lazzaro and A.G. Clark 2008. The evolutionary costs of immunological maintenance and deployment. BMC Evolutionary Biology, 8.

Mensah, B. A. and A. G. Gatehouse 1998. Effect of larval phase and adult diet on fecundity and related traits in *Spodoptera exempta*. Entomologia Experimentalis et Applicata, 86: 331–336.

Mitsui, J. and Y. Kunimi 1988. Effect of larval phase on susceptibility of the armyworm, *Pseudaletia separata* Walker (Lepidoptera, Noctuidae) to an entomogeneous Deuteromycete, *Nomuraea rileyi*. Japanese Journal of Applied Entomology and Zoology, 32: 129–134.

Møller, A. P., J. Erritzoe and N. Saino 2003. Seasonal changes in immune response and parasite impact on hosts. American Naturalist, 161: 657–671.

Moller, A. P., M. Martin-Vivaldi, S. Merino and J. J. Soler. 2006. Density-dependent and geographical variation in bird immune response. Oikos, 115: 463–474.

Moore, J. 1984 Altered behavioral responses in intermediate hosts—an acanthocephalan parasite strategy. American Naturalist, 123: 572–577.

Moret, Y. and P. Schmid-Hempel 2000. Survival for immunity: The price of immune system activation for bumblebee workers. Science, 290: 1166–1168.

Moret, Y. and M. T. Siva-Jothy 2003. Adaptive innate immunity? Responsive-mode prophylaxis in the mealworm beetle, *Tenebrio molitor*. Proceedings of the Royal Society of London Series B-Biological Sciences, 270: 2475–2480.

Nelson, R. J. 2004. Seasonal immune function and sickness responses. Trends in Immunology, 25: 187–192.

Nijhout, H. F. 1994. Insect Hormones. Princeton University Press.

Nijhout, H. F. and D. E. Wheeler 1982. Juvenile hormone and the physiological basis of insect polymorphisms. Quarterly Review of Biology, 57: 109–133.

Ogura, N. 1975. Hormonal control of larval coloration in the armyworm, *Leucania separata*. Journal of Insect Physiology, 21: 559–576.

Pener, M. P. 1991. Locust phase polymorphism and its endocrine relations. Advances in Insect Physiology, 23: 1–79.

Pie, M. R., R. B. Rosengaus, D. V. and J. F. A. Traniello 2003. Nest architecture, activity pattern, worker density and the dynamics of disease transmission in social insects. Journal of Theoretical Biology, 226: 45–51.

Pie, M. R., R. B. Rosengaus, D. V. Calleri II and J. F. A. Traniello 2005. Density and disease resistance in group-living insects: do eusocial species exhibit density-dependent prophylaxis? Ethology, Ecology and Evolution, 17: 41–50.

Rahman, M. M., G. Baggerman, M. Begum, A. De Loof and M. Breuer 2003a. Purification, isolation and search for possible functions of a phase-related 6080-Da peptide from the hemolymph of the desert locust, *Schistocerca gregaria*. Physiological Entomology, 28: 39–45.

Rahman, M. M., L. Vanden Bosch, G. Baggerman, E. Clynen, K. Hens, B. Hoste, K. Meylaers, T. Vercammen, L. Schoofs, A. De Loof and M. Breuer (2002) Search for peptidic molecular markers in hemolymph of crowd-(gregarious) and isolated-reared (solitary) desert locusts, *Schistocerca gregaria*. Peptides, 23: 1907–1914.

Rahman, M. M., A. Vandingenen, M. Begum, M. Breuer, A. De Loof and R. Huybrechts 2003b. Search for phase specific genes in the brain of desert locust, *Schistocerca gregaria*

(Orthoptera : Acrididae) by differential display polymerase chain reaction. Comparative Biochemistry and Physiology. A Molecular & Integrative Physiology, 135: 221–228.

Rantala, M. J., A. Vainikka and R. Kortet 2003. The role of juvenile hormone in immune function and pheromone production trade-offs: a test of the immunocompetence handicap principle. Proceedings of the Royal Society of London Series B-Biological Sciences, 270: 2257–2261.

Reeson, A. F., K. Wilson, J. S. Cory, P. Hankard, J. M. Weeks, D. Goulson and R. S. Hails 2000. Effects of phenotypic plasticity on pathogen transmission in the field in a Lepidoptera-NPV system. Oecologia, 124: 373–380.

Reeson, A. F., K. Wilson, A. Gunn, R. S. Hails and D. Goulson 1998. Baculovirus resistance in the noctuid *Spodoptera exempta* is phenotypically plastic and responds to population density. Proceedings of the Royal Society of London Series B-Biological Sciences, 265: 1787–1791.

Reiskind, J. 1970. Multiple mimetic forms in an ant-mimicking clubionid spider. Science, 169: 587–588.

Roessingh, P., A. Bouaichi and S. J. Simpson 1998. Effects of sensory stimuli on the behavioural phase state of the desert locust, *Schistocerca gregaria*. Journal of Insect Physiology, 44: 883–893.

Roessingh, P. and S. J. Simpson 1994. The time course of behavioral phase change in nymphs of the desert locust, *Schistocerca gregaria*. Physiological Entomology, 19: 191–197.

Roff, D. A. 2002. Life History Evolution. Sinauer Associates, Sunderland, Massachusetts.

Rolff, J. and M. T. Siva-Jothy 2002. Copulation corrupts immunity: A mechanism for a cost of mating in insects. Proceedings of the National Academy of Sciences of the United States of America.

Rogers, S. M., T. Matheson, K. Sasaki, K. Kendrick, S. J. Simpson and M. Burrows 2004. Substantial changes in central nervous system neurotransmitters and neuromodulators accompany phase change in the locust. Journal of Experimental Biology, 207, 3603–3617.

Rosengaus, R. B., M. R. Guldin and J. F. A. Traniello 1998. Inhibitory effect of termite fecal pellets on fungal spore germination. Journal of Chemical Ecology, 24: 1697–1706.

Rosengaus, R. B., C. Jordan, M. L. Lefebvre and J. F. A. Traniello 1999. Pathogen alarm behavior in a termite: A new form of communication in social insects. Naturwissenschaften, 86: 544–548.

Ryder J. J., K. M. Webberley, M. Boots and R. J. Knell. 2005. Measuring the transmission dynamics of a sexually transmitted disease. Proceedings of the National Academy of Sciences (USA), 102: 15140–15143.

Sasakawa, M. 1973. The influence of continuous contact on larval color in the larger pellucid hawk moth *Cephonodes hylas* L. (Lepidoptera: Sphingidae). Applied Entomology and Zoology, 8: 198–206.

Schmid-Hempel, P. 1998. Parasites in Social Insects. Princeton University Press.

Schmid-Hempel, P. 2003. Variation in immune defence as a question of evolutionary ecology. Proceedings of the Royal Society of London Series B-Biological Sciences, 270: 357–366.

Schmid-Hempel, P. 2005. Evolutionary ecology of insect immune defenses. Annual Review of Entomology, 50: 529–551.

Shelby, K. S., O. A. Adeyeye, B. M. Okot-Kotber and B. A. Webb 2000. Parasitism-linked block of host plasma melanization. Journal of Invertebrate Pathology, 75: 218–225.

Sheldon, B. C. and S. Verhulst 1996. Ecological immunology: costly parasite defences and trade-offs in evolutionary ecology. Trends in Ecology and Evolution, 11: 317–321.

Shudo, E. and Y. Iwasa 2001. Inducible defense against pathogens and parasites: Optimal choice among multiple options. Journal of Theoretical Biology, 209: 233–247.

Simonet, G., I. Claeys, B. Breugelmans, S. Van Soest, A. De Loof and J. V. Broeck 2004. Transcript profiling of pacifastin-like peptide precursors in crowd- and isolated-reared desert locusts. Biochemical and Biophysical Research Communications, 317: 565–569.

Simpson, S. J., E. Despland, B. F. Hagele and T. Dodgson 2001. Gregarious behavior in desert locusts is evoked by touching their back legs. Proceedings of the National Academy of Sciences of the United States of America, 98: 3895–3897.

Soller, M., M. Bownes and E. Kubli 1999. Control of oocyte maturation in sexually mature *Drosophila* females. Developmental Biology, 208: 337–351.

Sorrentino, R. P., Y. Carton and S. Govind 2002. Cellular immune response to parasite infection in the *Drosophila* lymph gland is developmentally regulated. Developmental Biology, 243: 65–80.

Stearns, S. C. 1992. The Evolution of Life Histories. Oxford University Press, Oxford.

Steinhaus, E. 1958. Crowding as a possible stress factor in insect disease. Ecology, 39: 503–514.

Steinhaus, E. 1963. Insect Pathology. Academic Press, New York.

Tanaka, S. 2000. Induction of darkening by corazonins in several species of Orthoptera and their possible presence in ten insect orders. Applied Entomology and Zoology, 35: 509–517.

Tawfik, A. I. and F. Sehnal 2003. A role for ecdysteroids in the phase polymorphism of the desert locust. Physiological Entomology, 28: 19–24.

Tawfik, A. I., A. Vedrova and F. Sehnal 1999. Ecdysteroids during ovarian development and embryogenesis in solitary and gregarious *Schistocerca gregaria*. Archives of Insect Biochemistry and Physiology, 41: 134–143.

Thomas, M. B., J. Klass and S. Blanford 2001. The year of the locust. Pesticide Outlook, 11: 192–195.

Tojo, S. 1991. Variation in phase polymorphism in the common cutworm, *Spodoptera litura* (Lepidoptera, Noctuidae). Applied Entomology and Zoology, 26: 571–578.

Traniello, J. F. A., R. B. Rosengaus and K. Savoie 2002. The development of immunity in a social insect: Evidence for the group facilitation of disease resistance. Proceedings of the National Academy of Sciences of the United States of America, 99: 6838–6842.

Uvarov, B. P. 1921. A revision of the genus *Locusta* (L.) (= *Pachytylus* Fieb.), with a new theory as to the periodicity and migration of locusts. Bulletin of Entomological Research, 12: 135–163.

Uvarov, B. P. 1966. Grasshoppers and Locusts. Vol. I. Cambridge University Press, London. 481 pp.

van Valen, L. 1973. A new evolutionary law. Evolutionary Theory, 1: 1–30.

Vilaplana, L., E.M. Redman, K. Wilson and J.S. Cory 2008. Density-related variation in vertical transmission of a virus in the African armyworm. Oecologia, 155: 237–246.

Watve, M. G. and M. M. Jog 1997. Epidemic diseases and host clustering: An optimum cluster size ensures maximum survival. Journal of Theoretical Biology, 184: 167–171.

Wedekind-Hirschberger, S., S. Sickold and A. Dorn 1999. Expression of phase-specific hemolymph polypeptides in a laboratory strain and field catches of *Schistocerca gregaria*. Journal of Insect Physiology, 45: 1097–1103.

White, K. A. J. and K. Wilson 1999. Modeling density-dependent resistance in insect-pathogen interactions. Theoretical Population Biology, 56: 163–181.

Wilson, K. 2000. How the locust got its stripes: the evolution of density-dependent aposematism. Trends in Ecology & Evolution, 15: 88–90.

Wilson, K. 2004. Evolutionary ecology of insect host-parasite interactions: an ecological immunology perspective. *In*: Rolff, J. and Fellowes, M. D. E. [eds]. Evolutionary Ecology of Insects. CABI (In press).

Wilson, K., S. C. Cotter, A. F. Reeson and J. K. Pell 2001. Melanism and disease resistance in insects. Ecology Letters, 4: 637–649.

Wilson, K., R. Knell, M. Boots and J. Koch-Osborne 2003. Group living and investment in immune defence: an interspecific analysis. Journal of Animal Ecology, 72: 133–143.

Wilson, K. and A. F. Reeson 1998. Density-dependent prophylaxis: Evidence from Lepidoptera-baculovirus interactions? Ecological Entomology, 23: 100–101.

Wilson, K., M. B. Thomas, S. Blanford, M. Doggett, S. J. Simpson and S. L. Moore 2002. Coping with crowds: Density-dependent disease resistance in desert locusts. Proceedings of the National Academy of Sciences of the United States of America, 99: 5471–5475.

Woods, S. A. and J. S. Elkinton 1987. Bimodal patterns of mortality from nuclear polyhedrosis virus in gypsy moth (*Lymantria dispar*) populations. Journal of Invertebrate Pathology, 50: 151–157.

Yagi, S. and K. Kuramochi 1976. The role of juvenile hormone in larval duration and spermiogenesis in relation to phase variation in the tobacco cutworm, *Spodoptera litura* (Lepidoptera: Noctuidae). Applied Entomology and Zoology, 11: 133–138.

Yerushalmi, Y. and M. P. Pener 2001. The response of a homochrome grasshopper, *Oedipoda miniata,* to the dark-color-inducing neurohormone (DCIN) of locusts. Journal of Insect Physiology, 47: 593–597.

Zhang, D. Q., D. L. Dahlman and D. B. Gelman 1992. Juvenile hormone esterase activity and ecdysteroid titer in *Heliothis virescens* larvae injected with *Microplitis croceipes* teratocytes. Archives of Insect Biochemistry and Physiology, 20: 231–242.

Zitnan D., S. I., Zitnanova I., Park Y. and Adams M. E. 2002. Multiple peptides released from the nervous system, gut and Inka cells control processes associated with insect ecdysis. *In* Abstracts of the Twenty-First Conference of European Endocrinologists, Bonn, Germany.

7

Natural Enemy-induced Plasticity in Plants and Animals

Douglas W. Whitman[1] and Leon Blaustein[2]

[1]*Department of Biological Sciences, Illinois State University, Normal, IL, USA 61790-4120, E-mail: dwwhitm@ilstu.edu*

[2]*Community Ecology Laboratory, Institute of Evolution, Department of Evolutionary and Environmental Biology, Faculty of Sciences, University of Haifa, Haifa 31905, Israel, E-mail: leon@research.haifa.ac.il*

Abstract

Natural enemy-induced plasticity in prey defense phenotype appears to be a generalized adaptation syndrome of considerable importance to ecology and evolution. Defense plasticity is evident across a wide range of animal, plant, and microbe taxa, and threatened prey can alter a wide range of morphological, physiological, behavioral, and life-history traits. By switching phenotypes in response to threats, prey can adaptively modify their relationships with natural enemies. Natural enemies can subsequently respond with phenotypic plasticity to prey plasticity. New prey phenotypes can alter biotic habitats, with sometimes profound consequences for communities. By changing phenotypes, prey come under new selective regimes, altering their evolutionary trajectories. The evolution and expression of defense plasticity are potentially influenced by nearly all aspects of the prey's physiology and ecology. Defense plasticity is expected to evolve when benefits outweigh costs; but, taxonomic, physiological, and ecological constraints may hinder evolution of plastic response. Natural enemy-induced phenotypic plasticity has a genetic basis and can evolve. Researchers are beginning to understand the mechanisms underlying natural enemy-induced plasticity in prey. This includes identification of the specific chemical, visual, and mechanical elicitors from natural enemies that cue prey, and the elicitor receptors, signal transduction pathways, and effector systems in prey that ultimately produce the altered phenotype. Signal transduction and effector systems vary from simple to complex, and can involve translational, transcriptional, enzymatic, hormonal, and neural regulation. Natural enemy induction of prey is important for agriculture and medicine. New technologies offer possibilities to manipulate prey induction systems for economic gain. This is a rapidly evolving field of science and many questions and hypotheses need to be addressed.

Introduction

We now know that prey can respond adaptively to predators via phenotypic plasticity[1]. When confronted with actual or potential predation, individual prey can alter their morphology, physiology, behavior and life history characteristics (Karban and Baldwin 1997, Agrawal et al. 1999, Tollrian and Harvell 1999, Werner and Peacor 2003, Benard 2004, Kusch et al. 2005, Mondor et al. 2008). Predator-induced plasticity is expressed across a wide array of taxonomic groups, including bacteria, fungi, protozoa, algae, plants, and numerous invertebrate, and vertebrate taxa, and is found in diverse habitats, including marine, aquatic, and terrestrial ecosystems (Kush 1999, Tollrian and Harvell 1999, Pavia and Toth 2000, Radman et al. 2003, Benard 2004, Duquette et al. 2005, Thoms et al. 2006, Vaughn 2007, Rohde and Wahl 2008). Many induced defenses are beneficial for prey in that they result in reduced attack by predators or increased prey survival (Traw et al. 2007, McCoy and Bolker 2008), and many appear to have been shaped by natural selection. Given its wide taxonomic and ecological expression, predator-induced plasticity appears to be a generalized adaptation syndrome of considerable importance in ecology and evolution.

Understanding predator-induced plasticity in prey is important for understanding the biology of individuals, populations, and species, and ecological relationships, such as predator-prey population dynamics, indirect effects, trophic webs, and community structure. Plasticity studies are also critical for understanding the production and maintenance of phenotypic, genetic, and interaction diversity, speciation, evolution, and co-evolution at the predator-prey interface (Agrawal 2001, Price 2002, Schlichting 2004, Strauss and Irwin 2004, Van Zandt and Agrawal 2004). Predator-induced changes in prey can have large ecosystem consequences (Agrawal 2005, Miner et al. 2005, Ohgushi 2005, Ohgushi et al. 2007). By inducing new prey phenotypes, predators alter biotic habitats, increasing habitat diversity, and alter interaction linkages, initiating new interactive cascades (Werner and Peacor 2003, Ohgushi 2005) that may favor instability (Abrams and Matsuda 1997) or stability (Vos et al. 2004, Miner et al. 2005). Predators induce changes in prey that can alter the availability of resources for, or otherwise influence, other species, and in this way, can function as indirect keystone species via trait-mediated indirect effects (Schmitz 2003,

[1] For the purpose of this chapter, we define "prey" as an organism whose live tissues are consumed by another organism, and "predator" as an organism that consumes prey. Hence, a prey may be a microbe, a plant, or an animal, and a predator may be an herbivore, parasite, parasitoid, pathogen, or a traditional predator.

Agrawal 2005, Ohgushi et al. 2007). For example, *Melanoplus femurrubrum* grasshoppers switch from feeding primarily on grasses to feeding primarily on forbs in the presence of predatory spiders (Schmitz 1998), a trait-mediated indirect effect. Such predator-induced defense could presumably impact both prey physiology (ingested nutrients and allelochemicals), and grasshopper, spider, and grass vs. forb species abundance, diversity, and ecosystem properties such as N-cycling (Schmitz 2006, 2008). If spiders switch from feeding on grasshoppers to other prey, this could initiate altered trophic cascades. Hence, a predator-induced response in a prey may alter community structure (Strauss and Irwin 2004, Ohgushi 2005).

Predator-induced phenotypic plasticity is currently an active area for physiology and biochemistry (Vasconsuelo and Boland 2007, Wittkopp 2007), development and morphology (Kishida and Nishimura 2006, Ishikawa et al. 2008), behavior (Bruce and Herberstein 2006, Korb and Fuchs 2006, Chapman et al. 2008), ecology (Agawala et al. 2006, Heil 2008), chemical ecology (Heil et al. 2008, Opitz et al. 2008), signal transduction (Maffei et al. 2007a, b, Pieterse and Dicke 2007), life history (Relyea 2007, Benard and McCauley 2008, Mikolajewski et al. 2008), genetic (Bakker et al. 2008, Gomez-Mestre et al. 2008), evolution (Hammill et al. 2008, Urban 2008), and modeling and theoretical research (Morris et al. 2006, Rinke et al. 2008). Because predator-induced phenotypic plasticity influences the relationships among agricultural plants, plant pests, and their natural enemies, and also between pathogens, disease vectors, and animal hosts, understanding phenotypic plasticity has high practical significance for agriculture and medicine (see Dicke and Grostal 2001, Dean and De Moraes 2006). Finally, the broad taxonomic, ecological, and mechanistic expression of predator-induced plasticity allows unlimited and flexible approaches and experimental designs to test numerous evolutionary and mechanistic hypotheses and evolutionary models. Many inducible traits are easily studied in the lab or field. As such, predator-induction serves as an ideal focus area for understanding phenotypic plasticity and predator-prey and multi-trophic interactions.

Evolution of Predator-induced Plasticity

What are the species characteristics and ecological parameters that select for plastic vs. canalized anti-predator defenses, and what form will plasticity take? Overall, plasticity should evolve when plasticity has higher fitness than canalization. But what factors might favor such an outcome?

(1) Strong selective pressure from predators (Riessen 1992). Without selective pressure from predators, no defense (one type of canalization) should evolve.
(2) Fluctuating predatory pressure in either degree or type, over time or space (Riessen 1992, Blaustein 1999, Van Buskirk 2002, Berrigan and Scheiner 2004, Karban and Nagasaka 2004). When predator level and type are uniform over time and space, there would be no selection or advantage for plastic defense.
(3) Induced defense is costly (Riessen 1992, Zangerl 2003). If the induced defense has little cost, then it should evolve to be constituently expressed.
(4) Induced defense is effective. If there is no benefit for the induced defense (but it has a cost), then it should be selected against. Hence, there should be overall fitness benefits for expression of the plastic defense when the predator is present, and costs when the predator is absent, and vice versa (Harvell 1990, Järemo et al. 1999, Zangerl 2003).
(5) Genetic variation for degree of plasticity. Without genetic variation, selection cannot occur (Agrawal et al. 2002, Berrigan and Scheiner 2004, Van Kleunen and Fischer 2005).
(6) Predators provide detectible and reliable cues (Riessen 1992, Blaustein 1999, Gabriel et al. 2005). If prey cannot detect the predator or detect cues that predict increased risk of predation, then they cannot respond (Karban et al. 1999, Van de Meutter et al. 2005), which would favor constitutive defenses (Berrigan and Scheiner 2004).

How often are these criteria met by species that have evolved predator-induced plasticity? Criteria 1 and 2 are nearly always met, because virtually all prey have multiple natural enemies, and because predatory pressure almost always varies in type and intensity, in space and time (Sih et al. 2000, Zangerl 2003). Indeed, prey can incur severe fitness loss from predators; in some cases, entire populations have been exterminated by natural enemies (Day 1989, Schoener et al. 2001). Moreover, a defense effective against one natural enemy may have little impact against another, or may actually benefit the other natural enemy (Salazar and Whitman 2001, Cipollini et al. 2004, Hoverman et al. 2005), creating a selection disadvantage against reliance on a single constitutive defense. In some cases, the degree of defense plasticity exhibited by different prey species correlates positively to the degree of predator variability in their natural habitat (Van Buskirk 2002) but not always (Karban and Nagasaka 2004, Van Buskirk and Arioli 2005). Note that for variation in space to favor evolution of plasticity, there would need to be suitable movement of individuals among sites with high and low predator-pressure (Kingsolver et al. 2002). However, gene flow between populations with different predator patterns could dampen evolution of predator-induction plasticity (Karban and Nagasaka 2004, Van Buskirk and Arioli 2005).

Induced defenses are often costly (Cipollini et al. 2003, Tian et al. 2003, Zavala et al. 2004, Heijari 2005, Walls et al. 2005), and costs can include both the cost to possess the inducible defense, and the cost of expressing the defense (DeWitt et al. 1998, Agrawal et al. 2002b, Cipollini et al. 2003, Zangerl 2003), including genetic costs (Agrawal et al. 2002a, Hoballah et al. 2004), ecological costs (Relyea and Auld 2004, Hoverman et al. 2005), and opportunity costs (Baldwin and Hamilton 2000, Zangerl 2003), expressed as direct or indirect costs (Järemo et al. 1999). However, induced defenses can be eminently effective (Benard 2004). Altered prey phenotypes have often, but not always (Nykanen and Koricheva 2004), been shown to harm predators (Kessler and Baldwin 2001, Traw and Dawson 2002, Viswanathan et al. 2005, Dugravot and Thibout 2006), suffer less damage (Traw and Dawson 2002, Heijari 2005), have greater survival (Dahl and Peckarsky 2002, Kishida and Nishimura 2005, Teplitsky et al. 2005), and leave more offspring (Heil 2004, Dietrich et al. 2005) than non-altered prey, when in the presence of predators (Karban and Baldwin 1997, Kolar and Wahl 1998). A major benefit of inducible vs. constitutive defenses may be in targeting specific predators and avoiding situations when constitutive defense to one natural enemy increases susceptibility to another (Salazar and Whitman 2001, Zangerl 2003).

Some studies have failed to find a cost or benefit to predator-induced defense (Johansson 2002, Zangerl 2003, Steiner and Van Buskirk 2008), and some authors suggest that costs are rare (DeWitt and Scheiner 2004). Furthermore, plasticity can be maladaptive (Langerhans and DeWitt 2002). However, measuring costs is problematic (Agrawal et al. 2002, Zangerl 2003, Agrawal 2005) and both costs and benefits might be expressed in any number of forms, which may or may not be identified or measured accurately (Strauss et al. 2002, Cipollini et al. 2003, Vonesh and Bolker 2005). For example, costs or benefits may extend to relatives (inclusive fitness) or into the next generation via maternal effects (Agrawal et al. 1999, Agrawal 2001, Mondor et al. 2005). Furthermore, both costs and benefits vary with ontogeny and current physiological and environmental conditions (Heil 2002, Dietrich et al. 2005), and hence their ratio changes constantly, making them hard to measure. Finally, selection for or against plasticity may be periodic or rare, but extremely strong, such that directional selection occurs, but is unlikely to be detected by short-term studies.

Predator-induced phenotypic plasticity has a genetic basis, varies among populations (Tollrian and Harvell 1999, Agrawal et al. 2002, Relyea 2002), and the degree and type of plasticity can be inherited (Relyea 2005). A capability for phenotypic plasticity or degree of response sometimes corresponds with predator sympatry, but not allopatry (Kats and Dill 1998, Stoks et al. 2003), suggesting its genetic basis and evolvability.

Finally, predator-induced plasticity is linked to detectible and specific predator-released cues or environmental stimuli that allow prey to reliably predict predatory risk, and respond with appropriate types and levels of plasticity (Karban et al. 1999, Van de Meutter et al. 2005), but see Langerhans and DeWitt (2002). Hence, the criteria given above for the evolution of predator-induced plasticity in prey are generally met in those species demonstrating predator-induced plasticity. However, many of these criteria are often present in non-plastic species. As such, there is a need for comprehensive comparisons of populations or species with vs. without plasticity.

Overall, facultative defenses are hypothesized to afford numerous benefits over constitutive (canalized) defenses (Agrawal and Karban 1999, Zangerl 2003) in that they can be flexible, specific (targeted), modulated, and not produced if not needed, and thus save energy and reduce trade-off costs. They can vary with ontogeny and in many cases are reversible (Relyea 2003). In this way, inducible defenses provide prey with greater flexibility to meet predatory challenges. Having a capability for predator-induced plasticity might expand niche breadth and geographic range, and preadapt organisms for changing environments such as new predators or dispersal into new habitats (Agrawal 2001, Stoks and McPeek 2003). However, it is assumed that phenotypic plasticity carries costs, in that it often requires energy to change or to maintain new phenotypes, or the induced forms suffer from constraints or trade-offs, including genetic or ecological costs, such as smaller body size, susceptibility to alternative natural enemies, and lower fitness in predator-free habitats (Dixon and Agarwala 1999, Hoverman et al. 2005, Relyea and Auld 2005).

Holistic View

The evolution of defense plasticity is not simply the interaction between a single defensive trait and a single natural enemy (Stinchcombe and Rausher

2002, Agrawal 2005, Hoverman et al. 2005, Ohgushi 2005). Plasticity evolution and expression are potentially influenced by nearly all aspects of a prey's physiology and ecology. The evolution of plasticity may be constrained or encouraged by pleiotropy and genetic correlations with other traits (Agrawal et al. 2002, Relyea 2002). Within a single individual, different traits are positively or negatively, directly or indirectly linked, and an induced change in one trait almost always requires or causes changes in other traits (Stratmann 2003, Zangerl 2003, Benard 2004, Teplitsky et al. 2005). For example, morphological and chemical defenses require energy, which could otherwise be used for growth, development, and reproduction (Han and Lincoln 1995, Traw and Feeny 2008, Kurashige and Agrawal 2005). Different inducible biochemical pathways may be antagonistic or synergistic (Li et al. 2004, Arimura et al. 2005, Huang et al. 2005). High titers of induced chemical defenses could disrupt other physiological pathways via autotoxicity (Baldwin and Callahan 1993, Gog et al. 2005). Defense allocation to one organ or structure may leave another less defended (Orians 2005). A strong immune response might carry costs, such as self-damage (Sadd and Siva-Jothy 2006). Immune response may be incompatible with predator-induced stress, because stress creates antigens, which require autoimmune suppression or immunoredistribution (Kurtz et al. 2000). Induced immunity could influence mating success and longevity (Jacot et al. 2004, 2005). Up-regulation of toxins in a single leaf may cause herbivore vectors to move, and thereby distribute a plant pathogen throughout the plant. Hiding by prey limits feeding, which alters nutrition, which may lower prey growth rate (Danner and Joern 2004, Wilder and Rypstra 2004, Preisser et al. 2005), and reduced activity or exposure could reduce thermoregulation or mating (Persons et al. 2002, Wilder and Rypstra 2004, Taylor et al. 2005), all of which could impact prey fecundity. Predator induced changes in life history can alter behavior or morphology (Benard 2004, Vonesh and Bolker 2005), such as when predators induce *Scytodes* spiders to hatch earlier, but smaller (Li 2002). Furthermore, taxonomic (i.e., physiological, morphological, functional, and ontogenetic) constraints limit or encourage the evolution of certain defenses (Orians 2005). For example, plants are limited in their ability to evolve behavioral defenses, and swimming or flying prey are constrained from evolving heavy morphological defenses. Hence, diverse tradeoffs and constraints influence the evolution of induced plasticity, yet are not always addressed by researchers.

The evolution of a specific plastic defense is likewise influenced by numerous other community members and their multitudinous interactions, producing negative genetic and ecological correlations that may limit the evolution of defense (Michimae and Hangui 2008). In wild radish, both petal pigment color and chemical defense are heritable, but are negatively genetically correlated such that individuals that express high levels of induced defensive glucosinolates also express a less attractive flower color for pollinators and vice versa (Strauss and Irwin 2004, Strauss et al. 2004). In other plants, chemical defenses leak into the nectar, deterring and in some cases killing pollinators (Barker 1978). Hence, pollinators may influence the evolution of induced defenses, and vice versa (Irwin et al. 2003, Strauss and Irwin 2004, Strauss et al. 2004). Defense may also conflict with competition (Cipollini 2004, Relyea and Auld 2004, 2005, Kurashige and Agrawal 2005, McGuire and Agrawal 2005). In contrast, positive genetic and ecological correlations can presumably favor the evolution of inducible defenses, such as when increased glucosinolates improve both defense and competition in *Brassica* (Siemens et al. 2002), or when prey gain a competitive advantage when predators switch to other, less defended prey.

Different predator species may exert antagonistic selection pressures for plasticity, and defense against one predator may leave prey more susceptible to the other. For example, some freshwater snails respond to crushing predators by strengthening their shells, but against invasion-predators by increasing shell volume, which allows defense-retraction (Hoverman et al. 2005). The two defenses are mutually exclusive. Either plastic response might be disadvantageous when both predators are present or when predator type changes rapidly. In tadpoles, dragonfly predators induce deeper tails, which reduces dragonfly predation, but makes the prey more susceptible to fish predators (Wilson et al. 2005). Likewise, increasing the level of induced direct or indirect chemical defense may deter generalist natural enemies, but might attract specialists that have evolved to orient to the defense chemical (Agrawal et al. 1999, Landholt et al. 1999, Carroll et al. 2006). For example, in mustard plants, glucosinulates are both induced and constitutively produced (Traw 2002; Strauss et al. 2004). These substances are deterrent and deadly to some non-adapted herbivores, but are attractive, and thus beneficial to some specialists (Agrawal and Sherriffs 2001, Aliabadi and Whitman 2001, Haribal and Renwick 2005). Similarly, many true bugs effectively repel some predators by ejecting noxious defensive sections (Aldrich 1988, Staples et al. 2002). However, specialist natural enemies use these same chemicals to locate the bugs (Aldrich 1994). In these cases, evolution of both constitutive defense and plasticity are presumably under bi-directional selection; it is unclear if the prey should evolve a greater or a lesser plastic response (Giamoustaris and Mithen 1995).

Another example is seen in plants that are fed on by both aphids and chewing insects. Aphid presence may benefit plants via attending ants, which attack various herbivores (Stadler and Dixon 2005). Normally, herbivore-induced chemical defenses might benefit a plant. However, higher chemical defense in plants may eliminate both aphids and ants (Mooney and Agrawal 2008), leaving the plants susceptible to other herbivore species. Hence, the evolution of both constitutive and induced defense in this case might hinge on the interaction of ants and their prey, or on aphids and their predators. This web becomes even more tangled in those aphid species where plant defense chemicals are sequestered and used by aphids as alarm pheromones or for their own defense (Dawson et al. 1987, Bowers 1990).

As suggested above, induced defenses in prey may serve to benefit or harm the natural enemies of predators (Zangerl 2003, Cory and Hoover 2006, Ode 2006). For herbivores that obtain their defense from their diet, defense titer, and thus defense efficacy is a plastic, environment-dependent trait (availability of plant toxins) (Jones et al. 1988, Whitman 1988, Blum et al. 1990, Bowers 1990). Thus, herbivore-carnivore interactions could influence the evolution of herbivore-induced plasticity in plants, when increased plastic response (increased levels of secondary compounds in plants) benefits the herbivore (Aliabadi et al. 2002). In this case, induction of chemical defenses can ultimately harm the plant, if it increases the herbivore's defense (Martinsen et al. 1998, Schappert and Shore 1999, Müller et al. 2006). However, consider when the major plant herbivore is also the primary pollinator, such as with the caltalpa sphinx moth, which acquires its iridoid glycoside chemical defense from the plant, but which also pollinates the plant (Stephenson 1982a, Bowers and Puttick 1986, Whitman 1988). Catalpa trees are defended by both toxic iridoid glycosides and extrafloral nectaries, the latter of which are induced by herbivore feeding and feed parasitoids of caterpillars (Stephenson 1982b, Whitman 1994). Given that plants benefit from pollination, should caterpillar feeding induce greater levels of defense in catalpa or not?

In tomato, damage from caterpillar feeding induces both non-volatile toxins and parasitoid-attractive volatiles. The former harm some herbivores; however, *Hyposoter* and *Cotesia* wasp parasitoids attracted by the induced volatiles develop poorly on *Spodoptera exigua* caterpillars reared on the induced plants (Thaler 1999, Rodriguez-Saona et al. 2005). Should the plant evolve more indirect or direct defense? Predicting how defense plasticity should evolve becomes very complex, when multiple interactions are included in the analysis (DeWitt et al. 1999, Tollrian and Dodson 1999, Siemens et al. 2002).

Finally, the evolution and expression of induced plasticity cannot be divorced from social, ecological, and life history constraints, such as time horizons. For example, populations in nutritionally stressed environments or those near the end of their season may evolve so as to ignore predator cues (Walker and Rypstra 2003, Sih 2004).

In sum, all organismal traits, including type and level of plasticity, should evolve to reach a compromise level that represents the highest fitness value considering all selective factors. The problem is that we will probably never know what these myriad factors are, nor can we determine all the costs and benefits of all direct and indirect consequences of induced plasticity, including compensatory responses (Zangerl 2003, Vonesh and Bolker 2005, Morris et al. 2006), transgenerational effects (Mousseau and Fox 1998, Agrawal et al. 1999), and inclusive fitness. In addition, selection pressures constantly change in time and space, and are thus difficult to study: a specific trait might be favored even if it is only selected for once every ten years. It is impossible to identify all the traits that are changed during induction, let alone know all the selective factors acting on a single plastic trait (Relyea 2004). However, precise and thorough measurements of multi-generational fitness approximate the evolutionary relevant sum of total costs and benefits, as opposed to measurements of single component of fitness such as growth rate or percent leaf eaten (see Sih 2004). On the other hand, focusing on individual traits may facilitate understanding of plasticity as an adaptation (DeWitt and Scheiner 2004).

Adding to the above complexity is the fact that in nature predators influence prey phenotype via multiple, interactive mechanisms, including selective culling of certain phenotypes during a single generation, selection over multiple generations, direct induction of traits, and indirect induction via processes such as reduced competition or altering abundance of other predator species (Relyea 2002b). Predators themselves are plastic, and such plasticity also influences the evolution of prey response (Agrawal 2001). For example, virulence or lethality of a natural enemy can vary with their environment or that of their parents (Taylor et al. 2006, Tseng 2006). Studies of predator-induced plasticity must accommodate and differentiate among these different effects.

In the same way that an entire community might influence trait evolution, a plastic trait might influence an entire community (Miner et al. 2005, Ohgushi 2005). Predator induced changes in prey have direct and indirect, multi-dimensional rippling effects throughout the community (Trussell et al. 2003, Werner and Peacor 2003, Strauss and Irwin 2004, Van Zandt and Agrawal 2004). Effects of plasticity can be separated in space and time, and can feed back to influence the original induced defense, such as when earlier defense induction or phenotypic change caused by predator species A influences subsequent feeding and success by predator species B on a prey (Wise and Weinberg 2002, Viswanathan et al. 2005). For example, root-feeding insects can induce systemic responses in plants that influence foliage feeders and vice versa (van Dam et al. 2003, Bezemer and van Dam 2005). Previous defense induction can either benefit or harm subsequent predators (Karban and Nagasaka 2004, Kessler and Baldwin 2004, De Vos et al. 2006, Stout et al. 2006). In some cases, predation induces not resistance, but increased susceptibility in prey (Nykänen and Koricheva 2004, Cui et al. 2005, Rodriguez-Saona et al. 2005). In willows, attack by a single herbivore, the spittle bug *Aphrophora pectoralis* induces compensatory shoot growth in the plant, which subsequently influences the abundance of more than 30 species of willow-associated insects (Nozawa and Ohgushi 2002, Nakamura and Ohgushi 2003, Ohgushi 2005). Presumably, these community changes feedback to influence the spittle bug. Likewise, plant species abundance might be altered when pollinators switch plants in response to presence of predators (Reader et al. 2006). Other ways that induced defenses could alter community structure, are when induced chemical defenses in plants alter population levels or distribution of generalist herbivores or carnivores (Underwood et al. 2005), or when induced extrafloral nectars attract generalist carnivores to the community, which then attack herbivores on other plant species (Rodriguez-Saona and Thaler 2005b). Predators may switch prey in response to induced prey defenses, which may eliminate competitors of the original prey (Van Dam et al. 2001). In contrast, reallocation of resources from competition to defense may benefit competitors (Van Dam and Baldwin 2001). Induced defenses have the capability to regulate populations because they operate in a density-dependent manner (Gardner and Agrawal 2002). On a longer time scale, the dramatic multi-year cycling of some herbivore populations might be due to population-wide defensive plasticity in trees (Haukioja 1991). Likewise, by creating new niches and dampening population fluctuation, plasticity fosters species diversity and stability, and may reduce local extinction (Miner et al. 2005). Such complexity in nature requires that we consider multiple effects simultaneously, and begs for more ecological realism in plasticity research (Miner et al. 2005).

Defense Plasticity as an Evolutionary Agent

Altered phenotypes, including the altered interactions among traits, places an individual in a different niche, with different selective forces, which can alter evolutionary trajectories for induced populations. For example, predator-induced changes in body form or mass, change mass and surface-to-volume relationships, altering metabolism, nutrition, thermoregulation, water-balance, osmoregulation, locomotion, and dispersal (Schmidt-Nielsen 1984, Padilla 1998). Defensive changes in color can influence body temperature via radiative absorption, and hence influence thermal-dependent metabolism, growth and development (Heinrich 1993). Predator-induced changes in prey might affect social and ecological relationships, such as mating, competition and resource defense, mutualism, dispersal, diet, behavioral time budgets, and predator type and intensity (Peters 1983, Heinrich 1993, Strauss and Irwin 2004). Escape dispersal or altered feeding patterns or sites exposes prey to new dangers (Ward et al. 1998, Kaitaniemi et al. 2004, Baverstock et al. 2005, Cory and Hoover 2006, Von Son and Thiel 2006). Plasticity can create new niches (niche construction), which subsequently alters selection pressures (Day et al. 2003, Donohue 2005). A predator-induced shift into a new refuge might initiate evolutionary changes such as when *Anolis sagrei* lizards shift from ground to trees following the introduction of a new ground predator into their community. In the new arboreal habitat, we would expect rapid evolution of shorter legs (Losos et al. 2000, 2004, 2006).

Induced defense might also influence the evolution of sexual vs. asexual reproduction, genome size (ploidy level) (Nuismer and Otto 2005), and macroevolution. Adaptive plasticity allows prey populations to survive environmental change and may allow prey to expand their geographic range or disperse into new habitats, and thus foster speciation (Agrawal 2001, Schlichting 2004). Predator-induced changes, including altered behavior, might induce rapid evolutionary jumps and speciation, via recurrence, genetic accommodation and assimilation (ten Cate 2000, West-Eberhard 2003). Conversely, phenotypic plasticity might inhibit macroevolution (de Jong 2005).

What Type of Plasticity Should Evolve?

Prey express a staggering diversity of predator-induced defenses (see below), and hence, presumably have "options" as to what type of plastic defense they can evolve. Although all inductions are ultimately physiological, plastic responses can manifest as chemical, physiological, morphological, developmental, behavioral, and/or as life-history changes. But which of many possible induced defenses should evolve or be expressed at any one time (De Meester et al. 1999, Berrigan and Scheiner 2004)? Each type of defense has advantages and disadvantages, including relative speed of induction, effectiveness, cost, specificity, breadth, persistence, reversibility, etc., which must be balanced over multiple natural enemies, and the general ecology and physiology of the organism over its lifetime (Zangerl 2003). And, as previously noted, genetic, taxonomic, or functional constraints limit or encourage the evolution of certain induced defenses in certain clades. For example, because they lack muscles, plants are expected to evolve chemical or morphological plasticity, whereas animals can evolve behavioral plasticity. In addition, defense possibilities, benefits and costs change with ontogeny (Boege and Marquis 2005). Some plastic defenses may be restricted to developmental windows (Relyea 2003b, Brodin et al. 2006). Transgenerational responses may circumvent ontogenetic constraints by waiting to apply the plastic defense to an inducible stage in the next generation (Agrawal et al. 1999). Different induced defenses may be antagonistic or synergistic (Baldwin and Preston 1999), such as direct and indirect induced-defenses, which may interfere with one another. However, overall, optimality theory predicts that the specific defense response should match the specific threat and type of natural enemy. Thus, we would expect induction speed to correspond with rate of change in predation risk (Von Buskirk 2002) and speed of damage, but, in general, rapid response may be better than a slow response (Padilla and Adolph 1996, Gabriel 1999, Duquette et al. 2005). Induced defenses vary in response speed from immediate to generational, along the approximate order of: *behavior > physiological/chemical > development/morphology > life history > transgenerational effects*. Likewise, the type of induced defense should change with different threats, and such flexibility is often observed in prey responses (Relyea 2003a, Kishida and Nishimura 2005). Type of induced defense should also change with ontogeny (Relyea 2003b, Boege and Marquis 2005). For example, young *Hyla* tree frog tadpoles respond to dragonfly predator cues by hiding, but older (larger) tadpoles respond morphologically with deeper tails (Relyea 2003b).

Reversible plasticity should evolve when predator impact is short compared to prey life-span, and induction should be rapid, but recovery slow (Mattiacci et al. 2001, Gabriel 2005, Gabriel et al. 2005). Indeed, behavioral, chemical, and physiological induced chemical defenses can be highly reversible (Fatouros et al. 2005), but not always (Van de Meutter et al.

2005). For example, in plants, induced defenses often remain only as long as stimuli or damage continues (Schultz 1988, Choh and Takabayashi 2006). However, even predator-induced morphology and life-history trajectory can be reversed when predator threat subsides (Young and Okello 1998, Relyea 2003b, Gabriel et al. 2005, Kishida and Nishimura 2006). In contrast, induced defenses should be persistent when predator threat is typically persistent, when there is low cost to maintain the defense, or when persistent morphology does not permit reversal (Relyea 2003b). Memory to natural enemies is an induced component of defense of extreme adaptive value. It can involve neural memory, such as when nymphal damselflies learn to recognize chemical cues from fish predators (Wisenden et al. 1997), or non-neural memory, such as immune response (Frost 1999, Schmid-Hempel 2005). For example, some trees can "remember" herbivore loads from the previous season (Young and Okello 1998, Haukioja 1991). Induced trans-generational changes in defense, such as are seen in plants, aphids, and locusts (Agrawal et al. 1999, Braendle et al. 2005, Mondor et al. 2005, Simpson and Sword, this volume), can also be viewed in terms of trans-generational non-neural memory of predation threat from the previous generation. If predator load for parents predicts that for offspring, then induced offspring may benefit by being defended at or soon after birth. Parasites and predators are also plastic and maternal and developmental conditions can alter their virulence (Tseng 2006). Finally, induced direct defenses may be metabolically more expensive than induced indirect defenses (Dicke and Sabelis 1992, Halitschke et al. 2000).

Theory assumes logic and order behind the evolution of plasticity; that is, the type and degree of plasticity exhibited by a population can be logically explained once we know all the costs, benefits, selective pressures and constraints, etc. (Van Buskirk and Arioli 2005). However, empirical studies do not always find clear patterns. For example, Benard (2004) reviewed predator-induced plasticity in animal life-history transitions. In 40 studies, virtually every type of prey response was observed in every combination (i.e., change in behavior, morphology, *or* time or size at metamorphosis, and often in ways counter to those predicted. Closely related species expressed diametrically opposite plastic responses (Benard 2004). Schmidt and Van Buskirk (2005) observed similar results with newts. Likewise, five species of larval odonates differed greatly in their behavioral responses to fish (Wohlfahrt et al. 2006), and there was low correlation between morphological and behavioral plasticity in 16 tadpole species (Van Buskirk 2002). Perhaps there are yet undiscovered logical reasons for such interspecific variability in type of plasticity exhibited, or perhaps there is a large random component. Maybe such variation simply represents alternative ways to solve one problem. Finally, some studies suggest that organisms are not unduly limited by costs or constraints in evolving multiple defenses (Koricheva et al 2004), and indeed, some prey respond to predators with combined behavioral, morphological, and life-history plasticity (Sih 2004, Hovermann et al. 2005). At this point, we do not always understand why one type of plastic defense evolves over another. In the end, stochasticity or complex genetic, physiological, or ecological constraints may determine type of defense, and the type of defense that evolves may not the best possible.

Genetics

The genetics of plasticity is under intense study (Kliebenstein et al. 2002, 2005, Berrigan and Scheiner 2004, Windig et al. 2004). Important parameters of predator-induced phenotypic plasticity such as the ability and level of induction, often vary among individuals and populations, are under genetic control, and are heritable (as shown by sibling, parent-offspring, or DNA studies), and thus can be selected for (Tian et al. 2003, Hoballah et al. 2004, Braendle et al. 2005, Duquette et al. 2005, Lass et al. 2005, Relyea 2005, Van Buskirk and Arioli 2005, Van Kleunen and Fischer 2005). Hence, predator-induced plasticity can evolve. Moreover, the ability to be induced often corresponds to predator sympatry, but not allopatry, implying genetic control (Huntingford et al. 1994, Kats and Dill 1998, Stoks and McPeek 2003). For example, in those *Acacia* that are protected by ant bodyguards, extra floral nectar production is inducible in facultatively guarded species, but constitutively expressed at high levels in ant-obligate species (Heil et al. 2004). Because phenotypic plasticity is often accomplished by environmentally induced up- or down-regulation of specific genes, expression of defense genes is a phenotypic plastic trait.

The identity and control of important genes underlying predator-induced phenotypic plasticity are being determined, in part, by using both forward and reverse genetics approaches as well as microarray analyses and bioinformatics (Arimura et al. 2005, Doss 2005, Kaloshian and Walling 2005, Mori et al. 2005). For example, although a single herbivore or pathogen can induce up to 1,500 defense-related genes in a plant, Schmidt et al. (2005) found through microarray analysis that only a 10% overlap exists in significantly regulated genes of two solanaceous plant species that were attacked by the same caterpillar species. Verhagen et al. (2004) examined the transcriptional response of over 8000 genes in *Arabidopsis* in response to rizobacterial colonization. Such studies are rapidly clarifying the genetic basis of plasticity.

In plants, effective induced defense relies on a matching of genes between the plant and the pathogen or herbivore. Only when both the plant gene for resistance (*R*) and the pathogen gene for avirulence (*avr*) are present, is there resistance. Accordingly, *R* products within the plant recognize *avr*-dependent cues from the phytophage, triggering signal transduction events that elicit a response (Nimchuk et al. 2003, Kaloshian 2004, Kruijt et al. 2005). Sequencing of the complete *Arabidopsis thaliana* genome and genetic screening has allowed workers to identify ~ 100 *R* loci distributed over all chromosomes (Dangl and Jones 2001, Dicke and Hilker 2003). Tian and coworkers (2003) asked why some plants are polymorphic for resistance genes, such that some individuals lacked inducible defenses, leaving them susceptible to pathogen attack. What prevented the resistance alleles from being driven to fixation? They demonstrated that in *A. thaliana*, maintaining the defense alleles carried a high cost; in the absence of the *Pseudomonas* pathogen, plants lacking the resistance gene had 9% greater seed production (Tian et al. 2003).

Most pathogen-induced defenses in plants are transcriptional, and careful biochemical analyses are revealing the specific transcription factors and the corresponding reading promoter regions and genes. Interestingly, the genes that code for the various enzymes that together produce a specific defense compound are often clustered, which facilitates their control and rapid and efficient transcription (Qi et al. 2004, Zhao et al. 2005).

Recognizing the Threat: Eliciting Cues

The first step in adaptive prey plasticity is recognition of predatory threat, and prey have evolved sophisticated mechanisms to detect predators and to translate such information into adaptive responses. In animals, the stimuli used by prey to detect risk of predation span the range of sensory modalities, and include chemical, visual, and mechanical cues (Dicke and Grostal 2001, Arimura et al. 2005, Schmitz 2005). If one includes anti-predator behaviors as induced defenses, then sound (moths vs. bats or caterpillars vs. wasps), substrate vibration (most terrestrial arthropods) (Li 2002, Bruschini et al. 2005, Castellanos and Barbosa 2006, Gish and Inbar 2006), pneumatic pressure (many insects) (Jablonski and Lee 2006), hydrostatic pressure (fish), tactile stimuli (most animals) (Williams and Wise 2003, Killian et al. 2006), electromagnetic stimuli (fish and amphibians) (Caputi and Budelli 2006, Fortune et al. 2006, Von der Emde 2006), and infra-red radiation (snakes) (Ebert and Westhoff 2006) can cue predator-induced defenses (Cocroff 1999, King and Leaich 2006). In contrast, plants respond to chemical and mechanical cues. Here, the term "elicitor" is often applied to substances from pathogens (exogenous elicitors) and compounds released from plants by the action of pathogens (endogenous elicitors) that induce plant defense (Montesano et al. 2003), however, this term can also designate any stimulus that initiates a plastic response in an organism.

Understanding the stimuli that trigger prey plasticity is complicated, because prey may respond to single or multiple cues from predators (Arimura et al. 2005, Mithöefer et al. 2005), or cues emanating directly from the predator in real time (such as visual or acoustic signals from a moving predator or mechanical damage) or cues that were previously deposited by a predator such as body scents, pheromones, silk, or fecal components (Hazlett 1999, Huryn and Chivers 1999, Ruzicka 2001, Li 2002, Fréchette et al. 2003, Wilder and Rypstra 2004, Bell et al. 2006). Prey may also respond to indirect cues that signal predator presence, such as body fluids leaked from injured prey, dead or fleeing conspecifics, and alarm signals released by con- or allospecifics (Grostal and Dicke 1999, 2000, Huryn and Chivers 1999, Venzon et al. 2000, Dicke and Grostal 2001, Schoeppner and Relyea 2005, Wong et al. 2005, Reader et al. 2006). Included here are visual, chemical, acoustic, and substrate vibratory intraspecific alarm or warning signals given by many prey species (Nuñez et al. 1997, Pijanowska 1997, Manzoli-Palma et al. 1998, Matsumoto et al. 1998, Shah et al. 1999, Kunert et al. 2005). Prey response also varies with predator diet. For example, some spiders, damselflies, and mayflies exhibit greater or faster induction when their predators have fed on con- vs. allospecific prey (Huryn and Chivers 1999, Li and Jackson 2005, Brodin et al. 2006). However, predator-released or predator-caused cues do not need to be present to induce prey defenses. In some cases, environmental factors that correlate strongly with probability of predator presence or risk of predation may serve as inducers. For example, many temporary pool organisms do not need to directly detect the predator, but may cue in on smaller or more ephemeral pools, which generally translates into fewer predators (Blaustein and Whitman, this volume).

Prey can also respond to presence of bodyguards or to environmental changes such as temperature, photoperiod, sun vs. shade, humidity, season, or environment background color that predict changing predation risk (Hairston et al. 1990, Karban et al. 1999). For example, some aphids increase soldier- or winged-dispersal forms in the absence of ant bodyguards (Kleinjan and Mittler 1975, Shingleton and Foster 2000). Certain Lepidoptera adjust larval or pupal color based on environment background color (Edmunds 1974, Brakefield and Frakino, this volume). Some grasshoppers and mantids change body color from green to brown or black under drought conditions or following fires; light, moisture level in food, or humidity serve as cues in different species (Burtt 1951, Otte and Williams 1972, Edmunds 1974). These environment-induced phenotypic changes are thought to increase prey defenses in the new environment (Lyytinen et al. 2004). In some *Anolis* lizards, leg length is plastic: individuals raised on broad substrates develop long legs and those raised on narrow surfaces develop short legs (Losos et al. 2000). Long legs enhance fast escape on flat surfaces, whereas short legs allow lizards to better escape along thin twisting branches.

In many predator-prey systems, prey mortality rates are a function of prey density. It is not surprising then that some prey species such as aphids and locusts switch anti-predator defenses based on some measure of their population densities (Shibao et al. 2004, Kunert et al. 2005, Simpson and Sword, this volume; Wilson and Cotter, this volume). In some termites, ants, and aphids, soldier production is suppressed by coexisting soldiers (Chapman 1998, Shibao et al. 2004). It is important to differentiate the eliciting cue (e.g., predator odor, change in season or prey density) which predicts the threat, from the selective factor (death or damage from predator). In some cases, they can be the same (Moran 1992).

Finally, alarm pheromones released by attacked prey can "prime" conspecifics, lowering thresholds for subsequent defense induction (de Bruijn et al. 2006) and naïve prey can learn to associate specific predator features with conspecifics alarm signals (Mirza et al. 2006, Ferrari et al., 2008). Sex or breeding status of conspecifics producing the alarm cues can influence the conspecifics' response (Pollock et al. 2006). Interestingly, social transfer of immunity may occur in social insects, but the mechanism is unknown (Traniello et al. 2002).

For arthropod prey, the specific inducing cues are mostly unknown; instead, the literature reports general stimuli such as "odors" from predators. However, in a few cases, the specific elicitors have been identified. In locusts, mechanical stimuli induce increased defense. Stroking locusts on the outside of their hind femora with a fine brush, which mimics tactile stimuli that would occur at high population density, can induce the change (Simpson et al. 2001, Rogers et al. 2003). In aphids, alarm pheromone, (E)-beta-farnesene, released during predator attack, appears to stimulate increased movement and tactile contact, which induces a winged, dispersal morph (Kunert et al. 2005). Aphid-parasitizing wasps move away from sites containing specific hydrocarbons deposited by predatory ladybeetles (Nakashima et al. 2006).

In contrast to animals, there has been rapid progress in identifying specific plant-inducing stimuli. In plants, both mechanical and chemical cues can initiate defensive responses, and plant taxa can differ in which elicitors are used (Mattiacci et al. 2001, Arimura et al. 2005). In many plants, mechanical damage alone is effective; different types of mechanical damage (cutting, hole-punching, abrasion, or single vs. multiple vs. continuous mechanical assault) can elicit similar or different levels or types of response depending on the plant species, developmental stage, and plant physiological state (Arimura et al. 2005, Mithöefer et al. 2005). Some plants are extraordinarily sensitive to mechanical stimuli, and even caterpillar footsteps can induce a response (Bown et al. 2002). Other plants and plant stages fail to respond to mechanical damage or respond slowly or only after repeated, continuous mechanical damage (Halitschke et al. 2000, Arimura et al. 2005, Ballhorn et al. 2006). Some plants respond to either mechanical damage or prey chemicals, but the highest response usually occurs when both are present. Sometimes, both types of cues must be present (Dicke et al. 2003, Arimura et al. 2005). For example, elms release volatiles attractive to wasp parasitoids that attack beetle eggs, only when both mechanical damage and chemical elicitors from ovipositing elm leaf beetles are present (Meiners and Hilker 2000). Beetle feeding alone does not induce the plant response (Meiners et al. 2005). Mechanical damage and chemical elicitors often induce different responses. In *Arabidopsis*, caterpillar feeding induces different genes than does mechanical damage (Reymond et al. 2000). Plants also actively cause phytophages to release elicitor compounds. For example, plants express various glucanases, some of which are induced, that attack the cell walls of invading pathogens, releasing cell wall fragments and cell contents of pathogens, which then serve as elicitors (Bishop et al. 2005). Plants also respond to the products or chemistry of mechanical or physiological damage (endogenous elicitors or plant-derived wound signals). Depolarization of plant cell membrane potential or plant cell debris such as oligosaccharide cell wall fragments can induce plant defense (Paré and Tumlinson 1997, Arimura et al. 2005). Plants also respond to touch. For example, simple touching induces an incredible 2.5% of *Arabidopsis* genes (Lee et al. 2005), including many involved in plant defense (Braam 2005).

A wide range of phytophage-derived chemicals induce plant defense, including enzymes and fatty acid-amino acid conjugates from insects, and peptides, proteins, glycopeptides, glycolipids, lipids, hydrocarbons, and carbohydrates, including saccharin, flagellin, chitosan, and chitin from pathogens (Boyle and Walters 2005, Arimura et al. 2005, Paré et al. 2005, Zhao et al. 2005). Plant elicitation is an active research field, and new plant elicitors and primers are discovered monthly (e.g., Bae et al. 2006, Charleston et al. 2006, Ortmann and Moerschbacher 2006).

Most known insect elicitors derive from regurgitates or oral secretions. For example, ß-glucosidase from the regurgitate of *Pieris brassicae* caterpillars induces the release of parasitoid-attracting volatiles from cabbage (Mattiacci et al. 1995). Likewise, a variety of fatty acid-amino acid conjugates (FAC) from caterpillar regurgitate or saliva elicits defense reactions in numerous plants (Arimura et al. 2005). Included here is *N*-(17-hydroxylinolenoyl)-L-glutamine (volicitin) from the beet armyworm, *Spodoptera exigua* (Alborn et al. 1997, Gomez et al. 2005). The tobacco hornworm, *Manduca sexta*, produces at least eight FAC, including N-acyl Gln/Glv (Halitschke et al. 2001, Roda et al. 2004). FAC may be synthesized by the insect (Lait et al. 2003) or by gut bacteria (Spiteller et al. 2000), and production is influenced by the plant (De Moraes and Mescher 2004, Peiffer and Felton 2005). Conceptually, chewing insects should not normally regurgitate (vomit) on their plant; and even if some regurgitant or saliva did contaminate a plant, these compounds would be removed during the next bite. Furthermore, such maladaptive behavior should be quickly selected against. Indeed Roda et al. (2004) noted that FAC were not present in *M. sexta* salivary or mandibular glands, but only in the gut and frass of feeding larvae, suggesting a digestive role for these substances. Furthermore, extracts of salivary and mandibular glands did not induce plant defense (Roda et al. 2004), and some insect salivas repress plant defenses (Musser et al. 2005, Bede et al. 2006). However, Truitt and Paré (2004) measured ~100 pmol of volicitin on damaged leaves after feeding by late 3rd instar *Spodoptera exigua* caterpillars. Hence, the extent to which saliva from chewing insects normally induces or suppresses plant defense in nature is still unclear (Musser et al. 2006).

In contrast, sucking insects (which inject digestive/solubilizing compounds), leaf miners and plant borers (which deposit fecal, cuticular, and respiratory compounds in plant tissues), and insects that oviposit into plants, might provide higher concentrations of chemical cues directly into plant tissues (Moraes et al. 2005). Indeed, strong elicitors have been found in the oviduct secretions of plant-ovipositing insects (Hilker et al. 2005, Schroder et al. 2007), including the bruchins (esterified long chain unsaturated diols) (Doss et al. 2000, Oliver et al. 2000, Doss 2005). Insect eggs, whether deposited in or on plant tissue, can induce strong plant defenses (Hilker and Meiners 2006). For example, *Pieris brassicae* caterpillar eggs glued to leaf surfaces appear to induce the release of parasitoid-arrestants in plants (Fatouros et al. 2005), and leaves with *Euphydras* butterfly eggs had 15 times more iridoid glycoside than adjacent, egg-free leaves (Peñuelas et al. 2006). Interestingly, some insect eggs and larvae contain high levels of jasmonic acid, a plant "defense" hormone (Tooker and De Moraes 2005). Feces deposited on plant wounds by externally feeding arthropods might also elicit plant responses.

Plants can also respond defensively to volatiles released from microorganisms and to volatile signals released by nearby damaged plants. The former includes butanediol and acetoin (Ping and Boland 2004, Ryu et al. 2004), whereas proposed intra- and interplant elicitors and priming agents include green leaf volatiles, terpenoids, methyl jasmonate and methyl salicylate (Farag et al. 2005, Paré et al. 2005, Kessler et al. 2006, Kost and Heil 2006, Engelberth et al. 2007, Frost et al. 2007). Plants may also communicate danger via root-to-root chemical signaling (Dicke and Dijkman 2001).

Because it is costly to respond to inappropriate cues, prey should evolve high discrimination: eliciting cues should be specific, detectible at low concentrations, timely, and reliable, which would allow prey to produce adaptive, graded, and predator-specific phenotypes (Zangerl 2003, Hoverman et al. 2005, Teplitsky et al. 2005, Ferrari et al. 2007). This may be the case (Alborn et al. 2003, Mori et al. 2003, Relyea 2003a, Voelckel and Baldwin 2004, Duquette et al. 2005). For example, *Baetis* mayfly naiads respond differently to cues from two different predators: fish kairomones induce hiding, whereas cues from stonefly predators induce drifting (Wooster and Shi 1995, McIntosh and Peckarsky 1999). Prey induction can vary with predator body size (Persons and Rypstra 2001, Hlivko and Rypstra 2003), and predator density (Duquette et al. 2005). *Bufo* tadpoles respond to cues from common predators, but not from uncommon predators (Kiesecker et al. 1996). In *Rana lessonae* tadpoles, predatory fish induced shallow, muscled tails, whereas dragonfly predators induced deeper, less muscled tails (Wilson et al. 2005). Deeper or colored tails stimulate dragonfly predators to attack the tail instead of the body, thus reducing mortality from dragonflies (Teplitsky et al. 2005, Touchon and Warkentin 2008). These cases suggest that prey discriminate among predators. In contrast, in a survey of prey life history plasticity, Benard (2004) found no differences in the way freshwater invertebrates responded to fish vs. invertebrate predator cues, suggesting perhaps a lack of discrimination. And, in some cases, prey respond to both dangerous and non-dangerous predator species (Langerhans and DeWitt 2002), which seems maladaptive.

In plants, response can vary with phytophage species, ontogenetic stage, type of plant damage, and plant part attacked (Hilker and Meiners 2002, Li et al. 2002, Dicke and Hilker 2003, De Vos et al. 2005, Leitner et al. 2005), and level of response is often directly correlated to amount of damage (Underwood 2000). Leaf miners induce lower phenolic levels in birch than do chewing caterpillars, and insect damage produces greater plant responses than does artificial mechanical damage (Hartley and Lawton 1991). Parsnip, *Pastinaca sativa*, expresses different furanocoumarins depending on if the attacking agent is an insect, a pathogenic fungus, or a nonpathogenic fungus (Berenbaum and Zangerl 1999). Pine species attacked by bark beetle-symbiotic fungi produce different responses depending on the particular species of fungus; the response is greatest when the normally associated species of fungus is present (Raffa and Berryman 1987, Raffa 1991). True bugs induce different volatile emissions in legumes than do caterpillars, and parasitoids of the bugs only orient to the former (Moraes et al. 2005). Bean plants emit different odors when attacked by different aphid species, and aphid parasitoids respond accordingly (Guerrieri et al. 1999). Other bean species release methyl salicylate and trimethyl,tridecaletraene, when herbivore mites, but not caterpillars, feed. The volatiles are attractive to predatory mites that can feed on the herbivore mites, but not the caterpillar (De Boer et al. 2004). Corn plants release different volatiles when attacked by early instar vs. late instar armyworm caterpillars (Takabayashi et al. 1995), and *Cotesia kariyai* parasitic wasps, which do better on young caterpillars, respond only to volatiles from the former. In this case, differentiation may be mediated by the herbivore's saliva (Turlings et al. 1993b, Mattiacci et al. 1994). Corn plants benefit more when young (vs. old) caterpillars are eliminated, and only regurgitate from younger armyworm instars induced the plants to release the attractive volatiles (Takabayashi et al. 1995). However, subsequent tests in other plant-arthropod systems failed to demonstrate this finding as a generality (Mattiacci and Dicke 1995, Gouinguene et al. 2003). In mustards, leaf trichomes and sinigrin levels are differentially induced by a variety of caterpillars and beetles (Agrawal 2000, Traw and Dawson 2002a,b). And, as previously mentioned, oviposition into elm tissues induces synomonal defense, whereas feeding by the same beetle herbivore does not (Meiners and Hilker 2000). Cabbage plants reduce parasitoid-attractive volatiles once their caterpillar herbivores are parasitized, allowing subsequent parasitoids to avoid competition (Fatouros et al. 2005b). In contrast, in many plants, there is little variation in induced response to different chewing insects (Kost and Heil 2006). Specificity and honesty in signaling by plants to the 3rd trophic level (i.e., communicating the true herbivore load to potential carnivores) might or might not benefit the plant (Zangerl 2003).

Detecting multiple cues from a predator or different predators, would be a better predictor of potential predation than a single cue, and should produce greater response than a single cue from a single predator (Hazlett 1999, Meiners and Hilker 2000a, Yamaguchi et al. 2002, Brown et al. 2004, Voelckel and Baldwin 2004). For example, induced hiding in fiddler crabs is strongest when crabs detect both the predator and fleeing conspecifics (Wong et al. 2005). Likewise, higher strengths or concentrations of specific elicitors induce greater induction. For example, wolf spiders show a graded response to increasing concentrations of predator kairomone (Persons and Rypstra 2001, Persons et al. 2001, 2002). Prey should distinguish among simultaneous predators and express the defense most suitable to the most dangerous predator species (McIntosh and Peckarsky 1999). Indeed, tadpoles react differently to different predators, but appear to perceive the risk of combined predators as being similar to that of the most dangerous predator, and not an average or sum of risks (Relyea 2003a, Teplitsky et al. 2004), whereas plastic responses to two simultaneous predator cues was additive in a marine snail (Trussell and Nicklin 2002). Similarly, prey should discriminate predators from competitors, and respond accordingly (Relyea and Auld 2005).

Finally, environmental conditions and predator variability influence elicitor concentration, apparency, and recognition, and thus induction, by prey. For example, water turbidity and pH alter visual- and chemical-elicitation, respectively, in fish (Leduc et al. 2004). Likewise, degree of predator hunger influences induction in spiders and clams (Bell et al. 2006, Smee and Weissburg 2006).

Physiological Mechanisms Underlying Plasticity

At present, we have a poor understanding of the physiology underlying predator-induced plasticity in animals, except for mammalian immune responses. Although we often know the elicitors, we generally do not know the receptors, signal transducers, signaling pathways, transcription factors, activated genes, enzyme and hormonal events and regulatory mechanisms that produce altered phenotypes. For example, in aphids, where increased density, predators, tactile stimulation, or alarm pheromones can induce the winged, dispersal form, we still do not know if hormones are involved in wing induction despite 45 years of research (Braendle et al. 2006). Likewise, the physiology of locust polyphenism remains cloudy after 50 years of research (Simpson and Sword, this volume). However, researchers are beginning to understand induced physiology in some animal systems. For example, in tadpoles, cell size and number determine tail musculature, a predator-influenced trait of proven defense value (Arendt 2006).

In plants, in contrast, rapid progress is illuminating the physiology underlying phytophage-induced plasticity—identifying the signaling molecules and biochemical, hormonal, cellular, and developmental processes that produce altered phenotypes, aided by modern metabol-, prote- and transcriptomics (Bezemer and Van Dam 2005). Induction can involve transcription, translation, and enzymatic and hormonal regulation (Arimura et al. 2005, Schmidt et al. 2005, Zhao et al. 2005), and positive and negative feedback. Interestingly, plants and animals exhibit many similarities in elicitor recognition, signal transduction, and defense-induction against pathogens and parasites (Lamotte et al. 2004, 2005, Ausubel 2005, Bothwell and Ng 2005, Schmid-Hempel 2005).

The current model of plant induction derives largely from studies of biochemical responses of plants to pathogens or pathogen-derived elicitors, as opposed to herbivores (Zhao et al. 2005). In this model, pathogen attack initiates numerous, simultaneous and sequential, interactive reactions and pathways, that ultimately culminate in diverse defense responses. The process starts with signal perception, when exogenous and/or endogenous elicitor molecules bind to specific elicitor-binding proteins, located primarily on the plasma membrane (Truitt et al. 2004, Kruijt et al. 2005). Next comes signal transduction and amplification, when G-proteins coupled to these receptors mediate depolarization and influx of H^+ and Ca^{++} into the cell, the latter serving as a secondary messenger (Mithöefer et al. 2005, Lecourieux et al. 2006). The resulting changes in pH and Ca concentrations activate various enzyme pathways involving peroxidases, NADPH oxidase, protein kinases (Ren et al. 2006), and phospholipases, which, in turn, produce cAMP, cADP ribose, ethylene, NO, and reactive oxygen species (ROS) (Malolepsza and Rozalska 2005, Torres et al. 2006) such as superoxide anion and H_2O_2, all of which serve as secondary messengers in a variety of pathways that ultimately lead to the production of transcription factors that result in a transcriptional reorganization from growth to defense, including activation of various defense genes (Léon et al. 2001, Orozco-Cárdenas et al. 2001, Lamotte et al. 2004, Wendehenne et al. 2004, Arimura et al. 2005). *Trans*-activating factors may coordinate this polygenetic response (Voelckel and Baldwin 2004). Rapid changes in protein phosphorylation assist these reactions or trigger additional transduction cascades. Transcription leads to biosynthesis of plant hormones such as jasmonic and salicylic acids, and upregulation of other enzyme pathways that culminate as chemical, morphological, and behavioral defenses (Léon et al. 2001, Montesano et al. 2003, Zhao et al. 2005). These cascading, interconnected events are organized temporally (Zhao et al. 2005). Oxidative burst can occur in seconds (Bown et al. 2002). G protein activation, phosphorylation, Ca influx, and cytoplasmic acidification can begin within minutes of pathogen attack or elicitor contamination. Nitrogen oxide production, and induction of "early response genes" proceeds within 1 hr, followed by production of ethylene and jasmonic acid, which can take hours, and finally, "late defense gene" expression and biosynthesis of secondary compounds up to a week later (Mattiacci et al. 2001). These events require nutrient input, reallocation, and increased metabolism (Arnold and Schultz 2002).

Mechanical damage alone, or application of exogenous chemical elicitors without mechanical damage, can sometimes induce many of the same reactions in plants, including immediate 4-aminobutyrate (GABA) synthesis, cell depolarization and proton and calcium influx, oxidative burst, and the production of numerous secondary signaling molecules including NO and ethylene, which, in turn, trigger transcription, translation, enzymatic activity, and defense (Bown et al. 2002, Maffei et al. 2004, Arimura et al. 2005, Zhao et al. 2005). As with clumping of associated defense genes (discussed previously), the enzyme chains that together produce specific defense compounds are often physically bound together on specific membranes, allowing rapid and efficient assembly-line biosynthesis of the end product (Burbulis and Winkel-Shirley 1999, Zhao et al. 2005).

In contrast, attack by phytophages often evokes different plant responses than does mechanical damage, and different types of phytophages often induce different primary induction pathways (Rose et al. 1996, Paré et al. 2005, Beckers and Spoel 2006, Halim et al. 2006), but with considerable cross-talk (De Vos et al. 2005). Communication among pathways allows plants to regulate responses and respond appropriately to the specific threat at hand (De Vos et al. 2006). Insect herbivores, particularly chewing insects, and necrotrophic pathogens, often activate the jasmonic acid-octadecanoid enzymatic pathway via lipoxygenation of linolenic acid, producing the plant hormone jasmonic acid (JA) and its methyl ester (MeJA) (Arimura et al. 2005, Glazebrook 2005). In many plants, wounding immediately activates protein kinase and other enzymes that elicit genes for JA synthesis (Seo et al. 1995, Kodama et al. 2000). JA, in turn, serves to activate various defense genes and stimulates numerous reactions including de novo synthesis of ethylene and a wide range of secondary plant compounds such as proteinase inhibitors, phenylpropanoids, polyphenol oxidase, alkaloids, steroid glycoalkaloids, glucosinolates, terpenoids, phenoloics, and flavinoids (Schmidt et al. 2005, Suzuki et al. 2005). The octadecanoid pathway also produces carnivore-attractive volatiles, and linolenic acid can serve as the starting point for other volatiles such as terpenoids and indol that are later released throughout the plant. The chemical pathways for volatile production are well known, and synthesis is both transcriptionally and enzymatically regulated (Mckay et al. 2003, Arimura et al. 2004, 2005). Sesquiterpenes (e.g., farnesene and caryophyllene) are produced in the cytosol via the mevalonate pathway, followed by an isopentenyl pyrophosphate intermediate. Mono- and diterpenes (e.g., ocimene and linalool) are derived from an alternative erythritol phosphate pathway localized in plastids, and then synthesized via an isopentenyl pyrophosphate intermediate (Lichtenthaler 1999, Rodríguez-Concepción and Boronat 2002). Indole and some other nitrogen-containing defense compounds are produced via the shikimic acid pathway, and putrescine and other polyamines from arginine decarboxylase (Cowley and Walters 2005). Volatile methanol is produced from pectin demethylation in the cell walls (Peñuelas et al. 2006). The octadecanoid pathway is also involved with induction of extrafloral nectar (Arimura et al. 2005), and JA increases trichome production (Traw and Bergelson 2003). Induction from insect herbivores via JA is often referred to as induced resistance (IR).

Plant biotrophic pathogens and sucking arthropods tend to induce the salicylic acid (SA) pathway, culminating in phytoalexin production or the hypersensitive response (programmed cell death) (Heidel and Baldwin 2004, Wendehenne et al. 2004, Glazebrook 2005, Montillet et al. 2005). Induced resistance from plant pathogens via the plant hormone SA is often referred to as systemic acquired resistance (SAR). Interestingly, phytophage attack can induce the production and release of volatile methyl salicylate, which attracts numerous carnivorous arthropods (Drukker et al. 2000, De Boer and Dicke 2004, James and Price 2004).

The volatile plant hormone ethylene can also serve a central role in defense induction (Huang et al. 2005, Hudgins et al. 2006, Schröder et al. 2007). This signaling molecule is synthesized in response to mechanical damage, herbivore feeding, necrotrophic pathogens, oligosaccharide elicitors, and surprisingly, in lima bean, by volatiles from mite-infested plants. It stimulates the induction of various defense genes and volatile and non-volatile defenses such as phytoalexins, phenolics, terpenoids, and GLV, often in conjunction with JA (Horiuchi et al. 2001, Arimura et al. 2002,

Huang et al. 2005, Xu et al. 2005). Ethylene can also antagonize JA-dependent pathways, inhibit the synthesis of some compounds, and, in some cases, sensitize tissues for SA production (Zhao et al. 2005, De Vos et al. 2006). Abscisic acid, another plant hormone involved in induction against insects, is thought to synergize JA pathways and antagonize SA pathways, and hence, may benefit some pathogens (Thaler and Bostock 2004).

Systemic (plant-wide) induction is still being worked out; however, both JA and SA pathways can lead to a systemic resistance via activation of defence genes and/or upregulation of defense substances in undamaged leaves or roots. Various compounds have been proposed to serve as mobile message molecules, including JA, MeJA, SA, peptides (including the systemins), proteins, and lipids (Pearce and Ryan 2003, Ryan and Pearce 2003, Arimura et al. 2005, Schilmiller and Howe 2005). Released volatiles may induce distant tissues (Engelberth et al. 2007, Frost et al. 2007), and electrical signals (Wildon et al. 1992, Herde et al. 1995) and hydrostatic pressure (Malone and Alarcón 1995) may also play a role in intraplant signaling. Shvetsova et al. (2001) have recorded plant action potentials traveling at 40 m/s.

Phytophage attack induces volatile release via several mechanisms. Mechanical damage alone can cause release of constitutively stored volatiles including green leaf volatiles (GLV), terpenoids, and aromatic compounds from glands and glandular trichomes (Paré and Tumlinson 1999, Degenhardt and Lincoln 2006). Additional GLV are immediately generated via autolytic oxidative breakdown of plant membrane lipids via the lipoxygenase pathway (Paré and Tumlinson 1999, Matsui et al. 2000). Besides their deterrent role, the GLV may also serve as plant elicitors (Farag and Paré 2002, Engelberth et al. 2007). Numerous other volatile and non-volatile toxins, such as cyanide, aldehydes, lactones, ketones, terpenoids, phenolics, nitriles, and sulfur compounds, including isothiocyanates can be immediately produced from preexisting precursors (Seigler 1991, Wittstock and Halkier 2002, Mewis et al. 2005, Peñuelas et al. 2005, Hudgins et al. 2006). Herbivory also induces plants to synthesize and release novel volatiles (Degenhardt and Lincoln 2006). Finally, pathogens induce plants to release numerous volatiles (Huang et al. 2003). All of these mechanisms may act sequentially or simultaneously in a single plant.

Volatiles from damaged plants can also prime (immunological memory) or induce defense genes and defense reactions in nearby undamaged plants (Arimura et al. 2001, Paré et al. 2005, Choh and Takabayashi 2006). Methyl jasmonate (Karban et al. 2000, 2004, Kessler et al. 2006), ethylene and terpenoids (Tscharntke et al. 2001), and GLV (Farag et al. 2005, Kishimoto et al. 2005, Engelberth et al. 2007) have been implicated, but see Preston et al. (2004). The physiology underlying such responses is currently being worked out. In lima beans, terpenoids from mite-damaged plants induce cell calcium influxes, protein-phosphorylation and dephosphorylation, and emission of ethylene in receiver plants (Arimura et al. 2002, 2005).

Induced plant defenses vary dramatically in response time, localization, and persistence. In many plants, mechanical damage induces an immediate release of green leaf, terpenoid, or other volatiles (Hatanaka 1993, Matsui et al. 2000). Wounded tobacco leaves initiate transcription of defense-related genes within 1 min, and increase JA and MeJA within 3 hrs (Seo et al. 1995). Some parasitoid-attractive green leaf volatiles decline quickly when feeding stops, whereas cotton plants release volatile terpenoids from throughout the plant via de novo synthesis only after 2 to 3 d after the start of feeding (Paré and Tumlinson 1997). Resin duct formation requires about a week in spruce (McKay et al. 2003). In contrast, induction of trichomes or spines requires weeks to grow, and remain for the life of the plant. Some *Acacia* continue to produce longer spines nearly 2 yrs after cessation of mammalian herbivory (Young and Okello 1998).

The physiology of plant-induction is actually much more complex than presented above, because phytophages cause mechanical damage, disrupt plant physiology, and shed chemicals that can be used by the plant as chemical elicitors to identify the phytophage. Mechanical damage releases cell wall fragments and plant cell contents that also serve as elicitors to initiate reaction cascades. These different stimuli can induce different or similar physiological responses in plants. In addition, some pathogens can activate JA, some insect herbivores can activate SA, and some phytophages activate neither (Fidantsef et al. 1999, Thomma et al. 2001). Also, there are multiple converging and diverging induction pathways and products, and plants vary somewhat in which pathways they use (Harms et al. 1995, Traw and Bergelson 2003, Schmidt et al. 2005, Zhao et al. 2005). Living systems represent biochemical webs, not linear pathways, and hence, a specific compound may derive from, and influence, any number of biochemical pathways (Walling 2000, Kliebenstein et al. 2002, Malolepsza and Urbanek 2002). Positive feedback and bi-directional reactions can make it difficult to track pathways, and there can be substantial antagonistic or synergistic cross-talk between pathways (Rojo et al. 2003, Arimura et al. 2005). Induction of one pathway can block the ability to defend via the alternative pathway, making the plant more susceptible to an alternate enemy (Thaler et al. 2002, Rojo et al. 2003, Traw and Bergelson 2003, Cipollini et al. 2004). Natural enemies also exhibit prey induced plasticity, and can undergo a similar processes as described above for plants. For example, specific leaf-surface compounds serve as elicitors to induce offensive genes in pathogenic fungi (Dickman et al. 2003).

Predator-induced plasticity usually involves simultaneous or sequential changes in multiple traits and their underlying pathways. All of these complex pathways must be coordinated via phenotypic integration to produce a compatible and cohesive new phenotype of high fitness (Pigliucci and Preston 2004, Hovermann et al. 2005, Canfield and Green, this volume; Simpson and Sword, this volume). In addition, those prey that produce different defense ensembles to different predators or over the course of development (Relyea 2003a,b) must maintain in their genetic closet different defensive wardrobes to be donned when needed.

In both plants and animals, prey physiological state and ecological conditions influence induced response; hence, predator-induced defenses can vary with age (Fréchette et al. 2003, Relyea 2003), body size (McIntosh et al. 1999), sex (Williams et al. 2001), previous experience (Fréchette et al. 2003, Jacot et al. 2005, Mirza et al. 2006), nutritional state (Glynn et al. 2003, Huang et al. 2003, Walls et al. 2005), number of predators species detected (McIntosh and Peckarsky 1999, Rostás et al. 2006), stress (Adamo and Parsons 2006), social context and crowding (Tseng 2004, Pollock et al. 2006), level of competition (Cipollini and Bergelson 2001, Relyea 2004), desiccation and osmotic balance (Takabayashi et al. 1994, Reymond et al. 2000, Mopper et al. 2004, Thaler and Bostock 2004), time of day or light intensity (McIntosh and Peckarsky 1999, McIntosh et al. 1999, Stratmann 2003, Arimura et al. 2004, Montillet et al. 2005), temperature (Thomas and Blanford 2003, Adamo and Parsons 2006), and water turbidity (Van de Meutter et al. 2005). Induced responses should be flexible to meet current physiological and ecological demands (McIntosh and Peckarsky 1999).

Predator Counter-measures to Prey Plasticity

Prey have evolved an astounding diversity of constitutive and induced anti-predator defenses. But predators are not passive bystanders in this evolutionary battle; they can evolve both constitutive and plastic counter-responses (reciprocal phenotypic change) to prey defenses. Predators may evolve plastic responses (prey-induced offence) to prey plasticity and vice versa — a type of phenotypic plasticity arms race (Adler and Grünbaum 1999, Agrawal 2001, Bishop et al. 2005). Hence, the evolution of predator-prey plasticity is shaped by the evolution of predator-prey plasticity (Wolf et al. 2004).

Predators have evolved multiple constitutive and induced counter-measures to prey defenses (Whitman and Nordlund 1994). For example, herbivorous insects may avoid induced defenses in plants by moving away from previously induced tissues or plants, or may reduce induction in plants by scattering their feeding among different plants (Rodríguez-Saona and Thaler 2005, Hilker and Meiners 2006). Some insects trench (vein-cut) or girdle their host-plant, which reduces transfer of plant signals and defenses across the gap, preempting some induced defenses (Chambers et al. 2007). Likewise, post-feeding petiole snipping or whole-leaf removal reduces intra-plant communication (Cipollini and Sipe 2001) and disrupts volatile attraction of carnivores when the chemical beacon (the damaged leaf) falls to the ground. In Monarch caterpillars, vein-cutting is plastic: it only occurs when latex pressure in the plant is high (Karban and Agrawal 2002). Scattering feeding throughout a tree spreads induced-plant volatile release throughout the tree, making it difficult for carnivores to locate herbivores. Some herbivores can detect predator-attractive volatiles released by plants, and avoid such (Shiojiri et al. 2002). In many herbivores, detoxification of plant allomones is inducible (Agrawal et al. 2002, Cianfrogna et al. 2002, Li et al. 2004). For example, proteinase inhibitors are commonly induced in plants by phytophages, and these compounds can harm feeding insects (Ryan 1990, Broadway et al. 1986, Broadway 1997). However, many caterpillars respond to elevated proteinase inhibitors in their diet by producing different proteinases that are less susceptible to the plant inhibitors (Jongsma et al. 1995, Mazumdar-Leighton and Broadway 2001). Perhaps this is why plants have evolved a high diversity of trypsin inhibitors (Haruta et al. 2001). Many insects detoxify plant poisons by up-regulating mixed-function oxidases (including p450s) (Feyeresen 1999, Mao et al. 2006). Not surprisingly, many plants have evolved herbivore-inducible compounds that inhibit these insect enzymes (Berenbaum and Neal 1985, Wen et al. 2006). Other phytophages may overcome host plants by blocking induction pathways (Bouarab et al. 2002, Cui et al. 2005) or exploiting the antagonistic interactions between competing defense pathways (Felton and Korth 2000, Musser et al. 2002). *Pieris rapae* caterpillars have evolved a novel gut protein that redirects the production of toxic plant isothiocyanates to less toxic nitriles (Wittstock et al. 2004). Diamondback moths possess a sulfatase enzyme that blocks the induced glucosinolate defenses of mustard plants (Ratzka et al. 2002). Some plant pathogens subvert induced plant defenses by releasing compounds that activate alternative, antagonistic defense pathways (Zhao et al. 2003, Brooks et al. 2005). Some tobacco-feeding caterpillars incorporate glucose oxidase (GOX) in their saliva, which may allow them to feed on this plant by inhibiting the induction of nicotine synthesis (Musser et al. 2002, 2005, 2006). *Spodoptera exigua* caterpillar saliva suppresses transcript levels of terpenoid defense genes in alfalfa (Bede et al. 2006). Plants may be countering this predator offence, since different plant species cause caterpillars to produce different levels of GOX (Merkx-Jacques and Bede 2005, Peiffer and Felton 2005).

Predators and prey, including animals, plants, and microbes, often use similar biochemistry. Hence, they might sense, respond, and evolve to each other's hormones and signaling molecules, resulting in "signal stealing" and "biochemical espionage" (Schultz 2002, Schultz and Appel 2004). Incredibly, the corn earworm counters induced plant defenses by monitoring the plant's signaling pathway: the plant hormones jasmonate and salicylate activate genes in the insect that rapidly produce P450 enzymes that detoxify plant toxins (Li et al. 2002). Some insects inoculate their hosts with plant pathogens, which then block induced plant defenses either directly, or indirectly by inducing a plant defense that is incompatible with an insect-targeted defense (Stout et al. 2006). Lepidopteran eggs and larvae contain substantial amounts of jasmonic acid, but it is not yet known if this influences plant-insect interactions (Tooker and De Moraes 2005).

Although the above examples concerns insect herbivores and plants, similar "games" are played by other predators and prey (Hammond et al. 2007). For example, pike may reduce their chemical conspicuousness by defecating away from their foraging areas (Brown et al. 1995). Predators change their behaviors to track predator-induced changes in prey foraging behavior or occupied habitat (Lima 2002). Some natural enemies evolved to "hide" their antigens or mimic host chemicals, making detection difficult (Elgar and Allen 2006). In the ant-colony parasite, *Myrmecaphodius excavaticollis*, chemical mimicry is an environment-dependent trait. Over time, each beetle acquires the odor of its specific host colony, reducing attack by the ants (Vander Meer and Wojcik 1982). Some pathogens and parasites over-stimulate the host's immune system, producing a disruptive physiological maelstrom destroying the host's ability for immune defense (Frost 1999). Others directly suppress the expression of immunity-related genes (Yang and Cox-Foster 2005). Likewise, parasites and pathogens usurp prey endocrine or biochemical systems to alter host physiology or behavior to their own advantage (Eberhard 2000, Moore 2001, Kamita et al. 2005, Bonds 2006, Pennacchio and Strand 2006, Roy et al. 2006). A good example of reciprocal phenotypic plasticity is seen in the ciliate *Lembadion* and its prey ciliate *Euplotes* (Kopp and Tollrian 2003). When exposed to predators, the latter develops "wings" that reduce predation. But *Lembadion* responds with an inducible offence: it increases gape size. Likewise, snails grow thicker shells in the presence of crab predators (Trussell and Smith 2000), and crabs grow larger, more powerful claws when reared on armored prey (Smith and Palmer 1994). Agrawal and Karban (1999) and Gardner and Agrawal (2002) suggest that prey can reduce the rate of predator counter-evolution by limiting induced defenses to only periods of high predation pressure, similar to our efforts to reduce unnecessary use of antibiotics in humans.

Predators and prey are caught in an endless cycle of co-evolutionary tit-for-tat, that over time should favor diversification of strategies and counter measures. Recognition of predators by prey often depends on specificity of both elicitors and receptors, whereas, for predators, escape from detection depends on altering detectable stimuli. Thus, prey and predators are always chasing one another as they continue to evolve novel attack and defense genes. Coevolution should result in different strategies of gene expression in predators vs. prey. Host resistance loci should evolve to co-express alleles, whereas pathogens and parasites should evolve so that individuals express single alleles, because multiple host alleles are more likely to recognize a diversity of parasites, whereas a parasite that expresses multiple elicitors would be more likely to be detected (Nuismer and Otto 2005). In addition, where predators are monophagous and predator-prey populations are isolated, we would expect local adaptation of both antagonists (Niemi et al. 2006) whereby plastic responses differed at each site.

The evolution of predator-prey counter-measures is constrained by physiological and ecological needs. For example, selection should eliminate the fatty acid-amino acid conjugates present in caterpillar guts and feces because these elicitors induce plant defenses (Halitschke et al. 2001). However, these compounds may serve an essential function in food assimilation (Roda et al. 2004), and thus possibly cannot be eliminated. Hence, caterpillars may be constrained from becoming stealthy to plants. However, perhaps caterpillars have found another way to solve this problem: many caterpillars fire their feces away from the plant, or house-clean (push feces with their head or remove feces with their mandibles or legs and drop the pellets off the edge of the leaf), and thus reduce possible elicitation (Salazar and Whitman 2001). An example of balancing ecological needs may be the interaction of tobacco plants and *Manduca sexta* caterpillars. The caterpillar's regurgitant can reduce levels of toxic plant nicotine, but not the release of parasitoid-attractive volatiles. However, lowered production of nicotine may be adaptive for the plant, because the parasitoids are harmed by high titers of nicotine in their caterpillar hosts (Kahl et al. 2000). Hence, there may be little selection pressure on the plant to evolve a counter-measure to the caterpillar's counter-measure.

Arthropod Predator-induced Defenses in Algae

Algal-feeding arthropods can induce a variety of defenses in algae, including dormancy, altered body shape, spines, and toxicity (Rengefors et al. 1998, van Donk et al. 1999, Pavia and Toth 2000). Exposure to water from feeding *Daphnia* induces unicellular *Scenedesmus* green algae to form colonies that are too large for the herbivore to ingest (Hessen and van Donk 1993, Lampert et al. 1994). Induction of chemical defenses in the brown alga *Eklonia* depends on season and species of grazer (Molis et al. 2006)

Phytophage-induced Defense in Plants

Plants respond to pathogens and vertebrate and invertebrate herbivores with a surprisingly diverse assortment of evolutionary tricks (Karban and Baldwin 1997, Agrawal et al. 1999, Dicke 1999, Arimura et al. 2005, Ohgushi 2005). *Acacia* browsed by rabbits, goats, or giraffes produce long thorns, *specifically* on those branches attacked (Young and Okello 1998, Milewski et al. 1991). In other *Acacia*, herbivores induce production of swollen thorn domatia that house ant bodyguards (Huntzinger et al. 2004). Caterpillars induce trichome production in mustards (Agrawal 2000, Traw and Dawson 2002). Leaf damage in broad bean increases the number of extrafloral nectaries (Mondor and Addicott 2003), and in cotton increases the number of defense-pigment glands (Agrawal and Karban 2000). Weevils induce resin ducts in Sitka spruce (Nagy et al. 2000). Pink bollworm invasion induces rapid cell proliferation in cotton plants, which crushes or smothers the caterpillar. Other plants respond to phytophage threat by adding lignin, glycoproteins and callose to cell walls, with neoplasms (growth of cells that trap, smother or crush the phytophage) (Doss et al. 2000, Hilker and Meiners 2002), abscission of parts (Williams and Whitham 1986, Simberloff and Stiling 1987), or rapid programmed cell/tissue death (hypersensitive necrosis), which isolates the phytophage and denies it nutrition (Hilker and Meiners 2002, Montesano et al. 2003, Greenberg and Yao 2004). Extrafloral nectars (EFN) are induced in numerous plants (Heil 2004, Heil et al. 2004, Choh and Takabayashi 2006). Caterpillar feeding on *Catalpa* induces elevated production of EFN, which feeds caterpillar-natural enemies, which then presumably attack the caterpillars (Stephenson 1982, Ness 2003). Wire worms feeding on roots doubled EFN flow in cotton leaves (Wäckers and Bezemer 2003). *Impatiens sultani* plants respond to stimulated herbivory by increasing amino acid content of extrafloral nectar (Smith et al. 1990). In *Macaranga*, EFN flow decreases in the absence of nectar feeders (Heil et al. 2000). Ant bodyguards induce food body production in *Cecropia* (Folgarait et al. 1994), and leaf damage induces higher ant density at the site (Agrawal and Dubin-Thaler (1999). Increased attendance by plant bodyguards following herbivore attack can be viewed as an induced response to herbivory (Agrawal and Dubin-Thaler 1999). Ant presence induces the ant-plant *Piper cenocladium* to reduce chemical defense and produce food bodies to feed its ant bodyguards; the elicitor is unknown. A clerid beetle parasitizes this relationship: not only can it induce food-body production, but the beetles feed on both food bodies and ants (Letourneau 1990, Dyer et al. 2004). Other plant responses to herbivory include altered cell structure, growth form, compensatory growth, reversion to juvenile stages, new meristems, leaf folding, climbing in vines, and changes in primary metabolites and nutrient allocation, including moving resources toward, or nutrients away from attacked tissues. Phytophage attack also induces changes in rate of photosynthesis, resource acquisition by roots, leaf toughness, maternal investment, sex ratio, phenology (bud, leaf, and flower appearance), and color, which might help hide the plant from herbivores, or signal visually hunting carnivores (Karban and Baldwin 1997, Fischer et al. 2004, Babst et al. 2005, Braam 2005, Gianoli and Molina-Montenegro 2005, Wise and Cummins 2006).

The most common induced response in plants is increased production of chemical or chemo-physical defenses, including saps, resins, latexes, gels, waxes, calcium oxalate, silica, and a vast array of secondary plant compounds (Bystrom et al. 1968, Karban and Baldwin 1997, Molano-Flores 2001, Traw 2002, Nykanen and Koricheva 2004, Ohgushi 2005). These include both non-volatile and volatile toxins, repellents, antifeedants, herbivore hormones and endocrine disrupters (Soriano et al. 2004, Ballhorn et al. 2006), but also herbivore-induced volatile synomones that attract arthropod predators and parasitoids, and even nematodes that subsequently attack the herbivores (Dicke et al. 2003, James and Price 2004, Turlings and Wackers 2004, Rasmann et al. 2005). The latter represents an induced indirect defense (Lou et al. 2005). Included here is a wide diversity of GLV, sulfur compounds, phenolics, including methyl salicylate, and terpenoids, including (E)-bata-caryopyllene, farnescene, indole, and methyl jasmonate (Van Poecke et al. 2001, James 2003, 2005, Arimura et al. 2005, Rasmann et al. 2005, Dugravot and Thibout 2006).

Herbivore-induced plant volatiles can also influence entomopathogenic fungi, suggesting the possibility that plants could employ microorganisms as bodyguards via induction (Hountondji et al. 2005). Plants can also vector herbivore diseases (Van Munster et al. 2005), and may have evolved plastic mechanisms to foster this interaction.

Individual plants may also communicate potential danger to adjacent plants—the still-debated phenomenon of "talking" or "listening trees" (Farmer and Ryan 1990, Karban et al. 2004, Kost and Heil 2006). Herbivore attack apparently induces some plants to release volatile pheromones that "alert" nearby conspecifics and prime them or induce them to increase allelochemical levels, produce EFN, or release their own synomones, to call in additional plant bodyguards (Arimura et al. 2000, Choh et al. 2004, Choh and Takabayashi 2006, Choh et al. 2006). Finally, phytophages can induce transgenerational defenses in some plants (Agrawal et al. 1999, Agrawal 2001).

Predator-induced Defense in Vertebrates

Fish respond to predator presence by altering foraging behavior, morphology, length of egg stage, and time of day of hatching (Parsons and Robinson 2006). Salmon eggs also respond differently to different predators (Jones et al. 2003). Amphibians respond to predator cues with spatial avoidance, increased hiding, reduced activity, and altered feeding patterns, tail form, musculature, color, body size at metamorphosis, and development time (Benard 2004, Wilson et al. 2005, Touchon and Warkentin 2008). Some amphibian eggs can sense predator presence and respond adaptively. Egg predators and egg pathogens induce rapid egg development and hatching, albeit at a smaller size (Chivers et al. 2001, Warkentin et al. 2001, Vonesh and Osenberg 2003), whereas larval predators induce later hatching, often at a larger size (Vonesh and Bolker 2005, Mandrillon and Saglio 2007). Interestingly, the direction of induced traits varies among species, and sometimes among populations of the same species. In other words, predators might induce larger *or* small sizes, or earlier *or* later development in different populations or under different conditions in the same prey species (Benard 2004). Sorting out the underlying ecological, taxonomic, and mechanistic reasons for divergent prey responses, and rectifying disparities between theoretical models and empirical results poses a challenge for researchers.

Mammals serve as the primary models for immune response, which represents pathogen-induced defense (Frost 1999). Interestingly, plant, invertebrate, and vertebrate immune systems share some common features (Schultz 2002, Schmid-Hempel 2005).

Predator-induced Morphological and Developmental Defenses in Arthropod Prey

In arthropods, predator presence can induce changes in development and morphology, including lengths of caudal filaments or spines in mayflies and dragonflies (Dahl and Peckarsky 2002, Johansson 2002), heavier exoskeletons in mayflies (Dahl and Peckarsky 2002), helmets, crests, conspicuousness, eye diameter, or length, size, or number of teeth or spines in *Daphnia* (Parejko and Dodson 1990, Tollrian and Dodson 1999), mucus sheets in *Holopedium* (Stenson 1987), and shell shape in barnacles (Lively 1999). Some social insects produce dispersing morphs when colonies are parasitized (Korb and Fuchs 2006, Hunt 2007). Parasitoids, predators, or lack of ant bodyguards can induce winged forms in aphids, which then disperse (Kleinjan and Mittler 1975, Sloggett and Weisser 2002, Kunert and Weisser 2003, Kunert et al. 2005). The aphid *Pseudoregma sundanica* rapidly produces more soldiers when ant bodyguards are absent (Shingleton and Foster 2000). In other aphids, larger colonies, which are more attractive to predators, produce proportionally more soldiers (Shibao et al. 2004). Gall-dwelling *Pemphigus* aphid soldiers extend the length of the soldier instar in older colonies, which increases defense investment for the clone (Pike et al. 2004), but reduce defensive behaviors and molt into reproductives after migrating to a new colony, possibly because there is little fitness benefit from defending non-kin (Abbot et al. 2001). Aphids also increase honeydew quality and quantity when ants are present (Fischer and Shingleton 2001, Yao and Akimoto 2002). This defense plasticity may increase ant guarding and aphid survival, but such a reallocation of nutrients may reduce aphid growth, size and fecundity (Yao et al. 2000, Stadler et al. 2002). Lycaenid caterpillars produce more attractive nectar when threatened by predators, causing ants to increase guarding (Agrawal and Fordyce 2000). Some spiders reduce conspicuousness of webs under predator threat (Li and Lee 2004).

Predators can alter arthropod phenology and life history (Benard 2004). For example, threat from egg predators induced earlier hatching in *Scytodes* spitting spiders (Li 2002), whereas predatory salamander larvae delay hatching in clam shrimp (Spencer and Blaustein 2001). In *Daphnia*, predators influence egg, clutch, and body size, egg diapause, and time to hatch and reproduce (Weider and Pijanowska 1993, Repka and Walls 1998,

Sakwinska 1998, Slusarczyk 1999, Lass et al. 2005). Other induced responses include smaller body size at metamorphosis in mayflies and damselflies (Johansson et al. 2001, Dahl and Peckarsky 2003); longer *or* shorter development time in mosquitoes, mayflies, midges, and damselflies (Ball and Baker 1996, Hechtel and Juliano 1997, Johansson et al. 2001, Dahl and Peckarsky 2003), and lower growth rate in mayflies, midges, and damselflies (Scrimgeour and Culp 1994, Ball and Baker 1996, Johansson et al. 2001). Induced responses can be trans-generational, such as the induced maternal effects in locusts (Simpson and Sword, this volume) and aphids (Mondor et al. 2005, Braendle et al. 2006), and when exposure to fish kairomones induces smaller body size or larger helmets in *Daphnia* offspring (Sakwinska 1998, Agrawal et al. 1999). Finally, numerous arthropods exhibit immune responses to pathogens or parasitoids (Beckage 2008) or to population density (density-dependent prophylaxis) (Wilson and Cotter, this volume; Pie et al. 2005).

Predator-induced Phenotypic Plasticity in Arthropod Behavior

Behavioral plasticity represents a type of phenotypic plasticity whereby expressed behavior of a genotype varies as a function of the environment (Dewitt and Scheiner 2004, Sih 2004). Under this definition, phenotypic plasticity in behavior could be considered to encompass the full range of behavioral mechanisms, including stimulus-response reactions, sensitization, habituation, imprinting, conditioning, repetitive learning, complex learning, and insight. Even an instinctive behavior could be considered as phenotypic plasticity if it is expressed in response to specific environmental stimuli. Thus, reactions such as escape behavior can, in a liberal sense, be construed as phenotypic plasticity.

Behavioral plasticity offers numerous advantages over transcriptionally controlled, non-reversible, morphological/developmental plasticity (Sih 2004). First, is that behavior is superbly responsive and flexible and can be immediately altered, rapidly reversed, and/or prolonged when needed, and thus finely modulated. This allows animals to closely track fluctuating environmental conditions in real time with immediate adaptive behavior of the appropriate type, strength, and duration (Laurila 2000). This is not possible with morphological phenotypic plasticity, which requires time to implement, and is usually non-reversible. Secondly, behavioral changes generally require less energy than morphological or physiological changes. Of course, the consequences of behavior changes, such as starvation due to hiding, could carry high costs. Behavioral plasticity is also advantageous in that it might be inducible at nearly any time during the active life stages of the individual, whereas most morphological, developmental, and life history defenses, and some physiological and chemical defenses are only inducible during certain ontological stages. For example, arthropods are, for the most part, unable to alter the exoskeleton once they reach the adult stage. Finally, behavioral plasticity can be highly adaptive. In general, behaviorally plastic responses are more rapid and less costly than induced morphological responses (e.g., Teplitsky et al. 2005), and can be highly sensitive, responsive, and rapidly reversible (Cooper 2006).

Arthropods exhibit a great diversity of plastic behaviors in response to risk of predation (Edmunds 1974, Evans and Schmidt 1990, Dicke and Grostal 2001, Salazar and Whitman 2001). Included are altered behaviors, such as escape, dispersal, or moving into alternative, adjacent microhabitats (Grostal and Dicke 1999, Shah et al. 1999, Williams and Wise 2003), drifting (Wooster and Sih 1995, McIntosh and Peckarsky 1999) avoidance (Schonewolf et al. 2006), depth selection or dial vertical migration (De Meester et al. 1999, Von Elert and Pohnert 2000), hiding (Pallini et al. 1998, Venzon et al. 2000, Van Son and Thiel 2006), altering feeding times (Stamp 1997), reducing activity levels or feeding (Snyder and Wise 2000, Williams and Wise 2003, Wilder and Rypstra 2004), autotomy (Maginnis 2006), increased wariness (De Meester and Pijanowska 1996), thanatosis or slower movement (Wilder and Rypstra 2004), settlement or colonization behavior (Welch et al. 1997, Pallini et al. 1999, Grostal and Dicke 2000), swarming (Pijanowska 1994, Kvam and Kleiven 1995), aggregation (Evans and Schmidt 1990), aggression or group defense (Evans and Schmidt 1990, Manzoli-Palma et al. 1998, Kunz et al. 2006), egg-guarding (Beal and Tallamy 2006), signaling (Nelson et al. 2006), digging deeper burrows (Hoelker and Stief 2005), fleeing from areas containing pathogens (Meyling and Pell 2006), increased hygiene (Schmid-Hempel and Schmid-Hempel 1998), behavioral-fever and self medication (Moore 2001, Thomas and Blanford 2003), entering resting states, and altering time of hatching or metamorphosis (Blaustein 1997, Li 2002).

As previously mentioned, learning can be considered a form of phenotypic plasticity that functions in predator-prey relationships (Papaj and Lewis 1993, Chivers et al. 1996, Chapman et al. 2008). For example, nymphal *Enallagma* damselflies can learn to express antipredatory responses to fish predators, if the damselflies are first exposed to stimuli from fish and injured conspecifics presented simultaneously (Wisenden et al. 1997). How imprinting, sensitization, and other forms of learning influence non-behavioral plasticity in prey is largely unexplored. Locust display trans-generational "learning" and induced maternal effects, when increased locust density (which predicts increased predator attack) induces females to alter their eggs such that the resulting hatchlings express different behavioral and morphological defenses (Simpson and Sword, this volume).

Because behavior, physiology, morphology, development, and life history are all interconnected, predator-induced changes in behavior can influence such traits, and vice versa (Benard 2004). Examples include when predator presence induces (1) reduced foraging, which causes smaller body size, smaller clutch size, or longer development, (2) a switch in diet, exposing prey to different nutrients or allelochemicals that result in altered growth, (3) movement into different habitats with different thermal characteristics, which subsequently alter growth or development rates or body color, or (4) altered clumping and dispersal, which alter competition, social interactions, etc. Likewise, predator-induced behaviors can alter trophic cascades (Schmitz et al. 2004).

Just like morphological and developmental plasticity, plastic behaviors have a genetic basis and individuals vary in expressed behavioral reaction norms (Foster 1999, Huntingford et al. 1994). Hence, behaviors are genotype-dependent, and the type and strength of predator-induced behaviors can be selected for. Plastic defensive behavior, including learning, is especially important because of its potential to foster rapid evolution and speciation (ten Cate 2000, Losos et al. 2000, 2004, 2006).

Predator-induced Plasticity in Oviposition Site Selection

Prey can also respond to perceived predation risk through altered oviposition site selection (Blaustein 1999, De Moraes et al. 2001, Resetarits 2005, Hilker and Meiners 2006). This has been observed in numerous arthropod groups and has been the focus of substantial experimental and theoretical attention during the past 15 years. For example, *Tetranychus* spider mites avoid ovipositing at sites contaminated by predatory mites or injured conspecifics (Grostal and Dicke 1999). The phytoseiid mite *Neoselius cucumeris* delays oviposition when an intraguild predator is present, but the embryo continues to develop inside the mother's body (Monserrat et al. 2007). *Chaoborus* midges refrain from ovipositing in water containing fish kairomones (Berendonk 1999), *Rhagoletis* fruit flies postpone oviposition in the presence of chemical cues from wasp parasitoids (Hoffmeister and Roitberg 1997), and *Aphidius* wasps disperse from areas containing odors of their hyperparasites (Höller et al. 1994). Minute pirate bugs lay fewer eggs in the presence of pathogenic *Beauveria* fungus (Meyling and Pell 2006). Some ladybeetles avoid laying near ant-attended aphids (Sloggett and Majerus 2003) or in the presence of cannibalistic con- or allospecifics (Nakashima et al. 2004, Laubertie et al. 2006). Gravid mosquito females respond to predator-related cues, avoiding ovipositing in water bodies containing these predators (Blaustein and Whitman, this volume). Allomones deter chrysopid oviposition (Ruzicka 1988). Numerous insects avoid ovipositing in the presence of conspecific alarm pheromones. However, ovipositing *Enallagma* damselfly Sometimes ignore fish kairomones (McGuffin et al. 2006).

Predictions, Questions and Future Directions

Predator-induced plasticity is still an emerging field. In this section we list some of the hypotheses, predictions and controversies that remain to be tested or solved.

- *Because plasticity evolves via selection on phenotypic values expressed in specific environments, plasticity must be controlled by loci expressed in those environments (allelic sensitivity model)* (Via 1994, Via et al. 1995, Kliebenstein et al. 2002) *vs. the gene regulation model, whereby plasticity and trait values are independent and thus genes that influence only plasticity must exist* (Schlichting and Pigliucci 1993).

- *Multiple predator species and negative genetic or ecological correlations among competing defensive traits.* Different predators require different defense strategies (Salazar and Whitman 2001), a defensive trait effective against one predator species may not be effective against another (Berenbaum et al. 1986, Agrawal et al. 1999). When there are negative genetic or ecological correlations among traits, we would expect plasticity to evolve. For example, when defense trait 1 is effective against predator 1, and trait 2 is effective against predator 2, but the two traits show a negative genetic correlation such that they cannot both be expressed simultaneously, then selection should favor the evolution of adaptive plasticity, whereby the most effective defense is expressed in the presence of the appropriate predator (but see Adler and Karban 1994). Hence, type of prey defense should change with different predators (Relyea 2003a). We observe negative genetic or ecological correlations between competing induced defensive traits in both plants (Berenbaum et al. 1986, Juenger and Bergelson 1998, Stinchcombe and Rausher 2001) and animals (Teplitsky et al. 2005).

- *Multiple vs. single major predator.* Either plastic or constitutive defenses could evolve under either single or multiple predators. For example, a single constitutive defense could evolve against a suite of different predator species, even if they attacked at different times, if the defense was inexpensive and effective against all of them.

- *Induction should be more effective against generalist predators than specialist predators* because specialists have coevolved with the prey. This prediction finds some empirical support (Agrawal and Sherriffs 2001, Van Zandt and Agrawal 2004).

- *Inducibility should be inversely proportional to fitness value of a particular tissue.* This is partly because induced responses usually require time to implement. During that lag period, the tissue may be less defended. Selection should remove genes that allow risk to vital tissues. Hence, for plants, tissues most related to fitness (reproductive tissues) should be constitutively defended, whereas tissues least linked to fitness (leaves and roots) should have inducible defenses (see Strauss et al. 2004, Traw and Feeny 2008). This prediction has been supported in plants, where reproductive tissues tend to be constitutively defended, and leaf defense tends to be inducible (van Dam et al. 1996, Zangerl and Rutledge 1996, Ohnmeiss and Baldwin 2000, Strauss et al. 2004).

- *Induced and constitutive defenses should be negatively associated among populations and species.* Populations or species with a high constitutive defense may not benefit from an induced defense, and vice versa (Traw 2002, Koricheva et al. 2004). However, some studies have not supported this prediction (Agrawal et al. 1999, Havill and Raffa 1999, Morris et al. 2006).

- *Restricted gene flow should result in genetic variability for predator-induced phenotypic plasticity* (Karban and Nagasaka 2004, Van Buskirk and Arioli 2005).

- *Migration and panmixis favor plasticity* (de Jong 1999, Tufto 2000, Sultan and Spencer 2002, Zhang 2006).

- *What was the ancestral state: constitutive, plastic, or no defense?* (Thaler and Karban 1997, Heil et al. 2004, Agrawal 2005).

- *r-Selected species should not exhibit defense plasticity, because they utilize rapid massive reproduction and dispersal as defensive strategies vs. r-selected should exhibit plasticity because they encounter greater predative diversity, because of greater dispersal* (Sultan and Spencer 2002).

- *K-Selected species should rely on constitutive defenses,* because longer lifespan, greater apparency and predictability, and lower fecundity require a greater commitment to defense (Feeny 1976).

- *Plasticity will evolve when the duration of environmental fluctuation is much less than the generation time, because genotype changes via mutation can only take place in generation time scales* (Bradshaw 1965).
- *Frequency of predative pressure vs. prey generation time* (Gomulkiewicz and Kirkpatrick 1992, Berrigan and Scheiner 2004). Organisms that rarely encounter predators should evolve inducible defenses, in part to avoid the costs of maintaining constitutive defenses (Lima and Bednekoff 1999, Sih et al. 2000, Traw 2002). Conversely, constitutive defenses should evolve in systems where high predative frequency is guaranteed.
- *Predictability of predator attack: Random vs. patterned variability in time of prey risk* (Bull 1987, Via 1994, Harvell and Tollrian 1999). If timing of predation is patterned, and thus predictable, then prey should evolve defenses that vary in synchrony with the threat, such as either variable constitutive defenses (ontogenetic change in defense, such as life history change) or (if predator cues or environmental cues that predict predator-threat are reliable) induced defense. Likewise, if predator attack is random, the prey could evolve constitutive defense, if it had low cost and was effective, or plastic defense if predators were detectable and plasticity was beneficial (Harvell and Tollrian 1999).
- *Plasticity is more likely to evolve in animals under temporal vs. spatial heterogeneity* (Moran 1992).
- *Short lag times for switching phenotypes are more adaptive than long lag times* (Padilla and Adolph 1996).
- *Reversible predator-induced plasticity should evolve when the period of predator attack is short compared to lifespan of prey and there is a cost to the induced defense* (Gabriel 2005).
- *Type of Plastic defenses should change with ontogeny* (Relyea 2003b).
- *Ontogeny: young organisms should evolve inducible defenses, because of the compounded costs of constitutive defenses over the lifetime, or because immatures often have higher developmental flexibility.* When immatures suffer low predation, but high mortality from other factors, such as competition or establishment loss, then most resources should be directed toward growth, competitive ability, and establishment. Early allocation toward defense would reduce allocation to growth and competitive ability, and thus reduce subsequent establishment, growth, competitive ability, or fecundity (Mondor and Roitberg 2003, Hoballah et al. 2004, Boege and Marquis 2005). In contrast, adults that have already established competitive dominance can afford constitutive defenses, and at this advanced point in development may

be unable to alter morphology and life history. Indeed, immature insects, cladocerans, and plants often show greater induced responses than older individuals (Brodin et al. 2006, but see Relyea 2003b). Predator-induced avoidance of oviposition sites should also decline with age for females, because of either egg load or because older females have less probability of finding an alternative patch before death (Frechette et al. 2004).

- *For organisms that undergo major developmental transitions (e.g., holometabolous insects and amphibians), time and size at life-history transition should depend on relative growth and mortality rates in the immature habitat* (Werner 1986, Abrams and Rowe 1996). Studies of time transitions have beencontridictory (Li 2002, Benard 2004). In a survey of 40 studies that examined predator-induced changes in animals with complex life cycles, Benard (2004) found only two in which individuals metamorphosed both earlier and smaller in response to predator threat during the immature stage.

- *Hosts should evolve to co-express resistance alleles and parasites should evolve such that individuals express only one of many possible elicitors* (Nuismer and Otto 2005). This appears to be the case in plants and their pathogens (Barbour and Restrepo 2000, Young 2000).

- *Constitutive defenses should evolve against predators that quickly attack and kill their prey, whereas inducible defenses should evolve against natural enemies that prolong their damage* (Järemo et al. 1999), in part because immediate death gives little chance to detect or respond. However, inducible defenses could still evolve when organisms can detect rapid predation on con- or allospecifics.

- *In plants, defenses that act prior to significant damage are more effective than those that act after tissue destruction* (Karban and Agrawal 2002). This would favor strong constitutive defenses that deter prior to attack, over induced defenses.

- *Plants should evolve local induction (i.e., only the damaged leaf increased defense) when herbivores are small and immobile, and systematic induction to large or mobile herbivores* (Järemo et al. 1999).

- *Induced resistance in plants causes aggregation vs. uniform distributions in herbivores* (Underwood et al. 2005).

- *Induction lag time determines whether or not predator-induced plasticity will dampen predator-prey population fluctuations* (Miner et al. 2005).

- *Plants should evolve to produce specific and honest signals to attract carnivorous bodyguards* (Zangerl 2003). However, some cheating (attracting a standing army of bodyguards when herbivores are absent) might also occur.

- *Constitutive defenses are more likely to evolve than inducible defenses.* This is because the latter involves coordinated evolution of receptors, internal signaling pathways, transcription factors and gene promoters or new biochemical pathways that in unison act to increase the defense, whereas, in the former, only the defense must be increased (Zangerl 2003).
- *If groups can, but individuals cannot, show a reaction norm, then a reaction norm as a distinct trait should evolve only by group selection* (Via et al. 1995, Sih 2004).

Perspectives

The interactions between prey and their natural enemies are much more complex than previously imagined. We now realize that understanding bitrophic relationships requires considering multi-trophic, multi-species, direct, and indirect effects, including predator and prey diets, associated competitors, mutualists, symbionts, and prey social factors (Dicke and Hilker 2003). Phenotypic plasticity changes the rules, because now all participants are moving targets—variable in short and long time scales, including across generations. Predator-induced phenotypic plasticity can alter species interactions, energy flow, and community structure. Hence, understanding species abundance and distribution, and untangling the interactions among species becomes much more difficult, albeit more interesting.

On a more practical scale, herbivore- and pathogen-induced plasticity has important consequences for agriculture, medicine, and industry. To maximize crop production, we need to understand plasticity in food plants, agricultural pests, and their natural enemies. Moreover, manipulation of phytophage-induction of plants offers intriguing possibilities for new forms of pest control (Agrawal 2005). For example, scientists are already vaccinating or inducing resistance in plants prior to predicted seasonal phytophage attack (Edreva 2004, Ping and Boland 2004, Heijari et al. 2005), or engineering plants to be hyper-sensitive or more resistant to crop pests (Dana et al. 2006). In the future, we will transfer entire genetic complexes coding for phytophage-induced plasticity or increased constitutive defenses to crop plants (Kappers et al. 2005). At present, we do not know what pitfalls lay ahead. For example, some elicitors, including SA, can be phytotoxic (Cohen et al. 1991, Kessmann et al. 1994). Perhaps because of energetic costs, we cannot genetically modify plants to always express high levels of induced defense. Likewise, negative cross-talk between induction-pathways or negative cross effects to different predators in plants may constrain the use of these techniques. Perhaps plants have already reached an optimal balance in inducibility, and our tinkering will only engender new problems. For example, biological control agents should quickly evolve resistance to artificial "false-signaling" (Drukker et al. 2000). In the same way, we do not know how genetic engineering, such as transferring *Bacillus thuringiensis* genes to plants will influence phytophage-induced effects (Dean and De Moraes 2006).

Phenotypic plasticity has important implications for human and veterinary medicine, and plant pathology. After all, both immune responses and pathogen counter-responses represent phenotypic plasticity. A better understanding of these processes will aid disease prevention and management. Some disease vectors might be amenable to vaccination against the vertebrate or plant pathogens they vector.

New molecular tools—cDNA-microarrays, quantitative PCR, QTL analysis, and proteomics, will allow us to identify specific genes and biochemical pathways that can be manipulated to our benefit (Kliebenstein et al. 2002, Von Rad et al. 2005). Genetic engineering, coupled with the use of appropriate elicitors, using whole plants or plant tissue culture may allow the mass production of a wide range of useful plant metabolites or engineered substances for use as pharmaceuticals, flavors, aromas, vitamins, dyes, pesticides, glues, etc. (Al-Tawaha et al. 2005, Jeong and Park 2005, Zhao et al. 2005). Industrial production of such substances may reduce pressure to harvest rare or endangered plant species. Furthermore, elicitation might be used for bioprospecting (Poulev et al. 2003, Li and Barz 2005). Remote monitoring of atmosphere levels and types of induced plant volatiles could be used to measure any number of ecosystem interactions (De Moraes et al. 2004).

Finally, recognition of induced defenses requires adjustments to theory and experimental methodologies (Vos et al. 2004, Agrawal 2005). If plants communicate and "eavesdrop" on one another, if prey routinely disperse away or hide from predators, and if all participants are constantly changing their phenotypes, then we need to redesign our experiments (Blaustein 1999, Spencer et al. 2002, Karban et al. 2004). We also need to examine more responses and more factors simultaneously (Benard 2004), including indirect effects, and examine interactions among factors. This added complexity makes it more difficult to understand or model predator-prey systems (Schmitz 1998). Evolutionary theory must likewise address predator-induced plasticity, because phenotypic change places individuals into different selective niches and produces new niches, by increasing interaction diversity, or via niche construction. In this way, plastic responses alter selection pressure and direction, and may (West-Eberhard 2003, Parsons and Robinson 2006) or may not (de Jong 2005) encourage evolution and speciation.

Acknowledgements

We thank Oswald Schmitz and John Sedbrook for editorial comments. This work supported by NSF CRUI Grant DBI-0442412 and Israel Science Foundation Grant 699-02.

References

Abbot, P., J. H. Withgott and N. A. Moran. 2001. Genetic conflict and conditional altruism in social aphid colonies. Proceedings of the National Academy of Sciences of the United States of America 98: 12068–12071.

Abrams, P. A. and H. Matusda. 1997. Prey adaptation as a cause of predator-prey cycles. Evolution 51: 1742–1750.

Abrams, P. A. and L. Rowe. 1996. The effects of predation on the age and size of maturity of prey. Evolution 50: 1052–1061.

Adamo, S. A. and N. M. Parsons. 2006. The emergency life-history stage and immunity in the cricket, *Gryllus texensis*. Animal Behavior 72: 235–244.

Adler, F. R. and D. Grünbaum. 1999. Evolution of forager responses to inducible defenses. In Tollrian, R. and C. D. Harvell (eds.). The Ecology and Evolution of Inducible Defenses. Princeton University Press, Princeton, NJ Pp. 257–285.

Adler, F. R. and R. Karban. 1994. Defended fortresses or moving targets? Another model of inducible defenses inspired by military metaphors. American Naturalist 144: 813–832.

Agrawal, A. A. 1999. Induced responses to herbivory in wild radish: effects on several herbivores and plant fitness. Ecology 80: 1713–1723.

Agrawal, A. A. 2000. Specificity of induced resistance in wild radish: causes and consequences for two specialist and two generalist caterpillars. Oikos 89: 493–500.

Agrawal, A. A. 2001a. Transgenerational consequences of plant responses to herbivory: an adaptive maternal effect? American Naturalist 157: 555–569.

Agrawal, A. A. 2001b. Phenotypic plasticity in the interactions and evolution of species. Science 294: 321–325.

Agrawal, A. A. 2005. Future directions in the study of induced plant responses to herbivory. Entomologia Experimentalis et Applicata 115: 97–105.

Agrawal, A. A. and J. A. Fordyce. 2000. Induced indirect defence in a lycaenid-ant association: the regulation of a resource in a mutualism. Proceedings of the Royal Society B. Biological Sciences 267: 1857–1861.

Agrawal, A. A. and R. Karban. 1999. Why induced defenses may be favored over constitutive strategies in plants. In Tollrian R and Harvell CD (eds.). The Ecology and Evolution of Inducible Defenses. Princeton University Press, Princeton, NJ Pp. 45–61.

Agrawal, A. A. and R. Karban. 2000. Specificity of constitutive and induced resistance: pigment glands influence mites and caterpillars on cotton plants. Entomologia Experimentalis et Applicata 96: 39–49.

Agrawal, A. A. and M. F. Sherriffs. 2001. Induced plant resistance and susceptibility to late-season herbivores of wild radish. Ann. Entomol. Soc. Am. 94: 71–75.

Agrawal, A. A. and B. J. Dubin-Thaler. 1999. Induced responses to herbivory in the neotropical ant-plant association between *Azteca* ants and *Cecropia* trees: response of ants to potential inducing cues. Behavioral Ecology & Sociobiology 45: 47–54.

Agrawal, A. A., P. M. Gorski and D. W. Tallamy. 1999. Polymorphism in plant defense against herbivory: constitutive and induced resistance in *Cucumis sativus*. J. Chem. Ecol. 25: 2285–2304.

Agrawal, A. A., C. Laforsch and R. Tollrian. 1999. Transgenerational induction of defences in animals and plants. Nature 401: 60–63.

Agrawal, A. A., S. Tuzun and E. Bent (eds.). 1999. Induced Plant Defenses Against Pathogens and Herbivores. Biochemistry, Ecology, and Agriculture. APS Press, St. Paul, MN, USA.

Agrawal, A. A, J. K. Conner, M. T. J. Johnson and R. Wallsgrove. 2002a. Ecological genetics of an induced plant defense against herbivores: additive genetic variance and costs of phenotypic plasticity. Evolution 56: 2206–2213.

Agrawal, A. A., A. Janssen, J. Bruin, M. A. Posthumus and M. W. Sabelis. 2002b. An ecological cost of plant defence: attractiveness of bitter cucumber plants to natural enemies of herbivores. Ecol. Lett. 5: 377–385.

Agrawal, A A., F. Vala and M. W. Sabelis. 2002c. Induction of preferences and performance afer acclimation to novel hosts in a phytophagous spider mite: adaptive plasticity? American Naturalist 159: 553–565.

Agrawal, A. A., J. A. Lau and P. A. Hamback. 2006. Community heterogeneity and the evolution of interactions between plants and insect herbivores Quart. Rev. Biol. 81: 349–376.

Agarwala, B. K., H. Yasuda and Y. Kajita. 2003. Effect of conspecific and heterospecific feces on foraging and oviposition of two predatory ladybirds: role of fecal cues in predator avoidance. J. Chem. Ecol. 29: 357–376

Alborn, H. T., T. C. J. Turlings, T. H. Jones, G. Stenhagen, J. H. Loughrin and J. H. Tumlinson. 1997. An elicitor of plant volatiles from beet armyworm oral secretion. Science 276: 945–949.

Alborn, H. T., M. M. Brennan and J. H. Tumlinson. 2003. Differential activity and degradation of plant volatile elicitors in regurgitant of tobacco hornworm (*Manduca sexta*) larvae. J. Chem. Ecol. 29: 1357–1372.

Aldrich, J. R. 1988. Chemical ecology of the Heteroptera. Annual Review of Entomology 33: 211–238.

Aldrich, J. R. 1994. Chemical communication in the true bugs and parasitoid exploitation. In Cardé RT and Bell WJ (eds.). Chemical Ecology of Insects. II. Chapman and Hall, New York Pp. 318–363.

Aliabadi, A. and D. W. Whitman. 2001. Semiochemistry of crucifers and their herbivores. Pp 71–94 in Ananthakrishnan TN (ed.). Insect and Plant Defence Dynamics. Science Publishers, Inc., Enfield, USA.

Aliabadi, A., J. A. A. Renwick and D. W. Whitman. 2002. Sequestration of glucosinolates by harlequin bug *Murgantia histrionica*. J. Chem. Ecol. 28: 1749–1762.

Al-Tawaha, A. M., P. Seguin, D. L. Smith and C. Beaulieu. 2005. Biotic elicitors as a means of increasing isoflavone concentration of soybean seeds. Annals of Applied Biology 146: 303–310.

Arendt, J. D. 2006. The cellular basis for phenotypic plasticity of body size in Western Spadefoot toad (*Spea hammondi*) tadpoles: patterns of cell growth and recruitment in response to food and temperature manipulations. Biological Journal of Linnean Society 88: 499–510.

Arimura, G., R. Ozawa, T. Shimoda, T. Nishioka, W. Boland and J. Takabayashi. 2000. Herbivory-induced volatiles elicit defence genes in lima bean leaves. Nature 406: 512–515.

Arimura, G., R. Ozawa, J. Horiuchi, T. Nishioka and J. Takabayashi. 2001. Plant-plant interactions mediated by volatiles emitted from plants infested by spider mites. Biochem. Syst. Ecol. 29: 1949–1061.

Arimura, G., R. Ozawa, T. Nishioka, W. Boland, T. Koch, F. Kühnemann and J. Takabayashi. 2002. Herbivore-induced volatiles induce the emission of ethylene in neighboring lima bean plants. Plant J. 29: 87–98.

Arimura, G., D. P. Huber and J. Bohlmann. 2004a. Forest tent caterpillars (*Malacosoma disstria*) induce local and systemic diurnal emissions of terpenoid volatiles in hybrid poplar (*Populus trichocarpa* × *deltoides*): cDNA cloning, functional characterization, and patterns of gene expression of (-)-germacrene D synthase, *PtdTPSI*. Plant J. 37: 603–616.

Arimura, G., R. Ozawa, S. Kugimiya, J. Takabayashi and J. Bohlmann. 2004b. Herbivore-induced defense response in a model legume. Two-spotted spider mites induce emission of (E)-beta-ocimene and transcript accumulation of (E)-beta-ocimene synthase in *Lotus japonicus*. Plant Physiology 135: 1976–1983.

Arimura, G., C. Kost and W. Boland. 2005. Herbivore-induced, indirect plant defences. Biochimica et Biophysica Acta 1734: 91–111.

Arnold, T. M. and J. C. Schultz. 2002. Induced sink strength as a prerequisite for induced tannin biosynthesis in developing leaves of *Populus*. Oecologia 130: 585–593.

Ausubel, F. M. 2005. Are innate immune signaling pathways in plants and animals conserved? Nat. Immunol. 6: 973–979.

Babst, B. A., R. A. Ferrieri, D. W. Gray, M. Lerdau, D. J. Schlyer, M. Schueller, M. R. Thorpe and C. M. Orians. 2005. Jasmonic acid induces rapid changes in carbon transport and artitioning in *Populus*. New Phytologist. 167: 63–72.

Bae, H., M. S. Kim, R. C. Sicher, H. Bae and B. A. Bailey. 2006. Necrosis- and ethylene-inducing peptide from *Fusarium oxysporum* induces a complex cascade of transcripts associated with signal transduction and cell death in Arabidopsis. Plant Physiology 141: 1056–1067.

Bakker, E. G., M. B. Traw, C. Toomajian, M. Kreitman and J. Bergelson. 2008. Low levels of polymorphism in genes that control the activation of defense response in *Arabidopsis thaliana*. Genetics 178: 2031–2043.

Baldwin, I. T. and P. Callahan. 1993. Autotoxicity and chemical defense—nicotine accumulation and carbon gain in solanaceous plants. Oecologia 94: 534–541.

Baldwin, I. T. and W. Hamilton. 2000. Jasmonate-induced responses of *Nicotiana sylvestris* results in fitness costs due to impaired competitive ability for nitrogen. J. Chem. Ecol. 26: 915–952.

Baldwin, I. T. and C. J. Preston. 1999. The ecophysiological complexity of plant responses to herbivores. Planta 208: 137–145.

Ball, S. L. and R. L. Baker. 1996. Predator-induced life history changes: antipredatory behavior costs or facultative life history shifts. Ecology 77: 1116–1124.

Ballhorn, D. J., M. Heil and R. Lieberei. 2006. Phenotypic plasticity of cyanogenesis in lima bean *Phaseolus lunatus*—activity and activation of â-glucosidase. J. of Chem. Ecol. 32(2): 261–275.

Barbour, A. G. and B. I. Restrepo. 2000. Antigenic variation in vector-borne pathogens. Emerg. Infect. Dis. 6: 449–457.

Barker, R. J. 1978. Poisoning by plants. In Morse RA (ed). Honey Bee Pests, Predators, and Diseases. Cornell University Press, Ithaca, NY Pp. 273–296.

Baverstock, J., P. G. Alderson and J. K. Pell. 2005. *Pandora neoaphidis* transmission and aphid foraging behaviour. J. Invert. Pathology 90: 73–76.

Beal, C. A. and D. W. Tallamy. 2006. A new record of amphisexual care in an insect with exclusive paternal care: *Rhynocoris tristis* (Heteroptera: Reduviidae). J. Ethol. 24: 305–307.

Beckage, N. E. 2008. *Insect Immunology*. Academic Press, San Deigo.

Beckers, G. J. M. and S. H. Spoel. 2006. Fine-tuning plant defence signalling: salicylate versus jasmonate. Plant Biol. 8: 1–10.

Bede, J. C., R. O. Musser, G. W. Felton and K. L. Korth. 2006. Caterpillar herbivory and salivary enzymes decrease transcript levels of *Medicago truncatula* genes encoding early enzymes in terpenoid biosynthesis. Plant Molecular Biol. 60: 519–531.

Bell, R. D., A. L. Rypstra and M. H. Persons. 2006. The effect of predator hunger on chemically mediated antipredator responses and survival in the wolf spider *Pardosa milvina* (Araneae: Lycosidae). Ethol. 112: 903–910.

Benard, M. F. 2004. Predator-induced phenotypic plasticity in organisms with complex life histories. Annu. Rev. Ecol. Evol. Syst. 35: 651–673.

Benard, M. F. and S. J. McCauley. 2008. Integrating across life-history stages: Consequences of natal habitat effects on dispersal. Am. Nat. 171: 553–567.

Berenbaum, M. R. 1999. Animal-plant warfare: molecular basis for cytochrome P450-mediated natural adaptation. In Puga A. and Wallace K.B. (eds.). Molecular Biology of the Toxic Response. Taylor & Francis, Philadelphia Pp. 553–571.

Berenbaum, M. R. 2002. Postgenomic chemical ecology: from genetic code to ecological interactions. J. Chemical Ecology 28: 873–896.

Berenbaum, M. R. and J. J. Neal. 1985. Synergism between myristicin and xanthotoxin, a naturally co-occurring plant toxicant. J. Chemical Ecol. 11: 1349–1358.

Berenbaum, M. R. and A. R. Zangerl. 1999. Coping with life as a menu option: inducible defenses of the wild parsnip. In Tollrian R and Harvell CF (eds.). The Ecology and Evolution of Inducible Defenses. Princeton University Press, Princeton, NJ Pp. 10–32.

Berenbaum, M. R., A. R. Zangerl and J. K. Nitao. 1986. Constraints on chemical coevolution: wild parsnips *Pastinaca sativa* and the parsnip webworm *Depressaria pastinacella*. Evolution 40: 1215–1228.

Berendonk, T. U. 1999. Influence of fish kairomones on the ovipositing behavior of *Chaoborus imagines*. Limnol. Oceanogr. 44: 454–458.

Berrigan, D. and S. M. Scheiner. 2004. Modeling the evolution of phenotypic plasticity. In DeWitt T.J. and Scheiner S.M. (eds.). Phenotypic Plasticity. Oxford University Press, Oxford Pp. 82–97.

Bezemer, T. M. and N. M. van Dam. 2005. Linking aboveground and belowground interactions via induced plant defenses. Trends in Ecology and Evolution 11: 617–624.

Blaustein, L. 1997. Non-cunsumptive effects of larval *Salamandra* on crustacean prey: can eggs detect predators? Oecologia 110: 212–217.

Blaustein, L. 1999. Oviposition habitat selection in response to risk of predation: Consequences for populations and community structure. In Wasser SP (ed.). Evolutionary Processes and Theory: Modern Perspectives. Kluwer Academic Publishers Pp. 441–456.

Blum, M. S., R. F. Severson, R. F. Arrendale, D. W. Whitman, P. Escoubas, O. Adeyeye and C. G. Jones. 1990. A generalist herbivore in a specialist mode: metabolic, sequestrative, and defensive consequences. J. Chem. Ecol. 16: 223–244.

Boege, K. and R. J. Marquis. 2005. Facing herbivory as you grow up: the ontogeny of resistance in plants. Trends Ecol. Evol. 20: 441–448.

Bonds, M. H. 2006. Host life-history strategy explains pathogen-induced sterility. American Naturalist 168(3): 281–293.

Bothwell, J. H. F. and C. K. Y. Ng. 2005. The evolution of Ca^{2+} signalling in photosynthetic eukaryates. New Phytologist 166(1): 21–38.

Bouarab, K., R. Melton, J. Peart, D. Baucombe and A. Osbourn 2002. A saponin-detoxifying enzyme mediates suppression of plant defences. Nature 418: 889–892.

Bowers, M. D. and G. M. Puttick. 1986. Fate of ingested iridoid glycosides in lepidopteran herbivores. J. Chem. Ecol. 12: 169–178.

Bowers, M. D. 1990. Recycling plant natural products for insect defense. In Evans D. L. and Schmidt J. O. (eds). Insect Defenses. State University of New York Press, Albany Pp. 353–386.

Bown, A. W., D. E. Hall and K. B. MacGregor. 2002. Insect footsteps on leaves stimulate the accumulation of 4-aminobutyrate and can be visualized through increased chlorophyll fluorescence and superoxide production. Plant Physiol. 129: 1430–1434.

Boyle, C. and D. Walters. 2005. Induction of systematic protection against rust infection in broad bean by saccharin: effects on plant growth and development. New Phytologist 167(2): 607–612.

Braam, J. 2005. In touch: plant responses to mechanical stimuli. New Phytologist 165: 373–377.

Bradshaw, W. E. 1965. Evolutionary significance of phenotypic plasticity in plants. Advances in Genetics 13: 115–155.

Braendle, C., I. Friebe, M. C. Caillaud and D. L. Stern. 2005. Genetic variation for an aphid wing polymorphism is genetically linked to a naturally occurring wing polymorphism. Proceedings of the Royal Society of Biological Sciences, Series B 272: 657–664.

Braendle, C., G. K. Davis, J. A. Brisson and D. L. Stern. 2006. Wing dimorphism in aphids. Heredity 97: 192–199.

Broadway, R. M. 1997. Dietary regulation of serine proteinases that are resistant to serine proteinase inhibitors. J. Insect Physiol. 43: 855–874.

Broadway, R. M., S. S. Duffey, G. Pearce and C. A. Ryan. 1986. Plant proteinase inhibitors: a defense against herbivorous insects? Entom. Exp. et Appl. 41: 33–38.

Brodin, T., D. J. Mikolajewski and F. Johansson. 2006. Behavioural and life history effects of predator diet cues during ontogeny in damselfly larvae. Oecologia 148: 162–169.

Brooks, D. M., C. L. Bender and B. N. Kunkel. 2005. The *Pseudomonas syringae* phytotoxin coronatine promotes virulence by overcoming salicylic acid-dependent defences in *Arabidopsis thaliana*. Mol. Plant Pathol. 6: 629–639.

Brown, G. E., D. P. Chivers and R. J. F. Smith. 1995. Localized defecation by pike: a response to labelling by cyprinid alarm pheromone? Behav. Ecol. Sociobiol. 36: 105–110.

Brown, G. E., J. F. Poirier and J. C. Adrian. 2004. Assessment of local predation risk: the role of subthreshold concentrations of chemical alarm cues. Behav. Ecol. 15: 810–815.

Brown, J. S. and B. P. Kotler. 2004. Hazardous duty pay and the foraging cost of predation . Ecology Letters 7: 999–1014.

Bruce, M. J. and M. E. Herbestein. 2006. The influence of predator cues on orb-web spider foraging behavior. Ethology, Ecology & Evolution 18: 91–98.

Bruschini, C., R. Cervo and S. Turillazzi. 2005. Defensive responses to visual and vibrational stimulations in colonies of the social wasp *Polistes dominulus*. Ethol. Ecol. & Evol. 17: 319–326.

Bull, J. J. 1987. Evolution of phenotypic variance. Evol 41: 303–305.

Burbulis, I. E. and B. Winkel-Shirley. 1999. Interactions among enzymes of the *Arabidopsis* flavonoid biosynthetic pathway. Proc. Nat. Acad. Sci. (USA). 96: 29–34.

Burtt, E. 1951. The ability of adult grasshoppers to change colour on burnt ground. Proc. R. ent. Soc, Lond. A 26: 45–48.

Bystrom, B. G., R. B. Glater, F. M. Scott and F. S. C. Bowler. 1968. Leaf surface of *Beta vulgaris*-electron microscope study. Bot. Gaz. 129: 133–138.

Caputi, A. A. and R. Budelli. 2006. Peripheral electrosensory imaging by weakly electric fish. J. Comp. Physiol. A. 192: 587–600.

Carroll, M. J., E. A. Schmelz, R. L. Meagher and P. E. A. Teal. 2006. Attraction of *Spodoptera frugiperda* larvae to volatiles from herbivore-damaged maize seedlings. J. Chem. Ecol. 32: 1911–1924.

Castellanos, I. and P. Barbosa. 2006. Evaluation of predation risk by a caterpillar using substrate-borne vibrations. Animal Behav. 72: 461–469.

Chambers, J. L., M. R. Berenbaum and A. R. Zangerl. 2007. Benefits of trenching behavior in the context of an inducible defense. Chemoecology 17: 125–130.

Chapman, B. B., L. J. Morrell, T. G. Benton and J. Krause. 2008. Early interactions with adults mediate the development of predator defenses in guppies. Behav. Ecol. 19: 87–93.

Chapman, R. F. 1998. *The Insects Structure and Function*. Cambridge University Press, Cambridge.

Charleston, D. S., R. Gols, K. A. Hordijk, R. Kfir, L. E. M. Vet and M. Dicke. 2006. Impact of botanical pesticides derived from *Melia azedarach* and *Azadirachta indica* plants on the emission of volatiles that attract parasitoids of the diamondback moth to cabbage plants. J. Chemical Ecology 32: 325–347.

Chivers, D. P. and R. J. F. Smith. 1998. Chemical alarm signaling in aquatic predator-prey systems: a review and prospectus. Ecoscience 5: 338–352.

Chivers, D. P., B. D. Wisenden and R. S. F. Smith. 1996. Damselfly larvae learn to recognize predators from chemical cues in the predator's diet.

Chivers, D. P., J. M. Kiesecker, A. Marco, J. DeVito, M. T. Anderson and A. R. Blaustein. 2001. Predator-induced life history changes in amphibians: egg predation induces hatching. Oikos 92: 135–142.

Choh, Y. and J. Takabayashi. 2006. Herbivore-induced extrafloral nectar production in lima bean plants enhanced by previous exposure to volatiles from infested conspecifics. J. Chemical Ecology 32: 2073–2077

Choh, Y., T. Shimoda, R. Ozawa, M. Dicke and J. Takabayashi. 2004. Exposure of lima bean leaves to volatiles from herbivore-induced conspecific plants results in emission of carnivore attractants: active or passive process? J. Chem. Ecol. 30: 1305–1317.

Choh, Y., S. Kugimiya and J. Takabayashi. 2006. Induced production of extrafloral nectar in intact lima bean plants in response to volatiles from spider mite-infested conspecifics plants as a possible indirect defense against spider mites. Oecologia 147: 455–460.

Cianfrogna, J. A., A. R. Zangerl and M. R. Berenbaum. 2002. Dietary and developmental influences on induced detoxification in an oligophage. J. Chem. Ecol. 28: 1349–1364.

Cipollini, D. 2004. Stretching the limits of plasticity: can a plant defend against both competitors and herbivores? Ecology 85: 28–37.

Cipollini, D. F. and J. Bergelson. 2001. Plant density and nutrient availability constrain constitutive and wound-induced expression of trypsin inhibitors in *Brassica napus*. J. Chem. Ecol. 27: 593–610.

Cipollini, D. F. and M. Sipe. 2001. Jasmonic acid treatment and mammalian herbivory differentially affect chemical defense expression and growth of *Brassica kaber*. Chemoecology 11: 137–143.

Cipollini, D. F., C. B. Purrington and J. Bergelson. 2003. Costs of induced responses in plants. Basic and Applied Ecology 4: 79–85.

Cipollini, D., S. Enright, M. B. Traw and J. Bergelson. 2004. Salicyclic acid inhibits jasmonic acid-induced resistance of *Arabidopsis thaliana* to *Spodoptera exigua*. Molecular Ecology 13: 1643–1653.

Cocroft, R. B. 1999. Parent-offspring communication in response to predators in a subsocial treehopper (Hemiptera: Membracidae: *Umbonia crassicornis*). Ethology 105: 553–568.

Cohen, Y., U. Gisi and E. Mosinger. 1991. Systemic resistance of potato plants against *Phytophthora infestans* induced by unsaturated fatty acids. Physiological and Molecular Plant Pathology 38: 255–263.

Cooper, W. E. 2006. Dynamic risk assessment: prey rapidly adjust flight initiation distance to changes in predator approach speed. Ethol. 112: 858–864.

Cory, J. S. and K. Hoover. 2006. Plant-mediated effects in insect-pathogen interactions. Trends in Ecology and Evolution 21(5): 278–286.

Cowley, T. and D. R. Walters. 2005. Local and systematic changes in arginine decarboxylase activity, putrescine levels and putrescine catabolism in wounded oilseed rape. New Phytologist 165(3): 807–811.

Cui, J., A. K. Bahrami, E. G. Pringle, G. Hernandez-Guzman, C. L. Bender, N. E. Pierce and F. M. Ausubel. 2005. *Pseudomonas syringae* manipulates systematic plant defenses against pathogens and herbivores. Proceedings of the National Academy of Sciences of the United States of America 102: 1791–1796.

Dahl, J. and B. L. Peckarsky. 2002. Induced morphological defenses in the wild: predator effects on a mayfly, *Drunella coloradensis*. Ecology 83: 1620–1634.

Dahl, J. and B. L. Peckarsky. 2003. Developmental responses to predation risk in morphologically defended mayflies. Oecologia 137: 188–194.

Dana, M. M., J. A. Pintor-Toro and B. Cubero. 2006. Transgenic tobacco plants over-expressing chitinases of fingal origin show enhanced resistance to biotic stress agents. Plant Physiology 142: 722–730.

Dangl, J. and J. D. G. Jones. 2001. Plant pathogens and integrated defence response to infection. Nature 411: 826–833.

Danner, B. J. and J. Joern. 2004. Development, growth, and egg production of *Ageneotettix deorum* (Orthoptera: Acrididae) in response to spider predation risk and elevated resource quality. Ecol. Entom. 29: 1–11.

Dawson, G. W., D. C. Griffiths, J. A. Pickett, L. J. Wadhams and C. M. Woodcock. 1987. Plant-derived synergists of alarm pheromone from turnip aphid, *Lipaphis* (*Hyadaphis*) *erysimi* (Homoptera, Aphidae). J. Chemical Ecology 13: 1663–1671.

Day, D. 1989. *Vanished Species*. Gallery Books, NY, USA.

Day, R. L., K. N. Laland and J. Odling-Smee. 2003. Rethinking adaptation: the niche-construction perspective. Perspectives in Biology and Medicine 46: 80–95.

Dean, J. M. and C. M. De Moraes. 2006. Effects of genetic modification on herbivore-induced volatiles from maize. J. Chem. Ecol. 32: 713–724.

De Boer, J. G. and M. Dicke. 2004. The role of methyl salicylate in prey searching behavior of the predatory mite *Phytoseiulus persimilis*. J. Chem. Ecol. 30: 255–271.

De Boer, J. G., M. A. Posthumus and M. Dicke. 2004. Identification of volatiles that are used in discrimination between plants infested with prey or nonprey herbivores by a predatory mite. J. Chem. Ecol. 30: 2215–2230.

De Bruijn, P. J. A., M. Egas, A. Janssen and M. W. Sabelis. 2006. Pheromone-induced priming of a defensive response in western flower thrips. J. Chem. Ecol. 32: 1599–1603.

Degenhardt, D. C. and D. E. Lincoln. 2006. Volatile emissions from an odorless plant in response to herbivory and methyl jasmonate exposure. J. Chemical Ecology 32: 725–743.

De Meester, L. and J. Pijanowska. 1996. On the trait-specificity of the response of *Daphnia* genotpes to the chemical presence of a predator. In Lenz PH, Hartline DK, Purcell JE and Macmillan DL (eds.). Sensory Ecology and Physiology. Gordon and Breach, Amsterdam Pp. 407–417.

De Meester, L., P. Dawidowicz, E. Van Gool and C. J. Loose. 1999. Ecology and evolution of predator-induced behavior of zooplankton: depth selection behavior and diel vertical migration. In Tollrian R and Harvell CD (eds.). The Ecology and Evolution of Inducible Defenses. Princeton University Press, Princeton, NJ Pp. 160–176.

De Moraes, C. M. and M. C. Mescher. 2004. Biochemical crypsis in the avoidance of natural enemies by an insect herbivore. Proceedings of the National Academy of Sciences of the United States of America 101(24): 8993–8997.

De Moraes, C. M., W. J. Lewis, P. W. Paré, H. T. Alborn and J. H. Tumlinson. 1998. Herbivore-infested plants selectively attact parasitoids. Nature 393: 570–573.

De Moraes, C. M., M. C. Mescher and J. H. Tumlinson. 2001. Caterpillar-induced nocturnal plant volatiles repel conspecific females. Nature 410: 577–580.

De Moraes, C. M., J. C. Schultz, M. C. Mescher and J. H. Tumlinson. 2004. Induced plant signaling and its implications for environmental sensing. J. Toxicology & Environmental Health 67: 819–834.

De Vos, M., V. R. Van Oosten, R. M. P. Van Poecke, J. A. Van Pelt, M. J. Pozo, M. J. Mueller, A. J. Buchala, J.-P. Metraux, L. C. Van Loon, M. Dicke and C. M. J. Pieterse. 2005. Signal signature and transcriptome changes of *Arabidopsis* during pathogen and insect attack. Molecular Plant-Microbe Interactions 18: 923–937.

De Vos, M., W. Van Zaanen, A. Koornneef, J. P. Korzelius, M. Dicke, L. C. Van Loon and C. M. J. Pieterse. 2006. Herbivore-induced resistance against microbial pathogens in *Arabidopsis*. Plant Physiology 142: 352–363.

DeWitt, T. J. and S. M. Scheiner (eds.). 2004. *Phenotypic Plasticity: Functional and Conceptual Approaches*. Oxford University Press, Oxford.

DeWitt, T. J. and S. M. Scheiner. 2004. Phenotypic variation from single genotypes: A primer. In: DeWitt T. J. and S. M. Scheiner (eds). Phenotypic plasticity: Functional and Conceptual Approaches. Oxford University Press. Pp. 1–9.

DeWitt, T. J., A. Shi and D. S. Wilson. 1998. Costs and limits of phenotypic plasticity. Trends Ecol. Evol. 13: 77–81.

DeWitt, T. J., A. Sih and J. A. Hucko. 1999. Trait compensation and cospecialization in a freshwater snail: size, shape and antipredator behaviour. Anim. Behav. 58: 397–407.

Dicke, M. 1999a. Evolution of induced indirect defense of plants. In Tollrian R and Harvell CJ (eds.). The Ecology and Evolution of Inducible Defenses. Princeton University Press, Princeton, NJ Pp 62–88.

Dicke, M. 1999b. Are herbivore-induced plant volatiles reliable indicators of herbivore identity to foraging carnivorous arthropods? Entomol. Exp. Appl. 91: 131–142.

Dicke, M. and H. Dijkman. 1992. Induced defense in detached uninfested plant leaves: effects on behavior of herbivores and their predators. Oecologia 91: 554–560.

Dicke, M. and H. Dijkman. 2001. Within-plant circulation of systemic elicitor of induced defence and release from roots of elicitor that affects neighbouring plants. Biochemical Systematics & Ecology 29: 1075–1087.

Dicke, M. and P. Grostal. 2001. Chemical detection of natural enemies by arthropods: an ecological perspective. Annu. Rev. Ecol. Syst. 32: 1–23.

Dicke, M. and M. Hilker. 2003. Induced plant defences: from molecular biology to evolutionary ecology. Basic and Applied Ecology 4: 3–14.

Dicke, M. and M. W. Sabelis. 1992. Costs and benefits of chemical information conveyance: proximate and ultimate factors. Pp. 31–60 in Roitberg B.S. and Isman M.B. (eds.). Insect Chemical Ecology: An Evolutionary Approach. Chapman and Hall, New York.

Dicke, M., R. M. P. Van Poecke and J. G. de Boer. 2003. Inducible indirect defence of plants: from mechanisms to ecological functions. Basic and Applied Ecology 4: 27–42.

Dickman, M. B., Y. S. Ha, Z. Yang, B. Adams and C. Huang. 2003. A protein kinase from *Colletotrichum trifolii* is induced by plant cutin and is required for appressorium formation. Molecular Plant-Microbe Interactions 16: 411–421.

Dietrich, R., K. Ploss and M. Heil. 2005. Growth responses and fitness costs after induction of pathogen resistance depend on environmental conditions. Plant Cell & Environment 28: 211–222.

Dixon, A. F. G. and B. K. Agarwala. 1999. Ladybird-induced life-history changes in aphids. Proc. R. Soc. London. B. 266: 1549–1553.

Donohue, K. 2005. Niche construction through phenological plasticity: life history dynamics and ecological consequences. New Phytologist 166: 83–92.

Doss, R. P. 2005. Treatment of pea pods with bruchin B results in up-regulation of a gene similar to MtN19. Plant Physiol. Biochem. 43: 225–231.

Doss, R. P., J. E. Oliver, W. M. Proebsting, S. W. Potter, S. R. Kuy, S. L. Clement, R. T. Williamson, J. R. Carney, and E. D. DeVilbiss. 2000. Bruchins: Insect-derived plant regulators that stimulate neoplasm formation. Proc. Natl. Acad. Sci. 97: 6218–6223.

Drukker, B., J. Bruin and M. W. Sabelis. 2000. Anthocorid predators learn to associate herbivore-induced plant volatiles with presence or absence of prey. Physiological Entomology 25: 260–265.

Dugravot, S. and E. Thibout. 2006. Consequences for a specialist insect and its parasitoid of the response of *Allium porrum* to conspecific herbivore attack. Physiological Entomology 31: 73–79.

Dukas, R. 1998. Ecological relevance of associative learning in fruit fly larvae. Behav. Ecol. Sociobiol. 19: 195–200.

Duquette, S. L., R. Altwegg and B. R. Anholt. 2005. Factors affecting the expression of inducible defences in *Euplotes*: genotype, predator density and experience. Functional Ecology 19: 648–655.

Dyer, L. A., D. K. Letourneau, C. D. Dodson, M. A. Tobler, J. O. Stireman and A. Hsu 2004. Ecological causes and consequences of variation in defensive chemistry of a neotropical shrub. Ecology 85: 2795–2803.

Eberhard, W. G. 2000. Spider manipulation by a wasp larvae. Nature 406: 255–256.

Ebert, J. and G. Westhoff. 2006. Behavioural examination of the infrared sensitivity of rattlesnakes (*Crotalus atrox*). J. Comp. Physiol. A 192: 941–947.

Edmunds, M. 1974. *Defense in Animals*. Longman, Harlow, UK.

Edreva, A. 2004. A novel strategy for plant protection: induced resistance. Cell and Molecular Biology 3: 61–69

Elgar, M. A. and R. A. Allan. 2006. Chemical mimicry of the ant *Oecophylla smaragdina* by the myrmecophilous spider *Cosmophasis bitaeniata*: Is it colony-specific? J. Ethology 24: 239–246.

Engelberth, J., I. Seidl-Adams, J. C. Schultz and J. H. Tumlinson. 2007. Insect elicitors and exposure to green leafy volatiles differentially upregulate major octadecanoids and transcripts of 12-oxo phytodienoic acid reductases in *Zea mays*. Mol. Plant-Microbe Interactions 20: 707–716.

Erbilgin, N., P. Krokene, E. Christiansen, G. Zeneli and J. Gershenzon. 2006. Exogenous application of methyl jasmonate elicits defenses in Norway spruce (*Picea abies*) and reduces host colonization by the bark beetle *Ips typographus*. Oecologia 128: 426–436.

Evans, D. L. and J. O. Schmidt (eds.). 1990. *Insect Defenses Adaptive Mechanisms and Strategies of Prey and Predators*. SUNY, Albany, NY.

Farag, M. A. and P. W. Paré. 2002. C_6-Green leaf volatiles trigger local and systemic VOC emissions in tomato. Phytochemistry 61: 545–554.

Farag, M., M. Fokar, H. A. H. Zhang, R. D. Allen and P. W. Pare. 2005. (Z)-3-Hexenol induces defense genes and downstream metabolites in maize. Planta 220: 900–909.

Farmer, E. E. and C. A. Ryan. 1990. Interplant communication: airborne methyl jasmonate induces synthesis of proteinase inhibitors in plant leaves. Proc. Natl. Acad. Sci. 87: 7713–7716.

Fatouros, N., K. G. Bukovinszkine, L. A. Kalkers, R. S. Gamborena, M. Dicke, and M. Hilker. 2005a. Oviposition-induced plant cues: do they arrest *Trichogramma* wasps during host location? Entom. Exp. Appl. 115: 207–215.

Fatouros, N., J. J. A. Van Loon, K. A. Hordijk, H. M. Smid and M. Dicke. 2005b. Herbivore-induced plant volatiles mediate in-flight host discrimination by parasitoids. J. Chem. Ecol. 31: 2033–2047.

Feeney, P. 1976. Plant apparancy and chemical defense. Recent Advances in Phytochemistry 10: 1–40.

Felton, G and K. Korth. 2000. Trade-offs between pathogen and herbivore resistance. Curr. Opin. Plant Biol. 3: 309–314.

Ferrari, M. C., F. Messier and D. P. Chivers. 2007. Variable predation risk and the dynamic nature of mosquito antipredator responses to chemical alarm cues. Chemoecology 17: 223–229.

Ferrari, M. C., F. Messier and D. P. Chivers. 2008. Threat-sensitive learning of predators by larval mosquitoes *Culex restuans*. Beh. Ecol. Soc. 62: 1079–1083.

Feyereisen, R. 1999. Insect P450 enzymes. Annu. Rev. Entomol. 44: 507–533.

Fidantsef, A. L., M. J. Stout and J. S. Thaler. 1999. Signal interactions in pathogen and insect attack: expression of lipoxygenase, proteinase inhibitor II, and pathogenesis-related protein P4 in the tomato, *Lycopersicon esculentum*. Physiol. Molec. Plant Path. 54: 97–114.

Fischer, M. K. and A. W. Shingleton. 2001. Host plant and ants influence the honeydew sugar composition of aphids. Functional Ecology 15(4): 544–550.

Fischer, S., J. Samietz, F. L. Wackers and S. Dorn. 2004. Perception of chromatic cues during host location by the pupal parasitoid *Pimpla turionellae* (L.) (Hymenoptera: Ichneumonidae). Environmental Entomology 33: 81–87.

Folgarait, P. J., H. L. Johnson and D. W. Davidson. 1994. Responses of *Cecropia* to experimental removal of Mullerian bodies. Functional Ecology 8: 22–28.

Fortune, E. S., G. J. Rosa and M. Kawasaki. 2006. Encoding and processing biologically relevant temporal information in electrosensory systems. J Comp. Physiol. A. 192: 625–635.

Foster, S. A. 1999. The geography of behaviour: an evolutionary perspective. TREE 14: 190–195.

Fréchette, B., C. Alauzet and J.-L. Hemptinne. 2003. Oviposition behavior of the two-spot ladybird beetle *Adalia bipunctata* (L.) (Coleoptera: Coccinellidae) on plants with conspecific larval tracks. Proc. 8th Inter. Symp. Ecol. Aphidophaga: Biology, Ecology and Behavior of Aphidiphagous Insects. Arquipélago—Life and Marine Science (Suppl. 5), 73–77.

Fréchette, B., A. F. G. Dixon, C. Alauzet and J.-L. Hemptinne. 2004. Age and experience influence patch assessment for oviposition by an insect predator. Ecological Entomology 29: 578–583.

Frost, S. D. W. 1999. The immune system as an inducible defense. In Tollrian R. and C. D. Harvell (eds.). The Ecology and Evolution of Inducible Defenses. Princeton University Press, Princeton, NJ. Pp. 104–126.

Frost, C. J., M. Appel, J. E. Carlson, C. M. De Moraes, M. C. Mescher and J. C. Schultz. 2007. Within-plant signaling via volatiles overcomes vascular constraints on systemic signaling and primes responses against herbivores. Ecol. Let. 10: 490–498.

Gabriel, W. 1999. Evolution of reversible plastic responses: inducible defenses and environmental tolerance. In Tollrian R. and Harvell C. D. (eds.). The Ecology and Evolution of Inducible Defenses. Princeton University Press, Princeton, NJ. Pp. 286–305.

Gabriel, W. 2005. How stress selects for reversible phenotypic plasticity. J. Evol. Biol 18: 873–883.

Gabriel, W., B. Luttbeg, A. Sih and R. Tollrian. 2005. Environmental tolerance, heterogeneity, and the evolution of reversible plastic responses. American Naturalist 166: 339–353.

Gardner, S. N. and A. A. Agrawal. 2002. Induced plant defence and the evolution of counter-defences in herbivores. Evolutionary Ecology Research 4: 1131–1151.

Giamoustaris, A. and R. Mithen. 1995. The effect of modifying the glucosinolate content of oilseed rape (*Brassica napus* ssp. *oleifera*) on its interaction with specialist and generalist pests. Ann. Appl. Biol. 126: 347–363.

Gianoli, E. and M. A. Molina-Montenegro. 2005. Leaf damage induces twining in a climbing plant. New Phytologist 167: 385–389.

Gish, M. and M. Inbar. 2006. Host location by apterous aphids after escape dropping from the plant. J. Insect Behavior 19: 143–153.

Glazebrook, J. 2005. Contrasting mechanisms of defense against biotrophic and necrotrophic pathogens. Ann. Rev. Phytopathol. 43: 205–227.

Glynn, C., D. A. Herms, M. Egawa, R. Hansen and W. J. Mattson. 2003. Effects of nutrient availability on biomass allocation as well as constitutive and rapid induced herbivore resistance in poplar. Oikos 101: 385–397.

Gog, L., M. R. Berenbaum, E. H. DeLucia and A. R. Zangerl. 2005. Autotoxic effects of essential oils on photosynthesis in parsley, parsnip, and rough lemon. Chemoecology 15(2): 115–119.

Gomez, S. K., M. M. Cox, J. C. Bede, K. Inoue, H. T. Alborn, J. H. Tumlinson and K. L. Korth. 2005. Lepidopteran herbivory and oral factors induce transcripts encoding novel terpene synthases in *Medicago truncatula*. Archives of Insect Biochemistry and Physiology 58(2): 114–127.

Gomez-Mestre, I., J. C. Touchon, V. L. Saccoccio and K. M. Warkentin. 2008. Genetic variation in pathogen-induced early hatching of toad embryos. J. Evol. Biol. 21: 791-800.

Gomulkiewicz, R. and M. Kirkpatrick. 1992. Quantitative genetics and the evolution of reaction norms. Evolution 46: 390–411

Gouinguene, S., H. Alborn and T. C. J. Turlings. 2003. Induction of volatile emissions in maize by different larval instars of *Spodoptera littoralis*. J. Chem. Ecol. 29: 145–162.

Greenberg, J. T. and N. Yao. 2004. The role and regulation of programmed cell death in plant-pathogen interactions. Cell. Microbiol. 6: 201–211.

Grostal, P. and M. Dicke. 1999. Direct and indirect cues of predation risk influence behavior and reproduction of prey: a case for acarine interactions. Behavioral Ecology 10: 422–427.

Grostal, P. and M. Dicke. 2000. Recognising one's enemies: a functional approach to risk assessment by prey. Behavioral Ecology & Sociobiology 47: 258–264.

Guerrieri, E., G. M. Poppy, W. Powell, E. Tremblay and R. Pennachio. 1999. Induction and systemic release of herbivore-induced plant volatiles mediating in-flight orientation of *Aphidius ervi*. J. Chem. Ecol. 25: 1247–1261.

Hairston, N. G., T. A. Dillon and B. T. DeStasio. 1990. A field test for the cues of diapause in a freshwater copepod. Ecology 71: 2218–2223.

Halim, V. A.., A. Vess, D. Scheel and S. Rosahl. 2006. The role of salicylic acid and jasmonic acid in pathogen defence. Plant Biol. 8: 307–313.

Halitschke, R., A. Kessler, J. Kahl, A. Lorenz and I. T. Baldwin. 2000. Ecophysiological comparison of direct and indirect defenses in *Nicotiana attenuata*. Oecologia 124: 408–417.

Halitschke, R., U. Schittko, G. Pohnert, W. Boland and L. T. Baldwin. 2001. Molecular interactions between the specialist herbivore *Manduca sexta* (Lepidoptera, Sphingidae) and its natural host *Nicotiana attenuata*: III. Fatty acid-amino amic conjugates in herbivore oral secretions are necessary and sufficient for herbivore-specific plant responses. Plant Physiol. 125: 711–717.

Hammill, E., A. Rogers and A. P. Beckerman. 2008. Costs, benefits and the evolution of inducible defenses: a case study with *Daphnia pulex*. J. Evol. Biol. 21: 705–715.

Hammond, J. I., B. Luttbeg and A. Sih. 2007. Predator and prey space use: dragonflies and tadpoles in an interactive game. Ecology 88: 1525–1535.

Han, K. and D. E. Lincoln. 1995. The evolution of carbon allocation to plant secondary metabolites: a genetic analysis of cost in *Diplacus aurantiacus* . Evolution 48: 1550–1563.

Haribal, M. and J. A. Renwick. 2005. Plasticity in insect response to variable chemistry of host plants. In Ananthakrishnan T.N. and Whitman D.W. (eds.). Insect Phenotypic Plasticity Diversity of Responses. Science Publishers, Enfield, NH, USA. Pp. 60–80.

Harms, K., R. Atzom, A. Brash, H. Kühn, C. Wasternack, L. Willmitzer, and H. Peña Cortés. 1995. Expression of a flax allene oxide synthase cDNA leads to increased endogenous jasmonic acid (JA) levels in transgenic potato plants but not to a corresponding activation of JA-responding genes. Plant Cell 7: 1645–1654.

Hartley, S. E. and J. H. Lawton. 1991. Biochemical aspects and significance of the rapidly induced accumulation of phenolics in birch foliage. In Tallamy D. W. and M. J. Raupp (eds.). Phytochemical Induction by Hervivores. John Wiley, New York, Pp. 105–132.

Haruta, M., I. T. Major, M. E. Christopher, J. J. Patton and C. P. Constabel. 2001. A Kunitz trypsin inhibitor gene family from trembling aspen (*Populus tremuloides* Michx.): cloning, functional expression, and induction by wounding and herbivory. Plant Molecular Biol. 46: 347–359.

Harvell, C. D. 1990. The ecology and evolution of inducible defenses. Q. Rev. Biol. 65: 323–340.

Harvell, C. D. and R. Tollrian. 1999. Why inducible defenses? Pp 3–9 in Tollrian R. and Harvell C.D. (eds.). The Ecology and Evolution of Inducible Defenses. Princeton University Press, Princeton, NJ.

Hatanaka, A. 1993. The biogeneration of green odour by green leaves. Phytochemistry 34: 1201–1218.

Haukioja, E. 1991. Induction of defenses in trees. Ann. Rev. Entomol. 36: 25–42.

Havill, N. P. and K. F. Raffa. 1999. Effects of elicitation treatment and genotypic variation on induced resistance in *Populus*; impacts on gypsy moth (Lepidoptera: Lymantriidae) development and feeding behavior. Oecologia 120: 295–303.

Hazlett, B. A. 1999. Responses to multiple chemical cues by the crayfish *Orconectes virilis*. Behaviour 136: 161–177.

Hechtel, L. J. and S. A. Juliano. 1997. Effects of a predator on prey metamorphosis: plastic responses of prey or selective mortality? Ecology 78: 838–851.

Heidel, A. J. and I. T. Baldwin. 2004. Microarray analysis of salicylic acid- and jasmonic acid- signalling in responses of *Nicotiana attenuata* to attack by insects from multiple feeding guilds. Plant Cell & Environment 27(11): 1362–1373.

Heijari, J., A.-M. Nerg, P. Kainulainen, H. Viiri, M. Vuorinen and J. K. Holopainen. 2005. Application of methyl jasmonate reduces growth but increases chemical defence and resistance against *Hylobius abietis* in Scots pine seedlings. Entomologia Experimentalis et Applicata 115: 117–124.

Heil, M. 2002. Ecological costs of induced resistance. Current Opinion in Plant Biology 5: 345–350.

Heil, M. 2004. Induction of two indirect defences benefits lima bean (*Phaseolus lunatus*, Fabaceae) in nature. J. Ecol. 92: 527–536.

Heil, M. 2008. Indirect defence via trophic interactions. New Phytologist 178: 41–61.

Heil, M., B. Fiala, B. Baumann and K. E. Linsenmair. 2000. Temporal, spatial and biotic variations in extrafloral nectar secretion by *Macaranga fanarius*. Funct. Ecol. 14: 749–757.

Heil, M., S. Greiner, H. Meimberg, R. Kruger, J. L. Noyer, G. Heubl, K. E. Linsenmair and W. Boland. 2004. Evolutionary change from induced to constitutive expression of an indirect plant resistance. Nature 430: 205–208.

Heil, M., U. Lion and W. Boland. 2008. Defense-inducing volatiles: in search of the active motif. J. Chem. Ecol. 34: 601–604.

Heinrich, B. 1993. *The Hot-Blooded Insects: Strategies and Mechanisms of Thermoregulation.* Harvard University Press, Cambridge.

Herde, O., H. Fuss, H. Peña-Cortés and J. Fisahn. 1995. Proteinase inhibitor II gene expression induced by electrical stimulation and control of photosynthetic activity in tomato plants. Plant Cell Physiology 36: 737–742.

Hessen, D. O. and E. van Donk. 1993. Morphological changes in *Scenedesmus* induced by substances released from *Daphnia*. Arch. Hydrobiol. 127: 129–140.

Hilker, M. and T. Meiners. 2002. Induction of plant responses to oviposition and feeding by herbivorous arthropods: a comparison. Entomol. Exp. Appl. 104: 181–192.

Hilker, M. and T. Meiners. 2006. Early herbivore alert: insect eggs induce plant defense. J. Chem. Ecol. 32: 1379–1397.

Hilker, M., C. Stein, R. Schröeder, M. Varama and R. Mumm. 2005. Insect egg deposition induces defence responses in *Pinus sylvestris*: characterisation of the elicitor. J. Exper. Biol. 208(10): 1849–1854.

Hlivko, J. T. and A. L. Rypstra. 2003. Spiders reduce herbivory: Nonlethal effects of spiders on the consumption of soybean leaves by beetle pests. Ann. of Entomol. Soc. of Amer. 96: 914–919.

Hoballah, M. E., T. G. Köllner, J. Degenhardt and T. C. J. Turlings. 2004. Costs of induced volatile production in maize. Oikos 105: 168–180.

Hoelker, F. and P. Stief. 2005. Adaptive behaviour of chironomid larvae (*Chironomus riparius*) in response to chemical stimuli from predators and resource density. Behav. Ecol. Sociobiol. 58: 256–263.

Hoffmeister, T. S. and B. D. Roitberg. 1997. Counterespionage in an insect herbivore-parasitoid system. Naturwissenschaften 84: 117–119.

Höller, C., S. G. Micha, S. Schulz, W. Francke and J. A. Pickett. 1994. Enemy-induced dispersal in a parasitic wasp. Experientia 50: 182–185.

Horiuchi, J., G. Arimura, R. Ozawa, T. Shimoda, J. Takabayashi and T. Nishioka. 2001. Exogenous ACC enhances volatiles production mediated by jasmonic acid in lima bean leaves. FEBS Lett. 509: 332–336.

Hountondji, F. C. C., M. W. Sabelis, R. Hanna and A. Janssen. 2005. Herbivore-induced plant volatiles trigger sporulation in entomopathogenic fungi: the case of *Neozgites tanajoe* infecting the cassava green mite. J. Chem. Ecol. 31: 1033–1021.

Hoverman, J. T., J. R. Auld and R. A. Relyea. 2005. Putting prey back together again: integrating predator-induced behavior, morphology, and life history. Oecologia 144: 481–491.

Huang, J., Y. J. Cardoza, E. A. Schmelz, R. Raina, J. Engelberth and J. H. Tumlinson. 2003a. Nitrogen deficiency increases volicitin-induced volatile emission, jasmonic acid accumulation, and ethylene sensitivity in maize. Plant Physiology 133: 295–306.

Huang, J., Y. J. Cardoza, E. A. Schmelz, R. Raina, J. Engelberth and J. H. Tumlinson. 2003b. Differential volatile emissions and salicylic acid levels from tobacco plants in response to different strains of *Pseudomonas syringae*. Planta 217: 767–775.

Huang, J., E. A. Schmelz, H. Alborn, J. Engelberth and J. H. Tumlinson. 2005. Phytohormones mediate volatile emissions during the interaction of compatible and incompatible pathogens: the role of ethylene in *Pseudomonas syringae* infected tobacco. J. Chem. Ecol. 31: 439–459.

Hudgins, J. W., S. G. Ralph and V. R. Franceschi. 2006. Ethylene in induced conifer defense: cDNA cloning, protein expression, and cellular and subcellular localization of 1-aminocyclopropane-1-carboxylate oxidase in resin duct and phenolic parenchyma cells. Planta 224: 865–867.

Hunt, J. H. 2007. *The Evolution of Social Wasps*. Oxford University Press, New York.

Huntingford, F. A., P. J. Wright and J. F. Tierney. 1994. Adaptive variation in antipredator behavior in threespine stickleback. In Bell M. A. and Foster S. A. (eds.). The Evolutionary Biology of the Threespine Stickleback. Oxford University Press, Oxford. Pp. 277–296.

Huntzinger, M., R. Karban, T. P. Young and T. M. Palmer. 2004. Relaxation of induced indirect defenses of acacias following exclusion of mammalian herbivores. Ecology 85: 609–614.

Huryn, A. D. and D. P. Chivers. 1999. Contrasting behavioral responses by detritivorous and predatory mayflies to chemicals released by injured conspecifics and their predators. J. Chem. Ecol. 25: 2729–2740.

Irwin, R. E., S. Y. Strauss, S. Storz, A. Emerson and G. Guibert. 2003. The role of herbivores in the maintenance of a flower-color polymorphism in wild radish. Ecology 84: 1733–1743.

Ishikawa, A., S. Hongo and T. Miura. 2008. Morphological and histological examination of polyphenic wing formation in the pea aphid *Acyrthosiphon pisum* (Hemiptera, Hexapoda). Zoomorphology 127: 121–133.

Jablonski, P. G. and S. D. Lee. 2006. Effects of visual stimuli, substrate-borne vibrations and air current stimuli on escape reactions in insect prey of flush-pursuing birds and their implications for evolution of flush-pursuers. Behaviour 143: 303–324.

Jacot, A., H. Scheuber, J. Kurtz and M. W. G. Brinkhof. 2004a. Costs of an induced immune response on sexual display and longevity in field crickets. Evolution 58: 2280–2286.

Jacot, A., H. Scheuber, J. Kurtz and M. W. G. Brinkhof. 2005b. Juvenile immune status affects the expression of a sexually selected trait in field crickets *Gryllus campestris*. J. Evol. Biol. 18: 1060–1068.

Jacot, A., H. Scheuber, J. Kurtz and M. W. G. Brinkhof. 2005. Juvenile immune system activation induces a costly upregulation of adult immunity in field crickets *Gryllus campestris*. Proc. Roy. Soc. Biol. Sci. Series B 272: 63–69.

James, D. G. 2003. Field evaluation of herbivore-induced plant volatiles as attractants for beneficial insects: methyl salicylate and the green lacewing, *Chrysopa nigricornis*. J. Chem. Ecol. 29: 1575–1583.

James, D. G. 2005. Further field evaluation of synthetic herbivore-induced plant volatiles as attractants for beneficial insects. J. Chem. Ecol. 31: 481–495.

James, D. G. and T. S. Price. 2004. Field-testing of methyl salicylate for recruitment and retention of beneficial insects in grapes and hops. J. Chem. Ecol. 30: 1613–1628.

Järemo, J., J. Tuomi and P. Nilsson. 1999. Adaptive status of localized and systemic defense response in plants. In Tollrian R. and C. D. Harvell (eds.). The Ecology and Evolution of Inducible Defenses. Princeton University Press, Princeton, NJ Pp. 33–44.

Jeong, G.-T. and D.-H. Park. 2005. Enhancement of growth and secondary metabolite biosynthesis: effect of elicitors derived from plants and insects. Biotechnology and Bioprocess Engineering 10: 73–77.

Johansson, F. 2002. Reaction norms and production costs of predator-induced morphological defences in a larval dragonfly (*Leucorrhinia dubia*: Odonata). Can. J. Zool. 80: 944–950.

Johansson, F., R. Stoks, L. Rowe and M. De Block. 2001. Life history plasticity in a damselfly: effects of combined time and biotic constraints. Ecology 82: 1857–1869.

Jones, M., A. Laurila, N. Peuhkuri, J. Piilronen and T. Seppa. 2003. Timing an ontogenetic niche shift: responses of emerging *Salmon alevins* to chemical cues from predators and competitors. Oikos 102: 155–163.

Jones, C. G., D. W. Whitman, P. J. Silk and M. S. Blum. 1988. Diet breadth and insect chemical defenses: a generalist grasshopper and general hypotheses. In Spencer K. C. (ed.). Chemical Mediation of Coevolution. Academic Press, San Diego Pp. 477–512.

de Jong, G. 1999. Unpredictable selection in a structured population leads to local genetic differentiation in evolved reaction norms. J. Evol. Biol. 12: 839–851.

de Jong, G. 2005. Evolution of phenotypic plasticity: patterns of plasticity and the emergence of ecotypes. New Phytologist 166: 101–118.

Jongsma, M. A., P. L. Bakker, J. Peters, D. Bosch and W. J. Stiekema. 1995. Adaptation of *Spodoptera exigua* larvae to plant proteinase inhibitors by induction of gut proteinase activity insensitive to inhibition. Proc. Natl. Acad. Sci. 92: 8041–8045.

Juenger, T. and J. Bergelson. 1998. Pairwise versus diffuse natural selection and the multiple herbivores of scarlet gilia, *Ipomopsis aggregata*. Evolution 52: 1583–1592.

Kahl, J., D. H. Siemens, R. J. Aerts, R. Gaebler, F. Kuehnemann, C. A. Preston and I. T. Baldwin. 2000. Herbivore-induced ethylene suppresses a direct defense but not a putative indirect defense against an adapted herbivore. Planta 210: 336–342.

Kaitaniemi, P., H. Vehvilainen and K. Ruohomaki. 2004. Movement and disappearance of mountain birch defoliators are influenced by the interactive effects of plant architecture and induced resistance. Ecological Entomology 29: 437–446.

Kaloshian, I. 2004. Gene-for-gene disease resistance: bridging insect pest and pathogen defense. J. Chem. Ecol. 30: 2419–2438.

Kaloshian, I. and L. L. Walling. 2005. Hemipterans as plant pathogens. Annual Review of Phytopathology 43: 491–521.

Kamita, S. G., K. Nagasaka, J. W. Chua, T. Shimada, K. Mita, M. Kobayashi, S. Maeda and B. D. Hammock. 2005. A baculovirus-encoded protein tyrosine phosphatase gene induces enhanced locomotory activity in a lepidopteran host. Proc. of the Natl. Acad. of Sci. USA 102: 2584–2589.

Kappers, I. F., A. Aharoni, van T. W. J. M. Herpen, L. L. P. Luckerhoff, M. Dicke and H. J. Bouwmeester. 2005. Genetic engineering of terpenoid metabolism attracts bodyguards to *Arabidopsis*. Science 309: 2070–2072, 2067.

Karban, R. and A. A. Agrawal. 2002. Herbivore offense. Annu. Rev. Ecol. Syst. 33: 641–664.

Karban, B. and T. Baldwin (eds.). 1997. *Induced Responses to Herbivory*. University of Chicago Press, Chicago.

Karban, R. and K. Nagasaka. 2004. Are defenses of wild radish populations well matched with variability and predictability of herbivory? Evolutionary Ecology 18: 283–301.

Karban, R., A. A. Agrawal, J. S. Thaler and L. S. Adler 1999. Induced plant responses and information content about risk of herbivory. TREE 14: 443–447.

Karban, R., I. T. Baldwin, K. J. Baxter, G. Laue and G. W. Felton. 2000. Communication between plants: Induced resistance in wild tobacco plants following clipping of neighboring sagebrush. Oecologia 125: 66–71.

Karban, R., M. Huntzinger and A. C. McCall. 2004. The specificity of eavesdropping on sagebrush by other plants. Ecology 85: 1846–1852.

Kats, L. B. and L. M. Dill. 1998. The scent of death: chemosensory assessment of predation risk by prey animals. Ecoscience 5: 361–394.

Kessler, A. and I. T. Baldwin. 2001. Defensive function of herbivore-induced plant volatile emissions in nature. Science 291: 2141–2144.

Kessler, A. and I. T. Baldwin. 2004. Herbivore-induced plant vaccination. Part I. The orchestration of plant defenses in nature and their fitness consequences fro the wild tobacco *Nicotiana attenuata*. Plant J. 38: 639–649.

Kessler, A., R. Halitshke, C. Diezel and I. T. Baldwin. 2006. Priming of plant defense responses in nature by airborne signaling between *Artemisia tridentata* and *Nicotiana attenuata*. Oecologia 148: 280–292.

Kessmann, H., T. Staub, C. Hofmann, T. Maetzke and J. Herzog. 1994. Induction of systemic acquired disease resistance in plants by chemicals. Annu. Rev. of Phytopathol. 32: 439–459.

Kiesecker, J. M., D. P. Chivers and A. R. Blaustein. 1996. The use of chemical cues in predator recognition by western toad tadpoles. Animal Behaviour 52: 1237–1345.

Killian, K. A., L. C. Snell, R. Ammarell and T. O. Crist 2006. Suppression of escape behaviour during mating in the cricket *Acheta domesticus*. Animal Behaviour 72: 487–502.

King, B. H. and H. R. Leaich. 2006. Variation in propensity to exhibit thnatosis in *Nasania vitripennis* (Hymenoptera: Pteromalidae). J. Insect Behavior 19: 241–247.

Kingsolver, J. G., D. W. Pfennig and M. R. Servedio. 2002. Migration, local adaptation and the evolution of plasticity. Trends Ecol. Evol. 17: 540–541.

Kishida, O. and K. Nishimura. 2005. Multiple inducible defences against multiple predators in the anuran tadpole, *Rana pirica*. Evolutionary Ecology Research 7: 619–631.

Kishida, O. and K. Nishimura. 2006. Flexible architecture of inducible morphological plasticity. J. Anim. Ecol. 75: 705–712.

Kishimoto, K., K. Matsui, R. Ozawa and J. Takabayashi. 2005. Volatile C6-aldehydes and allo-ocimene activate defense genes and induce resistance against *Botrytis cinera* in *Arabidopsis thaliana*. Plant & Cell Physiol. 46: 1093–1102.

Kliebenstein, D. J., A. Figuth and T. Mitchell-Olds. 2002. Genetic architecture of plastic methyl jasmonate responses in *Arabidopsis thaliana*. Genetics 161: 1685–1696.

Kliebenstein, D. J., J. Kroymann and T. Mitchell-Olds. 2005. The glucosinolate-myrosinase system in an ecological and evolutionary context. Current Opinion in Plant Biology 8: 264–271.

Kleinjan, J. E. and T. E. Mittler. 1975. A chemical influence of ants on wing development in aphids. Entomol. Exp. Applicata 18: 384–388.

Kodama, H., T. Nishiuchi, S. Seo, Y. Ohashi and K. Iba. 2000. Possible involvement of protein phosphorylation in the wound-responsive expression of *Arabidopsis* plastid *w*-3 fatty acid desaturase gene. Plant Science 155: 153–160.

Kolar, C. S. and D. H. Wahl. 1998. Daphnid morphology deters fish predators. Oecologia 116: 546–554.

Kopp, M. and R. Tollrian. 2003. Recirocal phenotypic plasticity in a predator-prey system: inducible offences against inducible defences? Ecol. Lett. 6: 742–748.

Korb, J. and A. Fuchs. 2006. Termites and mites-adaptive behavioural responses to infestation? Behaviour 143: 891–907.

Koricheva, J., H. Nykanen and E. Gianoli. 2004. Meta-analysis of trade-offs among plant antiherbivore defenses: Are plants jacks-of-all-trades, masters of all? American Naturalist (Electronic article) 163: E64–E75.

Kost, C. and M. Heil. 2006. Herbivore-induced plant volatiles induce an indirect defence in neighboring plants. J. of Ecology 94: 619–628.

Kruijt, M., M. J. D. De Kock and P. J. G. M. de Wit. 2005. Receptor-like proteins involved in plant disease resistance. Molecular Plant Pathology 6: 85–97.

Kunert, G. and W. W. Weisser. 2003. The interplay between density- and trait-mediated effects in predator-prey interactions: a case study in aphid wing polymorphism. Oecologia 135: 304–312.

Kunert, G. and W. W. Weisser. 2005. The importance of antennae for pea aphid wing induction in the presence of natural enemies. Bull. Entomol. Res. 95: 125–131.

Kunert, G., S. Otto, U. S. R. Roese, J. Gershenzon and W. W. Weisser. 2005. Alarm pheromone mediates production of winged dispersal morphs in aphids. Ecology Letters 8: 596–603.

Kunz, A. K., M. Ford and O. J. Pung. 2006. Behavior of the grass shrimp *Palaemonetes pugio* and its response to the presence of the predatory fish *Fundulus heteroclitus*. American Midland Naturalist 155: 286–294.

Kurashige, N. S. and A. A. Agrawal. 2005. Phenotypic plasticity to light competition and herbivory in *Chenopodium album*. American Journal of Botany 92: 21–26.

Kurtz, J., K. Reinhold and L. Engqvist. 2000. Immunosuppression under stress: necessary for condition-dependent signalling? TREE 15: 418–419.

Kusch, R. C., R. S. Mirza, M. S. Pollock, R. J. Tremaine and D. P. Chivers. 2005. The effects of predation on phenotypic and life history variation in an aquatic vertebrate. Chem. Sig. Vert. 10: 342–348.

Kush, J. 1999. Self-recognition as the original function of an amoeban defense-inducing kairomone. Ecology 80: 715–720.

Kvam, O. V. and O. T. Kleiven. 1995. Diel horizontal migration and swarm formation in *Daphnia* in response to *Chaoborus*. Hydrobiologia 307: 177–184.

Lait, C. G., H. T. Alborn, P. E. Teal and J. H. Tumlinson. 2003. Rapid biosynthesis of N-linolenoyl-L-glutamine, an elicitor of plant volatiles, by membrane-associated enzyme(s) in *Manduca sexta*. Proc. Natl. Acad. Sci. 100: 7027–7032.

Lamotte, O., K. Gould, D. Lecourieux, A. Sequeira-Legrand, A. Lebrun-Garcia, J. Durner, A. Pugin and D. Wendehenne. 2004. Analysis of nitric oxide signaling functions in tobacco cells challenged by the elicitor cryptogen. Plant Physiol. 135: 516–529.

Lamotte, O., L. Barnavon, A. Pugin and D. Wendehenne. 2005. Nitric oxide in plants: the biosynthesis and cell signaling properties of a fascinating molecule. Planta 211: 1–4.

Lampert, W., K. O. Rothhapt and E. von Elert. 1994. Chemical induction of colony formation in a green alga (*Scenedesmus acutus*) by grazers (*Daphnia*). Limnol. Oceanogr. 39: 1543–1505.

Landolt, P. J., J. H. Tumlinson and D. H. Alborn. 1999. Attraction of Colorado potato beetle (Coleoptera : Chrysomelidae) to damaged and chemically induced potato plants. Environmental Entomology 28: 973–978.

Langerhans, R. B. and T. J. DeWitt. 2002. Plasticity constrained: overgeneralized induction cues cause maladaptive phenotypes. Evolutionary Ecology Research 4: 857–870.

Lass, S., M. Vos, J. Wolinska and P. Spaak. 2005. Hatching with the enemy: *Daphnia* diapausing eggs hatch in the presence of fish kairomones. Chemoecology 15: 7–12.

Laugbertie, E., X. Martini, C. Cadena, M. Treilhou, A. F. G. Dixon and J. L. Hemptinne. 2006. The immediate source of the oviposition-deterring pheromone produced by larvae of *Adalia bipunctata* (L.) (Coleoptera, Coccinellidae). J. Insect Behavior 19: 231–240.

Laurila, A. 2000. Behavioural responses to predator's chemical cues and local variation in antipredator performance in *Rana temporaria* tadpoles. Oikos 88: 159–168.

Lecourieux, D., R. Ranjeva and A. Pugin. 2006. Calcium in plant defence-signalling pathways. New Pyhtologist 171: 249–269.

Leduc, A. O. H. C., M. C. O. Ferari, J. M. Kelly and G. E. Brown. 2004. Learning to recognize novel predators under weakly acidic conditions: the effects of reduced pH on acquired predator recognition by juvenile trout. Chemoecology 14: 107–112.

Lee, D., D. H. Polisensky and J. Braam. 2005. Genome-wide identification of touch- and darkness-regulated *Arabidopsis* genes: a focus on calmodulin-like and XTH genes. New Phytologist 165: 429–444.

Leitner, M., W. Boland and A. Mithoefer. 2005. Direct and indirect defences induced by piercing-sucking and chewing herbivores in *Medicago truncatula*. New Phytologist 167: 597–606.

Léon, J., E. Rojo and J. J. Sánchez-Serrano. 2001. Wound signaling in plants. J. Exper. Botany 52: 1–9.

Letourneau, D. K. 1990. Code of ant-plant mutualism broken by parasite. Science 248: 215–217.

Li, C., M. M. Williams, Y.-T. Loh, G. I. Lee and G. A. Howe. 2002. Resistance of cultivated tomato to cell content-feeding herbivores is regulated by the octadecanoid-signaling pathway. Plant Physiol. 130: 494–503.

Li, D. 2002. Hatching responses of subsocial spitting spiders to predation risk. Proc. R. Soc. Lond. B. 269: 2155–2161.

Li, D. and R. R. Jackson. 2005. Influence of diet-related chemical cues from predators on the hatching of egg-carrying spiders. Journal of Chemical Ecology 31: 333–342.

Li, D. and W. Lee. 2004. Predator-induced plasticity in web-building behaviour. Animal Behavior 67: 309–318.

Li, J., G. Brader and E. T. Palva. 2004. The WRKY70 transcription factor: a note of convergence for jasmonate-mediated and salicylate-mediated signals in plant defense. Plant Cell 16: 319–331.

Li, W., A. R. Zangerl, M. A. Schuler and M. R. Berenbaum. 2004. Characterization and evolution of furanocoumarin-inducible cytochrome P450s in the parsnip webworm, *Depressaria pastinacella*. Insect Molecular Biology 13: 603–613.

Li, W. W. and W. Barz. 2005. Biotechnological production of two new 8,4'-oxynorneolignans by elicitation of *Echinacea purpurea* cell cultures. Tetrahedron Lett. 46: 2973–2977.

Li, X., M. A. Schuler and M. R. Berenbaum. 2002. Jasmonate and salicylate induce expression of herbivore cytochrome P450 genes. Nature 419: 712–715.

Li, Y., G. Kiddle, R. N. Bennett and R. M. Wallsgrove. 1999. Local and systemic changes in glucosinolates in Chinese and European cultivars of oilseed rape (*Brassica napus* L.) after inoculation with *Sclerotinia sclerotiorum* (stem rot). Annals of Applied Biology 134: 45–58.

Lichtenthaler, H. K. 1999. The 1-deoxy-D-xylulose-5-phosphate pathway of isoprenoid biosynthesis in plants. Annu. Rev. Plant Physiol. Plant Mol. Biol. 50: 47–65.

Lima, S. L. 2002. Putting predators back into behavioral predator-prey interactions. Trends in Ecol. and Evol. 17: 70–75.

Lima, S. L. and P. A. Bednekoff. 1999. Temporal variation in danger drives antipredatory behavior: the predation risk allocation hypothesis. American Naturalist 153: 649–659.

Lively, C. M. 1999. Developmental strategies in spatially variable environments: Barnacle shell dimorphism and strategic models of selection. See Tollrian & Harvell 1999, pp. 245–258.

Losos, J. B., D. A. Creer, D. Glossip, R. Goellner, A. Hampton, G. Roberts, N. Haskell, P. Taylor and J. Ettling. 2000. Evolutionary implications of phenotypic plasticity in the hindlimb of the lizard *Anolis sagrei*. Evolution 54: 301–305.

Losos, J. B., T. W. Schoener and D. A. Spiller. 2004. Predator-induced behaviour shifts and natural selection in field-experimental lizard populations. Nature 432: 505–508.

Losos, J. B., T. W. Schoener, R. B. Langerhans and D. A. Spiller. 2006. Rapid temporal reversal in predator-driven natural selection. Science 314: 111.

Lou, Y.-G., B. Ma and J.-A. Cheng. 2005. Attraction of the parasitoid *Anagrus nilaparvatae* to rice volatiles induced by the rice brown planthopper *Nilaparvata lugens*. J. Chemical Ecology 31: 2357–2359.

Lyytinen, A., P. M. Brakefield, L. Lindstrom and J. Mappes. 2004. Does predation maintain eyespot plasticity in *Bicyclus anyana*? Proc. Royal Society Biological Sciences B. 271: 279–283.

Maffei, M., S. Bossi, D. Spiteller, A. Mithöfer and W. Boland. 2004. Effects of feeding *Spodoptera littoralis* on lima bean leaves: I. Membrane potentials, intrcellular calcium variations, oral secretions, and regurgitate components. Plant Physiol. 134: 1752–1762.

Maffei, M. E., A. Mithofer and W. Boland. 2007a. Insects feeding on plants: rapid signals and responses preceding the induction of phytochemical release. Phytochemistry 68: 2946–2959.

Maffei, M. E., A. Mithofer and W. Boland. 2007b. Before gene expression: early events in plant-insect interaction. Trends Plant Sci. 12: 310–316.

Maginnis, T. L. 2006. The costs of autonomy and regeneration in animals: a review and framework for future research. Behavioral Ecology 17: 857–872.

Malolepsza, U. and H. Urbanek. 2002. O-Hydroxyethylorutin-mediated enhancement of tomato resistance to *Botrytis cinerea* depends on a burst of reactive oxygen species. J. Phytopathology 150: 616–624.

Malolepsza, U. and S. Rozalska. 2005. Nitric oxide and hydrogen peroxide in tomato resistance—nitric oxide modulates hydrogen peroxide level in O-hydroxyethylorutin-induced resistance to *Botrytis cinerea* in tomato. Plant Phys. & Biochem. 43: 623–635.

Malone, M. and J. J. Alarcón. 1995. Only xylem-borne factors can account for systemic wound signaling in the tomato plant. Planta 196: 740–746.

Mandrillon, A. L. and P. Saglio. 2007. Effects of embryonic exposure to conspecific chemical cues on hatching and larval traits in the common frog (*Rana temporaria*). 2007. Chemoecology 17: 169–175.

Manzoli-Palma, M. F., N. Gobbi and M. S. Palma. 1998. Alarm pheromones and the influence of pupal odor on the aggressiveness of *Polybia paulista* (Ihering) (Hymenoptera: Vespidae). J. Venom. Anim. Toxins 4: 61–69.

Mao, W., M. A. Berhow, A. R. Zangerl, J. McGovern and M. R. Berenbaum. 2006. Chtychrome P450-mediated metabolism of xanthotoxin by *Papilio multicaudatus*. J. Chem. Ecol. 32: 523–.

Martinsen, G. D., E. M. Driebe and T. G. Whitham. 1998. Indirect interactions mediated by changing plant chemistry: beaver browsing benefits beetles. Ecology 79: 192–200.

Matsui, K., S. Kurishita, A. Hisamitsu and T. Kajiwara. 2000. A lipid-hydrolysing activity involved in hexenal formation. Biochem. Soc. Trans. 28: 857–860.

Matsumoto, K., M. Okamoto, H. Horikawa, K. Nakagawa, H. Yamamaura and Y. Kuwahara. 1998. The effects of the different environmental conditions on the dispersion of grain and house dust mites (Acari: Astigmata). Med. Entom. Zool. 49: 291–300.

Mattiacci, L. and M. Dicke. 1995. Host-age discrimination during host location by *Cotesia glomerata*, a larval parasitoid of *Pieris brassicae*. Entomol. Exp. Appl. 76: 37–48.

Mattiacci, L., M. Dicke and M. A. Posthumus. 1994. Induction of parasitoid attracting synomone in burssel sprouts plants by feeding of *Pieris brassicae* larvae: role of mechanical damage and herbivore elicitor. J. Chemical Ecology 20: 2229–2247.

Mattiacci, L., M. Dicke and M. A. Posthumus. 1995. β-Glucosidase: an elicitor of herbivore-induced plant odor that attracts host-searching parsitic wasps. Proc. Natl. Acad. Sci. USA 92: 2036–2040.

Mattiacci, L., A. B. Rocca, N. Scascighini, M. D'Alessandro, A. Hern and S. Dorn. 2001. Systematically induced plant volatiles emitted at the time of "danger". J. Chemical Ecology 27: 2233–2252.

Mazumdar-Leighton, S. and R. M. Broadway. 2001. Transcriptional induction of diverse midgut trypsins in larval *Agrotis ipsilon* and *Helicoverpa zea* feeding on the soybean trypsin inhibitor. Insect Biochem. & Molec. Biol. 31: 645–657.

McCoy, M. W. and B. M. Bolker. 2008. Trait-mediated interactions: influence of prey size, density and experience. J. Anim. Ecol. 77: 478–486.

McGuffin, M. A., R. Baker and M. R. Forbes. 2006. Detection and avoidance of fish predators by adult *Enallagma* damselfies. J. Insect Behavior 19: 77–91

McGuire, R. and A. A. Agrawal. 2005. Trade-offs between the shade-avoidance response and plant resistance to herbivores? Tests with mutant *Cucumis sativus*. Functional Ecology 19: 1025–1031.

McIntosh, A. R. and B. L. Peckarsky. 1999. Criteria determining behavioural responses to multiple predators by a stream mayfly. Oikos 85: 554–564.

McIntosh, A. R., B. L. Peckarsky and B. W. Taylor. 1999. Rapid size-specific changes in the drift of *Baetis bicaudatus* (Ephemeroptera) caused by alterations in fish odour concentration. Oecologia 118: 256–264.

McKay, S. A. B., W. L. Hunter, K. A. Godard, S. X. Wang, D. M. Martin, J. Bohlmann and A. L. Plant. 2003. Plant, insect attack and wounding induce traumatic resin duct development and gene expression of (-)-pinene synthase in Sitka spruce. Plant Physiol. 133: 368–378.

Meiners, T. and M. Hilker. 2000. Induction of plant synomones by oviposition of a phytophagous insect. Journal of Chemical Ecology 26: 221–232.

Meiners, T., N. K. Hacker, P. Anderson and M. Hilker. 2005. Response of the elm beetle to host plants induced by oviposition and feeding: the infestation rate matters. Entomologia Experimentalis et Applicata 115: 171–177.

Merkx-Jacques, M. and J. C. Bede. 2005. Influence of diet on the larval beet armyworm, *Spodoptera exigua*. J. Insect Science 5: 48.

Mewis, I., H. M. Appel, A. Hom, R. Raina and J. C. Schultz. 2005. Major signaling pathways modulate *Arabidopsis* glucosinolate accumulation and response to both phloem-feeding and chewing insects. Plant Physiology 138: 1149–1162.

Meyling, N. V. and J. K. Pell. 2006. Detection and avoidance of an entomopathogenic fungus by a generalist insect predator. Ecol. Entom. 31: 162–171.

Michimae, H. and J.-I. Hangui. 2008. A trade-off between prey- and predator-induced polyphenisms in larvae of the salamander *Hynobius retardatus*. Behav. Ecol. Sociobiol. 62: 699–704.

Mikolajewski, D. J., R. Stoks, J. Rolff and G. Joop. 2008. Predators and cannibals modulate sex-specific plasticity in life-history and immune traits. Functional Ecology 22: 114–120.

Milewski, A. V., T. P. Young and D. Madden. 1991. Thorns as induced defenses: experimental evidence. Oecologia 86: 70–75.

Miner, B. G., S. E. Sultan, S. G. Morgan, D. K. Padilla and R. A. Relyea. 2005. Ecological consequences of phenotypic plasticity. Trends in Ecol. & Evol. 20: 685–692.

Mirza, R. S., M. C. O. Ferrari, J. M. Kiesecker and D. P. Chivers. 2006. Responses of American toad tadpoles to predation cues: behavioural response thresholds, threat-sensitivity and acquired pedation recognition. Behaviour 143: 877–889.

Mithöefer, A., J. Ebel and H. H. Felle. 2005a. Cation fluxes cause plasma membrane depolarization involved in beta-glucan elicitor-signalling in soybean roots. Molecular Plant-Microbe Interactions 18: 983–990.

Mithöefer, A., G. Wanner and W. Boland. 2005b. Effects of feed *Spodoptera littoralis* on lima bean leaves: II. Continuous mechanical wounding resembling insect feeding is sufficient to elicit herbivory-related volatile emission. Plant Physiol. 137: 1160–1168.

Molano-Flores, B. 2001. Herbivory and calcium concentrations affect calcium oxalate crystal formation in leaves of *Sida* (Malvaceae). Ann. Bot. (Lond) 88: 387–391.

Molis, M., J. Korner, Y. W. Ko, et al. 2006. Inducible responses in the brown seaweed *Ecklonia cava*: the role of grazer identity and season. J. Ecology 94: 243–249.

Mondor, E. B. and J. F. Addicott. 2003. Conspicuous extra-floral nectaries are inducible in *Vicia faba*. Ecol. Lett. 6: 495–497.

Mondor, E. B. and B. D. Roitberg. 2003. Age-dependent fitness costs of alarm signaling in aphids. Can. J. Zool. 81: 757–762.

Mondor, E. B., J. A. Rosenheim and J. F. Addicott. 2005. Predator-induced transgenerational phenotypic plasticity in the cotton aphid. Oecologia 142: 104–108.

Mondor, E. B., J. A. Rosenheim and J. F. Addicott. 2008. Mutualist-induced transgenerational polyphenisms in cotton aphid populations. Functional Ecology 22: 157–162.

Monserrat, M., C. Bas, S. Magalhaes, et al. 2007. Predators induce egg retention in prey. Oecologia 150: 699–705.

Montesano, M., G. Brader and E. T. Palva. 2003. Pathogen derived elicitors: searching for receptors in plants. Molecular Plant Pathology 4: 73–79.

Montillet, J.-L., S. Chamnongpol, C. Rusterucci, J. Dat, B. van de Cotte and J.-P. Agnel. 2005. Fatty acid hydroperoxides and H_2O_2 in the execution of hypersensitive cell death in tobacco leaves. Plant Physiology 138: 1516–1526.

Mooney, K. A. and A. A. Agrawal. 2008. Plant genotype shapes ant-aphid interactions: Implications for community structure and indirect plant defense. Am. Nat. 171: E195-E205.

Moore, J. 2001. *Parasites and the Behavior of Animals*. Oxford University Press, Oxford.

Mopper, S., Y. Wang, C. Criner and K. Hasenstein. 2004. *Iris hexagona* hormonal responses to salinity stress, leafminer herbivory, and phenology. Ecology 85: 38–47.

Moraes, M. C. B., R. Laumann, E. R. Sujii, C. Pires and M. Borges. 2005. Induced volatiles in soybean and pigeon pea plants artificially infested with the neotropical brown stink bug, *Euschistus heros*, and their effect on the egg parasitoid, *Telenomus podisi*. Entom. Exp. et Appl. 115: 227–237.

Moran, N. A. 1992. The evolutionary maintenance of alternative phenotypes. American Naturalist 139: 971–989.

Mori, N., N. Yoshinaga and Y. Sawada. 2003. Identification of volicitin-related compounds from the regurgitant of lepidopteran caterpillars. Bioscience and Biotechnological Biochemistry 67: 1168–1171.

Mori, T., I. Hiraka, Y. Kurata, H. Kawachi, O. Kishida and K. Nishimura. 2005. Genetic basis of phenotypic plasticity for predator-induced morphological defenses in anuran tadpole, *Rana pirica*, using cDNA subtraction and microarray analysis. Biochemical & Biophysical Res. Commun. 330: 1138–1145.

Morris, W. F., M. B. Traw and J. Bergelson. 2006. On testing for a tradeoff between constitutive and induced resistance. Oikos 112: 102–110.

Mousseau, T. A. and C. W. Fox (eds.). 1998. *Maternal Effects as Adaptations*. Oxford University Press, New York.

Müller, M. S., S. R. McWilliam, D. Podlesak, J. R. Donaldson, H. M. Bothwell and R. L. Lindroth. 2006. Tri-trophic effects of plant defenses: chickadees consume caterpillars based on host leaf chemistry. Oikos 114: 507–517.

Musser, R. O., S. M. Hum-Musser, H. Eichenseer, M. Peiffer, G. Ervin, J. B. Murphy and G. W. Felton. 2002. Herbivory: caterpillar saliva beats plant defenses. Nature 416: 599–600.

Musser, R. O., D. E. Cipollini, S. M. Hum-Musser, S. A. Willians, J. K. Brown and G. W. Felton. 2005. Evidence that the caterpillar salivary enzyme glucose oxidase provides herbivore offense in Solanaceous plants. Archives of Insect Biochemistry and Physiology 58: 128–137.

Musser, R. O., E. Farmer, M. Peiffer, S. A. Willians and G. W. Felton. 2006. Ablation of caterpillar labial salivary glands: technique for determining the role of saliva in insect-plant interactions. J. Chemical Ecology 32: 981–992.

Nagy, N. E., V. R. Franceschi, H. Solheim, T. Krekling and E. Christiansen. 2000. Wound-induced traumatic resin duct development in stems of Norway spruce (Pinaceae): anatomy and cytochemical traits. Am. J. Bot. 87: 302–313.

Nakamura, M. and T. Ohgushi. 2003. Positive and negative effects of leaf shelters on herbivorous insects: linking multiple herbivore species on a willow. Oecologia 136: 445–449.

Nakashima, Y., M. A. Birkett, B. J. Pye, J. A. Pickett and W. Powell. 2004. The role of semiochemicals in the avoidance of the seven-spot ladybird, *Coccinella septempunctata* (Coleoptera: Coccinellidae) by the aphid parasitoid, *Aphidius ervi* (Hymenoptera: Braconidae). J. Chem. Ecol. 30: 1103–1116.

Nakashima, Y., M. A. Birkett, B. J. Pye and W. Powell. 2006. Chemically mediated intraguild predator avoidance by aphid parasitoids: interspecific variability in sensitivity to semiochemical trails of ladybird predators. J. Chemical Ecology 32: 1989–1998.

Nelson, X. J., R. R. Jackson and D. Li. 2006. Conditional use of honest signaling by a Batesian mimic. Behavioral Ecology 17: 575–580.

Ness, J. H. 2003. *Catalpa bignonioides* alters extrafloral nectar production after herbivory and attracts ant bodyguards. Oecologia 134: 210–218.

Niemi, L., A. Wennström, J. Hjältén, P. K. Waldmann and L. Ericson. 2006. Spatial variation in resistance and virulence in the host-pathogen system *Salix triandra-Melampsora amygdalinae*. J. Ecology 94: 915–921.

Nimchuk, Z., T. Eulgem, B. F. Holt and J. F. Dangl. 2003. Recognition and response in the plant immune system. Annu. Rev. Genetics 37: 579–609.

Nozawa, A. and T. Ohgushi. 2002. How does spittle-bug oviposition affect shoot growth and bud production in two willow species? Ecol. Res. 17: 535–543.

Nuismer, S. and S. Otto. 2005. Host-parasite interactions and the evolution of gene expression. PLoS Biology 3: 1283–1288.

Nuñez, J., L. Almeida, N. Balderrrama and M. Giurfa. 1997. Alarm pheromone induces stress analgesia via an apioid system in the honeybee. Physiol. Behav. 63: 75–80.

Nykanen, H. and J. Koricheva. 2004. Damage-induced changes in woody plants and their effects on insect herbivore performance: a meta-analysis. Oikos 104: 247–268.

Ode, P. J. 2006. Plant chemistry and natural enemy fitness: effects on herbivore and natural enemy interactions. Ann. Rev. Entomol. 51: 163–185.

Ohgushi, T. 2005. Indirect interaction webs: herbivore-induced effects through trait change in plants. Annu. Rev. Ecol. Evol. Syst. 36: 81–105.

Ohgushi, T., T. P. Craig and P. Price (eds.). 2007. *Ecological Communities: Plant Mediation in Indirect Interaction Webs*. Cambridge University Press, Cambridge.

Ohnmeiss, T. E. and I. T. Baldwin. 2000. Optimal defence theory predicts the ontogeny of an induced nicotine defence. Ecology 81: 1765–1783.

Oliver, J. E., R. P. Doss, R. T. Williamson, J. R. Carney and E. D. De Vilbiss. 2000. Bruchins—mitogenic 3-(hydroxypropanoyl) esters of long chain diols from weevils of the Bruchidae. Tetrahedron 56: 7633–7641.

Opitz, S., G. Kunert and J. Gershenzon. 2008. Increased terpenoid accumulation in cotton (*Gossypium hirsutum*) foliage is a general wound response. J. Chem. Ecol. 34: 508–522.

Orians, C. 2005. Herbivores, vascular pathways, and systemic induction: facts and artifacts. J. Chemical Ecology 31: 2231–2242.

Orozco-Cárdenas, M. L., J. Naváez-Vasquez and C. A. Ryan. 2001. Hydrogen peroxide acts as a second messenger for the induction of defense genes in tomato plants in response to wounding, systemin, and methyl jasmonate. Plant Cell 13: 179–191.

Ortmann, I. and B. M. Moerschbacher. 2006. Spent growth medium of *Pantoea agglemerans* primes wheat suspension cells for augmented accumulation of hydrogen peroxide and enhanced peroxidase activity upon elicitation. Planta 224: 963–970.

Otte, D. and K. Williams. 1972. Environmentally induced color dimorphisms in grasshoppers *Syrbula admirabilis*, *Dichromorpha viridis*, and *Chortophaga viridifasciata*. Annals Entom. Soc. Am. 65: 1154–1161.

Padilla, D. K. 1998. The effects of size and shape on passive dispersal potential in marine gastropods. American Zoologist 38: 184A.

Padilla, D. K. and S. C. Adolph. 1996. Plastic inducible morphologies are not always adaptive: the importance of time delays in a stochastic environment. Evolutionary Ecology 10: 105–117.

Pallini, A, A. Janssen and M. W. Sabelis. 1998. Predators induced interspecific competition for food in refuge space. Ecol. Lett. 1: 171–177.

Pallini, A., A. Janssen and M. W. Sabelis. 1999. Spider mites avoid plants with predators. Exp. Appl. Acarol. 23: 803–815.

Papaj, D. R. and A. C. Lewis. 1993. *Insect Learning*. Chapman & Hall, London.

Paré, P. W. and J. H. Tumlinson. 1997. De novo biosynthesis of volatiles induced by insect herbivory in cotton plants. Plant Physiol. 114: 1161–1167.

Paré, P. W. and J. H. Tumlinson. 1999. Plant volatiles as a defense against insect herbivores. Plant Physiology 121: 325–331.

Paré, P. W., M. A. Farag, V. Krishnamachari, H. Zhang, C.-M. Ryu and J. W. Kloepper. 2005. Elicitors and priming agents initiate plant defense responses. Photosynthesis Research 85: 149–159.

Parejko, K. and S. Dodson. 1990. Progress towards characterization of a predator/prey kairomone: *Daphnia pulex* and *Chaoborus americanus*. Hydrobiology 198: 51–59.

Parsons, K. J. and B. W. Robinson. 2006. Replicated evolution of integrated plastic responses during early adaptive divergence. Evolution 60: 801–813.

Pavia, H. and G. B. Toth. 2000. Inducible chemical resistance to herbivory in the brown seaweed *Ascophyllum nodosum*. Ecology 81: 3212–3225.

Pearce, G. and C. A. Ryan. 2003. Systemic signaling in tomato plants for defense against herbivores. J. Biol. Chem. 278: 30044–30050.

Peiffer, M. and G. W. Felton. 2005. The host plant as a factor in the synthesis and secretion of salivary glucose oxidase in larval *Helicoverpa zea*. Archives Insect Biochem. Physiol. 58: 106–113.

Pennacchio, F. and M. R. Strand. 2006. Evolution of developmental strategies in parasitic Hymenoptera. Annu. Rev. Entomol. 51: 233–258.

Peñuelas, J., I. Filella, C. Stefanescu and J. Llusia. 2005. Caterpillars of *Euphydryas aurinia* (Lepidoptera: Nymphalidae) feeding on *Succisa pratensis* leaves induce large foliar emissions of methanol. New Phytologist 167: 851–857.

Peñuelas, J., J. Sardans, C. Stefanescu, T. Paralla and I. Filella. 2006. *Lonicera implexa* leaves bearing naturally laid eggs of the specialist herbivore *Euphydryas aurinia* have dramatically greater concentrations of iridoid glycosides than other leaves. J. Chemical Ecology 32: 1925–1933.

Persons, M. H. and A. L. Rypstra. 2001. Wolf spiders show graded antipredator behavior in the presence of chemical cues from different sized predators. J. Chemical Ecology 27: 2493–2504.

Persons, M. H., S. E. Walker, A. L. Rypstra and S. D. Marshall. 2001. Wolf spider predator avoidance tactics and survival in the presence of diet-associated predator cues (Araneae: Lycosidae). Anim. Behav. 61: 43–51.

Persons, M. H., S. E. Walker and A. L. Rypstra. 2002. Fitness costs and benefits of antipredator behavior mediated by chemotactile cues in the wolf spider *Pardosa milvina*. Behavioral Ecology 13: 386–392.

Peters, R. H. 1983. *The Ecological Implications of Body Size*. Cambridge University Press, Cambridge.

Pie, M. R., R. B. Rosengaus, D. V. I. I. Calleri and J. F. A. Traniello. 2005. Density and disease resistance in group-living insects: do eusocial species exhibit density-dependent prophylaxis? Ethology Ecology & Evolution 17: 41–50.

Pieterse, C. M. and M. Dicke. 2007. Plant interactions with microbes and insects: from molecular mechanisms to ecology. Trends Plant Sci. 12: 564–569.

Pigliucci, M. and K. Preston. 2004. *Phenotypic Integration*. Oxford University Press, Oxford.

Pijanowska, J. 1994. Fish enhancement patchiness in *Daphnia* distribution. Verh. Int. Verein. Limmon. 25: 2366–2368.

Pijanowska, J. 1997. Alarm signals in *Daphnia*? Oecologia 112: 12–16.

Pike, N., C. Braendle and W. A. Foster. 2004. Seasonal extension of the soldier instar as a route to increased defence investment in the social aphid *Pemphigus spyrothecae*. Ecological Entomology 29: 89–95.

Ping, L. and W. Boland. 2004. Signals from the underground: bacterial volatiles promote growth in *Arabidopsis*. Trends in Plant Sci. 9: 263–266.

Pollock, M. S., R. G. Friesen, R. J. Pollock, R. C. Kusch and D. P. Chivers. 2005. The avoidance response of fathead minnows to chemical alarm cues: understanding the effects of donor gender and breeding condition. Chemoecology 15: 205–209.

Pollock, M. S., R. J. Pollock and D. P. Chivers. 2006. Social context influences the antipredatory behavior of fathead minnows to chemical alarm cues. Ethology 112: 801–806.

Poulev, A., J. M. O'Neal, S. Logendra, R. B. Pouleva and B. Timeva. 2003. Elicitation, a new window into plant chemodiversity and phytochemical drug discovery. J. Med. Chem. 46: 2542–2547.

Preisser, E. L., D. I. Bolnick and M. F. Benard. 2005. Scared to death? The effects of intimidation and consumption in predator-prey interactions. Ecology 86: 501–509.

Preston, C. A., G. Laue and I. T. Baldwin. 2004. Plant-plant signaling: application of *trans*- or *cis*-methyl jasmonate equivalent to sagebrush releases does not elicit direct defenses in native tobacco. J. Chem. Ecol. 30: 2193–2214.

Price, P. W. 2002. Species interactions and the evolution of biodiversity. In Herrera C. M. and O. Pellmyr (eds.). Plant-Animal Interactions: An Evolutionary Approach. Blackwell Science, Oxford Pp. 3–25.

Qi, X., S. Bakht, M. Leggett, C. Maxwell, R. Melton and A. Osbourn. 2004. A gene cluster for secondary metabolism in oat: implications for the evolution of metabolic diversity in plants. Proc. Natl. Acad. Sci. USA 101: 8233–8238.

Radman, R., T. Saez, C. Bucke and T. Keshavarz. 2003. Elicitation of plants and microbial cell systems. Biotechnol. Appl. Biochem. 37: 91–102.

Raffa, K. F. 1991. Induced defensive reactions in conifer-bark beetle systems. In Tallamy E. W. and Raupp M. J. (eds.). Phytochemical Induction by Herbivores. John Wiley, New York Pp. 245–276.

Raffa, K. F. and A. A. Berryman. 1987. Interacting selective pressures in conifer-bark beetle systems: a basis for reciprocal adaptations. American Naturalist 129: 234–262.

Rasmann, S., T. G. Koellner, J. Degenhardt, I. Hiltpold, S. Toepfer, U. Kuhlmann, J. Gershenzon and T. C. J. Turlings. 2005. Recruitment of entomopathogenic nematodes by insect-damaged maize roots. Nature 434: 732–737, 724.

Ratzka, A., H. Vogel, D. J. Kliebenstein, T. Mitchell-Olds and J. Kroymann. 2002. Disarming the mustard oil bomb. Proc. Natl. Acad. Sci. 99: 11223–11228.

Reader, T., A. D. Higginson, C. J. Bernard, F. S. Gilbert and The Behavioural Ecology Field Course. 2006. The effects of predation risk from crab spiders on bee faraging behavior. Behavioral Ecology 17: 933–939.

Relyea, R. A.. 2002a. Costs of phenotypic plasticity. American Naturalist 159: 272–282.

Relyea, R. A. 2002b. The many faces of predation: how induction, selection, and thinning combine to alter prey phenotypes. Ecology 83: 1953–1964.

Relyea, R. A. 2003a. Predators come and predators go: the reversibility of predator-induced traits. Ecology 84: 1840–1848.

Relyea, R. A. 2003b. How prey respond to combined predators: a review and an empirical test. Ecology 84: 1827–1839.

Relyea, R. A. 2004. Fine-tuned phenotypes: tadpole plasticity under 16 combinations of predators and competitors. Ecology 85: 172–179.

Relyea, R. A. 2005. The heritability of inducible defenses in tadpoles. J. Evol. Biol. 18: 856–866.

Relyea, R. A. 2007. Getting out alive: how predators affect the decision to metamorphose. Oecologia 152: 389–400.

Relyea, R. A. and J. R. Auld. 2004. Having the guts to compete: how intestinal plasticity explains costs of inducible defences. Eco. Lett. 7: 869–875.

Relyea, R. A. and J. R. Auld. 2005. Predator- and competitor-induced plasticity: how changes in foraging morphology affect phenotypic trade-offs. Ecology 86: 1723–1729.

Ren, D., K.-Y. Yang, G.-J. Li, Y. Liu and S. Zhang. 2006. Activation of Ntf4, a tobacco mitogen-activated protein kinase, during plant defense response and its involvement in hypersensitive response-like cell death. Plant Physiology 141: 1482–1493.

Rengefors, K., I. Karlsson and L. A. Hansson. 1998. Algal cyst dormancy: a temporal escape from herbivory. Proc. Royal Soc. London B. 265: 1353–1358.

Repka, S. and M. Walls. 1998. Variation in the neonate size of *Daphnia pulex*: the effects of predator exposure and clonal origin. Aquatic Ecol. 32: 203–209.

Resetarits, W. J. 2005. Habitat selection behaviour links local and regional scales in aquatic systems. Ecology Letters 8: 480–486

Reymond, P., H. Weber, M. Damond and E. E. Farmer. 2000. Differential gene expression in response to mechanical wounding and insect feeding in *Arabidopsis*. Plant Cell 12: 707–720.

Riessen, H. 1992. Cost-benefit model for the induction of an antipredator defense. American Naturalist 140: 349–362.

Rinke, K., S. Hulsmann and W. M. Mooij. 2008. Energetic costs, underlying resource allocation patterns, and adaptive value of predator-induced life-history shifts. Oikos 117: 273–285.

Roda, A., R. Halitschke, A. Steppuhn and I. T. Baldwin. 2004. Individual variability in herbivore-specific elicitors from the plant's perspective. Molecular Ecology 13: 2421–2433.

Rodríguez-Concepción, M. and A. Boronat. 2002. Elucidation of the methyl-erythritol phosphate pathway for ispprenoid biosynthesis in bacteria and plastids. A metabolic milestone achieved through genomics. Plant Physiol. 130: 1079–1089.

Rodríguez-Saona, C. and J. S. Thaler. 2005a. Herbivore-induced responses and patch heterogeneity affect abundance of arthropods on plants. Ecological Entomology 30: 156–163.

Rodríguez-Saona, C. and J. S. Thaler. 2005b. The jasmonate pathway alters herbivore feeding behavior: consequences for plant defences. Entomologia Experimentalis et Applicata 115: 125–134

Rodríguez-Saona, C., J. A. Chalmers, R. Sherosha and J. S. Thaler. 2005. Induced plant responses to multiple damagers: differential effects on an herbivore and its parasitoid. Oecologia 143: 566–577.

Rogers, S. M., T. Matheson, E. Despland, T. Dodgson, M. Burrows and S. J. Simpson. 2003. Mechanosensory-induced behavioural gregarization in the desert locust *Schistocerca gregaria*. J. Experimental Biology 206: 3991–4002.

Rohde, S. and M. Wahl. 2008. Antifeeding defense in Baltic macroalgae: induction by direct grazing versus waterborn cues. J. Phycology 44: 85–90.

Rojo, E., R. Solano and J. J. Sánchez-Serrano. 2003. Interactions between signaling compounds involved in plant defense. J. Plant Growth Regulation 22: 82–98.

Rose, U. S. R., A. Manukian and R. R. Heath. 1996. Volatile semiochemicals released from undamaged cotton leaves—A systemic response of living plants to caterpillar damage. Plant Physiol. 111: 487–495.

Rostás, M., J. Ton, B. Mauch-Mani and T. C. J. Turlings. 2006. Fungal infection reduces herbivore-induced plant volatiles of maize but does not affect naïve parasitoids. J. Chemical Ecology 32: 1897–1909.

Roy, H. E., D. C. Steinkraus, J. Eilenberg, A. E. Hajek and J. K. Pell. 2006. Bizarre interactions and endgames: entomopathogenic fungi and their arthropod hosts. Ann. Rev. Entomol. 51: 331–357.

Ruzicka, Z. 1998. Further evidence of oviposition-deterring allomone in chrysopids (Neuroptera: Chrysopidae). Eur. J. Entomol. 95: 35–39.

Ruzicka, Z. 2001. Oviposition responses of aphidophagous coccinellids to tracks of coccinellid (Coccinellidae) and chrysopid (Chrysopidae) larvae. European J. Entomol. 98: 183–188.

Ryan, C. A. 1990. Proteinase inhibitors in plants: genes for improving defenses against insects and pathogens. Ann. Rev. Phytopathol. 28: 425–449.

Ryan, C. A. and G. Pearce. 2003. Systemins: A functionally defined family of peptide signals that regulate defensive genes in Solanaceae species. Proc. Natl. Acad. Sci. 100: 14577–14580.

Ryu, C. M., M. A. Farag, C. H. Hu, M. S. Reddy, J. W. Kloepper and P. W. Pare. 2004. Bacterial volatiles induce systemic resistance in *Arabidopsis*. Plant Physiol. 134: 1017–1026.

Sadd, B. M. and M. T. Siva-Jothy. 2006. Self-harm caused by an insect's innate immunity. Proc. Royal Society B. 273: 2571–2574.

Sakwinska, O. 1998. Plasticity of *Daphnia magna* life history traits in response to temperature and information about a predator. Freshw. Biol. 39: 681–687.

Salazar, B. A. and D. W. Whitman. 2001. Defensive tactics of caterpillars against predators and parasitoids. In TN Ananthakrishnan (ed.). Insects and Plant Defence Dynamics. Science Publishers, Inc., Enfield, USA pp. 161–207.

Schappert, P. J. and J. S. Shore. 1999. Effects of cyanogenesis polymorphism in *Turnera ulmifolia* on *Euptoieta hegesia* and potential *Anolis* predators. J. Chem. Ecol. 25: 1455–1479.

Schilmiller, A. L. and G. A. Howe. 2005. Systemic signaling in the wound response. Current Opinion in Plant Biology 8: 369–377.

Schlichting, C. D. 2004. The role of phenotypic plasticity in diversification. Pp. 191–206 97 in DeWitt T. J. and Scheiner S. M. (eds.). Phenotypic Plasticity. Oxford University Press, Oxford.

Schlichting, C. D and Pigliucci. 1993. Control of phenotypic plasticity via regulatory genes. Amer. Nat. 142: 366–370.

Schmid-Hempel, P. 2005. Evolutionary ecology of insect immune defenses. Annu. Rev. Entomol. 50: 529–551.

Schmidt, B. R. and J. Van Buskirk. 2005. A comparative analysis of predator-induced plasticity in larval *Triturus* newts. J. Evol. Biol. 18: 415–425.

Schmidt, D. D. and I. T. Baldwin. 2006. Transcriptional responses of *Solanum nigrum* to methyl jasmonate and competition: a glasshouse and field study. Func. Ecol. 20: 500–508.

Schmidt, D. D., C. Voelckel, M. Hartl, S. Schmidt and I. T. Baldwin. 2005. Specificity in ecological interactions. Attack from the same lepidopteran herbivore results in species-specific transcriptional responses in two solanaceous host plants. Plant Physiology 138: 1763–1773.

Schmidt-Nielsen, K. 1984. *Scaling: Why is Animal Size so Important?* Cambridge University Press, Cambridge.

Schmitz, O. J. 1998. Direct and indirect effects of predation and predation risk in old-field interaction webs. American Naturalist 151: 327–342.

Schmitz, O. J. 2003. Top predator control of plant biodiversity and productivity in an old-field ecosystem. Ecology Letters 6: 156–163.

Schmitz, O. J. 2005. Behavior of predators and prey and links with population level processes. pp 256–278 In Barbosa P and Castellanos I (eds.). Ecology of Predator-Prey Interactions: Oxford University Press.

Schmitz, O. J. 2006. Predators have large effects on ecosystem properties by changing plant diversity, not plant biomass. Ecology 87: 1432–1437.

Schmitz, O. J. 2008. Effects of predator hunting mode on grassland ecosystem function. Science 31: 952–954.

Schmitz, O. J., V. Krivan and O. Ovadia. 2004. Trophic cascades: the primacy of trait-mediated indirect interactions. Ecology Letters 7: 153–163.

Schoener, T. W., D. A. Spiller and J. B. Losos. 2001. Predators increase the risk of catastrophic extinction of prey populations. Nature 412: 183–186.

Schoeppner, N. M. and R. A. Relyea. 2005. Damage, digestion, and defence: the roles of alarm cues and kairomones for inducing prey defences. Ecol. Lett. 8: 505–512.

Schonewolf, K. W., R. Bell, A. L. Rypstra and M. H. Persons. 2006. Field evidence of an airborne enemy-avoidance kairimone in wolf spiders. J. Chemical Ecology 32: 1565–1576.

Schroder, R., S. M. Cristescu, F. J. Harren and M. Hilker. 2007. Reduction of ethylene emission from Scots pine elicited by insect egg secretion. J. Exp. Biol. 58: 1835–1842.

Schultz, J. C. 1988. Plant responses induced by herbivores. TREE 3: 45–50.

Schultz, J. C. 2002. How plants fight dirty. Nature 416: 267.

Schultz, J. C. and H. M. Appel. 2004. Cross-kingdom cross-talk: hormones shared by plants and their insect herbivores. Ecology 85: 70–77.

Scrimgeour, G. J. and J. M. Culp. 1994. Foraging and evading predators: the effect of predator species on a behavioural trade-off by a lotic mayfly. Oikos 69: 71–79.

Seo, S., M. Okamoto, H. Seto, K. Ishizuka, H. Sano and Y. Ohashi. 1995. Tobacco MAP kinase: a possible mediator in wound signal transduction pathways. Science 270: 1988–1992.

Shah, P. A., J. A. Pickett and J. D. Vandenberg. 1999. Responses of Russian wheat aphid (Homoptera: Aphididae) to aphid alarm pheromone. Environ. Entomol. 28: 983–985.

Shibao, H., M. Kutsukake and T. Fukatsu. 2004a. Density-dependent induction and suppression of soldier differentiation in an aphid social system. J. Insect Physiology 50: 995–1000.

Shibao, H., M. Kutsukake and T. Fukatsu. 2004b. Density triggers soldier production in a social aphid. Proc. Royal Society Biological Sciences B 271: S71–S74.

Shingleton, A. W. and W. A. Foster. 2000. Ant tending influences soldier production in a social aphid. Proc. Royal Society Biological Sciences. B. 267: 1863–1868.

Shiojiri, K., T. Maeda, G.-I. Arimura, R. Ozawa, T. Shimoda and J. Takabayashi. 2002. Functions of plant infochemicals in tritrophic interactions between plants, herbivores and carnivorous natural enemies. Japanese Journal of Applied Entomology and Zoology 46: 117–133.

Shvetsova, T., J. Mwesigwa and A. G. Volkov. 2001. Plant electrophysiology: FCCP induces action potential and excitation waves in soybean. Plant Sci. 161: 901–909.

Siemens, D. H., S. H. Garner, T. Mitchell-Olds and R. M. Callaway. 2002. Cost of defense in the context of plant competition: *Brassica rapa* may grow and defend. Ecology 83: 505–517.

Sih, A. 2004. A behavioral ecological view of phenotypic plasticity. In: DeWitt, T. J. and Scheiner, S. M. (eds) Phenotypic Plasticity: Functional and Conceptual Approaches. Oxford University Press. pp. 112–125.

Sih, A., R. Ziemba and K. C. Harding. 2000. New insights on how temporal variation in predation risk shapes prey behavior. TREE 15: 3–4

Simberloff, D. and P. Stilling. 1987. Larval dispersion and survivorship in a leaf-mining moth. Ecology 68: 1647–1657.

Simpson, S. J., E. Despland, B. F. Hägele and T. Dodgson. 2001. Gregarious behavior in desert locusts is evoked by touching their back lags. Proc. Nat. Acad. Sci. (USA). 98: 3895–3897.

Sloggett, J. J. and M. E. N. Majerus. 2003. Adaptations of *Coccinella magnifica*, a myrmecophilous coccinellid to aggression by wood ants (*Formica rufa* group). II. Larval behaviour, and ladybird oviposition location. European J. Entomology 100: 337–344.

Sloggett, J. J. and W. W. Weisser. 2002. Parasitoids induce production of the dispersal morph of the pea aphid, *Acyrthosiphon pisum*. Oikos 98: 323–333.

Slusarczyk, M. 1999. Predator-induced diapause in *Daphnia magna* may require two chemical cues. Oecologia 119: 159–165.

Smee, D. L. and M. J. Weissburg. 2006. Hard clams (*Mercenaria mercenaria*) evaluate predation risk using chemical signals from predators and injured conspecifics. J. Chemical Ecology 32: 605–619.

Smith, L. D. and A. R. Palmer. 1994. Effects of manipulated diet on size and performance of brachyuran crab claws. Science 264: 710–712.

Smith, L. L., J. Lanza and G. C. Smith. 1990. Amino acid concentrations in extrafloral nectar of *Impatiens sultani* increase after stimulated herbivory. Ecology 71: 107–115.

Snyder, W. E., D. H. Wise. 2000. Antipredator behavior of spotted cucumber beetles (Coleoptera: Chrysomelidae) in response to predators that pose varying risks. Environ. Entomol. 29: 35–42.

Spencer, M. and L. Blaustein. 2001. Hatching responses of temporary pool invertebrates to signals of environmental quality. Israel J. Zoology 47: 397–417.

Spencer, M., L. Blaustein and J. E. Cohen. 2002. Oviposition habitat selection by mosquitoes (*Culiseta longiareolata*) and consequences for population size. Ecology 83: 669–679.

Spiteller, D., K. Dettner and W. Boland. 2000. Gut bacteria may be involved in interactions between plants, herbivores and their predators: microbial biosynthesis of N-aclglutamine surfactants as elicitors of plant volatiles. Biol. Chem. 381: 755–762.

Stadler, B. and A. F. G. Dixon. 2005. Ecology and evolution of aphid-ant interactions. Annu. Rev. Ecol. Evol. Syst. 36: 345–372.

Stadler, B., A. F. G. Dixon and P. Kindlmann. 2002. Relative fitness of aphids: effects of plant quality and ants. Ecol. Lett. 5: 216–222.

Stamp, N. E. 1997. Behavior of harassed caterpillars and consequences for host plants. Oikos 79: 147–154.

Staples, J. K., B. S. Krall, R. J. Bartelt and D. W. Whitman. 2002. Chemical defense in the plant bug *Lopidea robiniae* (Uhler). J. Chemical Ecology 28: 601–615.

Stav, G., L. Blaustein and J. Margalit. 1999. Experimental evidence for predation sensitive oviposition by a mosquito, *Culiseta longiareolata*. Ecological Entomology 24: 202–220.

Steiner, U. K. and J. Van Buskirk. 2008. Environmental stress and the costs of whole-organism phenotypic plasticity in tadpoles. J. Evol. Biol. 21: 97–103.

Stenson, J. A. E. 1987. Variation in the capsule size of *Holopedium gibberum* (Zaddach): a response to invertebrate predation. Ecology 68: 928–934.

Stephenson, A. G. 1982a. Iridoid glyucosides in the nectar of *Catalpa speciosa* are unpalatable to nectar theves. J. Chem. Ecol. 8: 1025–1034.

Stephenson, A. G. 1982b. The role of the extrafloral nectarines of *Catalpa speciosa* in limiting herbivory and increasing fruit production. Ecology 63: 663–669.

Stinchcombe, J. R. and M. D. Rausher. 2001. Diffuse selection on resistance to deer herbivory in the ivyleaf morning glory, *Ipomoea hederacea*. American Naturalist 158: 376–88.

Stinchcombe, J. R. and M. D. Rausher. 2002. The evolution of tolerance to deer herbivory: modifications caused by the abundance of insect herbivores. Proc. R. Soc. London Ser. B Biol. Sci. 269: 1241–46.

Stoks, R. and M. A. McPeek. 2003. Antipredator behavior and physiology determine *Lestes* species turnover along the pond-permanence gradient. Ecology 84: 3327–3338.

Stoks, R., M. A. McPeek and J. L. Mitchell. 2003. Evolution of prey behavior in response to changes in predation regime: damselflies in fish and dragonfly lakes. Evolution 57: 574–585.

Stout, M. J., J. S. Thaler and B. P. Thomma. 2006. Plant-mediated interactions between pathogenic microorganisms and herbivorous arthropods. Annu. Rev. Entomol. 51: 663–689.

Stratmann, J. 2003. Ultraviolet-B radiation co-ops defense signaling pathways. Trends in Plant Science 8: 526–533.

Strauss, S. Y. and R. E. Irwin. 2004. Ecological and evolutionary consequences of multispecies plant-animal interactions. Ann. Rev. Ecol. Evol. Syst. 35: 435–466.

Strauss, S. Y., J. A. Rudgers, J. A. Lau and R. E. Irwin. 2002. Direct and ecological costs of resistance to herbivory. TREE 17: 278–284.

Strauss, S. Y., R. E. Irwin and V. M. Lambrix. 2004. Optimal defence theory and flower petal colour predict variation in the secondary chemistry of wild radish. J. Ecology 92: 132–141.

Sultan, S. E. and H. G. Spencer. 2002. Metapopulation structure favors plasticity over local adaptation. American Naturalist 160: 271–283.

Suzuki, H., M. S. S. Reddy, M. Naoumkina, N. Aziz, G. D. May, D. V. Huhman, L. W. Sumner, J. W. Blount, P. Mendes and R. A. Dixon. 2005. Methyl jasmonate and yeast elicitor induce differential transcriptional and metabolic re-programming in cell suspension cultures of the model legume *Medicago truncatula*. Planta 220: 696–707.

Takabayashi, J., M. Dicke and M. A. Posthumus. 1994. Volatile herbivore-induced terpenoids in plant-mite interactions: variation caused by biotic and abiotic factors. J. Chem. Ecol. 20: 1329–1354.

Takabayashi, J., S. Takahashi, M. Dicke and M. A. Posthumus. 1995. Developmental stage of herbivore *Pseudaletia separate* affects production of herbivore-induced synomone by corn plants. J. Chem. Ecol. 21: 273–287.

Taylor, A. R., M. H. Persons and A. L. Rypstra. 2005. The effect of perceived predation risk on male courtship and copulatory behavior in the wolf spider *Pardosa milvina*. J. Arachnology 33: 76–81.

Taylor, P. D., T. Day, D. Nagy, G. Wild, J. André and A. Gardner. 2006. The evolutionary consequences of plasticity in host-pathogen interactions. Theoretical Population Biology 69: 323–331.

ten Cate. 2000. How learning mechanisms might affect evolutionary processes. TREE 15: 179–181.

Teplitsky, C., S. Plénet and P. Joly. 2004. Hierarchical responses of tadpoles to multiple predtors. Ecology 85: 2888–2894.

Teplitsky, C., S. Plénet and P. Joly. 2005a. Costs and limits of dosage response to predation risk: to what extent can tadpoles invest in anti-predator morphology? Oecologia 145: 364–370.

Teplitsky, C., S. Plénet, J. P. Léna, N. Mermet, E. Malet and P. Joly. 2005b. Escape behaviour and ultimate causes of specific induced defences in an anuran tadpole. J. Evol. Biol. 18: 180–190.

Thaler, J. S. 1999. Jasmonate-inducible plant defences cause increased parasitism of herbivores. Nature 399: 686–688.

Thaler, J. S. and R. M. Bostock. 2004. Interactions between abscisic-acid-mediated responses and plant resistance to pathogns and insects. Ecology 85: 48–58.

Thaler, J. and R. Karban. 1997. A phylogenetic reconstruction of constitutive and induced resistance in *Glossypium*. American Naturalist 149: 1139–1146.

Thaler, J. S., R. Karban, D. E. Ullman, K. Boege and R. M. Bostock. 2002. Cross-talk between jasmonate and salicylate plant defense pathways: effects on several plant parasites. Oecologia 131: 227–235.

Thomas, M. B. and S. Blanford. 2003. Thermal biology in insect-parasite interactions. TREE 18: 344–350.

Thomma, B. P. H. J., K. F. M. Tierens, I. A. M. A. Penninckx, B. Mauch-Mani, W. F. Broekaert and B. P. A. Cammue. 2001. Different micro-organisms differentially induce *Arabidopsis* disease response pathways. Plant Physiology & Biochemistry 39: 673–680.

Thoms, C., R. Ebel and P. Proksch. 2006. Activated chemical defense in *Aplysina* sponges revisted. J. Chemical Ecology 32: 97–123.

Tian, D., M. B. Traw, J. Q. Chen, M. Kreitman and J. Bergelson. 2003. Fitness costs of R-gene-mediated resistance in *Arabidopsis thaliana*. Nature 423: 74–77.

Tollrian, R. and S. I. Dodson. 1999. Inducible defenses in Cladocera: constraints, costs, and multipredator environments. See Tollrian & Harvell 1999, pp. 177–202.

Tollrian, R. and C. D. Harvell (Eds). 1999. *The Ecology and Evolution of Inducible Defenses*. Princeton, NJ: Princeton Univ. Press.

Tollrian, R. and C. D. Harvell. 1999. The Evolution of inducible defenses: current ideas. In Tollrian R. and Harvell C.D. (eds.). *The Ecology and Evolution of Inducible Defenses*. Princeton University Press, Princeton, NJ. Pp. 306–321.

Tooker, J. F. and De C. M. Moraes. 2005. Jasmonate in lepidopteran eggs and neonates. J. Chemical Ecology 31: 2753–2759.

Torres, M. A., J. D. G. Jones and J. L. Dangl. 2006. Reactive oxygen species signaling in response to pathogens. Plant Physiology 141: 373–378.

Touchon, J. C. and K. M. Warkentin. 2008. Fish and dragonfly nymph predators induce opposite shifts in color and morphology of tadpoles. Oikos 117: 634-640.

Traniello, J. F. A., R. B. Rosengaus and K. Savoiie. 2002. The development of immunity in a social insect: evidence for the group facilitation of disease resistance. Proc. Natl. Acad. of Sci. (USA) 99: 6838–6842.

Traw, M. B. 2002. Is induction response negatively correlated with constitutive resistance in black mustard? Evolution 56: 2196–2205.

Traw, M. B. and T.E. Dawson 2002a. Reduced performance of two specialist herbivores (Lepidoptera: Pieridae, Coleoptera: Chrysomelidae) on new leaves of damaged black mustard plants. Environmental Entomology 31: 714–722.

Traw, M. B. and T.E. Dawson 2002b. Differential induction of trichomes by three herbivores of black mustard. Oecologia 131: 526–532.

Traw, M. B. and P. Feeny. 2008. Glucosinolates and trichomes track tissue value in two sympatric mustards. Ecology 89: 763–772.

Traw, M. B. and J. Bergelson. 2003. Interactive effects of jasmonic acid, salicylic acid, and gibberellin on induction of trichomes in *Arabidopsis*. Plant Physiology 133: 1367–1375.

Traw, M. B., J. Kim, S. Enright, D. Cipollini and J. Bergelson. 2003. Negative cross-talk between salicylate- and jasmonate-mediated pathways in the Wassilewskija ecotype of *Arabidopsis thaliana*. Molecular Ecology 12: 1125–1135.

Traw, M. B., J. M. Kniskern and J. Bergelson. 2007. SAR increases fitness of *Arabidopsis thaliana* in the presence of natural bacterial pathogens. Evolution 61: 2444–2449.

Truitt, C. L. and P. W. Paré. 2004. In situ translocation of volicitin by beet armyworm larvae to maize and systemic immobility of the herbivore elicitor in plants. Planta 218: 999–1007.

Truitt, C. L., H. X. Wei and P. W. Paré. 2004. A plasma membrane protein from *Zea mays* binds with the herbivore elicitor volicitin. Plant Cell 16: 523–532.

Trussell, G. C. and M. O. Nicklin. 2002. Cue sensitivity, inducible defense, and trade-offs in a marine snail. Ecology 83: 1635–1647.

Trussell, G. C. and D. L. Smith. 2000. Induced defenses in response to an invading crab predator: an explanation of historical and geographic phenotypic change. Proc. Natl. Acad. of Sci. (USA) 97: 2123–2127.

Trussell, G. C., P. J. Ewanchuk and M. D. Bertness. 2003. Trait-mediated effects in rocky intertidal food chains: predator risk cues alter prey feeding rates. Ecology 84: 629–640.

Tscharntke, T., S. Thiessen, R. Dolch and W. Boland. 2001. Herbivory, induced resistance, and interplant signal transfer in *Alnus glutinosa*. Biochemical Systematics & Ecology 29: 1025–1047.

Tseng, M. 2004. Sex-specific response of a mosquito to parasites and crowding. Proceedings of the Royal Society B. Biological Sciences 271: S186–S188.

Tseng, M. 2006. Interactions between the parasite's previous and current environment mediate the outcome of parasite infection. American Naturalist 168: 565–571.

Tufto, J. 2000. The evolution of plasticity and nonplastic spatial and temporal adaptations in the presence of imperfect environmental cues. American Naturalist 156: 121–130.

Turlings, T. C. and R. Wackers. 2004. Recruitment of predators and parasitoids by herbivore injured plants. In Cardé RT and Millar JG (eds). Insect Chemical Ecology. Cambridge University Press, Cambridge Pp. 21–75.

Turlings, T. C. J., P. J. McCall, H. T. Alborn and J. H. Tumlinson. 1993. An elicitor in caterpillar oral secretions that induces corn seedlings to emit chemical signals attractive to parasitic wasps. J. Chemical Ecology 19: 411–425.

Underwood, N. 2000. Density dependence in induced plant resistance to herbivore damage: threshold, strength and genetic variation. Oikos 89: 295–300.

Underwood, N., K. Anderson and B. D. Inouye. 2005. Induced vs. constitutive resistance and the spatial distrubition of insect herbivores among plants. Ecology 26: 594–602.

Urban, M. C. 2008. The evolution of prey body size reaction norms in diverse communities. J. Anim. Ecol. 77: 346–355.

Van Buskirk, J. 2002. A comparative test of the adaptive plasticity hypothesis: relationships between habitat and phenotype in anuran larvae. American Naturalist 160: 87–102.

Van Buskirk, J. and M. Arioli. 2002. Dosage response of an induced defense: how sensitive are tadpoles to predation risk? Ecology 83: 1580–1585.

Van Buskirk, J. and M. Arioli. 2005. Habitat specialization and adaptive phenotypic divergence of anuran populations. J. Evol. Biol. 18: 596–608.

van Dam, N. H. and I. T. Baldwin. 2001. Competition mediates costs of jasmonate-induced defenses, nitrogen acquisition and transgenerational plasticity in *Nicotiana attenuata*. Functional Ecol. 15: 406–415.

van Dam, N. M., T. J. de Jong, Y. Iwasa and T. Kubo. 1996. Optimal distribution of defences: are plants smart investors? Functional Ecology 10: 128–136.

van Dam, N. M., U. Hermenau and I. T. Baldwin. 2001. Instar-specific sensitivity of specialist *Manduca sexta* larvae to induced defences in their host plant *Nicotiana attenuata*. Ecological Entomology 26: 578–586.

van Dam, N. M., J. A. Harvey, F. L. Wackers, T. M. Bezemer, W. H. van der Putten and L. E. M. Vet. 2003. Interactions between aboveground and belowground induced responses against phytophages. Basic & Applied Ecology 4: 63–77.

Van de Meutter, L. De Meester and R. Stoks. 2005. Water turbidity affects predator-prey interactions in a fish-damselfly system. Oecologia 144: 327–336.

Vander Meer, R. K. and D. P. Wojcik. 1982. Chemical mimicry in the myrmecophilous beetle, *Myrmecaphodius excavaticollis*. Science 218: 806–807.

van Donk, E., M. Lürling and W. Lampert. 1999. Consumer-induced changes in phytoplankton: inducibility, costs, benefits, and the impact on grazers. Pp. 90–103 in Tollrian R. and Harvell C. D. (eds.). The Ecology and Evolution of Inducible Defenses. Princeton University Press, Princeton, NJ.

van Kleunen, M. and M. Fischer 2005. Constraints on the evolution of adaptive phenotypic plasticity in plants. New Phytologist 166: 49–60.

Van Munster, M., A. Janssen, A. Clerivet and J. van den Heuvel. 2005. Can plants use an entomopathogenic virus as a defence against herbivores. Oecologia 143: 396–401.

Van Poecke, R. M. P., M. A. Posthumus and M. Dicke. 2001. Herbivore-induced volatile production by *Arabidopsis thaliana* leads to attraction of the parasitoid *Cotesia rubecula*: Chemical, behavioral, and gene-expression analysis. J. Chemical Ecology 27: 1911–1928.

Van Son, T. C. and M. Thiel. 2006. Multiple predator effects in intertidal food web. J. Animal Ecology 75: 25–32.

Van Zandt, P. A. and A. A. Agrawal. 2004a. Community-wide impacts of herbivore-induced plant responses in milkweed (*Asclepias syriaca*). Ecology 85: 2616–2629.

Van Zandt, P. A. and A. A. Agrawal. 2004b. Specificity of induced plant responses to specialist herbivores of the common milkweed *Asclepias syriaca*. Oikos 104: 401–409.

Vasconsuelo, A. and R. Boland. 2007. Molecular aspects of the early stages of elicitation of secondary metabolites in plants. Plant Science 172: 861-875.

Vaughn, D. 2007. Predator-induced morphological defenses in marine zooplankton: a larval case study. Ecology 88: 1030–1039.

Venzon, M., A. Janssen, A. Pallini and M. W. Sabelis. 2000. Diet of a polyphagous arthropod predator affects refuge seeking of its thrips prey. Anim. Behav. 60: 369–375.

Verhagen, B. W. M., J. Glazebrook, T. Zhu, H. Chang , L. C. Van Loon and C. M. J. Pieterse. 2004. The transcriptome of rhizobacteria-induced systemic resistance in *Arabidopsis*. Molecular Plant-Microbe Interactions 17: 895–908.

Via, S. 1994. The evolution of phenotypic plasticity: what do we really know? In Real LA (ed.). Ecological Genetics. Princeton University Press, Princeton, NJ Pp. 35–57.

Via, S., R. Gomulkiewicz, G. de Jong , S. M. Scheiner, C. D. Schlichting and P. H. Van Tienderen. 1995. Adaptive phenotypic plasticity: consensus and controversy. Trends Ecol. Evol. 10: 212–217.

Viswanathan, D. V., A. J. T. Narwani and J. S. Thaler. 2005. Specificity in induced plant responses shapes patterns of herbivore occurrence on *Solanum dulcamara*. Ecology 86: 886–896.

Voelckel, C. and I. T. Baldwin. 2004. Herbivore-induced plant vaccination. Part II. Array-studies reveal the transience of herbivore-specific transcriptional imprints and a distinct imprint from stress combinations. The Plant Journal 38: 650–663.

Von der Emde, G. 2006. Non-visual environmental imaging and object detection through active electrolocation in weakly electric fish. J. Comp. Physiol. A 192: 601–612.

Von Elert, E. and G. Pohnert. 2000. Predator specificity of kairomones in diel vertical migration of *Daphnia*: a chemical approach. Oikos 88: 119–28.

Vonesh, J. R. and B. M. Bolker. 2005. Compensatory larval responses shift trade-offs associated with predator-induced hatching plasticity. Ecology 86: 1580–1591.

Vonesh, J. R. and C. W. Osenberg. 2003. Multi-predator effects across life-history stages: non-additivity of egg- and larval-stage predation in an African treefrog. Ecol. Lett. 6: 503–508.

Von Rad, U., M. J. Mueller and J. Durner. 2005. Evaluation of natural and synthetic stimulants of plant immunity by microarray technology. New Phytologist 165: 191–202.

Vos, M., B. W. Kooi, D. L. DeAngelis and W. M. Mooij. 2004. Inducible defences and the paradox of enrichment. Oikos 105: 471–480.

Vos, M., A. M. Verschoor, B. W. Kooi, F. Wackers, D. L. DeAngelis and W. M. Mooij. 2004. Inducible defences and trophic structure. Ecology 85: 2783–2794.

Vrieling, K., W. Smit and E. Vandermeijden. 1991. Tritrophic interactions between aphids (*Aphis jocobaeae* Schrank), ant species, *Tyria jacobaeae* L., and *Senecio jacobaea* L. lead to maintenance of genetic variation in pyrrolizidine alkaloid concentration. Oecologia 86: 177–182.

Wäckers, F. L. and T. M. Bezemer. 2003. Root herbivory induces an above-ground indirect defence. Ecol. Lett. 6: 9–12.

Walker, S. E. and A. L. Rypstra. 2003. Hungry spiders aren't afraid of the big bad wolf spider. J. Arachnology 31: 425–427.

Walling, L. L. 2000. The myriad plant responses to herbivores. J. Plant Growth Regul. 19: 195–216.

Walls, R., H. Appel, M. Cipollini and J. Schultz. 2005. Fertility, root reserves and the cost of inducible defenses in the perennial plant *Solanum carolinense*. J. Chem. Ecol. 31: 2263–2288.

Ward, S. A., S. R. Leather, J. Pickup and R. Harrington. 1998. Mortality during dispersal and the cost of host-specificity in parasites: how many aphids find hosts? J. Anim. Ecol. 67: 763–773.

Warkentin, K. M., C. R. Currie and S. A. Rehner. 2001. Egg-killing fungus induces early hatching of red-eyed treefrog eggs. Ecology 82: 2860–2869.

Weider, L. J. and J. Pijanowska. 1993. Plasticity of *Daphnia* life histories in response to chemical cues from predators. Oikos 67: 385–392.

Welch, J. M., D. Rittschof, T. M. Bullock and R. B. Forward, 1997. Effects of chemical cues on settlement behavior of blue crab *Callinectes sapidus* postlarvae. Mar. Ecol. Progr. Ser. 154: 143–153.

Wen, Z., M. R. Berenbaum and M. A. Schuler. 2006. Inhibition of CYP6B1-mediated detoxification of xanthotoxin by plant allelochemicals in the black swallowtail (*Papilio polyxenes*). J. Chem. Ecol. 32: 507–522.

Wendehenne, D., J. Durner and D. F. Klessig. 2004. Nitric oxide: a new player in plant signalling and defence responses. Current Opinion in Plant Biology. 4: 449–455.

Werner, E. E. 1986. Amphibian metamorphosis: growth rate, predation risk and the optimal size at transformation. American Naturalist 128: 319–341.

Werner, E. E. and S. D. Peacor. 2003. A review of trait-mediated indirect interactions in ecological communities. Ecology 84: 1083–1100.

West-Eberhard M. J. 2003. *Developmental Plasticity and Evolution*. Oxford University Press, Oxford.

Whitman, D. W. 1988. Allelochemical interactions among plants, herbivore, and their predators. Pp 11–61 in Barbosa P. and Letourneau D. (eds.). Novel Aspects of Insect-Plant Interactions. Wiley, New York.

Whitman, D. W. 1994. Plant bodyguards: mutualistic interactions between plants and the third trophic level. In TN Ananthakrishnan (ed.) Functional Dynamics of Phytophagous Insects. Oxford & IBH, New Delhi Pp. 207–248.

Whitman, D. W. and D. A. Nordlund. 1994. Plant chemicals and the location of herbivorous arthropods by their natural enemies. In Ananthakrishnan TN (ed.). Functional Dymanics of Phytophagous Insects. Oxford & IBH, New Delhi Pp. 133–159.

Wilder, S. M. and A. L. Rypstra. 2004. Chemical cues from an introduced predator (Mantodea, Mantidae) reduce the movement and foraging of a native wold spider (Araneae, Lycosidae) in the laboratory. Environmental Entomology 33: 1032–1036.

Wildon, D. C., J. F. Thain and P. E. H. Minchin. 1992. Electrical signaling and systemic proteinase-inhibitor induction in the wounded plant. Nature 360: 62–65.

Williams, A. G. and T. G. Whitham. 1986. Premature leaf abscission: an induced plant defense against gall aphids. Ecology 67: 1619–1627.

Williams, J. L. and D. H. Wise. 2003. Avoidance of wold spiders (Araneae: Lycosidae) by striped cucumber beetles (Coleoptera: Chrysomelidae): laboratory and field studies. Environ. Entomol. 32: 633–640.

Williams, J. L., W. E. Snyder and D. H. Wise. 2001. Sex-based differences in antipredator behavior of the spotted cucumber beetle, *Diabrotica undecimpunctata howardi* (Coleoptera: Chrysomelidae). Environ. Entomol. 30: 327–332.

Wilson, R. S., P. G. Kraft and R. Van Damme. 2005. Predator-specific changes in the morphology and swimming performance of larval *Rana lessonae*. Functional Ecology 19: 238–244.

Windig, J. J., C. G. F. De Kovel and G. De Jong. 2004. Genetics and mechanics of plasticity. In DeWitt TJ and Scheiner SM (eds.). Phenotypic Plasticity. Oxford University Press, Oxford Pp. 31–97.

Wise, M. J. and J. J. Cummins. 2006. Strategies of *Solanum carolinense* for regulating maternal investment in response to foliar and floral herbivory. J. Ecology 94: 629–636.

Wise, M. J. and A. M. Weinberg. 2002. Prior flea beetle herbivory affects oviposition preference and larval performance of a potato beetle on their shared host plant. Ecological Entomology 27: 115–122.

Wisenden, B. D., D. P. Chivers and R. J. F. Smith. 1997. Learned recognition of predatory risk by *Enallagma* damselfly larvae (Odonata, Zygoptera) on the basis of chemical cues. Werner EE. 1986. Amphibian metamorphosis: growth rate predation risk and the optimal size at transformation. American Naturalist 128: 319–341.

Wittkopp, P. J. 2007. Variable gene expression in eukaryotes: a network perspective. J. Exp. Biol. 210: 1567–1575.

Wittstock, U. and B. A. Halkier. 2002. Glucosinolate research in the *Arabisopsis* era. Trends in Plant Science 7: 263–270.

Wittstock, U., N. Agerbirk, E. J. Stauber, C. E. Olsen, M. Hippler, T. Mitchell-Olds, J. Gershenzon and H. Vogel. 2004. Successful herbivore attack due to metabolic diversion of a plant chemical defense. Proc. Natl. Acad of Sci. (USA) 101: 4859–4864.

Wodarz, D. 2006. Ecological and evolutionary principles in immunology. Ecological Letters 9: 694–705.

Wohlfahrt, B., D. J. Mikolajewski, G. Joop and F. Suhling. 2006. Are behavioural traits in prey sensitive to the risk imposed by predatory fish? Freshwater Biology 51: 76–84.

Wolf, J. B., E. D. Brondie and M. J. Wade. 2004. The genotype-environment interaction and evolution when the environment contains genes. Pp. 173–190 in DeWitt T. J. and Scheiner S. M. (eds.). Phenotypic Plasticity. Oxford University Press, Oxford.

Wong, B. B. M., C. Bibeau, K. A. Bishop and G. G. Rosenthal. 2005. Response to perceived predation threat in fiddler crabs: trust thy neighbor as thyself? Behav. Ecol. Sociobiol. 58: 345–350.

Wooster, D. and A. Sih. 1995. A review of the drift and activity responses of stream prey to predator presence. Oikos 73: 3–8.

Xu, X., X. Hu, S. J. Neill, J. Fang and C. Weiming. 2005. Fungal elicitor induces singlet oxygen generation, theylene release and saponin synthesis in cultured cells of *Panax ginseng* C. A. Meyer. Plant Cell Physiology 46: 947–954.

Yamaguchi, T., Y. Maehara, O. Kodama, M. Okada, M. Matsumura and N. Shibuya. 2002. Two purified oligosaccharide elicitors, *N*-acetylchitohepatose and tetraglucosyl glucitol, derived from *Magnaporthe grisea* cell walls, syngeristically active biosynthesis of phytoalexin in suspension-cultured rice cells. J. Plant Physiology 159: 1147–1149.

Yang, X. and D. L. Cox-Foster. 2005. Impact of an ectoparasite on the immunity and pathology of an invertebrate: evidence for host immunosuppression and viral amplification. Proc. Acad. Natl. Sci. 102: 7470–7475.

Yao, I. and S. Akimoto. 2002. Flexibility in the composition and concentration of amino acids in honeydew of the drepanosiphid aphid *Tuberculatus quercicola*. Ecol. Entomol. 27: 745–752.

Yao, I., H. Shibao and S. Akimato. 2000. Cost and benefits of ant attendance to the drepanosiphid aphid *Tuberculatus quercicola*. Oikos 89: 3–10.

Young, N. D. 2000. The genetic architecture of resistance. Curr. Opin. Plant Biol. 3: 285–290.

Young, T. P. and B. D. Okello. 1998. Relaxation of of an induced defense after exclusion of herbivores: spines on *Acacia drepanolobium*. Oecologia 115: 508–513.

Zangerl, A. R. 2003. Evolution of induced plant responses to herbivores. Basic Appl. Ecol. 4: 91–103.

Zangerl, A. R. and C. E. Rutledge. 1996. The probability of attack and patterns of constitutive and induced defence: a test of optimal defence theory. American Naturalist 147: 599–608.

Zavala, J., A. Patankar, K. Gase and I. T. Baldwin. 2004. Constitutive and inducible trypsin protease inhibitor production incurs large fitness costs in *Nicotiana attenuata*. Proc. Natl. Acad. Sci. 101: 1607–1612.

Zhang, X. S. 2006. The phenotypic variance within plastic traits under migration-mutation-selection balance. Evolution 60: 1125–1136.

Zhao, Y., R. Thilmony, C. L. Bender, A. Schaller, S.-Y. He and G. A. Howe. 2003. Virulence systems of *Pseudomonas syringae* pv. *tomato* promote bacterial speck disease in tomato by targeting the jasmonate signaling pathway. Plant J. 36: 485–499.

Zhao, J., L. C. Davis and R. Verpoorte 2005. Elicitor signal transduction leading to production of plant secondary metabolites. Biotechnology Advances 23: 283–333.

8

Behavioral Plasticity to Risk of Predation: Oviposition Site Selection by a Mosquito in Response to its Predators

Leon Blaustein[1,2] **and Douglas W. Whitman**[3]

[1]*Community Ecology Laboratory, Institute of Evolution, Department of Evolutionary and Environmental Biology, Faculty of Sciences, University of Haifa, Haifa 31905, Israel.*

[2]*Center for Vector Biology, Rutgers University, New Brunswick, New Jersey 08901, USA. E-mail: leon@research.haifa.ac.il*

[3]*4120 Department of Biological Sciences, Illinois State University, Normal, IL 61790, USA. E-mail: dwwhitm@ilstu.edu*

Abstract

Oviposition habitat selection in response to risk of predation is an environmentally induced phenotypic plastic response. We suggest predator-prey characteristics for which such a response is more likely to evolve: high vulnerability of progeny to the predator; deposition of all eggs from a single reproductive event in a single site (i.e., inability to spread the risk spatially); few opportunities to reproduce (i.e., unable to spread the risk temporally); during habitat assessment by the gravid female, high predictability of future risk of predation for the period in which the progeny develop at the site; the predator is common but some sites are predator-free. We summarize work done on a particular system—the mosquito *Culiseta longiareolata* Macquart and its predators in pool habitats, a likely candidate system for oviposition habitat selection in response to risk of predation given these proposed characteristics.

Adult *C. longiareolata* females can chemically detect various predatory backswimmer species (Notonectidae) but do not appear to chemically detect odonate and urodele larvae. Evidence suggests however that ovipositing females may detect other predators by nonchemical cues.

C. longiareolata females can also detect and respond to larval food resource levels and conspecific larvae when choosing an oviposition site. When given only the two poor habitat choices of high conspecific larval densities or

Notonecta, overall oviposition rates drop and a larger proportion of the females oviposit in predator pools suggesting that females may operate in an ideal-free distribution manner. When comparing *Notonecta* versus non-predator pools, the proportion of egg rafts in predator-free pools increases when the proportion of predator-free pools increases. The most likely explanation among competing hypotheses is that the mosquito females that oviposit in predator pools are those that have encountered only predator pools when "sampling" the environment and ultimately choose to oviposit in a poor quality pool than no pool at all.

A growing body of literature suggests that predator avoidance when ovipositing is particularly prevalent among pool/pond species with a mobile adult stage (amphibians and insects). We suggest that this is because risk of predation during the period in which offspring will grow is predictable based on risk of predation at the time that the female assesses habitats. Nonrandom selection of oviposition sites in response to predators likely influences adult populations and community structure. In the case of *C. longiareolata,* a model suggests that this behavior, compared to nonrandom oviposition, results in a larger adult population. Nonrandom selection of oviposition sites in response to risk of predation also indicates that experiments that set up predator and nonpredator plots and then compare the abundance of prey immatures are overestimating the predator's effect on the prey population.

Introduction

Numerous studies over the past couple of decades have revealed a great diversity of induced responses by prey to risk of predation (see Whitman and Blaustein, this volume). The term "phenotypic plasticity" generally brings to mind developmental plasticity—e.g., altered growth rates (Peacor and Werner 2004), morphology (Trembath and Anholt 2001), life history (Blaustein 1997; Bernard 2004; Li 2002), or body color (Garcia et al. 2004). However, behavior, which is essentially an expressed phenotype of a genotype as a function of the environment, can also be considered a type of phenotypic plasticity (DeWitt and Scheiner 2004; Sih 2004). According to this definition, behaviors such as altered foraging (e.g., Kotler 1984; Brown and Kotler 2004) or altered oviposition site selection (Blaustein 1999; Resetarits 2005) in response to environmental conditions such as predation risk are examples of phenotypic plasticity. In general, behavioral plasticity allows a more rapid response and is generally less costly than induced morphological responses (e.g., Teplitsky et al. 2005). Moreover, behaviors are more easily reversible.

That gravid females could detect risk of predation and choose a site for oviposition based on risk of predation was rarely examined in the literature several decades ago. This has been explored considerably more recently and there is a quickly growing body of literature that this behavioral plasticity is far from rare. A number of examples have been demonstrated for terrestrial insects (see Whitman and Blaustein, this volume). But this behavioral phenotypic plasticity is apparently common and particularly strong in pond and pool habitats (Blaustein 1999, Resetarits 2005). In this chapter, we first consider predator and prey characteristics in which evolution of this behavior is more likely to evolve. We then review a specific case of this behavioral phenotypic plasticity—oviposition habitat selection by a mosquito in response to risk of predation to its progeny.

What Factors Favor the Evolution of Oviposition Plasticity in Prey?

A species may evolve in two ways to avoid ovipositing where predators are prevalent. There may be evolutionary responses to *predictors* of predation risk without detecting the predator itself. For example, larger and more permanent pools are more likely to have higher predator densities (Spencer et al. 1999, Wilcox 2001, Chase and Knight 2003) and the ovipositing female may cue in on the size of the pool without being able to detect the predator itself. A second way is where *detection of the predator itself* induces a behavioral oviposition response. Not all species respond in this way to their predators. In fact, probably most do not (Blaustein et al. 2004). Certain predator-prey properties that likely increase the probability that oviposition habitat selection will evolve are outlined by Blaustein (1999) and Blaustein et al. (2004) and are expanded here:

(1) *A gravid female lays all her eggs in a single patch*—i.e., she cannot spread the risk spatially. For example, the green toad, *Bufo viridis*, appears incapable of dividing its clutch and thus lays all its eggs in a single clutch while the tree frog *Hyla savignyi* may oviposit in several water bodies on the same night (Hill and Blaustein, unpublished data). Likewise, some mosquito species lay their entire clutch in a single egg boat whereas other mosquitoes may spread their eggs across several water bodies (Colton et al. 2003). A female that spreads her progeny across a number of sites is (by chance) likely to deposit eggs in both low- and high-quality habitats, and thus is likely to have at least some offspring survive. For the female that places all her eggs in a single site, site discrimination is more crucial to her fitness.

(2) *A female has few lifetime reproductive events (cannot spread risk temporally).* For example, some salamanders may live > 20 years, having at least one clutch per year (Warburg 1994). If a long-lived salamander deposits all of her progeny in a bad site during some years, it likely still has other opportunities.

(3) *High predator-induced mortality on immatures.* In resource-limited habitats, moderate predation rates may even be beneficial by reducing intraspecific competition (e.g. Wilbur 1997), so vulnerability needs to be high for selective pressure favoring oviposition habitat selection. We would predict then that there would not be strong selection pressure for evolving oviposition habitat selection for prey species that have evolved successful ways of evading the predator when they co-occur at the same habitat.

(4) *The female has multiple sites from which to choose to oviposit and there is predator heterogeneity among patches.* If suitable breeding patches are generally sufficiently distant, rare, or homogeneous, females may have evolved to oviposit into the first patch they encounter, without assessing patch quality. Otherwise, a female should search for a high quality site, though balancing costs of continued searching.

(5) *Predator abundance in a given site at time of oviposition predicts local predator abundance during subsequent larval development.* In many systems, this is not the case. For example, an insect might oviposit on a leaf devoid of its predators or devoid of predator-released kairomones (Nakashima et al. 2004). But if the predator is mobile, then the likelihood of future risk of predation may not be sufficiently different on this leaf than a leaf that presently has a predator. On the other hand, many predators of discrete aquatic habitats or parasitoid larvae already in a host cannot move from patch to patch. In these cases, the present distribution of predators among sites at the time a female is searching for an oviposition site should be a good predictor of future predation risk for her offspring.

(6) *Immigration of prey individuals from geographic areas devoid of predators into areas having these predators is "negligible."* If the range of a prey species extends beyond that of the predator range and there is sufficient immigration from the predator-free zone into the predator zone, this could flood out the natural selection for oviposition habitat selection. However, prey outside the zone of a particular predator species may also be cueing in on other predator species possessing the same cues. If the specific cue is the same for multiple predator species,

one might expect a response to predator "A" even if there are many prey individuals outside the predator "A" region dispersing into it.

(7) *The predator should be sufficiently common.* If predators are rare, there will not be strong selection to detect and avoid this predator. Like number 6, this is not the case if the prey does not detect a specific predator but can use the same cue to detect many predator species.

Why *Culiseta longiareolata* is a Likely Candidate to Evolve Oviposition Habitat Selection in Response to Risk of Predation

The mosquito *Culiseta longiareolata* is found throughout the Middle East, Southern Mediterranean, and Africa, where it is ubiquitous in small lentic aquatic habitats (van Pletzen and van der Linde 1981, Ward and Blaustein 1994). This species fits the characteristics suggested above favoring the evolution of predator-induced plasticity in oviposition:

- *All eggs are laid in a single patch.* Like all other congeners, it lays all eggs from each reproductive effort as a single clutch in a single pool (with rare exception, the laboratory: van Pletzen and van der Linde 1981). The eggs are deposited as an "egg raft" that floats on the water surface (Fig. 1a).
- *A female has few lifetime opportunities to reproduce.* After oviposition, a new blood meal, taken from a bird, is needed to develop a new egg batch. The period between blood acquisition and egg maturation varies from 4 -18 days in the laboratory (van Pletzen and van der Linde 1981). There are no published survival studies on adults of this species, but mosquito adults in general experience quite high mortality (~20 percent per day; Service 1993). Assuming similar adult mortality rates, most individuals that do succeed to lay a single batch will not live long enough to lay a second batch. Therefore, if habitat quality is highly variable, then where the female places her egg batch is crucial to her fitness.
- *Choice of oviposition patches is generally available with predator heterogeneity among patches.* Temporary pools in this region are often found in clusters along wadi bottoms and rock outcrops (Ward and Blaustein 1994, Spencer et al. 2002a) and are close enough that mosquitoes can visit a number of pools in a single night. Among these pools, predator densities vary greatly (Ward and Blaustein 1994, Blaustein et al. 1995, Spencer et al. 1999).
- *High predator mortality on immatures.* When vulnerability of *C. longiareolata* larvae was compared to other aquatic prey including

Fig. 1 (a) *Culiseta longiareolata* egg raft (photo credit: D. Arav); (b) The backswimmer, *Notonecta maculata*; (c) *Culiseta* larvae do not distance themselves from the dragonfly predator *Anax imperator*, and in fact, attempt to graze off the exoskeleton of this predator; (d) *Anax imperator* nymph consuming a *Culiseta* larva.

larvae of other mosquito species, *C. longiareolata* was always most vulnerable among species tested on many predator species (Blaustein and Margalit 1994a, Blaustein 1998, Eitam et al. 2002, Blaustein, unpubl. data). They are particularly vulnerable to notonectids (Fig. 1b) because both predator and prey are generally found in open, and not vegetated, habitats. Larvae of *C. longiareolata* do not shift to safer habitats when predators are present (Arav 2006). Figure 1c-d illustrates the lack of predator avoidance. The larvae do not distance themselves from predators such as nymphs of the dragonfly *Anax imperator*, and in fact even attempt to graze off the predator exoskeleton, making themselves easy prey.

- *Current among-pool predator distribution predicts future predator distribution.* Common predators of *C. longiareolata* include odonates, notonectids and urodeles. Adults of odonates and most urodeles are not predators of aquatic prey. They deposit their progeny in the water and with the exceptions of flash floods along wadi bottoms that may redistribute the larval predators among pools, they are restricted to the pool in which they were deposited until they metamorphose. Hemipteran (e.g., *Notonecta maculata*, Fig. 1b) and coleopteran adults can fly from pool to pool, and many of these adults may be predators of *C. longiareolata,* but the predatory nymphs, which are much more abundant than the adults, are restricted to the pool in which their egg was deposited. Thus, if *C. longiareolata,* and other temporary pool dipterans, are able to assess among-pool distributions of aquatic predators, this distribution is a fairly good predictor of risk of predation during the mosquito's larval period.
- *Some predators are abundant and ubiquitous across a wide radius from where we have conducted oviposition studies.* For example, our studies are conducted at various places in Israel. The backswimmer *Notonecta maculata* is distributed hundreds to thousands of kilometers in each direction. Therefore, it is very unlikely that *Culiseta* females would have immigrated into our study site areas from areas not containing the predator.

Despite the high vulnerability of this mosquito to many predators, *C. longiareolata* is the, or one of the, most abundant macroinvertebrate species inhabiting temporary pools in the region. This may be explained in part because these mosquitoes are abundant and ubiquitous early in the rainy season before most predators become abundant and disperse to a large fraction of the pools (Ward and Blaustein 1994). Another reason may be because predator and prey have different preferences for pools with respect to different pool characteristics, or at least the mosquito will colonize some types of pools that the predators will not colonize. For example, backswimmers and other predators are generally more abundant in larger pools (e.g. Wilcox 2001). While *Culiseta* oviposits more in intermediate sized pools than smaller ones (Blaustein et al. 2004, Arav 2006), the smaller pools may still provide predator-free space. Lastly, *C. longiareolata* may be so abundant because the females can assess predation risk to their progeny and oviposit accordingly.

Experimental Venue and Procedure for Oviposition Habitat Selection

We have conducted numerous studies on *C. longiareolata* oviposition behavior testing a number of hypotheses using a diversity of experimental designs and conditions. We have conducted experiments in different seasons and geographical areas, both in natural rock pools (Fig. 2a) and artificial mesocosms (Fig. 2b) with strikingly similar results (Blaustein et al. 2004). In some of our experiments, we have compared control (non-predator) pools versus ones having unconstrained predators. Females oviposit at night. We then count number of egg rafts on the water surface the following morning. If there are significantly fewer egg rafts in predator pools, this could indicate oviposition habitat selection in response to risk of predation. However, fewer egg rafts in predator pools does not necessarily demonstrate oviposition habitat selection. There could be fewer egg rafts because the predator has consumed females that alighted on the water surface to oviposit, or because the predator could have consumed or disrupted the egg rafts (Chesson 1984). When we cage predators (Fig. 2b), differences in egg raft numbers can only be attributed to differential oviposition responses of the mosquito females to the caged predators. When we compare predator-conditioned water without the predator to control water, oviposition differences can only be attributed to a chemical cue released by the predator.

Does Culiseta longiareolata Avoid Aquatic Predators When Ovipositing?: Experimental Evidence

We have assessed oviposition avoidance by *C. longiareolata* to a number of predators (Table 1). When comparing pools with unconstrained predators to control pools (no predators), egg raft abundance is not different for some predators (*Triturus vitattus* [newt] and *Lestes parvidens* [damselfly]) but lower for most (Table 1). Assuming predator discrimination by prey at fine

Fig. 2 Experimental venues to test for oviposition habitat selection by mosquito females in response to risk of predation: (a) natural rock pool (photo credit: O. Segev); (b) artificial plastic pool with cage for predator.

taxonomic resolution, it is not surprising that neither *T. vitattus* nor *L. parvidens* influence oviposition by *Culiseta*. Both are presently rare. *T. vitattus* used to be more common but neither this species nor *L. parvidens* were likely found in a high proportion of pools potentially available to *C. longiareolata*. When comparing egg raft abundance in nonpredator pools versus pools

Table 1 Experimental evidence for oviposition habitat selection by *Culiseta longiareolata* in response to various predators and various predator conditions (unconstrained predator, caged predator, and predator-conditioned water only). Predator treatments are compared below against non-predator controls. Yes = statistically fewer egg rafts in the predator treatment; NS= not statistically significant at p=0.05; — = not checked. Larval stages were used for all taxonomic groups except that hemipterans were adults or larvae.

Taxon	*Predator Species*	*Predator Distribution*	*Egg Raft reduction* Free predator	*Caged predator*	*Predator water only*	*Reference*
Hemiptera: Notonectidae	*Notonecta maculata*	Ubiquitous	**Yes**	**Yes**	**Yes**	Blaustein et al. 2004; Arav and Blaustein 2006
Hemiptera: Notonectidae	*Notonecta glauca*	Rare	**Yes**	**Yes**	—	Blaustein et al. unpubl.
Hemiptera: Notonectidae	*Anisops sardea*	Ubiquitous	**Yes**	**Yes**	—	Eitam et al. 2002
Hemiptera: Notonectidae	*Anisops debilis*	Limited distribution	**Yes**	—	—	Silberbush 2004
Hemiptera: Corixidae	*Sigara* sp.	?	—	NS reduction	—	Hill et al. unpubl.
Anisoptera: Aeshnidae	*Anax imperator*	Ubiquitous but not high densities	**Yes**	No	—	Stav et al. 1999; Stav et al. 2000
Anisoptera: Libellulidae	*Sympetrum fonscolumbi*	Ubiquitous	**Yes**	NS reduction	—	Eitam et al. unpubl.
Zygoptera: Lestidae	*Lestes parvidens*	Rare?	No	No	No	Hill et al. unpublished
Anisoptera: Libellulidae	*Trithemis arteriosa*	Ubiquitous	**Yes**	—	—	Blaustein, unpubl.
Anisoptera: Libellulidae	*Orthetrum chrysostigma*	Ubiquitous	**Yes**	—	—	Blaustein, unpubl.
Urodela: Salamandridae	*Salamandra salamandra*	Limited distribution	**Yes**	No	—	Eitam and Blaustein, unpubl.
Urodela: Salamandridae	*Triturus vitattus*	Limited distribution	No	No	—	Blaustein, unpubl.

containing odonate nymphs or fire salamander (*Salamandra salamandra*) larvae, egg raft abundance is much higher in nonpredator pools. In one documented case, a small fraction of reduced egg rafts in unconstrained predator pools can be attributed to egg raft predation or disruption (*Anax imperator*: Stav et al. 1999). In another predator (*Salamandra salamandra*; Blaustein, unpublished), most or all of the observed reductions can be attributed to egg raft predation or disruption. Like the newt, the fire salamander would be considered a weak natural selection agent for oviposition habitat selection as only a small proportion of the landscape used by *Culiseta* is a salamander-inhabited region. Some odonate species like *A. imperator* might not have high predation pressure on *Culiseta* even though *Culiseta* has been shown to be highly vulnerable to predation in an artificial pool experiment (Stav et al. 2000) because *Anax* prefers permanent to semi-permanent pools with macrophytes, and is found within the vegetation zone inside a pool. By contrast, *Culiseta* oviposits in pools with open areas and the larvae are found almost exclusively in the open areas (Blaustein and Margalit 1995).

Of the four notonectid species that we tested—*Notonecta maculata, N. glauca, Anisops sardea* and *A. debilis*—only *N. maculata* is widespread in and around Israel. However, for all four notonectid species, *Culiseta longiareolata* respond by avoiding caged-predator pools (e.g., Blaustein et al. 2004, Eitam et al. 2002, Silberbush 2004, Blaustein, unpubl. data). We have thus far tested *C. longiareolata* oviposition to predator-conditioned water only of *N. maculata* only and the mosquito avoids the predator-conditioned water (Fig. 3). This indicates that the cue is one or more predator-released kairomones and that the kairomone(s) is/are found generally in all notonectids. The chemical characterization of the kairomone(s) remains unknown but we do know something about the properties. It does not appear to be a chemical released by consumed conspecific prey as has been demonstrated in other prey species and other predation risk avoidance behaviors (Wisenden and Millard 2001); the oviposition avoidance occurs regardless of whether or not the predator has been fed mosquito larvae (Blaustein, unpublished). A density of one predator per 30 liters has produced the oviposition avoidance response and the necessary concentration of predators may be even lower (Eitam and Blaustein 2004). The kairomone has been shown to actively repel oviposition by *Culiseta* for 7-8 days in a field trial (Fig. 3; Blaustein et al. 2004). Both boiling the kairomone for 20 minutes or evaporating and then reconstituting the predator-conditioned water reduces the proportion of individuals that avoid predator-conditioned water, but does not eliminate the avoidance response entirely (Eitam et al., unpublished data). These

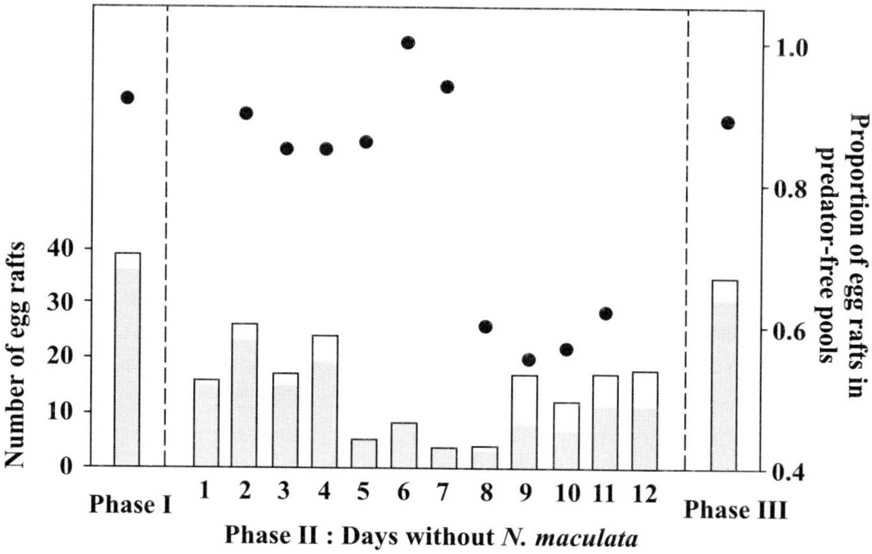

Fig. 3 Oviposition response (percent of total and number) of *Culiseta longiareolata* when offered control pools (no predator) *vs.* pools containing the predator *Notonecta maculata.* In the first period, *Notonecta* is introduced in a cage each day but only during the daylight hours. Oviposition by *Culiseta* occurs at night. In the second period, the pools remain *Notonecta*-free. The x-axis is the number of days since *Notonecta* was last present in the pool. In the third period, *Notonecta* was returned to the pools. From Blaustein et al. (2004), with kind permission of Springer Science and Business Media.

results suggest that a predator-released kairomone is not highly volatile. However, when mosquitoes are prevented from touching the predator-conditioned water, it still reduces the proportion avoiding the predator-conditioned water (Silberbush, unpublished data). These studies taken together suggest that there may be more than one kairomone—one fairly volatile and a second not very volatile.

Other *Notonecta*-Prey Combinations

We have assessed the oviposition response of other dipteran species to notonectids. These dipteran species, meet all the prey characteristics suggested earlier for oviposition habitat selection except that they range in vulnerability to the predator, and oviposition habitat selection is predicted based on vulnerability. The most vulnerable (*Culiseta*) and second most (*Culex laticinctus*) show oviposition avoidance while the less vulnerable, *Culex pipiens, Chironomus riparius* and *Forcipomya* sp. do not avoid ovipositing in the presence of backswimmers (Eitam et al. 2002, Kiflawi et al. 2003a, Blaustein et al. 2004, Arav and Blaustein 2006).

Why do Some *Culiseta longiareolata* Females Oviposit in Predator Pools?

C. longiareolata females cannot only detect predators but can also detect food levels for their larvae (Blaustein and Kotler 1993), interspecific competitors (Blaustein and Kotler 1993), and may also be able to detect the density of conspecific larvae irrespective of the resources (Kiflawi et al. 2003a). This being the case, is it possible that *Culiseta* can operate according to the ideal-free distribution in distributing their egg rafts among pools? As illustrated in Fig. 4, in the absence of predators, *C. longiareolata* progeny has increasing risk with increasing conspecific density due to intraspecific competition and cannibalism (Kiflawi et al. 2003a). Thus we would expect, and found, that females prefer to oviposit under conditions of point 1 (low conspecific density, no predator) than point 2 (high conspecific density, no predator) (Table 2). Likewise, we would expect, and found, that *C. longiareolata* females preferentially oviposit in pools with low conspecific density (point 1) over pools with *Notonecta* (point 2). When given two poor choices (points 2 and 3),

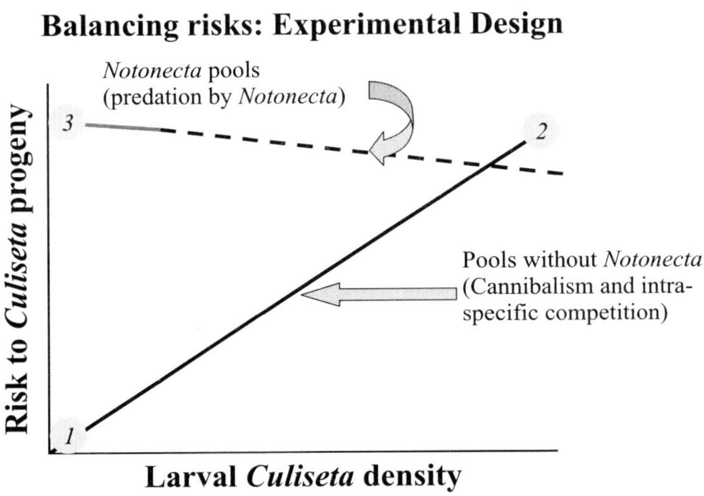

Fig. 4 Graphical representation of risk to *Culiseta* larvae given the absence and presence of *Notonecta* and a range of *Culiseta* larval densities. Point 1 is the condition of the absence or low density of *Culiseta* larvae in the absence of *Notonecta*. Point 2 is high density of *Culiseta* larvae in the absence of *Notonecta*. Point 3 is the presence of *Notonecta* with no or low densities of *Culiseta* larvae.

Table 2 Oviposition by *Culiseta longiareolata* under various paired conditions in an outdoor artificial pool experiment. The three treatments that were compared were low *Culiseta* density with no predator, high *Culiseta* density with no predator, and the predator *Notonecta*. The second column represents the total number of *Culiseta* egg rafts laid for that paired comparison. Binomial tests are one tailed. Last column is the total number of egg rafts laid in the experimental array per night.

Comparison	*Ratio*	*P(binom test)*	*No. nights*	*Egg rafts per night*
Low *Culiseta*: *Notonecta*	31:4 (89%)	< 0.0001	6	5.8
Low *Culiseta*: High *Culiseta*	50:8 (86%)	< 0.0001	7	8.2
High *Culiseta*: *Notonecta*	7:13 (35%)	0.132	8	2.5

the ratio of egg rafts oviposited in *Notonecta* pools increases dramatically which appears to support the ideal-free distribution hypothesis. However, this is open to interpretation. Given these two poor choices, the number compared to the low *Culiseta* larval density: *Notonecta* pairing, the number ovipositing total number of egg rafts deposited also drops instead of more evenly ovipositing all egg rafts to be expected at that night.

We have consistently found that when *C. longiareolata* females are given an experimental array of half nonpredator pools and half *N. maculata* pools, and all other factors are kept equal including conspecific density, a consistently small fraction of approximately 5-15 percent of the females still oviposit in the predator pools (Blaustein et al. 1995, Blaustein 1998, Spencer et al. 2002b, Kiflawi et al. 2003a, Blaustein, unpublished data). Why? Two plausible explanations are as follows: (1) genetic differences: a fraction of the population may not possess the ability to detect the predator and/or the larvae of females that oviposit in predator pools are less vulnerable to predation than the larvae arising from females that avoid predator pools; (2) searching for a "good" pool is limited and after a certain period of time or a certain number of pools assessed by the female, the small fraction of females that encounter only *Notonecta* pools oviposit in a *Notonecta* pool. Laboratory predation studies have shown that larvae arising from egg rafts deposited in predator pools are not less vulnerable than those arising from egg rafts deposited in nonpredator pools (Arav 2006). Unfortunately, our lab has thus far failed to maintain a sustained colony to address the other questions by observation and experimentation. However, to gain insight as to which of these was the more likely explanation, we ran an artificial pool field experiment in which, on different nights, we changed the ratio of control pools to *Notonecta* pools. On different nights, nonpredator:predator pool ratios were 3:9, 9:9, and 9:3 (Kiflawi et al. 2003 and unpublished data). We found an increase in the proportion of females using predator pools as the proportion of predator pool was increased (Fig. 5) which more strongly supports the second hypothesis.

Perspective

Behavioral responses to the environment are phenotypically plastic responses. The oviposition response of the mosquito *Culiseta longiareolata* to some of its predators provides one example. Oviposition habitat selection as a phenotypically plastic response to predation was, until recently, largely ignored as an important factor in explaining species distributions and community structure. Though examples of this behavior exist in terrestrial systems (e.g. Agarwala et al. 2003, Whitman and Blaustein, this volume), much of the literature documenting this behavior suggests that the oviposition avoidance is strongest and most prevalent for amphibians (e.g. Murphy 2003, Binckley and Resetarits 2003) and insects (e.g., Resetarits 2001, Blaustein et al. 2004) utilizing pool or pond habitats. The reason for this seems to be due to the predator-prey characteristics outlined in this chapter including predictability of future risk of predation when progeny

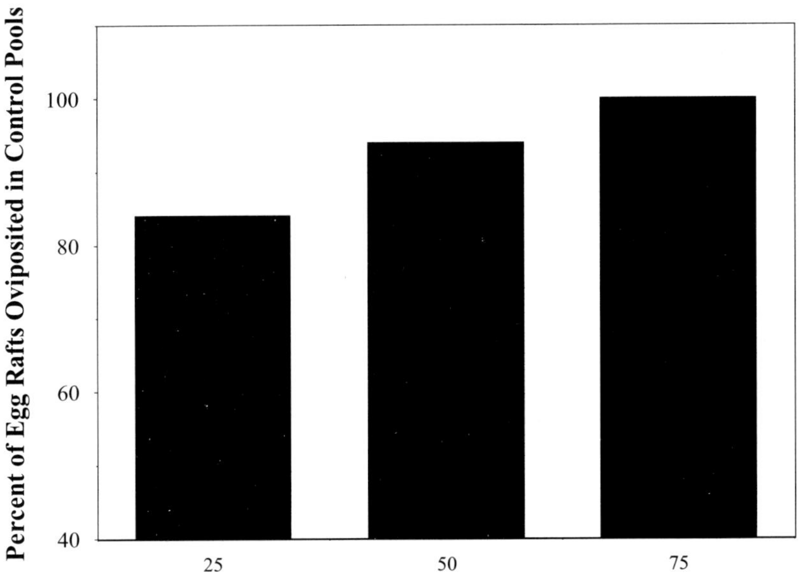

Fig. 5 The relationship between the percent of egg rafts laid in nonpredator pools with the changing ratio of nonpredator to *Notonecta* pools.

hatch. This behavior also has consequences for populations. In the case of *C. longiareolata* and backswimmers, a linear stage-structured model suggests that this behavior, compared to random oviposition, results in a larger adult mosquito population (Spencer et al. 2002b). Oviposition habitat selection in response to risk of predation also has consequences for communities (e.g. Resetarits 2005). In the specific system reported here, *C. longiareolata* larvae themselves are influential as competitors and predators, so how these larvae are distributed among pools should cause differences in animal and algal community structure (Blaustein et al. 1995, Blaustein and Margalit 1994b, 1996). This behavior also has consequences for how experiments should be designed to assess the true effect of the predator on prey populations (Blaustein 1999, Spencer et al. 2002b).

Acknowledgements

We thank collaborators on this specific research problem for their valuable discussion, namely Derah Arav, Jonathan Blaustein, Jonathan Chase, Joel E. Cohen, Avi Eitam, Naomi Hill, Moshe Kiflawi, Burt P. Kotler, Marc Mangel, Joel Margalit, Shai Markman, Jerrold Meinwald, Alon Silberbush, Mathew Spencer, Gil Stav and James Vonesh. The work was funded by Israel Science Foundation Grant 699-02 awarded to L. Blaustein.

References

Agarwala, B. K., H. Yasuda and Y. Kajita. 2003. Effect of conspecific and heterospecific feces on foraging and oviposition of two predatory ladybirds: Role of fecal cues in predator avoidance. Journal of Chemical Ecology 29: 357–376.

Arav, D. 2006. Oviposition and development responses of temporary pool dipterans to risks of predation and desiccation. MSc thesis, University of Haifa.

Arav, D. and L. Blaustein. 2006. Effects of pool depth and risk of predation on oviposition habitat selection by temporary pool dipterans. Journal of Medical Entomology 43: 443–497.

Benard, M. F. 2004. Predator-induced phenotypic plasticity in organisms with complex life histories. Annual Review of Ecology, Evolution and Systematics 35: 651–673.

Binckley, C. A. and W. J. Resetarits. 2003. Functional equivalence of non-lethal effects: generalized fish avoidance determines distribution of gray treefrog, *Hyla chrysoscelis*, larvae. Oikos 102: 623–629.

Blaustein, L. 1997. Nonconsumptive effects of larval *Salamandra* on its crustacean prey: can eggs detect predators? Oecologia 110: 212–217.

Blaustein, L. 1998. Influence of the predatory backswimmer, *Notonecta maculata*, on pool community structure. Ecological Entomology 23: 246–252.

Blaustein, L. 1999. Oviposition habitat selection in response to risk of predation: Consequences for populations and community structure. In: Evolutionary Processes and Theory: Modern Perspectives. Ed. by S. P. Wasser, Kluwer Academic Publishers. pp. 441–456.

Blaustein, L., M. Kiflawi, A. Eitam, M. Mangel and J. E. Cohen. 2004. Oviposition habitat selection in response to risk of predation: Mode of detection consistency across experimental venue. Oecologia 138: 300–305.

Blaustein, L. and B. P. Kotler. 1993. Oviposition habitat selection by *Culiseta longiareolata*: effects of immature conspecifics, tadpoles and food levels. Ecological Entomology 18: 104–108.

Blaustein, L., B. P. Kotler and D. Ward. 1995. Direct and indirect effects of a predatory backswimmer (*Notonecta maculata*) on community structure of desert temporary pools. Ecological Entomology 20: 311–318.

Blaustein, L. and J. Margalit. 1994a. Differential prey vulnerability of various mosquito species to the cyclopoid copepod, *Acanthocyclops viridis*. Israel Journal of Zoology 40: 55–60.

Blaustein, L. and J. Margalit. 1994b. Mosquito larvae (*Culiseta longiareolata*) compete with and prey upon toad (*Bufo viridis*) immatures. Journal of Animal Ecology 63: 841–850.

Blaustein, L. and J. Margalit. 1995. Spatial distributions of *Culiseta longiareolata* and *Bufo viridis* within and among desert pools. Journal of Arid Environments 29: 199–211.

Blaustein, L. and J. Margalit. 1996. Priority effects in temporary pools: Nature and outcome of mosquito larva-toad tadpole interactions depend on order of entrance. Journal of Animal Ecology 65: 77–84.

Brown, J. S. and B. P. Kotler. 2004. Hazardous duty pay and the foraging cost of predation. Ecology Letters 7: 999–1014.

Chase, J. M. and T. M. Knight. 2003. Drought-induced mosquito outbreaks in wetlands. Ecology Letters 6: 1017–1024.

Chesson, J. 1984. Effect of notonectids (Hemiptera, Notonectidae) on mosquitos (Diptera, Culicidae)—predation or selective oviposition. Environmental Entomology 13: 531–538.

Colton, Y. M., D. D. Chadee and D. W. Severson. 2003. Natural skip oviposition of the mosquito *Aedes aegypti* indicated by codominant genetic markers. Medical and Veterinary Entomology 17: 195–204.

Dewitt, T. J. and S. M. Scheiner. 2004. Phenotypic variation from single genotypes: A primer. In: DeWitt, T.J. and Scheiner, S.M. (eds) Phenotypic Plasticity: Functional and Conceptual Approaches. Oxford University Press. Pp. 1–9.

Eitam, A. and L. Blaustein. 2004. Oviposition habitat selection by mosquitoes in response to predator (*Notonecta maculata*) density. Physiological Entomology 29: 188–191.

Eitam, A., L. Blaustein and M. Mangel. 2002. Effects of *Anisops sardea* (Hemiptera: Notonectidae) on oviposition habitat selection by mosquitoes and other dipterans and community structure in artificial pools. Hydrobiologia 485: 183–189.

Garcia, T. S., J. Stacy and A. Sih. 2004. Larval salamander response to UV radiation and predation risk: Color change and microhabitat use. Ecological Applications 14: 1055–1064.

Kiflawi, M., L. Blaustein and M. Mangel. 2003a. Oviposition habitat selection in a mosquito, *Culiseta longiareolata* in response to risk of predation and conspecific density. Ecological Entomology 28: 168–173.

Kiflawi, M., L. Blaustein and M. Mangel. 2003b. Predation-dependent oviposition habitat selection by the mosquito *Culiseta longiareolata:* a test of competing hypotheses. Ecology Letters 6: 35–40.

Kotler, B. P. 1984. Risk of predation and the structure of desert rodent communities. Ecology 65: 689–701.

Li, D. Q. 2002. Hatching responses of subsocial spitting spiders to predation risk. Proceedings of the Royal Society of London Series B-Biological Sciences 269: 2155–2161.

Murphy, P. J. 2003. Context-dependent reproductive site choice in a Neotropical frog. Behavioral Ecology 14: 626–633.

Nakashima, Y., M. A. Birkett, B. J. Pye, J. A. Pickett and W. Powell. 2004. The role of semiochemicals in the avoidance of the seven-spot ladybird, *Coccinella septempunctata*, by the aphid parasitoid, *Aphidius ervi*. Journal of Chemical Ecology 30: 1103–1116.

Peacor, S. D. and E. E. Werner. 2004. Context dependence of nonlethal effects of a predator on prey growth. Israel Journal of Zoology 50: 139–167.

Resetarits, W. J. 2001. Colonization under threat of predation: avoidance of fish by an aquatic beetle, *Tropisternus lateralis* (Coleoptera : Hydrophilidae). Oecologia 129: 155–160.

Resetarits, W. J. 2005. Habitat selection behaviour links local and regional scales in aquatic systems. Ecology Letters 8: 480–486.

Service, M. W. 1993. *Mosquito Ecology. Field Sampling Methods*. 2nd edn. Elsevier Applied Sciences, London.

Sih, A. 2004. A behavioral ecological view of phenotypic plasticity. In: DeWitt, T.J. and Scheiner, S.M. (eds) Phenotypic Plasticity: Functional and Conceptual Approaches. Oxford University Press. pp. 112–125.

Silberbush, A. 2004. Effects of salinity and predation on dipteran and algal temporary pool community structure: physiological and behavioral mechanisms. MSc dissertation, Ben Gurion University.

Spencer, M., L. Blaustein, S. S. Schwartz and J. E. Cohen. 1999. Species richness and the proportion of predatory species in temporary pools: relationships with habitat size and permanence. Ecology Letters 2: 157–166.

Spencer, M., S. S. Schwartz and L. Blaustein. 2002a. Are there fine-scale spatial patterns in community similarity among temporary freshwater pools? Global Ecology and Biogeography 11: 71–78.

Spencer, M., L. Blaustein and J. E. Cohen. 2002b. Oviposition habitat selection by mosquitoes (*Culiseta longiareolata*) and consequences for population size. Ecology 83: 669–679.

Stav, G., L. Blaustein and J. Margalit. 1999. Experimental evidence for predation sensitive oviposition by a mosquito, *Culiseta longiareolata*. Ecological Entomology 24: 202–220.

Stav, G., L. Blaustein and Y. Margalit. 2000. Influence of nymphal *Anax imperator* (Odonata: Aeshnidae) on oviposition by the mosquito *Culiseta longiareolata* (Diptera:Culicidae). Journal of Vector Ecology 25: 190–202.

Teplitsky, C., S. Plenet and P. Joly. 2005. Costs and limits of dosage response to predation risk: to what extent can tadpoles invest in anti-predator morphology? Oecologia 145: 364–370.

Trembath, R. and B. R. Anholt. 2001. Predator-induced morphological and behavioral changes in a temporary pool vertebrate. Israel Journal of Zoology 47: 419–431.

Van Pletzen, R. and T. C. de K. Van Der Linde. 1981. Studies on the biology of *Culiseta longiareolata* (Macquart) (Diptera: Culicidae). *Bulletin of Entomological Research* 71: 71–79.

Warburg, M. R. 1994. Population ecology, breeding activity, longevity, and reproductive strategies of *Salamandra salamandra* during an 18-year long study of an isolated population on Mt. Carmel, Israel. Mertensiella, 4: 399–421.

Ward, D. and L. Blaustein. 1994. The overriding influence of flash floods on species:area curves in ephemeral Negev desert pools: a consideration of the value of island biogeography theory. Journal of Biogeography 21: 595–603.

Werner, E. E. and S. D. Peacor. 2003. A review of trait-mediated indirect interactions in ecological communities. Ecology 84: 1083–1100.

Wilbur, H. M. 1997. Experimental ecology of food webs: Complex systems in temporary ponds—The Robert H. MacArthur Award Lecture—Presented 31 July 1995 Snowbird, Utah. Ecology 78: 2279–2302.

Wilcox, C. 2001. Habitat size and isolation affect colonization of seasonal wetlands by predatory aquatic insects. Israel Journal of Zoology 47: 459–475.

Wisenden, B. D. and M. C. Millard. 2001. Aquatic flatworms use chemical cues from injured conspecifics to assess predation risk and to associate risk with novel cues. Animal Behaviour 62: 761–766.

9

Polyphenisms in Lepidoptera: Multidisciplinary Approaches to Studies of Evolution and Development

P.M. Brakefield[1] and W.A. Frankino[2]

[1]*Institute of Biology, Leiden University, P.O. Box 9516, 2300 RA Leiden, The Netherlands, E-mail: p.m.brakfield@biology.leidenuniv.nl*

[2]*Department of Biology and Biochemistry, University of Houston, Houston, TX. 77204–5007, USA, E-mail: frankino@uh.edu*

Abstract

Polyphenisms, where individuals can express one of two or more alternative, discrete phenotypes, offer a unique opportunity to study adaptation at different levels of biological organization. Here, we review research on polyphenisms in the Lepidoptera with the aim of illustrating how multidisciplinary, integrative approaches are providing new insights about the processes of adaptive evolution both within species and, using a comparative framework, among lineages. Many examples are associated with seasonal changes in the environment. Laboratory determinations of the pathways of induction of such seasonal polyphenisms in combination with field studies of natural selection have revealed how seasonal morph expression is often adaptive. Moreover, elucidation of the physiology and genetics of morph expression is leading to an ability to connect selection on the expression of the seasonal phenotypes to changes in the mechanisms that generate the alternative phenotypes. We review how work in evolutionary and developmental genetics is facilitating exploration of the contributions of developmental bias and generative constraints to the evolution of seasonal polyphenisms. In sum, integrative research on polyphenisms in the Lepidoptera is revealing how genetics, physiology, and development affect adaptive evolution on the micro-evolutionary scale, and how such polyphenisms can contribute to macro-evolutionary patterns. We close with a discussion of how research programs into the proximate and ultimate basis of polyphenisms could be directed to focus on the evolution of development itself.

Introduction

Polyphenisms represent an extreme case of phenotypic plasticity in which individuals express alternative, discrete phenotypes in different developmental environments. Even when data on the subject are scarce, polyphenisms are generally interpreted as being adaptive, because the maintenance of such dramatically distinct forms in a population suggests fitness benefits for the environment-specific expression of phenotypes (Lively 1986, Moran 1992). Alternative phenotypes are produced by control of potential variability in physiological and developmental pathways in response to environmental cues perceived at an earlier stage of ontogeny. Adaptive expression of a particular phenotype therefore requires exposure to inducing cues during a window of ontogeny sufficiently preceding the selective environment such that the appropriate phenotype can develop (Fig. 1) (Moran 1992). In addition to this form of developmental constraint, the evolution of polyphenisms may be limited by genetic correlations of trait values among alternatively expressed phenotypes (e.g. Via and Lande 1985). Lepidoptera exhibit polyphenisms in every life history stage, with some morphs being so distinct that they were historically misidentified as different species, e.g., larval morphs of *Nemoria arizonaria* (Greene 1989a) or adult morphs of *Araschnia levana* (Imms 1925). Because of their often striking appearance, apparent adaptive value, amenability to experimentation, and sheer life-historical and taxonomic diversity, polyphenisms in Lepidoptera are among the best studied of any insect order.

Lepidoptera have been used with great success to study the ecological conditions favoring the evolution and maintenance of polyphenisms. The physiological mechanisms regulating morph expression, and increasingly, the genetic mechanisms underlying the development of alternative trait states, have also been studied extensively. Such studies are of interest because they inform our understanding of morph determination and because they suggest generalities regarding the roles of development in the evolution of plastic responses, switch point evolution, and so on. The real power Lepidopteran systems offer, however, lies in combining these experimental approaches to understand how ecology, physiology, and development shape the evolution of polyphenic systems (Beldade and Brakefield 2002, Brakefield and Wijngaarden 2003). In particular, blending these approaches with comparative methodologies may offer insight into the general roles of developmental mechanisms in the evolution of polyphenisms specifically, and perhaps more generally, into the roles of development in shaping the evolution and diversification of morphology

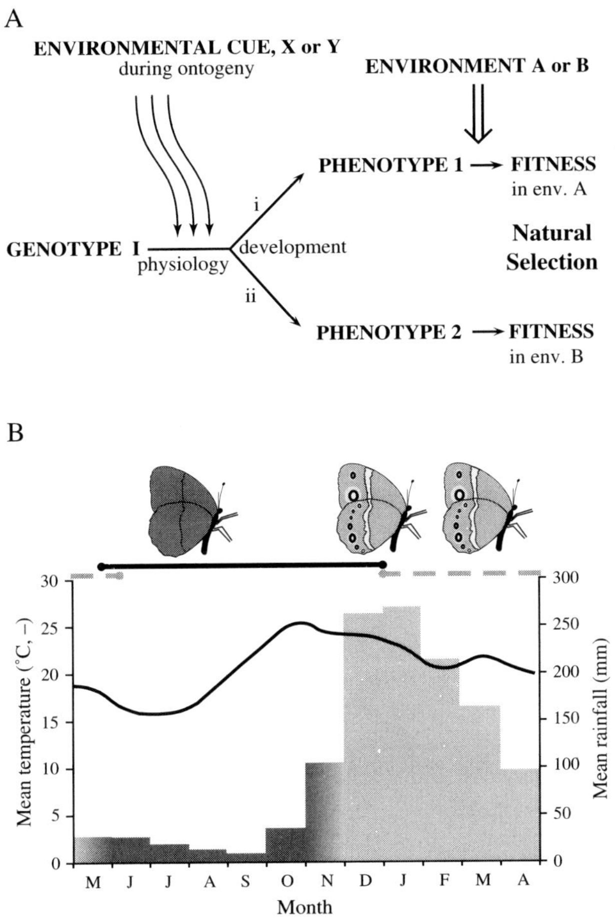

Fig. 1. (A) Scheme showing the basic components of developmental phenotypic plasticity in which two alternative seasonal forms (phenotypes 1 and 2) can be produced by modulation of development among individuals of a single genotype (I). An environmental cue predicting the environment in which selection will take place acts via physiological mechanisms to modulate the pathway of development (i or ii). Natural selection yields a higher relative fitness for each phenotype or morph in the environment (A or B) in which it spends most of its time. **(B)** A wet-dry seasonal cycle illustrated for *Bicyclus* butterflies in East Africa (see also Fig. 2). A dry season is followed by a wet season (dark and light grey bars, respectively). Two generations of the wet season form (WSF) with conspicuous eyespots occur in each rainy season. Larvae of both of these cohorts develop at high average temperatures (wavy line). The second WSF generation lays eggs before the grass food plants die out, and the larvae develop at progressively declining temperatures. This cohort produces the generation of the dry season form without eyespots that persists through the period of low rainfall. Periods of flight for each morph cohort are shown as bars along the X-axis and represented by the cartoons (redrawn from Brakefield and Reitsma, 1991).

and life history (Brakefield et al. 2003, Frankino and Raff 2004), and ultimately new species (West-Eberhard 2003). In the sections that follow, we review examples where such integrative approaches have lead to insights regarding the processes influencing the evolution of polyphenisms at different levels of biological organization and in different life history stages in Lepidoptera. We conclude by focusing on how studies placed in a comparative context can suggest general roles for development in the evolution of polyphenisms and morphological diversification.

Natural Selection, Fitness, and Seasonal Polyphenism

The seasonal morphs of adult butterflies are usually considered to be adapted to their matching environment, with each form having higher fitness during the season in which it flies (Shapiro 1976, Brakefield and Larsen 1984). However, few attempts have been made to test this hypothesis directly, and indeed even for some well known cases such as *Araschnia levana* (Nymphalidae) and *Precis coenia* (Nymphalidae) (see color images in Beldade and Brakefield 2002), we have only tenuous hypotheses addressing why differences in fitness may occur between seasonal morphs. There are, however, at least two well-worked examples where experiments provide both strong evidence for the expected seasonal differences in fitness between morphs and insights about the reasons why such differences exist. One of these examples involves thermoregulatory behavior, and the other, interactions with visually-hunting predators.

Melanization in *Pontia occidentalis*

A series of experiments by Joel Kingsolver represents an exceptionally thorough examination of how natural selection acts on the spring and summer forms of the western white butterfly, *Pontia occidentalis* (Pieridae), in North America, and the results provide a persuasive argument for the adaptive nature of the polyphenism. In this species, larvae developing under short photoperiods (spring) metamorphose into adults with a greater degree of melanization at the base of each dorsal wing surface and on the ventral hindwing than larvae reared under long-day (summer) conditions. Increased wing melanization and specific basking behaviors lead to more effective thermoregulation and elevated flight activity in the cooler conditions experienced by the spring form (Watt 1968, Kingsolver 1987). In combination with alternative patterns of thermoregulatory behavior, reduced wing-melanization in summer-flying *Pontia* butterflies lowers solar radiative heating and thereby decreases the likelihood of overheating. Kingsolver (1995a, b, 1996) used cohorts of bred or wild-collected butterflies in mark-release-recapture (MRR) experiments to examine patterns of relative survival of the two morphs across seasons (see also Kingsolver and Huey 1998).

Kingsolver's initial experiment examined directional selection on six wing traits in spring and in summer populations (Kingsolver 1995a). His results suggested that butterflies with more melanized wings survived longer within spring cohorts whereas paler individuals were favored in the summer. The pattern of selection on the degree of melaninization at the base of the dorsal hindwing suggested that some form of constraint may have been operating to restrict the amount of phenotypic plasticity exhibited across generations (see further discussion below). In a second experiment, Kingsolver released laboratory-reared butterflies of each morph to examine how morph relative fitness varied across seasons (Kingsolver 1995b). There were no differences between morphs in survival during the spring, perhaps because the weather was unseasonably warm that year or because selection is weaker during the spring. However, in the warmer, summer months, the paler summer morph had higher survival than the more melanized spring morph. This result is as expected but does not directly implicate reduced melanization of the summer morph as the target of selection. This issue was elegantly tackled in a third experiment using experimentally altered wing phenotypes (Kingsolver 1996). Butterflies collected in the wild were randomly assigned to two treatments: 1) *Blackened* in which case black pigment was added to the ventral hindwing veins, and 2) *Yellowed* in which the added pigment did not contrast with the natural background wing color. *Blackened* males showed substantially higher mortality relative to *Yellowed* males during the summer whereas *Blackened* and *Yellowed* females had equal survival. This sex-specific effect may result from a less marked change in female appearance relative to males (i.e., the manipulated female was closer to the wild-type phenotype than manipulated males were to wild-type males). Alternatively, the activity of males and females may differ such that males are subject to more intense selection by the thermal environment.

Kingsolver's results follow from the known relationship between melanin pattern on the wings and functional aspects of thermoregulation (Heinrich 1993) to demonstrate that each seasonal phenotype in *Pontia* is more frequent in environments in which its relative fitness is higher. Below, we discuss another seasonal polyphenism that has yielded similar results from cohort analyses in the field.

Eyespots in *Bicyclus anynana*

This satyrid butterfly occurs in wet-dry seasonal woodlands south of the equator in Africa (Brakefield and Reitsma 1991, Brakefield and Mazzotta 1995). They are active close to the ground. Adults of the wet season form (WSF) have a pale medial band and conspicuous, submarginal eyespots on their ventral wings (Fig. 2). Their marginal eyespots are exposed when at rest and probably function in deflecting some attacks of vertebrate predators away from the vulnerable body (Brakefield and Larsen, 1984, Lyytinen et al. 2003, 2004, Stevens 2005, Vlieger and Brakefield, 2007 for eyespots that are suddenly exposed to potential predators with a 'startling' or intimidating function, see Vallin et al. 2005). WSF butterflies are comparatively short-lived. They reproduce quickly and are active among green herbage that carpets the ground in the warm, wet season. In contrast, adults of the dry season form (DSF) are uniformly brown in color with highly reduced ventral eyespots (Fig. 2). DSF butterflies are inactive for much of their adult life and appear well camouflaged when at rest with wings closed amongst dead brown leaf litter (the green herbage having died away in the early dry season; Fig. 2a, b). They must survive several months before they can lay eggs at the beginning of the rainy season when larval food plants (grasses) produce fresh foliage.

Although temperature alone can induce the alternative seasonal forms of *Bicyclus*, in nature it is more likely that a combination of variables influence developmental time, which in turn is the actual morph-determining cue. Several observations support this hypothesis: 1) Populations artificially selected for fast pre-adult development exhibited more extreme WSF- ventral eyespots, whereas populations selected for slow pre-adult development evolved smaller eyespots in the direction of the DSF (Brakefield and Kesbeke 1997, Zijlstra et al. 2003, 2004); 2) larvae reared at high temperatures on a low-quality, artificial diet exhibited longer developmental times and DSF-like phenotypes (Holloway et al. 1992); 3) larvae reared at low temperature in the laboratory develop into butterflies that exhibit less extreme DSF phenotypes relative to those in nature, indicating additional roles for environmental factors such as the seasonal drying of larval food plants (Kooi et al. 1998). These results, in combination with the delayed reproduction, increased longevity, and alternative behaviors of the DSF suggest that it is a stress-tolerant morph (Brakefield 1997, Brakefield et al. 2007).

Cohort analyses using MRR experiments were performed with species of *Bicyclus* at Zomba in Malawi in the mid-1990s to test the hypothesis that seasonal morph expression is adaptive. Specifically, the hypothesis tested was that cryptic behavior and a uniform brown color is advantageous in the dry season, whilst camouflage is ineffective for active (brown) butterflies in the wet season when conspicuous marginal eyespots are favored (see Fig. 2).

Fig. 2 **(a)** late wet season and **(b)** mid dry season photographs taken at the same forest-edge site near Zomba, Malawi. Five species of *Bicyclus*, including *B. anynana*, *B. safitza* (c, e) and *B. cottrelli* (d) fly at this site. The thick layer of green herbage including larval food plants of grasses in a) has been replaced in b) by a carpet of dead foliage, leaves or bare ground (from Brakefield and Reitsma, 1991). **(c)** butterflies of the wet season form are highly active, rest on green herbage, and mate (the darker individual is the male) and oviposit soon after eclosion. **(d)** a female dry season form butterfly that has just alighted on a dead leaf. The conspicuous forewing eyespot will then be hidden by partial withdrawal of the forewings between the hindwings, producing effective camouflage. **(e)** females of the wet season form (left) and dry season form feeding on banana fruit. The butterflies are sisters reared in the same environment except for the final larval instar that was kept at 27°C or 20°C for the two individuals, respectively.

Experiments were performed in each season in several trials using lab-reared butterflies ranging continuously in phenotype from the extreme DSF through to the extreme WSF. Butterflies were released in a forest-edge environment containing a grid of about 40 fruit-baited traps (see Brakefield and Reitsma, 1991). Patterns of butterfly movement over the traps were similar for the different phenotypes. In the dry season, WSF *B. safitza* butterflies had a much lower survival (probability of recapture) than those of the DSF (Fig. 3A). In the wet season, the survival probability between seasonal forms reversed, but the difference in morph fitness was small (G. Engelhard, N. Reitsma and P.M. Brakefield, unpub. data). Thus, as with *P. occidentalis*, there is again a seasonal asymmetry in selection strength; extremely strong selection against eyespots in the dry season contrasts with weak selection in favor of eyespots in the wet season.

However, experimental releases of laboratory-reared individuals do not demonstrate unequivocally that it is the eyespots of the wet season morph that account for the lower survival of this morph in the dry season (cf. Lyytinen et al. 2004). Additional field experiments in Malawi used butterflies with experimentally manipulated eyespots to test the hypothesis that increased mortality results from larger eyespot size in the dry season (Fig. 3B). About 1800 wild-caught *B. safitza* of the DSF were marked individually on their dorsal wings (which are not exposed when at rest) and assigned randomly to one of three treatments: 1) no further treatment (control); 2) marked using felt-tip pens to have a full series of submarginal black-gold eyespots on each ventral wing surface (conspicuous eyespot treatment); or 3) marked in the same way as above but using pens of a similar color to the brown background of the wings (sham control with inconspicuous eyespots). All butterflies were released and recaptured over the trapping grid as in the previous MRR experiment. Unpainted control and sham control butterflies had similar survivorship curves whereas butterflies treated to have conspicuous submarginal eyespots similar to those of the WSF had much higher mortality rates (Fig. 3B). These results show unambiguously that the eyespots themselves contribute to the lower relative fitness of the WSF during the dry season. The reverse experiment (painting over eyespots of the WSF for releases in the wet season) proved more difficult, but the predicted switch in relative fitness occurred between seasonal morphs although with only a small difference in survival.

Recent studies in aviaries (Lyytinen et al. 2003; 2004) of bird predation on the seasonal forms of *B. anynana* also suggest a high premium on crypsis in the resting environment of brown leaf litter of the dry season, favoring the DSF, but only a small advantage favoring eyespots for deflective purposes

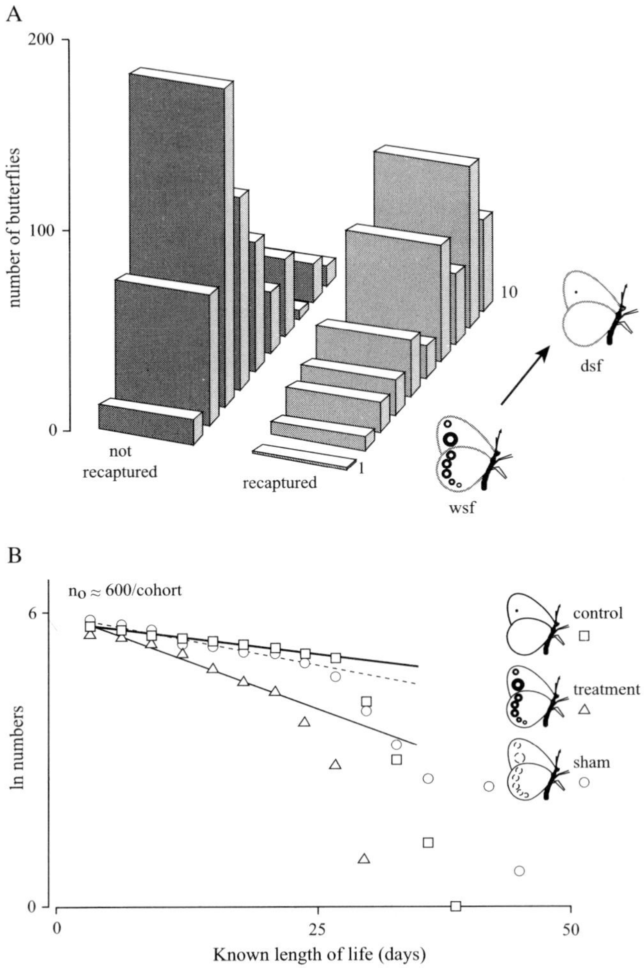

Fig. 3 Natural selection on *Bicyclus safitza* butterflies in the dry season illustrated by the results of representative experiments. A) Number of butterflies recaptured at least once or not at all over a grid of forest traps for releases of reared butterflies of ten phenotypic classes ranging from extreme wet season form (WSF with very large ventral eyespots) to extreme dry season form (DSF—no eyespots). The WSF shows much higher mortality. B) Survivorship curves over the same grid of traps for releases of DSF butterflies collected in another forest and divided among three treatments: unpainted controls (squares); painted with conspicuous eyespots (triangles); and painted with inconspicuous eyespots (circles). Butterflies with painted, conspicuous eyespots show a dramatically higher mortality ($P<0.001$) suggesting that the eyespots make them easier to find by predators in the dry season. Lines show periods of age-independent survival for each cohort. Previously unpublished data of N. Reitsma and P.M. Brakefield.

amongst the green herbage of the wet season. Marginal eyespots may only provide effective escape for some of the attacks made by naïve birds shortly after fledging, and not by birds that have already experienced butterfly prey (Lyytinen et al. 2004). Similar experiments demonstrated no influence of marginal eyespots on attacks by *Anolis* lizards (Lyytinen et al. 2003; Vlieger and Brakefield, 2007). In *Bicyclus*, there may also be some selective advantage to males with larger eyespots in the wet season resulting from female choice (see Breuker and Brakefield 2002; Robertson and Monteiro 2005); such sexual selection may contribute some additional advantage to the WSF (see Lyytinen et al. 2004).

These experiments on *Pontia* and *Bicyclus* demonstrate the adaptive aspects of seasonal morph expression and have begun to reveal why the observed seasonal patterns of relative fitness occur. These two systems illustrate how the adaptive significance of polyphenisms can be demonstrated in the field. However, considerable progress in understanding the evolution and maintenance of polyphenisms can also be made by studying how discrete morphological variation influences aspects of performance. Below, we discuss two examples where affects of morph on development and growth rate interact with differences in seasonal survival to maintain seasonal polyphenisms.

Color Polyphenisms in Eastern Black Swallowtail Larvae and Pupae

Hazel has investigated larval and pupal seasonal polyphenisms of *Papilio polyxenes* (Hazel 2002, Hazel and West 1979, 1983). This species is multivoltine, producing 2–4 generations each season before overwintering as a diapausing pupa. Larvae developing under warm, long day (summer) conditions are green with a series of narrow black bands centered within and between each of their segments. Larvae developing under cool, short day (autumnal) conditions have thickened bands, becoming almost fully black. This dark color allows more rapid heating and higher body temperatures, which in turn accelerates development. Rapid development is advantageous as late season larvae need to pupate before the first autumnal freeze to successfully over winter. Thus, induction of dark larval coloration late in the fall has a demonstrably adaptive explanation. However, the green and black coloration of the summer morphs is more difficult to explain. Summer morphs may be more cryptic in that season. Alternatively, greater melanization may confer physiological or size-based fitness costs manifested in other life history stages (Windig 1999, Talloen et al. 2004) or may be unnecessary for thermoregulation—even counterproductive—during the warm summer.

Papilio polyxenes also exhibits a pupal polyphenism that has a strong seasonal component. Larvae pupating on green (e.g. leafy) backgrounds tend to become green, whereas brown (e.g., woody) backgrounds tend to induce brown pupae. Interestingly, photoperiod and temperature affect the response to background color at the pupation site; short days foreshadow the loss of foliage during the autumn (Hazel and West 1979, 1983). Hence, brown pupae developing in the fall on green backgrounds will be cryptic over most of their pupal period. These dark individuals may also have increased fitness due to enhanced warming in the early spring months, hastening emergence.

Diet-induced Larval Polyphenism in the Emerald Moth

A striking seasonal polyphenism is exhibited in the bivoltine geometrid moth, *Nemoria arizonaria* (Greene 1989a; for other examples in geometrid moths see Kettlewell 1973). Spring broods feed on oak catkins (*Quercus* spp.), which are high in protein. These caterpillars develop into dramatic catkin mimics. Catkin morphs are bumpy, with red and yellow markings that make them highly cryptic on catkins. Summer broods feed on a lower-quality diet of oak leaves, and larvae develop a smooth, brown, twig-like phenotype. The two morphs differ in behavior; when provided a choice, each moves to the background on which it is cryptic. Finally, the morphs differ in feeding morphology in a way consistent with efficient processing of the different diets. Although morph is clearly diet-induced (Greene 1989a, Greene 1996), the exact dietary component cueing the switch (e.g., tannin or protein content, textural properties, etc.) is unclear (Faeth and Hammon 1989, Greene 1989b).

The factors favoring the maintenance of this polyphenism are probably complex (Greene 1989a). Presumably because of their rich diet, catkin morphs have greater development rates, survival rates to pupation, mean size at pupation, and fecundity relative to twig morphs. Parasitism and predation by birds can be intense, presumably favoring crypsis and rapid development. Fitness gains clearly favor production of a summer cohort, and since the catkins are present only in the spring, when selection by predators probably favors the seasonally cryptic twig morph. These strong verbal arguments regarding the benefits of seasonal morph expression, however, require testing. Predator/parasitoid exclusion, reciprocal transplantation of morphs, and measurements of growth rates of morphs fed typical and 'wrong' diets are necessary to fully elucidate the selective pressures acting on this fascinating system.

We will now move on from how natural selection shapes phenotypic variation in polyphenisms in the Lepidoptera to review some of what is known about how alternative phenotypes are generated.

Physiological Basis of Expression of Polyphenism

Hormonal regulation of polyphenism expression is well studied in Lepidoptera, but the patterns that emerge regarding physiological control are complex (Koch 1992). This is not surprising, considering the diversity of polyphenisms, even within a single developmental stage. For example, some adult polyphenisms result from differences in wing scale color intensity (e.g., the background coloration of *Bicyclus* morphs) whereas others result from differences in wing color patterning (e.g., seasonal variation in eyespot presence or size in *Bicyclus* morphs). Still others result from changes not just in wing color intensity or pattern, but also in wing shape (e.g., seasonal morphs of *Lycaena phlaeas* and *Melanitis leda*). At some level, morph expression is usually hormonally regulated. Hormones play a variety of roles in insect development, coordinating trait differentiation with the moulting cycle, coordinating ontogenies among traits themselves, and creating ontogenetic windows or sensitive periods during which plastic responses must be initiated developmentally (see reviews in Nijhout 1994, 1999a, b, and Nijhout and Davidowitz, this volume).

Cell differentiation and the molting cycle are controlled primarily by the interaction of two hormones, juvenile hormone (JH) and ecdysone, with target cells. Low titers of circulating JH are followed by prothoracicotropic hormone (PTTH) production, which induces ecdysone synthesis and release. Ecdysone is a steroid hormone that initiates the developmental cascade culminating in a molt. When ecdysone interacts with target cells and JH exceeds some threshold of cell senstivity, then the tissues retain their identity and a larval molt ensues. When JH is below the threshold, alternative developmental pathways may be expressed, changing cell fates and causing morphological differentiation. Such differentiation can take the form of ontogenetic shifts such as moulting from larva to pupa or they can result in switches among polyphenic morphs. In both cases though, the unfolding of different developmental pathways can be complex and involve many tissues across instars (see reviews in Nijhout 1994, 1999a, 1999b, Zera 1999).

The threshold for a given developmental response is dependent on the biological activity of the hormones and the sensitivity of the cell to the hormones at that time. Circulating hormone titers can be modifided by altering hormone production or the activity of degradatory enzymes (e.g., juvenile hormone esterase). Cell sensitivity can be modified in a variety of ways. Hormone receptor expression can vary over time, meaning that cells can alter sensitivity to endocrine signals via temporal modulation of receptor expression. Furthermore, receptors come in different 'flavors' which vary in their affinity for the same hormone, hence, modulating the composition of expressed receptors can also affect cell sensitivity to a hormone. In a growing number of examples, it appears that neuropeptides or other factors are released in response to environmental cues, affecting target cell sensitivity to hormonal signals as well. In this manner, polyphenism expression is tied to environmental cues through existing developmental events (see reviews in Nijhout 1994, 1999a, 1999b, Zera 1999).

Circulating titers of hormone and the various components that together determine cell sensitivity can change rapidly to achieve different physiological ends, and there is at least the potential for these mechanisms to vary evolutionarily (see review in Zera 1999, and this volume). Hence, in the signal-receiver aspect of the endocrine system alone, there lies a substantial amount of potential variation on which selection could act. In sum, endocrine regulation of trait ontogeny, integration, and control of the molting cycle produces the developmental windows before which environmental morph-inducing cues must be detected. These mechanisms also determine ontogenetic windows where morph expression is manifested. Finally, these mechanisms of endocrine regulation also represent a rich source of variation through which the expression of polyphenism can be shaped by selection. In the sections that follow, we review what is known about the endocrine basis of polyphenism in a few lepidopteran systems.

Physiological Control of Pupal Polyphenisms

Three lepidopteran families exhibit pupal color polyphenisms where individuals cryptically match the background color (green or brown) of their pupation site (see Hazel et al. 1998; Starnecker and Hazel 1999). In nymphalids and pierids, pupal melanization-reducing factor (PMRF) is released during the prepupal stage, producing green or yellow pupae. Failure to release this neuropeptide results in melanized pupae (Starnecker and Hazel 1999). Injections of PMRF induce yellow morphs on any background (Koch et al. 1990; Starnecker and Hazel 1999), demonstrating its causal role in inducing the lighter, non-melanized phenotypes. In many *Papilio* species, including *P. polyxenes*, a 'browning factor' is released during the prepupal stage that induces brown pupae. Failure to release this factor produces green pupae (Starnecker and Hazel 1999). Injection of *P. polyxenes* with PMRF-containing ganglionic extracts from *Inachis io* (Nymphalidae) produces the melanized pupal morph on any background. This result suggests a role for PMRF in pupal morph production in Papilionids, but in the opposite direction compared to PMRF action in Nymphalids and Pierids. Although interesting in their own right, these results gain additional significance when considered in a phylogenetic context, an idea to which we will return in a later section.

Physiological Control of Adult Polyphenisms

The European Map Butterfly, *A. levana* (Nymphalidae), exhibits one of the best-known adult seasonal polyphenisms of all Lepidoptera. Adults emerging from diapausing pupae in the spring have reddish brown or orange wings with black spots. Summer cohorts do not undergo diapause and have a dramatically different wing pattern, composed of a prominent white stripe and small patches of orange on a black background (see images in Beldade and Brakefield 2002). Moreover, the morphs differ in absolute size, relative wing size, and wing shape (reviewed in Fric et al. 2004). In a series of experiments, Koch and collogues investigated the endocrine basis of morph determination in this system (Koch and Buckmann 1985, 1987; reviewed in Koch 1992).

In *A. levana*, morph expression is linked to larval diapause. Two days after pupation, non-diapausing, summer individuals normally experience an increase in circulating levels of ecdysone, which in turn initiates adult development, producing the black-and-white summer morph. In contrast, diapausing individuals normally experience this ecdysone increase and initiate adult development after the winter, producing the reddish-brown spring morph. Injection of exogenous ecdysone into overwintering pupae interrupts diapause, prematurely resuming development. The resulting phenotype depends on the timing of ecdysone injection; the later in pupation the injection, the more 'spring-like' the adult. These results indicate that there is a developmental window of morph determination that occurs early in pupation, and that adult phenotype is tied to the mechanisms regulating diapause and adult development. The endocrine regulation of adult polyphenism expression in other systems, however, is more complicated.

Lycaena phlaeas (Lycaenidae) exhibits an adult polyphenism with spring (brown with red spots) and summer (more uniformly red) phenotypes. Morph determination occurs early in pupation, and is cued by photoperiod. A series of experiments combining environmental manipulations, ecdysone assays, treatment with exogenous ecdysone, and surgical removal and transplantation of a putative signaling organ—the brain—illustrate that a neuropeptide is involved in morph determination, affecting the response to signaling hormones (Endo and Kamata 1985). Similar experimental approaches revealed roles for neuropeptides in adult polyphenism expression in other butterflies (e.g., *Papilio xuthus*, Endo and Funatsu 1985), but showed no such role for a factor in determining *A. levana* morph expression (reviewed in Koch 1992).

These examples of the proximate basis of adult morph expression reveal complicated relationships between regulatory factors and timing of developmental events, and more data are needed from a variety of species before we can determine developmental or phylogenetic patterns in regulation of adult polyphenism. This is true even of seemingly tractable problems, such as the direct and indirect linking of pupal diapuase to morph expression or the physiology of pigment synthesis and wing scale maturation (Koch 1992, Koch et al. 2000). One informative avenue for investigating how these pathways evolve to affect morph expression has been to examine morph expression in mutants and artificially selected lineages.

The nymphalid butterfly, *Precis (= Junonia) coenia*, exhibits alternative seasonal forms differing in dorsal wing color. Rountree and Nijhout (1995a) found that response to photoperiod in the larval stage leads to differences in ecdysteroid titer in the early pupae, and thence to the divergent adult phenotypes. They also isolated a mutant that constitutively expresses only one phenotype (Rountree and Nijhout 1995b) and showed that the mutated gene does not affect the endocrine signal but alters the developmental response to the hormone in early pupae.

Ecdysteroid hormones mediate the development of the seasonal forms in *B. anynana* (Koch et al. 1996). The increase in ecdysteroid titer after pupation occurs at a later stage in pupae of the DSF of *B. anynana* than in those of the WSF. When animals, reared to produce the DSF, are microinjected as young pupae with ecdysone, the adult wing pattern is shifted towards developing the larger ventral eyespots and pale medial band characteristic of the WSF (Koch et al. 1996). Further research has begun to explore the genetic and developmental basis of the plasticity in these wing patterns.

Surveys of hormone titers have been made among artificially selected lineages of *B. anynana* (see below). Populations selected to express only one of the seasonal forms across all temperatures differ in the timing of the pupal ecdysteroid peak; the ecdysone profiles of these selected lineages when raised in a common environment is similar to that of wild-type butterflies developing in the corresponding DSF- or WSF-inducing environments (Koch et al. 1996, Brakefield et al. 1998). We can infer from this that responses to artificial selection for canalized production of each seasonal form result from changes, at least in part, in the same endocrine factors underlying natural expression of the morphs (see also Zijlstra et al. 2004).

Eyespot formation has been well studied in *B. anynana* (reviewed in Beldade and Brakefield 2002). Surgical experiments, gene expression studies, and analyses of wing pattern mutants have shown that eyespot development proceeds by the initial specification of a central organiser or 'focus', followed by signaling to the surrounding cells and their subsequent synthesis of specific color pigments. One of the developmental genes known to play a role in focus formation and signaling is *Distal-less* (*Dll*). The different seasonal phenotypes, whether genetically or environmentally determined, are associated with a divergent pattern of *Dll* gene expression in the wings of one-day-old pupae (Brakefield et al. 1996). This follows the time when surgical experiments indicate that the focus signals to specify the eyespot pattern. Understanding precisely how larval rearing temperature influences the secretion of ecdysteroids, and how the ecdysone titer in the early pupa then regulates eyespot development and expression of genes such as *Dll* (and only on the ventral wing surface, see Brakefield et al. 1998), are exciting challenges for the future.

These examples illustrate how morph expression is determined by complex relationships among environmental factors, timing of developmental events, and the neuroendocrine system. Changes in the expression of the physiological mechanisms underlying morph expression result from differential gene expression across inducing environments. Variation in these patterns of gene expression are then of course the ultimate target of selection, a topic we address in the next section.

Genetics of Seasonal Polyphenism and Evolutionary Change

Extensive studies of genetic variation for the ability to express seasonal polyphenism have been made in *B. anynana*. Field surveys of this species-rich genus in the wet-dry seasonal environments in Malawi suggested that temperature during larval development greatly influences expression of the adult seasonal forms (Fig. 1B; Brakefield and Reitsma 1991; Windig et al. 1994). Larvae are most sensitive to the effects of temperature on morph expression during the final and penultimate instars; high temperatures generate the WSF whereas low or declining temperatures yield adults of the DSF (Kooi and Brakefield 1999).

The analytical framework for much of the genetical analysis of phenotypic plasticity has been the 'norm of reaction', which describes the phenotypes a single genotype can produce across a range of environments. Reaction norms are typically depicted graphically as the mean phenotypic value of a genotype (or some surrogate, e.g., sibship, population, etc.) plotted against the environmental gradient (Schlichting and Pigliucci 1998). In diploid, sexual organisms such as butterflies, members of single families (full-sibs) provide individuals of similar genotype, and sibships split among different rearing environments are often used to estimate reaction norms for a given genotype. The different sibships sampled from a population constitute a bundle of reaction norms around some population average. Although field populations of *Bicyclus* spp. show a classical pattern of seasonal polyphenism with discrete forms (Windig et al. 1994), laboratory experiments demonstrate that the underlying reaction norms are continuous in form (Brakefield and Mazzotta 1995).

Quantitative variation in ventral eyespot size at a single rearing temperature in *B. anynana* has provided the basis to survey genetic variation available for the evolution of phenotypic plasticity in wing color pattern (Holloway and Brakefield 1995, Holloway et al. 1993). In artificial selection experiments during which rearing temperature was increased (Low Line) or decreased (High Line) progressively over several generations, the High Line eventually expressed only the WSF across all temperatures, although plasticity remained (i.e., higher temperatures yielded still larger eyespots). In sharp contrast, the Low Line produced only DSF at all temperatures (Brakefield et al. 1996). These selected lineages diverged dramatically in the mean intercept of their reaction norms (Fig. 4B) and effectively lost the ability to produce both seasonal forms, becoming canalized to produce a single seasonal morph at all temperatures. Analysis of crosses made between the High and Low lineages suggested that approximately 5 to 10 polymorphic genes contributed to evolution of these divergent phenotypes (Wijngaarden and Brakefield 2000). Extreme changes in the intercept or height of the reaction norm can therefore evolve rapidly. A response in degree or range of plasticity will, however, necessitate a change in reaction norm slope, which is only possible when there is genotype *x* environment interaction (i.e., when there is crossing of the slopes for reaction norms of different sibships). These selection experiments revealed strong positive genetic covariances across environments, suggesting that slope of reaction norms is unlikely to evolve as easily or as rapidly as the intercept. Low genetic variation for reaction norm slope could introduce limitations or constraints on the evolution in the range of phenotypic plasticity (see below).

We tested this hypothesis in a series of artificial selection experiments that directly targeted the slope or shape of the reaction norm for eyespot size on temperature in *B. anynana*. The first two experiments failed to yield either substantially steeper or shallower reaction norms, or ones of divergent shape (Wijngaarden and Brakefield 2001, Wijngaarden et al. 2002). However, a more recent study (P.M. Brakefield, unpub. data) has used a design with several advantages over previous work, namely: 1) a closer parallel to the life cycle in the field, 2) a much wider screening of genotypes generated by crossing and recombination, and 3) simultaneous selection for genotypes that produce favored phenotypes in both cool and warm temperature environments (Fig. 4A). There was no evolution towards a shallower or 'flat' reaction norm. However, there was a clear response towards steeper reaction norms with a larger difference between the mean phenotype expressed in the alternative temperature environments (Fig. 4B), as can be seen by examination of the range of variation found in the cohorts reared at each temperature (Fig. 4C). This result demonstrates that *Bicyclus* can evolve a change in reaction norm slope and degree of plasticity in at least some directions (see also Zijlstra et al. 2004). In particular, they should be

Fig. 4 The results of artificial selection on shape of the reaction norm for ventral eyespot size on rearing temperature in *Bicyclus anynana*. **(a)** Scheme of selection for two lines with target slopes for reaction norms of Flat or Steep, respectively. The experiment was started from a large F_2 population derived from a cross of the High and Low lines (see part B). Each diagram illustrates a single generation of larvae (*l*) and adults (*a*) at 20°C, or two consecutive generations at 27°C. Selected parents in each generation are indicated by the darker portions for each adult population (with eyespot size increasing along the *x*-axis). Note that reciprocal crosses ('×') were made after selection in each generation (cool, slow-developing environment with small eyespots) or in every other generation (warm environment with large eyespots). **(b)** Population reaction norms of eyespot size on temperature for the Flat and Steep selected lines after 12 cycles of selection. Note the apparent response in the Steep direction. The reaction norms for the unselected Stock, and also for the original High and Low lines (Brakefield et al 1996; and see text) are shown for comparison. All butterflies of each line were raised in several replicate groups within the same two temperature environments. Only data for females are shown, and the slope of the Steep reaction norm differs significantly from both Stock and Flat

Fig. 4 Contd. ...

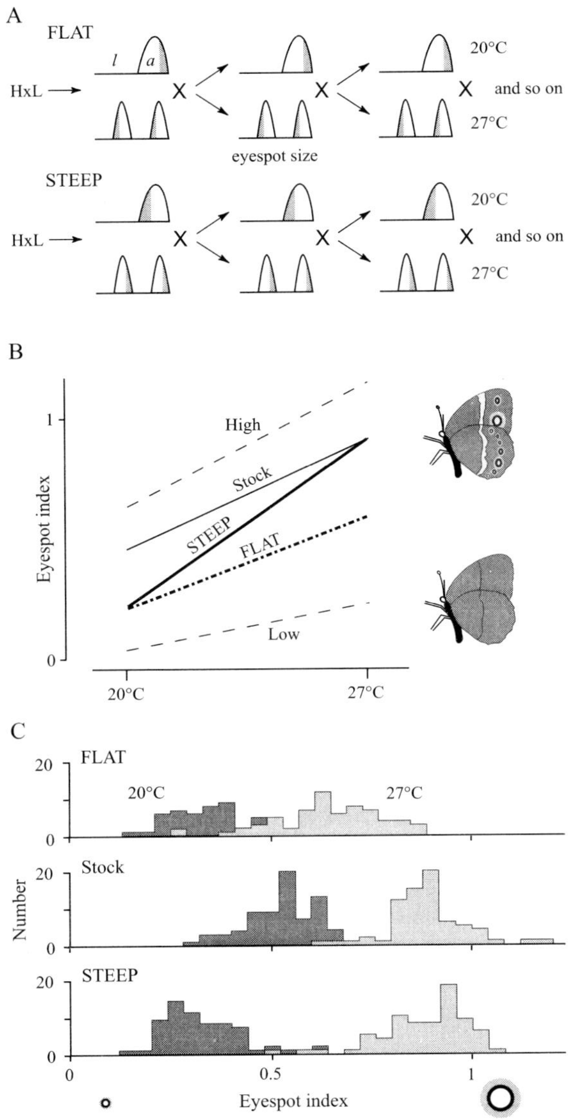

Fig. 4 Contd. ...

when tested by ANOVA (the latter two do not differ from each other). **(c)** Variation in ventral eyespot size for the cohorts of the female butterflies reared at two temperatures showing the wider difference for the two groups of Steep-line butterflies compared to both the unselected Stock and Flat-line. There has been a significant response ($P<0.01$) to artificial selection in the direction of an increase in the range of plasticity (but not in the direction of less plasticity). Unpublished data of P.M. Brakefield.

able track an environmental change towards a wider difference between the seasons.

This result is important because some field data for eyespot polyphenism in another satyrid butterfly, *Melanitis leda*, suggested that genetic constraints on reaction norm slope could inhibit the evolution of its seasonal morphs (N. Reitsma and P.M. Brakefield, unpublished data). *M. leda* exhibit a pattern of seasonal polyphenism very similar to *Bicyclus* in the same environments (Brakefield 1987, 1997). Samples of *M. leda* were obtained at Zomba in Malawi throughout several wet-dry seasonal cycles (cf Windig et al. 1994), and their eyespot size analyzed in the samples of the DSF. In each seasonal cycle, a single cohort of the DSF of *M. leda* ecloses in the early part of the dry season. This cohort shows quantitative variation in eyespot size from butterflies with no ventral marginal eyespots through to those with a series of small eyespots. However, after months of no recruitment at the end of the dry season, only individuals of the extreme DSF with no eyespots remain. The loss of *M. leda* butterflies possessing eyespots is consistent with the very strong directional selection observed in the natural selection experiments with *Bicyclus* butterflies (see Fig. 3). However, they leave open the question of why the individuals with small eyespots are produced in each dry season in the first place; theory predicts that genotypes producing the DSF with vestigial eyespots should be eliminated, resulting in a dry season cohort of extreme, spotless DSF. One possibility to explain the retention of the spotted DSF is that this cohort may result from cross-environmental genetic correlations for eyespot expression. The high fitness advantage of large eyespots during the wet season, combined with an inability of the slope of the reaction norm to evolve, may result in the production of spotted individuals in the dry season. However, evolution of reaction norm slope in response to artificial selection in *B. anynana* suggests that this is unlikely. Other possible constraints include variability in the seasonal climate from one cycle to another producing inaccuracies of environmental cues as a predictors of the coming selective environment, or genetic covariances among traits expressed in each morph that are 'mismatched' with the seasonal selection coefficients ('trade-offs'; see Zijlstra et al. 2004). More work will be necessary before the roles of such constraints in the evolution of seasonal polyphenism can be assessed properly.

Phylogenetic Pattern and Seasonal Polyphenism

In the preceding sections, we addressed instances of polyphenisms individually or among closely related groups of species. This approach is often necessary when conducting laboratory experiments designed to elucidate the proximate mechanisms that generate and maintain the polyphenism. However, there is also much to be gained by studying polyphenisms in a phylogenetic context. First, knowing that a polyphenism evolved as an adaptation *sensu* Gould and Lewontin (1979) requires identification of the ancestor-descendent relationship for a plastic response (Doughty 1995, Gotthard and Nylin 1995). Second, the general evolutionary importance of a plastic response may be suggested by its phylogenetic distribution. Clusters of homologous plastic responses within or among clades may suggest that the response facilitated adaptive radiation (Frankino and Raff 2004), and thus that plasticity represents a key innovation *sensu* Liem (1990). Frequent homoplasies (convergences, parallelisms, or other independent evolutionary origins) for particular plastic responses might indicate that the response is relatively easy to evolve when favored ecologically (Frankino and Raff 2004).

As discussed by Starnecker and Hazel (1999), pupal melanization-reducing factor (PMRF) or similar molecules have wide phylogenetic expression in Lepidoptera, including monomorphic species. This, coupled with PMRF expression at multiple points in ontogeny, suggests a wider role for PMRF than simply determining pupal color. Mapping the role of PMRF in pupal background color matching onto a phylogeny indicates that PMRF has been co-opted twice in the evolution of pupal color polyphenisms; once in the nymphalid-pierid clade and independently in the papilionids. This suggests that PMRF-expression occurs at an ontogenetic stage when developmental switches in melanization can occur easily. It would be extremely interesting to know whether the variable sensitivity among lineages to the cues influencing pupal color expression (e.g., background color or texture, temperature, photoperiod; Hazel and West 1979; Hazel et al. 1998) is associated with variation in PMRF-secretion, target cell sensitivity to PMRF, or some other part of the pathway.

The finding of variation in the developmental system underlying expression of polyphenism indicates that caution is necessary when interpreting the phylogenetic distribution of plastic responses. This is because development itself evolves, and therefore the proximate basis of polyphenism expression within a clade may change over time. This apparently has occurred in the case of wing loss in ant castes (Abouheif and Wray 2002). Derived, wingless castes are basal to ants, and although the same genetic pathway is interrupted among studied ant species to produce wingless individuals, the particular points at which the interruption occurs varies. These underlying changes in the wing gene network reflect variation among groups in the occurrence and morphology of residual wing discs. Such variation in development may result from drift, internal selection on the developmental system directly, or natural selection on life history or some other trait(s) which acts indirectly on the mechanisms controlling wing development (Abouheif and Wray 2002). In any case, this example illustrates how knowing the proximate basis of polyphenism expression across lineages can inform our understanding of its evolution. In the case of wing loss in ants, variation in the developmental data suggests that the polyphenism was modified among lineages and that it (or they) may still be evolving.

Conversely, understanding the proximate basis of polyphenism expression in a phylogenetic context may reveal that polyphenic responses among diverse taxa are more similar, perhaps resulting from parallel evolution, than might be apparent from examination of the phenotypes alone (Frankino and Raff 2004). For example, seasonally polyphenic wing color patterns in Holarctic *Pieris* and their South American *Tatochila* (Pieridae) counterparts have independent origins, but seem likely to share proximate mechanisms producing the morphs. Some species in these groups are seasonally polyphenic; adults with melanized wings (cool season) develop under cool/short day conditions whereas adults with greatly reduced melanization on the wings (immaculates) are produced during warmer/long days (cf *Pontia* butterflies above). Within these genera, other species are monomorphic, being canalized to develop either the melanized or immaculate morph.

In some lineages canalized to produce the immaculate morph, cold shock induces phenotypes similar to the cold season, melanized form naturally produced by other species (Shapiro 1976, 1980; reviewed in Nijhout 1991). Differences among lineages between the typically-expressed and experimentally-induced melanized morphs indicate that the two phenotypes are not homologous (Shapiro 1980). However, knowing the degree to which the proximate basis of morph determination is shared between the typically expressed seasonal morphs and the experimentally induced, novel alternative forms would suggest how easily such polyphenisms can appear in these highly successful butterflies (Frankino and Raff 2004). In particular, identification of the shared proximate basis of phenotype production would indicate if there is something in the mechanisms controlling the pattern of wing melanization that predisposes these lineages to the production of melanized scales around the veins, especially in cold environments (see True et al. 1999).

The phylogenetic approach has greatly aided our understanding of seasonal polyphenisms. For example, Figure 5 suggests that eyespot polyphenism has been gained and lost several times within the species-rich genus of *Bicyclus* (see Roskam and Brakefield 1996, 1999). Indeed we know that seasonal polyphenism involving the marginal eyespots is a common phenomenon within the whole subfamily of Satyrinae (see Brakefield and

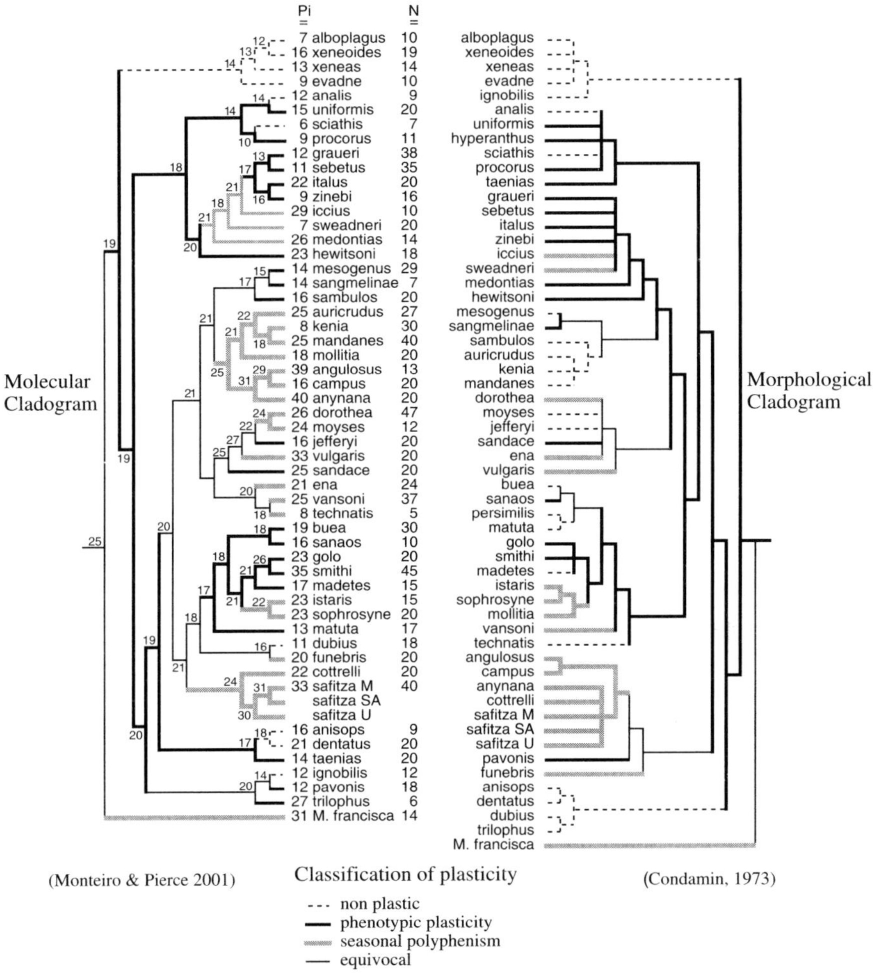

Fig. 5 *Bicyclus* clade trees highlighting species exhibiting seasonal polyphenisms. Both molecular-based and morphological cladograms are shown with a distinction between species showing discrete seasonal forms from others in which more continuous plasticity is found (unpublished work of J.C. Roskam using cladograms from Monteiro and Pierce 2001, and data from Condamin 1973).

Larsen 1984, Brakefield 1987, 1997, Braby 2002), and that a relationship between ventral wing eyespots and the environment during pre-adult development may be inherent to all species in the group (see Brakefield and Shreeve 1992). Polyphenisms involving black-and-white wing patterns has evolved repeatedly within nymphalid butterflies, and is ancestral to the *Araschnia*-clade (Fric et al. 2004). Hence, the appearance of the seasonal morphs predates the radiation of the genus (Fig. 6). Historical biogeography of *Araschnia* suggests that ecological differences associated with seasonally abundant and scarce rains—not seasonally variable temperature—favored the polyphenism in the common ancestor of the group. It will be fascinating to be able to make a full comparative analysis of the underlying mechanisms of a particular mode of polyphenism to tie down the extent of homology, and enable exploration of the extent to which differences at the phenotypic level reflect differences in adaptive evolution to different environments.

Perspectives

The evolutionary lability of seasonal polyphenism revealed by comparative analyses is fascinating. Such lability in phenotypic expression also suggests that plasticity may not only drive local adaptation, for example facilitating range extension, response to climate change or to environmental grain, but also may facilitate processes of divergence and speciation (Pfennig and Murphy 2000, West-Eberhard 2003). Determining how this lability occurs in terms of the underlying physiological, developmental and genetic mechanisms is an exciting challenge. Comparative work in evolutionary developmental biology is beginning to show that evolutionary lability of short protein-binding sequences in regulatory regions of key developmental genes might explain repeated patterns of disappearence and reappearance of particular traits such as dark wing markings across species of Drosophila (Gompel et al. 2005, Brakefield and French 2005). In the case of seasonal polyphenisms one can also imagine comparable genetic control points via the interactions with insect hormones, such as are involved in ecdysone modulation of *Distal-less* expression that underlies the polyphenism in *Bicyclus* eyespots (Brakefield et al. 1996).

Research on seasonal polyphenisms in Lepidoptera provides some of the most complete accounts of adaptive evolution, covering all levels of biological organization. Future application of molecular tools and genome-wide screens will identify the mechanisms underlying hormonal modulation of developmental pathways as well as the genes involved in

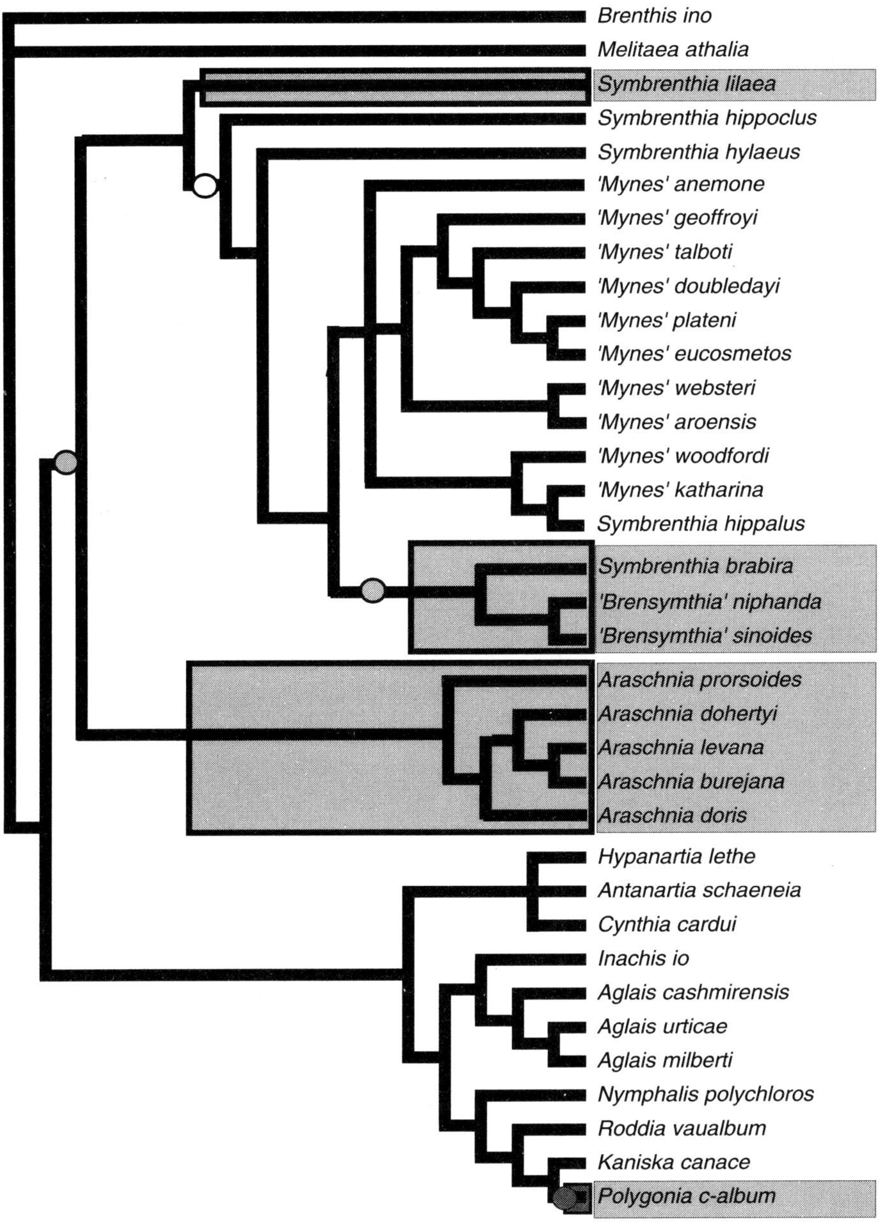

Fig. 6 *Araschnia* clade tree highlighting species exhibiting seasonal polyphenisms. The polyphenic species are shaded; circles within the tree or dark shaded sectors show possible scenarios for evolutionary change for polyphenism within the clade (from Fric et al. 2004).

polyphenism expression and evolution (Brakefield et al. 2003). Work has also begun to unravel the relationships among the morphological, life historical, and behavioral traits that covary among seasonal morphs. For example, it is clear that polyphenism in *Bicyclus* butterflies involves a wide suite of traits in addition to eyespot size (e.g. Brakefield 1997, Zijlstra et al. 2003, 2004). Understanding how proximate mechanisms produce co-ordinated changes across trait suites promises to elucidate the basis of trade-offs and how complex phenotypes evolve generally. Similarly, comparative analyses of the mechanisms underlying polyphenism expression at a deeper phylogenetic level will reveal much about the roles of development in phenotypic evolution. In sum, multi-disciplinary investigations of polyphenisms will continue to provide fascinating insights about the evolutionary process across levels of analysis.

Some described systems provide rich potential for further comparative analyses. For example, how does the seasonal polyphenism exhibited by *Nemoria arizonaria* fit into the wider phylogenic context of the group? Other geometrids show similarly dramatic larval polyphenisms (Kettlewell 1973); it would be interesting to know the degree of independence in evolutionary origins and proximate basis of expression across lineages. Do catkin or other rich diets induce catkin-like phenotypes in lineages that are monomorphic for the twig phenotype in nature? Induction by such environmental manipulations can inform much about the developmental basis of morph induction and the origins of the polyphenism (Frankino and Raff 2004). Also, are there fitness costs of this polyphenism? As in other examples, evolution of this polyphenism is presumably limited by some cost to maintaining the developmental machinery required to produce alternative phenotypes or restrictions of cross-environmental genetic correlations on morph expression.

It would also be informative to know where the proximate bases in sensitivity among lineages to the cues influencing morph expression lie. In particular, what are the the origins of quantitative genetic variation in response to cues or in the degree of morph expression? The insect endocrine system contains quantitative genetic variation underlying expression of polyphenisms, and this can respond to artificial selection (e.g., Fairbairn and Yadlowski 1997, Gu and Zera 1996, Zera and Bottsford 2001, Zera and Zhang 1995, Zijlstra et al. 2004). As the experiments performed on *B. anynana* demonstrate, such variation is the raw material that can be modified by natural selection to produce locally adapted populations. It may also be the source of genetic correlations which produce the tradeoffs among instars or morphs that can constrain the evolution of polyphenisms. It will be fascinating to unravel the origins of the variation among individuals in reponse to temperature, photoperiod, and light cues in determining pupal or larval color, or adult wing pattern in different systems. This will eventually provide insights about whether the proximate basis of developmental variation among individuals is also responsible for the variation in response to cues or morph expression documented among populations, species or lineages. Addressing these issues will require a challenging combination of manipulative and comparative approaches in different systems, but the payoffs will be large.

A multidisciplinary approach to studying polyphenisms offers an opportunity to understand the roles of development in evolution (Gilbert 2001, Beldade and Brakefield 2002, Brakefield et al. 2003, Frankino and Raff 2004). Questions regarding the roles of development in biasing the evolution of complex traits, facilitating or inhibiting diversification, and producing phylogenetic pattern in evolution can all be addressed by studying monomorphic and polyphenic lineages in a comparative context. Moreover, polyphenisms can be used as tools to study the evolution of development itself; the dramatic differences among alternative morphs in morphology and behavior offer a powerful experimental handle for examining the developmental regulation of trait differentiation and integration while avoiding the confounding effects of other differences among lineages. They also offer the opportunity to study how development responds to natural selection on a microevolutionary scale. Research on polyphenisms using integrative, multidisciplinary approaches such as those outlined above will continue to provide some of the deepest understanding of adaptive evolution at all levels of biological organization.

Acknowledgements

The research on *Bicyclus* butterflies has been supported by grants from the Dutch Organisation for Scientific Research and the Human Frontier Science Program. This work was supported by a 2001 Bioinformatics Postdoctoral Fellowship from the National Science Foundation to W.A.F. Vernon French kindly commented on a draft of this chapter, and we also thank Doug Whitman for his stimulating editorial work. We are extremely grateful to Hans Roskam for permission to use his unpublished drawings in Figure 5, and to Martin Brittijn who helped with all the figures.

References

Abouheif, E., and G. A. Wray. 2002. Evolution of the gene network underlying wing polymophism in ants. Science 297: 249–252.

Beldade, P., and Brakefield, P. M. 2002. The genetics and evo-devo of butterfly wing patterns. Nature Rev. Genet., 3: 442–452.

Braby, M. F. 2002. Life history strategies and habitat templets of tropical butterflies in north-eastern Australia. Evolutionary Ecology 16: 399–413.

Brakefield, P. M. 1987. Tropical dry and wet season polyphenism in the butterfly *Melanitis leda* (Satyrinae): Phenotypic plasticity and climatic correlates. Biol. J. Linn. Soc., 31: 175–191.

Brakefield, P. M. 1997. Phenotypic plasticity and fluctuating asymmetry as responses to environmental stress in the butterfly *Bicyclus anynana*. *In* Environmental Stress, Adaptation and Evolution. R. Bijlsma and V. Loeschke [eds]. pp. 65–78. Birkhäuser, Basel.

Brakefield, P. M., and French, V. 2005. How and why to spot fly wings. Nature 433: 466–467.

Brakefield, P. M., and Kesbeke, F. 1997. Genotype *x* environment interactions for insect growth in constant and fluctuating temperature environments. Proc. R. Soc. Lond., B, 264: 717–723.

Brakefield, P. M., and Larsen, T. B. 1984. The evolutionary significance of dry and wet season forms in some tropical butterflies. Biol. J. Linn. Soc., 22: 1–12.

Brakefield, P. M., and Reitsma, N. 1991. Phenotypic plasticity, seasonal climate and the population biology of *Bicyclus* butterflies (Satyridae) in Malawi. Ecological Entomology, 16: 291–303.

Brakefield, P. M., and Mazzotta, V. 1995. Matching field and laboratory environments: effects of neglecting daily temperature variation on insect reaction norms. J. Evol. Biol. 8: 559–573.

Brakefield P. M., and Shreeve T. 1992. Case studies in evolution. *In* The Ecology of Butterflies in Britain, R. L. H. Dennis [ed.]. Pp. 197–216. Oxford: Oxford University Press.

Brakefield, P. M., and Wijngaarden, P. J. 2003. Phenotypic plasticity. *In* Keywords & Concepts in Evolutionary Developmental Biology, pp. 288–297. B. K. Hall and W. M. Olson [eds]. Harvard University Press Reference Library.

Brakefield, P. M., French, V., and Zwaan, B. J. 2003. Development and the genetics of evolutionary change within insect species. Annu. Rev. Ecol. Evol.. Syst., 34: 633–660.

Brakefield, P. M., Kesbeke, F., and Koch, P. B. 1998. The regulation of phenotypic plasticity of eyespots in the butterfly *Bicyclus anynana*. Am. Nat., 152: 853–860.

Brakefield, P. M., Pijpe, J., and Zwaan, B. J. 2007. Developmental plasticity and acclimation both contribute to adaptive responses to alternative seasons of plenty and of stress in *Bicyclus* butterflies. J. Biosci. 32: 465–475.

Brakefield, P. M., Gates, J., Keys, D., Kesbeke, F., Wijngaarden, P. J., Monteiro, A., French, V., and Carroll, S. B. 1996. Development, plasticity and evolution of butterfly eyespot patterns. Nature 384: 236–242.

Breuker, C. J. and Brakefield, P. M. 2002. Female choice depends on size but not symmetry of dorsal eyespots in the butterfly *Bicyclus anynana*. Proc. Roy. Soc. Lond. B. 269: 1233–1239.

Doughty, P. E. 1995. Testing the ecological correlates of phenotyically plastic traits within a phylogenetic framework. Acta Oecologica 16: 519–524.

Endo, K., and Funatsu, S. 1985. Hormonal control of seasonal morph determination in the swallowtail butterfly, *Papilio xuthus* L. (Lepidoptera: Papilionidae). Journal of Insect Physiology 31: 669–674.

Endo, K., and Kamata, Y. 1985. Hormonal control of seasonal-morph determination in the small copper butterfly, *Lycaena phlaeas daimio* Seitz. Journal of Insect Physiology 31: 701–706.

Fairbairn, D. J., and Yadlowski, D.E. 1997. Coevolution of traits determining migratory tendency: correlated response of a critical enzyme, juvenile hormone esterase, to selection on wing morphology. Journal of Evolutionary Biology 10: 495–513.

Faeth, S. H., and Hammon, K. E. 1989. Caterpillars and polyphenisms. Science 246: 1639.

Frankino, W. A., and Raff, R. A. 2004. Evolutionary importance and pattern of phenotypic plasticity: Insights gained from development. Pp. 64–81. *In* T. J. DeWitt and S. M. Scheiner [eds]. Phenotypic Plasticity, Functional and Conceptual Approaches. Oxford Press.

Fric, Z., Konvicka, M., and Zrzavy, J. 2004. Red and black or black and white? Phylogeny of the *Araschnia* butterflies (Lepidoptera: Nymphalidae) and the evolution of seasonal polyphenism. Journal of Evolutionary Biology 17: 265–278.

Gilbert, S. F. 2001. Ecological developmental biology: Developmental biology meets the real world. Developmental Biology 233: 1–12.

Gompel, N., Prud'homme, B., Wittkopp, P. J., Kassner, V. A., and Carroll, S. B. 2005. Chance caught on the wing: *cis*-regulatory evolution and the origin of pigment patterns in *Drosophila*. Nature 433: 481–487.

Gotthard, K., and S. Nylin. 1995. Adaptive plasticity and plasticity as an adaptation: a selective review of plasticity in animal morphology and life history. Oikos 74: 3–17.

Gould, S. J., and Lewontin, R. C. 1979. The spandrels of San Marco and the Panglossian paradigm: a critique of the adaptationist programme. Proc. R. Soc. Lond. B 205: 581–598.

Greene, E. 1989a. A diet-induced developmental polymorphism in a caterpillar. *Science* 243: 643–646.

Greene, E. 1989b. Caterpillars and polymorphisms. Science 246: 1639–1640.

Greene, E. 1996. Effect of light quality and larval diet on morph induction in the polymorphic caterpillar *Nemoria arizonaria* (Lepidoptera: Geometridate). Biological Journal of the Linnean Society 58: 277–285.

Gu, X., and Zera, A. J. 1996. Quantitative genetics of juvenile hormone esteraste, juvenile hormone binding and general esterase activity in the cricket *Gryllus assimilis*. Heredity 76: 136–142.

Hazel, W. N. 2002. The environmental and genetic control of seasonal polyphenism in larval color and its adaptive significance in a swallowtail butterfly. Evolution 56: 342–348.

Hazel, W. N., Ante, S., and Stringfellow, B. 1998. The evolution of environmentally-cued pupal colour in swallowtail butterflies: natural selection of pupation site and pupal colour. Ecological Entomology 23: 41–44.

Hazel, W. N., and West, D. A. 1979. Environmental control of pupal colour in swallowtail butterflies (Lepidoptera: Papilioninae): *Battus philenor* (L.) and *Papilio polyxenes* Fabr. Ecological Entomology 4: 393–400.

Hazel, W. N., and West, D. A. 1983. The effect of larval photoperiod on pupal colour and diapause in swallowtail butterflies. Ecological Entomology 8: 37–42.

Heinrich, B. 1993. The Hot-bodied Insects. Harvard University Press, Cambridge, MA. 601 pp.

Holloway, G. J., Brakefield, P. M., Kofman, S., and Windig, J. J. 1992. An artificial diet for butterflies, including *Bicyclus* species, and its effect on development period, weight and wing pattern. J. Research on Lepidoptera, 30: 121–128.

Holloway, G. J., Brakefield, P. M., and Kofman, S. 1993. The genetics of wing pattern elements in the polyphenic butterfly, *Bicyclus anynana*. Heredity, 70: 179–186.

Holloway, G. J., and Brakefield, P. M. 1995. Artificial selection of reaction norms of wing pattern elements in *Bicyclus anynana*. Heredity 71: 91–99.

Imms, A. D. 1925. A General Textbook of Entomology. Methuen & Co.: London.

Kettlewell B. 1973. The Evolution of Melanism: The Study of a Recurring Necessity. Clarendon Press: Oxford.

Kingsolver, J. G. 1987. Evolution and coadaptation of thermoregulatory behavior and wing pigmentation pattern in pierid butterflies. Evolution 41: 472–490.

Kingsolver, J. G. 1995a. Viability selection on seasonally polyphenic traits: wing melanin pattern in western white butterflies. Evolution 49: 932–941.

Kingsolver, J. G. 1995b. Fitness consequences of seasonal polyphenism in western white butterflies. Evolution 49: 942–954.

Kingsolver, J. G. 1996. Experimental manipulation of wing pattern and survival in western white butterflies. Am. Nat. 147: 296–306.

Kingsolver, J. G., and Huey, R. B. 1998. Evolutionary analyses of morphological and physiological plasticity in thermally variable environments. Am. Zool. 38: 545–560.

Koch, P. B. 1992. Seasonal polyphenism in butterflies: A hormonally controlled phenomenon of pattern formation. Zool. Jb. Physiol. 96: 227–240.

Koch, P. B., Brakefield, P. M., and Kesbeke, F. 1996. Ecdysteroids control eyespot size and wing color pattern in the polyphenic butterfly *Bicyclus anynana* (Lepidoptera: Satyridae). J. Insect Physiol., 42: 223–230.

Koch, P. B., and Buckmann, D. 1985. The seasonal dimorphism of *Araschnia levana* L. (Nymphalidae) in relation to hormonal controlled development. Verh. dt. Zool. Ges 78: 260.

Koch, P. B., and Buckmann, D. 1987. Hormonal control of seasonal morphs by the timing of ecdysteroid release in *Araschnia levana* L. (Nymphalidae: Lepidoptera). Journal of Insect Physiology 33: 823–829.

Koch, P. B., Lorenz, U., Brakefield, P. M., and ffrench-Constant, R. H. 2000. Butterfly wing pattern mutants: developmental heterochrony and co-ordinately regulated phenotypes. Development, Genes, and Evolution 210: 536–544.

Koch, P. B., Starnecker, G., and Buckmann, D. 1990. Interspecific effects of the pupal melanization reducing factor on pupal coloration in different lepidopteran families. Journal of Insect Physiology 36: 159–164.

Kooi R. E., and Brakefield, P. M. 1999. The critical period for wing pattern induction in the polyphenic tropical butterfly *Bicyclus anynana* (Satyrinae) J. Insect Physiol., 45: 201–212.

Kooi, R. E., de Vries, M., and Brakefield, P. M. 1998. The effect of drought stress on food plants on the wing pattern induction of the polyphenic tropical butterfly *Bicyclus anynana*. Proc. Exper. & Appl. Entomol. 9: 217–222.

Liem, K. F. 1990. Key evolutionary innovations, differential diversity, and symecomorphosis. Pp. 147-170. *In* M. Nitecki [ed.]. Evolutionary Innovations. University of Chicago Press.

Lively, C. M. 1986. Canalization versus developmental conversion in a spatially variable environment. Amer. Natur. 128: 561–572.

Lyytinen, A., Brakefield, P. M., and Mappes, J. 2003. Significance of butterfly eyespots as an anti-predator device in ground-based and aerial attacks. Oikos, 100: 373–379.

Lyytinen, A., Brakefield, P. M., Lindstrom, L., Mappes, J. (2004). Does predation maintain eyespot plasticity in *Bicyclus anynana*? Proc. R. Soc. Lond. B. 271: 279–283.

Monteiro A., and Pierce N. E. 2001. Phylogeny of *Bicyclus* (Lepidoptera: Nymphalidae) inferred from COI, COII, and EF-1 alpha gene sequences. Mol. Phylogenet. Evol. 18: 264–281.

Moran, N. A. 1992. The evolutionary maintenance of alternative phenotypes. The American Naturalist 139: 971–989.

Nijhout, H. F. 1994. Insect Hormones. Princeton University Press, Princeton.

Nijhout, H. F. 1999a. Control mechanisms of polyphenic development in insects. BioScience 49: 181–192.

Nijhout, H. F. 1999b. Hormonal control in larval development and evolution—insects. Pp. 217–254. *In* B. K. Hall and M. H. Wake, [eds]. The Origin and Evolution of Larval Forms. Academic Press, San Diego.

Pfennig, D. W., and Murphy, P. J. 2000. Character displacement in polyphenic tadpoles. Evolution 54: 1738–1749.

Robertson, K. A., and Monteiro, A. 2005. Female *Bicyclus anynana* butterflies choose females on the basis of their dorsal UV-reflective eyespot pupils. Proc. R. Soc. Lond., B., 272: 1541–1546.

Roskam, J. C., and Brakefield, P. M. 1996. Comparison of temperature-induced polyphenism in African *Bicyclus* butterflies from a savannah-rainforest ecotone. Evolution 50: 2360–2372.

Roskam, J. C., and Brakefield, P. M. 1999. Seasonal polyphenism in *Bicyclus* (Lepidoptera: Satyridae) butterflies: Different climates need different cues. Biol. J. Linn. Soc. 66: 345–356.

Rountree, D. B., and Nijhout H. F. 1995a. Hormonal control of a seasonal polyphenism in *Precis coenia* (Lepidoptera, Nymphalidae). J. Insect Physiol. 41: 987–992.

Rountree, D. B., and Nijhout, H. F. 1995b. Genetic control of a seasonal morph in *Precis coenia* (Lepidoptera, Nymphalidae). J. Insect Physiol. 41: 1141–1145.

Schlichting, C. D., and Pigliucci, M. (1998). Phenotypic Evolution: A Reaction Norm Perspective. Sinauer Associates, Inc., Sunderland.

Shapiro A.M. 1976. Seasonal polyphenism. Evol. Biol. 9: 259–333.

Shapiro, A. M. 1980. Convergence in pierine polyphenisms (Lepidoptera). Journal of Natural History 14: 781–802.

Starnecker, G., and Hazel, W. 1999. Convergent evolution of neuroendocrine control of phenotypic plasticity in pupal colour in butterflies. Proc. R. Soc. Lond., B 266: 2409–2412.

Stevens, M. 2005. The role of eyespots as anti-predator mechanisms, principally demonstrated in the Lepidoptera. Biological Reviews 80: 573–588.

Talloen, W., Van Dyck, H., and Lens, L. 2004. The cost of melanization: butterfly wing coloration under environmental stress. Evolution 58: 360–366.

True, J. R., Edwards, K. A., Yamamoto, D., and Carroll, S. B. 1999. Drosophilia wing melanin patterns form by vein-dependent elaboration of enzymatic prepatterns. Current Biology 9: 1382–1391.

Vallin, A., Jakobsson, S., Lind, J., and Wiklund, C. 2005. Prey survival by predator intimidation: an experimental study of peacock butterfly defence against blue tits. Proc. R. Soc. Lond., B 272: 1203–1207.

Via, S., and Lande, R. 1985. Genotype-environment interaction and the evolution of phenotypic plasticity. Evolution 39: 505–522.

Vlieger, L., and Brakefield, P. M. 2007. The deflection hypothesis: eyespots on the margins of butterfly wings do not influence predation by lizards. Biol. J. Linn. Soc. (In press).

Watt, W. B. 1968. Adaptive significance of pigment polymorphisms in *Colias* butterflies. I. Variation in melanin pigment in relation to thermoregulation. Evolution 22: 437–458.

West-Eberhard, M. J. 2003. Developmental Plasticity and Evolution. Oxford University Press, Oxford, UK. 794 pp.

Windig, J. J. 1999. Trade-offs between melanization, development time and adult size in *Inachis io* and *Araschnia levana* (Lepidoptera: Nymphalidae). Heredity 82: 57–68.

Wijngaarden, P. J., and Brakefield, P. M. 2000. The genetic basis of eyespot size in the butterfly *Bicyclus anynana*: an analysis of line crosses. Heredity, 85: 471–79.

Wijngaarden, P. J., and Brakefield, P. M. 2001. Lack of response to artificial selection on the slope of reaction norms for seasonal polyphenism in the butterfly *Bicyclus anynana*. Heredity, 87: 410–420.

Wijngaarden, P. J., Koch, P. B., and Brakefield, P. M. 2002. Artificial selection on the shape of reaction norms for eyespot size in the butterfly *Bicyclus anynana*: direct and correlated responses. J. Evol. Biol., 15: 290–300.

Windig, J. J., Brakefield, P. M., Reitsma, N., and Wilson, J. G. M. 1994. Seasonal polyphenism in the wild: survey of wing patterns in five species of *Bicyclus* butterflies in Malawi. Ecological Entomology, 19: 285–298.

Zera, A. J. 1999. The endocrine genetics of wind polymorphism in *Gryllus*: Critique of recent studies and state of the art. Evolution 53: 973–977.

Zera, A. J., and Bottsford. J. 2001. The endocrine-genetic basis of life-history variation: The relationship between the ecdysteroid titer and morph-specific reproduction in the wing-polymorphic cricket *Gryllus firmus*. Evolution 55: 538–549.

Zera, A. J., and C. Zhang. 1995. Evolutionary endocrinology of juvenile hormone esterase in *Gryllus assimilis*: Direct and correlated responses to selection. Genetics 141: 1113–1124.

Zijlstra, W. G., Steigenga, M. J., Brakefield, P. M., and Zwaan, B. J. 2003. Simultaneous selection on two-components of life-history in the butterfly *Bicyclus anynana*. Evolution 57: 1852–1862.

Zijlstra, W. G., Steigenga, M. J., Koch, P. B., Zwaan, B. J., and Brakefield, P. M. 2004. Butterfly selected lines explore the hormonal basis of interactions between life histories and morphology. *Am. Nat.* 163: E76–E87.

10

Causes and Consequences of Phenotypic Plasticity in Body Size: The Case of the Yellow Dung Fly *Scathophaga stercoraria* (Diptera: Scathophagidae)

Wolf U. Blanckenhorn

Zoologisches Museum, Universität Zürich-Irchel, Winterthurerstrasse 190 CH-8057 Zürich, Switzerland, Phone: (+41) 44-635-4755, Fax: (+41) 44-635-4780, E-mail: wolfman@zm.uzh.ch

Abstract

Phenotypic plasticity occurs when a particular genotype assumes different phenotypes depending on its environment. In this article, I focus on phenotypic plasticity in body size and associated life history traits, primarily growth rate and development time, thus taking a trait-centered view. First, I discuss and review phenotypic plasticity in body size in insects in general, centering on its environmental causes and its behavioral and fitness consequences. Plasticity in body size can be caused by a variety of environmental variables, most notably food availability, predators, temperature, season length, photoperiod and humidity, producing recurrent patterns. Food limitation generally results in small body sizes. High temperatures often result in rapid growth but smaller body sizes (Bergmann's and the temperature-size rule). Short growing seasons limit growth and thus final body size, particularly in species with long development (the converse Bergmann rule). This is often proximately mediated by photoperiod. Predators also limit foraging and hence growth and final body size of prey species, and sometimes induce particular morphological adaptations (induced defenses). For an evolutionary ecologist, the central question is whether or not the phenotypic plasticity exhibited is adaptive. This requires demonstration of heritable variation of the trait and the existence of environment-specific fitness trade-offs. Based on these criteria, growth and body size plasticity in response to seasonal constraints and predators can often be shown to be adaptive, while size reductions due to food limitation typically represent merely the 'best of a bad situation'. Whether temperature-mediated size plasticity (i.e., Bergmann's rule) is adaptive remains unclear and contended.

Plasticity in adult body size influences plasticity in adult behavior. Small individuals are forced to use particular strategies in order to overcome their handicap and successfully compete against larger conspecifics to augment their fitness. This is most prominently seen in species with discrete size-dependent alternative foraging or reproductive strategies, but equally necessary when body size variation is continuous. Again, the question is addressed whether small size is merely the 'best of a bad situation,' or whether there are environments in which small individuals can match the fitness of larger conspecifics, in which case small size would be adaptive. Evidence suggests that in most cases it is the former. Behavioral plasticity can be classified according to the proximate mechanism, time scale, frequency of occurrence, and the variance component where variation may be detected.

In the second part of this chapter, I present a long-term case study of phenotypic plasticity in body size and associated traits of the yellow dung fly *Scathophaga stercoraria* (Diptera: Scathophagidae). I discuss various environmental causes and behavioral consequences, and conclude that although the extent of body size plasticity of yellow dung flies may not be untypical, the degree of **adaptive** plasticity perhaps is unusually high. This may be related to their particular ephemeral larval habitat (cow dung), the occurrence and amount of which is variable and unpredictable in space and time. Furthermore, high adult mobility promotes extensive gene flow, hampering local adaptation of populations. Long-term case studies of diverse species well suited for this purpose will elucidate which specific conditions foster the evolution of adaptive plasticities, be they morphological, life history, physiological or behavioral.

Introduction—Phenotypic Plasticity from an Evolutionary Ecologist's Perspective

Phenotypic plasticity occurs when a particular genotype assumes a different phenotype depending on its environment (Schmalhausen 1949, Roff 1992, Stearns 1989, 1992). Phenotypic plasticity is ubiquitous in plants and animals and affects diverse behavioral, physiological, morphological, and life history traits. Any variable environmental factor, in principle, can produce plastic responses in organisms, the most prominent factors being food availability, predators, temperature, season length, photoperiod and humidity.

For a quantitative trait, such as body size, the phenotypic variance of a trait (V_p) can, in the simplest case, be partitioned into three components due to genetic (V_g), environmental (V_e) and genotype-by-environment effects

($V_{g \times e}$): $V_p = V_g + V_e + V_{g \times e}$ (Falconer 1989, Roff 1997). This is illustrated in Fig. 1, where genotype 1 is larger than genotype 2 at any food level; thus, there is genetic variation in body size in the population (= V_g). In response to higher food availability, genotype 1 increases in size (= V_e); thus, this genotype is phenotypically plastic. As this is not the case for genotype 2, there is genetic variation in the response of genotypes to food availability in the population (= $V_{g \times e}$; Scheiner and Lyman, 1989). Phenotypic plasticity consequently has two components, a purely environmental (= V_e) and a heritable (= $V_{g \times e}$) component (Via and Lande 1985, Stearns et al. 1991). In general, relationships of phenotypic traits of genotypes (or populations) in response to any variable environmental factor, as depicted in Fig. 1, are called *reaction norms*. These reaction norms need not be linear and can be rather complex (van Noordwijk and de Jong 1986, Stearns and Koella 1986). When the reaction norms are not parallel, this is indicative of heritable genotype-by-environment interactions (as in Fig. 1; Via 1984, Via and Lande 1985).

From an evolutionary ecologist's point of view, the central question is whether or not phenotypic plasticity in response to a specific environmental factor is adaptive. The null hypothesis is, and should be, that it is not. For example, it should not be assumed that stunted growth and size of plant or animal species in harsh, high altitude alpine environments are adaptations (Franz 1979, Zettel and Zettel 1994, Sorci et al. 1996). While slow growth rates may be adaptive in stressful environments hampering growth (Arendt 1997), in many cases such plasticity is accompanied by fitness reductions, and, thus, organisms may simply be making the best of a bad situation in a limiting environment (cf. Dawkins 1980). However, sometimes plastic responses are adaptive, and occasionally stunningly so. There are three critical requirements to demonstrate that phenotypic plasticity is adaptive. First and foremost, it would have to be shown (using the above example) that stunted individuals have a fitness advantage in 'slow growth environments' (at high altitudes) *and* a disadvantage in 'fast growth environments' (at low altitudes) when compared to larger individuals (and *vice versa*; cf. Gotthard and Nylin 1995, Nylin et al. 1996, Gotthard 1998). Second, it has to be shown that the trait in question and/or the phenotypic plasticity exhibited have a heritable basis, i.e., there is V_g and/or $V_{g \times e}$ (Maynard-Smith et al. 1985). This is no trivial task, and few studies demonstrate this in its entirety. Indeed, heritability of plasticity ($V_{g \times e}$) is often ignored by ecologists (Nylin and Gotthard 1998). In this context it should be noted that there are two competing concepts about the heritability of phenotypic plasticity (Via and Lande 1985; Schlichting 1986, Scheiner 1993a,b,Via 1993a,b, de Jong 1995, Via et al. 1995). One school of thought posits that the reaction norm *per se* (i.e., the slope of the line in Fig. 1) is controlled by genes and can therefore evolve somewhat independently of the trait (Via and Lande 1985, Via 1993a,b). The other school of thought believes that only the trait itself (i.e., the points in Fig. 1) is controlled by genes, perhaps by different ones in different environments, and thus evolves. Thus, the reaction norm linking the environment-specific phenotypic values is merely an epiphenomenon (Scheiner 1993a,b). In either case, adaptive phenotypic plasticity *can* evolve, either directly or indirectly.

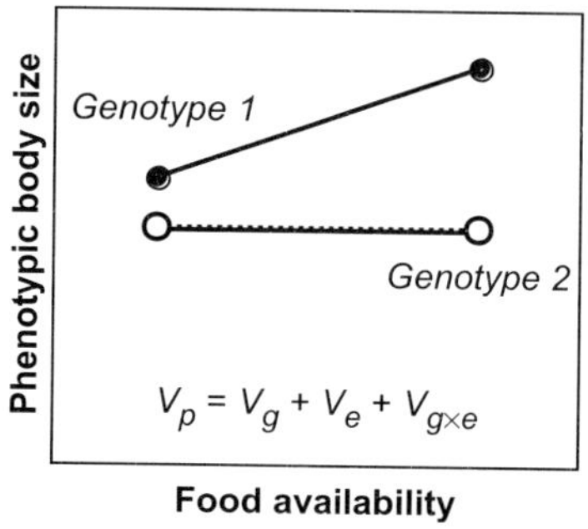

Fig. 1 Illustration of the variance partitioning of a phenotypic trait (V_p) into three components due to genetic (V_g), environmental (V_e) and genotype-by-environment effects ($V_{g\times e}$) using analysis of variance. Genotype 1 is larger than genotype 2 at any food level, so there is genetic variation in body size in the population (= V_g). In response to higher food availability genotype 1 increases in size (= V_e), so this genotype is phenotypically plastic. As this is not the case for genotype 2, there is genetic variation in the response of genotypes to food availability in the population (= $V_{g\times e}$). Phenotypic plasticity consequently has two components, a purely environmental (V_e) and a heritable ($V_{g\times e}$) component.

A third critical criterion is that the ability to show a plastic response bears a fitness benefit but also a cost, i.e., there is a trade-off. Without trade-offs or costs, all organisms would theoretically evolve to respond perfectly to all environments at all times. A good example is the occurrence of induced morphological plasticity in response to predators, such as wider tails in some frog tadpoles or neck spines in water fleas (Sell 2000, Van Buskirk 2000, 2002). The costs of phenotypic plasticity are obvious in this case because the morphological change is induced during the juvenile stage and fixed thereafter, even if the predator is no longer present. The costs and benefits of induced defenses can therefore be demonstrated experimentally by comparing the fitness of defended and undefended phenotypes in predator-free and predator-loaded environments: while defended phenotypes typically have higher survival when predators are present, they also often grow and/or forage more slowly when predators are absent (Van

Buskirk 2000, 2002, Van Buskirk and Schmidt 2000). While insects frequently cause induced plant defenses (e.g., Turlings et al. 1998, Wäckers and Bezemer 2003), induced morphological defenses of insect prey species analogous to those in anurans or *Daphnia* have not often been described (e.g., Arnqvist and Johansson 1998). However, the case of horned dung beetles, which respond to larval food (dung) quality and quantity, is a different example of such induced morphological plasticity (Eberhard 1982, Moczek et al. 2002, Moczek, this volume).

Induced defenses also illustrate the evolutionary causes of adaptive plasticity. Adaptive phenotypic plasticity is predicted to evolve by natural selection if genotypes are likely to encounter more than one environment and if the phenotype conferring highest fitness varies predictably between these environments (van Tienderen 1991, 1997, Newman 1992, Kawecki and Stearns 1993, West-Eberhard 2002). It is intuitively obvious that plasticity does not pay in a constant and predictable environment. If predators are present all the time, all individuals should bear the trait that reduces predation risk, so there is no need for plasticity. Only if predators are present in some habitats but not others, and provided the trait carries a cost, is a plastic, induced defense of selective advantage. This intuitive argument can be demonstrated by evolutionary models of plasticity (Van Tienderen 1991, 1997, de Jong 1995). Because the argument is so intuitive, it is also often argued *post-hoc*, i.e., after some research but without further rigorous proof that a particular plastic response is adaptive. But it is much more difficult in practice to specify or predict the conditions under which phenotypic plasticity is adaptive and should consequently evolve. This is successfully demonstrated only in rare cases, for example the induced defenses of frog tadpoles mentioned above (Van Buskirk and Schmidt 2000), or by means of time consuming and highly artificial experimental evolution in the laboratory (see Mery and Kawecki (2002) for an exceptional example on the evolution of learning in *Drosophila melanogaster*). However, it has to be borne in mind that all this only holds for *adaptive* phenotypic plasticity. As stated above, many plastic responses, such as stunted growth in poor environments, may not be advantageous at all and instead be viewed as making the best of a bad situation, due to physiological or developmental constraint.

Phenotypic Plasticity in Insect Body Size: Patterns, Causes and Consequences

Phenotypic plasticity has been classified in various ways, for example according to the traits in question (morphological, life history, physiological, behavioral, etc.), or according to the environmental variable causing the plasticity (food availability, predators, temperature, season length, photoperiod, humidity, etc.). A third scheme classifies plasticity according to the biological level of expression, differentiating between developmental (West-Eberhard 2002), physiological, or behavioral plasticity. And, as discussed above, plasticity can be classified as either adaptive or not. These classification schemes overlap considerably, because phenotypic plasticity of diverse traits, their various environmental causes, and the levels at which plasticity is expressed are all necessarily interconnected. Figure 2 diagrams how the development from genes of a functional phenotype in a variable environment is mediated by physiology and behavior. Physiological and behavioral mechanisms are involved in producing phenotypic plasticity, say in body size. But a particular body size, at the same time, entails further phenotypic plasticity in behavior and physiology as a consequence (Fig. 2). Entire syndromes can emerge. For example, the flexible induction of insect diapause can occur in response to photoperiod, temperature or humidity and involves physiological, behavioral and occasionally morphological changes (Tauber et al. 1986, Danks 1987). If it occurs at the pre-adult (larval or pupal) stage, as in the pitcher plant mosquito (Bradshaw

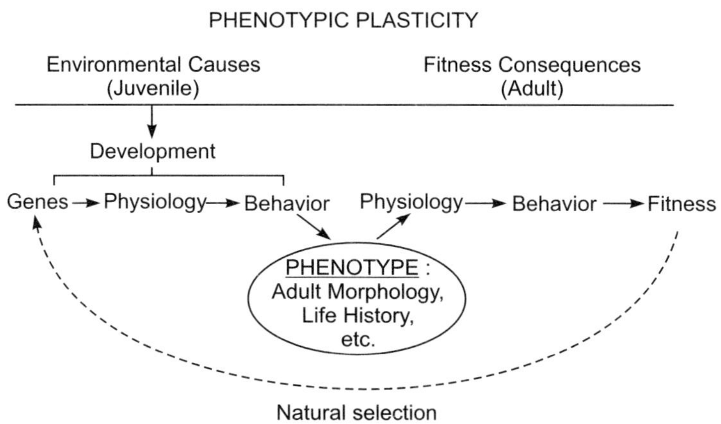

Fig. 2 Diagram showing the hypothetical adaptive nature of phenotypic plasticity. Development from genes is influenced by a variable environment, producing phenotypes that differ in physiology, morphology, behavior, and life history, within some limits set by environmental, physiological, developmental, genetic and phylogenetic constraints. Physiological and behavioral mechanisms are involved in producing phenotypic plasticity, say in body size, at the juvenile stage. A particular body size then produces further phenotypic plasticity in behavior and physiology at the adult stage. The resulting phenotypes differ in fitness. Selecting among the diverse phenotypes, natural selection alters population gene frequencies, increasing the proportion of genes that produce favorable phenotypic responses.

1976), it additionally involves growth and developmental plasticity. Furthermore, a trait such as body size can be regarded as both a morphological and a life history trait, as it is intimately intertwined with other life history traits like development time and growth rate (Nylin and Gotthard 1998). Hence, classification of phenotypic plasticity remains difficult due to the complex inter-connectedness of the various eliciting factors, processes, and outcomes.

In this chapter I focus on phenotypic plasticity in body size, thus taking a trait-centered view. I first briefly discuss variation in body size in insects in general, as it relates to developmental, physiological and behavioral plasticity (Fig. 2). In the second part I present a case study of the causes and consequences of phenotypic plasticity in body size and related traits in the yellow dung fly *Scathophaga stercoraria* (Diptera: Scathophagidae).

Body Size—An Important and Highly Plastic Trait

Body size is one of the most important quantitative traits of an individual. It greatly influences most physiological traits, such as metabolic expenditure or heart rate, producing strong but not necessarily well understood allometric patterns among organisms (Peters 1983, Calder 1984, Schmidt-Nielsen 1984, Reiss 1989, Roff, 1992). Fitness is generally also strongly positively correlated with body size. Fecundity increases with female size in most ectotherms including insects (reviewed by Wootton 1979, Shine 1988, Honek 1993, Andersson 1994). Similarly, mating and reproductive success typically increases with male size in many species, be it mediated by male-male competition, female choice or sexual conflict (reviewed by Andersson 1994, Kingsolver and Pfennig 2004). Even survival or lifespan is often a positive function of body size, at least when comparing species (Peters 1983, Calder 1984, Schmidt-Nielsen 1984), although this is much less clear and sometimes reversed within species (Blanckenhorn 2000a).

Life history, developmental plasticity, and body size

Virtually all species exhibit variation in body size. On average about 30–40% of this variation is heritable (i.e., V_g: Mousseau and Roff 1987), but most of it is phenotypic plasticity (i.e., V_e and $V_{g \times e}$). Multiple environmental factors generate variation in body size, the most prominent being temperature, season length, food availability and predators. Other environmental variables like humidity and photoperiod also systematically affect body size, but to a lesser extent. Most of the variation in body size is produced by variation in growth and development, as there are two ways to get large:

individuals can either grow faster (by increasing growth rate) or grow for longer time (by increasing development time). (We here ignore a third possibility, namely that the offspring are born larger to begin with.) Both strategies presumably incur costs (in terms of mortality) that have to be demonstrated (Nylin and Gotthard 1998, Blanckenhorn 2000a). Body size, growth rate and development time are thus intimately (even mathematically) interconnected and traded off against each other. These traits therefore cannot be investigated in isolation. Moreover, numerous life history models predict particular reaction norms for an organism's age and size at maturity in response to environmental variation (e.g., Roff 1980, 1992, Stearns and Koella 1986, Stearns 1992, Kozlowski 1992, Fraser and Gilliam 1992, Werner and Anholt 1993, Abrams et al. 1996, Niewiarowski 2001). These models imply that such phenotypic plasticity is adaptive. However, extrinsic environmental constraints are generally built into such models, and these cannot necessarily be regarded as adaptations (cf. above). Furthermore, plasticity requires time to evolve, during which the ecological pressures generating such plasticity may change, rendering evolution of adaptive plasticity more difficult (Padilla and Adolph 1996). Trade-offs also often require energy limitation to become apparent (van Noordwijk and de Jong 1986, Zera and Harshman 2001), but not all constraints are energetic or environmental in origin, as there also are developmental, physiological, genetic and phylogenetic constraints (Maynard-Smith et al. 1985, Ridley 1996). For insects, phenotypic plasticity in growth, development and body size has recently been reviewed by Arendt (1997) and Nylin and Gotthard (1998). These authors conclude that adaptive life history plasticity is common in insects, due to complex interactions of the organism with its environment. Both the costs of plasticity (Newman 1992, Padilla and Adolph 1996, Arendt 1997) and the underlying genetic mechanisms require further study before we can fully understand how adaptive plasticity in growth, development and body size can evolve and be maintained.

Some environmental factors produce well known evolutionary patterns of phenotypic plasticity in body size within species. In general, small body sizes are produced in environments that constrain growth (Berrigan and Charnov 1994). This definitely holds for **food limitation**, which increases intra-specific as well as inter-specific competition (for insect examples see Shorrocks 1970, Møller et al. 1989, Blanckenhorn 1994, 1998a, Bradshaw and Holzapfel 1996, Fox et al. 1996, Nylin and Gotthard 1998).

Predators also limit growth, development and ultimately body size: mortality risk typically increases with the number of predators present in the habitat, to which prey species often react with reduced activity and foraging success (e.g., Fraser and Gilliam 1992, Werner and Anholt 1993, Abrams et al. 1996). Insect examples are known from dragonflies (Martin et al. 1991), butterflies (Stamp and Bowers 1993) and mayflies (Scrimgeour and Culp 1994, Dahl and Peckarsky 2003).

A third environmental factor that can reduce body size is **season length**. The mechanism is straightforward and can be demonstrated easily by theory (Roff 1980, Rowe and Ludwig 1991): a short season limits the time available for growth and development and hence the body size that can be attained at maturity. Because season length systematically decreases towards the poles, the resulting pattern of decreasing body size with increasing latitude has been termed 'converse Bergmann rule' (see below) (Park 1949, Mousseau 1997). As latitudinal changes in season length are generally accompanied and mediated by systematic changes in **photoperiod** (Masaki 1972, Bradshaw 1976, Bradshaw and Lounibos 1977, Bradshaw and Holzapfel 2001), photoperiod can, at least indirectly, also systematically affect body size. In insects, latitudinal 'converse Bergmann' size variation has been demonstrated in beetles (Park 1949), crickets (Masaki 1967, 1972, Mousseau and Roff 1989, Mousseau 1997, Telfer and Hassall 1999), water striders (Blanckenhorn and Fairbairn 1995), butterflies (Nylin and Svärd 1991, Ayres and Scriber 1991), odonates (Johansson 2003) and ants (Elmes et al. 1999; reviewed by Blanckenhorn and Demont 2004). Analogous seasonal time constraints affecting body size can be caused by altitude (Blanckenhorn 1997a, Fischer and Fiedler 2002, Chown and Klok 2003). If high latitude or altitude populations of a given species compensate seasonal time constraints by evolving faster growth compared to their low latitude conspecifics, this is known as countergradient variation (Conover and Present 1990). Most of the insect species showing converse Bergmann size variation probably also show countergradient variation. However, demonstration of countergradient body size variation ultimately requires rearing various populations in common garden (laboratory) environments, which has been done only for a subset of the studies cited above (Ayres and Scriber 1991, Blanckenhorn and Fairbairn 1995, Elmes et al. 1999, Telfer and Hassall 1999).

Temperature is a notable exception to the rule that environments constraining growth always produce smaller body sizes. This even has been termed a life history puzzle (Berrigan and Charnov 1994), because low temperatures constrain growth but nevertheless result in larger body sizes (Taylor 1981, Atkinson and Sibly 1997). There are two related but conceptually different patterns. First, there is ample evidence from laboratory rearing that most ectotherms grow larger at lower temperatures;

this phenomenon is known as the temperature-size rule (Atkinson 1994, Angilletta and Dunham 2003, e.g., Blanckenhorn 1997a). The second, well-known pattern is 'Bergmann's rule,' the phenomenon that animals are larger in colder climates (Bergmann, 1847). Bergmann's rule originally referred to comparisons among endothermic (warm-blooded) species only, but later was (mostly) applied to comparisons among populations within species (Bergmann, 1847, Blackburn et al. 1999, Ashton et al. 2000). The adaptive explanation suggested was that larger individuals possess smaller surface-to-volume ratios more conducive to conserving heat. About one hundred years later, however, it transpired that Bergmann's rule extends to some ectothermic (cold-blooded) organisms, for which the cause must be different, because ectotherms do not conserve heat as endotherms do. Instead, their body temperatures closely follow that of the environment, and especially small ectotherms such as insects acclimate to the ambient temperature almost instantly (Stevenson 1985). A unifying explanation for Bergmann's rule is still lacking (Ray 1960, Atkinson 1994, Atkinson and Sibly 1997). There is still much debate about whether ectothermic 'Bergmann size clines' within species are adaptive or whether they are a consequence of physiological constraints at the cellular level (van der Have and de Jong 1996, Partridge and Coyne 1997, Huey et al. 2000, Blanckenhorn and Hellriegel 2002; arguments summarized in Atkinson and Sibly 1997, Blanckenhorn 2000b). Again, there is a genetic (V_g or $V_{g \times e}$) and an environmental component (V_e) to this temperature-mediated phenotypic plasticity. In insects, genetic Bergmann size clines (where body size in high latitude populations of a given species is larger than in lower latitude populations when they are reared in a common environment) occur in a number of small-bodied insect species such as *Drosophila* (Ray 1960, James et al. 1995, 1997, Huey et al. 2000), bees (Alpatov 1929, Daly et al. 1991), aphids (Sokal and Rinkel 1963) and ants (Heinze et al. 2003), but also in the relatively large ant lion (Arnett and Gotelli 1999; reviewed in Blanckenhorn and Demont 2004).

A recent review of the arthropod literature by Blanckenhorn and Demont (2004) shows that within-species Bergmann size clines, showing increased body size at higher latitudes, and converse Bergmann clines, showing decreased body size at higher latitudes, are about equally common. Moreover, the slope (i.e., the strength) of the latitudinal body size change varies in a continuous fashion among species. Which type of cline is evident, and how strong the effect is, depends crucially on the body size and/or development time of the species in question: larger species with typically longer development times tend to show converse Bergmann size clines, whereas smaller species with shorter development times tend to show Bergmann size clines. This lends empirical support to the original suggestion of Chown and Gaston (1999) that generation time relative to season length is a crucial parameter in determining which rule applies. That is, species with long development times relative to the season length, and consequently often only one generation per year, such as the water strider *Aquarius remigis* (Blanckenhorn and Fairbairn, 1995), are more prone to experience end of season time constraints than multivoltine species with short generation times and many generations per year, such as *Drosophila melanogaster* (James *et al.*, 1995). Thus, at least for arthropods, Bergmann's rule and the converse Bergmann rule, the two seemingly opposite rules long described, appear to be two ends of a continuum and not necessarily mutually exclusive, as they are driven by different causes (and presumably mechanisms): temperature and season length respectively (Chown and Gaston 1999, Blanckenhorn and Demont 2004).

Physiological plasticity and body size

Body size is a consequence of growth and development, and thus can be regarded as developmentally plastic (cf. West-Eberhard 2002). Developmental plasticity is mediated by physiological and behavioral plasticity at the juvenile stage (Fig. 2). It is important to understand the physiological mechanisms generating such plasticity, so as to link mechanisms and function. This call is often made, but rarely executed (Zera and Harshman 2001). As physiological plasticity is covered elsewhere in this book (Nijhout and Davidowitz, and also Zera), I shall only briefly discuss it here as it relates to phenotypic plasticity in body size.

In insects, growth entails a series of discrete larval instars before reaching the final and determinate adult size. The central questions of insect development are: which environmental cues trigger molting, and via which proximate mechanisms? Nijhout and coworkers (Nijhout 1994, 2003, D'Amico et al. 2001, Davidowitz et al. 2003, 2004) have investigated this, primarily in the moth *Manduca sexta*. In this species, but probably in insects in general, juvenile hormone and prothoracicotropic hormone regulate the duration of the developmental period and growth rate. Both hormones differentially respond to key environmental stimuli such as temperature or food limitation, either in their sensitivity or pattern of secretion. These changes in endocrine interactions result in alternative developmental pathways ultimately producing the body size adjustments in response to the various environmental stimuli described in the previous section (Davidowitz et al. 2004, Nijhout 2003).

One rare aspect of insect life history plasticity that has been well investigated in terms of physiological mechanisms are wing and flight polymorphisms (Zera and Denno 1997, Zera and Harshman 2001, Zera, this volume). Facultative production of winged individuals occurs in crickets (Orthoptera), planthoppers (Homoptera), ants (Hymenoptera) and bugs (Heteroptera), and occasionally in other insects (Roff and Fairbairn 1986, Zera and Denno 1997). This phenomenon is analogous to the induced defenses against predators discussed above. Typically, wing morph determination in a given species is partly genetic and partly environmentally induced. Again, the expression of wing polymorphism is proximately regulated by hormones, primarily juvenile hormone and ecdysone (Zera and Denno 1997, Zera and Harshman 2001). Winged individuals are typically produced only under particular environmental conditions conducive to dispersal (Dingle 1978, Vepsäläinen 1978, Roff 1986, Roff and Fairbairn 1986). This is because production of wings and wing muscles enabling flight and dispersal incurs costs that have to be traded off against other needs, such as reproduction (oocyte development), which can directly impact fitness (Roff 1986, Fairbairn and Desranleau 1987, Roff and Fairbairn 1986, Zera and Denno 1997). Wing polymorphism relates to body size plasticity, because generally winged individuals of a given species are distinctly larger than unwinged individuals (Zera and Denno 1997, e.g., Dingle et al. 1980, Fairbairn 1992). This is intuitive because the presence of wings and wing muscles requires at least a larger thorax. Because body size typically correlates positively with fecundity (Honek 1993) and mating success (Andersson 1994), one might expect greater reproductive success of winged individuals, but this is not so. Instead, the presence of wings can regularly be shown to trade off against other life history traits such as fecundity, development time or survivorship (Zera and Denno 1997, e.g., Zera 1984). These studies show that life history traits are intricately intertwined by trade-offs, resulting in complex syndromes involving many traits. In the case of wing polymorphisms, it is probably most parsimonious to conclude that large body size of winged individuals is not directly selected but instead a concomitant (physiological?) response to selection for wing production and the ability to disperse.

Behavioral plasticity and body size

Behavior can also be an environmentally induced plastic trait. Contrary to phenotypic plasticity in most morphological and life history traits, behavioral flexibility represents rapid, reversible, and ever-changing variation within, as opposed to between individuals (Fagen 1982, Clark and Ehlinger

1987). Nevertheless, some researchers may reject '**instantaneous**' or '**neural**' behavioral responses, such as acceptance or rejection of a mate or a prey item, seeking refuge from a predator, or switching between foraging patches or oviposition sites as examples of phenotypic plasticity. On the other hand, they will readily accept the above-mentioned **developmentally** induced defenses or flight syndromes as phenotypic plasticity *sensu strictu*, changes that generally occur only once, require considerable time to become fully expressed, and are typically not reversible (cf. Austad 1984, Caro and Bateson 1986, West-Eberhard 2002). What differs, in essence, are the time span and the immediate underlying proximate mechanism (Table 1). However, even this distinction is blurred, because some **physiologically** or **hormonally mediated plasticity** occurs at intermediate time scales (minutes to days), and may or not be fixed thereafter. Examples are the switch to territorial behavior in connection with sex change in some fish (Warner 1984), the (reversible) onset of migratory restlessness in many songbirds (Berthold 1971), or the (irreversible) resorption of flight muscles with its associated behavioural changes in water striders (Fairbairn and Desranleau 1987, Kaitala and Huldén 1990). At the other extreme, some behavioral attributes may be directly controlled by one or a few genes and therefore are fixed ('hard-wired'; = V_g), stereotypic, and not plastic at all. Prominent examples are nest-cleaning behavior in bees (Rothenbuhler 1964a,b), or 'roving' vs. 'sitting' foraging behavior of *Drosophila melanogaster* larvae (DeBelle and Sokolowski 1987). As mentioned previously, any classification scheme of phenotypic plasticity, such as the one in Table 1, will be limited in scope, because we are dealing with continuous variation.

Because growing larvae will sometimes forgo feeding to hide from predators, phenotypic plasticity in body size can be proximately mediated by behavioral plasticity (Fig. 2; e.g., *Drosophila* or butterfly larvae: DeBelle and Sokolowski 1987, Nylin et al. 1996, Gotthard 1998). Conversely, and

Table 1 Summary of a classification scheme of behavioral plasticity according to the proximate mechanism, time scale, frequency of occurrence, and the variance component where variation may be detected (examples in the text).

Proximate mechanism	*Flexibility*	*Time scale*	*Frequency of occurrence*	*Variance component*
genetic	none or low	cross-generational	None	V_G or fixed single allele
developmental	long-term	days to months	Once	$V_{G \times E}$
physiological	short-term	min to days	Repeated	$V_{G \times E}$ or V_E
neural	instant	sec	Frequent	V_E

perhaps more interestingly, phenotypic plasticity in body size can entail behavioral plasticity, as small individuals are forced to use different behavioral strategies in order to successfully compete against their larger conspecifics and augment their fitness. This is most prominently seen in species with alternative foraging or reproductive strategies (reviewed by Dawkins 1980, Rubenstein 1980, Austad 1984, Dominey 1984, Caro and Bateson 1986, Stephens and Krebs 1986,West-Eberhard 1986, Clark and Ehlinger 1987, Taborsky, 1994, 1998, 2001), which are often, but not always, associated with discontinuous variation in body size. Perhaps the most striking examples are those of bluegill sunfish and Pacific salmon (Dominey 1980, Gross and Charnov 1980, Gross 1985). In both these species, there are large males which typically compete for females by fighting for territories, and there are much smaller males ('jacks' or 'sneaks') which attempt to gain fertilizations by sneaking up to females without being detected by the large males. A number of examples of alternative mating or reproductive strategies involving morphological variation are also known from insects: scorpionflies (Thornhill 1979, 1981, 1984), earwigs (Tomkins 1999), crickets (Cade 1980), thrips (Terry and Dyreson 1996), dung beetles (Mozcek, this volume), and various wasp (Gadgil 1972, Evans and O'Neill 1978, Tsuchida et al. 2002), bee (Alcock 1997, Wuellner 1999) and ant species (Heinze and Keller 2000). A key question is whether such strategies are maintained by frequency dependent selection, i.e., whether at equilibrium the average fitness of all strategies (e.g., sneaks and fighters) is equal. Alternatively, some phenotypes, syndromes or strategies (e.g., sneaking) may be inherently inferior, 'best of a situation' strategies (Dawkins 1980), conferring partial but not complete fitness compensation to the bearer. This distinction may be clear in mating systems with discontinuous body size variation in which small males must, and do, employ discrete alternative reproductive strategies to gain mates. However, the necessity for small individuals to adjust their mating and foraging behavior to augment their fitness even holds, but more subtly so, when body size variation is continuous (e.g., Dunn et al. 1999). More subtle examples include facultative territorial, signaling or calling behavior in relation to body size and other traits in Orthopterans (e.g., Shelly and Greenfield 1991, Belovsky et al. 1996), Gerrids (Rubenstein 1984, Hayashi 1985) or Odonates (Nomakuchi 1992, Tsubaki and Ono 1986, Cordoba-Aguilar 2002). Because territoriality often influences increased mating success indirectly via increased foraging success, this relates directly to alternative foraging strategies (e.g., Spence and Wilcox 1986, Wilcox and Spence 1986, Belovsky et al. 1996; see also

DeBelle and Sokolowski 1987, Hoffmeister and Roitberg 1997, for examples of other alternative foraging strategies or tactics in insects).

The Yellow Dung Fly—Ubiquitous Phenotypic Plasticity in a Widespread Insect Species

In this section, I discuss phenotypic plasticity in the yellow dung fly as it relates to body size and associated traits. I follow Fig. 2 in that I discuss the environmental and evolutionary causes of phenotypic plasticity in body size, and then its behavioral, physiological and fitness consequences.

Distribution, Phenology, Life History and Reproductive Behavior of the Yellow Dung Fly

General biology and geographic distribution

The yellow dung fly, *Scathophaga stercoraria* (L.; Diptera: Scathophagidae; sometimes *Scatophaga*, formerly *Scopeuma*), occurs in north-temperate regions of the Old and New Worlds (Stone et al. 1965; Gorodkov 1984; Fig. 3). The larvae are coprophagous, feeding on the dung of large mammals, which

Fig. 3 Two guarding yellow dung fly pairs on a fresh cow pat, with a third pair in the background. Note the difference in size between the two males, and that the yellow males are larger than the green females. Photo by Peter Jann.

they thereby help to decompose, together with many other species of competing earthworms, beetles and flies (Hammer 1941, Holter 1979, Hanski 1980a,b, Hanski and Cambefort 1991, Ward and Wilhelm 1994). Adult yellow dung flies, in contrast, are sit-and-wait predators of small insects, but also imbibe nectar and fresh dung (Cotterell 1920, Hammer 1941, Foster 1967, Gibbons 1980a,b, Sasaki 1984, Failes et al. 1992). Adult flies are nutritionally anautogenous, and thus require protein and lipids from prey to produce eggs and sperm (Foster 1967). In the laboratory, adult flies can be maintained on just *Drosophila* and water (pers. observation), which indicates that individuals are able to extract sufficient sugars (without which they die) from prey and do not need to rely on nectar feeding.

Scathophaga stercoraria is a cool-climate species, and survives at high altitudes and latitudes (Gorodkov 1984, Sigurjónsdóttir and Snorrason 1995, Blanckenhorn 1997a). Its distribution appears limited in the south by hot temperatures (Hammer 1941, Parker 1970a, Gibbons 1987, Ward and Simmons 1990, Blanckenhorn 1998a, Blanckenhorn et al. 2001). In southern Europe yellow dung flies occur only at higher elevations, such as the Pyrenees, and this is probably also the case in southern North America, where they should be found primarily restricted to mountains (Stone et al. 1965). Yellow dung flies also occur in South Africa (http://www.museums.org.za/bio/insects/flies/scathophagidae/), but have not been reported from anywhere else in the southern hemisphere. In north-central Europe, *Scathophaga stercoraria* is clearly one of the most abundant and widespread insect species associated with cow dung. The species' widespread distribution may relate to human agricultural practices, although the degree to which the current distribution of yellow dung flies is caused by the expansion of animal husbandry in recent centuries remains to be investigated. While this species is considered a cow-dung specialist, it can successfully breed on dung of other large mammals such as sheep (Hirschberger and Degro 1996), horse, deer or wild boar (W. U. B. and D. Burkhard, unpublished data).

Reproductive behavior and life history

The reproductive behavior of yellow dung flies has been investigated in great detail by Parker and coworkers (see Parker 1978 and Simmons 2001 for summaries). Females spend most of their time foraging in the vegetation surrounding the pasture and only come to the dung to oviposit. Males are already waiting on and around fresh dung pats and immediately seize incoming females. Flies of both sexes are attracted to the dung pat by scent, as they approach them against the wind (Parker 1970b), and roughly in proportion to the surface area of the dung pat (Blanckenhorn et al. 2000). Copulation usually takes place in the surrounding grass or on the pat (Parker 1970b). Females show little behavioral choice or resistance in response to male mating attempts (e.g., they do not strongly shake or wing-flick to fend off unwanted males), perhaps because males are typically larger than females (Borgia 1981, 1982, Simmons and Ward 1991, Jann et al. 2000, Ding and Blanckenhorn 2002; Fig. 3). This is unusual for insects, where females are generally larger than males (Teder and Tammaru 2005). However, females appear to have subtle means of indirectly choosing particular males by arriving in a temporally dispersed manner (Parker 1970c, Reuter et al. 1998). Furthermore, Borgia (1981) reports that, at least at low densities when territories are economically defendable, larger, more dominant males may occupy strategic places on the dung pat, which females seem to prefer, as they apparently sneak up to the center of the dung pat, and thus avoid the smaller, peripheral males. During copulation, and the ensuing oviposition, the male guards the female against competitor males (Parker 1970b; Fig. 3). Defending females against other males is an important determinant of male reproductive success, since the last male fertilizes on average about 80% of the subsequent egg clutch (Parker 1970e, Simmons 2001). Violent struggles may ensue, during which all individuals (particularly the smaller females) may be harmed (Parker 1970d, Borgia 1980, Sigurjónsdóttir and Parker 1981). Competition among males for females is very strong, as the operational sex ratio at mating sites (Emlen and Oring 1977) is highly male biased (Borgia 1980, 1981, 1982; Parker 1970b,c,d, 1978; Blanckenhorn et al. 2000; Jann et al. 2000). A single dung pat may have as many as 400 males (Jann et al. 2000)

After copulation, which lasts about 30–40 minutes (Parker 1970e, Parker and Simmons 1991, 1994, 2000, Ward and Simmons 1991, Parker 1992, Simmons and Parker 1992, Parker et al. 1993, 1999, Nuyts 1994), the female lays about 30–70 eggs, partially submerged into the dung. Ovipositing females prefer ridges on the dung pat vs. crevices or sharp points, as egg mortality in the latter positions is higher due to drowning and desiccation, respectively (Ward et al. 1999). After oviposition, the female leaves the pat for further foraging in the vegetation, whereas the male waits for other females at the same pat or switches to another, fresher pat, as pats lose their attractiveness to this fresh-dung specialist within 1–2 h of being deposited (Parker 1970b, Blanckenhorn et al. 2000). As can be expected, males spend most of their time near oviposition sites in order to maximize their reproductive success. Males presumably must also forage in the vegetation, at least minimally (Blanckenhorn and Viele 1999), but, interestingly, only rarely forage on or around the dung pat, even though they could (Gibbons 1980b).

Larvae hatch from the eggs within 1–2 days, depending on temperature (Blanckenhorn 2000b), and immediately enter the dung to avoid desiccation or drowning. Thereafter they regularly surface for oxygen. Other sources of egg, larval and pupal mortality include numerous egg (e.g., staphylinid beetles) or larval (e.g., beetle larvae) predators, or larval and pupal parasitoids (Hammer 1941, Sowig et al. 1997); however, there is essentially no information about pre-adult mortality rates under natural conditions. At 20°C, larvae undergo three molts and grow exponentially and rapidly during the first 5 days of development; thereafter they require an additional 5 days to empty their guts and prepare for pupation during which no additional body mass is accumulated (Blanckenhorn 1999). Individuals pupate in the ground under the dung pat or near the encrusted parts of the dung. Pupal development takes an additional 10 days at 20°C (Blanckenhorn 1999).

In the case of direct development, adult flies emerge after a pre-adult (i.e., egg + larval + pupal) development time between 17 days (at 25°C) and 80 days (10°C and below) (Blanckenhorn 1997a,b, 1998a, Blanckenhorn et al. 2001, Blanchenhorn and Henseler 2005). The smaller females always emerge a few days earlier. Towards the end of the season, individuals regularly enter pupal winter diapause (Luvchiev and Tsankova 1982). Larvae that do not manage to pupate before the first winter frost die, (Blanckenhorn 1997b, 1998a, Teuschl et al. 2007). Pupal winter mortality is relatively low, albeit size and sex dependent: the larger male pupae are more likely to die, (Blanckenhorn 1998a, Teuschl et al. 2007). Post-winter emergence typically begins in March in the Swiss lowlands (Blanckenhorn 1997b, 1998a), long before the cows are released onto the pastures (in mid April). However, some individuals emerge during warm spells in winter, as early as February. Spring emergence is largely synchronized, although the earlier emergence of females is preserved (Blanckenhorn 1997b).

After emergence, adults forage for some time before they finally move to the pasture to reproduce (Gibbons 1980a). The adult pre-reproductive period depends on foraging success and temperature, and is, at minimum, 8–10 days for females and 3–6 days for males (Gibbons 1980a, Blanckenhorn and Henseler 2005). However, females may begin mating as early as 6 days after emergence, when confronted with a male. Thereafter, males can mate more or less continuously if their sperm stores are filled. Inter-clutch intervals are between 3–7 days in the laboratory, again depending on nutrition and temperature (Blanckenhorn 1997a), but probably considerably longer in the field (cf. Gibbons 1987). Under (benign) laboratory conditions, flies live on average for 1–2 months, but up to several months (Blanckenhorn 1997a). In the field, where predation and parasitism should be a serious threat, longevity is probably considerably lower, but data on this subject are scarce. Burkhard et al. (2002) used wing injuries to estimate sex- and size-specific adult field mortality, but found no clear trends. Using scars in the female reproductive tract indicating how many clutches a female had previously laid, Gibbons (1987) found a negative exponential distribution of female age: the majority of females were virgins (0 previous clutches), with the (exponential) average at about 3 and the maximum at 10 clutches. Using our laboratory estimates of pre-reproductive periods and inter-clutch intervals, mentioned above, this suggests that females can live up to 60 days in the field as well.

Phenology

Naturally, yellow dung fly phenology depends strongly on local climate. In central Europe, fly populations often exhibit a sharp population decline in summer (Hammer 1941, Parker 1970a, Gibbons 1987, Ward and Simmons 1990, Blanckenhorn 1997a, Blanckenhorn et al. 2001), subdividing the year into a spring and an autumn flight season. The summer decline is mediated by temperatures above 25°C, which tend to kill larvae, pupae and adults, and which the flies therefore avoid by moving into cooler microhabitats (Gibbons 1987, Ward and Simmons 1990, Blanckenhorn 1998a). In contrast, in the highlands (1500–2300 m) of the Swiss Alps, the flies are present only in summer (June-September; Blanckenhorn 1997a). The same appears to be the case in northern European countries such as Iceland (Sigurjónsdóttir and Snorrason 1995) and Finland (Otronen 1993, 1995a,b, 1996). In Britain, the summer decline occurs in some (warmer) years but not others (Gibbons 1987), and probably not in the north (i.e., Scotland; P. I. Ward, personal observation).

Accordingly, the number of generations per year varies with latitude and altitude. At our main study populations in Fehraltorf near Zurich, there are two to four (overlapping) generations per year (Jann et al. 2000, Blanckenhorn et al. 2001, Burkhard et al. 2002). As mentioned above, winter synchronizes emergence of the first generation in March (Blanckenhorn 1997b, 1998a). Their second-generation offspring start mating in the pastures about two months later at the earliest, i.e., in mid to late May. These individuals may reproduce (and then die) if born early enough and if the spring season lasts long enough (say until the beginning of July). In this case their third-generation adult offspring are expected to become quiescent to pass the summer in cooler locations (Blanckenhorn et al. 2001; see below), unless they die due to excessive heat, before they resume reproduction in early autumn (September). If born later, the second generation adults themselves should enter summer quiescence to re-emerge on the pasture in early autumn. These flies will then produce a fourth or third generation (respectively) which then overwinters as pupae and makes up the first generation of the following year. In the Swiss highlands, and in northern Europe, the season is shorter and thus only two (or even one) generation per year can be expected, but this has not yet been studied.

It is still a bit unclear where yellow dung flies are during the hot summer period of July and August. Blanckenhorn et al. (2001) found no evidence for pupal summer diapause. Adult flies also appear not to enter a proper summer diapause, but instead are quiescent and acclimate physiologically by suppressing reproduction and accumulating lipid reserves. Apparently the flies spend the summer in cooler, forested areas close to the pastures, where they may reproduce on deer, horse or wild boar dung, but this is unclear. This indicates a flexible, as opposed to a genetically fixed, life-history strategy (Blanckenhorn et al. 2001). The fact that lowland as well as highland flies disappear from the pastures on hot days, and the absence of the summer decline in cool years in England (Gibbons 1987), also support this interpretation. However, the possibility of incipient evolution of a genetic response (diapause or quiescence) at the southern fringes of their distribution, for which there are some indications in our Swiss populations, is intriguing and deserves further study.

The degree of population isolation, and thus the degree of potential local adaptation, depends on geographic location. In Switzerland, dairy farming is ubiquitous. There are numerous small farms of typically 10–30 cows throughout the country, and essentially all high altitude pasture land is populated during the summer months by cows transported up from the lowlands. At the same time, yellow dung flies are good and persistent flyers, as individuals have been shown to fly up to 70 km in a flight mill (Reim 2005). Accordingly, gene flow seems to be substantial, as indicated by the fact that Swiss populations are not significantly genetically differentiated based on six polymorphic enzyme loci (Kraushaar et al. 2002). In contrast, dairy farms in central Finland, for example, are fewer and somewhat isolated. Nevertheless, populations collected from Iceland, Finland, Sweden, Germany, England and Switzerland show only slight genetic differentiation based on 11 polymorphic microsatellite loci (Demont 2004). This intriguing result may be explained by high degrees of gene flow, even at this geographic scale, or by huge populations sizes not permitting genetic drift to result in population differentiation. Alternatively, it may be explained by a recent expansion of the species' range following expansion of dairy cow or cattle farming due to human migration within Europe during the past 3000 or so years (= 6000–9000 fly generations). Nevertheless, at the same scale within Europe, *S. stercoraria* populations revealed significant geographic differentiation in body size and development time (Blanckenhorn and Demont 2004; discussed below).

Causes of Phenotypic Plasticity in Body Size and Associated Life History Traits

As in most species, body size of yellow dung flies is extremely variable (Fig. 3). Body size varies latitudinally over long distances (Fig. 4; Blanckenhorn and Demont 2004), geographically over short distances (Kraushaar and Blanckenhorn 2002), altitudinally (Fig. 4; Blanckenhorn 1997a), seasonally (Fig. 5), between the sexes (Figs. 3–5 and 8), and even over the day (Fig. 5). What are the causes of this variation? In the introduction I already briefly discussed the main environmental factors that influence phenotypic body size in insects: juvenile nutrition (often mediated via competition), temperature, predators, photoperiod, humidity and genetic (as opposed to the former environmental) variation. Largely based on our own work, I now discuss these causes of plasticity in body size and associated traits in yellow dung flies, always asking whether it is adaptive or not.

Phenotypic plasticity in body size, development time and growth rate

Effects of larval nutrition. Much of the phenotypic variation in body size of yellow dung flies is due to variation in per capita food (i.e., dung) availability at the larval stage. This can be easily demonstrated experimentally by manipulating the amount of dung for a given number of larvae (Amano 1983, Sigurjónsdóttir 1984, Blanckenhorn 1998a; Fig. 6). As in other species, this effect it generally mediated by intra- as well as inter-specific competition for resources. However, in dung flies we cannot easily separate effects of larval competition from those of food availability. This is because the (fluid) dung is the larval habitat, much like a pond is for anuran tadpoles, and at the same time their food source. We suppose the main nutritional components consumed by the larvae are bacteria and fungi, which are presumably homogenously distributed in the dung, but this is not precisely known. Thus, in nature, and in practice in the laboratory, lower amounts of dung (i.e., smaller dung pats), more competitors, and pat drying

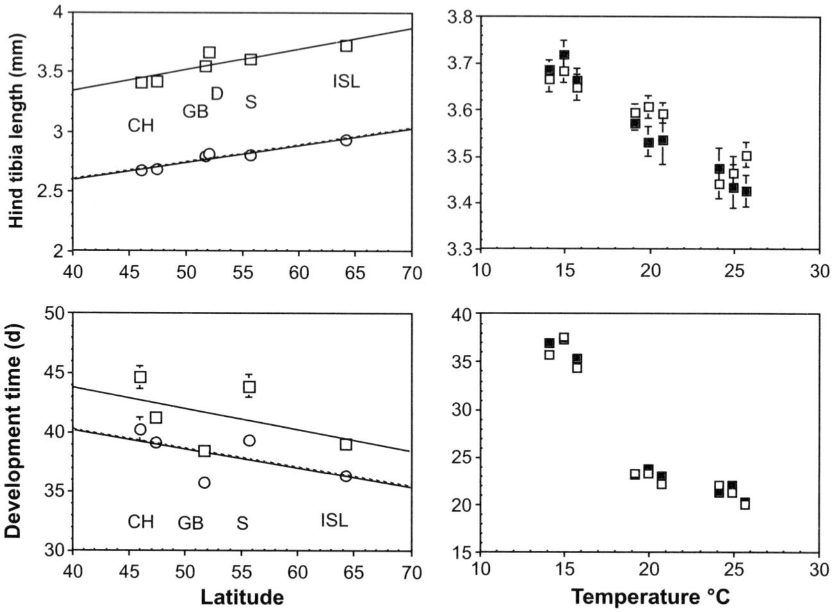

Fig. 4 Latitudinal variation (left panels) in body size (top) and development time (bottom) of yellow dung flies from Switzerland (CH), Britain (GB), Germany (D), Sweden (S) and Iceland (ISL) based on common garden laboratory rearing at 15°C (males: squares and unbroken lines, females: circles and broken lines; Blanckenhorn and Demont 2004). Body size (hind tibia length) increases while development time decreases with latitude. Analogous altitudinal variation (right panels) in body size (top) and development time (bottom): high altitude Swiss yellow dung flies from three populations (filled squares) are overall slightly smaller but do not develop faster than low altitude flies (open squares) from adjacent populations at three different temperatures (males only, but data for females are analogous; Blanckenhorn 1997a). Yellow dung flies clearly follow the temperature-size rule, as in the laboratory they grow larger at lower temperatures (top right).

(by the action of sun and wind) all have the same net effect of reducing per capita food availability (e.g., Blanckenhorn, 1998a; cf. discussion below). The latter occurs because larvae of this species can only process fluid dung.

Probably because the dung pat is habitat as well as food, yellow dung flies emerge smaller but earlier when dung is limited, a rare response exhibited by only few other organisms that are also regularly but unpredictably threatened by habitat depletion (Newman 1988a,b, 1992, Møller et al. 1989, Juliano and Stoffregen 1994, Fox 1998; Fig. 6). Most ectotherms exhibit (e.g., *Drosophila melanogaster*; Blanckenhorn 1999), and most life history models predict (e.g., Stearns and Koella 1986), later emergence at smaller body size because food restriction limits feeding,

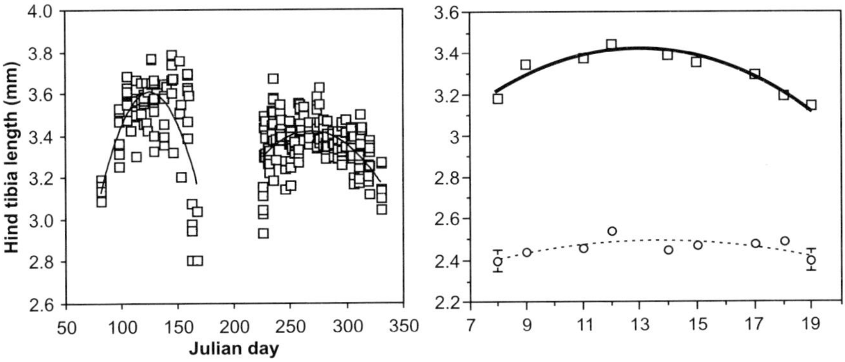

Fig. 5 Body size (hind tibia length) variation in one yellow dung fly population over two spring and fall seasons (left panel; males only, but the pattern for females is analogous; Jann et al. 2000) and over the day, based on means over two seasons (right panel; males: squares, females: circles; Jann 1997).

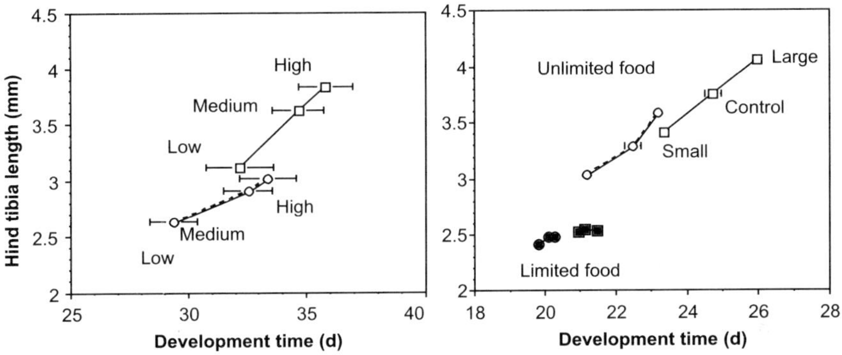

Fig. 6 Mean ± SE body size (hind tibia length) and development time of male (squares) and female (circles) full sib families reared in the field at three different larval food levels (low to high): at low food flies emerge smaller but sooner (left panels; data from Blanckenhorn 1998a). Mean ± SE body size and development time of yellow dung flies from three laboratory body size selection lines (small, control, large) at unlimited (top of right panel) and limited (bottom of right panel) larval food. At limited food, flies of both sexes and all selection lines emerge at roughly the same size, whereas clear differences are apparent at unlimited food, implying an increase in phenotypic plasticity in the large selection lines (Teuschl et al. 2007).

growth and thus final body size, which individuals compensate by growing for longer time to meet a presumed minimal target size. The plastic response of yellow dung flies is adaptive because dung pats regularly dry up as the larvae deplete the resource patch (in addition to dessication due to climate). The flies thus trade off faster development and smaller size against mortality, as they cannot switch to another patch and would certainly die if the pat were depleted before they can pupate (Blanckenhorn 1998a).

Temperature effects. For the yellow dung fly we have clear, albeit not universal, evidence for the temperature-size rule (cf. Atkinson 1994, Angilletta and Dunham 2003). When grown in the laboratory at constant temperatures, yellow dung flies grow smaller at warmer temperatures (Fig. 4, top right; Blanckenhorn 1997a). As argued in the general introduction above, the adaptive significance of Bergmann size clines and the temperature-size rule is unclear and disputed. One prominent alternative explanation is that physiological processes or constraints at the cellular level account for the temperature mediated body size variation (see Atkinson and Sibly 1997 or Blanckenhorn 2000b for summaries of the arguments). In agreement with the latter hypothesis, we found decreases in egg but not clutch size (Blanckenhorn 2000b; Fig. 7), as well as in wing cell and ommatidia size, with increasing temperature (Blanckenhorn and Llaurens 2005; Fig. 8). In contrast, sperm length increases with temperature, contrary to the temperature-size rule (Blanckenhorn and Hellriegel 2002; Fig. 7). Consequently, both

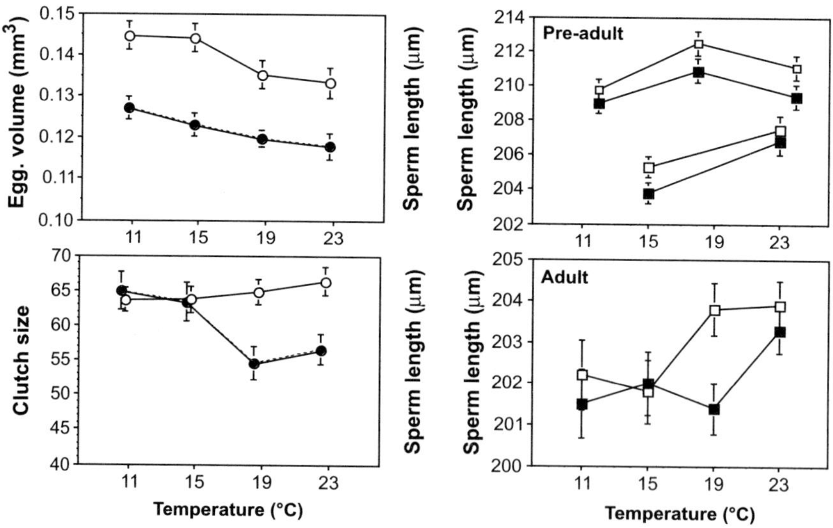

Fig. 7 Egg size (top left panel) of yellow dung flies decreases with holding temperature, thus following the temperature-size rule, at good (open symbols) and poor adult feeding (filled symbols) conditions, while clutch size (bottom left panel) does not change (data from Blanckenhorn 2000b). In contrast, sperm length increases as juvenile rearing (top right panel; data from two separate experiments) and adult holding temperature (bottom right panel) increases (data from Blanckenhorn and Hellriegel 2002).

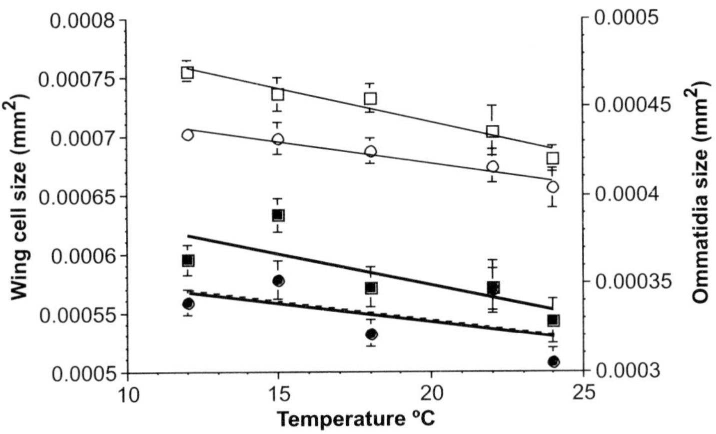

Fig. 8 Wing cell (open symbols and left scale) and ommatidia size (closed symbols and right scale) of yellow dung flies decrease with rearing temperature, following the temperature-size rule (males: squares and unbroken regression lines, females: circles and broken regression lines; Blanckenhorn and Llaurens 2005).

Bergmann (overall body size, egg and cell size) and converse Bergmann effects (sperm length and development time; discussed below; see Fig. 4) are evident in the yellow dung fly.

Genetic effects. The body size decrease in response to larval food limitation depicted in Fig. 6 (left panel) is an environmentally induced effect that represents V_e. However, because families were split among the three food environments tested (Blanckenhorn 1998a), the experiment also revealed genotype-by-environment interactions ($V_{g\times e}$), as sibships (i.e., genotypes) varied significantly in growth, development and body size in response to food limitation (Fig. 9; Blanckenhorn 1998a). There is of course also heritable variation (V_g). Broad sense heritabilities, based on full sibs and including dominance and common environment variance as well as maternal effects (Falconer 1989, Roff 1997), of body size and development time are high, often close to $h^2 = V_g / V_p = 1$ (Mühlhäuser et al. 1996, Blanckenhorn 2002). Narrow sense heritabilities, based on parent-offspring regression or half sibs and including only additive genetic variance, are considerably lower, ranging between $h^2 = 0.2–0.4$ (Simmons and Ward 1991, Blanckenhorn 2002). Realized heritabilites, based on artificial directional selection in the laboratory, are in the same range (Teuschl et al. 2007). Figure 6 (right panel) clearly shows the different body sizes and development times of various laboratory selection lines (Teuschl et al. 2007). Whatever the precise estimate, there is genetic variation in body size, growth rate and

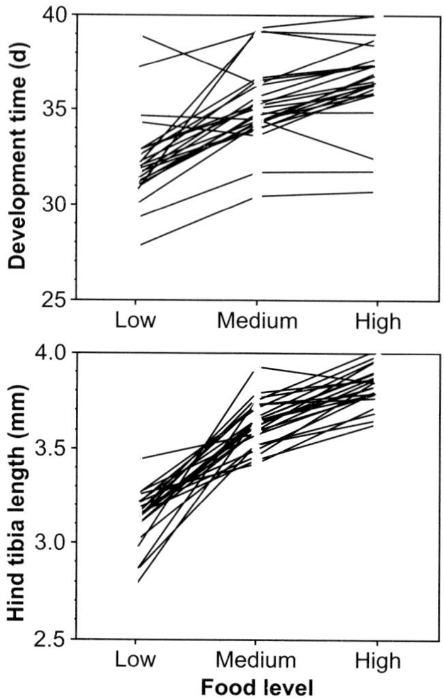

Fig. 9 Reaction norms (cf. Fig. 1) showing heritable phenotypic plasticity ($V_{g\times e}$, as indicated by non-parallel lines) of development time and body size for yellow dung fly full sib families reared in the field at three different larval food levels (males only; data from Blanckenhorn 1998a). The mean data for siblings of each family are linked by lines.

development time in the yellow dung fly, which was still present after over 20 generations of artificial selection on body size (Teuschl et al. 2007), but the largest fraction of the phenotypic variance represents heritable ($V_{g\times e}$) or environmentally induced (V_e) plasticity. This may sound extreme, but is not untypical: in their review of numerous species, Mousseau and Roff (1987) report average heritabilities of $h^2 = 0.3–0.4$ for morphologcial, life history and behavioral traits, implying that 60–70% of all variation is plasticity.

Other environmental effects. Photoperiod, humidity and predators. In addition to genetic and the prominent effects of food and temperature, other environmental variables can also affect growth, development and body size. Photoperiod is typically instrumental in triggering diapause in many insects (Masaki 1980, Tauber et al. 1986, Danks 1987). In yellow dung flies, however, temperature is at least as important as photoperiod (if not more so) in inducing and breaking winter diapause (Blanckenhorn 1998b; see below). Photoperiod *per se* has not yet been demonstrated to produce plastic responses of body size independent of temperature (Blanckenhorn 1998b and unpublished data).

An experiment specifically testing for effects of dung drying (or humidity) independent of temperature, was conducted by A. Heyland and myself (unpublished data). We expected that the drying of the dung would elicit earlier pupation at smaller size, analogous to dung limitation, because larvae can no longer feed on dry dung. While time to pupation (i.e., development time) was indeed prolonged by preventing dung drying, growth rate however significantly decreased, resulting in smaller body sizes, contrary to prediction. Our interpretation of this result is that preventing dung drying (at any temperature) leads to maladaptive plastic responses of yellow dung fly larvae because it is unnatural. Probably the larvae prefer a dry place to pupate, which they searched for in the humidity treatment but could not find in the end, thus expending considerable energy and time, ultimately reducing final body size.

As discussed in the introduction, predators typically have marked effects on prey mortality and feeding behavior, thus ultimately affecting their growth, development and body size. This has been clearly demonstrated in a number of species with aquatic juveniles such as fish, tadpoles, dragonflies and mayflies (e.g., Fraser and Gilliam 1992, Werner and Anholt 1993, Martin et al. 1991, Scrimgeour and Culp 1994, Dahl and Peckarsky 2003). Unfortunately, dung is a liquid medium too opaque to allow easy experimentation with larval predators, which definitely occur in the dung in nature (e.g. beetle larvae; Hammer 1941). So far, therefore, all our larval studies investigating growth, development and body size disregard predation, although we clearly expect some effects.

With this background knowledge, we can now attempt to understand the daily, seasonal, altitudinal and latitudinal variation in yellow dung fly body size depicted in Figs. 4 and 5.

Seasonal and daily variation. There is always considerable unsystematic body size variation in any dung fly population. This depends primarily on how many larvae, not only of this but several other species (Hammer 1941), compete for the dung in a given pat (the resource patch). Dung pats also vary tremendously in size. The degree of intra- and inter-specific competition, and consequently food availability, therefore varies unpredictably in time and space. However, there are some consistent patterns. Over both the spring and the fall seasons (cf. above), body size varies in a concave quadratic pattern, flies being largest in the middle of the season (Jann et al. 2000; Fig. 5). As larval food availability and competition do not vary systematically over the season, this can largely be explained by systematic temperature changes in combination with 'end-of-season' time constraints. In early spring, the overwintered flies of the previous fall generation emerge and appear at the mating site, smaller flies first, as these require less time to complete development (Blanckenhorn 1997b, 1998a) and presumably need to feed less before they can reproduce (Blanckenhorn and Viele 1999). Their offspring emerges successively smaller as spring progresses, due to warmer average temperatures and according to the temperature-size rule (Atkinson 1994, Angilletta and Dunham 2003). For the same reason, flies in early fall are small but their offspring emerge larger. Towards the end of the fall season, larvae face a time constraint and thus have to abbreviate their development to reach the pupal stage before the first frost (to avoid death). This implies some reduction in body size, although yellow dung flies can increase growth rates towards winter to attain large size despite shorter development periods (Blanckenhorn 1997b, 1998a). The latter can only be detected if development time is expressed in physiological time (i.e., degree days: Blanckenhorn 1999), as development in real time takes longer as it gets colder (Fig. 10). An analogous 'end-of-season' time constraint due to heat may operate in early summer, as yellow dung flies enter some sort of summer quiescence (cf. above). However, warm temperatures facilitate fast development anyway (Taylor 1981, Atkinson and Sibly 1997), so the increase in growth rate towards the end of the spring season is actually of little advantage, and expressing the data in degree days does not change the overall picture (Fig. 10). Thus, faster growth and/or shorter development (resulting in smaller size) before winter in order to meet a time constraint is an adaptive response (because a trade-off is involved: Blanckenhorn 1997b, 1998a), whereas faster growth and development and smaller sizes due to warmer temperatures towards the end of spring are not necessarily adaptive.

Yellow dung flies also show an interesting concave (quadratic) body size pattern on the pasture over the day, with flies being largest in the middle of the day (Fig. 5, right panel; Jann 1997). This pattern most probably reflects size-dependent behavior but is yet unexplained and remains to be explored. Smaller males are inferior competitors (Borgia 1980, 1981, 1982, Jann et al. 2000), and thus may choose to search for mates exclusively during marginal times of the day (early morning and late evening), while larger flies may populate the pasture primarily or exclusively during the middle of the day. However, there is no strong evidence yet that more females arrive at the dung to lay eggs during the middle of the day; in general female arrivals are very evenly dispersed (Parker 1970c, Reuter et al. 1998). A second hypothesis

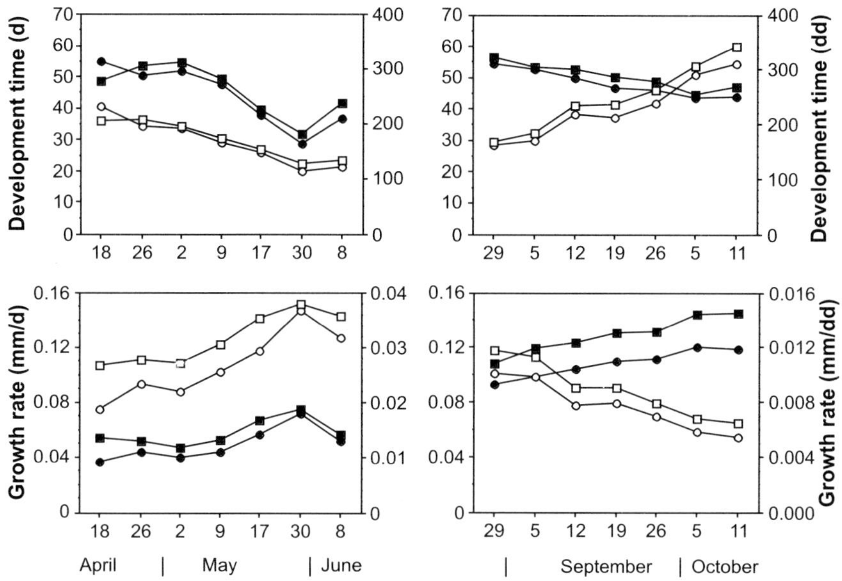

Fig. 10 Development time (top panels) and growth rate (bottom panels) as a function of laying date as yellow dung fly larvae approach the end of the spring (left panels) and fall seasons (right panels). In fall, development time increases and growth rate decreases towards winter in real time (left scale of all panels and open symbols), whereas the opposite becomes apparent when expressed in physiological time (i.e., degree days above 2°C: right scale of all panels and closed symbols). In spring there is no such effect because temperatures increase towards summer (males: squares, females: circles; data from Blanckenhorn 1997b, 1999, Blanckenhorn et al. 2001).

potentially explaining the daily body size pattern in Fig. 5 is that large flies of both sexes may better endure the warmer temperatures during the middle of the day than smaller flies. This consequently might limit the latter's activity during the hot parts of the day, although the converse is equally plausible (see Carroll and Quiring 1993, Dangerfield and Chipfunde 1995, Blanckenhorn 2000a). It must be noted that similar behavioral responses may of course also contribute to the seasonal body size pattern discussed in the previous paragraph (Fig. 5, left panel).

Latitudinal and altitudinal variation. Moving up the scale, yellow dung flies show an albeit slight body size increase with latitude, i.e., a Bergmann size cline (see also Blanckenhorn and Demont 2004 for another data set; Fig. 4). This cline is genetic, as it is expressed under common garden conditions when flies from various populations were reared in the laboratory. However, the corresponding development time data show a decrease with latitude, which is more consistent with the idea that at higher latitudes season length becomes limiting and populations consequently evolve faster growth rates. This at the same time demonstrates a converse Bergmann cline for development time (Roff 1980, Mousseau 1997) and countergradient growth rate variation (Conover and Present 1990; discussed above). We thus obtained a mixed pattern, supporting the hypothesis that Bergmann (due to temperature *per se*) and converse Bergmann clines (due to season length) are not necessarily mutually exclusive and may operate in conjunction, producing intermediate patterns (Chown and Gaston 1999; Blanckenhorn and Demont 2004; cf. above). While shorter development time at higher latitudes is clearly adaptive, as theory (e.g., Roff 1980, Rowe and Ludwig 1991) and other empirical studies demonstrate (e.g., water striders *Aquarius remigis*: Blanckenhorn 1991, 1994; Blanckenhorn and Fairbairn 1995), none of the Bergmann or temperature-size effects (Figs. 4, 7-8), nor the sperm length increase with temperature (Fig. 7), could thus far be demonstrated to be adaptive in yellow dung flies, despite some specific tests (e.g., Blanckenhorn 2000b).

Yellow dung flies also show analogous heritable body size variation with regard to altitude (Blanckenhorn 1997a; Fig. 4). High altitude populations are significantly smaller, but there is no difference in development time. This altitudinal body size cline is opposite to the larger scale latitudinal cline and thus consistent with a converse Bergmann cline due to seasonal constraints. However, while statistically significant, this effect is minute and thus probably of limited biological significance (Fig. 4). Altitudinal differentiation of course cannot be expected to be large: although high altitude dung fly populations are local, as indicated by emergence of the flies before the cows arrive at the high altitude pastures, their good flight capability (Reim 2005) necessarily generates considerable dispersal and hence gene flow counteracting local adaptation of populations over such short distances of only few kilometers (Kraushaar et al. 2002).

Phenotypic plasticity in other traits: diapause induction and energy reserves

As mentioned before, yellow dung flies diapause as pupae over winter (Blanckenhorn 1997a,b, 1998a,b) and are reproductively quiescent as adults during the hot summer (Blanckenhorn et al. 2001). In both situations, the flies show phenotypic plasticity with regard to the environmental cues triggering the response.

Analogous to the other life history traits (Blanckenhorn 1997a; Fig. 4), high and low altitude populations showed statistically significant genetic differentiation in diapause response in simulated laboratory environments

(Blanckenhorn 1998b), but these were slight and far from the differences predicted by their grossly different field phenologies: one spring and one fall season with a two month summer break in the Swiss lowlands vs. one summer season in the highlands (Blanckenhorn 1997a). Yellow dung flies thus enter winter diapause flexibly in response to temperatures close to freezing rather than photoperiod (Blanckenhorn 1998b). It is equally easy to break diapause: diapausing pupae brought from the field into the laboratory any time during winter will soon emerge (Blanckenhorn 1997b). Contrary to many insect species therefore (Tauber et al. 1986, Danks 1987, e.g., Bradshaw 1976, Bradshaw and Lounibos 1977), winter diapause is not strongly genetically fixed but highly phenotypically plastic, at least in our Swiss populations and probably other European populations as well. However, at the same time there is a clear and predictable (and hence adaptive) genetic latitidinal cline in that northern European populations are more likely to enter diapause and have longer diapause durations in a given photoperiod/temperature common garden laboratory environment (Blanckenhorn & Demont 2008). Crosses between Icelandic and Swiss populations further revealed dominant genetic factors in the control of diapause but few genetic maternal effects (Blanckenhorn & Demont 2008), although clearly a cool holding environment of the mother increases the diapause likelihood of her offspring, presumably mediated via as yet unspecified physiological compounds in the egg (W. U. B., unpublished data). Moreover, there are interesting interactive effects between larval food availability and temperature on winter diapause decisions (Blanckenhorn 1998a): food limited larvae destined to be small, as well as the smaller females, are more likely to enter diapause at a given time before winter. This is adaptive, as the smaller flies trade off higher pre-winter mortality against lower pupal mortality during winter (Blanckenhorn 1998a).

Phenotypic plasticity in relation to summer quiescence is indicated by temperature-dependent accumulation of physiological reserves, both in field flies at the end of the spring season and in the laboratory. At high temperatures above 24°C, mortality is generally increased (Ward and Simmons 1990; Blanckenhorn 1998a). In response, adult individuals accumulate lipids, and some, but not all, females but few males suppress reproduction (Blanckenhorn et al. 2001, Blanckenhorn and Henseler 2005). As no aestivation of pupae could be demonstrated, this presumably increases the survival of the flies during the hot summer. This suggests a highly flexible life history strategy, perhaps indicative of incipient evolution of summer diapause in this cold adapted fly.

Behavioral and Fitness Consequences of Phenotypic Plasticity in Body Size

Given the extensive plasticity in body size and associated traits, we can now ask what consequences this has for the bearer? Is it adaptive in certain situations or environments to be small, or are small flies merely making the 'best of a bad situation'? We already learned that emerging small in converse Bergmann end-of–season (Fig. 4), end-of-food or end-of-dung pat situations is adaptive because it is better to be small than dead (Amano 1983, Sigurjónsdóttir 1984, Blanckenhorn 1998a; Figs. 6, 9 and 10). In contrast, it is unclear (i.e., not necessarily adaptive) why it should generally be better to be small when it is warm (Bergmann's and the temperature-size rule: Atkinson and Sibly 1997).

It is well known that in many if not most species, large individuals on average have a fitness advantage (Andersson 1994, Blanckenhorn 2000a). This is also true for yellow dung flies. Clutch size of females increases with body size in the field, and so does the probability of mating for males (Jann et al. 2000; Fig. 11). Note that both fitness relationships are convex and increasing. This means that individuals of both sexes have accelerating fitness advantages as they get larger, and/or fitness advantages asymptote at very small sizes (these two interpretations cannot be discriminated without further evidence).

Size dependence of the third major component of individual fitness, larval or adult viability (or survival), is difficult to estimate in the field in yellow dung flies, and the evidence we have is equivocal. As it takes time to get large, a positive (genetic) correlation between body size and development time is often assumed in life history models (e.g., Roff 1980, 1992, 1997, Stearns 1992). We hence expected that large genotypes would die at higher rates when larval food (dung) is limited (cf. Fig. 6), because they require longer development and thus cannot terminate development before the food dries up. This was not generally the case (Blanckenhorn 1998a). However, when using flies selected for large size in the laboratory for up to 16 generations, we found such an effect, but only at very extreme environmental conditions (Teuschl et al. 2007). Growth plasticity in this species is thus very effective at avoiding death. Adult mortality in the field is even more difficult to estimate reliably, as yellow dung flies are small, fly well and disperse profusely. Direct mark-recapture estimates were unsuccessful, and indirect estimates via age-grading through wing injuries are imprecise and problematic in this species (Burkhard et al. 2002). The evidence we have does not strongly show longevity advantages of small flies; the data more likely suggest the converse (Burkhard et al. 2002).

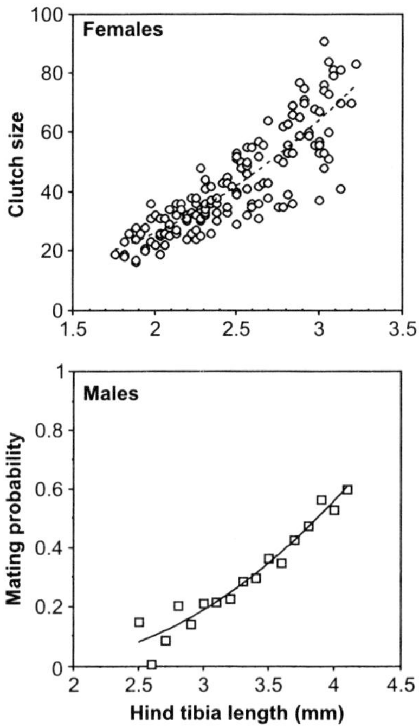

Fig. 11 Clutch size of female (top) and mating probability of male (bottom) yellow dung flies as a function of body size (hind tibia length) in the field (unpublished data from Jann et al. 2000).

However, the field estimates of Burkhard et al. (2002) and also those of Jann et al. (2000) depicted in Fig. 11 refer to the overall (i.e., average) field situation. It may be that under particular, stressful conditions, small flies can compensate their handicap and show relatively high fitness or, conversely, large flies have low fitness. This would indicate some, cryptic disadvantages of large body size, albeit only under restricted environmental conditions, but which are necessary to halt the evolution of ever larger individuals in any given species (Blanckenhorn 2000a). We have only very little evidence that this is the case, even after intensive search and experimentation. An obvious indication for this would be if at stressful conditions, say food limitation, small individuals would show the same or even higher fitness as large individuals. Again, this is typically not the case. Clutch size, egg size (Blanckenhorn and Heyland, 2005) and testis size (Hellriegel and Blanckenhorn 2002, but see Stockley and Seal 2001) are all smaller at limited food (Fig. 12). Interestingly, sperm are not (Fig. 12), but we do not have any strong evidence yet in this species, or many other species (Hosken 2003), that longer sperm are better and more likely to fertilize eggs (Hellriegel and Blanckenhorn 2002). We merely know that in yellow dung

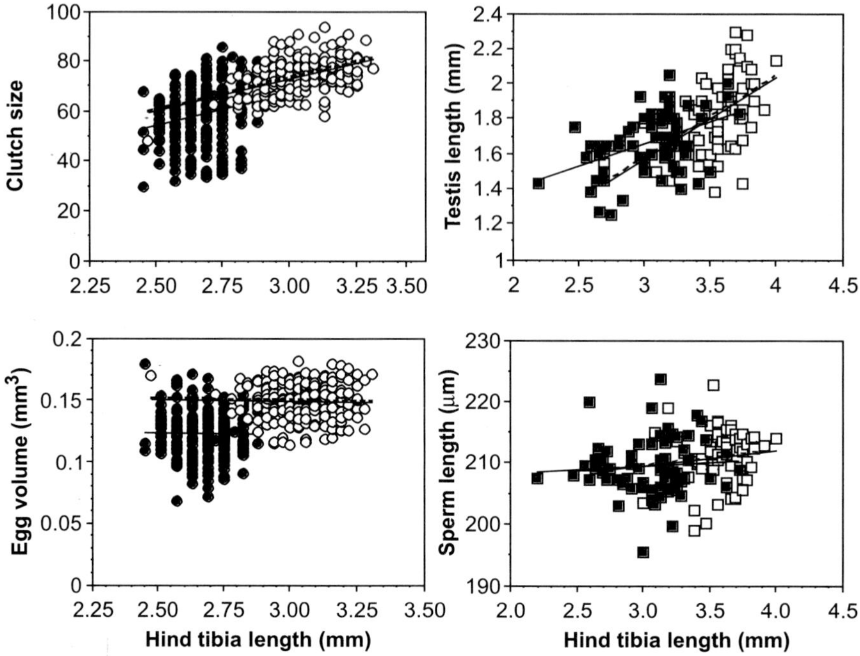

Fig. 12 Clutch size (top left panel) and egg size (bottom left panel) of yellow dung flies in relation to hind tibia length at unlimited (open symbols and broken regression lines) and limited (closed symbols and solid regression lines) adult food (data from Blanckenhorn 2000b), and testis length (top right panel) and sperm length (bottom right panel) in relation to hind tibia length at unlimited and limited larval food (data from Hellriegel and Blanckenhorn 2002). Egg volume and sperm length do not depend on body size, but clutch size and testis length do. At limited food all fitness components except sperm length are reduced.

flies sperm length varies strongly between males and is highly heritable (Ward and Hauschteck-Jungen 1993, Ward 2000b), to the extent that the length of sperm stored by females has been used to identify the father (e.g., Otronen et al. 1997, Hellriegel and Bernasconi 2000).

Another, more subtle indication of cryptic advantages of small individuals would be if the allometry between a fitness component and size changes in such a way that the advantage of large individuals disappears at stressful conditions. This appears to be the case for testis length, which is typically highly correlated with sperm numbers (Parker 1970e,f, Simmons 2001): the allometric relationship between testis length and hind tibia length in yellow dung fly males is steeper at abundant larval food (resulting in large flies) than at restricted larval food (compare the right with the left half of Fig. 12, top right panel; see also Stockley and Seal 2001). That is, small males have relatively large testes, presumably to stay competitive in sperm competition, for which the ability to transfer many sperm is crucial (Parker 1970e,f). However, while this is a plausible hypothesis, this cannot simply be inferred from a change in allometry (as depicted in Fig. 12) without further direct experimental evidence (cf. Fig. 11).

Figure 12 shows no change in the allometry between number of eggs (clutch size) and body size of yellow dung fly females in one of our laboratory experiments (Blanckenhorn 2000b, see also Blanckenhorn and Heyland 2005). However, other laboratory experiments have detected partial, or even full, fitness (component) compensation of small individuals in limited food environments. Jann and Ward (1999) showed that when prey was highly limited, the fecundity of small females was no longer less than that of large females. Reim et al. (2006) corroborated this effect using flies selected for large and small size (cf. Fig. 6), and further showed that this occurs because small females need less energy (primarily lipids, which are derived from prey) for maintenance and hence can invest relatively more into reproduction. For the same reason, small females and males are also able to start reproducing sooner at limited adult food (Reim et al. 2006), which may give them a fitness advantage in the field when populations are increasing, but there is no field evidence for this. Furthermore, energy reserve accumulation is heritable (Blanckenhorn and Hosken 2003) and also increases the chances of males to obtain a mate in the field (Otronen 1995a, Blanckenhorn et al. 2003a). In conclusion, there is some evidence for (cryptic) advantages of small body size at stressful environmental conditions, grounded largely in the fact that large individuals need higher absolute amounts of energy, rendering them less efficient or even not viable when resources are too limited (see Blanckenhorn et al. 1995; Reim et al. 2006).

There are other, more subtle behavioral consequences of body size to be discussed in this context. One example of compensatory behavior of small yellow dung flies concerning reproduction (and hence fitness) relates to sperm competition. Small males copulate for longer, as has been shown multiply in this well studied model species (Parker 1970e, 1992, Parker and Simmons 1991, 1994, 2000, Ward and Simmons 1991, Simmons and Parker 1992, Parker et al. 1993, 1999, Nuyts 1994). This is adaptive and predicted by theory (i.e., the marginal value theorem: Parker 1970e,f; summarized in Simmons 2001), because (1) smaller males have a lower probability of obtaining another mate (as they are competitively inferior; see above), and (2) small males presumably transfer fewer sperm per unit time (Simmons and Parker 1992, but see Hellriegel and Ward 1998). Small males consequently adjust copulation duration in such a way as to (presumably) equalizing the sperm investment of large males. Figure 13 shows an example of this effect from our own work (Blanckenhorn et al. 2003b).

Not least because males are considerably larger on average, female yellow dung flies have few behavioral means of choosing or rejecting mating partners (Ding and Blanckenhorn 2002). The degree of sperm competition is therefore high because females (have to) mate at least once every time they come to the dung to lay eggs. In males, this selects for high investment in sperm numbers and hence large testes (Hosken and Ward 2001, Simmons 2001). In females, this selects for internal morphological and physiological adaptations to exert cryptic female choice of sperm of particular males stored in different sperm storage compartments (so-called spermathecae) within the female reproductive tract (Ward 1993, 1998, 2000a, Hellriegel and Ward 1998, Hellriegel and Bernasconi 2000, Hosken and Ward 2000, 2001, Bernasconi et al. 2002, Hosken et al. 1999, 2001, 2002a,b), but this is controversial (Simmons et al. 1996, Stockley and Simmons 1998, Simmons 2001). Such cryptic choice benefits females because larvae of different phosphoglucomutase (PGM) genotypes grow better at certain temperatures, so females can match paternity with the conditions the larvae grow up in (Ward 1998, 2000a, Ward et al. 2002). Sperm storage patterns depend on the interaction of male and female body size: large size, implying large ducts, presumably not only allow males to pump in more sperm faster (Simmons and Parker 1992), but also allow females to better manipulate how much sperm gets stored in which spermatheca (Ward 1998, 2000a, Hellriegel and Ward 1998, Hosken et al. 1999, Hosken and Ward 2000, Hellriegel and Bernasconi 2000, Bernasconi et al. 2002).

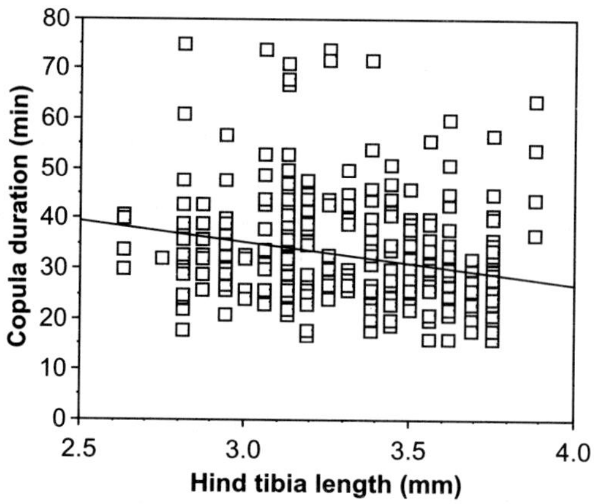

Fig. 13 Copula duration of yellow dung fly males as a function of body size (data from Blanckenhorn et al. 2003b).

A final example of compensatory behavior of small yellow dung flies relates not to reproduction but to foraging (and hence indirectly also to fitness). Laboratory studies indicate that although large yellow dung fly males (and females) extract nutrients more quickly, they do not catch prey faster (Blanckenhorn and Viele 1999). On the other hand, small individuals satiate more quickly and thus need to catch fewer prey items overall (Blanckenhorn and Viele 1999). Small males therefore can increase mating effort by minimizing foraging effort, thus enjoying a time budget advantage (Blanckenhorn et al. 1995). However, the degree to which prey hunting times in the laboratory reflect the natural situation is unclear.

Conclusions

I have shown that the yellow dung fly exhibits extensive phenotypic plasticity in body size, growth, development and other life history traits. A variety of common environmental factors cause this plasticity, most notably food availability, temperature, season length, photoperiod and humidity. I focused on the question whether the plasticity exhibited is adaptive or not. As outlined in the introduction, an adaptive response requires demonstration of both fitness trade-offs (i.e., costs and benefits) and genetic variation in the trait (V_g) and/or in the plasticity of the trait ($V_{g \times e}$). According to these criteria, the growth plasticity of yellow dung flies in response to food availability and season length can be considered adaptive (Figs. 4, left panel, 6, 9 and 10), whereas that in response to temperature and humidity cannot (Figs. 4, right panel, 7 and 8). I then discussed the fitness consequences of phenotypic plasticity in body size as mediated by behavior and physiology. In particular, I focused on the question whether small size in general confers fitness decrements, or whether there are environmental situations in which small individuals can behaviorally, physiologically or otherwise compensate for their handicap. Under most circumstances the former is the case, so small individuals are merely making the 'best of a bad situation'. However, the latter occurs too in particular stressful environments, largely due to the fact that large individuals need higher absolute amounts of energy, giving small males a time and energy budget advantage (see Blanckenhorn et al. 1995).

Do yellow dung flies exhibit particularly strong phenotypic plasticity in body size? Probably not. In their review of the literature, Mousseau and Roff (1987) found that for morphological traits genetic variation (V_g) averages about 30% of all phenotypic variation for a variety of species, the other 70% thus representing phenotypic plasticity (V_e and $V_{g \times e}$). This agrees roughly with what is found for yellow dung fly body size and development time (Ward and Simmons 1991, Blanckenhorn 2002). So do yellow dung flies exhibit particularly strong **adaptive** phenotypic plasticity in body size ($V_{g \times e}$)? Perhaps, although comparative data allowing quantitative comparisons with other species are not readily available. First, all our studies show relatively low genetic differentiation of local populations, and consequently high degrees of plasticity in body size and life history traits (Fig. 4; Blanckenhorn 1997a, Kraushaar and Blanckenhorn 2002, unpublished data). This agrees well with population genetic data (allozymes and microsatellites) equally showing very low genetic differentiation (Kraushaar et al. 2002), even when comparing European populations from Switzerland to Iceland (Demont 2004), and suggests extensive gene flow hampering local adaptation. In such situations, evolution of extensive phenotypic plasticity is the best strategy to cope with unpredictably changing environments. Second, as argued above, a large part of this plasticity is adaptive. The response of smaller but sooner pupation (and adult emergence) that yellow dung fly larvae show in situations of dung limitation / dung drying / high larval competition, is rare and clearly adaptive, as the flies eat up their habitat patch. This response is paralleled by those of treehole mosquitoes (Juliano and Stoffregen 1994) and desert frogs (Newman 1988a,b, 1992), which are also regularly threatened by habitat drying, and seed beetles (Møller et al. 1989; Fox 1998). Most ectotherms emerge smaller but later (Blanckenhorn 1999), and it is unclear whether this response is adaptive even though it is predicted by many life history models (e.g., Stearns and Koella 1986). So it appears that the peculiar ephemeral and unpredictable ecological situation selects for this particular type of adaptive growth plasticity. This implies that perhaps other dung inhabiting insects would show similar responses. However, this is not necessarily so. For example, the much smaller black scavenger fly *Sepsis cynipsea*, which also develops in fresh cow dung, shows an intermediate response, emerging smaller after the same development time (Blanckenhorn 1999). Unfortunately, there is little life history information on other species of the dung community (but see Moczek, this volume). Third, a major prerequisite for the evolution of adaptive phenotypic plasticity is the variable and unpredictable nature of the dung fly habitat combined with a predictable relationship between fitness and the environment inducing the particular plastic phenotype (van Tienderen 1991, 1997, Newman 1992, West Eberhard 2002). Although we have no systematic quantitative data on this, it is obvious that the size and spatio-temporal distribution of dung pats are variable and unpredictable. Therefore, even though there is some degree of 'ideal-free' patch size matching (Blanckenhorn et al. 2000), female dung flies cannot possibly predict the number of intra- and inter-specific competitors their offspring will face in any particular dung pat she has chosen to lay her eggs into, nor can the drying regime (i.e., the weather) be predicted. Admittedly, however, these are post-hoc arguments rather than a test of the theory.

The best and most direct test of the predicted conditions selecting for the evolution of phenotypic plasticity (van Tienderen 1991, 1997, Newman 1992, West Eberhard 2002) would be use of experimental evolution in the laboratory, along the lines of what was recently exemplified with regard to the evolution of learning (Mery and Kawecki 2002). This is difficult and tedious, and I am not aware of such work in the context of the evolution of phenotypic plasticity. Alternatively, comparative data could provide indirect evidence. However, this requires data of similar nature for many species, in this case on growth and body size plasticity. While some such case studies exist for particular (insect) species (e.g., bruchid beetles, pitcher plant mosquitoes or Satyrid butterflies: Møller et al. 1989, Gotthard et al. 1994, Bradshaw and Johnson 1995, Nylin et al. 1996, Bradshaw et al. 1997, Fox 1998, see Nylin and Gotthard 1998 for a review), we certainly need more. Case studies of particular 'model' species are crucial because only they provide the detailed information about a variety of life history traits in a variety of environmental contexts necessary to test the theory using comparative approaches (e.g., using meta analysis). In order to be general, we need to investigate adaptive plasticity in a number of traits (morphological, life history, physiological or behavioral), and relate it, quantitatively and in an abstract way, to the features, i.e., the variability and unpredictability of the particular environmental circumstances presumably causing the plasticity. It is the latter kind of quantitative data that are often missing (as is admittedly the case for the yellow dung fly), even though they are not necessarily difficult to gather (e.g., weather data; see Blanckenhorn 1997a). I believe that what we want to know in the end to evaluate the generality of any particular biological phenomenon, here the evolution phenotypic plasticity, is not whether it exists in any particular species (and why), but how often it exists and how general and thus how important it is in nature.

Acknowledgements

I thank the many colleagues and students who contributed to the work reviewed here, most of which are mentioned in the text (if the work is unpublished) or appear as co-authors of my work. I also thank the funding agencies, all above the Swiss National Fund, the University of Zürich and the Zoological Museum Zürich, for providing me with the means to conduct this research over the years. Thanks also to Gabriele Uhl and Martin Schäfer, Bonn University, for reviewing the chapter. Finally, I thank Drs. Ananthakrishnan and Whitman for giving me the opportunity to review our work on yellow dung fly life history plasticity after over ten years of intensive study. It was about time.

References

Abrams, P. A., O. Leimar, S. Nylin, and C. Wiklund. 1996. The effect of flexible growth rates on optimal sizes and development times in a seasonal environment. American Naturalist, 147: 381–395.

Alcock J. 1997. Competition from large males and the alternative mating tactics of small males of Dawson's burrowing bee (*Amegilla dawsoni*). Journal of Insect Behavior, 10: 99–113.

Alpatov, W. W. 1929. Biometrical studies on variation and races of the honey bee, *Apis mellifera*. Quarterly Review of Biology, 4: 1–58.

Amano, K. 1983. Studies on the intraspecific competition in dung breeding flies. I. Effects of larval density on the yellow dung fly. Japanese Journal of Sanitary Zoology, 34: 165–175.

Andersson, M. 1994. Sexual selection. Princeton University Press, Princeton, NJ, USA.

Angilletta, M. J. and A. E. Dunham. 2003. The temperature-size rule in ectotherms: simple evolutionary explanations may not be general. American Naturalist, 162: 332–342.

Arendt, J. D. 1997. Adaptive intrinsic growth rates: and integration across taxa. Quarterly Review of Biology, 72: 149–177.

Arnett, A. E and N. J. Gotelli. 1999. Geographic variation in life-history traits of the ant lion, *Myrmeleon immaculatus*: Evolutionary implications of Bergmann's rule. Evolution, 53: 1180–1188.

Arnqvist, G. and F. Johansson. 1998. Ontogenetic reaction norms of predator-induced defensive morphology in dragonfly larvae. Ecology, 79: 1847–1858.

Ashton, K. G., M. C. Tracy, and A. de Queiroz. 2000. Is Bergmann's rule valid for mammals? American Naturalist, 156: 390–415.

Atkinson, D. 1994. Temperature and organism size—A biological law for ectotherms? Advances in Ecological Research, 25: 1–58.

Atkinson, D. and R. M. Sibly. 1997. Why are organisms usually bigger in colder environments? Making sense of a life history puzzle. Trends in Ecology and Evolution, 12: 235–239.

Austad, S. N. 1984. A classification of alternative reproductive behaviours and methods for field-testing ESS models. American Zoologist, 24: 309–319.

Ayres, M. P. and J. M. Scriber., 1991. Local adaptation to regional climates in *Papilio canadensis*. Ecological Monographs, 64: 465–482.

Belovsky, G. E., J. B. Slade, and J. M. Chase. 1996. Mating strategies based on foraging ability: an experiment with grasshoppers. Behavioral Ecology, 7: 438–444

Bergmann, C. 1847. Über die Verhältnisse der Wärmeökonomie der Thiere zu ihrer Grösse. Göttinger Studien, 1: 595–708.

Bernasconi, G., B. Hellriegel, A. Heyland, and P. I. Ward. 2002. Sperm survival in the female reproductive tract in the fly *Scathophaga stercoraria* (L.). Journal of Insect Physiology, 48: 197–203.

Berrigan, D. and E. L. Charnov. 1994. Reaction norms for age and size at maturity in response to temperature: a puzzle for life historians. Oikos, 70: 474–478.

Berthold, P. 1971. Physiologie des Vogelzuges, pp. 257–299. *In* E. Schüz [ed.] 1971. Grundriss des Vogelzuges. Parey, Hamburg, Germany.

Blackburn, T. M., K. J. Gaston, and N. Loder. 1999. Geographic gradients in body size: a clarification of Bergmann's rule. Diversity and Distributions, 5: 165–174.

Blanckenhorn, W. U. 1991. Life history differences in adjacent water strider populations: phenotypic plasticity or heritable responses to water temperature? Evolution, 45: 1520–1525.

Blanckenhorn, W. U. 1994. Fitness consequences of alternative life histories in water striders, *Aquarius remigis*. Oecologia, 97: 354–365.

Blanckenhorn, W. U. 1997a. Altitudinal life history variation in the dung flies *Scathophaga stercoraria* and *Sepsis cynipsea*. Oecologia, 109: 342–352.

Blanckenhorn, W. U. 1997b. Effects of rearing temperature on growth, development and diapause in the yellow dung fly—against all rules? Oecologia, 111: 318–324.

Blanckenhorn, W. U. 1998a. Adaptive phenotypic plasticity in growth rate and diapause in the yellow dung fly. Evolution, 52: 1394–1407.

Blanckenhorn, W. U. 1998b. Altitudinal differentiation in diapause response in two species of dung flies. Ecological Entomology, 23: 1–8.

Blanckenhorn, W. U. 1999. Different growth responses to food shortage and temperature in three insect species with similar life histories. Evolutionary Ecology, 13: 395–409.

Blanckenhorn, W. U. 2000a. The evolution of body size: what keeps organisms small? Quarterly Review of Biology, 75: 385–407.

Blanckenhorn, W. U. 2000b. Temperature effects on egg size and their fitness consequences in the yellow dung fly. Evolutionary Ecology, 14: 627–643.

Blanckenhorn, W. U. 2002. The consistency of heritability estimates in field and laboratory in the yellow dung fly. Genetica, 114: 171–182.

Blanckenhorn, W. U. and M. Demont. 2004. Bergmann and converse Bergmann latitudinal clines in arthropods: two ends of a continuum? Integrative and Comparative Biology 44: 413–424.

Blanckenhorn, W. U. and D. J. Fairbairn. 1995. Life history adaptation along a latitudinal cline in water striders. Journal of Evolutionary Biology, 8: 21–41.

Blanckenhorn, W.U., J. Frei, and M. Birrer. 2003b. The effect of female arrivals on mate monopolization in the yellow dung fly. Behavioural Ecology and Sociobiology 54: 65–70.

Blanckenhorn, W. U. and B. Hellriegel. 2002. Against Bergmann's rule: fly sperm size increases with temperature. Ecology Letters, 5: 7–10.

Blanckenhorn, W.U. and C. Henseler, 2005. Temperature-dependent ovariole and testis maturation in the yellow dung fly. Entomologia Experimentalis et Applicata 116: 159-165.

Blanckenhorn, W. U., C. Henseler, D. U. Burkhard, and H. Briegel. 2001. Summer decline in populations of the yellow dung fly: diapause or quiescence? Physiological Entomology, 26: 260–265.

Blanckenhorn, W. U. and A. Heyland. 2004. The quantitative genetics of two life history trade-offs in the yellow dung fly in abundant and limited food environments. Evolutionary Ecology 18: 385–402.

Blanckenhorn, W. U. and D. J. Hosken. 2003. The heritability of three condition measures in the yellow dung fly. Behavioral Ecology, 14: 612–618.

Blanckenhorn, W. U., U. Kraushaar, and C. Reim. 2003a. Sexual selection on morphological and physiological traits and fluctuating asymmetry in the yellow dung fly. Journal of Evolutionary Biology, 16: 903–913.

Blanckenhorn, W. U. and V. Llaurens. 2005. The effects of temperature on cell size and number in the yellow dung fly *Scathophaga stercoraria*. Journal of Thermal Biology, 30: 213-219.

Blanckenhorn, W. U., C. Morf, and M. Reuter. 2000. Are dung flies ideal-free distributed at their mating and oviposition sites? Behaviour, 137: 233–248.

Blanckenhorn, W. U., R. F. Preziosi, and D. J. Fairbairn. 1995. Time and energy constraints and the evolution of sexual size dimorphism—to eat or to mate? Evolutionary Ecology, 9: 369–381.

Blanckenhorn, W. U. and S. T. N. Viele. 1999. Foraging in yellow dung flies: testing for a small male time budget advantage. Ecological Entomology, 24: 1–6.

Borgia, G. 1980. Sexual competition in *Scathophaga stercoraria*: size- and density related changes in male ability to capture females. Behaviour, 75: 185–206.

Borgia, G. 1981. Mate selection in the fly *Scathophaga stercoraria*: female choice in a male-controlled system. Animal Behaviour, 29: 71–80.

Borgia, G. 1982. Experimental changes in resource structure and male density: size-related differences in mating success among male *Scathophaga stercoraria*. Evolution, 36: 307–315.

Bradshaw, W. E. 1976. Geography and photoperiodic response in diapausing mosquito. Nature, 262: 384–386.

Bradshaw, W. E., C. M. Holzapfel, C. A. Kleckner, and J. J. Hard. 1997. Heritability of development time and protandry in the pitcher-plant mosquito, *Wyeomyia smithii*. Ecology, 78: 969–976.

Bradshaw, W. E. and C. M. Holzapfel. 1996. Genetic constraints to life history evolution in the pitcher plant mosquito, *Wyeomyia smithii*. Evolution, 50: 1176–1181.

Bradshaw, W. E. and C. M. Holzapfel. 2001. Genetic shift in photoperiodic response correlated with global warming. Proceedings of the National Academy of Sciences USA, 98: 14509–14511.

Bradshaw W. E. and K. Johnson. 1995. Initiation of metamorphosis in the pitcher-plant mosquito-effects of larval growth history. Ecology, 76: 2055–2065.

Bradshaw, W. E. and L. P. Lounibos. 1977. Evolution of dormancy and its photoperiodic control in pitcher-plant mosquitos. Evolution, 31: 546–567.

Burkhard, D. U., P. I. Ward, and W. U. Blanckenhorn. 2002. Using age-grading by wing injuries to estimate size-dependent adult survivorship in the field: A case study of the yellow dung fly *Scathophaga stercoraria*. Ecological Entomology, 27: 514–520.

Cade, W. H. 1980. Alternative male reproductive behaviors. Florida Entomologist 63: 30–45.

Calder, W. A. 1984. Size, Function, and Life History. Cambridge: Harvard University Press, Cambridge MA, USA.

Caro, T. M. and P. P. G. Bateson. 1986. Organization and ontogeny of alternative tactics. Animal Behaviour, 34: 1483–1499.

Carroll, A. L. and D. T. Quiring. 1993. Interactions between size and temperature influence fecundity and longevity of a Tortricid moth *Zeiraphera canadensis*. Oecologia, 93: 233–241.

Chown, S. L. and K. J. Gaston. 1999. Exploring links between physiology and ecology at macro-scales: the role of respiratory metabolism in insects. Biological Reviews, 74: 87–120.

Chown, S. L. and C. J. Klok. 2003. Altitudinal body size clines: latitudinal effects associated with changing seasonality. Ecography, 26: 445–455.

Clark, A. B. and T. J. Ehlinger. 1987. Pattern and adaptation in individual behavioural differences, pp. 1–47. *In* P. P. G. Bateson and P. H. Klopfer [eds.] 1987. Perspectives in Ethology, Vol. 7. Plenum Press, New York, NY, USA.

Conover, D. O. and T. M. C. Present. 1990. Countergradient variation in growth rate: compensation for length of the growing season among Atlantic silversides from different latitudes. Oecologia, 83: 316–324.

Cordoba-Aguilar, A. 2002. Wing pigmentation in territorial male damselflies, *Calopteryx haemorrhoidalis*: a possible relation to sexual selection. Animal Behaviour, 63: 759–766.

Cotterell, G. S. 1920. The life history and habits of the yellow dung-fly (*Scathophaga stercoraria*): a possible blow-fly check. Proceedings of the Zoological Society London, 1920: 629–647.

D'Amico, L. J., G. Davidowitz, and H. F. Nijhout. 2001. The developmental and physiological basis of body size evolution in an insect. Proceedings of the Royal Society of London B, 268: 1589–1593.

Dahl, J. and B. L. Peckarsky. 2003. Developmental responses to predation risk in morphologically defended mayflies. Oecologia, 137: 188–194.

Daly, H. V., K. Hoelmer, and P. Gambino. 1991. Clinal geographic variation in feral honey bees in California, USA. Apidologie, 22: 591–609.

Dangerfield, J. M. and L. Chipfunde. 1995. Stress tolerance and burrowing behaviour in the southern African millipede *Alloporus unicatus*. Journal of Zoology London, 236: 17–27.

Danks, H. V. 1987. Insect dormancy: an ecological perspective. Biological Survey of Canada, Ottawa, Canada.

Davidowitz, G., L. J. D'Amico, and H. F. Nijhout. 2003. Critical weight in the development of insect body size. Evolution and Development 5: 188–197.

Davidowitz, G., L. J. D'Amico, and H. F. Nijhout. 2004. The effects of environmental variation on a mechanism that regulates insect body size. Evolutionary Ecology Research, 6: 49–62.

Dawkins, R. 1980. Good strategy or evolutionary stable strategy?, pp. 331–367. *In* G. W. Barlow and J. Silverberg [eds] 1980. Sociobiology: beyond nature/nurture? Westview, Boulder, CO, USA.

DeBelle, D. S. and M. B. Sokolowski. 1987. Heredity of rover/sitter: alternative foraging strategies of *Drosophila melanogaster* larvae. Heredity, 59: 73–83.

de Jong, G. 1995. Phenotypic plasticity as a product of selection in a variable environment. American Naturalist, 145: 493–512.

Demont, M. 2004. Latitudinal Clinal Differentiation of Yellow Dung Fly Life History Traits across Europe: A Comparison of Molecular and Quantitative Genetic Data. Diploma Thesis, University of Zürich.

Demont, M. and W.U. Blanckenhorn. 2008. Genetic differentiation in diapause response along a latitudinal cline in European yellow dung fly populations. Ecological Entomology 33: 197–201.

Ding, A. and W. U. Blanckenhorn. 2002. The effect of sexual size dimorphism on mating behavior in two dung flies with contrasting dimorphism. Evolutionary Ecology Research, 4: 1–15.

Dingle, H. 1978. Migration and diapause in tropical, temperate, and island milkweed bugs, pp. 254–275. *In* H. Dingle [ed.] 1978. Evolution of Insect Migration and Diapause. Springer, Heidelberg, Germany.

Dingle, H., N. R. Blakley, and E. R. Miller. 1980. Variation in body size and flight performance in milkweed bugs (*Oncopeltus*). Evolution, 34: 371–385.

Dominey, W. J. 1980. Female mimicry in the male bluegill sunfish—a genetic polymorphism? Nature, 284: 546–548.

Dominey, W. J. 1984. Alternative mating tactics and evolutionary stable strategies. American Zoologist, 24: 385–396.

Dunn, D. W., C. S. Crean, C. L. Wilson, and A. S. Gilburn. 1999. Male choice, willingness to mate and body size in seaweed flies (Diptera: Coelopidae). Animal Behaviour, 57: 847–853.

Eberhard, W. G. 1982. Beetle horn dimorphism: making the best of a bad lot. American Naturalist, 119: 420–426.

Elmes, G. W., J. C. Wardlaw, M. G. Nielsen, V. E. Kipyakov, E. B. Lopatina, A. G. Radchenko, and B. Barr. 1999. Site latitude influences on respiration rate, fat content and the ability of worker ants to rear larvae: a comparison of *Myrmica rubra* (Hymenoptera: Formicidae) populations over their European range. European Journal of Entomology, 96: 117–124.

Emlen, S. T. and L. W. Oring. 1977. Ecology, sexual selection, and the evolution of mating systems. Science, 197: 215–223.

Evans, H. E. and K. M. O'Neill. 1978. Alternative mating strategies in digger wasp *Philanthus zebratus*. Proceedings of the National Academy of Sciences USA, 75: 1901–1903.

Fagen, R. 1982. Evolutionary issues in development of behavioural flexibility, pp. 365–383. *In* P. P. G. Bateson and P. H. Klopfer [eds] 1982. Perspectives in Ethology, Vol. 5. Plenum Press, New York, NY, USA.

Failes, E. S., J. W. Whistlecraft, and A. D. Tomlin. 1992. Predatory behaviour of *Scathophaga stercoraria* under laboratory conditions. Entomophaga, 37: 205–231.

Fairbairn, D. J. 1992. The origins of allometry: size and shape polymorphism in the common waterstrider, *Gerris remigis* Say (Heteroptera, Gerridae). Biological Journal of the Linnean Society, 45: 167–186.

Fairbairn, D. J. and L. Desranleau. 1987. Flight threshold, wing muscle histolysis, and alary polymorphism: correlated traits for dispersal tendency in the Gerridae. Ecological Entomology, 12: 13–24.

Falconer, D. S. 1989. Introduction to quantitative genetics, 3rd edn. Longman Scientific and Technical, Harlow, UK.

Fischer, K. and K. Fiedler. 2002. Reaction norms for age and size at maturity in response to temperature: a test of the compound interest hypothesis. Evolutionary Ecology, 16: 333–349.

Foster, W. 1967. Hormone-mediated nutritional control of sexual behavior in male dung flies. Science, 158: 1596–1597.

Fox, C. W. 1998. Genetic and maternal influences on body size and development time in the seed beetle Stator limbatus (Coleoptera: Bruchidae). Annals of the Entomological Society of America 91: 128-134.

Fox, C. W., J. D. Martin, M. S. Thakar, and T. A. Mousseau. 1996. Clutch size manipulations in two seed beetles: consequences for progeny fitness. Oecologia, 108: 88–94.

Franz, H. 1979. Oekologie der Hochgebirge. Ulmer Verlag, Stuttgart, Germany.

Fraser, D. F. and J. F. Gilliam. 1992. Nonlethal impacts of predator invasion: facultative suppression of growth and reproduction. Ecology, 73: 959–970.

Gadgil, M. 1972. Male dimorphism as a consequence of sexual selection. American Naturalist, 106: 574–584.

Gibbons, D. S. 1980a. Prey consumption, mating, and egg production in *Scathophaga* species (Dipt., Scathophagidae) in the laboratory. Entomologist's Monthly Magazine, 116: 25–32.

Gibbons, D. S. 1980b. The feeding habits of *Scathophaga stercoraria* (Dipt., Scathophagidae) on cow pats. Entomologist's Monthly Magazine, 116: 171.

Gibbons, D. S. 1987. The causes of seasonal changes in numbers of the yellow dung fly *Scathophaga stercoraria*. Ecological Entomology, 12: 173–185.

Gorodkov, K. B. 1984. Scathophagidae. pp. 11–41. *In* A .Soós and L. Papp [eds] 1984. Catalogue of Palaearctic Diptera. Elsevier, Amsterdam, The Netherlands.

Gotthard, K. 1998. Life history plasticity in the satyrine butterfly *Lasiommata petropolitana*: investigating an adaptive reaction norm. Journal of Evolutionary Biology, 11: 21–39.

Gotthard, K. and S. Nylin. 1995. Adaptive plasticity and plasticity as an adaptation: a selective review of plasticity in animal morphology and life history. Oikos, 74: 3–17.

Gotthard, K., S. Nylin, and C. Wiklund. 1994. Adaptive variation in growth rate: life history costs and consequences in the speckled butterfly. Oecologia, 99: 281–289.

Gross, M. 1985. Disruptive selection for alternative life histories in salmon. Nature, 313: 47–49.

Gross, M. and E. L. Charnov. 1980. Alternative male life histories in bluegill sunfish. Proceedings of the National Academy of Sciences USA, 77: 6937–6940.

Hammer, O. 1941. Biological and ecological investigations on flies associated with pasturing cattle and their excrement. Videnskabelige Meddelelser fra Dansk Naturhistorisk Forening, 105: 140–393.

Hanski, I. 1980a. Patterns of beetle successions in droppings. Annales Zoologicae Fennici, 17: 17–26.

Hanski, I. 1980b. The community of coprophagous beetles (Coleoptera, Scarabeidae and Hydrophilidae) in Northern Europe. Annales Entomologicae Fennici, 46: 57–74.

Hanski, I. and Y. Cambefort. 1991. Dung Beetle Ecology. Princeton University Press, Princeton, NJ, USA.

Hayashi, K. 1985. Alternative mating strategies in the water strider *Gerris oblongatus* (Heteroptera, Gerridae). Behavioural Ecology and Sociobiology, 16: 301–306.

Heinze, J., S. Foitzik, B. Fischer, T. Wanke, and V. E. Kipyatkov. 2003. The significance of latitudinal variation in body size in a holarctic ant, *Leptothorax acervorum*. Ecography, 26: 349–355.

Heinze, J. and L. Keller. 2000. Alternative reproductive strategies: a queen perspective in ants. Trends in Ecology and Evolution, 15: 508–512.

Hellriegel, B. and G. Bernasconi. 2000. Female-mediated differential sperm storage in a fly with complex spermathecae, *Scathophaga stercoraria*. Animal Behaviour, 59: 311–317.

Hellriegel, B. and W. U. Blanckenhorn. 2002. Environmental influences on the gametic investment of yellow dung fly males. Evolutionary Ecology, 16: 505–522.

Hellriegel, B. and P. I. Ward. 1998. Complex female reproductive tract morphology: Its possible use in postcopulatory female choice. Journal of Theoretical Biology, 190: 179–186.

Hirschberger, P. and H. N. Degro. 1996. Oviposition of the dung beetlw *Aphodius ater* in relation to abundance of yellow dung fly larvae (*Scathophaga stercoraria*). Ecological Entomology, 21: 352–357.

Hoffmeister, T. S. and B. D. Roitberg. 1997. To mark the host or the patch: Decisions of a parasitoid searching for concealed host larvae. Evolutionary Ecology, 11: 145–168.

Holter, P. 1979. Effect of dung-beetles (*Aphodius* spp) and earthworms on the disappearance of cattle dung. Oikos, 32: 393–402.

Honek, A. 1993. Intraspecific variation in body size and fecundity in insects: a general relationship. Oikos, 66: 483–492.

Hosken, D. J. 2003. Sperm biology: size indeed matters. Current Biology, 13: R355–R356.

Hosken, D. J., W. U. Blanckenhorn, and T. W. J. Garner. 2002a. Heteropopulation males have a fertilization advantage during sperm competition in the yellow dung fly (*Scathophaga stercoraria*). Proceedings of the Royal Society of London B, 269: 1701–1707.

Hosken, D. J., T. W. J. Garner, and P. I. Ward. 2001. Sexual conflict selects for male and female reproductive characters. Current Biology, 11: 489–493.

Hosken, D. J., E. P. Meyer, and P. I. Ward. 1999. Internal female reproductive anatomy and genital interactions during copula in the yellow dung fly *Scathophaga stercoraria*. Canadian Journal of Zoology, 77: 1975–1983.

Hosken, D. J., Uhía, E., and P. I. Ward. 2002b. The function of female accessory reproductive gland secretion and a cost to polyandry in the yellow dung fly. Physiological Entomology, 27: 87–91.

Hosken, D. J. and P. I. Ward. 2000. Copula in yellow dung flies (*Scathophaga stercoraria*): investigating sperm competition models by histological observation. Journal of Insect Physiology, 46: 1355–1363.

Hosken, D. J. and P. I. Ward. 2001. Experimental evidence for testis size evolution via sperm competition. Ecology Letters, 4: 10–13.

Huey, R. B., G. W. Gilchrist, M. L. Carlson, D. Berrigan, and L. Serra. 2000. Rapid evolution of a geographic cline in size in an introduced fly. Science, 287: 308–309.

James, A. C., R. B. R. Azevedo, and L. Partridge. 1995. Cellular basis and developmental timing in a size cline of *Drosophila melanogaster*. Genetics, 140: 659–666.

James, A. C., R. B. R. Azevedo, and L. Partridge. 1997. Genetic and environmental responses to temperature of *Drosophila melanogaster* from a latitudinal cline. Genetics, 146: 881–890.

Jann, P. 1997. Natürliche und sexuelle Selektion bei der Mistfliege *Scathophaga stercoraria*. Ph. D. dissertation, University of Zürich, Switzerland.

Jann, P., W. U. Blanckenhorn, and P. I. Ward. 2000. Temporal and microspatial variation in the intensities of natural and sexual selection in the yellow dung fly *Scathophaga stercoraria*. Journal of Evolutionary Biology, 13: 927–938.

Jann, P. and P. I. Ward. 1999. Maternal effects and their consequences for offspring fitness in the yellow dung fly. Functional Ecology, 13: 51–58.

Johansson, F. 2003. Latitudinal shifts in body size of *Enallagma cyathigerum* (Odonata). Journal of Biogeography, 30: 29–34.

Juliano, S. A. and T. L. Stoffregen. 1994. Effects of habitat drying on size at and time to metamorphosis in the tree hole mosquito *Aedes triseriatus*. Oecologia, 97: 369–376.

Kaitala, A. and L. Huldén. 1990. Significance of spring migration and flexibility in flight muscle histolysis in waterstriders. Ecological Entomology, 15: 409–416.

Kawecki, T. J. and S. C. Stearns. 1993. The evolution of life histories in spatially heterogeneous environments: optimal reaction norms revisited. Evolutionary Ecology, 7: 155–174.

Kingsolver, J. G. and D. W. Pfennig. 2004. Individual-level selection as a cause of Cope's rule of phyletic size increase. Evolution 58: 1608–1612.

Kozlowski, J. 1992. Optimal allocation of resources to growth and reproduction: implications for age and size at maturity. Trends in Ecology and Evolution, 7: 15–19.

Kraushaar, U., J. Goudet, and W. U. Blanckenhorn. 2002. Geographic and altitudinal population genetic structure of two dung fly species with contrasting mobility and temperature preference. Heredity, 89: 99–106.

Kraushaar, U. and W. U. Blanckenhorn. 2002. Population variation in sexual selection and its effect on body size allometry in two species of flies with contrasting sexual size dimorphism. Evolution, 56: 307–321.

Luvchiev, V. I. and R. N. Tsankova. 1982. On the hibernation of some coprobiont flies in Bulgaria (in Bulgarian). Ekologi, 10: 59–64.

Martin, T. H. D. M. Johnson, and R. D. Moore. 1991. Fish-mediated alternative life-history strategies in the dragonfly *Epitheca cynosura*. Journal of the North American Benthological Society, 10: 271–279.

Masaki, S. 1967. Geographic variation and climatic adaptation in a field cricket. Evolution, 21: 725–741.

Masaki, S. 1972. Climatic adaptation and photoperiodic response in the band-legged ground cricket. Evolution, 26: 587–600.

Masaki, S. 1980. Summer diapause. Annual Review of Entomology 25: 1–25.

Maynard-Smith, J., R. Burian, S. Kauffmann, P. Alberch, J. Campbell, B. Goodwin, R. Lande, D. Raup, and L. Wolpert. 1985. Developmental constraints and evolution. Quarterly Review of Biology, 60: 265–287.

Mery, F. and T. J. Kawecki. 2002. Experimental evolution of learning ability in fruit flies. Proceedings of the National Academy of Sciences USA, 99: 14274–14279.

Moczek, A. P., J. Hunt, D. J. Emlen, and L. W. Simmons. 2002. Threshold evolution in exotic populations of a polyphenic beetle. Evolutionary Ecology Research, 4: 587–601.

Møller, H., R. H. Smith, and R. M. Sibly. 1989. Evolutionary demography of a bruchid beetle. II. Physiological manipulations. Functional Ecology, 3: 683–691.

Mousseau, T. A. 1997. Ectotherms follow the converse Bergmann's rule. Evolution, 51: 630–632.

Mousseau, T. A. and D. A. Roff. 1987. Natural selection and the heritability of fitness components. Heredity, 59: 181–197.

Mousseau, T. A. and D. A. Roff. 1989. Adaptation to seasonality in a cricket: patterns of phenotypic and genotypic variation in body size and diapause expression along a cline in season length. Evolution, 43: 1483–1496.

Mühlhäuser, C., W. U. Blanckenhorn, and P. I. Ward. 1996. The genetic component of copula duration in the yellow dung fly, *Scathophaga stercoraria*. Animal Behaviour, 51: 1401–1407.

Newman, R. A. 1988a. Adaptive plasticity in development of *Scaphiopus couchii* tadpoles in desert ponds. Evolution, 42: 774–783.

Newman, R. A. 1988b. Genetic variation for larval anuran (*Scaphiopus couchii*) development time in an uncertain environment. Evolution, 42: 763–773.

Newman, R. A. 1992. Adaptive plasticity in amphibian metamorphosis. Bioscience, 42: 671–678.

Niewiarowski, P. H. 2001. Energy budgets, growth rates, and thermal constraints: toward an integrative approach to the study of life-history variation. American Naturalist, 157: 421–433.

Nijhout, H. F. 1994. Insect Hormones. Princeton University Press, Princeton, NJ, USA.

Nijhout, H. F. 2003. Development and evolution of adaptive polyphenisms. Evolution and Development, 5: 9–18.

Nomakuchi, S. 1992. Male reproductive polymorphism and form-specific habitat utilization of the damselfly *Mnais pruinosa* (Zygoptera, Calopterygidae). Ecological Research, 7: 87–96.

Nuyts, E. 1994. On the copulation duration of the yellow dung fly (*Scathophaga stercoraria*): the effects of optimisation per day. Acta Biotheoretica, 42: 271–279.

Nylin, S., K. Gotthard, and C. Wiklund. 1996. Reaction norms for age and size at maturity in *Lasiommata* butterflies: predictions and tests. Evolution, 50: 1351–1358.

Nylin, S. and K. Gotthard. 1998. Plasticity in life history traits. Annual Review of Entomology, 43: 63–83.

Nylin, S. and L. Svärd. 1991. Latitudinal patterns in the size of European butterflies. Holarctic Ecology, 14: 192–202.

Otronen, M. 1993. Size assortative mating in the yellow dung fly *Scathophaga stercoraria*. Behaviour, 126: 63–76.

Otronen, M. 1995a. Male distribution and mate searching in the yellow dung fly *Scathophaga stercoraria*: comparison between paired and unpaired males. Ethology, 100: 265–276.

Otronen, M. 1995b. Energy reserves and mating success in males of the yellow dung fly *Scathophaga stercoraria*. Functional Ecology, 9: 683–688.

Otronen, M. 1996. Effects of seasonal variation in the operational sex ratio and population density on the mating success of different sized and aged males in the yellow dung fly *Scathophaga stercoraria*. Ethology, Ecology and Evolution, 8: 399–411.

Otronen, M., P. Reguera, and P. I. Ward. 1997. Sperm storage in the yellow dung fly *Scathophaga stercoraria*: identifying the sperm from competing males in separate female spermathecae. Ethology, 103: 844–854.

Padilla, D. K. and S. C. Adolph. 1996. Plastic inducible morphologies are not always adaptive: The importance of time delays in a stochastic environment. Evolutionary Ecology, 10: 456–456.

Park, O. 1949. Application of the converse Bergmann principle to the carabid beetle, *Dicaelus purpuratus*. Physiological Zoology, 22: 359–372.

Parker, G. A. 1970a. The reproductive behaviour and the nature of sexual selection in *Scathophaga stercoraria* L. (Diptera: Scathophagidae). I. Diurnal and seasonal changes in population density around the site of mating and oviposition. Journal of Animal Ecology, 39, 185–204.

Parker, G. A. 1970b. The reproductive behaviour and the nature of sexual selection in *Scathophaga stercocaria* (Diptera: Scathophagidae). II. The fertilization rate and the spatial and temporal relationships of each sex around the site of mating and oviposition. Journal of Animal Ecology, 39: 205–228.

Parker, G. A. 1970c. The reproductive behaviour and the nature of sexual selection in *Scathophaga stercoraria* L. (Diptera: Scathophagidae). V. The female's behaviour at the oviposition site. Behaviour, 37: 140–168.

Parker, G. A. 1970d. The reproductive behaviour and the nature of sexual selection in *Scathophaga stercoraria* L. (Diptera: Scathophagidae). IV. Epigamic recognition and competition between males for the possession of females. Behaviour, 37: 113–139.

Parker, G. A. 1970e. Sperm competition and its evolutionary effect on copula duration in the fly *Scathophaga stercoraria.* Journal of Insect Physiology, 16: 1301–1328.

Parker, G. A. 1970f. Sperm competition and its evolutionary consequences in the insects. Biological Reviews, 45: 525–567.

Parker, G. A. 1978. Searching for mates, pp. 214–244. *In* J. R. Krebs and N. B. Davies [eds] 1978. Behavioural Ecology, 1st edition. Blackwell, Oxford, UK.

Parker, G. A. 1992. Marginal value theorem with exploitation time costs: diet, sperm, and optimal copula duration in dung flies. American Naturalist, 139: 1237–1256.

Parker, G. A. and L. W. Simmons. 1991. A model of constant and random sperm displacement during mating: evidence from *Scathophaga*. Proceedings of the Royal Society of London B, 246: 107–115.

Parker, G. A. and L. W. Simmons. 1994. Evolution of phenotypic optima and copula duration in dungflies. Nature, 370: 53–56.

Parker, G. A. and L. W. Simmons. 2000. Optimal copula duration in yellow dung flies: ejaculatory duct dimensions and size-dependent sperm displacement. Evolution, 54: 924–935.

Parker, G. A., L. W. Simmons, P. Stockley, D. M. McChristie, and E. L. Charnov. 1999. Optimal copula duration in yellow dung flies: effects of female size and egg content. Animal Behaviour, 57: 795–805.

Parker, G. A., L. W. Simmons, L.W., and P. I. Ward. 1993. Optimal copula duration in dungflies: effects of frequency dependence and female mating status. Behavioural Ecology and Sociobiology, 32: 157–166.

Partridge, L. and J. A. Coyne. 1997. Bergmann's rule in ectotherms: is it adaptive? Evolution, 51: 632–635.

Peters, R. H. 1983. The Ecological Implications of Body Size. Cambridge University Press, Cambridge, UK.

Ray, C. 1960. The application of Bergmann's and Allen's rules to the poikilotherms. Journal of Morphology, 106: 85–108.

Reim, C. 2005. Size-Dependent Energetics and their Life-History Consequences in the Yellow Dung Fly, *Scathophaga stercoraria.* Ph. D. Dissertation, University of Zürich.

Reim, C., Y. Teuschl and W. U. Blanckenhorn. 2006. Size-dependent effects of larval and adult food availability on reproductive energy allocation in the yellow dung fly. Functional Ecology, 20: 1012–1021.

Reiss, M. J. 1989. The Allometry of Growth and Reproduction. Cambridge University Press, Cambridge, UK.

Reuter, M., P. I. Ward, and W. U. Blanckenhorn. 1998. An ESS treatment of the pattern of female arrival at the mating site in the yellow dung fly *Scathophaga stercoraria*. Journal of Theoretical Biology, 195: 363–370.

Ridley, M. 1996. Evolution. Blackwell, Oxford, UK.

Roff, D. A. 1980. Optimizing development time in a seasonal environment: the 'ups and downs' of clinal variation. Oecologia, 45: 202–208.

Roff, D. A. 1986. The evolution of wing dimorphism in insects. Evolution, 40: 1009–1020.

Roff, D. A. 1992. The Evolution of Life Histories. Chapman and Hall, New York, NY, USA.

Roff, D. A. 1997. Evolutionary Quantitative Genetics. Chapman and Hall, New York, NY, USA.

Roff, D. A. and D. J. Fairbairn. 1986. Wing dimorphisms and the evolution of migratory polymorphisms among the Insecta. American Zoologist, 31: 243–251.

Rothenbuhler, W. C. 1964a. Behaviour genetics of nest-cleaning in honey bees. I. Responses of four inbred lines to disease-killed brood. Animal Behaviour, 12: 578–583.

Rothenbuhler, W. C. 1964b. Behaviour genetics of nest-cleaning in honey bees. IV. Responses of F1 and backcross generations to disease-killed brood. American Zoologist, 4: 111–123.

Rowe, L. and D. Ludwig. 1991. Size and timing of metamorphosis in complex life cycles: time constraints and variation. Ecology, 72: 413–427.

Rubenstein, D. I. 1980. On the evolution of alternative mating strategies, pp. 65–100. *In* Staddon, J. E. R. [ed] 1980. Limits to Action: The Allocation of Individual Behaviour. Academic Press, New York, NY, USA.

Rubenstein, D. I. 1984. Resource acquisition and alternative mating strategies in water striders, *Gerris remigis*. American Zoologist, 24: 345–353.

Sasaki, H. 1984. A comparative study of the ecology of yellow dung flies (Diptera: Scathophagidae). I. Predatory ability and predatory behaviour. Japanese Journal of Sanitary Zoology, 35: 325–331.

Scheiner, S. M. 1993a. Genetics and evolution of phenotypic plasticity. Annual Review of Ecology and Systematics, 24: 35–68.

Scheiner, S. M. 1993b. Plasticity as a selectable trait — reply. American Naturalist, 142: 371–373.

Scheiner, S. M. and R. F. Lyman. 1989. The genetics of phenotypic plasticity. I. Heritability. Journal of Evolutionary Biology, 2: 95–107.

Schlichting, C. D. 1986. The evolution of phenotypic plasticity in plants. Annual Review of Ecology and Systematics, 17: 667–693.

Schmalhausen, I. I. 1949. Factors of Evolution. Blakiston, Philadelphia, PA, USA.

Schmidt-Nielsen, K. 1984. Scaling. Why is Animal Size so Important? Cambridge University Press, Cambridge, UK.

Scrimgeour, G. J. and J. M. Culp. 1994. Feeding while evading predators by a lotic mayfly-linking short-term foraging behaviors to long-term fitness consequences. Oecologia, 100: 128–134.

Sell, A. F. 2000. Morphological defenses induced in situ by the invertebrate predator *Chaoborus*: comparison of responses between *Daphnia pulex* and *D. rosea*. Oecologia, 125: 150–160.

Shelly, T. E. and M. D. Greenfield. 1991. Dominions and desert clickers (Orthoptera, Acrididae)-influences of resources and male signaling on female settlement-patterns. Behavioural Ecology and Sociobiology, 28: 133–140.

Shine, R. 1988. Evolution of large body size in females: a critique of Darwin's fecundity advantage model. American Naturalist, 131: 124–131.

Shorrocks, B. 1970. Population fluctuations in the fruit fly (*Drosophila melanogaster*) maintained in the laboratory. Journal of Animal Ecology, 39: 229–253.

Sigurjónsdóttir, H. 1984. Food competition among *Scathophaga stercoraria* larvae with emphasis on its effects on reproductive success. Ecological Entomology, 9: 81–90.

Sigurjónsdóttir, H. and G. A. Parker. 1981. Dung fly struggles: evidence for assessment strategy. Behavioural Ecology and Sociobiology, 8: 219–230.

Sigurjónsdóttir, H. and S. S. Snorrason. 1995. Distribution of male yellow dung flies around oviposition sites: the effect of body size. Ecological Entomology, 20: 84–90.

Simmons, L. W. 2001. Sperm Competition and its Evolutionary Consequences in the Insects. Princeton University Press, Princeton, NJ, USA.

Simmons, L. W. and G. A. Parker. 1992. Individual variation in sperm competition success of yellow dung flies, *Scathophaga stercoraria*. Evolution, 46: 366–375.

Simmons, L. W., P. Stockley, R. L. Jackson, and G. A. Parker. 1996. Sperm competition or sperm selection: No evidence for female influence over paternity in yellow dung flies *Scathophaga stercoraria*. Behavioural Ecology and Sociobiology, 38: 199–206.

Simmons, L. W. and P. I. Ward. 1991. The heritability of sexually dimorphic traits in the yellow dung fly *Scathophaga stercoraria* (L.). Journal of Evolutionary Biology, 4: 593–601.

Sorci, G., J. Clobert, and S. Belichon. 1996. Phenotypic plasticity of growth and survival in the common lizard *Lacerta vivipara*. Journal of Animal Ecology, 65: 781–790.

Sokal, R. R. and R. C. Rinkel., 1963. Geographic variation of alate *Pemphigus populitransversus* in eastern North America. University of Kansas Scientific Bulletin, 44: 467–507.

Sowig, P., R. Himmelsbach, and W. Himmelsbach. 1997. Predator-prey relationship between insect larvae: growth of *Sphaeridium* larvae (Coleoptera: Hydrophilidae) under time constraints through predation on *Musca autumnalis* maggots (Diptera: Muscidae). Canadian Journal of Zoology, 75: 2069–2076.

Spence, J. R. and R. S. Wilcox. 1986. The mating system of two hybridizing species of water striders (Gerridae). II. Alternative tactics of males and females. Behavioural Ecology and Sociobiology, 19: 87–95.

Stamp, N. E. and M. D. Bowers. 1993. Presence of predatory wasps and stinkbugs alters foraging behavior of cryptic and non-cryptic caterpillars on plantain (*Plantago lanceolata*). Oecologia, 95: 376–384.

Stearns, S. C. 1989. The evolutionary significance of phenotypic plasticity. Bioscience, 39: 436–445.

Stearns, S. C. 1992. The Evolution of Life Histories. Oxford University Press, Oxford, UK.

Stearns, S. C., G. de Jong, and R. A. Newman. 1991. The effects of phenotypic plasticity on genetic correlations. Trends in Ecology and Evolution, 6: 122–126.

Stearns, S. C., and J. Koella. 1986. The evolution of phenotypic plasticity in life history traits: predictions of reaction norms for age and size at maturity. Evolution, 40: 893–914.

Stephens, D. W. and J. R. Krebs. 1986. Foraging Theory. Princeton University Press, Princeton, New Jersey.

Stevenson, R. D. 1985. Body size and limits to daily range of body temperature in terrestrial ectotherms. American Naturalist, 125: 102–117.

Stockley, P. and L. W. Simmons. 1998. Consequences of sperm displacement for female dung flies, *Scathophaga stercoraria*. Proceedings of the Royal Society of London B, 265: 1755–1760.

Stockley, P. and N. J. Seal. 2001. Plasticity in reproductive effort of male dung flies (*Scathophaga stercoraria*) as a response to larval density. Functional Ecology, 15: 96–102.

Stone, A., C. W. Sabrosky, W. W. Wirth, R. H. Foote, and J. R. Coulson., 1965. A catalog of the Diptera of America north of Mexico. US Department of Agriculture, Washington, DC.

Taborsky, M. 1994. Sneakers, satellites, and helpers: parasitic and co-operative behaviour in fish reproduction. Advances in the Study of Behaviour, 23: 1–100.

Taborsky, M. 1998. Sperm competition in fish: 'bourgeois' males and parasitic spawning. Trends in Ecology and Evolution, 13: 222–227.

Taborsky, M. 2001. The evolution of bourgeois, parasitic, and cooperative reproductive behaviors in fishes. Journal of Heredity, 92: 100–110.

Tauber, M. J., C. A. Tauber, and S. Masaki. 1986. Seasonal Adaptations of Insects. Oxford University Press, Oxford, UK.

Taylor, F. 1981. Ecology and evolution of physiological time in insects. American Naturalist, 117: 1–23.

Teder, T. and T. Tammaru. 2005. Sexual size dimorphism within species increases with body size in insects. Oikos, 108: 321–334.

Telfer, M. G. and M. Hassall. 1999. Ecotypic differentiation in the grasshopper *Chorthippus brunneus*: life history varies in relation to climate. Oecologia, 121: 245–254.

Terry, L. I. and E. Dyreson. 1996. Behavior of *Frankliniella occidentalis* (Thysanoptera: Thripidae) within aggregations, and morphometric correlates of fighting. Annals of the Entomological Society of America, 89: 589–602.

Teuschl, Y., Reim, C. and W. U. Blanckenhorn. 2007. Correlated responses to artificial body size selection in growth, development, phenotypic plasticity and juvenile viability in yellow dung flies. Journal of evolutionary Biology, 20: 87–103.

Thornhill, R. 1979. Adaptive female-mimicking behaviour in a scorpionfly. Science, 205: 412–414.

Thornhill, R. 1981. *Panorpa* (Mecoptera: Panorpidae) scorpionflies: systems for understanding resource-defense polygyny and alternative male reproductive efforts. Annual Reviews of Ecology and Systematics, 12: 355–386.

Thornhill, R. 1984. Alternative female choice tactics in the scorpionfly *Hylobittacus apicalis* (Mecoptera) and their implications. American Zoologist, 24: 367–383.

Tomkins, J. L. 1999. Environmental and genetic determinants of the male forceps length dimorphism in the European earwig *Forficula auricularia* L. Behavioral Ecology and Sociobiology, 47: 1–8.

Tsubaki, Y. and T. Ono. 1986. Competition for territorial sites and alternative mating tactics in the dragonfly *Nannophya pygmaea rambur* (Odonata, Libellulidae). Behaviour, 97: 234–252.

Tsuchida, K., N. Nagata, and J. Kojima. 2002. Diplold males and sex determination in a paper wasp, *Polistes chinensis antennalis* (Hymenoptera, Vespidae). Insectes Sociaux, 49: 120–124.

Turlings, T. C. J., U. B. Lengwiler, M. L. Bernasconi, and D. Wechlser. 1998. Timing of induced volatile emissions in maize seedlings. Planta, 207: 146–152.

Van Buskirk, J. 2000. The costs of an inducible defense in anuran larvae. Ecology, 81: 2813–2821.

Van Buskirk, J. 2002. A comparative test of the adaptive plasticity hypothesis: relationships between habitat and phenotype in anuran larvae. American Naturalist, 160: 87–102.

Van Buskirk, J. and B. R. Schmidt. 2000. Predator-induced phenotypic plasticity in larval newts: Trade-offs, selection, and variation in nature. Ecology, 81: 3009–3028.

van der Have, T. M. and de Jong, G. 1996. Adult size in ectotherms: temperature effects on growth and differentiation. Journal of Theoretical Biology, 18: 329–340.

van Noordwijk, A. J. and de Jong, G. 1986. Acquisition and allocation of resources: their influence on variation in life history tactics. American Naturalist, 128: 137–142.

van Tienderen, P. H. 1991. Generalists, specialists, and the evolution of phenotypic plasticity in sympatric populations of distinct species. Evolution, 51: 1372–1380.

van Tienderen, P. H. 1997. Evolution of generalists and specialists in spatially heterogeneous environments. Evolution, 45: 1317–1331.

Vepsäläinen, K. 1978. Wing dimorphism and diapause in *Gerris*: determination and adaptive significance, pp. 218–253. *In* H. Dingle [ed] 1978. Evolution of Insect Migration and Diapause. Springer, Heidelberg, Germany.

Via, S. 1984. The quantitative genetics of phytophagy in an insect herbivore. I. Genotype-environment interaction in larval performance on different host plant species. Evolution, 38: 881–895.

Via, S. 1993a. Adaptive phenotypic plasticity—target or by-product of selection in a variable environment. American Naturalist, 142: 352–365.

Via, S. 1993b. Regulatory genes and reaction norms. American Naturalist, 142: 374–378.

Via, S. and R. Lande. 1985. Genotype-environment interaction and the evolution of phenotypic plasticity. Evolution, 39: 505–522.

Via, S., R. Gomulkiewicz, G. de Jong, S. M. Scheiner, C. D. Schlichting, and P. H. van Tienderen. 1995. Adaptive phenotypic plasticity-consensus and controversy. Trends in Ecology and Evolution, 10: 212–217.

Wäckers, F. L. and T. M. Bezemer. 2003. Root herbivory induces an above-ground indirect defence. Ecology Letters, 6: 9–12.

Ward, P. I. 1993. Females influence sperm storage and use in the yellow dung fly *Scathophaga stercoraria*. Behavioural Ecology and Sociobiology, 32: 313–319.

Ward, P. I. 1998. Intraspecific variation in sperm size characters. Heredity, 80: 655–659.

Ward, P. I. 2000a. Cryptic female choice in the yellow dung fly *Scathophaga stercoraria*. Evolution, 54: 1680–1686.

Ward, P. I. 2000b. Sperm length is heritable and sex-linked in the yellow dung fly (*Scathophaga stercoraria*). Journal of Zoology, 251: 349–353.

Ward, P. I., M. Foglia, and W. U. Blanckenhorn. 1999. Oviposition site choice in the yellow dung fly *Scathophaga stercoraria*. Ethology, 105: 423–430.

Ward, P. I., and E. Hauschteck-Jungen. 1993. Variation in sperm length in the yellow dung fly *Scathophaga stercoraria* (L). Journal of Insect Physiology, 39: 545–547.

Ward, P. I. and L. W. Simmons. 1990. Short-term changes in the numbers of the yellow dung fly, *Scathophaga stercoraria*. Ecological Entomology, 15: 115–118.

Ward, P. I., and L. W. Simmons. 1991. Copula duration and testes size in the yellow dung fly, *Scathophaga stercoraria* (L.): the effects of diet, body size, and mating history. Behavioural Ecology and Sociobiology, 29: 77–85.

Ward, P. I., J. Vonwil, E. J. Scholte, and E. Knop. 2002. Field experiments on the distributions of eggs of different phosphoglucomutase (PGM) genotypes in the yellow dung fly *Scathophaga stercoraria* (L.). Molecular Ecology, 11: 1781–1785.

Ward, P. I. and A. Wilhelm. 1994. Degradation of pats of cow dung populated by different insect communities. Entomological Trends in Agricultural Sciences, 2: 89–94.

Warner, R. R. 1984. Mating behaviour and hermaphroditism in coral reef fish. American Scientist, 72: 128–136.

Werner, E. E. and B. R. Anholt. 1993. Ecological consequences of the trade-off between growth and mortality rates mediated by foraging activity. American Naturalist, 142: 242–272.

West-Eberhard, M. J. 1986. Alternative adaptations, speciation, and phylogeny (a review). Proceedings of the National Academy of Sciences USA, 83: 1388–1392.

West-Eberhard, M. J. 2002. Developmental Plasticity and Evolution. Oxford University Press, Oxford, UK.

Wilcox, R. S. and J. R. Spence. 1986. The mating system of two hybridizing species of water striders (Gerridae). I. Ripple signal functions. Behavioural Ecology and Sociobiology, 19: 79–85.

Wootton, R. J. 1979. Energy cost of egg production and environmental determinants of fecundity in teleost fishes. Symposia of the Zoological Society London, 44: 133–159.

Wuellner, C. T. 1999. Alternative reproductive strategies of a gregarious ground-nesting bee, *Dieunomia triangulifera* (Hymenoptera : Halictidae). Journal of Insect Behavior, 12: 845–863.

Zera, A. J. 1984. Differences in survivorship, development rate and fertility between the longwinged and wingless morphs of the waterstrider, *Limnoporus canaliculatus*. Evolution, 38: 1023–1032.

Zera, A. J. and R. F. Denno. 1997. Physiology and ecology of dispersal polymorphism in insects. Annual Review of Entomology, 42: 207–230.

Zera, A. J. and L. G. Harshman. 2001. The physiology of life histora trade-offs in animals. Annual Review of Ecology and Systematics, 32: 95–126.

Zettel, J. and Zettel, U. 1994. Adaptations to the alpine environment in *Isotomurus alticola* (Collembola, Isotomidae) in the Swiss alps. Acta Oecologica, 15: 93–104.

11

She Shapes Events As They Come: Plasticity in Female Insect Reproduction[1]

Jason Hodin
Hopkins Marine Station, Stanford University, Pacific Grove, CA USA 93950
E-mail: hodin@u.washington.edu

"Reproduction is what bugs do best. It's one of the reasons why they dominate the planet."

–from the IMAX film *Bugs! in 3D* (Slee and Aron 2003)

Abstract

The typical insect ovary has a modular arrangement, with the ovariole as its fundamental modular unit. In general, an increased ovariole number appears to correlate with total potential reproductive output, but other physiological characteristics of the ovary can theoretically influence the rate and timing of egg production as well, the rate of öocyte maturation being one such parameter. Nevertheless, it would be incorrect to imagine that an increased rate of egg production is the only relevant fitness parameter. While insects such as honeybees and drosophilid vinegar ("fruit") flies do seem to be characterized by a maximization of total egg production, there are clearly constraints (or trade-offs) involved even in these examples. The decreased reproductive output potential in worker *versus* queen honeybees, as well as interspecific variation in ovariole number within both of these taxa, suggests that maximization of reproductive output entails some physiological, ontogenetic and/or life history trade-off. More extreme examples are parasitic or viviparous insects (such as tsetse flies) that produce as few as one egg at a time. Furthermore, there is substantial variation across broad and narrow taxonomic groups of insects in the degree to which the rates (öocyte maturation, oviposition) and potential rates (ovariole number) of egg production are phenotypically plastic. Here I will review several well-documented cases of interspecific and/or phenotypically plastic variability in the rates and potential rates of egg production in a wide variety of insect

[1] Title derived from the *Tao te Ching* by Lao-Tzu (chapter 45, S. Mitchell translation 1992).

taxa. I will argue that developing a comprehensive theory of insect reproductive plasticity will require comparative phylogenetic approaches that take account of the interactions between ecological and ontogenetic factors, including developmental constraints. I will close by discussing the apparent similarities between the ecdysis and ovipositional behavioral networks.

TABLE OF CONTENTS

Introduction

It's probably fair to say that the decision of where, when, how much and how often to reproduce is the most important decision (from an evolutionary point of view) that most organisms will make in their lifetimes. But on what basis do they make this critical calculation? What external criteria influence this decision-making process? By what internal mechanisms are these external (environmental) signals integrated to alter development, physiology and/or behavior? And, what features of organisms (morphological, developmental, physiological, historical, etc.) might constrain or bias their responses to these environmental signals? Answering these questions has profound implications for, among other things, evolution, life history theory, conservation biology, and pest control.

Some General Thoughts on Plasticity

The "decision" referred to above is another way of describing plasticity in reproduction, which is generally understood as variation (morphological, physiological, behavioral) within genotypes resulting from environmental heterogeneity (reviewed in Schlichting and Pigliucci 1998; Kalisz and Kramer 2008). Two oft-discussed dichotomies bear consideration: continuous "versus" discontinuous plasticity, and adaptive versus non-adaptive plasticity. An example of continuous plasticity is the effect of rearing temperature on body size, while alternative spring/fall butterfly wing morphology (polyphenism) is an example of discontinuous plasticity. Adaptive plasticity is a plastic response that tends to yield an increase in fitness, such as the effect of day length on the likelihood of entering diapause. Non-adaptive plasticity, by contrast, indicates a plastic response that does not yield a predictable increase in fitness. It is merely, for example, a metabolic reality, such as the general correlation of temperature and developmental rate. Note, that such examples of non-adaptive plasticity are *not necessarily* maladaptive.

There seems to be a general impression [and Schlichting and Pigliucci (1998) as well as West-Eberhard (1989) appear to give this impression] that discontinuous plasticity is often indicative of adaptive plasticity. This seems reasonable for at least two reasons:

1. The discreteness of the morphologies that characterize discontinuous plasticity have obvious alternative fitness advantages in their discrete

environments, some of which have been demonstrated experimentally. Seasonal polyphenisms are famous examples.

2. There are convincing cases where continuously varying traits in basal lineages have been inferred to have become more discrete in derived lineages. The multiple, independent origins of insect eusociality/division of labor is one of the most widely cited examples (see Robinson 1992). Such a progression from continuous to discontinuous plasticity would strongly indicate molding by natural selection.

Less appreciated, though, is the fact that variation in the predictability of environmental signals that induce a discontinuous plastic response can result in any point on a continuum of plastic responses from adaptive to maladaptive in any given instance. For example, consider the use of a temperature threshold as a cue for breaking diapause, as in some ladybug beetles (Hodek 1996). Such a response would be quite maladaptive in the event of an unusually warm day in the middle of winter, since a cold snap the following week could lead to mass mortality. In such a situation, a graded (continuous) response to environmental conditions might be preferable.[2]

An underlying point here is that the adaptive nature of the response can only be judged in an appropriate ecological context. In the laboratory, while the advantage of controlling variables is obvious, the appropriateness of the ecological context can only be approximated. In the field, appropriate ecological contexts can be studied, but it is of course difficult to exclude factors which could not be controlled for (or, even worse, those which covary for trivial and even temporary reasons) as plausible alternative hypotheses to the adaptive hypothesis.

Thus, continuously varying traits may be adaptive in certain contexts, but not in others. Or, rather, the plasticity *per se* may be non-adaptive; for example, it may be a physiological constraint. However, the exact nature of the plasticity (the shape of the reaction norm itself) may be expected to have been tinkered with to provide the most adaptive solution within the context of the physiological constraint(s).

This suggests the hypothesis that most cases of adaptive plasticity may have, at their root, non-adaptive physiological responses which were ultimately molded by natural selection to produce either continuous or

[2] Indeed, such a graded response is found in ladybugs (Coleoptera: Coccinellidae), both within and among species (Hodek 1996), and may be one explanation for the evolutionary success of the group.

discontinuous adaptive plasticity.[3] These ideas are similar to those proposed previously by Schmalhausen (1949) and Matsuda (1987).

Such a situation may be particularly relevant for many instances of insect reproductive plasticity. The general correlation between food intake and ovarian growth is one possible example (see Labeyrie 1978, Wheeler 1996). Since the ovaries can grow in an adult insect in which the majority of structures have stopped growing, this correlation is completely expected, and would hardly be considered an adaptive response. However, modulations in the specific ways in which food induces growth of öocytes are probably good examples of adaptive plasticity. Öocyte growth up to resting stages in the sheep blowfly *Lucilia sericata* is one such example that I consider in some detail below.

Another confusing point is whether the observed continuity is individual or populational. In other words, if different individuals (for whatever reason, be it genetic or epigenetic) have different thresholds or otherwise differ in their environmental sensitivity to the plasticity cues, then a population may be seen to have a continuous response while each individual might have a predictably discontinuous response. Likewise, a discontinuous response may be an adaptive bet-hedging strategy from a populational perspective, while a given individual's response may strike the observer in that specific instance as being maladaptive.

Examples of insect reproductive plasticity cover the spectrum from discontinuous (e.g. soldier *vs.* reproductive castes in termites) to continuous (e.g. the effect of host plant availability on oviposition frequency or specificity, as per Mercander and Scriber, 2005). For the purposes of this chapter, I will focus mainly on instances where the plasticity is or is likely to be adaptive, whether in continuous or in discontinuous traits.

Providing an overview of insect reproductive plasticity that addresses not only the relevant ecological and evolutionary forces, but also the physiological and ontogenetic underpinnings, is a substantial and possibly unwieldy undertaking. From an ecological point of view, there are four major ways in which reproduction can be plastic: number of progeny, timing of reproduction, size/mass/quality of progeny, and place/timing of offspring release. On the other hand, from an ontogenetic point of view, one can think in terms of stages during ontogeny when particular aspects of insect

[3] This is an example of what Gould and Vrba (1982) termed an "exaptation": in this case, a non-adaptive feature of an organism later coopted for its current (adaptive) function. Such a "non-adaptive feature" is what Gould and Lewontin (1979) famously analogized to a cathedral's spandrel.

reproduction are subject to plasticity. In insects, these stages can be roughly divided into ovarian differentiation (the formation of an ovarian primordium from undifferentiated cells), ovarian maturation (also known as öogenesis), and oviposition (see Figure 1). To differing degrees, plasticity in each of these three ontogenetic processes influences the four ecological aspects of reproductive plasticity listed above (summarized at the end of the chapter in Figure 6). For example, I will describe instances in which plasticity in the timing of differentiation influences total number of progeny or the timing of reproductive maturity. Similarly, plasticity in maturation or oviposition can influence the timing, frequency and output of egg laying. As an organizational principle for this chapter, I will adopt an ontogenetic approach, focusing in turn on these three main ontogenetic processes. Within each process, I will attempt to show how plasticity for that process might influence plasticity in the four ecological aspects described above. I will end with a description of other aspects of reproductive plasticity not covered explicitly in the main body of the text.

Overview of Female Insect Reproductive Development

Embryo

In the majority of studied insects,[4] the germ cell primordia exist as groups of distinct, round "pole cells" at the posterior end of the early embryo (reviewed in Büning 1994). Subsequently, the pole cells migrate into the center of the embryo and release signals that recruit the mesodermal cells which will ultimately form the bulk of the ovary (reviewed in Santos and Lehmann 2004). While there has been a large body of research in *Drosophila melanogaster* on the signals and patterning molecules involved in these early events of ovarian development (see Santos and Lehmann 2004), the relationships between early embryonic development and plasticity in adult reproduction has been little explored. Also, different insects differ in their numbers of pole cells (Büning 1994); however, any relationship between pole cell number and adult reproductive plasticity is unknown. Still, it is theoretically possible that pole cell development may be a way in which the adults of some insects "communicate," via maternal effects, the state of the adult environment to their offspring.

[4] Honeybees (genus *Apis*) are a notable exception (reviewed in Büning 1994).

Larva/Nymph

Ovarian tissue proliferation and differentiation usually occurs during the larval/nymphal stage (Buning 1994). Grasshoppers are one exception, where ovarian differentiation is completed in the embryo stage (see Stauffer and Whitman 1997); Lepidoptera are another (see below and Figure 3). Variation in the stage at which differentiation occurs may influence later life-history events (e.g. whether or not the adults feed), as suggested by Büning (1994). I will focus here mainly on the vast majority of insects whose ovaries are subdivided into ovarioles (but see the section on paedogenesis below and Figure 5).

In almost all cases known, the mesodermal cells of the ovary proliferate during the earliest larval/nymph stages, usually without differentiation (reviewed in Büning 1994). In some cases, the germ cells also begin to divide at this time, sometimes forming into germ cell/nurse cell clusters. The stereotyped pattern of ovarian differentiation begins with the formation of ovarioles: öocyte maturation tubes that are the functional unit of the insect ovary (Figure 1). In *Drosophila melanogaster*, the process of ovarian differentiation is entirely regulated by the mesodermal cells of the ovary, as it can proceed in the complete absence of germ cells (Ashburner 1989). Ovariole differentiation in insects (King et al. 1968; Büning 1994) generally begins with the formation, in the anterior of the ovary, of pancake-like stacks of cells known as terminal filaments, which will ultimately cap each ovariole. Posterior to the terminal filament lies the germarium, a mixture of germ cells and mesoderm. These germ cells, when they divide, will produce the öocytes and nutritive nurse cells. Öocytes (with or without nurse cells, depending on the insect's ovariole type) will then be surrounded by follicle cells, forming the incipient "egg chambers." At the posterior end of the ovariole, a stack of cells somewhat analogous to the terminal filament (known as the basal stalk or pedicle) will form in some insects, later connecting to the oviduct (a structure not derived from the ovary anlagen). At this point, ovarian differentiation is complete. In the next phase, öocyte chambers grow into mature öocytes, in a process known as "öogenesis" or "öocyte maturation."

The distinction between ovarian differentiation and öocyte maturation is not merely a semantic one. The maximum number of ovarioles in all insects is fixed during pre-adult stages (coccids are the one confirmed exception; see below). This is significant, because ovariole number is correlated with potential fecundity (David 1970; Cohet and David 1978; Bouletreau-Merle et al. 1982; Stewart et al. 1991), as only one egg at a time can be matured from

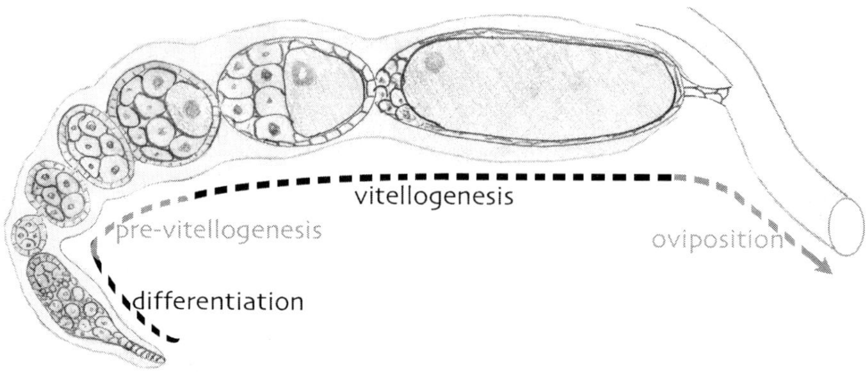

Fig. 1 Schematic drawing of a longitudinal section through an insect ovariole surrounded by its epithelial sheath. Anterior is to the left; the oviduct is the tubular structure at the far right. Ovarian differentiation is the process by which the terminal filaments (the stack of cells at the extreme anterior) form, the initial germ line stem cell divisions take place, and the first egg chambers begin to be surrounded by a monolayer of follicle cells. Note that ovarian differentiation is complete before pre-vitellogenesis begins, while the subsequent stages, in many insects, continue throughout the life of the adult female. Öocyte maturation (or öogenesis) is the growth of the egg chambers, six of which are indicated here. Maturation can be subdivided into a pre-vitellogenic, a vitellogenic, and a choriogenic (not shown) stage (see the text). Mature eggs then pass down the oviduct to be fertilized and oviposited. Öocytes that leave the body are termed eggs. Each of the indicated stages is subject to plasticity, as described in the text.

each ovariole. Hence, a possible determinant of fecundity becomes fixed early in the insect's life, and in some cases as early as the embryo. So although ovariole number is plastic in many insects (see more on this below), this plasticity does not extend into the adult phase. Öocyte maturation, by contrast, can be modulated in many insects during the adult stage (see below). Thus, there are sound functional reasons to consider the two stages of ovarian development—differentiation and maturation—as distinct from one another.

Adult

Insects are highly modular. The imaginal discs in some holometabolous insects are an extreme manifestation of this modularity, where the anlagen (primordia) for different appendages are spatially separated from one another, allowing the different structures the capacity to develop somewhat autonomously (but see Nijhout and Emlen 1998, Nijhout and Davidowitz this volume). In fact, the phenomenonal variety of insect morphologies can in some sense be accounted for by this modular organization. Still, there is a tremendous constraint on growth in all insects imposed by their hard cuticle and the presence of wings in adults: the transition to the adult stage is the terminal molt (with the single exception of mayflies). Therefore, no alterations in external morphology can be made beyond the adult molt.

The internal, soft structures in insects are not constrained in this way. Fat body can be built up and broken down, as can flight muscles, brain cells, and so on. Ovarian development, too, is extremely mutable during the adult stage. For example, öogenesis shuts down during adult diapause, and then resumes after diapause is broken. Still, as indicated above, there is one important apparent constraint on ovarian development in virtually all insects: maximum ovariole number is fixed during pre-adult stages, since no new ovarioles can be added after that time. This seems curious from a functional point of view. Ovariole number relates in some way to maximal egg production rate (e.g. David 1970; Cohet and David 1978; Bouletreau-Merle et al. 1982; Stewart et al. 1991; see below); thus, it would surely be advantageous for insects to be able to set their ovariole numbers after exploring their adult environment. The clearest example of such an advantage can be seen in insects whose pre-adult and adult habitats are different (as they often are, particularly in the Holometabola and in many aquatic insects), since their pre-adult habitats might well provide no reliable clues as to the state of their adult environment-to-be. The most obvious explanation is that there is some underlying constraint on ovariole differentiation in adults. We will explore this possibility below when discussing the scale insects (Hemiptera: Coccidae), the only insect taxon that is know to be able to substantially increase ovariole numbers as adults.[5]

Six key reproductive features can be modulated in the adult stage: öocyte maturation rate, the number of active ovarioles, egg size, the number of eggs held for oviposition, the timing of oviposition, and the place of oviposition.

Öocyte maturation (which, as mentioned above, often commences in the pre-adult stage; reviewed in Büning 1994) begins with the separation of egg

[5] A second group of insects—termites—are often cited as having this capacity as well. Many references are made to a study by Truckenbrodt and Amelung (1986) on *Odontotermes stercorivorus* (Termitidae) queens, showing that they increase their ovariole numbers by fission of pre-existing ovarioles. However, these authors only reported on ovarian growth in queen nymphs through the 5th stadium. Thus, although I consider it highly likely that this species can also add ovarioles as adults, this has not to my knowledge been technically shown.

chambers from the germarium (Figure 1). These egg chambers will then grow in size, taking up structural materials from the nurse cells and/or through metabolic processes in the öocytes themselves. This first stage of maturation is referred to as pre-vitellogenesis. In vitellogenesis (the 2nd stage), the öocyte begins to incorporate yolk proteins either from the hemolymph or, less often, from proteins produced in the follicle cells. This stage is typified by rapid öocyte growth. During choriogenesis (the 3rd stage of maturation), the chorion is deposited. The mature öocyte is then ovulated into the oviducts, fertilized, and oviposited (laid). The scenario outlined above is true for most insects, although there are exceptions, such as in parasitic and viviparous forms, where embryogenesis can precede the onset of vitellogenesis, and choriogenesis can be skipped entirely (reviewed in Wheeler 1996; Grbic 2003).

Generally, each mature ovariole contains a linear series of developing egg chambers, resulting in an anterior-posterior progression from germ cells, to incipient egg chambers, to mature öocytes (Figure 1). In many insects, the primary (posterior-most) öocyte in each ovariole matures in synchrony, so that each ovariole contributes one egg during each reproductive (gonotrophic) cycle. Other patterns include those species in which multiple öocytes within each ovariole undergo vitellogensis and mature simultaneously, while other species can have highly asynchronous öocyte maturation across ovarioles. Some insects oviposit batches of eggs into a single clutch, while others lay eggs singly. These features vary widely among insects, and are often plastic.

Öocyte maturation is famously plastic, and can be affected by food availability, the presence of males, oviposition site ("host") availability, temperature, humidity, day length, pathogens, parasitoids, and so on (reviewed in Labeyrie 1978, Wheeler 1996, Hopkins and Ekbom 1999, Tammaru and Javois 2000, Papaj 2000). Such plasticity will be a major topic of discussion below. Females can hold many mature öocytes awaiting an appropriate oviposition site. Developing öocytes, and even sometimes mature, unlaid eggs can also be resorbed, and the resources therein reallocated for other energetic needs or for future reproduction (Sundburg et al. 2001, Osawa 2005; reviewed in Bell and Bohm 1975). The numbers of mature eggs being held by the female is often referred to as "egg load," and has become an important characteristic in distinguishing reproductive patterns among insects (see Papaj 2000; Jervis and Ferns 2004, 2005).

Oviposition (the passage of eggs or embryos from the body to the environment) is itself subject to plasticity, as insects can judge the relative appropriateness of sites for the protection and/or growth of their offspring. The degree of clutch size and oviposition site plasticity, and the cues that elicit such plasticity can vary among species and populations (e.g. Fordyce 2005; Haribal and Renwick 2005; Mercander and Scriber 2005). Oviposition can also be modulated by many of the factors cited above that influence öocyte maturation (see above references and Hinton 1981), and facultative viviparity can lead to a decision to either oviposit or brood internally (see Schal et al. 1997). Finally, oviposition in many iteroparous (sequentially ovipositing) insects has been shown to modulate öocyte maturation, as well (reviewed in Papaj 2000).

I will discuss plasticity in a variety of insect groups in ovariole number, öocyte maturation rate (including reproductive diapause), clutch size, egg size, oviposition timing, and place of oviposition. I will consider how plasticity in these processes differ in solitary versus social insects, long-lived versus short-lived adults, r versus k selected species, and parasitic versus free living forms. I will then review some instances of plasticity in overall reproductive mode (viviparity, larval reproduction), and conclude by advocating a broad-based comparative strategy designed to integrate mechanistic (hormonal, genetic, cellular, biochemical), ecological and evolutionary approaches, to reach a holistic understanding of the startling variation in insect reproductive patterns.

Setting the Stage: Plasticity in Ovariole Number

Insect species vary widely in number of ovarioles per ovary, from one to several thousand. Likewise, individuals within populations vary in number of ovarioles, and in some cases this important reproductive characteristic has been shown to be plastic. Because all ovarioles can theoretically mature eggs simultaneously, maximum potential reproductive output correlates positively with ovariole number (David 1970; Cohet and David 1978; Bouletreau-Merle et al. 1982; Stewart et al. 1991; plus many examples discussed below). However, large ovaries can generate problems for lift and flight (Berrigan 1991), and also, rates of öogenesis may be inversely related to number of ovarioles and developing öocytes. These and other trade-offs suggest that ovariole number might be shaped by natural selection. Furthermore, differences in optimal ovariole numbers might be characteristic of different environmental conditions (see below), and selection should favor plasticity for this character in insect populations existing in fluctuating environments.

Several aspects of the pre-adult environment can influence the numbers of ovarioles in the adult, including temperature, food quality, food abundance and crowding (e.g. Saviliev 1928, Robertson 1957, Hinton 1981, Rhamhalinghan 1986; Grenier and Nardon 1994; Delpuech et al. 1995; Morin et al. 1997; Moreteau et al. 1997, Hodin and Riddiford 2000a, Tu and Tatar 2003). In general, higher quality, abundant food assimilated in uncrowded conditions leads to an increase in the ovariole number. The temperature effect on ovariole number, by contrast, is a bell-shaped function, with a certain moderate temperature (which varies widely among populations and species) leading to a maximal ovariole number (see below).

It is tempting to consider the effects of pre-adult feeding on ovariole number to be adaptive and anticipatory, whereby females use current conditions to predict future conditions. It would follow, for example, that during a "poor" reproductive season, it may be advantageous for a females to reduce her number of ovarioles, and instead direct more resources into simply staying alive. Still, a purely correlative explanation for such a pattern cannot be excluded. For example, ovariole number clearly correlates with body size (e.g. Stewart et al. 1991; Gasser et al 2000, Tu and Tatar 2003; reviewed in Honek 1993), so ovariole number differences resulting from differential feeding *per se* are by no means indicative of adaptive plasticity. As I stressed above, phenotypic plasticity does not have to be adaptive, nor to have undergone selection for the plastic response.

Likewise, temperature effects on ovariole number might also be, in essence, a non-adaptive bio-physical plastic response, with the optimum temperature merely representing the metabolic optimum for the molecules involved in terminal filament formation. Still, comparative studies among drosophilid vinegar ("fruit") flies (Diptera: Drosophilidae) demonstrate predictable geographic differences in temperature optima for ovariole number (e.g. Delpuech et al. 1995; Moreteau et al. 1997, Morin et al. 1997, Karan et al 1999, 2000; Gibert et al. 2004; Wayne et al. 2005). Thus, the bell shape of the reaction norm might be a purely physiological reality, whereas the particular nature of the reaction norm (e.g. the optimum temperature and the steepness of the curve; e.g. Gibert et al. 2004) may shift under different selection conditions. In this way, non-adaptive plasticity might give way to an exaptation (see footnote 3, above) allowing for adaptive evolutionary shifts in the mean numbers of ovarioles in different populations or species.

One way of testing this hypothesis is to compare the mechanisms underlying the plastic response with those underlying genetically-fixed differences among related populations or species. We (Hodin and Riddiford

2000a) undertook such a test, comparing food and temperature-induced plasticity to intra- and inter-specific variation among members of the *melanogaster* species group of *Drosophila* (Figure 2). We reasoned that since maximal ovariole number is set before metamorphosis in vinegar flies, any differences in ovariole number between two flies must be either due to ontogenetic differences in the processes of ovarian differentiation in larval stages, or to subsequent cell death and removal of differentiated ovarioles. We used this reasoning to compare trajectories of ovarian development within and across species, as well as in flies raised under a variety of temperature and food conditions. We showed that the ontogenetic mechanisms underlying within- and across-species variation in ovariole number were broader than the mechanistic underpinnings of the plastic responses in *D. melanogaster* (Table 1). In other words, only a subset of the mechanisms underling genetically-based differences (among populations and species) demonstrated plasticity under a variety of food and temperature conditions.

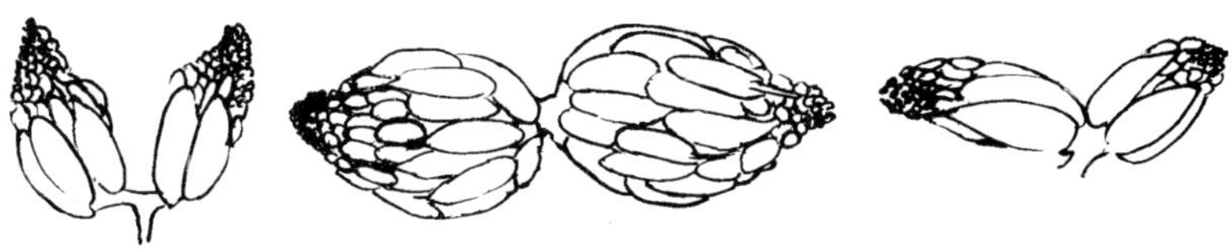

Fig. 2 Gross similarity in ovarian development in *D. melanogaster* when reared in food-limiting conditions, and its sister species *D. sechellia* when fed *ad libitum*. See Hodin and Riddiford (2000a) for details.

At first glance, these results seem to refute the exaptation hypothesis outlined above. However, our plasticity experiments with *D. melanogaster* larvae demonstrate that ovarian growth and ovariole differentiation are processes that can be decoupled under a variety of environmental conditions. The fact that these processes are not *necessarily* correlated suggests the possibility that they can be independently acted upon by natural selection, ultimately yielding the variety of mechanisms underlying ovariole number differences across populations and species noted above. Hence, the exaptation scenario might indeed be valid. Further tests with a diversity of insect groups would help resolve this situation.

For example, comparative studies on Hawai'ian drosophilids (a sub-family separated by at least 60 million years from the branch containing the

Table 1 The array of mechanisms underlying within- and across-species differences in ovariole number in the *melanogaster* species group of the genus *Drosophila* are more broad than the mechanisms underlying plasticity for ovariole number in *Drosophila melanogaster.* Control conditions were the same in all cases : 25°C rearing temperature, uncrowded conditions, full amounts of food. Food reductions were half rations. Mechanisms: **a**=smaller ovarian primordium, 2nd larval stage ("instar"); **b**=slower ovarian growth, 2nd instar; **c**=slower ovarian growth, 3rd instar; **d**=delayed onset of terminal filament (TF) formation; **e**=reduced rate of TF formation. Note: none of the observed differences in ovariole number was due to increased rates in larval ovarian cell death, reduction of ovariole number in the pupal or adult stage or early completion of TF formation. Data compiled from Hodin and Riddiford (2000a).

species	*population / locality*	*rearing condition*	*mean ovariole number*	*mechanism(s) underlying ovariole number difference (all relative to Sevelen control unless indicated)*
melanogaster	Sevelen / Sevelen, Switzerland	control	21.1	
melanogaster	Sevelen / Sevelen, Switzerland	low temp. (15°C)	11.8	**d** and/or **e**
melanogaster	Sevelen / Sevelen, Switzerland	high temp. (30°C)	17.7	**d** and/or **e**
melanogaster	Sevelen / Sevelen, Switzerland	food reduction	18.2	**e**
melanogaster	Capitol Hill / Seattle, USA	control	21.3	no significant difference
melanogaster	Capitol Hill / Seattle, USA	food reduction	18.9	**d** and/or **e**
melanogaster	Nahal / "Evolution Canyon", Israel	control	18.6	**c**, **d**
simulans	"hond" / Zamorano, Honduras	control	18.9	**d**
simulans	"st" / Florida City, USA	control	15.5	**b**, **e?** (relative to Sevelen and *simulans* Honduras)
mauritiana	Riviere Noire, Mauritius	control	12.9	**b**, **d**
sechellia	Cousin Island, Seychelles Islands	control	8.5	**a**, **c**, **d**, **e**
yakuba	Cote d'Ivoire	control	12.9	**a?**, **b**, **c**, **d**, **e**

melanogaster species group) have revealed an extreme range of mean ovariole numbers in different species (from 1 to 50 per ovary; Kambysellis and Heed 1971). Furthermore, this ovariole number variation correlates with profound ecological differences among the various species. For example, species that oviposit many eggs at once under bark have high ovariole numbers, whereas those that oviposit only one egg at a time on decaying leaves are characterized by lower ovariole numbers. An adaptive explanation for the former seems obvious: higher potential öocyte maturation rate. An explanation for the latter might be found in experiments suggesting that ovary weight is negatively correlated with lift production in a flesh fly (Berrigan 1991), and that wing to thorax ratio (but, importantly, not body size) shows a significant positive correlation with ovariole number across species of the *obscura* group (Drosophilidae; Moreteau et al. 2003). Furthermore, these differences in ovariole number/oviposition strategy correlate with phylogeny (Kambysellis et al. 1995), as there is a general progression among Hawai'ian drosophilids from ancestral decaying leaf laying specialists (1–4 ovarioles per ovary), to decaying stem-laying specialists (5–11 ovarioles per ovary), to highly derived, decaying bark laying generalists (12–50 ovarioles per ovary).[6] In any case, this range of ovariole numbers among Hawai'ian species would provide an independent case with which to explore the mechanistic underpinnings of ovariole number plasticity and variation, and to test the exaptation scenario introduced above, in a phylogenetic context. Plasticity for ovariole number has not been examined for any Hawai'ian species of which I am aware.

Whereas the Drosophilidae might contain the best documented examples of ovariole number disparity (Pappas and Engstrom 1974; Mahowald and Kambysellis 1980), ovariole number variability is found in almost every insect group that has been examined. In Table 2, I organized ranges of reported ovariole numbers (previously compiled by Büning 1994 and Robertson 1961) according to insect order (I have excluded the social insects, which I consider separately below). The first trend that is obvious is that many groups of insects show substantial interspecific variation in ovariole number, notably the Diptera, Neuroptera and Orthoptera. Interestingly, the Lepidoptera and most Hemiptera show surprisingly little variation in ovariole number. The Lepidoptera represent a striking case, in which all nine genera examined have the same number of ovarioles (4). This is particularly intriguing given the vast differences among butterflies and moths in body size, and their ecological and reproductive patterns (reviewed in Ramaswamy et al. 1997), from those that do not feed as adults

[6] While this pattern (from specialist ancestors to generalist descendants) might seem counterintuitive, Van Valen (1965) accounted for such a pattern of expanded "niche width" in derived, island populations.

Table 2 Variation in ovariole numbers in six different orders of insects. The Coleoptera data are from Robertson (1961), and the Orthoptera data from Stauffer and Whitman (1997). All other data are from Büning (1994), with four extra lepidopteran species added [the hawk moth *Manduca sexta* (Nijhout and Riddiford, 1974), the monarch butterfly *Danaus plexippus* (Urquhart 1960), the apple coddling moth *Cydia pomonella* (Benz 1969), and the yucca moths *Tegetigula* spp. (Nielsen and Kristensen 1989)] in order to further validate the striking pattern among the Lepidoptera. I have only included those orders for which Büning (1994) listed data for genera from at least 3 different families. In the Hemiptera, the coccids and the psyllids may be special cases, as described in the text, and were thus excluded. *Meloe*, which is reported to have approximately 1000 ovarioles per ovary, is such an extreme outlier for the Coleoptera that it was excluded here (see the text). I treat the social insects and their close relatives (the Hymenoptera and the Dictyoptera + Isoptera) separately (see below). Clearly, there are many genera with reported ovariole numbers that were not included in Büning's review, so this list should not by any means be considered comprehensive. I merely intend to demonstrate the broad, inter-ordinal trends here.

order	*number of genera examined*	*number of families represented*	*range of ovariole numbers per ovary*
Orthoptera	33	7	2–150
Hemiptera (except coccids and psyllids)	15	11	4–15
Coleoptera (except *Meloe*)	223	31	1–70
Neuroptera	4	4	10–160
Lepidoptera	9	9	4
Diptera	10	8	1–150

and eclose with their full complement of mature eggs (such as the silk moth *Bombyx mori*, where only oviposition is plastic in adults), to those with an extended adult phase that eclose with their ovaries in a completely pre-vitellogenic state (including the monarch butterfly *Danaus plexippus*, where all reproductive plasticity is manifest in the adult stage).

Within orders, we can tentatively distinguish a few consistent patterns. For example, in the Diptera, ovariole numbers are lowest in specialist taxa that brood their offspring (such as the tsetse fly), and highest in generalists that exploit rich, plentiful and ephemeral food resources (such as decaying fruit). Beetles that oviposit on dependable resources of borderline nutritional value (k-selected taxa, such as the flour beetle *Tribolium castaneum*) have few ovarioles (4 ovarioles per ovary), while those that feed on episodic, high nutrition resources (r-selected taxa, such as aphid-feeding ladybugs in the genus *Coccinella*) have many more ovarioles (from 10–60 or more ovarioles per ovary). Non-nesting dung beetles (Coleoptera: Scarabeidae) of the sub-family Aphodiinae have multiple ovarioles per ovary (seven, for example, in *Aphodius fossor*), while advanced nesting dung beetles of the sub-family Scarabeinae all have only one ovary which has exactly one ovariole (Halffter and Edmonds 1982). Still, I would caution the reader from making any firm conclusions based upon these results. With the exception of very few studies, such as that previously described for the Hawai'ian drosophilids, such ovariole number comparisons have not been subjected to rigorous phylogenetic analyses. It is critical that such a strict comparative analysis be done, on a broad assemblage of insect taxa, before we can hope to paint a complete picture of the relationship between reproductive ecology and ovariole number. As I will argue later, such analyses would be one way to identify the constraints (be it developmental, physiological or phylogenetic) that so clearly interact with insect ecology to mold the evolution of insect reproduction.

What are we to make of the non-conformist taxa indicated in Table 2? I argue that the explanations are both ecological and ontogenetic. Take the case of the blister beetles (family Meloidae) from the genus *Meloe*. This fascinating group of insects is perhaps best known for their hypermetamorphoses, in which several distinct larval morphologies are produced in turn, each one specialized for egg predation, bee parasitization, mimicry, overwintering and so on (reviewed in Gillott 1995). Less well known is their massive reproductive potential: *Meloe proscarabaeus* is reported to have approximately 1000 ovarioles per ovary (Büning 1994). This high reproductive potential is translated into enormous bouts of egg laying, in which the beetles dig a hole and therein deposit their immense pile of eggs (as many as 4218 eggs oviposited in one location in *M. cicatricosus*; Fabre 1857). How do these females handle such a tremendous egg load? Apparently, the gravid adult females, which had fattened themselves within the colonies of mason bees (genus *Anthophora*), simply drop to the ground and find an appropriate oviposition location by walking (Fabre 1857). The mostly sedentary lifestyle of these adults is associated with their greatly reduced wings. Flightlessness and enlarged abdomens (physogastry) are seen in other meloid species, sugesting that these insects trade-off wings for ovary development, a topic that we will return to below. Thus, *Meloe* females may have relieved themselves of the trade off between ovariole number/ ovary size and foraging/dispersal function by largely avoiding the need to expend energy in either foraging or dispersal.

Interestingly, there is substantial variation in egg production within the family Meloidae, and this variation appears to be associated with proximity of hosts for the larvae. In those meloids, for example, that attack locust egg pods, females deposit their clutches in the general vicinity of such egg pods. The total numbers of eggs deposited by these meloids is 10 or more times lower than in those bee parasitic meloids, whose larvae wait in flowers for a bee visitation so that they can hitch a ride back to the bee nest (Hinton, 1981). Since *M. cicatricosus* larvae must encounter and attach to a mason bee, each larva presumably has a fairly low probability of survival to reproduction (Robertson 1961). Hence, females should produce many larvae. Recently, cooperative behavior has been described for clutches of synchronously-hatching *M. franciscanus* larvae, who can mimic the female pheromone of their mason bee hosts, *Habropoda pallida* (Saul-Gershenz and Millar 2006). The pheromone attracts males, which the groups of larvae respond to by quickly arranging themselves in such a way as to increase the chance that one (or usually many) of their number will be able to attach to the male bee. These lucky larvae then transfer to the female bee during mating attempts, and finally get transported to that females nest to feed on pollen, nectar and eggs. In sum, the particular level of reproductive output of meloid beetles is consistent with life history theory, which predicts, all other factors being equal, a negative correlation between chance of survival to adulthood and egg number (Stearns 1992). Hinton further provides evidence that such egg production strategies correlate more with ecology than taxonomy. A rigorous phylogenetic approach would assist greatly in deciding this issue.

A second non-conformist group is the scale insects (Hemiptera: Coccidae). I excluded this group from Table 2 as they are the only insects[7] known to add substantial numbers of ovarioles during the adult stage (Weglarska 1961, Büning 1994). As indicated previously, this ability would seem to be highly advantageous to any insect whose adult habitat differs from their pre-adult habitat, since they would then be able to optimize ovariole number after assessing the quality of their adult environment. Long lived adults, which may be expected to experience temporal shifts in their adult environment during the adult period, would also be predicted to gain finer control towards reproductive optimization if they had the ability to add ovarioles after adult eclosion. Ovarian development in the ovoviviparous scale insect *Quadraspidiotus ostraeformis* has been studied in the most detail, and ovarioles are added by budding off ovarioles from undifferentiated cells in their tubular gonad (Weglarska 1961). In fact, this process is so divergent

[7] see footnote 5, above

from the mode of ovariole differentiation in typical insects that Weglarska suggests that *Q. ostraeformis* has no ovarioles *per sé*. However, since the follicles do develop along progressions of stages in these buds, they are certainly similar to ovarioles, and for clarity I will continue to use that term here. The epithelial sheath degenerates as ovarioles form in these coccids, which is presumably one of the reasons why the continual process of ovariole differentiation is mechanistically possible in the group. The permanently ensheathed ovarioles of typical insects (see Figure 1), and their apparent lack of pluripotent ovarian cells (see Kirilly and Xie 2007), would seem to explain why they do not retain the ability to add ovarioles as adults (via some sort of constraint). The nature of this constraint presumably has to do with the process of oviposition. The sheaths in typical insect ovaries separates the ovarioles from one another and from the body cavity, and the ovarioles are connected to the lateral oviduct in order to conduct the eggs there in preparation for ovipostion (see Figure 1). By contrast, adult *Q. ostraeformis* females have no such sheaths surrounding their individual ovarioles. Instead, they only have thin peritoneum surrounding the entire ovary, within which they brood their young (Weglarska 1961). As a result, the ovarioles can project any which way into this peritoneal cavity. Such a release from the constraints inherent in typical ovarian morphology would, thus, have resulted in a highly modified process of ovarian morphogenesis.

An analogous situation is found in the paedogenetic (larvally-reproductive) gall midges, which truly lack ovarioles altogether (see Figure 5 and below for more on this group). Indeed, even non-paedogenetic gall midge species lack true ovarioles, as their egg tubes form by secondary fusion (Matuszewski 1968), rather than the typical assembly-line process described in the introduction and illustrated in Figure 1. This derived pattern of ovarian morphogenesis in the non-paedogenetic gall midges may have been one of the features that preadapted[8] that taxon for the evolution of paedogenesis, something which occurred at least twice independently in the group (Hodin and Riddiford 2000b).

The third non-conformist group is the jumping plant lice (Hemiptera: Psyllidae), which have been reported to contain up to 100 ovarioles per ovary (Büning 1994), 10 times the number found in typical hemipterans (see Table 2). Many psyllids are major crop pests, including the Asian citrus psyllid *Diaphorina citri*. This species can survive for months as adults awaiting appropriate oviposition conditions: young, furled leaves. When

[8] *sensu* Gould (1984): features adapted for one function, that are fortuitously suited for another.

such conditions arise, the insects can utilize their high reproductive potential to lay as many as 800 eggs in a few days (Mead 2002), an ability undoubtedly enhanced by the high ovariole numbers characteristic of psyllids. The fact that the adults can feed on mature leaves, while their offspring require young leaves, has two important consequences: 1) it allows the adults to mature eggs while awaiting the appearance of the young leaves for their offspring; 2) it allows adults with high egg loads to remain on their host plant, reducing the need to disperse to find an oviposition site. As in the case of *Meloe* discussed above, such a situation obviates the typical trade off between ovariole number/ovary size and foraging/dispersal function by largely avoiding the need to expend energy in foraging and dispersal. Again, the explanation for their deviant numbers of ovarioles appears to be both ecological and ontogenetic.

The observation that many of the aforementioned trends in ovariole number variation apply across different insect orders makes the situation in the Lepidoptera all the more striking. How can we account for the apparent total lack of variation in ovariole numbers in the Lepidoptera? The patterns that we have described above, where different ecological parameters correlate with differences in ovariole number across several insect orders, indicate that ovariole numbers are subject to natural selection. However, these same selective criteria would presumably apply to Lepidoptera as well. The nine families of Lepidoptera noted in Table 2 include species with a broad range of ecologies and life histories, from r to k strategists, unitvoltine to multivoltine, generalists to specialists, migratory to non-migratory species, short lived to long lived adults, tropical to temperate forms, small to large body size, and so on. Also, they include representatives from at least two relatively basal lepidopteran groups, the leafroller moths (*C. pomonella* from the family Tortricidae) and the yucca moths (family Prodoxidae), in addition to several highly derived families (Kristensen and Skalski 1999). Still, all have exactly 4 ovarioles per ovary.

Such a situation represents a perfect candidate for a developmental constraint (see Hodin 2000). But what is the nature of this constraint? Very little is known concerning the ontogenetic processes underlying ovariole differentiation in the Lepidoptera (Büning 1994). Interestingly, ovariole number determination in female Lepidoptera takes place in the embryo, and is coincident with sperm follicle tube formation in males; male embryos also form four gonadal tubes per gonad (Figure 3; Grünberg 1903). By contrast, testis and ovarian differentiation follow quite distinct ontogenetic routes in most non-lepidopteran insect species (reviewed in Büning 1994), again suggesting an additional and unique level of constraint on lepidopteran

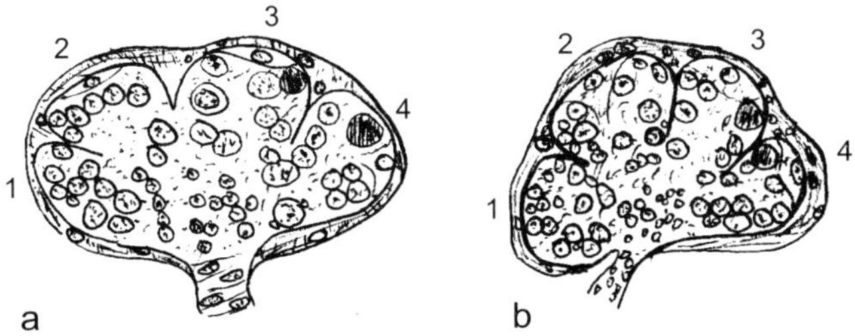

Fig. 3 Cross sections through the embryonic gonads in *Bombyx mori*, redrawn from Grünberg (1903, Figures 1 and 18). (a) embryonic ovary; (b) embryonic testis. The four ovarioles (numbered in a) are already evident at this stage (see also Beckemeyer and Shirk 2004), much earlier than is the case in most insects (reviewed in Büning 1994). Homologous gonadal tubes are differentiating in the testis as well at this stage (numbered in b). In typical insects, ovarian differentiation follows a quite distinct ontogenetic trajectory from testis differentiation (reviewed in Büning 1994). This early ovariole differentiation and the similar male and female ontogenetic trajectories in these moths may help to explain the apparent total lack of variation in ovariole number across the Lepidoptera.

gonadal development. In other words, it is possible that the tightly coupled (canalized?) processes of male and female gonadal formation in Lepidoptera have constrained their evolutionary potential.

One useful avenue for investigating the nature of such (presumed) constraints is to examine the analogous processes in an unconstrained outgroup. The caddis flies (order Trichoptera) are widely accepted as the sister group to the Lepidoptera (e.g. Kristensen 1984; Wheeler et al. 2001). Since the net-spinning caddis fly *Parasthenopsyche sauteri* has approximately 130 ovarioles per ovary (Matsuzaki 1972), caddis flies do not appear to be similarly constrained. Comparative studies on ovarian differentiation in caddis flies and lepidopterans might, therefore, yield some insight into the presumed constraints on ovariole determination in lepidopterans. Despite the wide geographic distribution, ease of collection and high population densities of larval caddis flies, ovarian differentiation remains mostly unstudied in the group (Büning 1994), so comparative data are lacking.

Dividing Up Who Goes into Labor: Ovariole Number Plasticity in Social Insects

By far the most extreme examples of ovariole number plasticity are the queen-worker differences in ovariole numbers in many social insects (Keller 1993). Although data in some taxa suggest a genetic component to queen-worker determination (Michener 1974), by far the predominant mechanism involves differential feeding of larvae (reviewed in Wheeler 1986). So, when a female honeybee larva is born, she has the potential of developing either as a queen or a worker, depending upon the quantity and quality of food that she receives during larval development (Beetsma 1979). If she gets the queen ration, then she ecloses with up to 200 ovarioles in each ovary, and a fully functional reproductive system with sperm storage organs and reproductive fluid ducts. If, by contrast, this larva only receives the meager worker food allotment, then she ecloses with less than 10 ovarioles per ovary, an underdeveloped sperm storage organ, and an inability to mate (Michener 1974).

The multiple independent origins of sociality within the Hymenoptera, and the entirely independent origin of sociality in the termites (order Isoptera), provide fertile ground for testing evolutionary hypotheses regarding, for example, the mechanisms underlying plastic and genetically-fixed differences in ovariole numbers. Caste differences in ovariole numbers between queens and workers have evolved independently in many social taxa (Keller 1993). This finding, in and of itself, indicates two important points: first, that there is strong selection pressure for increases in ovariole number to maximize reproductive output (in queens/reproductives); and second, that these high ovariole numbers involve trade-offs with non-reproductive tasks (in workers vs. queens/reproductives). Thus, the often lower ovariole numbers found in reproductive females from non-social species when compared to their social counterparts indicates that the former are trading-off potential reproductive output against non-reproductive functions, such as flight, foraging, defense, somatic growth and so on. The extreme specialization of queens for reproduction in many social species releases them from such trade-offs.

An examination of the situation in queenless social insects provides some potent support for the presence and importance of such trade-offs in social colonies, though in this case with respect to öocyte maturation rather than ovariole number. For example, the queenless ponerine ant *Pachycondyla (=Ophthalmopone) berthoudi* has variable proportions of reproductives (gamergates), depending on colony conditions (i.e. they are polygynous). Sledge et al. (1999) compared several such colonies, examining the behavioral profiles of individual ants, and then dissecting them to examine their state of öocyte maturation. The gamergates in colonies with high proportions of reproductives were much like workers in their colony task profiles, and had only moderately mature ovaries. By contrast, gamergates in colonies with a low proportion of reproductives performed no colony labor tasks, and had many more mature ovaries. Ito and colleagues (1996) obtained similar results for the East Asian ponerine ant *Odontomachus rixosus*, another species with scores of gamergates in each colony. In *Ectatomma tuberculatum*, a Central and South American ponerine species with facultative polygyny (Cook 1905), there is a strong, age-independent correlation between colony duties and reproduction: nurse ants had more mature ovaries and laid trophic eggs (to be fed to larvae), while forager ants had degenerate ovaries and well-developed poison glands (Féneron et al. 1996). Thus, the plasticity of queenless/polygynous ants for reproductive versus worker tasks is functionally related to ovarian developmental plasticity. An examination of the hormonal mechanisms underlying such a trade-off would be extremely edifying.

One widespread misconception concerning reproduction in social insects is that workers are "sterile." This is only technically true in a handful of highly derived social insects, such as the stingless bee (Apidae: Meliponini) species *Frieseomelitta varia* and *Trigona minangkabau* (Cunha et al. 1986; Suka and Inoue 1993), the Ceylon Black Termite (Isoptera: Termitidae) *Hospitalitermes (=Eutermes) monoceros* (Bugnion 1909) and the fire ant (Hymenoptera: Formicidae) *Solenopsis invicta* (Hölldobler and Wilson 1990). In the vast majority of social insects, workers have functional ovaries, but full öocyte maturation and oviposition is repressed (to varying degrees in different taxa) by the presence of a queen or queens (e.g. Michener 1974, Hölldobler and Wilson 1990, Noirot 1990). Thus insect sociality is not the just-so story of queens who have all of the offspring, and workers who toil away their whole lives for the sole inclusive fitness advantage inherent in the reproductive potentials of their little sisters and brothers. It is true that workers in many social hymenopteran taxa are incapable of mating (though there are some notable exceptions that I will discuss below), but they still retain the potential of laying unfertilized (male-producing) eggs that can develop and ultimately mate with conspecific queens. The varying degrees to which reproduction in such workers is held in check by the presence of the queen is a topic that I will now consider.

Let's begin by reviewing the reproductive potential of workers and queens/reproductives in two different social taxa: basal termites and honeybees. Later I will return to these same taxa, and also include a discussion of the stingless bees and the multi-queen (polygynous) ants. If the reader's favorite social insect is not among these four groups, he or she can find information on other social taxa in one of several excellent reviews (Wheeler 1986; Engels and Imperatriz-Fonseca 1990; Peeters 1991; Keller 1993; Peeters 1993; West-Eberhard 1996; Robinson and Vargo 1997; Thorne 1997; O'Donnell 1998; Reeve and Keller 2001; Thorne and Traniello 2003; Hartfelder and Emlen 2004; Schwarz et al. 2007).

Isoptera

All termites are social, and are thought to have arisen from a cockroach-like ancestor. Phylogenetic data, as well as the incipient sociality in some cockroach taxa, provides strong evidence for this evolutionary scenario (reviewed in Thorne and Traniello 2003). Two key features distinguish sociality in the Isoptera and the Hymenoptera. First, termites are hemimetabolous insects, whose pre-adults are not the helpless grubs characterizing the social Hymenoptera. As a result, immature termites can function as workers. Furthermore, these immatures not only have the potential to follow several different ontogenetic routes-to workers, soldiers, reproductives and so on-but some basal taxa exhibit an amazing plasticity in that they can backtrack under the appropriate conditions, by molting to earlier, less differentiated forms, and then continue development along totally altered trajectories (Noirot 1990). Thus, worker termites have the potential to undergo a regressive molt and then begin to develop as a secondary reproductive. The second major distinguishing feature of the Isoptera is that they do not have the haplo-diploid sex determination system found in all Hymenoptera. Thus, all termite reproductives can lay both male and female eggs, and sisters and brothers are equally related. In the Hymenoptera, by contrast, many worker females cannot mate, and can therefore only lay male eggs. Also, if their queens only mated once, then the workers share, on average, 75% relatedness to their sisters. For these reasons, the evolutionary dynamics within termite and hymenopteran colonies are predicted to be quite different (Hamilton 1964).

Still, caste differences in ovariole number are features that many termites share with the majority of social hymenopterans. And, as in many hymenopterans, caste in termites is determined environmentally rather then genetically (Noirot 1990). The termite family Termopsidae is often proposed as the prototypical ancestral termite (Thorne and Traniello 2003), despite the fact that phylogenetic analyses robustly place the families Mastotermitidae and Hodotermitidae, respectively, as the most basal taxa within the Isoptera (e.g. Eggleton 2001; though one topology places the Hodotermitidae as the sister group to the Termopsidae). The justification for this unconventional character analysis is that the Mastotermitidae and Hodotermitidae are widely considered to be highly derived in their social organization (Thorne and Traniello 2003). For example, the Mastotermitidae and Hodotermitidae have more rigid castes than do the Termopsidae and other presumed basal families (reviewed in Thorne 1997). I will tentatively follow Thorne and

Traniello's suggestion, while expressing reservations about the cladistic relevance of their hypothesis.

Female termite workers develop along two different pathways: alate (the sexual forms, in which wings or wing buds are present) and non-alate (workers with no wing buds). Termopsids are characterized by extreme plasticity in reproductive tasks, even within non-alates. All castes, except possibly soldiers, retain the capacity to develop as either workers or reproductives (reviewed in Thorne 1997). And in *Archotermopsis*, soldiers have gonads that are as fully developed as mature alates (Imms 1920). If a termite colony has healthy primary reproductives, then they suppress reproduction in other castes through pheromones. A non-alate termite that becomes reproductive is known as a secondary reproductive.

A unique feature of secondary reproductives, at least in the termopsid *Zootermopsis angusticollis*, is that they increase their ovariole numbers as they begin to develop along the reproductive trajectory. [Note that these reported increases in ovariole number were judged at a stage when maturation had begun; thus the authors cannot distinguish between an actual increase in ovariole number and an activation of pre-existing "filamentous" ovarioles.] In one experiment (Brent and Traniello 2001b), secondaries increased their numbers of ovarioles by approximately 25% in 30 days (from 26 to about 32 ovarioles) when housed with non-alate workers and one reproductive female in experimental colonies. Primary (adult stage) reproductives placed in the same conditions underwent no change in ovariole number (approximately 31 ovarioles throughout), as would be expected for insect adults in general. In an even more dramatic experiment (Brent and Traniello 2001a), secondaries housed with one male and 6 non-alate workers increased their ovariole numbers by almost 50% in 60 days (to 38 ovarioles), while primaries again underwent virtually no change in ovariole number. Still, the primaries had greater egg laying rates under all conditions, suggesting a greater rate of öocyte maturation (see more on this below). In any case, these experiments demonstrate that termites can adjust their ovariole numbers in a seemingly adaptive fashion depending on colony condition.

Interestingly, termopsids have small colonies inhabiting decaying wood and do not leave their nest to forage (Thorne and Traniello 2003). These features may or may not be ancestral features for termites, but certainly provide a plausible explanation for their lack of caste rigidity. Once again, we see here a situation where a presumed relaxation of the ovarian growth *vs.* foraging function trade-off leads to an increase in reproductive potential.

And as West-Eberhard has pointed out (1978, page 853), "as long as a female has 'hope' of laying eggs...her participation in the worker tasks can be viewed as possibly or partially an investment in her own reproductive future." Clearly, the smaller the colony, the greater the chance that such a hope will be fulfilled (Thorne and Traniello 2003). This argument not only provides a compelling account for termopsid workers' acceptance of their non-reproductive status, despite their state of reproductive readiness, but also indicates a most plausible scenario for how sociality could have arisen in the first place from non-social ancestors. In fact, there is good reason to believe that such a scenario has played itself out repeatedly in the evolution of many of the independently derived, social taxa.

Honeybees

Honeybees (family Apidae, group Apinae) have some of the most spectacular instances of queen-worker differences in ovariole number (reviewed in Michener 1974, Ruttner 1988). These differences are entirely due to phenotypic plasticity: depending on the food allotment, a given female larva can develop as either a queen (with up to two hundred ovarioles per ovary) or a worker (generally with fewer than 10 ovarioles per ovary). In addition, workers are not endowed with the full reproductive system of queens, and thus can only lay unfertilized (male-determined) eggs. As I will discuss below, maturation in worker ovaries is repressed by the presence of a queen, but workers still have the capacity to reproduce. Still, even in the absence of a queen, a reproductive worker is quite limited in her reproductive output relative to queens, owing to her greatly reduced number of ovarioles.

The mechanisms underlying queen worker differences in ovariole number have recently begun to be elucidated. In the fourth instar larvae, in both queens and workers of the European honeybee *Apis mellifera carnica*, the ovaries contain over one hundred incipient ovarioles. But during the final stage and the lead up to metamorphosis, worker ovaries undergo massive cell death, which ultimately removes the great majority of these incipient ovarioles (Hartfelder and Steinbrück 1997; Reginato and Cruz-Landim 2002, 2003). Hormonal studies indicate that juvenile hormone levels in larval honeybees, which are higher in incipient queens than in workers, may underlie these morphogenetic differences (Rachinsky et al. 1990; Schmidt-Capella and Hartfelder 1998, 2002; reviewed in Hartfelder and Engels 1998).

Although this queen-worker ovariole number disparity holds for all honeybees, there is substantial variation in worker ovariole numbers in different honeybee species and races (Table 3). Perhaps the most intriguing situation in honeybee reproduction comes from the cape honeybee, *Apis mellifera capensis*. It has been known for some time that this bee is unique among honeybees (and indeed all social Hymenoptera) in that the workers are capable of laying female eggs by thelytokous parthenogenesis (Onions 1912). Clearly such a situation creates a different social dynamic within the hive, since workers have the potential of winning the reproductive jackpot by giving birth to a future queen. Still, this ability has until recently only warranted a footnote in accounts of honeybee reproduction, as this unique bee is restricted to a small region in the very southern tip of South Africa. In fact, at one point, it was suggested that the dominant, aggressive southern African honeybee race, *Apis mellifera scutellata* (the bee race that begot the Africanized bee when transported to Brazil), was destined to overrun the poor, docile *capensis* bees (Ruttner 1977). Ironically, the reverse has happened. In recent years, a single clonal lineage of *capensis* bees has arisen that has the capacity to invade *scutellata* colonies, as described in more detail later in the chapter.

In queenright capensis hives, the queens are able to maintain some semblance of reproductive dominance. But even here, the workers are more uppity than in a typical *A. mellifera* colony, laying substantial numbers of eggs even in the presence of the queen pheromone (Moritz et al. 1999, Pirk et al. 2002). The possibility that the larger ovaries, themselves, are responsible for this difference in capensis is an intriguing one that remains untested. Interestingly, the trend in other honeybee species provides some support for this suggestion. As Table 3 shows, the queen-worker differences in ovariole number are far less dramatic in *A. cerana* than in typical *A. mellifera* races, and as we shall see, *A. cerana* workers have greater reproductive capacity than typical members of the genus. Furthermore, *A. dorsata* have the least profound worker-queen ovariole number differences among honeybees (Table 3), and have been described as having "the least pronounced caste dimorphism" in the genus (Engels and Imperatriz-Fonseca 1990, page 212). Recent work on Africanized honeybees in Brasil showed that worker bees with greater ovariole numbers and more active ovaries came from identifiable, genetic "patrilines" that dominated in new drone rearing after queen removal (Makert et al. 2006). A second study with *A. mellifera* showed a strong connection between progression through worker tasks (such as age at first foraging), pollen *versus* nectar foraging, and both ovariole number and degree of ovarian maturation state (Amdam et al. 2006). Furthermore, endocrinological variation among genotypes underlies these differences

Table 3 Differences in ovariole numbers between queens and workers in honeybee species, and in different races of *Apis mellifera*. Indicated are mean numbers of ovarioles per ovary (reported ranges in parentheses). Asterisks indicate range data for workers that was calculated differently from the other races/species: shown here are ranges of mean ovariole number values across the geographic ranges of *A. m. scutellata* and *A. m. capensis* in South Africa, as reported by Hepburn and Radloff (2002). Note that the numbers throughout the table were obtained in a variety of different conditions (temperature, seasonal, wild versus cultivated colonies, etc.), and not all of the *Apis mellifera* data are from colonies in their native area. Thus, the reader is cautioned from making too much of particular, small differences among races and species. Data from Alpatov 1938, Velthuis 1970, Michener 1974, Woyke et al. 1974, Weiss 1975, Buys 1988, Ruttner 1988, Koeniger et al. 1990, Dedej et al. 1998 and Hepburn and Radloff 2002. *A. andreniformis* data also from N. Koeniger pers. comm. Locality information from Ruttner (1988) and Wongsiri et al. (1996).

species	*native locality*	*queen ovariole number*	*worker ovariole number*
Apis mellifera adansonii	Central Africa	?	(1-11)
Apis mellifera carnica	Austrian Alps, northern Yugoslavia, Danube valley	175 (146-204)	(2-11)
Apis mellifera caucasica	Caucasus region	175 (169-181)	?
Apis mellifera lingustica	Italy	172.5 (155-190)	3.3 (1-24)
Apis mellifera mellifera	Central to southern Russia (Balkans to NE Mongolia)	162 (127-183)	5.3 (1-12)
Apis mellifera scutellata	Southern Africa	140 (136-149)	(2-5)*
Apis mellifera capensis	Cape region, South Africa	139 (127-151)	(9-18)*
Apis florea	Red Sea region, India, Bangladesh, Burma, Yunnan Province (China), southern Vietnam, Cambodia, Lao, Thailand	?	?
Apis andreniformis	NE India, Yunnan Province (China), Thailand, Lao, southern Vietnam, Malaysia, Sumatra, Java, Borneo, Palawan	48 (36-52)	2?
Apis cerana	Afghanistan, India, Pakistan, SE Asia, China, Korea, Japan, Indonesia (west of New Guinea), Philippines	73	8.6 (4-21)
Apis dorsata	India, Pakistan, SE Asia, Indonesia (west of New Guinea), Philippines	130	33 (17-60)

(Amdam et al. 2007). The implications of these data are profound: the size of the ovary, which is at center stage in the queen-worker social distinction in honeybees, is also intimately connected to the endocrinological status of the adult workers through genetic factors acting on ovary development during the larval phase [a connection also suggested by Sledge and colleagues' (1999) experiments on Ponerine ants described above, as well as Cepeda's (2006) studies on stingless bees (Apidae: Meliponini)]. We will revisit these studies below when considering the multiple evolutionary origins of eusociality among the Hymenoptera.

Which Stage-hands Set the Stage? The Genetic Determination of Ovariole Number

Not surprisingly, almost all work on the genetic determination of ovariole number comes from studies with *Drosophila melanogaster* and its close relatives. Because of the substantial (up to two-fold) intra-specific variability in ovariole numbers within *D. melanogaster* (Pappas and Engstrom 1974), and the even greater scope of interspecific variability within the melanogaster species group (see Table 1; Figure 2), this approach has proved and will continue to prove fruitful. Intraspecific hybridization experiments in *D. melanogaster* suggest that loci affecting ovariole number are concentrated on both of the large autosomes (chromosomes 2 & 3) rather than the X chromosome (Coyne et al. 1991; Chakir et al. 1995). Jones' (2004) interspecific hybridization studies showed that all chromosomes (especially chromosome 2 and the X chromosome) contribute to the lower fecundity of *D. sechellia*[9] when compared to *D. simulans*. However, unlike the Coyne et al. and Chakir et al. studies, Jones mapped differences in egg production, thus identifying factors involved in the lower öocyte maturation and/or oviposition rates in *D. sechellia*. So, it seems, there is not a tremendous amount of overlap between the genes involved in ovariole number and those involved in öocyte maturation/oviposition. Given the substantial dissociability among these different processes, both evolutionarily and developmentally, as I will describe throughout this chapter, this seems hardly surprising.

Initial QTL mapping studies indicated that a thankfully small number of loci might contribute the vast share of the heritability component in ovariole number differences among laboratory-selected lines of *D. melanogaster*

[9] *D. sechellia* are endemic to the Seychelles Islands, and are specialists on a fruit (*Morinda citrifolia*) that is toxic to other drosophilids. They also have the lowest ovariole number in the *melanogaster* species group (see Table 1; Figure 2).

(Wayne and Mackay 1998; recent studies might suggest otherwise, Bergland et al. submitted). The chromosomal regions with the predominant effects on ovariole number were further narrowed (Wayne et al. 2001), and 34 candidate loci in that region have now been identified (Wayne and McIntyre 2002). Two of the most noteworthy genes among these loci are *yellow-f* and *Actin87E*. The predicted *yellow-f* protein sequence is only similar to one known group of metazoan proteins: the honeybee (*Apis mellifera*) royal jelly proteins (approximately 30% amino acid identity; Maleszka and Kucharski 2000; Malecová et al. 2003). Because royal jelly seems to be important for the development of caste specific (including ovariole number) differences between queen and worker honeybees (Beetsma 1979), this finding for *Drosophila* raises the exciting possibility that the royal jelly proteins may have a common function in ovarian development in these divergent insect taxa. Still, that function would have been substantially modified in honeybees, where the proteins are obtained by the larvae through feeding by nurse bees, and are apparently involved in differential cell death in worker ovarioles (see Hartfelder and Steinbrück 1997), a process not seen in *Drosophila* ovarioles (Hodin and Riddiford 2000a). Actin, too, has been implicated in queen-worker ovariole number differences in honeybees (Schmidt-Capella and Hartfelder 2002), and *Actin87E* is one of several *Drosophila* actin genes. Further comparative studies are clearly warranted to determine whether these similarities are due to parallel evolution or common ancestry of ovariole determination mechanisms.

Simple experiments [using techniques such as those that we presented in Hodin and Riddiford (2000a)] with these 34 candidate loci from the laboratory lines would substantially increase the likelihood of identifying given genes that are involved in ovariole number variability in natural populations, or even differences due to plasticity. In other words, if mutations in these candidate loci phenocopy specific ontogenetic differences (either plastic or genetically-fixed) that we identified in 2000 (see Table 1), then those loci would be of particular interest. For example, partial loss of function mutations in the ecdysone receptor (*EcR*) and ultraspiracle (*usp*) genes, whose protein products dimerize to form the insect ecdysteroid receptor, phenocopy one of the mechanisms (mechanism "**d**" in Table 1) accounting for the lower ovariole numbers (relative to the Sevelen strain of *D. melanogaster*) in *D. sechellia, D. mauritiana, D. yakuba,* the Honduras strain of *D. simulans* and the Nahal Canyon strain of *D. melanogaster* (Hodin and Riddiford 1998, 2000a). Indeed, one of the epistatic QTLs identified by Bergland et al. (submitted) includes the EcR gene (A. Bergland, pers. comm.).

Intriguingly, Malecová et al. (2003) identified USP binding sites in the 5' untranslated regions of all five of the honeybee royal jelly protein genes, directly upstream of their translational start sites. Such USP binding sites have been proposed as possible JH-mediated USP regulatory sequences, distinct from the classic EcR/USP-mediated ecdysteroid binding sites (Jones and Sharp 1997). Indeed, methyl farnesoate (MF), a biosynthetic precursor of JH, binds USP 150x more strongly than does JHIII (Jones et al. 2006). This finding raises the exciting possibility that MF could be a bona fide ligand for USP. USP, in the absence of EcR, is expressed at high levels in the differentiating terminal filament cells of the larval ovary in *D. melanogaster* (and its sister species as well; Hodin and Riddiford 2000a), and alterations in this terminal filament expression of USP results in defects in ovariole morphogenesis and reduction in ovariole number (Hodin and Riddiford 1998). These results could provide the first evidence of a direct link between the nutritional (yellow/royal jelly proteins) and hormonal (ecdysteroid/JH/MF) regulation of ovariole number. Recent studies suggest that insulin-like signaling may also be directly involved in this apparent link between nutrition and hormonal regulation (Tu and Tatar 2003; Flatt et al. 2005), and a recent QTL study of ovariole number differences between *D. simulans* and *D. sechellia* found that the genomic region of largest effect contains the *insulin receptor* (*InR*) ortholog in *D. simulans* (Orgogozo et al. 2006). Bergland et al. (submitted) have also identified QTL's with large effect on ovariole number that contain genes involved in insulin and related signaling pathways. Indeed, mutations in insulin pathway genes in *D. melanogaster* result in reduced ovariole numbers (Tu and Tatar 2003; Richard et al. 2003).

Thus, with these recent developments, we are tantalizingly close to being able to determine if the genetic factors that are involved in inter- and intra-specific ovariole number differences are also involved in ovariole number plasticity. Not only would such information add tremendously to our understanding of reproductive plasticity in insects, but it would be groundbreaking for our general understanding of the evolution of phenotypic plasticity as well.

So Ovariole Numbers Are Set: Now What? Plasticity in Öocyte Maturation

In all insects except coccids (see above), maximal ovariole number is fixed by adult eclosion, and sometimes much earlier. However, plasticity in reproductive potential hardly ends here. In fact, modulation in the rate of öocyte maturation is probably the predominant mechanism by which insects attempt to optimize their reproductive output to suit environmental conditions (Labeyrie 1978). Many different external stimuli are known to modulate the rates of öocyte maturation in different insects: temperature, food availability, day length, the presence of mates, mating, the availability of ovipositional resources, oviposition itself, the presence of dominant reproductives (in social insects), and so on (reviewed in Labeyrie 1978, Wheeler 1996, Hopkins and Ekbom 1999, Tammaru and Javois 2000, Papaj 2000, Jervis et al. 2005). This is eminently sensible. Many organisms, including insects, are known to trade off current versus future reproduction (reviewed in Hopper 1999) and to vary substantially in degree of parental investment per offspring (e.g. Halffter and Edmonds 1982, Tallamy 1984). This particular trade-off can be more serious than just a timing of reproduction issue: reproduction in a diversity of insects induces a direct cost to life expectancy (Partridge et al. 1987, Tatar and Carey 1995, Sgro and Partridge 1999; Herman and Tatar 2001, Jervis et al. 2005, Flatt and Kawecki 2007) *via* a nutrient allocation mechanism (Tatar and Carey 1995). Also, such allocation trade-offs vary within taxonomic groups among species with differing life histories (e.g. Stevens et al. 2000, Emlen 2001) suggesting that such trade-offs are moldable by natural selection. Furthermore, not only is the production of eggs energetically costly, but the extra weight imposed by a fully mature ovary can have substantial mechanistic consequences as well, such as speed of anti-predator escape or lift in flying insects (Berrigan 1991). Therefore, insects under variable conditions would be expected to adjust their processes of öocyte maturation according to the suitability of the environment. The goal, of course, would be for a female to produce the maximum number of eggs in her lifetime that have the maximum chance of surviving to adulthood (within the parameters of ontogenetic and other constraints, of course). There are many excellent examples that suggest that this is precisely what insects attempt to do.

Plasticity in the timing of öocyte maturation has been adequately reviewed on numerous occasions (see above). Still, most (though not all) of these reviews have focused on mainly one of two areas: ecological stimuli influencing öocyte maturation, or the mechanisms (hormonal and otherwise) controlling the maturation process. My purpose here is not to attempt a comprehensive review of this unwieldy subject. Instead, I intend to illustrate, with specific examples, four points: 1) different insects use different environmental cues to modulate öocyte maturation in a seemingly adaptive fashion; 2) öocyte maturation entails trade offs with somatic functions (the most famous is flight, the so-called "ovary flight syndrome"); 3) the internal mechanisms regulating öocyte maturation are numerous, and vary even within closely related groups; and 4) a thorough understanding of plasticity in insect reproduction will only come from a holistic approach, combining ecology and ontogenetic mechanisms in a phylogenetic context. I will begin this section by examining specific examples of plasticity in öocyte maturation in a wide diversity of insect taxa. I will conclude the section with some generalized thoughts concerning patterns (or lack thereof) in the control of öocyte maturation in insects.

Perhaps the most typical, and intuitively obvious, influence on öocyte maturation is nutrition (Labeyrie 1978). Many insects eclose with substantial stored materials left over from pre-adult development. In such cases, it is not uncommon to find adults that eclose with öocyte maturation underway, or even complete. We can make several predictions concerning selective regimes that might favor such a reproductive strategy: 1) plentiful food resources in the larval environment; 2) lack of larval competition; 3) low larval predation and/or parasitization rates; 4) lack of adult feeding and/or short lived adults; 5) typically poor or ephemeral adult food resources; 6) typically heavy predation pressures on adults; 7) intense competition for oviposition sites; 8) high probabilities of finding a mate quickly; and 9) oviposition sites nearby (or identical with) the larval habitat that the individual just left. Several of these hypotheses have not, to my knowledge, been rigorously tested, though examples of others abound (reviewed in Labeyrie 1978, Papaj 2000).

Blowflies Need Protein

For those insects that neither eclose with substantial reserve materials, nor have mature ovaries at eclosion (as well as those insects, such as mosquitoes, that will mature multiple batches of eggs in succession), adult feeding is often the stimulus for öocyte maturation (reviewed in Wheeler 1996, Papaj 2000). In such cases, proteins are often limiting, since excess proteins are needed for the synthesis of vitellogenins (the major class of yolk proteins produced by the fat body and transferred to the öocytes during vitellogenesis). Thus, protein intake (feeding) is often the direct trigger for öocyte maturation.

Adult sheep blowflies, *Lucilia sericata* (Diptera: Calliphoridae), are short-lived (less than a week, though they can survive longer in captivity) and cannot initiate vitellogenesis until they find a source of protein, such as a live sheep, carrion, or even manure. Barton Browne et al. (1979) and Wall et al. (2002) fed females liquid liver in different concentrations, and noticed that öocyte maturation seemed to pass through at least two protein threshold stages. At the lowest protein levels, vitellogenesis began in all ovarioles (stage 1) but was not completed.[10] At moderate levels, vitellogenesis continued (stage 2), but many follicles arrested (their development could be reinitiated) or were resorbed, allowing the materials to be reused by the female, perhaps for somatic maintenance. At high levels, the majority of the follicles matured, although some were still resorbed. The authors suggest that these responses, where the flies continually assess their nutritional state at multiple stages, and modify öocyte development accordingly, are bet-hedging strategies regarding possibilities of finding further protein resources. In protein limited individuals, it would be disadvantageous to attempt to mature too many eggs, since if no further protein sources are found, effective reproduction may be impossible in these short lived adults. Moderately fed individuals can remain at a later resting stage, to take full advantage of possible additional resources, if they are found. Another (not mutually exclusive) explanation is that the bet-hedging strategy will allow them to mature the maximum number of eggs at one time under any given feeding regime, since the adults will hold and then lay essentially all of their eggs at once (independent of the egg load) once an appropriate host is located (Barton Browne et al. 1990). And for *Lucilia*, that would likely be their one and only opportunity for oviposition.

In accord with this ecological account for the multi-stage maturation process in *Lucilia* is the observation that this species apparently starts to mature all of its primary öocytes simultaneously in every ovariole (Barton Browne et al. 1979). While this ability clearly allows for the rapid, synchronous maturation of a complete batch of eggs, it may also impose a constraint upon öocyte maturation, such that a miscalculation of maturational timing could be disastrous for their chances of rearing successful offspring. Thus, the multi-stage maturational process of *Lucilia* might be best understood as a bet-hedging life history strategy in the context of constraints relating to their particular mechanics of synchronic öocyte maturation.

[10] Such maturational synchrony across all ovarioles is a common feature among several disparate so-called "anautogenous" species of dipterans (Barton Browne 2001). Such a situation, which is not by any means universal among insects as a whole, may be considered a developmental constraint in *L. sericata* that could help explain their threshold pattern of öocyte maturation.

Lubber Grasshoppers and the Limits of Food-induced Plasticity

Lubber grasshoppers, *Romalea microptera* (=*guttata*) (Orthoptera: Acrididae), eclose as adults with immature ovaries, containing only early-stage öocytes in each of its ~ 65 ovarioles. Well-fed adults oviposit for the first time about one month later, while those with a reduced food ration (87% reduction in amount offered) delay first oviposition by an additional 2 weeks (Moehrlin and Juliano, 1998). Surprisingly, a switch from high to low food 14 or 21 days after eclosion did not result in a significant lengthening of time to oviposition. Thus, the reproductive cycle for lubbers can be described in terms of a plastic (early) and canalized (late) phase (Juliano et al. 2004). The switch from the plastic to the canalized phase corresponds to a transient rise in the titers of juvenile hormone (JH) (Hatle et al. 2000). Although no direct connection has been demonstrated between this rise in JH and the reproductive events that follow, these results suggest that the release of JH may initiate a chain of events involved in the control of reproductive timing, which can no longer be modulated by external conditions. By contrast, number of öocytes being developed remains plastic throughout the reproductive cycle (see below), so the reproductive process as a whole is not canalized, just the timing of reproduction.

In sum, plasticity in timing seems to be constrained late in the oviposition cycle in lubbers, though at different times in different populations (Hatle et al. 2002). A parallel process of canalization in the timing of the last molt in lubbers (Hatle et al. 2003b) might indicate that there is something in general about the hormonal control of development in this species that constrains timing of developmental events (the similarities in the ovipositional and ecdysis behavioral networks in insects may be related to this apparent constraint; see Figure 7, below)

This canalized (or constrained) ontogenetic trajectory contrasts sharply with the highly plastic reproductive trajectory in the sheep blowfly (Diptera: Calliphoridae) discussed above. What is the nature of this difference? Does the apparent use of JH as an öogenic regulator in lubbers impose this constraint in timing? If that is the case, then one would need to account for the extreme variation in the functions of hormones in insect reproduction in different insects, including orthopterans (Strambi et al. 1997). Interestingly, vitellogenesis in black blowflies (*Phormia regina*, also in the family Calliphoridae), as in most dipterans, seems to be under the control of ecdysteroids rather than JH (Yin and Stoffolano 1997), while ecdysteroids are probably not involved in vitellogenesis in lubbers (Hatle et al. 2003a). Indirect evidence suggests that ecdysteroids might also be responsible for the threshold stage progression described above for the sheep blowflies (A.D. Clift 1972 Ph.D. thesis, cited in Barton Browne at al. 1979), as has been shown at least in part for other so-called "anautogenous" dipterans: those in which their öocytes mature synchronically, such as *P. regina*, mosquitoes (family Culicidae) and the house fly (family Muscidae) *Musca domestica* (Barton Browne 2001). Still, it would be far too cavalier at this stage to suggest any connection between these differences in JH/ecdysteroid control and relative plasticity in ovary maturational processes in blowflies and lubbers.

Several key pieces of information, including detailed studies on the ovarian events occurring during the plastic and canalized phases in lubbers, as well as more detailed hormonal studies on both lubbers and sheep blowflies, would go a long way towards an understanding of the mechanistic underpinnings of these differences in reproductive patterns. Furthermore, selection experiments could allow one to distinguish between ontogenetic constraints and stabilizing selection on these reproductive processes. Finally, detailed intra- and interspecific comparative studies on related groups of grasshoppers (extending on Hatle et al 2002) and blowflies, as well as other anautogenous and autogenous dipterans, would offer a much needed comparative focus.

Cross-continental Insect Migrations and the Ovary-flight Syndrome

In examining various instances of reproductive plasticity in insects, one pattern has appeared repeatedly: an apparent trade-off between flight and reproduction known as the "ovary-flight syndrome." Whether this trade off is energetic (i.e. conflicting metabolic demands) or biomechanic (i.e. lift production) in nature (or, more likely, both), the trade-off clearly imposes substantial constraints on insect life histories. A cogent example for the negative relationship between dispersal and reproduction is the process of absconding in honey bees, typically a plastic response to deteriorating conditions (due to weather, resource availability, parasites, etc.). When absconding, the entire colony abandons its nest site, leaving behind mainly empty combs, and re-establishes at a new, suitable location. In preparation for this transition, the mature queen has to regress her ovaries in order to fly. While absconding is rare among the European races and the European-derived honey bees introduced to North America, it is commonplace among honey bees of the tropics and subtropics (Spivak et al. 1991).[11] Thus, for

[11] In fact, this propensity to abscond may be considered a preadaptation (see footnote 8, above) for the invasive behavior of the infamous Africanized bees of Brasil.

example, a given African honey bee queen from a particularly mobile colony might mature and regress her ovaries several times throughout her lifetime.

The classic examples of the ovary-flight syndrome, however, involve migratory insects that disperse at most once in their lifetimes, and generally at a defined stage. Several excellent reviews have described the ovary-flight syndrome in particularly well-studied migratory taxa, such as solitary and gregarious locusts (*Locusta migratoria*; Applebaum et al. 1997), crickets (*Gryllus* spp.; Zera and Harshman 2001), the army worm *Pseudaletia unipunctata* (McNeil et al. 1996), the soap-berry bug *Jadera haematoloma* (Dingle and Winchell 1997) and the water strider *Limnoporus canaliculatus* (Zera 1985). Here I will focus on what is perhaps the most famous of all migrating insects, the monarch butterfly *Danaus plexippus*. Clearly any insect that accomplishes a yearly migration from Canada to Mexico must have nearly optimized both energetics and lift production, and thus would make an excellent case study for investigating the ovary-flight syndrome. Furthermore, while many papers have been written on monarch migration, ontogeny, physiology and reproduction, I am aware of no attempts to synthesize these varied aspects of monarch biology in the explicit context of reproductive-flight trade-offs.

Although the spectacular southern migration of monarchs in autumn alluded to above is accomplished by individual insects that fly the entire route, the northward migration the following spring is more of a stepping-stone process (reviewed in Brower and Malcolm 1991). The eastern North American monarchs (the population that stretches from the Rocky mountains to the Atlantic ocean) migrate south and overwinter in the highlands north of Mexico City. In spring, these same individuals migrate north to their summer breeding grounds in the southern United States, and reproduce and die there. Several weeks later, their offspring continue the migration north. In the Mexican overwintering sites, the ovaries of female monarchs remain in an immature state (Barker and Herman 1976). So during their northward migration, the ovary is small, and energy is instead invested in the flight muscles and energetic stores. By the springtime, photoperiod and temperature interact synergistically to promote ovarian development; overwintering monarchs transferred to long day, warmer incubators will begin to mature öocytes (Barker and Herman 1976; Herman et al. 1989). Still, excessive temperatures inhibit reproduction, and by summertime, larvae are not found in the southern part of their range. Furthermore, the photoperiod and temperature levels which maximally promote öocyte maturation in monarchs are also the same levels that cue growth of their primary host plant, the milkweed *Asclepias syriaca* (Barker and Herman 1976). All of these factors presumably work together, resulting in a continued northward migration into summer: bouts of reproduction are followed by sub-optimal reproductive conditions, leading to ovarian regression and further migration north.

Throughout the long days of summer, several non-migratory generations of monarchs are produced in the northern US and southern Canada. In fact, the lack of migration during this phase is not the only manifestation of the female monarch's ovary-flight trade-off. Activity patterns of male and female breeding monarchs show a striking difference: while males engage in longer flights, patrolling mating grounds and searching for females, female flights are limited mainly to shorter, less-frequent and slower plant-to-plant foraging trips (Zalucki and Kitching 1982). Thus, during the period of most intense monarch reproduction, öocyte maturation correlates with low flight activity.

Monarch adults eclose with only immature öocytes. Under winter conditions, the ovaries remain in this immature state, but summertime photoperiod conditions induce öocyte maturation within five days after emergence (Herman et al. 1981). This surge in öocyte growth under summer conditions is preceded by a rise in the JH titers in the hemolymph of newly eclosed adults (Lessman et al. 1989), which is required for öocyte development to proceed (Barker and Herman 1973, Herman 1975, Lessman et al. 1982).

Levels of JH also vary seasonally (Lessman and Herman 1983), with high female JH titers in June and July (reproductive generations), moderate titers in August to October (migrating generation) and low titers in November to March (overwintering generation).[12] Indeed, the low JH titers in overwintering monarchs has been recently proposed to be related to longevity, as JH injections into overwintering monarchs lower their life expectancy in much the same way as transfer of such butterflies to artificial summer-like conditions (Herman and Tatar 2001). High JH titers promote öocyte maturation in a wide variety of insects (see below), while moderate JH titers have been shown to induce migration in the milkweed bug *Oncopeltus fasciatus* as well (Rankin and Riddiford 1978). Removal of the corpora allata results in a drastic reduction in long distance flights in tethered monarchs, further supporting the involvement of moderate JH levels in monarch migration. Finally, increasing JH levels (by injection) in migratory adults can initiate öocyte maturation (Herman 1975).

[12] Note that these data are combined from two distinct populations of monarchs: the eastern population (wild caught in Wisconsin and Minnesota) in May to September, and the western population (from overwintering sites near San Francisco, California) in October to March. Note also that these latter titers were relative values derived from a *Galleria* wax test.

Intriguingly, a direct mechanistic connection has been suggested between flight activity and reduced öocyte maturation in monarchs. Monarchs injected with radioactively-labeled JH showed a rapid increase in JH metabolites following a 40-minute tethered flight relative to unflown controls (Lessman and Herman 1981). Increased thoracic temperature had the same effect, leading to the fascinating proposal that the increased thoracic temperature resulting from flight activity directly causes a reduction in JH activity (Lessman and Herman 1981). These authors suggested that the thorax acts as a "JH gauntlet" which the hormone passes through on the way to the ovary-containing abdomen. Under high activity conditions, JH would be broken down on the way to the abdomen, and öocyte maturation would fail to be activated.[13] Hence, in monarchs, physical activity may feed back to control reproduction via hormones. This idea that environmental temperatures directly control the concentration, timing, and effects of hormones, has great importance for phenotypic plasticity.

Finally, Rankin (1986) showed an inverse relationship between the number of mature eggs being held by a female monarch and the length and likelihood of continuous, tethered flight by the same female. Furthermore, monarchs with their corpora allata removed flew very little, but flew about two fold more when injected with JH. In sum, these results support the existence of a direct ovary-flight trade-off in monarchs, and further suggest that JH lies at the center of this trade-off.

Few attempts have been made to apply modern techniques in insect biochemistry, physiology and development to further test these and other hypotheses regarding the ovary-flight syndrome in monarchs. Recent work with the cricket *Gryllus firmus* has demonstrated a direct link between fatty acid metabolism and reproductive versus flight energetics (Zera and Zhao 2003a, Zera, this volume), and long wing and short wing *G. firmus* morphs have different energetic profiles reflecting their different propensities for migration *versus* reproduction (Zera and Zhao 2003b). An application of these approaches to long-distance migratory species such as *D. plexippus* would not only allow us to ascertain the generalities of the underlying mechanisms (such as those uncovered by Zera and Zhao) and their associated trade-offs, but would also aid in our comprehension of how the spectacular cross-continental migration of monarchs is physiologically possible.

[13] Incidentally, this negative relationship between flight activity and circulating JH titers has apparently been circumvented in the long-distance migrating grasshopper *Melanoplus sanguinipes*, where tethered flights to exhaustion are followed by a rapid rise in JH titers and rapid onset of öocyte maturation (Min et al. 2004). Such rises in JH were not found in unflown or briefly-flown controls.

Parasitoid Wasps: How Important Is Numbers of Eggs Being Held?

Life histories in parasitoid wasps (Hymenoptera: Apocrita) have been intensively studied, largely due to their usefulness in insect "pest" control. Substantial research has focused on understanding the conditions leading to egg maturation and successful reproduction in this group. Not surprisingly, a vast array of reproductive strategies is employed in the various parasitoid taxa. Jervis and colleagues (2001, 2003) conducted a broad comparative study of parasitoids, and made the surprising observation that body size is inversely correlated with "ovigeny index" (the proportion of eggs that are mature at adult eclosion), intraspecifically as well as interspecifically (thus it is likely a plastic response as well). Why would smaller wasps tend to have a greater proportion of mature ovaries at eclosion? Jervis and colleagues suggest the possibility that small size correlates with higher mortality rates (and thus shorter life expectancies) in the field due to increased risks of predation and/or desiccation. The relevant data on these points, however, are limited and equivocal.

Wang and Messing (2003) examined the stimulant for öocyte maturation in the braconid wasp *Fopius arisanus* (Hymenoptera: Braconidae), a fruit fly (Diptera: Tephritidae) generalist with a very low ovigeny index. During the days following eclosion, öocyte maturation proceeds independently of host stimuli, food and access to males in any combination. However, after this first batch of öocytes matured, the rate of maturation in subsequent öocytes was substantially increased only in females given access to host stimuli, food and males. An even greater increase in öocyte maturation was seen in wasps provided with the host and the host fruit (and were, thus, allowed to oviposit for the first time), even when starved. Thus, the first round of öocyte maturation seems to be an inherent process, largely independent of external cues. Further maturation, though, was most effective in response to oviposition, a maturational cue found in many parasitoid wasps and other insects (reviewed in Papaj 2000).

These results have three important implications. First, the apparent inverse proportionality between longevity and reproductive effort suggests that forgoing additional öocyte development allows for longer survival (reviewed in Papaj 2000). Second, the finding that oviposition was the predominant maturational trigger begs the following question: how does oviposition transmit the signal to initiate öocyte maturation? We will return to this point near the end of the chapter. Third, these results appear to contradict a hopeful assumption on the part of many parasitoid researchers:

that a fully mature ovary's "egg load" (the numbers of eggs held) is a general stimulus to oviposition in parasitoids (e.g. Mangel 1989). If this were true, it would greatly simplify pest management strategies, as releasing wasps with full egg loads would then enhance their success in bio-control. Unfortunately, the explanations for reproductive decisions of parasitoids will only be found, it seems, in the context of a detailed understanding of the underlying öocyte maturation mechanisms, which are likely to be quite variable across taxa (see also Jervis and Ferns 2004).

In another example, Rivero-Lynch and Godfray (1997) have shown that *Leptomastix dactylopii* parasitoids (Hymenoptera: Encyrtidae), when provided with plentiful numbers of their mealybug hosts, and thus unlimited opportunities to oviposit, have egg loads equivalent to host-limited and host-deprived wasps. It seems that the presence of plentiful hosts and/or oviposition leads to an increase in maturation rates (see Alonso-Pimentel et al. 1998 for an ingenious method used to distinguish among similar possibilities in walnut flies). In any case, egg load seems largely irrelevant to the reproductive biology of these encyrtid wasps as well.

Tephritid Fruit Flies: Comparative Approaches

For many of the same reasons indicated above for parasitoids, tephritid fruit flies (Diptera: Tephritidae), which are major fruit pests around the world, have been the focus of much research into reproductive life histories. Happily, many of these studies have involved a comparative approach, something which is decidedly lacking from the majority of studies in insect life histories.

One set of studies (Aluja et al. 2001; Díaz-Fleischer and Aluja 2003) involved a comparison of the reproductive biology of two species from the genus *Anastrepha*. *A. obliqua* oviposits on a wide variety of fruits, mostly from the family Anacardiaceae, which includes the cashew, mango and pistachio. *A. ludens* is more of a specialist on plants in the rue family (Rutaceae), including the yellow chapote (*Sargentia greggii*) and the white sapote (*Casimiroa edulis*). *A. obliqua* tends to lay a single egg on its high quality host fruits, which are, for the most part, abundant, synchronous and highly ephemeral. The lower quality host fruits of *A. ludens* are comparatively less numerous, less synchronous and less ephemeral, and *A. ludens* tends to lay eggs in batches of up to 40 eggs (Aluja et al. 2001). *A. ludens* are larger in size than *A. obliqua*, and have fewer ovarioles (22–25 vs. 30–33) (F. Díaz-Fleischer, personal communication).

Under similar dietary conditions, *A. obliqua* consistently matured more than twice as many öocytes than did *A. ludens*. In addition, the presence of host fruit volatiles led to substantial increases in öocyte maturation only in *A. obliqua* (Aluja et al. 2001). A second study (Díaz-Fleischer and Aluja 2003) examining lifetime oviposition in the two species can help account for these results. When offered a low, high or variable availability of hosts, *A. ludens* maintained a strikingly constant egg oviposition pattern. *A. obliqua*, by contrast, modulated its oviposition patterns to match host availability. It seems that the high egg loads are maintained in *obliqua* (and aided by their higher ovariole numbers) in order to take advantage of rare but high abundance host patches. *A. ludens* females are relatively non-selective about their oviposition choices, since their hosts tend not to appear in blooms in nature. While the mechanisms underlying the two different maturation trajectories remain to be elucidated, work on other insects suggests some plausible hypotheses, which we will return to at the end of this section.

A second comparative approach involving tephritids has focused on flies in the genus *Bactrocera* (formerly *Dacus*). The genus includes species with a wide range of reproductive ecologies, from specialists (e.g. the olive fly *B. oleae* and the solanum fruit fly *B. cacuminatus*) to generalists (e.g. the Queensland fruit fly *B. tryoni*) to intermediates (the cucumber fly *B. cucumis* and the Jarvis' fruit fly *B. jarvisi*). Such diversity within a restricted taxonomic group allows for a fairly rigorous test of a commonly-held hypothesis (e.g. Labeyrie 1978): namely, that generalist taxa (in which oviposition possibilities are abundant) tend to have less specific control mechanisms for initiating öocyte maturation than specialist taxa (where hosts are limited). The example cited above of the two *Anastrepha* species tends to support this hypothesis. In *B. oleae*, the ovaries are activated in the presence of olives, which, on the Hellenic island of Corfu, are available in May and late July, but not June through early July (Fletcher et al. 1978). *B. oleae* ovaries regressed in the field in July, and lab experiments showed that presence of fruits, as well as temperature and humidity conditions mimicking those in late July, lead to öocyte maturation (Fletcher et al. 1978).[14] So, it seems, this specialist fly has fairly specific (environmental)

[14] A correlated adaptive explanation here is that immature ovaries in the absence of appropriate hosts aid in dispersal by decreasing wing loading. Such a situation has been nicely demonstrated for the potato tuberworm *Phthorimaea operculella* (Lepidoptera: Gelechiidae), where development on tomato (a sub-optimal host) leads to lower rates of egg maturation and a greater tendency to fly, while development on potato (an optimal host) leads to higher maturation rates and a reduced tendency to fly (Coll and Yuval 2004).

control mechanisms that match its maturational timing with the presence of its preferred host fruit. But does this pattern hold in related species with different reproductive patterns?

Fitt (1986) deprived groups of *B. cacuminatus, cucumis, jarvisi* and *tryoni* of hosts for up to 16 days, and then offered either previously unacceptable or highly unpreferred fruit hosts (determined in choice experiments; different fruits for different species). He recorded numbers of eggs laid in oviposition tests conducted on various days after host deprivation when compared to undeprived controls. All flies were well fed on a sugar/yeast mixture. The specialist *B. cacuminatus* and intermediate *B. cucumis* laid virtually no eggs on unpreferred hosts even after 16 days of host deprivation. *B. jarvisi* did lay eggs on unpreferred hosts, but there were no differences between deprived and undeprived flies, and time since deprivation had no effect on numbers of eggs laid on these unpreferred hosts. By contrast, the generalist *B. tyroni* readily laid on an unpreferred host (the wild tobacco *Solanum mauritianum*), but only when host-deprived for 4 days or more.

These differences in oviposition under host-deprived condition were mirrored by egg load measurements (Fitt 1986). *B. cacuminatus* (a specialist) showed no difference in egg load between deprived and undeprived flies, deprived *B. cucumis* and *B. jarvisi* (intermediates) had double the egg load of undeprived flies, and deprived *B. tryoni* (a generalist) had 3–5 times the egg load as undeprived flies. Interestingly, the egg loads in the three non-generalist species never exceeded one egg per ovariole, while egg loads in *B. tyroni* were up to twice the ovariole number. Given the positive correlation between ovariole number and number of host plant species across 14 *Bactrocera* species (Fitt 1990), it seems that generalist taxa use several tricks to increase their potential reproductive output. However, the full significance of these results will only be clear when more specialist and generalist *Bactrocera* (and *Anastrepha*) species (there are hundreds) are examined in a phylogenetic context.

Hormones and Insect Reproductive Variability and Plasticity: Nothing New Under the Sun

It is fair to say that the major lesson learned from a half-century of comparative studies on insect development is that hormones are involved in just about every ontogenetic process you can think of (Nijhout 1994). Still, this ubiquity of hormonal involvement in insect development is paradoxical: how is the extreme variability in insect life cycles, morphology, physiology and behavior controlled by what is largely a highly stereotyped pattern of ontogenetic hormonal profiles? One answer comes from the apparent modularity in hormonal response indicated by tissue- (and cell-type-) specific patterns of hormone receptor expression. Evidence for such evolutionary variation has been presented (e.g. Hodin and Riddiford 2000b) or is suggested (e.g. Rountree and Nijhout 1985a,b) by studies in a wide variety of insects. A second answer involves the observation that hormonal profiles are much more variable in adults than in pre-adult stages (Nijhout 1994). A third answer may relate to the fact that hormonal release in some cases is directly triggered by nutritional inputs (reviewed in Adams 1999; Barton Browne 2001), demonstrating the physiological basis for the connection between nutritional intake and reproductive plasticity. Still, the extreme variability in hormonal control of reproduction across insects defies an easy characterization. I'll begin by presenting some evidence for this extreme variability, and end with some speculation concerning evolutionary scenarios that could have generated such a variable system.

Table 4 gives an indication for some of this variability. While JH promotion of vitellogenesis and öocyte maturation is fairly widespread (and has, thus, been often proposed to be an ancient function in insects; e.g. Bellés 1998), this pattern is by no means universal.[15] For example, JH inhibits vitellogenesis and/or vitellogenin (VG) synthesis in the sweet potato weevil *Cylas formicarius*, the gypsy moth *Lymantria dispar* (where VG synthesis apparently depends on falling JH titers) and, possibly, the western tent caterpillar *Malacosoma pluviale*. There are also many taxa in which JH seems to have no role at all (or, at least, a drastically reduced function; Barchuk et al. 2002) in vitellogenesis (such as most ants and the eusocial honeybees and stingless bees, as well as the silk moth *Hyalophora cecropia*), and those in which JH effects seem to be restricted to late- or post-vitellogenic stages (such as the tobacco hawk moth *Manduca sexta* and the apple codling moth *Cydia pomonella*). As is obvious from Table 4, the reproductive functions of ecdysteroids are, if anything, more variable than those for JH. This table is not a comprehensive list of all insects for which hormonal effects on reproduction have been studied, but gives a flavor for the diversity in hormonal control mechanisms in insects. As such, it overemphasizes the

[15] I use the term "JH" here and throughout to refer collectively to all of the different forms of juvenile hormone (JHI, JHII, JHIII, JH-bis-epoxide, etc.) that have been shown to be active in various insects groups. I also have not distinguished here among experiments in which, for example, corpora allata have been removed versus perhaps less convincing hormone manipulation methods. My apologies to my former advisor and all other JH enthusiasts for this oversimplification. Interested readers can find more detailed information in the papers to which I have referred herein.

Table 4 Vitellogenesis control mechanisms in selected insects. Information comes from a wide variety of sources and experimental evidence. Question marks indicate mostly correlative data (such as a correlation of whole body hormone titers with the timing of vitellogenesis), while mechanisms without question marks derive from manipulative experiments (either pharmacological or manual removal of hormone sources and/or ectopic hormonal applications). Blanks mean that the functions of this hormone in insect reproduction have not, to my knowledge, been investigated. References (some of which are reviews, others are experimental studies) as follows: [a]Taub-Montemayor et al. 1997; [b]Ram et al 1988; [c]Scott et al. 2001; [d]Rankin et al. 1997; [e]Stay et al. 1980; [f]Schal et al. 1997; [g]Yin and Stoffolano 1997; [h]Klowden 1997; [i]Bownes 1989; [j]Audit-Lamour and Busson 1981; [k]Hodin and Riddiford 1998; [l]Davey 1997; [m]Hartfelder et al. 2002; [n]Bloch et al. 2002; [o]Sommer and Hölldobler 1995; [p]Robinson and Vargo 1997; [q]Ramaswamy et al. 1997; [r]Webb et al. 1999; [s]Fescemeyer et al. 1992; [t]Zeng et al. 1997; [u]Shaaya et al. 1993; [v]Nijhout and Riddiford 1974, 1979; [w]Applebaum et al. 1997; [x]Strambi et al. 1997; [y]Bradley et al. 1995; [z]Bellés 1998.

Order	*Family*	*Species*	*Function of JH in female reproduction*	*Function of ecdysteroids in female reproduction*	*Ref.*
Coleoptera	Chrysomelidae	*Leptinotarsa decemlineata*	promotes but not necessary for VG synthesis or oviposition	no effect on VG synthesis	a
Coleoptera	Coccinellidae	*Coccinella septempunctata*	promotes VG synthesis		a
Coleoptera	Curculionidae	*Anthonomus grandis*	promotes VG synthesis in pre-reproductive but not in diapausing females; not involved in yolk uptake or oviposition	no effect on VG synthesis in diapausing females; not involved in oviposition	a
Coleoptera	Curculionidae	*Cylas formicarius*	inhibits vitellogenesis?		b
Coleoptera	Silphidae	*Nicrophorus orbicollis*	promotes öocyte maturation?; inhibits oviposition; promotes maternal care?		c
Dermaptera	Carcinophoridae	*Euborellia annulipes*	promotes reproductive behavior; promotes maturation; inhibits brood protection	inhibit brood protection	d

(Contd.)

Table 4 (Contd.)

Order	*Family*	*Species*	*Function of JH in female reproduction*	*Function of ecdysteroids in female reproduction*	*Ref.*
Dictyoptera	Blaberidae	*Diploptera punctata*	promotes VG synthesis and öocyte maturation; inhibits ovulation	Inhibit JH production preceding ovulation	e
Dictyoptera	Blattellidae	*Blatella germanica*	promotes VG synthesis; inhibits choriogenesis?; inhibits oviposition? induces receptivity; inhibits brood protection?	promote choriogenesis, oviposition?	f
Diptera	Calliphoridae	*Phormia regina*	promotes sexual receptivity; not necessary for VG synthesis; promotes VG uptake	Promote VG synthesis	g
Diptera	Culicidae	*Aedes aegypti*	promotes pre-vitellogenic growth, competency for vitellogenesis, sexual receptivity	promote VG synthesis; deposition of vitelline membrane	h
Diptera	Drosophilidae	*Drosophila melanogaster*	promotes vitellogenesis and yolk uptake	promote pre-vitellogenic ovarian differentiation; promote yolk protein synthesis; inhibit vitellogenesis	i,j,k
Hemiptera	Lygaeidae	*Oncopeltus fasciatus*	no effect on VG-A synthesis; induces conversion of VG-A to mature form; promotes öocyte maturation	inhibit vitellogenesis?	l
Hemiptera	Reduviidae	*Rhodnius prolixus*	promotes but not *necessary* for VG synthesis and öocyte maturation; promotes VG uptake (patency)	inhibit vitellogenesis?; promote ovulation and oviposition	l
Hymenoptera	Apidae	*Apis mellifera*	not involved in maturation or later stages	probably not involved in adults	m,n

(Contd.)

Table 4 (Contd.)

Order	*Family*	*Species*	*Function of JH in female reproduction*	*Function of ecdysteroids in female reproduction*	*Ref.*
Hymenoptera	Apidae	*Bombus terrestris*	promotes VG synthesis and öocyte maturation	promote öocyte maturation	n
Hymenoptera	Apidae	*Melipona quadrifasciata*	not involved in maturation or later stages	probably not involved in adults	m
Hymenoptera	Formicidae	*Diacamma* (unnamed sp.)	not detectable in reproductive workers		n
Hymenoptera	Formicidae	*Lasius niger*	reduces egg output; apparently uninvolved in dominance interactions		o
Hymenoptera	Formicidae	*Solenopsis invicta*	promotes VG synthesis; higher levels promote VG uptake?; promotes egg output		p
Hymenoptera	Vespidae	*Polistes dominulus*	promotes VG synthesis, öocyte maturation and reproductive dominance	promote reproductive dominance	n,p
Lepidoptera	Bombycidae	*Bombyx mori*	possibly none; certainly not required	declining titers promotes VG synthesis (and patency?)	q
Lepidoptera	Danaidae	*Danaus plexippus*	promotes VG synthesis, patency, choriogenesis	no effect on VG synthesis	q
Lepidoptera	Lasiocampidae	*Malacosoma pluviale*	suppresses VG synthesis?	promotes VG synthesis?	q,r
Lepidoptera	Lymantriidae	*Lymantria dispar*	suppresses VG synthesis?	promotes VG synthesis?	q,s
Lepidoptera	Noctuidae	*Helicoverpa zea*	promotes VG synthesis, patency, choriogenesis	none?	q

(Contd.)

Table 4 (Contd.)

Order	*Family*	*Species*	*Function of JH in female reproduction*	*Function of ecdysteroids in female reproduction*	*Ref.*
Lepidoptera	Noctuidae	*Heliothis virescens*	promotes VG synthesis, patency, choriogenesis	none?	q,t
Lepidoptera	Noctuidae	*Pseudaletia unipuncta*	promotes VG synthesis, patency, choriogenesis	none?	q
Lepidoptera	Nymphalidae	*Nymphalis antiopa*	promotes VG synthesis, patency, choriogenesis	no effect on VG synthesis	q
Lepidoptera	Nymphalidae	*Polygonia c-aureum*	promotes VG synthesis, patency, choriogenesis	no effect on VG synthesis	q
Lepidoptera	Nymphalidae	*Vanessa cardui*	promotes VG synthesis, patency, choriogenesis	no effect on VG synthesis	q
Lepidoptera	Pieridae	*Pieris brassicae*	promotes VG synthesis, patency, choriogenesis	no effect on VG synthesis	q
Lepidoptera	Pyralidae	*Diatraea grandiosella*	induces choriogenesis	declining titers promotes VG synthesis (and patency?)	q
Lepidoptera	Pyralidae	*Plodia interpunctella*		declining titers promotes VG synthesis	u
Lepidoptera	Saturniidae	*Hyalophora cecropia*	not necessary		q
Lepidoptera	Sphingidae	*Manduca sexta*	induces choriogenesis; increases the rate of (but not necessary for) VG synthesis; necessary only for late stages of vitellogenesis	VG synthesis initiated in absence of ecdysteroids; addition of ecdysteroids to pharate adults suppresses JH promotion of öocyte maturation	q,v,
Lepidoptera	Tortricidae	*Cydia pomonella*	induces choriogenesis, represses oviposition		r

(Contd.)

Table 4 (Contd.)

Order	*Family*	*Species*	*Function of JH in female reproduction*	*Function of ecdysteroids in female reproduction*	*Ref.*
Orthoptera	Acrididae	*Locusta migratoria*	promotes VG synthesis and öocyte maturation		w
Orthoptera	Gryllidae	*Acheta domesticus*	promotes VG synthesis and öocyte maturation	low levels promote maturation; high levels inhibit	x
Orthoptera	Gryllidae	*Gryllus bimaculatus*	promotes but not *necessary* for VG synthesis	more prominent involvement in maturation than JH?	x
Orthoptera	Gryllidae	*Gryllus campestris*	promotes but not *necessary* for VG synthesis	more prominent involvement in maturation than JH?	x
Orthoptera	Gryllidae	*Teleogryllus commodus*	promotes but not *necessary* for VG synthesis	more prominent involvement in maturation than JH?	x
Phasmatodea	Heteronemiidae	*Carausius morosus*	unnecessary for VG synthesis; involved in VG uptake		y
Thysanura	Lepismatidae	*Thermobia domestica*	necessary for vitellogenesis	probably not involved	z

cases in which, for example, JH control has deviated from its canonical function in promoting vitellogenesis and öocyte maturation.

Lepidopteran (moths and butterflies) endocrinology has been particularly well studied (reviewed in Ramaswamy et al. 1997). Figure 4 is one fairly robust phylogenetic hypothesis for the Lepidoptera (based upon Kristensen and Skalski 1999) in which I have only included species for which vitellogenesis control mechanisms have been studied (not all such species are included). Next to the species names, I have indicated the stage at which mature öocytes (chorionated eggs ready to be laid) are produced, and if JH and/or ecdysteroids are involved in VG synthesis. Clearly, both reproductive features are highly variable. And while some of these character states show a strong phylogenetic signal (e.g. adult reproductive maturation among the butterflies—*P. brassicae, N. antiopa, D. plexippus*), others are most likely homoplasious (e.g. ecdysteroid regulation of vitellogenesis in the pyralid moth *P. interpunctella* and the silk moth *B. mori*). Interestingly, those taxa in which JH promotes vitellogenesis are also the only taxa among those considered that undergo öocyte maturation in the adult stage (the three butterflies included plus the tobacco budworm *H. virescens*). This finding raises the fascinating possibility that the reproductive control mechanisms are less associated with phylogeny than they are with selection on the timing of reproductive maturation (see below).

I have purposefully avoided attempting to map ancestral character states on the phylogeny in Figure 4. Since homoplasy would be extensive under any evolutionary scenario that one might favor, such an exercise would seem to be guesswork. Furthermore, as I mentioned previously, the taxon sampling here may not be completely random. It is intriguing, though, that the "typical" JH function in promoting VG synthesis is only found among the Macrolepidoptera, the most derived lepidopteran taxon, and specifically in only those macrolepidopterans in which öocyte maturation occurs entirely in the adult stage. Let us allow the assumption (and there is substantial evidence to support this assumption; see Ramaswamy et al. 1997) that the stage of öocyte maturation is intimately connected to the life history of the organism. The conclusion that follows from the apparent correlation between öocyte maturation stage and hormonal control mechanism is, therefore, that the hormonal mechanisms are coopted to follow suit with selection on insect life histories. Testing this hypothesis rigorously would necessitate more independent contrasts than presented in this limited phylogenetic data set. In any case, it seems clear that some caution is warranted in any attempts to reconstruct ancestral states for such characters, either in restricted insect taxa or among insects as a whole.

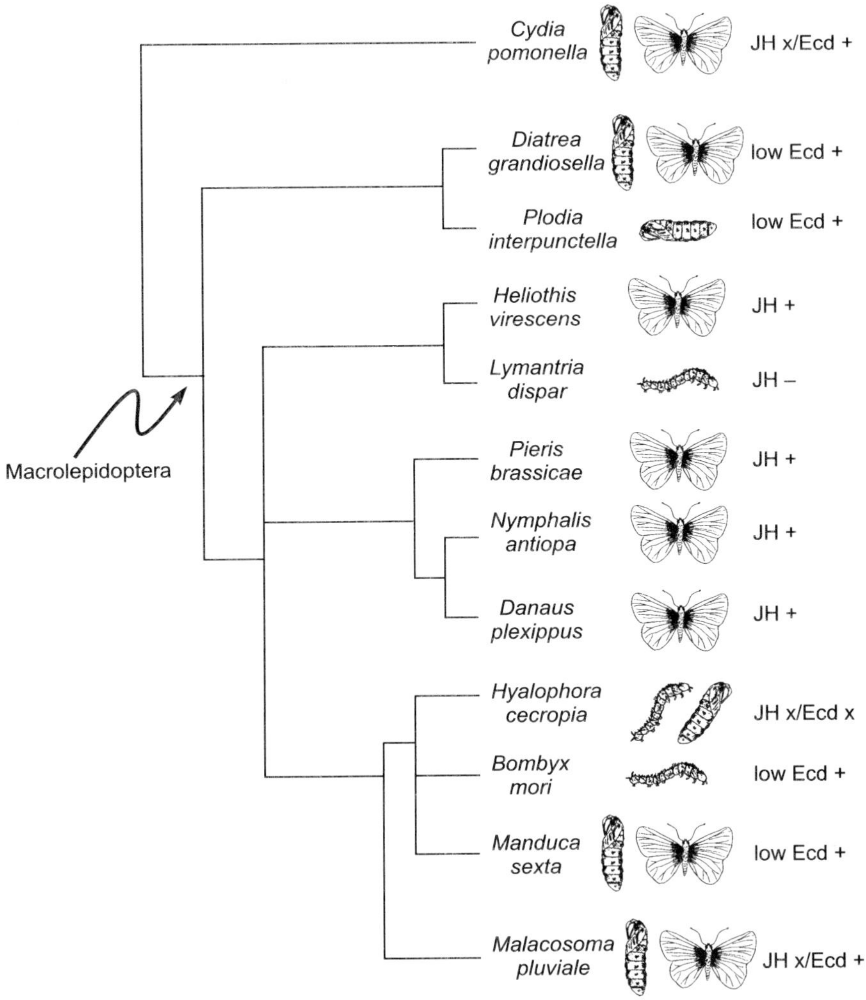

Fig. 4 Data from the Lepidoptera in Table 4 mapped onto a current phylogenetic hypothesis for the order (after Kristensen and Skalski 1999). Pictures indicate the stage at which öocyte maturation occurs, whether larval (e.g. *B. mori*), larval/pupal (*H. cecropia*), pupal (*P. interpunctella*), pupal/adult (e.g. *M. sexta*) or adult (e.g. *D. plexippus*). + indicates promotion of vitellogenesis by that hormone, - indicates repression, and **x** indicates that the hormone is apparently not involved in/required for vitellogenesis. See Table 4 for references.

As we will see in the next section, attempts to reconstruct ancestral character states have not only been popular in developing ideas regarding control mechanisms underlying the evolution of eusociality in insects, but, indeed, seem to have led to some circular reasoning, where supposed similarities in control mechanisms are cited as evidence for close relationships among certain highly eusocial hymenopteran taxa.

Extreme Reproductive Plasticity: Queen-worker Dimorphism in Social Insects

Sociality has been hypothesized to have arisen numerous times independently: in termites, thrips (order Thysanoptera), aphids (order Hemiptera), beetles and multiple times in the Hymenoptera (ants, wasps and bees) (reviewed in Lin and Michener 1972; Halffter and Edmonds 1982; Stern and Foster 1996; Schwarz et al. 2007). While the close association between the evolution of sociality and a haplo-diploid mechanism of sex determination has long been cited (Hamilton 1964) as a critical correlate of sociality in the Hymenoptera (and the thrips as well), beetles and termites have the more typical diploid sex determination mechanism, and aphid sociality is related to clonality (reviewed in Stern and Foster 1996). These findings then beg the question of what factors have led to the evolution of sociality in insects in general, where certain members of a colony (queens, reproductive workers) have a higher reproductive potential than others (reviewed in Keller 1993; Reeve and Keller 2001). I will return to this point later. I will begin, however, with a discussion of some of the mechanisms underlying plasticity in öocyte maturation in the termites and a few groups of Hymenoptera. In the section on oviposition, I will revisit some of these same social insect groups, and consider some general questions regarding the evolution of eusociality in termites and hymenopterans.

Termites: Fixating on Nitrogen

As mentioned previously, hymenopterans and basal termites differ in at least one important respect: the termites have much more plasticity, as they can switch among reproductive and non-reproductive tasks depending on colony conditions.

I will return to a consideration of reproductive plasticity in the Termopsidae, the relatively basal family of termites that I introduced earlier. As is the case with ovariole number determination, termopsids are also quite plastic with respect to öocyte maturation. Secondary reproductives of *Zootermopsis angusticollis*, when raised with attendant workers, began laying eggs about 10 days earlier than unattended controls (Brent and

Traniello 2001a). Still, their egg laying rate, while enhanced in the presence of attendants, remained about 3-fold lower than that in primary reproductives under any condition. Interestingly, when the termites' diets (sawdust) included nitrogen (sawdust + uric acid) the secondaries achieved the same rate of egg laying as primaries, who did not show a change in rate with nitrogen supplementation (Brent and Traniello 2002).[16] Also, many termite species exhibit plasticity in cannibalism, consuming colony mates when nitrogen levels are low (Whitman et al. 1994). It seems, then, that nitrogen is limiting for secondary reproductives, and is a potent stimulator of öocyte maturation in termites. In sum, the dual ability of secondaries to increase their potential reproductive output by increasing ovariole number and maturation rate under certain conditions gives them the potential to attain primary reproductive-like levels of reproduction. As we shall see, this option, for several reasons, is not available to the majority of hymenopterans.

Ponerine Ants: The Workers Control the Means of Production (Sometimes)

Ponerines (Formicidae: Ponerinae) are a relatively basal lineage of ants, having less division of labor than most ants, and a class of workers (the so-called "gamergates") that can attain substantial levels of reproduction in the absence of their queen (reviewed in Peeters 1991). Some ponerine species lack queens altogether (see below). And unlike most hymenopteran workers, ponerine gamergates can mate with males. This is key, because it allows them the potential to raise female daughters, both workers and queens.

In the ponerine ant *Gnamptogenys menadensis*, worker-run colonies are the norm: Gobin and colleagues (1998) observed only two queenright colonies out of 37 examined in Sulawesi. There are several gamergates (between 1 and 14; mean=5) in each of the worker-run colonies. When all of the gamergates are removed, a subset of the virgin workers ("dominants") begin dominance interactions, involving agonistic behaviors (antennal boxing and biting; Gobin et al. 2001). Non-reproductive workers ("subordinates") did not engage in such behaviors, but occasionally ganged up on individual dominant worker, inevitably converting such a worker to a subordinate. After several weeks or months, some of the dominants began sexual "calling" behavior (i.e. adopting a pheromone-releasing posture), and from that time forward, dominance interactions were no longer observed.

[16] Nitrogen contents were determined in all samples by an elemental analysis assay sensitive to within 0.003%.

I previously mentioned the South African ant, *Pachycondyla (=Ophthalmopone) berthoudi*, as one of the 100 or so ponerine species that are permanently queenless. Their colonies can have up to one hundred gamergates (mean = 38) who never stray from the nest, display no agonistic interactions and do not get excess food shares when compared to non-reproductive workers (Peeters and Crewe 1985; Sledge et al. 2001).

Thus, in *G. menadensis* and *P. berthoudi* colonies (as is true in many ponerine species, including, for example, *Odontomachus rixosus*; Ito et al. 1996), reproductive division of labor is established among groups of workers, where the majority have repressed öocyte maturation and a minority have mature ovaries. Furthermore, except during initial founding of the dominant class in the case of *G. menadensis*, agonistic interactions are rare. These results may seem surprising given the expected low relatedness of offspring of the gamergates to their caretakers (the subordinate workers). Certain peculiarities of ponerine ant colonies have been hypothesized to account for these findings (Peeters 1991): (1) most ponerines have small colony sizes (100 or fewer workers; rarely more than a few hundred); (2) colonies typically have low division of labor; (3) reproductives mate nearby the nest; (4) colony founding is by fission (i.e. workers and gamergates/queens found new colonies together); (5) even in species with queens, ovariole number in queens is comparable to that in workers; (6) the workers retain spermathecae, so can mate. Therefore, the potential reproductive output of workers and queens is roughly equivalent. The plasticity of worker reproduction expressed in the adult stage in ponerines is thus understandable as an efficient way to respond to variable surroundings, maximizing colony efficiency by making conflicts less likely (Monnin and Ratnicks 2001), and facilitating colony fission.

The Cape Honeybee: Revenge of the South African Black Bees

In honeybees, a single, long-lived queen dominates reproduction, releasing pheromones that keep the vast majority of workers in a state of repressed ovarian differentiation. Furthermore, there is substantial age-based division of labor, with younger bees performing tasks in the hive, and older bees foraging. For these reasons, honeybees are considered highly social (Michener 1974).

It would be incorrect, however, to conclude that workers have zero direct reproductive potential. After a queen dies, honeybee colonies desperately try to raise one of the last female offspring of the departed queen as her heir. But, in addition, a certain portion of the workers begin to mature their ovaries, as the queen's repressive pheromone is no longer present. Still, worker honeybees are incapable of mating. As a result, de-repressed reproducing workers can only lay unfertilized, and, hence, male-determined eggs. In queenright honeybee colonies, worker reproduction is extremely rare, with less than 0.1% of the male brood attributable to worker reproduction in *Apis mellifera, A. cerana, A. dorsata* and *A. florea* (reviewed in Barron et al. 2001). Of these, *A. cerana* may have the highest capacity for worker reproduction (Oldroyd et al. 2001), since many *A. cerana* queenright workers have vitellogenic ovaries, and the delay between loss of the queen and first worker brood is relatively short (just a few days). This enhanced reproductive capacity is associated with a relatively low rate of egg production in queens (approximately 1/4 that of *A. mellifera* races under similar conditions; reviewed in Ruttner 1988), and corresponding relatively high worker and low queen ovariole numbers (see Table 3). Also, the rate at which *A. cerana* eggs are removed by police is substantially lower than in other *Apis* species. This sluggish rate of policing, according to a recent hypothesis, may indicate that *A. cerana* workers have recently evolved the ability to mimic the queen's egg marking pheromone (Nanork et al. 2007). These authors further suggest (p. 1513) that the "existence of reduced policing efficiency reduces the incentive for self-restraint by workers...[which] may explain why we observe quite high levels of ovary activation in *A. cerana*." Nevertheless, even in *A. cerana,* worker policing of eggs results in a situation where few to no males are derived from worker brood in queenright colonies (Oldroyd et al. 2001).

The major exception to the "male brood only, and only in the absence of the queen" rule in honeybees is found in the cape honeybee *Apis mellifera capensis.* These dark-colored bees are restricted to the southern tip of South Africa, and are unique among honeybees in that the workers, when they lay eggs, give rise to female offspring through thelytokous parthenogenesis (by fusion of the egg pronucleus with one of its non-sister polar nuclei; and since there is apparently no meiotic recombination in capensis workers, these offspring are clones of the parent; Moritz and Haberl 1994), a trait recently found to be conferred by a single locus recessive allele (Lattorff et al. 2005). The possibility that such worker-derived brood have a chance to become queens gives *capensis* workers a much higher potential individual (i.e. non-inclusive) fitness than typical *A. mellifera* workers.

Queen exclusion experiments (where the queen is excluded from certain parts of the hive) show that substantial numbers of *capensis* worker-derived eggs (between 0 and 23 eggs per day) are laid even in the presence of queen inhibitory pheromone, though most of these are removed by other workers (Moritz et al. 1999, Pirk et al. 2002; it is not known what percentage of brood is worker derived under normal, queenright conditions). As we shall see below, this situation is more akin to what happens in many stingless bee species, where queenright workers often produce eggs, many of which are inviable trophic eggs consumed by the queen. In addition, *capensis* workers have many more ovarioles per ovary on average than typical *A. mellifera* workers (Table 3). Thus, through their higher ovariole numbers, mature ovaries, and capacity to lay female eggs even in the presence of a queen, *capensis* workers are more "queen-like" than typical honeybee workers.

In 1992, the situation became even more complicated. Possibly through migratory bee-keeping practices (Allsopp 1992), *capensis* bees were introduced to regions of South Africa in which another honeybee, the African honeybee *Apis mellifera scutellata*, is native. Beekeepers noticed that some of their *scutellata* colonies had become populated and were eventually over-run by black bees, and such colonies eventually and inevitably crashed. What had happened, and continues to happen in a growing portion of *scutellata*'s range, was that a single clonal lineage of *capensis* workers (Kryger 2001; Martin et al. 2002; Dietemann et al. 2007) arose, with the capacity to enter *scutellata* colonies, and commenced laying eggs.[17] Furthermore, the *scutellata* workers attend such laying *capensis* workers as if they were queens, probably due to their queen-like pheromone cocktail (Wossler 2002; Dietemann et al. 2007), and feed their offspring disproportionately (reviewed in Calis et al. 2002). These offspring hatch into laying workers themselves, with more ovarioles than they would have had if reared by *capensis* nurse bees (a mean of 24 versus 14 ovarioles per ovary, respectively; Wossler 2002). Eventually, the colony is over-run by reproductives, the *scutellata* queen disappears for unknown reasons, and then the colony structure collapses (Allsopp 1993).

Presumably, the capacity of *capensis* workers to undertake this colonial parasitism is related to the reasons cited above for their greater reproductive capacity (*vis-à-vis* typical *A. mellifera* workers) in their own colonies, as well as their ability to lay female-producing eggs. In other words, their unique reproductive biology can be considered an evolutionary preadaptation[18] for the recent parasitic behavior (see also Neumann and Hepburn 2002).

[17] See Lopez-Vaamonde et al. (2004) for a fascinatingly similar instance of "social parasitism" in the bumble bee *Bombus terrestris*, though in this case only male offspring are produced by the invading workers.

[18] see footnote 8, above

Importantly, these traits may increase the inter-racial competitiveness of *capensis*, resulting in competitive displacement of *scutella* in some areas.

Stingless Bees: Behavioral Reproductive Control By the Queen in a Highly Eusocial Bee Group

It has been proposed that there is a trend in the evolution of the highly eusocial Hymenoptera in which behavioral dominance by the queen over worker reproduction predominates in basal social taxa (akin to what we described for ponerines, or more properly, what is seen in bumble bees and pollistine wasps), while pheromonal dominance is the means of control in the more derived, more highly social taxa (such as honeybees). Stingless bees are a fascinating exception to this rule, where highly ritualized behavioral interactions between queens and workers keeps reproduction in check, with varying degrees of success in different species (reviewed in Zucchi 1993).

Stingless bees (family Apidae; group Meliponini) are found throughout the old and new world tropics and sub-tropics. There has been much speculation concerning the evolution of high sociality (extreme division of labor) among the bees, and it has been often assumed, in the absence of solid phylogenetic evidence, that stingless bees are the sister group to the honeybees (e.g. Michener 1974; Schultz et al. 1999). As alluded to previously, the lack of JH control of vitellogenesis in honeybees and stingless bees, and its apparent cooption for regulating worker polyethism (progression of worker tasks) in the two taxa are strong uniting features of the two taxa. The implication, therefore, is that high sociality evolved once in the Apidae, and that bee evolution is a progression of sorts from non-social, to primitively eusocial, through higher levels of sociality (bumble bees), to the highest sociality in the stingless bees and honeybees (Lin and Michener 1972). Such a progressive view of social evolution has been commonly postulated for other social taxa (wasps, ants, termites) as well (e.g. Lin and Michner 1972; Thorne and Traniello 2003; Bloch et al. 2002). Two pieces of recent evidence have challenged this view. First, the oldest known fossil bee (65 million years old) is a stingless bee (now known as *Creotrigona*) that has been recently redescribed as having affinities to the most highly derived stingless bee taxon (Engel 2000)! Furthermore, the single fossil specimen is a female with a very worker-like abdominal morphology. These results seem to imply that extreme sociality was already present in the earliest known members of the Apidae. If so, and if bee evolution was indeed a progression as postulated, then this progression must have occurred fairly rapidly during the initial diversification of the group. Second, a recent phylogenetic analysis of multiple molecular data sets for the Apidae has also questioned the progressive evolution view (Cameron and Mardulyn 2001). In this analysis, stingless bees are the sister group to the moderately eusocial Bombini (bumble bees), while honeybees are more closely related to the mainly solitary Euglossini (orchid bees). Thus, it seems, high sociality has probably evolved at least twice independently within the Apidae. By contrast, Engel's (2001) morphological phylogeny of the Apidae (which includes fossil taxa) yields the more traditional view of monophyly of the highly social bee groups, as does a recent cladistic analysis of the sting apparatus (Cardinal and Packer 2007). The combined morphological and molecular approach of Schultz et al. (1999) tended to favor the traditional view, but not unequivocally so, while a comparable analysis by Cameron and Mardulyn (2001)—with a different, and larger, morphological data set—tended to support the opposite view.[19] Furthermore, the suprisingly extensive differences in gene order of the mitochondrial tRNA genes of *Melipona bicolor* and *Apis mellifera* (50% are in different positions in the two taxa; Silvestre and Arias 2006) are difficult to reconcile with the tradition phylogenetic view, as mitochondrial gene order tends to be very conservative in closely related taxa. Indeed, a recent analysis of mitochondrial genes in the bumblebee *Bombus ignitus* points to a closer relationship between *Melipona* and *Bombus* (Cha et al. 2007), concordant with the hypothesis of Cameron and Mardulyn regarding independent origins of eusociality. Still, given the substantial differences in the mechanics of sociality in the stingless bees and honeybees, a conclusion of independent origins of sociality in honeybees and stingless bees would hardly seem surprising. Indeed, recent studies I described above (p. 393) on the connection between honeybee worker ovarian development, endocrinological state, pollen *versus* nectar foraging, and progression through worker tasks (Makert et al. 2006; Amdam et al. 2006, 2007) as well as some similar results from stingless bees (Cepeda 2006), bears on this issue. These findings suggest a persistent signal in modern honeybees and stingless bees of preadaptation for social evolution in their solitary ancestors: solitary bees in non-reproductive cycles have immature ovaries, low vitellogenin levels and focus on nectar collection; those in reproductive cycles have mature ovaries, high vitellogenin levels, and focus on pollen collection and hoarding. These patterns are still seen in modern honeybees, and to some extent in stingless bees, and underlie the social organization of their hives (see Cepeda 2006, Amdam et al. 2007, and references therein). If

[19] Cameron and Mardulyn note, however, that combined morphological and molecular phylogenies from such incongruous data sets as those for the bees is a questionable undertaking at best (Cunningham 1997)

this analysis is correct, then it suggests that solitary bees are not only preadapted for the evolution of eusociality, but indeed would be predicted to evolve parallel patterns of eusociality in independent lineages (*via* a type of developmental constraint; see Hodin 2000), in line with the evolutionary hypothesis of Cameron and Mardulyn (2001).

One major difference between honeybee and stingless bee reproductive biology is that while honeybee queens are typically multiply-mated, stingless bees in general are thought to mate only once (for exceptions see Imeratriz-Fonseca et al. 1998, Paxton 2000).[20] Hence, workers in stingless bee colonies tend to be full sibs, while workers in honeybee colonies are often only half sibs (reviewed in Peters et al. 1999, Tóth et al. 2004). Thus, when a worker lays a male egg in a honeybee colony, such offspring have relatively low relatedness to their aunt workers, so the inclusive fitness inherent in rearing nephews is lower than rearing brothers (i.e. sons of the queen). In stingless bees, by contrast, workers are actually more closely related to their nephews than to their brothers. Thus, it is argued, stingless bee workers will be more likely to tolerate reproduction by their sisters than would be the case for honeybees (see, for example, Engels and Imperatriz-Fonseca 1990; but see also Drumond et al. 2000, Green and Oldroyd 2002, Tóth et al. 2004).

Another major difference between stingless bees and honeybees is that the former mass provision brood cells, and close them upon oviposition (Roubik 1992). In honeybees, the larvae are continually fed by workers throughout the larval period. Also, there is evidence that honeybee queens mark eggs with a pheromone that identifies them as her offspring (reviewed in Oldroyd et al. 2002; but see Katzav-Gozansky et al. 2002), while little such evidence exists for egg marking in stingless bees (but see Koedam et al. 2007).[21] Furthermore, the enormous caste differences in ovariole number described above for honey bees are not found in stingless bees, where both the queens and workers in most species have four ovarioles per ovary (Engels and Imeratriz-Fonseca 1990). All of these differences are correlated with a much higher level of öocyte maturation in stingless bee workers than in their honeybee counterparts.

Indeed, egg laying by workers in stingless bees is quite common. Many species show evidence for substantial drone production by stingless bee workers in queenright colonies (e.g. Koedam et al. 1999, 2001; Tóth et al.

[20] It may well be, however, that single mating was the ancestral condition for the honeybee lineage as well (see Korb and Heinze 2004)

[21] Pheromonal egg marking by queens has also been recently confirmed for the Florida carpenter ant *Camponotus floridanus* (Endler et al. 2004).

2002a,b, 2004), though the majority of workers within a colony apparently do not lay eggs, and have relatively undeveloped ovaries. In most species, workers also produce what are considered trophic eggs, as they differ in morphology from functional eggs, and they are, in fact, often consumed by the queen (e.g. Engels and Imperatriz-Fonseca 1990; Koedam et al. 1999, 2001; Toth et al. 2002a, 2004).

Trophic eggs can be considered a plastic mechanism by which workers channel nutrition to the queen. Indeed, under conditions of high colony nutrition, the numbers of trophic eggs laid increases, and this is also correlated with a higher rate of queen ovarian development (Grosso et al. 2000, Capeda 2006). Still, a second (not mutually exclusive) possibility is that many stingless bee colonies function right on the verge of queen control over worker reproduction (Sakagami and Zucchi 1974; Wenseleers et al. 2004). Indeed, in some species, worker derived males appear to be the norm (*Tetragona clavipes, Paratrigona subnuda* and several *Melipona* species; Toth et al. 2002a,b, 2004).

Are we to consider stingless bees, therefore, to be less highly social than honeybees due to this blurry distinction between queens and workers in terms of egg production? Not at all. The fact that worker reproduction is positively correlated with colony nutrition is eminently sensible. Under reduced food conditions, worker reproduction decreases, and the queen lays mostly female eggs—an investment in future colony productivity (Moo-Valle et al. 2001). Once again, it seems that ecological realities dictate reproductive choices in insects. This is hardly surprising.

Now or Later, and How Many? Ovipositional Plasticity in Insects

Much of the information on insect reproductive plasticity comes from studies of oviposition. This bias has a partly pragmatic explanation: unlike ovariole number and öocyte maturation, oviposition can be readily observed in many insect species, both in the lab in the field, without having to kill the insect. The propensity of studies on oviposition also has an important ecological justification: a poor decision on when and how many eggs to oviposit can have disastrous consequences for an insect's fecundity (number of offspring surviving to reproductive age).

As expected, a multitude of environmental factors, from day length, nutrition, availability of hosts, temperature, the presence of mates and mating, can all influence oviposition, to varying degrees in different insects (reviewed in Labeyrie 1978, Hinton 1981, Papaj 2000). And, as mentioned previously, oviposition can itself be the trigger for further öocyte maturation, resetting the reproductive cycle. These mechanisms have all been extensively reviewed for many individual insect groups and for insects as a whole, and I have already touched on many of these points in the preceding sections, since maturation and oviposition are intricately connected (Papaj 2000). Thus, I will concentrate here on just a few additional examples which illustrate some of the evolutionary pressures that might select for one particular ovipositional strategy over another. I will also address the mechanisms by which oviposition behavior is connected to the öocyte maturation cycle.

Should I Lay or Should I Go? Host Preference and Oviposition Rates in Pollen Beetles

The pollen beetle *Meligethes aeneus* feeds and oviposits on various brassica (family Brassicaceae), and are important pests of seed crops. Because the adults feed on pollen and nectar in open flowers whereas the larvae consume pollen inside flower buds there is not a perfect relationship between adult food and ovipositional host availability (Hopkins and Ekbom 1996). Perhaps as a result of this discontinuity, these beetles maintain a high degree of plasticity with respect to oviposition rate depending upon the quality of available hosts. For example, choice experiments identify oilseed rape as a high quality host and mustard as one of low quality (Hopkins and Ekbom 1999). Host switching experiments resulted in alterations in ovipositional rates, demonstrating a substantially modifiable ovipositional response to the immediate presence of host plants of differing quality (Hopkins and Ekbom 1999). The long life span, low adult mortality, and high lifetime egg laying potential of these beetles (Hopkins and Ekbom 1996) may indicate why they can afford to be particularly choosy in the presence of low quality hosts: chances are that a higher quality host will come along in their lifetimes.

Contrast these findings to those for the sheep blowfly (*Lucilia cuprina*; see above): extreme specialists that will lay all of the eggs that they have upon locating an appropriate host. Clearly, the ecological parameters under which reproduction in these two disparate insects occur provides a logical explanation for the extreme ovipositional plasticity in one species and its complete absence in the other. At the end of this chapter, I will discuss the implications for evolvability of such extreme differences in plasticity capacity.

"Sunlight will Increase Your Reproductive Potential," and Other Sound Advice from Grasshoppers

Stauffer and Whitman (1997) provided an excellent and comprehensive review of grasshopper oviposition biology. They note a marked variability among grasshopper species in ovariole number, egg mass, clutch size, time to first reproduction and inter-ovipositional interval, and so on (all of which correlate with mean body mass of these species). These results beg the following question: are certain grasshopper lineages constrained in their ovipositional ecology by the dynamics of their ovarian differentiation and/ or maturation processes? The suite of ovipositional diversity cited above is, thus, crying out for a phylogenetic analysis.

Stauffer and Whitman (1997) also point out that grasshopper oviposition is modulated—to varying degrees in different species—by most of the environmental factors that are known to influence ovipositional biology in insects as a whole. Temperature is particularly important for determining the timing of öocyte maturation and oviposition. Developmental rates and egg production decline dramatically under cold or cloudy conditions, and this influences biogeography and life history (Dearn 1977). For example, *Locusta migratoria* is univoltine in cool northern Japan, bivoltine in middle Japan, and multivoltine in warm southern Japan (Tanaka et al. 1993). Because body temperature is so important, most grasshoppers spend substantial portions of their daily cycle solar-basking (thermomaximizing) (Chappell and Whitman 1990). Others exhibit phenotypic plasticity in body color, becoming darker in cool seasons, which allows them to better asorb solar radiation and maintain warm body temperatures (Key and Day 1954, Colvin and Cooter 1995). Basking grasshoppers can increase their body temperature by as much as 18°C above ambient (Kemp 1986). Many species seek out warm microhabitats, such as sun-facing slopes, elevated ant mounds, or open, sunny ground, in which to oviposit (Stauffer and Whitman 1997).

The environmental factors known to modulate oviposition in grasshoppers include: nutrition (affecting the pre-ovipositional and inter-ovipositional periods), temperature (through metabolic effects on öocyte maturation, activity rates, indirectly through food quality, etc.), exposure to solar radiation (to overcome the physiological constraints of the metabolic effects of temperature), rainfall (through, for example, specific triggers of oviposition by softening ovipositional substrates), photoperiod, crowding (which can enhance or suppress oviposition, depending on the species), finding the correct substrate (Mefferd et al. 2005), and mating stimulation (reviewed in Stauffer and Whitman 1997; Walker et al. 1999). These factors likely vary from largely metabolic realities (such as the effects of temperature and nutrition) to specific environmental clues that it is a good time to oviposit (such as the effect of rainfall and photoperiod). But it would be incorrect to assume that, for insects in general, the metabolic realities are simply a non-adaptive plastic response, as two examples will demonstrate. First, because the specific temperatures that enhance oviposition show geographic differences (e.g. reviewed in Hinton 1981 for a wide range of insects); and second, because the degree to which nutritional factors modulate oviposition varies among populations and species (Wheeler 1996). These observations indicate that insects fine-tune their plastic responses within the framework of metabolic and environmental realities, a highly parsimonious evolutionary solution. The regulation of insect oviposition is, thus, a prime example of how phenotypic plasticity arises at the crossroads of constraints (ontogenetic, phylogenetic, physical), selection and the environment.

Night Time is the Right Time: Circadian Egg Laying Cycles in Kissing Bugs

Most of the factors described above involved permissive conditions for ovipositional cycling or oviposition itself. Still, environmental modulation of egg laying is often controlled at an even finer scale, so that eggs are laid preferentially only at certain times of day. So, for example, apple infesting tephritid fruit flies will oviposit late in the day (Hinton 1981), thus perhaps minimizing the possibility that their eggs will be over-heated before the larvae hatch and have the ability to actively avoid the hottest parts of the fruit. The most likely explanation for such behavior is that oviposition is modulated by the circadian cycle, and experiments with the primary Chagas disease vector *Rhodnius prolixus* (Hemiptera: Reduviidae), among other insects, have clearly demonstrated a circadian timing of oviposition. Under a wide variety of light-dark cycle lengths, *R. prolixus* females tended to concentrate their ovipositional bouts right around the time of lights out, and such cycles persist in total darkness, indicating an entrained circadian control (Ampleford and Davey 1989).

Insect circadian cycling has been well-documented to be regulated by daily cycles of active protein products of the *period* gene (reviewed in Ashmore and Sehgal 2003). A myotropic peptide hormone released from neurosecretory cells of the brain increases the power of ovipositional contractions in *Rhodnius* (reviewed in Davey 1997; studies in other insects suggest that this hormone could be either CCAP, proctolin or a FMRFamide family protein; see below). Thus, an examination of *period* protein cycling in these neurosecretory cells, as well as in neurons that interact with these cells, might offer the potential of understanding the entire network from environmental signal to behavioral output. Such an approach might eventually be applied to a diversity of species that differ in the particular circadian timing (or lack of timing) of their ovipositional behavior.

Note that not all apparently "circadian" oviposition patterns are in fact photoperiod-controlled. In grasshoppers, for example, temperature largely determines the hour of laying. On cold days grasshoppers lay at midday if at all, while on hot days some species lay in the morning and late afternoon, but not at mid-day. On extremely hot days, some species lay primarily at night (Stauffer and Whitman 1997, 2007).

Miscellaneous Strategies to Modulate Reproductive Output

I have thus far mostly concentrated on plasticity in three aspects of reproductive development and physiology: ovariole number, öocyte maturation, and timing of oviposition. Still, there are several other ways in which insects modulate reproduction that are worth mentioning, if only briefly.

Why Don't We Do It in the Road? Plasticity in Place of Oviposition

Many insects are extreme specialists on certain ovipositional hosts, such as rabbit bot flies on rabbits, fig wasps on figs, and gall forming midges on specific plant species. Such insects are highly canalized for host species, but can still exhibit high adaptive ovipositional plasticity in response to different host individuals or host body parts (Souza et al. 2005). By contrast, other insect taxa exhibit varying degrees of plasticity for ovipositional location, from extreme generalists like some drosophilid vinegar flies to what might be called "hierarchical specialists": those that exhibit preferences for certain ovipositional hosts, but will lay on suboptimal hosts if the preferred host is unavailable. I discussed one example of such hierarchical specialization above when describing maturational plasticity and egg load in tephritid fruit flies of the genus *Bactrocera*. When deprived of preferred hosts, some generalist-leaning taxa like *Bactrocera tyroni* readily lay on suboptimal hosts, whereas specialist taxa like *B. cucumis* are more likely to hold their eggs (Fitt 1986). Presumably, the adults make calculations based on likely larval survival/quality on the suboptimal hosts and the likelihood of their own continued survival to find a more optimal host plant (see Papaj 2000 for a general review).

The Australian plague locust *Chortoicetes terminifera* exhibits an adaptive plasticity in egg pod shape and location. Females that experience high temperatures and a long photoperiod lay egg pods deep in the ground, whereas females exposed to low temperatures and a short photoperiod lay shallow egg pods. This is adaptive in that the deep pods are laid in summer, when heat and drying would kill shallow eggs (Wardhaugh 1977).

Extreme Plasticity in Reproductive Timing

Reproduction is generally limited to the adult stage in insects. This seems strange *a priori*, particularly given the fact that other arthropods routinely reproduce at very early growth stages, and that hemimetabolous insects, at least, have already adopted an adult-like morphology at hatching. These observations indicate some constraint on precocious reproduction in most insects. Still, there are a few notable cases of insects that have circumvented this constraint, including certain gall midges, aphids, beetles, and hover flies (reviewed in Hodin and Riddiford 2000b; Achterkamp et al. 2000; Normark 2003). The best studied cases are paedogenesis in gall midges and telescoped generations in aphids. A commonality among these disparate examples is that alternative reproductive strategies are triggered by the quality and/or the quantity of available food. While the physiological basis for the connection between the nutritional input and the physiological/morphogenetic output is not understood for any such paedogenetic taxa, recent work on a variety of insects has indicated a direct nutritional control of reproduction via hormonal release (reviewed in Adams 1999, Barton Browne 2001), possibly involving insulin-like signaling (Tu and Tatar 2003). It is likely that a modification of such a situation is at work in the paedogenetic taxa as well.

All larvally reproductive (paedogenetic) gall midges are fungus feeders, and good quality fungal resources encountered in the few hours after hatching induces precocious ovarian differentiation and maturation in early stage larvae (reviewed in Went 1979). This precocious maturational process is correlated with an ecdysteroid peak (Went et al. 1984), and increased levels of ecdysteroid receptor proteins specifically in the cells of the immature ovary (Hodin and Riddiford 2000b). When newly hatched larvae encounter poor food conditions, receptor expression is delayed, the larvae proceed through metamorphosis, and the winged adults disperse to find a new mushroom patch (Figure 5). This pattern holds for two gall midge

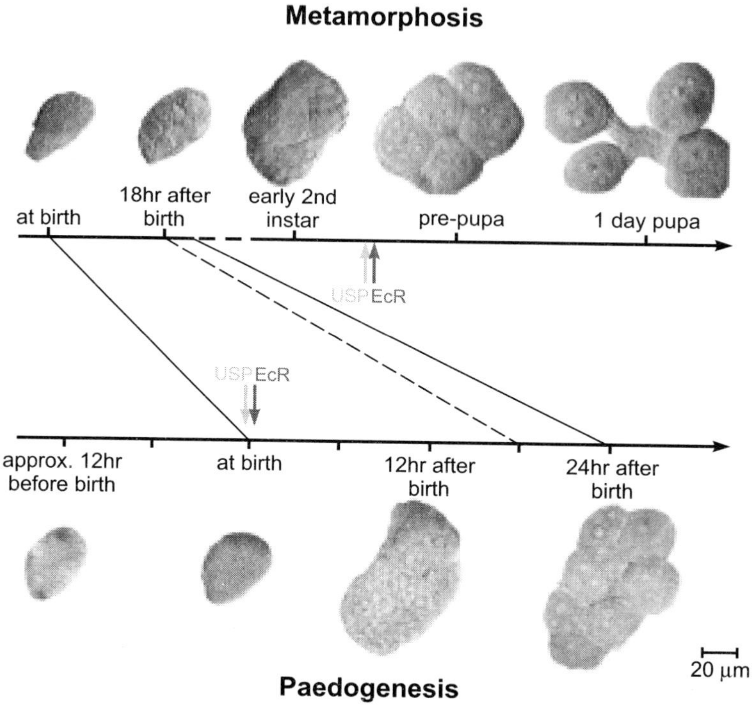

Fig. 5 Comparison of ovarian morphogenesis during metamorphic (upper panels) and paedogenetic (=larvally reproductive; lower panels) development in the gall midge *Mycophila speyeri* (Diptera: Cecidomyiidae). Ovarian development is greatly accelerated in paedogenetic larvae; for example, paedogenetic ovaries 24 hours after birth are at about the same stage of öocyte maturation as pre-pupal metamorphic ovaries. This acceleration in the paedogenetic life cycle is correlated with early appearance of the Ultraspiracle (USP) and Ecdysone Receptor (EcR) proteins in ovarian cells (arrows). These two proteins constitute the functional insect ecdysteroid receptor. The independently-evolved paedogenesis in the gall midge *Heteropeza pygmaea* follows an almost identical pattern (Hodin and Riddiford 2000b; see this reference for life cycles and other details in both taxa). We took these photomicrographs using DIC (differential interference contrast) optics.

species (*Mycophila speyeri* and *Heteropeza pygmaea*) in which paedogenesis has apparently arisen independently (according to the phylogenetic analysis in Jaschof 1998), suggesting that standard hormonal control mechanisms need to be altered in specific ways in order for paedogenesis to evolve (Hodin and Riddiford 2000b). Furthermore, a suite of other preadaptations[22] (including parthenogenesis, modified öogenesis in the absence of ovarioles, early germ cell proliferation, and substantial

[22] see footnote 8, above

nutritional plasticity in reproduction) are also presumably involved. It seems likely that the explanation for the restricted distribution of paedogenesis among insects lies in the rarity with which this particular suite of preadaptations is in place in different insect lineages.

Telescoped generations in aphids represents an even more bizarre situation (reviewed in Simon et al. 2002), where apterous (wingless) generations of early nymph stage aphids can not only brood offspring within their ovaries (again via parthenogenetic activation of öocytes), but these brooded children can themselves have embryos developing within their gonads! Again, idiosyncratic characteristics of aphid ovarian development in general (such as parthenogenesis and extremely early germ cell proliferation; reviewed in Büning 1994) might have pre-adapted the group for this highly modified life cycle. Furthermore, such a situation (in gall midges and aphids) affords these insects the opportunity to communicate specific information[23] to their offspring concerning the likely environment that they will face upon emergence from their mother. Maternal feeding conditions are also known to strongly influence the phenotype (metamorphic *versus* paedogenetic) of brooded gall midge larvae (reviewed in Went 1979). Such examples of maternal (and, as in the case of aphids, grandmaternal) effects on insect reproduction is my next topic for consideration.

Mother Knows Best: Maternal Influences on Insect Reproduction

As mentioned previously, in variable environments, it might be predicted that developing insects would benefit from maternal information about the quality of the habitat that they would be likely to experience following adult eclosion. Such maternal information would seem most useful for insects that: 1) differ in their pre-adult *versus* adult habitats; 2) produce multiple overlapping generations; 3) develop relatively quickly from hatching to adult eclosion; and/or 4) feed on rare but rich and/or long-lived resources. Unfortunately, too little comparative information is available concerning maternal effects on insect reproduction to be able to adequately evaluate these hypotheses. Still, several notable examples, where maternal conditions (such as crowding, nutritional availability and temperature)

[23] Rapidly accumulating data in a variety of insects (e.g. Broughton et al. 2005, Wu and Brown 2006, and citations therein) suggests to me that insulin-like signaling may be the pathway by which such information on external food conditions is communicated to brooded offspring.

have predictable effects on offspring reproductive characters, suggest that such effects may be more widespread than is generally appreciated. This class of plasticity has been particularly well-studied in grasshoppers (reviewed in Stauffer and Whitman 1997, Hartfelder and Emlen 2004), where even grandmaternal effects have been reported (e.g. Smith 1972). Recently, a rare example where such plasticity seems demonstrably adaptive has been elucidated for the soft scale insect (Hemiptera: Coccidae) *Saissetia coffeae* (Spitzer 2004). For example, *S. coffeae* mothers raised in an environment where fungal infection is more likely had offspring that showed slight increases in resistance to fungal attack.

Each Child Is Precious: Reproduction in Tsetse Flies

Tsetse flies of the genus *Glossina* (Diptera: Glossinidae) are major disease-causing vectors in Africa, as they transmit trypanosome parasites to vertebrates (African sleeping sickness in humans; nagana in cattle) through blood-feeding. Interestingly, this group of flies is fairly unique among insects in brooding their young almost all of the way to pupation (an extreme form of larviparous reproduction). As a result, they mature only one egg at a time from highly reduced ovaries (only two ovarioles per ovary; reviewed in Tobe and Langley 1978). In another eerie parallel of the reproductive biology of one of their nutritional hosts (humans), maturation of öocytes occurs in an alternating pattern, where the first egg matures in left ovariole 1, then right ovariole 1, then left ovariole 2, then right ovariole 2, and so on. The speed of maturation is related to blood feeding, which also increases receptivity to mating. While pre-vitellogenesis and vitellogenesis may be regulated by similar mechanisms elucidated for mosquitoes, the ovipositional (or, in *Glossina*, the comparable "larvipositional") process seems to be more intensively controlled in *Glossina* (reviewed in Tobe and Langley 1978). Ablation of the brain's median neurosecretory cells (a source of many neurohormones) or the corpora cardiaca (a source of a variety of hormones), but not the corpora allata (where JH biosynthesis occurs), inhibits larviposition behavior. After spending so much energy protecting and feeding their offspring, they are certainly more invested than, say, mosquitoes in choosing the appropriate time and place to larviposit; this could account for the apparently enhanced larvipositional network in tsetse flies. Extending this analysis to ovipositional control mechanisms in a diversity of blood feeding dipterans that either brood (such as louse flies; family Hippoboscidae) or don't brood (such as black flies; family Simuliidae) would allow a rigorous, comparative test of this hypothesis.

Why Do Today What You Could Put Off Until Tomorrow? Reproductive Diapause

Many insect taxa occupy habitats where the conditions during part of the year are too harsh (too hot, too cold, too dry, poor nutrition, etc.) to support continued growth and reproduction. To deal with these temporary extreme conditions, such insects often go into diapause, a situation not unlike hibernation. Diapause occurs in different taxa at different stages of development: embryo, larva, pupa or adult. In adult female diapause, the insect typically regresses her ovaries, and remains in a relatively dormant state for the period unsuitable for reproduction. Then, when conditions improve, the ovaries reinitiate maturational processes, and reproduction ensues. Examples of reproductive diapause abound, including the ladybird beetles that invade houses in temperate latitudes in the wintertime and the absconding honeybee queens I described above. The physiological, ecological and evolutionary bases of insect reproductive diapause have been extensively reviewed elsewhere, so I will merely point the reader to one of several excellent reviews for more information on the topic (Tauber et al. 1986, Danks 1987, Leather et al. 1993, Hopper 1999, Hodek 2003).

Does Size Matter? The Egg Size/Egg Number Trade-off

Much life history theory centers on the apparent trade-off between egg size and egg number (reviewed in Stearns 1992). The general presumption is that increased maternal investment in individual eggs (as evidenced by the egg mass and/or egg volume) leads to increased chance of survival to reproduction, via any number of mechanisms, for example, more rapid development to initial mobility (e.g. to avoid egg dessication or egg predators) or to reproductive maturity. Those individuals that lay more eggs of lower quality, then, are playing a numbers game, where a decreased chance of survival of an individual offspring is balanced by a greater number of offspring.

This appears to be the case in insects, as egg size/number trade offs have been noted in mayflies (Corkum et al. 1997), grasshoppers (reviewed in Stauffer and Whitman 1997), parasitic bees (Wcislo and Canne 1996), piophilid cervid flies (Bonduriansky and Brooks 1999; Bonduriansky 2003), lycaenid butterflies (Fischer and Fiedler 2001), calliphorid flies (reviewed in Wall et al. 2002), ladybug beetles (Stewart et al. 1991) and so on. An egg size/ovariole number trade-offs has been described for Hawai'ian and mushroom-eating flies in the family Drosophilidae (Starmer et al. 2002). Hinton (1981) argues that, in general, sub-social taxa (such as coprophagous dung beetles) lay fewer, larger eggs than their non-social counterparts, owing to their greater reproductive investment per egg. Likewise, r selected taxa are predicted to invest in many small eggs, while k selected taxa might generally invest more into individual eggs, thus resulting in fewer larger eggs by comparison (see Pianka 1970). Such hypotheses are certainly tempting and plausible, but await rigorous phylogenetic analyses.

Egg size plasticity in insects is less known than plasticity in reproductive timing or clutch size. However, there are some good examples, including the classic study by Fox and coworkers (1997) who showed that female seed beetles, *Stator limbatus* (Coleoptera: Bruchidae), adjust the size of their eggs to match seed quality of their hosts in the family Fabaceae. Larger eggs were laid on seeds of the palo verde tree (*Cercidium floridum*), as smaller eggs almost never survive on such hosts. By contrast, these beetles consistently laid smaller eggs on cat claw acacia seeds (*Acacia greggii*), since egg size was not related to survival on these seeds, and when beetles laid smaller eggs they could lay more of them. Some grasshoppers also alter egg size in response to environmental stimuli. Late-maturing *Chortophaga viridifasciata* and *Arphia sulphurea* grasshoppers produce fewer, larger eggs at shorter interoviposition periods than do females that mature early in the season (Landa 1992a,b).

The tephritid fly "*Anastrepha fraterculus*" (which is actually two cryptic species) produces different sized eggs, though the embryos themselves are similarly sized. The excess yolk in larger eggs is extruded and later consumed by the larva shortly before hatching (Selivon et al. 1997), indicating a constraint on embryo size which these flies have circumvented in a novel way. Although life history theory (e.g. Stearns 1992) predicts consequences for larval and/or post larval performance, survival etc. resulting from such differently-sized eggs, such consequences have not yet been explored in this case. Indeed, very few insect studies have rigorously tested the hypothesis that increased egg size would lead to an increase in fitness in the offspring (*via* faster development, greater resistance to starvation as larvae, increased reproductive capacity, etc.; the Fox et al. study outlined above is a notable exception). For example, Honek (1996) reviews the literature on such studies for coccinellid beetles, and notes a few instances where species with larger eggs showed greater starvation resistance or shorter development times than those species with larger eggs. However, in the absence of an explicitly phylogenetic approach, one cannot exclude possible confounding effects of (i.e. correlations with) phylogeny rather than the noted egg size differences.

The ideal situation for testing the egg size/number trade-off would be examining within-species variation, and rearing the larvae derived from different sized eggs in a common garden (e.g. Fox et al. 1997; see above). Such an experiment would be difficult for coccinellids, since the vast majority of their intraspecific variation in reproductive output is due to differences in egg number rather than size (Stewart et al. 1991). By contrast, in the African squinting bush brown butterfly *Bicyclus anynana* (Lepidoptera: Nymphalidae), oviposition temperature influences egg size: 27°C oviposition results in a greater than 20% reduction in egg volume than 20°C oviposition (Fischer et al. 2003). However, the resultant larvae reared under identical conditions showed no detectable fitness differences. Thus, caution is certainly warranted in drawing premature conclusions regarding the life history consequences of egg size differences within and among insect taxa.

It may be that egg size *per sé* is not a particularly good predictor of egg energy across insects generally, and it is egg energy that would presumably have the direct effects on the fitness of subsequent life stages. Indeed, a similar debate has surrounded the use of egg size as a proxy for egg energy among marine invertebrates as well (see McEdward and Morgan 2001).

One final topic worth mentioning related to energy investment in eggs is the production of sterile eggs as food for other conspecifics. The production of such so-called "trophic eggs" are particularly widespread among the social stingless bee workers (family Apidae; group Meliponini), as described above.

Can't Buy Me Love? Nuptial Gifts and Copulation Duration in Dance Flies

Although the vast majority of this chapter has only considered female reproduction, a complete analysis of insect reproductive potential (in sexual species) must consider the contributions of the male as well. While a full analysis of male reproductive plasticity is well beyond the scope of this review, one particular example nicely illustrates the interplay between male and female investment in reproduction: nuptial gifts by males to females before mating. Such offerings are positively related to female fitness in fireflies (Coleoptera: Lampyridae; in the form of spermatophores in this case,

Rooney and Lewis 2002), and thus would benefit the male as well. These benefits are reciprocated most directly, though, in the case of empidid dance flies, where the size of the nuptial gift shows a strong positive correlation with time spent copulating (Svensson et al. 1990). Because copulation duration correlates with quantities of sperm transferred, longer copulations would be predicted to increase male's reproductive success.

In the Orthoptera, öocyte development rates, time of oviposition and number of eggs laid are plastic in response to males and mating (see Loher and Dambach 1989; Brown and Gwynne 1997; Walker et al 1999). For example, copulating male *Teleogryllus commodus* crickets pass prostaglandin synthetase to the females in their spermatophores, inducing in the female the production of prostaglandins and the rapid release of several hundred eggs, followed by dramatically higher JH levels and further rapid egg development (Loher and Dambach 1989; Loher and Zaretsky 1989). Also, there is evidence in some species (such as the tobacco budworm *H. virescens*; Park et al. 1998) for direct transfer of JH from males to females, which can then act to promote öocyte maturation. In *Drosophila melanogaster*, ovulation-inducing "sex peptides" are transferred by males during copulation (Chen et al. 1988). Still, there is evidence for costs to the female (in terms of longevity) of these copulatory substances (Wolfner 1997, Wigby and Chapman 2005, Gage 2005), perhaps *via* immune suppression (e.g. Fedorka et al. 2004, 2007). Thus, additional studies are required before we can say with some confidence that more time spent copulating is definitively better.

Conclusions and Speculations

Plasticity at What Level? Patterns and Constraints

I have described examples where insects modulate reproductive output during immature phases, *via* ovariole number determination or larval reproduction, and adult phases, *via* öocyte maturational and ovipositional regulation (Figure 6). Why this variability? Why are honeybees characterized by such extreme queen-worker dimorphism in ovariole number, while stingless bees express queen-worker differences mainly at the level of rates of öocyte maturation? Why are some insect classes (such as the Diptera and Orthoptera) characterized by such extreme variation in ovariole number, while other classes (such as most Hemiptera and all Lepidoptera) show little to no such variation?

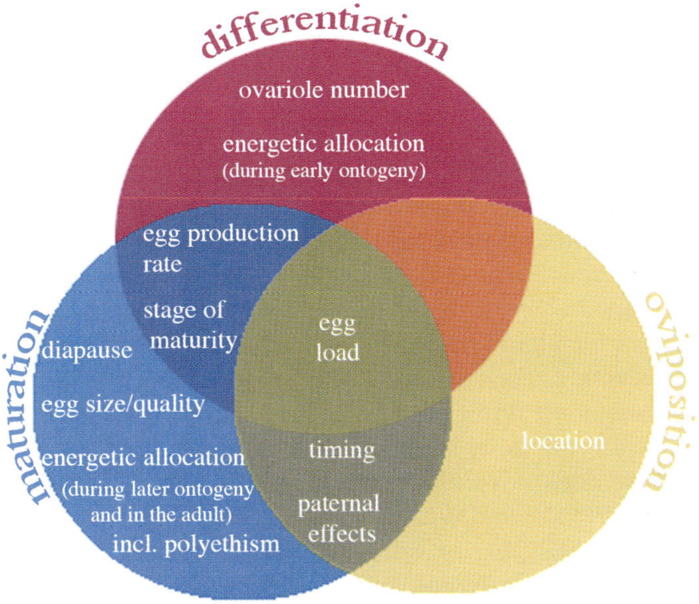

Fig. 6 A pictorial representation of the integration of an ontogenetic and ecological focus with respect to insect reproductive plasticity. The three circles represent the three ontogenetic phases of reproduction subject to plasticity: ovarian differentiation, ovarian maturation and oviposition (see also Figure 1). Inside of the circles are the ecologically relevant parameters subject to plasticity; their locations indicate the ontogenetic phase(s) during which each ecological parameter can be modified. So, for example, location of egg deposition is modulated only during the ovipositional phase, whereas the stage at which reproductive maturity is reached can be modulated during either the differentiation or the maturation phase (or both). Egg load (numbers of mature eggs being held) is the one ecological parameter that can be influenced by differentiation, maturation and/or oviposition.

I have endeavored to show that the answers to these questions lie at the intersection of ecological factors and developmental (and other) constraints (Figure 6). Most of the examples that I have discussed concerning reproductive plasticity can be seemingly accounted for by ecological factors, such as the differences in maturational plasticity in generalist versus specialist tephritid fruit flies. Indeed, much of the published literature on reproductive plasticity has adopted such an ecological focus. Still, without a consideration of proximate mechanisms, such an ecological approach will always be incomplete. Indeed, life history theorists are beginning to appreciate as much, as Stephen Stearns recently wrote (2000, p. 484):

> *When we understand the phylogenetic origins of developmental constraints on the adult phenotype, we should be in the position to explain some of the major puzzles in life history evolution—as well as many other things.*

Developmental constraint is an elusive concept that often seems to either be overused (e.g. where any lack of variation is considered evidence of constraint) or underappreciated. I (Hodin 2000) as well as others (e.g. Beldade et al. 2002; Richardson et al. 2003; Brakefield and Roskam 2006) have reexamined constraints in a modern context in an attempt to define useful criteria for identifying and understanding the phenomenon. One example relevant to the topic at hand is ovariole number determination among grasshoppers. While there is substantial genetically-based and some plasticity-based variability in this characteristic among grasshoppers, pre-adult feeding does not influence numbers of ovarioles (Stauffer and Whitman 1997, Taylor and Whitman in prep.) as it does in other taxa (such as in drosophilids; e.g. Hodin and Riddiford 2000a). Ecological explanations, such as hypothesizing the absence of environmental conditions that would favor such variability, cannot account for this lack of plasticity. Indeed, ovariole number **is** subject to plasticity in locusts, but only as a result of environmental effects in the maternal, rather than the embryonic/nymphal environment (reviewed in Stauffer and Whitman 1997). This observation demonstrates both a capacity for and probable selective advantage of ovariole number plasticity in the group. Furthermore, grasshopper nymphs, like many hemimetabolous insects, exist in a similar environment to what they will likely experience as adults. Thus, it would seem, ovariole number plasticity resulting from exposure to the pre-adult environment would be even more selectively advantageous than such plasticity in holometabolous insects like *Drosophila melanogaster*, whose adult environment could differ substantially from the one in which they existed as pre-adults (larvae).

So, if the ecological pressure for ovariole number plasticity is present in grasshoppers, and is even manifest through maternal effects, then why on earth do grasshopper nymphs themselves not express such plasticity? The answer appears to be a simple developmental constraint: maximal ovariole number is determined in all known grasshoppers in the embryonic stage (reviewed in Stauffer and Whitman 1997). Therefore, by the time the nymphs hatch and explore their habitat, it is too late to have any positive influence on ovariole number. As a result, grasshoppers have seemingly evolved the only mechanism available to them to influence ovariole numbers in a plastic manner: they modulate ovariole numbers through maternal effects.[24]

I relate this example to highlight the pitfalls inherent in any attempts to over-generalize explanations for the presence, absence, degree and/or stage specificity of plasticity, or of the utilization of a particular plasticity-inducing environmental cue. A full understanding of reproductive plasticity in a given instance necessitates not only a thorough understanding of life history, ecology, and biology of the insect in question, but also an appreciation of the physiological and ontogenetic peculiarities of ovarian development in that particular insect taxon. And, it is only through a comparative approach that such taxon-specific trends will become apparent, allowing us to take tentative steps in the direction of developing a predictive theory of insect reproductive plasticity.

She's Leaving Home: Holometabolous *versus* Hemimetabolous Insects

If immature insects have the ability to predict the quality of their adult habitat, then one would expect that they would harbor greater reproductive plasticity in their immature stage. For example, if a larva could predict that food resources would be scarce in their adult environment, then it would seem to be a waste of energy for them to differentiate a large number of ovarioles.

While different insects are likely to vary widely in their opportunities to predict the quality of their adult habitat in this manner, hemimetabolous nymphs might be predicted, on the whole, to have a greater insight into their likely adult habitat than their holometabolous counterparts (Wheeler 1996; Papaj 2000). Because a major way in which immature insects modulate their reproductive potential is through ovariole number, the prediction is that hemimetabolous nymphs would tend to demonstrate greater ovariole

[24] Perhaps related to this constraint, grasshoppers have an ingenious method to match reproductive effort to current environmental conditions. This method is rapid, efficient, and, most importantly, reversible. Grasshoppers are exceedingly adept at adjusting the number of developing oocytes via oocyte resorption, to keep pace with current conditions (Sundberg et al 2001). For example, migratory locusts, *Locusta migratoria*, can resorb virtually all of their developing oocytes within days, when the environment fails (Launois 1972). Presumably, the oocyte nutrients are recycled for somatic maintainance, allowing the female to live to breed again when conditions improve. These adaptations can be seen as methods of circumventing the ovariole number determination constraint. Such a process, where constraints lead to novel evolutionary adaptations to circumvent the constraints, have been described in a wide variety of organisms and evolutionary contexts (reviewed in Hodin 2000).

number plasticity than sympatric holometabolous insect larvae that feed on the same food resources. To my knowledge, no such comparative studies have been attempted.

Still, some work on ovariole number plasticity in social insects provides tangential support for this hypothesis. Social hymenopteran (holometabolous) larvae tend to be helpless little grubs that do little but feed and grow. By contrast, some termite (hemimetabolous) nymphs perform colony tasks as nymphs. Thus, termite nymphs have the ability to more closely assess the colony's needs than their hymenopteran counterparts. Indeed, termite nymphs often show great degrees of flexibility in caste differentiation routes, and can even increase ovariole numbers under conditions favoring the development of secondary reproductives (reviewed in Thorne and Traniello 2003). By contrast, hymenopteran workers are all adults, and by this stage, caste and ovariole numbers are more rigidly fixed. Whether such differences between hemimetabolous and holometabolous insects would hold across non-social taxa as well remains to be seen.

Predictable Cooption? Ecdysis, Oviposition and Settlement of Marine Invertebrate Larvae

Growth by ecdysis is a synapomorphy (shared derived feature) for the Ecdysozoa, a proposed clade of protostome invertebrate phyla that includes the Arthropoda and Nematoda, and some lesser known groups including the Nematomorpha, Kinoryncha, Tardigrada and Onycophora (Aguinaldo et al. 1997). While ecdysteroid regulation of molting is also likely found in all of these taxa (though strong evidence for it is lacking for most of these groups), the neurophysiological events regulating ecdysis behavior itself is only well known for the insects, and in particular the hawk moth *Manduca sexta* (Figure 7; reviewed in Fuse and Truman 2002, Zitnan et al. 2007). The peptide prothoracicotropic hormone (PTTH) from the brain stimulates ecdysteroid synthesis, causing deposition of new cuticle. Falling levels of ecdysteroids lead to the release of ecdysis triggering hormone (ETH) from the epitracheal cells, initiating pre-ecdysis behavior (the loosening of the cuticle in preparation for ecdysis). This results in a positive feedback loop where eclosion hormone (EH) is released from the brain, triggering more ETH release, then more EH release and so on. This ultimately leads to the release of crustacean cardioactive peptide (CCAP) from the ventral nerve chord *via* an increase in levels of the second messenger guanosine 3',5'-cyclic monophosphate (cGMP) in the CCAP cells (Gammie and Truman 1999). CCAP release suppresses the pre-ecdysis ETH activity and induces ecdysis behavior (shedding of the old cuticle).

Recent evidence suggests that a remarkably similar regulatory network is involved in insect oviposition behavior as well (Figure 7). In mosquitoes, ovarian ecdysteroidogenic hormone (OEH) released from the brain stimulates the pre-vitellogenic ovary to produce ecdysteroids (Hagedorn et al. 1979; Brown et al. 1998). These ecdysteroids then act on the fat body cells to produce vitellogenin and other yolk proteins, leading to vitellogenesis and öocyte maturation (reviewed in Klowden 1997). The peak of ecdysteroid levels correlates with vitelline membrane production, and falling levels then terminate vitellogenin production by the fat body. At the same time, an unidentified factor ("hormone Q" in Figure 7; this could be a FMRFamide family neuropeptide or possibly octopamine-see the legend to Figure 7) is released into the hemolymph, promoting oviposition behavior (reviewed in Klowden 1997). In *Manduca sexta*, CCAP is present in nerves that innervate the oviduct muscles, and has been shown in vitro to stimulate oviduct contractions (Marshall and Reynolds 1998), providing strong evidence that CCAP may be an endogenous oviposition inducer.[25] The neuroendocrine control of reproduction in insects has been recently reviewed by Simonet and colleagues (2004).

Do these similarities in hormonal networks imply that oviposition is somehow homologous to ecdysis? Not necessarily: the interaction of morphogenetic hormones (such as ecdysteroids) with neuropeptides and neuromodulators (such as EH, CCAP and cGMP) may be a common mechanism by which complex developmental transitions are controlled (Bishop and Hodin 2001, Hodin 2006). Indeed, metamorphosis and settlement in marine invertebrates is a similarly complex developmental transition (Figure 7). In echinoids (sand dollars and sea urchins from the phylum Echinodermata), juvenile morphogenesis during metamorphosis is under the control of thyroid hormones (THs), which are also involved in the attainment of competence to undergo the planktonic-to-benthic settlement behavior (Heyland and Hodin 2004). Once competence is reached, settlement is actively repressed by a NO/cGMP regulatory network until a settlement cue is encountered, at which time the regulatory network is de-repressed, and settlement occurs (Bishop and Brandhorst 2003).

Thus, networks involving interactions between morphogenetic hormones and neuromodulators seem to have been independently coopted in a wide variety of life history transitions. Much is known about how environmental signals (such as day length) can modulate the ecdysis

[25] I described another proposed similarity between oviposition and ecdysis regulation earlier with respect to food induced plasticity and canalization in lubber grasshoppers.

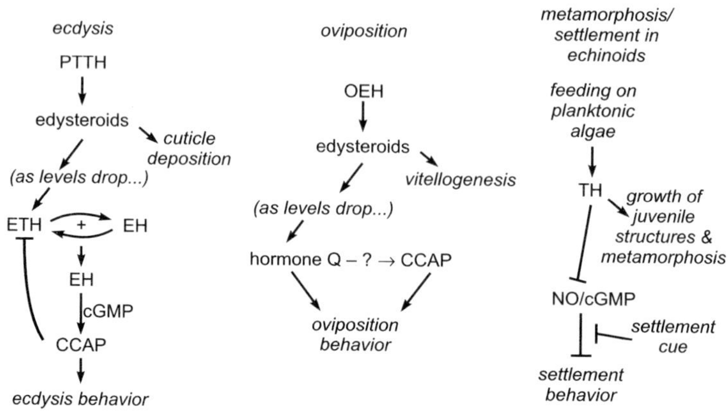

Fig. 7 Similarities in three different complex morphogenetic/behavioral networks. The ecdysis network data is mainly from experiments with the moth *Manduca sexta* (Lepidoptera). The ovipositional network is pieced together here from work on mosquitoes (Diptera) and *M. sexta*. "Hormone Q" is how I refer to the unidentified blood-borne factor that promotes oviposition behavior in mosquitoes (see the text). The blood-feeding reduviid *Rhodnius prolixus* (Hemiptera) appears to have a slightly altered ovipositional network from the one diagrammed here: ovarian ecdysteroids promote the release of a FMRFamide family neuropeptide that promotes oviposition behavior (reviewed in Davey 1997). In *Locusta migratoria* (Orthoptera), FMRFamide-like peptides also induce oviduct muscle contractions, and neurons containing these peptides innervate the oviduct. There is also evidence in *L. migratoria* that an antagonistic interaction between the neuropeptides octopamine and proctolin stimulates oviposition behavior via modulations in cAMP activity (reviewed in Lange and Da Silva 2006), possibly via interactions with CCAP signaling (Donini and Lange 2002). In *Drosohila*, apparent interference with octopamine signaling results in egg retention (Lee et al. 2003). In *D. melanogaster*, oviposition is also regulated by cAMP via an octopamine-proctolin antagonistic interaction (Lee et al. 2003; Rodríguez-Valentín et al. 2006). It is, thus, possible that "hormone Q" may in fact be a series of neuropeptides, the cocktail of which may differ among taxa. The identification of a FMRFamide-like peptide in the genome of the African malaria mosquito *Anopheles gambiae* (Riehle et al. 2002) should facilitate comparative studies to flesh out the nature of this ovipositional network. TH control of metamorphosis has been established for a wide variety of echinoids (sea urchins, sand dollars and their relatives; reviewed in Heyland and Hodin 2004), while the NO/cGMP control of settlement is best studied in the sea urchin *Lytechinus pictus* (Bishop and Brandhorst 2003; Bishop et al. 2006). A more complete analysis of the network can be found in Hodin (2006). Abbreviations are defined in the text. Note tnat this does *not* necessarily imply homologies among these disparate processes. Indeed, the similarities between insects and echinoids are almost certainly due to convergent evolution (see Hodin 2000). The similarities in the ovipositional and ecdysis networks, however, could be indicative of a serial homology-like process, where the presumably more ancient ecdysis network was coopted for regulation of insect oviposition behavior. Subsequently, the details of the two networks would have diverged.

behavioral network (reviewed in Myers 2003). Parallel investigations into the modulation of the analogous oviposition behavioral network, by the environmental signals discussed previously, would greatly enhance our understanding of the mechanistic basis of insect reproductive plasticity. If these patterns do, indeed, turn out to be parallel at a meaningful level, we will have taken one large step towards a predictive science of life history evolution.

Acknowledgements

I would first like to acknowledge all of the glorious insects that never cease to be inspiring. My heartfelt gratitude to Douglas Emlen, Klaus Hartfelder and Doug Whitman for reading an earlier version of this chapter in its entirety, and for their extremely useful comments and suggestions. Thanks also to Lynn Riddiford, Alan Bergland, Corbin Jones and James Baker for their comments on individual sections, and to Otto Pellmyr, Mike Allsopp, Jerome Regier and Tracey Cole for help tracking down obscure references. A special thanks to Jasson at the Seattle Mac Store for data recovery, and Dave's Computer in Santa Cruz, CA for resuscitating my computer. I am most grateful to Joseph Wible and the staff of the Miller Library at Hopkins for invaluable help obtaining resources, and to Hopkins Marine Station (and in particular David Epel) for providing facilities for the completion of this chapter. Funding was provided by the Washington State Employment Security Department. The chapter title is from the *Tao te Ching* by Lao-Tzu (chapter 45, S. Mitchell translation 1992).

References

Achterkamp, B., M. Ottenheim, L. Beukeboom and P. Brakefield. 2000. Paedogenesis in *Eristalis arbustorum* (Diptera: Syrphidae). Proceedings Experimental and Applied Entomology, N.E.V. Amsterdam, 11: 83–97

Adams, T. S. 1999. Hematophagy and hormone release. Annals of the Entomological Society of America, 92: 1–13.

Aguinaldo, A. M. A., J. M. Turbeville, L. S. Linford, M. C. Rivera, J. R. Garey, R. A. Raff and J. A. Lake. 1997. Evidence for a clade of nematodes, arthropods and other moulting animals. Nature, 387: 489–493.

Allsopp M. H. 1992. The *capensis* calamity. South African Bee Journal, 64: 52–55.

Allsopp M. H. 1993. Summarized overview of the *capensis* problem. South African Bee Journal, 65: 127–136.

Alonso-Pimentel, H., J. B. Korer, C. Nufio and D. R. Papaj. 1998. Role of colour and shape stimuli in host-enhanced oogenesis in the walnut fly, *Rhagoletis juglandis*. Physiological Entomology, 23: 97–104.

Alpatov, W. W. 1938. Contribution to the study of variation in the honey bee. V. Number of egg tubules in ovaries of worker bees in the g. *Apis* as related to the evolution of polymorphism. Zoologicheskii Zhurnal, 17: 241–245 [in Russian with English summary].

Aluja, M., F. Díaz-Fleischer, D. R. Papaj, G. Lagunes and J. Sivinski. 2001. Effects of age, diet, female density, and the host resource on egg load in *Anastrepha ludens* and *Anastrepha obliqua* (Diptera: Tephritidae). Journal of Insect Physiology, 47: 975–988.

Amdam, G. V., A. Csondes, M. K. Fondrk, R. E. Page Jr. 2006. Complex social behaviour derived from maternal reproductive traits. Nature, 439: 76–78.

Amdam, G. V., K.A. Nilsen, K. Norberg, M. K. Fondrk and K. Hartfelder. 2007. Variation in endocrine signaling underlies variation in social life history. American Naturalist, 170: 37-46.

Ampleford, E. J. and K. G. Davey. 1989. Egg-laying in the insect *Rhodnius prolixus* is timed in a circadian fashion. Journal of Insect Physiology, 35: 183–187.

Applebaum, S. W., E. Avisar and Y. Heifetz. 1997. Juvenile hormone and locust phase. Archives of Insect Biochemistry and Physiology, 35: 375–391.

Ashburner, M. 1989. Drosophila: A Laboratory Handbook. Cold Spring Harbor Laboratory, Cold Spring Harbor, N.Y.

Ashmore, L. J. and A. Sehgal. 2003. A fly's eye view of circadian entrainment. Journal of Biological Rhythms, 18: 206–216.

Audit-Lamour, C. and D. Busson. 1981. Oogenesis defects in the *Ecd-1* mutant of *Drosophila melanogaster*, deficient in ecdysteroid at high temperature. Journal of Insect Physiology, 27: 829–837.

Barchuk, A. R., M. M. G. Bitondi and Z. L. P. Simões. 2002. Effects of juvenile hormone and ecdysone on the timing of vitellogenin appearance in hemolymph of queen and worker pupae of *Apis mellifera*. Journal of Insect Science, 2: 1–8.

Barker, J. F. and W. S. Herman. 1973. Neuroendocrine control of ovarian development in monarch butterfly. Journal of Experimental Zoology, 183: 1–9.

Barker, J. F. and W. S. Herman. 1976. Effect of photoperiod and temperature on reproduction of monarch butterfly, *Danaus plexippus*. Journal of Insect Physiology, 22: 1565–1568.

Barron, A. B., B. P. Oldroyd and F. L. W. Ratnieks. 2001. Worker reproduction in honey-bees (*Apis*) and the anarchic syndrome: a review. Behavioral Ecology and Sociobiology, 50: 199–208.

Barton Browne, L. 2001. Quantitative aspects of the regulation of ovarian development in selected anautogenous Diptera: integration of endocrinology and nutrition. Entomologia Experimentalis et Applicata, 100: 137–149.

Barton Browne, L., A. C. M. Vangerwen and K. L. Williams. 1979. Oocyte resorption during ovarian development in the blowfly *Lucilia cuprina*. Journal of Insect Physiology, 25: 147–153.

Barton Browne, L., A. C. M. Vangerwen and W. G. Vogt. 1990. Readiness to lay in the blowfly *Lucilia cuprina* is unaffected by oviposition site deprivation or egg-load. Entomologia Experimentalis et Applicata, 55: 33–40.

Beckemeyer, E. F. and P. D. Shirk. 2004. Development of the larval ovary in the moth, *Plodia interpunctella*. Journal of Insect Physiology, 50: 1045-1051.

Beetsma, J. 1979. Process of queen-worker differentiation in the honeybee. Bee World, 60: 24–39.

Beldade, P., K. Koops and P. M. Brakefield. 2002. Developmental constraints versus flexibility in morphological evolution. Nature, 416: 844–847.

Bell, W. J. and M. K. Bohm. 1975. Oosorption in insects. Biological Review, 50: 373–396.

Bélles, X. 1998. Endocrine effectors in insect vitellogenesis, pp. 71–90. *In* G. M. Coast and S. G. Webster [eds]. Recent Advances in Arthropod Endocrinology, vol. 65. Cambridge University Press, Cambridge.

Benz, G. 1969. The influence of the presence of individuals of the opposite sex, and some other stimuli on sexual activity, oogenesis and oviposition in five lepidopterous species, pp. 175–206. Colloques Internationaux du Centre National de la Recherche Scientifique No. 189: L'Influence des Stimuli Externes sur la Gamétogenèse des Insectes. CNRS, Paris.

Berrigan, D. 1991. Lift production in the flesh fly, *Neobellieria (= Sarcophaga) bullata* Parker. Functional Ecology, 5: 448–456.

Bishop, C. and J. Hodin. 2001. Metamorphosis. Developmental Biology, 235: 242.

Bishop, C. D. and B. P. Brandhorst. 2003. On nitric oxide signaling, metamorphosis, and the evolution of biphasic life cycles. Evolution & Development, 5: 542–550.

Bishop, C. D., M. Huggett, A. Heyland, J. Hodin and B. P. Brandhorst. 2006. Interspecific variation in metamorphic competence in marine invertebrates: the significance for comparative investigations of regulatory systems. Integrated and Comparative Biology, 46: 662–682.

Bloch, G., D. E. Wheeler and G. E. Robinson. 2002. Endocrine influences on the organization of insect societies, pp. 195–235. *In* D. W. Pfaff [ed]. Hormones, Brain, and Behavior. Volume II, Non-Mammalian Hormone-Behavior Systems. Academic Press, Boston.

Bonduriansky, R. 2003. Layered sexual selection: a comparative analysis of sexual behaviour within an assemblage of piophilid files. Canadian Journal of Zoology-Revue Canadienne De Zoologie, 81: 479–491.

Bonduriansky, R. and R. J. Brooks. 1999. Reproductive allocation and reproductive ecology of seven species of Diptera. Ecological Entomology, 24: 389–395.

Boulétreau-Merle, J., R. Allemand, Y. Cohet, and J. R. David. 1982. Reproductive strategy in *Drosophila melanogaster*: significance of genetic divergence between temperate and tropical populations. Oecologia 53: 323–329.

Bownes, M. 1989. The roles of juvenile hormone, ecdysone and the ovary in the control of *Drosophila* vitellogenesis. Journal of Insect Physiology, 35: 409–413.

Bradley, J. T., M. Masetti, A. Cecchettini and F. Giorgi. 1995. Vitellogenesis in the allatectomized stick insect *Carausius morosus* (Br) (Phasmatodea, Lonchodinae). Comparative Biochemistry and Physiology B-Biochemistry & Molecular Biology, 110: 255–266.

Brady, S. G., S. Sipes, A. Pearson and B. N. Danforth. 2006. Recent and simultaneous origins of eusociality in halictid bees. Proceedings of the Royal Society B-Biological Sciences, 273: 1643–1649.

Brakefield, P. M. and J. C. Roskam. 2006. Exploring evolutionary constraints is a task for an integrative evolutionary biology. American Naturalist, 168: S4–S13.

Brent, C. S. and J. F. A. Traniello. 2001a. Social influence of larvae on ovarian maturation in primary and secondary reproductives of the dampwood termite *Zootermopsis angusticollis*. Physiological Entomology, 26: 78–85.

Brent, C. S. and J. F. A. Traniello. 2001b. Influence of sex-specific stimuli on ovarian maturation in primary and secondary reproductives of the dampwood termite *Zootermopsis angusticollis*. Physiological Entomology, 26: 239–247.

Brent, C. S. and J. F. A. Traniello. 2002. Effect of enhanced dietary nitrogen on reproductive maturation of the termite *Zootermopsis angusticollis* (Isoptera : Termopsidae). Environmental Entomology, 31: 313–318.

Broughton, S. J., M. D. W. Piper, T. Ikeya, T. M. Bass, J. Jacobson, Y. Driege, P. Martinez, E. Hafen, D. J. Withers, S. J. Leevers and L. Partridge. 2005. Longer lifespan, altered metabolism, and stress resistance in *Drosophila* from ablation of cells making insulin-like ligands. Proceedings of the National Academy of Sciences of the United States of America, 102: 3105–3110.

Brower, L. P. and S. B. Malcolm. 1991. Animal migrations—endangered phenomena. American Zoologist, 31: 265–276.

Brown, M. R., R. Graf, K. M. Swiderek, D. Fendley, T. H. Stracker, D. E. Champagne and A. O. Lea. 1998. Identification of a steroidogenic neurohormone in female mosquitoes. Journal of Biological Chemistry, 273: 3967–3971.

Brown, W. D. and D. T. Gwynne. 1977. Evolution of Mating in Crickets, Katydids and Wetas (Ensifera), pp. 281–314 *In* S. K. Gangwere, M. C. Muralirangan and M. Muralirangan [eds]. The Bionomics of Grasshoppers, Katydids, and Their Kin. CAB International, New York.

Bugnion, W. 1909. Le Termite noir de Ceylan, *Eutermes monoceros*, Koen. Annales de la Société Entomologique de France, 78: 271–281, plates 8–10 [in French].

Büning, J. 1994. The Insect Ovary: Ultrastructure, Previtellogenic Growth, and Evolution. New York, London.

Buys, B. 1988. Artificial insemination of the Cape honeybee *Apis mellifera capensis* Esch. and the influence of various factors on success. M.Sc. Thesis. University of Stellenbosch, South Africa.

Calis, J. N. M., W. J. Boota, M. H. Allsopp and M. Beekman. 2002. Getting more than a fair share: nutrition of worker larvae related to social parasitism in the Cape honey bee *Apis mellifera capensis*. Apidologie, 33: 193–202

Cameron, S. A. and P. Mardulyn. 2001. Multiple molecular data sets suggest independent origins of highly eusocial behavior in bees (Hymenoptera : Apinae). Systematic Biology, 50: 194–214.

Cardinal, S. and L. Packer. 2007. Phylogenetic analysis of the corbiculate Apinae based on morphology of the sting apparatus (Hymenoptera: Apidae). Cladistics, 23: 99–118.

Cepeda, O. I. 2006. Division of labor during brood production in stingless bees with special reference to individual participation. Apidologie, 37: 175–190.

Chakir, M., J. R. David, E. Pla and P. Capy. 1995. Genetic basis of some morphological differences between temperate and equatorial populations of *Drosophila melanogaster*. Experientia, 51: 744–748.

Chappell, M. and D. Whitman. 1990. Grasshopper Thermoregulation, pp. 143–172. *In* R. Chapman and A. Joern [eds]. Biology of Grasshoppers. Wiley, New York.

Chen, P. S., E. Stummzollinger, T. Aigaki, J. Balmer, M. Bienz and P. Bohlen. 1988. A male accessory gland peptide that regulates reproductive behavior of female *Drosophila melanogaster*. Cell, 54: 291–298.

Cha, S. Y., H. J. Yoon, E. M. Lee, M. H. Yoon, J. S. Hwang, B. R. Jin, Y. S. Han and I. Kim. 2007. The complete nucleotide sequence and gene organization of the mitochondrial genome of the bumblebee, *Bombus ignitus* (Hymenoptera: Apidae). Gene, 392: 206–220.

Cohet, Y. and J. David. 1978. Control of adult reproductive potential by pre-imaginal thermal conditions—study in *Drosophila melanogaster*. Oecologia, 36: 295–306.

Coll, M. and B. Yuval. 2004. Larval food plants affect flight and reproduction in an oligophagous insect herbivore. Environmental Entomology, 33: 1471–1476.

Colvin, J. and R. J. Cooter. 1995. Diapause induction and coloration in the Senegalese grasshopper, *Oedaleus senegalensis*. Physiological Entomology 20: 13–17.

Cook, O. F. 1905. The social organization and breeding habits of the cotton-protecting 'kelep' of Guatemala. Technical Series, US Department of Agriculture Bureau of Entomology, 10: 1–55.

Corkum, L. D., J. J. H. Ciborowski and R. G. Poulin. 1997. Effects of emergence date and maternal size on egg development and sizes of eggs and first-instar nymphs of a semelparous aquatic insect. Oecologia, 111: 69–75.

Coyne, J. A., J. Rux and J. R. David. 1991. Genetics of morphological differences and hybrid sterility between *Drosophila sechellia* and its relatives. Genetical Research, 57: 113–122.

Cunha, M. A. S., G. M. Gomes and L. A. O. Campos, 1986. Desenvolvimento ovariano em operárias adultas de *Frieseomelitta silvestri languida* (Hymenoptera, Apidae) sob condições normais e de orfandade. Ciência e Cultura (São Paulo), 38: 1725–1731 [in Portugese].

Cunningham, C. W. 1997. Can three incongruence tests predict when data should be combined? Molecular Biology and Evolution, 14: 733–740.

Danks, H. V. 1987. Insect Dormancy: An Ecological Perspective. Biological Survey of Canada (Terrestrial Arthropods), Ottawa.

Davey, K. G. 1997. Hormonal controls on reproduction in female Heteroptera. Archives of Insect Biochemistry and Physiology, 35: 443–453.

David, J. R. 1970. Le nombre d'ovarioles chez *Drosophila melanogaster*: relation avec la fécondité et valeur adaptative. Archives de Zoologie Experimentale et Génerale. 111: 357–370.

Dearn, J. M. 1997. Variable life history characteristics along an altitudinal gradient in three species of Australian grasshopper. Oecologia 28: 67–85.

Dedej S., K. Hartfelder, P. Aumeier, P. Rosnekaranz and W. C. Engels. 1998. Caste determination is a sequential process: effect of larval age at grafting on ovariole number, hindleg size and cephalic volatiles in the honey bee (*Apis mellifera carnica*). Journal of Apicultural Research, 37: 183–190.

Delpuech, J. M., B. Moreteau, J. Chiche, E. Pla, J. Vouidibio and J. R. David. 1995. Phenotypic plasticity and reaction norms in temperate and tropical populations of *Drosophila melanogaster*-ovarian size and developmental temperature. Evolution, 49: 670–675.

Díaz-Fleischer, F. and M. Aluja. 2003. Behavioural plasticity in relation to egg and time limitation: the case of two fly species in the genus *Anastrepha* (Diptera : Tephritidae). Oikos, 100: 125–133.

Dietemann, V., P. Neumann, S. Härtel, C. W. W. Pirk, R. M. Crewe. 2007. Pheromonal dominance and the selection of a socially parasitic honeybee worker lineage (*Apis mellifera capensis* Esch.). Journal of Evolutionary Biology, 20: 997–1007.

Dingle, H. and R. Winchell. 1997. Juvenile hormone as a mediator of plasticity in insect life histories. Archives of Insect Biochemistry and Physiology, 35: 359–373.

Donini, A. and A. B. Lange. 2002. The effects of crustacean cardioactive peptide on locust oviducts are calcium-dependent. Peptides, 23: 683–691.

Drumond, P. M., B. P. Oldroyd and K. Osborne. 2000. Worker reproduction in *Austroplebeia australis* Friese (Hymenoptera, Apidae, Meliponini). Insectes Sociaux, 47: 333–336.

Eggleton, P. 2001. Termites and trees: a review of recent advances in termite phylogenetics. Insectes Sociaux, 48: 187–193.

Emlen, D. J. 2001. Costs and the diversification of exaggerated animal structures. Science, 291: 1534–1536.

Endler, A., J. Liebig, T. Schmitt, J. E. Parker, G. R. Jones, P. Schreier and B. Holldobler. 2004. Surface hydrocarbons of queen eggs regulate worker reproduction in a social insect. Proceedings of the National Academy of Sciences of the United States of America, 101: 2945–2950.

Engel, M. S. 2000. A new interpretation of the oldest fossil bee (Hymenoptera: Apidae). American Museum Novitates, 3296: 1–11.

Engel, M.S. 2001. Monophyly and extensive extinction of advanced eusocial bees: Insights from an unexpected Eocene diversity. Proceedings of the National Academy of Sciences of the United States of America, 98: 1661–1664.

Engels, W. and V. L. Imperatriz-Fonseca. 1990. Caste development, reproductive strategies and control of fertility in honeybees and stingless bees, pp. 167–230. *In* W. Engels [ed.]. Social Insects: An Evolutionary Approach to Castes and Reproduction. Springer Verlag, New York.

Fabre, J. H. 1857. Mémoire sur l'hypérmétamorphose Et les moeurs des Méloides. Annales de Sciences Naturelles, Zoologie, 7: 299–365.

Fedorka, K. M., M. Zuk and T. A. Mousseau. 2004. Immune suppression and the cost of reproduction in the ground cricket, *Allonemobius socius*. Evolution, 58: 2478–2485.

Fedorka, K. M., J. E. Linder, W. Winterhalter and D. Promislow. 2007. Post-mating disparity between potential and realized immune response in *Drosophila melanogaster*. Proceedings of the Royal Society B-Biological Sciences, 274: 1211–1217.

Féneron, R., J. L. Durand and P. Jaisson. 1996. Relation between behaviour and physiological maturation in a ponerine ant. Behaviour, 133: 791–806.

Fescemyer, H. W., E. P. Masler, R. E. Davis and T. J. Kelly. 1992. Vitellogenin synthesis in female larvae of the gypsy moth, *Lymantria dispar* (L)-suppression by juvenile hormone. Comparative Biochemistry and Physiology B-Biochemistry & Molecular Biology, 103: 533–542.

Fischer, K. and K. Fiedler. 2001. Egg weight variation in the butterfly *Lycaena hippothoe*: more small or fewer large eggs? Population Ecology, 43: 105–109.

Fischer, K., A. N. M. Bot, P. M. Brakefield and B. J. Zwaan. 2003. Fitness consequences of temperature-mediated egg size plasticity in a butterfly. Functional Ecology, 17: 803–810.

Fitt, G. P. 1986. The Influence of a shortage of hosts on the specificity of oviposition behavior in species of *Dacus* (Diptera, Tephritidae). Physiological Entomology, 11: 133–143.

Fitt, G. P. 1990. Variation in ovariole number and egg size of species of *Dacus* (Diptera, Tephritidae) and their relation to host specialization. Ecological Entomology, 15: 255–264.

Flatt, T. and T. J. Kawecki. 2007. Juvenile hormone as a regulator of the trade-off between reproduction and life span in *Drosophila melanogaster*. Evolution, 61: 1980–1991.

Flatt, T., M-P. Tu and M. Tatar. 2005. Hormonal pleiotropy and the juvenile hormone regulation of *Drosophila* development and life history. BioEssays, in press.

Fletcher, B. S., S. Pappas and E. Kapatos. 1978. Changes in ovaries of olive flies (*Dacus oleae* Gmelin) during summer, and their relationship to temperature, humidity and fruit availability. Ecological Entomology, 3: 99–107.

Fordyce, J. A. 2005. Clutch size plasticity in the Lepidoptera, pp. 125–144. *In* T. Ananthakrishnan and D. Whitman [eds]. Insect Phenotypic Plasticity: Diverstiy of Responses. Science Publishers, Inc., Enfield, NH, USA.

Fox, C. W., M. S. Thakar, and T. A. Mousseau. 1997. Egg size plasticity in a seed beetle: an adaptive maternal effect. American Naturalist 149: 149–163.

Fuse, M. and J. W. Truman. 2002. Modulation of ecdysis in the moth *Manduca sexta*: the roles of the suboesophageal and thoracic ganglia. Journal of Experimental Biology, 205: 1047–1058.

Gage, M. J. G. 2005. Reproductive biology: Direct delivery of costly sex peptides. Current Biology, 15: R124–R126.

Gammie, S. C. and J. W. Truman. 1999. Eclosion hormone provides a link between ecdysis-triggering hormone and crustacean cardioactive peptide in the neuroendocrine cascade that controls ecdysis behavior. Journal of Experimental Biology, 202: 343–352.

Gasser, M., M. Kaiser, D. Berrigan and S. C. Stearns. 2000. Life-history correlates of evolution under high and low adult mortality. Evolution, 54: 1260–1272.

Gibert, P., P. Capy, A. Imasheva, B. Moreteau, J. P. Morin, G. Pétavy and J. R. David. 2004. Comparative analysis of morphological traits among *Drosophila melanogaster* and *D. simulans*: genetic variability, clines and phenotypic plasticity. Genetica, 120: 165–179.

Gillott, C. 1995. Entomology. Plenum Press, New York.

Gobin, B., C. Peeters and J. Billen. 1998. Production of trophic eggs by virgin workers in the ponerine ant *Gnamptogenys menadensis*. Physiological Entomology, 23: 329–336.

Gobin, B., J. Billen and C. Peeters. 2001. Dominance interactions regulate worker mating in the polygynous ponerine ant *Gnamptogenys menadensis*. Ethology, 107: 495–508.

Gould, S. J. 1984. Challenges to neo-Darwinism and their meaning for a revised view of human consciousness. Tanner Lectures on Human Values. Clare Hall, Cambridge University, Cambridge, UK.

Gould, S. J. and E. S. Vrba. 1982. Exaptation—a missing term in the science of form. Paleobiology, 8: 4–15.

Gould, S. J. and R. C. Lewontin. 1979. Spandrels of San-Marco and the Panglossian paradigm—a critique of the adaptationist program. Proceedings of the Royal Society of London Series B-Biological Sciences, 205: 581–598.

Grbic, M. 2003. Polyembryony in parasitic wasps: evolution of a novel mode of development. International Journal Of Developmental Biology, 47: 633–642.

Green, C. L. and B. P. Oldroyd. 2002. Queen mating frequency and maternity of males in the stingless bee *Trigona carbonaria* Smith. Insectes Sociaux, 49: 196–202.

Grenier, A. M. and P. Nardon. 1994. The genetic control of ovariole number in *Sitophilus oryzae* L. (Coleoptera, Curculionidae) is temperature-sensitive. Genetics Selection Evolution, 26: 413–430.

Grosso, A. F., L. R. Bego and A. S. Martinez. 2000. The production of males in queenright colonies of *Tetragonisca angustula angustula* (Hymenoptera, Meliponinae). Sociobiology, 35: 475–485.

Grünberg, K. 1903. Untersuchungen über die Keim- und Nährzellen in den Hoden und Ovarien der Lepidopteren. Zeitschrift für wissenschaftliche Zoologie, 74: 327–395, plates 16–18 [in German].

Hagedorn, H. H., J. P. Shapiro and K. Hanaoka. 1979. Ovarian ecdysone secretion is controlled by a brain hormone in an adult mosquito. Nature, 282: 92–94.

Halffter, G. and W. D. Edmonds. 1982. The Nesting Behavior of Dung Beetles (Scarabaeinae): An Ecological and Evolutive Approach. Instituto de Ecología, México, D. F.

Hamilton, W. D. 1964. Genetical evolution of social behaviour 2. Journal of Theoretical Biology, 7: 17–52.

Haribal, M. and J. A. Renwick. 2005. Plasticity in insect responses to variable chemistry of host plants, pp. 59–80. *In* T. Ananthakrishnan and D. Whitman [eds]. Insect Phenotypic Plasticity: Diverstiy of Responses. Science Publishers, Inc., Enfield, NH, USA.

Hartfelder, K. and D. G. Emlen. 2004. Endocrine control of insect polyphenism. *In* L. I. Gilbert, K. Iatrou and S. Gill [eds]. Comprehensive Insect Physiology, Biochemistry Pharmacology and Molecular Biology. Elsevier, Oxford (in press).

Hartfelder, K. and W. Engels. 1998. Social insect polymorphism: Hormonal regulation of plasticity in development and reproduction in the honeybee, pp. 45–77. Current Topics in Developmental Biology, Vol. 40.

Hartfelder, K. and G. Steinbrück. 1997. Germ cell cluster formation and cell death are alternatives in caste-specific differentiation of the larval honey bee ovary. Invertebrate Reproduction & Development, 31: 237–250.

Hartfelder, K., M. M. G. Bitondi, W. C. Santana and Z. L. P. Simoes. 2002. Ecdysteroid titer and reproduction in queens and workers of the honey bee and of a stingless bee: loss of ecdysteroid function at increasing levels of sociality? Insect Biochemistry and Molecular Biology, 32: 211–216.

Hatle, J. D., M. C. Crowley, A. L. Andrews and S. A. Juliano. 2002. Geographic variation of reproductive tactics in lubber grasshoppers. Oecologia, 132: 517–523.

Hatle, J. D., S. A. Juliano and D. W. Borst. 2000. Juvenile hormone is a marker of the onset of reproductive canalization in lubber grasshoppers. Insect Biochemistry and Molecular Biology, 30: 821–827.

Hatle, J. D., S. A. Juliano and D. W. Borst. 2003a. Hemolymph ecdysteroids do not affect vitellogenesis in the lubber grasshopper. Archives of Insect Biochemistry and Physiology, 52: 45–57.

Hatle, J. D., W. A. Miller and D. W. Borst. 2003b. Canalization of development and ecdysteroid timing during the last instar in lubber grasshoppers. Journal of Insect Physiology, 49: 73–80.

Hepburn, R. and S. E. Radloff. 2002. *Apis mellifera capensis*: an essay on the subspecific classification of honeybees. Apidologie, 33: 105–127.

Herman, W. S. 1975. Endocrine regulation of posteclosion enlargement of the male and female reproductive glands in monarch butterflies. General and Comparative Endocrinology, 26: 534–540.

Herman, W. S. and M. Tatar. 2001. Juvenile hormone regulation of longevity in the migratory monarch butterfly. Proceedings of the Royal Society of London Series B-Biological Sciences, 268: 2509–2514.

Herman, W. S., C. A. Lessman and G. D. Johnson. 1981. Correlation of juvenile hormone titer changes with reproductive tract development in the post-eclosion monarch butterfly. Journal of Experimental Zoology, 218: 387–395.

Herman, W. S., L. P. Brower and W. H. Calvert. 1989. Reproductive tract development in monarch butterflies overwintering in California and Mexico. Journal of the Lepidopterists Society, 43: 50–58

Heyland, A. and J. Hodin. 2004. Heterochronic developmental shift caused by thyroid hormone in larval sand dollars and its implications for phenotypic plasticity and the evolution of non-feeding development. Evolution, 58 (in press).

Hinton, H. E. 1981. Biology of Insect Eggs. Pergamon Press, Oxford.

Hodek, I. 1996. Dormancy, pp. 239–315. *In* I. Hodek and A. Honek [eds]. Ecology of Coccinellidae. Kluwer, Boston.

Hodek, I. 2003. Role of water and moisture in diapause development (A review). European Journal of Entomology 100: 223–232.

Hodin, J. 2000. Plasticity and constraints in development and evolution. Journal of Experimental Zoology Part B, Molecular Development Evolution, 288: 1–20.

Hodin, J. 2006. Expanding networks: Signaling components in and a hypothesis for the evolution of metamorphosis. Integrated and Comparative Biology, 46: 719–742.

Hodin, J. and L. M. Riddiford. 1998. The ecdysone receptor and ultraspiracle regulate the timing and progression of ovarian morphogenesis during *Drosophila metamorphosis*. Development Genes and Evolution, 208: 304–317.

Hodin, J. and L. M. Riddiford. 2000a. Different mechanisms underlie phenotypic plasticity and interspecific variation for a reproductive character in drosophilids (Insecta : Diptera). Evolution, 54: 1638–1653.

Hodin, J. and L. M. Riddiford. 2000b. Parallel alterations in the timing of ovarian Ecdysone Receptor and Ultraspiracle expression characterize the independent evolution of larval reproduction in two species of gall midges (Diptera : Cecidomyiidae). Development Genes and Evolution, 210: 358–372.

Hölldobler, B. and E. O. Wilson. 1990. The Ants. Belknap Press of Harvard University Press, Cambridge, Massachusetts, USA.

Honek, A. 1993. Intraspecific variation in body size and fecundity in insects—a general relationship. Oikos, 66: 483–492.

Honek, A. 1996. Life history and development, pp. 61–93. *In* I. Hodek and A. Honek [eds]. Ecology of Coccinellidae. Kluwer, Boston.

Hopkins, R. J. and B. Ekbom. 1996. Low oviposition stimuli reduce egg production in the pollen beetle *Meligethes aeneus*. Physiological Entomology, 21: 118–122.

Hopkins, R. J. and B. Ekbom. 1999. The pollen beetle, *Meligethes aeneus*, changes egg production rate to match host quality. Oecologia, 120: 274–278.

Hopper, K. R. 1999. Risk-spreading and bet-hedging in insect population biology. Annual Review of Entomology, 44: 535–560.

Imms, A. D. 1920. On the structure and biology of *Archotermopsis*, together with descriptions of new species of intestinal protozoa, and general observations on the Isoptera. Phil. Trans. Roy. Soc. Lond. B, 209: 75–180.

Imperatriz-Fonseca, V. L., E. T. Matos, E. Ferreira and H. H. W. Velthuis. 1998. A case of multiple mating in stingless bees (Meliponinae). Insectes Sociaux, 45: 231–233.

Ito, F., N. R. Yusoff and A. H. Idris. 1996. Colony composition and queen behavior in polygynous colonies of the Oriental ponerine ant *Odontomachus rixosus* (Hymenoptera Formicidae). Insectes Sociaux, 43: 77–86.

Jaschof, M. 1998. Revision der "Lestremiinae" (Diptera, Cecidomyiidae) der holarktis. Studia Dipterologica Supplement, 4: 1–552 [in German].

Jervis, M. A. and P. N. Ferns. 2004. The timing of egg maturation in insects: ovigeny index and initial egg load as measures of fitness and of resource allocation. Oikos, 107: 449–460.

Jervis, M. A., G. E. Heimpel, P. N. Ferns, J. A. Harvey and N. A. C. Kidd. 2001. Life-history strategies in parasitoid wasps: a comparative analysis of 'ovigeny'. Journal of Animal Ecology, 70: 442–458.

Jervis, M. A., P. N. Ferns and G. E. Heimpel. 2003. Body size and the timing of egg production in parasitoid wasps: a comparative analysis. Functional Ecology, 17: 375–383.

Jervis, M. A., C. Boggs and P. N. Ferns. 2005. Egg maturation strategy and its associated trade-offs: a synthesis focusing on Lepidoptera. Ecological Entomology, 30: 359–375.

Jones, C. D. 2004. Genetics of egg production in *Drosophila sechellia*. Heredity, 92: 235–241.

Jones, G. and P. A. Sharp. 1997. Ultraspiracle: An invertebrate nuclear receptor for juvenile hormones. Proceedings of the National Academy of Sciences of the United States of America, 94: 13499–13503.

Jones, G., D. Jones, P. Teal, A. Sapa and M. Wozniak. 2006. The retinoid-X receptor ortholog, ultraspiracle, binds with nanomolar affinity to an endogenous morphogenetic ligand. FEBS Journal, 273: 4983–96.

Juliano, S. A., J. R. Olson, E. C. Murrell, and J.D. Hatle. 2004. Plasticity and canalization of insect reproduction: testing alternative models of lifehistory transitions. Ecology, 85: 2986–2996.

Kalisz, S. and E. M. Kramer. 2008. Variation and constraint in plant evolution and development. Heredity, 100: 171–177.

Kambysellis, M. P. and W. B. Heed. 1971. Studies of oogenesis in natural populations of Drosophilidae. I. Relation of ovarian development and ecological habitats of Hawaiian species. American Naturalist, 105: 31–49.

Kambysellis, M. P., K. F. Ho, E. M. Craddock, F. Piano, M. Parisi and J. Cohen. 1995. Pattern of ecological shifts in the diversification of Hawaiian *Drosophila* inferred from a molecular phylogeny. Current Biology, 5: 1129–1139.

Karan, D., B. Moreteau and J. R. David. 1999. Growth temperature and reaction norms of morphometrical traits in a tropical drosophilid: *Zaprionus indianus*. Heredity, 83: 398–407.

Karan, D., S. Dubey, B. Moreteau, R. Parkash and J. R. David. 2000. Geographical clines for quantitative traits in natural populations of a tropical drosophilid: *Zaprionus indianus*. Genetica, 108: 91–100.

Katzav-Gozansky, T., V. Soroker and A. Hefetz. 2002. Evolution of worker sterility in honey bees: egg-laying workers express queen-like secretion in Dufour's gland. Behavioral Ecology & Sociobiology, 51: 588–589.

Keller, L. [ed.] 1993. Queen Number and Sociality in Insects. Oxford University Press, New York.

Kemp, W. P. 1986. Thermoregulation in three rangeland grasshopper species. Canadian Entomologist, 118: 335–343.

Key, K. H. and M. F. Day. 1954. A physiological mechanism of colour change in the grasshopper *Kosciuscola tristis* Sjost. (Orthoptera: Acrididae). Australian Journal of Zoology, 2: 340–363.

King, R. C., S. K. Aggarwal and U. Aggarwal. 1968. Development of female *Drosophila* reproductive system. Journal of Morphology, 124: 143–166.

Kirilly, D., and T. Xie. 2007. The *Drosophila* ovary: an active stem cell community. Cell Research, 17: 15–25.

Klowden, M. J. 1997. Endocrine aspects of mosquito reproduction. Archives of Insect Biochemistry and Physiology, 35: 491–512.

Koedam, D., F. A. L. Contrera and V. L. Imperatriz-Fonseca. 1999. Clustered male production by workers in the stingless bee *Melipona subnitida* Ducke (Apidae, Meliponinae). Insectes Sociaux, 46: 387–391.

Koedam, D., H. H. W. Velthuis, M. R. Dohmen and V. L. Imperatriz-Fonseca. 2001. The behaviour of laying workers and the morphology and viability of their eggs in *Melipona bicolor bicolor*. Physiological Entomology, 26: 254–259.

Koedam, D., O. I. Cepeda and V. L. Imperatriz-Fonseca. 2007. Egg laying and oophagy by reproductive workers in the polygynous stingless bee *Melipona bicolor* (Hymenoptera, Meliponini). Apidologie, 38: 55–66.

Koeniger, G., N. Koeniger, M. Mardan, R. W. K. Punchihewa and G. W. Otis. 1990. Numbers of spermatozoa in queens and drones indicate multiple mating of queens in *Apis andreniformis* and *Apis dorsata*. Apdiologie, 21: 281–286.

Korb, J. and J. Heinze. 2004. Multilevel selection and social evolution of insect societies. Naturwissenschaften, 91: 291–304.

Kristensen, N. P. 1984. Studies on the morphology and systematics of primitive Lepidoptera (Insecta). Steenstrupia, 10: 141–191.

Kristensen, N. P. and A. W. Skalski. 1999. Phylogeny and paleontology, pp. 7–25. *In* N. P. Kristensen [ed.]. Handbook of Zoology. Vol. 4, part 35: Lepidoptera, Moths and Butterflies. de Gruyter, New York.

Kryger, P. 2001. An obligate social parasite in honeybees: the pseudo-clone of *Apis mellifera capensis*. Proceedings of the 13th Entomological Congress of South Africa, p. 38.

Labeyrie, V. 1978. Significance of environment in control of insect fecundity. Annual Review of Entomology, 23: 69–89.

Landa, K. 1992a. Seasonal declines in offspring fitness and selection for early reproduction in nymph-overwintering grasshoppers. Evolution 46: 121–135.

Landa, K. 1992b. Adaptive seasonal variation in grasshopper offspring size. Evolution 46: 1553–1558.

Lange, A. B. and R. da Silva. 2007. Neural and hormonal control of muscular activity of the spermatheca in the locust, *Locusta migratoria*. Peptides, 28: 174–184.

Lao-Tzu. 1992. Tao te Ching (S. Mitchell, trans.). Harper Collins, New York.

Lattorff, H. M. G., R. F. A. Moritz and S. Fuchs. 2005. A single locus determines thelytokous parthenogenesis of laying honeybee workers (*Apis mellifera capensis*). Heredity, 94: 533–537.

Launois, M. 1972. Contribution à l'etude du fonctionnement ovarien du criquet migrateur, *Locusta migratoria capito* Saussure dans la nature. Annales de Zoologie et Ecologie Animale 72: 55–116.

Leather, S. R., K. F. A. Walters and J. S. Bale. 1993. The Ecology of Insect Overwintering. Cambridge University Press, Cambridge.

Lee, H. G., C. S. Seong, Y. C. Kim, R. L. Davis and K. A. Han. 2003. Octopamine receptor OAMB is required for ovulation in *Drosophila melanogaster*. Developmental Biology, 264: 179–190.

Lessman, C. A. and W. S. Herman. 1981. Flight enhances juvenile hormone inactivation in *Danaus plexippus plexippus* L. (Lepidoptera, Danaidae). Experientia, 37: 599–601.

Lessman, C. A. and W. S. Herman. 1983. Seasonal variation in hemolymph juvenile hormone of adult monarchs (*Danaus p. plexippus*, Lepidoptera). Canadian Journal of Zoology-Revue Canadienne de Zoologie, 61: 88–94.

Lessman, C. A., L. Rollins and W. S. Herman. 1982. Effects of juvenile hormone-I, hormone-II, and hormone-III on reproductive tract growth in male and female monarch butterflies. Comparative Biochemistry and Physiology A-Physiology, 71: 141–144.

Lessman, C. A., W. S. Herman, D. A. Schooley, L. W. Tsai, B. J. Bergot and F. C. Baker. 1989. Detection of juvenile hormone-I, hormone-II and hormone-III in adult monarch butterflies (*Danaus plexippus*). Insect Biochemistry, 19: 431–433.

Lin, N. and C. D. Michener. 1972. Evolution of sociality in insects. The Quarterly Review of Biology, 47: 131–159.

Loher, W. and M. Zaretsky. 1989. Structure and Function of the Endocrine System, pp. 114–146. *In* H. Huber, T. Moore, and W. Loher [eds]. Cricket Behavior and Neurobiology. Cornell University Press, Ithaca, NY.

Loher, W. and M. Dambach. 1989. Reproductive Behavior, pp. 43–82 *In* H. Huber, T. Moore, and W. Loher [eds]. Cricket Behavior and Neurobiology. Cornell University Press, Ithaca, NY.

Lopez-Vaamonde, C., J. W. Koning, R. M. Brown, W. C. Jordan and A. F. G. Bourke. 2004. Social parasitism by male-producing reproductive workers in a eusocial insect. Nature, 430: 557–560.

Mahowald, A. P. and M. P. Kambysellis. 1980. Oogenesis, pp. 141–224. *In* M. Ashburner and T. R. F. Wright [eds]. The Genetics and Biology of *Drosophila*. Academic Press, London.

Malecová, B., J. Ramser, J. K. O'Brien, M. Janitz, J. Judova, H. Lehrach and J. Simuth. 2003. Honeybee (*Apis mellifera* L) *mrjp* gene family: computational analysis of putative promoters and genomic structure of *mrjp1*, the gene coding for the most abundant protein of larval food. Gene, 303: 165–175.

Maleszka, R. and R. Kucharski. 2000. Analysis of Drosophila *yellow-B* cDNA reveals a new family of proteins related to the Royal Jelly proteins in the honeybee and to an orphan protein in an unusual bacterium *Deinococcus radiodurans*. Biochemical and Biophysical Research Communications, 270: 773–776.

Mangel, M. 1989. Evolution of host selection in parasitoids—does the state of the parasitoid matter? American Naturalist, 133: 688–705.

Makert, G. R., R. J. Paxton and K. Hartfelder. 2006. Ovariole number—a predictor of differential reproductive success among worker subfamilies in queenless honeybee (*Apis mellifera* L.) colonies. Behavioral Ecology and Sociobiology, 60: 815–825.

Marshall, A. K. and S. E. Reynolds. 1998. Control of the insect oviduct: the role of the neuropeptide CCAP in the tobacco hornworm, *Manduca sexta*, pp. 334–353. *In* G. M. Coast and S. G. Webster [eds]. Recent Advances in Arthropod Endocrinology, vol. 65. Cambridge University Press, Cambridge.

Martin, S., T. Wossler and P. Kryger. 2002. Usurption of African *Apis mellifera scutellata* colonies by parasitic *Apis mellifera capensis* workers. Apidologie, 33: 215–231.

Matsuda, R. 1987. Animal Evolution in Changing Environments with Special Reference to Abnormal Metamorphosis. John Wiley and Sons, New York.

Matsuzaki, M. 1972. Oogenesis in adult net-spinning caddisfly, *Parasthenopsyche sauteri* (Trichoptera, Stenopsychidae), as revealed by election microscopic observation. Science Reports—Faculty of Education Fukushima University, 22: 27–40.

Matuszewski, B. 1968. Regulation and growth of nurse nuclei in the development of egg follicles in Cecidomyiidae (Diptera). Chromosoma, 25: 429–469.

McEdward, L. R. and K. H. Morgan. 2001. Interspecific relationships between egg size and the level of parental investment per offspring in echinoderms. Biological Bulletin, 200: 33–50.

McNeil, J. N., M. Laforge, C. Bedard and M. Cusson. 1996. Juvenile hormone production and sexual maturation in true armyworm, *Pseudaletia unipuncta* (HAW) (Lepidoptera: Noctuidae): A comparison of migratory and non-migratory populations. Archives of Insect Biochemistry and Physiology, 32: 575–584.

Mead, F. W. 2002. Asiatic citrus psyllid, *Diaphorina citri* Kuwayama. UF/IFAS Featured Creatures. EENY-33. http://creatures.ifas.ufl.edu/citrus/acpsyllid.htm

Mefferd, C., W. Hatch, R. Burries, and D. Whitman. 2005. Plasticity in the length of the ovulation-oviposition interval in the lubber grasshopper *Romalea microptera*. Journal of Orthopteran Research 14: in press.

Mercader, R. J. and J. M. Scriber. 2005. Phenotypic Plasticity of Host Selection in Adult Tiger Swallowtail Butterflies, *Papilio glaucus* L. (Lepidoptera: Papilionidae), pp. 25–57 *In* T. Ananthakrishnan and D. Whitman [eds] Insect Phenotypic Plasticity: Diversity of Responses. Science Publishers, Inc., Enfield, NH, USA.

Michener, C. D. 1974. The Social Behavior of the Bees; A Comparative Study. Mass. Belknap Press of Harvard University Press, Cambridge.

Min, K. J., N. Jones, D. W. Borst and M. A. Rankin. 2004. Increased juvenile hormone levels after long-duration flight in the grasshopper, *Melanoplus sanguinipes*. Journal of Insect Physiology, 50: 531–537.

Moehrlin, G. S. and S. A. Juliano. 1998. Plasticity of insect reproduction: testing models of flexible and fixed development in response to different growth rates. Oecologia, 115: 492–500.

Monnin, T. and F. L. W. Ratnieks. 2001. Policing in queenless ponerine ants. Behavioral Ecology and Sociobiology, 50: 97–108.

Moo-Valle, H., J. J. G. Quezada-Euan and T. Wenseleers. 2001. The effect of food reserves on the production of sexual offspring in the stingless bee *Melipona beecheii* (Apidae, Meliponini). Insectes Sociaux, 48: 398–403.

Moreteau, B., J. P. Morin, P. Gibert, G. Petavy, E. Pla and J. R. David. 1997. Evolutionary changes of nonlinear reaction norms according to thermal adaptation: a comparison of two *Drosophila* species. Comptes Rendus De L Academie Des Sciences Serie Iii-Sciences De La Vie-Life Sciences, 320: 833–841.

Moreteau, B., P. Gibert, G. Petavy, J. C. Moreteau, R. B. Huey and J. R. David. 2003. Morphometrical evolution in a *Drosophila* clade: the *Drosophila obscura* group. Journal of Zoological Systematics and Evolutionary Research, 41: 64–71.

Morin, J. P., B. Moreteau, G. Petavy, R. Parkash and J. R. David. 1997. Reaction norms of morphological traits in *Drosophila*: Adaptive shape changes in a stenotherm circumtropical species? Evolution, 51: 1140–1148.

Moritz R. F. A. and M. Haberl. 1994. Lack of meiotic recombination in thelytokous parthenogenesis of laying workers of *Apis mellifera capensis* (the Cape honeybee). Heredity, 73: 98–102.

Moritz R. F. A., P. Kryger and M. H. Allsopp. 1999. Lack of worker policing in the Cape honeybee (*Apis mellifera capensis*). Behaviour, 136: 1079–1092.

Myers, E. M. 2003. The circadian control of eclosion. Chronobiology International, 20: 775–794.

Nanork, P., S. Wongsiri and B. P. Oldroyd. 2007. Preservation and loss of the honey bee (*Apis*) egg-marking signal across evolutionary time. Behavioral Ecology and Sociobiology, 61:1509–1514

Neumann P. and H. R. Hepburn. 2002. Behavioural basis for social parasitism of Cape honeybees (*Apis mellifera capensis* Esch.). Apidologie, 33: 165–192.

Nielsen, E. S. and N. P. C. A. C. Kristensen. 1989. Primitive Ghost Moths : Morphology and Taxonomy of the Australian Genus Fraus Walker (Lepidoptera: Hepialiadae s. lat.). CSIRO, East Melbourne, Australia.

Nijhout H. F. 1994. Insect Hormones. Princeton University Press, Princeton.

Nijhout, H. F. and D.J. Emlen. 1998. Competition among body parts in the development and evolution of insect morphology. Proceedings of the National Academy of Sciences of the United States of America, 95: 3685–3689.

Nijhout, M. M. and L. M. Riddiford. 1974. Control of egg maturation by juvenile hormone in tobacco hornworm moth, *Manduca sexta*. Biological Bulletin, 146: 377–392.

Nijhout, M. M. and L. M. Riddiford. 1979. Juvenile hormone and ovarian growth in *Manduca sexta*. International Journal of Invertebrate Reproduction, 1: 209–219.

Noirot, C. 1990. Sexual castes and reproductive strategies in termites, pp. 5–37. *In* W. Engels [ed.]. Social Insects: An Evolutionary Approach to Castes and Reproduction. Springer Verlag, New York.

Normark, B. B. 2003. The evolution of alternative genetic systems in insects. Annual Review of Entomology, 48: 397–423.

O'Donnell, S. 1998. Reproductive caste determination in eusocial wasps (Hymenoptera : Vespidae). Annual Review of Entomology, 43: 323–346.

Oldroyd, B. P., F. L. W. Ratnieks and T. C. Wossler. 2002. Egg-marking pheromones in honey-bees *Apis mellifera*. Behavioral Ecology & Sociobiology, 51: 590–591.

Oldroyd, B. P., L. A. Halling, G. Good, W. Wattanachaiyingcharoen, A. B. Barron, P. Nanork and S. Wongsiri. 2001. Worker policing and worker reproduction in *Apis cerana*. Behavioral Ecology and Sociobiology, 50: 371–377.

Onions G. W. 1912. South African 'fertile worker bees'. South African Agricultural Journal 1: 720–728.

Orgogozo, V., K. W. Broman and D. L. Stern. 2006. High-resolution quantitative trait locus mapping reveals sign epistasis controlling ovariole number between two *Drosophila* species. Genetics, 173: 197–205.

Osawa, N. 2005. The effect of prey availability on ovarian development and oosorption in the ladybird beetle *Harmonia axyridis* (Coleoptera: Coccinellidae). European Journal of Entomology, 102: 503–511.

Papaj, D. R. 2000. Ovarian dynamics and host use. Annual Review of Entomology, 45: 423–448.

Pappas, N. and L. Engstrom. 1974. Survey of ovariole numbers in wild-type *Drosophila melanogaster*. Drosophila Information Service, 51: 27.

Park, Y. I., S. Shu, S. B. Ramaswamy and A. Srinivasan. 1988. Mating in *Heliothis virescens*: Transfer of juvenile hormone during copulation by male to female and stimulation of biosynthesis of endogenous juvenile hormone. Archives of Insect Biochemistry and Physiology, 38: 100–107.

Partridge, L., A. Green and K. Fowler. 1987. Effects of egg-production and of exposure to males on female survival in *Drosophila melanogaster*. Journal of Insect Physiology, 33: 745–749.

Paxton, R. J. 2000. Genetic structure of colonies and a male aggregation in the stingless bee *Scaptotrigona postica*, as revealed by microsatellite analysis. Insectes Sociaux, 47: 63–69.

Peeters, C. 1991. The occurrence of sexual reproduction among ant workers. Biological Journal of the Linnean Society, 44: 141–152.

Peeters, C. 1993. Monogyny and polygyny in ponerine ants with or without queens, pp. 234–261. *In* L. Keller [ed.]. Queen Number and Sociality in Insects. Oxford University Press, New York.

Peeters, C. and R. Crewe. 1985. Worker reproduction in the ponerine ant *Ophthalmopone berthoudi*—an alternative form of eusocial organization. Behavioral Ecology and Sociobiology, 18: 29–37.

Peters, J. M., D. C. Queller, V. L. Imperatriz-Fonseca, D. W. Roubik and J. E. Strassmann. 1999. Mate number, kin selection and social conflicts in stingless bees and honeybees. Proceedings of the Royal Society of London Series B-Biological Sciences, 266: 379–384.

Pianka, E. R. 1970. On "*r*" and "*K*" selection. American Naturalist, 104: 592–597.

Pirk, C. W. W., P. Neumann and H. R. Hepburn. 2002. Egg laying and egg removal by workers are positively correlated in queenright Cape honeybee colonies (*Apis mellifera capensis*). Apidologie, 33: 203–211.

Rachinsky, A., C. Strambi, A. Strambi and K. Hartfelder. 1990. Caste and metamorphosis-hemolymph titers of juvenile-hormone and ecdysteroids in last instar honeybee larvae. General and Comparative Endocrinology, 79: 31–38.

Ram, G. M., B. K. Rao, S. S. Thakur, S. Ashok and L. S. Rao. 1988. Histological changes in the ovaries of sweet potato weevil *Cylas formicarius* F. (Coleoptera: Curculionidae) caused by the juvenoid methoprene. Journal of Animal Morphology and Physiology, 35: 1–6.

Ramaswamy, S. B., S. Q. Shu, Y. I. Park and F. R. Zeng. 1997. Dynamics of juvenile hormone-mediated gonadotropism in the Lepidoptera. Archives of Insect Biochemistry and Physiology, 35: 539–558.

Rankin, M. A. 1986. The oogenesis-flight syndrome revisited, pp. 27–48. *In* W. Danthanarayana [ed.]. Insect Flight: Dispersal and Migration. Springer Verlag, New York.

Rankin, M. A. and L. M. Riddiford. 1978. Significance of hemolymph juvenile hormone titer changes in timing of migration and reproduction in adult *Oncopeltus fasciatus*. Journal of Insect Physiology, 24: 31–38.

Rankin, S. M., J. Chambers and J. P. Edwards. 1997. Juvenile hormone in earwigs: Roles in oogenesis, mating, and maternal behaviors. Archives of Insect Biochemistry and Physiology, 35: 427–442.

Ratnieks, F. L. W., K. R. Foster and T. Wenseleers. 2006. Conflict resolution in insect societies. Annual Review of Entomology, 51: 581–608.

Reeve, H. K. and L. Keller. 2001. Tests of reproductive-skew models in social insects. Annual Review of Entomology, 46: 347–385.

Reginato, R. D. and C. Cruz-Landim. 2002. Morphological characterization of cell death during the ovary differentiation in worker honey bee. Cell Biology International, 26: 243–251.

Reginato, R. D. and C. Cruz-Landim. 2003. Ovarian growth during larval development of queen and worker of *Apis mellifera* (Hymenoptera: Apidae): a morphometric and histological study. Brazilian Journal of Biology 63: 121–127.

Rhamhalinghan, M. 1986. Seasonal variations in ovariole numbers/ovary in *Coccinella septempunctata* L. (Coleoptera: Cooccinellidae). Proceedings of the Indian National Science Academy, Part B: Biological Sciences, 52: 619–623.

Richard, D. S., R. Rybczynski, T. G. Wilson, Y. Wang, M. L. Wayne, Y. Zhou, L. Partridge, L. G. Harshman. 2005. Insulin signaling is necessary for vitellogenesis in *Drosophila melanogaster* independent of the roles of juvenile hormone and ecdysteroids: female sterility of the *chico1* insulin signaling mutation is autonomous to the ovary. Journal of Insect Physiology, 51: 455–464.

Richardson, M. K. and A. D. Chipman. 2003. Developmental constraints in a comparative framework: a test case using variations in phalanx number during amniote evolution. Journal of Experimental Zoology Part B, Molecular Development Evolution, 296: 8–22.

Riehle, M. A., S. F. Garczynski, J. W. Crim, C. A. Hill and M. R. Brown. 2002. Neuropeptides and peptide hormones in *Anopheles gambiae*. Science, 298: 172–175.

Rivero-Lynch, A. P. and H. C. J. Godfray. 1997. The dynamics of egg production, oviposition and resorption in a parasitoid wasp. Functional Ecology, 11: 184–188.

Robertson, F. W. 1957. Studies in quantitative inheritance. X. Genetic variation of ovary size in *Drosophila*. J. Genet. 55: 410–427.

Robertson, J. G. 1961. Ovariole numbers in Coleoptera. Canadian Journal of Zoology-Revue Canadienne De Zoologie, 39: 245–263.

Robinson, G. E. 1992. Regulation of division-of-labor in insect societies. Annual Review of Entomology, 37: 637–665.

Robinson, G. E. and E. L. Vargo. 1997. Juvenile hormone in adult eusocial hymenoptera: Gonadotropin and behavioral pacemaker. Archives of Insect Biochemistry and Physiology, 35: 559–583.

Rodríguez-Valentín, R., I. López-González, R. Jorquera, P. Labarca, M. Zurita and E. Reynaud. 2006. Oviduct contraction in *Drosophila* is modulated by a neural network that is both, octopaminergic and glutamatergic. Journal of Cellular Physiology, 209, 183–198.

Rooney, J. and S. M. Lewis. 2002. Fitness advantage from nuptial gifts in female fireflies. Ecological Entomology, 27: 373–377.

Roubik, D. W. 1992. Ecology and Natural History of Tropical Bees. Cambridge University Press, New York.

Rountree, D. B. and H. F. Nijhout. 1995a. Hormonal control of a seasonal polyphenism in *Precis coenia* (Lepidoptera, Nymphalidae). Journal of Insect Physiology, 41: 987–992.

Rountree, D. B. and H. F. Nijhout. 1995b. Genetic control of a seasonal morph in *Precis coenia* (Lepidoptera, Nymphalidae). Journal of Insect Physiology, 41: 1141–1145.

Ruttner, F. 1977. Problem of Cape Bee (*Apis mellifera capensis* Escholtz)-parthenogenesis-size of population-evolution. Apidologie, 8: 281–294.

Ruttner, F. 1988. Biogeography and Taxonomy of Honeybees. Springer Verlag, Berlin.

Sakagami, S. F. and R. Zucchi. 1974. Oviposition behavior of two dwarf stingless bees, *Hypotrigona (Leurotrigona) muelleri* and *H. Trigonisca duckei*, with notes on the temporal articulation of oviposition process in stingless bees. Journal of the Faculty of Science Hokkaido University, Series VI Zoology, 20: 647–690.

Santos, A. C. and R. Lehmann. 2004. Germ cell specification and migration in *Drosophila* and beyond. Current Biology, 14: R578–R589.

Saul-Gershenz, L. S. and J. G. Millar. 2006. Phoretic nest parasites use sexual deception to obtain transport to their host's nest. PNAS, 103: 14039–14044.

Saviliev, V. I. 1928. Manifold effects of the gene *vestigial* in *Drosophila melanogaster*. Trudy Leningradskogo obshchestva estestvoispytatelei, 58: 65–88.

Schal, C., G. L. Holbrook, J. A. S. Bachmann and V. L. Sevala. 1997. Reproductive biology of the German cockroach, *Blattella germanica*: Juvenile hormone as a pleiotropic master regulator. Archives of Insect Biochemistry and Physiology, 35: 405–426.

Schlichting, C. and M. Pigliucci. 1998. Phenotypic Evolution: A Reaction Norm Perspective. Sinauer, Sunderland, MA.

Schmalhausen, I. I. 1949. Factors of Evolution: The Theory of Stabilizing Selection. Blakiston, Philadelphia.

Schmidt-Capella, I. C. and K. Hartfelder. 1998. Juvenile hormone effect on DNA synthesis and apoptosis in caste-specific differentiation of the larval honey bee (*Apis mellifera* L.) ovary. Journal of Insect Physiology, 44: 385–391.

Schmidt-Capella, I. C. and K. Hartfelder. 2002. Juvenile hormone-dependent interaction of actin and spectrin is crucial for polymorphic differentiation of the larval honey bee ovary. Cell and Tissue Research, 307: 265–272.

Schultz, T. R., M. S. Engel and M. Prentice. 1999. Resolving conflict between morphological and molecular evidence for the origin of eusociality in the "corbiculate" bees (Hymenoptera: Apidae): a hypothesis-testing approach. University of Kansas Museum of Natural History Special Publication, 24: 125–138.

Schwarz, M. P., M. H. Richards and B. N. Danforth. 2007. Changing paradigms in insect social evolution: insights from halictine and allodapine bees. Annual Review of Entomology, 52: 127–50.

Scott, M. P., S. T. Trumbo, P. A. Neese, W. D. Bailey and R. M. Roe. 2001. Changes in biosynthesis and degradation of juvenile hormone during breeding by burying beetles: a reproductive or social role? Journal of Insect Physiology, 47: 295–302.

Selivon, M., J. S. Morgante and A. L. P. Perondini. 1997. Egg size, yolk mass extrusion and hatching behavior in two cryptic species of *Anastrepha fraterculus* Wiedemann (Diptera, Tephritidae). Brazilian Journal of Genetics, 20 (4).

Sgro, C. M. and L. Partridge. 1999. A delayed wave of death from reproduction in *Drosophila*. Science, 286: 2521–2524.

Shaaya, E., P. D. Shirk, G. Zimowska, S. Plotkin, N. J. Young, H. H. Rees and D. L. Silhacek. 1993. Declining ecdysteroid levels are temporally correlated with the initiation of vitellogenesis during pharate adult development in the indianmeal moth, *Plodia interpunctella*. Insect Biochemistry and Molecular Biology, 23: 153–158.

Silvestre, D. and M. C. Arias. 2006. Mitochondrial tRNA gene translocations in highly eusocial bees. Genetics and Molecular Biology, 29: 572–575.

Simon, J. C., C. Rispe and P. Sunnucks. 2002. Ecology and evolution of sex in aphids. Trends in Ecology & Evolution, 17: 34–39.

Simonet, G., J. Poels, I. Claeys, T. Van Loy, V. Franssens, A. De Loof and J. Vanden Broeck. 2004. Neuroendocrinological and molecular aspects of insect reproduction. Journal of Neuroendocrinology, 16: 649–659.

Sledge, M. F., C. Peeters and R. M. Crewe. 1999. Fecundity and the behavioural profile of reproductive workers in the queenless ant, *Pachycondyla (= ophthalmopone) berthoudi*. Ethology, 105: 303–316.

Sledge, M. F. ,C. Peeters and R. M. Crewe. 2001. Reproductive division of labour without dominance interactions in the queenless ponerine ant *Pachycondyla (=Ophthalmopone) berthoudi*. Insectes Sociaux, 48: 67–73

Slee, M. and A. Aron. 2003. Bugs! in 3D. Principal Large Format film and Image Quest 3-D, UK.

Smith, D. S. 1972. Crowding in grasshoppers. II. Continuing effects of crowding on subsequent generations of *Melanoplus sanguinipes* (Orthoptera: Acrididae). Environmental Entomology, 1: 314–317.

Sommer, K. and B. Holldobler. 1995. Colony founding by queen association and determinants of reduction in queen number in the ant *Lasius niger*. Animal Behaviour, 50: 287–294.

Souza, A. L. T., M. O. Tanaka, and R. P. Martins. 2005. Adaptive allometric response of galling insects to availability of ovipositing sites, pp. 173–202. *In* T. N. Ananthakrishnan and D. Whitman [eds]. Insect Phenotypic Plasticity Diversity of Responses. Science Publishers, Inc. Enfield, NH. USA.

Spitzer, B. W. 2004. Maternal effects in the soft scale insect *Saissetia coffeae* (Hemiptera : Coccidae). Evolution, 58: 2452–2461.

Spivak, M., D. J. C. Fletcher and M. C. Brees [eds]. 1991. The "Africanized" Honey Bee. Westview Press, San Francisco, CA.

Starmer, W. T., M. Polak, J. S. F. Barker, S. McEvey, S. Pitnick and L. L. Wolf. 2002. Phylogenetic, geographic, and temporal analysis of female reproductive trade-offs in *Drosophila*. Evolutionary Biology, 33: 139–171.

Stauffer, T. W. and D. W. Whitman. 1997. Grasshopper oviposition, pp. 231–280. *In* S. K. Gangwere, M. C. Muralirangan and M. Muralirangan [eds]. The Bionomics of Grasshoppers, Katydids, and Their Kin. CAB International, New York.

Stauffer, T. W. and D. W. Whitman. 2007. Divergent oviposition behavious in a desert *vs* a marsh grasshopper. J. Orthoptera Research 16: 103–114.

Stay, B., T. Friedel, S. S. Tobe and E. C. Mundall. 1980. Feedback-control of juvenile hormone synthesis in cockroaches-possible role for ecdysterone. Science, 207: 898–900.

Stearns, S. C. 1992. The Evolution of Life Histories. Oxford University Press, Oxford.

Stearns, S. C. 2000. Life history evolution: successes, limitations, and prospects. Naturwissenschaften, 87: 476–486.

Stern, D. L. and W. A. Foster. 1996. The evolution of soldiers in aphids. Biological Reviews of the Cambridge Philosophical Society, 71: 27–79.

Stevens, D. J., M. H. Hansell and P. Monaghan. 2000. Developmental trade-offs and life histories: strategic allocation of resources in caddis flies. Proceedings of the Royal Society of London Series B-Biological Sciences, 267: 1511–1515.

Stewart, L. A., J. L. Hemptinne and A. F. G. Dixon. 1991. Reproductive tactics of ladybird beetles-relationships between egg size, ovariole number and developmental time. Functional Ecology, 5: 380–385.

Strambi, A., C. Strambi and M. Cayre. 1997. Hormonal control of reproduction and reproductive behavior in crickets. Archives of Insect Biochemistry and Physiology, 35: 393–404.

Suka, T. and T. Inoue. 1993. Nestmate recognition of the stingless bee *Trigona (Tetragonula) minangkabau* (Apidae, Meliponinae). Journal of Ethology, 11: 141–147.

Sundberg, S. V., M. H. Luong-Skovmand, and D. W. Whitman. 2001. Morphology and development of oocyte and follicle resprption bodies in the lubber grasshopper, *Romalea microptera* (Beauvois). Journal of Orthoptera Research. 10: 39–51.

Svensson, B. G., E. Petersson and M. Frisk. 1990. Nuptial gift size prolongs copulation duration in the dance fly *Empis borealis*. Ecological Entomology, 15: 225–229.

Tallamy, D. W. 1984. Insect parental care. Bioscience, 34: 20–24.

Tammaru, T. and J. Javois. 2000. Responses of ovipositing moths (Lepidoptera: Geometridae) to host plant deprivation: life-history aspects and implications for population dynamics. Environmental Entomology 29: 1002–1010.

Tanaka, S., T. Hakomori, and E. Hasegawa. 1993. Effects of daylength and hopper density on reproductive traits in a Japanese population of the migratory locust, *Locusta migratoria* L. Journal of Insect Physiology 39:571–580.

Tatar, M., Carey, J. R. 1995. Nutrition mediates reproductive costs in the beetle *Callosobruchus maculatus*. Ecology. 76: 2066–2073.

Taub-Montemayor, T. E., J. O. Palmer and M. A. Rankin. 1997. Endocrine regulation of reproduction and diapause in the boll weevil, *Anthonomus grandis* Boheman. Archives of Insect Biochemistry and Physiology, 35: 455–477.

Tauber, M. J., C. A. Tauber and M. Sazaki. 1986. Seasonal Adaptations of Insects. Oxford University Press, New York.

Thorne B. L. 1997. Evolution of eusociality in termites. Annual Review of Ecology and Systematics, 28: 27–54

Thorne, B. L. and J. F. A. Traniello. 2003. Comparative social biology of basal taxa of ants and termites. Annual Review of Entomology, 48: 283–306.

Tobe, S. S. and P. A. Langley. 1978. Reproductive physiology of *Glossina*. Annual Review of Entomology, 23: 283–307.

Tóth, E., D. C. Queller, V. L. Imperatriz-Fonseca and J. E. Strassmann. 2002a. Genetic and behavioral conflict over male production between workers and queens in the stingless bee *Paratrigona subnuda*. Behavioral Ecology and Sociobiology, 53: 1–8.

Tóth, E., J. E. Strassmann, P. Nogueira-Neto, V. L. Imperatriz-Fonseca and D. C. Queller. 2002b. Male production in stingless bees: variable outcomes of queen-worker conflict. Molecular Ecology, 11: 2661–2667.

Tóth, E., D. C. Queller, A. Dollin and J.E.Strassmann. 2004. Conflict over male parentage in stingless bees. Insectes Sociaux, 51: 1–11

Truckenbrodt, W. and U. Amelung. 1986. Postembryonic development of the queens ovary of *Odontotermes stercorivorus* (Isoptera, Termitidae). Entomologia Generalis, 11: 237–248.

Tu, M. P. and M. Tatar. 2003. Juvenile diet restriction and the aging and reproduction of adult *Drosophila melanogaster*. Aging Cell, 2: 327–333.

Urquhart, F. A. 1960. The Monarch Butterfly. University of Toronto Press, Toronto.

van Valen, L. 1965. Morphological variation and the width of the ecological niche. American Naturalist, 99: 377–390.

Velthuis, H. H. 1970. Ovarian development in *Apis mellifera* worker bees. Entomologia Experimentalis et Applicata, 13: 377–394.

Walker, M., C. Lewis, and D. Whitman. 1999. Effects of males on the fecundity and fertility of female *Romalea microptera* grasshoppers. Journal of Orthoptera Research 8: 277–283.

Wall, R., V. J. Wearmouth and K. E. Smith. 2002. Reproductive allocation by the blow fly *Lucilia sericata* in response to protein limitation. Physiological Entomology, 27: 267–274.

Wang, X. G. and R. H. Messing. 2003. Egg maturation in the parasitoid *Fopius arisanus* (Hymenoptera : Braconidae): Do host-associated stimuli promote ovarian development? Annals of the Entomological Society of America, 96: 571–578.

Wardhaugh, K. G. 1977. The effects of temperature and photoperiod on the morphology of the egg-pod of the Australian plague locust (*Chortoicetes terminifera* Walker, Orthoptera: Acrididae). Australian Journal of Ecology 2: 81–88.

Wayne, M. L. and L. M. McIntyre. 2002. Combining mapping and arraying: An approach to candidate gene identification. Proceedings of the National Academy of Sciences of the United States of America, 99: 14903–14906.

Wayne, M. L. and T. F. C. Mackay. 1998. Quantitative genetics of ovariole number in *Drosophila melanogaster*. II. Mutational variation and genotype-environment interaction. Genetics, 148: 201–210.

Wayne, M. L., J. B. Hackett, C. L. Dilda, S. V. Nuzhdin, E. G. Pasyukova and T. F. C. MacKay. 2001. Quantitative trait locus mapping of fitness-related traits in *Drosophila melanogaster*. Genetical Research, 77: 107–116.

Wayne, M. L., A. Korol and T. F. C. Mackay. 2005. Microclinal variation for ovariole number and body size in *Drosophila melanogaster* in 'Evolution Canyon.' Genetica, 123: 263–270.

Wcislo, W. T. and J. H. Cane. 1996. Floral resource utilization by solitary bees (Hymenoptera: Apoidea) and exploitation of their stored foods by natural enemies. Annual Review of Entomology, 41: 257–286.

Webb, T. J., S. Q. Shu, S. B. Ramaswamy and S. Dorn. 1999. Influence of juvenile hormone and mating on oogenesis and oviposition in the codling moth, *Cydia pomonella*. Archives of Insect Biochemistry and Physiology, 41: 186–200.

Weglarska, B. 1961. Oogenesis in the ovoviviparous scale insect *Quadraspidiotus ostraeformis* (Curt.) (Homoptera, Coccidae, Aspidiotini). Part I. Morphological and cytological investigations. Zoologica Poloniae, 11: 267–294, plates 1–4.

Weiss, K. 1975. Zur kastenspezifischen Ernährung der weiblichen Bienenlarve (*Apis mellifica* L.). Apidologie, 6: 95–120 [in German].

Wenseleers, T., A. G. Hart, F. L. W. Ratnieks and J. J. G. Quezada-Euan. 2004. Queen execution and caste conflict in the stingless bee *Melipona beecheii*. Ethology, 110: 725–736.

Went, D. F. 1979. Pedogenesis in the dipteran insect *Heteropeza pygmaea*—an interpretation. International Journal of Invertebrate Reproduction, 1: 21–30.

Went, D. F., V. Gentinetta and B. Lanzrein. 1984. Ecdysteroid titers during larval reproduction of the dipteran insect *Heteropeza pygmaea*. Experientia, 40: 998–1000.

West-Eberhard, M. J. 1978. Polygyny and the evolution of social behavior in wasps. Journal of the Kansas Entomological Society, 51: 832–856.

West-Eberhard, M. J. 1989. Phenotypic plasticity and the origins of diversity. Annual Review of Ecology and Systematics, 20: 249–278.

West-Eberhard, M. J. 1996. Wasp societies as microcosms for the study of development and evolution, pp. 290–317. *In* S. Turillazzi and M. J. West-Eberhard [eds]. Natural History and Evolution of Paper-Wasps. Oxford University Press, Oxford.

Wheeler, D. E. 1986. Developmental and physiological determinants of caste in social hymenoptera-evolutionary implications. American Naturalist, 128: 13–34.

Wheeler, D. E. 1996. The role of nourishment in oogenesis. Annual Review of Entomology, 41: 407–431.

Wheeler, W. C., M. Whiting, Q. D. Wheeler and J. M. Carpenter. 2001. The phylogeny of the extant hexapod orders. Cladistics-the International Journal of the Willi Hennig Society, 17: 113–169.

Whitman, D. W., M. S. Blum, and F. S. Slansky, Jr. 1994. Carnivory in phytophagous insects, pp. 161–205. In T. Ananthakrishnan [ed.]. Functional Dynamics of Phytophagous Insects. Oxford and IBH, New Delhi.

Wigby, S. and T. Chapman. 2005. Sex peptide causes mating costs in female *Drosophila melanogaster*. Current Biology, 15: 316–321.

Wolfner, M. F. 1997. Tokens of love: Functions and regulation of *Drosophila* male accessory gland products. Insect Biochemistry and Molecular Biology, 27: 179–192.

Wongsiri, S., C. Lekprayoon, R. Thapa, K. Thirakupt, T. E. Rinderer, H. A. Sylvester, B. P. Oldroyd and U. Booncham. 1996. Comparative biology of *Apis andreniformis* and *Apis florea* in Thailand. Bee World, 77: 23–35.

Wossler, T. C. 2002. Pheromone mimicry by *Apis mellifera capensis* social parasites leads to reproductive anarchy in host *Apis mellifera scutellata* colonies. Apidologie, 33: 139–163.

Woyke, J., Z. Jasinski and B. Smagowska. 1974. [Comparison of reproductive organs and the effect of artificial and natural insemination of bees of different races and their hybrids.]. Pszczelnicze Zeszyty Naukowe, 18: 53–75 [in Polish].

Wu, Q. and M. R. Brown. 2006. Signaling and function of insulin-like peptides in insects. Annual Review of Entomology, 51: 1–24.

Yin, C. M. and J. G. Stoffolano. 1997. Juvenile hormone regulation of reproduction in the cyclorrhaphous Diptera with emphasis on oogenesis. Archives of Insect Biochemistry and Physiology, 35: 513–537.

Zalucki, M. P. and R. L. Kitching. 1982. Movement pattern in *Danaus plexippus* L. Behaviour, 80: 174–198.

Zeng, F., S. Shu, Y. I. Park and S. B. Ramaswamy. 1997. Vitellogenin and egg production in the moth, *Heliothis virescens*. Archives of Insect Biochemistry and Physiology, 34: 287–300.

Zera, A. J. 1985. Wing polymorphism in waterstriders: Mechanism of morph determination and adaptive significance, pp. 674–686. *In* M. A. R. Rankin [ed.] Migration: Mechanisms and Adaptive Significance. Marine Science Institute, The University of Texas at Austin, Port Aransas, TX.

Zera, A. J. and L. G. Harshman. 2001. The physiology of life history trade-offs in animals. Annual Review of Ecology and Systematics, 32: 95–126.

Zera, A. J. and Z. W. Zhao. 2003a. Life-history evolution and the microevolution of intermediary metabolism: Activities of lipid-metabolizing enzymes in life-history morphs of a wing-dimorphic cricket. Evolution, 57: 586–596.

Zera, A. J. and Z. W. Zhao. 2003b. Morph-dependent fatty acid oxidation in a wing-polymorphic cricket: implications for the trade-off between dispersal and reproduction. Journal of Insect Physiology, 49: 933–943.

Zitnan, D., Y. J. Kim, I. Zitnanová, L. Roller and M. E. Adams. 2007. Complex steroid-peptide-receptor cascade controls insect ecdysis. General and Comparative Endocrinology, 153: 88–96.

Zucchi, R. 1993. Ritualized dominance, evolution of queen-worker interactions and related aspects in stingless bees (Hymenoptera, Apidae), pp. 207–249. *In* T. Inoue and S. Yamane [eds]. Evolution of Insect Societies. Hakuhinsha, Tokyo.

12

Temperature Dependence of Development Rate, Growth Rate and Size: From Biophysics to Adaptation

Gerdien de Jong[1] **and Tom M. van der Have**[2]

[1]*Evolutionary Population Biology, Utrecht University, Padualaan 8, NL-3584 CH Utrecht, The Netherlands, tel: +31 30 253 2246, fax: +31 30 253 2837 E-mail: g.dejong@bio.uu.nl*

[2]*Resource Ecology Group, Wageningen University and Research Centre, Bornsesteeg 69, NL-6708 PD Wageningen, The Netherlands, tel: +31 317 483806 E-mail: Tom.vanderhave@wur.nl*

Abstract

In insects, temperature strongly influences phenotype, and insects differ in sensitivity to temperature. The viable temperature range might be narrow or wide, and body size often changes with development temperature. Insect development rate depends strongly on temperature, while insect growth rate does so to a lesser extent. Development rate and growth rate show a more or less triangular shape with temperature, rising slowly and almost linearly with temperature to a maximum rate at a fairly high temperature, and decreasing steeply after at higher temperatures. Insect size in many species decreases with increasing temperature over a large part of the viable temperature range. The decrease in adult size indicates that development rate is more temperature sensitive than growth rate. The decrease might be very slight, almost amounting to temperature compensation.

Many models have been proposed to describe the temperature dependence of development rate, from the degree-day summation to a model based upon biophysics. We present the degree-day model and the biophysical Sharpe-Schoolfield model. We prefer the latter model as it has a clear biophysical base, and provides an accurate description of the temperature dependence of biological rates.

We detail the possibilities of the Sharpe-Schoolfield model. (i) It can be used to describe phenotypic plasticity in development rate, growth rate and insect size. (ii) Any change in parameters in the model immediately explains why genetic variation for phenotypic plasticity can be found. (iii) The optimal temperature for organismal functioning is part of the model. This optimal temperature proves not to be identical to the temperature of highest development rate or highest growth rate. (iv) Some of the parameters in the model can be held to describe the boundaries of the viable temperature range. The Sharpe-Schoolfield model can be used to specify in how far the boundaries of the viable temperature range and the temperature dependence of the development rate could be determined by the same biophysical parameters.

By way of the Sharpe-Schoolfield model, biophysics can be used to explain size differences over temperatures and geographical clines in adult body size. Selection on development rate or growth rate would translate into selection on the parameters of the model. So would selection for enzyme efficiency or enzyme stability. The Sharpe-Schoolfield model can therefore be used to link adaptation at the physiological level to phenotypic plasticity in body size. We can see why phenotypic plasticity is adaptive, or not, what traits are the prime movers of adaptation and what traits might be easily observed but not be adaptive themselves.

Introduction

Temperature has pervasive effects on all biological systems, and it strongly influences phenotypic plasticity. The basis of this is that all biochemical reactions are sensitive to temperature, increasing in rate with increasing temperature (Hochachka and Somero 1984). In ectothermic organisms, all biochemical reactions have to be very precisely balanced and coordinated to enable the organism to perform over a range of temperatures: very different temperature sensitivities for biochemical reactions essential for life functions would lead to a speedy surmise of an organism at any temperature change. Thus, temperature coefficients of life functions have to be balanced over a range of temperatures to ensure proper functioning of organisms when temperature varies (Hochachka 1991). In ectotherms, adaptation to ambient temperatures is a physiological, behavioral, and evolutionary task, and temperature coefficients are shaped by natural selection (Clarke 2003).

Temperature is, from pole to pole, a major determinant of life histories: of lifetime patterns of development, differentiation, growth, storage of reserves, age of maturity, fecundity patterns and longevity (Atkinson 1996, Nylin and Gotthard 1998). Life histories summarize selection patterns. Development time, fecundity, and longevity are all temperature dependent. Adaptation to any environmental temperature will, therefore, be reflected in an integrated life history pattern for that temperature (Gotthard and Nylin 1995). To understand temperature adaptation it is necessary to know how proximate mechanisms influence organism performance and fitness (Pigliucci 1996, Angilletta et al. 2003).

Through its influence on the individual organism's life history, temperature influences not only population density (Huey et al. 1999), but also community structure (Petchey et al. 1999). The outcome of virtually all organismal interactions, including herbivore-plant, host-parasite, and prey-predator relationships and others, depends, in part, on temperature (Harrington et al. 1999). The influence of temperature on density is especially of importance in damage by insects to crops. Temperature influences plant growth rate and insect development time, and, hence, the damage insects can do to crops. Successful biological pest management, therefore, requires a detailed understanding of the temperature sensitivities of the plant, insect pests, insect predators and parasitoids and subsequent selection regimes (Gilbert and Raworth 1996).

Observations on Temperature Dependence of Viability, Development Rate and Adult Size

Viability, development time and adult size of insects and ectotherms in general are plastic, and strongly depend upon temperature. The natural viable range of temperatures allowing survival and development can differ strongly between species, and this correlates with habitat. Hence, alpine grylloblattids survive quite nicely at around 0°C, but will succumb when temperatures exceed 12°C (Morrisey and Edwards 1979), whereas some desert-inhabiting insects are active at 50°C, but relatively sluggish at 20°C (Chapman 1982). Tropical insects with low climatic variability appear to have a narrower developmental temperature range than temperate insects living in areas with high climatic variability. This pattern has been found in amphibians (Snyder and Weathers 1975) and fish (Brett 1970). Data on temperature ranges of insects from different latitudinal ranges are, however, surprisingly difficult to come by. Within the genus *Drosophila*, for example, species differ in upper and lower temperatures that allow full development (Table 1). Temperate species seem to have shifted their viable range to a wider and lower temperature region. The tropical species seem to have a higher lower boundary temperature, but not an appreciably higher upper temperature to the viable range. As a consequence, the viable temperature range is narrower in tropical than in temperate and cosmopolitan species (Honěk 1996). What genetic basis such differences might have is unknown. A wider thermal range of the fruitfly *Dacus tryoni* evolved in possibly less

Table 1 Temperature range for development in *Drosophila* species

Species	*Temperature range*	*Ecology*
D. melanogaster	12-32[a]	cosmopolitan
D. simulans	12-32[a]	cosmopolitan
D. yakuba	13-31[a]	tropical (Africa)
D. ananassae	16-31[b]	tropical (India)
D. iri	17-32[a]	tropical (Africa)
D. frabura	16-28[a]	tropical (Africa)
D. willistoni	15-29[b]	tropical (Brazil)
D. funebris	10-29[b]	temperate (France)
D. subobscura	6-26[b]	temperate (France)

a (David et al. 1983b)
b (Gibert and de Jong 2001b)

than hundred years after spreading from tropical regions to the temperate climate of southern Australia, but this wider thermal range was attributed to hybridization (Lewontin and Birch 1966).

Over the viable temperature range, a graph of development time versus temperature has almost the shape of a mirrored and very inclined capital-J. Development time increases steeply when the temperature drops. In the middle range of temperatures the reaction norm of development time changes relatively slowly but significantly with temperature. At the highest temperatures, development time increases again. Development rate is as often quoted as development time; by definition, development rate equals 1/(development time). When development rate rather than development time is studied, relatively small differences in development time at fast development emerge as relatively large differences in development rate, and the large differences in development time near the lower border of the viable temperature range emerge as small differences. In Fig. 1, data for development rate and development time of the fly *Dacus cucurbitae* (Coq.) are plotted. The reaction norm of development rate shows the triangular shape with temperature that is very pervasive over biological rates (Huey and Kingsolver 1989). The left, increasing leg of the reaction norm rises slowly and almost linearly (Honěk and Kocourek 1988). After a maximum, the decline in development rate is precipitous, and often very short due to failure to survive. The downturn in the reaction norm and low viability are generally related and must be due to the same temperature sensitive physiology.

In ectotherms, increased developmental temperature often results in reduced adult body size (Atkinson 1996). In 91 out of 109 studies, adult

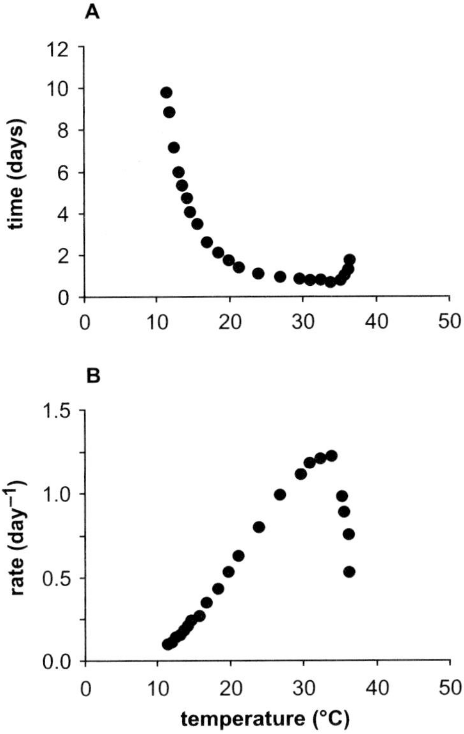

Fig. 1 Egg development as a function of constant development temperature for *Dacus cucurbitae* (Coq.). Fig. 2 from Wagner et al. (1991), after data from Messenger and Flitters (1958). A. Development time. B. Development rate.

ectotherm body size decreased with temperature (Atkinson 1996). It has been widely realized that the ratio of growth rate to development rate has to be altered if rearing temperature alters adult size (Atkinson 1996). Both growth rate (the rate of biomass increase), and development rate (the rate of tissue differentiation) increase with temperature. However, if development rate increases faster with increasing temperature than growth rate, adult size decreases.

To date, satisfactory explanations are lacking for this general phenomenon. Explanations are of two types: physiological and evolutionary: physiological explanations range from Von Bertalanffy's explanation (see Atkinson 1996, Atkinson and Sibly 1997) to the influence of temperature on cell size. Von Bertalanffy's explanation involves weight derived from different rates of anabolism and catabolism, where anabolism is less temperature sensitive than catabolism. Von Bertalanffy's explanation is a special case of the hypothesis of a growth constraint due to temperature. Such a constraint would result from a shortage of resources that is exacerbated by an increase in size or an increase in temperature. In an experimental study with *Drosophila melanogaster*, Frazier et al. (Frazier et al. 2001) found that temperature and oxygen availability interacted in the determination of adult size. Hyperoxia increased adult body size at higher rearing temperatures, but at lower rearing temperatures hyperoxia had a very small effect on body size. Flies reared in hypoxic conditions were generally smaller, had longer eclosion times, slower growth rates, and reduced survival—all signs of adverse conditions. At cooler temperatures, hypoxia had relatively modest or non-significant effects on development, while at higher temperature the effects of hypoxia were large. Whether different geographical populations of *Drosophila melanogaster*, with different body size and different temperature sensitivity, all react this way has not been established, but Frazier et al.'s (2001) study is the most explicit on the subject of growth constraints by temperature. Another physiological approach is the study of cell size. Cell size in invertebrates increases at lower temperature (Van Voorhies 1996, de Moed et al. 1997a, French et al. 1998, Blanckenhorn and Llaurens 2005). This appears to be a main mechanism causing larger adult size in insects; larger cell size per se remains unexplained but might be a consequence of limited change in cell number together with an increase in biomass at lower temperatures.

An evolutionary explanation of larger adult size at lower development temperature is as problematic, despite the progress claimed on the physiological problem (Berrigan and Charnov 1994). Low temperature selects for larger adult size if food and length of season are not limiting, as is the case in *Drosophila melanogaster* in laboratory studies (Partridge et al. 1994, Bochdanovits and de Jong 2003). In the cosmopolitan *Drosophila melanogaster* multivoltine temperate populations have larger body size than multivoltine tropical populations, on all continents (Robinson and Partridge 2001, reviewed in de Jong and Bochdanovits 2003). Not only are adult *D. melanogaster* from temperate populations larger at any developmental temperature, but the change in size with temperature is larger too (Noach et al. 1996). The higher temperature sensitivity at lower temperatures is part of the life history puzzle.

Temperature therefore presents three problems for insect size: why larger at lower developmental temperature, why larger in more temperate geographic regions and why more sensitive to developmental temperature if originating from a temperate population.

Is Temperature Dependence of Development Rate Adaptive or a Constraint?

The fact that all biochemical rates and most development rates are highly sensitive to temperature has been interpreted as a constraint on development rates by fundamental thermodynamics. Gillooly et al. (2002) and Charnov and Gillooly (2003) propose that development rate vs. temperature in their mean environment follows an exponential relationship among populations and species when corrected for body size. Charnov and Gillooly (2003) extend this model to an exponential development rate with temperature within a species. However, most thermal reaction norms of insects are linear at intermediate temperatures, not exponential. Genetic variation in thermal sensitivity has repeatedly been found (Gilbert and Raworth 1996) indicating scope for selection and adaptation rather than absolute rule by thermodynamics (Clarke 2004). Rather than describing a universal exponential function, Gillooly et al. (2002) describe an average temperature dependence of development rate where the most interesting biology is found in the deviations around the mean.

Biologists are tempted to regard the temperature at which the maximum development rate is reached as the optimal temperature for the organisms. This might be so in some performance characters (Bennett and Huey 1990, Angilletta et al. 2002a, Angilletta et al. 2002b), but it is not inherent in the physiology of the organism and development in particular. In *Drosophila melanogaster*, maximum development rate is reached at 28-29°C, but the maximum in the reaction norm for the number of ovarioles is realized at 22.5°C (David et al. 1983, Delpuech et al. 1995). The latter temperature might be nearer to the optimal functioning of the organism.

Is Temperature Dependent Adult Size Adaptive, or Mostly Temperature Compensation?

Despite higher growth and development rates at higher temperatures, adult size generally decreases with increasing temperature (Atkinson 1994, 1996). The negative slope of the thermal size reaction norm might be the result of approximate temperature compensation and other proximate mechanisms, or an adaptation to selection such as increased predation on juvenile stages at higher temperatures (Sibly and Atkinson 1994). Temperature compensation is the response of organism to a change in temperature such that they can maintain homeostatic physiological functions (Clarke 2003) or maintain a specific character nearly constant. An increasing number of studies suggest that similar thermal reaction norms can be realized with very different proximate mechanisms, influenced by different tradeoffs (Angilletta et al. 2003). These tradeoffs can occur during allocation or acquisition of resources (e.g. increased predation risk during increased intensity of foraging), or can be related to the decrement in performance within one range of the thermal environment resulting from an increase in performance within another range.

We suggest that only field and laboratory experiments, including reciprocal transplantations which measure fitness of different clones or populations with different thermal reaction norms in different thermal environments, can determine the real opportunity for adaptation (Gotthard and Nylin 1995).

Questions for the Chapter

In this chapter we focus on larval development to adult size, and review several models to describe the temperature dependence of development rate and growth rate. We regard adult size as a governed by the ratio of growth rate to development rate; as both rates are temperature dependent, the temperature dependence of adult size will follow. First we examine the importance of the linearity of the thermal reaction norm for development. Then we discuss the application of a biophysical model to growth rate and development rate, and therefore to adult size. We use this biophysical model of development rate to model the viability boundaries of the thermal range. We demonstrate how variation in the model parameters can be used to model genetic variation within and between populations. A choice of the parameter values exemplifies geographical and between-species patterns in rates and size.

Models of Development Rate

Many mathematical models have been used to describe development rate as a function of temperature. For surveys of the performance of these models, see Wagner et al. (1991) and Kontodimas et al. (2004). Many of these models are empirical and statistical: they aim at an accurate description of possible data on insect development rate, without attempting to explain development rate in terms of biological mechanisms. Deductive and explanatory models however seek to understand the causes of particular development rates from physiological mechanisms that might themselves be understood in terms of lower level processes.

Of all existing descriptive models we will mention only one, the degree-day model. The degree-day model is the simplest model for development rate, over the more or less linear middle part of the temperature range where development rate increases (Honěk 1996, Honěk and Kocourek 1990). However, the model potentially has an interesting biological background, in terms of physiological time. And even more interesting, the degree-day model might relate with the molecular genetics of cell growth and cell proliferation, through the link provided by physiological time.

In biology, mechanistic explanations are usually given in terms of lower level biological, cellular, and/or biochemical processes. However, in the case of temperature dependent development rate, the lower level explanation of biological rate is in terms of a biophysical explanatory model, not a biochemical model. One model dominates this field of explanatory models based on biophysical reaction rates: the Sharpe-Schoolfield model. This model, developed by Sharpe and DeMichele (1977) and Schoolfield et al (1981), provides an accurate description of biological rates based upon plausible biophysics. Not only does the Sharpe-Schoolfield model accurately describe temperature-dependent biological rates (Kontodimas et al. 2004), it provides a framework of fruitful thinking on the causation of such rates.

The Sharpe-Schoolfield model closely mimics empirical data, yielding a fairly linear increase in development rate over the middle part of the viable temperature range, as does the degree-day model. The parameters of the Sharpe-Schoolfield model have to conform to constraints for this to be true. We will show how these two models can be made to connect.

Degree-day Model

Empirical observations on ectothermic plants and animals show that development can often be accurately characterized by a 'temperature sum.' This was realized as early as 1735: de Réaumur (1735), as cited by Wang (1960). The rather surprising observation was that plants or insects needed a total amount of heat to reach any particular state. This amount of heat is constant over a large part of the viable temperature range—in particular, over the mid-region where development rate increases. The actual temperature of development does not matter. The threshold temperature for an insect to develop might, for instance, be 12°C. If development takes 20 days at 17°C, the temperature sum equals $(17 - 12) \times 20 = 100$ degree-days. In the same species, development at 22°C would take about 10 days, and imply an identical temperature sum of $(22 - 12) \times 10 = 100$ degree-days. A constant temperature sum, that is, a constant number of degree-days, leads to the prediction that at 27°C development would take $100/(27 - 12) = 6.66$ days. Such simple predictions are widely used in applied entomology.

Empirical Data

Constancy of the number of degree-days needed to complete development, whatever the actual development rate and temperature, depends upon a very simple feature of development rate. Development rate increases with temperature over the middle region of the viable range in a linear manner, to a very good approximation (Figs. 2 & 3). A linear regression line can be fitted through this part of the curve. Where the regression line representing development rate crosses the temperature axis (the x axis) is the threshold point h for the development rate. It is taken to indicate a zero growth rate, but is in reality, a temperature that is slightly above the biological temperature of zero growth, but quite sufficiently near it for practical purposes. The equation for development rate r_d over its middle part can be written as: $r_d = c(T - h) = cx$, where T is temperature in °C, $x = T - h$ is the difference between the actual temperature and the threshold temperature, and c is the slope of the linear regression line representing development rate. Development time equals $t_d = 1/r_d = 1/cx$. Therefore, the total number of degree-days $t_d x$ equals $1/c$. As long as the development rate is a linear function of temperature, the number of degree-days will be constant, and equal to the inverse of the slope of development rate with temperature. Once we know that development rate is a linear function of temperature, this constant temperature sum based on the temperature (above threshold) and development time, is a mathematical certainty.

In the literature, many degree-day values can be found, as the method is actively used in predicting development in insect pests (for instance, Honěk and Kocourek 1988, 1990). In the journal *Environmental Ecology* many articles reporting on insect development mention degree-day values or average degree-day values. For instance, Judd et al. (Judd et al. 1991) give both an average degree-day value and a standard error (602 ± 13 degree-days) in their study of development in the pepper maggot. Empirical degree-day values are not absolutely constant, even if the feeding environment and all other environmental influences are the same. Environmental differences other than temperature, such as food, population density, and allospecifics, influence the number of degree-days, and, in addition, development rates sometimes deviate slightly from linearity. However, the number of degree-days to reach maturity is constant enough to be of biological interest. It implies that the linearity of development rate as a function of temperature is more than a statistical first approximation: it seems a biological property. Therefore, it is a biological question how linearity of development rate is caused.

Physiological Time

Van Straalen (1983) noted that the concept of degree-days or of a temperature sum could be regarded as a transformation from physical time to physiological time. Physiological time is very useful in biology, as it can be used to construct models that are time-invariant. A physiological time-scale can be defined as follows (van Straalen 1983):

> *A physiological time-scale for a specified biological process is a time-scale obtained by transforming a physical time-scale so that the rate of the process becomes time-invariant in physiological time.*

The relationship between physiological time, physical time, and temperature, is very important for understanding animal development, ultimate phenotype, and constraints. If different traits possess the same physiological time scale under temperature change, the relation between those traits remains the same for all temperatures. Two traits that have the same physiological time over temperature change are temperature-invariant relative to each other; hence, their ratio remains constant at all temperatures. Obviously, the great multitude of biochemical/physiological processes that underlie an organism's development must possess a similar temperature-specificity, or be relatively temperature-invariant, in order to produce individuals that function properly when exposed to different temperatures. However, few traits, especially physiological traits, are truly temperature invariant; circadian rhythm might be one of the few temperature invariant traits. Temperature invariance might be difficult to achieve for physiologically dependent traits, and most traits are ultimately physiologically dependent.

The argument of van Straalen (1983) demonstrates the relation between the linear portion of the development rate, the constant temperature sum and physiological time. Van Straalen's mathematical argument is presented in verbal form in Box 1. The argument concerns when a transformation between physical time and physiological time can be defined. A linear development rate implies a constant temperature sum for total development. A constant temperature sum implies similar development under a physiological time scale. Development is time-invariant under a physiological time scale under the condition of a constant temperature sum, i.e., of a development rate that is linear with temperature.

Box 1 Linear Development Rate and Physiological Time

We explore when a conversion from varying physical time to constant physiological time might be possible, assuming that insect development time depends upon temperature. The development rate r from egg to adult depends upon temperature, but is constant, at a specific temperature, over the developmental period.

The question is whether it is possible to define a physiological time scale such that the rate of the developmental process is constant over physiological time and independent of temperature. This temperature-independent physiological rate would be given by k if it exists; $1/k$ therefore has the unit time, and is the unit in which physiological time is measured.

The starting point of the conversion between physical time and physiological time is that the same amount of development is taking place at both time scales (as it is the same developing organism). This amount of development can be visualized as a length of time × the development rate over that time. Total development over a length of physiological time $\Delta\tau$ × the physiological development rate k equals $k\Delta\tau$. Total development over an equivalent length of physical time Δt × the physical development rate r equals $r\,\Delta t$. Since both expressions concern the same development, $k\Delta\tau$, $r\Delta t$, and the potential conversion of physical time to physiological time can be given as $\Delta\tau = r\Delta t/k$. Here r and Δt both depend upon temperature, but the temperature dependence should cancel in the conversion if a temperature independent rate k is possible in physiological time.

The amount of development has to be proportional to the temperature sum in physical time. This is the essence of temperature dependence. Total developmental time t_a from egg to adult equals $1/r$ in physical time. The temperature sum of completed development therefore equals $t_a(T-h) = \frac{1}{r}(T - h)$, where h °C equals the minimum temperature for development. For a conversion from physical time to constant physiological time to be possible, the temperature sum $\frac{1}{r}(T-h)$ should be independent of temperature for all temperature dependent rates r.

The temperature sum over total development is constant if the temperature dependent rate equals $r = c\,(T-h)$. In that case, the temperature sum equals $1/c$ and is constant. A constant temperature sum implies that development is independent of temperature. Therefore rt_a is constant, independent of temperature, as is each $r\Delta t$, and as $k\Delta\tau = r\Delta t$, a constant physiological time exists. Over total development, τ_a is defined to equal 1.

Contd.

Contd.

Physiological time is measured as $1/k$. Development is time-invariant under a physiological time scale.

The linearity of development rate with temperature implies a scaling of development rate to constant physiological time; only a linear development rate has this property (van Straalen 1983).

Growth rate and development rate might both be linear with temperature. If so, both growth rate and development rate can be scaled to a temperature-independent physiological time scale, if not directly to the same time scale. However, different temperature-independent physiological time scales $1/k_g$ and $1/k_d$ are related to each other by a simple conversion factor. The temperature sums for growth rate and development rate must be constant, as otherwise no physiological time scales can be defined. If so, the division of the temperature sums is temperature independent and equal to c_d/c_g. This constant gives a conversion factor between time scales.

Body size, as the division of growth rate by temperature rate, will be constant if the minimum h °C is identical for the two rates. This will be clear from the expression for size, as size $= \dfrac{c_g(T - h_g)}{c_d(T - h_d)}$. At equal minimum temperature, size $= c_g/c_d$. Temperature independent body size therefore implies a direct relation of the physiological time scales of the two rates.

The relative duration of stages or instars are often found to be independent of temperature (Jarošík et al. 2002). If development rate is linear with temperature for each instar i according to $r_i = c_i(T - h_i)$, a constant temperature sum is present for each instar, and a separate physiological time scale can be defined for each instar. The division of the temperature independent temperature sums for two instars i and j is temperature-independent and equal to c_i/c_j, and the relative duration of the instars is constant over temperature. If the instars have identical minimal temperatures h, the relative duration of the instars equals the division of their development rates, c_i/c_j.

Biology of Physiological Time

The scaling of the linear development rate to constant physiological time is biologically interesting. What might correspond, in organismal time, to the physiological time implied by the temperature sum? A possible answer to this question is number of cell divisions in the developing organism.

This possible answer derives from detailed investigations into the components of morphological phenotypic plasticity in insects, in particular *Drosophila melanogaster*. This species shows the classical tilted triangular shape of development rate as a function of temperature (Fig. 2).

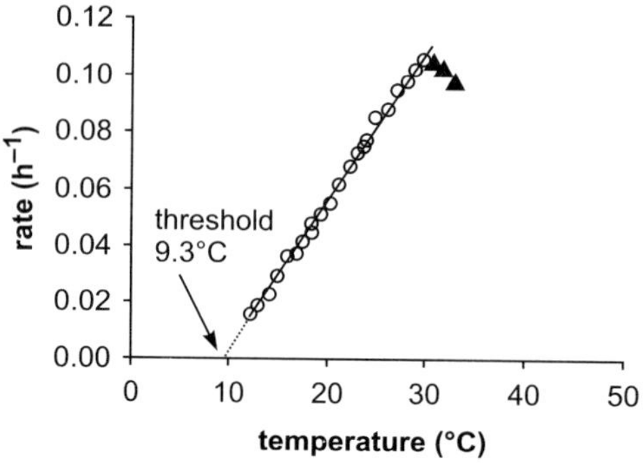

Fig. 2 Development rate of prepupae of *Drosophila melanogaster.* Fig. 1 from Gilbert and Raworth (1996), using data from Bliss (1926).

D. melanogaster adult body size has been extensively investigated for phenotypic plasticity. The most phenotypically plastic trait is wing size; thorax size might be more generally indicative of body size and has lower but still appreciable phenotypic plasticity (Noach et al. 1996, Karan et al. 1998, Karan et al. 1999). The *Drosophila* wing lends itself easily to the determination of cell size and cell number. Geographic variation in wing size is mostly but not exclusively determined by changes in cell number, cell number being higher in flies from temperate latitudes (De Moed et al. 1997b, Zwaan et al. 2000). Larger wing size, due to lower rearing temperature, proved to be mainly determined by larger cell size, with little or no effect of cell number (e.g. De Moed et al. 1997a, b, French et al. 1998). This effect of cell size might be wing-specific or population-specific. Therefore, Azevedo and coworkers (2002) surveyed phenotypic plasticity in cell size of the wing, the basitarsus of the leg and the cornea of the eye of *Drosophila melanogaster* from two populations at opposite ends of the South American latitudinal cline in body size. They found that lower rearing temperature increases wing size, leg length and eye size, through an effect on epidermal cell size, but without a significant change in cell number.

The explanation of the environmental effect on body size through cell size might be found in the growth characteristics of the imaginal discs. In *Drosophila*, the epidermis of the adult head and of the adult thoracic segments is formed by separate imaginal discs that grow and differentiate inside the developing larva. Given full larval nutrition and constant temperature, the cell proliferation of these imaginal discs might be intrinsic: that is, they might grow to a predetermined number of cells, at least for a specific *Drosophila* strain (Bryant and Simpson 1984, Bryant and Levinson 1985). Cell size can therefore be regulated and potentially selected apart from cell number. Cell size seems mainly determined by growth temperature. Constant cell number in imaginal discs (within a line) seems independent of temperature. Therefore, cell number makes a good candidate as deriving from organismal physiological time during development. In other words, the same number of cell divisions might be needed to complete animal development, independent of temperature, at least in the mid-region of the viable temperature range. This suggests that all biochemical processes of cell division have rates that increase linearly with temperature over this range. Of course, the rates of all chemical and biochemical reactions are inherently temperature sensitive (Hochachka and Somero 1984). This brings us to the other type of model of development rate, a deductive and explanatory model based upon the biophysics of reaction rates.

Sharpe-Schoolfield Model

The Sharpe-Schoolfield model is based upon two principles: (i) temperature dependent reaction kinetics per active enzyme molecule; (ii) reversible inactivation of enzyme molecules at high or low temperature. Both the reaction kinetics and the inactivation of molecules use classical models of biophysical reaction rates. We will use these models without more explanation of biophysical detail than is necessary for understanding the biology. The simplest version of the theory of biochemical reactions can be found in Willmer et al. (2000). Slightly more advanced treatments can be found in the discussion of biochemical adaptation by Hochachka and Somero (1984) or in Lowry and Richardson (1987). Watt (1968) gives a clear ecological introduction for practical biologists.

Model Based Upon Biophysics

Temperature dependence of reaction rates has traditionally been described by an empirical equation due to Arrhenius (around 1900) and a more theoretical equation due to Eyring (1935). Sharpe and DeMichele (1977) developed a model for development rate based upon the Eyring equation. The parameters of this model conformed to the usage in the biophysics of chemical reactions, and had no direct biological interpretation. The model made a number of assumptions.

The first assumption is that development is regulated by a single control-enzyme whose reaction rate determines the development rate of the organism. This assumption might be less restrictive than it seems at first look. Sharpe and DeMichele (1977) show how several limiting enzymes, for parts of the temperature range, leave the overall impression of determination of the development rate by one controlling enzyme over all the range. Hence, although the physiological reality might be several enzymes, the model can proceed as if one enzyme is in control of development over all the range. Moreover, development rate is proportional to the product of the concentration of active molecules of the control enzyme and their temperature-dependent reaction rate. For each enzyme molecule in the active state, the reaction rate exponentially increases with temperature.

The second assumption is that the control-enzyme can exist in two temperature-dependent reversible inactivation states as well as in an active state; one reversible inactivation state pertains to high temperature, the other to low temperature.

Schoolfield et al. (1981) modified the model of Sharpe and DeMichele to make it more convenient for biological interpretation. They introduced a reference reaction rate that absorbed physical constants and the entropy of activation, combined other parameters and succeeded in writing the Sharpe and DeMichele model in a way that can be easily understood and now has parameters that all have clear biological relevance.

The classical biophysical Eyring equation describes a reaction rate's exponential increase with temperature. The Eyring equation describes temperature sensitivity of reaction rates without enzyme inactivation. Schoolfield et al (1981) give the Eyring equation in the original parameters (their Equation 2). In the parameters, as defined by Schoolfield et al. (1981), the Eyring equation reads:

$$r(T) = \rho \frac{T}{T_{ref}} \exp\left[\frac{H_A}{R}\left(\frac{1}{T_{ref}} - \frac{1}{T}\right)\right] \qquad \text{(Eq. 1)}$$

The ratio $r(T + 10)/r(T)$ is known as Q_{10}, a widely used rough and ready guide to the temperature sensitivity of reactions.

In the Eyring equation, reaction rate $r(T)$ as a function of temperature T is given as a modification of a reference reaction rate ρ at a reference temperature T_{ref} (in °K). The dependence of the reaction rate on temperature is given by the temperature sensitivity coefficient H_A (in J mol^{-1}), officially the enthalpy of activation of any reaction that is catalyzed by a rate-controlling enzyme (Hochachka and Somero 1984, pg 379–380, notation simplified); R is the universal gas constant (8.314 J K^{-1} mol^{-1}). The exponential factor in Equation 1 yields a dimensionless scaling to temperature of the reference reaction rate ρ. The exponential factor is larger than 1 if temperature is higher than the reference temperature and the reaction rate is then higher than the reference reaction rate. Conversely, the exponential factor is less than 1 if the environmental temperature is lower than the reference temperature, and the reaction rate is then lower than the reference reaction rate.

What rate the equation refers to is specified by the units of the reference rate ρ. If the biological rate described by Equation 1 refers to development rate, ρ has the unit time^{-1}. If the considered rate refers to growth rate, then ρ's units are biomass per time. The same equation applies to any intended temperature sensitive rates that are determined by biochemical reactions.

Biological rates show this exponential increase at most across a very limited temperature range. At middle temperatures, rates are almost linear. At higher temperatures, development rate usually sharply decreases. At low temperatures, development rate slowly approaches zero. Any feasible model designed to explain all of development rate has to include the high and low temperature behavior of biological rates, and accommodate the linearity. A very useful assumption is that the sharp downturn in development rate at high temperature is due to enzyme inactivation. This is a particularly plausible assumption as the downturn in development rate often is near the high end of the viable range for development (Fig. 3).

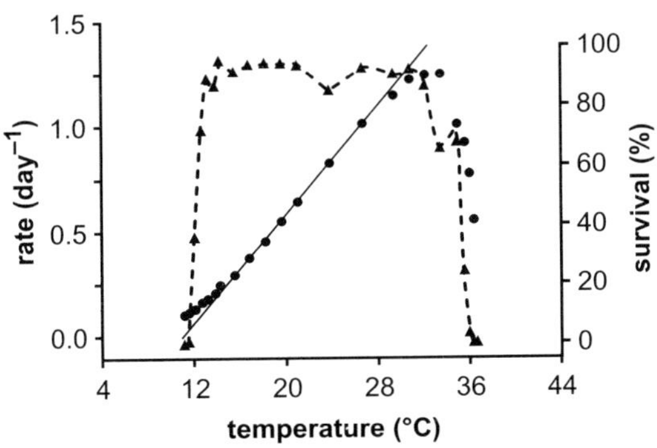

Fig. 3 Development rate (circles) and survival (triangles) for eggs from *Dacus cucurbitae* (Coq.). Fig. 4 from Wagner et al. (1991), after data from Messenger and Flitters (1958).

Sharpe and DeMichele (1977) therefore proposed to modify the Eyring equation in order to describe biological rates better. The Eyring equation can be modified by the probability that any enzyme molecule is active at a given temperature. Any enzyme molecule is supposed to be subject to reversible inactivation at both high and low temperature; the reversible inactivation processes or sensitivities at the two temperature extremes are supposed to be independent. The probability for the enzyme to be active can be thought of as proportional enzyme efficiency over temperatures. This enzyme efficiency is characterized by two parameters at low temperature and by two parameters at high temperature. At low temperature, the parameters are: (1) the temperature at which 50% of all molecules of a rate-limiting protein or enzyme are reversibly inactive due to cold, the lower boundary temperature T_L, in °K; and (2) the specific sensitivity to cold inactivation, the cold inactivation coefficient H_L. The sensitivity to heat inactivation at low temperature is formally the change in enthalpy associated with low temperature inactivation of the enzyme, and expressed in J mol^{-1}. At high temperature, the parameters are: (1) the temperature at which 50% of all molecules of a rate-limiting protein or enzyme are reversibly inactive due to heat, the upper boundary temperature T_H, in °K; and (2) the specific sensitivity to heat inactivation, the heat inactivation coefficient H_H. The sensitivity to heat inactivation at high temperature is formally the change in enthalpy associated with high temperature inactivation of the enzyme, and again expressed in J mol^{-1}.

The enzyme efficiency, as probability P_T for the enzyme to be active as a function of temperature, is given by:

$$1/P_T = 1 + \exp\left[\frac{H_L}{R}\left(\frac{1}{T_L} - \frac{1}{T}\right)\right] + \exp\left[\frac{H_H}{R}\left(\frac{1}{T_H} - \frac{1}{T}\right)\right] \quad \text{(Eq. 2)}$$

Incorporation of the probability of the enzyme to be active yields the Sharpe-Schoolfield equation for any biological rate as a function of temperature:

$$r(T) = \rho \frac{TP_T}{T_{ref}} \exp\left[\frac{H_A}{R}\left(\frac{1}{T_{ref}} - \frac{1}{T}\right)\right] \quad \text{(Eq. 3)}$$

The six parameters of the Sharpe-Schoolfield equation can be used in paired combinations.

The reference rate ρ and the temperature sensitivity coefficient H_A refer to the total temperature range. These two parameters by themselves might be sufficient to describe a short middle region of development rate, in a two-parameter version of the Sharpe-Schoolfield model (Fig. 4A). Two parameters, T_L and H_L, are specific for the low-temperature range, and two parameters, T_H and H_H are specific for the high temperature range. Two versions of a four-parameter model exist, one at low temperatures with reference rate ρ and temperature sensitivity coefficient H_A together with lower boundary temperature T_L and cold inactivation coefficient H_L (Fig. 4B), and one with parameters ρ and H_A together with upper boundary temperature T_H and heat inactivation coefficient H_H, at high temperatures

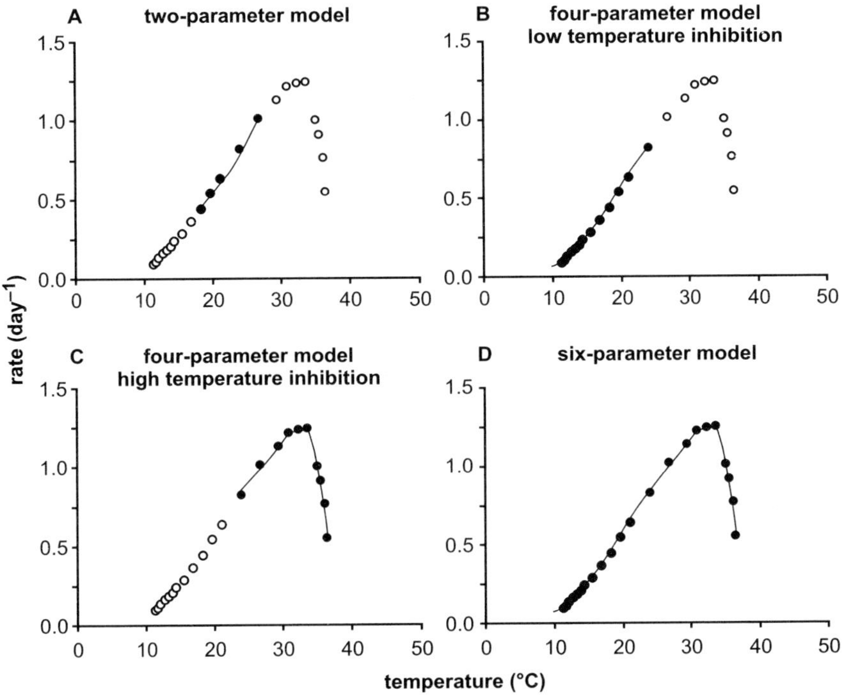

Fig. 4 Sharpe-Schoolfield model fitted to the data from Messenger and Flitters (1958). A. Two-parameter model over the middle temperature range using r and H_A. B. Four-parameter model over the low temperature using ρ and H_A, as well as H_L and T_L. C. Four-parameter model over the high temperature using r and H_A, as well as H_H and T_H. D. Six-parameter model for the entire data set, ρ and H_A, as well as H_L and T_L, as H_H and T_H. Fig. 5 from Wagner et al. (1991).

(Fig. 4C). Over the total temperature range, the full six-parameter model can be used (Fig. 4D). These four figures are given by Wagner et al. (1991) as examples of the four sub-models and their good fit to data on insect development.

Parameter Variation

The Sharpe-Schoolfield equation possesses in total six parameters, two from the Eyring equation and four from the description of temperature dependent reversible enzyme activation. Together, these parameters describe development rates that are overall similar in shape but might be very different in detail.

Different shapes of the probability P_T for the control enzyme to be active represent biologically different cases of enzyme adaptation to temperature. The probability of the rate-determining enzyme to be active as a function of temperature decreases at both high and low temperature; the probability might reach a value of 1, but this is not necessary. High absolute values of the inactivation coefficients H_L and H_H at the lower and upper boundary temperatures T_L and T_H imply a wider plateau and a faster decrease of the probability that the rate-determining enzyme is active. In Fig. 5A, two enzymes with different parameters are shown as examples. One enzyme has parameters H_L and H_H that are large in absolute value; therefore the enzyme activity declines steeply both at high and low temperature. The other enzyme has very low values for the parameters H_L and H_H, and, as a consequence, the probability to be active increases and decreases slowly with temperature. In this case, the enzyme need not be fully active at any temperature. In Fig. 5B, two enzymes are shown that are a high and low temperature specialist, due to one temperature inactivation that goes very rapid and another temperature inactivation that is very gradual. The maximum in the probability P_T for the control enzyme to be active is at high or low temperature. This contrast provides a possibility to model enzyme specialization to high or low temperature (Hochachka and Somero1984).

The Eyring equation describes an exponential increase of a biological rate with temperature; higher temperature sensitivity coefficient H_A implies a faster increase at temperatures higher than the reference temperature (Fig. 5C) and higher reference rate ρ implies both a higher rate at the reference temperature and a faster increase with temperature (Fig. 5D). Changing parameter combinations lead to different development rates. The steepness of the rates depends highly on the inactivation parameters H_L and H_H (Figs. 5E, F). The maximum of the rates is for a large part dependent upon H_A and ρ, larger H_A implying higher development rate (Figs. 5E, F).

Arrhenius Plot and Parameter Estimation

The six parameters of the Sharpe-Schoolfield model have to be estimated in order to investigate the validity and sufficiency of the model in describing empirical data on development rate. The high temperature range of development rate often shows a sudden down-curve. Likewise, the low temperature range of development might show a development rate that approaches zero asymptotically. Many data points are necessary to get accurate estimates in both of these regions. However, data collection at low temperatures represents a disproportionate investment in effort compared to high temperatures, because of the greater time to complete development, and the problems associated with keeping insects healthy at low temperatures.

For statistical reasons, seven data points are minimally necessary to estimate the six parameters of the Sharp-Schoolfield model. However, it

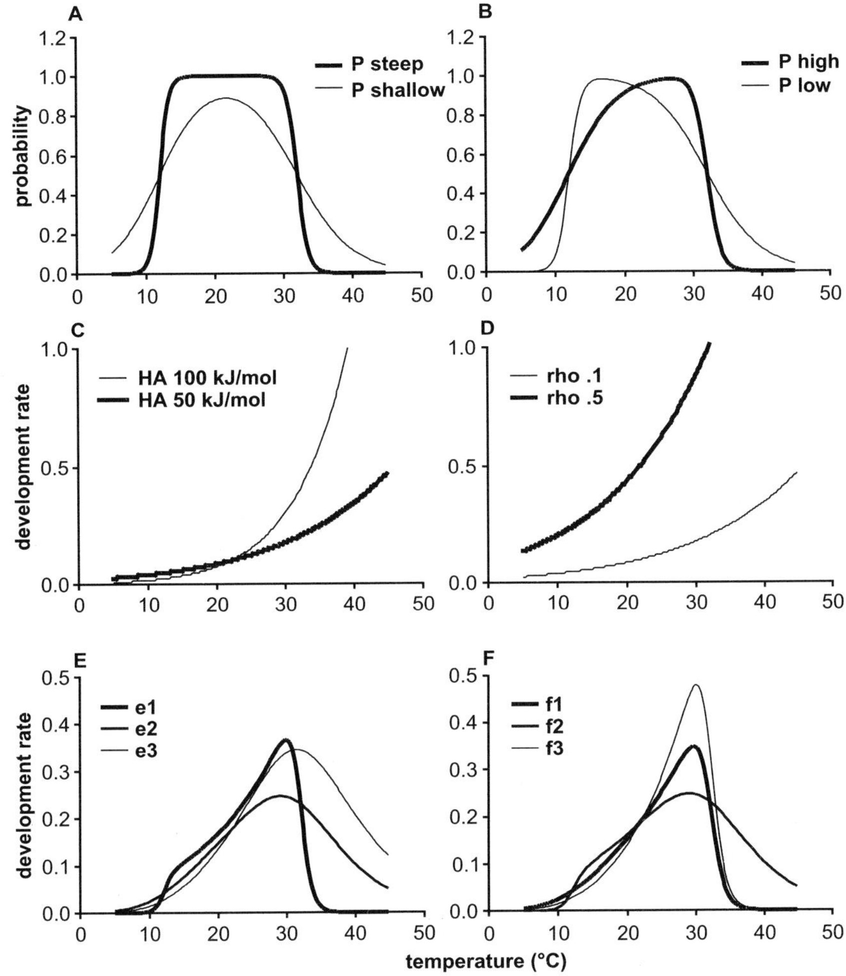

Fig. 5 Influence of parameter variation in the Sharpe-Schoolfield model on development rate. In all cases, T_{ref} = 295°K = 2 1.8°C, T_H = 305°K = 3 1.8°C, and T_L = 285°K = 1 1.8°C. A. Parameter variation in denominator P: H_H = 1000 kJ mol^{-1}, H_L = −1000 kJ mol^{-1} (steep) versus H_H = 200 kJ mol^{-1}, H_L = −200 kJ mol^{-1} (shallow). B. Parameter variation in denominator P: H_H = 800 kJ mol^{-1} and H_L = −200 kJ mol s^{-1} (skewed to high temperature) versus H_H = 200 kJ mol^{-1} and H_L = −800 kJ mol^{-1} (skewed to low temperature). C. Variation in H_A: 100 kJ mol^{-1} and 50 kJ mol^{-1}: note identity at reference temperature at ρ = 0.1. D. Variation in ρ: ρ = 0.1 t^{-1}, ρ = 0.5 t^{-1} at H_A = 50 kJ mol^{-1}. E. Variation in rate: (e1) H_H = 1000 kJ mol^{-1}, H_L = −1000 kJ mol^{-1} with H_A = 60 kJ mol^{-1} and ρ = 0.2 ; (e2) H_H = 200 kJ mol^{-1}, H_L = − 200 kJ mol^{-1} with H_A = 60 kJ mol^{-1} and ρ = 0.2; (e3) H_H = 200 kJ mol^{-1}, H_L = −200 kJ mol^{-1} with H_A = 90 kJ mol^{-1} and ρ = 0.2. F. Variation in rate: (f1) H_H = 800 kJ mol^{-1}, H_L = −200 kJ mol^{-1} with H_A = 60 kJ mol^{-1} and ρ = 0.2 t^{-1} ; (f2) H_H = 200 kJ mol^{-1}, H_L = − 800 kJ mol^{-1} with H_A = 60 kJ mol^{-1} and ρ = 0.2 t^{-1}; (f3) H_H = 800 kJ mol^{-1}, H_L = −200 kJ mol^{-1} with H_A = 90 kJ mol^{-1} and ρ = 0.2 t^{-1}.

should be remembered that while seven points will estimate a curve with six parameters, the reliability of the estimates depends on the density of points along the low- and high-temperature inflections. In statistical programs like SAS or SPSS, a general non-linear regression curve-fitting module using the Marquardt Method is present and can be used. Such programs need initial parameter values. If only seven points are available, initial parameter values have to be found in a range of values published in the literature. If the number of data points is optimized for curve fitting rather than the minimum needed for statistics, an Arrhenius plot might help to estimate initial parameter values (Schoolfield et al. 1981, Wagner et al. 1984).

In an Arrhenius plot, the logarithm of development rate is plotted against the inverse of temperature, in °K. In order to make clear how estimation proceeds, we will not start with data but with known theoretical values for the six parameters of the Sharpe-Schoolfield model (Schoolfield et al. 1981). In Fig. 6, theoretical development rates are plotted both as rate versus temperature (Fig. 6A) and as the logarithm of the rate versus the inverse of the temperature (Figs. 6C, D). Moreover the probability for the enzyme to be active is plotted (Fig. 6B). Note that the probability for the enzyme to be active is higher than 0.9 between 16°C and 28°C. The logarithm of the development rate plotted against the inverse of temperature $1/T$ shows a more or less linear middle part, and curves down at the left (high temperature) and right (low temperature). A linear middle part corresponds to temperatures where the probability for the rate-controlling enzyme to be active equals 100%. Only then will we find linearity in an Arrhenius plot. In such a middle part of high enzyme activity, development rate can be sufficiently described by the Eyring equation. High probabilities for the enzyme to be active, greater than 90% or so, lead to an approximately linear part in the Arrhenius plot. This linear part can be used to estimate the reference rate ρ and the temperature sensitivity coefficient H_A, as the slope of the straight line equals $-\frac{H_A}{R}$ and the intercept $\ln \rho + \frac{H_A}{R}\frac{1}{T_{ref}}$.

Model Fit to Empirical Pattern

The Sharpe-Schoolfield model is excellently suited to describe biological rates (Kontomidas et al. 2004). Of course, any model with six parameters might be expected to perform well, but the Sharpe-Schoolfield model is better able than most alternatives to describe the fine detail of development rate, especially the sudden decrease at high temperature. The initial curve fittings by Sharpe and DeMichele (1977) showed the suitability of the model.

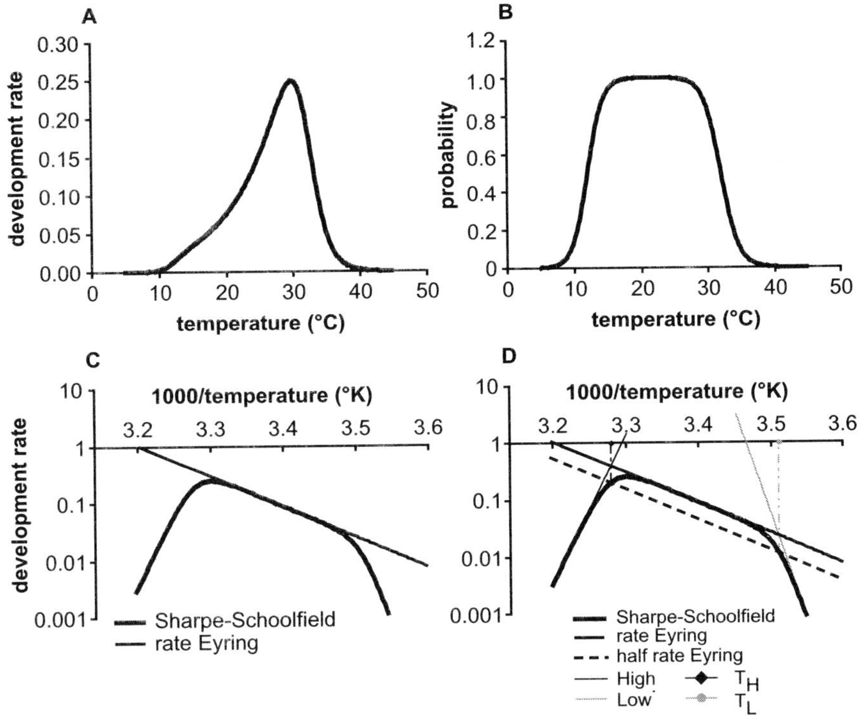

Fig. 6 Graphical estimation of parameter values. The graphs are drawn according to the parameter values: $H_A = 100$ kJ mol^{-1}, $\rho = 0.1\ t^{-1}$, $H_H = 600$ kJ mol^{-1}, $H_L = -600$ kJ mol^{-1}, $T_{ref} = 295°$ K = 2 1.8°C, $T_H = 305$°K = 3 1.8°C, and $T_L = 285$ °K = 1 1.8°C. Figs. 6A and 6B give the data. Figs. 6C and 6D the estimation. A. Rate. B. Probability enzyme active. C. Ln rate versus $1/T$: estimation of H_A and ρ from the 17°C to 26°C temperature range as $H_A = 102.46$ kJ mol^{-1} and $\rho = 0.099t^{-1}$. D. Ln rate versus $1/T$: estimation of H_H, H_L, and T_H, T_L as $H_H = 599.9$ kJ mol^{-1}, $H_L = -599.9$ kJ mol^{-1}, and $T_H = 305$°K, $T_H = 285$°K.

Wagner and co-authors (1984) applied the model successfully to data from a range of species, and compared the performance of the Sharpe-Schoolfield model with a range of alternative models common to entomology (Wagner et al. 1991).

For *Drosophila*, some data are available for development rate (Table 2) and growth rate (Table 3). Van der Have and de Jong (1996) estimated the six parameters for *Drosophila melanogaster* development rate and growth rate on the base of data from David and Clavel (1967). Both rates are fairly linear with temperature over the middle range (Fig. 7A). In the developmental rate case, the rate-controlling enzyme is at no temperature more than 90% active, for growth rate the maximum probability for the enzyme to be active reaches

Box 2 Parameter Estimation from an Arrhenius Plot

Parameter estimation from an Arrhenius plot is detailed in Schoolfield et al. (1981) and Wagner et al. (1984). Taking logarithms, the Eyring equation (Equation 1) becomes

$$\ln \rho(T) = \ln \rho + \ln \frac{T}{T_{ref}} + \frac{H_A}{R}\frac{1}{T_{ref}} - \frac{H_A}{R}\frac{1}{T} \qquad \text{(Eq. 4)}$$

Over physiological ranges from some 10°C to some 30°C, the term $\frac{T}{T_{ref}}$ will be very small and can be neglected. In a plot of $1/T$ on the x-axis and $\ln \rho$ on the y-axis, the slope of the straight line will then be $-\frac{H_A}{R}$, and the intercept $\ln \rho + \frac{H_A}{R}\frac{1}{T_{ref}}$. The slope in an Arrhenius plot can therefore be used to estimate H_A. After estimation of H_A, the intercept can be used to estimate ρ. In Fig. 6C, the curve represents Sharpe-Schoolfield model, and the straight line its numerator, the Eyring equation.

Estimation of H_H proceeds by dividing the Eyring equation by the term from the probability of enzyme inactivation that is specific for high temperature (Schoolfield et al. 1981, Wagner et al. 1984) and taking logarithms:

$$\frac{\rho \frac{T}{T_{ref}} \exp\left[\frac{H_A}{R}\left(\frac{1}{T_{ref}} - \frac{1}{T}\right)\right]}{\exp\left[\frac{H_H}{R}\left(\frac{1}{T_H} - \frac{1}{T}\right)\right]} = \rho \frac{T}{T_{ref}} \exp\left[\frac{H_A}{R}\left(\frac{1}{T_{ref}} - \frac{1}{T}\right) - \frac{H_H}{R}\left(\frac{1}{T_H} - \frac{1}{T}\right)\right]$$

$$\ln \rho + \ln \frac{T}{T_{ref}} + \left(\frac{H_A}{R}\frac{1}{T_{ref}} - \frac{H_H}{R}\frac{1}{T_H}\right) - \left(\frac{H_A}{R} - \frac{H_H}{R}\right)\frac{1}{T} \qquad \text{(Eq. 5)}$$

The result is again linear if the logarithm of this division is plotted against $1/T$, now with slope $-(H_A - H_H)/R$ (Fig. 6D, line High). Similarly, estimation of H_L proceeds by dividing the Eyring equation by the term from the probability of enzyme inactivation that is specific for low temperature. The resulting is linear in a plot of the logarithm of this division against $1/T$ (Fig. 6D), now with slope $-(H_A - H_L)/R$. In both the high and low temperature cases, accuracy of estimation would require many data points near the temperature extremes for viability.

Estimation of T_H and T_L can proceed in two ways. On the one hand, the property that the probability for the enzyme to be active is exactly one half at both temperatures can be used. This implies that the numerator from the Sharpe-Schoolfield model can be divided by 2 to obtain a straight line that crosses the curve of the full Sharpe-Schoolfield model at T_H and T_L (Fig. 6D; (Schoolfield et al. 1981, Wagner et al. 1984)). On the other hand, in the logarithmic plot the straight line High crosses the straight line Eyring at T_H, and the straight line Low crosses the straight line Eyring at T_L (compare Equations 4 and 5; Fig. 6D).

Table 2 Parameter estimates of development rate in *Drosophila* species.

Species[1]	*Sex, population*	$\rho_{\cdot}^{2}$	H_A	T_H	H_H	T_L	H_L
D. melanogaster	David & Clavel	0.567	66064	32.3	220632	12.3	−176707
D. melanogaster	male, Houten	0.487	73580	31.6	355904	11.3	−318001
D. melanogaster	male, Adana	0.530	86868	30.3	244182	13.5	−1138734
D. melanogaster	female, Houten	0.498	74274	31.9	318775	11.6	−366259
D. melanogaster	female, Adana	0.510	82772	30.8	272655	13.5	−1114203
D. simulans	male, Houten	0.507	52262	32.4	371974	13.1	−210782
D. simulans	male, Adana	0.584	83441	30.6	271521	11.0	−365652
D. simulans	female, Houten	0.500	49204	32.1	556844	13.1	−238509
D. simulans	female, Adana	0.586	82508	30.8	300512	11.1	−382167
D. ananassae	male	0.3378	80290	31.5	341733		
D. ananassae	female	0.3400	77163	31.8	362784		
D. willistoni	male	0.3258	86402	28.9	339119		
D. willistoni	female	0.3440	86239	28.8	345762		
D. funebris	male	0.2985	99242	26.2	260893		
D. funebris	female	0.2900	96628	26.7	267376		
D. subobscura	male	0.2560	69957				
D. subobscura	female	0.2550	70244				

[1] First row: *D. melanogaster* as estimated by van der Have and de Jong (1996) on data from David and Clavel (1967). *D. melanogaster* and *D. simulans* Houten, The Netherlands, and Adana, Turkey, populations data by Jeroen Bohré (unpublished data student graduation project, G. de Jong's lab); estimations based upon 10 vials. *D. ananassae, D. willistoni, D. funebris* and *D. subobscura* from (Gibert and de Jong 2001); estimation based upon 20 vials. All data apart from David and Clavel's on same fly food in the same lab.

[2] $\rho_{\cdot}$ in 10^{-2} h^{-1}; H_A, H_H, and H_L in J mol^{-1}; T_H and T_L in °C.

0.975 (Fig. 7B). The middle part of the Arrhenius plot is not really linear: as the probability for the rate-controlling enzyme to be active has no plateau at 100%, the line of the Eyring equation corresponding to the numerator only approaches the curve for the Sharpe-Schoolfield model (Fig. 7C). As the rates are fairly linear with the temperature range, the number of degree-days can be estimated. Linear regression on temperature for the temperature range 11°C to 30°C yields a threshold of 10.2°C for development rate, and 9.6°C for growth rate. The actual number of degree-days is computed as the product of the development time at a specific temperature as found from the Sharpe-Schoolfield model × the degrees over the threshold. The number of degree-days is not constant (as the development rate is not perfectly linear) but is restrained to a fairly narrow region over the feasible range of development temperatures (Fig. 7D).

Estimates of all six parameters are available for a further two *D. melanogaster* populations and two *D. simulans* populations, from the same

Table 3 Parameter estimates of growth rate in *Drosophila* species.

Species[1]	*Sex, population*	$\rho.$	H_A	T_H	H_H	T_L	H_L
D. melanogaster	David & Clavel[2]	0.667	39603	32.4	578105	14.3	−248771
D. melanogaster	male, Houten[2]	0.00400	59446	31.5	448098	12.4	−351301
D. melanogaster	male, Adana[2]	0.00400	66760	31.2	288458	13.7	−860398
D. melanogaster	female, Houten[2]	0.00690	71906	31.4	435743	12.6	−335092
D. melanogaster	female, Adana[2]	0.00610	72459	30.9	471114	13.7	−719711
D. simulans	male, Houten[2]	0.00424	73484	31.3	340266	10.2	−345749
D. simulans	male, Adana[2]	0.00430	77136	30.5	297395	10.4	−437998
D. simulans	female, Houten[2]	0.00590	69099	31.5	726660	11.3	−290767
D. simulans	female, Adana[2]	0.00730	99638	30.5	358431	11.6	−613751
D. ananassae	thorax male[3]	3.4930	83054	30.4	339826		
D. ananassae	thorax female[3]	3.8483	81790	30.8	34652		
D. willistoni	thorax male[3]	2.8473	83770	28.6	357922		
D. willistoni	thorax female[3]	3.4401	92798	27.8	325489		
D. funebris	thorax male[3]	4.1904	103583	25.3	285400		
D. funebris	thorax female[3]	4.6264	101733	25.4	27145		
D. subobscura	thorax male[3]	2.7515	66302				
D. subobscura	thorax female[3]	3.0947	66563				
D. ananassae	wing male[3]	4.8448	73526	30.8	343128		
D. ananassae	wing female[3]	5.4173	72606	30.0	354770		
D. willistoni	wing male[3]	4.4513	75354	28.5	326061		
D. willistoni	wing female[3]	5.1449	768 17	28.3	389725		
D. funebris	wing male[3]	6.5630	95011	26.5	281633		
D. funebris	wing female[3]	7.1903	94747	25.5	271604		
D. subobscura	wing male[3]	4.5094	61830				
D. subobscura	wing female[3]	4.4958	62267				

[1] First row: *D. melanogaster* as estimated by van der Have and de Jong (1996) on data from David and Clavel (1967). *D. melanogaster* and *D. simulans* Houten and Adana populations data by Jeroen Bohré (unpublished data student graduation project, G. de Jong's lab); estimations based upon 10 vials. *D. ananassae, D. willistoni, D. funebris* and *D. subobscura* from (Gibert and de Jong 2001); estimation based upon 20 vials. All data apart from David and Clavel's on same fly food in the same lab.

[2] $\rho.$ in mg h^{-1}; H_A, H_H, and H_L in J mol^{-1}; T_H and T_L in °C.

[3] $\rho.$ in 10^{-3} mm h^{-1}; H_A, H_H, and H_L in J mol^{-1}; T_H and T_L in °C.

temperate (Houten, The Netherlands) and Mediterranean (Adana, Turkey) locations. Over the populations, no species differences in development rate or growth rate were found (Figs. 8A, B). The probabilities for the enzyme to be active never reached 100% (Figs. 8C, D). The high temperature four-parameter version of the Sharpe-Schoolfield model was applied to development rate and growth rate of four other *Drosophila* species, two tropical species and two temperate (Tables 2 & 3). In all these cases, the

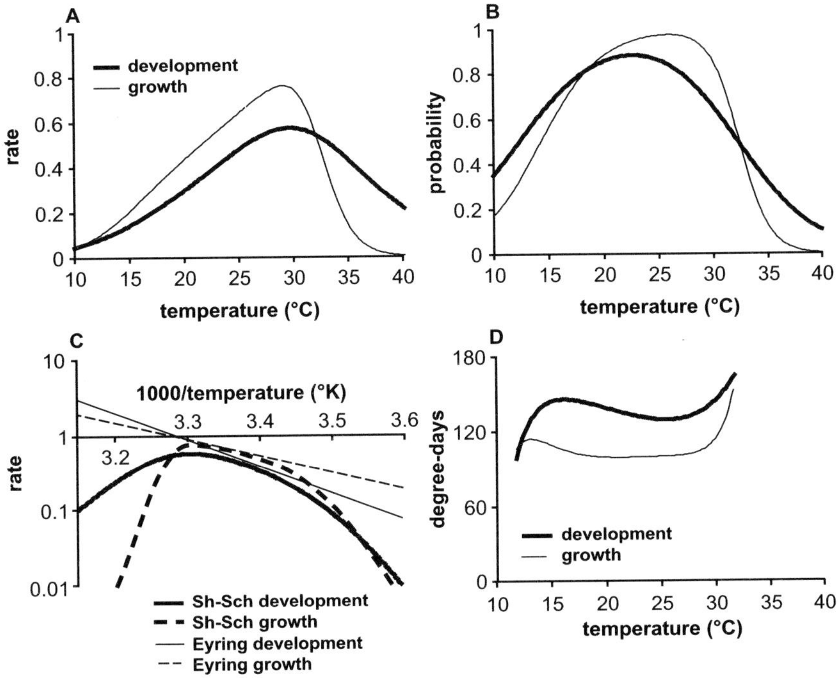

Fig. 7 Rates and probabilities in *Drosophila melanogaster.* Data David & Clavel (1967). Estimates of parameters for development rate are given in Table 2, for growth rate in Table 3. A. Development and growth rates and temperature. B. Probabilities for developmental and growth enzyme to be active. C. Arrhenius plot of ln rate versus $1/T$, both according to the Sharpe-Schoolfield equation (Sh-Sch) under estimated parameters, and according to the Eyring equation under estimated parameters. D. Degree-day plots for development and growth.

experiment itself has not been replicated, and experimental error is therefore unknown. Nevertheless, the order of magnitude of the parameter values is now established for *Drosophila*.

One clear result is that whenever all six parameter values can be estimated, the probability for the rate-controlling enzyme to be active proved never to reach 100%; a maximum enzyme activity might be near to 100% but is not consistently maintained over a range of temperatures (Figs. 8C, D). According to these data there is no temperature at which all enzyme molecules are active; reversible enzyme inactivation is present at all temperatures, and the biological rates are never completely ruled by the Eyring equation and exponentially increasing. Accordingly, development rates seem fairly linear over the middle temperature range in these data, whereas linearity of development rate is impossible with full enzyme activity over a large part of the temperature range: a probability of 100% for

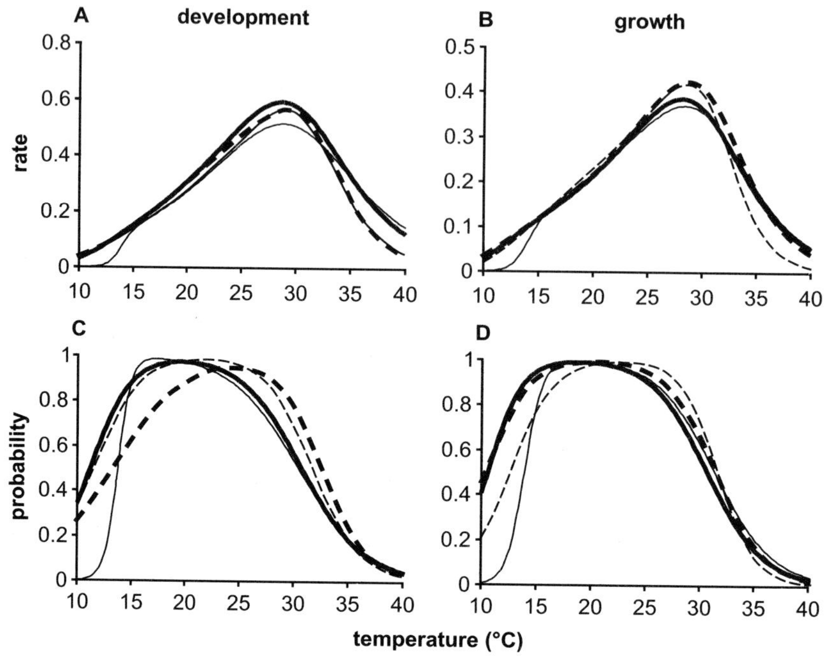

Fig. 8 Rates and probabilities in *Drosophila melanogaster* (thin lines) and *D. simulans* (thick lines), for the Adana (Turkey, continuous lines) and Houten (The Netherlands, broken lines) populations (data by Jeroen Bohré). Estimates of parameters for development rate are given in Table 2, for growth rate in Table 3. A. Development rate. B. Growth rate. C. Probability for the enzyme to be active: development. D. Probability for the enzyme to be active: growth.

the control enzyme to be active results in an exponential increase in development rate. As the development rate approaches linearity over the middle temperature range, the degree-day model is approximately valid. For *D. simulans*, degree-days over 15°C to 27°C range from 121 – 136 (mean 130) in the population from Adana to 132 – 145 (mean 138) in the population from Houten. For *D. melanogaster*, the degree-days between 15°C to 27°C vary from 121 – 136 (mean 131) in the Adana population to 130 – 146 (mean 140) in the Houten population. The conclusion is that the degree-day model and the Sharpe-Schoolfield model are compatible over the middle temperature range of the reaction norm for development rate.

Sharpe-Schoolfield Model and Degree-Days

Development rates might seem fairly linear over the middle temperature range, and the degree-days model is sufficiently experimentally supported, although any increase in development rate might seem linear to a sufficient approximation due lack of statistical power. On the other hand, the Sharpe-Schoolfield model provides a good fit of development rate over the total temperature range. The question arises whether and why empirical parameter values found for the Sharpe-Schoolfield model give rise to approximate linearity over the middle temperature range, and therefore to the degree-day model. One possible approach is to look at the degree-days that would result from the observed parameter estimates in *Drosophila* spp. (Table 2).

Development rates, as in the Houten and Adana populations of *D. melanogaster* and *D. simulans*, were given in Fig. 8A. As the development rate is known for each temperature, the number of degree-days can be computed. The number of degree-days is relatively constant, at about 90 to 100 dependent upon the *Drosophila* line. Variation in number of degree-days between 15°C and 27°C leads to a coefficient of variation of an order of magnitude (4%) that might pass muster in experimental studies as indicating validity of the degree-day model. Perhaps the *D. simulans* model-fitting and *D. melanogaster* Adana model-fitting show as good a linear increase in development rate as any set of experimental data. The model for the *D. melanogaster* Houten population shows a curve that looks somewhat more exponential. These curves are estimated from development times and show a middle part as linear as the Sharpe-Schoolfield model seems capable of. At least, it is difficult to find better linearity by adjusting the model.

Both the Sharpe-Schoolfield model and the degree-day model might be regarded as curve fitting exercises rather than a description of some biological reality. If we take the parameters in the Sharpe-Schoolfield model as representing biological reality, we have to face the question why the actual parameter combinations lead to an increase in development rate with temperature that is near linear. The estimated parameter values for H_H and H_L were relatively low in absolute value, leading to a gradual change in probability that the rate-controlling enzyme would be active. Less enzyme inactivation and a larger range of full enzyme activity is possible with higher absolute values for H_H and H_L. We have to face the conclusion that maximal enzyme activity over a fairly extended temperature range is not present in these data. The prevalence of linearity in the development rate and the applicability of the degree-day model argue against 100% enzyme activity. Maximal enzyme activity over an extended temperature range would lead to an exponential development rate.

If linearity in development rate rather than maximal enzyme activity is selected for, selection might be for constant physiological time. Van Straalen

(1983) noted that linearity of development rate with temperature and constant number of degree-days implied a direct transformation from physical time to physiological time. Above, we have argued that a possible biological equivalent of physiological time might be cell number. Functioning of the organism at all temperatures might well involve selection to keep cell number as constant as possible in development. Selection on enzyme inactivation would be a consequence of selection on developmental homeostasis, rather than a cause of differences in development rate.

We make two steps in reasoning here. Both steps start with the observed quasi linearity in development rate. The first step concerns the Sharpe-Schoolfield model. The estimated parameters of the Sharpe-Schoolfield model show that approximate linearity of increase in development rate with temperature corresponds to probabilities of less than one for the control enzyme to be active, throughout. The second step concerns the relation between linearity of development rate and physiological time (van Straalen 1983, Box 1). If we take the two steps together, physiological time is kept constant if an organism performs at less than 100% enzyme activity at all temperatures, and rates are never constrained by the Eyring equation. If so, organisms adapt to temperature but neutralize its effect by manipulating enzyme inactivation. At 100% enzyme activity, temperature sensitivities of reaction rates rule organismal properties. With enzyme inactivation, the organismal properties can prevail over environmental temperature influences.

Rates and Size: from Sharpe-Schoolfield Model to van der Have Model

The Sharpe-Schoolfield model is usually applied to development rate. In the model description, the temperature dependent rate $r(T)$ is specified by the reference rate ρ. The units used in the definition of r determine what the rate is about—the units of the rate $r(T)$ are those of the reference rate ρ. For development rate, ρ has the unit per time (per hour in Table 2, per day in van der Have and de Jong 1996). The Sharpe-Schoolfield model can be immediately applied to growth in biomass, if ρ is defined as biomass per time, or to growth in length, if ρ is defined as length per time. The temperature sensitivity coefficient H_A, temperature inactivation coefficients H_H and H_L, and boundary temperatures T_H and T_L are numerically different between growth rate and development rate, but not in units used (Joule per mol, and degree Kelvin). The parameters of growth rate and development rate are indicated by the index G for growth rate and index D for development rate. The parameters for growth rate and the parameters for development rate are assumed to be fully independent. That is, the biological processes for development rate and biomass increase are supposedly different.

The discussion of the properties of the model will take this independence as its starting point, but estimation of parameters introduces inevitably a link. Growth rate can only be estimated from biomass accrued over time, and the time itself is therefore implicated in growth rate.

Van der Have and de Jong (1996) posited that adult size could be found as the ratio of growth rate and development rate. Dividing a temperature dependent growth rate (units mass/time) by a temperature dependent development rate (unit 1/time) yields a temperature dependent adult body size:

$$m(T) = \frac{\rho_G}{\rho_D}\frac{P_G}{P_D}\exp\left[\frac{H_{AG} - H_{AD}}{R}\left(\frac{1}{T_{ref}} - \frac{1}{T}\right)\right] \qquad \text{(Eq. 6)}$$

The temperature dependence of body size depends upon two features: upon the relation between the temperature sensitivity coefficients H_{AG} and H_{AD}, and upon the relation between the probabilities for the enzymes for growth and development to be active. The relative temperature sensitivities of the rates if the control enzymes are active and the probabilities for the control enzymes to be active both influence body size. At high probability of the control enzymes to be active or at equal probability of the control enzymes to be active, body size depends upon the difference in the temperature coefficients H_{AG} and H_{AD}. If P_G and P_D both equal one, the temperature dependence of body size fully depends upon the difference in the temperature sensitivity coefficients H_{AG} and H_{AD}. Lower probabilities for the reaction-limiting enzyme to be active might lead to complications, if these probabilities P_G and P_D differ. If growth rate is more temperature sensitive than development rate, the difference $H_{AG} - H_{AD}$ is positive, and body size increases with temperature. If development rate is more temperature sensitive than growth rate, the difference $H_{AG} - H_{AD}$ is negative, and body size decreases with temperature. Constancy of body size over temperature implies that the temperature sensitivity coefficients H_{AG} and H_{AD} are equal, while the probabilities for the enzyme to be active P_G and P_D are equal too.

Equality of the corresponding parameters in growth rate and development rate might mean one of two things. Development and growth are either biologically identical, or growth is strongly temperature compensated. The first possibility would imply that the increase in biomass and the developmental program of the animal are identical: that development has no features other than biomass increase. This cannot be so, as development entails differentiation next to biomass increase. The second possibility is very interesting. Temperature compensation implies that biomass is selected to remain constant over temperature: the temperature sensitivity of growth rate would be selected to compensate the temperature sensitivity of development rate, leading to a very similar organism at all temperatures. In the data of Gibert and de Jong (2001), the differences between H_{AG} and H_{AD} for thorax size are very small (compare Table 2 with Table 3), and thorax size for the four *Drosophila* species is nearly temperature-compensated.

A decrease of body size with temperature might also result from a decrease in the ratio P_G/P_D (Equation 6). Depending upon the parameter values, this might prevail even if the difference $H_{AG} - H_{AD}$ is positive. In the data of Gibert and de Jong (2001), the difference $H_{AG} - H_{AD}$ is always negative for wing size, and negative for thorax size in male *D. subobscura* and *D. willistoni* (Fig. 9A), but this difference is positive for *D. funebris* (Fig. 9B) and *D. ananassae* (Fig. 9C). The lines marked 'Eyring' in Figs. 9A, B, C indicate how thorax size (or wing size) would change with temperature if the ratio of the probabilities P_G/P_D equaled 1. Actually, all these ratios differ from 1 at higher temperatures (at lower temperature probabilities could not be estimated: this is a six-parameter version of the model). The decreases in the ratios P_G/P_D are shown in Fig. 9D. The actual thorax size decreases with temperature, but this is completely the consequence of the decrease of the ratios P_G/P_D for these characters and the three species. The lines marked 'Sharpe-Schoolfield' indicate actual thorax size (or wing size), and these lines mirror the probability ratios. In fact, in *D. funebris* (Fig. 9B) and *D. ananassae* (Fig. 9C), the decrease in the ratio overrides the increase in thorax size due to the positive value of $H_{AG} - H_{AD}$. Maximum thorax size at intermediate temperature might well indicate approximate temperature compensation in $H_{AG} - H_{AD}$, and a slight difference in the ratio of P_G and P_D towards higher temperature. The decrease of wing size with temperature is a consequence both of negative $H_{AG} - H_{AD}$, and of a decrease in the ratio P_G/P_D (Fig. 9A, B, C). A decrease in P_G/P_D implies that $P_G < P_D$. This works out to be the case in all three species, both for the probabilities for the enzyme to be active estimated for thorax size and for wing the probabilities for the enzyme to be active estimated for wing size.

The actual body size—wing size, thorax size, weight—seems as much or more dependent upon the ratio of the probabilities for the responsible control enzymes to be active than upon the difference between the temperature sensitivity coefficients without enzyme inactivation. Size follows P_G/P_D as much as $H_{AG} - H_{AD}$, in the few data that are available. In Fig. 10, the body weight is given according to the Sharpe-Schoolfield model

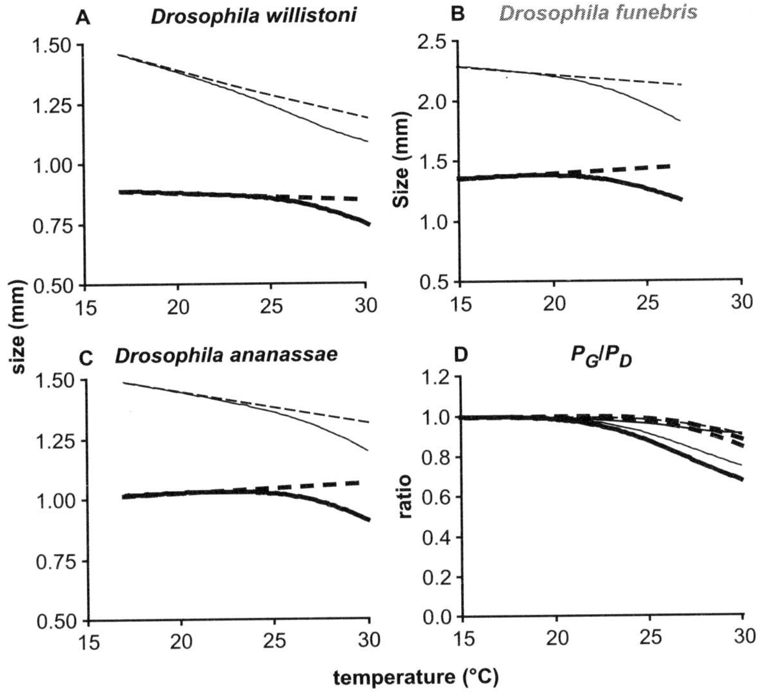

Fig. 9 Body size according to the model of van der Have and de Jong (1996) is given for *D. willistoni, D. funebris* and *D. ananassae*. Comparison of the Sharpe-Schoolfield (continuous lines) with the Eyring model (broken lines) shows the influence of the ratio of probabilities, P_G/P_D, on body size, as compared with only the difference $H_{AD} - H_{AG}$ for the Eyring model. The influence of the various parameters differs between species, and between thorax length (thick lines) and wing length (thin lines). All parameter estimates are from Gibert & de Jong (2001), where the observed data for wing length and thorax length are shown. Estimates of parameters for development rate are given in Table 2, for growth rate in Table 3. A. *D. willistoni* wing length and thorax length, in mm. B. *D. ananassae* wing length and thorax length, in mm. C. *D. funebris* wing length and thorax length, in mm. D. Ratio P_G/P_D of the probabilities for the rate determining enzymes for growth and development to be active. In *D. funebris*, the influence of this ratio is high compared with *D. willistoni* and *D. ananassae*.

including both P_G/P_D and $H_{AG} - H_{AD}$, and according to the Eyring model just involving $H_{AG} - H_{AD}$, for the Houten and Adana population of *D. melanogaster* and *D. simulans*. As all reaction norms for development rate and growth rate show quite a linear increase over the middle temperature range, indicating that the probabilities P_G and P_D do not possess a plateau at 100%, this prevalence of P_G/P_D over $H_{AG} - H_{AD}$ is consistent. It is however fairly surprising in its effect. The estimates for the *D. simulans* population from Houten actually indicate a strong conflict between P_G/P_D and $H_{AG} - H_{AD}$

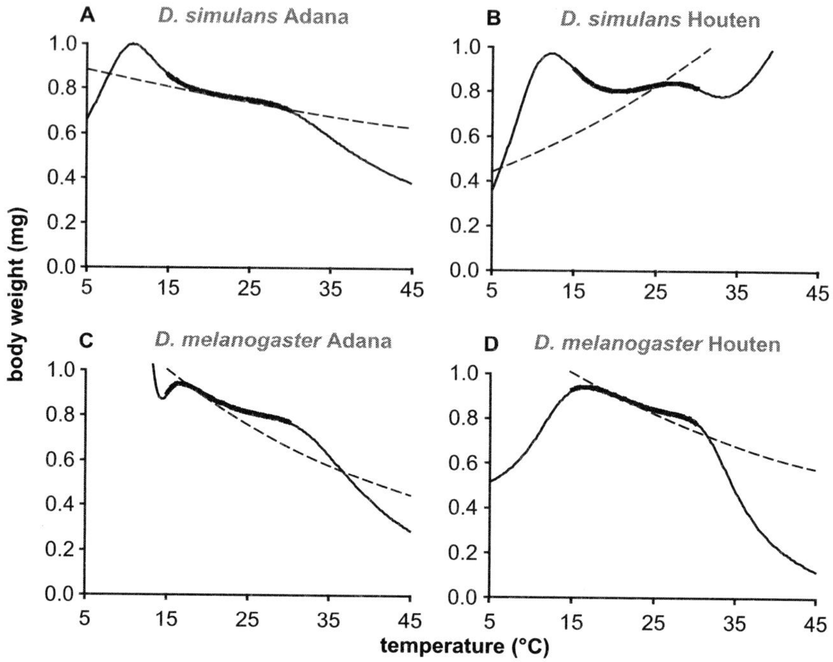

Fig. 10 The model of van der Have-de Jong (1996) is fitted to body size of *Drosophila melanogaster* and *D. simulans*, for the Adana (Turkey) and Houten (The Netherlands) populations. The van der Have-de Jong (1996) model (continuous line) closely follows the data (body weight in mg). Comparison with the Eyring model (broken lines) shows the influence of the ratio of probabilities, P_G/P_D, on body size, as compared with only the difference $H_{AD} - H_{AG}$ for the Eyring model. The influence of the various parameters differs between the species and the populations. Over the viable range 15°C to 30°C (thick lines) the influence of the ratio P_G/P_D is less than outside this range. All parameter estimates are from Jeroen Bohré. Estimates of parameters for development rate are given in Table 2, for growth rate in Table 3. Compare Fig. 8 that gives rates and probabilities for the same estimated parameter values for these populations. A. *D. simulans* Adana. B. *D. simulans* Houten. C. *D. melanogaster* Adana. D. *D. melanogaster* Houten.

over increase or decrease in size. A decrease in size is not one phenomenon. Rather, the causes of phenotypic plasticity, in terms of the model's parameters, differ with the trait and the temperature range (Angilletta et al. 2003).

This raises the question as to what actually might be under selection—at least, if we trust the biological reality of the model. If body size itself is under selection, any body size could be reached by any number of parameter combinations. If the parameter values were under selection, body size would follow. This might lead to interesting biological phenomena.

Biological Patterns

We have seen that the Sharpe-Schoolfield model fits data for development rate, and can be used to fit data on development rate and body size. Van der Have and de Jong (1996) and Gibert and de Jong (2001) estimated parameters, and additional estimates are found in Tables 2 and 3. The model fit works well, but would need additional data, especially data from replicated experiments to determine the experimental error rate.

We will proceed on the assumption that the Sharpe-Schoolfield model for rates and the van der Have-de Jong model for size perform well, in order to examine the possible properties of the model. We know that the upper temperature limit to viability is near the down-curve in development rate at high temperature. In the model this down-curve is caused by enzyme inactivation. Any potential link between the viability boundaries and the inactivation boundaries needs therefore further examination.

Another point is whether the variation in model parameters can reflect genetic variation in body size within a population. Moreover, we would like to see patterns of body sizes that can conceivably represent a geographic cline in body size within a species, or patterns of body sizes for species with different temperature ranges. In all these cases we only attempt to show the possibilities of the model.

Viability Boundaries

Temperature-viability reaction norms of ectotherms, including insects, grown at constant temperatures generally have an inverted u-shape (Figs. 3 & 11). The thermal limits of development resemble sharply defined thresholds at high and low temperatures and are symmetrical around the median temperature of viability. The permissive temperature range of embryonic development is usually much narrower compared to the tolerance range of adult physiology like respiration, metabolism in general, or derived performance parameters like running speed or flight speed [fish, Brett (1970); anura, van der Have (2002); *Drosophila*, David et al. (1983)]. It should be noted that the upper thermal limit of development is also much lower than temperatures at which proteins denature irreversibly. To date, few attempts have been made to explain the threshold character or shape of

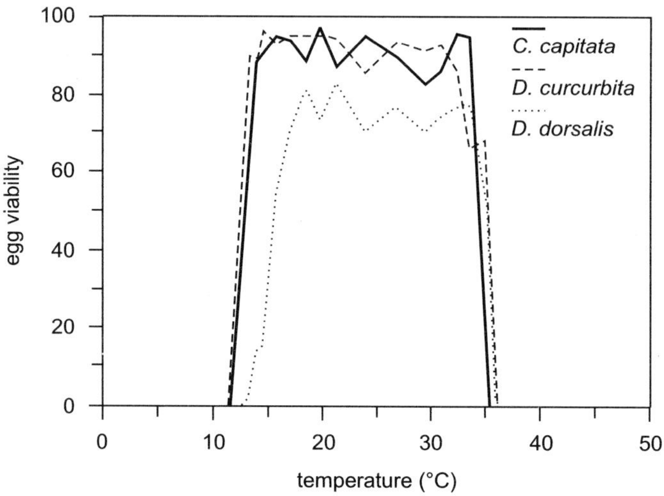

Fig. 11 Temperature viability curves of *Dacus dorsalis, D. cucurbita* and *Ceratitis capitata* illustrate the general pattern of viability boundaries in developing ectotherms. Data are taken from Messenger and Flitters (1958).

the viability boundaries and the difference between adult and embryonic performance.

One obvious difference in ectotherms between the adult stage and the embryonic and larval stages in ectotherms is the relative intensity of cell division and differentiation. During development most cells are actively dividing, while in the adult stages cell division occurs mainly in regenerating processes and reproductive tissue not directly linked to performance of the whole organism. This suggests that temperature-induced conformational changes of proteins involved in cell cycle regulation may block cell division and by implication determine the thermal limits of development.

The development of a multicellular organism from zygote to the adult stage proceeds through a series of cell divisions. Cell growth and differentiation are closely co-ordinated with cell division during the larval stage, but are dissociated during embryogenesis. Overall, development can be considered as the interaction between differentiation and growth. Development rate (time^{-1}) is assumed to be primarily determined by the cell division rate (van der Have and de Jong 1996).

In this section, a proximate model is presented which shows that temperature inactivation of cell cycle proteins may interact with their regulation and subsequently can predict the temperature tolerance limits of ectothermic development (van der Have 2002). The analysis suggests that reversible temperature inactivation at high and low temperatures has a symmetrical, inhibiting effect on the balance between synthesis and degradation of cell cycle proteins, resulting in sharp thresholds at the high and low temperatures, above and below which the cell cycle becomes arrested and development blocked. Observed viability boundaries will be compared to thermal limits predicted by the model and derived from differentiation rate-temperature reaction norms in fourteen insect species taken from literature data.

Reversible inactivation of cell cycle proteins During division, each cell proceeds through a sequence of well-defined stages, together known as the cell cycle. The cell cycle is regulated by the interactions of the subunits Cdc2 and cyclin of the heterodimer MPF (Maturation Promoting Factor) and various cell cycle enzymes (Murray 1992, 1994, Tyson 1991). One of the main goals of this tight regulation is to ensure that the DNA is duplicated exactly once and only once. Several detailed mathematical models of the cell cycle have been developed (Goldbeter 1991, Novak and Tyson 1993a, Novak and Tyson 1993b, Norel and Agur 1991). These quantitative models can quite precisely explain the oscillator phenomena in early embryos and switch mechanisms in growth-controlled cell cycles.

Reversible inactivation of cell cycle enzymes will slow cell division down at low temperatures as well as decrease it at high temperatures. Furthermore, when all enzymes involved will be reversibly inactivated, a gradual response of the whole system can be expected, not the switch-like behaviour of the developmental tolerance limits we are pursuing to explain.

A simple model of derepression as a control mechanism for the cell cycle in eukaryotes was developed by Tyson and Sachsenmaier (1979). They showed how a genetic control system can account for the periodic synthesis of a mitotic activator by sequential dosage changes of an early-replicated repressor and a late-replicated inducer. These dosage changes result in periodic switching of the operon from the derepressed to the repressed state and the activator synthesis respectively off and on at the beginning and end of *S*. The Tyson and Sachsenmaier model is relatively simple and involves both protein-DNA (repressor-operon) and protein-protein (repressor-inducer) binding. It therefore fulfils the above stated prerequisite to serve as a starting point for the analysis of the effects of temperature inactivation on proteins regulating the cell cycle.

The following simplifying assumptions have been made for the thermodynamics of the regulation of the cell cycle. (1) The inducer and repressor are proteins, which are assumed to occur in three energy states: active or reversibly inactive at high or low temperature (Sharpe and DeMichele 1977). At high and low temperatures these proteins undergo a conformational transition which renders them inactive with respect to binding properties (Somero 2002). (2) The repressor is more thermolabile than the inducer (Polyak et al. 1994), so that the temperature inactivation of the repressor can be ignored over the temperature range at which inactivation of the inducer occurs. (3) The reversible inactivation of the inducer follows Equation 8. (4) The temperature dependencies of all transition rates and equilibrium constants are assumed to be similar, i.e., the rate of inducer degradation is temperature-independent.

Under these assumptions it can be shown (van der Have 2002) that DNA-replication, and, as a result, cell division will become (reversibly) blocked at the temperature at which the inducer is only half active, while the potential range of biological activity is much wider (Fig. 12). Temperature inactivation of the inducer, therefore, mimics the decrease in inducer concentration resulting from gene dosage changes during the cell cycle. Temperature inactivation of the inducer brings the connection between this model for cell division and the Sharpe-Schoolfield model.

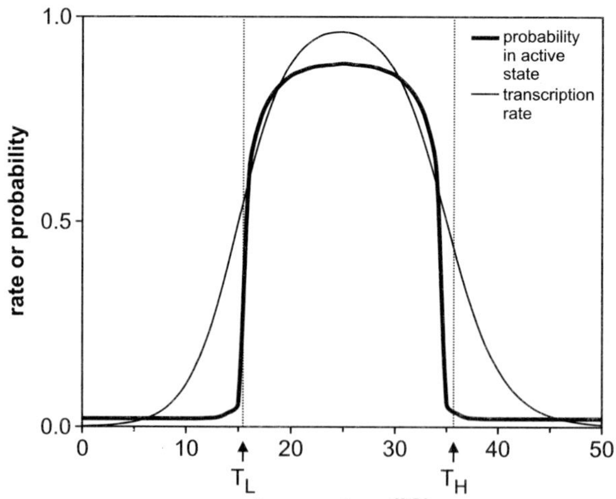

Fig. 12 The probability that the inducer is in active state (P_a) (thick continuous line) and the transcription rate of the genetic operon (thin continuous line) at different temperatures. A steep decrease from near maximum transcription rate to zero transcription rate occurs at temperatures when the inducer (rate-limiting developmental protein) is approximately half active and half inactive (broken line). Consequently, cell division becomes blocked and development is arrested at the upper (T_H) and lower (T_L) thermal limit.

Testing the theory: comparison of predictions with observed tolerance limits The theory that thermal limits of development are determined by reversible inactivation of cell cycle regulatory proteins can be tested as follows. The temperatures, T_L and T_H, at which 50% of all molecules of a rate-limiting protein or enzyme are reversibly inactivated, can be estimated from temperature-development reaction norms for development rate (see Table 2 for *Drosophila* estimates).

Published datasets of development rates and viability of 14 insect species were used to estimate the thermodynamic parameters from the development rate reaction norm. The estimated values of the 50% inactivation temperatures T_L and T_H from development rate are independent of viability. These temperatures were compared with observed viability curves. An important condition was that the experimental temperatures should cover the full range of viable development (the whole thermal window) for both development rate (embryonic and/or larval) and viability (embryonic or egg-to-adult). The datasets which fulfilled these conditions included eight species of *Drosophila* (Cohet et al. 1980; Gibert and de Jong, 2001), three species of *Dacus* fruitflies (Messenger and Flitters 1958), the southern pine beetle *Dendroctonus frontalis* (Wagner et al. 1984), and two Homoptera [aphids] *Myzus persicae* and *Lipaphis erysimi* (Liu and Meng 1989).

When the estimated temperatures at which the developmental enzyme has equal probability to be active or inactive at low and high temperatures (T_L and T_H) were compared with the observed thermal limits, 23 out of 25 comparisons (92%) fell closely together (Fig. 13). The correspondence at high temperatures is remarkably close in all species. The observed lower tolerance limits (T_L) in *Dacus dorsalis* and *Lipaphis erysimi* do not agree with the observed lower thermal limits, but it should be noted that in these species the estimates for H_A were also outliers. Loss of viability seems therefore the result of too much temperature inactivation of crucial enzymes in cell division—the same enzymes in cell division that are responsible for the temperature dependent development rate.

Phenotypic Plasticity

The van der Have & de Jong model, i.e., Sharpe-Schoolfield model as applied to body size, is eminently suitable to describe phenotypic plasticity of organismal size, if the temperature of larval development determines organismal size. A full model for body size requires estimation of 12 parameters, 6 for development rate and 6 for growth rate. Development rate

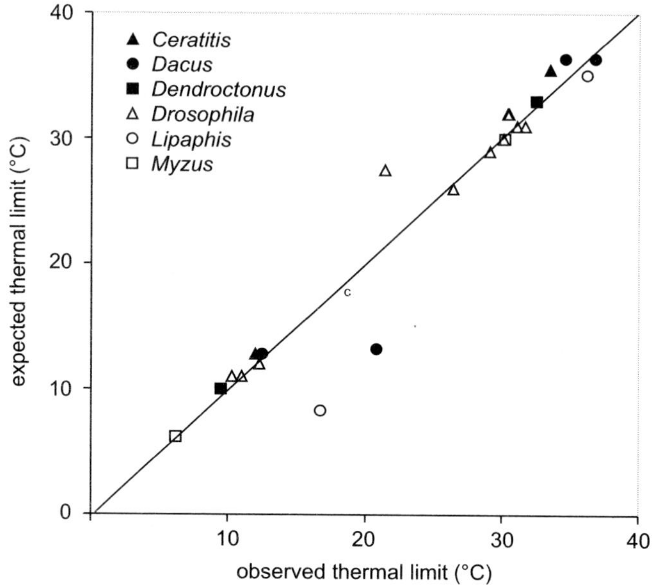

Fig. 13 Observed thermal limits of viability are compared with expected thermal limits of development, T_L and T_H, in fourteen insect species (*Ceratitis*, *Dacus*, *Dendroctonus*, *Drosophila*, *Lipaphis* and *Myzus*) estimated with the Sharpe-Schoolfield model. The equality line (y=x) is drawn.

and growth rate have a roughly triangular shape unless parameter values are very deviant—unless the probabilities for the control enzyme to be active are very low due to very low absolute values of H_H and H_L. Within the range of parameter values that lead to the classical triangular shape of the reaction norms for growth rate and development rate, many patterns of phenotypic plasticity in body size are possible. The patterns in phenotypic plasticity mostly originate from the ratio of the probabilities for the control enzymes of growth and development to be active; only if this ratio P_G/P_D equals 1 (that includes the case that both probabilities equal 1) does the difference between the temperature sensitivity coefficients of the control enzymes, $H_{AG} - H_{AD}$, exert a major influence.

In Fig. 14, body sizes are given where the ratio P_G/P_D is changed, by changing either H_{HG} or H_{LG}, or H_{HD} or H_{LD}, while the temperature sensitivity coefficient of growth H_{AG} is kept equal to the temperature sensitivity coefficient of development H_{AD}. This gives a survey of plasticity types for change in one temperature sensitivity parameter superimposed upon a constant body size. The effect of an underlying constant or decreasing body size is given in Fig. 15, where moreover both the high temperature sensitivity

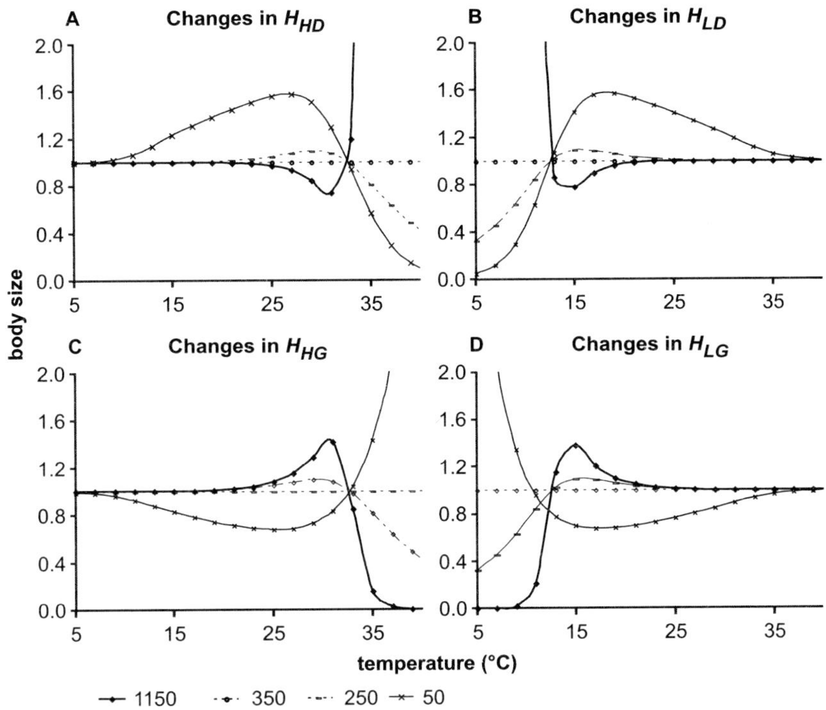

Fig. 14 The influence of the ratio P_G/P_D on body size if either H_{HD}, H_{LD}, H_{HG}, or H_{LG} are changed. In all cases $H_{AD} = H_{AG} = 70$ kJ mol^{-1} and $\rho_D = 0.4$ t^{-1}, $\rho_G = 0.4$ mg t^{-1} : as a consequence, no variation in body size with temperature is expected. The expected body size equals 1 over all temperatures. In all cases $T_{ref} = 295.7$°K = 22.5°C, $T_H = 305.7$°K = 32.5°C, and $T_L = 285.7$°K = 12.5°C. H_{HD}, H_{LD}, H_{HG}, and H_{LG} equal 350 kJ mol^{-1} if H_{HD} or H_{LD} are varied. H_{HD}, H_{LD}, H_{HG}, and H_{LG} equal 250 kJ mol^{-1} if H_{HG} or HL_G are varied between 50 and 1150 kJ mol^{-1}. A. Changes in H_{HD}. B. Changes in H_{LD}. C. Changes in H_{HG}. D. Changes in H_{LG}.

parameter and the low temperature sensitivity parameter are changed at the same time.

Body size follows the probability ratio P_G/P_D as much as it follows the differences of the temperature sensitivity coefficients $H_{AG} - H_{AD}$. Any increase or decrease of body size with temperature can have very different combinations of underlying parameters. This means that the possibilities for alternative outcomes in parameter values are very large if body size itself would be under selection. But if the parameters values themselves are under selection—or rather, the biology they might stand for—body size would in many respects be an epiphenomenon to physiological functioning at different temperatures.

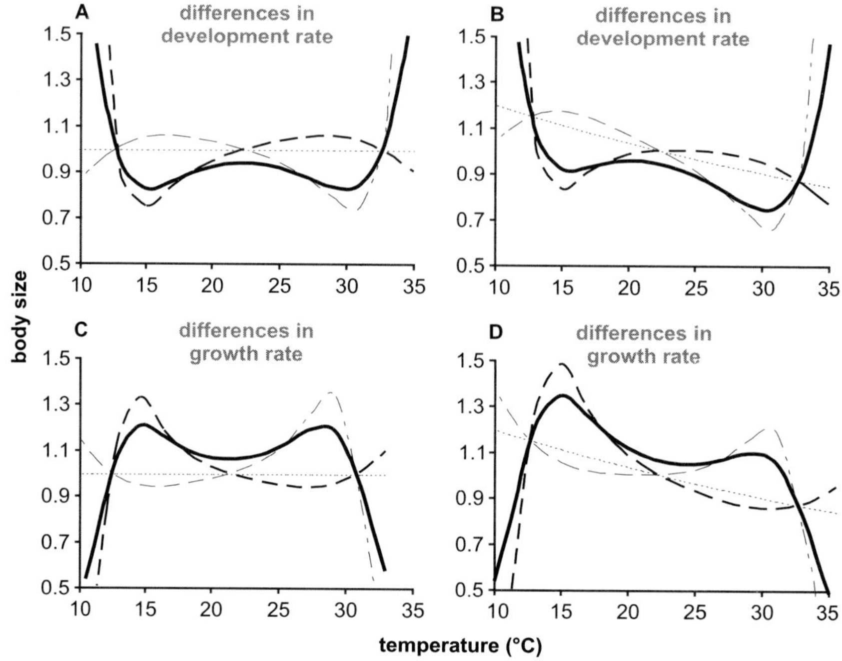

Fig. 15 The influence of the ratio P_G/P_D on body size if H_{HD}, H_{LD}, H_{HG}, or H_{LG} are changed and compared with the Eyring model (dotted lines). For basic body weight temperature invariant, $H_{AD} = H_{AG} = 70$ kJ mol^{-1} and $\rho_D = 0.4$ t^{-1}, $\rho_G = 0.4$ mg t^{-1}. If body weight basically decreases with temperature, $H_{AD} = 80$ kJ mol^{-1} and $H_{AG} = 70$ kJ mol^{-1}. In all cases T_{ref} = 295.7°K = 22.5°C, T_H = 305.7°K = 32.5°C, and T_L = 285.7°K = 12.5°C. Three combinations are presented: H_{HD} or H_{HG} = 200 kJ mol^{-1} and H_{LD} or H_{LG} = 800 kJ mol^{-1} (broken lines); H_{HD} or H_{HG} = 500 kJ mol^{-1} and H_{LD} or H_{LG} = 500 kJ mol^{-1} (continuous lines); H_{HD} or H_{HG} = 800 kJ mol^{-1} and H_{LD} or H_{LG} = 200 kJ mol^{-1} (broken and dotted lines). **A.** $H_{AD} = H_{AG}$, temperature invariant body size, H_{LD} and H_{HD} changed. **B.** $H_{AD} > H_{AG}$, body size decreases with temperature, H_{LD} and H_{HD} changed. **C.** $H_{AD} = H_{AG}$, temperature invariant body size, H_{LG} and H_{HG} changed. **D.** $H_{AD} > H_{AG}$, body size decreases with temperature, H_{LG} and H_{HG} changed.

Genetic Variation

Genetic variation in the parameter values is necessary for selection to have any evolutionary effect. A first indication of the effect of genetic variation in the reference rate ρ, the temperature sensitivity coefficient H_A, and the temperature inactivation coefficients H_H and H_L on (development) rate is given in Fig. 15. Genetic variation in the inactivation temperatures T_H and T_L would lead to a lateral translation of the curves. From genetic variation in parameter values we can infer genetic variation in growth rate, development rate and body size.

A major question is how selection would act. This question has two components. The first question is on what organismal property selection acts: on body size itself, on development rate, on growth rate or on enzymatic properties as temperature sensitivity or inactivation parameters. Often, we cannot ascertain what selection is actually acting on-body size itself, or development rate, or temperature sensitivity of enzymes. This is the question of the nature of the relevant selection pressure. But when thinking in the Sharpe-Schoolfield and van der Have-de Jong modeling approach, selection on body size translates into selection on parameter values, and selection on parameter values translates into selection on body size. The second question is how selection pressure is translated in a selection response. In general the selection response of any quantitative trait depends upon direct selection mediated by the trait's genetic variance and indirect selection mediated by the genetic covariances with all other traits (Box 3).

Higher genetic variance or covariance implies faster change in mean phenotypic trait value for identical selection pressures. In the modeling approach, this question translates into the question after the appropriate genetic variances and covariances. What the appropriate genetic variances and covariances are depends partly upon our interest. On the one hand, we might be interested in simultaneous selection at all or a number of temperatures. On the other hand, we might be interested selection at one temperature on body size or some of the model parameters. In all these cases we need to know what the genetic variances and covariances are, specific for each case. In Box 3, selection is explained, but the selection equations are far more general than the model of body size we are dealing here with and are not necessary for understanding the properties of the biophysical model of insect size.

We will consider how genetic variance in body size is caused by genetic variation in parameter values. Genetic variances in development rate, growth rate and body size will be temperature dependent if genetic variation in the parameters of the Sharpe-Schoolfield model exists in natural populations. Additive genetic variation in parameter values will not only lead to additive genetic variation in development rate, growth rate and body size, but will generate non-additive genetic variation at each temperature too. Both the temperature dependence of the genetic variance and the appearance of non-additive genetic variance are a consequence of the non-linear transformations between parameter variation and rate and body size variation.

The temperature dependence of the genetic variance and its subdivision in additive, and non-additive genetic variance components can be studied

Box 3 Selection on Body Size, Selection on Parameters

The selection response of a quantitative trait is given by

$$\Delta \bar{z}_i = \frac{1}{\bar{w}} \left(\sigma_i^2 \beta_i + \sum_{j \neq i} \sigma_{ij} \beta_j \right) \qquad \text{(Eq. 7)}$$

where $\Delta \bar{\mathbf{z}}_i$ stands for the change in mean phenotype of trait i, σ_i^2 for the additive genetic variance in trait i, σ_{ij} for the additive genetic covariance of traits i and j, and $\beta_i = \frac{\partial \bar{w}}{\partial \bar{z}_i}$ for the selection gradient of mean fitness $\bar{w}$ towards mean phenotypic value $\bar{z}_i$ of trait i (Lynch and Walsh 1997).

Simultaneous selection on n quantitative traits can be described by

$$\Delta \bar{\mathbf{z}} = \frac{1}{\bar{w}} \mathbf{G_A} \boldsymbol{\beta} \qquad \text{(Eq. 8)}$$

The column vector $\Delta \bar{z}$ contains the changes $\Delta \bar{z}_i$ in mean phenotype of the traits, $\mathbf{G_A}$ is the additive genetic variance-covariance matrix with the additive genetic variances σ_i^2 in the traits on the diagonal and the other elements, the additive genetic covariances σ_{ij} of traits i and j, and the column vector $\boldsymbol{\beta}$ has as elements the selection gradients $\frac{\partial \bar{w}}{\partial \bar{z}_i}$ of mean fitness $\bar{w}$ towards mean phenotypic value $\bar{z}_i$ of trait i (Lynch and Walsh 1997). Note that the direct selection response in a trait from its additive genetic variance and its direct selection gradient need not have the same sign as the indirect selection response due to the genetic covariances between traits and the selection gradients on all other traits.

Equation 8 is the basic equation. Selection at n different temperatures has the same form as selection on n different traits; the only complication is that mean fitness over all temperatures is now weighted by how often each temperature occurs. The genetic variance-covariance matrix contains the genetic variance at each temperature, and the genetic covariance between temperatures. Here, we only show how genetic variance in body size depends upon the temperature. In de Jong and Imasheva (2000), a genetic covariance between temperatures is shown too.

The selection response in body size m at any constant temperature T can be predicted by

$$\Delta \bar{m} = \frac{1}{\bar{w}} \sigma^2 \beta \qquad \text{(Eq. 9)}$$

Here σ^2 is additive genetic variance in body size, and $\beta = \frac{\partial \bar{w}}{\partial \bar{m}}$ is the selection gradient of mean fitness $\bar{w}$ towards mean body size. The selection response in body size can be expressed in the selection responses for the values of all 12 parameters. Call the parameters d_1 to d_{12}. Define three column vectors each

Contd.

Contd.

with 12 elements. The column vector $\overline{\boldsymbol{m}}$ has as elements the change in mean body size with mean parameter value, $\frac{\partial \overline{m}}{\partial \overline{d}_i}$; the column vector $\Delta \overline{d}$ contains the selection responses in mean parameter $\Delta \overline{d}_i$; and the column vector $\boldsymbol{\beta}_{wd}$ has as elements the selection gradients towards the mean parameter values, $\frac{\partial \overline{w}}{\partial \overline{d}_i}$. The additive genetic variance-covariance matrix $\mathbf{G_A}$ is here the matrix of additive genetic variances and covariances in the parameter values, a 12 × 12 matrix.

The predicted selection response in the mean parameter values is

$$\overline{\mathbf{d}} = \frac{1}{\overline{w}} \mathbf{G_A} \boldsymbol{\beta}_{wd'} = \frac{1}{\overline{w}} \mathbf{G_A} \overline{\boldsymbol{m}} \beta \qquad \text{(Eq. 10)}$$

with $\beta = \frac{\partial \overline{w}}{\partial \overline{m}}$ as before. This tells us how a selection gradient β on body size is distributed over the various parameters, and how this selection on parameter values themselves leads to a selection response in the parameters. On the other hand, the selection response in body size depends on the selection responses in the parameter values, as

$$\Delta \overline{m} = \overline{\boldsymbol{m}}^t \, \overline{\boldsymbol{d}} \qquad \text{(Eq. 11)}$$

where t stands for transpose (i.e. a row vector). Taking Equation 10 and Equation 11 together, we see how change in mean body size derives from selection on the parameter values:

$$\Delta \overline{m} = \overline{\boldsymbol{m}}^t \, \overline{\boldsymbol{d}} = \frac{1}{\overline{w}} \overline{\boldsymbol{m}}^t \, \mathbf{G_A} \boldsymbol{\beta}_{w,d} \qquad \text{(Eq. 12)}$$

In Equation 12, the genetic variance-covariance matrix $\mathbf{G_A}$ between the parameter values does not depend upon temperature, but all other quantities do.

using a two-locus model. In a two-locus model, additive effects, dominance effects per locus and interaction effects between loci can be estimated if two homozygote lines would be present, and a set crosses involving these lines made. The effects can be estimated from the homozygote line phenotypic values P1 and P2, the mean phenotypic values of the first filial cross F1 and second filial cross F2, and of the backcrosses B1, F1*P1, and B2, F1*P2. The method involves standard quantitative genetics statistical techniques, and is explained in its application to the Sharpe-Schoolfield model by de Jong and Imasheva (2000).

De Jong and Imasheva (2000) showed how the genetic variance in development rate and body size over temperature resulted from genetic variation in the inactivation parameters T_L and H_L, and T_H and H_H of development rate. The same method is applied here to make clear possible effects of genetic variation in the parameters of growth rate. In each case, a two-locus model is employed. The parameter values chosen are variations on the values found by van der Have and de Jong (1996) for *Drosophila melanogaster*, but otherwise the values are arbitrarily chosen to demonstrate the ranges of possible behaviors of the model.

In Fig. 16, the effect of additive genetic variation in temperature sensitivity parameters ρ_G and H_{AG} is shown, as well as the effect of genetic variation in both parameters (Figs. 16A, C, and E, respectively). The resulting graphs show that the genotypes differ in size and the additive genetic variance to be temperature dependent (Fig. 16B,D,F). Non-additive genetic variance is absent if variation is only in the reference rate ρ_G (Fig. 16B) but appears with genetic variation in temperature sensitivity H_{AG} (Figs. 16D, F). Differences in H_{AG} lead to higher genotype by environment interaction than differences in ρ_G: compare Fig. 16C with Fig. 16D. Combination of genetic variation in both parameters leads to a fairly naturally looking set of genotypic values for body size (Fig. 16E), with much higher additive genetic variance than results from variation in just one parameter (Fig. 16F).

Genetic variation in the temperature inactivation parameters T_{LG} and H_{LG}, and T_{HG} and H_{HG} leads to more changes in the shape of the reaction norm of body size, and a more pronounced presence of non-additive genetic variance. Here too natural looking patterns of genetic variation in body size can be modeled. Fig.17A gives the effect of additive genetic variation in the temperature boundaries T_{LG} and T_{HG}, Fig.17B shows the effect of additive genetic variation in the inactivation coefficient H_{LG} and H_{HG} and in Fig.17C variation in all four parameter values is depicted. Of course, genetic variation is mainly apparent at low and high temperatures at the edges of the viable range. The impression would be of increased genetic variation in unfavorable environments (Hoffmann and Merilä 1999). Variation in the inactivation parameters leads to more non-additive genetic variance than variation in the sensitivity parameters. This will be the consequence of the position of the inactivation parameters in the denominator of the Sharpe-Schoolfield equation for growth rate. The higher amount of non-additive genetic variance found by de Jong and Imasheva (2000) for genetic variation in the parameters of development rate rather than the parameters of growth rate must be due to the presence of the development rate in the denominator of body size in the model.

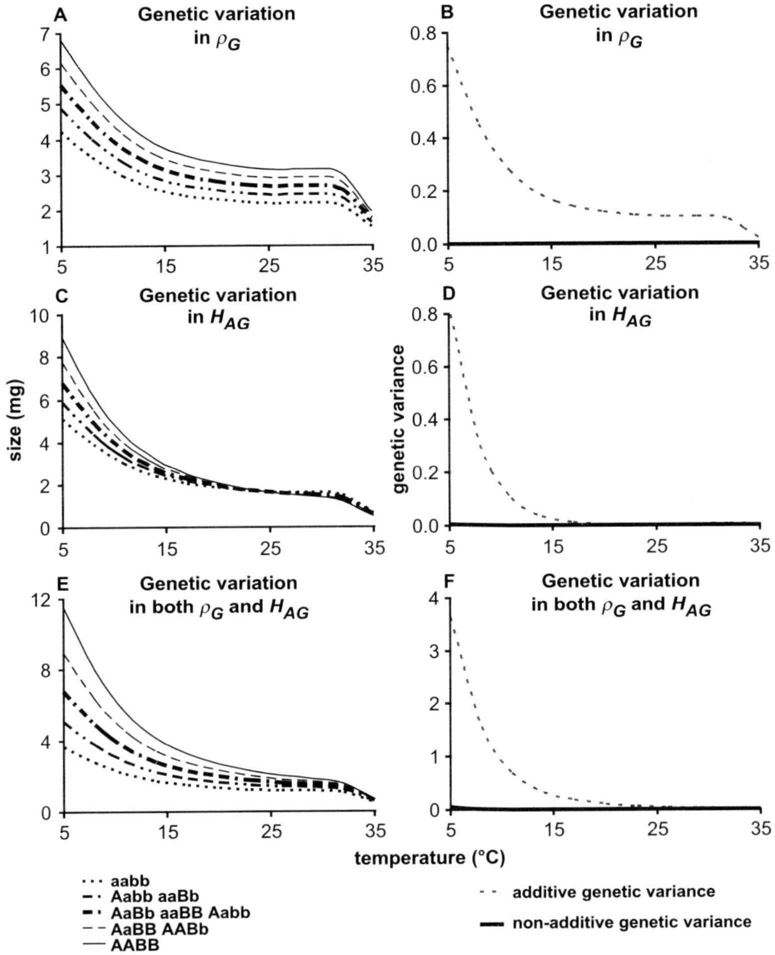

Fig. 16 The effect of genetic variation in reference rate r_G and temperature coefficient H_{AG} of growth rate: sizes and genetic variance in body size. Mostly additive genetic variance results from additive genetic variation in the reference rate and temperature coefficient. The parameter values are based upon the parameter estimates by van der Have and de Jong (1996), based upon *Drosophila melanogaster* data from David & Clavel (1967). See Table 2 for estimates of parameters for development rate and Table 3 for parameter for growth rate. Only mentioned parameter values change, all other parameter values are kept identical. A. Nine genotypes at two loci additively differing in ρ_G : ρ_G varies additively from 0.15 mgt^{-1} to 0.27 mgt^{-1}. B. Genetic variance resulting if the allele frequencies at the two loci are $p_1 = 0.6$ and $p_2 = 0.3$. C. Nine genotypes at two loci additively differing in H_{AG} : H_{AG} varies additively from 15 to 35 kJ mol^{-1}. D. Genetic variance resulting if the allele frequencies at the two loci are $p_1 = 0.6$ and $p_2 = 0.3$. E. Nine genotypes at two loci additively differing in ρ_G and H_{AG}: ρ_G varies additively from 0.15 to 0.27, H_{AG} varies from 15 to 35 kJ mol^{-1}. F. Genetic variance resulting if the allele frequencies at the two loci are $p_1 = 0.6$ and $p_2 = 0.3$.

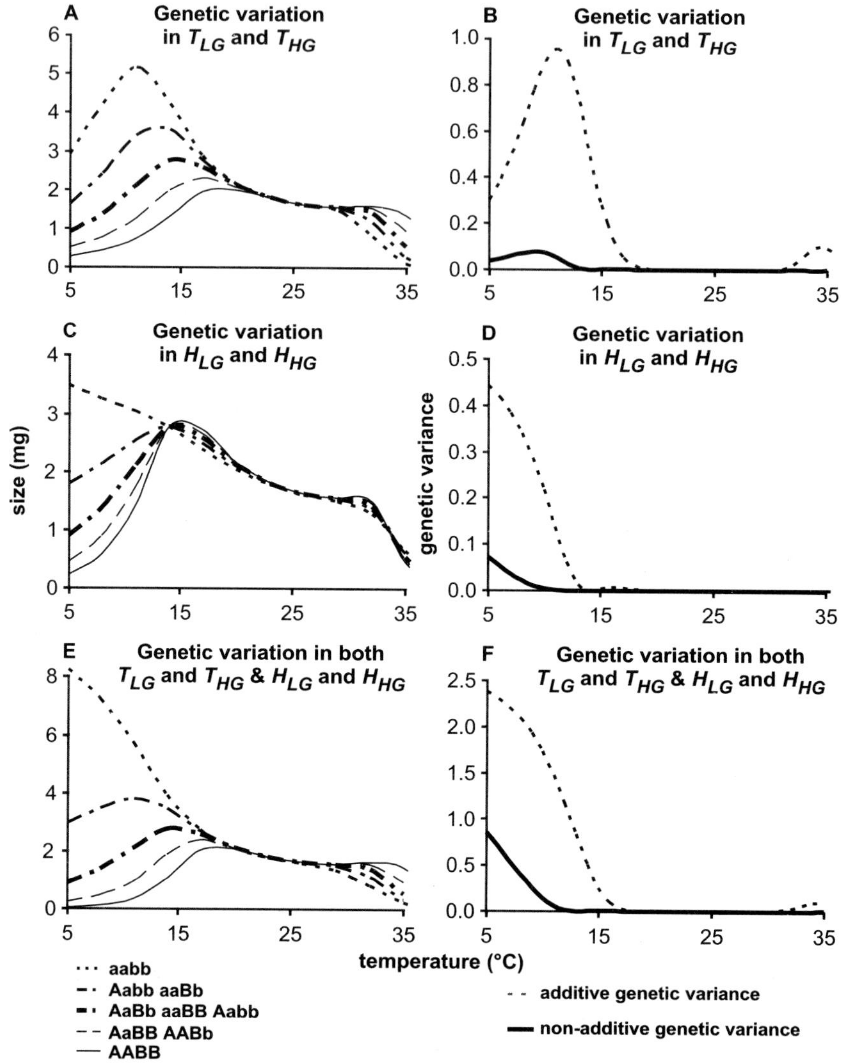

Fig. 17 The effect of genetic variation in temperature boundaries T_{LG} and T_{HG} and temperature sensitivities H_{LG} and H_{HG} of growth rate: sizes and genetic variance in body size. Additive genetic variance results from additive genetic variation in the temperature boundaries and temperature sensitivities, but much non-additive genetic variation is present. The parameter values are based upon the parameter estimates by van der Have and de Jong (1996), based upon *Drosophila melanogaster* data from David & Clavel (1967). See Table 2 for estimates of parameters for development rate and Table 3 for parameter for growth rate. Only mentioned parameter values change, all other parameter values are kept identical. **A.** Nine genotypes at two loci additively differing in T_{LG} and T_{HG}: T_{LG} and T_{HG} vary in concert additively from 285 °K /

Fig. 17 Contd. ...

Geographical Variation

Geographical variation between populations of the same species implies genetic differences in some of the parameters between the populations. A very interesting source of genetic variation in body size might be genetic variation in the parameters that control the probability for the enzyme to be active, as these parameters are candidates to represent direct adaptation to the environment in enzymatic properties. We might assume that a population that is adapted to a fairly cold environment needs a high probability for the enzyme determining rates to be active at fairly low temperature, but at the cost of higher enzyme inactivation at high temperature. In such a population, a larger absolute value of H_L would lead to a high probability of the enzyme to be active at low temperature. Relatively low H_H would lead to a certain measure of high temperature inactivation at moderate temperatures. A population adapted to high temperatures might show the opposite pattern in probability for the enzyme to be active—a fairly low absolute value for H_L and a higher H_H. The question is whether such patterns in the parameters are able to mimic actual patterns in body size.

In *Drosophila melanogaster*, body size in temperate populations is larger than in tropical populations (Noach et al. 1996, Zwaan et al. 2000). The development time shows little or no difference (James et al. 1995, James et al. 1997) but growth rate differs between temperate and tropical populations (De Moed et al. 1998) and increases under cold temperatures in experimental evolution experiments (Robinson and Partridge 2001, Bochdanovits and de Jong 2003).

In keeping with this, we chose an example in which the probabilities for the growth enzyme to be active differ between populations, but everything that has to do with development is the same for the different populations. Five possibilities potentially representing a temperature cline in enzyme parameters within one species are given. The difference between the

---

Fig. 17 Contd. ...

305 °K (aabb) to 289 °K / 309 °K (AABB); T_{ref} does not vary. B. Genetic variance resulting if the allele frequencies at the two loci are $p_1 = 0.6$ and $p_2 = 0.3$. C. Nine genotypes at two loci additively differing in H_{LG} and H_{HG}: H_{LG} and H_{HG} vary in concert additively from 900 kJ mol^{-1} / –500 kJ mol^{-1} (AABB) to 500 kJ mol^{-1} / –300 kJ mol^{-1} (aabb). D. Genetic variance resulting if the allele frequencies at the two loci are $p_1 = 0.6$ and $p_2 = 0.3$. E. Nine genotypes at two loci additively differing in T_{LG} and T_{HG} and H_{LG} and H_{HG}: T_{LG} and T_{HG} vary in concert additively from 285 °K / 305 °K (aabb) to 289 °K / 309 °K (AABB), H_{LG} and H_{HG} vary in concert additively from 900 kJ mol^{-1} / –500 kJ mol^{-1} (AABB) to 500 kJ mol^{-1} / –300 kJ mol^{-1} (aabb). F. Genetic variance resulting if the allele frequencies at the two loci are $p_1 = 0.6$ and $p_2 = 0.3$.

populations is only in H_{LG} and H_{HG}. Figure 18 shows the corresponding probabilities for the enzymes to be active (Fig. 18A), growth rates (Fig. 18B), enzyme probability pertaining to development and development rate (Fig. 18C) and body size (Fig. 18D) for the different populations.

The assumed pattern of genetic differences in the temperature inactivation parameters H_{LG} and H_{HG} of growth rate leads to different body sizes in the populations. A higher probability for the enzyme to be active at lower temperature leads to larger body size at lower temperature. The pattern of body size shows a satisfying resemblance to the actual pattern of body sizes laboratory experiments with tropical and temperate populations (for instance, Noach et al. 1996).

An important point here is that a parameter representing enzymatic properties is varied, and body size differences follow. This demonstrates that body size itself is not necessarily selected on. The differences in enzymatic properties might be primarily under selection, and body size differences might follow. Alternatively, changes in enzymatic properties might be an efficient way to adapt to the environment if body size itself would be the selected trait.

Of course, clines in body size between tropical and temperate populations might be due to changes in other parameters than the enzyme inactivation coefficients H_{LG} and H_{HG} of growth rate. Genetic variation in the temperature sensitivity coefficient H_{AG} or genetic variation in the reference growth rate ρ_G too might cause a cline. However, changes in H_{LG} and H_{HG} between populations mediate a more intuitively obvious connection to the environment.

Between Species Variation

Between species differences in body size might result from many different changes in parameter values, but we will concentrate upon one set of parameters that by itself already changes the body sizes—a set of parameters that might not be regarded at first sight as directly influencing body size. In Table 1, the temperature ranges of a number of *Drosophila* species are given. We will concentrate upon the question whether just changing the temperature range but not the sensitivity coefficients or the inactivation coefficients is by itself sufficient to cause a change in body size.

We will model a set of species that are identical in their temperature sensitivity coefficients H_{AG} and H_{AD} for their growth rate and development rate. Moreover, the species are identical in their temperature inactivation coefficients H_{HG}, H_{LG}, and H_{HD}, H_{LD}. The half-way inactivation temperatures

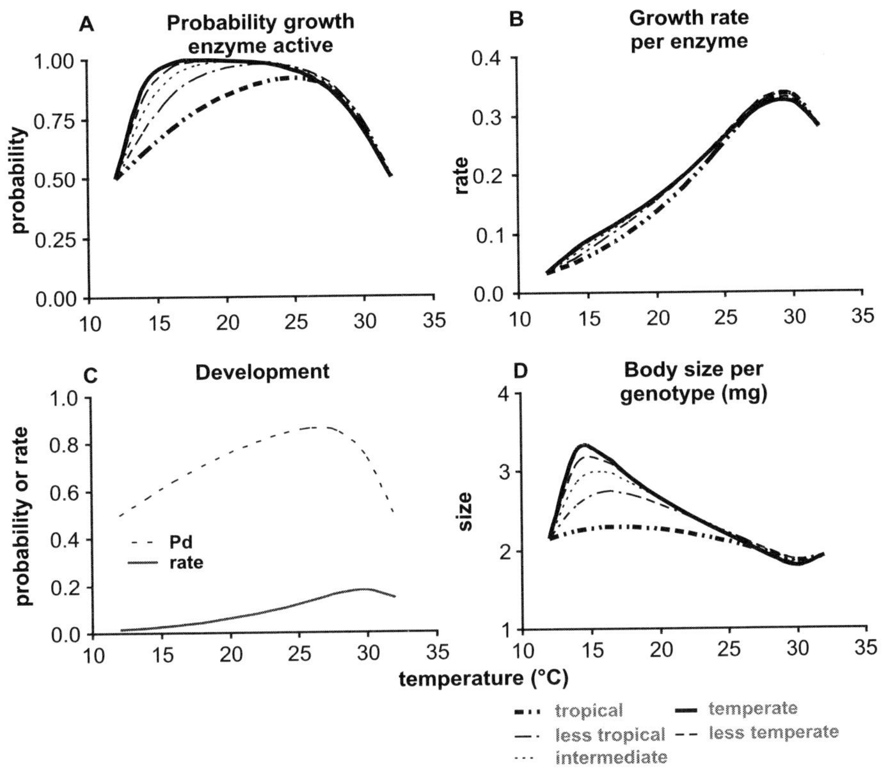

Fig. 18 A cline in body size can be caused by differences in the enzyme inactivation coefficients between populations. Five populations are depicted that differ in the temperature inactivation parameters of growth rate, H_{LG} and H_{HG}. H_{LG} and H_{HG} differ additively and consistently from H_{LG} =–700 kJ mol^{-1} for the temperate population to H_{LG} =–150 kJ mol^{-1} for the tropical population, and H_{HG} =325 kJ mol^{-1} for the temperate population to H_{HG} =400 kJ mol^{-1} for the tropical population. All other parameters are identical between genotypes for the different populations: H_{AD}= 80 kJ mol^{-1}and H_{AG}= 75 kJ mol^{-1}, ρ_D= . 1, ρ_G= 0.2. In all cases T_{ref} = 295°K = 21.8°C, T_H= 305°K = 31.8°C, and T_L= 285°K = 11.8°C. A. Probability for the growth limiting enzymes to be active. B. Growth rates. C. Probability for the development limiting enzymes to be active and development rate. D. Body sizes.

T_{HG} and T_{LG}, and T_{HD} and T_{LG}, differ between the species, but are identical for growth and development: $T_{HD} = T_{HG} = T_H$ and $T_{LD} = T_{LG} = T_L$. The reference temperature for each species is chosen exactly at the midpoint between T_L and T_H. The reference rates ρ_G and ρ_D for each species are found from a reference Eyring equation with given H_{AD} or H_{AG}, and ρ_D =0. 1 and ρ_G=0.2 at T_{ref}=295°K. This was done in order to avoid using a reference rate at a reference temperature that is actually outside the enzyme activation range of the species (the equations might not suffer but this way seems more biologically realistic). The growth rates and development rates over the species therefore refer back to one Eyring equation for growth rate and one Eyring equation for development rate. All differences between the species derive from different temperature ranges of enzyme inactivation.

Two different sets of species are compared, both consisting of a graded range from a high temperature adapted species to a low temperature adapted species. In the first comparison, the range $T_H - T_L$ for the enzymes to be active is the same (20°C) for the five species. In the second comparison, the range $T_H - T_L$ is larger for the cold adapted species than for a species adapted to higher temperature.

Five species differ in temperature boundaries but not in the length of their temperature range. The difference between T_H and T_L equals 20°C, for both development and growth. The ends of the ranges differ by 1.5°C between the species. The probabilities for the enzyme to be active are identical, only horizontally shifted by 1.5°C to different temperatures (Fig. 19A development, Fig. 19B growth). Combined with an identical Eyring equation at different temperature ranges - combining different parameter values for Equation 2 with an identical values in Equation 1 to Equation 3 –, a family of growth rates and a family of development rates originates (Figs. 19C, D). Due to the Eyring equation (Fig. 19E), these rates are not just horizontally translated, but differ in maximum height and slightly in shape. The resulting body sizes are appreciably different, with the species that possesses the lowest temperature range obtaining the highest maximum body size while being smallest at the temperature range the species have in common (Fig. 19F). Maximum body size seems to be related linearly to temperature. The body sizes are translated horizontally by 1.5°C, and the downward slope over the main temperature range is slightly steeper in the lower temperature species.

A similar but more pronounced pattern is found when the temperature range $T_H - T_L$ is larger for the cold adapted species than for the species adapted to warmer temperatures. This is supposed to be caused by a faster decrease in T_L than in T_H. The probabilities for the enzyme to be active now change in shape, as the width of the temperature range decides whether some sort of plateau in the probability for the enzyme to be active will occur. A low temperature species with a wider temperature range reaches higher enzyme activities (Figs. 20A, B). Development rate and growth rate show more difference between the species, in increase and downturn (Figs. 20C, D). The body sizes differ more than in the case of equal temperature range. Again the species that possesses the range of lowest temperatures obtains

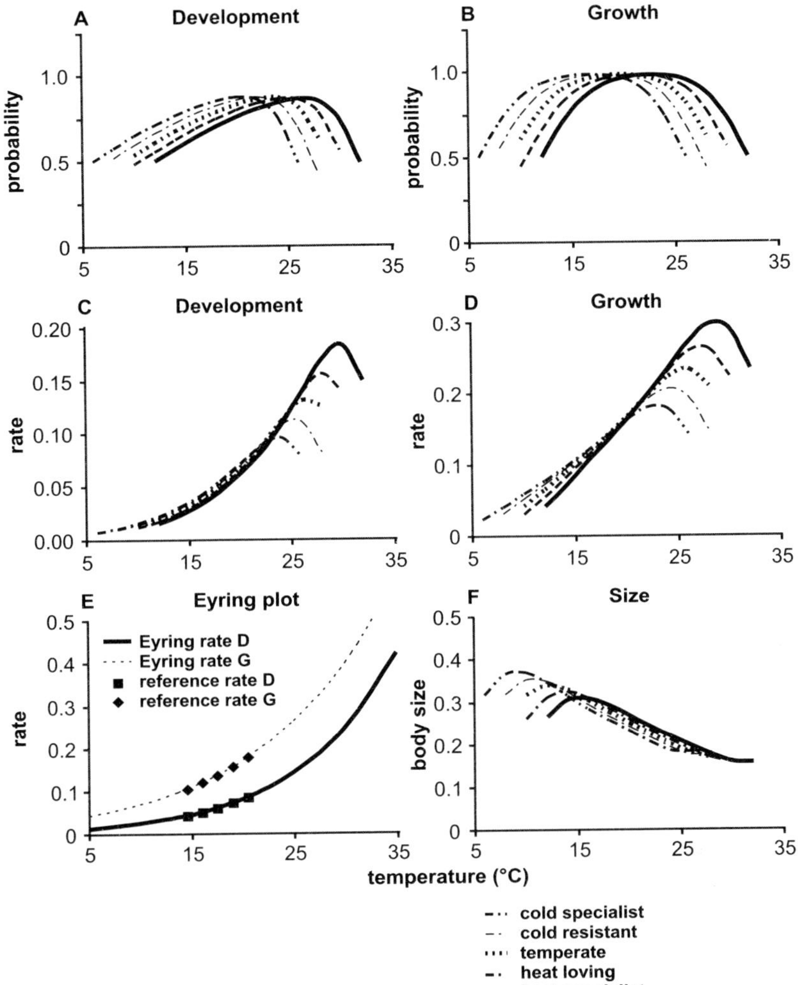

Fig. 19 Between species variation in T_{LD} and T_{HD}, and in T_{LG} and T_{HG}: the range between T_L and T_H equals 20°C. Five species are depicted ranging from cold specialist to heat specialist. The highest temperature range is 12°C to 32°C, the lowest 6°C to 26°C. The difference between the species is a 1.5°C interval. The per species reference temperature is found at the midpoint of the range. The per species reference rate is read from an Eyring equation with $\rho_D = 0.1\,t^{-1}$ and $\rho_G = 0.2\,mg\,t^{-1}$ at $T_{ref} = 295°K$. Change of the temperature interval therefore does not imply a change in the Eyring equation in the numerator of the Sharpe-Schoolfield equation. All other parameters are identical between genotypes for the different populations: $H_{AD} = 80$ kJ mol^{-1} and $H_{AG} = 60$ kJ mol^{-1}, $H_{HD} = 500$ kJ mol^{-1} and $H_{HG} = 38\,1.25$ kJ mol^{-1} and $H_{AD} = -100$ kJ mol^{-1} and $H_{AG} = -287.5$ kJ mol^{-1}. A. Probability developmental enzyme is active. B. Probability growth enzyme is active. C. Development rates. D. Growth rates. E. Eyring plot to find ρ_D and ρ_G. F. Body size.

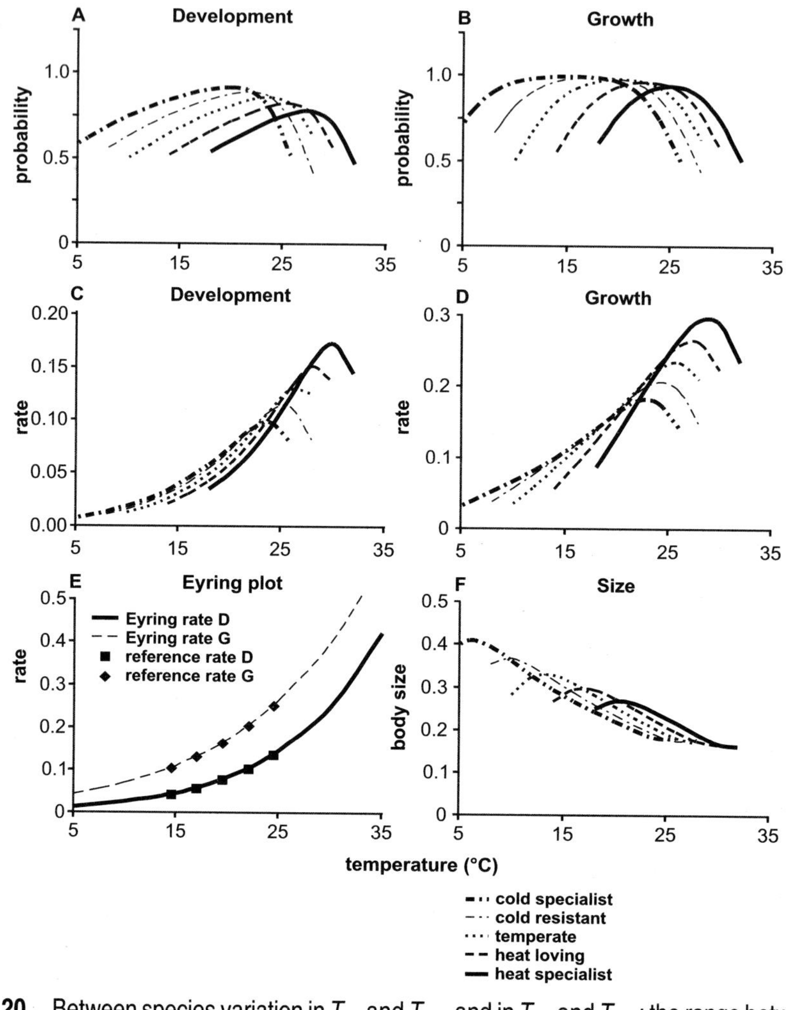

Fig. 20 Between species variation in T_{LD} and T_{HD}, and in T_{LG} and T_{HG}: the range between T_L and T_H increases if the midpoint is at a lower temperature. Five species are depicted. The highest temperature range is 17°C to 32°C, the lowest 3°C to 26°C. The difference between the species is a 1.5°C interval for T_H, a 2.5°C interval for the species specific T_{ref}, and a 3.5°C interval for T_L. The per species reference temperature is found at the midpoint of the range. The per species reference rate is read from an Eyring equation with $\rho_D = 0.1\,t^{-1}$ and $\rho_G = 0.2$ mg t^{-1} at $T_{ref} = 295$°K. Change of the temperature interval therefore does not imply a change in the Eyring equation in the numerator of the Sharpe-Schoolfield equation. All other parameters are identical between genotypes for the different populations: $H_{AD} = 80$ kJ mol^{-1} and $H_{AG} = 60$ kJ mol^{-1}, $H_{HD} = 500$ kJ mol^{-1} and $H_{HG} = 38\ 1.25$ kJ mol^{-1} and $H_{AD} = -100$ kJ mol^{-1} and $H_{AG} = -287.5$ kJ mol^{-1}. A. Probability developmental enzyme is active. B. Probability growth enzyme is active. C. Development rates. D. Growth rates. E. Eyring plot to find ρ_D and ρ_G. F. Body size.

the highest maximum body size while being smallest at the temperature range the species have in common (Fig. 20F). The downward slope of body size over the main temperature range is steeper in the lower temperature species. In total, the pattern is similar to but more pronounced than in the case of species with equal width of temperature range.

The importance of this example is as follows: differences in body size are caused by the temperature range, not by any parameter that has to do with growth, development or temperature inactivation. The non-linearity of the Eyring equation implies that no body size based upon a temperature range can be identical to another body size based upon a different temperature range—even if all parameters are other wise equal. One Eyring equation and horizontally translated but identical probabilities for the enzyme to be active do not lead to horizontally translated body sizes of the same shape, but to different reaction norms of body size, of different shapes.

Discussion

How Valid or Interpretable are Biophysical Models?

The temperature dependence of biological rates might have been known even before agriculture started, but the earliest cited studies are from 1735 (de Réaumur, as cited by Wang 1960) and 1855 (de Candolle, as cited by Sharpe and DeMichele 1977). Both de Réaumur and de Candolle described the degree-day rule. Biophysical descriptions of reaction rates started with the empirical description by Arrhenius. The theoretical derivation by Eyring (1935) and Johnson et al. (1974) remains the basis of virtually all further biological descriptions of reaction rates. In Eyring's formulation, any reaction rate scales with temperature T in °K by a factor $\exp\left[-\frac{H_A}{RT}\right]$, where H_A is the enthalpy of activation and R the gas constant. Eyring's approach must be accepted as standard (Hochachka and Somero 1984, pg 379 – 380), and has been indicated as the basis of the temperature dependence of biological rates (Watt 1968, Johnson et al. 1974).

Universal Temperature Dependence Lately, Gillooly and co-authors (Gillooly et al. 2001) have developed a theory of the general temperature dependence of biological rates. They aim to document a general exponential relation between rates and temperature within and between species, with one underlying temperature coefficient, in what they call "Universal

Temperature Dependence" of biological rates. Gillooly and co-authors write $\exp\left[-\frac{E}{kT}\right]$, with E the Arrhenius activation energy of a biological reaction in eV, and k Boltzmann's constant (Eyring 1935). The Arrhenius activation energy is an approximation of the enthalpy of activation, here called the temperature sensitivity coefficient H_A.

Clarke (Clarke 2004, Clarke and Fraser 2004) criticized an exponential increase of biological rates over species and Universal Temperature Dependence on the basis of the biochemistry of reactions. For one thing, the enthalpy of activation is not the only factor involved in the activation energy, and between species patterns in temperature not only depend upon the enthalpy of activation. For another thing, Gillooly and co-authors' "Universal Temperature Dependence" does not provide any explanation. A number of intertwined patterns were found to be present in the data, but why "Universal Temperature Dependence" should exist is a different question. An explanation in terms of biophysics differs from a statistical description, even if it has the same mathematical form.

We emphasize the importance of parameter variation, and do not subscribe to "Universal Temperature Dependence" as a fundamental constraint. But of more importance here is whether Clarke's criticism of the between species approach of Gillooly and co-authors applies to within species patterns too. The important point is whether biophysical descriptions are compatible with biochemical knowledge of how reactions proceed. Clarke argues that biological reaction rates cannot be exponential with temperature as rates of enzyme catalyzed reactions depend not only on substrate and product but on conformational changes in the enzymatic complex as well. We hope the modification of the Eyring equation to the Sharpe-Schoolfield equation represents the properties of enzymatic complexes adequately for within species purposes. If so, Clarke's criticism is not applicable to the Sharpe-Schoolfield model as used here. The Sharpe-Schoolfield model provides a sufficient statistical description of observed patterns, but we use it mostly as if its parameters provide explanatory variables. Therefore, we have to be concerned with biochemical adequacy, even if our names for the parameters (temperature sensitivity coefficient, temperature inactivation coefficient) might not be biochemically appropriate.

Linearity of biological rates has been extensively documented (cf Figs. 1–3). Charnov and Gillooly (2003) proposed that this linearity could be regarded as a linear approximation of an exponential increase with temperature. If so, a linear approximation at the reference temperature of the

Eyring equation (that is, the numerator of the Sharpe-Schoolfield equation), would have a slope of $\frac{\rho H_A}{RT_{ref}^2}$, in the present notation and in agreement with Charnov & Gillooly (2003). The number of predicted degree-days would be $\frac{RT_{ref}^2}{\rho H_A}$. In the *Drosophila* data, this prediction is nowhere near the actual number of degree-days. In all insect data we have, the linearity of development rate with temperature is clearly not an approximation to an exponential function (Figs. 1–3). The degree-day model depends strictly upon linearity of development rate with temperature. The success of the degree-day model argues against the validity of Charnov & Gillooly (2003)'s approximation: an exponential increase of development rate with temperature gives not even approximately a constant number of degree-days.

The implications of the degree-day model Linearity of development rate requires and implies that in the Sharpe-Schoolfield model the 'probability for the enzyme to be active' never reaches 100%. For direct empirical data, we possess only a limited data set of parameter values. In studies of *Drosophila*, the probability for the enzyme to be active never reaches 100%, and in an Arrhenius plot, a region of strict linearity does not appear. The same observation has been made in other studies using the Sharpe-Schoolfield model, starting with data on plants and insects by Sharpe and DeMichele (1977). Van Straalen (2001) presents data on springtails, and observes the same non-linearity in an Arrhenius plot. The temperature sensitivity coefficient, the enthalpy of activation H_A as used in Sharpe-Schoolfield model, therefore never rules the temperature dependence of biological rates in insects.

Actually, the best evidence for this is the applicability of the degree-day model. The Eyring equation by itself would lead to development rates that are exponential with temperature. Linear development rates imply other processes, and in a model of temperature sensitivity this would require additional parameters. But the most important implication of the linearity of development rate with temperature is the presence of a temperature independent physiological time (van Straalen 1983). As argued above, this physiological time might well be interpreted as constancy of the number of cell divisions. The formal reversible enzyme inactivation in the Sharpe-Schoolfield model would represent the existence of a temperature invariant physiological time, and the regulation of the number of cell divisions. Note though, that the applicability of the biophysical models to temperature sensitivity of biological rates does not depend upon any interpretation of invariant physiological time as a fixed number of cell divisions. Similarly, the van der Have-de Jong model of temperature dependent body size does not depend upon an interpretation of development rate as involving cell number and growth rate as involving cell size.

From Biophysical Parameters to Adaptive Patterns in Rate and Size

Biophysical parameters are not universal constants that rule biology; they are not unavoidable constraints. Temperature dependence abounds, but not equally among species or populations within a species. Yet, given a set of parameter values, temperature dependence of rates is unavoidable. A species' environment will indicate what combination of development rate and growth rate are optimal and in concord with evolved enzymatic parameters. One environment might allow several combinations of rates, sizes and biochemistry to evolve, either as equivalent solutions to the same life history, or as alternative options defining different niches. The temperature dependence of development rate can be used to adjust emergence towards a specific date. Consider for instance a univoltine species with a short mating season in early summer. Eggs might have hatched at different times in spring. The temperature dependent development rate should be selected to synchronize adult emergence and thereby enhance efficient reproduction (Gilbert and Raworth 1996). To function in this way, development rate needs to be temperature dependent, within any specific latitude and between latitudes, but the temperature dependence must be under genetic and evolutionary control.

Generation time relative to season length will decide whether a latitudinal cline in body size will show larger body at higher latitudes or lower body size at higher latitudes (Chown and Gaston 1999). A strictly univoltine species might well show smaller body size at higher latitude as a consequence of shorter season length. Small multivoltine insect species with relatively fast development show larger body size at higher latitudes (Blanckenhorm and Demont 2004). Over species, growth rate and development rate have to respond independently to account for this diversity of observed patterns in insect latitudinal clines.

Adult size is widely regarded as the ratio of growth rate to development rate (Gilbert and Raworth 1996). Temperature compensation in adult size occurs when the effect of temperature dependence disappears due to the division—that is, when the temperature sensitivity coefficients of development and growth H_{AD} and H_{AG} are equal and when the probabilities for the enzyme to be active are equal. In the *Drosophila* data, H_{AD} and H_{AG} differ significantly between species but for thorax, H_{AD} and H_{AG} do not differ significantly. Thorax length might be the most obvious indicator for adult size, and the evidence seems to be that *Drosophila* species almost compensate for temperature dependence in body size. The generally observed decrease in adult size with temperature (Atkinson 1994, 1996) might be restricted to the region of high temperature enzyme inactivation—or be a consequence of experimenter choice of character to measure. Wing size, in contrast to thorax size, decreases steadily with temperature, but this might be as much an adaptation to flight at different temperatures as indicative of a general decrease in body size with temperature (Petavy et al. 1997), though the actual decrease in wing size depends upon the population (Noach et al. 1996).

In *Drosophila*, the differences in the temperature sensitivity coefficient for development H_{AD} are almost significant (data in Table 2, P=0.087 over 6 species) but the differences in the temperature sensitivity coefficients for growth H_{AG} are significant (data in Table 3, P=0.013 over 6 species for thorax and P<0.001 over 4 species for wing). Larger H_A occurs in the species *D. willistoni* and *D. funebris*. No obvious ecological correlate is evident.

Van Straalen (2001) applied the Sharpe-Schoolfield model to temperature dependent development time in springtails. The model described the data very well. Over 38 species, parameter values for development rate at 15°C and for temperature sensitivity coefficient H_A were compiled. Springtails can be grouped in ecological classes depending on their place in the leaf litter and soil surface. The epigeon, the group of species that lives on the soil surface, has highest H_A values, significantly different from the values in the euedaphon, the group of species that lives in the soil. Hemiedaphon species that live in the litter layer have intermediate values for H_A. In van Straalen's study temperature sensitivity of development and ecological niche are clearly related.

From Biophysical Parameters to Adaptive Phenotypic Plasticity

Different theoretical possibilities for phenotypic plasticity emerge just by changing a few of all the parameters in the model. The most obvious parameters to change to obtain a change in phenotypic plasticity of adult body size are the temperature sensitivity coefficients H_{AD} and H_{AG} (the enthalpy of activation of enzymatic reactions for development and growth). The difference between H_{AD} and H_{AG} is a main factor in deciding plasticity of body size towards temperature, as shown in Fig. 15. But, as is clear too in Fig. 15, it is quite possible to observe phenotypic plasticity in body size of opposite sign to that indicated by the difference between H_{AD} and H_{AG}. The temperature inactivation coefficients (H_{HD} and H_{HG}, H_{LD} and H_{LG}) decide as much about phenotypic plasticity as the temperature sensitivity coefficients.

The patterns of phenotypic plasticity we demonstrate are therefore not derived from difference in the temperature sensitivity coefficients H_{AD} and H_{AG}. Without changing the temperature sensitivity coefficients we simulated temperature dependence of body size. The parameter values we changed have been chosen to indicate how actual patterns of phenotypic plasticity might be formally caused, and we took our clue from observations in *Drosophila*. We concentrated on patterns of differences in phenotypic plasticity that were actually observed, and tried to find biologically plausible but simple ways to generate such patterns. In *D. melanogaster*, clines in allozyme frequency have repeatedly been described (Eanes 1999). In such clines, allozymes with higher enzyme activity might predominate in more temperate populations. For some enzymes, an allozyme with high enzyme activity at crucial points in the metabolic network has relatively higher activity at low temperature and reaches high frequency in temperate populations (Verrelli and Eanes 2001). Relatively high enzyme activity at lower temperature might lead to larger adult flies (Bijlsma-Meeles and Bijlsma 1988). The observation is for the enzyme alcohol dehydrogenase; its generality is unknown.

In Fig. 18, a cline in enzyme activity was simulated, taking our clue from the enzyme clines in *Drosophila melanogaster*. In Fig. 18, a situation is modeled where enzyme activity changes from a maximum at low temperature to a maximum at high temperature. Body size changes with enzyme activity, leading to a larger body size at low temperature for the parameter set with the maximum enzyme activity for growth at low temperature. Of course, many other parameter values can be chosen to represent differences in enzyme activity. Changing from enzyme activity for growth with a low maximum at low temperature to high enzyme activity for growth over a much wider range, for instance results in almost parallel reaction norms for body size over the *D. melanogaster* viable range. We have been able to reproduce the observed patterns of reaction norm comparisons in natural populations by changing the probability of the enzyme for growth to be active.

Larger size at lower temperature resulted from the model equivalent of a higher enzyme activity at lower temperature (Fig. 18). This leaves open what selection pressures would operate in a natural population. Selection could be for larger adult size at lower temperatures, without any selection pressure for populations to differ at higher temperatures. The selection pressure on body size itself could be translated into a selection pressure on enzyme activity at low temperature (see Box 3 how such translation works). Or, selection could directly be on enzymatic functioning, on enzymes that perform better at lower temperatures, and any change in body size itself could be a correlated response. The formal description in terms of the biophysical model would be the same.

Other changes in body size and of phenotypic plasticity in body size can be derived from changing the temperature range of enzyme activity without changing the temperature sensitivity coefficients (H_{AD} and H_{AG}) or the temperature inactivation coefficients (H_{HD} and H_{HG}, H_{LD} and H_{LG}). This is a particularly interesting observation, as it indicates that many patterns in temperature dependence of adult body size in insects might derive from the temperature boundaries rather than from temperature sensitivity over the viable range itself.

Biophysical models like the Sharpe-Schoolfield model are a fertile field for biological explanation of temperature related plasticity in development rate and body size in insects. At the moment, the number of available parameter estimates is very low, and this prevents us from having insight into the ecological and evolutionary patterns that might be associated with biophysical parameters. We hope it will be clear that the models are applicable and can be used to categorize temperature related ecological differences, as exemplified in the observations of van Straalen (2001) on springtails.

Acknowledgements

We thank Nico van Straalen for comment on the manuscript, and Marjolein Kortbeek for the definitive version of the figures.

References

Angilletta, M. J., T. Hill, and M. A. Robson. 2002b. Is physiological performance optimized by thermoregulatory behavior?: a case study of the eastern fence lizard, *Sceloporus undulatus*. Journal of Thermal Biology, 27: 199–204.

Angilletta, M. J., P. H. Niewiarowski, and C. A. Navas. 2002a. The evolution of thermal physiology in ectotherms. Journal of Thermal Biology, 27: 249–268.

Angilletta, M. J., R. S. Wilson, C. A. Navas, and R. S. James. 2003. Tradeoffs and the evolution of thermal reaction norms. Trends in Ecology & Evolution, 18: 234–240.

Atkinson, D. 1994. Temperature and organism size—a biological law for ectotherms. Advances in Ecological Research, 25: 1–58.

Atkinson, D. 1996. Ectotherm life-history responses to developmental temperature. pp. 183–204. *In* I. Johnston and A. F. Bennett [eds] 1996. Animals and Temperature: Phenotypic and Evolutionary Adaptation. Cambridge University Press, Cambridge, UK.

Atkinson, D. and R. M. Sibly. 1997. Why are organisms usually bigger in colder environments? Making sense of a life history puzzle. Trends in Ecology & Evolution, 12: 235–239.

Azevedo, R. B. R., V. French, and L. Partridge. 2002. Temperature modulates epidermal cell size in *Drosophila melanogaster*. Journal of Insect Physiology, 48: 231–237.

Bennett, A. and R. B. Huey. 1990. Studying the evolution of physiological performance. Oxford Surveys of Evolutionary Biology, 7: 251–284.

Berrigan, D. and E. L. Charnov. 1994. Reaction norms for age and size at maturity in response to temperature—a puzzle for life historians. Oikos, 70: 474–478.

Bijlsma-Meeles, E. and R. Bijlsma. 1988. The alcohol-dehydrogenase polymorphism in *Drosophila melanogaster*: fitness measurements and predictions under conditions with no alcohol stress. Genetics, 120: 743–753.

Blanckenhorn, W. U. and M. Demont. 2004. Bergmann and converse Bergmann latitudinal clines in arthropods: two ends of a continuum? Integrative and Comparative Biology, 44: 413–424.

Blanckenhorn, W. U. and V. Llaurens. 2005. Effects of temperature on cell size and number in the yellow dung fly *Scathofaga stercoraria*. Journal of Thermal Biology, 30: 213–219.

Bochdanovits, Z. and G. de Jong. 2003. Experimental evolution in *Drosophila melanogaster*: Interaction of temperature and food quality selection regimes. Evolution, 57: 1829–1836.

Brett, J. R. 1970. Fishes (Temperature). pp. 515–573. *In* O. Kinne [ed.] 1970. Marine Ecology: A Comprehensive, Integrated Treatise on Life in Oceans and Coastal Waters. Wiley, London.

Bryant, P. J. and P. Levinson. 1985. Intrinsic growth control in the imaginal primordia of *Drosophila*, and the autonomous action of a lethal mutation causing overgrowth. Developmental Biology, 107: 355–363.

Bryant, P. J. and P. Simpson. 1984. Intrinsic and extrinsic control of growth in developing organs. Quarterly Review of Biology, 59: 387–415.

Chapman, R. F. 1982. The Insects Structure and Function. Harvard University Press, Cambridge, MA. 919 pp.

Charnov, E. L. and J. F. Gillooly. 2003. Thermal time: body size, food quality and the 10 degrees C rule. Evolutionary Ecology Research, 5: 43–51.

Chown, S. L., and K. J. Gaston. 1999. Exploring links between physiology and ecology at macro-scales: The role of respiratory metabolism in insects. Biological Review, 74: 87–120.

Clarke, A. 2003. Costs and consequences of evolutionary temperature adaptation. Trends in Ecology & Evolution, 18: 573–581.

Clarke, A. 2004. Is there a Universal Temperature Dependence of metabolism? Functional Ecology, 18: 252–256.

Clarke, A. and K. P. P. Fraser. 2004. Why does metabolism scale with temperature? Functional Ecology, 18: 243–251.

Cohet, Y. 1980. Thermal tolerance and geographic distribution: a comparison of cosmopolitan and tropical endemic *Drosophila* species. Journal of Thermal Biology, 5: 69–74.

David, J. R. and M. F. Clavel. 1967. Influence de la température subie au cours du développement sur divers caractères biometrique des adultes de *Drosophila melanogaster* Meigen. Journal of Insect Physiology 13: 717–729.

David, J. R., R. Allemand, J. Van Herrewege, and Y. Cohet. 1983. Ecophysiology: abiotic factors. pp. 105–170. *In* M. Ashburner, H. L. Carson, and J. N. Thompson. [eds] 1983b. The Genetics and Biology of Drosophila 3d. Academic Press, London, New York.

de Jong, G. and Z. Bochdanovits. 2003. Latitudinal clines in *Drosophila melanogaster*: body size, allozyme frequencies, inversion frequencies, and the insulin-signalling pathway. Journal of Genetics, 82: 207–223.

de Jong, G. and A. Imasheva. 2000. Genetic variance in temperature dependent adult size deriving from physiological genetic variation at temperature boundaries. Genetica, 110: 195–207.

de Moed, G. H., G. de Jong, and W. Scharloo. 1997a. Environmental effects on body size variation in *Drosophila melanogaster* and its cellular basis. Genetical Research, 70: 35–43.

de Moed, G. H., G. de Jong, and W. Scharloo. 1997b. The phenotypic plasticity of wing size in *Drosophila melanogaster*: the cellular basis of its genetic variation. Heredity, 79: 260–267.

de Moed, G. H., G. de Jong, and W. Scharloo. 1998. The energetics of growth in *Drosophila melanogaster*: Effect of temperature and food conditions. Netherlands Journal of Zoology, 48: 169–188.

de Réaumur, R. 1735. Observation du thermomètre, faites à Paris pendant l'année 1735, comparée avec celles qui ont été faites sous la ligne, à l'Isle de France, à Alger et en quelques-unes de nos isles de l'Amérique. Mémoire de l'Académie des Sciences de Paris.

Delpuech, J., B. Moreteau, J. Chice, E. Pla, J. Vouidibio, and J. R. David. 1995. Phenotypic plasticity and reaction norms in temperate and tropical populations of *Drosophila melanogaster*: ovarian size and developmental temperature. Evolution, 49: 670–675.

Eanes, W. F. 1999. Analysis of selection on enzyme polymorphisms. Annual Review of Ecology and Systematics, 30: 301–326.

Eyring, H. 1935. The activated complex in chemical reactions. Journal of Chemical Physics, 3: 107–115.

Frazier, M. R., H. A. Woods, and J. F. Harrison. 2001. Interactive effects of rearing temperature and oxygen on the development of *Drosophila melanogaster*. Physiological and Biochemical Zoology, 74: 641–650.

French, V., M. Feast, and L. Partridge. 1998. Body size and cell size in *Drosophila*: the developmental response to temperature. Journal of Insect Physiology, 44: 1081–1089.

Gibert, P. and G. de Jong. 2001. Temperature dependence of development rate and adult size in *Drosophila* species: biophysical parameters. Journal of Evolutionary Biology, 14: 267–276.

Gilbert, N. and D. A. Raworth. 1996. Insects and temperature—a general theory. Canadian Entomologist, 128: 1–13.

Gillooly, J. F., J. H. Brown, G. B. West, V. M. Savage, and E. L. Charnov. 2001. Effects of size and temperature on metabolic rate. Science, 293: 2248–2251.

Gillooly, J. F., E. L. Charnov, G. B. West, V. M. Savage, and J. H. Brown. 2002. Effects of size and temperature on developmental time. Nature, 417: 70–73.

Goldbeter, A. 1991. A minimal cascade model for the mitotic oscillator involving cyclin and cdcc kinase. Proceedings of the National Academy of Sciences of the United States of America, 88: 9107–9111.

Gotthard, K. and S. Nylin. 1995. Adaptive plasticity and plasticity as an adaptation: a selective review of plasticity in animal morphology and life history. Oikos, 74: 3–17.

Harrington, R., I. Woiwod, and T. Sparks. 1999. Climate change and trophic interactions. Trends in Ecology & Evolution, 14: 146–150.

Hochachka, P. W. 1991. Temperature: the ectothermy option. pp. 313–322. *In* P. W. Hochachka and T. P. Ommsen [eds] 1991. Biochemistry and Molecular Biology of Fishes. Elsevier Science Publishers, Amsterdam.

Hochachka, P. W. and G. N. Somero. 1984. Biochemical Adaptation. Princeton University Press, Princeton, NJ.

Hoffmann, A. A. and J. Merilä. 1999. Heritable variation and evolution under favourable and unfavourable conditions. Trends in Ecology & Evolution, 14: 96–101.

Honěk, A. 1996. Geographical variation in thermal requirements for insect development. European Journal of Entomology, 93: 303–312.

Honěk, A. and F. Kocourek. 1988. Thermal requirements for development of aphidophagous Coccinellidae (Coleoptera), Chrysopidae, Hemerobiidae (Neuroptera), and Syrphidae (Diptera)—some general trends. Oecologia, 76: 455–460.

Honěk, A., and F. Kocourek. 1990. Temperature and development time in insects: a general relationship between thermal constants. Zoologische Jahrbücher, Abteilung für Systematik, 117: 410–439.

Huey, R. B., D. Berrigan, G. W. Gilchrist, and J. C. Herron. 1999. Testing the adaptive significance of acclimation: A strong inference approach. American Zoologist, 39: 323–336.

Huey, R. B. and J. G. Kingsolver. 1989. Evolution of thermal sensitivity of ectotherm performance. Trends in Ecology & Evolution, 4: 131–135.

James, A. C., R. B. R. Azevedo, and L. Partridge. 1995. Cellular basis and developmental timing in a size cline of *Drosophila melanogaster*. Genetics, 140: 659–666.

James, A. C., R. B. R. Azevedo, and L. Partridge. 1997. Genetic and environmental responses to temperature of *Drosophila melanogaster* from a latitudinal cline. Genetics, 146: 881–890.

Jarošík, V., A. Honěk, and A. F. G. Dixon. 2002. Developmental rate isomorphy in insects and mites. American Naturalist, 160: 497–510.

Johnson, F. H., H. Eyring, and B. J. Stover. 1974. The Theory of Rate Processes in Biology and Medicine. Wiley, New York.

Judd, G. J. R., G. H. Whitfield, and H. E. L. Maw. 1991. Temperature dependent development and phenology of pepper maggots (Diptera, Tephritidae) associated with pepper and horsenettle. Environmental Entomology, 20: 22–29.

Karan, D., J. P. Morin, E. Gravot, B. Moreteau, and J. R. David. 1999. Body size reaction norms in *Drosophila melanogaster*: temporal stability and genetic architecture in a natural population. Genetics Selection Evolution, 31: 491–508.

Karan, D., J. P. Morin, B. Moreteau, and J. R. David. 1998. Body size and developmental temperature in *Drosophila melanogaster*: Analysis of body weight reaction norm. Journal of Thermal Biology, 23: 301–309.

Kontodimas, D. C., P. A. Eliopoulos, G. J. Stathas, and L. P. Economou. 2004. Comparative temperature-dependent development of *Nephus includens* (Kirsch) and *Nephus bisignatus* (Boheman) (Coleoptera : Coccinellidae) preying on *Planococcus citri* (Risso) (Homoptera : Pseudococcidae): Evaluation of a linear and various nonlinear models using specific criteria. Environmental Entomology, 33: 1–11.

Lewontin, R. C. and L. S. Birch. 1966. Hybridization as a source of variation for adaptation to new environments. Evolution, 20: 315–336.

Liu, S.-S., and X.-D. Meng. 1989. The change pattern of development rates under constant and variable temperatures in *Myzus persicae* and *Lipaphis erysimi*. Acta Ecologica Sinica 9: 182–190.

Lowry, T. H. and K. S. Richardson. 1987. Mechanism and Theory in Organic Chemistry. HarperCollins Publishers.

Lynch, M. and J. B. Walsh. 1997. Genetics and Analysis of Quantitative Traits. Sinaure, Sunderland MA.

Messenger P. S. and N. E. Flitters. 1958. Effect of constant temperature environments on the egg stage of three species of Hawaiian fruit flies. Annals of the Entomological Society of America, 51: 109–119.

Morrisey, R. and J. S. Edwards. 1979. Neural function in an alpine grylloblattid: A comparison with the house cricket, *Acheta domesticus*. Physiological Entomology 4: 241–250.

Murray, A. W. 1992. Creative blocks-cell cycle checkpoints and feedback controls. Nature, 359: 599–604.

Murray, A. W. 1994. Cell-cycle—a rum tale of replication. Nature, 367: 219–220.

Noach, E. J. K., G. de Jong, and W. Scharloo. 1996. Phenotypic plasticity in morphological traits in two populations of *Drosophila melanogaster*. Journal of Evolutionary Biology, 9: 831–844.

Norel, R. and Z. Agur. 1991. A model for the adjustment of the mitotic clock by cyclin and MPF levels. Science, 251: 1076–1078.

Novak, B. and J. J. Tyson. 1993a. Numerical analysis of a comprehensive model of M-phase control in *Xenopus*-oocyte extracts and intact embryos. Journal of Cell Science, 106: 1153–1168.

Novak, B. and J. J. Tyson. 1993b. Modeling the cell-division cycle-M-phase trigger, oscillations and size control. Journal of Theoretical Biology, 165: 101–134.

Nylin, S. and K. Gotthard. 1998. Plasticity in life-history traits. Annual Review of Entomology, 43: 63–83.

Partridge, L., B. Barrie, K. Fowler, and V. French. 1994. Evolution and development of body size and cell size in *Drosophila melanogaster* in response to temperature. Evolution, 48: 1269–1276.

Petavy, G., J. P. Morin, B. Moreteau, and J. R. David. 1997. Growth temperature and phenotypic plasticity in two *Drosophila* sibling species: probable adaptive changes in flight capacities. Journal of Evolutionary Biology, 10: 875–887.

Petchey, O. L., P. T. McPhearson, T. M. Casey, and P. J. Morin. 1999. Environmental warming alters food-web structure and ecosystem function. Nature, 402: 69–72.

Pigliucci, M. 1996. How organisms respond to environmental changes: From phenotypes to molecules (and vice versa). Trends in Ecology & Evolution, 11: 168–173.

Polyak, K., J. Y. Kato, M. J. Solomon, C. J. Sherr J. Massague, J. M. Roberts, and A. Koff. 1994. P27(KIP1), a cyclin-cdk inhibitor, links transforming growth-factor-beta and contact inhibition to cell-cycle arrest. Genes & Development, 8: 9–22.

Robinson, S. J. W. and L. Partridge. 2001. Temperature and clinal variation in larval growth efficiency in *Drosophila melanogaster*. Journal of Evolutionary Biology, 14: 14–21.

Schoolfield R.M., Sharpe P.J.H., and Magnuson C.E. 1981. Non-linear regression of biological temperature-dependent rate models based on absolute reaction-rate theory. Journal of Theoretical Biology, 88: 719–731.

Sharpe, P. J. H. and D. DeMichele. 1977. Reaction kinetics of poikilotherm development. Journal of Theoretical Biology, 64: 649–670.

Sibly, R. M. and D. Atkinson. 1994. How rearing temperature affects adult size in ectotherms. Functional Ecology, 8: 486–493.

Snyder, G. K. and W. W. Weathers. 1975. Temperature adaptations in amphibians. American Naturalist, 109: 93–101.

Somero, G. N. 2002. Thermal limits to life: underlying mechanisms and adaptive plasticity. Integrative and Comparative Biology, 42: 1316–1316.

Tyson, J. J. 1991. Modeling the cell-division cycle-cdc2 and cyclin interactions. Proceedings of the National Academy of Sciences of the United States of America, 88: 7328–7332.

Tyson, J. and W. Sachsenmaier. 1979. Derepression as a model for control of the DNA-division cycle in eukaryotes. Journal of Theoretical Biology, 79: 275–280.

van der Have, T. M. 2002. A proximate model for thermal tolerance in ectotherms. Oikos, 98: 141–155.

van der Have, T. M. and G. de Jong. 1996. Adult size in ectotherms: Temperature effects on growth and differentiation. Journal of Theoretical Biology, 183: 329–340.

van Straalen, N. 1983. Physiological time and time invariance. Journal of Theoretical Biology, 104: 349–357.

van Straalen, N. 2001. Adaptive significance of temperature response in Collembola. Acta Zoologica Fennica, 195: 135–142.

Van Voorhies, W. A. 1996. Bergmann size clines: A simple explanation for their occurrence in ectotherms. Evolution, 50: 1259–1264.

Verrelli, B. C. and W. F. Eanes. 2001. The functional impact of *Pgm* amino acid polymorphism on glycogen content in *Drosophila melanogaster*. Genetics, 159: 201–210.

Wagner, T. L., R. L. Olson, and J. L. Willers. 1991. Modeling arthropod development time. Journal of Agricultural Entomology, 8: 251–270.

Wagner, T., H. Wu, P. J. H. Sharpe, R. Schoolfield, and R. N. Coulson. 1984. Modeling insect development rates: a literature review and application of a biophysical model. Annals of the Entomological Society of America, 77: 208–225.

Wang, J. 1960. A critique of the heat unit approach to plant science response studies. Ecology, 41: 785–790.

Watt, K. 1968. Ecology and Resource Management: A Quantitative Approach. McGraw-Hill , New York.

Willmer, P., G. Stone, and I. Johnston. 2000. Environmental Physiology of Animals. Blackwell Science, Oxford.

Zwaan, B. J., R. B. R. Azevedo, A. C. James, J. van 't Land, and L. Partridge. 2000. Cellular basis of wing size variation in *Drosophila melanogaster*: a comparison of latitudinal clines on two continents. Heredity, 84: 338–347.

13

The Developmental-Physiological Basis of Phenotypic Plasticity

H. Frederik Nijhout* and Goggy Davidowitz**
**Department of Biology, Duke University, Durham, NC 27708, USA*
***Department of Ecology and Evolutionary Biology, University of Arizona, Tucson, AZ 85721, USA*

Abstract

Research on the developmental physiology of animals has greatly expanded our understanding of the nature of phenotypic plasticity. We are no longer constrained to hypothesize "genes for plasticity" because we are beginning to understand how the different parts of the mechanism that generates the phenotype respond to specific environmental variables. What we observe as phenotypic plasticity is due to the plasticity of a broad diversity of developmental processes that underlie the phenotype. Here we outline recent advances in our understanding of the developmental and physiological determinants of complex traits and how these determinants contribute to plastic variation of traits. We highlight the control of allometry, body size, and polyphenism, and conclude with a brief discussion of a physiological perspective on the relationships between homeostatic mechanisms and phenotypic plasticity.

Introduction: The Role of Physiology

Until recently, phenotypic plasticity was viewed strictly in terms of genes and environments. Genes coded for the phenotype, and sensitivity of gene expression to environmental variation led to plastic variation of the phenotype. The mechanisms and pathways by which genes affected phenotypes and by which environments altered gene expression were generally not a part of the picture. Today we know that phenotypes are affected by the simultaneous, as well as sequential, interaction of many genes and environmental variables, so that the specific effect of any one gene or environmental factor is not always easy to interpret. A strict one-to-one relationship between a mutation and a specific phenotypic defect is often assumed, but the vast majority of mutations have variable penetrance and expressivity, which indicates that both genetic background and environment have a large effect on the way any given gene affects the phenotype (Nijhout, 2002, 2003).

Physiology and development provide the mechanisms through which genetic and environmental factors affect the expression of the phenotype. Understanding the details of these mechanisms is critical for a full understanding of the genetic and functional properties of complex traits. In the absence of such understanding we are inevitably forced to explain the inheritance and evolution of complex traits by assuming hypothetical mechanisms and genetic interactions that may be unrealistic and non-existent. Different investigators may favor different hypothetical mechanisms, and this can lead to unnecessary and unproductive disagreement and confusion. For instance, West-Eberhard (2003, page 55) has argued that the disagreement and controversy over the existence and role of "genes for" plasticity is largely due to a failure to understand, and take account of, the mechanistic basis of phenotypic plasticity.

Phenotypic plasticity results from variation in developmental, physiological, biochemical, and behavioral processes that are sensitive to environmental variables. Most traits that have been of interest in the study of phenotypic plasticity (such as body size, allometry, fecundity) are established during post-embryonic development and are therefore unlikely to be directly influenced by the genetic processes that regulate early embryonic development (Nijhout, 1999). Postembryonic development is regulated largely by hormonal mechanisms whose dynamics depend on many genetic and environmental factors. Hence, the control of postembryonic development is a largely physiological process. Studies of developmental physiology have revealed many features of phenotypic regulation that are relevant to an understanding of phenotypic plasticity.

Below we will outline some of the areas where an understanding of the developmental physiology of a trait helps us to better understand the nature and control of its plasticity. We will focus on recent developments in our understanding of the control of allometry, body size and polyphenism. We conclude with a brief discussion of the relationships between homeostatic mechanisms and phenotypic plasticity.

Allometry and Tradeoffs

Allometry

Allometry refers to the covariation of the sizes of different characters. The relative size of different body parts change throughout development either because different parts grow at different rates, or because the timing and duration of their growth periods differ. This change over time is referred to as *ontogenetic allometry*, and can be measured as a time series on a single individual, or as the means of groups of synchronized individuals. In addition, at a given stage in development, there can be individual variation in the relative sizes of body parts. For instance, variation in development results in adults of different body sizes, and the allometric variation of different individuals is called *static allometry*. If the variation in adult size is environmentally determined, as in the case of the beetle, *Onthophagus taurus* (Emlen, 1994, 1997b; Moczek, this volume), then body size is a plastic trait, and the range of the allometry (though not its shape) is also a plastic trait that is mediated through body size. Insofar as body size variation has a strong environmental component in most species, static allometry is a common manifestation of phenotypic plasticity. Finally, species differ characteristically in the relative size of body parts, and the allometric variation of different species is called *evolutionary allometry* (Cock, 1966; Klingenberg and Zimmerman, 1992).

The most common assumption underlying our understanding of allometric relationships is that different body parts grow at different rates during development. Thus if part X grows as

$$dX/dt = aX,$$

and part Y grows as

$$dY/dt = bY,$$

then the variation of part X with respect to part Y can be found by solving

$$dX/dY = aX/bY,$$

which gives

$$X = cY^{a/b},$$

where c is a constant. This is the fundamental allometry equation. It can also be written in logarithmic form

$$\log X = a/b * \log Y + c \quad \text{(Eq. 1)}$$

In this formulation, logX varies linearly with logY. In other words, the graph that relates the size of X to the size of Y is a straight line in a double-logarithmic plot.

In spite of the substantial simplifying assumptions used to derive the allometry equation (i.e., that body parts grow exponentially, and that they grow simultaneously), a large number of traits exhibit linear allometric relationships in double logarithmic plots. In addition, structures that initiate or terminate their growth at different times during ontogeny, such as wings and body in many insects, fore- and hindlimbs in tetrapods, and forceps and thorax in earwigs, also tend to exhibit nicely linear allometries. Even structures that regrow periodically, such as the antlers of deer, show a linear allometry with body size (Huxley 1932). These observations suggest that differences in growth rate may not be the critical determinant of allometry, but that the sizes of specific body parts are somehow cued by, or limited by, overall body size.

Imaginal Disks

One of the most extreme deviations from the assumption of simultaneous growth occurs in the holometabolous insects (insects with compete metamorphosis). In these animals, the imaginal disks that will give rise to various adult structures (legs, wings, genitalia, eyes antennae, mouthparts, head horns), remain small during most of larval life, and undergo a period of rapid growth after the larva has stopped feeding and enters the prepupal stage in preparation for metamorphosis (Svacha, 1992; Nijhout and Wheeler, 1996; Kremen and Nijhout, 1998). The imaginal disks thus do most of their growth after the body has stopped growing, at the expense of reserves and other tissues accumulated during larval life. As a consequence, the final sizes of these structures often bear complex relationships to overall body size (Nijhout and Wheeler, 1996).

The allometric relationships of imaginal disk-derived structures in holometabolous insects are often non-linear, curving upward or downward with increasing body size, or sigmoidal (Fig. 1), although many other non-linear shapes have been described (Eberhard, 1982; Eberhard end Gutierrez, 1991; Emlen and Nijhout, 2000; Rowland, 2003). This diversity of shapes can be explained by variation in the relative degree and timing of growth of imaginal disks within the non-growing body of the prepupa (Nijhout and Wheeler, 1996). The prepupa is effectively a nutritionally closed system, and growth of some internal parts must necessarily be at the expense of other parts. Body size determines the amount of tissue and reserves that are available to support the growth of the imaginal disks. Small bodies can therefore pose different limits on growth than do large bodies, because they have smaller amounts of reserves. The interaction between the gradual depletion of reserves (which is at the cost of body mass that is not imaginal

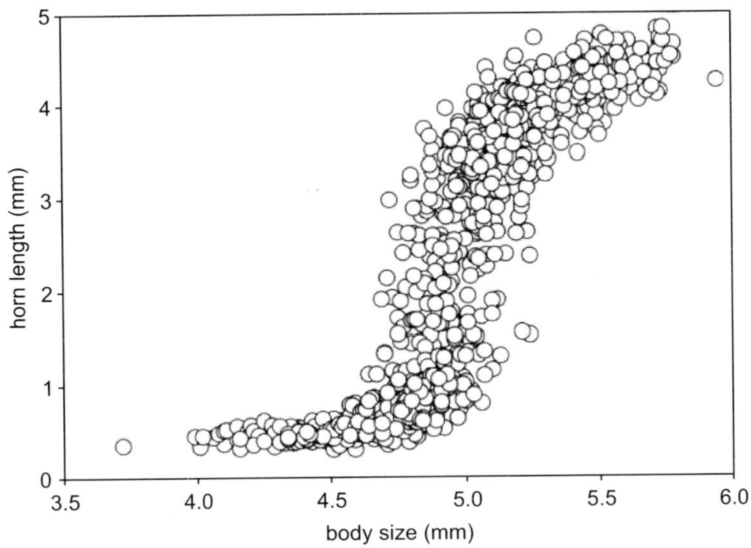

Fig. 1 Non-linear allometry of horn length in *Onthophagus taurus.* (After Moczek and Nijhout, 2003)

disks), and variation in disk growth rate can give rise to non-linear variation in the size of the imaginal disks relative to body size (Fig. 2). The complex non-linear allometries that result can therefore be understood as direct consequences of the mechanism by which imaginal disks and bodies of holometabolous insects grow.

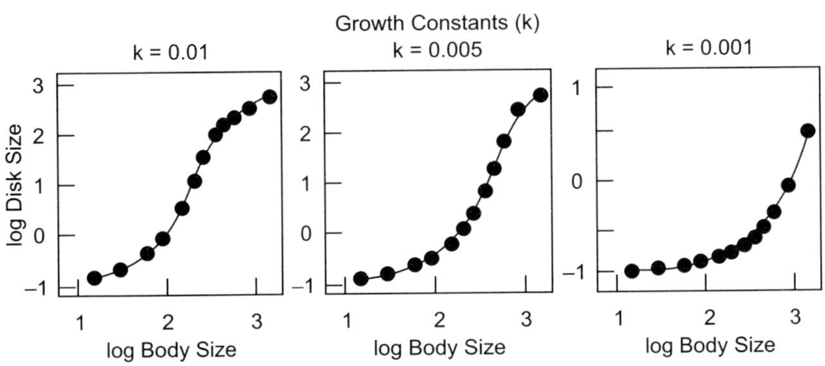

Fig. 2 Allometric relationships between disk size and body size predicted by a mathematical model that assumes disks grow within the nutritionally closed system of the prepupa. These panels illustrate the effect of different assumptions about the growth rate of disks (k represents the rate constant and is equivalent to a in Equation (1)). Each point is the endpoint of development of a disk in a body of a given size. (After Nijhout and Wheeler, 1996)

Hence, the existence of non-linear allometries does not require a special evolutionary explanation, because such allometries could simply be a mechanical consequence of the developmental physiology of the animal (Nijhout and Wheeler, 1996; Nijhout and Emlen, 1998; Stern and Emlen, 1999). Before an evolutionary explanation is attempted, the underlying physiological mechanism should be investigated. Whether a particular allometry confers a positive fitness advantage can, of course, only be determined through evolutionary studies. If the allometry is advantageous, then evolution could, of course, act to stabilize or further modify it (Nijhout and Emlen, 1998; West-Eberhard, 2003). The developmental/physiological mechanism accounts for the origin of the allometry, and provides the foundation on which further evolutionary modifications could take place. Whether such evolutionary modifications have indeed occurred would be difficult to deduce.

Tradeoffs

An interesting consequence of the fact that the imaginal disks of holometabolous insects grow rapidly within the closed environment of the prepupa is that they can potentially compete with each other for limiting resources for growth (Nijhout and Wheeler, 1996). Nijhout and Emlen (1998) and Moczek and Nijhout (2004) have shown that when one imaginal disk is removed, other disks can grow disproportionally large. Because growth of imaginal disks occurs in a closed system, competition will occur whenever resources for growth cannot be supplied or mobilized as fast as competing growing organs need or utilize them.

The sizes of imaginal disks that are in competition with each other during development will be negatively correlated because excessive growth of one disk must necessarily be accompanied by a decreased growth of the competing disk. A negative genetic correlation (or a negative additive genetic covariance) between two traits can result in divergent evolution of those traits because directional selection on one trait will result in a negatively correlated response to selection of the other trait. If the correlated response to selection of the second trait results in a maladaptive phenotype, then this developmental resource tradeoff among imaginal disks will act as a constraint on their respective evolutionary trajectories. For instance, in *Onthophagus taurus*, an increase in the size of the head horns in males is associated with a decrease in the size of the eyes (Nijhout and Emlen, 1998; Emlen and Nijhout, 2000), and in rhinoceros beetles, an increase in the thoracic horns is associated with a decrease in the size of the wings

(Kawano, 1995). Although larger horn size makes a male more effective in competing for access to females, it is conceivable that smaller eyes or wings would have a negative effect on fitness. If this negative effect is sufficiently strong it could inhibit the evolutionary enlargement of horns.

Examples of negative genetic correlations abound in the literature on quantitative genetics. Whenever a negative genetic correlation among body parts or physiological functions is encountered, some kind of developmental or physiological tradeoff should be suspected (Zera and Harshman, 2001; Zera, this volume). Rather than hypothesize the existence of genes or alleles for negative correlations, as is typically done, we can obtain a deeper understanding of the evolutionary potential of a system if we attempt to discover the underlying developmental-physiological mechanism that gives rise to the correlation.

Allometry and size tradeoffs among body parts are closely associated with the control of growth and body size. The mechanisms that control growth and size must therefore account for a broad spectrum of plastic traits, and understanding the mechanisms by which animals regulate the growth and final size of their body parts, and of their body as a whole, will be essential for developing a deep understanding of plasticity. We discuss some recent developments in our understanding of the control of growth and size next.

Growth and Body Size

Body size is among the characteristic features of species, and changes in body size are among the most common and widespread signals in evolution. Although body size and growth rate tend to be relatively stable within a species, they are also among the best-known plastic traits. In most animals, growth and body size are sensitive to both nutrition and temperature (de Jong and van der Have, this volume). Nutrition affects growth and body size in an indirect and interesting way. Tissues generally do not appear to respond directly to circulating nutrients: although nutrients are necessary for growth, they do not appear to be a sufficient condition for growth. In all animals that have been investigated, the growth of cells and tissues is controlled by specific growth factors or growth hormones, such as the many insulin-related peptides and insulin-like growth factors (e.g. Cross and Dexter, 1991; Böhni et al., 1999; Britton et al., 2002). These growth factors are typically neurosecretory hormones, or hormones whose secretion is controlled via tropic hormones by the central nervous system (CNS). It is the CNS that senses the availability of nutrients during development and controls the secretion of growth factors and thus, indirectly, the growth of cells and tissues. Often the CNS responds to a specific class of nutrients, such as amino acids or carbohydrates (Nijhout, 2003a).

Plasticity in growth rate due to variation in nutrition can thus be mediated by mechanisms that control the production and secretion of growth factors, as well as by the mechanisms that control the uptake and assimilation of nutrients. Genetic variation in plasticity can therefore be due to variation in the sensitivity of either or both of these processes to nutrients. For instance, if the rate of nutrient uptake typically runs at or near saturation, then variation in nutrients will have little or no effect on that part of the mechanism, and plasticity of growth rate will be largely due to nutrient effects on the hormone signaling mechanism.

Although growth must obviously occur in order for animals to achieve their final body size, it turns out that body size depends only indirectly on the growth rate. Adult body size in animals with determinate growth is primarily determined by the mechanisms that cause growth to stop at a particular time in development, or when a particular size has been attained; in such species, variation in growth rate typically only affects the time required to reach this endpoint (Blanckenhorn, 1999; Nijhout, 1994; 2003b; Shafiei et al., 2001; Davidowitz et al., 2003, 2004; Davidowitz and Nijhout, 2004). We know little about the mechanisms that terminate growth. However, in at least one insect group, the Hemiptera, larval growth stops when abdominal stretch receptors become stretched to a critical degree (Nijhout, 1979, 1984). These receptors then send action potentials to the brain that initiate a cascade of endocrine events that culminates in the secretion of the molting hormone, ecdysone. The secretion of ecdysone causes the insect to stop feeding and initiates the sequence of events that culminates in a molt to the adult. Hence, stretch-induced molting in the last larval stage, in effect, determines the body size of the adult.

In the tobacco hornworm, *Manduca sexta*, the control of adult body size is a bit more complex (Fig. 3). In about the middle of the last larval instar, there is a critical size that causes the initiation of a well-defined endocrine cascade. When a larva reaches this critical weight it stops secreting juvenile hormone (JH) and initiates the synthesis of JH-esterase, an enzyme that breaks down circulating JH (Nijhout, 1994; Browder et al., 2001; D'Amico et al., 2001; Davidowitz et al., 2003, 2004; Davidowitz and Nijhout 2004). As long as JH is present, it represses secretion of the prothoracicotropic hormone (PTTH) and the molting hormone, ecdysone, so no molt takes place. Once JH has been eliminated this inhibition is relieved, and the larva secretes PTTH and ecdysone during the following photoperiodic gate

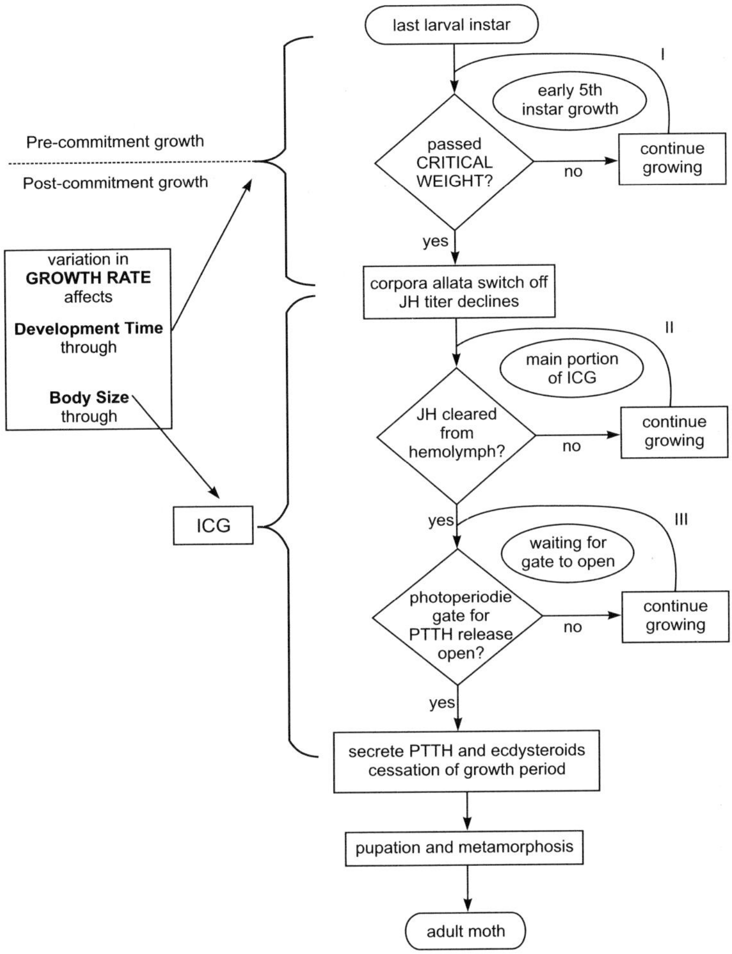

Fig. 3 Flow diagram for the physiological control of body size and development time in *Manduca sexta*. Growth in this species is exponential, and 90% of mass accumulates during the last (5th) larval instar, depicted here. In the right portion of the diagram, diamonds indicate decision points, rectangles indicate processes, ellipses are descriptors, and rounded rectangles indicate beginning and end stages. The commitment to stop growing and pupate is made when larvae reach the critical weight. Development time from that point onward no longer depends on nutrition and growth but is determined entirely by a sequence of endocrine events (decay of JH and secretion of PTTH and ecdysteroids). The interval between attainment of the critical weight and the secretion of PTTH and ecdysteroids is called the Interval to Cessation of Growth (ICG). During this period larvae can roughly double their mass, so the duration of this period has a major effect on body size. The box on the left indicates the mechanisms through which variation in growth rate (either due to variation in temperature or nutrition) affects body size and development time. (Redrawn from Davidowitz and Nijhout, 2004).

(Truman and Riddiford, 1974). As in the case of the Hemiptera, the secretion of ecdysone causes the larva to stop feeding and initiate metamorphosis. Hence, adult size is determined, in effect, by the size that the larva has achieved at the time ecdysone secretion begins. In female *M. sexta*, variation in the peak larval size explains about 64% of the variation in adult body mass (Davidowitz et al., 2004). The time interval between the achievement of the critical size and the actual secretion of ecdysone (the interval to cessation of growth, ICG) is one to three days (depending on the genetic strain), and all this time the larva continues to grow on its normal exponential trajectory. During this interval a larva can double its body size, so any variation in the processes that affect this interval has a profound effect on the final size of the adult.

Plasticity of body size in *Manduca* is determined by the degree to which environmental factors affect these physiological events. Nutrition and temperature can alter the growth rate and therefore the total mass that accumulates during the interval between the achievement of the critical weight and the final secretion of ecdysone. Plasticity of body size in response to nutrition is due to the effect of nutrition on the growth rate and on the value of the critical weight (Davidowitz et al., 2004). Temperature can also affect the rate at which JH breaks down, and therefore the total time available for growth. Accordingly, plasticity of body size in response to variation in temperature (Fig. 4) is due to the effect of temperature on both the overall growth rate, and on the delay time between the critical weight and the secretion of PTTH and ecdysone (Davidowitz et al., 2004).

Thus, nutrition and temperature have common effects on some processes that control body size and unique effects on others. Understanding the physiology of size regulation has turned out to be critical in dissecting the mechanism by which these two environmental variables affect plasticity (Davidowitz et al. 2004).

Polyphenisms

What are Polyphenisms?

Polyphenisms are extreme forms of phenotypic plasticity in which individuals can develop one of several discrete alternative phenotypes, depending on the environment they experience during development. Polyphenisms are distinguished from polymorphisms in that in the latter the alternative phenotypes are due to allelic (genetic) differences between individual, whereas in the former they are due to environmental differences.

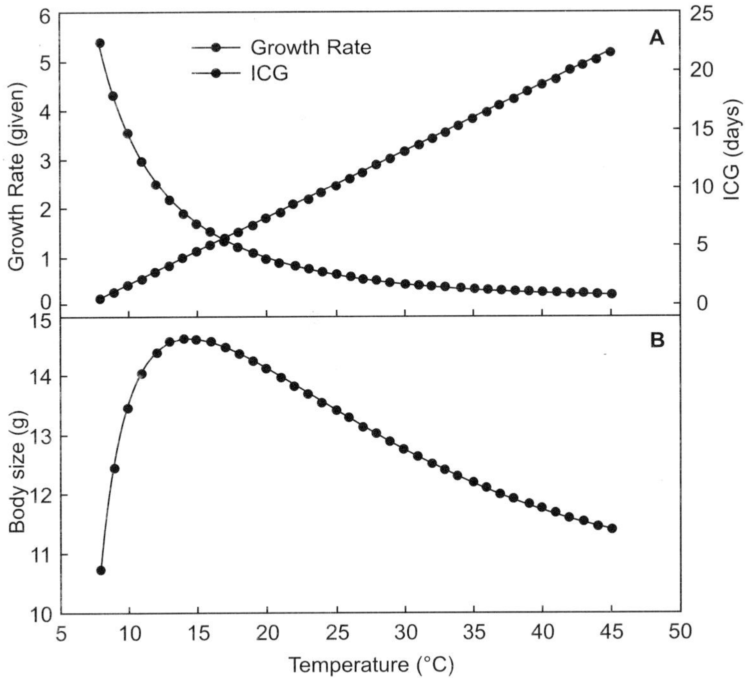

Fig. 4 Plasticity of body size with temperature in *Manduca sexta*. In many insects, body size decreases with increasing temperature. The mechanism that produces this non-intuitive effect can be understood by examining the temperature-dependence of the underling physiological mechanism. Panel A illustrates the empirical experimental data and shows that the growth rate increases linearly with temperature over a broad range of temperatures. The ICG is the interval to cessation of growth which is, in effect, the time required to break down juvenile hormone and disinhibit the secretion of PTTH and ecdysone, after the critical weight is reached (see Figure 3). The ICG decreases non-linearly with increasing temperature. Panel B illustrates the predicted body size, obtained by: Body Size = Critical Weight + (Growth Rate * ICG); where Critical Weight = 7 grams. The physiological mechanism predicts that over a small range of very low temperatures body size should increase with temperature, but that over a broad range of physiological temperatures body size will decrease with increasing temperatures. (Redrawn from Davidowitz and Nijhout, 2004).

Best known among the polyphenisms are the caste polyphenism of social insects (queens, workers, soldiers), the phase polyphenism of migratory locusts (gregarious migratory and solitary sedentary phases), and the seasonal polyphenisms (spring/summer, summer/autumn, dry season/ wet season) that are common throughout the insects. Examples of seasonal polyphenisms are the wing-length polyphenisms of many Orthoptera, Hemiptera and Homoptera, and the color and wing shape polyphenisms of the Lepidoptera and other insects (Fig. 5; Nijhout, 1994, 1999; Brakefield and Frankino, this volume).

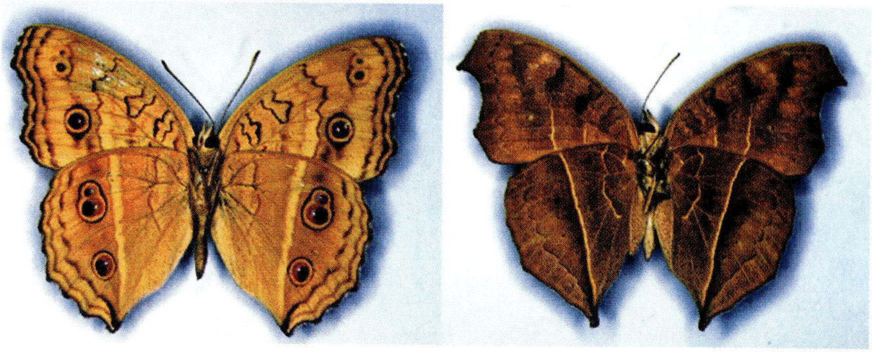

Fig. 5 Seasonal polyphenism in the tropical nymphalid butterfly, *Precis almana*. The dry season form has a more angular wing shape, and its color pattern is a dead-leaf mimic.

Seasonal polyphenisms are very common among the insects, and are highly evolved adaptations to predictable environments (Shapiro, 1976; Moran, 1992). They are almost always cued by token stimuli (i.e., a stimulus such as photoperiod that itself is not the environmental variable to which the polyphenism is an adaptation, but is predictive of that environment), and the environment-sensitive period usually precedes the development of the phenotype by many days, weeks or months (Nijhout, 1999, 2003c). Seasonal polyphenisms are typically cued by the photoperiod, and occasionally by moderate temperature changes or changes in food quality. Caste polyphenisms in social insects are triggered by pheromones or particular nutrients (Wilson, 1971; Nijhout and Wheeler, 1982). Body size polyphenisms occur in some mating systems in insects in which males above (or below) a particular body size develop special morphologies that aid in mating success (Eberhard, 1985; Emlen, 1997a; Emlen and Nijhout, 2000). Body size polyphenisms are manifest as allometries with more-or-less sharply sigmoid relationships between the relevant trait and body size.

Developmental/Genetic Switching

The developmental and physiological regulation of insect polyphenisms is relatively well understood. In all cases studied so far, there is a developmental switch at the time of a molt (usually the metamorphic molt that produces the adult insect) that affects the phenotype of the following stage, and this switch is controlled by hormones. In insects, the hormones that control metamorphosis, ecdysone and JH, are most commonly involved in polyphenic switching, but there are also a number of neurosecretory hormones that can play an important role in some systems (Nijhout, 1994; Zera and Denno, 1997; see also Zera, this volume). These developmental hormones control the pattern of gene expression in specific tissues during discrete critical periods in development (Evans and Wheeler, 2000). Different patterns of gene expression, and therefore different developmental trajectories are taken depending on whether a hormone is present (or above some threshold) or absent during a given sensitive critical period. Each tissue has a different pattern of hormone-sensitivity, so that a particular pulse of hormone secretion affects some tissues but not others (Jindra et al., 1996; Nijhout, 1999, 2003c).

Environmental Input

Developmental switching in polyphenisms is thus compartmentalized in time and space by temporal variation in hormone secretion and by spatial and temporally varying patterns of hormone responsiveness at the tissue level. The pattern of hormone secretion is controlled by the CNS, and the CNS appears to provide the principal mechanism through which environmental stimuli affect gene expression, and ultimately phenotype. Environmental experiences such as temperature, photoperiod, pheromones, and nutrition reprogram the brain's neurosecretory activity, and the altered pattern of hormone secretion activates alternative gene expression pathways in the target tissues. There is usually a more-or-less well-defined sensitive period during which the inductive signal must be received. This sensitive period occurs days to weeks before the actual developmental switch takes place, and in many cases a prolonged exposure to the inducing environment is required to get a normal developmental switch. Thus, the reprogramming of the CNS appears to require the integration of environmental signals that accumulate over a period of several days or weeks. The physiological mechanism by which specific environmental input is integrated, how this information is stored, and how it results in an altered pattern of hormone secretion by the CNS remains unknown and largely unexplored.

Evolution

Polyphenisms can evolve by changes in the physiological mechanism that controls the developmental switch. Changes in this mechanism would be manifest as changes in the sensitivity to the environmental signal, changes in the pattern of hormone secretion, or changes in the genetic response to the hormonal signal. The diversity of the control mechanisms of polyphenic development (Fig. 6) may indicate that the evolution of a specific mechanism is contingent on previous taxon-specific evolutionary changes in that particular mechanism.

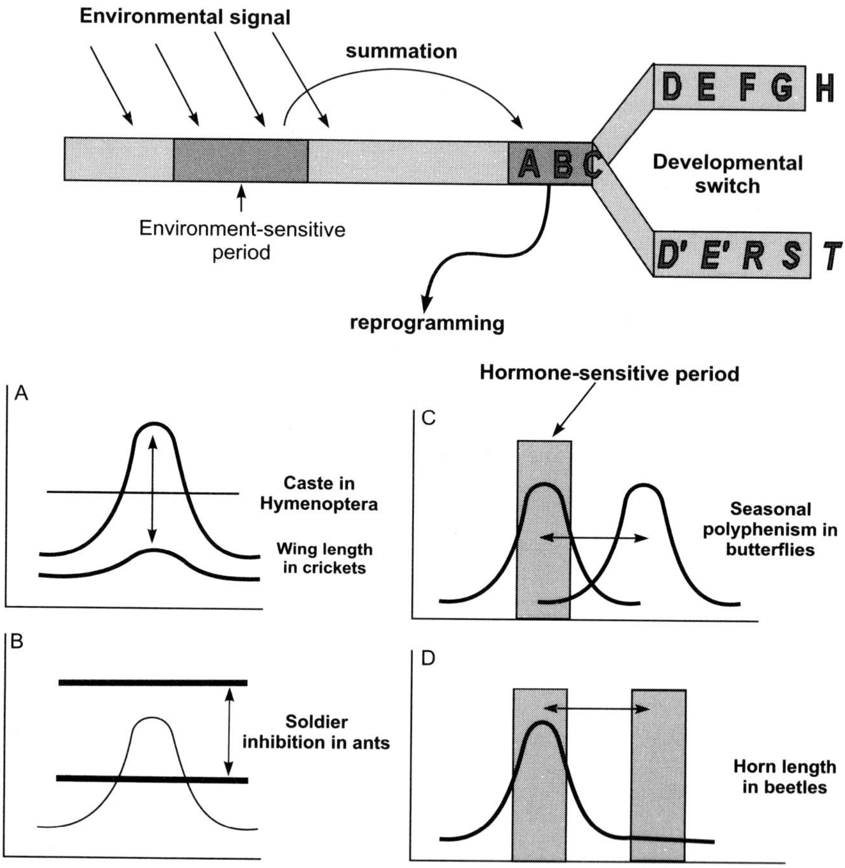

Fig. 6 Control of polyphenic development. Upper diagram represents a developmental trajectory with a switchpoint resulting in two alternative phenotypes **H** and ***T***. Sometime during development there is an environment-sensitive period. Environmental signals received during this critical period are integrated and result in a switch in the developmental program so that one of the two alternative pathways is chosen. Reprogramming occurs via alteration in a hormone signaling system. Four possible endocrine switch mechanisms have been found: A, Variation in hormone level relative to a threshold; B, variation in the threshold of sensitivity to a hormone; C, variation in the timing of hormone secretion relative to a hormone-sensitive period; D, variation in the timing of a hormone sensitive period. Next to each diagram we give an example of a polyphenism that is controlled by such a mechanism. (Redrawn from Nijhout, 2003c).

Although in most cases we do not know how a specific mechanism evolved, we do have a nice example of rapid evolution in the horn polyphenism of the dung beetle, *Onthophagus taurus* (Moczek and Nijhout, 2002, 2003; Moczek, this volume). Horns in this beetle develop in males that grow larger than a threshold body size. Body size in this species is not heritable, but is entirely determined by nutrition during the larval stage (Emlen, 1994; Shafiei et al., 2001). After introduction into Australia, *O. taurus* rapidly evolved a much larger threshold body size, presumably in response to increased interference competition for access to females (Moczek and Nijhout, 2002, 2003). Horn development is controlled by a brief JH-sensitive period that occurs during the late larval stage (Emlen and Nijhout, 1999, 2001). High JH during this sensitive period leads to horn development. The evolutionary change in the threshold has been shown to be due to a shift in this JH-sensitive period (Moczek and Nijhout, 2003) .

Because polyphenic development is highly compartmentalized by the temporal pattern of hormone secretion and the spatial pattern of tissue responsiveness, the development of a given trait can be switched while imposing little or no correlated changes in other traits. It is possible that this compartmentalization inherent in the developmental physiology of insects has facilitated the evolution of the great diversity of polyphenic responses to the environment.

Homeostasis

Perhaps the greatest contribution that physiology has made to biology is a mechanistic understanding of homeostasis. A century of physiological studies have shown that all life processes obey the laws of physics and chemistry and are therefore sensitive to environmental variables such as temperature, pH, osmolarity, ionic strength, and hydration. In addition, biochemistry tells us that metabolism is exquisitely sensitive to variation in substrate availability and micronutrients (such as vitamins that are needed as co-enzymes). Likewise, toxins in chemically defended food items, such as toxic plants, can disrupt many physiological regulatory mechanisms and alter metabolic pathways. Deficiencies or excesses of various ingested substances can lead to a broad diversity of physiological and developmental abnormalities.

Homeostatic physiology is the set of mechanisms that evolved to make the function and development of animals largely insensitive to environmental variation. While homeostasis represents a type of canalization, its existence is, interestingly, due to an underlying physiological plasticity, whereby the organism's physiology (or behavior) responds plastically to a suboptimal internal environment. The pervasiveness of homeostatic mechanisms poses a special problem for the evolution of phenotypic plasticity. Do homeostatic mechanisms constrain the evolution of phenotypic plasticity? Is phenotypic plasticity merely the amount of phenotypic variation that is "tolerated" by the homeostatic mechanisms? Does phenotypic plasticity occur preferentially in traits for which no homeostatic mechanism has evolved? Does adaptive phenotypic plasticity involve the evolution of mechanisms that specifically circumvent or eliminate certain homeostatic mechanisms?

A wide range of homeostatic mechanisms modulate body temperature, osmolarity, pH, hunger regulation (including salt hunger, fat hunger and sugar hunger), and various aspects of metabolism. In each of these mechanism there is a sensor that detects the level of a particular variable, and a response system that brings the relevant variable back to a given *set-point*. The sensors, and the feedback pathways by which the set-point is restored have been widely studied and are generally well understood (Schmidt-Nielsen, 1997). Interestingly, little or nothing is known about the nature of the set-points (Schmidt, Nielsen, 1994). Exactly how 37 degrees Celsius is represented in human physiology, for instance, remains an unsolved mystery, as does the nature of every other homeostatic set-point.

The existence of physiological homeostatic mechanisms, coupled with the observation that the end-product of development is stable and relatively insensitive to many environmental perturbations, has been thought by many to imply the existence of developmental homeostatic mechanisms. These hypothetical mechanisms go by various terms such as canalization, robustness and developmental homeostasis (Nijhout and Davidowitz, 2003). These terms are today primarily used metaphorically, and while many hypotheses exist, no mechanism of developmental homeostasis has yet been described.

It is difficult to imagine how the constancy of the phenotype is achieved in the face of extreme environmental and experimental perturbation except through the action of some kind of plastic mechanism that targets a final set-point. A number of investigators have observed that the morphology of developing embryos is much more variable than the final adult phenotype. Errors accumulated during development do not necessarily add up to an ever-increasing variability with time but are somehow eliminated or compensated as development progresses (Klingenberg, 1996; Hallgrimsson et al., 2003). Development gives the appearance of being somehow "goal oriented," and, in spite of initial variability, development is convergent on a characteristic phenotype.

Convergent development is postembryonic, and is therefore likely to be controlled in the same way that other aspects of postembryonic development are controlled, namely physiologically, through the interaction of hormones, growth factors and other cytokines. But whether developmental homeostasis operates by mechanisms that are analogous to those of physiological homeostasis is not known at present. It is possible that physiological homeostasis does not provide a good model for understanding developmental homeostasis, and it seems likely that this quandary will not be resolved until we understand the nature of the set-points, or targets, of ontogeny. Here is where studies on regeneration may shed some light, since regenerating tissues develop to the same endpoint as tissues that developed normally, but obviously do so from very different starting points.

Postscript

Recent findings on the developmental physiology of insects have greatly expanded our understanding of the nature of phenotypic plasticity. We are no longer constrained to hypothesize "genes for plasticity" that are somehow sensitive to the "environment" but are beginning to understand how the different parts of the mechanism that generates the phenotype respond to specific environmental variables. Phenotypic plasticity can now be explained in terms of plasticity of specific developmental processes that give rise to the phenotype.

Among the significant findings of the developmental-physiological approach is that postembryonic development is coordinated by the central nervous system via the secretion of hormones. Involvement of the CNS provides an entirely new way for the environment to affect development of the phenotype. Not only can cells and tissues be directly sensitive to say, temperature, but temperature experienced during development can also alter the pattern of secretion of developmental hormones, and these can affect the development of tissues that themselves are relatively temperature-insensitive. Hormone-mediated responses to the environment should thus be able to evolve independently from the direct sensitivities of tissues to environmental variables, and may counteract the negative effects of the environmental variables.

Another significant finding is that the control of postembryonic development is spatially and temporally compartmentalized by the patterns of developmental hormone secretion and by the patterns of tissue responsiveness to those hormones. This is especially true in insects that display both extreme metamorphosis, and extreme organ-specific developmental ontogeny (imaginal disks). This compartmentalization of control allows substantial uncoupling of variation and evolution of different body parts, and allows different features of an organism to evolve different sensitivities to environmental variation. A consequence of this compartmentalization is the widespread evolution of polyphenisms and complex allometries among the insects.

Acknowledgments

The work in our laboratories was supported by grants IBN-9975168 and IBN-0212621 from the National Science Foundation.

References

Blanckenhorn, W. U. 1999. Different growth responses to temperature and resource limitation. Evol. Ecol. 13: 395–409.

Böhni, R., Riesgo-Escovar, J. Oldham, S., Brogiolo, W., Stocker, H., Andruss, B. F. Beckingham, K. and Hafen, E. 1999. Autonomous control of cell and organ size by CHICO, a *Drosophila* homolog of vertebrate IRS1–4. Cell 97: 865–875.

Britton, J. S., Lockwood, W. K., Li, L., Cohen, S. M. and Edgar, B. A. 2002. *Drosophila* insulin/PI3 kinase pathway coordinates cellular metabolism with nutritional conditions. Dev. Cell 2: 239–249.

Brogiolo, W. Stocker, H., Ikeya, T., Rintelen, F., Fernandez, R. and Hafen, E. 2001. An evolutionarily conserved function for the *Drosophila* insulin receptor and insulin-like peptides in growth control. Curr. Biol. 11: 213–221.

Browder, M. H., D'Amico, L. J. and Nijhout, H. F. 2001. The role of low levels of juvenile hormone esterase in the metamorphosis of *Manduca sexta*. J. Insect Science 1(11): 1–5.

Cock, A. G. 1966. Genetical aspects of metrical growth and form in animals. Quart. Rev. Biol. 41: 131–190.

Cross, M. and Dexter, T. M. 1991. Growth factors in development, transformation, and tumorigenesis. Cell 64: 271–280.

D'Amico, L. J., Davidowitz, G. and Nijhout, H. F. 2001. The developmental and physiological basis of body size evolution in an insect. Proc. Roy. Soc. London, Ser. B 268: 1589–1593.

Davidowitz, G., D'Amico. L. J. and Nijhout, H. F. 2003. Critical weight in the development of insect body size. Evolution & Development 5: 188–197.

Davidowitz, G., D'Amico, L. J. and Nijhout, H. F. 2004. The effects of environmental variation on the mechanism that regulates insect body size. Evolutionary Ecology Research 6: 49–62.

Davidowitz, G. and H. F. Nijhout, 2004. The physiological basis of reaction norms: The interaction between growth rate, the duration of growth and body size. Integrative and Comparative Biology 44: 443–449.

Eberhard, W. G. 1985. Beetle horn dimorphism: making the best of a bad lot. Amer. Nat. 119: 420–426.

Eberhard, W. G. and Gutierrez, E. E. 1991. Male dimorphisms in beetles and earwigs and the question of developmental constraint. Evolution 45: 18–28.

Emlen, D. J. 1994. Environmental control of horn length dimorphism in the beetle *Onthophagus acuminatus* (Coleoptera: Scarabaeidae). Proc. Roy. Soc. London Ser. B 256: 131–136.

Emlen, D. J. 1997a. Alternative reproductive tactics and male dimorphism in the horned beetle *Onthophagus acuminatus*. Behav. Ecol. Sociobiol. 41: 335–341.

Emlen, D. J. 1997b. Diet alters male horn allometry in the beetle *Onthophagus acuminatus* (Coleoptera: Scerabaeidae). Proc. Roy. Soc. London Ser. B 264–567–574.

Emlen, D. J. and Nijhout, H. F. 2000. The development and evolution of exaggerated morphologies in insects. Ann. Rev. Entomol. 45: 661–708.

Emlen, D. J. and Nijhout, H. F. 1999. Hormonal control of horn length dimorphism in the dung beetle *Onthophagus taurus* (Coleoptera: Scarabaeidae). J. Insect Physiol. 45: 45–53.

Emlen, D. J. and Nijhout, H. F. 2001. Hormonal control of horn length dimorphism in *Onthophagus taurus* (Coleoptera: Scarabaeidae): a second critical period of sensitivity to juvenile hormone. J. Insect Physiol. 47: 1045–1054.

Evans, J. D. and Wheeler, D. E. 2000. Expression profiles during honeybee caste determination. Genome Biology 2(1): research0001.1–0001.6.

Hallgrimsson, B., Miyake, T., Wilmore, K. and Hall, B. K. 2003. Embryological origins of developmental stability: size, shape and fluctuating asymmetry in prenatal random bred mice. J. Exp. Zool. (Mol. Dev. Evol.) 296B: 40–57.

Huxley, J. S. 1932. Problems of Relative Growth. Methuen, London.

Jindra, M., Malone, F., Hiruma, K. and Riddiford, L. M. 1996. Developmental profiles and ecdysteroid regulation of the mRNAs for two ecdysone receptor isoforms in the epidermis and wings of the tobacco hornworm, *Manduca sexta.* Dev. Biol. 180: 258–272.

Kawano, K. 1995. Horn and wing allometry and male dimorphism in giant rhinoceros beetles (Coleoptera: Scarabaeidae) of tropical Asia and America. Ann. Entomol. Soc. Amer. 88: 92–99.

Klingenberg, C. P. 1996. Individual variation of ontogenies: a longitudinal study of growth and timing. Evolution 50: 2412–2428.

Klingenberg, C. P. and Zimmerman, M. 1992. Static, ontogenetic, and evolutionary allometry: A multivariate comparison of nine species of waterstriders. *Amer. Nat.* 140: 601–620.

Kremen, C. and Nijhout, H. F. 1998. Control of pupal commitment in the imaginal disks of *Precis coenia* (Lepidoptera: Nymphalidae). J. Insect Physiol. 44: 287–296.

Moczek, A. P. and Nijhout, H. F. 2002. Developmental mechanism of threshold evolution in a polyphenic beetle. Evolution & Development 4: 252–264.

Moczek, A. P. and Nijhout, H. F. 2003. Rapid evolution of a polyphenic threshold. Evolution & Development 5: 259–268.

Moczek, A. P. and Nijhout., H. F. 2004. Trade-offs during the development of primary and secondary sexual traits in a horned beetle. Amer. Nat. 163: (in press).

Moran, N. A. 1992. The evolutionary maintenance of alternative phenotypes. Amer. Nat. 139: 971–989.

Nijhout, H. F., 1979. Stretch induced molting in *Oncopeltus fasciatus.* J. Insect Physiol. 25: 277–281.

Nijhout, H. F. 1984. Abdominal stretch reception in *Dipetalogaster maximus* (Hemiptera: Reduviidae). J. Insect Physiol. 30: 629–633.

Nijhout, H. F., 1990, Metaphors and the role of genes in development. BioEssays 12: 44–446.

Nijhout, H. F. 1994. Insect Hormones. Princeton University Press, Princeton, NJ.

Nijhout, H. F. 1999. Control mechanisms of polyphenic development in insects. BioScience 49: 181–192.

Nijhout, H. F. 2002. The nature of robustness in development. BioEssays 24: 553–563.

Nijhout, H. F. 2003a. The control of growth. Development. 130: 5863–5867.

Nijhout, H. F. 2003b. The control of body size in insects. Developmental Biology 261: 1–9.

Nijhout, H. F. 2003c. The development and evolution of adaptive polyphenisms. Evolution & Development 5: 9–18.

Nijhout, H. F. 2003d. The importance of context in genetics. Amer. Sci. 91: 416–423.

Nijhout, H. F. and Davidowitz, G. 2003. Developmental perspectives on phenotypic variation, canalization, and fluctuating asymmetry. *In* Developmental Instability: Causes and Consequences (M. Polak, Ed.), pp 3–13. Oxford Univ. Press, Oxford, UK.

Nijhout, H. F. and Emlen, D. J. 1998. Competition among body parts in the development and evolution of insect morphology. Proc. Nat. Acad. Sci. 95: 3685–3689.

Nijhout, H. F. and D. E. Wheeler, 1982. Juvenile hormone and the physiological basis of insect polymorphisms. Quart. Rev. Biol. *57*: 109–133.

Nijhout, H. F. and Wheeler, D. E. 1996. Growth models of complex allometries in holometabolous insects. Amer. Nat. 148: 40–56.

Rowland, J. M. 2003. Male horn dimorphism, [phylogeny and systematics of rhinoceros beetles of the genus *Xylotrupes* (Scarabeidae: Coleoptera). Austral. J. Zool. 51: 213–258.

Schmidt-Nielsen, K. 1994. How are control systems controlled? Amer. Sci. 82: 38–44.

Schmidt-Nielsen, K. 1997. Animal Physiology: Adaptation and Environment. Cambridge Univ. Press, Cambridge, UK.

Shafiei, M., A. P. Moczek and H. F. Nijhout. 2001. Food availability controls the onset of metamorphosis in the dung beetle *Onthophagus taurus* (Coleoptera: Scarabeidae). Physiol. Entomol. 26: 173–180.

Shapiro, A. M. 1976. Seasonal polyphenism. Evol. Biol. 9: 259–278.

Stern, D. L. 2001. Body size evolution: how to evolve a mammoth moth. Curr. Biol. 11: R917–R919.

Stern, D. L. and Emlen, D. J. 1999. The developmental basis for allometry in insects. Development 126: 1091–1101.

Svacha, P. 1992. What are and what are not imaginal disks: reevaluation of some basic concepts (Insecta: Holometabola). Dev. Biol. 154: 101–117.

Truman, J. W. and Riddiford, L. M. 1974. Physiology of insect rhythms: III. the temporal organization of the endocrine events underlying pupation of the tobacco hornworm. J. Exp. Biol. 60: 371–382.

Via, S. and Lande, R. 1985. Genotype-environment interaction and the evolution of phenotypic plasticity. Evolution 39: 505–523.

West-Eberhard, M. J. 2003. Developmental Plasticity and Evolution. Oxford Univ. Press, Oxford, UK.

Wilson, E. O. 1971. The Insect Societies. Harvard Univ. Press, Cambridge, MA, USA.

Zera, A. J. and Denno, R. F. 1997. Physiology and ecology of dispersal polymorphism in insects. Ann. Rev. Entomol. 42: 207–231.

Zera, A. J. and Harshman, L. G. 2001. The physiology of life history trade-offs in animals. Ann. Rev. Ecol. Syst. 32: 95–126.

14

Wing Polymorphism in *Gryllus* (Orthoptera: Gryllidae): Proximate Endocrine, Energetic and Biochemical Mechanisms Underlying Morph Specialization for Flight vs. Reproduction

Anthony J. Zera[1]

[1]*School of Biological Sciences, University of Nebraska, Lincoln, NE 68588-0118*
fax: 402-472-2083; phone: 402-472-2768, E-mail: azera1@unlnotes.unl.edu

Abstract

Wing polymorphism is a common feature of insects, and consists of a flight-capable morph that delays reproduction, and one or more flightless morphs that exhibit substantially elevated early-age fecundity. Despite considerable speculation for over five decades, until recent studies were undertaken on wing-polymorphic crickets of the genus *Gryllus*, little detailed information was available concerning the endocrine or biochemical mechanisms that control morph development or specialization for flight vs. reproduction. Titers of two key morphogenetic hormones, juvenile hormone (JH) and ecdysone, and the activity of juvenile hormone esterase, an enzyme that degrades and partially regulates the JH titer, differ dramatically between nascent morphs of two species of *Gryllus*. Evidence points to a role for either or both of these hormones in regulating important aspects of morph development. In the adult stage, an unexpected morph-specific circadian rhythm for the hemolymph JH titer was observed in *G. firmus*, in both the laboratory and field. The JH titer changes 10–50 fold daily in the flight-capable morph, rising above and falling below the titer in the flightless morph, which exhibits a relatively constant JH titer. The cyclic change in the JH titer likely regulates some aspect of nocturnal flight in the dispersing morph. By contrast, the ecdysteroid titer is consistently higher in the more fecund flightlesss morph, and possibly plays an important role in regulating the enhanced egg production of that morph. Morph-specific differences in JH and ecdysteroid titer regulation are currently known in greater detail for wing polymorphism in *Gryllus* than for any other case of ecologically-important, complex (multi-trait) polymorphism (e.g. phase, wing, or flight-muscle polymorphism).

During adulthood, the dispersing morph accumulates considerably more total lipid and triglyceride, its main flight fuel, compared with its flightless counterpart; increased lipid accumulation may be an important cause of reduced egg production in the dispersing morph. Morphs consume and assimilate nearly the same amount of nutrients, ruling out increased acquisition of nutrients as the cause of elevated lipid in the dispersing morph. *In vivo* radiotracer studies have directly documented that the dispersing morph biosynthesizes more fatty acid, triglyceride, and total lipid, and oxidizes less fatty acid for energy production than the flightless morph. Morph-specific differences in the flow of nutrients through pathways of lipid metabolism, in turn, result from dramatic morph-specific differences in the activities of enzymes that comprise these pathways. These studies currently represent the most detailed analysis of the biochemical-metabolic basis of a life history trade-off, underlying morph-specialization in a complex polymorphism. Alterations of intermediary metabolism, that play a key role in morph specialization for flight vs. reproduction, appear to have evolved by modification of the endocrine control of intermediary metabolism.

Introduction

Many insects exhibit complex polymorphism, in which phenotypes (morphs) differ qualitatively in a diverse array of traits and are specialized for functions such as flight, reproduction, defense, offense, or crypsis (Nijhout, 1994, 1999). These polymorphisms result from a variety of causes: alternate morphs may be encoded by different genotypes (genetic polymorphism), induced by different environments (environmental polyphenism), or produced by variation in both genetic and environmental factors (referred to simply as polymorphism if the specific cause of morph production is not specified). Examples include phase polyphenism in locusts, caste polyphenism in social insects, seasonal polyphenism in butterflies, and dispersal (wing) polymorphism in a wide variety of insect groups (Fig. 1) (Hardie and Lees 1985; Zera and Denno, 1997; Pener and Yerushalmi 1998; Nijhout 1994, 1999; West-Eberhard, 2003; Zera, 2004, 2005). These polymorphisms often play an integral role in the life cycle of the species in which they are found, for example, producing dispersing or reproductive phenotypes adapted to a particular season or habitat.

For decades, the proximate physiological mechanisms that control the expression of traits that comprise alternate phenotypes have been intensively studied by insect physiologists and, more recently, by evolutionary biologists. These physiological studies have provided important insights

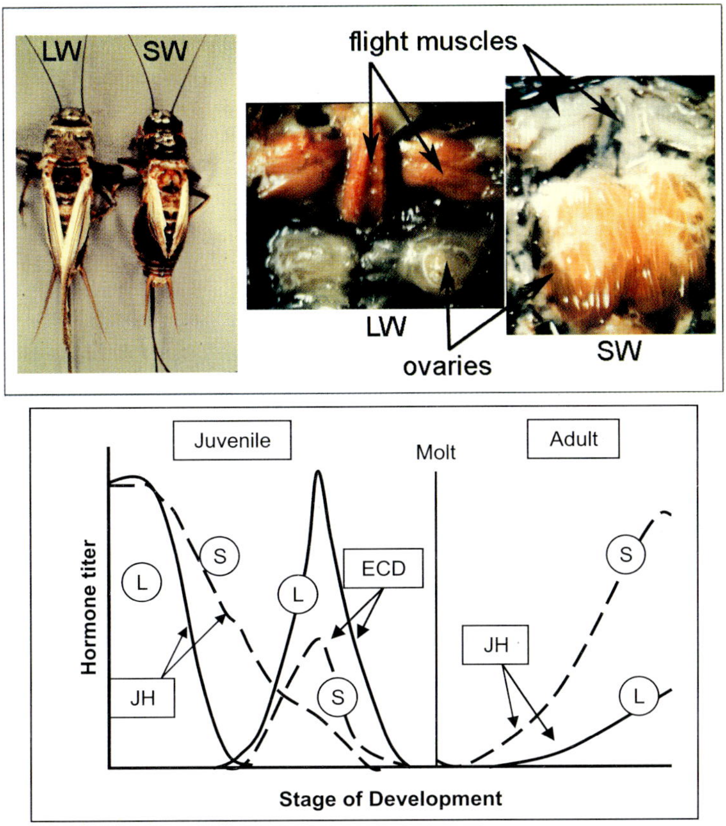

Fig. 1 Top panels: Flight-capable, long-winged, (LW) and flightless, short-winged (SW) female morphs of *Gryllus rubens* of the same age (day 5 of adulthood). In the top left panel, the fore wings have been removed to show variation in the hind wings. The middle and right panels illustrate dissections of same-aged morphs showing much larger and functional flight muscles, but much smaller ovaries in the LW females, and substantially-underdeveloped flight muscles but much larger ovaries in SW females. Bottom panel: The "classical model" of the endocrine control of wing morph development and reproduction. This panel illustrates hypothetical variation in the juvenile hormone (JH) and ecdysteroid (ECD) titers that potentially regulate differences in development and reproduction between long-winged (L) and short-winged (S) morphs during juvenile and adult stages (see text for explanation of hypotheses).

into a number of important evolutionary topics such as the mechanistic basis of adaptive phenotypic plasticity and genetic polymorphism, the role of development in evolution, and the physiological basis of life history evolution (Nijhout, 1994, 1999; Zera and Harshman, 2001; West-Eberhard, 2003; Zera, 2004, 2005).

One of the most intensively studied complex polymorphisms has been wing polymorphism (Wigglesworth 1961; Lees 1966; Gould 1977; Nijhout and Wheeler 1982; Hardie and Lees 1985; Roff 1986; Socha 1993; Tanaka 1993; Nijhout, 1999), especially in crickets of the genus *Gryllus* (Zera and Denno 1997; Zera and Harshman 2001; Zera and Zhao 2003a, b; Zhao and Zera 2002, 2004a; Zera, 2004, 2005; Zera et al. 2007a). During the past two decades work undertaken on crickets of the genus *Gryllus* has provided fundamental information on the endocrine control of morph development, the endocrine regulation of morph specializations for flight vs. reproduction, and the biochemical and physiological underpinnings of morph-specific adaptations for flight vs. reproduction. Indeed, wing polymorphism in *Gryllus* now constitutes one of the most intensively studied complex polymorphisms with respect to the physiological and biochemical underpinnings of morph specialization. In this review I will mainly discuss work undertaken in my laboratory during the past two decades on the proximate hormonal, energetic and biochemical mechanisms regulating morph development, reproduction, and flight capability. Although this chapter is contained in a book on phenotypic plasticity in insects, most of the endocrine and biochemical work reviewed here deals with genetically-based wing polymorphism. This focus was necessitated since most of the detailed endocrine and biochemical work has been obtained from studies of genetic stocks that produce alternate morphs. However, it is likely that similar endocrine and biochemical mechanisms underlie both environmentally-induced polyphenism and genetically-based polymorphism for dispersal capability (see below for a specific, documented example of this point, and West-Eberhard (2003) for a general discussion of this topic).

Background on Wing Polymorphism

Before discussing the details of the endocrine and biochemical bases of wing polymorphism, it is essential to give some background information on the polymorphism itself. Dispersal and reproduction are important organismal features that commonly trade-off (i.e. are negatively associated; Dingle 1996; Zera and Denno 1997; Zera and Harshman 2001). The most dramatic expression of this trade-off is manifest as dispersal polymorphism in which qualitatively discrete phenotypes are produced, that are adapted for dispersal at the expense of reproduction and vice-versa (Fig. 1). The most extensively studied type of dispersal polymorphism, wing polymorphism, is common in many insect groups, most notably the Hemiptera/Homoptera (waterstriders, planthoppers and aphids), Coleoptera (beetles), and

Orthoptera (crickets and grasshoppers) (Harrison 1980; Roff 1986; Masaki and Shimizu 1995; Zera and Denno 1997). While often referred to as "wing polymorphism," the polymorphism actually consists of morphs that differ qualitatively in numerous important aspects of flight capability (e.g. morphology of wings and flight muscles, production of flight fuel) and reproduction. Typically, one morph has fully-developed wings, large functional flight muscles, and a high concentration of flight fuel, but delays ovarian growth (termed LW = long-winged morph). The alternate, obligately flightless morph has underdeveloped wings [either shortened (SW) or absent (APT = apterous)] and underdeveloped flight muscles, but begins reproducing earlier and has substantially-elevated early-age fecundity (Roff 1986; Zera and Denno 1997; Zera and Harshman 2001). Elevated reproduction, especially during early adulthood, is one of the hallmarks of the flightless morph in wing-polymorphic insects (Roff 1986; Zera and Denno, 1997).

Many species, including many *Gryllus* species, have another flightless morph that has fully developed wings but reduced, non-functional flight muscles, and is derived from the flight-capable morph by histolysis of flight muscles during adulthood (Zera and Denno 1997; Zera *et al.* 1997). The long-winged flightless (i.e. flight-muscle-polymorphic) morph is not considered in this review (see discussions in Zera and Denno 1997; Zera et al. 1997; Zera and Larsen 2001; Zera and Cisper 2001; and references therein). Hence, in this chapter, "LW" refers to a morph with both fully-developed wings *and* fully-developed flight muscles. In addition to these morphological and reproductive differences, a wide variety of biochemical, physiological, and behavioral traits differ between the morphs, and contribute importantly to morph-specialization for flight vs. reproduction (Tanaka 1993; Sula et al. 1998; Zera et al. 1997, 1999; Zera and Denno 1997; Zera and Harshman 2001; Zera and Zhao 2003, 2004; Zhao and Zera 2002, 2004).

Depending upon the species, wing polymorphism may be exclusively a genetic polymorphism (e.g. some beetles) or an environmental polyhenism (some aphids). Most commonly (e.g. waterstriders, planthoppers, and crickets), the polymorphism results from variation in both genetic factors (typically polygenic control), and environmental factors (most commonly density, photoperiod, temperature, and food quality) (Roff and Fairbairn 1991; Morooka and Tojo 1992; Masaki and Shimizu 1995; Zera and Denno 1997). Sensitivity to environmental signals allows the production of phenotypes that are adapted to specific environments, such as the production of the dispersing morph in aphids as a function of increased crowding, or during a particular season (Lees 1966; Dixon 1985). Because clones of aphids can be easily obtained and reared, proximate environmental signals that control wing polyphenism have been particularly well-studied in this group (Lees 1966; Dixon 1985; Hardie and Lees 1985).

Ecological aspects of wing polymorphism are reviewed in Vepsalainen (1978), Roff (1990 1994), and Zera and Denno (1997), and only a very brief summary will be given here. Extensive theory predicts that dispersal should be diminished in persistent habitats, because of the various physiological and fitness costs of flight and flight-capability (e.g., production and maintenance of the flight apparatus; Roff 1994). Conversely, flight is essential in many insects for exploiting or escaping temporary habitats and for tracking changing resources or mates. Thus, environmental heterogeneity should favor the evolution of wing polymorphism. The most detailed ecological studies of wing-polymorphism have been undertaken in species of waterstriders, and especially planthoppers (Vepsalainen 1978; Denno 1994; Denno et al. 1980, 1985, 1991; Longellatto and Denno 2001); these studies have provided strong empirical support for the theoretical ideas mentioned above. For example, strong positive correlations have been documented between habitat persistence and degree of flightlessness (e.g. frequency of the flightless morph) in both intraspecific and interspecific comparisons. Especially significant are the detailed studies of Denno and colleagues on the influence of environmental heterogeneity on wing-polyphenism in salt-marsh planthoppers (Denno 1994; Denno et al. 1980, 1985, 1991, 2000; Longellatto and Denno 2001). These studies provide the most detailed account of the specific ecological factors that favor flight-capable or flightless morphs, and the various environmental cues (crowding, host plant nutrition) involved in morph induction in the field.

Endocrine Control of Wing Polymorphism: The Classical Model

In the early 1960's insect physiologists and ecologists used newly obtained information on the endocrine control of metamorphosis to formulate an endocrine model of wing polymorphism (Wigglesworth 1961; Southwood 1961; Lees 1966). This model has been the focus of experimentation and thinking about the hormonal mechanisms that regulate wing polymorphism up to the present (Lees 1966; Gould 1977; Nijhout and Wheeler 1982; Hardie and Lees 1985; Matsuda 1987; Zera and Denno 1997; Zera and Harshman 2001). Although numerous hormones affect metamorphosis and reproduction, two hormones, ecdysone and juvenile hormone, are especially important in regulating these processes (Nijhout 1994, 1999; Klowden, 2002), and have been the almost exclusive foci of physiological studies of wing polymorphism.

Ecdysone has two important functions: it induces the molt and causes the expression of genes that specify adult characteristics (metamorphosis; Fig. 1). Juvenile hormone (JH), on the other hand, antagonizes the metamorphic action of ecdysone, while allowing the molt to occur. A high JH concentration causes a juvenile-to-juvenile molt, whereas a drop in the JH titer to a low or imperceptible level during the early last juvenile instar, allows ecdysone-induced gene expression to occur resulting in a metamorphic molt [i.e., a molt from a juvenile to an adult in hemimetabolous insects (e.g., crickets, aphids, waterstriders), or to a pupa in holometabolous insects (beetles, butterflies, moths, flies)]. During the adult stage, JH, and/or ecdysone, takes on a new role as a gonadotropin, regulating, among other things, the synthesis of vitellogenin or yolk protein, and the uptake of these molecules into the developing oocyte. Finally, JH is thought to control a variety of behaviors involved in reproduction and flight in adult morphs. The hormonal regulation of metamorphosis, and reproduction, are, of course, much more complex than depicted in the brief outline given above, and many other hormones are involved in the regulation of these processes (Nijhout 1994; Klowden, 2002; Gilbert et al. 2005). Furthermore, temporal and tissue-specific expression of hormone receptors play critically-important roles in metamorphosis and reproduction, by defining the periods of time ("sensitive stages") during which tissues are sensitive to metamorphic or reproductive hormones (Nijhout 1994; 1999).

The classical JH-wing-polymorphism model (Fig. 1) has focused primarily on variation in the developmental timing, duration, and height of JH and ecdysteroid titer peaks, as the physiological causes of expression of morph-specific differences in morphology and reproduction. Variation in the expression of JH receptors has not been considered in this model, because the JH nuclear receptor has yet to be identified (see below). A variety of possible changes in the profiles of JH and ecdysone titers during development might underlie alternate morph development, only two of which are illustrated in Fig. 1. The most common expectation of the classical model is that the JH titer should be higher in nascent short-wing-destined individuals during some "sensitive" stage in development. In Fig. 1, this is depicted as an elevated (delay in the decrease of) JH titer during the last juvenile stadium, sufficient to block the metamorphic effect of the rising ecdysteroid titer. Other possibilities would include an elevated JH titer during earlier instars or within the mother (in species with pre-natal morph determination), as proposed for some aphids (Hardie and Lees 1985). Alternatively, a reduced ecdysteroid titer (Fig. 1), a change in the timing of the ecdysteroid peak relative to the JH titer, or decreased expression of ecdysteroid receptors (not shown in Fig. 1) could result in ecdysteroid titers or tissue sensitivity that are insufficient to allow full growth and differentiation of wings and flight muscles, resulting in a short-winged morph with underdeveloped flight muscles. A significant point here, is that, given the complexity of the endocrine control of wing and flight muscle development, any of the numerous components of the endocrine regulatory network (hormone levels, timing of release, expression of tissue receptors) could potentially be modulated by either the environment or natural selection to alter the expression of wings or flight muscles.

In adults, an earlier rise in the titer of JH (Fig. 1), the most widely studied gonadotropin in insects (Nijhout 1994; Klowden, 2002), has been the most common hypothesis put forward to account for the earlier ovarian growth of the flightless morph, which occurs in virtually all wing polymorphic species (Harrison 1980; Roff 1986; Zera and Denno 1997). JH also appears to negatively affect many aspects of flight capability. For example, topical application of JH often prevents the accumulation of lipid flight fuel and causes degeneration of flight muscles in the flight-capable morph (Zera and Denno 1997; Zera and Harshman 2001). Thus, the JH titer is expected to be reduced in the flight-capable morph. However, JH also positively affects flight behaviors (e. g. increases propensity to fly) in some insects (Rankin 1991). This suggests that expression of traits in LW vs. SW morphs might be regulated by a more complex mechanism than difference in the JH titer above vs. below a single threshold.

Testing the Model: Aphids and Planthoppers

One of the striking aspects of endocrine studies of wing polymorphism is the paucity of hard data on endocrine traits such as hormone titers, hormone receptors or activities of hormone regulating enzymes in adult or juvenile morphs. Indeed, although the JH-regulation of morph development and reproduction has attained the status of near dogma in some textbooks, and reviews (e.g. Hartfelder and Emlen, 2005), this conclusion is based, in most cases, on gross-level hormone manipulation experiments (for critique see Zera 2007; Zera et al. 2007a). To a large extent, this situation has resulted from severe technical limitations imposed by insects species, such as aphids and planthoppers, which have, until very recently, been the primary experimental models in physiological studies of wing polymorphism. Aphids were the predominant model used during the 1960s, 70s and early 80s to investigate the endocrine control of wing polymorphism (mainly density- or photoperiodically-mediated wing polyphenism; Hales 1976;

Rankin and Singer 1984; Hardie and Lees 1985; Mittler 1991; Hardie et al. 1995; Zera and Denno, 1997; Zera 2004). Only a brief summary of the aphid work will be given here since it is already well reviewed (see above references). In a nutshell, despite decades of intensive work, the importance of JH (or ecdysone) as a regulator of wing polymorphism in aphids remains uncertain. This uncertainty has resulted from a number of unfortunate technical problems inherent in aphid endocrine physiology. For example, the very small size of aphids has precluded the use of standard techniques such as the surgical removal of endocrine glands, or the *in vitro* measurement of juvenile hormone biosynthesis. The major JH in this group has yet to be identified (there are several structurally-similar juvenile hormones in insects; Nijhout 1994; Klowden, 2002). The lack of information on the specific JH in aphids (or planthoppers), together with the small size of these insects, have precluded reliable, direct measurement of the JH titer (Hardie and Lees 1985; Hardie et al. 1985; Zera and Denno 1997).

Because of these limitations, workers have been forced to infer morph-regulating endocrine mechanisms almost exclusively from results of experiments involving application of exogenous hormones, agonists, or antagonists. However, only very limited conclusions can be drawn from such an approach when it is used as the sole or primary experimental method in endocrine studies of morph development (Zera and Denno 1997; Zera, 2007). For example, topical application of JH or JH analogues alters many *in vivo* endocrine traits, such as hormone titers (e.g. ecdysteroids) and release of neuropeptides (e.g., Smith and Nijhout 1981; Stay et al., 1994). Thus, in the absence of additional data (e.g. direct measures of *in vivo* hormone titers), one can never be certain that an effect of an exogenous hormone is due to the applied hormone itself as opposed to an induced regulator. Furthermore, no information on endocrine mechanisms (e.g. do hormone titers differ between nascent morphs?) can be obtained from topical application experiments alone. Simply stated, strong support for a particular hormone or hormonal mechanism regulating morph development must include direct documentation that key components of the regulatory mechanism differ between alternate morphs. Key components would include the *in vivo* hormone titer, activities of regulators of the hormone titer (e.g. rates of hormone biosynthesis, or activities of biosynthetic or degradative enzymes), or degree of receptor expression. The problematic nature of the data on hormone application experiments in aphids, and indeed with this experimental approach in general, has been discussed in detail in a number of reviews (Rankin and Singer 1984; Hardie and Lees 1985; Hardie et al. 1995; Gao and Hardie 1996; Zera and Denno 1997; Zera and Cisper, 2001; Zera 2004, 2007). Unfortunately, there currently is a resurgence in the use of topical application of hormones, hormone analogues or inhibitors as the sole or primary approach to infer endocrine mechanisms controlling insect polyphenisms (e.g., Dingle and Winchell 1997; Emlen and Nijhout 1999, 2001; Moczek and Nijhout 2002). Endocrine mechanisms purportedly regulating morph specialization reported in these studies should be viewed with caution [for critiques see Zera (2007) and Zera et al. (2007a)].

The strongest case for a regulatory role for JH in aphid wing polymorphism is in photoperiodically-mediated wing polyphenism in *Aphis fabae* where topically applied JH strongly redirects morph development from the long-winged to the short-winged morph (Hardie 1980; Hardie and Lees 1985; Hardie et al. 1995). However, it is unclear whether this is a special case with resepct to the endocrine control of wing polymorphism (Hardie et al., 1995), or whether JH itself is even involved, given the uncertainties inherent in topical-application experiments discussed above. A number of hormone-manipulation studies of planthoppers, especially *Nilaparvata lugens* (Ayoade et al. 1999; Bertuso et al. 2002; and references therein) also implicate juvenile hormone as a potential regulator of morph induction and development. Use of molecular approaches will likely revolutionize studies of wing polymorphism in aphids (Brisson et al, 2007).

Testing the Classical Model: Crickets (*Gryllus*)

Morph-specific Difference in Endocrine Titers and Titer Regulators

Endocrine control of wing polymorphism has been most intensively studied in crickets of the genus *Gryllus*. Indeed, results of investigations over the past 20 years, provide some of the most detailed information on the endocrine mechanisms that potentially regulate morph-specific development for any case of complex polymorphism (Zera 2004, 2006, Zera et al. 2007a). In addition, recent studies on the JH titer in adult morphs have yielded unexpected results, which indicate that the endocrine control of morph-specific traits is much more complex in wing polymorphic insects that previously suspected (Zera and Cisper 2001; Zhao and Zera 2004a, Zera et al. 2007b). Advantages in using species of *Gryllus* as a model in endocrine studies of wing polymorphism include their large size, which allows direct quantification of hormone titers and titer regulators in the hemolymph (blood) of single individuals, various surgical manipulations, and *in vivo* studies of hormone metabolism (discussed below). Furthermore, the specific JH in crickets (and Orthoptera in general; Nijhout, 1994) has been identified, and several sensitive and specific radioimmunoassays have been developed for this JH (Zera et al. 1989; Zera and Tobe 1990; Goodman et al. 1993; Zera and Cisper 2001; Zhao and Zera 2004a).

Endocrine studies have been undertaken on genetic stocks of two cricket species (*Gryllus rubens* and *G. firmus*) that produce primarily flight-capable (>85%) or flightless (<15%) morphs under standard rearing conditions. Shifting presumptive long-winged (LW) *G. rubens* from standard to high densities as late as the penultimate and early last juvenile instar redirected development to the flightless, short-winged (SW) morph. Juvenile hormone applied during this time to nascent LW individuals had the same SW-promoting effect in *Gryllus rubens* (Zera and Tiebel 1988) and *Modicogryllus confirmatus* (Zera and Tanaka 1996). These results suggested that development of alternate morphs may be regulated by modulation of the JH titer/receptors during the last two juvenile instars.

Possibly the most important finding with respect to the endocrine control of wing polymorphism in *Gryllus* was the dramatic (3- to 6-fold) elevation in the hemolymph activity of juvenile hormone esterase (JHE) during the last juvenile instar in nascent LW vs. SW morphs in two *Gryllus* species (Zera and Tiebel 1989; Roff et al. 1997; Zera and Huang 1999; Fig. 2b, d). JHE degrades juvenile hormone and is thought to regulate its titer in a number of insects (Hammock 1985; Wyatt and Davey 1996: Roe and Venkatesh, 1990). The activity of JH-epoxide hydrolase, which degrades JH in non-hemolymph compartments of an insect, did not differ between morphs of *G. firmus* (Zera and Huang 1999).

Several pieces of evidence strongly suggest that hemolymph JHE is involved in the development of alternate LW and SW morphs in *Gryllus* (Figs. 2 and 3). First, as expected from the classical JH-wing-morph hypothesis, JHE activity is substantially reduced in the nascent SW morph, which is expected to result in an elevated JH titer in that morph (Fig 2b, d; Zera et al., 1989; Zera and Huang 1999). Moreover, reduced JHE activity occurred during the same period of time during which topical application of JH redirected development from the LW to the SW morph (Zera and Tiebel 1988). Second, high JHE activity exhibits a nearly prefect co-segregation with long wings (LW morph) in crosses and backcrosses between LW and SW lines of two *Gryllus* species (Zera and Tiebel 1989; unpublished data). Indeed, hemolymph JHE activity exhibits the tightest correlation with wing morph for any endocrine regulator investigated thus far in any case of insect polymorphism or polyphenism. Third, morph-specific differences in JHE

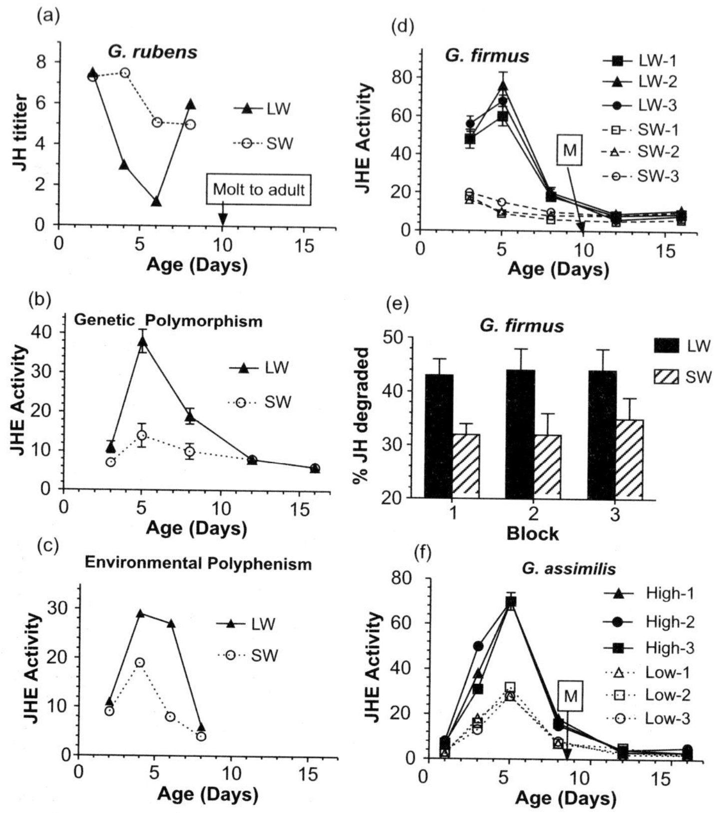

Fig. 2 Morph-specific differences in various endocrine traits in *Gryllus* species. Panels "a" and "b" illustrate hemolymph juvenile hormone (JH) titers and hemolymph juvenile hormone esterase (JHE) activities in LW and SW genetic stocks of *G. rubens* reared under the same, standard environmental conditions. Panel "c" illustrates JHE activities in LW and SW morphs of *G. rubens* that had been produced from one and the same LW stock when individuals were reared under standard or high density; high density caused about 65% of individuals of the LW stock to molt into SW adults. Panels "d" and "f" illustrate JHE activities in LW and SW genetic stocks of *G. firmus*, or in lines of the wing-monomorphic *G. assimilis*, artificially-selected for high or low hemolymph JHE activity. Panel "e" illustrates the elevated rate of *in vivo* JH degradation in LW- vs. SW stocks of *G. firmus*. Age refers to days since molt to the last juvenile instar. Data are from Zera et al. (1989), Zera and Zhang (1995), and Zera and Huang (1999). JH titer is nM, JHE activities are nmol JH-acid min^{-1} ml hemolymph^{-1}, and rate of JH degradation is percentage injected, radiolabeled JH converted to metabolites during a standard assay period in day-5 crickets. Note the reciprocal relationship between the JH titer and JHE activity (panels "a" and "b"), and the similarity between genetic polymorphism and environmental polyphenism for JHE activity (panels "b" and "c"). Note also the similarities between the morph-specific developmental profiles of JHE activities in lines of *Gryllus* species selected for wing morph (panels "b" and "d") or selected for high or low JHE activity (panel "f").

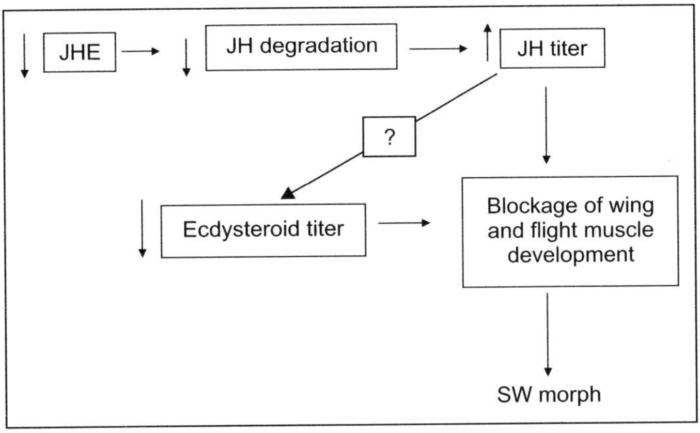

Fig. 3 Endocrine model illustrating the regulation of development of LW vs. SW morphs of *Gryllus* species by modulation of juvenile hormone esterase activity, the juvenile hormone titer, and the ecdysteroid titer. See text for explanation.

activity, measured *in vitro,* are positively correlated with morph-specific differences in whole-organism JH degradation measured *in vivo* (Fig 2e; Zera and Huang 1999). This association is critically important because there are numerous examples in which variation in the activity of a particular enzyme, measured *in vitro,* has no observed physiological effects *in vivo* (Zera and Huang, 1999). By contrast, JHE activity appears to be sufficiently elevated within the nascent LW morph to increase the rate of JH degradation in that morph relative to the SW morph, and thus to produce morph-specific differences in the JH titer. Preliminary studies have identified no significant differences between nascent morphs in JH binding in the hemolymph (Zera and Holtmeier, 1992) indicating that differential hormone binding is less likely to regulate morph development.

Thus far, the *in vivo* JH titer has been compared between nascent flight-capable and flightless morphs of only one *Gryllus* species, *G. rubens* (Zera et al. 1989). No difference in the hemolymph JH titer was observed between morphs during the penultimate stadium. However, consistent with the JH-wing-morph hypothesis, a slightly elevated JH titer was found in nascent SW vs. LW morphs during the last stadium (Fig. 2a; Zera et al. 1989). Importantly, a reciprocal relationship was observed between the median hemolymph JH titer (higher in SW) and median hemolymph JHE activity (lower in SW; Fig. 2a, b), consistent with the hypothesis that differences between morphs in the JH titer result from differences in JHE activity.

Despite this striking reciprocal relationship, only small differences in the JH titer have thus far been observed between nascent morphs during the last juvenile instar (Fig. 2). Thus, even in this species, the role of JH in regulating alternate morph development is not firmly established. On the one hand, titer differences between morphs may have been underestimated because of the large experimental error associated with JH titer measurements during the last juvenile stadium, when JH levels are very low (see discussion in Zera et al. 1989). Thus, *in vivo* JH titers may actually differ to a greater degree between morphs than is indicated by current titer measurements. Alternatively, since the JH titer affects many aspects of metamorphosis, there may be a severe constraint on the degree to which the JH titer can differ between the morphs, even if this hormone is an important regulator of morph development. Thus, only subtle JH titer differences might exist between morphs, but these differences could be functionally important by inducing the differential expression of JH receptors in nascent LW and SW morphs. In amphibians, a low thyroxine titer is functionally important since it induces expression of the thyroxine receptor (Tata 1996). Furthermore, alternate morph development in crickets may be regulated by combined differences in the titers of several hormones, only one of which is JH (see below). Finally, it is possible that morph development may be regulated by mechanisms different from those typically considered, in which JHE plays a more important role than JH (discussed below).

Hemolymph JHE was directly selected for high or low activity during the last juvenile stadium of the wing-*mono*morphic congener, *Gryllus assimilis* (i.e., a species that produces only LW individuals; Zera and Zhang, 1995; Fig. 2f). Interestingly, selected lines of *G. assimilis* produced JHE activity developmental profiles that were very similar to those observed in the wing-polymorphic congeners. That is, activities differed substantially between selected lines during the last juvenile instar, but did not differ significantly between lines during the adult stage (Fig 2b, d, f). Many other parallels were observed with respect to JHE activity differences between selected lines of *G. assimilis* and selected lines of *G. firmus* (Zera and Huang, 1999). For example, in high activity lines of *G. assimilis* and LW (high activity) lines of *G. firmus*, a much greater proportion of whole-organism JHE activity was found in the hemolymph vs. the rest of the body compared with the low activity (*G. assimilis*) or SW (*G. firmus*) lines [see Table 4 of Zera and Huang (1999) for other examples]. These data suggest that genetically-variable factors that were selected to evolve wing polymorphism in natural populations are similar or identical to genetically-variable factors that are currently segregating in wing-monomorphic species.

Another important finding, with respect to the endocrine control of polymorphism, was that morphs of *G. rubens* differed to a much greater degree in aspects of the ecdysteroid titer than the JH titer (Zera et al. 1989; Fig. 4). During the penultimate stadium, the ecdysteroid titer rose later and remained elevated for a shorter period of time in the SW compared with the LW morph. No difference was observed between the morphs in the height of the ecdysteroid peak. By contrast, during the last stadium, the timing of the rise, and the duration of elevation of the ecdysteroid titer were similar in both morphs. However, the height of the ecdysteroid titer peak was substantially

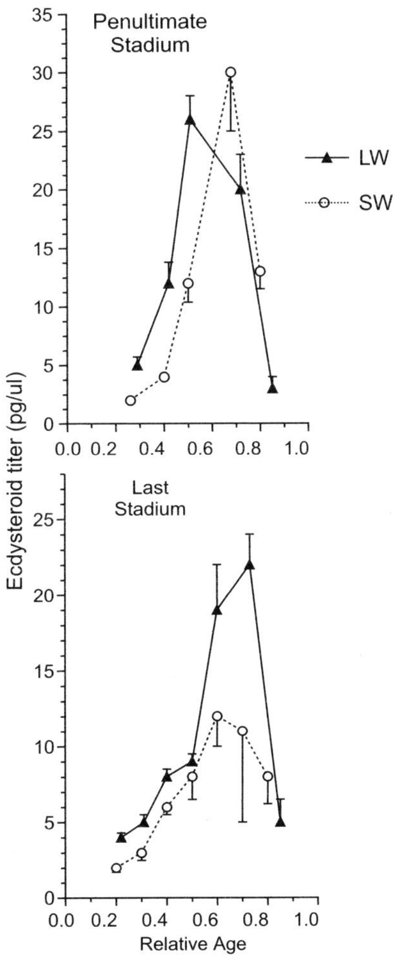

Fig. 4 Differences in the hemolymph ecdysteroid titer between LW and SW morphs of *G. rubens* during the penultimate and last juvenile instars. Note the delayed rise in the ecdysteroid titer during the penultimate instar and lower ecdysteroid titer during the last juvenile instar in nascent SW vs. LW morphs. Data are from Zera et al. (1989).

lower in the SW morph. These various differences in the ecdysteroid titer are consistent with a role for ecdysteroids in morph determination (as discussed above; Fig. 3). Alternatively, the combined effect of an elevated JH titer and a reduced ecdysteroid titer during the last juvenile instar may be the key factor that inhibits wing and flight-muscle development in the SW morph, rather than variation in the titer of either hormone alone. At present, no information is available on morph-specific differences in tissue or developmental expression of ecdysteroid receptors in *Gryllus*. Since the ecdysteroid receptor has been cloned in a number of insects (see below), such a study would be feasible and highly informative in *Gryllus*.

Relationship between Mechanisms Regulating Genetic Polymorphism and Environmental Polyphenism

Although many species exhibit both genetically-specified polymorphism and environmentally-induced polyphenism, few data exist on the relationship between endocrine mechanisms controlling these two types of polymorphism (e.g., Zera et al. 1989; Roundtree and Nijhout 1995a, b; Evans and Wheeler 2001). In *G. rubens*, individuals from a LW-selected line were either reared under standard density (producing primarily the LW morph) or under higher density, which diverted development of most (~ 65%) individuals to the SW morph. JHE activity profiles differed between these wing-polyphenic morphs from the same genetic stock in a similar manner to profiles between LW individuals from a LW-selected line and SW individuals from a SW-selected line raised under the same environmental conditions (Fig. 2). Thus, the endocrine mechanisms that underlie genetically-specified and environmentally-induced wing polymorphism in *Gryllus* appear to share some common components.

In these studies we used environmental manipulation solely as an experimental tool to investigate the endocrine regulation of genetically-based wing polymorphism. It is therefore uncertain whether the short-wing-promoting effect of high density is an adaptive response or a laboratory response which does not normally occur under field conditions. If this is solely a laboratory-induced polyphenism, these data are still interesting because they indicate that *G. rubens* has the capacity to exhibit a stress-induced polyphenism using regulatory elements (juvenile hormone esterase) normally employed to produce an adaptive genetic polymorphism. However, there are numerous examples in which increased density or other stressful conditions induce the production of the flightless morph in crickets (e.g., Shimizu and Masaki 1993). Moreover, flightless *G. firmus* survive stressful, nutrient-poor conditions much better than flight-capable individuals (A. J. Zera, unpublished data). This is probably because the flightless morph is more energy-efficient (Mole and Zera, 1993; Zera and Harshman, 2001) since it does not have to divert resources into production of the massive flight muscles, and other components of the flight apparatus. Thus, it is likely that stress-induced flightlessness, modulated by juvenile hormone esterase, is an adaptive polyphenism. Interestingly, the production of a phenotype in *Gryllus* which is better able to tolerate stress, is very different from the typical response to stress by aphids, which is to produce a dispersing morph which can escape from the local, stressful environment by flight (Lees, 1966; Hardie and Less, 1985).

Physiological, Biochemical, and Molecular Causes of JHE Activity Differences between Morphs

As a step towards identifying specific genes involved in the regulation of JHE and wing polymorphism in *Gryllus*, JHE enzymes from LW and SW morphs of *G. rubens*, or from genetic stocks of the congener *G. assimilis* selected for high or low JHE activity, were compared. The JHEs did not differ in any biochemical characteristic such as Michaelis constant, inhibition, or thermostability (Zera et al. 1992; Zera and Zeisset 1996). These and additional physiological studies (Zera and Huang 1999) point to variation in as yet unidentified loci that regulate the concentration of the JHE enzyme, and the degree to which it is exported into the hemolymph, as the most likely cause of the substantial differences in hemolymph JHE activity between LW and SW morphs. The JHE from *G. assimilis* has been purified to homogeneity (Zera et al. 2002), and a full-length cDNA from *G. assimilis* has been isolated and sequenced (Crone et al. 2007). Molecular probes designed from the cDNA sequences have recently been used to demonstrate that JHE activity differences between selected lines of G. assimilis are correlated with line differences in JHE transcript abundance (Zera et al. 2007a).

JH Titer Differences Between Adult Morphs

The first direct comparison of hemolymph JH titers between adult morphs of a wing-polymorphic insect was undertaken in *G. firmus* (Cisper and Zera 2000; Zera and Cisper 2001). Because the JH titer is much higher in adults than in juveniles, titers can be measured in single individuals and with much greater accuracy than in juveniles. Ovaries grow considerably faster (100–400%; Fig. 1; see below) during the first week of adulthood in SW vs. LW *G. firmus* (Zera and Cisper 2001; Zera and Bottsford 2001; Zera and

Larsen 2001). Since JH is a major gonadotropin in orthopterans such as *Gryllus* (Nijhout 1994), the classic JH-wing-morph model predicts that the JH titer should rise earlier in the SW vs. the LW morph. Consistent with this expectation, topical application of JH or JH analogues to LW females of *G. firmus*, or other cricket species, during early adulthood, caused them to increase ovarian growth to a level seen in SW females (Tanaka 1994; Zera et al., 1998; Zera and Cisper 2001; unpublished data)

However, direct titer measurements have revealed a much more complex and interesting situation in *Gryllus*: the JH titer cycles on a daily basis in the LW morph, with the titer rising and dropping 10–100 fold between the late photophase and early scotophase. By contrast, in the SW morph, the hemolymph JH titer is relatively constant, during the day and night, on each of the first 8 days of adulthood (Fig. 5; Zera and Cisper 2001; Zhao and Zera 2004a). The large amplitude, morph-specific daily cycle in the JH titer appears to be specific for JH, since comparable morph-specific changes were not seen for the hemolymph ecdysteroid titer (Fig. 6; Zhao and Zera, 2004a).

The morph-specific JH titer cycle persists under constant darkness indicating that it is a true circadian cycle (unpublished data). Furthermore, a similar morph-specific daily cycle for the JH titer was seen in *G. firmus* laboratory stocks raised in the field and in individuals of *G. firmus* from natural populations sampled in the field (Zera et al. 2007b). Thus, the

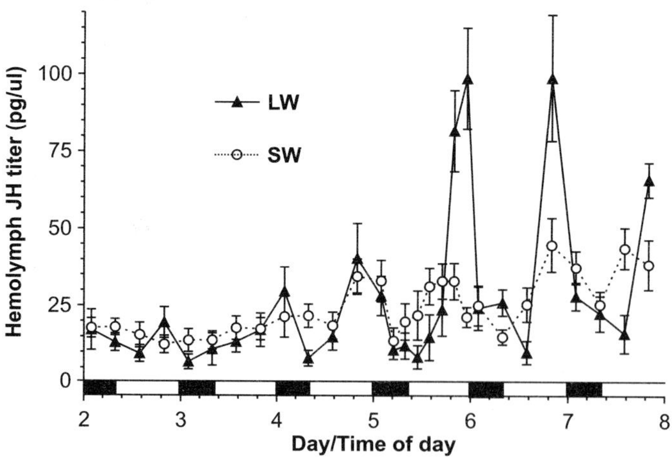

Fig. 5 Differences in the hemolymph juvenile hormone titer (pg/ul) between *adult* female LW and SW *G. firmus*. Numbers on x-axis refer to day since molt to adulthood and light and dark bars refer to photophase (light) and scotophase (dark) of photoperiod. Note the large-amplitude daily cycle in the JH titer in LW females but relatively constant titer in SW females. Data are from Zhao and Zera (2004a).

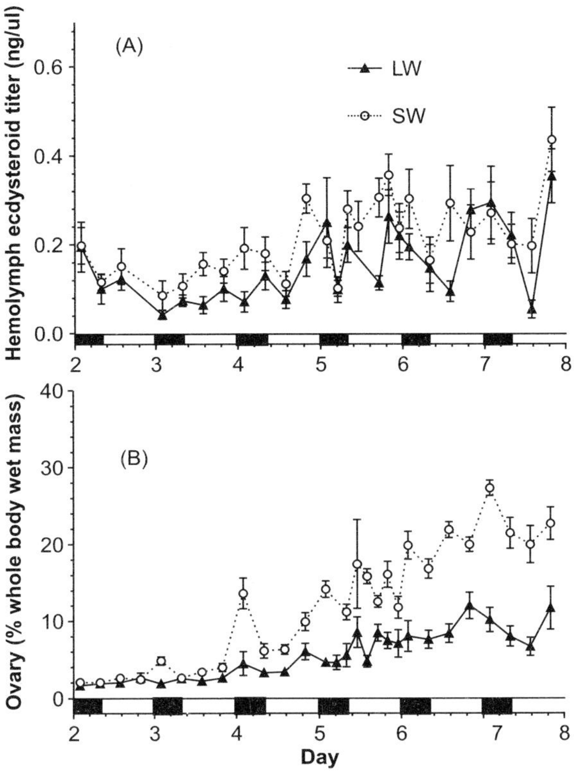

Fig. 6 Panel A. Hemolymph ecdysteroid titer in LW and SW adult female *G. firmus*. Note the absence of a large-amplitude daily cycle in either morph (a small amplitude cycle is seen in both morphs due to diel change in hemolymph volume; see Discussion in Zhao and Zera, 2004a). The ecdysteroid titer was approximately 30% higher in SW than in LW females over days 2-7. Panel B. Elevated ovarian growth in SW (open circles) vs. LW (closed triangles) females during the first week of adulthood. Data are from Zhao and Zera (2004a).

morph-dependent daily cycle in the JH titer in *G. firmus* does not appear to be a laboratory artifact. Finally, rate of juvenile hormone biosynthesis in *G. firmus* also cycles in a morph-specific manner, and the phases of the cycles of JH biosynthesis and the JH titer are co-incident (Zhao and Zera 2004b). These strongly covarying cycles suggest that the JH titer differences between morphs (presence or absence of a daily cycle) are regulated, at least in part, by differences in rates of hormone biosynthesis. Mathematical modeling indicates that estimated values of rates of JH biosynthesis and in vivo JH degradation can quantitatively explain the dynamics of the JH titer daily cycle (timing and amplitude of titer peaks) in the LW morph reasonably well (unpublished data).

These findings clearly demonstrate that, at least for crickets, JH titer differences between flight-capable and flightless morphs are very different from those expected by the "classical model" (Fig. 1). Furthermore, these intriguing results raise questions concerning the functional significance of the morph-specific daily cycle in the JH titer. One possibility is that the daily rise in the JH titer in the LW morph activates (sensu Elekonich and Robinson 2000a) flight behavior in that morph. This hypothesis is supported by a number of lines of evidence. First, the large-amplitude diurnal cycle only occurs in the flight-capable morph, and begins on day 4 of adulthood (Zhao and Zera 2004a), precisely when the LW morph first attains flight ability (Zera et al. 1999). Second, as is the case with many other wing polymorphic insects, some LW adult *G. firmus* histolyze their flight muscles and become flightless (Zera et al. 1997). Loss of the daily cyclicity in the JH titer occurs co-incident with flight muscle histolysis in the LW morph (Zera and Cisper 2001). As discussed previously, topical application of JH or JH analogues is known to increase flight propensity in many insects (Rankin 1991). Thus, it is not unexpected that a rise in the JH titer late in the photophase might be involved in releasing noctural flight behaviors.

The above hypothesis raises an interesting paradox. Since experimental elevation of the JH titer in *Gryllus* (and many other species) causes histolysis of flight muscles and ovarian growth (Tanaka 1994; Zera et al. 1998; Zera and Cisper 2001), why doesn't the daily rise in the JH titer in the LW morph, which is several-fold higher than that in the SW morph, cause flight muscle histolysis and initiate ovarian growth in the LW morph? One possible solution to this paradox is that the rise in the JH titer in the LW morph may be of sufficient duration to release flight behavior but of insufficient duration to initiate flight-muscle histolysis and ovarian growth. If this is the case, then the differential expression of morph-specific traits may be regulated by a novel mechanism: variation in the length of time required for an elevated titer of a single hormone to affect the expression of antagonistic traits. Alternatively, counteracting production of JH inhibitors, or down-regulation of JH receptors may occur in the LW morph, which may obviate the effects of the increased JH titer on certain tissues. In addition, the ecdysteroid titer is consistently elevated in the SW vs. the LW(f) morph independent of time-of-day, either when measured in the laboratory or in the field (Fig. 6; Zera and Bottsford 2001; Zhao and Zera 2004a; Zera et al. 2007b). Indeed, the magnitude of elevation of the ecdysteroid titer in SW compared with LW females is much greater in the field (200–300%) than in the lab (30%). Thus, this hormone may play a more important role than JH in regulating morph-specific differences in ovarian growth, or may mitigate the effect of JH on flight muscles and ovaries. Finally, other, as yet undiscovered factors might play a role. Whatever the mechanisms involved, the endocrine regulation of morph-specific traits in adults in *Gryllus* is clearly more complex than envisioned by the classical model. Importantly, these unexpected JH titer differences between morphs could only have been identified by direct *in vivo* titer measurements. Not only would these titer differences have been completely missed if studies had only consisted of topical hormone application, we would have erroneously inferred that the JH titer is consistently higher in SW vs. LW females.

Physiology and Biochemistry of Internal Resource Allocation to Flight Capability versus Reproduction in Morphs of *Gryllus*

A long-standing topic of research on wing polymorphism has been the physiological mechanisms responsible for the differential allocation of internal resources to organs of flight vs. reproduction in adult morphs (Zera and Denno 1997). For many decades, the increased fecundity of the flightless morph has been assumed to result from the "freeing up" of resources that are required to build and maintain various components of a functional flight apparatus in the LW morph (e.g., growth and maintenance of large flight muscles; accumulation of large flight fuel reserves [(Mole and Zera 1993; Zera et al. 1997; Zera and Denno 1997; Zera and Harshman 2001; see below)]. Note that the key issue here is the functional cause of the fitness gain in the flightless morph due to reduction of costs involved in flight capability *in the absence of flight*. Despite the widespread acceptance of the aforementioned hypothesis, only during the past decade have detailed physiological studies begun to adequately investigate (1) whether internal trade-offs of resources actually exist, (2) the importance of such trade-offs with respect to morph-specialization for flight vs. reproduction, and (3) the proximate physiological mechanisms involved (Mole and Zera 1993; Zera and Denno 1997; Zera and Harshman 2001). Biochemical processes underlying the trade-off between flight-capability and fecundity have only been studied in detail during the past few years (Zhao and Zera 2001, 2002; Zera and Zhao 2003a, b; 2004, 2006; Harshman and Zera, 2007). These physiological or biochemical studies have primarily been undertaken in *Gryllus* (Tables 1 and 2; see above references), and a few related cricket species (*Modicogryllus*; Tanaka, 1993, 1994). Proximate physiological, and biochemical-metabolic mechanisms underlying morph specializations for flight vs. reproduction are now best understood in wing-polymorphic species of *Gryllus* (Zera and Harshman 2001; Zhao and Zera 2002; Zera and

Table 1 Physiological differences between flight-capable (LW) and flightless (SW) morphs of *Gryllus* related to internal resource allocation[1]

Trait	*Differences between morphs*	*References*
Organ mass		
Flight muscles	LW > SW	A[2], B, C
Fat body	LW > SW	E
Ovaries	SW > LW	A, B, C, D
Nutrient consumption	Equivalent	B, C, F
Nutrient assimilation	Equivalent	B, C, F, G
Nutrient conversion to biomass	SW>LW	B, C, G
Whole organism respiration	LW>SW	B, C, G, H
Flight muscle respiration	LW>SW	A, B
Ovarian respiration	SW > LW	A, B
Flight muscle + ovarian respiration	LW > SW	B
Activities of enzymes in flight muscles	LW > SW	A, I
Whole-organism lipid content	LW > SW	D
" " Triglyceride	LW > SW	B, D
" " Phospholipid	SW > LW	D

[1]Characteristics of individuals reared on the "standard" diet [see Zera and Larsen (2001) for diet composition]. Flight-capable = long-winged individuals with large pink (functional) flight muscles; flightless = short-winged individuals with underdeveloped flight muscles. Flightless, long-winged individuals (with histolyzed flight muscles) are not considered here (see Zera et al., 1997, 1999).

[2]References: A = Zera et al., 1997; B = Zera et al., 1998; C = Zera and Mole, 1993; D = Zera and Larsen, 2001; E = Zera and Zhao, 2003a; F = unpublished data; G = Zera and Brink, 2000; H = Crnokrak and Roff, 2002; I = Zera et al., 1999.

Zhao 2003a, b; 2004, 2006; Zera 2005), although many fundamental aspects of this problem remain unstudied.

Detailed feeding studies of various *Gryllus* and *Modicogryllus* species provided the first direct information on the relative importance of nutrient acquisition vs. allocation in flight-capable and flightless morphs (Table 1; Mole and Zera 1993; Tanaka 1993; Zera and Denno 1997; Zera et al. 1998; Zera and Harshman 2001). Naturally-occurring or hormonally-induced flightless morphs of *G. rubens*, or *G. assimilis* consumed and assimilated the same amount of nutrients as their flight-capable counterpart (Mole and Zera 1993; Zera et al. 1998; A. J. Zera, unpubl). Substantially greater nutrient consumption was originally reported in flight-capable vs. flightless female *G. firmus* (Mole and Zera 1994). However, that result has subsequently been shown to be an experimental artifact, and morphs of *G. firmus* consume and assimilate nearly the same amount of nutrients (within 10%) when fed a standard diet (A. J. Zera, unpublished data). Thus the large differences in the

Table 2 Differences in lipid metabolism between flight-capable and flightless morphs of *G. firmus*

Aspect of metabolism	*Flight-capable morph relative to flightless morph*	*References*
Biosynthesis of total lipid	Higher	A, B
Absolute or relative biosynthesis of triglyceride	Higher	A
Absolute or relative biosynthesis of phospholipid	Lower	A
Oxidation of fatty acid	Lower	C
Activities of lipogenic enzymes	Higher	B, D
Conversion of amino acids into lipid	Higher	E
Oxidation of amino acid	Higher	E
Activity of transaminases	Higher	D, B
Biosynthesis of ovarian protein from amino acids	Lower	E
Allocation of biosynthesized lipid to soma vs. ovaries		
Triglyceride	Higher	A
Phospholipid	Lower	A
Assimilation of lipid from diet	Equivalent	F

[1]Characteristics of individuals reared on the "standard" diet [see Zera and Larsen (2001) for diet composition]. Flight-capable = long-winged individuals with large pink (functional) flight muscles; flightless = short-winged individuals with underdeveloped flight muscles. Flightless, long-winged individuals (with histolyzed flight muscles) are not considered here.

References: A = Zhao and Zera, 2002; B = Zhao and Zera, 2001; C = Zera and Zhao, 2003a; D = Zera and Zhao, 2003b; E = Zera and Zhao, 2006; F = Zera and Brink, 2000.

masses of organs of flight and reproduction (e. g., 200–300% heavier ovaries in the flightless vs. flight-capable morph of each species) must have been produced by the differential allocation of internal nutrients to the growth and maintenance of these organs, and not by the differential intake of nutrients. Although numerous studies have investigated internal trade-offs, very few have quantified or controlled nutrient input (Zera and Harshman 2001, Harshman and Zera 2007). Thus, the extent to which life history differences between morphs, or life history trade-offs in general, are actually caused by morph-specific differences in internal nutrient allocation (i.e., an internal trade-off), as opposed to variation in nutrient consumption or assimilation, remains a largely open issue (Harshman and Zera 2007).

A number of studies have identified physiological costs of flight-capability that potentially reduce egg production in the LW morph. For

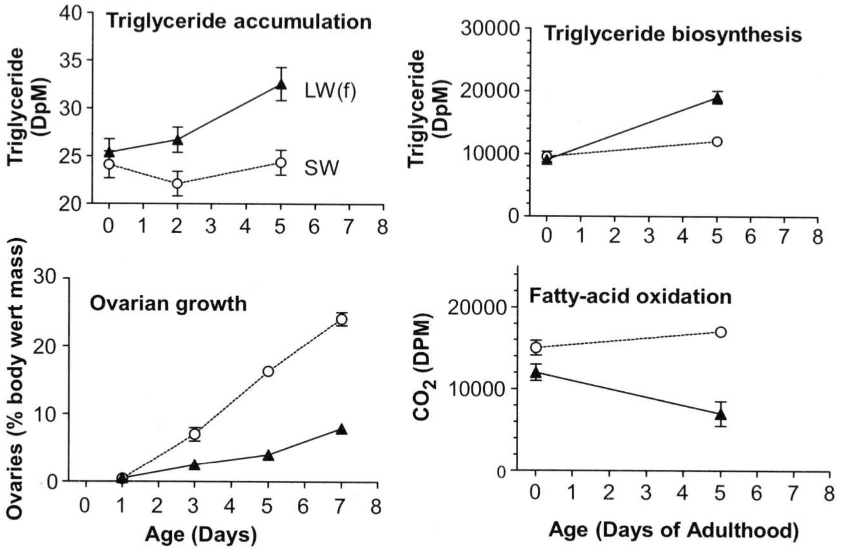

Fig. 7 Relationships among triglyceride accumulation and biosynthesis, fatty acid oxidation, and ovarian growth in LW and SW adult *G. firmus*. Data are from Zera and Larsen (2001), Zhao and Zera (2002) and Zera and Zhao (2003a).

example, whole-organism respiration is elevated in the flight-capable vs. flightless morph of several *Gryllus* species, especially on low-nutrient diets (Table 1; Zera and Mole 1993; Zera and Denno 1997; Zera et al. 1997, 1998; Crnocrak and Roff 2002). Basal respiration of the large flight muscles and production of large quantities of lipid flight fuel also have been implicated as important energetic costs of flight capability in *Gryllus* species (Zera et al. 1997, 1998, 1999; Zera and Larsen 2001; Zera and Harshman 2001; Zhao and Zera 2002, Harshman and Zera 2007). Triglyceride is the main flight fuel in *Gryllus* (Zera et al. 1999), as in most orthopterans (Beenakkers et al. 1985; and references therein). A substantially greater amount of triglyceride (30–40%), and total lipid, accumulates in the LW morph during early adulthood, similar to the situation in dispersing forms of other wing- or phase-polymorphic insects (Zera and Larsen 2001; Zera and Denno 1997). Importantly, this occurs precisely during the time when ovarian growth is substantially reduced in the LW compared with the SW morph (Fig. 7), and suggests that enhanced lipid accumulation may directly constrain ovarian growth and vice versa in morphs of *G. firmus*. Studies of nutrient consumption and assimilation in *G. firmus* have eliminated the possibility that the increased accumulation of lipid in the LW morph is due to increased nutrient consumption or assimilation (Zera and Brink 2002; unpublished data).

The trade-off between triglyceride accumulation and ovarian growth could result from a variety of causes, such as limited internal nutrients which require that increased allocation of nutrients to "reproduction" be matched by decreased allocation to "the soma"—the "classical' explanation, proposed or assumed by most workers (Zera and Denno 1997; Zera and Harshman 2001). Alternatively, space within the organism may be insufficient to accommodate large ovaries and a large amount of flight-fuel reserves. Finally, although not typically considered, the trade-off might result from a non-energetic "regulatory constraint" (Mole and Zera 1993; Zera and Harshman 2001; Harshman and Zera 2007; discussed below). That is, enhanced biosynthesis of triglyceride might require a hormonal environment that precludes an elevated biosynthesis of yolk protein. The negative-pleiotropy of juvenile hormone, which positively affects yolk protein biosynthesis and negatively affects lipid biosynthesis in many insects, including *Gryllus* (Zera et al. 1998; Zhao and Zera 2002; Zera and Zhao 2004), may set the stage for such a "hormonal constraint". Thus, even in the presence of abundant nutrients, ovarian growth and flight-fuel production may be negatively correlated. Although differential allocation of internal resources clearly underlies the trade-off between flight-capability and reproduction in insects, the key constraint that causes this trade-off is presently unknown (Zera and Harshman 2001, Harshman and Zera 2007).

Elevated lipid is an important component of morph-specialization in many insect polymorphisms, and in various life history adaptations (e.g. phenotypes exhibiting increased longevity or stress-resistance; Ricklefs 1996; Rose and Bradley 1998; Zera and Denno 1997; Harshman and Schmidt 1998; Zera and Harshman 2001; Zera and Zhao 2004; Zera 2005; Harshman and Zera 2007). However, very little is known about the specific aspects of lipid metabolism that have been modified to allow increased lipid accumulation. Most work has focused on total lipid as a somatic energy reserve, the enhanced accumulation of which is thought to reduce nutrients allocated to reproduction (see above references). However, the situation is likely to be much more complex, because lipid is a heterogeneous class of molecules whose components have a variety of somatic and reproductive functions. For example, phospholipid, the second most abundant lipid class in insects, is the major lipid component of vitellogenin and vitellins ("yolk proteins"), and is abundant in eggs, where it is used during embryonic development (Lipsitz and McFarlane 1970; Bailey 1975; Beenakkers et al. 1985; Gropes et al. 1998). Indeed, phospholipid accumulates to a greater degree in the reproductive flightless vs. the flight-capable morph of *G. firmus* during the first week of adulthood (Zera and Larsen 2001), and is found to a greater degree in ovaries vs. non-ovarian tissues (unpublished data). Furthermore, triglyceride, in addition to being an important somatic energy reserve, also is found in high concentration in eggs where it plays a reproductive function, namely as an energy reserve for the developing embryo (Beenakkers et al. 1985). Thus a variety of trade-offs involving specific lipid classes are expected to occur between flight-capable and flightless morphs of a wing polymorphic species, in addition to the general trade-off between whole-body lipid and ovarian mass. *A priori*, one would expect whole-body triglyceride to be elevated in LW females, while phospholipid should be elevated in SW females. A greater allocation of whole organism triglyceride to ovaries vs. the fat body (soma) also would be expected in the SW vs the LW morph. The existence and biochemical basis of each of these types of trade-off have recently been investigated in *G. firmus* (Zhao and Zera 2002; Zera and Zhao, 2003a, b).

Radiotracer studies have documented that the rate of biosynthesis of total lipid (= total fatty acid) and triglyceride are significantly elevated in the flight-capable, LW morph. Moreover, enhanced lipid biosynthesis in the LW morph occurs during the period of adulthood when total lipid and triglyceride accumulate to a greater degree in that morph (Figs. 7–9; Table 2; Zhao and Zera 2001, 2002). Thus an important biochemical cause of the greater accumulation of total lipid/triglyceride in the LW morph, appears to be increased rate of biosynthesis of these compounds. In addition to this enhancement of total lipid biosynthesis in the LW morph, there is a downstream trade-off in the proportional conversion of fatty acid into

Table 3 Trade-off in the biosynthesis of phospholipid vs. triglyceride in LW and SW *G. firmus* (modified from Zhao and Zera, 2002). Values are mean±SEM percentage total biosynthesized total lipid (= triglyceride + phospholipid) that was phospholipid in three pairs (blocks) of LW- and SW-selected lines. Note the greater proportional biosynthesis of phospholipid in the SW vs. the LW morph

Morph	*Block*		
	1	*2*	*3*
Acetate			
LW	15.8±0.8 (26)***[a]	11.7±0.5 (23)***	13.7±0.8 (24)***
SW	24.0±1.3 (23)	17.6±1.9 (24)	20.9±1.1 (24)

[a]Percentages differed between LW and SW lines within each block (*** = P < 0.005; t-test of arc-sine transformed values)

Data are the same as those presented in Fig. 9, and are illustrated graphically in Fig. 10.

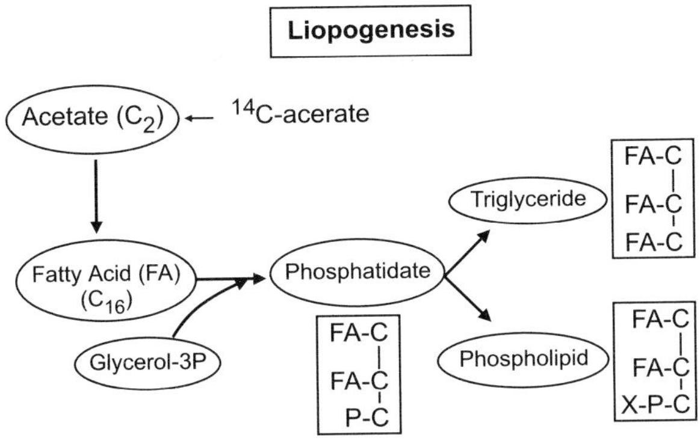

Fig. 8 Simplified pathway of *de novo* biosynthesis of triglyceride and phospholipid from acetate. Glycerol-3P = glycerol-3-phosphate. Box under phosphatidate indictes that this compound is composed of two fatty acids (FA) linked to glycerol-3-phosphate. Boxes next to triglyceride and phospholipid indicate that these compounds are produced from phosphatidate by removal of the phosphate group and addition of another fatty acid (triglyceride), or modification of the phosphate group (phospholipid). ^{14}C-acetate next to acetate indicates that radiotracer studies of lipid biosynthesis were conducted by injection of radiolabeled acetate into whole crickets and subsequent quantification of amount of radiolabel in either triglyceride or phospholipid. See Zhao and Zera (2001, 2002) for experimental details.

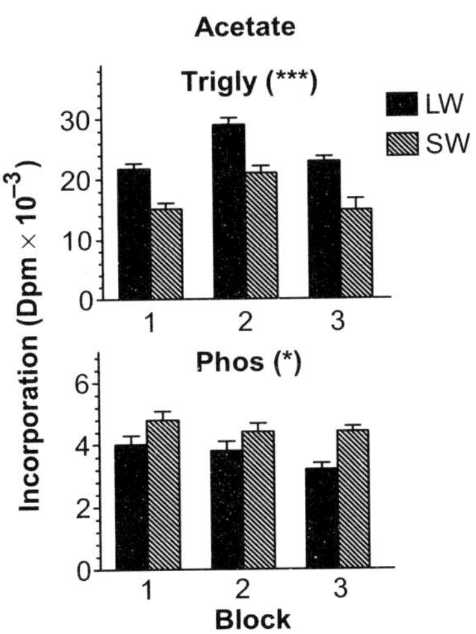

Fig. 9 Rate of biosynthesis [DPM (radiolabel) incorporation/7 hours] of triglyceride and phospholipid by LW and SW morphs on day 5 of adulthood. Histograms refer to means (±SEM) of LW or SW genetic stocks from 3 independent selection trials (Blocks) (data from Zhao and Zera, 2002). Asterisks refer to results of paired t-tests on LW and SW line means (*** = $P < 0.005$, * = $P < 0.05$). Note the greater biosynthesis of triglyceride but lesser biosynthesis of phospholipid in LW vs. SW *G. firmus*.

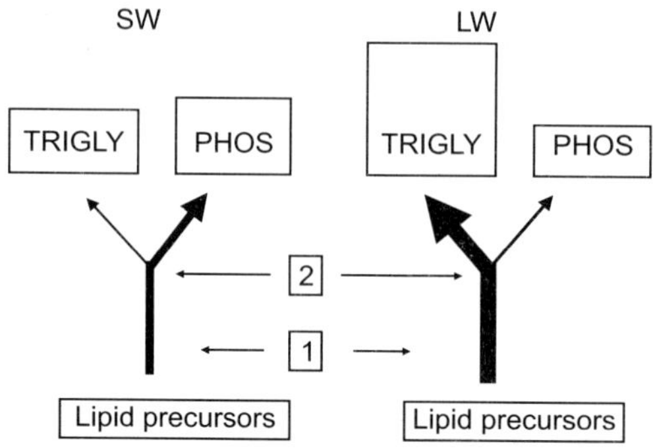

Fig. 10 Trade-offs in lipid biosynthesis in *G. firmus* identified in radiotracer studies of Zhao and Zera (2002). "Y" diagrams illustrate relative flow of ^{14}C through *de novo* pathway of lipid biosynthesis in LW and SW females. Total lipid biosynthesis (= total fatty acid biosynthesis; denoted by width of base of "Y") is greater in LW than in SW females. Total and relative biosynthesis of triglyceride is greater in LW than in SW females, while total and relative biosynthesis of phospholipid is greater in SW than in LW females. Widths of diagrams, which are only meant to illustrate rank-order and not quantitative differences between morphs, are based on data from Table 3 and Zhao and Zera (2002)

triglyceride vs. phospholipid in alternate morphs (Table 3, Figs. 8–10). A greater absolute and relative amount of fatty acid is allocated to the production of triglyceride and less to the production of phospholipid in the LW morph, while the opposite occurs in the SW morph. Indeed, although SW females produce a lower amount of total lipid, they produce a greater absolute and relative amount of phospholipid, than LW females. This morph-specific trade-off in the biosynthesis of triglyceride vs. phospholipid accounts for the elevated accumulation of triglyceride and decreased accumulation of phospholipid in LW vs. SW morphs during the first week of adulthood (Zera and Larsen 2001). The radiotracer study of Zhao and Zera (2002) represents the first demonstration that morphs of a polymorphic or polyphenic species differ in flux through a specific pathway of intermediary metabolism that produces end products (e.g. phospholipid vs. triglyceride) important for morph specialization.

Other functionally-important differences in lipid metabolism occur between morphs of *G. firmus* (Table 2). For example, not only does the LW morph biosynthesize more total lipid (i.e., fatty acid), it also utilizes (oxidizes) less fatty acid as an energy source during the first week of adulthood (Fig. 7; Zera and Zhao 2003a). Thus, the greater accumulation of total lipid and triglyceride in the LW morph occurs by both decreased utilization, as well as by increased biosynthesis. Furthermore, the SW morph converts a greater proportion of amino acids into ovarian protein than does the LW morph, which both oxidizes and converts a greater proportion of amino acids into lipid (Table 2; Zera and Zhao, 2006). In other words, intermediary metabolism in LW females is adapted for (1) the production and retention of triglyceride, presumably for flight, and (2) utilization of amino acids as an energy source, probably to biosynthesize triglyceride. By contrast, SW females allocate a greater proportion of amino acids to reproduction (increased biosynthesis of ovarian protein), and use a greater amount of lipid than LW females as an energy source, probably to fuel the biosynthesis of yolk protein (Zera and Zhao, 2006). Thus, morph-specializations for flight vs. reproduction in *G. firmus* result from a remarkable, global alteration of lipid and amino acid metabolism.

In addition to morph-specific trade-offs in the whole-organism production of total lipid, triglyceride, and phospholipid, morphs also differ in the relative allocation of these biosynthesized lipid components to organs of flight vs. reproduction (Zhao and Zera 2002). For example, a greater proportion of biosynthesized triglyceride is allocated to the fat body (soma) than to ovaries in the LW morph, while the opposite occurs in the SW morph. These studies of organ-specific allocation are important because they provide a more accurate assessment of the energetic costs of reproduction and flight capability. For example, if the morph-specific difference in the allocation of triglyceride to soma vs. ovaries is not taken into account, the energetic cost of flight capability is substantially underestimated in the LW morph, while the energetic cost of reproduction is substantially underestimated in the SW morph (Zhao and Zera, 2002). Most physiological studies of insect polymorphism and life history trade-offs have only measured whole-organism total lipid, and have classified total lipid as a somatic energy reserve (Zhao and Zera, 2002; Zera, 2005).

Differences between morphs in activities of lipogenic enzymes measured *in vitro* have corroborated the morph-specific differences in *in vivo* rates of lipid biosynthesis discussed above (Figs. 11–12; Table 2; Zhao and Zera 2001; Zera and Zhao 2003b). Activities of all studied enzymes involved in lipid biosynthesis and in the conversion of amino acids into lipid (transaminases) are elevated in the flight-capable morph, relative to the flightless morph. These enzymatic differences further underscore the dramatic, global modification of lipid metabolism in morphs of *G. firmus*.

What co-ordinates the expression of the numerous aspects of lipid metabolism that differ between flight-capable and flightless morphs of *G. firmus*? Endocrine regulation of wing polymorphism has typically focused on the hormonal control of growth, development, and degeneration of

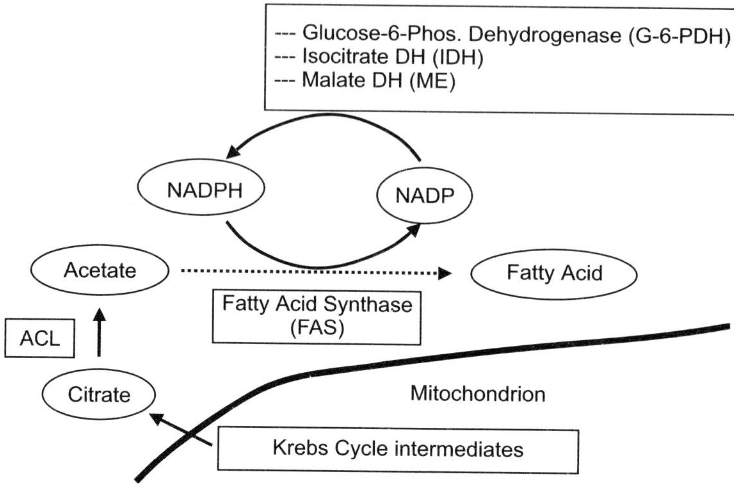

Fig. 11 Location of enzymes in the pathway of fatty-acid biosynthesis whose activities were compared between LW and SW morphs. ACL (ATP-citrate lyase) converts citrate, transported outside the mitochondrion, into acetate (= acetyl CoA). Fatty acid synthase (FAS) is an enzyme complex that converts acetyl CoA into a 16 carbon fatty acid through a series of chemical reactions denoted by the dotted line. Reducing equivalents (NADPH) required for fatty acid biosynthesis are produced from NADP by the enzymes glucose-6-phosphate dehydrogenase, isocitrate dehydrogenase, and malate dehydrogenase.

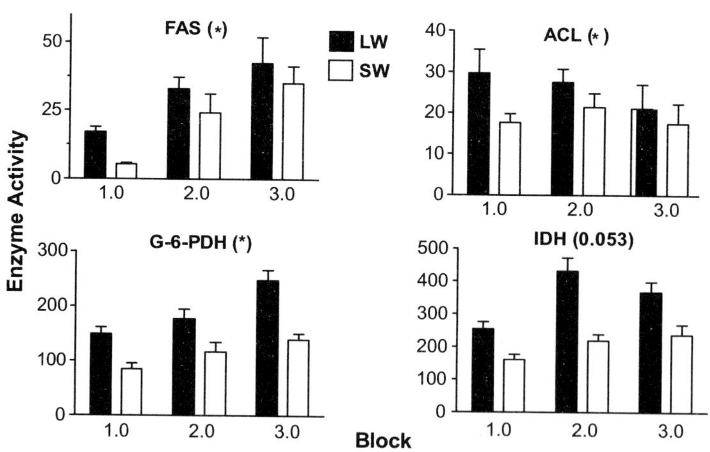

Fig. 12 Elevated activities of lipid-biosynthesizing enzymes in LW vs. SW lines *G. firmus* measured on day 5 of adulthood (data from Zera and Zhao, 2003b). See legend of Figure 11 for names of enzymes. See Figure 9 for definition of "Block" and asterisks.

organs of flight and reproduction, in adults and juveniles. However, hormones play an important role in regulating and co-ordinating various aspects of intermediary metabolism (Granner and Pilkis 1990; Sul and Wang 1998). Thus, morph-specific differences in the hormonal regulation of metabolism also are expected to play a key role in regulating metabolic aspects of morph specialization for flight vs. reproduction. As mentioned previously, juvenile hormone (JH) is a prime candidate regulator of lipid and protein metabolism in the LW and SW morphs, since it positively affects yolk protein biosynthesis, and inhibits the accumulation of lipid flight fuel in many insects (Zera and Zhao 2004). Thus, a long-duration elevation of the JH titer or increased sensitivity of fat body and ovaries to JH in SW vs. LW females would account for both their increased ovarian growth and reduced biosynthesis of total lipid and triglyceride. Evidence supporting this hypothesis was obtained from a series of experiments in which the JH titer was artificially elevated in the flight capable morph of *G. firmus* by topical application of a JH analogue (methoprene). This manipulation produced a remarkable SW phenocopy with respect to numerous reproductive, anatomical, and biochemical traits (Fig. 13; Zera and Zhao 2004). For example, LW females treated with methoprene during adulthood had larger ovaries, decreased flight-muscle mass, decreased rate of triglyceride biosynthesis, decreased activities of lipogenic enzymes, increased rate of fatty acid oxidiation, and increased rate of phospholipid biosynthesis, all characteristics of the SW morph. Remarkably, even such traits as the *proportional* biosynthesis of triglyceride vs. phospholipid in LW treated females was altered to values seen in unmanipulated SW females (Zera and Zhao 2004).

Although morph-specific differences in lipid metabolism appear to be caused by variation in endocrine regulation, general aspects of the mechanisms involved, and the specific the role of JH itself, are unclear. As discussed previously (see pp. 563-564), the mechanism by which JH influences the expression of traits in adult morphs is complicated by the existence of a dramatic (10–100-fold) daily cycle in the JH titer in the LW morph (Fig. 5). However, while short-term, circadian variation in the JH titer may regulate the expression of some traits (e.g., possibly flight behavior in the LW morph; see Previous section) other key morph-specific traits appear to be unaffected. For example, the relative increase in ovarian mass in SW vs. LW morphs remains relative constant throughout a 24 h cycle (Fig. 6B). A similar situation exists for morph-specific differences in aspects of lipid metabolism: activities of lipogenic enzymes, and rates of lipid biosynthesis, are elevated in LW vs. SW morphs, to a similar level throughout the

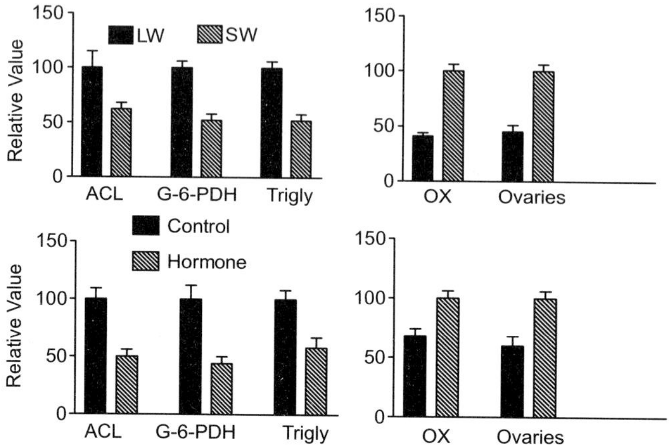

Fig. 13 Production of a SW biochemical and reproductive phenocopy from LW *G. firmus* by topical application of a juvenile hormone analogue. Top panels: Relative depression in activities of lipid biosynthetic enzymes and triglyceride biosynthesis (left), and relative increase in fatty acid oxidation and ovarian mass (right) in unmanipulated SW vs. LW females on day 5 of adulthood. Bottom panels: Relative depression in activities of lipid biosynthetic enzymes and triglyceride biosynthesis (left), and relative increase in fatty acid oxidation and ovarian mass (right) in JH-treated LW vs. control (untreated) LW females on day 5 of adulthood. Note the parallel increase or decrease in aspects of lipid metabolism and ovarian mass in hormone-treated vs. untreated LW crickets and untreated SW vs. untreated LW crickets. All comparisons between LW and SW morphs or control or treated LW individuals were statistically significant ($P < 0.01$). Data are from Zera and Zhao (2004). See legend of Figure 11 for names of enzymes. Trigly and OX = rate of triglyceride and fatty acid oxidation, respectively; Ovaries = relative ovarian mass.

photophase (AJZ and Z. Zhao, unpublished data). Thus, the expression of increased ovarian growth and decreased lipid biosynthesis in the SW morph may require a longer duration elevation in the JH titer or tissue sensitivity to this hormone; the short-duration spike (< 6 h) in the JH titer (Fig. 5) may be insufficient to affect the expression of these traits. Consistent with this idea, the JH manipulation that produced the SW biochemical phenocopy in the study of Zera and Zhao (2004), involved a long-duration elevation of the JH titer (over several days, which could also have mimiced a long-duration elevation in tissue sensitivity to JH) (Zera and Zhao 2004). Alternately, other hormones, such as ecdysteroids, may be the primary regulators of morph-specific differences in metabolism and reproduction. As mentioned previously, the ecdysteroid titer is consistently higher in SW vs. LW adult females during early adulthood (Zera and Bottsford, 2001; Zhao and Zera, 2004a; Fig. 6). Topical application of JH and JH analogues is known to alter the ecdysteroid titer (Smith and Nijhout 1981), and, thus, the JH manipulation of Zera and Zhao (2004) may also have caused changes in lipid metabolism by altering the ecdysteroid titer. Although the specific mechanism is unknown, variation in endocrine regulation appears to be an important cause of morph-specific differences in lipid metabolism.

Ongoing work on wing polymorphism and lipid metabolism is focusing on the regulation and enzymology of lipogenic enzymes. Several of these enzymes (IDH, G-6-PDH), ACL; see Fig. 11) have been purified, antibodies raised, and their genes cloned and sequenced. The key issue being investigated is the extent to which higher enzyme activities in the LW morph are due to enhanced gene expression or enhanced catalytic efficiency.

Summary, Synthesis and Future Directions

Endocrine Regulation of Wing Morphs

During the past two decades, considerable progress has been made in identifying the endocrine mechanisms that regulate aspects of wing polymorphism. However, because of the complexity of this problem, only the broad outlines of a few *potential* regulatory mechanisms can be seen at present. Most significantly, morph-specific differences in hemolymph hormone titers, and activities of titer regulators (juvenile hormone esterase) have been directly identified in a wing polymorphic insect.

In some cases, results of endocrine studies in *Gryllus* are consistent with expectations of the "classical model" (Figs. 1 and 3). Most notable is the very strong correlation between reduced activity of juvenile hormone esterase (JHE) during the last stadium in nascent SW individuals in two wing-polymorphic cricket species. A wide variety of data are consistent with the hypothesis that reduced JHE activity specifies the development of the SW morph (Figs. 2 and 3). Indeed, studies of JHE activity in wing-polymorphic crickets represent one of the most detailed investigations of the role of a specific endocrine regulator in morph development for any case of complex polymorphism in insects. The ecdysteroid titer has also been implicated in regulating morph development in *Gryllus*, either in concert with, or independent of JH (Figs. 3 and 4). Indeed variation in the ecdysteroid titer between morphs of *G. rubens* is greater than variation in the JH titer raising the possibility that ecdysteroids may be more important than JH in regulating wing polymorphism in *Gryllus*.

Ironically, the weak link in the JHE-JH-wing-morph hypothesis in *Gryllus*, is the JH titer itself, which differs only subtly between morphs. As discussed above, there are reasons to expect that only small differences in the JH titer between nascent morphs during the last juvenile instar can be tolerated without disrupting metamorphosis, and that these titer differences are functionally important. On the other hand, recent studies raise the possibility that JHE may affect morph development in unexpected ways. JHE hydrolyzes compounds other than JH (Zera et al. 2002) and thus might influence morph development by metabolizing a non-JH regulator. Numerous JH-like compounds of unknown function exist in insects (Darrouzet et al. 1998; Davey 2000), and a diverse array of lipid molecules are involved in cell signaling (Karp 2002). Furthermore, some evidence suggests that JH-acid (JHE converts juvenile hormone into juvenile hormone acid), rather than being an inactive metabolite is a hormone required for metamorphosis. That is, JH may be a prohormone that is converted into the active hormone, JH-acid, by JHE (Ismail et al 1998, 2000). This might explain why JHE activity is strongly associated with wing morph, while the JH titer is much less so (Zera, 2004). The important point is that even fundamental aspects of the endocrine regulation of morph development (and JH mode-of-action; Wheeler and Nijhout 2003) are not yet firmly established and may be very different than those originally envisioned. Indeed, the first direct measures of the JH titer in *adult* wing morphs resulted in the completely unexpected finding that the JH titer cycles dramatically in the flight-capable morph, but not in the flightless morph (*G. firmus*; Zera and Cisper 2001; Zhao and Zera 2004).

It is also important to emphasize that despite significant advances during the past two decades, the endocrine mechanisms that regulate wing polymorphism in insects as a whole remain largely an open issue. Hormone titers and titer regulators have only been measured in detail in two cricket species, and even these basic pieces of endocrine information are lacking in aphids and planthoppers, despite decades of study. Many central aspects of morph regulation remain completely unstudied (e.g., hormone receptors; see below). The situation is not restricted to wing-polymorphim. For example, although the role of JH in insect phase polyphenism has been the subject of experimentation and discussion since the 1950's, only very recently have detailed measures of the JH titer been reported for locust phases during adulthood (Tawfik et al. 2000), and comparable JH titer data still are lacking for juvenile phases!

Future studies on the endocrine regulation of wing polymorphism should focus on four areas. First, basic studies on the role of juvenile hormone and ecdysteroids in regulating morph development and reproduction are still important, given that direct estimates of hormone titers and titer regulators are only available for a few cricket species. Second, no published information is available on morph-specific differences in any hormone receptor for any wing-polymorphic insect. Thus, a major aspect of the endocrine regulation of morph development and reproduction remains virtually unstudied. Investigations of JH receptors in wing polymorphic insects must await the identification of the nuclear JH receptor in insects [see Truman and Riddiford (2002) and Wilson (2004) for state-of-the-art in JH receptor studies], if, indeed, JH operates through a traditional nuclear receptor (Wheeler and Nijhout, 2003). On the other hand, the ecdyteroid receptor has been cloned in several insects (Fujiwara et al. 1995; Riddiford et al. 2001), including a wing polymorphic aphid (A. Pawlak-Skrzecz, G. N. Hannan, D. F. Hales and R. J. Hill; pers. comm.). Third, detailed investigations of the mechanisms by which hormones differentially regulate the development of target organs (e.g. wings and flight muscles) are just beginning (e.g. Kobayashi and Ishikawa 1994; Niitsu 2001) and will also likely expand considerably during the next decade. Microarray studies of aphid wing polymorphism show great promise for investigating, in detail, mechanisms of morph development (Brisson et al. 2007).

Finally, studies of the endocrine regulation of wing polymorphism thus far have been narrowly focused on juvenile hormone and ecdysone. Numerous neurohormones regulate various aspects of reproduction and development in insects (Raabe 1989; Nijhout 1994; Bendena et al. 1999; Klowden, 2002). Locust phases differ in various neurohormones (Pener and Yerushalmi 1998), and, numerous studies have implicated neurohromones in morph induction in aphids (Hardie and Lees 1985; Zera and Denno 1997). Neurohormonal regulation of morph development and reproduction will likely become one of the most prominent foci of future work on the endocrine control of wing polymorphism in insects.

Physiology and Biochemistry of Resource Allocation in Morphs

Detailed physiological and biochemical studies of resource allocation in flight-capable and flightless morphs began much later than endocrine studies of morph development and reproduction (Zera and Denno 1997). However, we currently have a more detailed understanding of the mechanisms underlying morph-specific resource allocation, albeit still in only a few species. The past decade has seen the first direct measures of morph-specific nutrient consumption, assimilation, and allocation in wing-polymorphic insects (Mole and Zera 1993; Tanaka 1993; Zera and Denno 1997; Zera and Harshman 2001). These studies provide the best, and thus far the only direct, evidence that an internal trade-off of nutrients is responsible for important aspects of morph specialization. Insect feeding studies have rarely been used to investigate the causes underlying life history aspects of morph specialization. Yet such studies are indispensable because they provide the only way of quantifying the extent to which morph specializations result from an internal trade-off of resources as opposed to the differential acquistion of nutrients from the diet. Additional important advances on the topic of resource allocation have been the identification of important energetic requirements of flight capability that negatively impact egg production, most notably, increased maintenance metabolism in the flight-capable morph due to the presence of large flight muscles, and biosynthesis of large quantities of lipid flight fuel (Zera et al. 1997; Zera and Harshman 2001; Zhao and Zera 2002; Zera 2005). Wing polymorphism in crickets is now the best studied example of insect polymorphism with respects to energetic aspects of morph-specialization. However, the ultimate cause of these internal trade-offs of nutrients (e.g., constraints due to limited energy or space within the organism; antagonistic effects of regulators on different aspects of metabolism), remains an open issue. In addition, as is the case for endocrine studies of morph development, only a few studies have investigated detailed energetics aspects of morph specialization in a few species of crickets. Thus, the extent to which results obtained for these cricket species can be generalized to other wing-polymorphic insects is unknown.

Recent studies of lipid metabolism in morphs of *G. firmus* represent the first detailed investigations of the biochemical mechanisms underlying morph-specific patterns of resource allocation in an insect polymorphism (Zhao and Zera 2002; Zera and Zhao 2003a, b, 2004). These studies illustrate the remarkable global restructuring of intermediary metabolism that underlies morph specialization for flight vs. reproduction. Most notably, studies in *G. firmus* have directly documented that the differential flow of metabolites through a specific pathway in intermediary metabolism underlies an internal, resource-based trade-off important in morph specialization. That is, the SW and LW morphs differ in the degree to which fatty acids are partitioned into pathways of triglyceride and phospholipid biosynthesis (Zhao and Zera 2002). This partitioning results in enhanced production of triglyceride flight fuel in the LW(f) morph, and enhanced production of phospholipid in the SW morph, which is important in reproduction. This branch point in lipid biosynthesis will likely serve as a useful model to investigate the molecular basis of resource-allocation trade-offs (Harshman and Zera, 2007).

Finally, studies in *Gryllus* are beginning to identify the endocrine mechanisms that modulate intermediary metabolism leading to morph-specific specializations for flight vs. reproduction (Zera and Zhao 2004). These studies illustrate the importance of viewing regulation and energetics as important complementary aspects of morph specialization rather than alternative explanations (Harshman and Zera 2005).

Proximate Endocrine Mechanisms and Phenotypic Plasticity

Phenotypic plasticity is a broad topic that includes many physiological, ecological, and evolutionary aspects. The present chapter has focused primary on the proximate mechanisms that underlie expression of phenotypic variation in wing polymorphic crickets. There is an increasing appreciation that information on proximate mechanisms of phenotypic expression is critically important for understanding the evolution of adaptive phenotypic plasticity. This argument has recently been cogently stated by West-Eberhard (2003, p. 11): "For evolutionary biology, proximate mechanisms represent more than just different levels of analysis or research styles. They are *the* causes of variation upon which selection acts......Among the consequences of neglect of mechanisms in modern evolutionary biology are the problems that arise when the black box of mechanism is filled with imaginary devices." Clearly, understanding evolutionary aspects of phenotypic plasticity requires a deep understanding of the mechanisms responsible for variation in phenotypic expression. Yet detailed data on these mechanisms are available for only a few species, and current models and discussions of phenotypic plasticity are still often filled with the "imaginary devices" alluded to above (Zera 1999, 2007). Recent in-depth studies of proximate mechanisms controlling alternate morph development and specializations for flight and reproduction in wing polymorphic crickets are helping to fill in this gap in our understanding of phenotypic plasticity. Most importantly, investigations of wing polymorphic *Gryllus* are beginning to identify specific endocrine control mechanisms that are modulated to produce phenotypes that differ in a diverse array of traits which adapt them for flight vs. reproduction. It is remarkable that relatively slight variations in endocrine regulators appear to direct nascent morphs along very different developmental trajectories. Equally remarkable is the endocrine orchestration of morph-specific expression of genes controlling intermediary metabolism during adulthood, which is a key component of morph specialization for flight vs. reproduction. Results of these studies of *Gryllus*, together with studies of other polymorphic insects (e.g., Nijhout, 1994, 1999; Suzuki and Nijhout 2006), indicate that modulation of endocrine regulation plays a cardinal, causal role in phenotypic plasticity.

In addition, studies of *Gryllus* have documented that phenotypic plasticity and genetic polymorphism for wing length are both associated with modulation of the same endocrine regulator, juvenile hormone esterase (Fig. 2). These data support the hypothesis of West-Eberhard (2003) on the interchangability of environmental and genetic influences on phenotypic expression (also see Suzuki and Nijhout 2006). The proximate control of phenotypic plasticity is a topic that will occupy physiologists and evolutionary biologists for decades to come. Results obtained on the endocrine control of polymorphism in *Gryllus* provide a firm foundation for future studies of this complex and fascinating topic.

Acknowledgements

AJZ gratefully acknowledges the National Science Foundation which has supported his work on wing polymorphism in *Gryllus* during the past 20 years (most recently grants IOS-0516973, IBN-0212486, IBN-0130665, and IBN-9808249).

References

Ayoade, O., S. Morooka, and S. Tojo. 1999. Enhancement of short wing formation and ovarian growth in the genetically defined macropterous strain of the brown planthopper, *Nilaparvata lugens*. Journal of Insect Physiology, 45: 93–100.

Bailey, E. 1975. Biochemistry of insect flight. Part 2. Fuel supply. *In* D. J. Candy and B. A. Kilby [eds] 1975. Insect Biochemistry and Function. Chapman and Hall, London.

Beenakkers, A. M. T., D. J. Van der Horst, and W. J. A. Van Marrewijk. 1985. Insect lipids and lipoproteins, and their role in physiological processes. Progress in Lipid Research, 24: 19–67.

Bendena, W. G., B. C. Donly and S. S. Tobe 1999. Allatostatins: A growing family of neuropeptides with structural and functional diversity. Annals of the New York Academy of Science, 897: 311–329.

Bertuso, A., S. Morooka, and S. Tojo. 2002. Sensitive periods for wing development and precocious metamorphosis after precocene treatment of the brown planthopper, *Nilaparvata lugens*. Journal of Insect Physiology, 48: 221–229.

Brisson, J. A., G. K. Davis, and D. L. Stern. 2007. Common genome-wide patterns of transcript accumulation underlying the wing polyphenism and polymorphism in the pea aphid (*Acrythosiphon pisum*). Evol. Dev. 9: 338–346.

Crnokrak, P. and D. Roff. 2002. The trade-off to flight capability in *Gryllus firmus*: the influence of whole-organism respiration rate on fitness. Journal of Evolutionary Biology, 15: 388–398.

Crone, E., A. J. Zera, A. Anand, J. G. Oakeshott, T. Sutherland, R. J. Russell, L. G. Harshman, F. G. Hoffmann and C. Claudianos. 2007. JHE in Gryllus assimilis: cloning, sequence activity associations and phylogeny. Insect Biochem. Molec. Biol. (In Press).

Davey, K. G. 2000. The modes of action of juvenile hormones: some questions we ought to ask. Insect Biochemistry and Molecular Biology, 30: 663–669.

Dingle, H. 1996. Migration: the Biology of Life on the Move. Oxford University Press, Oxford.

Darrouzet, E., F. Rossignol, and F. Couillaud. 1998. The release of isoprenoids by locust corpora allata in vitro. Journal of Insect Physiology, 44: 103–111.

Denno, R. F. 1994. The evolution of dispersal polymorphism in insects: the influence of habitats, host plants, and mates. Researches on Population Biology, 36: 127–135.

Denno, R. F., L. Douglass, and D. Jacobs. 1985. Crowding and host plant nutrition: environmental determinants of wing-form in *Proklesia marginata*. Ecology, 66: 1588–1596.

Denno, R. F., D. Hawthorne, B. Thorne, and C. Gratton. 2000. Reduced flight capability in British Virgin Island populations of a wing-dimorphic insect: role of habitat isolation, persistence, and structure. Ecological Entomology, 26: 25–36.

Denno, R. F., M. Raupp, D. Tallamy, and C. Reichelderfer. 1980. Migration in heterogeneous environments: differences in habitat selection between wing-forms of the dimorphic planthopper *Proklesia marginata* (Homoptera:Delphacidae).

Denno, R. F., G. Roderick, K. Olmstead, and H. Dobel. 1991. Density-related migration in planthoppers (Homoptera:Delphacidae): role of habitat persistence. American Naturalist, 138: 1513–1541.

Dingle, H. 1996. Migration: the Biology of life on the move. Oxford University Press, Oxford.

Dingle, H. , and R. Winchell. 1997. Juvenile hormone as a mediator of plasticity in insect life histories. Archives of Insect Biochemistry and Physiology, 35: 359–373.

Dixon, A. F. G. 1985. Aphid Ecology. Chapman and Hall, London.

Elekonich, M. M., and G. E. Robinson. 2000. Organizational and activational effects of hormones on insect behavior. Journal of Insect Physiology, 46: 1509–1515.

Emlen, D. and H. F. Nijhout. 1999. Hormonal control of male horn length dimorphism in the dung beetle *Onthophagus taurus* (Coleoptera:Scarabaeidae). Journal of Insect Physiology, 45: 45–53.

Emlen, D. J. and H. F. Nijhout. 2001. Hormonal control of male horn dimorphism in *Onthophagues taurus* (Coleoptera:Scarabaeidae): a second critical period. Journal of Insect Physiology 47: 1045–1054.

Evans, J., and D. E. Wheeler. 1999. Differential gene expression between developing queens and workers in the honey bee, *Aphis mellifera*. Proceedings of the National Academy of Sciences USA, 96: 5575–5580.

Evans, J., and D. E. Wheeler. 2001. Gene expression and the evolution of insect polyphenisms. Bioessays, 23: 62–68.

Fujiwara, H., M. Jindra, R. Newitt, S. R. Palli, K. Hiruma, and L. M. Riddiford. 1995. Cloning of an ecdysone receptor homolog from *Manduca sexta* and the developmental profile of its mRNA in wings. Insect Biochemistry and Molecular Biology, 25: 845–856.

Gao, N., and J. Hardie. 1996. Pre- and post-natal effects of precocenes on aphid morphogenesis and differential rescue. Archives of Insect Biochemistry and Physiology, 32: 503–510.

Gilbert, L. I., I. Kostos, and S. Gill 2005. Comprehensive molecular Insect Science. Vol. 3. Endocrinology Elsevier. Amsterdam.

Goodman, W. G., Z.-Y. Huang, G. E. Robinson, A. Strambi, and C. Strambi. 1993. A comparison of two juvenile hormone radioimmunoassays. Archives of Insect Biochemistry and Physiology, 23: 147–152.

Gould, S. J. 1977. Ontogeny and Phylogeny. Belknap Harvard Press, Cambridge.

Granner, D. and S. Pikis. 1990. The genes of hepatic glucose metabolism. Journal of Biological Chemistry, 265: 10173–10176.

Grapes, M., P. Whiting, and L. Dinan. 1989. Fatty acid and lipid analysis of the house cricket, *Acheta domesticus*. Insect Biochemistry, 19: 767–774.

Hales, D. F. 1976. Juvenile hormone and aphid polymorphism, pp. 105–115. *In* M. Lucher [ed.] 1976. Phase and Caste Determination in Insects. Endocrine Aspects. Pergamon Press, Oxford.

Hammock, B. D. 1985. Regulation of the juvenile hormone titer: Degradation, pp. 431–472. *In* G. A. Kerkut and L. I. Gilbert [eds] 1985. Comprehensive Insect Physiology, Biochemistry and Pharmacology. Pergamon Press, New York.

Hardie, J. 1980. Juvenile hormone mimics the photoperiodic apterization of the alate gynopera of the aphid, *Aphis fabae*. Nature, 286: 602–604.

Hardie, J., F. C. Baker, G. C. Jamieson, A. D. Lees, and D. A. Schooley. 1985. The identification of an aphid juvenile hormone, and its titre in relation to photoperiod. Physiological Entomology, 10: 297–302.

Hardie, J., K. Honda, T. Timar, and L. Varjas. 1995. Effects of 2,2-dimethylchromene derivatives on wing determination and metamorphosis in the pea aphid, *Acyrthosiphon pisum*. Archives of Insect Biochemistry and Physiology, 30: 25–40.

Hardie, J., and A. D. Lees. 1985. Endocrine control of polymorphism and polyphenism, pp. 441–489. *In* G. A. Kerkut and L. I. Gilbert [eds] 1985. Comprehensive Insect Physiology, Biochemistry and Pharmacology. Pergamon Press, New York.

Harrison, R. G. 1980. Dispersal polymorphisms in insects. Annual Review of Ecology and Systematics, 11: 95–111.

Harshman, L. G. and J. L. Schmidt. 1998. Evolution of starvation resistance in *Drosophila melanogaster*: Aspects of metabolism and counter-impact selection. Evolution, 52: 1679–1685.

Harshman, L. G. and A. J. Zera. 2007. The cost of reproduction: the devil in the details. Trend Ecol. Evol. 22: 80–86.

Hartfelder, K. and D. J. Emlen. 2005. Endocrine control of insect polyphenism. pp 651–703. *In* L. I. Gilbert, I. K. Iatrou and S. Gill [eds]. Comprehensive Molecular Insect Science. Vol. 3. Endocrinology. Elsevier, Amsterdam.

Ismail, S. M., K. Satyanaryama, J. Y. Bradfield, K. H. Dahm, G. Bhaskaran. 1998. Juvenile hormone acid; evidence for a hormonal function in induction of vitellogenin in larvae of *Manduca sexta*. Archives of Insect Biochemistry and Physiology, 37: 305–314.

Ismail, S. M., C. Goin, K Muthumani, M. Kim, K. H. Dahm, G. Baskaran. 2000. Juvenile hormone acid and ecdysteroid together induce competence for metamorphosis of the Verson's gland in *Manduca sexta*. Journal of Insect Physiology, 46: 59–68.

Karp, G. 2002. Cell and Molecular Biology. Concepts and Experiments. Third Edition. John Wiley & Sons, New York.

Klowden, M. J. 2002. Physiological Systems in Insects. Academic Press, Amsterdam.

Kobayashi, M. and H. Ishikawa. 1994. Involvement of juvenile hormone and ubiquinone-dependent proteolysis in flight muscle breakdown of altate aphid. Journal of Insect Physiology, 40: 107–111.

Langellotto, G. and R. F. Denno. 2001. Benefits of dispersal in patchy environments: Mate location by males of a wing-dimorphic insect. Ecology, 82: 1870–1878.

Lees, A. D. 1966. The control of polymorphism in aphids. Advances in Insect Physiology 3: 207–277.

Lipsitz, E. Y. and J. E. McFairlane. 1970. Total lipid and phsopholipid during the life cycle of the house cricket, *Acheta domesticus* (L.). Comparative Biochemistry and Physiology, 34: 699–704.

Masaki, S. and T. Shimizu. 1995. Variability in wing form in crickets. Researches in Population Ecology, 37: 119–128.

Matsuda, R. 1987. Animal Evolution in Changing Environments with Special Reference to Abnormal Metamorphosis. Wiley-Interscience, New York.

Mittler, T. E. 1991. Juvenile hormone and aphid polymorphism, pp. 453–474. *In* A. P. Gupta [ed.] 1991. Morphogenetic Hormones of Arthropods,. Rutgers University Press, New Brunswick.

Moczek, A. and H. F. Nijhout. 2002. Developmental mechanisms of threshold evolution in a polyphenic beetle. Evolution and Development, 4: 252–264.

Mole, S., and A. J. Zera. 1993. Differential allocation of resources underlies the dispersal-reproduction trade-off in the wing-dimorphic cricket, *Gryllus rubens*. Oecologia, 93: 121–127.

Mole, S. and A. J. Zera. 1994. Differential resource consumption obviates a potential flight-fecundity trade-off in the sand cricket (*Gryllus firmus*). Functional Ecology, 8: 573–580.

Morooka, S. and S. Tojo. 1992. Maintenance and selection of strains exhibiting specific wing form and body color under high density conditions in the brown planthopper, *Nilaparvata lugens* (Homoptera:Delphacide). Applied Entomology and Zoology 27: 445–454.

Nijhout, H. F. 1994. Insect Hormones. Princeton University Press, Princeton.

Nijhout, H. F. 1999. Control mechanisms of polyphenic development in insects. Bioscience 49: 181–192.

Nijhout, F. J., and D. Wheeler. 1982. Juvenile hormone and the physiological basis of insect polymorphism. Quarterly Review of Biology, 57: 109–133.

Niitsu, S. 2001. Wing degeneration due to apotosis in the female of the winter moth *Nyssiodes lefuarius* (Lepidoptera, Geometridae). Entomological Science 4: 1–7.

Pener, M. P., and Y. Yerushalmi. 1998. The physiology of locust phase polymorphism. Journal of Insect Physiology, 44: 365–377.

Raabe, M. 1989. Recent Developments in Insect Neurohormones. Plenum Press, New York.

Rankin, M. A. 1991. Endocrine effects on migration. American Zoologist 31: 217–230.

Rankin, M. A., and M. C. Singer. 1984. Insect movement: Mechanisms and effects, pp. 185–216. *In* C. B. Huffaker and R. L. Rabb [eds] 1984. Ecological Entomology. John Wiley & Sons, New York.

Ricklefs, R. 1996. Avain energetics, ecology and evolution, pp. 1–30. *In* C. Cary [ed.] 1996. Avian Energetics. Chapman, New York.

Riddiford, L. M., P. Cherbas, and J. W. Truman. 2001. Ecdysone receptors and their biological actions. Vitamins and Hormones 60: 1–73.

Roe, R. M., and K. Venkatesh. 1990. Metabolism of juvenile hormones: degradation and titer regulation, pp. 126–179. *In* A. P. Gupta [ed.] 1990. Morphogenetic Hormones of Arthropods. Rutgers University Press, New Brunswick.

Roff, D. A. 1986. The evolution of wing dimorphism in insects. Evolution, 40: 1009–1020.

Roff, D. A. 1990. The evolution of flightlessness in insects. Ecological Monographs, 60: 389–421.

Roff, D. A. 1994. Habitat persistence and the evolution of wing dimorphism in insects. American Naturalist, 144: 772–798.

Roff, D. A., and D. J. Fairbairn. 1991. Wing dimorphisms and the evolution of migratory polymorphisms among insects. American Zoologist, 31: 243–251.

Roff, D. A., G. Stirling, and D. J. Fairbairn. 1997. The evolution of threshold traits: A quantitative genetic analysis of the physiological and life-history correlates of wing dimorphism in the sand cricket. Evolution, 51: 1910–1919.

Rose, M. R., and T. J. Bradley. 1998. Evolutionary physiology of the cost of reproduction. Oikos, 83: 443–451.

Roundtree, D. B., and H. F. Nijhout. 1995a. Genetic control of a seasonal polyphenism in *Precis coenia* (Lepidoptera: Nymphalidae). Journal of Insect Physiology, 41: 1141–1145.

Roundtree, D. B., and H. F. Nijhout. 1995b. Hormonal control of a seasonal polyphenism in *Precis coenia* (Lepidoptera: Nymphalidae). Journal of Insect Physiology, 41: 987–992.

Shimizu, T. and S. Mazaki. 1993. Injury causes microptery in the ground cricket, *Dianemobius fascipes*. Journal of Insect Physiology, 39: 1021–1027.

Smith, W. and H. F. Nijhout. 1981. Effects of a juvenile hormone analogue on duration of the fifth instar in the milkweed bug *Oncopeltus fasciatus*. Journal of Insect Physiology, 27: 169–173.

Southwood, T. R. E. 1961. A hormonal theory of the mechanism of wing polymorphism in Heteroptera. Proceedings of the Royal Entomological Society, London, 36: 63–66.

Stay, B., J. A. S. Bachman, C. A. Stoltzman, S. E. Fairbairn, C. G. Yu, and S. S. Tobe. 1994. Factors affecting allatostatin release in a cockroach (*Diploptera punctata*): Nerve section, juvenile hormone analog and ovary. Journal of Insect Physiology, 40: 365–372.

Socha, R. 1993. *Pyrrhocoris apterus* (Heteroptera) - an experimental model species: a review. European Journal of Entomology 101: 539–545.

Sul, H., and D. Wang. 1998. Nutritional and hormonal regulation in fat synthesis: Studies of fatty acid synthase and mitochondrial glycerol-3-phosphate acyltransferase gene transcription. Annual Review of Nutrition, 18: 331–351.

Sula, J., R. Socha, and R. Zemek. 1998. Wing morph-related physiological differences in adults of temperate population of *Pyrrhocoris apterus* (L.) (Heteroptera). Comparative Biochemistry and Physiology (A), 121: 365–373.

Suzuki, Y. and H. F. Nijhout. 2006. The evolution of a polyphenism by genetic accommodation. Science 311: 650–652.

Tanaka, S. 1993. Allocation of resources to egg production and flight muscle development in a wing dimorphic cricket, *Modicogryllus confirmatus*. Journal of Insect Physiology, 39: 493–498.

Tanaka, S. 1994. Endocrine control of ovarian development and flight muscle histolysis in a wing dimorphic cricket, *Modicogryllus confirmatus*. Journal of Insect Physiology, 40: 483–490.

Tata, J. 1996. Hormonal interplay and thyroid hormone receptor expression during amphibian metamorphosis, pp. 465–503. *In* L. I. Gilbert, J. R. Tata and B. G. Atkinson [eds] 1996. Metamorphosis: Postembryonic Reprogramming of Gene Expression in Amphibian and Insect Cells. Academic Press, New York.

Tawfik, A. I., K. Treiblmayr, A. Hassanali, and E. O. Osir. 2000. Time-course haemolymph juvenile hormone titres in solitarious and gregarious adults of *Schistocerca gregaria*, and their relation to pheromone emission, CA volumetric changes and oocyte growth. Journal of Insect Physiology, 46: 1143–1150.

Truman, J. W. and L. M. Riddiford. 2002. Endocrine insights into the evolution of metamorphosis in insects. Annual Review of Entomology, 47: 467–500.

Vepsalainen, K. 1978. Wing dimorphism and diapause in *Gerris*: determination and adaptive significance, pp. 218–253. *In* H. Dingle [ed.] 1978. Evolution of Insect Migration and Diapause. Springer-Verlag, New York.

West-Eberhard, M. J. 2003. Developmental Plasticity and Evolution. Oxford University Press, Oxford.

Wheeler, D. E. and H. F. Nijhout. 2003. A perspective for understanding the modes of juvenile hormone action as a lipid signalling system. BioEssays, 25: 994–1001.

Wigglesworth, V. B. 1961. Insect polymorphism—a tentative synthesis. *In* J. S. Kennedy [ed.]. Insect Polymorphism, pp. 103–113. Royal Entomological Society, London.

Wilson, T. B. 2004. The molecular site of action of juvenile hormone and juvenile hormone insecticides during metamorphosis: how these compounds kill insects. Journal of Insect Physiology 50: 111–121.

Wyatt, G. R., and K. G. Davey. 1996. Cellular and molecular actions of juvenile hormone. II. Roles of juvenile hormone in adult insects. Advances in Insect Physiology, 26: 1–155.

Zera, A. J. 2004. The endocrine regulation of wing polymorphism: State of the art, recent surprises, and future direction. Integrative and Comparative Biology, 43: 607–616.

Zera, A. J. 2005. Intermediary metabolism and life history trade-offs: Lipid metabolism in lines of the wing-polymorphic cricket, *Gryllus firmus*, selected for flight capability vs. early-age reproduction. Integrative and Comparative Biology, 45: 511–524.

Zera, A. J. 2006. Evolutionary genetics of juvenile hormone and ecdysteroid regulation in *Gryllus*. a case study in the microevolution of endocrine regulation. Comp. Biochem. Physiol. 144A: 365–79.

Zera, A. J. 2007. Endocrine analysis in evolutionary-developmental studies of insect polymorphism: Use and misuse of hormone manipulation. Evol. Dev. (In Press).

Zera, A. J., and J. Bottsford. 2001. The endocrine-genetic basis of life-history variation: Relationship between the ecdysteroid titer and morph-specific reproduction in the wing-polymorphic cricket, *Gryllus firmus*. Evolution, 55: 538–549.

Zera, A. J. and T. Brink. 2000. Nutrient absorption and utilization by wing and flight-muscle morphs of the cricket *Gryllus firmus*: implications for the trade-off between flight capability and early reproduction. Journal of Insect Physiology 46: 1207–1218.

Zera, A. J., and G. Cisper. 2001. Genetic and diurnal variation in the juvenile hormone titer in a wing-polymorphic cricket: Implications for the evolution of life histories and dispersal. Physiological and Biochemical Zoology, 74: 293–306.

Zera, A. J., and R. F. Denno. 1997. Physiology and ecology of dispersal polymorphism in insects. Annual Review of Entomology, 42: 207–231.

Zera, A. J., X. Gu, and M. Zeisset. 1992. Characterization of juvenile hormone esterase from genetically-determined wing morphs of the cricket, *Gryllus rubens*. Insect Biochemistry and Molecular Biology, 22: 829–839.

Zera, A. J. and L. G. Harshman. 2001. The physiology of life-history trade-offs in animals. Annual Review of Ecology and Systematics, 32: 95–126.

Zera, A. J., L. G. Harshman, and T. D. Williams. 2007a. Evolutionary Endocrinology: The developing synthesis between endocrinology and evolutionary genetics. Annu. Rev. Ecol. Evol. Syst. 38: 793–817.

Zera, A. J. and C. L. Holtmeier. 1992. In vivo and in vitro degradation of juvenile hormone-III in presumptive long-winged and short-winged *Gryllus rubens*. J. Insect Physiol. 38: 61–74.

Zera, A. J. and Y. Huang. 1999. Evolutionary endocrinology of juvenile hormone esterase: functional relationship with wing polymorphism in the cricket, *Gryllus firmus*. Evolution, 53: 837–847.

Zera, A. J., and A. Larsen. 2001. The metabolic basis of life history variation: Genetic and phenotypic differences in lipid reserves among life history morphs of the wing-polymorphic cricket, *Gryllus firmus*. Journal of Insect Physiology, 47: 1147–1160.

Zera, A. J., J. Potts, and K. Kobus. 1998. The physiology of life history trade-offs: experimental analysis of a hormonally-induced life history trade-off in *Gryllus assimilis*. American Naturalist, 152: 7–23.

Zera, A. J., J. Sall, and K. Grudzinski. 1997. Flight-muscle polymorphism in the cricket *Gryllus firmus*: Muscle characteristics and their influence on the evolution of flightlessness. Physiological Zoology, 70: 519–529.

Zera, A. J., J. Sall, and K. Otto. 1999. Biochemical aspects of flight and flightlessness in *Gryllus*: flight fuels, enzyme activities, and electrophoretic profiles of flight muscles from flight-capable and flightless morphs. Journal of Insect Physiology, 45: 275–285.

Zera, A. J., T. Sanger, J. Hanes, and L. Harshman. 2002. Purification and characterization of hemolymph juvenile hormone esterase from the cricket, *Gryllus assimilis*. Archives of Insect Biochemistry and Physiology, 49: 41–55.

Zera, A. J., C. Strambi, K. C. Tiebel, A. Strambi, and M. A. Rankin. 1989. Juvenile hormone and ecdysteroid titers during critical periods of wing morph determination in *Gryllus rubens*. Journal of Insect Physiology, 35: 501–511.

Zera, A. J., and S. Tanaka. 1996. The role of juvenile hormone and juvenile hormone esterase in wing morph determination in *Modicogryllus confirmatus*. Journal of Insect Physiology, 42: 909–915.

Zera, A. J., and K. C. Tiebel. 1988. Brachypterizing effect of group rearing, juvenile hormone-III, and methoprene on winglength development in the wing-dimorphic cricket, *Gryllus rubens*. Journal of Insect Physiology, 34: 489–498.

Zera, A. J. and K. C. Tiebel. 1989. Differences in juvenile hormone esterase activity between presumptive macropterous and brachypterous *Gryllus rubens*: implications for the hormonal control of wing polymorphism. Journal of Insect Physiology, 35: 7–17.

Zera, A. J. and S. S. Tobe. 1990. Juvenile hormone-III biosynthesis in presumptive long-winged and short-winged *Gryllus rubens*: implications for the hormonal control of wing polymorphism. Journal of Insect Physiology, 36: 271–280.

Zera, A. J., and M. Zeisset. 1996. Biochemical characterization of juvenile hormone esterases from lines selected for high or low enzyme activity in *Gryllus assimilis*. Biochemical Genetics, 34: 421–435.

Zera, A. J. and C. Zhang. 1995. Evolutionary endocrinology of juvenile hormone esterase in *Gryllus assimilis*: Direct and correlated responses to selection. Genetics, 141: 1125–1134.

Zera, A. J. and Z. Zhao. 2003a. Morph-dependent fatty-acid oxidation in a wing-polymorphic cricket: Implications for morph specialization for dispersal vs. reproduction. Journal of Insect Physiology, 49: 933–943.

Zera, A. J., and Z. Zhao. 2003b. Life history evolution and the microevolution of intermediary metabolism: Activities of lipid metabolizing enzymes in life history morphs of a wing-dimorphic cricket. Evolution, 57: 568–596.

Zera, A. J. and Z. Zhao. 2004. Effect of a juvenile hormone analogue on lipid metabolism in a wing-polymorphic cricket: Implications for the biochemical basis of the trade-off between reproduction and dispersal. Biochemical and Physiological Zoology, 77: 255–266.

Zera, A. J. and Z. Zhao. 2006. Intermediary metabolism and life history trade-offs: Differential metabolism of amino acids underlies the dispersal-reproduction trade-off in a wing-polymorphic cricket. American Naturalist, 167: 889–900.

Zera, A. J., Z. Zhao, and K. Kaliseck. 2007b. Hormones in the field: Evolutionary endocrinology of juvenile hormone and ecdysteroids in field populations of the wing-dimorphic cricket *Gryllus firmus*. Physiol. Biochem. Zool. (In Press).

Zhao, Z. and A. J. Zera. 2001. Enzymological and radiotracer studies of lipid metabolism in the flight-capable and flightless morphs of the wing-polymorphic cricket, *Gryllus firmus*. Journal of Insect Physiology, 47: 1337–1347.

Zhao, Z., and A. J. Zera. 2002. Differential lipid biosynthesis underlies a tradeoff between reproduction and flight capability in a wing-polymorphic cricket. Proceedings of the National Academy of Sciences USA, 99: 16829–16834.

Zhao, Z. and A. J. Zera. 2004a. The JH titer exhibits a high-amplitude, morph-dependent, diurnal cycle in the wing-polymorphic cricket, *Gryllus firmus*. Journal of Insect Physiology, 50: 93–102.

Zhao, Z. and A. J. Zera. 2004b. A morph-specific daily cycle in the rate of JH biosynthesis underlies a morph-specific daily cycle in the hemolymph JH titer in a wing-polymorphic cricket. Journal of Insect Physiology, 50: 965–973.

15

Evolution of Homeostatic Physiological Systems

H. Arthur Woods
Division of Biological Sciences, University of Montana, Missoula, MT 59812 USA;
E-mail: art.woods@mso.umt.edu

Abstract

Physiological homeostasis represents a specialized kind of trait canalization that stems from phenotypic plasticity in other underlying physiological, morphological, and behavioral traits. Although homeostatic systems perform much of the physiological work that keeps organisms alive and functioning, how these systems evolve is unknown. Here I develop a set of predictions—focusing on rates at which physiological systems accumulate genetic variation, likely patterns of genetic correlation across character states, and conditions under which system components experience selection. Driving these predictions is the distinction between physiological components doing work and those processing information. Because these functions are supported by interacting but distinct organs, tissues, molecular structures, and sets of genes, the two functions likely evolve in substantially different ways.

Introduction

Because homeostatic systems consist of interacting physiological parts, research on them has, not surprisingly, fallen to physiologists. But these systems should be of interest also to evolutionary biologists, as they underlie all higher-level performance; how they evolve is fundamentally important. Recent efforts have begun to explore the evolution of homeostasis across broad time scales (e.g., evolution of osmotic stress signaling over the past 1–1.5 billion years; Kultz and Burg 1998) and to examine the *in silico* origin of homeostatic systems in simulation models (Stern 1999). Moreover, the introduction of homeostasis into evolutionary theory (and *vice versa*) is natural, as the idea shares common ground with two other evolutionary ideas—canalization and plasticity. In particular, homeostasis *is* physiological canalization stemming from plasticity in other physiological, morphological, and behavioral traits. What we lack still, however, is a theoretical framework suitable for generating predictions about how microevolutionary processes shape homeostatic systems. Below I develop such a framework, focusing on rates at which physiological systems accumulate genetic variation, likely patterns of genetic correlation across character states, and conditions under which system components experience selection.

Homeostatic systems are universal features of living organisms because they regulate, within certain bounds, the concentrations of molecules, ions, and heat (Bernard 1872; Cannon 1932; Eckert and Randall 1983), though obviously not all organisms regulate all, or even a fraction, of possible substances. The field has generated a complicated terminology. If an organism's concentration of a specific factor is constant despite environmental variation (Fig. 1A, line a), the organism is said to exhibit strict homeostasis. In contrast, simple non-homeostasis occurs when a variable's

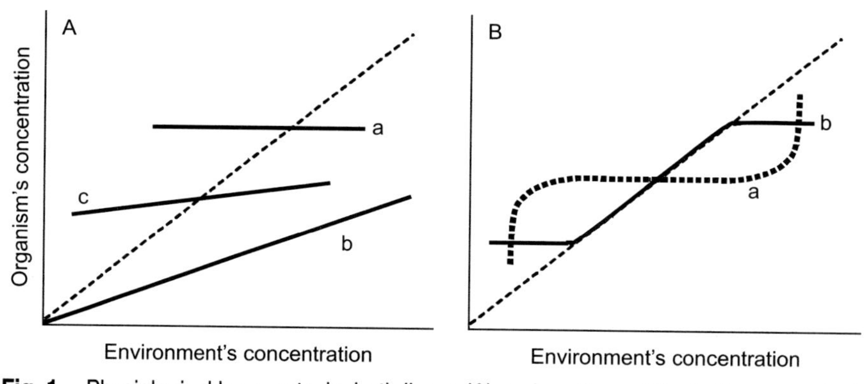

Fig. 1 Physiological homeostasis, both linear (A) and nonlinear (B). A. If an organism's concentration (of any substance of interest, or of heat) always matches the environment's concentration, the organism is not homeostatic. The simplest alternative case is perfect homeostasis (line a), in which the organism's concentration is invariant across environments. Slopes between 0 and 1 can represent either of two situations. Line b shows non-homeostasis arising from simple proportionality rules—e.g., if a caterpillar always absorbs 20% of the nitrogen in ingested leaf pieces. Line c, by contrast, represents imperfect homeostasis, arising either from inability, or lack of desire, to control the variable; or from a variable set point that changes continuously across environments. The lines in panel B show more complicated homeostasis—either strong homeostasis in intermediate environments but failure outside some range (line a) or *vice versa* (line b). See Sterner and Elser (2002).

concentration is identical in organism and environment (Fig. 1A, dashed line). Between the extremes lie a multitude of other possibilities, a few of which are illustrated in Fig. 1. These concepts have been elaborated further by those concerned with how set points change and whether organisms make anticipatory changes (Mrosovsky 1990; Schulkin, 2003).

Homeostatic systems move perturbed internal states toward set points. They may do so by comparing the actual physiological state to some reference signal or set point (Mrosovsky 1990) (Fig. 2A). Information about internal state, obtained from sensors, is processed by the genome or, in the case of labile physiology traits (those that are rapid and reversible), by endocrine, nervous, or other pre-built processing structures (see Weng et al. 1999). The processing structures then produce a signal proportional to the magnitude of the internal state's deviation from its set point. This signal, possibly amplified by other biological systems, is inverted and fed back into the system such that system components (enzymes, organs, tissues) perform work to counteract the perturbation. Physiological systems may also achieve regulation by the use of opposing subsystems, without reference to specific 'reference signals' or set points (Mrosovsky 1990) (Fig. 2B & C). These systems, however, *behave* as if using reference signals, and, for the purposes of this chapter, I will treat them as identical.

Homeostasis arises from plasticity in physiological parts; more specifically, the stability of internal concentrations (a kind of canalization) arises from plasticity in the activity of other, labile physiological components that themselves vary as a function of the environment. One could object to calling labile physiological traits 'plastic' (Piersma and Lindström (1997) prefer the term 'phenotypic flexibility'). And, indeed, labile physiological plasticity is distinct from developmental plasticity (e.g., the antipredator defenses studied by Harvell 1998). In physiological plasticity, both genotypes and individuals can exhibit plasticity across a range of character states. In developmental plasticity, by contrast, genotypes are the units exhibiting plasticity; individuals may exhibit only one or a few character states. This difference likely affects how these traits evolve, but at present is theoretically underdeveloped.

Below I will refer to labile physiological traits as 'plastic,' because the evolutionary issues fit best into the frameworks already developed for developmental plasticity. Physiological plasticity also falls within Bradshaw's (1965) operational definition of plasticity and satisfies most of his additional conditions (summarized by Smith-Gill 1983, p. 48): "(1) [plasticity] is specific for a specific character; (2) it is specific in relation to

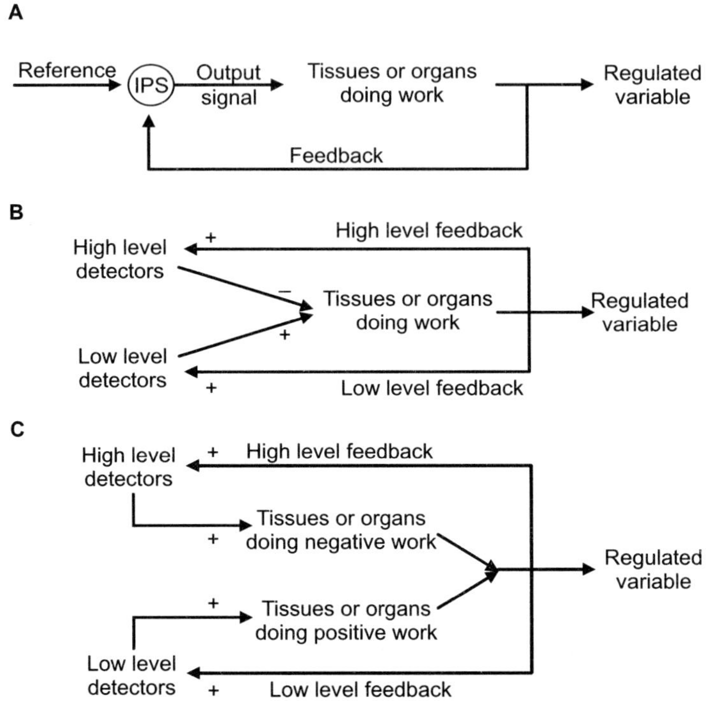

Fig. 2 Three kinds of physiological regulation. A. Control system that uses a reference signal or set point. The information processing system (IPS) compares the physiological state to the set point, and adjusts its output so that tissues doing work counteract the perturbation. B. Control system that exhibits behavior similar to that of A, but without use of a refence signal. Two separate detectors monitor the physiological state. When the state value is low, the target tissue or organ is activated to do more work and, when it is high, to do less work. C. Control system similar to the one above (and exhibiting the same behavior as both A and B), but with different wiring between detectors and effectors. One detector responds to high state values, activating an organ or tissue that depresses the state value; the other detector responds to low state values, activating a separate organ or tissue that increases the state value. The behavioral similarity exhibited by the three systems emphasizes that mechanistic details matter less than does the common division between components doing work and processing information (A and B are adapted from Mrosovsky 1990).

particular environmental influences; (3) it is specific in direction; (4) it is under direct and specific genetic control; (5) it is able to be radically altered by selection." The condition least often met in physiology is the requirement for direct and specific genetic control, especially in systems labile over minutes or less. For the present discussion, therefore, I relax this requirement to include physiological traits that are specified by genes but not necessarily under their minute-to-minute control.

How homeostatic systems evolve likely hinges on a dichotomy between components doing *work* and components devoted to *information processing* (see Fig. 2).

Below I explore the consequences of this distinction for: (1) levels of genetic variation; (2) genetic correlations across character states; and (3) selection on, and responses of, homeostatic systems. Because the work-information dichotomy is one of the chapter's central ideas, it is worth illustrating in a real physiological system.

Consider insect water balance. Many terrestrial insects are subjected, at various points in their lives, to significant risk of dehydration (Edney 1977; Hadley 1994; Tauber et al. 1998); others, such as nectar or blood feeders, experience infrequent but large water loads that require rapid elimination (Maddrell 1963). How do individuals solve these problems? The *work* of water balance falls to a common set of organs found across insect taxa: the Malpighian tubules and the hindgut, particularly the rectum (Figure 3). Malpighian tubules produce primary urine by pumping ions (ultimately energized by V-type ATPases; Wieczorek et al 1999) from the hemolymph into the tubule lumen. Water follows osmotically. This primary urine flows into the gut near the junction between the midgut and ileum. The rectum is specialized for water absorption, which depends on vigorous ion pumping (Phillips et al. 1986) to establish steep osmotic gradients. Thus, the work components in this system are the Malpighian tubules and rectum, or, at a molecular level, the V-ATPases and ion exchangers.

How much work is done, and its timing, is decided by *information processing systems* (IPSs). For insect attempting water homeostasis, the basic rules are standard: when dehydrated, increase water absorption by the hindgut and decrease water secretion by the Malpighian tubules; when over-hydrated, decrease absorption from the rectum and increase rates of secretion by the tubules. How these rules are physiologically manifest is less clear. *A priori*, we might suppose that an IPS consists of four parts (see Fig. 2): (i) sensors that monitor internal state, (ii) some logical processing unit that decides how to interpret signals from the sensors and what to do in response, (iii) the signal that this unit sends out, and (iv) the receivers associated with the work components that listen for and interpret the signal. In the context of water balance, the signals sent by processing units are known best (reviewed by Phillips et al. 1998; O'Donnell and Spring 2000; Coast et al. 2002). In particular, water loss (diuresis) is promoted by a well-known set of peptide hormones (and serotonin), which stimulate secretion

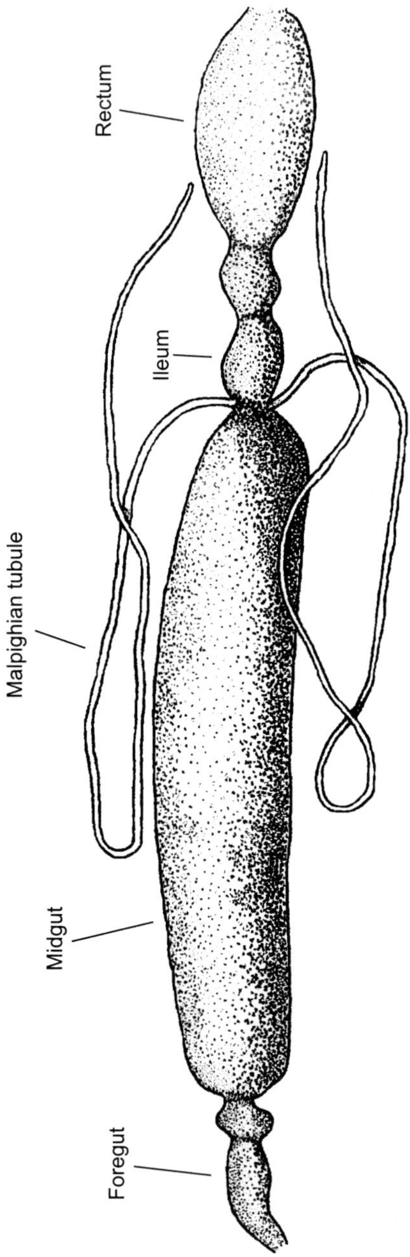

Fig. 3 Insect tissues important in water balance. The Malpighian tubules produce primary urine, which flows into the hindgut. Water (and ions) may be resorbed by the rectal epithelium.

of primary urine by the Malpighian tubules. Water conservation (antidiuresis) is promoted by inhibiting tubule secretion, or by increasing rates of water recovery from the hindgut and, possibly, the midgut.

Both kinds of factors—diuretic and antidiuretic—are released into the hemolymph by neurosecretory cells (Maddrell and Nordmann 1979), the most important of which are in the corpora cardiaca and corpora allata, associated with the brain (Predel and Manfred 2000). But how much hormone should neuroscretory cells manufacture, and how rapidly should they do it? How many nerve terminals should release the factor? How do the nerve terminals decide when to release (and to stop releasing) the factor? Answers must involve the structure and sensitivity of sensors, the wiring of sensors to neurohaemal organs, and communication among and within neurosecretory cells themselves. The details of these IPS components are poorly known. However, in the hemipteran *Rhodnius prolixus*, large blood meals stretch the abdomen dorsoventrally and activate stretch receptors in the tergosternal muscles (Maddrell 1964), which then signal the mesothoracic ganglionic mass to release diuretic hormone (Maddrell 1966).

Pre-signal information processing is not the only problem: tissues downstream must decide whether they are targets of a particular signal and, if so, how strong it is and how to respond. The action of diuretic peptides, for example, seems to be mediated by G protein-coupled receptors (Coast 1996; Vanden Broeck 2001). To date, however, only a few receptors have been cloned and characterized. The receptors activate second messenger systems which themselves stimulate the components doing work. In the locust rectum, for example, cAMP simulates the apical Cl^- pump and opens K^+ and Cl^- channels in both apical and basolateral membranes (Hanrahan and Phillips 1984).

Table 1 partitions insect water-balance physiology into components devoted to work versus information processing. The obvious point is that tissue- or organ-level traits are easily sorted into groups. Less obvious, but more important for evolution, is that *the sets of genes underlying work and information components are distinct*. That is, genes specifying the construction of sensors, of CNS wiring, and of peptide hormones are not the same as those shaping the development of the Malpighian tubules or hindgut, or controlling the density and kinetics of ion pumps and V-type ATPases. The main exception occurs in the genes encoding pumps. Pumps doing work are under the proximate control of second messengers. To respond to these messengers, the pumps (doing work) must also contain domains involved in information processing.

Table 1 List of work and information components in physiological systems devoted to water balance in insects

Work components	*Information components*
Malpighian tubules and hindgut	Hormone
Size	Chemical structure
Number (of tubules)	Concentration
Metabolic parameters: supply of blood and metabolic substrates, tracheation	Timing of release
Supply of blood/metabolic substrate	Physiological sensor (stretch- or osmo-receptors)
Transporter density and kinetics	Sensitivity
Permeability	Gain
	Saturability
	Tissue receptors
	Binding kinetics
	Second messenger system + interactions
	Specificity

The framework developed above applies to more than just water balance: *in general*, homeostatic systems process information and do work. The phenotypic manifestations (enzymes, tissues, organs) of these two components are distinct, as are the sets of genes that specify them. Such a dichotomy has important implications for how homeostatic systems evolve.

Genetic Variation in Homeostatic Systems

A trait's short-term evolutionary trajectory depends on the kind and amount of genetic variation underlying it. For two reasons, the physiological components *doing work* should exhibit higher levels of genetic variation than components *processing information*.

First, for most taxa, most of the time, environmental conditions will be 'normal' or 'expected' (more on extreme environments below). In such environments, the state of a regulated internal variable is dictated by its corresponding IPS. The actual capacity of work components may be largely irrelevant (excepting null mutants), because system outputs do not approach their limits—i.e., feedback from the IPS directs more or less work to be done until the target, not necessarily optimal, state is achieved, regardless of actual tissue capacity. Thus, mutations in capacity may often have neutral fitness consequences. The concept of regulation by IPSs is related to West-Eberhard's (2003) ideas about phenotypic accommodation ('adaptive mutual adjustment among variable parts during development without genetic change', p. 51).

Mutations affecting IPS components, by contrast, may less often be neutral. First, novel mutations in IPSs may undermine the process of accommodating unusual work components. Second, if a mutant IPS encodes a physiological set-point that is distant from the optimal physiological state, the organism carrying it may suffer negative consequences even in normal or benign environments. Such variants may be rapidly weeded out. Hence, regulatory adjustments (plastic responses) of IPS systems can buffer genetic changes in work systems, but not *vice versa*.

The net outcome of the processes above should be greater accumulation of genetic variation in work than in feedback components. Is this so? Human homeostatic systems are a useful starting point for discussion. Anecdotally, many such systems appear to show greater variability among structural than among information components (set points). For example, normal human oral temperatures range between only 35.5–37.5°C (Rising et al. 1992), despite tremendous variation in body components contributing heat (relative masses of organs, fat, and muscles) and in organ systems involved in offloading heat (e.g., density of sweat glands, dermal capillary density). Likewise, salt loads or deficits induce hormonal (IPS) responses (atrial natriuretic peptide, vasopressin, aldosterone) and can alter levels of thirst. The main organ of sodium control is the kidney, which absorbs sodium in hyponatremia and excretes it in hypernatremia. Clearly, individuals reduced to one kidney by donating the other can live into old age, even if the surgery puts them at higher risk in situations where extreme dehydration or unusual salt conditions are possible. This is true even though the normal plasma levels in the sodium (136–145 mM) are close to the upper level beyond which individuals can suffer seizures or death (152 mM), a balancing act accomplished by adjustment of the remaining kidney's function by the sodium IPS (Mimran et al. 1993). Although neither example above demonstrates greater *genetic* variation *per se* in organ function than in its corresponding IPS, they do show that substantial variation in organ function can be accommodated.

A complementary approach is to ask whether genetic variation in work or information components is more strongly associated with disease. Loktionov (2003), for example, reviews the genetics of energy homeostasis. Genetic disturbance in this system is associated with obesity, cardiovascular disease, and cancer. Although factors leading to disease are complex, multifactorial, and (therefore) poorly understood, Loktionov's

(2003) review shows clearly that most of the important polymorphisms suspected in disease states involve genes whose products participate in information processing—including ghrelin (appetite-promoting peptide), cholecystokinin (a short-term satiety signal produced by the small intestine), peptide $YY_{3\text{-}36}$ (longer-term satiety signal from the colon), and leptin (a peptide expressed primarily in adipocytes believed to be important in fat metabolism). Genetic variation in receptors for many of the signaling peptides appears also to be associated with disease states (Loktionov 2003).

A priori, such a distribution of disease-associated variation (associated with IPSs) could result from either of two patterns of genetic variation. First, work components may contain so little variation that what there is cannot be reliably associated with disease. The opposite extreme is that they exhibit vast genetic variation, but that the variation is unrelated to disease states (and therefore is ignored by the medical community)—i.e., the variation is usually accommodated by an IPS. Almost certainly, the latter is the case. Surgical removal of parts or all of organs (e.g., the kidney donation mentioned above) often has minimal effects on organismal function. Moreover, human organs often show substantial natural variation. For example, Anson (1951) catalogued variation in stomach size and shape in humans, finding astonishingly high levels of variation (though again how much of the variation is genetic is unknown).

A different approach is afforded by progress in large-scale analysis of DNA sequences and expression patterns. In particular, publication of the human genome sequence (Lander et al. 2001; Venter et al. 2001), together with efforts to catalogue the frequency and nature of single nucleotide polymorphisms (SNPs) (Chakravarti 1999), promises to reveal the complex genetics of normal and disease states. Whether such techniques will be immediately applicable to understanding the evolution of homeostatic systems is less certain, as suggested by several caveats. For example, one could potentially partition genes into structural versus regulatory genes (à la Jacob and Monod 1961) and simply catalog relative density of SNPs in the two classes. This approach, though appealing, is inadequate. First, the initial assumption—equating structural versus regulatory genes to work versus information genes—is unlikely to be valid: many regulatory genes participate in building or modifying the organs and tissues that do work. Second, data currently available focus on variation in coding regions. However, much of the important variation within populations, and between species, stems from changes in the timing, location, and intensity of gene expression (King and Wilson 1975)—i.e., is due to variation in regulatory regions upstream of ORFs. Another caveat is that the genetics of homeostatic systems often will be polygenic, involving many genes of small effect and frequent epistatic interactions. In the long run, as better data on genome-wide variation become available and the genetic basis of complex traits is increasingly well understood, genomics are increasingly likely to shed significant light on this problem.

Genetic Correlations in Homeostatic Systems

The outcome of selection also depends on whether phenotypic states are genetically correlated across environments. By analogy to thermal performance curves (Huey and Kingsolver 1989), the output of a physiological system—either a flux or a concentration—can be described as a continuous function of an environmental factor (Fig. 4A, B). The exact shape is unknown in most systems. Nevertheless, system output will often change monotonically, if not linearly, over much of its factor's range. Fecal water content of *Manduca sexta* caterpillars, for example, decreases with declining environmental water availability or increasing rates of transcuticular water loss (Woods and Bernays 2000; Woods and Harrison 2001).

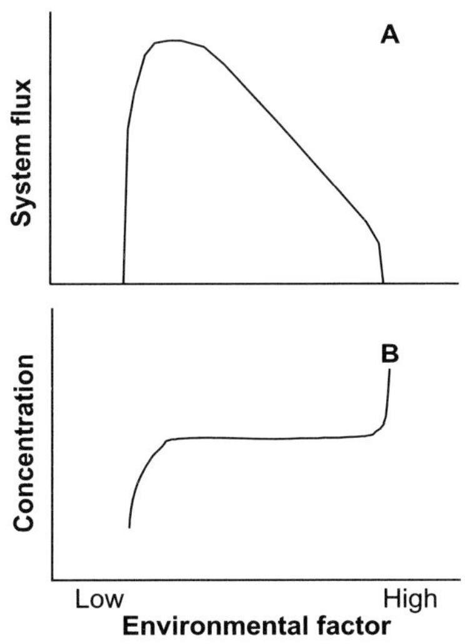

Fig. 4 Physiological performance curves for the same factor, focusing on either system flux (A) or concentration (B). Fluxes fall in extreme environments as organismal integrity is progressively compromised; beyond these extremes, concentrations change much more rapidly with the environment.

Consider selection, in one particular environment, for higher system flux (Fig. 5). If appropriate genetic variation exists and fluxes are weakly correlated across environments, evolutionary change in the performance curve will be local (Fig. 5B). Strong genetic correlations, however, may lead to change in system outputs at other sections of the curve (in other environments). For example, output at one end of the curve could be inversely correlated with output at the other; if so, the entire curve may shift in response to selection (Fig. 5C). Yet another possibility is that system fluxes at intermediate and extreme environments tradeoff (Fig. 5D), a kind of 'jack-of-all-trades-is-master-of-none' genetic structure. Other more generalized frameworks have been designed to analyze continuously variable genetic correlations and responses to selection *functions* (Gomulkiewicz and Kirkpatrick 1992; Kingsolver et al. 2001).

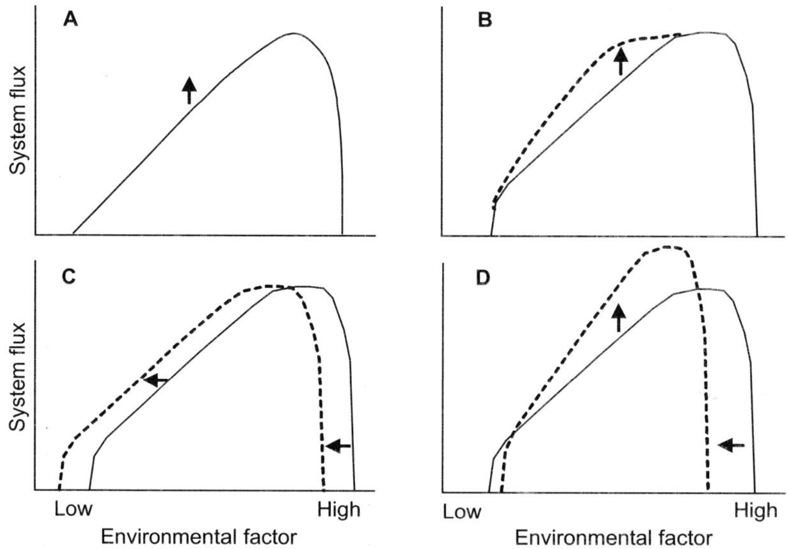

Fig. 5 How different kinds of genetic correlation affect response to selection (after Huey and Kingsolver 1993). **(A)** The arrow represents strong selection, in one particular environment, for higher system flux. **(B)** If fluxes are weakly correlated across environmental states, evolutionary change will be local (post-selection curves represented by dashed lines). Nearby states will never be completely uncorrelated (taken to an extreme, this would imply that a state infinitesimally close to another would be uncorrelated); but the correlation could fall off rapidly with environmental distance. Alternatively, the output at one end of the curve could exhibit a strong inverse correlation with output at the other end. In this situation, selection at one end may shift the entire curve **(C)**. Another alterative is a 'jack-of-all-trades-is-master-of-none' genetic structure, in which system fluxes at intermediate and extreme environments tradeoff **(D)**.

The dichotomy in homeostatic systems, between components devoted to work and to information processing, suggests specific forms of correlation and constraint. First, the existence of information processing modules, whatever their biological form, will lead to genetic correlations. Correlations arise because genes do not make independent decisions about outputs in each of all possible environments. Rather, information processing structures encode sets of response rules. In general, genetic alterations to these rules are likely to be simple—either linear offsets or changes in gain. If so, a population's response to strong selection in one environment may affect system outputs across all environments. This idea is related to those of Kingsolver et al. (2001) about function-valued traits. They discussed an important constraint in function-valued traits: imagine two phenotypes in two different environments; as the distance between environmental states goes to zero, the phenotypes must converge. Correlations stemming from IPSs are an information-specific form of this kind of constraint.

By contrast, genetic correlations between system outputs at one environmental extreme versus the other may be small—especially in systems where two or more tissues contribute to homeostasis (most systems). For example, in insects the main route of water loss, urine production by Malpighian tubules, is functionally decoupled from water recovery across the rectal epithelium. Selection for better water conservation may not directly affect the ability to produce urine rapidly; and, conversely, selection for faster urine production may not affect the ability of the rectum to absorb water. This conclusion likely depends on the strength of tradeoffs for energy and material costs (Piersma and Lindström 1997)—i.e., if better water conservation requires reallocating material from Malpighian tubules to rectal tissues, a more diffuse tradeoff arising from tissues drawing on common pools of energy and materials (see Nijhout and Emlen 1998) may result in genetic correlations.

Do any data bear on these predictions? Essentially no homeostasis-specific data are available, but work on thermal performance curves provide an interesting view of how such data might be analyzed and what outcomes we might expect. Kingsolver et al. (2001) extended Lande and Arnold's (1983) multivariate model of evolution to traits that vary continuously as a function of the environment—so called 'function-valued traits.' Using 21 full-sib families of caterpillars, they collected data on feeding and growth rate across a range of temperatures and estimated a genetic variance-covariance *function* from a standard variance-covariance matrix. To identify significant axes of variation within this function, they performed principal components on the covariance function, which essentially decomposes the variance-covariance function into a set of eigenfunctions. The main eigenfunction, representing 62% of the total variance, had zero loading at low temperatures and high loading at high temperatures. This indicates that, especially at high temperatures, growth rates across temperatures were positively correlated. Like growth rates, homeostatic characters (either set points or component fluxes and concentrations) are function-valued and promise to be amenable to this kind of analysis. Moreover, the participation of IPSs suggests that the main eigenfunction for homeostatic traits should indicate strong positive correlations across states, especially in the 'normal' range (see next section) of environmental variation.

The second prediction—that work done in the extremes at opposite ends of an environmental continuum will often be genetically uncorrelated—appears also to be supported, though again only by data on thermal performance curves. Huey and Kingsolver (1993) analyzed data on the thermal dependence of running speed in 19 species of iguanid lizards. Analogous to the scenario proposed above, performance at extreme high temperatures may be decoupled from performance at extreme low temperatures if limits at high and low temperatures stem from different structures or processes. Huey and Kingsolver explicitly tested for correlations between critical thermal maximum and minimum temperatures, and indeed found none. A recent review (Angilleta et al. 2002) concluded that, in general, genetic constraints (or tradeoffs) play a minor role in the evolution of thermal performance physiology. Unfortunately such data on homeostasis *per se* are unavailable. Perhaps the predictions could be tested by examining, for example, the structure and function of Malpighian tubules and rectal epithelia in insect populations artificially selected for desiccation resistance (e.g., *Drosophila*).

Selection on Homeostatic Systems

How selection acts on homeostatic systems is contingent on whether environmental conditions are benign or extreme. The distinction between kinds of environments is not sharp: benign conditions for one species may be extreme for another (and *vice versa*). Nonetheless, it is possible, in principle, to identify the timing and duration of the most severe of environmental conditions, e.g., the 5% of most dehydrating environments experienced by a population of caterpillars. Kingsolver et al. (2001) have explored specifically how environmental frequency affects the outcome of selection on continuous traits.

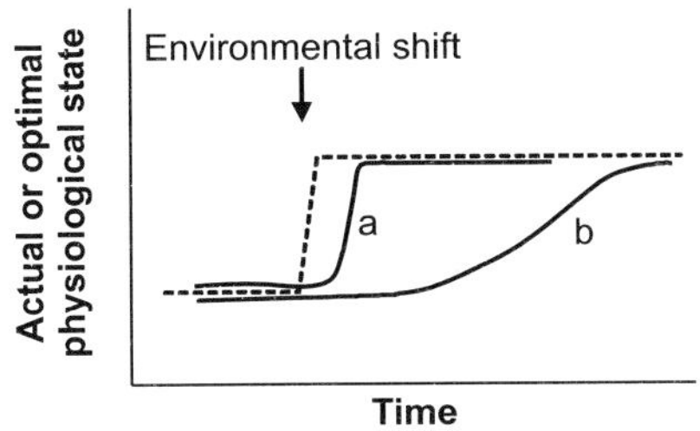

Fig. 6 Selection on work capacity, even in benign environments. The environment changes such that the optimal physiological state assumes a new value (dashed line). A physiological system with substantial built-in capacity for work will be able to track environmental change rapidly (assuming desired and optimal states are equivalent), and therefore will spend little time in non-optimal physiological states. By contrast, a system with little capacity for work will respond slowly, and the organism will spend more time away from the optimum.

Benign conditions should exert selection primarily on information processing—because it, rather than tissue capacity, determines the distance between actual and desired (possibly optimal) physiological states. If, however, an environmental factor changes rapidly from one 'benign' state to another, work components may also experience selection. The organism's feedback system will sense the change and signal tissues to alter fluxes, as a means of achieving a new desired internal concentration (Fig. 6, dashed line). Arriving there, however, requires that work be done—and an organism with more capacity for work (Fig. 6, line a) will arrive more rapidly than one with less (Fig. 6, line b). All else equal, selection will favor the organism with greater tissue capacity, because its internal state will be closer to the desired state for more of the time. Therefore, even if feedback is well tuned, rapid environmental change may select on tissue capacity. Of course, all else may not be equal: maintaining and operating tissues capable of higher rates of work likely incurs energy and materials costs. Such costs can significantly constrain the evolution of optimal phenotypes (DeWitt et al. 1998).

Over longer intervals of time, environmental extremes are inevitable. It is these extremes that likely exert strong selection on components devoted to work, particularly when insufficient work is done despite intense signaling from information processing components. Moreover, evolutionary responses to selection at extremes should be rapid, because reservoirs of genetic variation accumulated in the usual range of environmental states will suddenly be expressed (see Gerhart and Kirschner 1997).

A classic, non-physiological example of such 'revealed' genetic variation, and its response to selection, is Waddington's (1953) work on genetic assimilation in *Drosophila melanogaster*. In wings of normal adults, two major crossveins link longitudinal veins. Individual flies may exhibit gaps in, or absence of, one or both crossveins, although these mutants are rare in benign environments. Waddington (1953) induced much higher incidence of disruption by exposing pupae to heat shock (4 hours, 40°C) and choosing adults that expressed unusual (in the 'upward' line) or normal (in the 'downward' line) crossvein morphologies. By generation 18 in the 'upward' line, the crossveinless phenotype came to dominate the population (> 95% of flies). Moreover, after additional selection, up to 95% of flies in the 'upward' line were crossveinless even at normal (non heat shock) temperatures (25°C). In other words, exposure to heat shock revealed hidden genetic variation in an arbitrarily chosen trait, which then responded rapidly to selection.

Physiological systems likely are shaped by analogous processes: extreme environments both expose genetic variation and impose strong selection (West-Eberhard 2003), leading to rapid evolution. Importantly, the environment that exposes variation often will also impose selection; in Waddington's experiments, by contrast, the selected trait (wing vein morphology) was arbitrary in that heat itself did not select for the crossveinless phenotype. In the wild, the coupling of revealed genetic variation and selection likely has profound effects; indeed, selection in extreme environments—rather than the amount or duration of selection in benign environments—may be the main force shaping the organization and function of work components in physiological systems (see Hoffmann and Parsons 1997).

Conclusions

This chapter has examined dichotomies in homeostatic systems, between physiological components that do work and those that process information. This division leads to four testable predictions about how such systems evolve:

1. Work components will, in general, exhibit more genetic variation than will information processing components. This pattern arises because new, functional work variants are likely to be accommodated or regulated, whereas new information processing variants often will be weeded out because they result in sub-optimal physiological states across a broad range of environments.
2. Because IPSs encode simple rules, homeostatic systems will exhibit strong genetic correlations across normal environments (with the caveat that 'normal' is defined with respect to particular characters and taxa). By contrast, system performance at one environmental extreme should often be uncorrelated with performance at the opposite extreme, because the parts used at the two extremes usually differ.

3. Benign environments will select primarily on IPSs, because they, rather than work components, determine actual physiological states. Selection for rapid physiological responses is the exception; such situations may select for greater work capacity as a means of responding more rapidly to signals from IPSs. Extreme environments, by contrast, exhibit selection particularly on components doing work.
4. As a consequence of 1–3, work and information components will evolve in markedly different ways. Work components—with high levels of genetic variation, weak genetic correlations between parts operating at opposite extremes on environmental continuums, and experiencing episodic strong selection—will evolve rapidly in short bursts bracketing long periods of stasis. Information components—with low levels of genetic variation and strong genetic correlations, and experiencing persistent weak selection—will evolve slowly but more continuously. The framework does not predict which kind of component will exhibit greater change on macroevolutionary time scales.

Testing the predictions will be a challenge. In the genes-to-phenotype hierarchy, physiological systems occupy a difficult middle ground: too emergent for easy genetic analysis yet too atomized for clean fitness measurements. Such a conclusion could be discouraging. These systems, however, underlie all physiological performance traits, and understanding how they evolve should be a high priority. Progress will be most rapid when physiological views about homeostasis are integrated into broader evolutionary ideas about plasticity and canalization.

Acknowledgments

I thank David Berrigan, Jim Bull, Creagh Breuner, Tom Clark, Mark Kirkpatrick, Sue Nicolson, and Doug Whitman for comments and discussion that substantially clarified this chapter. Its writing was supported by a grant from the National Science Foundation (IBN 0213087) and The University of Texas at Austin.

References

Angilleta, M. J., P. H. Niewiarowski, and C. A. Navas. 2002. The evolution of thermal physiology in ectotherms. Journal of Thermal Biology 27: 249–268.

Bernard, C. 1872. Physiologie Générale. Paris, Hatchette.

Bernays, E. A. 1990. Water regulation. *In* R. F. Chapman, A. Joern, eds, Biology of Grasshoppers. John Wiley & Sons, New York, pp. 129–141.

Bradshaw, A. D. 1965. Evolutionary significance of phenotypic plasticity in plants. Advances in Genetics 13: 115–155.

Cannon, W. B. 1932. The Wisdom of the Body. Norton Press, New York.

Chakravarti, A. 1999. Population genetics—making sense out of sequence. Nature Genetics 21 (supplement), 56–60.

Coast, G. M. 1996. Neuropeptides implicated in the control of diuresis in insects. Peptides 17: 327–336.

Coast, G. M. I. Orchard, J. E. Phillips, and D. A. Schooley 2002. Insect diuretic and antidiuretic hormones. Advances in Insect Physiology 29: 279–409.

DeWitt, T. J., Sih A., Wilson D. S. 1998. Costs and limits of phenotypic plasticity. Trends in Ecology and Evolution 13: 77–81.

Eckert, R., Randall D. 1983. Animal Physiology: Mechanisms and Adaptations, 2nd ed. W. H. Freeman and Company, New York.

Edney, E. B. 1977. Water Balance in Land Arthropods. Springer-Verlag, Berlin.

Gerhart, J. and M. Kirschner 1997. Cells, Embryos, and Evolution: Toward a Cellular and Developmental Understanding of Phenotypic Variation and Evolutionary Adaptability. Blackwell, Malden, MA.

Gomulkiewicz, R. and M. Kirkpatrick 1992. Quantitative genetics and the evolution of reaction norms. Evolution 46: 390–411.

Hadley, N. F. 1994. Water Relations of Terrestrial Arthropods. Academic Press, Inc., San Diego, CA.

Hanrahan, J. W. and J. E. Phillips 1984. KCl transport across an insect epithelium. II. Electrochemical potentials and electrophysiology. Journal of Membrane Biology 80: 27–47.

Harrison, J. F. 2001. Insect acid-base physiology. Annual Review of Entomology 46: 221–250.

Harvell, C. D. 1998. Genetic variation and polymorphism in the inducible spines of a marine bryozoan. Evolution 52: 80–86.

Hoffmann, A. A. and P. A. Parsons 1997. Extreme Environmental Change and Evolution. Cambridge University Press, Cambridge.

Huey, R. B. and J. G. Kingsolver 1989. Evolution of thermal sensitivity of ectotherm performance. Trends in Ecology and Evolution 4. 131–135.

Huey, R. B. and J. G. Kingsolver 1993. Evolution of resistance to high temperature in ectotherms. American Naturalist 142: S21–S46.

Jacob, F. and J. Monod 1961. Genetic regulatory mechanisms in the synthesis of proteins. Journal of Molecular Biology 3: 318–356.

King, M. C. and A. C. Wilson 1975. Evolution at two levels in humans and chimpanzees. Science 188: 107–116.

Kingsolver, J. G., R. Gomulkiewicz, and P. A. Carter 2001. Variation, selection and evolution of function-valued traits. Genetica 112–113, 87–104.

Kultz, D. and M. Burg 1998. Evolution of osmotic stress signaling via map kinase cascades. Journal of Experimental Biology 201: 3015–3021.

Lande, R. and S. J. Arnold 1983. The measurement of selection on correlated characters. Evolution 37: 1210–1226.

Lander E. S., Linton L. M., Birren B., et al. 2001. Initial sequencing and analysis of the human genome. Nature 409: 860–921.

Loktionov A. 2003. Common gene polymorphisms and nutrition: emerging links with pathogenesis of multifactorial chronic diseases. Journal of Nutritional Biochemistry 14: 426–451.

Maddrell S. H. P. 1963. Excretion in the blood-sucking bug, *Rhodnius prolixus* Stål. I. The control of diuresis. Journal of Experimental Biology 40: 247–256.

Maddrell S. H. P. 1964. Excretion in the blood-sucking bug, *Rhodnius prolixus* Stål. III. The control of the release of the diruetic hormone. Journal of Experimental Biology 41: 459–472.

Maddrell S. H. P. 1966. The site of release of the diuretic hormone in *Rhodnius*—a new neurohaemal system in insects. Journal of Experimental Biology 61: 155–171.

Maddrell S. H. P. and J. J. Nordmann 1979. Neurosecretion. John Wiley and Sons, New York.

Mimram A., G. Mourad, and J. Ribstein 1993. Early systemic and renal responses to nephrectomy in normotensive kidney donors. Nephrology Dialysis Transplantation 8: 448–453.

Mrosovsky, N. 1990. Rheostasis: The Physiology of Change. Oxford, Oxford University Press.

Nijhout H. F. and Emlen D. J. 1998. Competition among body parts in the development and evolution of insect morphology. Proceedings of the National Academy of Sciences USA 95: 3685–3689.

O'Donnell M. J. and J. H. Spring 2000. Modes of control of insect Malpighian tubules: synergism, antagonism, cooperation and autonomous regulation. Journal of Insect Physiology 46: 107–117.

Phillips J. E., Hanrahan J., Chamberlin M., Thomson B. 1986. Mechanisms and control of reabsorption in insect hindgut. Advances in Insect Physiology 19: 329–422.

Phillips J. E., J. Meredith, N. Audsley, N. Richardson, A. Macins, and M. Ring 1998. Locust ion transport peptide (ITP): A putative hormone controlling water and ionic balance in terrestrial insects. American Zoologist 38: 461–470.

Piersma, T. and Å. Lindström 1997. Rapid reversible changes in organ size as a component of adaptive behaviour. Trends in Ecology and Evolution 12: 134–138.

Predel R. and E. Manfred 2000. Neurosecretion: peptidergic systems in insects. Naturwissenschaften 87: 343–350.

Rising, R., A. Keys, E. Ravussin, and C. Bogardus 1992. Concomitant interindividual variation in body temperature and metabolic rate. Am J Physiol 263: E730–E734.

Schulkin, J. 2003. Rethinking Homeostasis: Allostatic Regulation in Physiology and Pathophysiology. The MIT Press, Cambridge, MA.

Smith-Gill, S. J. 1983. Developmental plasticity: developmental conversion *versus* phenotypic modulation. American Zoologist 23: 47–55.

Stern M. D. 1999. Emergence of homeostasis and "noise imprinting" in an evolution model. Proceedings of the National Academy of Sciences USA 96: 10764–10751.

Sterner, R. W. and J. J. Elser 2002. Ecological Stoichiometry. Princeton University Press, Princeton.

Tauber M. J., C. A. Tauber, J. P. Nyrop, M. G. Villani. 1998. Moisture, a vital but neglected factor in the seasonal ecology of insects: hypotheses and tests of mechanisms. Environmental Entomology 27: 523–530.

Vanden Broeck, J. 2001. Insect G protein-coupled receptors and signal transduction. Archives of Insect Biochemistry and Physiology 48: 1–12.

Venter J. C., Adams M. D., Myers E. W., et al. 2001. The sequence of the human genome. Science 291: 1304–1351.

Waddington, C. H. 1953. Genetic assimilation of an acquired character. Evolution 7: 118–126.

Weng, G., U. S. Bhalla, and R. Iyengar 1999. Complexity in biological signaling systems. Science 284: 92–96.

West-Eberhard M. J. 2003. Developmental Plasticity and Evolution. Oxford University Press.

Wieczorek H., D. Brown, S. Grinstein, J. Ehrenfeld and W. R. Harvey 1999. Animal plasma membrane energization by proton motive V-ATPases. BioEssays 21: 637–648.

Woods, H. A. and E. A. Bernays 2000. Water homeostasis by wild larvae of *Manduca sexta*. Physiological Entomology 25: 82–87.

Woods, H. A. and J. F. Harrison 2001. The beneficial acclimation hypothesis versus acclimation of specific traits: physiological change in water-stressed *Manduca sexta* caterpillars. Physiological and Biochemical Zoology 74: 32–44.

16

Acclimation

Douglas W. Whitman
4120 Department of Biological Sciences, Illinois State University, Normal, IL, 61790 USA, Phone: (309) 438-5123. E-mail: dwwhitm@ilstu.edu

Abstract

Acclimation refers to a physiological change in an individual stimulated by exposure to a different, often stressful, environment. As such it represents physiological phenotypic plasticity. This chapter reviews both early (1900 – 1960) and current research on arthropod acclimation, including: definitions, abiotic and biotic elicitors, types of acclimatory responses, tolerance and capacity acclimation, persistence and speed of response, confounding factors, including different experimental designs and metrics, graphic models, underlying physiological mechanisms, and possible adaptive value. Current acclimation research emphasizes molecular biology, environment-induced gene activation, passive vs. active responses, ecological and fitness consequence of acclimation, and its costs, adaptiveness, and evolution. Current studies attempt to integrate acclimation from genes-to-ecology, and relate acclimation to homeostatic physiology, phenotypic plasticity and stress studies. Understanding acclimation has numerous practical benefits.

Introduction

Everything old is new again

The current literature on phenotypic plasticity often proclaims the novelty of this exciting research area. However, like many fields of science, phenotypic plasticity actually has a long and diverse history, some of which has been nearly forgotten. During the early and mid 20th Century, as geneticists and evolutionary biologists worked to develop the initial ideas about phenotypic plasticity (Baldwin 1896, 1902, Morgan 1896a,b, Osborn 1897, Woltereck 1909, Johannsen 1911, Nilsson-Ehle 1914, Dobzhansky 1937, Clausen et al. 1940, Goldschmidt 1940, Waddington 1942, Schmalhausen 1949,

Bradshaw 1965), another group of scientists labored, largely beyond their view, on a sub-discipline of phenotypic plasticity: acclimation and acclimatization. Publishing in different journals, these physiologists, entomologists, and horticulturalists, produced a substantial body of literature that is underappreciated by modern workers on phenotypic plasticity (Pigliucci 1996). In this chapter, I hope to draw the attention of researchers to this prior work.

Acclimation and Acclimatization

Acclimation and acclimatization are forms of phenotypic plasticity. Acclimation refers to a change in the physiological phenotype of an individual following exposure to one or two well-defined environmental parameters such as temperature, osmolarity, or O_2 concentration, usually under controlled laboratory conditions. The analogous phenomenon in nature is acclimatization, which refers to a change in the physiological phenotype of an individual after exposure to different natural conditions.

It is interesting that despite the fact that around 10,000 papers have been published on acclimation, no one actually knows what it is (Lagerspetz 2006). Acclimatization was first applied to plants and insects that became more tolerant to freezing temperatures following pre-exposure to cold. As such, acclimatization implied the following:

1. The elicitor is a stressful (harmful) environmental factor.
2. The elicitor is an abiotic factor.
3. The elicitor stimulates a physiological change (as opposed to a morphological or behavioral change).
4. Acclimatization is beneficial (increases fitness) in that it allows the organism to better function or survive in the new environment. Hence, acclimatization allows individuals to adjust to a changing environment.
5. The elicitor is also the selective agent, i.e., the organism becomes more tolerant to the *same* factor that elicited the physiological change.
6. The response is anticipatory in that it prepares the animal for future environmental stress.
7. The response is gradual or delayed.
8. The response is medium- to long-lasting.
9. The effect is reversible.
10. The capacity for response is gene-regulated and has undergone natural selection.

Over the years, authors have offered numerous definitions of acclimation and acclimatization, few of which include all of the above criteria:

> *Habituation of an organism to different environmental conditions.*
>
> —Gordh 2001

> *A reversible physiological or morphological change evinced by one individual in response to some alterations in its environment.*
>
> —Morris 1992.

> *Acclimatization is the progressive physiological adjustment or adaptation by an organism to a change in an environmental factor, such as temperature . . . The adjustment can take place immediately or over a period of days or weeks . . . short-term responses include shivering or sweating in warm-blooded animals.* —Rittner and McCabe 2004.

> *. . . the changes which take place in a process in an organism up to the time that the steady level is reached, after the organism has been transferred suddenly from one temperature to another within viable limits . . .*
>
> —Grainger 1958.

> *Physiological, emotional, and behavioral adjustments by an individual to changes in the environment.* —Geller 2003.

> *Physiological or behavioral changes occurring within an organism, which reduces the strain or enhances endurance of strain caused by experimentally induced stressful changes in particular climatic factors.* —IUPS (2001).

> *A physiological response involving sensory mechanisms detecting an environmental change and effecting a gene-regulated change in phenotypic expression.* —Wilson and Franklin 2002b.

According to Prosser (1973), acclimation includes some morphological and behavioral changes, but not diapause, and represents "*. . . compensatory change in an organism under maintained deviation of a single environmental factor . . .*"

Williams et al. (2005) separate acclimation (short term) from longer-term phenotypic or developmental plasticity, such as when raising rodents under water-stress alters kidney structure or function.

Huey and Berrigan (1996) suggest that acclimation is reversible, may be anticipatory *or* reactive, and may be induced by temperature and photoperiod, in time frames of minutes to months. They differentiate acclimation from labile (acute) effects, cross-generational effects, and developmental switches, in which, " *. . . the phenotype is fixed irreversibly by* *environmental conditions experienced during a critical phase of development."* Examples of developmental switches might include permanent changes in adult body size, color, and bristle number.

As can be seen from the above, definitions of acclimation/acclimatization vary greatly and tend to be ambiguous and contradictory, leading to confusion. Does acclimation need to be adaptive? Must acclimatization be anticipatory (e.g., acclimatization of cold-tolerance *before* the onset of winter) or can it be responsive (e.g., induction of heat-shock proteins *after* heat-shock)? Can it be passive (i.e., not regulated by the organism) (e.g., Pigliucci 1996)? Must it be regulated by genes that are activated by the environment? What about cases where environmental stresses bypass genes and accomplish acclimation by *directly* influencing hormones or enzyme cofactors? Must acclimation be reversible? Are immediate physiological responses such as sweating included? How does acclimation differ from acute homeostasis? Can behavior, emotion, morphology, or life-history acclimate? Indeed, over the years, these terms have come to assume such breadth and imprecision, that they can encompass nearly all responses to the environment, such as the photoperiodic entrainment of circadian rhythms or the induction of aphids alates by crowding. However, restrictive definitions can be just as problematic. This is because, whether an environment-induced change is regulated or passive, immediate or delayed, permanent or reversible, physiological or behavioral, elicited by a stressful factor (such as cold) or an innocuous factor (such as photoperiod), such a change (1) still represents a modified phenotype and, hence, phenotypic plasticity, (2) places those individuals into a different selective regime, altering their evolutionary trajectory, (3) may be adaptive, and (4) may eventuate via similar physiological mechanisms (Schlichting and Smith 2002). In addition, whether an individual responds to cold by immediate shivering, moving into sunlight, erection of fur or feathers, vasoconstriction, thermogenesis, migration, diapause, gradual production of cold-adapted enzymes or lipids, or increased pelage, all of these altered phenotypes are similar in that they represent homeostasis or homeokinesis. Hence, a broader definition of acclimation can encompass these diverse and biologically relevant concepts, and this is the approach I use in this chapter. I will not attempt to provide a precise definition of acclimation, but will leave that problem to future workers.

Because acclimation and acclimatization are expressions of different physiological phenotypes in a single genotype after being exposed to different environments, they represent physiological phenotypic plasticity

(Huey and Berrigan 1996). As such they offer fertile ground for exploring all aspects of plasticity, including elicitors, signal transduction, effector systems, genetic control, population, geographic, and genetic variability of response, adaptive value, and evolution. Because physiologically plastic responses are typically more rapid than morphological responses, and because there are innumerable physiological systems, acclimation offers many experimental advantages to researchers. Because we now have a fair understanding of physiological control in organisms, the genetics underlying such physiology, and powerful new molecular tools, acclimation offers the opportunity to finally begin to understand what is happening during phenotypic modification *inside* the organism, which was previously considered a "black box." Laboratory acclimation studies have the advantage of precise control of all variables. In contrast, field acclimatization can be difficult to study, because conditions in nature are always changing, because individuals within natural populations vary in age, experience, and condition, and because of possible interactions among numerous factors. However, acclimatization has the advantage that it often can be directly linked to real fitness benefits in nature, such as when acclimatized insects withstand winter temperatures, but non-acclimatized insects die. Finally, understanding acclimation is essential for all experimental biologists, because both within- and trans-generational acclimation might influence their results (Falconer 1989, Garland and Adolph 1991).

Initial interest in acclimation and acclimatization originated for practical reasons: the need to induce winter hardiness or vernalization of seed germination or blooming in commercial plants, and the desire to understand the winter biology of pest insects. Hence, much of the initial research in this area was undertaken by applied entomologists and horticulturists. However, the field was soon subsumed by basic researchers (e.g., Davenport and Castle 1896, Parhon 1909). As a result, by mid-Century, scientists already had a good understanding of the environmental factors that triggered acclimation, the co-factors that influenced the response, the types and degree of responses possible, and the presumably adaptive nature of acclimation in a great diversity of taxa, including humans. The early workers in this area produced substantial observational and experimental data and theory (see Bělehrádek 1935, Prosser 1950, 1958a, 1961, 1973, H. Precht 1951, Bullock 1955, Fry 1958, Ingrid Precht 1967, H. Precht et al. 1973), and some simple graphical models (see H. Precht 1951, Smith 1951, H. Precht et al. 1973, Prosser 1973), much of which remains of value to current scientists.

Elicitors

By the 1960's researchers had a good understanding of the environmental factors that elicited acclimation in animals. These included temperature (Robinson 1928, Bullock 1955), oxygen level (Wigglesworth 1954), osmotic pressure and salinity (Haas and Strenzke 1957, Kinne 1958, Waddington 1959, Oglesby 1965), humidity (Breitenbrecher 1918, Giersberg 1928), water pressure and density (Bardenfleth and Ege 1916, Damant 1924), food type (Knox et al. 1956, Knox 1958), food quality (Gambaro 1954, Hughes 1960, de Wilde 1962), food quantity (Dreyer 1932), food moisture level (Okay 1953, 1956), and, if one ignores Criteria 1 and 5 (above), photoperiod (Yeates 1954, Hoar and Robertson 1959, de Wilde 1962, Müller 1970), light intensity (Herrstroem 1949, Lees 1955), light color (Kogure 1933), and crowding (Iwao 1962, Uvarov 1966, Rowell 1970) (see also Precht 1951, Bullock 1955, Prosser 1961, Rowell 1971) (Table 1). Subsequently, toxins, pH, CO_2, lunar cycles, and tidal cues were shown to induce long-lasting changes in physiological state (Husain and Mathur 1936a, Prosser 1973, Hill and Wyse 1976, Nicolas and Sillans 1989).

In most cases, exposure to an elicitor stimulates the organism to alter its physiology such that it subsequently shows a different physiological response to that *same* factor, such as when exposure to cold temperature induces low-temperature tolerance or alteres temperature-specific metabolic rates in individuals (Bullock 1955, Prosser 1958a). In nature, these same eliciting factors often serve as selective agents, capable of directly or indirectly lowering fitness. In other cases, the eliciting factor induces a change in physiological response to a *different* factor, such as when changes in photoperiod induce improved cold-tolerance (Danilevskii 1965), or when changes in humidity alter body color in grasshoppers and walking sticks (Giersberg 1928, Faure 1932, 1933, Key 1954), which alters thermoregulation, and hence, thermal physiology (Buxton 1924, Hill and Taylor 1933), as well as antipredatory defense (Rowell 1971). In such cases, the elicitor is not the selective agent, but serves as a proxy (token) stimulus for that agent and cues the individual to adaptively alter its physiology such that it is more resistant to the selective agent. In some cases the elicitor can be far removed from the harmful factor. Hughes (1960) noted a two-step elicitation process, whereby photoperiod altered plant chemistry, which subsequently triggered pupal diapause in feeding cabbage root fly larvae.

In some cases, such as with photoperiod, the insect has evolved specific structures to detect the elicitor and transfer that information to effector systems. In other cases, there may be no formal sensing of the environmental

Table 1 Examples of the progressive steps during acclimation. Examples include physiological, physio-morphological, and physio-behavioral responses, and their assumed fitness consequences. HSPs = heat shock proteins. SCP = supercooling point.

Step 1 Elicitor	Step 2 Physiological response	Step 3 Physio-ecological outcome	Step 4 Fitness consequence	Reference
Cold	↑ Glycerol titer	↓ Supercooling point	↑ Survival in cold	Salt 1958, 1959
Cold	Change in enzymes	↑ Metabolic rate and activity	↑ Survival in cold	Marzusch 1952
Heat	↑ Heat-shock proteins	↑ Thermal protein stability	↑ Survival in heat	Feder and Hofmann 1999
Low O_2	↑ Hemoglobin titer	↑ Tissue O_2 concentration	↑ Survival at low O_2 levels	Wigglesworth 1938
Low O_2	↑ Tracheal growth	↑ Tissue O_2 concentration	↑ Survival at low O_2 levels	Locke 1958
High salinity	↑ Papillae growth	↑ Osmoregulation	↑ Survival at high salinity	Waddington 1959
Short photoperiod	↑ Diapause	↑ SCP or ↑ freeze tolerance	↑ Winter survival	de Wilde 1962
		↓ Metabolic rate		Danilevskii 1965
Thermoperiod	↑ HSPs and polyols	↑ Protein stability in cold	↑ Survival in cold	Wang et al. 2006
Low humidity	↓ Temperature preference	↓ Body temperature and	↑ Survival in dry	Breitenbrecher 1918
	↓ Phototropism	↓ Desiccation		Franenkel and Gunn 1961
Low humidity	↑ Unsaturated lipids	↑ Membrane fluidity in cold	↑ Survival in cold	Holmstrup et al. 2002
Food toxins	↑ P450 enzymes	↑ Diet detoxification	↑ Growth and survival on diet	Berenbaum 2002

factor. Instead, a new phenotype is produced via direct, passive, biophysical effects.

What Changes during Acclimation?

Acclimation can alter physiology, morphology, behavior, and/or life history. However, because the latter traits are simply extensions of physiology, all acclimation is physiological.

Early studies documented a great variety of acclimating traits (Fig. 1, Table 1), including biochemical traits such as enzyme concentrations (Carlsen 1953, Knox et al. 1956, Knox 1958), membrane permeability (Oglesby 1965), O_2-binding capacity of blood, synthesis of thermal-adaptive lipids (homeoviscous adaptation) (Fraenkel and Hopf 1940, Munson 1953, Fast 1970), Q_{10} (Bĕlehrădek 1935, Agrell 1947, Edwards 1958), and enzyme activity rates (Mutchmor and Richards 1961, Prosser 1962), with resulting alterations in concentrations of numerous organic molecules (Fox 1955, Precht et al. 1955, Mews 1957). Performance traits could also acclimate, including: supercooling points, the ability to survive freezing, nerve output and conductance, heartbeat rate (Thompson 1937, Perttunen and Lagerspetz 1956), buoyancy (Damant 1924), digestive functions (Applebaum et al. 1964), metabolic rate (Fig. 2) (Sayle 1928, Agrell 1947, Lühmann and Drees 1952, Dehnel and Segal 1956), blood coagulation (Numanoi 1938, Dean and Vernberg 1966), and tolerance to extreme temperatures, salinities, humidities, and oxygen levels (Bodenheimer and Klein 1930, Wigglesworth 1933, 1938, Bĕlehrădek 1935, Prosser 1950, Baldwin 1954, Baldwin and House 1954, Benthe 1954, Bullock 1955, Fox 1955, Kerkut and Taylor 1957, 1958, Maynard Smith 1957, Straub 1957, Tribe and Bowler 1968) (Table 1; Fig. 1). For example, a short exposure to 37.5°C triples survival time of *Drosophila* pupae at 42.5°C (Milkman 1962).

Fig. 1 Each graph represents a reaction norm showing acclimation (phenotypic plasticity) when insects of the same genotype are maintained under two or more different environments. Examples include acclimation to temperature (a-d), O_2 level (e), and osmolarity (f), as evinced by changes in survival (tolerance) (a & b), supercooling point (c), behavior (d), chemical titer (e), and morphology (f). a) Mortality of *Anagasta* (*Ephestia*) *kuhniella* moth pupae when exposed to –15°C, after pre-conditioning for 4 h at various temperatures. After Atwal (1960a). b) Mortality of adult *Dahlbominus fuscipennis* wasps when exposed to 43°C for 3 h, with or without a prior 2-h heat shock at 36°. After Baldwin (1954). c) Supercooling points for *Bracon cephi* wasp larvae pre-conditioned at various temperatures in the laboratory. Control larvae collected outside in August. After Salt (1959). d) Percentage of male *Blattella germanica*

Fig. 1 Contd. ...

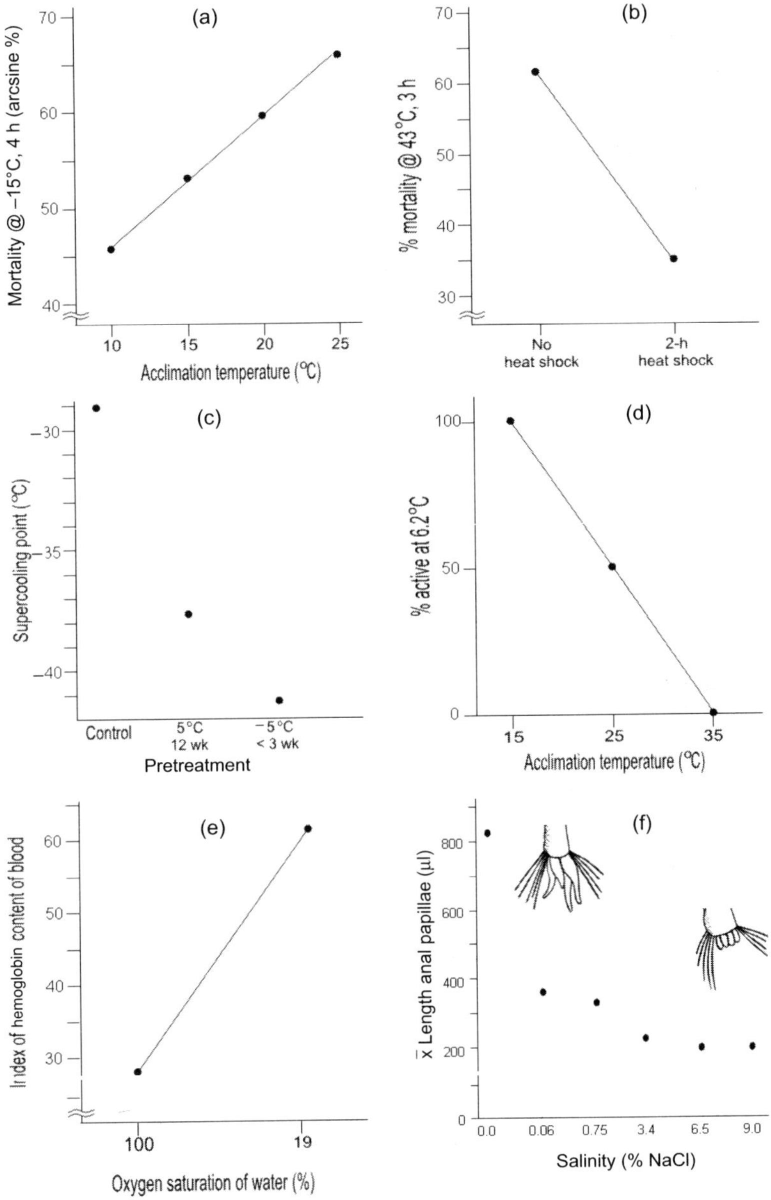

Fig. 1 Contd. ...

roaches active (not in cold-torpor) at 6.2°C after pre-conditioning at various temperatures. After Colhoun 1954). e) Relative hemoglobin content of the blood of *Chironomus* midge larvae from well-aerated and poorly aerated water. After Fox (1955). f) Length of anal papillae of *Culex pipiens* mosquito larvae reared in water of different salinities. After Wigglesworth (1938).

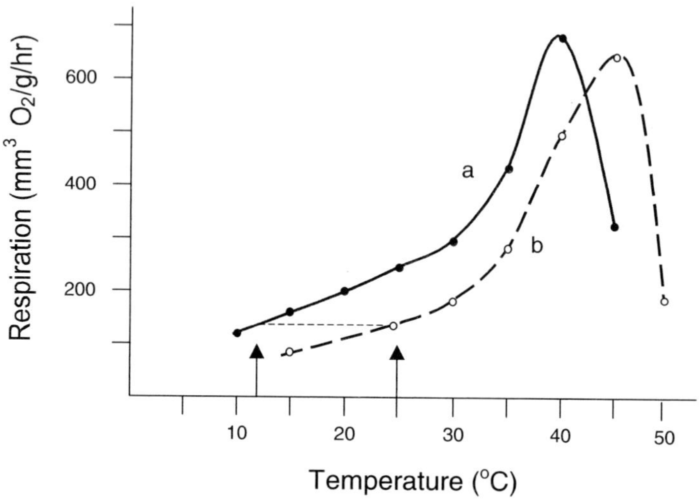

Fig. 2 Metabolic rates of *Melasoma populi* chrysomelid beetles tested at various acute temperatures, for a) insects acclimated to 12°C, and b) insects acclimated to 25°C. Horizontal dotted line at points where acclimation temperature = test temperature, shows temperature compensation in O_2 consumption. After Marzusch (1952).

Environmentally induced diapause and estivation (as opposed to developmentally programmed) are forms of acclimatization of great importance to insects, and are largely responsible for their success and distribution. Various environmental cues trigger diapause and estivation, inducing manifold coordinated physiological changes that make the individual more resistant to the temperature-, food-, and water-stresses of seasonal environments (Lees 1955, Maynard Smith 1957, Salt 1961, Müller 1970). Diapause dramatically alters physiological phenotype. For example, diapausing cecropia silkworm pupae (Saturniidae) consume 1.4% of the O_2 of non-diapausing caterpillars (Schneiderman and Williams 1953), and diapausing *Bracon* wasp larvae can survive temperatures as low as –47°C (Salt 1959).

Many morpho-physiological traits also acclimate, such as number and size of tracheae and tracheoles in *Rhodnius* and *Tenebrio* (Wigglesworth 1954, Locke 1958), diameter of nephridial canals in polychaete worms (Jones 1967), size of anal papillae in midges, mosquitoes, and *Drosophila* (Pagast 1936, Wigglesworth 1938, Harnisch 1951, Haas and Strenzke 1957, Waddington 1959) (Fig. 1f), surface area in hydra (Kinne 1958), buoyancy and size of air sacs and swimbladders in aquatic and marine organisms (Bardenfleth and Ege 1916, Damant 1924, Prosser 1973), body color in insects (Knight 1924, Giersberg 1928, 1929, Atzler 1930, Faure 1932, Husain and Mathur 1936a,b, Key and Day 1954a,b, Ergene 1956), number of mitochondria in cold-acclimated roaches (Thiessen and Mutchmor 1967), hemocyte number in *Galleria* wax moths (Marek 1970), and, in vertebrates, increased pelage or fat (Yeates 1954, Hill and Wyse 1976).

Behavio-physiological traits can also acclimate (Thorpe 1963), including preferred temperatures in ants (Herter 1924), beetles (Bodenheimer and Schenkin 1928, Miller 1969), mites (Henschel 1929), ticks (Totze 1933), and wireworms (Campbell 1937), preferred humidity in roaches (Krijgsman 1930, Gunn and Cosway 1938), preferred light intensity in ground beetles (Herrstroem 1949), low and high temperature thresholds for activity (Bĕlehrádek 1935, Mellanby 1939, Colhoun 1954, McLeese and Wilder 1958), circadian rhythms (Hoffmann 1957, Papi and Parrini 1957), feeding rates (Breitenbrecher 1918, Chapman 1957), and rates of locomotion (Gunn and Hopf 1942, Fry and Hart 1948). For example, nymphal snow scorpionflies, *Boreus hiemalis*, prefer 34°C in summer, but 10°C in winter (Herter 1953). In many insects, including some cockroaches, grasshoppers, beetles, earwigs, lice, flies, and Thysanura, pretreatment with low humidity or dry food alters temperature- or light-preference (Herter 1924, Henke 1930, Bodenheimer 1931, Gunn 1931, 1934, Jack 1939, van Heerdt 1946, Jakovlev and Krüger 1954, Heeg 1967). For example, Colorado potato beetles, *Leptinotarsa*, kept in low humidity for 8 d, refused to feed, became positively geotactic and negatively phototactic, burrowed into the soil, and remained there until it rained (Breitenbrecher 1918). Those kept at high humidity remained active, negatively geotactic and positively phototactic. Photoperiod alters salinity preferences in sticklebacks (Baggerman 1957). Even learning may acclimate to temperature, as suggested when adult *Tenebrio molitor* beetles became more adept at learning after acclimating to a cold temperature (Alloway 1969). Likewise, life history properties such as rate of reproduction (Dick 1937), growth and development rate (Parker 1930, Uvarov 1966), number of instars (Hunter-Jones 1958, Farrow 1975), and time to oviposit can be said to acclimate (Ryan 1941, Elens 1953, Pantyukhov 1962). Finally, Kinne (1958) and Uvarov (1966) give examples of apparent trans-generational acclimation in amphipods and locusts, respectively.

One problem with the early rush to document acclimating traits was a failure to differentiate cause and effect. The process of acclimation proceeds through at least four steps (Table 1): (1) An environmental factor (elicitor) stimulates (2) a complex series of physiological changes that (3) alters an ecophysiological relationship, which (4) alters fitness. For example, exposure to cold and/or short day lengths may induce antifreeze production, which prevents nucleation and lowers freezing temperature, which increases cold-tolerance and hence, winter survival. These steps are mechanistically linked, but not conceptually equivalent. For example, physiologists claim that acclimation alters capacity *or* tolerance (resistance) (Precht 1958, Prosser 1961). But capacity acclimation usually refers to changes in rates such as changes in rate of synthesis, nerve firing, heartbeat, metabolism, behavior, etc., and hence represents Steps 2 or 3 (Table 1). Tolerance usually refers to the ability to survive a specific environmental condition, such as an extremely high or low temperature or osmolarity, and hence represents Step 4. Furthermore, tolerance normally applies to extreme conditions, whereas capacity normally refers to adjustments within normal or mid-ranges of environmental factors (Prosser 1967). However, tolerance and capacity are linked in complex ways. A desynchronization of compensation across all functions may reduce tolerance, and intolerance of a single function or organ can alter lower- or higher-level compensation.

A second major problem with early acclimation studies was insufficient knowledge of the transcriptional, biochemical, and physiological mechanisms underlying acclimation (Step 2). Although researchers identified bits and pieces of the complex physiological cascades effecting acclimation (i.e., concentration changes in specific hormones, enzymes, organic compounds, etc.), most early research documented only inputs (the eliciting factors) and outputs (i.e., changes in capacity or tolerance). What happened in-between often remained clouded. For example, referring to rudimentary flow-diagrams explaining acclimation, Knox (1958) stated, "*... they permit us to express a relationship while remaining ignorant of the nature of the relation and even the nature of the parts affected.*"

Although early workers lacked the tools to elucidate complete pathways, they recognized that acclimation represented not just changes in single traits, but complex cascades involving numerous branching and interacting physiological pathways (see Mechanisms, below). Because complex physiological traits, such as metabolic rate, represented aggregates of multiple underlying systems and processes, it was clear that acclimation proceeded at multiple levels and in multiple systems simultaneously. As such, adaptation was apparent at the enzyme, membrane, organelle, cell, tissue, and whole-animal levels (Bullock 1955, Precht 1958, Prosser 1958a, 1961, Thiessen and Mutchmor 1967, Spencer-Davis and Tribe 1969). Each different level, organelle or tissue might or might not acclimate tolerance or capacity in a different way. Thus, demonstrating acclimation in whole tissues or animals is quite different from knowing which of many possible underlying traits actually acclimated to produce the complete effect (Das and Singh 1974). Likewise, interaction of multiple simultaneous processes confounded understanding. For example, studying high-temperature tolerance in *Drosophila*, Maynard-Smith (1957) found evidence of two kinds of acclimation (developmental and physiological) occurring simultaneously in the same individual. Acclimation was also sometimes confused with developmental changes in physiology (Davison 1971).

It also became evident that numerous traits could be altered by a single elicitor (Anderson and Mutchmor 1971, Precht 1973a,b). For example, a pre-exposure to high temperature can not only adaptively alter metabolic and behavioral capacity (compensation), but can improve heat tolerance and resistance to oxygen, salinity, and toxic stresses (Precht 1973b). In some cases of acclimation, entire suites of co-adaptive traits are altered in synchrony, an example of phenotypic integration (see Canfield and Greene, this Volume). Diapause and estivation represent phenotypic integration, whereby much of the insect's physiology and behavior are altered, starting with temperature, light, and geotactic preferences, and alterations in concentrations of numerous biomolecules, such as enzymes, polyols, amino acids, peptides, proteins, charbohydrates, and lipids, leading to changes in growth, development, metabolic rate, and resistance to temperature extremes, dehydration, lowered O_2, and starvation (Schneiderman and Williams 1953, Lees 1955, Prosser 1961, 1973).

The manifold effects of single environmental factors are seen in the response of some grasshoppers to moisture. In response to humidity, *Locusta* grasshoppers alter both body color and starvation resistance (Rowell 1971). Dry conditions induce brown forms that are more starvation-resistant in dry conditions, and moist conditions induce green forms that are more resistant in moist conditions (Albrecht 1964, 1965). Likewise, darker cuticle is generally more desiccation-resistant (Needham 1974) and shields against harmful UV radiation (Burtt 1981). Both low RH and food moisture can lower temperature preference in grasshoppers and thus reduce evaporative water loss (Jakovlev and Krüger 1954, Laudien 1973). Even activity levels can be altered, given that the brown morph of *Locustana pardalina* is more active than the green morph (Pick and Lea 1970).

Early researchers noted that insect species and stages varied greatly in the ability to acclimate, and many species showed no acclimation (Dreyer 1932, Mellanby 1940a, Salt 1956, Atwal 1960a,b). For example, some Orthoptera alter body color in response to temperature, humidity, food moisture, background color, crowding, etc. (Rowell 1971). However, in regard to color-polymorphic Tetrigidae, Nabours (1929) frustratingly proclaimed, "...*neither excessive humidity, temperature, aridity, acidity, salinity, sunlight through glass or direct, darkness, color of soil, food, excreta, starvation, fungus disease, parasitism, nor any other observable feature of the environment has ever changed color pattern to any appreciable extent.*" Keister and Buck (1974) even suggested that insects in general had poor ability for respiratory compensation in comparison to other taxa.

Early researchers quickly discovered that different traits varied greatly in pattern of acclimation (Fig. 3). Many traits show no acclimation (Nabours 1929, Dryer 1932, Atwal 1960a, Keister and Buck 1961), and some traits

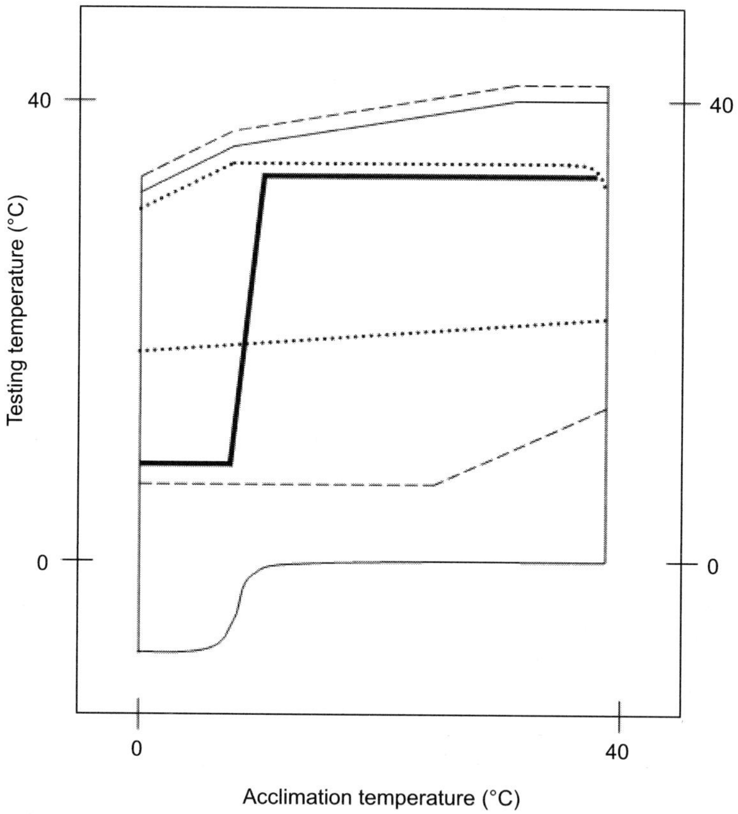

Fig. 3 Hypothetical tolerance polygon showing the potential complexity of acclimation in a genetically identical population of male or female adult insects. Horizontal axis gives temperature at which animal was acclimated. Vertical axis gives temperature at which animal was tested, subsequent to acclimation. Graph shows four functions, each acclimating differently. Areas bounded by similar lines represent zone of tolerance (ability to tolerate specific combinations of acclimation and test temperatures). **Thin solid line:** 1-d survival at test temperature. **Dashed**

Fig. 3 Contd. ...

exhibit reverse acclimation (Precht 1951, Marzusch 1952, Kirberger 1953, Roberts 1953). Different enzyme systems, tissues, and functions acclimate differently under the same experimental treatment (Edwards 1957, Precht 1958, Free and Spencer-Booth 1960, Prosser 1961, Anderson and Mutchmor 1971, Sømme 1972, Das and Singh 1974). For example, in the same animal, anabolic and energy-producing enzyme systems often acclimate to cold, whereas catabolic enzyme systems do not (Marzusch 1952, Mutchmor 1967, Hazel and Prosser 1970). In eels, metabolic rate of head muscles acclimated differently to temperature than that of tail muscle (Schultze 1965). In *Periplaneta americana*, different regions of the nervous system respond differently to changes in temperature (Kerkut and Taylor 1957). Even different regions of individual cells can apparently acclimate: Chatfield and coworkers (1953) examined compensatory increases in nerve conduction in the long nerve that runs the length of the leg in herring gulls, acclimated to cold vs. warm temperatures. In cold-acclimated gulls, the distal portion of the nerve (in the thin distal portion of the leg) became cold-acclimated, but the proximal portion of the nerve (in the thicker, partially feathered and warmer tibial region of the leg) did not.

Speed of Change

Early work on acclimation demonstrated inter- and intraspecific variation in rate of acclimation to the same conditions, and great variation within

---

Fig. 3 Contd. ...

line: Ability to move (= torpor temperature) (1-min exposure). **Dotted line:** Temperature allowing oogenesis (or other function such as spermogenesis or molting). **Bold solid line:** Temperature preference. Note that animals cannot survive below 0°C or above 38°C, unless first acclimated. Also note that each function (survival, activity, and oogenesis) exhibits a different thermal capacity, with survival possible over a wide temperature range, behavior over a medium temperature range, and oogenesis over a narrow temperature range. Hence, at extreme temperatures, activity and survival are possible, but not oogenesis. Each function acclimates differently, and, in this example, the high and low limits for each function acclimate differently. For example, low-temperature (< 0°C) survival only acclimates following exposure to low temperatures. In contrast, high-temperature survival generally rises gradually across all acclimation temperatures. Survival acclimates beneficially to both heat and cold. In contrast, there is little beneficial acclimation for oogenesis, suggesting that there are overriding thermal constraints for that function. Indeed, acclimation at high temperatures reduces the subsequent ability to complete oogenesis at high temperatures (downward curved line). Note high-temperature survival and activity change gradually and in unison with increasing acclimation temperature, but temperature preference and low-temperature survival switch dramatically at about 10°C acclimation temperature.

individuals in the rates of acclimation to different factors (Mellanby 1939, Prosser 1958a). Each different enzyme system, tissue, and function acclimated at a different speed (Mellanby 1939, 1940a,b). For example, Meats (1973) claimed that *Dacus tryani* fruit flies could acclimate "immediately" to changes in temperature. A 2-h-long, 36°C high-temperature shock immediately doubles the amount of time that *Dahlbominus* wasps can withstand 43°C (Baldwin 1954, Baldwin and Riordan 1956). *Periplaneta* roaches begin to acclimate metabolic rates to heat in 4 h (Dehnel and Segal 1956). In *Blattella* roaches, the same cold-exposure increases cold-activity within a few hours, but lowers super-cooling only after several days (Colhoun 1960). *Tribolium confusum* requires only a few hours to increase cold-resistance (survival) (Sømme 1968), ~ 60 h to acclimate O_2 consumption, ~ 17 d for full behavioral acclimation (locomotion) to cold, and 38 d for ATPase acclimation (Anderson and Mutchmor 1971). Brine shrimp, *Artemia*, may take up to three weeks to acclimate to low O_2 levels by increasing hemoglobin levels (Bowen et al. 1969).

Because acclimation often involves numerous physiological systems, the rate at which the whole tissue or organism can acclimate is limited by the acclimation speed of the slowest step (Bullock 1955). In addition, speed of acclimation varies with direction of change and high vs. low levels of a particular environmental factor (Prosser 1973). For example, some animals acclimate faster to warm than to cold temperatures (Prosser 1973). Also, as in the case of *Tribolium* (above), resistance acclimation often proceeds faster than capacity acclimation, and species from variable environments often acclimate faster than those from stable environments (Precht 1973b). Interestingly, acclimation, which is often considered a delayed reaction, can be much faster than "physiological regulation." For example, osmoregulation is generally not considered acclimation, yet, steelhead trout required 80 to 160 h to osmotically adjust transfer from fresh to sea water (Houston 1959). This is an order of magnitude longer than the initiation of thermal acclimation in trout (Werner et al. 2006), and 100 times longer than some rapid cold- or heat-hardening (see below). Finally, early workers realized that acclimation is not a change that continues indefinitely, i.e., at some point, additional conditioning produces no additional change in physiology (Fig. 4) (Atwal 1960a, Sømme 1968), and, in some cases, additional conditioning appears to reduce acclimation (Baldwin and Riordan 1956) (Fig. 5).

Permanency of Change

Historically, acclimation was considered a delayed and medium- to long-lasting, but reversible, adaptive change in physiology in response to

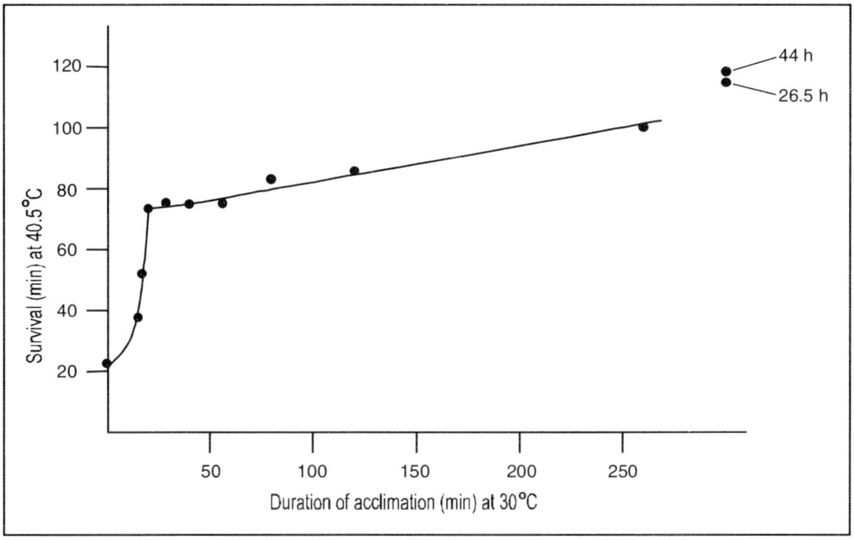

Fig. 4 Relationship between duration of acclimation period (time spent at 30°C) and degree of tolerance acclimation (survival at 40.5°C) for *Artemia* brine shrimp. Most acclimation occurs within the first 20 min. After Grainger (1958).

external conditions. As such, permanency of acclimation fell between highly flexible, acute, immediate, and labile regulation and irreversible developmental switches (Huey and Berrrigan 1996). However, acclimation grades into acute regulation on one hand and developmental switches on the other, in regard to speed, reversibility, and underlying physiological mechanisms (Prosser 1958b). Similarly, there are no clear criteria separating rapid-acting hardening and heat-shock from traditional, slower-acting acclimation (Precht 1973b), but see Bowler (2005).

Many acclimating traits, such as some metabolic traits, activity rates, and even morphological traits are rapidly reversible (Levins 1969, Precht 1973a). For example, although most temperature-, humidity-, and light-induced changes in grasshopper body color are permanent (Rowell 1971), *Kosciuscola* grasshoppers can rapidly (~1 h), change their color back-and-forth in response to temperature (Key and Day 1954a,b). In *Ephestia* moth pupae, acclimation is quickly reversible. Half-day-old pupae acclimate to cold in only 2-3 h, but can reverse cold acclimation after only 2 h under warm conditions (Atwal 1960a). Other acclimations and acclimatizations are permanent, such as when lowered oxygen level induces greater tracheation in *Rhodnius* and *Tenebrio* (Wigglesworth 1954, Locke 1958) or lowered salinity induces larger anal papillae in aquatic Diptera larvae (Fig. 1f) (Pagast 1936, Wigglesworth 1938, Harnisch 1951, Haas and Strenzke 1957).

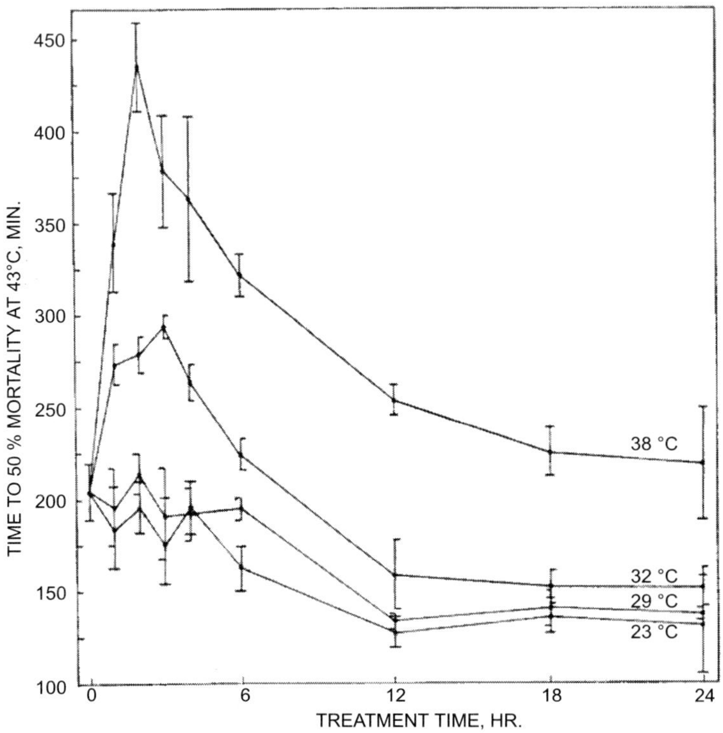

Fig. 5 Influence of acclimation temperature and length of acclimation on tolerance to high temperature (43°C) in the chalcidoid wasp, *Dahlbominus fuscipennis*. Higher acclimation temperatures induce greater acclimation. Maximal tolerance occurs after only 2 to 3 h of acclimation. Tolerance declines with increasing acclimation time, possibly because general heat resistance declines with adult development. From Baldwin and Riordan (1956), courtesy Canadian Journal of Zoology & NRC Research Press.

In *Drosophila suboscura*, high-temperature acclimation is semi-permanent. Rearing larvae at high temperatures increases high-thermal tolerance in adults, suggesting that larval acclimation is carried through the pupal stage (Maynard Smith 1957). A long-acting trans-generational acclimation is seen in some locusts, where nymphal density determines phase state of adult females, which then determines the type of egg diapause, and color and phase state of subsequent hatchling (Matthe, 1950, Uvarov 1966, Simpson and Sword, this Volume). Also, it must be remembered that acclimation is a continuing dynamic process. During acclimation, physiological parameters often under- or overshoot before settling at a new constant value (Precht et al. 1973, Prosser 1973). Because in nature conditions constantly change, acclimatizing individuals may never reach steady state.

Factors that Influence Acclimation

By the middle of the last century, scientists had identified numerous co-factors and variables that influenced the speed, strength, and permanency of acclimation. These included body size, sex, age, circadian phase and activity levels, molting and reproductive stage, hormones, diapause-, estivation-, and desiccation-state, and co-occurring environmental factors such as temperature, salinity, oxygen level, relative humidity, density and crowding, nutrition, time of day, photoperiod, light intensity, and season (Bodenheimer 1931, Gunn 1931, 1934, Lühmann and Drees 1952, Baldwin 1954, Dehnel and Segal 1956, Okay 1956, Edwards 1958, Kinne 1958, Prosser 1958a, Atwal 1960a, Dehnel 1960, Todd and Dehnel 1960, Hunter-Jones 1962, de Wilde 1962, Danilevskii 1965, McWhinnie and O'Connor 1967, Rowell 1970, Tauber and Tauber 1976, Takeda and Masaki 1979). For example, nymphal *Periplaneta americana* cockroaches show a greater acclimation response than do adults to temperature, and small adults acclimate better than large adults (Dehnel and Segal 1956). *Homarus americanus* lobsters can acclimate to temperature, salinity, *or* oxygen level, but acclimation to any two of these factors strongly alters tolerance to the third (McLeese 1956). Likewise, photoperiod strongly influences thermal-acclimation of metabolism in sunfish (Roberts 1964), and photoperiod and temperature interact in triggering diapause in many insects (Fig. 6) (Danilevskii 1965).

Experimental variables, such as speed or intensity of environmental change and time spent under the acclimating environment greatly influence acclimation (Figs. 4 and 5) (Nicholson 1934, Kennedy 1939, Bullock 1955, Precht et al. 1955, Baldwin and Riordan 1956, Grainger 1958, Atwal 1960a). For example, in *Ephestia* moth pupae, a 4-h cold-shock greatly improves subsequent cold-tolerance, but a 16-h cold-shock does not (Atwal 1960a). One confounding problem is that general physiology changes with development and age, and that high acclimation temperature accelerates such changes (Bowler 1967, Davison 1969, 1971). For example, because they are immobile, pupae may have evolved high thermal tolerances (Burnette 1957, Hollingsworth and Bowler 1966). Such tolerance may rapidly decline after eclosion (e.g., Fig. 5). Also, acclimating to a constant temperature or oxygen level may produce a different effect than acclimating to a cycling variable with the same mean as the constant variable (Rogers 1929, Salt 1956, Richards and Suanraksa 1962, Matthews 1976). Greater acclimation may be seen in nature than in the lab, possibly because multiple synergistic factors may operate in nature (Salt 1956). Likewise, the degree of acclimation

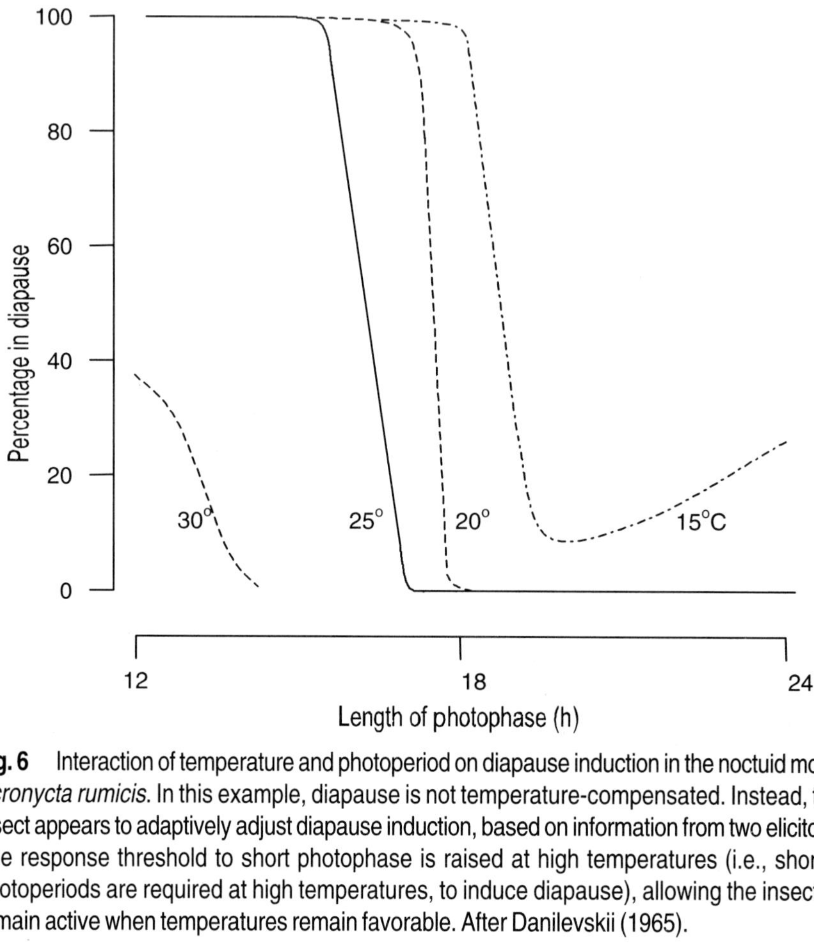

Fig. 6 Interaction of temperature and photoperiod on diapause induction in the noctuid moth, *Acronycta rumicis*. In this example, diapause is not temperature-compensated. Instead, the insect appears to adaptively adjust diapause induction, based on information from two elicitors. The response threshold to short photophase is raised at high temperatures (i.e., shorter photoperiods are required at high temperatures, to induce diapause), allowing the insect to remain active when temperatures remain favorable. After Danilevskii (1965).

recorded by a researcher varies with both the level of the environmental factor during acclimation and during subsequent measurement. Acclimation often occurs at one temperature, oxygen, or light level, but not at another (Prosser 1958a, Rowell 1971). Colhoun (1960) noted rapid (3 h) cold-acclimation when *Blattella* roaches were transferred from 25 to 15 °C, but slow acclimation (> 24 h) from 35 to 25 °C. Despite having to delay their publication because of "*. . . injuries sustained by one of us during an air raid,*" Gunn and Hopf (1942) emphasized that speed and direction of changing test conditions influenced the expression of acclimation. Perhaps Gunn and Hopf came to study rapid acclimation as a result of a pressing need to rapidly complete their experiments.

Overall, organisms tend to acclimate better and exhibit better compensatory acclimation at viable vs. extreme environmental values (Kinne 1958, Buffington 1969). For example, cold-acclimation is often induced by cool temperatures, but not extreme cold temperatures (Mellanby 1940a). Thresholds for elicitation vary greatly among traits. In some cases, only a brief exposure to the elicitor stimulates rapid and strong acclimation (Meats 1973). A 2-hr-long high-temperature shock increases heat-tolerance in *Dahlbominus* wasps, but this response declines as the length of the heat-shock increases, even at relatively low heat-shock temperatures of 29 and 32°C (Baldwin 1954, Baldwin and Riordan 1956), probably due to developmental changes in base line tolerance (Davison 1971) (Fig. 5). *Ephestia* flour moth pupae require only a 4-h exposure to cool temperatures to increase survival at 15 °C (Atwal 1960a). In contrast, thermal acclimation in *Drosophila subobscura* is transitory unless the insects are conditioned at the high temperature for a long period (Maynard Smith 1957). Hence, to fully understand acclimation, it is essential to acclimate organisms to, and subsequently measure them, at a wide range of any variable (Precht et al. 1955, Prosser 1958b).

Acclimation researchers should be aware of confounding factors. For example, temperature-induced change in activity rates (escape response, torpor) and thermal pathologies may alter O_2 use, which may be mistaken for metabolic acclimation (Anderson and Muchmor 1971). Likewise, unrecognized physiological damage that may occur under extreme conditions may underlie the inability of some animals to acclimate to extreme vs. moderate conditions (e.g., Edwards 1958, Atwal 1960a). Acclimation to high temperature may really represent increased resistance to desiccation at high temperatures, and not temperature-resistance, per se (Maynard Smith 1957). Finally acclimation triggered by environmental factors must be separated from developmental physiological changes; under warm (as opposed to cold) treatments, insects continue to age, which might confound interpretation of acclimation data (Bowler 1967, Davison 1971).

Modeling Acclimation

Early workers produced numerous descriptive and graphical models of acclimation (Agrell 1947, Precht 1951, 1958, 1973b, Smith 1951, Zerbst et al. 1966, Brett 1971, Prosser 1973), and these remain instructive, today. Both Precht (1951, 1958) and Prosser (1958b) emphasized how acclimation could alter rate curves by displacing them up, down, left, or right (translation), or via rotation (clockwise or counterclockwise), or by altering their length or shape (Precht 1951, Bullock 1955) (Fig. 7). Changes in length or shape, and rotation of rate curves altered Q_{10} (Fig. 7). Precht (1949) proposed five types of acclimation responses (Fig. 8). Although he stressed thermal acclimation, his model is appropriate for acclimation to other factors as well. In Fig. 8, each line represents one possible way that an individual might acclimate a rate (such as metabolic rate) to three different levels of a particular environmental factor such as temperature. Line 2 represents an organism with perfect compensation (i.e., the organism acclimates such that it maintains a similar rate in all environments. Line 4 shows no acclimation. It also represents the acute rates that would occur immediately following transfer to a new environment, *before* acclimation occurred. Because, in this case, the acute rates do not change after a long period at the new environment, there is no acclimation. Line 1 represents excess compensation, Line 3 represents incomplete or partial acclimation, and Line 5 represents inverse acclimation. One might not expect inverse acclimation, but it occurs (Gunn and Hopf 1942, Garside and Tait 1958, Sømme 1968, Davison 1971). For example, some crab and crayfish muscles show Type 5 acclimation to temperature (Bowler 1963, Vernberg and Vernberg 1967a,b, Jungreis and Hooper 1968), and *Tribolium* reared in cold temperatures for a month survive better at hot temperatures than those reared at hot temperatures (Edwards 1958). However, as Prect and Prosser pointed out, each pattern of acclimation, including inverse acclimation, may (or may not) be adaptive under the appropriate environmental and physiological conditions. For example, inverse acclimation of a specific enzyme might aid general homeostasis if it moderated the effects of harmful rate changes in other enzymes (Precht et al. 1973).

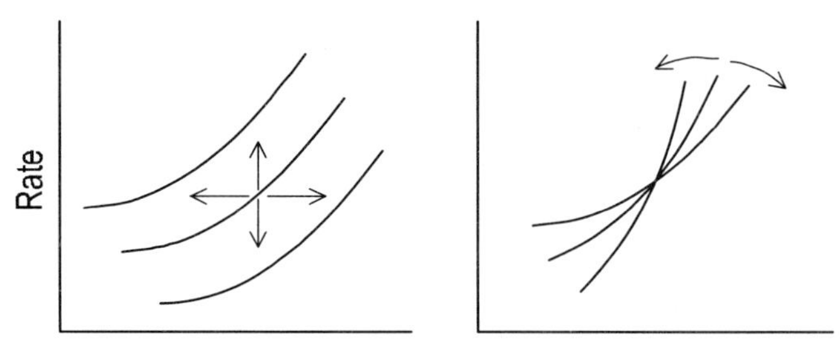

Fig. 7 Possible theoretical changes in biological rates during acclimation as per Precht (1951) and Prosser (1958b). Left graph shows translation. Right graph shows rotation, which alters Q_{10}.

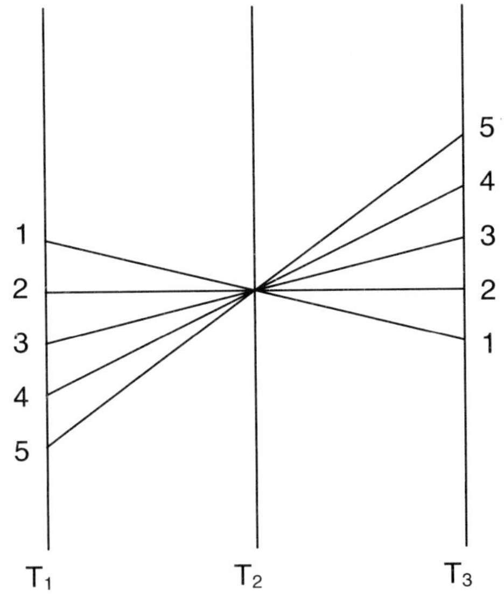

Fig. 8 Five patterns (lines 1-5) of acclimation response for animals acclimated to, and tested at, three different temperatures, as per Precht (1949) and Prosser (1973). Vertical axis represents physiological response (rate, tolerance, titer of metabolic product, etc.) after acclimation. T_1, T_2, and T_3 represent low, medium, and high acclimation and test temperatures, respectively. Line 2 represents perfect compensation, whereby the physiological response does not change when the individual is successively acclimated to and tested at three different temperatures. See text for further explanation.

Adaptive Value of Acclimatization

Early workers touted the adaptive value of acclimatization (Davenport and Castle 1985), which seemed obvious in nature, when non-acclimatized individuals died and acclimatized ones lived. Adaptive evolution was also implied by the fact that acclimatization was often integrated and anticipatory, whereby organisms altered their physiology in complex ways *before* harmful conditions occurred. Adaptive arguments were especially strong for those insects that responded to token elicitors that were not themselves harmful, but which served as proxies for those that were, such as when photoperiod triggered beneficial acclimation to harmful temperatures (Hoar and Robertson 1959). It was hypothesized that because some harmful environmental factors could change too fast for animals to prepare, organisms had evolved to respond to proxy elicitors that predicted future dangers. Hence, acclimatization was seen as a type of phenotypic plasticity that permitted individuals to adaptively alter their physiology in response to environmental cues that signal changing conditions. In some cases, acclimatization allowed organisms to extend tolerances or limits (Baldwin 1954, Mellanby 1954, Salt 1959). In other cases it permitted compensation such that an individual maintained the same physiological rate or function (such as metabolism, heartbeat, ventilation, activity, etc.) in the face of changing environmental conditions (Parhon 1909, Sayle 1928, Lühmann and Drees 1952) (Figs. 2 and 9). Trans-generational acclimatization allowed parents to adaptively alter their offspring's physiology, based on the parent's knowledge of the current environment, which predicted the offspring's environment (Levins 1968). However, acclimation could also be responsive, such as when chironomid midges increase hemoglobin *after* being moved to a low-O_2 environment (Fox 1955).

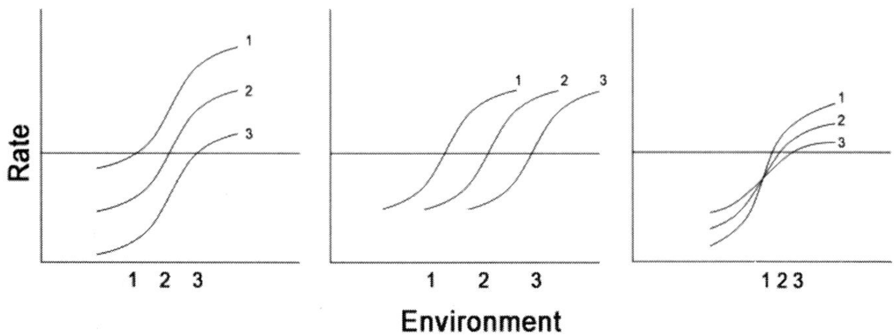

Fig. 9 Three ways that change in rates during acclimation can result in compensation. Curves 1, 2 and 3 show acute biological rates following acclimation to three different environments (1, 2 and 3), when tested at various levels (horizontal axis) of a given factor (temperature, osmolarity, O_2 concentrration, etc.). Points 1, 2, and 3 on horizontal axis represent when testing conditions equal acclimation conditions. Note that in each case, rate is identical (i.e., perfect compensation) at each of the three acclimation levels.

Acclimatization may be beneficial in that it permits organisms to continuously adjust their physiology, throughout their lives (Fig. 10). This presumably allows organisms to become more competitive and to inhabit broader ecological niches and wider seasonal and geographical ranges (Bullock 1955). Prosser (1958b) suggested that many physiological races actually represent acclimatization, and not genetically distinct populations.

However, acclimatization may carry physiological costs (Hoffmann 1995) and does not always easily lead to an adaptional explanation (Tantawy and Mallah 1961, Huey and Berrigan 1996, Huey et al. 1999. Gilbert et al. 2001). This is especially true in cases of reverse acclimation (see

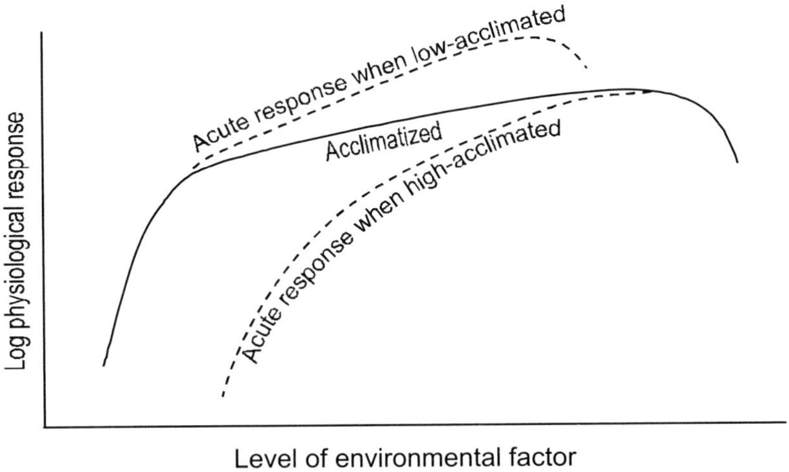

Fig. 10 Hypothetical effects of acclimation on acute physiological response. Solid line shows acute physiological response for an individual acclimatized to each tested level of a given factor (temperature, salinity, O_2 concentration, etc.). Dotted lines show acute physiological response at various factor levels for individuals acclimated to a low level of a given environmental factor (top curve) vs. a high level (bottom curve). Note that, in this example, acclimatization improves both tolerance and compensation. After Bullock (1955).

above). In some cases, "acclimation" makes insects more susceptible to a stressful environmental factor, such as when acclimation to 38°C made *Tribolium* beetles less tolerant to 40°C than beetles acclimated at 18°C (Edwards 1958). However, pretreatments at extreme conditions may produce lingering damage, masking or disrupting acclimation, or making individuals less viable, and high temperatures may speed aging and developmental changes in physiology (Davison 1971). Although Laudien (1973) cautioned that lab-derived conclusions about adaptive value should be tested in the field, the laboratory physiologists who studied acclimation generally did not do so, nor did they consider lifetime fitness consequences of specific acclimations.

The benefits of tolerance adaptation can be illustrated by a tolerance polygon, which shows the tolerance of individuals acclimated to all possible levels of a particular environmental factor. Tolerance polygons derived from empirical data often show that acclimated organisms can tolerate a wider range of conditions and thus extend their ecological niche. Figure 3 shows a hypothetical tolerance polygon, illustrating complexities when considering different traits. In this hypothetical example, different traits have dramatically different tolerance thresholds, different traits acclimate differently, individual traits acclimate differently to high vs. low levels of a given environmental factor during acclimation, and tolerance to low extremes show a different pattern than tolerance to high extremes. Actual tolerance polygons derived from real animals (Fig. 11) show the great diversity of responses present in nature, and reaffirm that, in biology, anything is possible, and that not all acclimation may be beneficial.

Early researchers assumed genetic control for the ability to acclimate, and that populations had undergone selection for both an ability to acclimatize, and degree of response (Key 1954, Prosser 1955, Rowell 1971). Breeding, hybridization, and selection experiments on high-temperature acclimation (Maynard Smith 1956, Bowler and Hollingsworth 1965), diapause induction (Danilevskii 1965), environmentally induced color morphs (Fuzeau-Braesch 1960, Nel 1968), and locust phase characteristics (Gunn and Hunter-Jones 1952) supported these beliefs. After 21 generations of selection for high salt tolerance, *D. melanogaster* pupae not only had larger anal papillae (used in osmoregulation) and greater survival in salty media, but had evolved greater plasticity for papillae size than non-selected flies, demonstrating that plasticity can be selected (Waddington 1959).

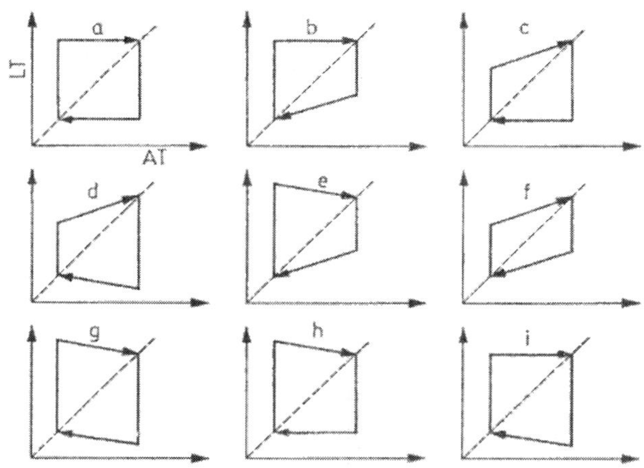

Fig. 11 Tolerance polygons for interaction of acclimation temperature (AT) vs lethal temperature (LT), showing possible acclimation patterns among different species. a) No acclimation. b) Cold acclimation only. c) Heat acclimation only. d) Heat acclimation, but paradoxical (reverse) cold acclimation. e) Cold acclimation, but paradoxical heat acclimation. f) Both cold and heat acclimation. g) Paradoxical or reverse acclimation to both cold and heat. h) No cold acclimation, but paradoxical heat acclimation. i) No heat acclimation, but paradoxical cold acclimation. Precht (1973) gives actual species examples for each type. Reprinted from Precht (1973), Fig. 2, p. 429. With kind permission of Springer Science and Business Media.

Geographic and habitat differences in acclimation abilities further supported genetic-adaptational hypotheses (Heart 1952, Precht et al. 1955, Prosser 1955). Indeed, related species inhabiting highly variable environments often (but not always) showed better compensatory acclimation capability than those inhabiting less-variable environments (Payne 1926, Marzusch 1952, Hunter 1968, Levins 1969, Anderson and Mutchmor 1971). For example, temperate fiddler crabs exhibited greater thermal acclimation than tropical ones (Vernberg and Vernberg 1966, Vernberg 1969). Furthermore, species that remain active over a wide range of environmental conditions (temperature, salinity, oxygen) are generally better acclimators than those that relocate, hibernate, estivate, or enter dormancy in response to variable conditions (Prosser 1973).

For an organism to function, all biochemical and physiological processes must be coordinated and balanced. But, in some cases, acclimation of one function interferes with another (Clark 1967, Hoffmann and Stockmeier 1975). As such, the difficulty of integrating acclimation (plasticity) among innumerable physiological systems and functions was hypothesized to be the factor that ultimately limited adaptation in organisms (i.e., why all individuals cannot perfectly acclimatize to all environmental conditions, and why populations and species must specialize or are restricted to specific environments). At extreme environmental conditions, it becomes harder for organisms to maintain steady rates (to compensate). Hence, compensatory acclimation generally only occurs over a certain medium range (Fig. 10) (Platzer 1967). For example, metabolic rates remain relatively steady for *Culex* mosquitoes acclimated between 15 and 25°C, but deviate beyond that range (Buffington 1969). Acclimation may speed speciation, when sympatric races acclimate differently or possess different environmental sensitivities (thresholds to elicitors). Genetic divergence may increase in locals where hybrids produce inappropriate acclimation responses to the local environment. As previously mentioned, many species and many traits show no acclimation (Nabours 1929, Dryer 1932, Keister and Buck 1961), and in fact, only a few species have been shown to undergo simultaneous capacity and tolerance acclimation (Precht 1958, Lagerspetz 2006). Answering why some species, functions, and traits acclimate and others do not, and understanding the consequences of such differences presents major challenges to evolutionary biology.

Mechanisms Underlying Acclimation

Early studies on acclimation laid the framework for how current workers analyze acclimation mechanisms (Fox 1936, 1939). For example, Prosser

(1950, 1961), Bullock (1955), Precht (1955, 1964), Knox (1958), and Clarke (1966) discuss inductors, sensing elements, signal detection, effectors, controllers, information-control, servo-loops, adapting systems, and the possibility of enzymatic, cellular, and hormonal regulation of plastic responses. As such, these workers presaged the elicitors, receptors, and signal transduction pathways that are discussed in the current literature. Altered enzyme activity was an early focus, be it changes in conformation (Milkman 1967), concentration (Knox 1958, Rao 1967), or isozymes (Baldwin and Hochachka 1970). Hochachka (1967) proposed that acclimation initiated alternative pathways, and Gordon (1972) suggested that some animals compensate by changing one set of rate-determining reactions for another. Hence, organisms store in their genetic closets, different biochemical ensembles, to be employed under the appropriate environmental conditions. Prosser (1958b) even discusses the possibility of environmental factors inducing acclimation by acting directly on the DNA of individuals, and Ritossa (1962) and Tissieres et al. (1974) showed that heat shock induced both transcription and translation. Hence, by 1970, workers understood that acclimation was accomplished by regulation of hormones, enzymes, transcription, translation, and concentrations of specific compounds (Marzusch 1952, Carlson 1953, Precht 1953, Prosser 1958a, Siminovitch et al. 1967, Somero 1969, Weiser 1970).

Although pioneering physiologists lacked modern molecular methods, some still made impressive progress, elucidating partial physiological pathways for acclimation. Thus, Fox (1955) showed that acclimation to low O_2 level in chironomid midge larvae (blood worms) and *Daphnia* was effected by up-regulation of oxygen-transport pigments such as hemoglobin, and (in *Daphnia*) increased cytochrome. Thermal acclimation of metabolic rate in whole *Sarcophaga* or *Calliphora* (Diptera) was mirrored by that in isolated mitochondria (Spencer-Davis and Tribe 1969, Danks and Tribe 1979). Likewise, researchers quickly determined that winter acclimation in insects was often accomplished by simultaneously decreasing water and increasing polar, polyhydric solutes such as glycerol, sorbitol, and mannitol. The organic solutes inhibit ice nucleation, and both processes lower freezing point by increasing osmolarity (Salt 1961, Prosser 1973). Hence, increasing hemolymph glycerol concentration to 5 M (25% of fresh mass) allowed some *Bracon* wasp larvae to supercool to –47 °C (Salt 1958, 1959, 1964). Similarly, winter acclimation in tent caterpillar (*Malacosoma*) eggs raised glycerol concentration to 35% (dry wt), giving them the ability to supercool to –41°C. Pre-acclimation eggs could supercool to only –21°C

(Hanec 1966). Other insects acclimated to winter by becoming freeze tolerant (Salt 1962, Sømme 1964), with some surviving freezing as low as −80°C (Miller 1982).

By the 1950's, researchers already had a fair understanding of how environment induced color change in insects, whereby known wavelengths of light, perceived via the eyes or individual epidermal cells of the integument, induced production of specific pigments, sometimes under endocrine control (Knight 1924, Schlottke 1926, Kaestner 1931, Ergene 1954, Key 1954a,b, Rowell 1971). Similarly, scientists discovered that animals often acclimate to temperature by altering the saturation of membrane lipids (homeoviscous adaptation). Saturated lipids are less fluid, which stabilizes membranes and lipoproteins at high temperatures. Temperature and humidity can elicit this response (Fraenkel and Hopf 1940, Precht et al. 1955, Holmstrup et al. 2002), but so does feeding on saturated lipids (House et al. 1958). Note that all three examples use the same mechanisms (altered lipids) to produce the same consequences (increased thermal tolerance or compensation), but initiate via different factors (temperature, humidity, or diet). This illustrates the difficulty in defining acclimation: despite the similarities in all three examples, change in lipid saturation via diet would be considered a passive process, and not acclimation.

Early workers realized that acclimation to different factors required different underlying mechanisms, that different traits acclimated differently (Fig. 3) (Precht 1951, 1958, Precht et al. 1955), and that different mechanisms operated during exposure to high vs. low states of a given environmental factor (Fig. 3) (Brett 1946, Prosser 1950). Rearing insects at high temperatures often increased thermal-death points, but had relatively small effects on cold torpor temperatures, and vice versa (Edwards, 1957, 1958), suggesting independence of the mechanisms effecting high- vs. low-temperature acclimation (Mellanby 1954, Precht 1958). In contrast, Salt (1961) opined that cold-acclimation could be considered as the loss of heat-acclimation and vice versa.

Initial researchers also articulated the need to maintain overall internal physiological balance – Precht (1958) used the phrase, *"harmony of cell metabolism."* Hence, for acclimatization to be beneficial and adaptive, the thousands of individual biochemical reactions and physiological rates and functions, which in total comprise a viable organism, all must work in harmony and synchrony *after* a phenotypic change (phenotypic integration). Separate traits must acclimate in lock-step.

Current Studies of Acclimation

Acclimation continues as an exciting and active research area. In this section, I briefly highlight some current trends. Good treatments of various topics relating to acclimation include: Huey and Berrigan 1996, Johnston and Bennett 1996, Kingsolver and Huey 1998, Bale 2002, Hoffmann et al. 2003, Wang et al. 2003, Chown and Nicholson 2004, Morris and Vosloo 2004, Bowler 2005, Danks 2005, Sinclair and Roberts 2005, Gimenz 2006, Harrison et al. 2006, Lagerspetz 2006, Lagerspetz and Vainio 2006.

The greatest change in acclimation research has been the advent of modern molecular biology and bioinformatics, which has dramatically transformed physiological research (Cossins et al. 2006, Gracey 2007, Kültz et al. 2007, Quackenbush 2007, Wittkopp 2007, Shiu and Borevitz 2008), providing powerful tools for understanding how phenotypes are made, maintained, and altered (but see Feder and Walser 2005, Barrett et al. 2007). For the first time, transcriptome, proteome and metabolome open-screens facilitate system-wide assessment of acclimation responses (Malmendal et al. 2006), aiding understanding of mechanisms and regulation at the molecular level, allowing identification of candidate genes, and, importantly, generating new hypotheses (Liang et al. 2004, Storey 2004, Kayukawa et al. 2005, Cossins et al. 2006, Malmendal et al. 2006, Sonoda et al. 2006, Mathias et al. 2007). Mutant and transgenic lines aid this process (Raushenbach et al. 2004, Nielsen et al. 2005). We now know that regulation of gene expression underlies much acclimation, and that the environment can influence phenotype by *directly* turning on or off specific genes (Buckley et al. 2006, Henry et al. 2006, Nielsen et al. 2006, Sonoda et al. 2007). For example, a 1-h heat shock altered expression in 1222 *D. melanogaster* genes (Sorensen et al. 2005). The environment can also directly stimulate hormones (Gade 2004, Schooley et al. 2005), and hormones can turn on or off genes (Clever 1961, Thissieres et al. 1974, Raikhel et al. 2005).

Debate continues as to the definition of acclimation (Bowler 2005, Loeschcke and Sørensen 2005, Sinclair and Roberts 2005, Lagerspetz 2006) and how adaptive acclimation differs from passive responses (Pigliucci 1996, Wilson and Franklin, 2002a,b). In adaptive acclimation, the capacity for, and mechanism to produce, the physiological change are presumed to have undergone natural selection. The response is considered active, specific, coordinated, and beneficial. That is, the organism has evolved to respond adaptively to environmental elicitors via specific sensors, signal transduction pathways, effector systems, and integration, all controlled by the coordinated activities of numerous regulatory genes and feedback loops. In contrast, passive responses are assumed to include unavoidable, direct, inevitabilities, biophysical effects, pathologies, physiological damages, and environment-induced physiological or developmental constraints. Such responses may or may not be detrimental. These categorizations, however, are problematic, because it is not only difficult to document their defining qualities, but at some point, passive and active physiological responses grade into one another. In addition, natural selection chooses among phenotypes, not genes. Thus, as previously mentioned, whether an environmentally induced change in phenotype is passive or gene-regulated, beneficial or detrimental, it still is a change in phenotype, and thus places that individual into a different selective regime. Recurrence of those particular environmental conditions, leading to recurrence of the new phenotype, allows the environment to select for genes that produce a beneficial plastic response (given genetic variability and heritability for plasticity) (West-Eberhard 2003). In this sense, an unavoidable, passive physiological pathology may be the first stage of adaptive acclimatization evolution. Thus, environmental susceptibilities and highly evolved, gene-regulated acclimatization responses may represent two points on a continuum.

Definitions of acclimation and related phenomena will become less importation in the future. As modern molecular physiology clarifies the pathways and mechanisms underlying physiological change, the traditional categories (homeostasis, heat shock, hardening, acclimation, developmental switches, etc.) will be distinguished or replaced by mechanistic categories. Hence, homeostasis, acclimation, and developmental switches (in different species) that utilize the same genes, mechanisms, and pathways, will be grouped together.

A lively discussion centers around the question of the adaptiveness of acclimation (Huey and Berrigan 1996, Huey et al. 1999, Hoffmann and Hewa-Kapuge 2000, Thompson et al. 2001, Woods and Harrison 2001, 2002, Loeschcke and Hoffmann 2002, Wilson and Franklin 2002a,b, Deere and Chown 2006, Lagerspetz 2006). This has spurred articulation of methods to test adaptive hypotheses (Huey and Berrigan 1996, Kingsolver and Huey 1998, Garland and Kelly 2006), and research on the costs and benefits of acclimation (Kingsolver 1995, 1996, Krebs and Holbrook 2001, Loeschcke and Hoffmann 2002, Hoffmann et al. 2003, Stillwell and Fox, 2005), especially in the field (McMillan et al. 2005, Kristenson et al. 2007, Loeschcke and Hoffmann 2007). Realization that the single, rapid, large-step environmental changes applied in the lab do not accurately represent nature (Wang et al. 2006) has stimulated a call for more ecological relevance in testing (Bale 1987, Feder 1997, Feder et al. 1997, Kelty and Lee 2001, Sinclair 2001).

Although we know that a single environmental factor can induce manifold changes in gene expression, and numerous subsequent biochemical and physiological cascades, we need to sort out which of these myriad phenotypic changes represent adaptive physiological adjustments, and which are developmental pathologies and other non-beneficial pleiotropic and physiological by-products (Fischer et al. 2003). Likewise, we need to take a more holistic view when assessing costs, benefits, and tradeoffs of acclimation (Angilletta et al. 2003, Seebacher and Wilson 2006, Loeschcke and Hoffmann 2007), including considering long-term effects (Layne and Peffer 2006). Acclimatization might provide any number of unrecognized benefits such as enhanced mating or fighting ability (Lagerspetz 2000, Seebacher and Wilson 2006). In contrast, acclimatization might greatly improve a specific physiological function, but still lower overall fitness when it reduces competitiveness, disease resistance, longevity, etc. (Zwaan et al. 1992, Zamudio et al. 1995, Wilson et al. 2007). A stress factor may elicit an adaptive beneficial acclimation response, and simultaneously induce physiological damage (Woods and Harrison 2001, 2002, Wilson and Franklin 2002a). The two opposite effects are difficult to disentangle.

Some have called into question the existence of, or at least the correct measurement of, acclimation. For example, Clark (1993) suggests that seasonal acclimation of metabolic rate has no useful biological meaning. This is because temperature influences nearly all physical properties of organisms (lipid fluidity, diffusion rates, water density, viscosity, solubility, pH, ionization, etc.), and we do not know how these combined factors affect metabolic rate. Also, growth, activity, and reproduction often slow in winter, because of seasonal changes in resource availability, and these changes greatly influence metabolism. Hence, observed changes in metabolic rate may or may not have anything to do with direct thermal adaptation per se. Other authors have pointed out similar problems with Q_{10} (Johnson et al. 1974). Acclimation researchers have yet to adequately reply to these criticisms. Sinclair and Roberts (2005) point out that interpreting acclimation studies is difficult given the great diversity of treatments applied (seconds to months), and responses measured. Also, supercooling points may not be a reliable metric for estimating cold hardiness or survival (Wang and Kang 2005).

Despite the above arguments, documentation of acclimation continues at a furious pace, with weekly reports of new species or traits showing one-or-another form of acclimation (Hawes et al. 2007, Jensen et al. 2007, Slabber et al. 2007, Sonoda et al. 2007), and some showing no acclimation (Pitts and Wall, 2006, Terblanche and Chown 2007). For example, many additional insects have been found to acclimate lipids to environmental temperature (Kostal et al. 2003, Overgaard et al. 2006, Tomcala et al. 2006), such as *Melanoplus sanguinipes* grasshoppers, which synthesize higher-melting-point *n*-alkanes when reared at high temperatures (Gibbs and Mousseau 1994). Drought, short cold-shock, and photoperiod, induce altered lipids in Collembola (Holmstrup et al. 2002), *Sarcophaga* flies (Michaud and Denlinger 2006), and *Pyrrhocoris* bugs (Hodkova et al. 2002), respectively. In Collembola, drought acclimation improves cold-tolerance (Bayley et al. 2001, Holmstrup et al. 2002). Both polyols and heat shock proteins can be induced by cold *or* heat shock, and both apparently aid in thermal tolerance (Wolfe et al. 1998, Salvucci et al. 2000). Caterpillars and some other insects may acclimate to dry conditions by altering cuticular transpiration and rectal water absorption (Martin and Van't Hof 1988, Reynolds and Bellward 1989, Woods and Bernays 2000, Woods and Harrison 2001). Rapid cold hardening can be induced by both high temperatures (Goto and Kimura 1998, Sinclair and Chown 2006) and anoxia (Coulson and Bale 1991). In tsetse flies, thermal acclimation strongly affects rates of water loss (Terblanche et al. 2006). Thermoperiod influences cold hardening in locust eggs (Wang et al. 2006) and beet armyworms (Kim and Song 2000). Additional arthropods have been shown to acclimate to hypoxia by increasing tracheal diameter or branching (Loudon 1989, Jarecki et al. 1999, Henry and Harrison 2004, Harrison et al. 2006), or to salinity (Henry et al. 2006, Mendonca et al. 2007). Salinity acclimation alters salt uptake in freshwater shrimp, *Gammarus zaddachi*. *Culex tarsalis* mosquito larvae acclimate to salty water by increasing concentrations of two organic osmolytes; trehalose increases 2-fold, and hemolymph proline 50-fold (Patrick and Bradley 2000). Other mosquito larvae acclimate to pH (Clark et al. 2004). Caterpillars respond to thermal stress by lightening body color (Nice and Fordyce 2006). Cold acclimation allows *Periplaneta japonica* roaches to walk on ice (Tanaka 2002). In Colorado potato beetles, chilling or temporary freezing, followed by transfer to 24°C alters behaviors, causing the beetles to burrow into the soil and remain there for 3-4 weeks, a response possibly mediated via JH (Hiiesaar et al. 2001). We now know that insects can acclimate to dietary toxins by up-regulating P450 and other enzymes (Mazumdar-Leighton and Broadway 2001, Agrawal et al. 2002, Berenbaum

2002, Cianfrogna et al. 2002, Li et al. 2002, 2004), and to pollutants by producing heat-shock proteins (Lee et al. 2006). Insects can acclimate to tough and fibrous food by altering head, mandible, and muscle morphology (Thompson 1992). Insects can "acclimate" to novel foods (Agrawal et al. 2002), and time to oviposit (Scott 1996, Batemann et al. 2005), and even ability to reproduce can be said to acclimate (Scott et al. 1997). There is also growing interest in transgenerational acclimation (Huey et al. 1995, Crill et al. 1996, Watson and Hoffmann 1996, Magiafoglou and Hoffmann 2003, Guan and Wang 2006a,b, Rako and Hoffmann 2006, Simpson and Sword, this Volume).

Particularly important for entomologists who use CO_2 as a narcotizing agent, is that this molecule can induce numerous, profound, and long-term physiological changes in insects. CO_2 can alter metabolic pathways, changing titers and types of proteins, lipids, carbohydrates, and hormones (Nicolas and Sillans 1989). It also changes nerve function, behavior, development, growth, body size, and reproduction (Nicolas and Sillans 1989). Such effects were realized over 50 years ago. For example, a single exposure to high CO_2 alters temperature preferences and causes individual honey bees to switch from hive work to foraging (Ribbands 1950). It also influences JH titers (Buhler et al. 1983) and initiates vitellogenin production and oviposition in queen bees (Mackensen 1947, Engels et al. 1976). CO_2 induces permanent changes in body color in some insects (Rowell 1971). In *Helix* snails, it influences estivation (Michaelidis et al. 2007). However, at this time, it is not known if such responses represent pathologies or beneficial acclimation. Interestingly, some midges, ants, weevils, and tiger beetles switch to anaerobic respiration under hypoxia or high CO_2 levels (Hoback and Stanley 2001, Nielsen and Christian 2007). Nilson et al. (2006) believe that short CO_2 exposure does not significantly influence cold hardiness in *D. melanogaster*.

The study of stress proteins, including heat shock proteins (HSPs), has become a major research industry unto itself (Korsloot et al. 2004, Henderson and Pockley 2005, Yin et al. 2006, Asea and DeMaio 2007, Calderwood et al. 2007, Pappas et al. 2007, Wang et al. 2007). Stress proteins are induced by various environmental factors, including heat, cold, desiccation, crowding, heavy metals, organic toxins, toxic gases, UV, anoxia, salinity, parasites, disease, and mechanical injury (Kulz 1996, Nepple and Bachofen 1997, Feder and Hofmann 1999, Tammariello et al. 1999, Bayley et al. 2001, Sørensen and Loeschcke 2001, Rinehart et al. 2002, Hoffmann et al. 2003, Williams et al., this Volume). They are often mediated by cytoplasmic stress-induced heat-shock factors that bind to promoter regions of HSP genes, initiating transcription. HSPs repair or protect enzymes from stress-damage, and play a vital role in increasing thermal or cold tolerance, and thus appear to function in acclimation. They can be induced within minutes of stress. During and following heat shock, normal protein synthesis can be largely replaced by production of heat-shock proteins (Chown and Nicolson 2004), demonstrating that environmental stress can redirect much of the cell's biochemistry. A critical goal is to determine if HSP-production is only reactive or if it can be anticipatory (i.e., can be induced by harmless levels of factors or harmless elicitors that predict changing environments).

Researchers continue to explore the many factors that influence acclimation, including exposure time (Jian et al. 2005, Mahroof 2005), rate of environmental change (Wang and Kang 2005, Overgaard et al. 2006, Powell and Bale 2006), fluctuating conditions (Colinet et al. 2006, Wang et al. 2006, Lalouette et al. 2007, Hawes et al. 2008), repetitive exposure (Woods et al. 2001, Hawes 2007), age and ontogeny (Miyazaki et al. 2006, Terblanche and Chown 2006, Jensen et al. 2007, Pappas et al. 2007), season (Lee et al. 2006, Ma et al. 2006), diapause state (Izumi et al. 2005, Kayukawa et al. 2005, Teixeira and Polavarapu 2005, Cho et al. 2007), and interactions among multiple, simultaneous environmental factors (Hernandez et al. 2006, Kandori et al. 2006, Tomcala et al. 2006).

Recent studies have uncovered new physiological mechanisms and pathways (Gorr 2004, Michaud and Denlinger 2004, Storey 2004, Marjanovic et al. 2005, Marden 2008), and consequences of acclimation (Gulevsky et al. 2006, Kim et al. 2006, Sonoda et al. 2006, Lalovette et al. 2007, Tsai and Lin 2007). For example, reports continue to suggest an important role for insect hormones in stress response and acclimation (Gruntenko et al. 2000a,b). Both cold and heat acclimation are much more complicated than previously thought (Bale 2002). Some Collembola dramatically lower supercooling points by actively dehydrating (Worland et al. 1998). Other insects may track daily temperature cycles by constantly resetting their thermal thresholds, (Kelty and Lee 2001, Powell et al. 2004). Locust hoppers become cold hardened after only 2 h of cold shock, greatly increasing their survival at –7 °C; however, they quickly lose this protective physiology if returned to 30°C for 2 h (Wang and Kang 2003). Long-term acclimation is both differentiated from, and related to, rapid cold-hardening and rapid heat shock (Hoffmann et al. 2003, Sinclair and Chown 2003, Bowler 2005, Loeschcke and Sørensen 2005, Powell and Bale 2005, Sinclair and Roberts 2005). In *Drosophila*, thermal acclimation influences ethanol tolerance by changing membrane lipids (Montouth et al. 2006), in *Heliocoverpa* caterpillars it influences susceptibility to pathogens (Chandrashekar et al.

2005), and in *Cherax* crayfish, thermal acclimation improves fighting ability (Seebacher and Wilson 2006).

Researchers are just beginning to trace the temporal patterns of gene, proteome, and metabolome expression during and following acclimation (Hayward et al. 2005, Sorensen et al. 2005, Malmendal et al. 2006, Sonoda et al. 2007), but in the next few years, we will have compete pathways, including their temporal expression, for some acclimations. This will help answer whether acclimation is top-down (centrally controlled via hormones or the CNS), vs. bottom-up (i.e., produced via an aggregate of individual biochemical and cell responses) (Bowler 2005). Cuculescu et al. (1999) and Pearson et al. (1999) employed a clever method to test central vs. local control of acclimation (Fig. 12). By differentially heating different body regions, they showed that thermal acclimation in crabs was relatively independent of both CNS and endocrine control. Indeed, isolated cells can acclimate (Schmidt et al. 1984).

There is accelerated effort in trying to understand the hierarchal physiological relationships that produce acclimation (Garland and Kelly 2006, Gracey 2007, Wittkopp 2007) – how numerous lower-level transcriptional, translational, and enzymatic responses combine to shape complex traits, such as metabolic rate, growth rate, body size, etc., which

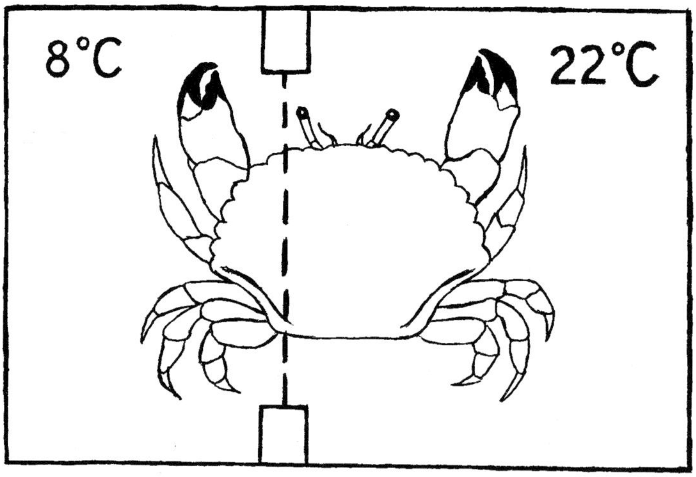

Fig. 12 Apparatus to produce heterothermal acclimation. Crab is suspended by a tight-fitting rubber diaphragm (dotted line) between two chambers of different temperature. Diaphragm is positioned such that the CNS and most of the endocrine system is on one side. Pearson et al. (1999) and Cuculescu et al. (1999) used this method to determine that thermal acclimation in crabs was under local, not central, control.

determine life history. This is important, because genes and gene × environmental interaction control the lower-level processes, yet, natural selection may occur primarily on complex traits (Swallow et al. 2005), partially because complex traits may be more closely related to fitness (Garland and Kelly 2007). Because complex traits are aggregates of dozens of underlying, basic traits, which are themselves controlled by hundreds, if not thousands of genes, selection on a single complex trait must influence many genes. Furthermore, natural selection may act on numerous complex traits simultaneously (Arnold 2003, Ghalambor et al. 2003, Sinervo and Calsbeek 2003). Thus there is a great need to understand hierarchical interactions, including pleiotrophy and epistasis (Swallow et al. 2005). How is integrative physiology maintained under change? Adaptive analyses of acclimation are difficult, when each level of biological organization has its own tradeoffs and constraints, and when there are incalculable, unrecognized interactions (Woods and Harrison 2001, Angilletta et al. 2003).

Numerous researchers are exploring the broader ecological consequences of acclimation – how acclimation influences abundance, distribution, and interspecific interactions (Slabber et al. 2007, Terblanche et al. 2007). Acclimation may allow individuals to enter new niches or expand geographic ranges. Plasticity can determine the strength and direction of ecological interactions, including both con- and allospecific interactions (Fordyce 2006). Acclimation has traditionally emphasized abiotic factors, but individuals may also acclimate to biotic factors, such as other organisms or organism-influenced factors. An exciting new idea concerns reciprocal plastic responses, where two individuals interact reciprocally to one another (Agrawal 2001, Fordyce 2006). We would expect to observe reciprocal acclimation among symbiotic organisms such as mutualists or parasitoids and hosts.

Overall, acclimation research is becoming increasingly integrated, combining genetic, molecular, endocrine, systems physiology, ecological, phyletic, biogeographical, and evolutionary studies, and each benefits from the other (Chown and Nicholson 2004, Loeschcke et al. 2004, Tian et al. 2004, Kellett et al. 2005, Angilletta et al. 2006, Almaas 2007). To a certain extent, acclimation has been subsumed into phenotypic plasticity and phenotypic plasticity into stress studies (Bijlsma and Loeschcke 2005). There is greater awareness of interaction effects and the connections, tradeoffs and constraints among different functional modules (Angilletta et al. 2003, Korsloot et al. 2004, Gimenez 2006, Rako and Hoffmann 2006, Quackenbush 2007). For example, behavioral plasticity in choice, can place animals into different habitats (climatic, nutrition, etc.) which then influences their subsequent physiology (Price et al. 2003, Garland and Kelly 2006). Swallow et al. (2005) propose "self-induced adaptive plasticity" for cases where a plastic behavior results in a physiological plastic change, which subsequently feeds back to enhance the ability to perform that behavior. Examples would be when an animal switches habitat preferences, subjecting it to a different temperature, salinity, O_2 level, etc., or switches diet. The former might lead to beneficial acclimation to those stressful environmental factors, and the latter might (1) alter preference, which increases feeding, and (2) induce new digestive enzymes that increase tolerance to dietary toxins, resulting in greater fitness for those insects with greater behavioral plasticity (Agrawal et al. 2002). In the above cases, physiological acclimation cannot be separated from behavioral plasticity.

Because acclimation is now accepted as phenotypic plasticity, it is now included in modeling and theoretical studies of plasticity evolution (Gabriel 2005, Borenstein et al. 2006, Garland and Kelly 2006). A growing literature demonstrates the genetic basis of acclimation, its heritability, and its ability to be selected (Harshman et al. 1991, Cavicchi et al. 1995, Krebs and Loeschcke 1996, Loeschcke and Krebs 1996, Lerman et al. 2001, Scheiner 2002, Hoffmann et al. 2003, Wang and Kang 2005). Plasticity may preserve genetic diversity (Lagerspetz 2006) or drive evolution (Schlichting and Smith 2002, Price et al. 2003), and directional selection on a trait mean can increase plasticity (Waddington 1959, Garland and Kelly 2006).

Theoreticians and empiricists are examining the broad patterns that favor the evolution of acclimation vs. other types of homeostatic mechanisms. Clearly, rapid dramatic environmental changes require rapid homeostatic adjustments. Slow-changing marine and aquatic environments might favor the evolution of enzymatic acclimation whereas rapid-changing land environments might favor behavioral plasticity and acute physiological homeostasis. Land is more spatially variable, which allows individuals to rapidly move among habitats, therefore favoring behavioral responses, such as microhabitat shifts, thermoregulation, shelter building, migration, etc., over physiological responses. In contrast, many deep caves and marine environments are thermal- and osmo-stable, and, hence, there might be no need for acclimation (Willmer et al. 2005). Likewise, length of stress periods (Gabriel 2005, Garland and Kelly 2006), cue reliability and predictability of environmental changes (Deere and Chown 2006, Deere et al. 2006), latitude (Ishiguro et al. 2007), altitude (Sorensen et al. 2005), lifespan (Lee et al. 2006), and phylogenetic constraints (Deere and Chown 2006, Terblanche et al. 2007), should all influence evolution of acclimation.

Global warming appears to already be selecting for altered acclimatory responses (Bradshaw and Holzapfel 2006). There are multiple mechanistic pathways to achieve the same functional endpoint. Understanding why one species evolves one solution and another species a different solution is an important focus of current evolutionary physiology (Angilletta et al. 2003).

Scientists have come to realize that acclimation represents just one segment along a gradation of available options that individuals can employ to counter the negative effects of habitat variation. These options include behaviors, acute physiological adjustments, longer term acclimations, and developmental switches (Chown and Nicolson 2004). Parents can also influence offspring physiology via trans-generational effects. Finally, natural selection acts on altered phenotypes, favoring those with genomes that produce beneficial physiological responses to environmental variation. These various responses grade into one-another and interact, blurring divisions (Price et al. 2003). In sum, acclimation is just one of many adaptation strategies, all of which are, more-or-less, interconnected.

Acclimation, of course, has a practical, applied side. Understanding insect acclimation may allow us to predict species ranges (Chen and Kang 2004, Klok and Chown 2005, Régnière and Bentz 2007) and establishment of invasive species (Bale 2002, Slabber et al. 2007). It can also aid pest control (Bean et al. 2007, Jagdale and Grewal 2007, Jian et al. 2007, Kaliyan et al. 2007, Luczynski et al. 2007, Pereira et al. 2007, Terblanche et al. 2008), and ecological monitoring. We may be able to assess relative levels of environmental stress, such as pollutants, by measuring the degree of stress acclimation in insects (De Coen et al. 2006, Lee et al. 2006). Understanding of acclimation may help us predict species responses to changes in Earth's ecology, such as global warming (Bradshaw and Holzapfel 2006, 2008, Gienapp et al. 2008). In the future, we will select or design organisms for economically beneficial acclimation (Collier et al. 2006). Acclimation is also important in animal husbandry and dairy, poultry, and fisheries production (Collier et al. 2006). Finally, a general knowledge of acclimation helps us understand the fascinating topic of human acclimation (Beal 2001, Bae et al. 2006, Flück 2006, Geurts et al. 2006, Swynghedauw 2006, Asea and DeMaio 2007, Calderwood et al. 2007, Radakovic et al. 2007), including human stress and disease response, and pharmacology (Nedeau and Topol 2006).

Lastly, returning to the theme that current researchers can learn from past acclimation studies, I end with a caution from Mellanby (1940b). Although he addressed temperature, his warning is apropos for nearly all environmental factors: *At every temperature, some acclimatization is probably going on in a living tissue, and the extent of the acclimation depends partly on the length of the exposure. This appears to me to mean that at any particular temperature there may be no absolute rate for a biological process, and that all conditions previously experienced by the animal must be considered.*

Acknowledgements

I thank Rachel Bowden and Ken Bowler for editorial comments, and Leigh Whitman and Ebony Murrell for figures and illustrations. Supported by NSF CROI grant DBI–0442412.

References

Agrawal AA. 2001. Phenotypic plasticity in the interactions and evolution of species. Science 294: 321–326.

Agrawal AA, Vala F and Sabelis MW. 2002. Induction of preference and performance after acclimation to novel hosts in a phytophagous spider mite: adaptive plasticity? American Naturalist 159: 553–565.

Agrell I. 1947. Some experiments concerning thermal adjustments and respiratory metabolism in insects. Arkiv. Zool., Stockholm 39A: 1–48.

Albrecht FO. 1964. Etat hygrométrique, coloration et résistance chez l'imago de *Locusta migratoria migratorioides* R. and F. Experientia 20: 1–3.

Albrecht FO. 1965. Influence du groupement, de l'état hygrométrique et de la photopériode sur la résistance au jeûne de *Locusta migratoria migratorioides* R. et F. (Orthoptére acridien). Bull. biol. Fr. Beig. 99: 287–339.

Alloway TM. 1969. Effects of low temperature upon acquisition and retention in the grain beetle, *Tenebrio molitor*. J. Comp. Physiol. Psychol. 69: 1–8.

Almaas E. 2007. Biological impacts and context of network theory. J. Exp. Biol. 210: 1548–1558.

Anderson RL and Mutchmor JA. 1971. Temperature acclimation in *Tribolium* and *Musca* at locomotory, metabolic and enzyme levels. J. Insect Physiol. 17: 2205–2219.

Angilletta MJ, Wilson RS, Navas CA and James RS. 2003. Tradeoffs and the evolution of thermal reaction norms. Trends in Ecology and Evolution 18: 234–240.

Angilletta MJ, Bennett AF, Guderley H, Navas CA, Seebacher F and Wilson RS. 2006. Coadaptation: A unifying principle in evolutionary thermal biology. Physiological and Biochemical Zoology 79: 282–294.

Applebaum SW, Janković M, Grozadanović J, and Marinković D. 1964. Compensation for temperature in the digestive metabolism of *Tenebrio molitor* larvae. Physiol. Zool. 37: 90–95.

Asea AAA and DeMaio A (eds). 2007. Heat Shock Proteins: Potent Mediators of Inflammation and Immunity. Springer, New York.

Atwal AS. 1960a. Influence of temperature, duration of conditioning, and age of *Anagasta* (*Ephestia*) *kuhniella* (Zell.) (Lepidoptera:Pyralidae) on acclimation to a sub-zero temperature. Can. J. Zool. 38: 131–141.

Atwal AS. 1960b. Influence of temperature and duration of conditioning on oxygen consumption and specific gravity of the haemolymph of *Anagasta* (*Ephestia*) *kuhniella* (Zell.) (Lepidoptera:Pyralidae). Can. J. Zool. 38: 143–148.

Atzler M. 1930. Untersuchungen uber den morphologischen und physiologischen Farbwechsel von *Dixippus* (*Carausius*) *morosus*. Z. vergl. Physiol., Berlin 13: 505–533.

Bae JS, Lee JB, Matsumoto T, Othman T, Min YK and Yang HM. 2006. Prolonged residence of temperate natives in the tropics produces a suppression of sweating. Pflugers Archiv – European Journal of Physiology 453: 67–72.

Baggerman B. 1957. An experimental study of the timing of breeding and migration in the three-spined stickleback (*Gasterosteus aculeatus* L.). Arch. neerl. zool. 12: 105–318.

Baldwin JM. 1896. A new factor in evolution. American Naturalist 30: 441–451.

Baldwin JM. 1902. Development and Evolution. Macmillan, New York.

Baldwin J and Hochacka PW. 1970. Functional significance of isoenzymes in thermal acclimation: acetylcholinesterase from trout brain. Biochem. J. 116: 883–887.

Baldwin WF. 1954. Acclimation and lethal high temperatures for a parasitic insect. Canadian J. Zool. 32: 157–171.

Baldwin WF and House HL. 1954. Studies on effect of thermal conditioning in two species of sawfly larvae. Canadian J. Zool. 32: 9–15.

Baldwin WF and Riordan DF. 1956. Acclimation times in *Dahlbominus fuscipennis* (Zett.) (Hymenoptera: Chalcidoidea). Canadian J. Zool. 34: 565–567.

Bale JS. 2002. Insects and low temperatures: from molecular biology to distributions and abundance. Phil. Trans. R. Soc. Lond. B 357: 849–862.

Bardenfleth KS and Ege R. 1916. On the anatomy and physiology of the air-sacs of the larva of *Corethra plumicornis*. Kobenhavn Nath. Medd. 67: 25–42.

Barrett T, Troup DB, Wilhite SE, Ledoux P, Rudnev D, Evangelista C, Kim IF, Soboleva A, Tomashevsky M and Edgar R. 2007. NCBI GEO: mining tens of millions of expression profiles – database and tools update. Nucleic Acids Res. 35: D760–D765.

Bateman PW, Verburgt L and Ferguson JWH. 2005. Exposure to male song increases rate of egg development in the cricket *Gryllodes sigillatus*. African Zoology 40: 323–326.

Bayley M, Petersen SO, Knigge T, Kohler H-R and Holmstrup M. 2001. Drought acclimation confers cold tolerance in the soil collembolan *Folsomia candida*. J. Insect Physiol. 47: 1197–1204.

Beal CM. 2001. Adaptations to altitude: a current assessment. Annu. Rev. Anthropol. 30: 423–456.

Bean DW, Dudley TL and Keller JC. 2007. Seasonal timing of diapause induction limits the effective range of *Diorhabda elongata deserticola* (Coleoptera: Chrysomelidae) as a biological control agent for tamarisk (*Tamarisk* spp.). Environ. Entomol. 36: 15–25.

Beck SD (ed.). 1980. Insect Photoperiodism. Academic Press, New York.

Bělehrádek J. 1935. Temperature and Living Matter. Protoplasma Monographien (Borntraeger, Berlin) 8: 1–277.

Benthe HF. 1954. Über die Temperaturabhängigkeit neuromuskulärer Vorgänge. Z. vergl. Physiol. 36: 327–351.

Berenbaum MR. 2002. Postgenomic chemical ecology: from genetic code to ecological interactions. J. Chem. Ecol. 28: 873–869.

Bijlsma R and Loeschcket V. 2005. Environmental stress, adaptation and evolution: an overview. J. Evol. Biol. 18: 744–749

Bodenheimer FS. 1931. Die Beziehung der Vorzugstemperatur zur Luftfeuchtigkeit der Umgebung. Z. vergl. Physiol. 13: 740–747.

Bodenheimer FS and Klein HZ. 1930. Uber die Temperatur Abhangigkeiten von Insekten. II. Die abhangigkeit der aktivitat bei der Ernteameise, *Wessor semirufus*, E. Andre von Temperatur und anderen Faktoren. Z. vergl. Physiol. 11: 345–385.

Bodenheimer FS and Schenkin D. 1928. Die temperaturabhängigkeiten einiger Insekten. Z. vergl. Physiol. 8: 1–15.

Borenstein E, Meilijson I and Ruppin E. 2006. The effect of phenotypic plasticity on evolution in multipeaked fitness landscapes. Journal of Evolutionary Biology 19: 1555–1570.

Bowen ST, Lebherz HG, Poon M, Chow VHS and Grigliatti TA. 1969. The hemoglobins of *Artemia salina* I. Determination of phenotype by genotype and environment. Comp. Biochem. Physiol. 31: 733–747.

Bowler K. 1963. A study of the factors involved in acclimatization to temperature and death at high temperatures in *Astacus pallipes*. I. Experiments on intact animals. J Cell Comp. Physiol. 62: 119–146.

Bowler K. 1967. Changes in temperature tolerance with adult age in *Tenebrio molitor*. Ent. exp. Appl. 10: 16–22.

Bowler K. 2005. Acclimation, heat shock and hardening. J. Thermal Biology 30: 125–130.

Bowler K and Hollingsworth MJ. 1965. The effect of inbreeding on temperature acclimatizaton in *Drosophila subobscura*. Genet. Res. 6: 1–12.

Bradshaw AD. 1965. Evolutionary significance of phenotypic plasticity in plants. Advances in Genetics 13: 115–155.

Bradshaw WE and Holzapfel CM. 2006. Evolutionary response to rapid climate change. Science 312: 1477–1478.

Bradshaw WE and Holzapfel CM. 2008. Genetic response to rapid climate change: it's seasonal timing that matters. Molecular Ecology 17: 157–166.

Breitenbrecher JK. 1918. The relation of water to the behavior of the potato beetle in a desert. Pub. Carnegie Inst. Washington 263: 343–348.

Brett JR. 1946. Some lethal temperature relations of Algonquin Park fishes. Univ. Toronto Studies in Biology no. 53: 5–49.

Brett JR. 1971. Energetic responses of salmon to temperature. A study of some thermal relations in the physiology and freshwater ecology of sockeye salmon (*Oncorhynchus nerkd*). Amer. Zool. 11: 99–113.

Buckley BA, Gracey AY and Somero GN. 2006. The cellular response to heat stress in the goby *Gillichthys mirabilis*: a cDNA microarray and protein-level analysis. J. Exp. Biol. 209: 2660–2677.

Buffington JD. 1969. Temperature acclimation of respiration in *Culex pipiens* (Diptera: Culicidae) and the influence of seasonal selection. Comp. Biochem. Physiol. 30: 865–878.

Buhler A, Lanzrein B and Wille H. 1983. Influence of temperature and carbon dioxide concentration on juvenile hormone titre and dependent parameters of adult worker honey-bees (*Apis milifera* L.). J. Insect Physiol. 29: 885–893.

Burnett GC. 1957. The relation between age and the cold resistance of tsetse flies and the value of chilling when transporting tsetse for experiments. Proc. R. ent. Soc. Lond. Ser. A. 32: 53–58.

Bullock TH. 1955. Compensation for temperature in the metabolism and activity of poikilotherms. Biological Reviews 30: 311–342.

Burtt EH Jr. 1981. The adaptiveness of animal colors. Bioscience 31: 723–729.

Buxton PA. 1924. Heat, moisture, and animal life in the deserts. Proc. R. Soc. London, B. 96: 123–131.

Calderwood SK, Sherman, MY and Ciocca DR (eds). 2007. Heat Shock Proteins in Cancer. Springer, New York.

Campbell RE. 1937. Temperature and moisture preferences of wireworms. Ecology 18: 479–489.

Carlsen H. 1953. Der cocarboxylasegehalt des Aales (*Anguilla vulgaris* L.). Journal of Comparative Physiology A: Neuroethology, Sensory, Neural, and Behavioral Physiology 35: 199–208.

Carlsen H. 1953. Uber Ruhestadien erwachsener Insekten. II. Aneurenbestimmungen bei Kartoffelkäfern (*Leptinotarsa decemlineata* Say). Journal of Comparative Physiology A: Neuroethology, Sensory, Neural, and Behavioral Physiology 35: 344–352.

Cavicchi S, Guerra D, La Torre V and Huey RB. 1995. Chromosomal analysis of heat-shock tolerance in *Drosophilia melanogaster* evolving at different temperatures in the laboratory. Evolution 49: 676–684.

Chandrashekar K, Kumari A, Kalia V and Gujar GT. 2005. Baseline susceptibility of the American bollworm, *Helicoverpa armigera* (Hubner) to *Bacillus thuringiensis* Berl var. *kurstaki* and its endotoxins in India. Current Science 88: 167–175.

Chapman RF. 1957. Observations on the feeding of adults of the red locust (*Nomadacris septemfasciata* (Serville). Brit. J. Anim. Behav. 5: 60–73.

Chatfield PO, Lyman CP and Irving L. 1953. Physiological adaptation to cold of peripheral nerve in the leg of the herring gull (*Larus argentatus*). American J. Physiology 172: 639–644.

Chen B and Kang L. 2004. Variation in cold hardiness of *Liriomyza huidobrensis* (Diptera: Agromyzidae) along latitudinal gradients. Environ. Entomol. 33: 155–164.

Chen C-P, Lee RE and Denlinger DL. 1991. Cold shock and heat shock: a comparison of the protection generated by brief pretreatment at less severe temperatures. Physiol. Entomol. 16: 19–26.

Cho JR, Lee M, Kim HS, Kim Y and Boo KS. 2007. Cold hardiness in the black rice bug, *Scotinophara lurida*. Physiol. Entomol. 32: 167–174.

Chown SL and Nicolson SW (eds). 2004. Insect Physiological Ecology Mechanisms and Patterns. Oxford University Press, Oxford.

Cianfrogna JA, Zangler AR and Berenbaum MR. 2002. Dietary and developmental influences on induced detoxification in an oligophage. J. Chem. Ecol. 28: 1349–1364.

Clarke A. 1993. Seasonal acclimatization and latitudinal compensation in metabolism: do they exist? Functional Ecology 7: 139–149.

Clark KU. 1966. Histological changes in the endocrine system of *Locusta migratoria* L. associated with the growth of the adult under different temperature regimes. J. Insect Physiol. 12: 163–170.

Clark KU. 1967. Insects and temperature. Pp. 293–352 in Rose AH (ed.). Thermobiology. Academic, London.

Clark TA, Flis BJ and Remold SK. 2004. pH tolerances and regulatory abilities of freshwater and euryhaline Aedine mosquito larvae. J. Exp. Biol. 207: 2297–2304.

Clausen J, Keck D and Hiesey WM 1940. Experimental studies on the nature of plant species. I. Effect of varied environment on western North American plants. Carnegie Institution, Washington, DC.

Clever U. 1961. Genaktivitaten in den Riesenchro mosomen von *Chironomus tentans* und ihre Beziehungen zur Entwicklung. I. Genaktivierungen durch Ecdyson. Chromosoma, Berlin 12: 607–675.

Colhoun EH. 1954. Temperature acclimatization in insects. Nature 173–582.

Colhoun EH. 1960. Acclimation to cold in insects. Entomologia exp. appl. 3: 27–37.

Colinet H, Renault D, Hance T and Vernon P. 2006. The impact of fluctuating thermal regimes on the survival of a cold-exposed parasitic wasp, *Aphidius colemani*. Physiological Entomology 31: 234–240.

Collier RJ, Dahl GE and VanBaale MJ. 2006. Major advances associated with environmental effects on dairy cattle. Journal of Dairy Science 89: 1244–1253.

Cossins A, Fraser J, Hughes M and Gracey A. 2006. Post-genomic approaches to understanding the mechanisms of environmentally induced phenotypic plasticity. J. Exp. Biol. 209: 2328–2336.

Coulson SJ and Bale JS. 1991. Anoxia induces rapid cold hardening in the housefly *Musca domestica* (Diptera, Muscidae). J. Insect Physiol. 37: 497–501.

Crill WD, Huey RB and Gilchrist GW. 1996. Within- and between-generation effects of temperature on the morphology and physiology of *Drosophila melanogaster*. Evolution 50: 1205–1218.

Cuculescu M, Pearson T, Hyde D and Bowler K. 1999. Heterothermal acclimation: an experimental paradigm for studying the control of thermal acclimation in crabs. Proc. Natl. Acad. Sci. USA 96: 6501–6505.

Damant GCC. 1924. The adjustment of the buoyancy of the larva of *Corethra plumicornis*. J. Physiol. Cambridge 59: 345–356.

Danilevskii AS. 1965. Photoperiodism and Seasonal Development of Insects (English translation). Oliver and Boyd, Edingburgh.

Danks HV. 2005. Key themes in the study of seasonal adaptations in insects I. Patterns of cold hardiness. Applied Entomol. and Zool. 40: 199–211.

Danks SM and Tribe MA. 1979. Biochemical changes in blowfly flight muscle mitochondria following temperature acclimation. J. Therm. Biol. 4: 183–195.

Das AB and Singh SP. 1974. Thermal acclimation in the rate of endogenous oxygen consumption of the fat body and the coxal muscle of *Periplaneta americana* (Linn.). Environ. Physiol. Biochem. 4: 139–146.

Davenport CB and Castle WE. 1895. Studies in morphogenesis. III. On the acclimatization of organisms to high temperatures. Arch. Entwickelungsmechanik der Organismen 2: 227–249.

Davison TF. 1969. Changes in tolerance during the life cycle of *Calliphora erythrocephala*. J. Insect Phys. 15: 977–988.

Davison TF. 1971. The effect of temperature on oxidative phosphorylation in isolated flight muscle sarcosomes. Comp. Biochem. Physiol. 38B: 21–34.

Davison TF. 1971. The relationship between age, acclimatization temperature, and heat death point in adult *Calliphora erythrocephala*. J. Insect Phys. 17: 575–585.

Dean JM and Vernberg FJ. 1965. Variations in the blood glucose level of Crustacea. Comp. Biochem. Physiol. 14: 29–34.

Dean JM and Vernberg FJ. 1966. Hypothermia and blood of crabs. Comp. Biochem. Physiol. 17: 19–22.

De Coen W, Robbens J and Janssen C. 2006. Ecological impact assessment of metallurgic effluents using in situ biomarker assays. Environmental Pollution 141: 283–294.

Deere JA and Chown SL. 2006. Testing the beneficial acclimation hypothesis and its alternatives for locomotor performance. American Naturalist 168: 630–644.

Deere JA, Sinclair BJ, Marshall DJ and Chown SL. 2006. Phenotypic plasticity of thermal tolerances in five oribatid mite species from sub-Antarctic Marion Island. Journal of Insect Physiology 52: 693–700.

Dehnel PA. 1960. Effect of temperature and salinity on the oxygen consumption of two intertidal crabs. Biol. Bull. 118: 215–249.

Dehnel PA and Segal E. 1956. Acclimation of oxygen consumption to temperature in the American cockroach (*Periplaneta americana*). Biol. Bull. 111: 53–61.

Dick J. 1937. Oviposition in certain Coleoptera. Ann. Appl. Biol. 24: 762–796.

Dobzhansky T (ed.). 1937. Genetics and the Origin of Species. Columbia University Press, New York.

Dow JAT. 2007. Integrative physiology, functional genomics and the phenotype gap: a guide for comparative physiologists. J. Exp. Biol. 210: 1632–1640.

Dryer WA. 1932. The effect of hibernation and seasonal variation of temperature on the respiratory exchange of *Formica ulkei* Emery. Physiol. Zool. 5: 301–331.

Edwards DK. 1957. Temperature-respiration curve of flour beetles exposed to nonoptimal temperatures. Science 125: 651–652.

Edwards DK. 1958. Effects of acclimatization and sex on respiration and thermal resistance in *Tribolium* (Coleoptera: Tenebrionidae). Can. J. Zool. 36: 363–382

Elens AA. 1953. Agricultura (Louvain) 1: 19–32.

Engels W, Goncalves LS and Engels E. 1976. Effects of carbon dioxide on vitellogenin metabolism in unmated queen honeybees. J. Apic. Res. 15: 3–10.

Ergene S. 1950. Untersuchungen über Farbanpassung und Farbwechesel bei *Acrida turrita*. Z. vergl. Physiol. 32: 530–551.

Ergene S. 1952. Farbanpassung entsprechend der jeweiligen Substratfärbung bei *Acrida turrita*. Z. vergl. Physiol. 34: 69–74.

Ergene S. 1953. Homochrome Farbanpassungen bei *Mantis religiosa*. Z. vergl. Physiol. 35: 36–41.

Ergene S. 1954. Homochromer Farbwechsel bei *Oedaleus*-Imagines. Zoo. Anzeiger 153: 110–113.

Ergene S. 1955. Über die Faktoren die Grünfärbung bei *Acrida* bedingen. Z. vergl. Physiol. 37: 221–225.

Ergene S. 1956. Homochromer Farbwechsel bei geblendeten Oedaleus decorus. Z. vergl. Physiol. 38: 311–316.

Falconer DS. 1989. Introduction to Quantitative Genetics. Longman, London.

Farrow RA. 1975. The African migratory locust in its main outbreak area of the Middle Niger: quantitative studies of solitary populations in relation to environmental factors. Locusta 11: 1–198.

Fast PG. 1970. Insect lipids. Pp. 181–242 in Holman RT (ed). Progress in the Chemistry of Fats and Other Lipids. Vol. II, Part 2. Pergamon, New York.

Faure JC. 1932. The phases of locusts in South Africa. Bull. Ent. Res. 23: 293–424.

Faure JC. 1933. The phases of the Rocky Mountain Locust *Melanoplus mexicanus* (Saussure). J. Econ. Ent. 26: 706–718.

Feder ME. 1997. Necrotic fruit: a novel model system for thermal ecologists. J. Thermal Biol. 22: 1–9.

Feder ME and Hofmann GE. 1999. Heat-shock proteins, molecular chaperones, and the stress response: evolutionary and ecological physiology. Ann. Rev. Physiol. 61: 243–282.

Feder ME and Walser JC. 2005. The biological limitations of transcriptomics in elucidating stress and stress responses. J. Evol. Biol. 18: 901–910.

Feder ME, Blair N and Figueras H. 1997. Natural thermal stress and heat-shock protein expression in *Drosophila* larvae and pupae. Functional Ecol. 11: 90–100.

Fischer K, Eenhoorn E, Bot ANM, Brakefield PM and Zwaan BJ. 2003. Cooler butterflies lay larger eggs: developmental plasticity versus acclimation. Proc. R. Soc. Biol. Sci. Series B. 270: 2051–2056.

Flück M. 2006. Functional, structural and molecular plasticity of mammalian skeletal muscle in response to exercise stimuli. J. Exp. Biol. 209: 2239–2248.

Fordyce JA. 2006. The evolutionary consequences of ecological interactions mediated through phenotypic plasticity. J. Exp. Biol. 209: 2377–2383.

Fox HM. 1936. The activity and metabolism of poikilothermal animals in different latitudes. I. Proc. zool. Soc. Lond for 1936 (1937): 945–955.

Fox HM. 1939. The activity and metabolism of poikilothermal animals in different latitudes. V. Proc. zool. Soc. London A 109: 141–156.

Fox HM. 1955. The effect of oxygen on the concentration of haem in invertebrates. Proc. Roy. Soc. Lond. B. 143: 203–214.

Fox HM and Taylor AER. 1955. The tolerance of oxygen by aquatic invertebrates. Proc. Roy. Soc. B. 143: 214–225.

Fraenkel GS and Gunn DL. 1961. The Orientation of Animals. Dover, New York.

Fraenkel G and Hopf HS. 1940. The physiological action of abnormally high temperatures on poikilothermic animals. I. Temperature adaptation and the degree of saturation of the phosphates. Biochem. J., Cambridge 34: 1085–1092.

Free JB and Spencer-Booth Y. 1960. Chill-coma and cold death temperatures of *Apis mellifera*. Entomol. Exp. Appl. 3: 222–230.

Fry FEJ. 1958. Temperature compensation. Annu. Rev. Physiol. 20: 207–224.

Fry FEJ and JS Hart. 1948. Cruising speed of goldfish in relation to water temperature. J. Fish. Res. Board Canada 7: 169–175.

Fuzeau-Braesch S. 1960. Etude biologique et biochimique de la pigmentation d'un insecte, *Gryllus bimaculatus* de Geer (Gryllidae, Orthoptère). Bull. bioil. Fr. Belg. 94: 525–627.

Gabriel W. 2005. How stress selects for reversible phenotypic plasticity. J. Evol. Biol. 18: 873–883.

Gade G. 2004. Regulation of intermediary metabolism and water balance of insects by neuropeptides. Annu. Rev. Entomol. 49: 93–113.

Gambaro P. 1954. L'importanza del fattore "alimentazione" nella determinazione della diapausa di *Carpocapsa pomonella* L. Bull. Zool., Naples 21: 163–169.

Garland T Jr and Adolph ST. 1991. Physiological differentiation of vertebrate populations. Annu. Rev. Ecol. Syst. 22: 193–228.

Garland T Jr and Kelly SA. 2006. Phenotypic plasticity and experimental evolution. J. Exp. Biol. 209: 2344–2361.

Garside ET and Tait JS. 1958. Preferred temperature of rainbow trout (*Salmo gairdneri* Richardson) and its unusual relationship to acclimation temperature. Canadian J. Zool. 36: 563–567.

Geller E (ed). 2003. McGraw-Hill Dictionary of Scientific and Technical Terms. McGraw-Hill, New York.

Geurts CLM, Sleivert GG and Cheung SS. 2006. Local cold acclimation during exercise and its effect on neuromuscular function of the hand. Applied Physiology, Nutrition, and Metabolism = Physiologie Appliquee, Nutrition et Metabolisme 31: 717–725.

Ghalambor CK, Walker JA and Reznick DN. 2003. Multi-trait selection, adaptation, and constraints on the evolution of burst swimming performance. Integr. Comp. Biol. 43: 431–438.

Gibbs AG and Mousseau TA 1994. Thermal acclimation and genetic variation in cuticular lipids of the lesser migratory grasshopper (*Melanoplus sanguinipes*); effects of lipid composition on biophysical properties. Physiological Zoology 67: 1523–1543.

Gienapp P, Teplitsky C, Alho JS, Mills JA and Merila J. 2008. Climate change and evolution: disentangling environmental and genetic responses. Molecular Ecology 17: 167–178.

Giersberg H. 1928. Über den morphologischen und physiologischen Farbwechsel der Stabheuschrecke *Dixippus* (*Carausius morsus*). Z. vergl. Physiol. 7: 657–695.

Giersberg H. 1929. Die Färbung der Schmetterlinge. I. Z. vergl. Physiol. 9: 523–552.

Gilbert P, Huey RB and Gilchrist GW. 2001. Locomotor performance of *Drosophilia melanogaster:* interactions among developmental and adult temperatures, age, and geography. Evolution 55: 205–209.

Gimenez L. 2006. Phenotypic links in complex life cycles: conclusions from studies with decapod crustaceans. Integrative and Comparative Biology 46: 615–622.

Goldschmidt RB (ed.). 1940. The Material Basis of Evolution. Yale University Press, New Haven.

Gordh G. 2001. A Dictionary of Entomology. CABI Publishing, New York.

Gorr TA. 2004. *Daphnia* and *Drosophila*: two invertebrate models for O_2 responsive and HIF-mediated regulation of genes and genomes. International Congress Series 1275: 55–62.

Goto SG and Kimura MT. 1998. Heat- and cold-shock responses and temperature adaptations in subtropical and temperate species of *Drosophila*. J. Insect Physiol. 44: 1233–1239.

Gracey AY. 2007. Interpreting physiological responses to environmental change through gene expression profiling. J. Exp. Biol. 209: 1584–1592.

Grainger JNR. 1958. First stages in the adaptation of poikilotherms to temperature change. Pp. 79-91 in Prosser CL (ed). Physiological Adaptation. American Physiological Soc. Washington, DC.

Gruntenko NE, Khlebodarova TM, Vasenkova IA, Sukhanova MJh, Wilson TG and Rauschenbach IY. 2000a. Stress-reactivity of a *Drosophila melanogaster* strain with impaired juvenile hormone action. J. Insect Physiol. 46: 451–456.

Gruntenko NE, Wilson TG, Monastirioti M and Rauschenbach IY. 2000a. Stress-reactivity and juvenile hormone degradation in *Drosophila melanogaster* strains having stress-related mutations. Insect Biochem and Molecular Biol. 30: 775–783.

Guan R and Wang WX. 2006a. Comparison between two clones of *Daphnia magna*: Effects of multigenerational cadmium exposure on toxicity, individual fitness, and biokinetics. Aquatic Toxicology 76: 217–229.

Guan R and Wang WX. 2006b. Multigenerational cadmium acclimation and biokinetics in *Daphnia magna*. Environmental Pollution 141: 343–352.

Gulevsky AK, Relina LI and Grishchenkova YA. 2006. Variations of the antioxidant system during development of the cold-tolerant beetle, *Tenebrio molitor*. Cryo-Letters 27: 283–290.

Gunn DL. 1931. Temperature and humidity relations of the cockroach. Nature 128: 186–187.

Gunn DL. 1934. Temperature and humidity relations of the cockroach. II. Temperature preference. Z. vergl. Physiol. 20: 617–625.

Gunn DL and Cosway CA. 1938. The temperature and humidity relations of the cockroach. V. Humidity preference. J. Exp. Biol. 15: 555–563.

Gunn DL and Hopf HS. 1942. The biology and behaviour of *Ptinus tectus* Boie. (Coleoptera: Ptinidae), a pest of stored products. II. The amount of locomotory activity in relation to experimental and to previous temperatures. J. Exp. Biol. 18: 278–289.

Gunn DL and Hunter-Jones P. 1952. Laboratory experiments on phase differences in locusts. Anti-Locust Bull., London 12: 1–29.

Haas H and Strenzke K.1957. Experimentelle Untersuchungen über den Einflus der ionalen Zusammensetzung des Mediums auf die Entwicklung der Analpapillen von *Chironomus thummi*. Biologisches Zentralblatt (Leipzig) 76: 513–528.

Hanec WJ. 1966. Cold-hardiness in the forest tent caterpillar, *Malacosoma disstria* Hübner (Lasiocampidae, Lepidoptera). J. Insect Physiology 12: 1443–1449.

Harnisch O. 1951. Hydrophysiologie der Tiere. Elemente zu ihren Aufbau. Binnengewasser. 19: 1–299.

Harrison J, Frazier MR, Henry JR, Kaiser A, Klok CJ and Rascon B. 2006. Responses of terrestrial insects to hypoxia or hyperoxia. Respiratory Physiology and Neurobiology 154: 4–17.

Harshman LG, Ottea JA and Hammock BD. 1991. Evolved environment-dependent expression of detoxication enzyme activity in *Drosophila melanogaster*. Evolution 45: 791–795.

Hart JS. 1952. Geographic variations of some physiological and morphological characters in certain fresh-water fish. Publ. Ont. Fish. Res. Lab. 721: 1–79.

Hawes TC. 2006. Re-crystallisation temperature (T-rc) in an Antarctic springtail. Cryo-Letters 27: 333–340.

Hawes TC, Bale JS, Worland MR and Convey P. 2007. Plasticity and superplasticity in the acclimation potential of the Antarctic mite *Halozetes belgicae* (Michael). Journal of Experimental Biology 210: 593–601.

Hawes TC, Bale JS, Worland MR and Convey P. 2008. Trade-offs between microhabitat selection and physiological plasticity in the Antarctic springtail, *Cryptopygus antarcticus* (Willem). Polar Biology 31: 681–689.

Hayward SAL, Pavlides SC, Tammariello SP, Rinehart JP and Denlinger DL. 2005. Temporal expression patterns of diapause-associated genes in flesh fly pupae form the onset of diapause through post-diapause quiescence. J. Insect Physiol. 51: 631–640.

Hazel J and Prosser CL. 1970. Interpretation of inverse acclimation to temperature. Z. vergl. Physiol. 67: 217–228.

Heeg J. 1967. Studies on Thysanura. II. Orientation reactions of *Machiloides delanyi* Wygodzinsky and *Ctenolepisma longicaudata* Escherich to temperature, light and atmospheric humidity. Zool. Afr. 3: 43–57.

Henderson B (ed.). 2005. Molecular Chaperones and Cell Signaling. Cambridge University Press, London.

Henke K. 1930. Die Lichtorientierung und die Bedingungen der Lichtstimmung bei *Armadillidium*. Z. vergl. Physiol. 13: 535–626.

Henry RJ and Harrison JF. 2004. Plastic and evolved responses of larval tracheae and mass to varying atmospheric oxygen content in *Drosophila melanogaster*. J. Exp. Biol. 207: 3559–3567.

Henry RP, Thomason KL and Towle DW. 2006. Quantitative changes in branchial carbonic anhydrase activity and expression in the euryhaline green crab, *Carcinus maenas*, in response to low salinity exposure. Journal of Experimental Zoology Part A-Comparative Experimental Biology 305A: 842–850.

Henschel J. 1929. Reizphysiologische Untersuchung der Kasemilbe *Tyrolichus casei* (Oudemans). Z. vergl. Physiol. 9: 802–837.

Hernandez RM, Buckle RF, Palacios E and Baron SB. 2006. Preferential behavior of white shrimp *Litopenaeus vannamei* (Boone 1931) by progressive temperature-salinity simultaneous interaction. Journal of Thermal Biology 31(7): 565–572.

Herrstroem G. 1949. Illumination of preferendum and adaptation experiments with *Agonum dorsale* Pont. (Col., Carabidae). Oikos 1: 48–55.

Herter K. 1924. Temperatursinn einiger Insekten. Z. vergl. Physiol. 1: 221–280.

Herter K. 1953. Der Temperatursinn der Insekten. Duncker & Humblot, Berlin.

Hiiesaar K, Kuusik A, Joudu J, Metspalu L and Hermann P. 2001. Laboratory experiments on cold acclimation in overwintering Colorado potato beetles, *Leptinotarsa decemlineata* (Say). Norwegian J. Entomol. 48: 87–90.

Hill L and Taylor HJ. 1933. Locusts in sunlight. Nature 132: 276.

Hill RW and Wyse GA. 1976. Animal Physiology. Harper and Row, New York.

Hoar WS and Robertson GB. 1959. Temperature resistance of goldfish maintained under controlled photoperiods. Canadian J. Zool. 37: 419–428.

Hoback WW and Stanley DW. 2001. Insects in hypoxia. J. Insect Physiol. 47: 533–542.

Hochachka PW. 1967. Organization of metabolism during temperature compensation. Pp. 177-203 in Prosser CL (ed.). Molecular Mechanisms of Temperature Adaptation. AAAS, Washington, DC.

Hochachka PW and Somero GN. 1973. Strategies of Biochemical Adaptation. Saunders, Philadelphia.

Hodkova M, Berkova P and Zahradnickova H. 2002. Photoperiodic regulation of the phospholipid molecular species composition in thoracic muscles and fat body of *Pyrrhocoris apterus* (Heteroptera) via an endocrine gland, corpus allatum. J. Insect Physiol. 48: 1009–1019.

Hoffman K. 1957. Angeborene Tages-periodik bei Eidechsen. Naturwiss. 44: 358–360.

Hoffmann AA. 1995. Acclimation: increasing survival at a cost. Trends in Ecology and Evolution 10: 1–2.

Hoffmann AA and Hewa-Kapuge S. 2000. Acclimation for heat resistance in *Trichogramma* nr. *brassicae*: can it occur without costs? Functional Ecology 14: 55–60.

Hoffmann KH and Stockmeier E. 1975. Wirkung von vollsynthetischer Nahrung und Temperature auf Gesamtfettsäuresektrum und Neutralfette von *Gryllus bimaculatus* (Orthopt.). Verh Dtsch. Zool. Ges. 1974: 286–291.

Hoffmann AA, Sorensen JG and Loeschcke V. 2003. Adaptation of *Drosophila* to temperature extremes: bringing together quantitative and molecular approaches. J. Thermal Biology 28: 175–216.

Hollingsworth MJ and Bowler K. 1966. The decline in ability to withstand high temperature with increase in age in *Drosophila subobscura*. Exp. Geront. 1: 251–257.

Holmstrup M, Hedlund K and Boriss H. 2002. Drought acclimation and lipid composition in *Folsomia candida*: implications for cold shock, heat shock and acute desiccation stress. J. Insect Physiol. 48: 961–970.

House HL, Riordan DF and Barlow JS. 1958. Effects of thermal conditioning and of degree of saturation of dietary lipids on resistance of an insect to a high temperature. Canadian J. Zool. 36: 629–632.

Houston AH. 1959. Osmoregulatory adaptation of steelhead trout (*Salmo gairdneri* Richardson) to sea water. Canadian J. Zool. 37: 729–748.

Huey RB and Berrigan D. 1996. Testing evolutionary hypotheses of acclimation. Pp. 205-237 in Johnston IA and Bennett AF (eds.). Animals and Temperature. Phenotypic and Evolutionary Adaptation. Cambridge University Press, London.

Huey RB, Wakefield T, Crill WD and Gilchrist GW. 1995. Within- and between-generation effects of temperature on early fecundity of *Drosophila melanogaster*. Heredity 74: 216–223.

Huey RB, Berrigan D, Gilchrist GW and Herron JC. 1999. Testing the adaptive significance of acclimation: a strong inference approach. Amer. Zool. 39: 323–336.

Hughes RD. 1960. Induction of diapause in *Erioishia brassicae* Bouché (Diptera: Anthomyidae). J. Exp. Biol. 37: 218–223.

Hunter AS. 1968. Effects of temperature on *Drosophila*. IV. Adaptation of *D. immigrans*. Comp. Biochem. Physiol. 37: 218–223.

Hunter-Jones P. 1958. Laboratory studies on the inheritance of phase characters in locusts. Anti-Locust Bull. 29: 1–32.

Hunter-Jones P. 1962. Coloration of the desert locust (*Schistocerca gregaria* Forskal) reared in isolation. Entomologist's Monthly Magazine 98: 89–92.

Husain MA and CB Mathur. 1936a. Studies on *Schistocerca gregaria* Forsk. VIII. Influence of carbon dioxide on development of black pigmentation in *Schistocerca gregaria* Forsk. Indian J. Agric. Sci. 6: 1005–1030.

Husain MA and CB Mathur. 1936b. Studies on *Schistocerca gregaria* Forsk. V. Pigmentation and physical exertion. Indian. J. Agric. Sci. 6: 591–623.

Ishiguro S, Li Y, Nakano K, Tsumuki H and Goto M. 2007. Seasonal changes in glycerol content and cold hardiness in two ecotypes of the rice stem borer, *Chilo suppressalis*, exposed to the environment in the Shonai district, Japan. J. Insect Physiol. 53: 392–397.

IUPS (The Commission for Thermal Physiology of the International Union of Physiological Sciences). 2001. Glossary for terms for thermal physiology. Japanese J. Physiol. 51: 245–280.

Iwao S. 1962. Studies on the phase variation and related phenomena in some lepidopterous insects. Mem. Coll. Agric. Kyoto 84: 1–80.

Izumi Y, Anniwaer K, Yoshida H, Sonoda S, Fujisaki K and Tsumuki H. 2005. Comparison of cold hardiness and sugar content between diapausing and nondiapausing pupae of the cotton bollworm, *Helicoverpa armigera* (Lepidoptera: Noctuidae). Physiol. Entomol. 30: 36–41.

Jack RW. 1939. Studies in the physiology and behaviour of *Glossina morsitans*, Westw Mem. Dept. Agric. N. I. S. Rhodesia. No. 1: 1–203.

Jagdale GB and Grewal PS. 2007. Storage temperature influences desiccation and ultra violet radiation tolerance of entomopathogenic nematodes. Journal of Thermal Biology 32: 20–27.

Jakovlev V and Krüger F. 1954. Untersuchungen über die Borzugstemperatur einiger Acrididen. Bilogisches Zentralblatt. 73: 633–650.

Jarecki J, Johnson E and Krasnow MA. 1999. Oxygen regulation of airway branching in *Drosophila* is mediated by Branchless FGF. Cell 99: 211–220.

Jensen D, Overgaard J and Sorensen JG. 2007. The influence of developmental stage on cold shock resistance and ability to cold-harden in *Drosophila melanogaster.* Journal of Insect Physiology 53: 179–186.

Jian FJ, Jayas DS and White NDG. 2005. Effects of temperature acclimation and age on movement of *Cryptolestes ferrugineus* (Coleoptera: Laemophloeidae) adults in response to temperature gradients. Can. Entomol. 137: 71–82.

Jian FJ, Jayas DS, White NDG and Fields PG. 2007. A distributed-delay model to predict ageing and survival rates of adults of *Cryptolestes ferrugineus* (Stephens) (Coleoptera : Laemophloeidae) in granaries filled with wheat. Ecological Modelling 200: 412–420.

Johannsen, W. 1911. The genotype conception of heredity. American Naturalist 45: 129–159.

Johnson FH, Eyring H and Stover BJ. 1974. The Theory of Rate Processes in Biology and Medicine. Wiley, New York.

Johnston IA and Bennett AF (eds.). 1996. Animals and Temperature. Cambridge University Press, London.

Jones ML. 1967. Morphology of nephridia of *Nereis*. Biol. Bull. 132: 362–380.

Jungreis AM and Hooper AB. 1968. Physiological properties of cold resistance adaptation in the fresh-water crayfish *Orconectes virilis*. Comp. Biochem. Physiol. 26: 91–100.

Kaestner H. 1931 Temperature and morphological color change in insects. Arch. Entw.-Mech. Org. 124: 1–16

Kaliyan N, Morey RV, Wilcke WF, Carrillo MA and Cannon CA. 2007. Low-temperature aeration to control Indian meal moth, *Plodia interpunctella* (Hubner), in stored grain in twelve locations in the United States: a simulation study. Journal of Stored Products Research 43: 177–192.

Kandori I, Kimura T, Tsumuki H and Sugimoto T. 2006. Cold tolerance of the sweet potato weevil, *Cylas formicarius* (Fabricius) (Coleoptera : Brentidae), from the southwestern islands of Japan. Applied Entomology and Zoology 41: 217–226.

Kayukawa T, Chen B, Miyazaki S, Itoyama K, Shinoda T and Ishikawa Y. 2005. Expression of mRNA for the t-complex polypeptide-1, a subunit of chaperonin CCT, is upregulated in association with increased cold hardiness in *Delia antiqua*. Cell Stress and Chaperones 10: 204–210.

Keister M and Buck J. 1961. Respiration of *Phormia regina* in relation to temperature and oxygen. J. Insect Physiol. 7: 51–72.

Keister M and Buck J. 1974. Respiration: some exogenous and endogenous effects on rate of respiration. Pp. 470-509 in Rockstein M (ed). The Physiology of Insecta. Vol VI. Academic Press, New York.

Kellett M, Hoffmann AA and McKechine SW. 2005. Hardening capacity in the *Drosophila melanogaster* species group is constrained by basal thermotolerance. Functional Ecology 19: 853–858.

Kelty JD and Lee RE. 1999. Induction of rapid cold hardening by cooling at ecologically relevant rates in *Drosophila melanogaster*. J. Insect Physiol. 45: 719–726.

Kelty JD and Lee RE Jr. 2001. Rapid cold-hardening of *Drosophila melanogaster* (Diptera: Drosophilidae) during ecologically based thermoperiodic cycles. J. Exp. Biol. 204: 1659–1666.

Kennedy JS. 1939. The behaviour of the desert locust (*Schistocerca gregaria* Forsk.) (Orthopt.) in an outbreak centre. Trans. Roy. Entomol. Soc. Lond. 89: 385–542.

Kerkut GA and Taylor BJR. 1957. A temperature receptor in the tarsus of the cockroach, *Periplaneta americana*. J. Exp. Biol. 34: 486–493.

Kerkut GA and Taylor BJR. 1958. The effect of temperature changes on activity of poikilotherms. Behavior. 13: 259–279.

Key KHL. 1954. The Taxonomy, Phases, and Distribution of the Genera *Chortoicetes* Brunn. and *Austroicetes* Uv. (Orthoptera: Acrididae). CSIRO, Canberra.

Key KHL and Day MF. 1954a. A temperature-controlled physiological colour response in the grasshopper, *Kosciuscola tristis* Sjöst. (Orthoptera, Acrididae). Australian J. Zool. 2: 309–339.

Key KHL and Day MF. 1954b. The physiological mechanism of colour change in the grasshopper, *Kosciuscola tristis* Sjöst. (Orthoptera, Acrididae). Australian J. Zool. 2: 340–363.

Kim M, Robich RM, Rinehart JP and Denlinger DL. 2006. Upregulation of two actin genes and redistribution of actin during diapause and cold stress in the northern house mosquito, *Culex pipiens*. Journal of Insect Physiology 52: 1226–1233.

Kim YG and Song WR. 2000. Effect of thermoperiod and photoperiod on cold tolerance of *Spodoptera exigua* (Lepidoptera: Noctuidae), Environ. Entomol. 29: 868–873.

Kingsolver JG. 1995. Fitness consequences of seasonal polyphenism in western white butterflies. Evolution 49: 942–954.

Kingsolver JG. 1996. Experimental manipulation of wing pigment pattern and survival in western white butterflies. American Naturalist 147: 296–306.

Kingsolver JG. 1999. Experimental manipulation of wing pigment pattern and survival in western white butterflies. American Naturalist 147: 296–306.

Kingsolver JG and Huey RB 1998. Evolutionary analyses of morphological and physiological plasticity in thermally variable environments. Amer. Zool. 38: 545–560.

Kinne O. 1958. Adaptations to salinity variations – some facts and problems. Pp. 92-106 in Prosser CL (ed). Physiological Adaptation. Amer. Physiol. Soc., Washington DC.

Kirberger C. 1953. Untersuchungen über die temperaturabhängigkeit von Lebensprozessen bei verschiedenen Wirbellosen. Z. vergl. Physiol. 35: 175–198.

Klok CJ and Chown SL. 2005. Inertia in physiological traits: *Embronopsis halticella* caterpillars (Yponomeutidae) across the Antarctic polar frontal zone. J. Insect Physiol. 51: 87–97.

Knight HH. 1924. On the nature of the color patterns in Heteroptera with data on the effects produced by temperature and humidity. Ann. Entom. Soc. Amer. 17: 258–272.

Knox WE. 1958. Adaptive enzymes in the regulation of animal metabolism. Pp. 107–125 in Prosser CL (ed). Physiological Adaptation. Amer. Physiol. Soc., Washington DC.

Knox WE, Auerbach VH and Lin ECC. 1956. Enzymatic and metabolic adaptations in animals. Physiol. Rev. 36: 164–254.

Kogure M. 1933. The influence of light and temperature on certain characters of the silkworm, *Bombyx mori*. J. Dep. Agric. Kyushu Imp. Univ., Fukuoka 4: 1–93.

Korsloot A, van Gestel CAM and van Straalen NM (eds). 2004. Environmental Stress and Cellular Response in Arthropods. CRC Press, Boca Raton, Florida.

Kostal B, Berkova P and Simek P. 2003. Remodeling of membrane phospholipids during transition to diapause and cold-acclimation in the larvae of *Chymomyza costata* (Drosophilidae). Comp. Biochem. & Physiol. Part B Biochem. & Molecul. Biol. 135B: 407–419.

Krebs RA and Holbrook SA. 2001. Reduced enzyme activity following Hsp70 overexpression in *Drosophila melanogaster*. Biochemical Genetics 39: 73–82.

Krebs RA and Loeschcke B. 1996. Acclimation and selection for increased resistance to thermal stress in *Drosophilia buzzatii*. Genetics 142: 471–479.

Krijgsman BJ. 1930. Reizphysiologische Untersuchungen an *Stomoxys calcitrans*. Z. verg. Physiol. 11: 702–729.

Kristensen TN, Loeschcke V and Hoffmann AA. 2007. Can artificially selected phenotypes influence a component of field fitness? Thermal selection and fly performance under thermal extremes. Proc. R. Soc. B 274: 771–778.

Kültz D. 1996. Plasticity and stressor specificity of osmotic and heat-shock responses of *Gillichthys mirabilis* gill cells. Cell Physiol. 40C: 1181–1193.

Kültz D, Fiol D, Valkova N, Gomez-Jimenez S, Chan SY and Lee J. 2007. Functional genomics and proteomics of the cellular osmotic stress response in 'non-model' organisms. J. Exp. Biol. 210: 1593–1601.

Lagerspetz KYH. 2000. Thermal avoidance and preference in *Daphnia magna*. J. Therm. Biol. 25: 405–410.

Lagerspetz KYH. 2006. What is thermal acclimation? Journal of Thermal Biology 31: 332–336.

Lagerspetz KYH and Vainio LA. 2006. Thermal behaviour of crustaceans. Biological Reviews of the Cambridge Philosophical Society 81: 237–258.

Lalouette L, Kostal V, Colinet H, Gagneul D and Renault D. 2007. Cold exposure and associated metabolic changes in adult tropical beetles exposed to fluctuating thermal regimes. FEBS Journal 274: 1759–1767.

Laudein H. 1973. IV. Activity, behavior, etc. Pp. 441-469 in Precht H et al. (eds). Temperature and Life. Springer-Verlag, Berlin.

Layne JR and Peffer BJ. 2006. The influence of freeze duration on postfreeze recovery by caterpillars of *Pyrrharctia isabella* (Lepidoptera : Arctiidae): When is survival enough to qualify as recovery? Journal of Experimental Zoology Part A – Comparative Experimental Biology 305A: 570–575.

Lee RE, Elnitsky MA, Rinehart JP, Hayward SAL, Sandro LH and Denlinger DL. 2006. Rapid cold-hardening increases the freezing tolerance of the Antarctic midge *Belgica antarctica*. Journal of Experimental Biology 209: 399–406.

Lee SM, Lee SB, Park CH and Choi J. 2006. Expression of heat shock protein and hemoglobin genes in *Chrionomus tentans* (Diptera, Chironomidae) larvae exposed to various environmental pollutants: A potential biomarker of freshwater monitoring. Chemosphere 65: 1074–1081.

Lees AD (ed.). 1955. The Physiology of Diapause in Arthropods. Cambridge University Press, London.

Lerman DN and Feder ME. 2001. Laboratory selection at different temperatures modifies heat-shock transcription factor (HSF) activation in *Drosophilia melanogaster*. J. Exp. Biol. 204: 315–323.

Levins R (ed.) 1968. Evolution in Changing Environments. Princeton Univ. Press, Princeton.

Levins R. 1969. Thermal acclimation and heat resistance in *Drosophilia* species. Amer. Nat. 103: 483–499.

Li W, Zangerl AR, Schuler MA and Berenbaum MR. 2004. Characterization and evolution of furanocoumarin-inducible cytochrome P450s in the parsnip webworm, *Depressaria pastinacella*. Insect Molecular Biology 13: 603–613.

Li X, Schuler MA and Berenbaum MR. 2002. Jasmonate and salicylate induce expression of herbivore cytochrome P450 genes. Nature 419: 712–715.

Liang M, Cowley A and Greene A. 2004. High throughput gene expression profiling: a molecular approach to integrative physiology. J. Physiol. 554: 22–30.

Locke M. 1958. The coordination of growth in the tracheal system of insects. Q. J. Microsc. Sci. 99: 373–391.

Loeschcke V and Hoffmann AA. 2002. The detrimental acclimation hypothesis. Trends in Ecol. and Evol. 17: 407–408.

Loeschcke V and Hoffmann AA. 2007. Consequences of heat hardening on a field fitness component in *Drosophila* depend on environmental temperature. American Naturalist 169: 175–183.

Loeschcke V and Krebs RA. 1996. Selection for heat-shock resistance in larval and in adult *Drosophila buzzatii*: comparing direct and indirect responses. Evolution 50: 2354–2359.

Loeschcke V and Sørensen JG. 2005. Acclimation, heat shock and hardening – a response from evolutionary biology. J. Thermal Biol. 30: 255–257.

Loeschcke V, Sørensen JG and Kristensen TN. 2004. Ecologically relevant stress resistance: from microarrays and quantitative trait loci to candidate genes – A research plan and preliminary results using *Drosophila* as a model organism and climatic and genetic stress as model stresses. J. Biosciences 29: 503–511.

Loudon C. 1989. Tracheal hypertrophy in mealworms: design and plasticity in oxygen supply systems. J. Exp. Biol. 147: 217–235.

Luczynski A, Nyrop JP and Shi A. 2007. Influence of cold storage on pupal development and mortality during storage and on post-storage performance of *Encarsia formosa* and *Eretmocerus eremicus* (Hymenoptera : Aphelinidae). Biological Control 40: 107–117.

Lühmann M and Drees O. 1952. Uber die Temperaturabhängigkeit der Atmung sommerschlafender Blattkäfer (Ein Beitrag zum Adaptationsproblem) Zool Anz. 148: 13–22.

Ma RY, Hao SG, Tian J, Sun JH and Kang L. 2006. Seasonal variation in cold-hardiness of the Japanese pine sawyer *Monochamus alternatus* (Coleoptera : Cerambycidae). Environmental Entomology 35: 881–886.

Mackensen O. 1947. Effect of carbon dioxide on initial oviposition of artificially inseminated and virgin queen bees. J. Econ. Entomol. 40: 344–349.

Magiafoglou A and Hoffmann AA. 2003. Cross-generation effects due to cold exposure in *Drosophila serrata*. Functional Ecology 17: 664–672.

Mahroff R, Zhu KY and Subramanyam B. 2005. Changes in expression of heat shock proteins in *Tribolium castaneum* (Coleoptera: Tenebrionidae) in relation to developmental stage, exposure time, and temperature. Ann. Entomol. Soc. Am. 98: 100–107.

Malmendal A, Overgaard J, Bundy JG, Sorensen JG, Nielsen NC, Loeschcke V and Holmstrup M. 2006. Metabolomic profiling of heat stress: hardening and recovery of homeostasis in *Drosophila*. American Journal of Physiology – Regulatory Integrative and Comparative Physiology 291: R205–R212.

Marden JH. 2008. Quantitative and evolutionary biology of alternative splicing: how changing the mix of alternative transcripts affects phenotypic plasticity and reaction norms. Heredity 100: 111–120.

Marek M. 1970. Effect of stress of cooling on the synthesis of immunoglobulins in haemolymph of *Galleria mellonella* L. Comp. Biochem. Physiol. 35: 615–622.

Marjanovic M, Lawrence, BA, Wright NC, Costanzo JP and Lee RE. 2005. Possible role of phosphodiesters in cold tolerance. FASEB Journal 19 (5, Suppl. S, Part 2). A1653.

Martin MM and Van't Hof HM. 1988. the cause of reduced growth of *Manduca sexta* larvae on a low-water diet: increased metabolic processing costs or nutrient limitation. J. Insect Physiol. 34: 515–525.

Marzusch K. 1952. Untersuchungen über die Temperaturabhängigkeit von Lebensprozessen bei Insekten unter besonderer Berüksichtigung winterschlafender Kartoffelkäfer. Z. vergl. Physiol. 34: 75–92.

Mathias D, Jacky L, Bradshaw WE and Holzapfel CM. 2007. Quantitative trait loci associated with photoperiodic response and stage of diapause in the pitcher-plant mosquito, *Wyeomyia smithii*. Genetics 176: 391–402.

Matthee JJ. 1950. The eggs of the brown locust. Fmg. S. Afr. 25: 255–257.

Matthews EG (ed.) 1976. Insect Ecology. University of Queensland Press, St. Lucia.

Maynard Smith J. 1956. Acclimatization to high temperatures in inbred and outbred *Drosophila subobscura*. J. Genet. 54: 497–505.

Maynard Smith J. 1957. Temperature tolerance and acclimatization in *Drosophila subsobscura*. J. Exp. Biol. 34: 85–96.

Mazumdar-Leighton S and Broadway R. 2001. Transcriptional induction of diverse midgut trypsins in larval *Agrotis ipsilon* and *Helicoverpa zea* feeding on the soybean trypsin inhibitor (STI). Insect Biochem. Molec. Biol. 31: 645–657.

McLeese DW. 1956. Effects of temperature, salinity and oxygen on the survival of the American lobster. J. Fish. Res. Board. Canada 13: 247–272.

McLeese DW and Wilder DG. 1958. The activity and catchability of the lobster (*Homarus americanus*) in relation to temperature. J. Fisheries Res. Board Canada 15: 1345–1354.

McMillan DM, Fearnley SL, Rank NE and Dahloff EP. 2005. Natural temperature variation affects larval survival, development and Hsp70 expression in a leaf beetle. Functional Ecol. 19: 844–852.

McWhinnie MA and O'Connor JD. 1967. Metabolism and low temperature acclimation in the temperate crayfish, *Orconectes virilis*. Comp. Biochem. Physiol. 20: 131–145.

Meats A. 1973. Rapid acclimatization to low temperature in the Queensland fruit fly, *Dacus tryoni*. J. Insect Physiol. 19: 1903–1911.

Mellanby K. 1939. Low temperature and insect activity. Proceedings of the Royal Society B 127: 473–487.

Mellanby K. 1940a. The activity of certain artic insects at low temperatures. J. Anim. Ecol. 9: 296–301.

Mellanby K. 1940b. Temperature coefficients and acclimatization. Nature 146: 165–166.

Mellanby K. 1954. Acclimatization and the thermal death point in insects. Nature 173: 582–583.

Mendonca NN, Masui DC, McNamara JC, Leone FA and Furriel RPM. 2007. Long-term exposure of the freshwater shrimp *Macrobrachium olfersii* to elevated salinity: Effects on gill (Na+,K+)-ATPase alpha-subunit expression and K+-phosphatase activity. Comparative Biochemistry and Physiology. Part A, Molecular and Integrative Physiology 146: 534–543.

Mews HH. 1957. Temperaturadaptation der Sekretion von Verdauungsferment en und deren Hitzesistenz. Journal of Comparative Physiology A: Neuroethology, Sensory, Neural, and Behavioral Physiology 40: 345–355.

Michaelidis B, Vavoulidou D, Rousou J and Portner HO. 2007. The potential role of CO_2 in initiation and maintenance of estivation in the land snail *Helix lucorum*. Physiological and Biochemical Zoology 80: 113–124.

Michaud MR and Denlinger DL. 2004. Molecular modalities of insect cold survival: current understanding and future trends. International Congress Series 1275: 32–46.

Michaud MR and Denlinger DL. 2006. Oleic acid is elevated in cell membranes during rapid cold-hardening and pupal diapause in the flesh fly, *Sarcophaga crassipalpis*. Journal of Insect Physiology 52: 1073–1082.

Milkman RD. 1962. Temperature effects on day-old *Drosophila* pupae. J. Gen. Physiol. 45: 777–799.

Milkman RD. 1967. Kinetic analysis of temperature adaptation in Drosophila pupae. Pp. 147–162 in Prosser CL (ed.). Molecular Mechanisms of Temperature Adaptation. AAAS, Washington, DC.

Miller IK. 1982. Cold-hardiness strategies of some adult and immature insects overwintering in interior Alaska. Comp. Biochem. Physiol. A 73: 595–604.

Miller LK. 1969. Freezing tolerance in an adult insect. Science 166: 105–106.

Miner BG, Sultan SE, Morgan SG, Padilla DK and Relyea RA. 2005. Ecological consequences of phenotypic plasticity. Trends Ecol. Evol. 20: 685–692.

Miyazaki S, Kayukawa T, Chen B, Nomura M and Ishikawa Y. 2006. Enhancement of cold hardiness by acclimation is stage-specific in the non-diapausing pupae of onion maggot *Delia antiqua* (Diptera : Anthomyiidae). European Journal of Entomology 103: 691–694.

Montooth KL, Siebenthall KT and Clark AG. 2006. Membrane lipid physiology and toxin catabolism underlie ethanol and acetic acid tolerance in *Drosophila melanogaster*. Journal of Experimental Biology 209: 3837–3850.

Morgan CL. 1896a. On modification and variation. Science 4: 733–740.

Morgan CL. 1896b. Habit and Instinct. Arnold, London

Morris C (ed.). 1992. Academic Press Dictionary of Science and Technology. Academic Press, San Deigo.

Morris S and Vosloo A (eds.). Animals and Environments. Proceedings of the Third International Conference of Comparative Physiology and Biochemistry. International Congress Series. Elsevier. Vol. 1275: 1–374.

Müller HJ. 1970. Formen der Dormanz bei Insekten. Nove Acta Leopold N F 35 191: 7–27.

Munson SC. 1953. Some effects of storage at different temperatures on the lipids of the American roach and on the resistance of this insect to heat. J. Econ. Entomol. 46: 657–666.

Mutchmor JA. 1967. Temperature adaptation in insects. Pp. 165-176 in Prosser CL (ed). Molecular Mechanisms of Temperature Adaptation. Amer. Assoc. Advan. Sci. Washington, DC.

Mutchmor JA and Richards AG. 1961. Low-temperature tolerance of insects in relation to the influence of temperatutre on muscle apyrase activity. J. Insect Physiol. 7: 141–158.

Nabours RK. 1929. "The Genetics of the Tettigidae." Berlag M. Nijhoff, den Haag.

Nadeau JH and Topol EJ. 2006. The genetics of health. Nat. Genet. 38: 1095–1098.

Needhan AE (ed.). 1974. The Significance of Zoochromes. Springer, Berlin.

Nel MD. 1968. Selection at a high humidity for green and brown *solitaria* hopper coloration in *Locustana pardalina* (Walker). S. Afr. J. Agric. Sci. 11: 163–172.

Neple BB and Bachofen R. 1997. Induction of stress proteins in the phototophic bacterium *Rhodobacter sphaeroides*. FEMS Microbiol. Lett. 153–180.

Nice CC and Fordyce JA. 2006. How caterpillars avoid overheating: behavioral and phenotypic plasticity of pipevine swallowtail larvae. Oecologia 146: 541–548.

Nicholson AJ. 1934. Influence of temperature on the activity of sheep blowflies. Bull. Ent. Res. 25: 85–99.

Nicolas G and Sillans D. 1989. Immediate and latent effects of carbon dioxide on insects. Ann. Rev. Entomol. 34: 97–116.

Nielsen MG and Christian KA. 2007. The mangrove ant, *Camponotus anderseni*, switches to anerobic respiration in response to elevated CO_2 levels. J. Insect Physiol. 53: 505–508.

Nielsen MM, Overgaard J, Sorensen JG, Holmstuup M, Justesen J and Loeschcke V. 2005. Role of HSF activation for resistance to heat, cold and high-temperature knock-down. J. Insect Physiol. 51: 1320–1329.

Nielsen MM, Sorensen JG, Kruhoffer M, Justesen J and Loeschcke V. 2006. Phototransduction genes are up-regulated in a global gene expression study of *Drosophila melanogaster* selected for heat resistance. Cell Stress and Chaperones 11: 325–333.

Nilson TL, Sinclair BJ and Roberts SP. The effects of carbon dioxide anestesia and anoxia on rapid cold-hardening and chill coma recovery in *Drosophila melanogaster*. J. Insect Physiol. 52: 1027–1033.

Nilsson-Ehle, H. 1914. Vilka erfarenheter hava hittills vunnits rörande möjligheten av växters acklimatisering. Kunglig Landtbruks-Akad. Handlinger och Tidskrift 53: 537–572.

Numanoi H. 1938. On crustacean blood coagulation. Jap. J. Zool. 7: 613–641.

Oglesby LC. 1965. Steady-rate parameters of water and chloride regulation in "estuarine" Nereid polychaetes. Comp. Biochem. Physiol. 14: 621–640.

Okay S. 1953. Formation of green pigment and colour changes in Orthoptera. Bull. Entomol. Res. 44: 299–315.

Okay S. 1956. The effect of temperature and humidity on the formation of the green pigment in *Acrida bicolor* (Thunb.). Arch. Int. Physiol. Biochim. 64: 80–91.

Osborn HF. 1897. Organix selection. Science 15: 583–587.

Overgaard J, Sorensen JG, Petersen SO, Loeschcke V and Holmstrup M. 2006. Reorganization of membrane lipids during fast and slow cold hardening in *Drosophila melanogaster*. Physiological Entomology 31: 328–335.

Pagast F. 1936. Ueber Bau und Funktion der Analpapillen bei *Aedes aegypti* L. (fasciatus Fabr.). Zool. Jahrb. Abt. Physiol. 58: 183–218.

Pantyukhov GA. 1962. The effects of positive temperatures on different gegoraphic populations of the European gold tail (*Euproctis chryorrhea* L.) and the gypsy moth (*Lymantria dispar* L. – Lepidoptera, Orgyidae). Entomol. Rev. 14: 169–175.

Papi F, Serretti L and Parrini S. 1957. Nuove ricerche sull' orientamento e il senso del tempo di *Arctosa perita* (Latr.) (Araneae Lycosidae). Z. vergl. Physiol. 39: 531–561.

Pappas C, Hyde D, Bowler K, Loeschcke V and Sorensen JG. 2007. Post-eclosion decline in 'knock-down' thermal resistance and reduced effect of heat hardening in *Drosophila melanogaster*. Comparative Biochemistry and Physiology. Part A, Molecular and Integrative Physiology 146: 355–359.

Parhon M. 1909. Les échanges nutritifs chez les abeilles pendant les quatre saisons. Ann. Sci. Nat. (ser. 9), Zool. 9: 1–58.

Parker JR. 1930. Some effects of temperature and moisture upon *Melanoplus mexicanus mexicanus* Saussure and *Camnula pellucida* Scudder (Orthoptera). Univ. Montana Agri. Exp. Sta. Bull. 223: 1–132.

Patrick ML and Bradley TJ. 2000. Salt-lovin' skeeters: the osmocon-forming strategy of mosquito larvae. Amer. Zool. 40: 1166.

Payne NM. 1926. The effect of environmental temperature upon insect freezing points. Ecology 7: 99–106.

Pearson T, Hyde D and Bowler K. 1999. Heterologous acclimation: a novel approach to the study of thermal acclimation in the crab *Cancer pagurus*. Am. J. Physiol. 277: R24–R30.

Pereira R, Silva N, Quintal C, Abreu R, Andrade J and Dantas L. 2007. Effect of acclimation to outdoor conditions on the sexual performance of mass-produced medflies (Diptera : Tephritidae). Florida Entomologist 90: 171–174.

Perttunen V and Lagerspetz K. 1956. On the temperature acclimation of the heart beat and of the onset of muscle tremor in *Corethra plumicornis* (Diptera, Culicidae) larvae. Arch. Soc. Vanamo, Helsinki 11: 65–70.

Pick FE and Lea A. 1970. Field observations on spontaneous movements of solitarious hoppers of the brown locust, *Locustana pardalina* (Walker), and behavioural differences between various colour forms. Phytophlactica 2: 203–210.

Pigliucci M. 1996. How organisms respond to environmental changes: from phenotypes to molecules (and vice versa). Trends in Ecol. and Evol. 11: 168–173.

Pitts KM and Wall R. 2006. Cold shock and cold tolerance in larvae and pupae of the blow fly, *Lucilia sericata*. Physiol. Entomol. 31: 57–62.

Platzer I. 1967. Untersuchungen zur Temperaturadaptation der tropischen Chironomidenart *Chironomus strenskei* Fittkau (Diptera). Z. vergl. Physiol. 54: 58–74.

Powell SJ and Bale JS. 2004. Cold shock injury and ecological costs of rapid cold hardening in the grain aphid *Sitobion avenae* (Hemiptera: Aphididae). J. Insect Physiol. 50: 277–284.

Powell SJ and Bale JS. 2005. Low temperature acclimated populations of the grain aphid *Sitobion avenae* retain ability to rapidly cold harden with enhanced fitness. J. Exp. Biol. 208: 2615–2620.

Powell SJ and Bale JS. 2006. Effect of long-term and rapid cold hardening on the cold torpor temperature of an aphid. Physiological Entomology 31: 348–352.

Precht H. 1949. Die Temperat-urabhuangigkeit von Lebensprozessen. Z. Naturforsch Wiesbaden 4b: 26–35.

Precht H. 1951. Der Einfluss der temperature auf fermentsystem. Berh. dtsch. zool. Ges. p. 179.

Precht H. 1953. Über Ruhestadien erwachsener Insekten. I. Versuche an Kartoffelkäfern (*Leptinotarsa decemlineata* Say). Z. Vergl. Physiol. 35: 326–343.

Precht H. 1958. Concepts of the temperature adaptation of unchanging reaction systems of cold-blooded animals. Pp. 50-78 in Prosser CL (ed). Physiological Adaptation. Amer. Physiol. Soc., Washington, DC.

Precht H. 1964. Anpassungen wechselwarmer Tiere im normalen Temperaturbereich und ihre Ursachen. Naturw. Rundschau 11: 438–442.

Precht H. 1964. Über die Bedeuntung des Blutes für die Temperaturadaptation von Fischen. Zool. Jb. Physiol. 71: 313–327.

Precht H. 1973a. Constant systems. Pp. 302-352 in Precht H, Christophersen J, Hensel H and Larcher W (eds). Temperature and Life. Springer-Verlag, Berlin.

Precht H. 1973b. Limiting temperatures of life functions. Pp. 400-440 in Precht H, Christophersen J, Hensel H and Larcher W (eds). Temperature and Life. Springer-Verlag, Berlin.

Precht H, Christophersen J and Hensel H (eds). 1955. Temperatur und Leben. Springer-Verlag, Berlin.

Precht H, Christophersen J, Hensel H and Larcher W (eds). 1973. Temperature and Life. Springer-Verlag, Berlin.

Precht I. 1967. Untersuchungen über Diapause, Leistungsadaptation und Temperaturresistenz einiger Insekten und Schnecken. Z. Wiss. Zool. 176: 122–172.

Price TD, Qvarnstöm A and Irwin DE. 2003. The role of phenotypic plasticity in driving genetic evolution. Proceedings of the Royal Society Biological Sciences Series B 270: 1433–1440.

Prosser CL. 1950. Temperature: metabolic aspects and perception. Pp. 341–380 in Prosser CL (ed). Comparative Animal Physiology. Saunders, Philadelphia.

Prosser CL. 1955. Physiological variation in animals. Biological Reviews 30: 229–262.

Prosser CL. 1958a. Physiological Adaptation. American Physiological Soc. Whasington, DC.

Prosser CL. 1958b. General summary. Pp 167–180 in Prosser CL (ed). Physiological Adaptation. American Physiological Soc. Whasington, DC.

Prosser CL. 1961. Temperature. Pp. 227-244 in Prosser CL and Brown FA Jr (eds). Comparative Animal Physiology. Saunders, Philadelphia.

Prosser CL. 1962. Acclimation of poikilothermic vertebrates to low temperatures. Pp. 1-44 in Hannon JP and Biereck E (eds.). Comparative Physiology of Temperature Regulation. Part III. Artic Aeromedical Laboratory, Fort Wainwright, AK.

Prosser CL (Ed.). 1967. Molecular Mechanisms of Temperature Adaptation. Amer. Assoc. Adv. Sci., Washington, DC.

Prosser CL. 1973. Comparative Animal Physiology. W.B. Saunders, Philadelphia.

Quackenbush J. 2007. Extracting biology from high-dimentional biological data. The Journal of Experimental Biology 210: 1507–1517.

Radakovic SS, Maric J, Surbatovic M, Radjen S, Stefanova E, Stankovic N and Filipovic N 2007. Effects of acclimation on cognitive performance in soldiers during exertional heat stress. Military Medicine 172: 133–136.

Raikhel AS, Brown MR, and Belles X. 2005. Hormonal control of reproductive processes. Pp. 433–491 in Gilbert LI, Iatrou K and Gill SS (eds.). Comprehensive Molecular Insect Science Vol. 3: Endocrinology. Elsevier Pergamon, Amsterdam.

Rako L and Hoffmann AA. 2006. Complexity of the cold acclimation response in *Drosophila melanogaster*. Journal of Insect Physiology 52: 94–104.

Rao KP. 1967. Biochemical correlates of temperature acclimation. Pp. 227–244 in Prosser CL (ed.). Molecular Mechanisms of Temperature Adaptation. AAAS, Washington, DC.

Raushenbach, IY, Gruntenko NE, Bownes M, Adonieva NV, Terashima J, Karpova EK, Daddeeva NV and Chentsova NA. 2004. The role of juvenile hormone in the control of reproductive function in *Drosophila virilis* under nutritional stress. J. Insect Physiol. 50: 323–330.

Régnière J and Bentz B. 2007. Modeling cold tolerance in the mountain pine beetle, *Dendroctonus ponderosae*. J. Insect Physiol. 53: 559–572.

Reynolds SE and Bellward K. 1989. Water balance in *Manduca sexta* caterpillars: water recycling from the rectum. J. Exp. Biol. 141: 33–45.

Ribbands CR. 1950. Changes in the behavior of honey-bees following their recovery from anesthesia. J. Exp. Biol. 27: 302–310.

Richards AG and Suanraksa S. 1962. Energy expenditure during embryonic development under constant versus variable temperatures (*Oncopeltus fasciatus* (Dallas)). Ent. Exp. Appl. 5: 167–178.

Rinehart JP, Denlinger DL, and Rivers DB. 2002. Upregulation of transcripts encoding select heat shock proteins in the flesh fly *Sarcophaga crassipalpis* in response to venom from the ectoparastoid wasp *Nasonia vitripennis*. J. Invert. Pathol. 79: 62–63.

Ritossa FM. 1962. A new puffing pattern induced by temperature shock and DNP in *Drosophila*. Experientia 18: 571–573.

Rittner D and McCabe TL. 2004. Encyclopedia of Biology. Facts On File, Inc. New York.

Robb R, Hammond R and Bieber LL. 1972. Temperature-dependent changes in sphingosine composition of fatty acids of glycerophosphatides from *Musca domestica* larvae. Insect Biochem. 2: 131–136.

Roberts JL. 1953. Studies on thermal acclimatization in the lined shore crab, *Pachygrapsus crassipes* Randall. XIXth Int. Physiol. Congr. pp. 706–707.

Roberts JL. 1964. Metabolic responses of fresh-water sunfish to seasonal photoperiods and temperatures. Helgoländer Wiss. Meeresuntersusch 9: 459–473.

Robinson W. 1928. Response and adaptation of insects to external stimuli. Ann. Entomol. Soc. Am. 21: 407–417.

Rogers E. 1929. The effect of temperature on the oxygen consumption of an insect, *Melanoplus differentialis*. Physiol. Zool. 2: 275–283.

Rowell CHF. 1970. Environmental control of coloration in an acridid, *Gastrimargus africanus* (Saussure). Anit-Locust Bull. 47: 1–48.

Rowell CHF. 1971. Variable coloration of the Acridoid grasshoppers. Adv. Insect Physiol. 8: 145–198.

Ryan FJ. 1941. Temperature change and the subsequent rate of development. J. Exp. Zool. 88: 25–54.

Salt, RW. 1953. The influence of food on cold hardiness of insects. Can. Entomol. 85: 261–269.

Salt RW. 1956. Influence of moisture content and temperature on cold-hardiness of hibernating insects. Can. J. Zool. 34: 283–294.

Salt RW. 1958a. Role of glycerol in producing abnormally low supercooling and freezing points in an insect, *Bracon cephi* (Gahan). Nature 181: 1281.

Salt RW. 1958b. Proc. 10th International Congress Entomol. 2: 73–77.

Salt RW. 1959. Role of glycerol in the cold-hardening of *Bracon cephi* (Gahan). Can. J. Zool. 37: 59–69.

Salt, RW. 1961. Principles of insect cold-hardiness. Annu. Rev. Entomol. 6: 55–74.

Salt RW. 1962. Intracellular freezing in insects. Nature 193: 1207–1208.

Salt, RW. 1964. Terrestrial animals in cold: arthropods. Handbook of Physiology. American Physiology Society 4: 349–359.

Salvucci ME, Stecher DS and Henneberry TJ. 2000. Heat shock proteins in whiteflies, an insect that accumulates sorbitol in response to heat stress. J. Thermal Biol. 25: 363–371.

Saunders DS. 1982. Insect Clocks. Pergamon, Oxford.

Sayle MH. 1928. Factors influencing the rate of metabolism of *Aeschna umbrosa* nymphs. Biological Bulletin, Woods Hole 54: 212–229.

Scheiner SM. 2002. Selection experiments and the study of phenotypic plasticity. J. Evol. Biol. 15: 889–898.

Schlichting CD and Smith H. 2002. Phenotypic plasticity: linking molecular mechanisms with evolutionary outcomes. Evolutionary Ecology 16: 189–211..

Schmalhausen II (ed). 1949. Factors of Evolution: The Theory of Stabilizing Selection. University of Chicago Press, Chicago.

Schmidt J, Laudien H, and Bowler K. 1984. Acute adjustments to high temperature in FHM cells from *Pimephales promelas* (Pices, Cyprinidae). Comp. Biochem. Physiol. A Physiol. 78: 823–828.

Schneiderman JA and Williams CM 1953. The physiology of insect diapause. VII. The respiratory metabolism of the cecropia silkworm during diapause and development. Biol. Bull. (Woods Hole). 105: 320–334.

Schooley DA, Horodyski FM and Coast GM. 2005. Hormones controlling homeostasis in insects. Pp. 493-550 in Gilbert LI, Iatrou K and Gill SS (eds.). Comprehensive Molecular Insect Science, Vol 3: Endocrinology. Elsevier Pergamon, Amsterdam.

Schultze D. 1965. Beitrage zur Temperaturadaptation des Aales (*Anguilla vulgaris* L.) II. Zeitschrift Wissenschaftliche Zoologie 172: 104–133.

Scott MP. 1996. Communal breeding in burying beetles. American Scientist 84: 376–382.

Scott M, Berrigan D and Hoffmann AA. 1997. Costs and benefits of acclimation to elevated temperature in *Trichogramma carverae*. Entom. Exp. et Appl. 85: 211–219.

Seebacher F and Wilson RS. 2006. Fighting fit: thermal plasticity of metabolic function and fighting success in the crayfish *Cherax destructor*. Functional Ecology 20: 1045–1053.

Shiu SH and Borevitz JO. 2008. The next generation of microarray research: applications in evolutionary and ecological genomics. Heredity 100: 141–149.

Siminovitch D, Rheaume B and Sachar R. 1967. Seasonal increase in protoplasm and metabolic capacity in tree cells during adaptation to freezing. Pp. 3-40 in Prosser CL (ed.). Molecular Mechanisms of Temperature Adaptation. Amer. Assoc. Adv. Sci., Washington, DC.

Sinclair BJ. 2001. Field ecology of freeze tolerance: interannual variation in cooling rates, freeze-thaw and thermal stress in the microhabitat of the alpine cockroach *Celatoblatta quinquemaculata*. Oikos 93: 286–293.

Sinclair BJ and Chown SL. 2003. Rapid responses to high temperature and desiccation but not to low temperature in the freeze tolerant sub-Antarctic caterpillar *Pringleophaga marioni* (Lepidoptera, Tineidae). J. Insect Physiol. 49: 45–52.

Sinclair BJ and Chown SL. 2006. Rapid cold-hardening in karo beetle, *Afrinus* sp. Physiol. Entomol. 31: 98–101.

Sinervo B and Calsbeek R. 2003. Physiological epistasis, ontogenetic conflict and natural selection on physiology and life history. Integr. Comp. Biol. 43: 419–430.

Slabber S, Worland MR, Leinaas HP and Chown SL. 2007. Acclimation effects on thermal tolerances of springtails from sub-Antarctic Marion Island: Indigenous and invasive species. Journal of Insect Physiology 53: 113–125.

Slachta M, Vambera J, Zahradnickova H and Kostal V. 2002. Entering diapause is a prerequisite for successful cold-acclimation in adult *Graphosoma lineatum* (Heteroptera: Pentatomidae). J. Insect Physiol. 48: 1031–1039.

Smith CL. 1951. The temperature pulse rate curve of the isolated frog's heart, *Rana temporaria*. J. Exp. Biol. 28: 141–146.

Somero GN. 1969. Enzymic mechanisms of temperature compensation. American Naturalist 103: 517–530.

Sømme L. 1964. Effects of glycerol on cold hardiness in insects. Can. J. Zool. 42: 89–101.

Sømme L. 1968. Acclimation to low temperatures in *Tribolium confusum* Duvat (Coleopt., Tenebrionidae). Norsk. Entomologisk Tidsskrift 15: 134–136.

Sømme L. 1972. The effects of acclimation and low temperatures on enzyme activities in larvae of *Ephestia kuehniella* Zell. (Lep., Pyralidae). Entomol. Scand. 3: 12–18.

Sonoda S, Fukumoto K, Izumi Y, Ashfaq M, Yoshida H and Tsumuki H. 2006. Methionine-rich storage protein gene in the rice stem borer, *Chilo suppressalis*, is expressed during diapause in response to cold acclimation. Insect Molecular Biology 15: 853–859.

Sonoda S, Fukumoto K, Izumi Y, Yoshida H and Tsumuki H. 2006. Cloning of heat shock protein genes (hsp90 and hsc70) and their expression during larval diapause and cold tolerance acquisition in the rice stem borer, *Chilo suppressalis* Walker. Archives of Insect Biochemistry and Physiology 63: 36–47.

Sonoda S, Fukumoto K, Izumi Y, Ashfaq M, Yoshida H and Tsumuki H. 2007. Expression profile of arylphorin gene during diapause and cold acclimation in the rice stem borer, *Chilo suppressalis* Walker (Lepidoptera : Crambidae). Applied Entomology and Zoology 42: 35–40.

Sørensen JG and Loeschcke V. 2001. Larval crowding in *Drosophila melanogaster* induces Hsp70 expression, and leads to increased adult longevity and adult thermal stress resistance. J. Insect Physiol. 47: 1301–1307.

Sørensen JG, Nielsen MM, Kruhoffer M, Justesen J and Loeschcke V. 2005. Full genome gene expression analysis of the heat stress response, in *Drosophila melanogaster*. Cell Stress and Chaperones 10: 312–328.

Sørensen JG, Norry FM, Scannapieco AC and Loeschcke V. 2005. Altitudinal variation for stress resistance traits and thermal adaptation in adult *Drosophila buzzatii* from the New World. J. Evol. Biol. 18: 829–837.

Spencer-Davis P and Tribe MA. 1969. Temperature dependence of metabolic rate in animals. Nature 224: 723–724.

Stillwell RC and Fox CW. 2005. Complex patterns of phenotypic plasticity: interactive effects of temperature during rearing and oviposition. Ecology 86: 924–934.

Story KB. 2004. Gene regulation in physiological stress. International Congress Series 1275: 1–13.

Straub M. 1957. Weitere Untersuch-ungen zur Temperaturadaptation der Sauerstoffbindung des Blutes von *Rana esculenta* L. Z. vergl. Physiol. 39: 507–523.

Swallow JG, Rhodes JS and Garland T Jr. 2005. Phenotypic and evolutionary plasticity of organ masses in response to voluntary exercise in house mice. Integr. Comp. Biol. 45: 426–437.

Swynghedauw B. 2006. Phenotypic plasticity of adult myocardium. Molecular mechanisms. J. Exp. Biol. 209: 2320–2327.

Takeda M and Masaki S. 1979. Asymmetric perception of twilight affecting diapause induction by the fall webworm, *Hyphantria cunea*. Entomol. Exp. Appl. 25: 317–327.

Tammariello SP, Rinehart JP and Denlinger DL 1999. Desiccation elicits heat shock protein transcription in the flesh fly, *Sarcophaga crassipalpis*, but does not enhance tolerance to high or low temperatures. J. Insect Physiol. 45: 933–938.

Tanaka S. 2002. Temperature acclimation in overwintering nymphs of a cockroach, *Periplaneta japonica*: walking on ice. J. Insect Physiol. 48: 571–583.

Tantawy AO and Mallah GS. 1961. Studies on natural populations of *Drosophila*. I. Heat resistance and geographical variation in *Drosophila melanogaster* and *D. simulans*. Evolution 15: 1–14.

Tauber MJ and Tauber CA. 1976. Insect Seasonality: Diapause Maintenance, Termination, and Post-Diapause Development. Annu. Rev. Entomol. 21: 81–107.

Teixeira LAF and Polavarapu S. 2005. Expression of heat shock protein 70 after heat stress during pupal diapause in *Rhagoletis mendax* (Diptera: Tephritidae). Ann. Entomol. Soc. Am. 98: 966–972.

Terblanche JS and Chown SL. 2006. The relative contributions of developmental plasticity and adult acclimation to physiological variation in the tsetse fly, *Glossina pallidipes* (Diptera, Glossinidae). Journal of Experimental Biology 209: 1064–1073.

Terblanche JS, Klok CJ, Krafsur ES and Chown SL. 2006. Phenotypic plasticity and geographic variation in thermal tolerance and water loss of the tsetse *Glossina pallidipes* (Diptera : Glossinidae): Implications for distribution modelling. American Journal of Tropical Medicine and Hygiene 74: 786–794.

Terblanche JS, Marais E and Chown SL. 2007. Stage-related variation in rapid cold hardening as a test of the environmental predictability hypothesis. J. Insect Physiol. 53: 455–462.

Terblanche JS, Clusella-Trullas S, Deere JA and Chown SL. 2008. Thermal tolerance in a south-east African population of the tsetse fly *Glossina pallidipes* (Diptera, Glossinidae): implications for forecasting climate change impacts. J. Ins. Physiol. 54: 114–127.

Thiessen CI and Mutchmor JA. 1967. Some effects of thermal acclimation on muscle apyrase activity and mitochondrial number in *Periplaneta americana* and *Musca domestica*. J. Insect. Physiol. 13: 1837–1842.

Thompson DB. 1992. Consumption rates and the evolution of diet-induced plasticity in the head morphology of *Melanoplus femurrubrum* (Orthoptera: Acrididae). Oecologia 89: 204–213.

Thompson LJ, Robinson M and Hoffmann AA. 2001. Field and laboratory evidence for acclimation without costs in an egg parasitoid. Functional Ecology 15: 217–221.

Thompson V. 1937. Effects of temperature on movements of embryos (Acrididae, Orthoptera). Physiol. Zool. 10: 21–30.

Thorpe WH. 1963. Learning and Instinct in Animals. Methuen, London.

Tian Q, Stepaniants SB, Mao M, Weng L, Feetham MC, Doyle MJ, Yi EC, Dai H, Thorsson V, Eng J, et al. 2004. Integrated genomic and proteomic analysis of gene expression in mammalian cells. Mol. Cell. Proteomics 3: 960–969.

Tissieres A, Mitchell HK, and Tracy UM. 1974. Protein synthesis in salivary glands of *Drosophila melanogaster*: Relation to chromosome puffs. J. Mol. Biol. 84: 189–198.

Todd ME and Dehnel PA. 1960. The influence of temperature and salinity on heat tolerance in two grapsoid crabs, *Hemigrapsus nudus* and *Hemigrapsus oregonensis*. Biol. Bull. 118: 150–172.

Tomcala A, Tollarova M, Overgaard J, Simek P and Kostal V. 2006. Seasonal acquisition of chill tolerance and restructuring of membrane glycerophospholipids in an overwintering insect: triggering by low temperature, desiccation and diapause progression. Journal of Experimental Biology 209: 4102–4114.

Totze R. 1933. Beitrage zur sinnesphysiologie der Zecken. Z. vergl. Physiol. 19: 110–161.

Tribe, MA and Bowler K. 1968. Temperature dependence of "standard metabolic rate" in a poikilotherm. Comp. Biochem. Physiol. 25: 427–436.

Tsai JR and Lin HC. 2007. V-type H+-ATPase and Na+,K+-ATPase in the gills of 13 euryhaline crabs during salinity acclimation. Journal of Experimental Biology 210: 620–627.

Uvarov B (ed.). 1966. Grasshoppers and Locusts. Cambridge Univ. Press, London.

van Heerdt PF. 1946. (Physiology and Ecology of Forficula). Thesis, Utrecht.

Vernberg FJ. 1969. Acclimation of intertidal crabs. Amer. Zool. 9: 333–341.

Vernberg WB and Vernberg FJ. 1966. Studies on the physiological variation between tropical and temperate zone fiddler crabs of the genus *Uca*. 5. Effect of temperature on tissue respiration. Comp. Biochem. Physiol. 17: 363–374.

Vernberg WB and Vernberg, FJ. 1967a. Physiological diversity in metabolism in marine and terrestrial Crustacea. Amer. Zool. 8: 449–458.

Vernberg WB and Vernberg FJ. 1967b. Studies on the physiological variation between tropical and temperate zone fiddler crabs of the genus *Uca*-10. The influence of temperature on cytochrome-C oxidase activity. Comp. Biochem. Physiol. 26: 499–508.

Waddington CH. 1942. Canalization of development and the inheritance of acquired characters. Nature 150: 563–565.

Waddington CH. 1959. Canalization of development and genetic assimilation of acquired characters. Nature 183: 1654–1655.

Wang HS and Kang L. 2005. Effect of cooling rates on the cold hardiness and cryoprotectant profiles of locust eggs. Cryobiology 51: 220–229.

Wang HS, Zhou CS, Guo W and Kang L. 2006. Thermoperiodic acclimations enhance cold hardiness of the eggs of the migratory locust. Cryobiology 53: 206–217.

Wang HS, Wang XH, Zhou CS, Huang LH, Zhang SF, Guo W and Kang L. 2007. cDNA cloning of heat shock proteins and their expression in the two phases of the migratory locust. Insect Molecular Biology 16: 207–219.

Wang XH and Kang L. 2003. Rapid cold hardening in young hoppers of the migratory locust *Locusta migratoria* L. (Orthoptera: Acrididae). CryoLetters 24: 331–340.

Wang XH and Kang L. 2005. Differences in egg thermotolerance between tropical and temperate populations of the migratory locust *Locusta migratoria* (Orthoptera: Acrididae). J. Insect Physiol. 51: 1277–1285.

Wang XH, Qi XL and Kang L. 2003. Rapid cold hardening process of insects and its ecologically adaptive significance. Progress in Natural Science 13: 641–647.

Watson MJO and Hoffmann AA 1996. Acclimation, cross-generation effects, and the response to selection for increased cold resistance in *Drosophila*. Evolution 50: 1182–1192.

Weiser CJ. 1970. Cold resistance and injury in woody plants. Science 169: 1269–1278.

Werner EE and Peacor SD. 2003. A review of trait-mediated indirect interactions in ecological communities. Ecology 84: 1083–1100.

Werner I, Viant MR, Rosenblum ES, Gantner AS, Tjeerdema RS and Johnson ML. 2006. Cellular responses to temperature stress in steelhead trout (*Oncorhynchus mykiss*) parr with different rearing life histories. Fish Physiol. and Biochem. 32: 261–273.

West-Eberhard MJ. 2003. Developmental Plasticity and Evolution. Oxford University Press, Oxford.

Wheeler WM (ed.). 1910. Ants. Columbia University Press, New York.

Wigglesworth VB. 1933. The adaptation of mosquito larvae to salt water. J. Exp. Biol. 10: 27–37.

Wigglesworth VB. 1938. The regulation of osmotic pressure and chloride concentration in the haemolymph of mosquito larvae. J. Exp. Biol. 15: 235–247.

Wigglesworth VB. 1954. Growth and regeneration in the tracheal system of an insect, *Rhodnius prolixus* (Hemiptera). Q. J. Microsc. Sci. 95: 115–137.

de Wilde J. 1962. Photoperiodism in insects and mites. Annu. Rev. Entomol. 7: 1–26.

Willmer P, Stone G and Johnston I. 2005. Environmental Physiology of Animals, 2nd ed. Blackwell, Malden, MA.

Wilson RS and Franklin CE. 2002a. Testing the beneficial acclimation hypothesis. Trends in Ecology and Evolution 17: 66–70.

Wilson RS and Franklin CE. 2002b. The detrimental acclimation hypothesis. Trends in Ecology and Evolution 17: 408.

Wilson RS, Hammill E and Johnston IA. 2007. Competition moderates the benefits of thermal acclimation to reproductive performance in male eastern mosquitofish. Proceedings of the Royal Society of London – Series B: Biological Sciences 274: 1199–1204.

Wittkopp PJ. 2007. Variable gene expression in eukaryotes: a network perspective. J. Exp. Biol. 210: 1567–1575.

Wolfe GR, Hendrix DL and Salvucci ME. 1998. A thermoprotective role for sorbitol in the silverleaf whitefly, *Bemisia argentifolii*. J. Insect Physiol. 44: 597–603.

Woltereck R. 1909. Weitere experimentelle Untersuchungen über Artveränderung, speziell über das Wesen quantitativer Artunterschiede bei Daphniden. Verhandlungsbericht der Deutschen Zoologischen Gesellschaft 19: 110–172.

Woods HA and Bernays EA. 2000. Water homeostasis by wild larvae of *Manduca sexta*. Physiol. Entomol. 25: 82–87.

Woods HA and Harrison JF. 2001. The beneficial acclimation hypothesis versus acclimation of specific traits: physiological change in water-stressed *Manduca sexta* caterpillars. Physiol. Biochem. Zool. 74: 32–44.

Woods HA and Harrison JF. 2002. Interpreting rejections of the beneficial acclimation hypothesis: when is physiological plasticity adaptive? Evolution 56: 1863–1866.

Worland MR, Grubor-Lajsic G, and Montiel PO. 1998. Partial desiccation induced by sub-zero temperatures as a component of the survival strategy of the Arctic collembolan *Onychiurus arcticus* (Tullberg). J. Insect Physiol. 44: 211–219.

Yeates NTM. 1954. Environmental control of coat changes in cattle. Nature 174: 609–610.

Yin X, Wang S, Tang J, Hansen JD and Lurie S. 2006. Thermal conditioning of fifth-instar *Cydia pomonella* (Lepidoptera: Tortricidae) affects HSP70 acumulation and insect mortality. Physiol. Entomol. 31: 241–247.

Zamudio KR, Huey RB and Crill WD. 1995. Bigger isn't always better; body size, development and parental temperature, and territorial success in *Drosophila melanogaster*. Animal Behaviour 49: 671–677.

Zerbst E, Schmidt U and Müller E. 1966. Pflügers Archiv für die Gesamte Physiologie des Menschen und der Tierre. 289: R.26.

Zwaan BJ, Bijlsma R and Hoekstra RF. 1992. On the developmental theory of ageing. II. The effect of developmental temperature on longevity in relation to adult body size in *D. Melanogaster*. Heredity 68: 123–130.

17

Heat Shock Proteins and their Role in Generating, Maintaining and Even Preventing Alternative Insect Phenotypes

Jason B. Williams, Stephen P. Roberts and Michelle M. Elekonich

School of Life Sciences, University of Nevada, Las Vegas, 4505 S. Maryland Parkway, Las Vegas, NV 89154-4004, USA, E-mails: jason.williams@unlv.edu; sroberts@unlv.edu; michelle.elekonich@unlv.edu

Abstract

Heat shock proteins (Hsps) are part of a larger suite of molecular chaperones that play diverse roles in protein conformation-dependent processes in both unstressed and stressed cells. Hsps and their encoding genes (*hsps*) are nearly universal in organisms, highly-conserved, and assigned to families on the basis of sequence homology and typical molecular weight. Members of several Hsp families differ in inducibility by stressors, intracellular localization, and function. Hsps interact with other proteins that are in non-native conformations (whether due to protein-denaturing stress or because the peptides they comprise are not fully mature) to promote refolding, minimize their aggregation or target them for degradation and removal from the cell. In doing so, Hsps play key roles in regulating organismal development and protecting its proteomic underpinnings from denaturing stressors including extremes of temperature, cellular energy depletion, and concentrations of ions, other osmolytes, gases, and various toxic substances. In this chapter we explore how the activities of Hsps play key roles in novel manifestations of alternative insect phenotypes (phenocopies, behavioral phenotypes, and cryptic mutants). These three examples suggest that Hsps may play a more complex role in brokering individual adaptation and speciation in addition to the protective homeostatic effects of the cellular stress response.

Heat Shock Proteins

Proteins are central to all cellular and higher-order biological processes and their function is highly dependent on proper folding (Hochachka and

Somero 2002, Somero 1995). The maintenance of native protein structure is conferred in large part by the activity of heat shock proteins (Hsps), which are part of a larger suite of molecular chaperones that play diverse roles in successful folding, assembly, intracellular localization and trafficking, secretion, regulation, and degradation of proteins in both unstressed and stressed cells (Fig. 1). As such, Hsps affect a wide variety of cellular functions including signal transduction, apoptosis, antigen presentation and cell signaling (Feder and Hofmann 1999, Feder et al. 1995, Gething 1997, Gething and Sambrook 1992, Hartl 1996, Morimoto et al. 1994, Parsell and Lindquist 1993, Pratt and Toft 2003). Hsps and their encoding genes (*hsps*) are nearly universal in organisms, highly-conserved and assigned to families on the basis of sequence homology and typical molecular weight. These families include but are not limited to *hsp110, hsp100, hsp90, hsp70, hsp60, hsp40, hsp10,* and small *hsp* families (Gething 1997). In eukaryotes, many Hsp families include both constituitively expressed and inducible forms. Family members may also differ in intracellular localization, and function (Gething and Sambrook 1992, Hartl 1996, Morimoto et al. 1994).

The best studied of these is Hsp70, which is a generalist Hsp and stabilizes a wide range of nascent polypeptide chains or unfolded proteins by recognizing hydrophobic stretches of four or five amino acid residues (Flynn et al. 1991, Rudiger et al. 1997). Hsp70 induction leads to enhanced cell survival and occurs in response to a variety of stresses, including extremes of temperature, cellular energy depletion, concentrations of ions, other osmolytes, and gases, dessication, changes in salinity and exposure to various toxic substances. Hsp70 induction is not limited to inducible forms; significant stressors can cause temporary increases in constitutive forms as well (Feder and Hofmann 1999, Lindquist 1986, Palter et al. 1986, Ritossa 1996).

Hsp Expression in Nature

There is ample evidence that Hsp induction is not simply a laboratory based phenomenon but we still do not know how frequently wild organisms routinely express inducible Hsps (Feder and Hofmann 1999). Mobile organisms likely minimize exposure to Hsp-inducing stress by exploiting equable microrefugia in otherwise stressful habitats. Furthermore, organisms use other physiological and behavioral mechanisms to protect against the deleterious effects of environmental extremes. Even so, that wild organisms express Hsps in response to natural and anthropogenic stress exposure is firmly established. Far less is known about the frequency and

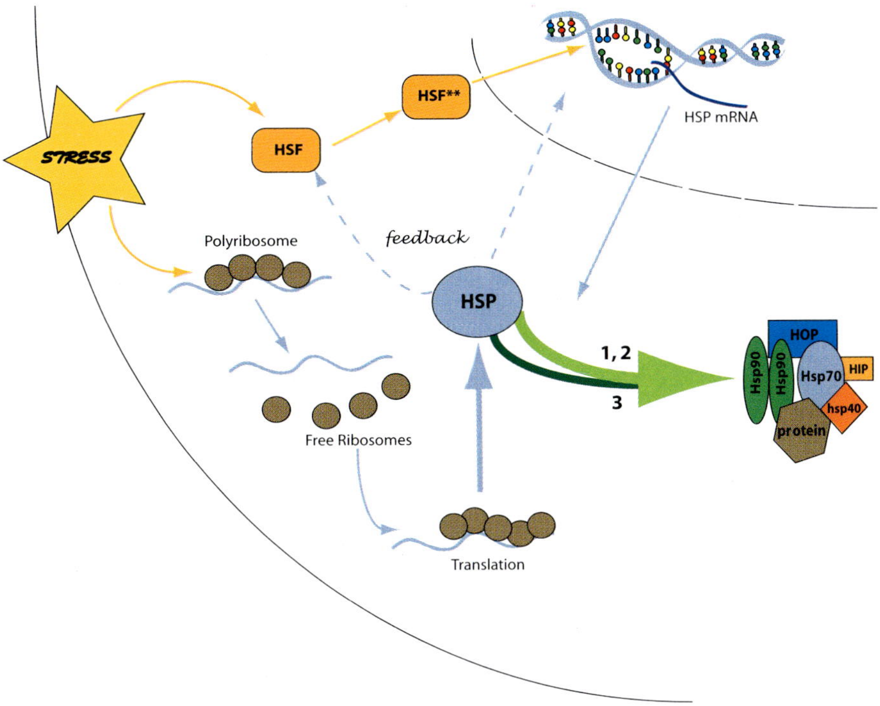

Fig. 1 Multiple roles of heat shock proteins (Hsps) in stressed and unstressed cells. During development (1), constitutively expressed Hsps regulate folding and degradation of developmental proteins such as during honey bee behavioral development. During exposure to a stressor, Hsp transcription is upregulated through the transcription factor HSF which dimerizes and translocates from the cytosol to the nucleus. Hsp translation is also upregulated in the cytosol. Hence, more are made Hsps available to combat stress-induced changes in protein structure and protein aggregation. Constitutively-expressed Hsps are also reallocated (2) to offset stress effects, perhaps compromising their ability to buffer hidden genetic variation as is the case with HSP90. Elevated levels of inducible Hsps (such as Hsp70) apparently provide protection to stress-sensitive developmental programs and may reduce the likelihood of teratogenesis (3).

intensity of Hsp expression throughout the lifespan of an organism in nature as well as the specific life-history and environmental correlates of Hsp expression.

Studies of sessile organisms such as plants and marine intertidal species have shown that the level of Hsp induction, the threshold for Hsp induction in the organism and in specific tissues, and the inducing stressor can all vary with the season and time of day (Alamillo et al. 1995, Buckley and Hofmann 2004, Burke et al. 1985, Colombo et al. 1995, Hamilton et al. 1996, Helmuth and Hofmann, 2001, Hendershot et al. 1992, Hernandez and

Vierling 1993, Hofmann and Somero 1996, Kimpel and Key 1985, Nguyen et al. 1994). Far less data exist for insects and other terrestrial animals. Sub-adult *Drosophila* encounter temperatures exceeding 40°C if the necrotic fruit they infest is sunlit, and express Hsp70 (the primary inducible Hsp in this species) in response (Feder 1997, Feder et al. 1997). However, adult *Drosophila* rarely express Hsp70 even on very hot days, presumably because their mobility and small size allows them to exploit thermally neutral microfugia. Body temperatures of the desert ants *Cataglyphis bombycina* and *Cataglyphis bicolor* exceed 50°C while they are foraging. In this case, Hsp70 proteins increase prior to foraging bouts, perhaps reflecting an anticipatory response (Gehring and Wehner 1995). Even terrestrial vertebrates are occasionally hyperthermic, such as during intense physical activity or fever. In birds and mammals, such hyperthermia activates HSF1 (the primary Hsp transcription factor) and increases the level of Hscs (Hsp cognates, constitutively-expressed Hsps) and Hsps (Brown and Rush 1996, Di et al. 1997, Locke and Noble 1995). Naturally occurring hypothermic events (such as diapause, overwintering in exposed sites, hibernation, and sometimes outright freezing) are also coincident with Hsp expression (Denlinger 2002).

Hsps and Phenotypic Plasticity

Hsp/Hsc expression during such naturally occurring environmental hyperthermic and hypothermic events could be thought of as phenotypic plasticity under the most inclusive definition, often stated as: *the expression of different phenotypes in a single genotype when subjected to different environments* (Ananthakrishnan 2005). However numerous reviews that expertly detail the nature of the stress response (and signature upregulation of Hsps), its role in homeostasis and the evolution of Hsps have been published and we refer the reader to the ecologically relevant review by Hoffmann and Feder (1999) and the classic references cited therein. Thus, as this territory is well-covered we wish to consider a more restrictive definition of phenotypic plasticity used by other researchers which limits phenotypic plasticity to adaptive, permanent, and discrete alternative developmental programs that alter morphology (for a complete discussion see West-Eberhard 2003). Because Hsp-induction is a graded, reversible, cellular phenomenon with temporal kinetics on the order of minutes/hours, it falls outside the latter, restrictive definition. In fact, to our knowledge, none of the numerous reviews of Hsp induction and function explicitly links the activities of Hsps with developmental programs and discrete alternative phenotypes in insects or any other taxa.

Thus we focus this chapter on the roles of Hsps in phenotypic variation arising from the effects of different external or internal environments on individuals of the same (in the case of *Drosophila*) or similar (in the case of the honey bee) genotypes *sensu* Schlichting and Pigliucci (1998). This sort of phenotypic plasticity may arise as an adaptation to spatial or temporal environmental heterogeneity or as an outcome of environmental influences on the individual organism's physiology (Levins 1963, Bradshaw 1965).

Specifically, in this chapter we provide a discussion of how Hsps play key roles in three novel manifestations of insect phenotypic plasticity (Fig. 1). We first discuss the role of Hsps to limit phenotypic plasticity by mitigating teratogenesis in laboratory and wild populations of *Drosophila*. Next we discuss how Hsps support the maintenance of behavioral phenotypes in the honey bee *Apis mellifera*. Finally, we discuss how Hsps limit or permit phenotypic plasticity as their action or impairment hides or reveals expression of genotypic variation affecting developmental signaling and adult morphogenesis in *Drosophila*. In this context, Hsps allow cryptic genetic variation to accumulate in populations of phenotypically similar individuals and act putatively as capacitors of morphological evolution. During periods of stress, Hsps are redistributed and their buffering of cryptic genetic variation is compromised, which in turn increases the expression of morphogenetic variation and perhaps evolvability. These three examples suggest that Hsps may play a more complex role in brokering individual adaptation and speciation in addition to the protective homeostatic effects of the cellular stress response.

Hsps can Prevent Insect Teratogenesis

Development in insects and other complex organisms is exceedingly sensitive to environmental perturbations, which can significantly alter developmental trajectories to yield a wide variety of adult phenotypes. Over a century ago Merrifield (1890, 1893) showed in butterflies that a temperature shock imposed during pupation disrupts wing pigmentation pattern. Soon thereafter Standfuss (1896) and Goldschmidt (1938) demonstrated that butterfly color patterns induced by pupal temperature shock sometimes mimic the normal genetically-controlled patterns of related races or species living at different temperatures, inspiring Goldschmidt to coin the term "phenocopy" to describe environmentally produced phenotypes that mimic genetically produced phenotypes. Research since then (mostly on *Drosophila*) establishes that numerous agents in addition to temperature are insect teratogens and that insect development is sensitive to the identity of the teratogen, the duration and intensity of exposure to it, and the developmental stage at which it is applied. For example, laboratory heat shock of wild-type *Drosophila melanogaster* can induce specific external cuticular phenocopies depending on the developmental stage at which the heat shock is imposed (e.g., abdominal segmentation in embryos; Maas 1948, Welte et al. 1995, wing defects in larvae: Goldschmidt 1935, wing defects in pupae: Goldschmidt 1935, Milkman 1962, Mitchell and Lipps 1978, Mitchell and Petersen 1982).

Again, these abnormalities often resemble defined phenotypes of known genetic mutations (yet occur against wild type genetic backgrounds), represent stress-induced disruptions of otherwise normal developmental programs, and do not appear in non-stressed offspring of affected individuals. While many phenocopies of well-characterized mutations have been identified, there are only a few descriptions of how environmental stress targets specific genes and/or gene products to disrupt insect developmental programs. For example, embryonic heat shock disrupts abdominal segmentation in *D. melanogaster* by delaying turnover of the protein encoded by the gene *fushi tarazu*, which regulates *bithorax*-complex genes that encode body segment patterning (Welte et al. 1995). Likewise, disruptions of *Drosophila* eye development, including phenocopies of the *eyeless* and *bar* mutations, can be induced by larval exposure to the mitotic poisons vinblastine and colchesine (Clayton and Francoeur 1971, Wolsky 1983, Drozdovskaya and Rapoport 1988, Isaenko et al. 1994, Isaenko and Shvartsman 1999) which disrupt microtubules and microtubule-dependent processes (Liang and Satya-Prakash 1985). Vinblastine induces depolymerization and aggregation of tubulin polymers (Bensch and Malawista 1969), while colchesine inhibits polymerization of tubulin by site-specific binding to β-monomers of tubulin within dimers (Margolis and Wilson 1977, Uppuluri et al. 1993).

Insect teratogenesis is not just a laboratory phenomenon, although far fewer examples of this phenomenon have been documented in wild insect populations. Roberts and Feder (1999) have shown that the natural counterpart to phenocopying occurs in orchard populations of *D. melanogaster* along gradients of temperature stress. In these wild populations, up to 15% of individuals surviving peak summer temperatures exhibit substantial wing and abdominal abnormalities (Fig. 2). Necrotic fruit temperatures in orchards and natural areas are extremely variable and can reach 45°C during peak sun exposure, causing a dramatic increase in the mortality of indwelling larvae (Feder et al. 1997, Roberts and Feder 1999). In nature, wild flies eclosing from frequently sunlit fruit exhibit wing and

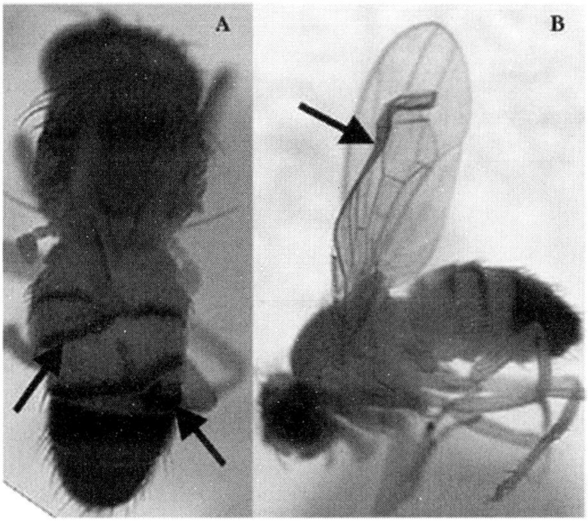

Fig. 2 Developmental disruption of abdominal tergites (A) and a wing (B) in *Drosophila melanogaster* due to heat stress during sub-adult stages. Certain heat shock proteins protect developing insects against some (but not all) of such injuries (see text). Photo adapted from Roberts and Feder (1999).

abdominal segment deformities 2-3 times more frequently than wild flies eclosing from fruits collected from deeply shaded areas (Roberts and Feder 1999). When necrotic fruit from natural sites are brought into the laboratory and the indwelling *Drosophila* eclose, the temporal sequence of eclosing wild flies with abnormalities resembles that predicted by laboratory phenocopy induction (abdominal segment deformities in flies presumably exposed as embryos and wing abnormalities in flies presumably exposed as larvae and pupae). However, the proportion of wild flies exhibiting defects is lower than that reported in studies of laboratory temperature-induced phenocopies, which can approach 100% (see Williams et al. 2003 and above references); this difference could be due to the differing thermal regimes in natural heat stress (characterized by large spatial and temporal thermal variation) *vs.* controlled laboratory thermal treatments. Even so, the frequency of phenocopying in wild *Drosophila* populations is sufficiently high to predict strong selection favoring mechanisms to prevent such injuries.

Numerous laboratory studies suggest that Hsps are a significant mechanism of protection against insect developmental defects. Capacity for Hsp expression and vulnerability to heat damage are generally inversely correlated during *Drosophila* development, with developmental stages lacking Hsp expression particularly sensitive to heat damage and/or death (Michaud et al. 1997, Elefant and Palter 1999). Exposure of sub-adult *Drosophila* to a Hsp-inducing heat pretreatment prior to a teratogenic heat-shock significantly reduces some (but not all) developmental defects (Milkman 1962, Milkman and Hille 1966, Mitchell et al. 1979, Petersen and Mitchell 1981, 1991). Heat pretreatment has also been shown to mitigate many developmental and mutagenic events in *Drosophila* caused by vinblastine and colchesine (Isaenko and Shvartsman 1999, Shvartsman and Isaenko 1999a, 1999b).

Genetic or pharmacological manipulations of Hsps provide direct proof of their role in protecting insect development. Mutagenesis of the *hsp83* gene and inhibition of its product with geldanomycin increase the incidence of developmental abnormalities in *Drosophila melanogaster* (Rutherford and Lindquist 1998). When compared to control flies that express normal amounts of Hsp70, heat pretreatment in a *Drosophila* line engineered for heat-inducible overexpression of Hsp70 yields a greater reduction in the frequency and severity of vinblastine-induced eye disruptions, including nuclear mispositioning in third-instar eye imaginal disks (with accompanying eye disk distortion) and ommatidia reduction/distortion and trichome doubling in adults (Isaenko et al. 2002). Furthermore, flies from this same Hsp70 overexpression line are more resistant to wing deformities following larval exposure to natural regimes of hyperthermia in field conditions (Roberts and Feder, 1999) and adult walking impairment following laboratory heat-shock of pupae (Roberts et al. 2003) than flies control lines. However, Hsp70 overexpression lines and control lines are equally well-protected by heat pretreatment against wing deformities and flight impairment following laboratory heat-shock of pupae (Roberts et al. 2003, Williams et al. 2003), indicating that other yet-to-be-described inducible mechanisms/chaperones in addition to Hsp70 offer protection against developmental disruption.

Hsps Aide in Maintaining Specific Behavioral Phenotypes

By protecting proper developmental outcomes Hsps limit phenotypic plasticity. Hsps can also support the maintenance of multiple existing phenotypes by maintaining the protein population necessary for a specific phenotype or creating conditions which ameliorate inherent stresses associated with a phenotype. One such place where Hsps appear to support the maintenance of alternative phenotypes is the switch from hive work to foraging in the honey bee. Because of their experimental tractability, sequenced genome and well understood biology, honey bees are an ideal model system for integrating molecular, genetic, physiological and sociobiological perspectives to advance understanding of phenotypic plasticity. As detailed below, changes in the transcriptome and proteome coincide with changes in systemic physiology and behavior as bees age and move from primarily non-flying tasks in the hive to foraging for the colony. Honey bees exhibit a form of adult behavioral development termed "temporal polyethism", moving through a series of behaviorally defined life history stages in an age-related fashion. For the first 2-3 weeks of life, adult workers perform tasks inside the hive such as brood care ("nursing") and hive maintenance. Typically at about 3 weeks of age, workers transition to performing tasks outside the hive such as foraging. While foraging, bees do not eat but collect nectar and pollen from flowering plants. Upon the foragers' return to the colony the nectar and pollen are made into honey and "bee bread" from which all adult members of the colony are fed. Foraging bees are typically the oldest workers in the hive. The rate of movement through these behavioral stages varies in response to colony demography and colony needs. In fact individual bees can accelerate or slow their rate of behavioral development and even return to previous behavioral stages (Huang and Robinson 1992, 1996, Leoncini et al. 2004a,b, Page et al. 1992, Pankiw 2004, Robinson et al. 1989, 1992).

Physiological changes with the transition to foraging include regression of the hypopharyngeal glands which produce the food fed to the developing larvae and increased production of enzymes for processing nectar, increased juvenile hormone levels, decreased body mass and water content, and increased metabolic and flight capacity (Fluri et al. 1982, Harrison 1986, Huang et al. 1994, Ohashi et al. 1996, 1999, Pontoh and Low 2002, Robinson and Vargo 1997, Winston 1987). The increases in flight capacity in particular are spectacular. Adult honey bees go from being unable to fly during the first day following eclosion to generating spectacular rates of metabolism and aerodynamic power (up to 0.8 W g^{-1} and 0.2 W g^{-1} respectively; Roberts and Harrison 1999) that enable later foraging and undertaking (removal of corpses from the hive) during which they travel up to 8 km from the hive and carry loads equivalent to their body mass. Foragers may increase Hsp expression in their thoraces due to thermogenesis in the flight muscles and/or extreme protein degradation, repair, maturation and replacement likely needed by the heavily-taxed flight muscles. Indeed, insect flight muscles have one of the highest metabolic rates of any animal tissue (Harrison and Roberts 1999). Bee muscles are conservatively estimated to contract over 4 million times per day based on an average of 5 hours of flight per day and 240 wing beats/sec (Harrison et al. 1996, Winston 1987).

Equally large differences in metabolism between hive bees and foragers are also observed. Flight metabolic rates, thoracic enzyme levels and thoracic glycogen levels remain relatively constant over the 1–3-week period when the bees work within the hive. Then, at the onset of foraging (14-21 days post-ecolosion) there is an approximate 15% increase in agitated flight metabolic rate, coincident with an approximate doubling of thoracic glycogen levels and dramatic increases in thoracic pyruvate kinase and citrate synthase activies (Fewell and Harrison 2001, Harrison 1986, Harrison and Fewell 2002, Moritz 1988, Neukirch 1982).

Structural and regulatory proteins of the flight muscle may also change as honey bees age and transition to flight-dependent behaviors. For example, Troponin-T is the tropomyosin-binding protein of the calcium-regulated troponin complex of striated muscle, and honey bees express different troponin-T (TnT) isoforms in their thoraces at 1 day *vs.* 5 days post-eclosion over which time they become flight capable (Domingo et al. 1998). This result suggests that honey bees are altering their calcium-dependent regulation of muscle contraction in an age-specific manner consistent with the acquisition of functional flight capability. Studies of *Drosophila* mutants suggest numerous other genes which may be upregulated during the transition to foraging including those coding for myosin regulatory light chain (Moore et al. 2000, Tohtong et al. 1995), flightin (Ayer and Vigoreaux 2003, Henkin et al. 2004, Reedy et al. 2000), paramyosin/miniparamyosin (Maroto et al. 1996), calcineurin (Gajewski et al. 2003), kettin (Kulke et al. 2001) and tropomodulin (Mardahl-Dumesnil and Fowler 2001).

In addition to changes in behavior and activity level, honey bees foraging outside of the hive are subject to several spatial and temporal gradients of heterothermia. Thoracic/flight muscle must reach a minimum temperature of 27°C to fly, but foragers typically warm up to ~37°C prior to departing the hive. Despite the ability of an individual bee to thermoregulate in flight by varying evaporative heat loss and metabolic heat production, the temperature of flying/foraging honey bees strongly varies with air temperature which in turn is subject to strong diurnal and seasonal variation. During flight at nearly all air temperatures there is strong heterothermia *within* an individual bee, with thoracic (flight muscle) temperatures higher than head or abdominal temperatures. During flight at warmer air temperatures each body segment can exceed 40°C (Roberts and Harrison 1999). At the cellular level, muscle proteins, metabolic enzymes and resultant function are protected from exercise and heat-damage in part by the expression of the heat shock proteins.

Previous studies demonstrated increased Hsp expression in honey bees in response to foulbrood (*Paenibacillus*) infection (Gregorc and Bowen 1999), and heat shock of either tissue or whole bees (Chacon-Almeida et al. 2000, Severson et al. 1990). Recent data using microarrays generated from honey bee ESTs found significant differences in brain Hsp mRNA levels between foragers and age-matched hive bees (Whitfield et al. 2003), suggesting hsps might play a role in the transition from in-hive tasks to foraging. However, our data suggest a more complex Hsp *vs.* tissue-type interaction with Hsp70 proteins and antioxidants playing an even greater role in the heavily used flight muscles in the thorax than in the brain (Williams et al. 2008).

We measured Hsp70 and other members of the 70-kD family of molecular chaperones in honey bee heads and thoraces as a function of age/behavior (newly emerged "day old" bees, vs. "nurse" bees vs. older foragers returning to the hive). Foragers expressed more Hsp70 protein in their thoraces than nurse bees, although there was no significant difference in head Hsp70 expression between the two groups (Roberts and Elekonich 2005, Fig. 3).

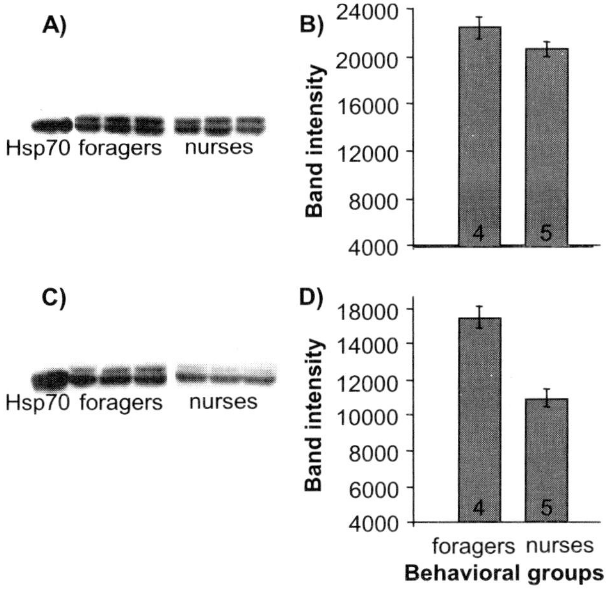

Fig. 3 Hsp70 expression in honey bee nurses (hive bees) and foragers. (A) and (B) Hsp70 expression in heads; (C) and (D) Hsp70 expression in thoraces. Hsp70 family proteins were isolated on Western blots labeled with a Hsp70 mouse monoclonal antibody (Sigma H5147) and an anti-mouse secondary antibody conjugated to HRP (Sigma). Proteins were visualized with ECL and quantified on a Typhoon Phosphorimager (GE Healthcare). The Hsp70 standard is 0.5 μg of purified Hsp70 protein (from Bovine brain, Sigma H9776). Numbers in bars indicate sample size. (Adapted from Roberts and Elekonich 2005).

One explanation for this result may be that foragers have hotter thoraces, but not heads, than hive bees. While this is true in some circumstances (Stabentheiner 2001), it is also possible that elevated Hsp70 expression in forager thoraces may be due to extreme protein degradation, repair, maturation and replacement needed by the heavily taxed forager flight muscles. Exercise related damage is well known in a number of animals including race horses and humans and is correlated with increased Hsp70 expression (e.g. Clarkson et al. 2005, Kinnunen et al. 2005, McArdle et al. 2004, Thompson et al. 2003).

Hsp70 family proteins increase in honey bee brains and thoraces following heat exposure in the laboratory; but induction is relatively modest and occurs only after several hours or at high temperatures. Groups of bees exposed to various temperatures above normal hive temperature showed very modest increases in Hsp70 proteins above 40°C with maximal induction of proteins in both the head and thorax at 48°C (Fig. 4). This

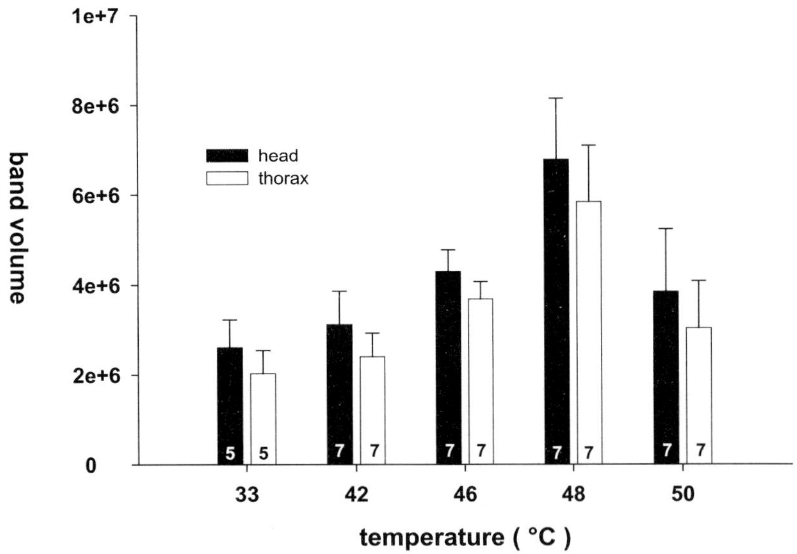

Fig. 4 Hsp70 protein expression as a function of laboratory heat-shock in 9-day-old honey bees. Different groups of 30 bees were exposed for one hour to 33, 42, 46, 48 or 50°C. Hsp70 family proteins were isolated on Western blots labeled with a Hsp70 mouse monoclonal antibody (Sigma H5147) and an anti-mouse secondary antibody conjugated to HRP (Sigma). Proteins were visualized with ECL and quantified on a Typhoon Phosphorimager (GE Healthcare). Hsp70 proteins varied significantly with temperature, due to maximal induction at 48°C ($F = 5.313$, $p < 0.001$; Tukey's HSD $p < 0.05$ for all comparisons to 48°C, but no other post-hoc comparisons). Data are means ± S.E. Sample size in bars. Black bars = heads; white bars = thoraces.

response occurs at a higher temperature and is much more attenuated than in other organisms. For comparison, even the normal hive temperature of 33°C approaches maximal Hsp70 induction temperatures in ectotherms such as *Drosophila* where normal induction causes as much as a 100 fold increase in transcript (e.g. Krebs and Feder 1997a,b). Given that the typical forager's flight lasts less than 30 minutes (Winston 1987), it is unlikely that even high air temperatures would induce Hsp expression during a normal foraging bout suggesting that use-related stress to the flight muscles during foraging is responsible for the increased expression in foragers.

When we compared *hsp70/hsc70* mRNA expression in hive workers and foragers we saw significant differences between the behavioral groups in levels of both the inducible form, *hsp70,* and the constitutive form, *hsc70,* transcript in the brain, but only the inducible form was upregulated in the thorax (Fig. 5). During these collections daytime temperatures ranged from 12°C to 33°C. Since bees typically warm up to ~37°C prior to flight the foragers' thoraces would have exceeded air temperature in all cases. Thus, it is unlikely that the the increased expression of Hsps in foragers' thoraces is due to environmentally induced heat stress. It is more likely that increased Hsps in forager's thoraces are due to exercise related muscle damage and heat production by the flight muscle as it undergoes millions of contractions a day (Harrison et al. 1996). It may be that increased expression of Hsps protects flight muscles against overwhelming damage during the day and facilitates repair of damage during the night. Thus, Hsp70 expression may help maintain the forager phenotype. Our laboratories are currently investigating the interaction between air temperature and flight on Hsp expression and Hsp mitigation of protein damage in honey bees. With the annotation of the honey bee genome now completed we will be able to track expression of all the Hsp/Hsc70 isoforms under varying regimes of flight and heat stress.

These studies suggest that Hsps help maintain alternative behavioral phenotypes such as hive work and foraging in the honey bee. Hsps may be important where the behavior involved is stressful in some way such as alternative mating tactics, or the maintenance of dominance behaviors in hierarchical groups just to suggest a few possibilities.

Hsps can Reveal Phenotypic Variation

Hsps may also uncover genetically based variation to provide the means to develop alternative behavioral or physical phenotypes. Our previous examples have focused on effects of Hsp70 during periods of stress which

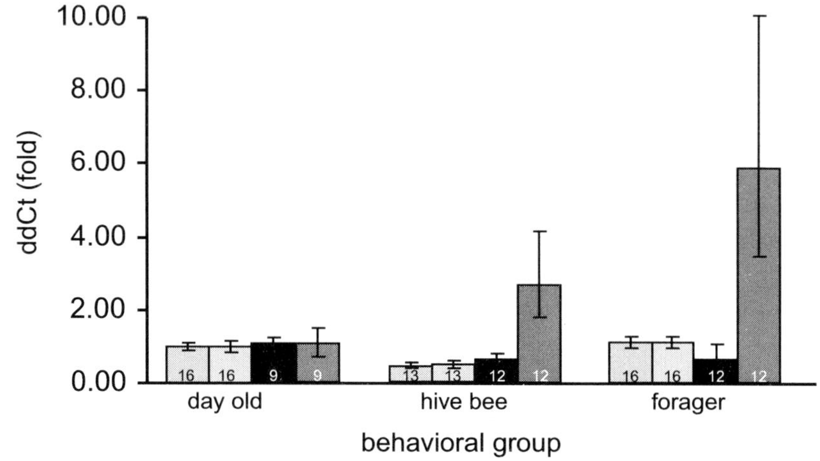

Fig. 5 *Hsp70* and *hsc70* mRNA expression in the brains and thoraces of 1-day-old, hive bees and foragers. Following dissection of the brain from the head capsule, RNA was extracted with the RNEasy kit (Qiagen). Thoraces (90% flight muscle) were processed without dissection. Primers for *hsp70* and *hsc70-4* and for the control gene rp49 were made from sequences in the honey bee EST set. Expression of *hsp70*, *hsc70-4* and the control gene *rp49* were quantified with quantitative real time pcr using either Fam-Tamra or SyberGreen labeled probes/primers. Graphical presentation follows the $2^{\Delta\Delta Ct}$ method (ABI User Bulletin #2). Levels of *hsp70* and *hsc70* mRNA were significantly higher in the brains of day olds and foragers than in those of hive bees ($F_{2,39}$ = 11.28, p<0.001; $F_{2,39}$ = 17.63, p<0.001 respectively, Tukey's HSD for pairwise comparisons p<0.05). Levels of *hsp70* mRNA were higher in the thoraces of foragers and hive bees than newly emerged day old bees ($F_{2,27}$ = 4.24, p<0.05; Tukey's HSD p<0.05). Although levels of *hsp70* in the thorax were not significantly different between foragers and hive bees, the increased variance in the forager group likely reflects differences in individual foragers' flight histories. *hsc70-4* levels in the thorax did not differ significantly between groups. Mean ± SE, sample sizes in bars. Brains in light grey (*hsp70*) and white (*hsc70*). Thoraces in black (*hsp70*) and dark grey (*hsc70*).

may promote protein denaturation or aggregation. In contrast, Hsp90 recognizes metastable folds in a specific, yet diverse, set of proteins, many of which are associated specifically with cellular proliferation and embryonic development (Rutherford and Zucker 1994, Nathan et al. 1997). Hsp90 plays diverse roles for signal transduction complexing with over 100 known client proteins (Pratt and Toft 2003). Hsp90 function is highly conserved and abundant in the cytosol of eukaryotes even in the absence of stress (Picard et al. 1990, Buchner 1999, Young et al. 2001, Picard, 2002).

Hsp90 plays a central role in regulating normal unstressed growth and development by stabilizing inherently unstable proteins so those proteins can recognize and respond to upstream regulatory components (Rutherford and Zucker 1994, Nathan and Lindquist 1995). During stress, however, Hsp90 is redeployed from its normal targets to other denatured proteins (Jakob et al. 1995, Nathan et al. 1997, Ali et al. 1998, Zou et al. 1998). When this occurs, the original signal transduction targets may lose activity and sensitize those developmental pathways to cryptic genetic variation and altered phenotypes (Rutherford and Lindquist 1998, Queitsch et al. 2002). When Hsp90 function is redirected due to pharmacological inhibition, genetic manipulation, or environmental stress, a small proportion of *D. melanogaster* display a wide range of phenotypes including malformed wings, antennae, eyes, and bristle patterns (Fig. 6; Rutherford and Lindquist 1998). It is through this means that Hsp90 may uncover additional genetic variation to provide previously unavailable targets for natural selection and thus the basis for new manifestations of phenotypic plasticity.

The altered phenotypes expressed by *D. melanogaster* following Hsp90 inhibition are dependant upon the individual's genetic background and the signal transduction protein targets that interact with Hsp90. In a series of selection experiments Rutherford and Lindquist (1998) demonstrated that the Hsp90 dependant eye and wing deformations were heritable. In addition, after repeated crosses the flies displayed a pattern of heritability that produced similar defects suggesting the defect was the result of genetic background and the traits were polygenic, arising from multiple genetic

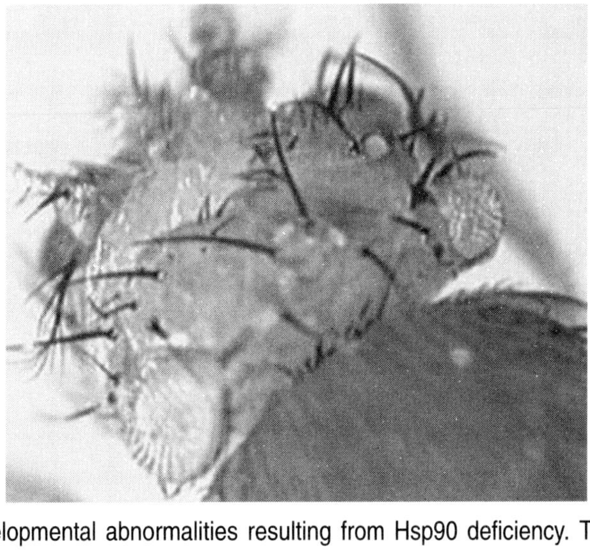

Fig. 6 Developmental abnormalities resulting from Hsp90 deficiency. The *Drosophila melanogaster* individual pictured has defects in the color and shape of the eye and abnormal bristles (see Rutherford and Lindquist 1998 for a full description).

determinates. Remarkably, selecting for a deformed eye phenotype in flies containing a mutant Hsp90 gene for only a few generations resulted in 80-90% of the offspring displaying the deviant trait. In addition, after the repeated selections the resulting offspring no longer had the mutant Hsp90 gene (Rutherford and Lindquist 1998). Thus, temporarily redirecting Hsp90 from its typical function is sufficient to allow an altered phenotype to become fixed.

Several lines of evidence suggest that the Hsp90 buffering system proposed by Rutherford and Lindquist may produce phenotype variation in nature. For instance, impairing Hsp90 function in the plant *Arabidopsis thaliana* also elicits altered phenotypes affecting all parts of the plant (Queitsch et al. 2002) and pharmacological inhibition of Hsp90 in the parasite *Leishmania donovani* triggers the transition from the promastigote sand fly-inhabiting to the amastigote mammal-inhabiting life stage (Wiesgigl and Clos 2001), indicating the Hsp90 may act as a buffer to phenotype change in many different types of organisms. Furthermore, naturally occurring stress, such as high temperature, limits the buffering capacity of Hsp90, producing altered phenotypes in both *Drosophila* and *Arabidopsis*, regardless of pharmacological inhibition or genetic manipulation of Hsp90. When flies from a *D. melanogaster* line with high-expression for an eye abnormality but normal Hsp90 function was allowed to develop at its normal rearing temperature of 18°C, the eyeless trait remained unexpressed and silent (Rutherford and Lindquist 1998). However, when these flies were reared at temperatures above 25°C there was a dramatic increase in expression of the eyeless trait, with greater than 15% of the flies having abnormal eyes when raised at 30°C. Studies of flies in nature indicate that 30°C is the maximum sustained temperature at which *D. melanogaster* can grow and reproduce (Hoffmann and Parsons 1997); yet developing larvae can experience temperatures greater than 40 °C in sunlit necrotic fruit (Roberts and Feder 1999). These extremely high temperatures may reduce the buffering capacity of Hsp90 to allow cryptic genetic variation to be expressed.

It is important to note that alteration of phenotype in nature would occur only when the Hsp90 buffering capacity is extremely taxed. For instance, if typical levels of stress were to routinely lower the buffering capacity of Hsp90, the "hidden" genetic variation would frequently be revealed. Consequently, the relatively large amount of genetic variation revealed in wild collected *D. melanogaster* and *A. thaliana* would not have accumulated (Rutherford and Lindquist 1998, Queitsch et al. 2002, Sangster et al. 2004). This is because such altered phenotypes are usually maladaptive

(Rutherford and Lindquist 1998, Queitsch et al. 2002), and would be selected against and not maintained in the genome. Thus, if, in nature, phenotypes are altered due to a reduction in Hsp90 function, it likely occurs infrequently and under abnormal circumstances.

The effect of phenotypic plasticity on accelerating or retarding evolution is a widely debated topic (see discussion and references in West-Eberhard, 2003, Görür 2005, Whitman, this volume). Extremely plastic variation that produces a wide range of possible phenotypes or reaction norms, such as those involved in learning, likely retards the rates of evolution as selection cannot act on multiple non-recurring traits. In contrast, when phenotypic plasticity is involved in switching between alternative traits, such as those produced due to a reduced Hsp90 buffering capacity, evolution can be greatly accelerated. As mentioned above, Rutherford and Lindquist (1998) selected for and increased the percentage of Hsp90 mutant fruit flies that had deformed eyes from essentially 0 to 80% of the population in just a few generations. This suggests that if an altered, yet beneficial phenotype were produced due to a reduced Hsp90 buffering capacity in nature, it could be selected for and incorporated in the population, thus speeding evolution.

The studies by Rutherford and Lindquist (1998) and Queitsch et al. (2002) revealed that a large amount of potentially potent genetic variation exists in normal populations. When functioning normally, Hsp90 can maintain developmental pathways to prevent the expression of such variation in the genetic code. However, when organisms are exposed to environmental stress, Hsp90 may be redirected to other functions resulting in the expression of certain altered phenotypes. As mentioned above, the consequent altered phenotypes in *D. melanogaster* and *A. thaliana* were maladaptive. But occasionally phenotypic changes due to altered Hsp90 function may produce a novel adaptive phenotype that could be assimilated genetically at a rapid rate. How frequently traits arise via this mechanism is still unknown and remains open to debate (Patridge and Barton 2000, Sniegowski et al. 2000, Rutherford 2003, Masel and Bergman 2003, Sangster et al. 2004).

Directions for Future Research

In addition to their classical role in the stress response we have provided evidence that Hsps can inhibit, maintain or promote phenotypic plasticity in insects. Given that Hsps are involved in a wide range of organismal and cellular functions including stress, apoptosis, signal transduction, membrane conductance, cell cycle control, regulation of transcription, protein shuttling and folding, and the immune response, it is likely that the examples presented here are only a few of the cases where Hsps are involved in mediating insect phenotypic plasticity at the cellular, physiological, behavioral, ecological and evolutionary levels.

Although we are starting to understand the connections between evolution and ecological, physiological and cellular genetic levels of analysis, we are only beginning to elucidate the details of how the environment and the genotype interact to produce a phenotype and phenotypic plasticity. The environment appears to play a dual role to induce variation and via natural selection to winnow that variation. While Hsps appear to be involved in both the response of a particular phenotype to environmental change and in the presentation of previously hidden genetic variation as a target for natural selection, a number of questions regarding Hsps remain to be answered:

- What is the role of Hsps in uncovering phenotypic plasticity in natural populations?
- How do Hsps support the spectacular variety of insect behavioral transitions and maintain ongoing behavior?
- Can the role of Hsps during early development be separated from their role in phenotypic plasticity?
- Is Hsp support of behavior and physiological capacity compromised during senescence or with high levels of exposure to stressors?
- What developmental programs, and the specific genes and proteins operating within them, are especially sensitive to teratogens? How do teratogens' target genes/proteins/programs and their sensitivities explain the age-specificity of specific teratogens?
- What teratogens are ecologically relevant for insects? What insect taxa or ecological communities are especially prone to teratogen exposure?
- Does evolutionary variation in Hsp expression ability in natural and laboratory insect populations correspond predictably with exposure and resistance to teratogens?

This list is by no means exhaustive, but we hope that these questions and others will (a) fuel continued interest in the complex mechanisms and relationships underlying phenotypic plasticity and (b) serve as inspiration to a new generation of biologists who integrate functional, genetic, evolutionary and ecological approaches out of the conviction that all of these components are necessary for a rigorous analysis of phenotypic plasticity and other fundamental biological phenomena.

Acknowledgments

We would like to thank Gene Robinson for his generous and enthusiastic support of the early stages of the honey bee Hsp work while MME was a post-doctoral researcher in his laboratory. We thank Suzanne Rutherford for graciously providing the photo for Fig. 6. We thank Karen Pruiett of the University of Illinois Bee Research Facility, Rodney Mehring of the Southern Nevada Local Honey Company for assistance with beekeeping and Kate Shen of the University of Nevada Las Vegas (UNLV) Genomics Center for technical support. This work is supported by a National Science Foundation Grant (IBN-0517635) to MME and SPR, a Nevada Applied Research Initiative Award to SPR and a UNLV New Investigator Award to MME.

References

Alamillo, J., C. Almoguera, D. Bartels and J. Jordano. 1995. Constitutive expression of small heat shock proteins in vegetative tissues of the resurrection plant *Craterostigma plantagineum*. Plant Molecular Biology 29: 1093–1099.

Ali A., S. Bharadwaj, R. O'Carroll and N. Ovsenek. 1998. HSP90 interacts with and regulates the activity of heat shock factor 1 in *Xenopus* oocytes. Molecular and Cellular Biology 18: 4949–4960.

Ananthakrishnan, T. N. 2005. *Insect Phenotypic Plasticity: Diversity of Responses*. Science Publishers, Inc. Enfield, NH, USA.

Ayer, G. and J. O. Vigoreaux. 2003. Flightin is a myosin rod binding protein. Cell Biochemistry and Biophysics 38: 41–54.

Bensch, K. G. and S. E. Malawista. 1969. Microtubular crystals in mammalian cells. The Journal of Cell Biology 40: 95–107.

Bradshaw, A. D. 1965. The evolutionary significance of phenotypic plasticity in plants. Advances in Genetics 13: 115–155.

Brown, I. R. and S. J. Rush. 1996. In vivo activation of neural heat shock transcription factor HSF1 by a physiologically relevant increase in body temperature. Journal of Neuroscience Research 44: 52–57.

Buchner J. 1999. Hsp90 & Co.—a holding for folding. Trends in Biochemical Science 24: 136–141.

Buckley, B. A. and G. E. Hofmann. 2004. Magnitude and duration of thermal stress determine kinetics of hsp gene regulation in the goby *Gillichthys mirabilis*. Physiological and Biochemical Zoology 77: 570–581.

Burke, J. J., J. L. Hatfield, R. P. Klein and J. E. Mullet. 1985. Accumulation of heat shock proteins in field-grown cotton. Plant Physiology 78: 394–398.

Chacon-Almeida, V. M. L., Z. L. P Simões and M. M. G. Bitondi. 2000. Induction of heat shock proteins in the larval fat body of *Apis mellifera* L. bees. Apidologie 31: 487–501.

Clarkson, K., J. D. Kieffer and S. Currie. 2005. Exhaustive exercise and the cellular stress response in rainbow trout, *Oncorhynchus mykiss*. Comparative Biochemistry and Physiology A 140: 225–232.

Clayton, F. C. and R. T. Francoeur. 1971. Some teratogenic effects of vinblastine on the external morphology of *Drosophila melanogaster*. Oncology 25: 188–192.

Colombo, S. J., V. R. Timmer, M. L. Colclough and E. Blumwald. 1995. Diurnal variation in heat tolerance and heat shock protein expression in black spruce (*Picea mariana*). Canadian Journal of Forest Research 25: 369–375.

Denlinger, D. L. 2002. Regulation of diapause. Annual Review of Entomology 47: 93–122.

Di, Y. P, E. A. Repasky and J. R. Subjeck. 1997. Distribution of HSP70, protein kinase C, and spectrin is altered in lymphocytes during a fever-like hyperthermia exposure. Journal of Cellular Physiology 172: 44–54.

Domingo, A., J. Gonzalez-Jurado, M. Maroto, C. Diaz, H. Vinos, C. Carrasco, M. Cervera and R. Marco. 1998. Troponin-T is a calcium-binding protein in insect muscle: *in vivo* phosphorylation, muscle-specific isoforms and developmental profile in *Drosophila melanogaster*. Journal of Muscle Research and Cell Motility 19: 393–403.

Drozdovskaya, L. N. and I. A. Rapoport. 1988. Influence of colchicine on *Drosophila morphogenesis*. Izvestiia Akademii Nauk SSSR 1: 125–128.

Elefant, F. and K. B. Palter. 1999. Tissue-specific expression of dominant negative mutant *Drosophila* HSC70 causes developmental defects and lethality. Molecular Biology of the Cell 10: 2101–2117.

Feder, M. E. 1997. Necrotic fruit: a novel model system for thermal ecologists. Journal of Thermal Biology 22: 1–9.

Feder, M. E. and G. E. Hofmann. 1999. Heat-shock proteins, molecular chaperones, and the stress response: evolutionary and ecological physiology. Annual Review of Physiology 61: 243–282.

Feder, M. E., D. A. Parsell and S. Lindquist. 1995. The stress response and stress proteins. *In* J.J. Lemasters and C. Olivers (Eds.). *Cell Biology of Trauma*. CRC Press, Boca Raton, Florida Pp. 177–191.

Feder, M. E., N. Blair and H. Figueras. 1997. Natural thermal stress and heat-shock protein expression in *Drosophila* larvae and pupae. Functional Ecology 11: 90–100.

Fewell, J. H. and J. F. Harrison. 2001. Variation in worker behavior of African and European honey bees. *In* R.E. Page and E. Erickson (Eds.). Proceedings of the Second International Congress on Africanized Bees and Bee Mites.

Fluri, P., M. Lüsher, H. Willie and L. Gerig. 1982. Changes in weight of the pharyngeal gland and haemolymph titers of juvenile hormone, protein and vitellogenin in worker honey bees. Journal of Insect Physiology 28: 61–68.

Flynn, G. C., J. Pohl, M. T. Flocco and J. E. Rothman. 1991. Peptide-binding specificity of the molecular chaperone BiP. Nature 353: 726–730.

Gajewski, K., J. B. Wang, J. D. Molkentin, E. H. Chen, E. N. Olson and R. A. Schulz. 2003. Requirement of the calcineurin subunit gene canB2 for indirect flight muscle formation in *Drosophila*. Proceedings of the National Academy of Science 100: 1040–1045.

Gehring, W. J. and R. Wehner. 1995. Heat shock protein synthesis and thermotolerance in *Cataglyphis*, an ant from the Sahara desert. Proceedings of the National Academy of Science 92: 2994–2998.

Gething, M. J. ed. 1997. *Guidebook to Molecular Chaperones and Protein Folding Catalysts*. Oxford University Press, Oxford.

Gething, M. J. and J. Sambrook. 1992. Protein folding in the cell. Nature 355: 33–45.

Goldschmidt, R. B. 1938. *Physiological Genetics*. McGraw-Hill, New York.

Goldschmidt, R. 1935. Gen und Aubeneigenschaft (Untersuchungen and Drosophila) I. Z. Indukt. Abstamm. Vererbungsl. 69: 38–69.

Görür, G. 2005. The importance of phenotypic plasticity in herbivorous insect speciation. *In* T.N. Ananthakrishnan and D. Whitman (Eds.). Insect Phenotypic Plasticity. Science Publishers, Inc., Enfield, NH, USA Pp. 145–171.

Gregorc, A. and I. D. Bowen. 1999. *In situ* localization of heat-shock and histone proteins in honey bee (*Apis mellifera* L.) larvae infected with *Paenibacillus* larvae. Cell Biology International 23: 211–218.

Hamilton, E. W., S. A. Heckathorn, C. A. Downs, T. E. Schwarz, J. S. Coleman and R. L. Hallberg. 1996. Heat shock proteins are produced by field grown naturally occurring plants in the summer in the temperate northeast U.S. Bulletin of the Ecological Society of America 77, Suppl. Part 2: 180.

Harrison, J. M. 1986. Caste specific changes in honeybee flight capacity. Physiological Zoology 59: 175–187.

Harrison, J. F. and J. H. Fewell. 2002. Environmental and genetic influences on flight metabolic rate in the honey bee, *Apis mellifera.* Comparative Biochemistry and Physiology A 133: 323–333.

Harrison, J. F., J. H. Fewell, S. P. Roberts and H. G. Hall. 1996. Achievement of thermal stability by varying metabolic heat production in flying honeybees. Science 274: 88–90.

Hartl, F. U. 1996. Molecular chaperones in cellular protein folding. Nature 381: 571–580.

Helmuth, B. S. T. and G. E. Hofmann. 2001. Microhabitats, thermal heterogeneity, and patterns of physiological stress in the rocky intertidal zone. The Biological Bulletin 201: 374–384.

Hendershot, K. L., J. Weng and H. T. Nguyen. 1992. Induction temperature of heatshock protein synthesis in wheat. Crop Science 32: 256–261.

Henkin, J. A., D. A. Maughan and J. O. Vigoreaux. 2004. Mutations that affect flightin expression in *Drosophila* alter the viscoelastic properties of flight muscle fibers. American Journal of Physiololgy—Cell Physiology 286: C65–C72.

Hernandez, L. D. and E. Vierling. 1993. Expression of low molecluar weight heatshock proteins under field conditions. Plant Physiology 101: 1209–1216.

Hochachka, P. W. and G. N. Somero. 2002. *Biochemical Adapatation.* Princeton University Press, Princeton, New Jersey.

Hofmann, G. E. and G. N. Somero. 1996. Protein ubiquitination and stress protein synthesis in *Mytilus trossulus* occurs during recovery from tidal emersion. Molecular Marine Biology and Biotechnology 5: 175–184.

Hoffmann, A. A. and P. A. Parsons. 1997. *Extreme Environmental Change and Evolution.* Cambridge University Press, Cambridge.

Huang, Z. Y. and G. E. Robinson. 1992. Honey bee colony integration: Worker-worker interactions mediate hormonally regulated plasticity in division of labor. Proceedings of the National Academy of Science 89: 11726–11729.

Huang, Z. Y. and G. E. Robinson. 1996. Regulation of honey bee division of labor by colony age demography. Behavioral and Ecological Sociobiology 39: 147–158.

Huang, Z. Y., G. E. Robinson and D. W. Borst. 1994. Physiological correlates of division of labor among similarly aged honey bees. Journal of Comparative Physiology A 174: 731–739.

Isaenko, O. A., T. B. Romashkina, P. Y. Shvartsman and L. F. Shelomova. 1994. Analysis of the mutagenic and teratogenic effects of griseofulvin in the mutagen-sensitive line mus(2)201G1 of *Drosophila melanogaster*. Genetika 30: 796–800.

Isaenko, O. A. and P. Y. Shvartsman. 1999. Genetic effects of mitotic poisons and their modification by heat-shock in strains of *Drosophila melanogaster* having defective adaptive response. Russian Journal of Genetics 35: 518–527.

Isaenko, O. A., T. L. Karr and M. E. Feder. 2002. Hsp70 and thermal pretreatment mitigate developmental damage due to mitotic poisons in *Drosophila*. Cell Stress Chaperones 7: 297–308.

Jakob U., H. Lilie, I. Meyer and J. Buchner. 1995. Transient interaction of Hsp90 with early unfolding intermediates of citrate synthase. Implications for heat shock in vivo. Journal of Biological Chemistry 270: 7288–7294.

Kimpel, J. A. and J. L. Key. 1985. Presence of heat shock mRNAs in field grown soybeans. Plant Physiology 79: 672–678.

Kinnunen, S., S. Hyppa, J. Lappalainen, N. Oksala, M. Venojarvi, C. Nakao, O. Hanninen, C.K. Sen and M. Atalay. 2005. Exercise-induced oxidative stress and muscle stress protein responses in trotters. European Journal of Applied Physiology 93: 496–501.

Krebs, R. A. and M. E. Feder. 1997a. Natural variation in the expression of the heat-shock protein Hsp70 in a population of *Drosophila melanogaster* and its correlation with tolerance of an ecologically relevant thermal stress. Evolution 51: 173–179.

Krebs, R. A. and M. E. Feder. 1997b. Tissue specific in Hsp70 expression and thermal damage in larval *Drosphila melanogaster* larvae. The Journal of Experimental Biology 200: 2007–2015.

Kulke, M., C. Neagoe, B. Kolmerer, A. Minajeva, H. Hinssen, B. Bullard and W. A. Linke. 2001. Kettin, a major source of myofibrillar stiffness in *Drosophila* indirect flight muscle. The Journal of Cell Biology 154: 1045–1057.

Leoncini, I., G. E. Robinson and Y. Le Conte. 2004a. Worker-worker inhibition of honey bee behavioural development independent of queen and brood. Insectes Sociaux 51: 392–394.

Leoncini, I., Y. Le Conte, G. Costagliola, E. Plettner, A. L. Toth, M. Wang, Z. Huang, J-M. Bécard, D. Cruaser, K. N. Slessor and G. E. Robinson. 2004b. Regulation of behavioral maturation by a primer pheromone produced by adult worker honey bees. Proceedings of the National Academy of Science 101: 17559–17564.

Levins R. 1963. Theory of fitness in a heterogenous environment. II. Developmental flexibility and niche selection. American Naturalist 47: 75–90.

Liang, J. C. and K. L. Satya-Prakash. 1985. Induction of aneuploidy by mitotic arrestants in mouse bone marrow. Mutation Research 155: 61–70.

Lindquist, S. 1986. The heat shock response. Annual Review of Biochemistry 55: 1151–1191.

Locke, M. and E. G. Noble. 1995. Stress proteins: the exercise response. Canadian Journal of Applied Physiology 20: 155–167.

Masel J. and A. Bergman. 2003. The evolution of the evolvability properties of the yeast prion [PSI+]. Evolution 57: 1498–1512.

Maas, H. 1948. Uber die Auslosbarkeit von Temperaturmodifikationen wahrend der Embryonal Entwicklung von *Drosophila melanogaster* Meigen. Wilhelm Roux' Archir fur Entwicklungsmechanick der Organism 143: 515–572.

Mardahl-Dumesnil, M. and V. M. Fowler. 2001. Thin filaments elongate from their pointed ends during myofibril assembly in *Drosophila* indirect flight muscle. The Journal of Cell Biology 155: 1043–1053.

Margolis, R. L. and L. Wilson. 1977. Addition of colchicine–tubulin complex to microtubule ends: the mechanism of substoichiometric colchicine poisoning. Proceedings of the National Academy of Science 74: 3466–3470.

Maroto, M., J. Arredondo, D. Goulding, R. Marco, B. Bullard and M. Cervera. 1996. *Drosophila* paramyosin miniparamyosin gene products show a large diversity in quantity, localization, and isoform pattern: A possible role in muscle maturation and function. The Journal of Cell Biology 134: 81–92.

McArdle, A., W. H. Dillmann, R. Mestril, J. A. Faulkner and M. J. Jackson. 2004. Overexpression of HSP70 in mouse skeletal muscle protects against muscle damage and age-related muscle dysfunction. The FASEB Journal 18: 355–357.

Merrifield, F. 1890. Systematic temperature experiments on some Lepidoptera in all their stages. Transactions of the Entomological Society of London, pp. 131–159.

Merrifield, F. 1893. The effects of temperature in the pupal stage on the colouring of *Pieris napi, Vanessa atalanta, Chrysophanus phloeas*, and *Ephyra punctaria.* Transactions of the Entomological Society of London, pp. 425–438.

Michaud, S., R. Marin and R. M. Tanguay. 1997. Regulation of heat shock gene induction and expression during *Drosophila* development. Cellular and Molecular Life Sciences 53: 104–113.

Milkman, R. D. 1962. Temperature effects on day old *Drosophila* pupae. The Journal of General Physiology 45: 777–799.

Milkman, R. D. and B. Hille. 1966. Analysis of some temperature effects on *Drosophila* pupae. The Biological Bulletin 131: 331–345.

Mitchell, H. K. and L. S. Lipps. 1978. Heat shock and phenocopy induction in *Drosophila.* Cell 15: 907–918.

Mitchell, H. K. and N. S. Petersen. 1982. Developmental abnormalities in *Drosophila* induced by heat shock. Developmental Genetics 3: 91–102.

Mitchell, H. K., G. Moller, N. S. Petersen and L. Lipps-Sarmiento. 1979. Specific protection from phenocopy induction by heat shock. Developmental Genetics 1: 181–192.

Moore J. R., Dickinson M. H., Vigoreaux J. O. and D. W. Maughan. 2000. The effect of removing the N-terminal extension of the *Drosophila* myosin regulatory light chain upon flight ability and the contractile dynamics of indirect flight muscle. Biophysical Journal 78: 1431–1440.

Morimoto, R. I., A. Tissieres and C. Georgopoulos (Eds.). 1994. *Heat Shock Proteins: Structure, Function and Regulation*. Cold Spring Harbor Lab. Press, Cold Spring Harbor, New York.

Moritz, R. F. A. 1988. Biochemical changes during honeybee flight muscle development. *In* W. Nachtigall (Ed.). *The Flying Honeybee; Aspects of Energetics*. Gustav Fischer, New York Pp. 51–65.

Nathan D. F. and S. Lindquist. 1995. Mutational analysis of Hsp90 function: interactions with a steroid receptor and a protein kinase. Molecular and Cellular Biology 15: 3917–3925.

Nathan D. F., M. H. Vos and S. Lindquist. 1997. In vivo functions of the *Saccharomyces cerevisiae* Hsp90 chaperone. Proceedings of the National Academy of Science 94: 12949–12956.

Neukirch, A. 1982. Dependence of the life-span of the honeybee (*Apis mellifera*) upon flight performance and energy consumption. Journal of Comparative Physiology B 146: 35–40.

Nguyen, H. T, C. P. Joshi, N. Klueva, J. Weng, K. L. Hendershot and A. Blum. 1994. The heat-shock response and expression of heat-shock proteins in wheat under diurnal heat stress and field conditions. Australian Journal of Plant Physiology 21: 857–867.

Ohashi, K., M. Sawata, H. Takeuchi, S. Natori and T. Kubo. 1996. Molecular cloning of cDNA and analysis of expression of the gene for alpha-glucosidase from the hypopharyngeal gland of the honeybee *Apis mellifera* L. Biochemical and Biophysical Research Communications 221: 380–385.

Ohashi, K., S. Natori and T. Kubo. 1999. Expression of amylase and glucose oxidase in the hypopharyngeal gland with an age-dependent role change of the worker honeybee (*Apis mellifera* L.). Eururopean Journal of Biochemistry 265: 127–133.

Page, R. E., Jr., G. E. Robinson, D. S. Britton and M. K. Fondrk. 1992. Genotypic variability for rates of behavioral development in worker honey bees (*Apis mellifera* L.). Behavioral Ecology 3: 173–180.

Palter, K. B., M. Watanabe, L. Stinson A. P. Mahowald and E. A. Craig. 1986. Expression and localization of *Drosophila melanogaster* hsp70 cognate proteins. Mol. Cell. Biol. 6: 1187–1203.

Pankiw, T. 2004. Worker honey bee pheromone regulation of foraging ontogeny. Naturwissenschaften 91: 178–181.

Parsell, D. A. and S. Lindquist. 1993. The function of heat shock proteins in stress tolerance: degradation and reactivation of damaged proteins. Annual Review of Genetics 27: 437–496

Partridge L. and N. H. Barton. 2000. Evolving evolvability. Nature 407: 457–458.

Petersen, N. S. and H. K. Mitchell. 1981. Recovery of protein synthesis after heat shock: prior heat treatment affects the ability of cells to translate mRNA. Proceedings of the National Academy of Science 78: 1708–1711.

Petersen, N. S. and H. K. Mitchell. 1991. Environmentally induced development defects in *Drosophila.* Results and Problems in Cell Differentiation 17: 29–43.

Picard, D. 2002. Heat-shock protein 90, a chaperone for folding and regulation. Cellular and Molecular Life Science 59: 1640–1648.

Picard D., B. Khursheed, M. J. Garabedian, M. G. Fortin, S. Lindquist and K.R.Yamamoto. 1990. Reduced levels of hsp90 compromise steroid receptor action in vivo. Nature 348: 166–168.

Pontoh, J. and N. H. Low. 2002. Purification and characterization of beta-glucosidase from honey bees (*Apis mellifera*). Insect Biochemistry and Molecular Biology 32: 679–690.

Pratt W. B. and D. O. Toft. 2003. Regulation of signaling protein function and trafficking by the hsp90/hsp70-based chaperone machinery. Experimental Biology and Medicine 228: 111–133.

Queitsch C., T. A. Sangster and S. Lindquist. 2002. Hsp90 as a capacitor of phenotypic variation. Nature 417: 618–624.

Reedy, M. C., B. Bullard and J. O. Vigoreaux. 2000. Flightin is essential for thick filament assembly and sarcomere stability in *Drosophila* flight muscles. The Journal of Cell Biology 151: 1483–1499.

Ritossa, F. 1996. Discovery of the heat shock response. Cell Stress Chaperones 1: 97–98.

Roberts, S. P. and M. E. Feder. 1999. Natural hyperthermia and expression of the heat-shock protein Hsp70 affect development in *Drosophila melanogaster*. Oecologia 121: 323–329.

Roberts S. P. and J. F. Harrison. 1999. Mechanisms of thermal stability during flight in the honeybee *Apis mellifera*. The Journal of Experimental Biology 202: 1523–1533.

Roberts, S. P. and M. M. Elekonich. (in press). Commentary: Behavioral Development and the Ontogeny of Flight Capacity in Honey Bees. The Journal of the Experimental Biology.

Roberts, S. P., M. E. Feder and J. H. Marden. 2003. Dropping like flies: Environmentally-induced impairment and protection of locomotor performance in adult *Drosophila melanogaster*. Physiological and Biochemical Zoology 75: 615–621.

Robinson, G. E. and E. L. Vargo. 1997. Juvenile hormone in adult eusocial Hymenoptera: Gonadotropin and behavioral pacemaker. Archives of Insect Biochemistry and Physiology 35: 559–583.

Robinson, G. E., R. E. Page, C. Strambi and A. Strambi. 1989. Hormonal and genetic control of behavioral integration in honeybee colonies. Science 246: 109–112.

Robinson, G. E., R. E. Page, C. Strambi and A. Strambi. 1992. Colony integration in honey bees: mechanisms of behavioural reversion. Ethology 90: 336–350.

Rudiger, S., L. Germeroth, J. Schneider-Mergener and B. Bukau. 1997. Substrate specificity of the DnaK chaperone determined by screening cellulose-bound peptide libraries. The EMBO Journal 16: 1501–1507.

Rutherford, S. L. 2003. Between genotype and phenotype: protein chaperones and evolvability. Nature Reviews Genetics 4: 263–274.

Rutherford, S. L. and C. S. Zucker. 1994. Protein folding and the regulation of signaling pathways. Cell 79: 1129–1132.

Rutherford, S. L. and S. Lindquist. 1998. Hsp90 as a capacitor for morphological evolution. Nature 396: 336–342.

Sangster T. A., S. Lindquist and C. Queitsch. 2004. Under cover: causes, effects and implications of Hsp90-mediated genetic capacitance. Bioessays 26: 348–362.

Schlichtling C. D. and M. Pigliucci. 1998. *Phenotypic Evolution, a Reaction Norm Perspective*. Sinauser, Massachusetts, Sunderland.

Severson, D. W., E. H. Erickson, Jr., J. L. Williamson and J. M. Aiken. 1990. Heat stress induced enhancement of heat shock protein gene activity in the honey bee (*Apis mellifera*). Experientia 46: 737–739.

Shvartsman, P. Y. and O. A. Isaenko. 1999a. Effect of heat shock and microtubule disorganization on the frequency of embryonic lethality in different *Drosophila melanogaster* strains. Russian Journal of Genetics 35: 601–604.

Shvartsman, P. Y. and O. A. Isaenko. 1999b. Effect of heat-shock and microtubule disorganization on the frequency of sex chromosome nondisjunction in various *Drosophila melanogaster* strains. Russian Journal of Genetics 35: 760–765.

Sniegowski P. D., P. J. Gerrish, T. Johnson and A. Shaver. 2000. The evolution of mutation rates: separating causes from consequences. Bioessays 22: 1057–1066.

Somero, G. N. 1995. Proteins and temperature. Annual Review of Physiology 57: 43–68.

Stabentheiner, A. Thermoregulation of dancing bees: thoracic temperature of pollen and nectar foragers in relation to profitability of foraging and colony need. Journal of Insect Physiology 47: 385–392.

Standfuss, M. 1896. *Handbuch der palearctischen Gross-Schmetterlinge für Forscher und Sammler*. Gustav Fischer, Jena.

Thompson, H. S., E. B. Maynard, E. R. Morales and S. P. Scordilis. 2003. Exercise-induced HSP27, HSP70, and MAPK responses in human skeletal muscle. ACTA Physiologica Scandinavica 178: 61–72.

Tohtong, R., H. Yamashita, M. Graham, J. Haeberle, A. Simcox and D. Maughan. 1995. Impairment of muscle function caused by mutations of phosphorylation sites in myosin regulatory light-chain. Nature 374: 650–653.

Uppuluri, S., L. Knipling, D. L. Sackett and J. Wolff. 1993. Localization of the colchicine-binding site of tubulin. Proceedings of the National Academy of Science 90: 11598–11602.

Weisgigl, M., J. Clos. 2001. Heat shock protein 90 homeostasis controls stage differentiation in *Leishmania dovoni*. Mol. Cell. Biol. 12: 3307–3316.

Welte, M. A., I. Duncan and S. Lindquist. 1995. The basis for a heat-induced developmental defect: defining crucial lesions. Genes and Development 9: 2240–2250.

West-Eberhard, M. J. 2003. *Developmental Plasticity and Evolution*. Oxford University Press, New York.

Whitfield, C. W., A. M. Cziko and G. E. Robinson. 2003. Gene expression profiles in the brain predict behavior in individual honey bees. Science 302: 296–299.

Williams, K. D., A. B. Helin, J. Posluszny, S. P. Roberts and M. E. Feder. 2003. Effect of heat shock, pretreatment, and *hsp70* copy number on wing development in *Drosophila melanogaster*. Molecular Ecology 12: 1165–1177.

Williams, J. B., S. P. Roberts and M. M. Elekonich. 2008. Age and natural metabolically-intensive behavior affect oxidative stress and antioxidant mechanisms, Experimental Gerontology 43: 538–549.

Winston, M. L., 1987. *The Biology of the Honey Bee*. Harvard University Press, Cambridge, Massachusetts.

Wolsky, A. 1983. Abnormal eye development in *Drosophila* after treatment with vinblastine. Experimental Cell Biology 51: 345–351.

Young J.C., I. Moarefi and F. U. Hartl. 2001. Hsp90: a specialized but essential protein-folding tool. The Journal of Cell Biology 154: 267–273.

Zou, J., Y. Guo, T. Guettouche, D. F. Smith and R. Voellmy. 1998. Repression of heat shock transcription factor HSF1 activation by HSP90 (HSP90 complex) that forms a stress-sensitive complex with HSF1. Cell 94: 471–480.

18

Learned Host Preferences

Andrew B. Barron
Centre for the Integrative Study of Animal Behaviour, Macquarie University, Sydney, NSW 2109, Australia, E-mail: andrew.barron@mq.edu.au

Abstract

Learning, a form of behavioural phenotypic plasticity, can have significant fitness benefits. In this chapter I discuss the evidence that insects can learn features of their hosts, and consider whether learned changes in behaviour are restricted to a single life history stage, or whether a learned behavioural change acquired in the preimaginal stages can influence imaginal behaviour. I then discuss which evolutionary scenarios promote learning over innate responses to hosts, and how learned changes in behaviour can interact with natural selection in the evolution of host races.

Insect learning

Behaviour is a component of phenotype, and environmentally induced changes in behaviour represent behavioural phenotypic plasticity. Learned behavioural changes are clearly a subset of behavioural phenotypic plasticity, but defining the boundaries of what constitutes learning is difficult. Thorpe (1956) defined learning as "that process which manifests itself by adaptive changes in behaviour as a result of experience", and this definition (or something similar) has been used widely in the animal behaviour literature (Alcock, 1989; Lorenz, 1965; Manning and Dawkins, 1992). Even so, this definition has been criticised because writing adaptation into the definition of learning excludes examples of behavioural plasticity that do not fit easily into an adaptive paradigm (Jermy, 1987). A dramatic example of this is the tendency of some caterpillars, having sampled one host plant, to refuse a second acceptable host species to the point of starvation (Hanson, 1976; Jermy, 1987). For this reason many authors prefer a simpler definition of learning as: a change in behaviour with experience (Stephens, 1993; Vet and Groenewold, 1990), and this is the definition adopted for this review.

Learning is widespread in insects; it is documented in representatives of every order studied to date (Jermy, 1987). Learning abilities have been explored in greatest detail in the honey bee (*Apis mellifera*). The honey bee performs extremely well in elementary associative learning tasks, and also demonstrates advanced 'cognitive' learning abilities (that rival the performance of vertebrates) including generalization, categorization and cross-modal recall (Giurfa et al., 2001; Menzel and Giurfa, 2001; Reinhard et al., 2004; Srinivasan et al., 1998; Zhang et al., 2005). At present, a lack of comparative studies makes it difficult to judge to what extent other insects share the tremendous 'cognitive powers' of the honey bee.

More often, studies of insect learning involve simple bioassays of food preferences or oviposition behaviour and compare behaviour of naive insects with those given controlled exposure to a stimulus. Changes in behaviour are often referred to as 'conditioned' or 'induced' behavioural changes. These neutral terms are used deliberately, since it is often unclear whether the learned behavioural change is driven by a Pavlovian associative process, sensitisation, habituation, or any of the many specific forms of learning. Induction may represent several different learning processes (Bernays, 1996; Papaj and Prokopy, 1989).

Imaginal and Preimaginal Conditioning of Adult Behaviour

Experience can alter the feeding, foraging, or oviposition preferences of adult Orthoptera (Muralirangan et al., 1997), Lepidoptera (Cunningham et al., 1998; Karowe, 1989; Rojas and Wyatt, 1999), Diptera (Jaenike, 1988; McCall and Kelly, 2002; Prokopy et al., 1982; Prokopy et al., 1986), Coleoptera (Phillips, 1977; Wallin and Raffa, 2002), and Hymenoptera (Bjorksten and Hoffmann, 1998; Vet et al., 1995; Whitman, 1988). Insects can also learn during the larval stages (Kennedy et al., 1987; Tully et al., 1994a; Yamamoto, 1974). That experience within a life history stage can induce a learned change in behaviour is not in doubt, but there has been a great deal of debate over whether experience in the preimaginal stages can induce a behavioural change that persists through metamorphosis to influence the behaviour of the adult.

In holometabolous insects the reorganisation of the nervous system during metamorphosis is so dramatic that it is difficult to see how memory with a neural basis could persist. This is especially true of the higher

Diptera. In *Drosophila*, adult sensory neurons and interneurons are formed *de novo* from imaginal disks (embryonic neuroblasts that persist into larval life) during metamorphosis (Carlson, 1991; Truman, 1996). The mushroom bodies are the regions of the brain most often implicated in learning and memory, and in *Drosophila* fibres of intrinsic neurons in the mushroom bodies degenerate during early stages of metamorphosis. The cell bodies remain and most regenerate new adult fibres, which make new connections (Truman, 1990, 1996).

Compelling demonstrations of preimaginal conditioning of adult behaviour in holometabolous insects are rare, and are greatly outnumbered by studies refuting the possibility. Several authors found no effect of larval feeding experience on adult feeding and oviposition preferences in Lepidoptera (Palmiter, 1966; Rojas and Wyatt, 1999; Thompson and Parker, 1928) and Diptera (Jaenike, 1982), and much of the evidence purporting to show preimaginal conditioning has been criticised for failing to completely isolate emergent adults from the preimaginal environment (Corbet, 1985; Jaenike, 1982). Corbet (1985) discusses how many cases of apparent preimaginal learning could be adult learning of chemical cues from the larval environment carried on the pupa or cocoon that are encountered by the emergent adult (so called chemical legacies). For *Drosophila*, traces of the larval diet on the surface of the puparium are sufficient to induce a change in the responses of emerging adults to the larval diet (Barron and Corbet, 1999). Many parasitoid wasps emerge on, or from within, the body of their larval host, and exposure to host cues during this early imaginal period has been shown to be of particular importance in determining adult oviposition preferences, whereas there is little support for preimaginal conditioning in parasitoid wasps (Caubet et al., 1992; Kester and Barbosa, 1991; van Emden et al., 1996; Vet and Groenewold, 1990). Since traces of the larval diet can be found within the body of the imago, or within the pupal cuticle as well as on the surface of the pupa or cocoon (Corbet, 1985), preimaginal conditioning of adult behaviour is extremely difficult to prove.

Despite this difficulty, there remain a few studies that provide evidence in support of preimaginal conditioning of adult behavioural responses. *Hyssopus pallidus* (family Eulophidae) is a gregarious larval ectoparasitoid of the codling moth, *Cydia pomonella*, which infests apples. Gandolfi et al. (2003) found that exposure of *H. pallidus* as young larvae to an odour induced a strong change in the oviposition response of adults to the same odour. This effect was observed for both apple fruit extracts and menthol, a pure compound that would not normally be encountered in the parasitoid wasp's life cycle (Gandolfi et al., 2003). This finding is exciting because

Gandolfi et al's (2003) methods control for the possibility of significant chemical legacies, and these authors also report that preimaginal conditioning of adult behaviour was most effective when young parasitoid larvae are exposed to odours immediately post hatching. This implies the development of a persistent behavioural change that endured through all larval moults, as well as metamorphosis, to influence the oviposition behaviour of the adult.

A study of *Drosophila* provides, perhaps, the best evidence for preimaginal conditioning of adult behaviour to date (Tully et al., 1994a). Tully et al. used a Pavlovian conditioning procedure to study associative learning and memory retention in *Drosophila* larvae (Tully et al., 1994a). Larvae were exposed to two odours and trained to associate one with delivery of a powerful electric shock. This resulted in a persistent conditioned avoidance of the shock-associated odour that could still be observed in adult flies eight days after training (Tully et al., 1994a). While the effect of preimaginal conditioning on the behavioural responses of the adults was not strong (compared to similar conditioning of adult flies), Tully's evidence argues for the retention of memory of overt negative reinforcement through metamorphosis. Whether memories that endure through metamorphosis are neurobiologically similar to memories acquired during the adult stage is an open question. Of relevance here is the remarkable report that an induced host preference in house fly (*Musca domestica*) larvae could be transferred between flies by injecting dissociated nervous tissue from trained larvae into the heads of untrained larvae immediately prior to pupariation (Ray, 1999). Emergent adult flies that received grafts of neural tissue from larvae exposed to novel odours showed an induced behavioural change that reflected the experience of the donor larvae (Ray, 1999). Whether the grafted tissue integrated with the host nervous system, or whether the induced behavioural change was mediated by humoral factors is not clear.

Evolutionary Costs and Benefits of Learning

For parasitoid wasps, learning reduced the search-time taken to locate hosts thereby increasing rates of parasitism (Olson et al., 2003; Steidle, 1998; Vet et al., 1995) and fitness (Dukas and Duan, 2000), but there are costs associated with learning. Mery and Kaweki (2004b; 2005) have shown that in *Drosophila*, repeated memory formation reduces fecundity, and that formation of long-term memory reduces survival in harsh environments. Long-term memory formation requires protein synthesis (Tully et al., 1994b), and Mery and Kawecki's studies seem to show that these metabolic costs are not trivial. There may even be a constitutive fitness cost to possessing a learning ability, whether it is utilised or not, since selection for improved learning ability in replicated experimental *Drosophila* populations has been consistently associated with a decline of larval competitive ability (Mery and Kawecki, 2003).

Regardless of any intrinsic costs to learning and memory, learning (and other forms of phenotypic plasticity) is often considered to be disadvantageous in stable environments. It is argued that in a stable world, an innate host preference would be preferable over a learned one, because individuals would not have to waste time and resources learning what to do (Dukas, 1998; Laverty and Plowright, 1988; Stephens, 1993). For these reasons it is often stated that learning has evolved as a response to an unpredictable environment (Papaj and Prokopy, 1989; Shettleworth, 1984; Thorpe, 1956). Similarly, generalists are often assumed to exhibit greater learning abilities than specialists since the range of microenvironments encountered and behavioural responses available to generalists are assumed to be greater than those for specialists. This seems to hold true for some parasitoids (Geervliet et al., 1998; Poolman Simons et al., 1992; Vet, 1983; Vinson et al., 1977), but for phytophagous species the evidence is mixed (Papaj and Prokopy, 1989).

Stephens (1991; 1993) distinguished between two timescales on which environmental unpredictability could operate: timescales greater than an animal's generation time (between-generation predictability), and timescales less than an animal's generation time (within-generation predictability). Environmental stability could be different at these two timescales for a number of reasons. For example, if offspring are dispersed widely from the parental habitat and are then relatively sedentary as they grow up then the environment would be less predictable between generations than within one. Predictions from a mathematical model suggested that learning was most favoured when the environment was unpredictable between generations but relatively constant within a generation (Table 1). Under these conditions it is impossible to evolve a fixed optimal behaviour, but there are the greatest returns for learning by an individual.

Stephen's simulation model has shown that there is still a benefit to learning in an apparently fixed environment. One shift in the environment every fifty generations or so was enough to give learning a selective advantage over fixed behaviour (Stephens, 1993).

Table 1 (Stephens, 1991). The influence of between and within generation predictability on the evolution of learning. If the environment is unpredictable over the short term (within generations) then there is little selective advantage to learning, since learning could not be predictive in this circumstance. If the environment is very stable and predictable both within and between generations, then a fixed behavioural pattern would be more advantageous than having to relearn the same behaviour each generation. The greatest benefit to learning is found when the environment is predictable within a generation but variable between generations. Under these circumstances a single fixed optimal behaviour cannot evolve and there is a strong benefit to learning.

		within-generation predictability	
		low	*high*
between-generation predictability	low	ignore experience	learn
	high	ignore experience	ignore experience

Papaj (1994) has taken a different approach to investigate the evolutionary advantages of learning. His model explores the different ways in which an animal could adapt to a stable environment. Papaj assumes that there is an optimal behavioural response that is composed of an instinctive component and a learned component. The learned component is assumed to converge on the optimal behaviour with progressive experience. A species could adapt to a stable environment over evolutionary time either by increasing its instinctive behavioural response until it matches the optimum, or by increasing its rate of learning of the optimal behaviour, or by a combination of both strategies. Papaj's simulations suggest that when both processes are possible, changes in learning rate seem to dominate. In simulations, learning rate increased to a point at which further changes in the size of the instinctive behavioural response were effectively halted so that at the end point of the simulations learning was rapid, but innate responses were relatively weak. This leads to the counterintuitive conclusion that an evolutionary response to a stable environment could often be the evolution of rapid learning (Papaj, 1994).

The take-home message of these two models is that learning is selectively advantageous in a broad variety of environmental conditions. Perhaps the only environment that would not foster learning abilities is one of complete unpredictability. Clearly, testing hypotheses for the evolution of learning is extremely difficult, but a series of ambitious experiments selecting for innate and learned host preferences in *Drosophila* lends some support to Papaj's prediction that faster learning can evolve as a response to a stable, predictable, selection pressure (Mery and Kawecki, 2004a).

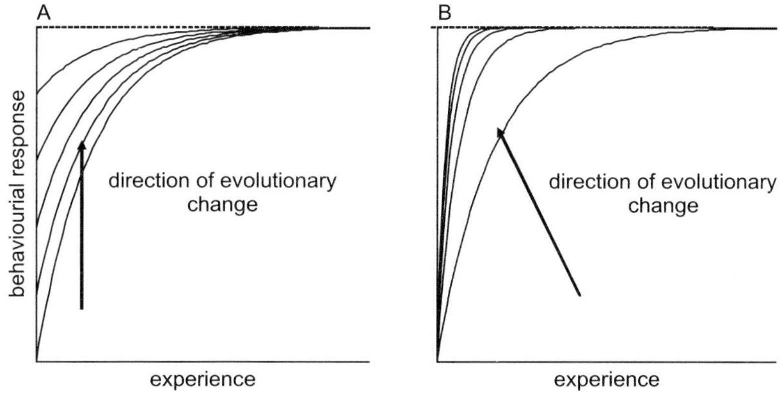

Fig. 1 Alternative evolutionary responses to a stable environment. A behavioural response has an instinctual component, which is fixed, and a learned component that varies with the experience according to the learning coefficient. If the learning coefficient is zero animals cannot learn, but for any learning coefficient greater than zero behaviour will eventually converge on the optimal response (denoted in each graph by the dotted line) with experience. The higher the learning coefficient the faster an animal learns. An animal can adapt to a stable environment either by increasing the instinctive component of the behavioural response until it matches the optimum (A) or by increasing the learning coefficient to increase the rate of learning of the optimal response (B). Adapted from Papaj, 1994.

Learning and the Evolution of Host Preferences

Would learning accelerate or delay the evolution of a novel host preference in an insect? The effect of learning (and behavioural plasticity in general) on evolution has long been debated, but still remains controversial. Some of the most influential arguments stem from Baldwin (1896), Lloyd Morgan (1896) and Osborn (1896) who independently developed ideas about 'organic selection', which suggests that behavioural plasticity facilitates the evolution of innate behaviour. Their arguments (often described as 'The Baldwin effect', and summarised by Bateson, 2004) imagine a population encountering a new environment. Selection acts initially on behavioural, physiological, and morphological phenotypic plasticity such that those who are able to modify themselves to the new environment survive, while those that do not perish. Initially phenotypic plasticity allows the population to persist in the new environment, but the population is under selective pressures imposed by the change in environment. Each generation adapts via plasticity to meet the challenges of the new environment and those that fail to adapt die off, but the adaptations are not passed on genetically.

This process repeats generation on generation. If there is genetic variation for the ease with which the modified characters are expressed, then individuals that express the modifications (the environmentally induced plastic response) most readily will be selected for in the population. As a consequence an inherited predisposition to express the beneficial modifications will evolve. The result is that over time there is an evolutionary change in the likelihood of expressing a plastic modification in behaviour or physiology, which, taken to the extreme, could lead to permanent expression of the originally plastic modification, in virtually all environments. This could then be recognised as the evolution of a new trait or the expression of a new instinct (West-Eberhard, 2003), and appear as genetically fixed in the population. Hence, in theory, adaptive environment-induced plastic responses can become genetically fixed traits, over time. These authors were not suggesting the inheritance of acquired characters, rather their argument considers the role of natural selection on genes that regulate the expression of plastic traits (Bateson, 2004). Natural selection can act to increase the frequency of genes that aid the expression of a beneficial plastic response in the population, and eliminate genes that stimulate a disadvantageous phenotypic plastic response.

Another view states that because plasticity allows individuals to adapt to a new situation irrespective of its genotype, learning can buffer some of the harshness of natural selection and slow the rate of evolution of an adaptive character (Ancel, 2000; Papaj, 1994). For example, if an insect could learn to avoid the parts of a novel host that contained the highest concentration of a defensive allelochemical, this could reduce the selection pressure of the allelochemical on the insect population, and perhaps slow the evolution of a detoxification enzyme for a novel plant allelochemical.

These two arguments appear to give contradictory predictions, but they address different processes. The Baldwin effect considers how natural selection can adjust the rate of existing plastic adaptation whereas the second view considers the evolution of a novel adaptive character. Thus, a particular selective force might, over time, directly alter a morphological or physiological trait, or might alter the capacity for an adaptive phenotypic plastic response. In reality, it is likely that both processes will occur as an evolutionary response to a new situation, and the effect of learning on the rate of evolution of a novel fixed character can vary case by case (Ancel, 2000; French and Messinger, 1994). Empirical studies have shown that even within the same experiment, learning sometimes accelerates the evolution of a preference for a novel host and sometimes delays it (Mery and Kawecki, 2004a).

Leaving aside theoretical considerations of learning and the rates of evolution, there is strong evidence that learned host preferences can influence population structure and gene flow in insect populations, and therefore might contribute to host race formation. Learning can accelerate behavioural adaptation to a new host, and can reinforce genetically based differences in host preference (Jaenike and Papaj, 1992). Host race formation depends strongly on assortative mating between members of different host races (Bush, 1994), which could occur if matings took place predominantly on or near the host. Host fidelity greatly reduces gene flow between host races of the apple maggot fly *Rhagoletis pomonella* (Feder et al., 1994). The effects of host preference on gene flow will be reinforced if the alternative hosts occur in different microhabitats, or if developmental times of the insect are different on the different hosts, which could result in adults from the different host races emerging out of phase (Feder and Filchak, 1999; Leclaire and Brandl, 1994; Payne and Berlocher, 1995). Hence, behavioural phenotypic plasticity (in our case, learning), through its effects on gene flow and host race formation, can stimulate speciation and effect the course of evolution (Gorur, 2005).

Summary

Learning, a type of behavioural phenotypic plasticity, has been observed in all insect orders studied so far. For most insects, analyses of learning capabilities have focussed on changes in feeding or ovipositional behaviour with experience, but in honey bees far more advanced forms of learning have been documented. Learning has been demonstrated in adult and juvenile forms of both holometabolous and hemimetabolous insects. The evidence that learned changes in behaviour can persist between life history stages is mixed, but this does appear to be possible for some insects and some forms of learning. Models suggest that learning has an advantage over invariable innate behaviour in most evolutionary scenarios. The few experimental studies that have tested these models support the suggestion that learning has an advantage even in a relatively fixed environment. Whether learning accelerates or slows the evolution of host preferences is contentious, but empirical studies have shown that by reinforcing host fidelity learning can influence gene flow in populations of phytophagous insects, and assist speciation.

Acknowledgements

I thank Sally Corbet and Doug Whitman for critical reviews of this manuscript.

References

Alcock, J. 1989. Animal behavior: an Evolutionary Approach. Sinauer Associates. 596 pp.

Ancel, L. W. 2000. Undermining the Baldwin expediting effect: does phenotypic plasticity accelerate evolution? *Theoretical Population Biology*, 58: 89–101.

Baldwin, J. M. 1896. A new factor in evolution. American Naturalist, 30: 441–451, 536–553.

Barron, A. B. and S. A. Corbet. 1999. Pre-imaginal conditioning in *Drosophila* revisited. Animal Behaviour, 58: 621–628.

Bateson, P. 2004. The active role of behaviour in evolution. Biology and Philosophy, 19: 283–298.

Bernays, E. A. 1996. Selective attention and host-plant specialization. Entomologia Experimentalis et Applicata, 80: 125–131.

Bjorksten, T. A. and A. A. Hoffmann. 1998. Persistence of experience effects in the parasitoid *Trichogramma* nr. *brassicae*. Ecological Entomology, 23: 110–117.

Bush, G. L. 1994. Sympatric speciation in animals: new wine in old bottles. Trends in Ecology & Evolution, 9: 285–288.

Carlson, J. R. 1991. Olfaction in *Drosophila*: genetic and molecular analysis. Trends in Neuroscience, 14: 520–524.

Caubet, Y., P. Jaisson and A. Lenoir. 1992. Preimaginal induction of adult behaviour in insects. Quarterly Journal of Experimental Psychology, 44B: 165–178.

Corbet, S. A. 1985. Insect chemosensory responses: a chemical legacy hypothesis. Ecological Entomology, 10: 143–153.

Cunningham, J. P., M. F. A. Jallow, D. J. Wright and M. P. Zalucki. 1998. Learning and host selection in *Helicoverpa armigera* (Hübner) (Lepidoptera: Noctuidae). Animal Behaviour, 55: 227–234.

Dukas, R. 1998. Evolutionary ecology of learning. *In* Dukas, R., ed. Cognitive Ecology, pp. 129–174. University of Chicago Press, Chicago.

Dukas, R. and J. J. Duan. 2000. Potential fitness consequences of associative learning in a parasitoid wasp. Behavioral Ecology, 11: 536–543.

Feder, J. L. and K. E. Filchak. 1999. It's about time: the evidence for host plant-mediated selection in the apple maggot fly, *Rhagoletis pomonella*, and its implications for fitness trade-offs in phytophagous insects. Entomologia Experimentalis et Applicata, 91: 211–225.

Feder, J. L., S. B. Opp, B. Wlazlo, K. Reynolds and W. Go. 1994. Host fidelity is an effective premating barrier between sympatric races of the apple maggot fly. Proceedings of the National Academy of Science USA, 91: 7990–7994.

French, R. and A. Messinger. 1994. Genes, phenes and the Baldwin effect. *In* Brooks, R. and P. Maes, [eds]. Artificial Life IV, pp. 277–282. MIT Press, Cambridge MA.

Gandolfi, M., L. Mattiacci and S. Dorn. 2003. Preimaginal learning determines adult response to chemical stimuli in a parasitic wasp. Proceedings of the Royal Society of London Series B-Biological Sciences, 270: 2623–2629.

Geervliet, J. B. F., A. I. Vreugdenhil, M. Dicke and L. E. M. Vet. 1998. Learning to discriminate between infochemicals from different plant-host complexes by the parasitoids *Cotesia glomerata* and *C. rubecula*. Entomologia Experimentalis et Applicata, 86: 241–252.

Giurfa, M., S. Zhang, A. Jenett, R. Menzel and M. V. Srinivasan. 2001. The concepts of 'sameness' and 'difference' in an insect. Nature, 410: 930–933.

Gorur, G. 2005. The importance of phenotypic plasticity in herbivorous insect speciation. *In* Ananthakrishnan, T. and D. Whitman, [eds]. Insect Phenotypic Plasticity Diversity of Responses, pp. 145–171. Science Publishers, Inc, Enfield.

Hanson, H. E. 1976. Comparative studies on induction of food choice preferences in lepidopterous larvae. *In* Jermy, T., [ed.]. The host plant in relation to insect behaviour and reproduction, pp. 71–77. Plenum Press, New York.

Jaenike, J. 1982. Environmental modification of oviposition behavior in *Drosophila*. American Naturalist, 119: 784–802.

—. 1988. Effects of early adult experience on host selection in insects: some experimental and theoretical results. Journal of Insect Behavior, 1: 3–15.

Jaenike, J. and D. R. Papaj. 1992. Behavioural plasticity and patterns of host use by insects. *In* Roitberg, B. D. and M. B. Isman, eds., Insect Chemical Ecology: An Evolutionary Approach, pp. 245–264. Chapman & Hall, London.

Jermy, T. 1987. The role of experience in the host selection of phytophagous insects. *In* Chapman, R. F., E. A. Bernays and J. G. Stoffolano, [eds]. Perspectives in Chemoreception and Behavior, pp. 143–157. Springer-Verlag, New York.

Karowe, D. N. 1989. Facultative monophagy as a consequence of prior feeding experience: behavioural and physiological specialisation in *Colias philodice* larvae. Oecologia, 78: 106–111.

Kennedy, G. G., R. R. Farrar and M. R. Riskallah. 1987. Induced tolerance of neonate *Heliothis zea* to host plant allelochemicals and carbaryl following incubation of eggs on foliage of *Lycopersicon hirsutum* f. *glabratum*. Oecologia, 73: 615–620.

Kester, K. M. and P. Barbosa. 1991. Postemergence learning in the insect parasitoid, *Cotesia congregata* (Say) (Hymenoptera: Braconidae). Journal of Insect Behaviour, 4: 727–742.

Laverty, T. M. and R. C. Plowright. 1988. Flower handling by bumblebees: a comparison of specialists and generalists. Animal Behaviour, 36: 733–740.

Leclaire, M. and R. Brandl. 1994. Phenotypic plasticity and nutrition in a phytophagous insect: consequences of colonising a new host. Oecologia, 100: 379–385.

Lloyd Morgan, C. 1896. On modification and variation. Science, 4: 733–740.

Lorenz, K. Z. 1965. Evolution and modification of behaviour. University of Chicago Press, Chicago. pp.

Manning, A. and M. S. Dawkins. 1992. An Introduction to Animal Behaviour. Cambridge University Press, Cambridge. 196 pp.

McCall, P. J. and D. W. Kelly. 2002. Learning and memory in disease vectors. Trends in Parasitology, 18: 429–433.

Menzel, R. and M. Giurfa. 2001. Cognitive architecture of a mini-brain: the honeybee. Trends in Cognitive Science, 5: 62–71.

Mery, F. and T. J. Kawecki. 2003. A fitness cost of learning ability in *Drosophila melanogaster*. Proceedings of the Royal Society of London Series B-Biological Sciences, 270: 2465–2469.

—. 2004a. The effect of learning on experimental evolution of resource preference in *Drosophila melanogaster*. Evolution, 58: 757–767.

—. 2004b. An operating cost of learning in *Drosophila melanogaster*. Animal Behaviour, 68: 589–598.

—. 2005. A cost of long-term memory in *Drosophila*. Science, 308: 1148–1148.

Muralirangan, M. C., M. Muralirangan, and P. D. Partho. 1997. Feeding behaviour and host selection strategies in Acridids. *In* Gangwere, S., M. C. Muralirangan, and M. Muralirangan, eds. The Bionomics of Grasshoppers, Katydids and their Kin, pp. 163–182. CAB International, Wallingford, UK.

Olson, D. M., T. A. Hodges and W. J. Lewis. 2003. Foraging efficacy of a larval parasitoid in a cotton patch: influence of chemical cues and learning. Journal of Insect Behavior, 16: 613–624.

Osborn, H. F. 1896. Ontogenic and phylogenic variation. Science, 4: 786–789.

Palmiter, R. D. 1966. Absence of olfactory conditioning in an oligophagous insect *Heliothis zea* (Boddie). Animal Behaviour, 14: 236–238.

Papaj, D. R. 1994. Optimizing learning and its effect on evolutionary change in behaviour. *In* Real, L. A., [ed.]. Behavioral Mechanisms in Evolutionary Ecology, pp. 133–156. University of Chicago Press, Chicago.

Papaj, D. R. and R. J. Prokopy. 1989. Ecological and evolutionary aspects of learning in phytophagous insects. Annual Review of Entomology, 34: 315–350.

Payne, J. A. and S. H. Berlocher. 1995. Phenological and elctophoretic evidence for a new blueberry-infesting species in the *Rhagoletis pomonella* sibling species complex. Entomologia Experimentalis et Applicata, 75: 183–187.

Phillips, W. M. 1977. Modification of feeding 'preference' in the flea beetle *Haltica lythri* (Coleoptera: Chrysomelidae). Entomologia Experimentalis et Applicata, 21: 71–80.

Poolman Simons, M. T. T., B. P. Suverkropp, L. E. M. Vet and G. de. Moed. 1992. Comparison of learning in related generalist and specialist eucoilid parasitoids. Entomologia Experimentalis et Applicata, 64: 117–124.

Prokopy, R. J., A. L. Averil, S. S. Cooley and C. A. Roitberg. 1982. Associative learning in egg laying site selection by apple maggot flies. Science, 218: 76–77.

Prokopy, R. J., D. R. Papaj, S. S. Cooley and C. Kallet. 1986. On the nature of learning in oviposition site acceptance by apple maggot flies. Animal Behaviour, 34: 98–107.

Ray, S. 1999. Survival of olfactory memory through metamorphosis in the fly *Musca domestica*. Neuroscience Letters, 259: 37–40.

Reinhard, J., M. V. Srinivasan and S. W. Zhang. 2004. Olfaction: scent-triggered navigation in honeybees. Nature, 427: 411–411.

Rojas, J. C. and T. D. Wyatt. 1999. The role of pre- and post-imaginal experience in the host-finding and oviposition behaviour of the cabbage moth. Physiological Entomology, 24: 83–89.

Shettleworth, S. J. 1984. Learning and behavioural ecology. *In* Krebs, J. R. and N. B. Davies, [eds]. Behavioural Ecology: An Evolutionary Approach, pp. 170–194. Blackwell Scientific Publications, London.

Srinivasan, M. V., S. W. Zhang and H. Zhu. 1998. Honeybees link sights to smells. Nature, 396: 637–638.

Steidle, J. L. M. 1998. Learning pays off: influence of experience on host finding and parasitism in *Lariophagus distinguendus*. Ecological Entomology, 23: 451–456.

Stephens, D. W. 1991. Change, regularity and value in the evolution of animal learning. Behavioral Ecology, 2: 77–89.

—. 1993. Learning and behavioral ecology: incomplete information and environmental predictability. *In* Papaj, D. R. and A. C. Lewis, [eds]. Insect learning: ecological and evolutionary perspectives, pp. 195–218. Chapman & Hall, New York.

Thompson, W. R. and H. L. Parker. 1928. Host selection by *Pyrausta nubilais* Hubn. Bulletin of Entomological Research, 18: 359–564.

Thorpe, W. H. 1956. Learning and Instinct in Animals. Methuen, London. 493 pp.

Truman, J. W. 1990. Metamorphosis of the CNS of *Drosophila*. Journal of Neurobiology, 21: 1072–1084.

—. 1996. Metamorphosis of the insect nervous system. *In* Gilbert, L. I., J. R. Tata and B. G. Atkinson, [eds]. Metamorphosis Postembryonic Reprogramming of Gene Expression in Amphibian and Insect Cells, pp. 283–320. Academic Press, San Diego.

Tully, T., V. Cambiazo and L. Kruse. 1994a. Memory through metamorphosis in normal and mutant *Drosophila*. Journal of Neuroscience, 14: 68–74.

Tully, T., T. Preat, S. C. Boynton and M. D. Vecchio. 1994b. Genetic dissection of consolidated memory in *Drosophila*. Cell, 79: 35–47.

van Emden, H. F., B. Sponagl, E. Wagner, T. Baker, S. Ganguly and S. Douloumpaka. 1996. Hopkins' 'host selection principle', another nail in its coffin. Physiol Entomol, 21: 325–328.

Vet, L. E. M. 1983. Host-habitat location through olfactory cues by *Leptopilina clavipes* (Hartig) (Hym.: Eucoilidae), a parasitoid of fungivorous *Drosophila*: the influence of conditioning. Netherlands Journal of Zoology, 33: 225–248.

Vet, L. E. M. and A. W. Groenewold. 1990. Semiochemicals and learning in parasitoids. Journal of Chemical Ecology, 16: 3119–3133.

Vet, L. E. M., W. J. Lewis and R. T. Cardé. 1995. Parasitoid foraging and learning. *In* Cardé, R. T., [ed]. Chemical Ecology of Insects 2, pp. 65–101. Chapman & Hall, New York.

Vinson, S. B., C. S. Barfield and R. D. Henson. 1977. Oviposition behaviour of *Bracon mellitor*, a parasitoid of the boll weevil (*Anthonomus grandis*). II. Associative learning. Physiol Entomol, 2: 157–164.

Wallin, K. F. and K. F. Raffa. 2002. Prior encounters modulate subsequent choices in host acceptance behaviour by the bark beetle *Ips pini*. Entomologia Experimentalis et Applicata, 103: 205–218.

West-Eberhard, M. J. 2003. Developmental Plasticity and Evolution. Oxford University Press, Oxford. 794 pp.

Whitman, D. W. 1988. Plant natural products as parasitoid cuing agents. *In* Cutler, H., [ed.]. Biologically Active Natural Products, pp. 386–396. American Chemical Soc. Washington, DC.

Yamamoto, R. T. 1974. Induction of host plant specificity in the tobacco hornworm, *Manduca sexta*. Journal of Insect Physiology, 20: 641–650.

Zhang, S., F. Bock, A. Si, J. Tautz and M. V. Srinivasan. 2005. Visual working memory in decision making by honey bees. Proceedings of the National Academy of Science USA, 102: 5250–5255.

19

Adaptive Maternal Effects: A Case Study of Egg Size Plasticity in a Seed-Feeding Beetle

Charles W. Fox[1] and Mary Ellen Czesak[2]

[1]*Department of Entomology, University of Kentucky, Lexington, Kentucky, USA 40546-0091, E-mail: Fox@uky.edu*

[2]*Department of Biology, Vassar College, Poughkeepsie, NY 12604, USA E-mail: maczesak@vassar.edu*

Abstract

When the maternal environment is predictive of the environment that her offspring will encounter, females have an opportunity to manipulate the phenotype of their offspring to prepare them for the conditions predicted by the cue. We briefly review examples of this transgenerational (cross-generational) phenotypic plasticity. We then focus on a case study of egg size plasticity in a seed-feeding beetle, *Stator limbatus*. Females adjust egg size in response to the plant species (seeds) on which they lay their eggs, in a manner consistent with the variation in selection (the relationship between offspring size and fitness varies depending on the host species). We explore the genetics underlying evolutionary responses in egg size and plasticity. We find that genetic relationships among traits (egg size, body size and fecundity) depend on the host upon which females oviposit. The evolutionary dynamics of resource allocation strategies will thus vary substantially among the environments in which beetles lay eggs, not just because selection varies among environments but also because the genetic relationships among traits vary among environments. We also find that the evolution of egg size plasticity is not predictable from a cross-environment genetic correlation estimated from a sib-breeding experiment. Realized cross-environment genetic correlations (a measure of the observed correlated response to selection) depend on the direction of selection (Up vs. Down) and on the environment (host species) in which selection is imposed. We describe a simple two-locus model that is adequate to explain the observed correlated responses to selection and evolution of plasticity. This model posits two types of loci, one that affects the trait (egg size) in both environments and one that affects the trait in only one environment. Lastly, we explore some ecological lessons learned from studies of how egg size plasticity affects beetle colonization of alien plants.

Organisms live in spatially and temporally variable environments. Variation in selection resulting from this environmental variation can maintain genetic variation in populations (e.g., maintain polymorphism) or can favor generalist genotypes that do well across a range of environmental conditions. Phenotypic plasticity is a mechanism by which a single genotype can respond adaptively to a variety of environments (Roff 2002; Ghalambor et al. 2007). Most often, organisms respond to early developmental conditions by modulating their own development, and thus manipulate their own phenotype, to prepare them for the environmental conditions they expect to encounter. However, when the maternal environment is predictive of the environment that her offspring will encounter, females have an opportunity to manipulate the phenotype of their offspring to prepare them for the environmental conditions predicted by the cue (Fox and Mousseau 1998; Mousseau and Fox 1998a, b).

There are numerous well studied examples of maternally-mediated phenotypic plasticity, commonly called transgenerational (or cross-generational) phenotypic plasticity (Fox and Mousseau 1998; Mousseau and Fox 1998a; Marshall and Uller 2007; Rasanen and Kruuk 2007; Marshall 2008). Offspring dormancy, including diapause in insects, is frequently under maternal control (Mousseau and Dingle 1991; Donahue and Schmitt 1998; Evans and Dennehy 2005). Flight polymorphisms in many aphids, some grasshoppers and a variety of other taxa, are mediated in part or full by maternal environmental conditions (Fox and Mousseau 1998; Simpson and Miller 2007; Mondor et al. 2008; Chapter 5, this volume). There are also many examples by which mothers more subtly influence the phenotype of their offspring. For example, herbivory or parasitism on mothers can affect resistance of their offspring to herbivores/parasites (Agrawal 2002; Gallizzi et al. 2008), and maternal parasite exposure to resistant hosts can affect offspring susceptibility to that resistance (see below). It is not surprising that mothers can have such large effects on ecologically-significant traits of their offspring since mothers determine the protein and RNA composition of eggs, and can thus manipulate early gene expression in eggs and embryos, including genes that affect major developmental trajectories (Johnstone and Lasko 2001; Bettegowda et al. 2008). Mothers can also make epigenetic changes to chromatin to significantly modulate gene expression in an egg or embryo (Bossdorf et al.

2008). However, while the genetics and physiology of maternal influences on development have been well studied in model systems, they are poorly understood for traits that have ecological significance (but see Miller et al. 2008).

In addition to controlling developmental switches in offspring, mothers also control the quantity and quality of resources provided to their offspring; i.e., they can manipulate offspring size (e.g., egg or seed size) and quality. Offspring size is an evolutionarily interesting trait because, even when selection favors large offspring, females necessarily encounter a constraint making their offspring large; for any fixed allocation to reproduction mothers that make larger offspring will have lower fecundity (Smith and Fretwell 1974). For females to produce larger offspring without experiencing a fecundity cost, they must increase total reproductive allocation (Winkler and Wallin 1987), which necessarily must come at the cost of allocation to other functions (such as growth or maintenance). Alternatively, they need to increase their total resource pool (Sakai and Harada 2005), which comes with costs such as delayed maturation or increased foraging risk.

The size of offspring that are favored by selection varies substantially with environmental conditions (Fox and Czesak 2000). This generates variation in offspring size among taxa and across space and time within species (Marshall and Keough 2008a, b). However, because females frequently have reliable information on the environmental conditions their offspring will encounter (such as when they can assess the condition of the environment in which they are laying their eggs), there is an opportunity for females to manipulate reproductive allocation to produce offspring of a size appropriate for the environmental conditions that the offspring are expected to encounter. Indeed, organisms from a wide diversity of taxa manipulate offspring size in response to environmental cues. For example, *Daphnia* females lay larger eggs when food limited (Glazier 1998). This may improve offspring survival during food stress experienced immediately after hatching, though some studies do not support this hypothesis (Guinnee et al. 2007). In some mites, cladocerans, copepods and bryozoans, mothers produce larger offspring at higher densities (Cleuvers et al. 1997; Plaistow et al. 2007; Allen et al. 2008), likely because large offspring have significant advantages when competing with conspecifics (Marshall et al. 2006) or dealing with food stress (Cleuvers et al. 1997).

Of course, not all effects of the maternal environment on offspring size are adaptive (Marshall and Uller 2007). For example, food-stressed females commonly lay smaller eggs (opposite of the pattern described above), especially at older ages (Fox 1993), probably an unavoidable consequence of nutritional stress (Fox and Czesak 2000). The adaptive significance of other plastic responses is unclear. For example, in most insects, females reared at low temperature lay larger eggs (Ernsting and Isaaks 2000; Fischer et al. 2006), though there are exceptions (McKee and Ebert 1996; Stillwell and Fox 2005). Some studies have found that temperature affects the fitness consequences of egg size (Azevedo et al. 1996; Fischer et al. 2003b; Hassall et al. 2006) but others have not (Ernsting and Isaaks 1997; Blanckenhorn 2000; Fischer et al. 2003a). In some cases, temperature mediated plasticity may prepare insects for seasonal shifts in other environmental variables, rather than temperature per se (Seko and Nakasuji 2006). More often, though, temperature effects on egg size probably result from non-adaptive effects of temperature on the physiology of oogenesis (e.g., shift in the rate of oocyte production relative to the rate of oocyte growth) (Ernsting and Isaaks 1997; Steigenga and Fischer 2007).

Many herbivorous insects manipulate egg size in response to the host plant on which they are ovipositing (Fox et al. 1997b; Mizumoto and Nakasuji 2007). In some cases the observed plasticity is opposite that predicted by patterns of selection, such that its adaptive significance is unclear (Ekbom and Popov 2004). In others, the plasticity matches predictions based on how selection on egg size varies among host species (Braby 1994; Fox and Mousseau 1996; Fox and Czesak 2000; Agosta 2008). In this chapter, we describe a case study for the evolution of phenotypic plasticity: egg size in a seed-feeding beetle, *Stator limbatus*. In *S. limbatus* the relationship between offspring size and fitness varies depending on the plant species. Females adjust egg size in response to the host species on which they lay their eggs, in a manner consistent with the variation in selection. We discuss the sources of selection that generate this egg size plasticity, and the genetics underlying evolutionary responses in egg size and plasticity. We then explore some ecological lessons learned from studies of how egg size plasticity affects beetle colonization on alien plants.

Plasticity in the Size of Eggs Laid by *Stator limbatus*

Stator limbatus (family Chrysomelidae, sub-family Bruchinae) (Figure 1) is a seed-feeding beetle that is distributed from northern South America to the southwestern United States. It is a relative generalist in that it feeds on seeds of > 70 species of legumes, substantially more than other species in the genus *Stator* and more than most other bruchines (Morse and Farrell 2005b, a). In the Sonoran Desert of the southwestern United States, *S. limbatus* uses seeds

Fig. 1 *Stator limbatus* on seeds of cat-claw acacia, *Acacia greggii.*

of primarily three host species. Two of these are congeneric caesalpinoid legumes (blue paloverde, *Parkinsonia florida*, and foothill or small-leaf paloverde, *P. microphylla*), both of which vary from large shrubs to trees, and one shrubby mimosoid legume (cat-claw acacia, *Acacia greggii*). Blue paloverde is primarily limited to lower elevations and in or around desert washes. Foothill paloverde is much more widespread but generally found at higher elevations than is blue paloverde. Cat-claw acacia is the most widespread of the three hosts, common throughout the Sonoran, Mojave and Chihuahuan deserts (aka *Acacia wrightii* in the Chihuahuan Desert). Despite large regions of allopatry, the three species are commonly sympatric, and the two paloverde species frequently hybridize (Siemens et al. 1994; Fox et al. 1997c). There is thus substantial spatial variation in the host species available to beetles.

There is also temporal variation among the three species in the phenology of fruit production and thus in seed availability to *S. limbatus*. Though both paloverde species flower in April or May in the Sonoran Desert, and begin fruiting in May or June (Turner 1963), the availability of their fruits to beetles varies substantially. Foothill paloverde fruits are dehiscent and seeds are dropped shortly after maturation to be gathered quickly by rodents (Siemens et al. 1992). In contrast, blue paloverde fruits are indehiscent and stay attached to the maternal plant for many months, with seeds available into late fall. Because blue paloverde fruits are indehiscent, beetles can only access seeds inside of fruits that have been damaged by other animals, including other seed beetles (such as *Mimosestes* species) and rodents (Mitchell 1977). Cat-claw acacia generally fruits later than either paloverde species, especially at higher elevations where they can flower as late as October, and produces indehiscent fruits that are only available to beetles when damaged by another animal (though fruits also crack along seams, providing access to seeds).

Most significantly for the ecology of *S. limbatus* egg size is that these three legume species differ substantially in their suitability for beetle development. Cat-claw acacia is generally the highest quality host; larval survival is high (frequently >95% in laboratory experiments) and development fast on this host (Fox et al. 1994, 1995). At the other extreme is blue paloverde, on which larval survival is poor (often <50%) and larval development slow compared to development on cat-claw acacia. On this host, hard seed coats and seed coat chemical defenses cause high mortality in first instars (Siemens et al. 1992; Siemens et al. 1994). Indeed, pre-hatch embryonic beetle mortality is much greater for eggs on seeds of blue paloverde than on seeds of the other hosts (suggesting that the defense is dissolving into the eggs), and extracts of paloverde seed surfaces are toxic to larvae. Though most larval mortality occurs at the seed surface, larval mortality is also higher and larval development slower inside blue paloverde seeds than inside the other Sonoran hosts. However, beetle larvae reared on blue paloverde ingest less seed mass per mg of beetle mass gained during development. Thus, it appears that blue paloverde seeds are nutritionally better resources than are other Sonoran hosts but that larvae are coping with non-nutritional stresses inside of these seeds, such as resistance mechanisms. The chemical defense on the surface of blue paloverde seeds does not appear to be present on foothill paloverde seeds – larval survival on foothill paloverde is intermediate to survival on blue paloverde and cat-claw acacia, and mortality on foothill paloverde rarely occurs as first instar larvae are penetrating the seed-coat.

Selection on Egg Size Varies Within and Among Host Species

The high larval mortality on blue paloverde seeds generates substantial selection on females to produce offspring that can overcome the seed's physical and chemical defenses. A variable, under maternal control, that has a large effect on offspring survival is egg size. Larvae hatching from large eggs are better able to penetrate the seed coat of blue paloverde seeds, generating substantial natural selection favoring large eggs on this host (Fox and Mousseau 1996). However, resistance of seeds to larval penetration varies substantially among individual blue paloverde trees (but varies little within trees because the seed coat is maternal tissue)(Siemens and Johnson 1990). This variation in resistance among trees generates substantial variation in selection on egg size; selection favoring large eggs is greater on more resistant trees (Figure 2)(Fox 2000). This positive relationship between seed-coat resistance and selection on egg size has been found in all populations of blue paloverde that we have studied, and occurs both within populations (selection favoring large eggs is greatest on the most resistant individual trees) and among populations (selection favoring large eggs is the greatest in populations that have, on average, the most resistant trees)(Fox et al. 2001).

Of course, beetles encounter more than just blue paloverde trees in nature. There are many locations where blue paloverde is sympatric with either foothill paloverde and/or cat-claw acacia. Beetles at such locations switch hosts through the season as fruit availability of the three species changes. Individual trees also vary substantially from year to year in whether and how much they fruit, such that beetles must disperse to find new hosts, and frequently switch hosts.

On seeds of cat-claw acacia, larval mortality is low and uncorrelated with egg size – larvae hatching from small or large eggs survive equally well (Fox and Mousseau 1996) (Figure 3A). Offspring hatching from large eggs have slightly shorter development time, but the effect is small (Fox 1997). Females can thus lay small eggs with minimal fitness cost (minimal reduction in offspring survival) but substantial fitness benefit since females that lay

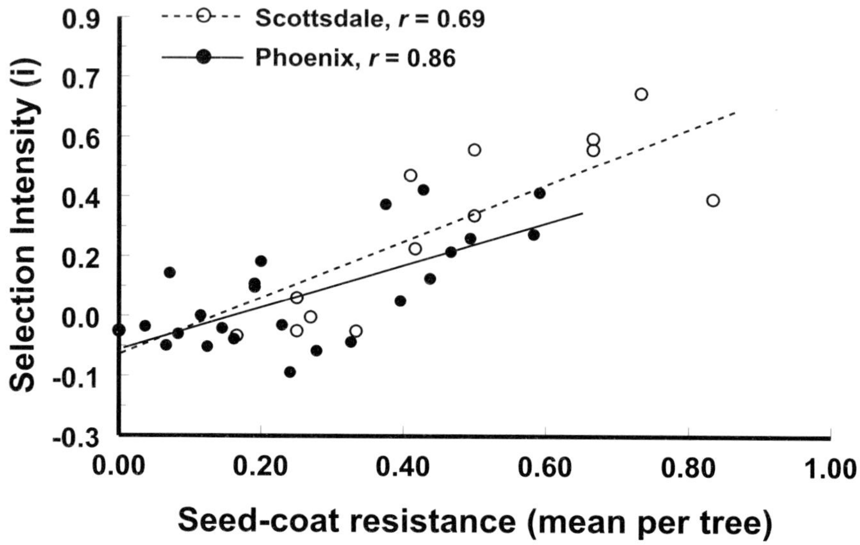

Fig. 2 The resistance of blue paloverde (*Parkinsonia florida*) seed coats affects the intensity of selection on *Stator limbatus* egg size. Each point represents the intensity of selection occurring on a single tree, and the regression is the relationship among trees within a population (Scottsdale or Phoenix). Seed coat resistance is defined as the probability of mortality of a larval beetle as it attempts to penetrate the seed coat. Selection intensity, *i*, is the standardized selection differential, $i = (z_A - z_B) / s$, where z_A is the mean size of egg that produced a surviving offspring (i.e., mean *after* selection), z_B is the mean size of all eggs laid (i.e., mean *before* selection), and *s* is the standard deviation in egg size before selection.

smaller eggs can have increased fecundity. Females that lay small eggs on blue paloverde seeds also lay more eggs, but this benefit is balanced by the fitness cost of laying small eggs (Figure 3B). Thus, fecundity selection favors laying small eggs regardless of host (for increased fecundity), but the degree to which selection via offspring survival balances this fecundity selection varies among hosts, and thus the egg size that maximizes female fitness is much larger when females lay eggs on blue paloverde than when they lay eggs on cat-claw acacia (Figure 4) (see also modeling on pp. 433-438 of Roff 2002).

Egg Size Plasticity

Stator limbatus has responded to host-specific selection on egg size by evolving egg size plasticity – females lay larger eggs when ovipositing on blue paloverde seeds than when ovipositing on cat-claw acacia seeds. Figure 5 shows an example in which females are allowed to mature their

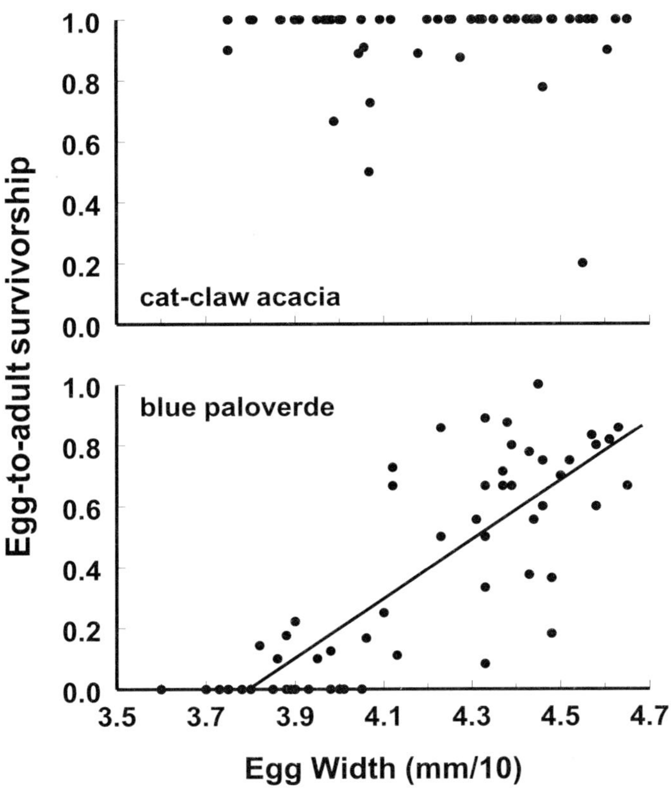

Fig. 3 The relationship between egg size and egg-to-adult survival in the seed beetle, *Stator limbatus*, when larvae are reared on seeds of (top) cat-claw acacia (*Acacia greggii*) or (bottom) blue paloverde (*Parkinsonia florida*). Data from Fox and Mousseau (1996).

eggs while in contact with seeds of blue paloverde (dashed lines) or cat-claw acacia (solid lines) but then were switched to the alternate host immediately after they start laying eggs (Time 0). Note first that females mature different size eggs on the two hosts (egg size at Time 0 in Figure 5) – they start laying large eggs when in the presence of blue paloverde seeds and small eggs when in the presence of cat-claw acacia seeds. The difference in egg size is quite large – paloverde-size eggs are ~25-30% larger than acacia-sized eggs. After the switch to a new host it takes females ~48-72 hours to adjust egg size to their new oviposition substrate (Fox et al. 1997b; Savalli and Fox 2002). This difference in egg size between host species has a large effect on female fecundity; females lay 40% more eggs when they oviposit on cat-claw acacia than when they oviposit on blue paloverde (Fox et al. 1997b; Czesak and Fox 2003a).

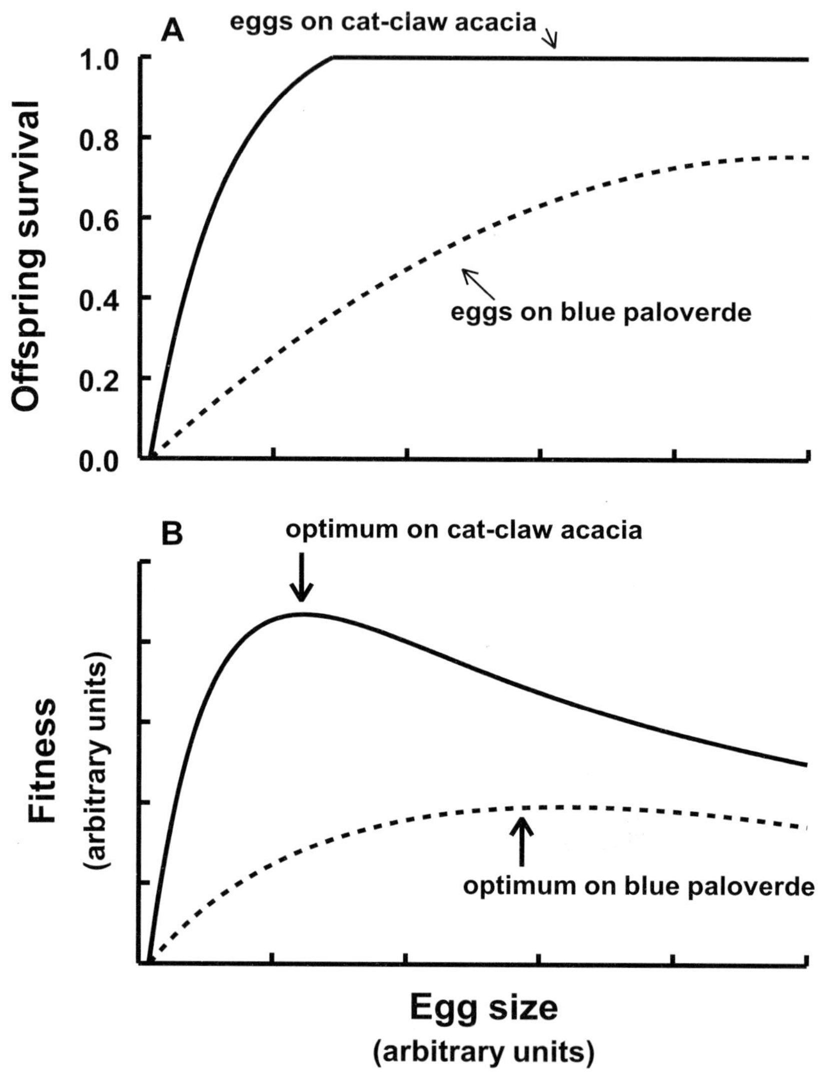

Fig. 4 A simple optimality model demonstrating how female fitness (the number of surviving offspring produced) is influenced by egg size on cat-claw acacia and blue paloverde. (A) The relationship between egg size and offspring survival on cat-claw acacia (solid line) or blue paloverde (dashed line). We assume fitness is unaffected by egg size above some threshold but drops substantially below a threshold (if we assume that there is no effect of egg size on survival for all egg sizes then the smallest physiological possible egg size is the optimal egg size). (B) The fitness of a female laying on each host as a function of egg size, assuming no other selection and a simple trade-off between egg size and number (in which x resources are divided equally among n eggs). Note that the optimal egg size is much larger on blue paloverde.

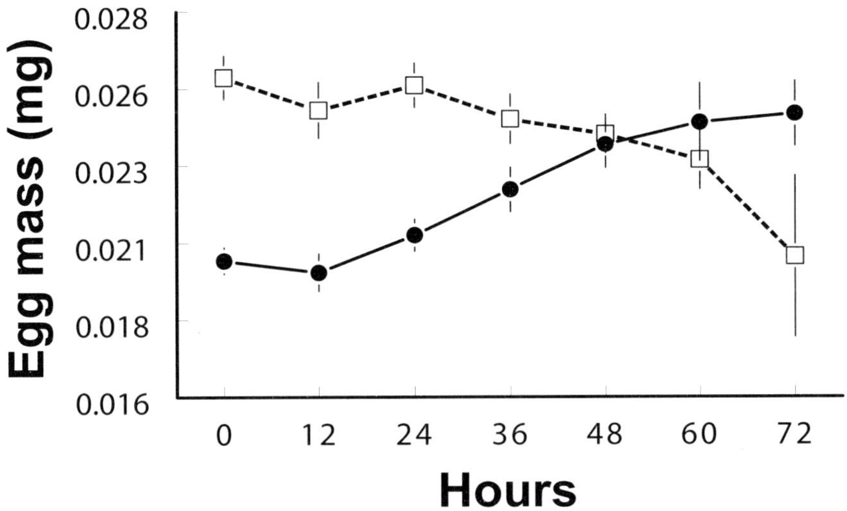

Fig. 5 The effect of oviposition host on egg size in *Stator limbatus*, and the change in egg size following a switch to an alternate host. Dashed lines are females that start laying on blue paloverde and are switched to cat-claw acacia (switched at time 0); solid lines are females that start laying on cat-claw acacia and are switched to blue paloverde. Redrawn from Savalli and Fox (2002).

Is this plasticity adaptive? Possibly blue paloverde is a non-preferred host and thus females delay oviposition, such that eggs recruit more resources (e.g., more yolk), and thus get larger, simply because they are retained longer in the female. We can reject this hypothesis because females that are forced to retain eggs (by confining them on non-preferred hosts or without hosts) lay small eggs immediately upon encountering a suitable host. Instead, it appears that females control their egg size, and that plasticity in egg size is an adaptive response to the seed type. There are substantial fitness costs to females that do not lay eggs of a size appropriate for the seed on which they are ovipositing. Females that lay small eggs on blue paloverde produce few or no surviving offspring, and those that lay unnecessarily large eggs on cat-claw acacia have reduced fecundity.

Interestingly, females not provided access to any host seeds during egg maturation, or those provided only novel hosts during egg maturation, mature eggs of a size determined by their natal host – females raised on blue paloverde mature larger eggs than do females raised on cat-claw acacia (Savalli and Fox 2002)(this study controls for evolution during the experiment). However, even females reared on blue paloverde will lay relatively small eggs if given access to a cat-claw acacia seed immediately after emergence rather than allowed 48h before given an acacia seed. Females thus appear to emerge from their rearing host with small eggs in their oviduct, after which they assess their environment. If they find host seeds quickly, they begin adjusting their eggs to match the size appropriate for the seed they find. Until they find a seed, however, they begin adjusting their eggs to match the size that would be appropriate for their natal host. This is likely an adaptive strategy for females because it is likely that early emergence experiences are good predictors of the host species to be encountered.

The cue to which females respond is currently unknown. Females need to be in contact with seeds to correctly identify the species and lay the appropriate sized egg (Savalli and Fox 2002). The cue is thus on the seed surface and not volatile. Females do not use the difference in seed size between catclaw acacia and blue paloverde as the cue for adjusting egg size (Savalli and Fox 2002). Also, the cue to which females respond is not the same as the seed trait that confers resistance against larvae; in a paloverde hybrid swarm, in which trees vary substantially from low to high resistance, seed coat resistance and the cue stimulating the production of large eggs appear to assort independently (Fox et al. 1997c).

Though our studies of egg size plasticity have focused primarily on the three most common host species of *S. limbatus* in the Sonoran Desert, we are only scratching the surface of the potential complexity in plasticity that exists in nature. We have recently studied two *S. limbatus* populations in Colombia that are genetically substantially divergent from the Arizona populations and which may represent a different species (Morse 2003; Morse and Farrell 2005b). Those Colombian populations also exhibit egg size plasticity. They lay large eggs on a native Colombian host, igua (aka guachapeli or cenicero, *Pseudosamanea guachapele*) but small eggs on some other hosts (including cat-claw acacia, *A. greggii*) (Amarillo-Suarez and Fox 2006). Curiously, though, igua does not have significant seed-coat defenses, and larvae hatching from large eggs have no obvious survival advantage over larvae hatching from small eggs (in laboratory studies). However, igua seeds are quite small and larval densities on these seeds are often quite high. Could larvae hatching from large eggs have a fitness advantage during larval competition? Regardless of the explanation, that egg size plasticity is present in such diverse branches of the *S. limbatus* phylogeny indicates that egg size plasticity is ancestral in this species, likely pre-dating (and possibly facilitating) the diet expansion of *S. limbatus* onto the wide diversity of hosts that it currently uses (Amarillo-Suarez and Fox 2006).

Many other interesting patterns have yet to be explained. For example, Arizona populations of *S. limbatus* lay larger eggs on igua than on cat-claw acacia (but smaller eggs than they lay on blue paloverde; A. Amarillo-Suárez, unpublished), despite no close relative of igua being native in the Sonoran desert (Amarillo-Suarez and Fox 2006). This suggests that the cues stimulating the plastic response to blue paloverde and igua are similar despite the hosts being phylogenetically unrelated (different subfamilies of legumes), and even though the sources of selection maintaining egg size plasticity in the two localities must be quite different. Likewise intriguing is the observation that Colombian and Texas populations of *S. limbatus* lay large eggs on blue paloverde, despite not having access to blue paloverde in their native ranges. This suggests that either gene flow is high among geographically quite distant populations (consistent with the phylogeography)(Morse 2003; Morse and Farrell 2005b) and that plasticity genes confer little cost to beetles in the absence of a host for which it is needed, or that plasticity is maintained by selection associated with different ecological conditions in different parts of the beetle's distribution. Clearly much about the ecology, biogeography and evolutionary history of egg size plasticity is still a mystery.

Selection on Egg Size Mediates Selection on Adult Body Size

The variation among host species in selection on egg size of *S. limbatus* has some interesting consequences for how selection affects other traits. For example, as is typical in insects, female body size has a moderate effect on egg size and a large effect on fecundity, such that larger females lay both larger and more eggs (see Chapter 11, this book). Direct selection on either egg size or fecundity thus causes indirect selection on female body size. Using a simple path analysis (Scheiner et al. 2000) we examined how oviposition host affects the relative magnitude of indirect selection on body size through these two different paths (fecundity vs. egg size)(Figure 6) (Fox and Czesak 2006). When females lay eggs on blue paloverde, selection on female body size is influenced by both selection through the fecundity path (body size → fecundity → fitness), because larger females lay more eggs, and selection through the egg size path (body size → egg size → fitness), because larval survival is positively correlated with egg size (Figure 6). In contrast, when females lay eggs on cat-claw acacia, selection affects female body size only through the fecundity path. This is because egg size does not affect larval survival when eggs are laid on cat-claw acacia, so there is no selection on female size through the egg size path.

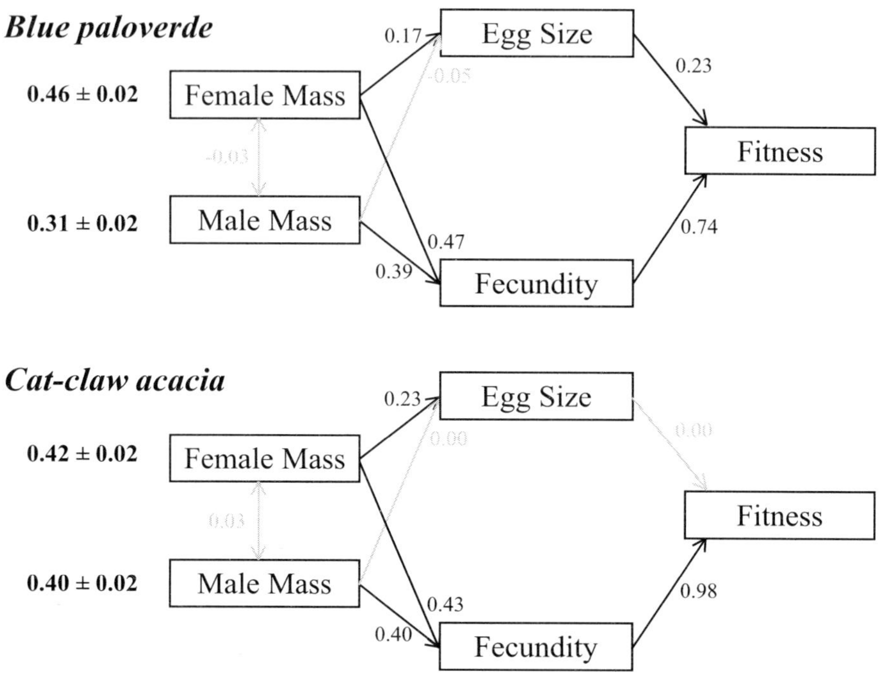

Fig. 6 Path analysis demonstrating how oviposition host can affect the magnitude of indirect selection on body size in the seed beetle *Stator limbatus*. Path coefficients are standardized. Black paths are statistically significant and grey paths are non-significant. 'Fitness' is the number of larvae produced that successfully survive until completely inside their host seed. Numbers to the far left (± standard errors) are total indirect selection on male or female body size. Modified from Fox and Czesak (2006).

The story becomes more intriguing when we consider selection on male body size. Male *S. limbatus* provide large nuptial gifts to females (as large ejaculates) which in turn affects female fecundity (Savalli and Fox 1998; Moya-Larano and Fox 2006). Because larger males produce larger nuptial gifts, there is fecundity selection on male size that is roughly similar in magnitude to the fecundity selection on female size (Savalli and Fox 1998; Fox and Czesak 2006). However, male size has no effect on the size of eggs laid by females (Czesak and Fox 2003b), such that the only indirect selection on male size, regardless of oviposition host, is through the fecundity path (in our simple model; Figure 6). This has an important implication for the relative magnitude of selection on male versus female size and thus the evolution of sexual size dimorphism. Specifically, when eggs are laid on cat-claw acacia, the indirect selection on male and female size is only through the fecundity path (fecundity selection), resulting in a fairly similar amount of selection on both male and female size. In contrast, when eggs are laid on blue paloverde there is indirect selection on female body size through both the egg size and fecundity paths, whereas indirect selection on male body size is still only through the fecundity path. Thus, when eggs are laid on blue paloverde there is greater indirect selection on female body size than on male body size, whereas when eggs are laid on cat-claw acacia indirect selection on body size is of similar intensity on both sexes. This shift in relative selection on males versus females is a consequence of where females lay eggs. It is independent of any changes in male or female investment into reproduction, variation in sexual selection, or any other direct effects on adult beetles, and is not caused by differential mortality of males versus females, large versus small beetles, or any direct effect of male size on fitness.

Though our study was performed in the lab, the difference in the intensity of indirect selection on male body size between the two hosts is similar in magnitude to the median total amount of directional selection observed in nature in studies of morphological traits (Kingsolver et al. 2001). We thus suspect that variation among hosts in the fitness consequences of egg size is a major source of variation in selection on *S. limbatus* body size in nature.

The Evolutionary Genetics of Egg Size

Selection on the size of eggs that females lay affects a variety of resource allocation decisions in females, including how much to allocate to reproduction (reproduction versus growth and maintenance) and how to divide total reproduction among eggs (larger versus more eggs) (Fox and Czesak 2000). However, because the genetic architecture of traits commonly depends on the environmental conditions in which those traits are expressed the degree to which these various allocation relationships co-evolve could likewise be dependent on environmental conditions (Roff 1997). To explore how host species affects these evolutionary relationships, we imposed artificial selection on egg size and quantified the correlated evolutionary responses of fecundity and a variety of other traits (e.g., body size). We created replicate lines selected for increased egg length ("Up" lines; top 20% each generation) and decreased egg length ("Down" lines; bottom 20%), each paired with unselected controls. Selection was imposed separately using blue paloverde and cat-claw acacia seeds (Czesak and Fox 2003a). To control for natural selection on egg size in the paloverde lines, all larvae were reared to adult on seeds of cat-claw acacia (after scoring the size of eggs females laid on blue paloverde).

After nine generations of selection egg size had diverged substantially among the selected lines. As expected based on the previously demonstrated trade-off between egg size and number, females selected to lay larger eggs on blue paloverde evolved to have lower fecundity, whereas females selected to lay smaller eggs evolved higher fecundity (Figure 7A). In contrast, fecundity did not evolve in lines selected to lay large or small eggs on seeds of cat-claw acacia. Though a surprising result, this matched results from a half-sib breeding experiment run just before the selection experiment; when females laid eggs on blue paloverde, egg size and fecundity were negatively genetically correlated ($r_A = -0.51$) (i.e., there was a substantial genetically based trade-off between egg size and number), whereas when females laid eggs on cat-claw acacia the genetic correlation was not significantly different from 0 (r_A = -0.13). Initially, this appears inconsistent with an allocation trade-off; how can females lay larger eggs without the cost of reduced fecundity? The answer is that females selected to lay larger eggs on cat-claw acacia evolved to be larger beetles that allocated more total resources to reproduction, whereas those selected to lay smaller eggs evolved to be smaller beetles that allocated fewer resources to reproduction (Figure 7B)(Czesak and Fox 2003a). Again, this was consistent with results from a half-sib breeding experiment; when females laid eggs on cat-claw acacia, egg size and body size were positively genetically correlated ($r_A = -0.32$) whereas when females laid eggs on blue paloverde egg size and body size were uncorrelated ($r_A = -0.07$). Recent studies of egg size in other insects have found quite similar results, that the genetic relationships among egg size, body size and fecundity are very environmentally-dependent (Steigenga et al. 2005; Seko et al. 2006)

This experiment illustrates two key points of relevance to understanding evolutionary dynamics of traits in complex environments. First, genetic variances for traits, and genetic covariances underlying relationships among traits, are highly dependent on the environment in which traits are expressed. This environmental dependence of genetic (co)variances has long been expected from theory (Roff 1997), and has been demonstrable in studies comparing genetic (co)variance matrices. However, more recent studies demonstrate that this environmental effect on genetic (co)variance structure in populations is more than just subtle changes in magnitude of the various genetic parameters; the changes in magnitude are often large and sometimes even in direction (Hoffmann et al. 1995; Sgro and Hoffmann 2004; Charmantier and Garant 2005; Sultan 2007). This leads to our second key conclusion, which is that the large environmental dependence of genetic (co)variances can have substantial ecological implications. In our study, we

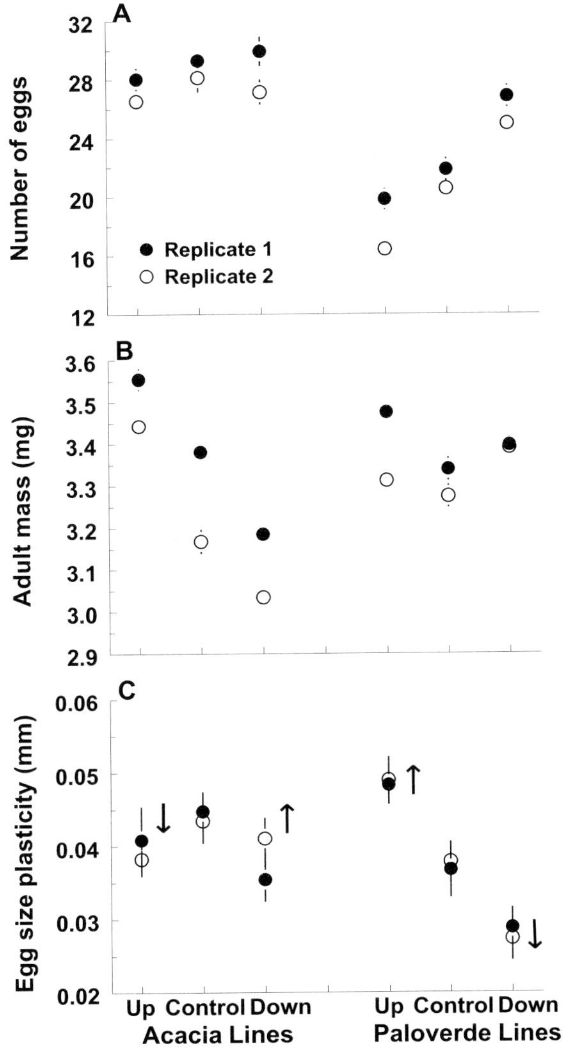

Fig. 7 Correlated responses to selection on *Stator limbatus* egg size; (A) Fecundity evolved in response to selection on the size of eggs laid on blue paloverde but not in response to selection on cat-claw acacia, whereas (B) body mass evolved in response to selection on the size of eggs laid on cat-claw acacia but not in response to selection on blue paloverde; (C) Phenotypic plasticity evolved as predicted in response to selection on blue paloverde, but not as predicted in response to selection on cat-claw acacia. Plasticity = egg size on paloverde – egg size on acacia. The arrows in C are the predicted correlated responses based on the genetic correlation estimated from a half-sib breeding design. Closed and open circles are replicates 1 and 2, respectively.

see that the evolution of resource allocation strategies (e.g., to growth versus reproduction and to large versus many eggs) should vary substantially among the environments in which beetles lay eggs, not just because selection varies among environments but also because the genetic relationships among traits vary among environments.

The Evolutionary Genetics of Egg Size Plasticity

Selection on a trait within one environment affects not only other traits expressed in that same environment but also the same trait when it is expressed in alternate environments. This is because traits (including egg size) can be affected by the same genes, or at least a significantly overlapping set of genes, in all of the environments in which they are expressed. However, some of the loci that affect the trait of interest are expressed in only one or a subset of environments, whereas other loci are expressed in all environments but affect the trait differently in each environment. That genes have effects on traits in multiple environments means that selection on a trait in one environment will affect expression of that trait in other environments. However, the environmental-dependence of gene expression and genetic effects means the evolution of the trait in that alternate environment (the correlated response to selection) will generally be smaller than the direct response to selection. In some cases, loci can even have opposite effects in different environments such that the correlated response to selection can be opposite the direct response.

A common way to quantify the degree to which a trait expressed in one environment is influenced by selection occurring on that same trait in a different environment is to calculate a *cross-environment genetic correlation* (Falconer 1952; Via 1994). This is analogous to a genetic correlation between any two traits, such as body size and fecundity, except that the traits are defined according to the environment in which they are expressed. For example, we can measure a genetic correlation between the size of eggs that females lay on cat-claw acacia and the size of eggs that they lay on blue paloverde (Fox et al. 1999). A high cross-environment genetic correlation indicates that gene expression is similar in different environments, such that selection for a trait in one environment will have similar effects on the trait in the other environment. Cross-environment genetic correlations can constrain the adaptive divergence of traits across environments. The higher the genetic correlation, the slower the divergence of the two traits will be when they are selected for different optima in each environment. However, though high genetic correlations affect the rate and trajectory of evolution of correlated traits (Fry 1996) they do not constrain traits from ultimately attaining their 'optimal' trait values in different environments (Lande 1980b; Reeve and Fairbairn 2001) unless the correlation = 1 and the genetic and phenotypic variances are the same for the trait in each environment.

Experimental Evolution of Egg Size Plasticity in *Stator limbatus*

For *S. limbatus*, the size of eggs that females lay on cat-claw acacia seeds is genetically correlated with the size of eggs that they lay on blue paloverde seeds, r_A (the additive genetic correlation) > 0.6 when calculated from half-sib breeding experiments (Fox et al. 1999; Czesak et al. 2006). Thus, selection on egg size on cat-claw acacia seeds should have a large effect on the size of eggs laid on blue paloverde, and vice versa. However, when the genetic correlation estimates are less than 1.0 (as for egg size), or the genetic and phenotypic variances for egg size differ between the two oviposition environments, the correlated responses to selection will be different than the direct responses to selection and plasticity will evolve.

To better understand how cross-environment genetic correlations influence the evolution of egg size plasticity in *S. limbatus*, we used the artificial selection experiment described above to test (a) whether the cross-environment additive genetic correlation estimated from our half-sib quantitative genetic breeding designs accurately predicted the correlated evolution of egg size in one environment to selection in the alternate environment (and vice versa), and (b) whether the evolution of phenotypic plasticity is predictable from this cross-environment genetic correlation. Plasticity should increase (relative to control lines) when we selected for increased egg size on blue paloverde or decreased egg size on cat-claw acacia. In contrast, plasticity should decrease when we selected for decreased egg size on blue paloverde or increased egg size on cat-claw acacia. This is because the cross-environment genetic correlation is less than 1.0 and thus the correlated response to selection is less than the direct response to selection. When we select on one trait away from the mean of the other trait (e.g., selection for larger eggs on paloverde or smaller eggs on acacia) plasticity should increase, whereas when we select towards the mean of the other trait (for smaller eggs on paloverde or larger eggs on acacia) plasticity should decrease. This is commonly called the Jinks-Connolly rule (Falconer 1952; Jinks and Connolly 1973; Falconer 1990) and has been observed in selection studies (Gavrilets and Scheiner 1993; Perez and Garcia 2002), though specific combinations of genetic variances and covariances can lead to exceptions (e.g., when correlated responses are greater than direct responses to selection)(Falconer 1990).

In our artificial selection experiment, selection on the size of eggs that females laid on cat-claw acacia did indeed lead to the correlated evolution of the size of eggs laid on blue paloverde, and vice versa. To quantify the magnitude of the correlated responses, we calculated *realized* cross-environment genetic correlations. These are calculated from the observed direct and correlated responses to selection, as

$$r_A = \frac{CR_Y}{\frac{1}{2}(h_X h_Y) i \sigma_Y}$$

where CR_Y is the correlated response of trait Y estimated as the difference between control and selected lines, h_X and h_Y are the square roots of the narrow-sense heritabilities of each trait (estimated from half-sib analysis), i is the selection intensity, and σ_Y is the standard deviation of the distribution in trait Y estimated from half-sib analysis (note that the equation contains ½ in the denominator because selection was applied to one sex only, egg length is a trait of females). These estimates were then compared among the selection lines, and compared to the estimated genetic correlation from a half-sib analysis conducted immediately before we imposed selection. The realized cross-environment genetic correlations were positive for all selected lines (Table 1). However, they varied substantially among the selection lines, both within and between host species (Table 1). When selection was imposed on the size of eggs laid on cat-claw acacia, the realized cross-environment genetic correlation was highly asymmetrical; r_A was ≥ 0.71

Table 1 Estimates of the realized genetic correlations (r_A; ± standard error) between the length of eggs laid by female *Stator limbatus* on cat-claw acacia (*Acacia greggii*) and on blue paloverde (*Parkinsonia florida*) calculated after nine generations of selection for egg length. "Acacia lines" refers to lines selected to lay large (Up lines) or small (Down lines) eggs on cat-claw acacia, and paloverde lines refers to lines selected to lay large (Up lines) or small (Down lines) eggs on blue paloverde. The table is modified from Czesak et al (2006).

	Realized r_A Up Lines	*Realized r_A Down Lines*
Acacia lines		
Replicate 1	0.45 ± 0.08	0.91 ± 0.04
Replicate 2	0.39 ± 0.08	0.71 ± 0.06
Paloverde lines		
Replicate 1	0.75 ± 0.06	0.77 ± 0.06
Replicate 2	0.97 ± 0.02	0.71 ± 0.06

when we selected for small eggs but $r_A \leq 0.47$ when we selected for large eggs. In contrast, when selection was imposed on the size of eggs laid on blue paloverde seeds, the realized cross-environment genetic correlations were less asymmetrical. Thus, the realized genetic correlations varied depending on the environment in which selection was imposed and the direction of selection. In other words, the realized genetic correlation between two traits, egg size on acacia (trait X) and egg size on paloverde (trait Y), depended on the direction of selection on trait X (but not on the direction of selection on trait Y) and depended on which trait (X or Y) was under selection.

The evolution of phenotypic plasticity in egg size also depended on the environment in which selection was imposed – selection on the size of eggs laid by females on paloverde changed phenotypic plasticity in the predicted direction whereas selection on the size of eggs laid on acacia did not (Figure 7C). Previous studies have also found that the evolution of phenotypic plasticity can be dependent on the environment in which selection is imposed (Scheiner and Lyman 1991; Matsumura 1996; Noach et al. 1997; Noach et al. 1998). Complex responses for plasticity evolution appear to be the norm rather than the exception in selection experiments.

Genetic architecture affects the evolution of egg size plasticity

The results of our selection experiment – that realized genetic correlations were asymmetrical and dependent on the trait on which selection was imposed (egg size on paloverde versus egg size on acacia) – go against the common notion in quantitative genetics (based largely on the assumptions of the Gaussian infinitesimal model) that a genetic correlation can be used to predict the evolutionary response to selection over multiple generations regardless of the direction of selection and the trait on which selection acts. Similarly, other studies have found that observed correlated responses to selection do not agree with predicted responses based on genetic correlation estimates (Palmer and Dingle 1986; Gromko 1995; Worley and Barrett 2000), and some have observed variation in the correlated responses among replicate selection lines (Gromko et al. 1991; Hillesheim and Stearns 1991) or between divergent (up vs. down) selection lines (Wilkinson et al. 1990; Hillesheim and Stearns 1991; Worley and Barrett 2000). Stochastic effects, such as random variation in which loci contribute to the selection response (Gromko 1995) likely explain results of some experiments such as those for which correlated responses are inconsistent among replicates (Hillesheim and Stearns 1991). However, stochasticity is not adequate to explain the repeatable asymmetrical correlated responses to selection seen in our experiment.

Asymmetrical correlated responses to selection are a common outcome of genetic models (Roff 1997). There are only a limited set of conditions under which an asymmetrical correlated response to selection is not predicted. These include the extreme cases of when there is only a single locus with additive effects on a pair of traits or when traits are affected by an infinite number of loci and population sizes are infinite (Lande 1980a; Bulmer 1985). In the former case (single locus additive inheritance) the genetic correlation cannot evolve, whereas in the latter case the genetic variances and covariances remain approximately constant due to the large (infinite) number of loci. However, for all cases that fall between the extremes of the single locus additive model and the Gaussian infinitesimal model, some degree of asymmetrical correlated response to selection is common (Bohren et al. 1966). Thus, the constancy of genetic correlations predicted by the architecture of quantitative genetic variation depends critically upon the assumptions of the genetic model.

To gain insight into the type of inheritance that can explain the patterns of experimental evolution observed in our selection experiment, including both the observed pattern of realized cross-environment additive genetic correlations and the observed evolution of plasticity, we constructed a simple two locus model analogous to the multi-locus model of genotype-by-environment interactions presented in de Jong (1990). We consider a single trait expressed in two environments as two separate traits (traits X and Y) having phenotypic values z_X and z_Y respectively. We define the trait 'plasticity' (P) as the difference between the values of z_X and z_Y for a given genotype (i.e., $z_P = z_Y - z_X$). We thus consider plasticity an emergent property of the two traits, in contrast to some previous plasticity models that incorporate plasticity as a parameter in the model (Scheiner 1998). We assume that there are two unlinked loci (A and B) in linkage equilibrium with two alleles at each locus. We consider a two-locus model because it is simple but also flexible and produces general results that hold when we add together multiple additive loci. At each locus we have 2 alleles, A_1, A_2 at the A locus and B_1, B_2 at the B locus. With two traits, X and Y, we have four additive effects: a_{AX}, a_{BX}, a_{AY}, and a_{BY}, where the subscripts denote the locus being considered (A or B) and the trait affected (X or Y); these are the additive effects of locus i on trait j, where one allele has the genotypic value of $+a_{ij}$ and the other allele $-a_{ij}$ (Table 2A). The four alleles have the frequencies p_1, p_2, q_1 and q_2 for A_1, A_2, B_1 and B_2 respectively. We assume a large randomly mating diploid sexual population with discrete generations. The details of the model, including calculation of genetic variances, covariances and the

Table 2 The (A) generalized two-locus model and (B) the variant of the model that generates the evolutionary dynamics observed in our selection experiment (which we call *asymmetrical genetic architecture*). For comparison with the experimental results, Trait X is the size of eggs that females lay on cat-claw acacia, and trait Y is the size of eggs that females lay on blue paloverde. The asymmetrical genetic architecture model differs from the generalized two locus-model in that the affect of locus A is the same for both Trait X and Y (i.e., $a_{AY} = a_{AX}$) and locus B affects only one trait (Trait Y; $a_{BX} = 0$ but $a_{BY} \neq 0$). Note that the model includes no epistasis or dominance. The model is described in more detail in Czesak et al. (2006).

Table A – generalized two-locus model

		Effect on Trait X	*Effect on Trait Y*
Locus A			
	Allele 1	$+a_{AX}$	$+a_{AY}$
	Allele 2	$-a_{AX}$	$-a_{AY}$
Locus B			
	Allele 1	$+a_{BX}$	$+a_{BY}$
	Allele 2	$-a_{BX}$	$-a_{BY}$

Table B – asymmetrical genetic architecture

		Effect on Trait X	*Effect on Trait Y*
Locus A			
	Allele 1	$+a_{A}$	$+a_{A}$
	Allele 2	$-a_{A}$	$-a_{A}$
Locus B			
	Allele 1	0	$+a_{B}$
	Allele 2	0	$-a_{B}$

evolutionary dynamics of traits X, Y and plasticity are presented in Czesak et al. (2006) and thus not described here.

The evolutionary patterns observed in our selection experiment can be predicted by one particular genetic architecture – that is where one locus affects egg size only on blue paloverde while the other locus has approximately equal effects on egg size on both hosts (i.e., either $a_{AY} = a_{AX}$ and $a_{BX} = 0$, $a_{BY} \neq 0$ or $a_{BY} = a_{BX}$ and $a_{AX} = 0$, $a_{AY} \neq 0$ (Table 2B)(Czesak et al. 2006). The evolutionary dynamics of plasticity and the cross-environment genetic correlations are shown in Figure 8. Note in particular two key results. First, this genetic architecture lead to the evolution of plasticity in egg size only when selection is imposed on the size of eggs laid on blue paloverde, and not when selection is imposed on the size of eggs laid on cat-claw acacia. When selection is imposed on trait *X* (egg size on cat-claw acacia) only alleles at locus *A* will evolve. This is because locus *B* has no effect on Trait *X* (Table 2B). However, locus *A* has identical effects on Trait *Y* (egg size on blue paloverde), such that the correlated response to selection (i.e., evolution of Trait *Y*) will be identical to the direct response (evolution of Trait *X*) and thus plasticity does not evolve (Figure 8A). In contrast, selection on Trait *Y* affects allele frequencies at both loci, *A* and *B*. Because only one of these two loci (locus *A*) affects Trait *X*, the correlated response to selection will be smaller than the direct response and plasticity will evolve in the direction observed in the experiment (Figure 8A).

Second, the model predicts conditions under which the realized genetic correlations will be asymmetrical and differ between hosts in the manner observed in the experiment. Figure 8B presents an example of initial allele frequencies that would generate the observed realized genetic correlations. Note that selection on Trait *Y* (the size of eggs laid on blue paloverde) changes allele frequencies at both loci and the cross-environment genetic correlation stays fairly constant (the evolutionary trajectory is along an isoclines). In contrast, selection on Trait *X* (the size of eggs laid on cat-claw acacia) affects only allele frequencies at locus *A*; considering the hypothetical case where the starting allele frequency as are in Figure 8B, we see that selection for an increase in Trait *X* leads to a decline in the genetic correlation (the genetic correlation evolves across the isoclines), whereas selection for a decrease in Trait *X* leads to an increase in the genetic correlation, exactly as observed in the selection experiment. More generally, the genetic correlation changes much more rapidly as a function of allele frequencies at locus *A* compared to locus *B*, and generally becomes smaller as allele frequency evolve away from intermediate allele frequencies at locus *A*. This is because locus A contributes to both the numerator and denominator of the correlation (i.e., to both the genetic covariance and genetic variances) while locus *B* contributes only to the denominator (one genetic variance) and thus contributes a component of variance that does not change as the covariance changes. The model also leads to a variety of observations about genetic and environmental variances that are consistent with the empirical observations (Czesak et al. 2006).

Our simple two locus model of asymmetrical genetic architecture can thus produce the complex evolutionary dynamics observed in our selection experiment. This is despite the fact that the model is strictly additive – we assume no dominance or epistasis. Epistatic interactions among loci do influence the evolution of genetic variances and covariances (Schlichting

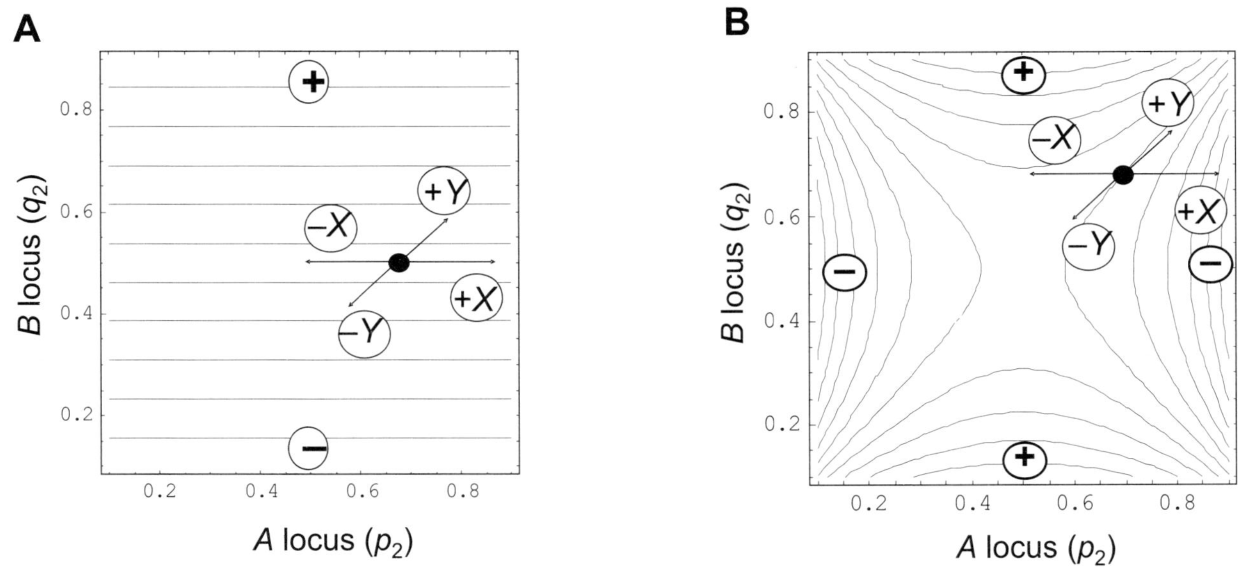

Fig. 8 (A) Contour plots showing the means of phenotypic plasticity as a function of the frequencies of alleles with positive effects at a pair of loci (*A* and *B*). Values were calculated using the genetic model under the asymmetrical genetic architecture, as in Table 2B, in which locus *A* affects both traits *X* and *Y* and locus *B* only affects trait. Contour lines are isoclines of equal value. The relative elevation on the surface is indicated with a + or –. Overlaid onto each surface is an evolutionary trajectory of a population experiencing directional selection for either larger or smaller values of trait *X* or *Y* (the arrow indicates the direction of response). (B) Contour plot of the additive genetic cross-environment correlation (r_{XY}) as a function of allele frequencies at a pair of loci (*A* and *B* locus). Contour lines are isoclines. All values are positive and relative elevation on the surface is indicated with a + or –. Overlaid onto the surface is an example of the evolutionary trajectory of a population experiencing directional selection for either larger or smaller values of trait *X* or *Y* (the arrow indicates the direction of evolution). Figures modified from Czesak et al (2006).

and Pigliucci 1993) and can contribute to the evolution of phenotypic plasticity (Pigliucci 2005), but they are not necessary to generate complex correlated responses to selection or complex evolutionary dynamics for plasticity (Bohren et al. 1966; Scheiner and Lyman 1991). Also, though other forms of genetic effects can have large influences on evolutionary dynamics, short-term evolutionary dynamics should be dominated by additive effects. Additive effects are thus likely to explain the pattern of response to selection seen in *S. limbatus*. However, we do not argue that the true genetic architecture underlying egg size evolution in *S. limbatus* is really this simple – affected by only two loci with strictly additive inheritance. The same basic results seen in this simple model hold for an arbitrary (n) number of loci (where $n > 1$) as long as one set of loci have similar effects on both traits (egg size on both hosts) while a second set of loci affect only Trait Y (egg size only on blue paloverde).

This pattern of allelic effects, where the effect of a locus is sensitive to the environment and the effects of another locus or loci are not, was proposed by Jinks and colleagues (Brumpton et al. 1977; Jinks et al. 1977) and has been called an 'epistasis model' by Scheiner and colleagues (Scheiner and Lyman 1991; Scheiner 1998; Berrigan and Scheiner 2004) because the final phenotype depends on environmental interactions across locus types. However, because our model and that of Scheiner and colleagues include only additive allelic effects, we called this *asymmetrical genetic architecture* to avoid confusion on the use of the term 'epistasis' which more correctly refers to within genome interactions between genotypes at different loci (Czesak et al. 2006). That loci have environment-specific effects is well documented, as discussed above. For *S. limbatus*, our model suggests that one or more loci affect the size of eggs laid on both cat-claw acacia and blue paloverde, whereas at least one additional locus affects only the size of eggs laid on blue paloverde. We propose that females default to laying "small" eggs except in the presence of specific host stimuli and that at least one gene mediates a shift in egg size when encountering this stimulus; this gene may be a regulatory control gene (Schlichting and Pigliucci 1993; Schlichting and Smith 2002). Experimental studies are consistent with this. Though females respond to their natal host plant in the absence of other host stimuli, even females emerging from blue paloverde seeds start by laying smaller eggs; they need time after emergence to increase their eggs to a size appropriate for blue paloverde (Savalli and Fox 2002). However, the details of the genetic architecture underlying egg size plasticity in *S. limbatus* are in need of much further research.

Egg Size Plasticity and the Colonization of Novel Environments

Stator limbatus is a broad generalist that has incorporated > 20 alien species into its diet. Some of these are ornamentals and others are invasives. Some are widespread enough that *S. limbatus* has expanded its geographic range by using these species as hosts. The use of Texas ebony (*Ebenopsis ebano*) is particularly intriguing. This tree is not used as a host by *S. limbatus* in locations where the beetle and plant naturally co-occur (southern Texas and the gulf coast of Mexico); it is used by *S. limbatus* only where the tree is non-native. However, a daughter species to *S. limbatus* (*S. beali*) is a specialist on Texas ebony in the tree's natural range. The speciation event that split *S. limbatus* and *S. beali* was associated with beetle colonization of Texas ebony ~1.2 mya (Morse 2003; Morse and Farrell 2005b, a). *S. beali* has since diverged from *S. limbatus* in a variety of life history traits directly related to the large size of Texas ebony seeds (Nilsson & Johnson, 1993; Fox & Mousseau, 1995; Fox et al., 1996). The current expansion of *S. limbatus* onto Texas ebony may thus provide insights into the ecological and evolutionary processes that occurred prior to the speciation of *S. beali*.

When reared in the laboratory, egg-to-adult survival of *S. limbatus* larvae is quite poor on Texas ebony seeds, generally < 5% (Fox et al. 1997a). However, in the field in central Arizona (where Texas ebony is commonly planted as an ornamental) survival of larvae is much greater, generally ~10-15%. This discrepancy is due to female experiences during egg maturation. Our main study populations in central Arizona are surrounded by blue paloverde and the majority of females ovipositing on Texas ebony immigrate from blue paloverde seeds (Fox 2006). These females immigrating from blue paloverde lay larger eggs on Texas ebony than they would have laid had they matured on Texas ebony or immigrated from cat-claw acacia. Females do not recognize Texas ebony as a host on which they should lay large eggs, even though egg size has a large effect on the egg-to-adult survival of larvae on Texas ebony (Fox and Savalli 2000). Thus, it is the maternal effect induced by exposure to blue paloverde seeds that allow larvae to have higher survival on Texas ebony seeds in nature than would be possible were Texas ebony the only host. When females encounter blue paloverde during egg maturation their larvae have up to 10-fold higher survival on seeds of Texas ebony than do offspring hatching from eggs laid by mothers that never encounter paloverde (Fox and Savalli 2000). Maternal experiences with blue paloverde also appear to facilitate colonization of the invasive Mexican paloverde (*P. aculeata*) (Fox et al. 2006). Like Texas ebony, Mexican paloverde is native in parts of the distribution of *S. limbatus*, but is not used as a normal host of *S. limbatus* in regions where the plant is native.

Later experiments have demonstrated that the improved survival on Texas ebony (and probably Mexican paloverde) is only partly due to the effect of maternal experience on the size of eggs that she lays; females also manipulate egg content in response to paloverde seeds, and these changes in egg composition are the major factor allowing offspring to survive on Texas ebony seeds (Fox and Savalli 2000; Fox 2006). Females are clearly either putting into eggs, or inducing expression inside of eggs, something that dramatically improves larval survival on both blue paloverde and Texas ebony. It is this egg composition effect that is the most exciting new development in the *S. limbatus* egg size plasticity story.

The role of phenotypic plasticity in facilitating colonization of new habitats, and in adaptation to a global change, is a topic of extensive research and debate (Ghalambor et al. 2007; Hulme 2008; Visser 2008). Plasticity likely facilitates colonization of new environments by increasing environmental tolerances and by allowing non-genetic "acclimation" to the novel environment to occur in a single generation (Sultan 2005). Plasticity may even facilitate adaptation to those new environments, and possibly speciation (West-Eberhard 1989; Agrawal 2001; Badyaev and Oh 2008), if traits that are originally plastic become canalized. However, experimental results on the importance of plasticity for colonization of novel environments, and facilitating invasiveness, have been mixed (Hulme 2008). Our *S. limbatus* results demonstrate that phenotypic plasticity can indeed directly influence the ability of offspring to survive in a novel environment. However, the colonization of that novel environment, and subsequent adaptation to that environment, can be influenced by the species composition of the local community and/or the composition of the community from which the colonists migrated. In this case, blue paloverde provides a stepping stone for colonization of Texas ebony. In the absence of blue paloverde, or when blue paloverde is at low density in the community, larval survival will be dramatically reduced on seeds of these alien species, and that should limit the ability of *S. limbatus* to colonize and possibly adapt to these alien species.

Lastly, we believe that phenotypic plasticity in egg size has been an important factor in the history of diet evolution of *S. limbatus*. Most species in the genus *Stator* are extreme specialists, using seeds of just one species or genus (Morse and Farrell 2005a). Only four species of *Stator* use hosts from more than one genus. *Stator limbatus* uses >50 native species, in 18 genera, plus a large number of alien species (>20), as hosts. Is this extreme generalization in part due to egg size plasticity? As discussed above, egg size plasticity appears to be an ancestral trait in *S. limbatus*. We have proposed that plasticity facilitates diet expansion by increasing larval survival on otherwise low quality hosts, reducing the selection against use of those hosts and providing populations time to respond evolutionarily to their new host before going extinct (Amarillo-Suarez and Fox 2006; Carroll and Fox 2007). Continued study of *S. limbatus* populations that have incorporated alien species into their diet should allow us to test this hypothesis and develop a better understanding of the role of plasticity in facilitating diet evolution in insects and, more generally, expansion of organisms into novel environments.

Acknowledgements

We thank Dustin Marshall and Angela Amarillo for constructive comments on an earlier draft of this manuscript. The research described here was funded by the US National Science Foundation (DEB-94-03244, DEB-98-07315, DEB-99-96371, DEB-01-10754 and DEB-02-71929), Kentucky EPSCoR, the Kentucky Agricultural Experiment Station, the University of Kentucky, and Fundación para la Promoción de la Investigación y la Tecnología (in Colombia).

References

Agosta, S. J. 2008. Selection on offspring size varies within and among families in relation to host nutritional quality. Evolutionary Ecology 22: 71–83.

Agrawal, A. A. 2001. Phenotypic plasticity in the interactions and evolution of species. Science 294: 321–326.

Agrawal, A. A. 2002. Herbivory and maternal effects: Mechanisms and consequences of transgenerational induced plant resistance. Ecology 83: 3408–3415.

Allen, R. M., Y. M. Buckley and D. J. Marshall. 2008. Offspring size plasticity in response to intraspecific competition: An adaptive maternal effect across life-history stages. American Naturalist 171: 225–237.

Amarillo-Suarez, A. R. and C. W. Fox. 2006. Population differences in host use by a seed-beetle: local adaptation, phenotypic plasticity and maternal effects. Oecologia 150: 247–258.

Azevedo, R. B. R., V. French and L. Partridge. 1996. Thermal evolution of egg size in *Drosophila melanogaster*. Evolution 50: 2338–2345.

Badyaev, A. V., and K. P. oh. 2008. Environmental induction and phenotypic retention of adaptive maternal effects. BMC Evolutionary Biology 8: 3.

Berrigan, D. and S. M. Scheiner. 2004. Modeling the evolution of plasticity *in* T. J. DeWitt and S. M. Scheiner, eds. Phenotypic plasticity: functional and conceptual approaches. Oxford University Press, New York.

Bettegowda, A., K. B. Lee and G. W. Smith. 2008. Cytoplasmic and nuclear determinants of the maternal-to-embryonic transition. Reproduction Fertility and Development 20: 45–53.

Blanckenhorn, W. U. 2000. Temperature effects on egg size and their fitness consequences in the yellow dung fly *Scathophaga stercoraria*. Evolutionary Ecology 14: 627–643.

Bohren, B. B., W. G. Hill and A. Robertson. 1966. Some observations of asymmetrical correlated responses to selection. Genetical Research 7: 44–57.

Bossdorf, O., C. L. Richards and M. Pigliucci. 2008. Epigenetics for ecologists. Ecology Letters 11: 106–15.

Braby, M. F. 1994. The significance of egg size variation in butterflies in relation to host plant quality. Oikos 71: 119–129.

Brumpton, R. J., H. Boughey and J. L. Jinks. 1977. Joint selection for both extremes of mean performance and of sensitivity to a macro-environmental variable. 1. Family selection. Heredity 38: 219–226.

Bulmer, M. G. 1985. The mathematical theory of quantitative genetics. Clarendon, Oxford.

Carroll, S. P. and C. W. Fox. 2007. Dissecting the evolutionary impacts of plant invasions: bugs and beetles as native guides. Global Change Biology 13: 1644–1657.

Charmantier, A. and D. Garant. 2005. Environmental quality and evolutionary potential: lessons from wild populations. Proceedings of the Royal Society B-Biological Sciences 272: 1415–1425.

Cleuvers, M., B. Goser and H. T. Ratte. 1997. Life-strategy shift by intraspecific interaction in *Daphnia magna*: Change in reproduction from quantity to quality. Oecologia 110: 337–345.

Czesak, M. E. and C. W. Fox. 2003a. Evolutionary ecology of egg size and number in a seed beetle: Genetic trade-off differs between environments. Evolution 57: 1121–1132.

Czesak, M. E. and C. W. Fox. 2003b. Genetic variation in male effects on female reproduction and the genetic covariance between the sexes. Evolution 57: 1359–1366.

Czesak, M. E., C. W. Fox and J. B. Wolf. 2006. Experimental evolution of phenotypic plasticity: How predictive are cross-environment genetic correlations? American Naturalist 168: 323–335.

Donahue, K. and J. Schmitt. 1998. Maternal environmental effects in plants. pp. 137–158 *in* T. A. Mousseau, and C. W. Fox, eds. Maternal Effects as Adaptations. Oxford University Press, New York.

Ekbom, B. and S. Y. A. Popov. 2004. Host plant affects pollen beetle (*Meligethes aeneus*) egg size. Physiological Entomology 29: 118–122.

Ernsting, G. and A. Isaaks. 2000. Ectotherms, temperature, and trade-offs: Size and number of eggs in a carabid beetle. American Naturalist 155: 804–813.

Ernsting, G. and J. A. Isaaks. 1997. Effects of temperature and season on egg size, hatchling size and adult size in *Notiophilus biguttatus*. Ecological Entomology 22: 32–40.

Evans, M. E. K. and J. J. Dennehy. 2005. Germ banking: Bet-hedging and variable release from egg and seed dormancy. Quarterly Review of Biology 80: 431–451.

Falconer, D. S. 1952. The problem of environment and selection. American Naturalist 86: 293–298.

Falconer, D. S. 1990. Selection in different environments: effects on environmental sensitivity (reaction norm) and on mean performance. Genetical Research 56: 57–70.

Fischer, K., S. S. Bauerfeind and K. Fiedler. 2006. Temperature-mediated plasticity in egg and body size in egg size-selected lines of a butterfly. Journal of Thermal Biology 31: 347–354.

Fischer, K., A. N. M. Bot, P. M. Brakefield and B. J. Zwaan. 2003a. Fitness consequences of temperature-mediated egg size plasticity in a butterfly. Functional Ecology 17: 803–810.

Fischer, K., P. M. Brakefield and B. J. Zwaan. 2003b. Plasticity in butterfly egg size: Why larger offspring at lower temperatures? Ecology 84: 3138–3147.

Fox, C. W. 1993. The influence of maternal age and mating frequency on egg size and offspring performance in *Callosobruchus maculatus* (Coleoptera: Bruchidae). Oecologia 96: 139–146.

Fox, C. W. 1997. Egg-size manipulations in the seed beetle *Stator limbatus*: consequences for progeny growth. Canadian Journal of Zoology-Revue Canadienne De Zoologie 75: 1465–1473.

Fox, C. W. 2000. Natural selection on seed-beetle egg size in nature and the laboratory: Variation among environments. Ecology 81: 3029–3035.

Fox, C. W. 2006. Colonization of a new host by a seed-feeding beetle: Genetic variation, maternal experience, and the effect of an alternate host. Annales Zoologici Fennici 43: 239–247.

Fox, C. W. and M. E. Czesak. 2000. Evolutionary ecology of progeny size in arthropods. Annual Review of Entomology 45: 341–369.

Fox, C. W. and M. E. Czesak. 2006. Selection on body size and sexual size dimorphism differs between host species in a seed-feeding beetle. Journal of Evolutionary Biology 19: 1167–1174.

Fox, C. W., M. E. Czesak and R. W. Fox. 2001. Consequences of plant resistance for herbivore survivorship, growth, and selection on egg size. Ecology 82: 2790–2804.

Fox, C. W., M. E. Czesak, T. A. Mousseau and D. A. Roff. 1999. The evolutionary genetics of an adaptive maternal effect: Egg size plasticity in a seed beetle. Evolution 53: 552–560.

Fox, C. W., D. M. Gordon and P. Bojang. 2006. Genetic and environmental sources of variation in survival on nonnative host species in the generalist seed beetle, *Stator limbatus*. Southwestern Naturalist 51: 490–501.

Fox, C. W. and T. A. Mousseau. 1996. Larval host plant affects fitness consequences of egg size variation in the seed beetle *Stator limbatus*. Oecologia 107: 541–548.

Fox, C. W. and T. A. Mousseau. 1998. Maternal effects as adaptations for transgenerational phenotypic plasticity in insects. pp. 159-177 *in* T. A. Mousseau, and C. W. Fox, eds. Maternal effects as adaptations. Oxford University Press, New York.

Fox, C. W., J. A. Nilsson and T. A. Mousseau. 1997a. The ecology of diet expansion in a seed-feeding beetle: Pre-existing variation, rapid adaptation and maternal effects? Evolutionary Ecology 11: 183–194.

Fox, C. W. and U. M. Savalli. 2000. Maternal effects mediate host expansion in a seed-feeding beetle. Ecology 81: 3–7.

Fox, C. W. M. S. Thakar and T. A. Mousseau. 1997b. Egg size plasticity in a seed beetle: An adaptive maternal effect. American Naturalist 149: 149–163.

Fox, C. W., K. J. Waddell, J. DesLauriers and T. A. Mousseau. 1997c. Seed beetle survivorship, growth and egg size plasticity in a paloverde hybrid zone. Ecological Entomology 22: 416–424.

Fox, C. W., K. J. Waddell and T. A. Mousseau. 1994. Host-associated fitness variation in a seed beetle (Coleoptera: Bruchidae): Evidence for local adaptation to a poor quality host. Oecologia 99: 329–336.

Fox, C. W., K. J. Waddell and T. A. Mousseau. 1995. Parental host plant affects offspring life histories in a seed beetle. Ecology 76: 402–411.

Fry, J. D. 1996. The evolution of host specialization: Are trade-offs overrated? American Naturalist 148: S84–S107.

Gallizzi, K., L. Gern and H. Richner. 2008. A flea-induced pre-hatching maternal effect modulates tick feeding behaviour on great tit nestlings. Functional Ecology 22: 94–99.

Gavrilets, S. and S. M. Scheiner. 1993. The genetics of phenotypic plasticity. VI. Theoretical predictions for directional selection. Journal of Evolutionary Biology 6: 49–68.

Ghalambor, C. K., J. K. McKay, S. P. Carroll and D. N. Reznick. 2007. Adaptive versus non-adaptive phenotypic plasticity and the potential for contemporary adaptation in new environments. Functional Ecology 21: 394–407.

Glazier, D. S. 1998. Does body storage act as a food-availability cue for adaptive adjustment of egg size and number in *Daphnia magna*? Freshwater Biology 40: 87–92.

Gromko, M. H. 1995. Unpredictability of correlated response to selection: Pleiotropy and sampling interact. Evolution 49: 685–693.

Gromko, M. H., A. Briot, S. C. Jensen and H. H. Fukui. 1991. Selection on copulation duration in *Drosophila melanogaster*: predictability of direct response versus unpredictability of correlated response. Evolution 45: 69–81.

Guinnee, M. A., A. Gardner, A. E. Howard, S. A. West, and T. J. Little. 2007. The causes and consequences of variation in offspring size: a case study using Daphnia. Journal of Evolutionary Biology 20: 577–587.

Hassall, M., R. J. Walters, M. Telfer and M. R. J. Hassall. 2006. Why does a grasshopper have fewer, larger offspring at its range limits? Journal of Evolutionary Biology 19: 267–276.

Hillesheim, E. and S. C. Stearns. 1991. The response of *Drosophila melanogaster* to artificial selection on body weight and its phenotypic plasticity in two larval food environments. Evolution 45: 1909–1923.

Hoffmann, A. A., C. M. Sgro and S. H. Lawler. 1995. Ecological population genetics: The interface between genes and the environment. Annual Review of Genetics 29: 349–370.

Hulme, P. E. 2008. Phenotypic plasticity and plant invasions: is it all Jack? Functional Ecology 22: 3–7.

Jinks, J. L. and V. Connolly. 1973. Selection for specific and general response to environmental differences. Heredity 30: 33–40.

Jinks, J. L. N. E. M. Jayasekara and H. Boughey. 1977. Joint selection for both extremes of mean performance and of sensitivity to a macro-environmental variable. 2. Single seed descent. Heredity 39: 345–355.

Johnstone, O. and P. Lasko. 2001. Translational regulation and RNA localization in *Drosophila* oocytes and embryos. Annual Review of Genetics 35: 365–406.

Kingsolver, J. G., H. E. Hoekstra, J. M. Hoekstra, D. Berrigan, S. N. Vignieri, C. E. Hill, A. Hoang, P. Gibert and P. Beerli. 2001. The strength of phenotypic selection in natural populations. American Naturalist 157: 245–261.

Lande, R. 1980a. The genetic covariance between characters maintained by pleiotropic mutations. Genetics 94: 203–215.

Lande, R. 1980b. Sexual dimorphism, sexual selection, and adaptation in polygenic characters. Evolution 34: 292–305.

Marshall, D. J. 2008. Transgenerational plasticity in the sea: Context-dependent maternal effects across the life history. Ecology 89: 418–427.

Marshall, D. J., C. N. Cook and R. B. Emlet. 2006. Offspring size effects mediate competitive interactions in a colonial marine invertebrate. Ecology 87: 214–225.

Marshall, D. J., and M. J. Keough. 2008a. The relationship between offspring size and performance in the sea. American Naturalist 171: 214–224.

Marshall, D. J., and M. J. Keough. 2008b. The evolutionary ecology of offspring size in marine invertebrates. Advances in Marine Biology 53: 1–60.

Marshall, D. J. and T. Uller. 2007. When is a maternal effect adaptive? Oikos 116: 1957–1963.

Matsumura, M. 1996. Genetic analysis of a threshold trait: Density-dependent wing dimorphism in *Sogatella furcifera* (Horvath) (Hemiptera: Delphacidae), the whitebacked planthopper. Heredity 76: 229–237.

McKee, D. and D. Ebert. 1996. The interactive effects of temperature, food level and maternal phenotype on offspring size in Daphnia magna. Oecologia 107: 189–196.

Miller, G. A., M. S. Islam, T. D. W. Claridge, T. Dodgson, and S. J. Simpson. 2008. Swarm formation in the desert locust *Schistocerca gregaria:* isolation and NMR analysis of the primary maternal gregarizing agent. Journal of Experimental Biology 211: 370–376.

Mitchell, R. 1977. Bruchid beetles and seed packaging by palo verde. Ecology 58: 644–651.

Mizumoto, M. and F. Nakasuji. 2007. Egg size manipulation in the migrant skipper, *Parnara guttata guttata* (Lepidoptera: Hesperiidae), in response to different host plants. Population Ecology 49: 135–140.

Mondor, E. B., J. A. Rosenheim, and J. F. Addicott. 2008. Mutualist-induced transgenerational polyphenisms in cotton aphid populations. Functional Ecology 22: 157–162.

Morse, G. E. 2003. Ecological and evolutionary diversification in seed beetles (Coleoptera: Bruchinae). . PhD Dissertation, Department of Organismic and Evolutionary Biology, Harvard University.

Morse, G. E. and B. D. Farrell. 2005a. Ecological and evolutionary diversification of the seed beetle genus *Stator* (Coleoptera: Chrysomelidae: Bruchinae). Evolution 59: 1315–1333.

Morse, G. E. and B. D. Farrell. 2005b. Interspecific phylogeography of the *Stator limbatus* species complex: The geographic context of speciation and specialization. Molecular Phylogenetics and Evolution 36: 201–213.

Mousseau, T. A. and H. Dingle. 1991. Maternal effects in insect life histories. Annual Review of Entomology 36: 511–534.

Mousseau, T. A. and C. W. Fox. 1998a. The adaptive significance of maternal effects. Trends in Ecology & Evolution 13: 403–407.

Mousseau, T. A. and C. W. Fox. 1998b. Maternal effects as adaptations. Oxford University Press, New York.

Moya-Larano, J. and C. W. Fox. 2006. Ejaculate size, second male size, and moderate polyandry increase female fecundity in a seed beetle. Behavioral Ecology 17: 940–946.

Noach, E. J. K., G. De Jong, and W. Scharloo. 1998. Locomotor activity in *Drosophila melanogaster* selected for different wing lengths. Netherlands Journal of Zoology 48: 27–37.

Noach, E. J. K., G. DeJong and W. Scharloo. 1997. Phenotypic plasticity of wings in selection lines of *Drosophila melanogaster*. Heredity 79: 1–9.

Palmer, J. O. and H. Dingle. 1986. Direct and correlated responses to selection among life history traits in milkweed bugs (*Oncopeltus fasciatus*). Evolution 40: 767–777.

Perez, A. and C. Garcia. 2002. Evolutionary responses of *Drosophila melanogaster* to selection at different larval densities: changes in genetic variation, specialization and phenotypic plasticity. Journal of Evolutionary Biology 15: 524–536.

Pigliucci, M. 2005. Evolution of phenotypic plasticity: where are we going now? Trends in Ecology & Evolution 20: 481–486.

Plaistow, S. J., J. J. H. St Clair J. Grant, and T. G. Benton. 2007. How to put all your eggs in one basket: Empirical patterns of offspring provisioning throughout a mother's lifetime. American Naturalist 170: 520–529.

Rasanen, K. and L. E. B. Kruuk. 2007. Maternal effects and evolution at ecological time-scales. Functional Ecology 21: 408–421.

Reeve, J. P. and D. J. Fairbairn. 2001. Predicting the evolution of sexual size dimorphism. Journal of Evolutionary Biology 14: 244–254.

Roff, D. A. 1997. Evolutionary Quantitative Genetics. Chapman and Hall, New York.

Roff, D. A. 2002. Life History Evolution. Sinauer Associates, Inc., Sunderland, MA.

Sakai, S. and Y. Harada. 2005. Production of offspring using current income and reserves: size-number trade-off and optimal offspring, size. Journal of Theoretical Biology 233: 65–73.

Savalli, U. M. and C. W. Fox. 1998. Sexual selection and the fitness consequences of male body size in the seed beetle *Stator limbatus*. Animal Behaviour 55: 473–483.

Savalli, U. M. and C. W. Fox. 2002. Proximate mechanisms influencing egg size plasticity in the seed beetle *Stator limbatus* (Coleoptera: Bruchidae). Annals of the Entomological Society of America 95: 724–734.

Scheiner, S. M. 1998. The genetics of phenotypic plasticity. VII. Evolution in a spatially-structured environment. Journal of Evolutionary Biology 11: 303–320.

Scheiner, S. M. and R. F. Lyman. 1991. The genetics of phenotypic plasticity. II. Response to selection. Journal of Evolutionary Biology 4: 23–50.

Scheiner, S. M., R. J. Mitchell and H. S. Callahan. 2000. Using path analysis to measure natural selection. Journal of Evolutionary Biology 13: 423–433.

Schlichting, C. D. and M. Pigliucci. 1993. Control of phenotypic plasticity via regulatory genes. American Naturalist 142: 366–370.

Schlichting, C. D. and H. Smith. 2002. Phenotypic plasticity: linking molecular mechanisms with evolutionary outcomes. Evolutionary Ecology 16: 189–211.

Seko, T., T. Miyatake, S. Fujioka and F. Nakasuji. 2006. Genetic and environmental sources of egg size, fecundity and body size in the migrant skipper, *Parnara guttata guttata* (Lepidoptera: Hesperiidae). Population Ecology 48: 225–232.

Seko, T. and F. Nakasuji. 2006. Adaptive significance of egg size plasticity in response to temperature in the migrant skipper, *Parnara guttata guttata* (Lepidoptera: Hesperiidae). Population Ecology 48: 159–166.

Sgro, C. M. and A. A. Hoffmann. 2004. Genetic correlations, tradeoffs and environmental variation. Heredity 93: 241–248.

Siemens, D. H. and C. D. Johnson. 1990. Host-associated differences in fitness within and between populations of a seed beetle (Bruchidae) - Effects of plant variability. Oecologia 82: 408–413.

Siemens, D. H., C. D. Johnson and K. J. Ribardo. 1992. Alternative seed defense mechanisms in congeneric plants. Ecology 73: 2152–2166.

Siemens, D. H., B. E. Ralston and C. D. Johnson. 1994. Alternative seed defense mechanisms in a palo verde (Fabaceae) hybrid zone - effects on bruchid beetle abundance. Ecological Entomology 19: 381–390.

Simpson, S. J. and G. A. Miller. 2007. Maternal effects on phase characteristics in the desert locust, *Schistocerca gregaria*: A review of current understanding. Journal of Insect Physiology 53: 869–876.

Smith, C. C. and S. D. Fretwell. 1974. Optimal balance between size and number of offspring. American Naturalist 108: 499–506.

Steigenga, M. J. and K. Fischer. 2007. Ovarian dynamics, egg size, and egg number in relation to temperature and mating status in a butterfly. Entomologia Experimentalis Et Applicata 125: 195–203.

Steigenga, M. J., B. J. Zwaan, P. M. Brakefield and K. Fischer. 2005. The evolutionary genetics of egg size plasticity in a butterfly. Journal of Evolutionary Biology 18: 281–289.

Stillwell, R. C. and C. W. Fox. 2005. Complex patterns of phenotypic plasticity: Interactive effects of temperature during rearing and oviposition. Ecology 86: 924–934.

Sultan, S. E. 2005. An emerging focus on plant ecological development. New Phytologist 166: 1–5.

Sultan, S. E. 2007. Development in context: the timely emergence of eco-devo. Trends in Ecology & Evolution 22: 575–582.

Turner, R. M. 1963. Growth in four species of Sonoran Desert trees. Ecology 44: 760–765.

Via, S. 1994. The evolution of phenotypic plasticity: what do we really know? pp. 35-57 *in* L. A. Real (ed.), Ecological Genetics. Princeton University Press, Princeton, New Jersey.

Visser, M. E. 2008. Keeping up with a warming world; assessing the rate of adaptation to climate change. Proceedings of the Royal Society B-Biological Sciences 275: 649–659.

West-Eberhard, M. J. 1989. Phenotypic plasticity and the origins of diversity. Annual Review of Ecology and Systematics 20: 249–278.

Wilkinson, G. S., K. Fowler and L. Partridge. 1990. Resistance of genetic correlation structure to directional selection in *Drosophila melanogaster*. Evolution 44: 1990–2003.

Winkler, D. W. and K. Wallin. 1987. Offspring size and number: a life history model linking effort per offspring and total effort. American Naturalist 129: 708–720.

Worley, A. C. and S. C. H. Barrett. 2000. Evolution of floral display in *Eichhornia paniculata* (Pontederiaceae): Direct and correlated responses to selection on flower size and number. Evolution 54: 1533–1545.

20

On the Origins of Insect Hormone Signaling

Jason Hodin
Hopkins Marine Station, Stanford University, Pacific Grove, CA USA 93950
E-mail: hodin@u.washington.edu

Abstract

Hormones are generally thought of as endogenously produced substances, yet many key animal hormones are actually derived intact or as complex precursors from external sources, such as plants. Ecdysteroid hormones in insects are one such example, since insects are incapable of synthesizing steroids endogenously. Therefore, the numerous phenotypic effects of ecdysteroids in insects are actually examples of phenotypic plasticity, where an environmental signal (an ingested sterol) ultimately regulates flexible responses such as molting, oviposition and wing pattern polyphenisms. Here I present the hypothesis that other key insect hormones, notably the juvenile hormones (JH), originated from externally produced substances, initially ingested by herbivorous proto-insects. The plant hormone abscisic acid (ABA) is the most likely candidate: it is ubiquitous among plants, is chemically related to JH, and has juvenilizing effects on some insects. One key argument against this hypothesis is the close similarity of the biosynthetic pathways for JH in insects and a similar hormone in crustaceans, methyl farnesoate (MF). This similarity between JH and MF has led to the commonly-held belief that the aquatic ancestor of insects and crustaceans already had the ability to produce JH, MF or a related substance endogenously. To help distinguish among these possibilities (i.e. an endogenous versus an exogenous origin for JH in insects), I present a sequence comparison of the recently isolated genes encoding crustacean and insect methyltransferase, an enzyme involved in both JH and MF biosynthesis. The lack of orthology between crustacean and insect methyl transferase lends support to my hypothesis of an exogenous origin for JH signaling in insects.

Abbreviations: ABA (abscisic acid), FA (farnesoic acid), FPP (farnesyl pyrophosphate), HGT (horizontal gene transfer), JH (juvenile hormone), Ma (million years ago), MF (methyl farnesoate), MT (methyltransferase).

Hormonal signaling molecules regulate just about every physiological, behavioral and developmental process in insects[1] (reviewed in Nijhout 1994; De Loof 2008). There are several probable explanations for this widespread usage. First, complex organismal functions (such as ecdysis, oviposition and migration) require precise temporal coordination and control. A hormone released into the circulation at a given time can reach all of the cells in the body in a relatively short period, thus allowing for coordination among tissues and cells in different parts of the organism. Second, in addition to this potential for synchrony, morphogenetic hormones—mainly via their binding to nuclear receptors—are well-suited for orchestrating temporally and spatially complex processes such as metamorphosis (Truman et al. 1994). By binding to different receptor isoforms in different cells and tissues at different times, one broad hormonal peak can coordinate a series of events occurring over a period of days or even longer. And third, by placing the release of morphogenetic and other hormones under the control of neurohormones and other neuroregulatory molecules, the timing and amounts of hormones released can be modified by environmental inputs acting on the nervous system[2]. Taken together, this view of hormones suggests the existence of a communication network, in which external and internal signals integrate across a wide range of both spatial and temporal scales. Or, put another way, hormonally-regulated events are notably amenable to phenotypic plasticity (reviewed in Hatle 2003).

Dominant among insect hormones are the juvenile hormones and the ecdysteroids. These two families of insect hormones are unique among chemical messengers in their fundamental importance to insects (reviewed in Nijhout 1994, Flatt et al. 2005). Not only are they required for the successful completion of all insect life cycles, but they function specifically in a wide range of organismal processes in insects, including embryonic development, ecdysis (≈molting) and growth, reproduction, metamorphic transformations, foraging behavior, social caste differentiation, and numerous other polyphenisms (reviewed in Nijhout 1994).

What is the evolutionary history of these functionally diverse yet critical groups of hormones? Ecdysteroids (a class of steroids) and juvenile hormones (JH, a group of sesquiterpenes) are so integral to the biology of the

[1] I use the term "insect" in the broad sense here, including related hexapod groups such as collembolans.

[2] The same logic can apply to plants and fungi, whose plastic responses are not often thought of as being relayed by "nervous systems" *per sé* (but see Wildon et al. 1994; Li et al. 2002).

widest range of living insect groups that their uses are presumably ancient, perhaps dating back to the origin of the Ecdysozoa[3] or even before. As far as we know, all insects and crustaceans utilize ecdysteroids and sesquiterpenes (JH in insects; methyl farnesoate in crustaceans) as regulators of both somatic growth and some aspect of reproduction (Laufer and Biggers 2001, Nijhout 1994). It is thus tempting to consider the internal use of ecdysteroids and sesquiterpenes to be a plesiomorphy (ancestral characteristic) for the arthropods, and possibly for all ecdysozoans (e.g. Tobe and Bendena 1999). I refer to this as the **ancient hormone hypothesis**.

Ecdysteroids are chemically derived from cholesterol or other sterols, and insects are incapable of producing cholesterol or any sterol endogenously (reviewed in Robbins et al. 1971; Behmer and Nes 2003). Thus, it is quite curious that insects rely so heavily on hormones that they cannot produce themselves without first obtaining complex precursors derived from plants or other organisms on which they feed (see Robbins et al. 1971; Svoboda et al. 1975; Pennock 1977; Behmer and Nes 2003). Roundworms (Nematoda) and crustaceans are also apparently incapable of synthesizing sterols endogenously, and must obtain cholesterol or other complex precursors from their diet (Rothstein 1968, Kanazawa 2001). In fact, there is, to my knowledge, no published evidence for endogenous ecdysteroid (or any sterol) production in a single member of the proposed Ecdysozoa clade! This leads to the perhaps surprising conclusion that ecdysteroid signaling in insects is actually an example of phenotypic plasticity, since the myriad cellular, developmental and behavioral processes regulated by ecdysteroids are dependent on an environmental cue: ingested sterols.

More curious still is the fact that the ability to produce sterols internally from relatively simple precursors (e.g. mevalonate or acetate) is apparently widespread among other groups of 'non-ecdysozoan' animals (reviewed in Kanazawa 2001), and the evolutionary pattern among animal phyla seems to indicate multiple gains and/or multiple losses of this trait. In fact, internal sterol production is the rule rather than the exception in all eukaryotes (reviewed in Behmer and Nes 2003). Even ecdysteroids themselves are

[3] The Ecdysozoa (Aguinaldo et al. 1997) are a proposed monophyletic group that would unite a diverse array of molting animals. This super-phylum grouping includes what are arguably the two most dominant animal taxa on the planet at this time: the Arthropoda (insects, spiders, crabs, etc.) and the Nematoda (roundworms). The Ecdysozoa also includes less known groups, such as water bears (Tardigrada), velvet worms (Onycophora), horsehair worms (Nematomorpha), fanghead worms (Priapulida, with probable relatives in the Burgess Shale) and awlhead worms (Kinoryncha or Echinoderida).

synthesized by organisms in a wide array of kingdoms, including fungi, plants and 'non-ecdysozoan' animals (see Dinan 2004). But how is it possible that such globally dominant and ancient groups as arthropods and nematodes are unable make a common hormone that they absolutely need in order to grow? Why have some insect taxa evolved such elaborate mechanisms of utilizing particular sterols to produce ecdysteroids (Behmer and Nes 2003), and none have apparently figured out what so many other living forms have: how to make the hormone themselves? Did the ancestors of arthropods and nematodes originally have this capacity, and then lose it? Or, did they never have it? In any event, lack of ability to synthesize this essential compound from simple precursors apparently has not held back the insects and nematodes in their ascendancy to taxonomic dominance.

The origin of juvenile hormones (JH) in insects is also of interest. These sesquiterpene hormones may exceed even ecdysteroids in the breadth of their involvement in various biological processes (reviewed in Nijhout 1994, Flatt et al. 2005). A similar sesquiterpene, methyl farnesoate (MF), has been isolated from several crustaceans, and has apparent roles in reproduction (reviewed in Laufer and Biggers 2001). The diverse and critical functions for JH across insects, the extremely close chemical similarity of JH and MF, and the fact that both JH and MF are involved in reproduction seems to support the ancient hormone hypothesis.

Still, a closer look at the role of juvenile hormones across insects uncovers some curious patterns. First, a comparative analysis of the precise reproductive functions of JH in insects reveals substantial differences across insect taxa. Even vitellogenesis (the synthesis of egg yolk), the process often described as the most fundamental role of JH in insect reproduction (e.g. Wyatt 1997), does not appear to depend upon JH in several notable insect groups. Indeed, in the lepidopterans (moths and butterflies), the canonical vitellogenic role for JH is only found among a derived group of butterflies, though taxon sampling there is still low (see Chapter 11: Table 4 and Figure 4, this volume). These puzzling patterns are rarely mentioned in print when discussing JH evolution. Furthermore, the precise roles for MF in crustacean reproduction are, relative to JH, very poorly studied. The glands that produce JH in insects (the corpora allata; henceforth CA) and MF in crustaceans (the mandibular organ; henceforth MO) are nearly universally described as homologous. Nevertheless, evidence for this characterization is rarely presented or analyzed (e.g. Stay and Tobe 2007). In those few papers that do carefully consider the evidence for homology, the argument mainly rests upon similar ultrastructural features of their respective glandular cells and similar embryonic origin (as well as the production of MF by the MO). While similar embryonic origins of the respective glands are indeed found across the majority of insects and crustaceans examined, within-class variations exist, casting possible doubt on the homology argument.[4] In the case of the ultrastructural similarities, the cellular features that are cited to suggest homology (such as the ER ultrastructure and prominence of vacuoles) are, in fact, common characteristics of steroid/lipid secreting cells across animals. Indeed, the two initial papers suggesting an endocrine function of the MO (Hinsch & Hajj 1975; Byard et al. 1975) both noted the strong ultrastructural similarity to vertebrate steroid secreting glands. Surely, such similarities in crustacean and vertebrate glandular ultrastructure are due to homoplasy, so one wonders why CA-MO similarities could not be due to homoplasy as well. In fact, comparative analyses of MO ultrastructures within decapod crustaceans reveal rather dramatic differences between species (Byard et al. 1975). Finally, as Tobe and Bendena (1999) have noted, the innervation patterns of the MF and CA are quite different. In sum, the common presumption of homology for endocrine signaling in insects and crustaceans overlays a more complex story than is often acknowledged. Do the comparative observations cited above regarding ecdysteroid and sesquiterpene signaling cast doubt on the veracity of the ancient hormone hypothesis?

I noted above that all insects obtain their sterols from the environment, and that therefore, ecdysteroid signaling represents a form of phenotypic plasticity. Like steroids, structurally diverse sesquiterpenes have been isolated from a vast array of plants, fungi and algae (see below), suggesting that sesquiterpenes were probably abundant at the time that the first insect ancestors invaded land. Were the earliest insects likely exposed to a consistent source of sesquiterpenes, as was surely the case with steroids? If so, is it possible that the history of both of these hormones in insects began as plastic responses to highly active, externally produced compounds?

An analysis of the substantial fossil record of insects along with comparative biochemistry and genomics can help to provide some answers to these questions. Insects appear in the fossil record during the late Silurian

[4] Although the embryonic origin of the corpora allata in many insects is from ectodermal invaginations in or around the mandibular pouch, it is different in others (reviewed in Kobayashi and Ando 1983), including the rice weevil *Calandra oryzae*, whose corpora allata originate from the antennal mesodermal sacs in the embryo (Tiegs and Murray 1938). Also, in the Collembola (springtails) and other basal insect/hexapod orders, the position of the corpora allata in the adults is suboesophageal and/or lateral, rather than the supraesophageal position seen in most insects (Cassagnau and Juberthie 1983).

to early Devonian, starting at around 410 million years ago ("410 Ma"). And while there is some controversy concerning their original food source (whether live or dead/decaying plant material), the earliest evidence of insect live plant feeding is from spore protoplasts from around 400 Ma, in the late Silurian and early Devonian (Edwards et al. 1995; Habgood 2004; also see Labandeira 1998, 2002a,b). Subsequently, the earliest leaf damage is recorded on seed ferns about 326 Ma, from the late Mississippian (Labandeira, personal communication). Thus, it seems that when live feeding on plants began in earnest, particularly on foliar tissues, gymnospermous plants[5] had largely diversified (Niklas 1997). Indeed, recent evidence suggests that the spectacular radiation of the Coleoptera, which has led them to become by far the most speciose taxon on earth, began on gymnosperms, possibly preadapting[6] them for their multiple, independent moves to angiosperms (Farrell 1998). Therefore, the major events of insect diversification took place when insects were feeding mainly on gymnosperms (Labandeira 1998, 2002b). Phylogenetic comparisons among terpene biosynthetic systems in various plant groups, as well as the near ubiquitous presence of sesquiterpenes in modern land plants (e.g. Asakawa et al. 2001), indicate that these ancient gymnosperms probably had well-developed sesquiterpene biosynthetic capacity (Martin et al. 2004; Bohlmann et al. 1998). Indeed, terpenoids are the largest and most diverse group of plant organic compounds and sesquiterpenes are the largest group of known terpenoids (Ryan 2002). The original function of such sesquiterpenes has been hypothesized as anti-microbial, anti-fungal and insecticidal (reviewed in Bohlmann et al. 2000). Furthermore, algae and even fungi can produce sesquiterpenes (e.g. Anke and Sterner 1991; Smyrniotopoulos et al. 2003), and JH analogs are abundant in a vast array of plants (reviewed in Bowers 1997, Eales 1997). In fact, many plants produce JH intermediates (see below), and sedges (*Cyperus iria*) can actually synthesize *bona fide* JH-III, the active form of JH in most insects (Toong et al. 1988; Bede et al. 2001).

Abscisic acid (ABA) is one noteworthy sesquiterpene found in all groups of land plants, as well as algae and fungi (Jolivet et al. 1991, Oritani and Kiyota 2003, Kroemer et al. 2004). Like JH in insects, ABA is a hormone involved in an extremely wide array of ontogenetic and physiological pro-

[5] The term "gymnosperm" here and throughout refers essentially to the non-flowering seed plants, both living (e.g. cycads, ginkgoes and conifers) and extinct (e.g. seed ferns).

[6] *sensu* Gould (1984): features adapted for one role that are fortuitously suited for another.

cesses in plants, including dessication tolerance, seed dormancy, root growth, root-shoot ratios, cold tolerance, leaf polyphenisms ("heterophylly") and pathogen resistance (reviewed in Davies and Jones 1991, Oritani and Kiyota 2003, Wheeler and Nijhout 2003, Taylor et al. 2004). The similarities between JH and ABA apparently extend to their molecular action: the searches for the JH and ABA receptors have been impeded by a similar series of difficulties, indicating to Wheeler and Nijhout (2003) that each of these sesquiterpenes may act through low affinity interactions with several different endogenous receptor molecules, both nuclear and membrane-based. Furthermore, both JH and ABA are derived from the same sesquiterpenoid precursor: farnesyl pyrophosphate (FPP; see below and Figure 2) (reviewed in Oritani and Kiyota 2003). Given these similarities, the obvious question is: does exposure to ABA have JH-like effects on insects?

Several studies have shown that plants upregulate ABA production in response to insect damage (e.g. Peña-Cortés and Willmitzer 1995), indicating a function for ABA in protection against herbivory (Thaler and Bostock 2004). Still, ABA treatments of adult insects through their food give somewhat complicated results when they are compared across species. In general, leaf-feeding adult insects show a decrease in fecundity when exposed to excess ABA, whereas seed-feeding, carnivorous and detritivorous adult insects tend to show the opposite effect (reviewed[7] in Visscher 1983). Although few studies have been done to determine the effects of ABA exposure on preadult insects, the unpublished results of Carroll Williams suggest that feeding ABA to nymphs of the fire bug *Pyrrhocoris apterus* (Hemiptera: Pyrrhocoridae) can have JH-mimicking effects on cuticle deposition (cited in Visscher 1983), a similar result to that reported by Eidt and Little (1970) with ABA injections into pupae of the mealworm beetle, *Tenebrio molitor* (Coleoptera: Tenebrionidae).

In sum, these various results suggest that ABA can either act as an antagonist or an agonist of JH signaling, depending on the ontogenetic stage and the feeding mode of the insect in question. Why do we see such variability in ABA effects across insects? Perhaps it is unrealistic to expect a uniform effect of such a potent and potentially detrimental plant hormone among insects as a whole. Put simply, 400 million years of co-evolution between insects and plants have undoubtedly resulted in a wide variety of insect responses to ABA, as well as plant counter-responses to these insect

[7] There have been a mere handful of additional studies on this topic since 1983, including Bur 1985; Yesilada and Bozcuk 1995, 1996. The results of these studies are essentially consistent with those reported in and reviewed by Visscher (1983).

responses, etc. (see Fraenkel 1959, Herout 1970 for more general discussions of this idea). Therefore, in order to infer anything reliable about the evolutionary history of ABA-insect interactions, we need more data on the roles of ABA in basal plant taxa, as well as its effects on basal hexapod orders.

In the mean time, it seems reasonable to assume that the earliest insect feeders were exposed to highly bioactive sesquiterpenoid compounds, including ABA and possibly other JH-like chemicals (see Slama et al. 1974, for a review). What if the biological effect of these ancient sesquiterpenes was similar to juvenoids in modern insects: namely, in repressing adult differentiation? Presumably, then, successful ancient gymnosperm-feeding insects eventually developed a degree of resistance (or biochemical accommodation) to such sesquiterpenes, ultimately allowing for adult differentiation to occur on such host plants. At this point, the first juvenile hormonal "control" of development would have been in place, and the source of the hormone would have been exogenous, a situation similar to that of ecdysteroids in insects today[8]. Such a process is not unlikely; indeed, many animal hormones have exogenous sources (reviewed in Heyland et al. 2005). Furthermore, an external source of metamorphically-active JH-like compounds has been proposed for marine polychaetes as well (Biggers and Laufer 1992, 1999).

In fact, this idea that insect hormones may have been evolutionarily derived from plant compounds was proposed as early as 1979 by Karel Sláma (page 683), who hypothesized that "some compounds that are present for various other reasons in plants may accidentally fit certain structural requirements and consequently act as animal hormones. Like some other secondary plant substances, the hormonally active compounds in plants could have been a factor of natural selection that modulated the co-evolutionary relationships between plants and their insect herbivores."[9] Abscisic acid and other JH-mimicking sesquiterpenoids in plants are likely examples.

An exogenous origin for sesquiterpenes in insects implies an alternative to the ancient hormone hypothesis outlined above. If the original source of a

[8] The difference is that ecdysteroids are generally not the ingested form in insects; instead, complex sterols are ingested, which the insects then convert to ecdysteroids.

[9] Such ideas were directly descended from Sláma and Williams' (1966) identification of the JH-mimicking "paper factor." The active compound in the "paper factor" is juvabione from balsam fir (*Abies balsamea*), another FPP-derived sesquiterpene with very close chemical similarity to ABA (Bowers et al. 1966; see Figure 2 and legend).

JH-like compound in stem group insects[10] was in fact exogenous, then how was sesquiterpenoid biosynthesis internalized in the ancestors of modern insects, all of whom synthesize juvenile hormones in specialized structures in the anterior of the animal called the corpora allata (Nijhout 1994)?

There is an evolutionary mechanism that can account for this internalization that should be familiar to all aficionados of phenotypic plasticity: genetic assimilation. This term was introduced by Conrad Waddington (1961 for review), who selected heterogeneous populations of *Drosophila melanogaster* vinegar flies for an induced bithorax phenotype (the conversion of the haltere balancer organs into wing tissue) resulting from environmental stress (ether treatment of embryos). Amazingly, some of the flies in the eighth generation of selection showed the bithorax phenotype in the absence of the ether treatment. In other words, what began as an environmentally induced response quickly became a constitutive response: the induced bithorax condition had somehow "assimilated" into the genome, becoming a genetically fixed condition.

Recent selection experiments using either *D. melanogaster* or the mustard *Arabidopsis thaliana* have revealed one possible cellular mechanism for this process of assimilation of environmental effects: inhibition of the normal functioning of the molecular chaperone HSP90, a protein involved in maintaining protein cellular functions under conditions of stress, uncovers cryptic variation that can subsequently be selected for and ultimately fixed genetically (reviewed in Sangster et al. 2004).

The general principle here is that a response dependent on an environmental stimulus in an ancestral state can become assimilated in a derived state. Such an evolutionary pattern was actually proposed more than a half century before Waddington by James Mark Baldwin (1896, 1902), who wrote that "heredity provid[es] for the modification of its own machinery. Heredity not only leaves the future free for modifications, it also provides a method of life in operation of which modifications are bound to come" (1896, page 552). Baldwin's idea of "future modifications" that are "bound to come," clearly prefigures Waddington's bithorax results. Thus, I adopt the formulation '**Baldwinian assimilation**' throughout this chapter to describe the internalization of environmentally induced phenotypes. In the context of the origin of insect hormones, a Baldwinian assimilation hypothesis could account for the possible internalization of sesquiterpene biosynthetic capacity in stem group insects, as an alternative to the ancient hormone hypothesis presented above.

[10] A "stem group" is an extinct taxonomic group that branched off before the all of the living representatives of a given clade appeared. Thus, a "stem group insect" would hold a phylogenetic position basal to all of the living insect orders.

Some recent work on bark beetles in the genus *Ips* (Coleoptera: Scolytidae) might provide a second example where terpene biosynthetic ability evolved *de novo* in insects via Baldwinian assimilation. Conifers produce monoterpene volatiles, such as α-pinene in pines, in response to insect grazing (reviewed in Harborne 1991). Bark beetles, though, are not only resistant to this feeding deterrent, they utilize the hosts α-pinene as a precursor for one of their sex and aggregation pheromones, the monoterpene cis-verbenol (Byers 1981, 1983, 1989). Some *Ips* species (such as *I. paraconfusus*) seem to engage the services of gut symbionts for this production of cis-verbenol, as well as for the synthesis of other monoterpenoid pheromonal components (Brand et al. 1975, Byers and Wood 1981). In addition, this same species and other bark beetles actually have the ability to synthesize their terpenoid pheromones endogenously, and thus independently of pine terpenoid precursors (Byers and Birgersson 1990, Seybold et al. 1995, Hall et al. 2002). In fact, these beetles are the only insects (indeed the only metazoans) known to have an endogenous monoterpene synthase (Martin et al. 2003). Therefore, the most obvious evolutionary scenario is one where an ancestral bark beetle taxon, possibly with the aid of gut symbionts, converted the plant's own monoterpenes into aggregation pheromones. Then, later, some male bark beetles evolved the ability to synthesize the compounds themselves, and could thus initiate aggregation responses and attract mates independent of host-produced volatiles. As in the JH biosynthesis mechanism discussed above, the evolutionary acquisition of this novel terpene biosynthesis mechanism in bark beetles is a perfect candidate for Baldwinian assimilation.

Thus, Baldwinian assimilation might explain the evolution of phenotypic plasticity in response to a wide range of exogenous hormones, pheromones and other chemicals (Figure 1). In the first stage, the animal is exposed to such a chemical, which may induce a harmful, or even a neutral or (less likely) a beneficial response. In the second stage, the insect acquires resistance to any ill effects of the chemical. Not only would this allow the animal to exploit the chemical-producing resource more efficiently, but it would also allow for the environmental signal to be co-opted to induce a particular, selectively favorable (phenotypically plastic) reaction in the animal.[11] During this stage, specific insect receptor molecules, which may

---

[11] The idea here is that organisms cue their life cycle transitions and other responses to detectable, reliable environmental signals. Such reliable signals (for example: day length, volatile compounds produced by hosts, rainfall, etc.) are precisely what organisms respond to in cases of adaptive phenotypic plasticity. Furthermore, there is substantial precedent in animals for the utilization of potent chemicals derived from food sources as both hormonal regulators of life cycle transitions, and as cues for adaptive plasticity (e.g. Pfennig 1992, Heyland and Hodin 2004; reviewed in Heyland et al. 2005).

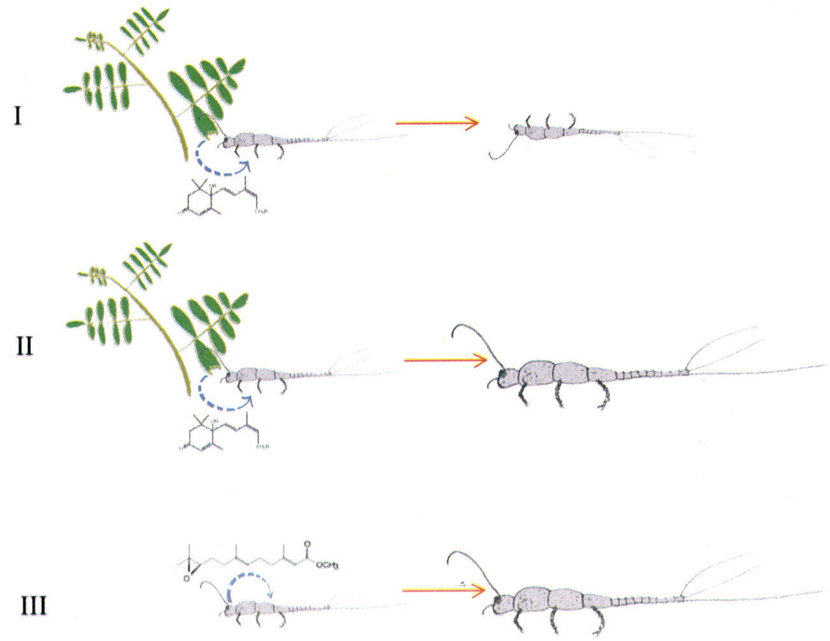

Fig. 1 Baldwinian assimilation hypothesis for the origin of hormonal signaling in insects. In the first stage (**I**), a potent compound [abscisic acid (ABA) is shown here] produced by a plant induces a detrimental plastic response in the proto-insect feeding on that plant. In the second stage (**II**), the proto-insect acquires resistance to the potent compound. Thus, the proto-insect can now safely feed on this plant. Ultimately, the potent compound is used as a signaling molecule/plasticity cue involved in life stage transitions (red arrow in II) in the proto-insect. At this stage, the insect's genome has adapted so that the external compound produces a beneficial plastic response. In the final stage (**III**), the proto-insect has acquired the ability to synthesize a chemically related compound endogenously [juvenile hormone III (JH-III) is shown here]. Now, the proto-insect can complete the same life stage transitions in the absence of that particular food source. The insect has co-opted and internalized ("assimilated") a formerly environmentally-dependent process.

have started out at low affinity for the external signal, are gradually modified through natural selection to attain higher and higher affinities (see for example Tallamy et al. 1999). Such a receptor molecule could either be something akin to a detoxification enzyme, or it could be a member of a pre-existing signaling cascade that induces a particular physiological or neurophysiological reaction to the presence of the novel plant signal (see Baker 2005). In the third stage, the animal is primed for internal synthesis of the chemical (or a similar substance), as the "favorable reaction" could then occur in a wider diversity of environmental contexts, where and when the specific exogenous chemical is absent. Stage three, then, represents a fertile condition for the Baldwinian "internalization" of a synthetic mechanism for the exogenous chemical by genetic assimilation.[12]

There is a second (albeit not entirely mutually exclusive) evolutionary mechanism that could account for the acquisition of plant chemical synthetic enzymes by herbivorous insects: **horizontal gene transfer** (HGT). For example, there is quite strong evidence that several bacterial genes involved in manipulating plant chemistry and physiology, including cellulases, have been acquired by plant parasitic nematodes via HGT (reviewed in Bird and Koltai 2000). And despite widespread understanding that pieces of DNA have the ability to cross species boundaries (as in yearly flu virus outbreaks as well as HIV evolution), the possibility remains underappreciated that modern organismal genomes may be, in reality, patchwork mosaics of DNA derived both horizontally and vertically. Indeed, Palmer et al. (2004) reached the shocking conclusion that the "true" tree of all living forms may never be knowable, due to the exceedingly high rates of HGT among bacteria and cyanobacteria.

These findings have at least two implications for the Baldwinian assimilation hypothesis delineated above. First, the insect receptor molecules that I postulate to have some fortuitous affinity for the plant chemical ("stage two" above) could have been acquired by HGT from a bacterium (or other microbe), one that, perhaps, had a longer term association with the plant than the insect in question. In this way, in a literal evolutionary instant, the ability to bind the plant chemical could have been acquired. Second, the internalization of the chemical synthetic mechanism ("stage three" above) could have also been a HGT event, either from the plant itself, or from some plant-associated microbe.

Thus we have three working hypotheses to explain the acquisition of plant signaling systems by insects: the Baldwinian assimilation hypothesis, the horizontal gene transfer (HGT) hypothesis, and the ancient hormone

[12] The scenario I present here can be seen as broadly consistent with Waddington's (1960, Figure 9) evolutionary outline for assimilation, where genetic change involves four different but overlapping "subsystems" operating somewhat sequentially. The first two subsystems ('exploitive' and 'epigenetic') involve environmental variation and behavioral responses to them, and can be called "phenotypic plasticity" in modern parlance. The third ('natural selective') subsystem involves novel combinations of alleles yielding new phenotypes from standing variation. In the final ('genetic') subsystem, changes result from mutation.

hypothesis.[13] The main prediction of the Baldwinian assimilation hypothesis is that different lineages exposed to the same chemical would evolve internal synthesis by quite different mechanistic (convergent) routes (see Hodin 2000). Thus, for example, the mechanisms of methyl farnesoate (MF) synthesis in crustaceans would be substantially different from the mechanisms of JH biosynthesis in insects. In other words, the biosynthetic enzymes would have been independently co-opted in crustaceans and insects, and would not necessarily be orthologs. The HGT hypothesis predicts that the enzymatic proteins involved in these hormone synthesis pathways would be more similar to specific plant or microbe proteins than they are to any proteins in other ecdysozoans or other animals. The null hypothesis here is the ancient hormone hypothesis: namely, that the JH and MF biosynthetic pathways are homologous, in that sesquiterpene biosynthesis was present in the last common ancestor of insects and crustaceans. In this case these enzymes would be expected to be orthologous in crustaceans and insects, and would also be found in other animals.

Distinguishing among these hypotheses requires detailed comparative biochemistry and genomics. In insects, MF is a JH precursor (Figure 2). Do insects and crustaceans use similar or different biosynthetic enzymes to produce MF? The key steps in the insect JH and crustacean MF biosynthetic pathways involve conversion of farnesol to MF via oxidase, dehydrogenase and methyltransferase activities (Figure 2). Do crustaceans use orthologous enzymes here? What about those plants and other organisms that produce JH or related compounds? Are these independently-evolved biosynthetic pathways at all similar to the pathways in arthropods?

The starting point for JH biosynthesis in insects is the same as for cholesterol synthesis in other eukaryotes, as well as for abscisic acid (ABA) synthesis in plants and fungi: farnesyl pyrophosphate (FPP; see Figure 2).[14] The enzyme responsible for catalyzing the formation of FPP is a FPP synthase (not shown in Figure 2), and, not surprisingly, FPP synthase orthologs are found in plants and animals (Poulter and Rilling 1981; Bellés et al. 2005). Therefore, with FPP we have an example where the same

[13] The former two (assimilation and HGT) together might be examples of "phylogenetic espionage" hypotheses *sensu* Schultz (2002). The scenario outlined above for the evolutionary acquisition of monoterpene production in bark beetles as sex and aggregation pheromones is another plausible example. Shultz (2002) and Schultz and Apel (2004) provide several additional and striking examples of parallel uses of hormones and other chemical signaling molecules in plants and herbivorous animals.

[14] FPP is also well-known for its involvement in protein modifications in plants and animals (prenylation), as shown in Figure 2 (see also Poulter and Rilling 1981).

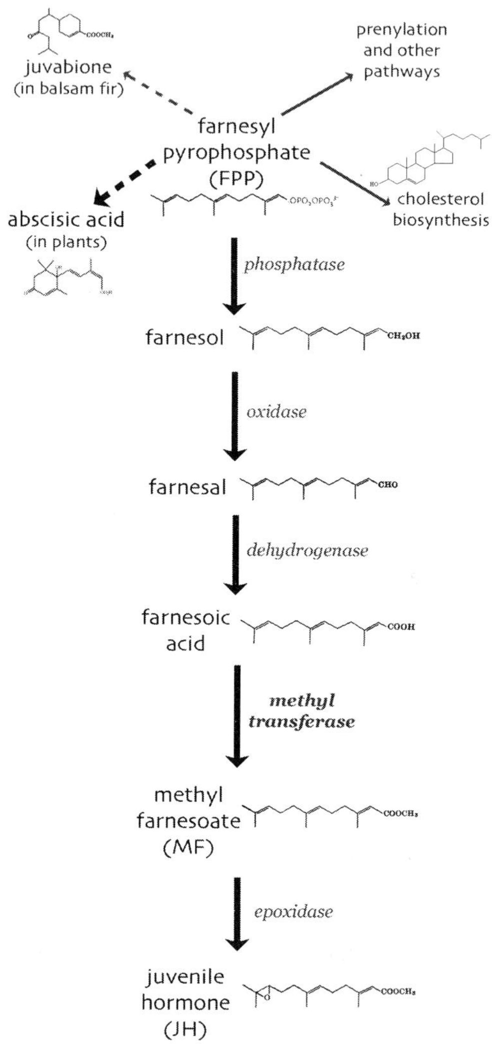

Fig. 2 The MF/JH biosynthetic pathway in arthropods, and related pathways in plants. Juvenile hormone III (JH-III), the active form in most insects, is shown here. In italics are enzymes that catalyze the different steps in the MF/JH pathway. Each of these enzymes has been identified by their activity, but the genes have, to this date, only been isolated for the crustacean and insect methyltransferases (see the text and Figure 3) as well as the insect epoxidase (cytochrome P450). Crustaceans are not known to make JH. FPP is involved in multiple pathways in various organisms, as indicated (see also the text). Juvabione (the "paper factor") from balsam fir (*Abies balsamea*) is a potent JH-mimic in hemipteran bugs in the family Pyrrhocoridae (see the text). Cholesterol biosynthesis from FPP is conspicuously absent in all arthropods and nematodes.

molecule is produced by an apparently conserved biosynthetic pathway across a wide array of organisms.

Now what about JH/MF biosynthesis in arthropods, a pathway unknown from any other animal group (Figure 2)? Unfortunately, the oxidase and dehydrogenase genes responsible for the conversion of farnesol to farnesoic acid have not yet been identified (Bellés et al. 2005). Recent work, though, has identified methyltransferases (MTs) from a variety of insects and crustaceans. To begin to distinguish among the above hypotheses, I performed sequence alignment comparisons of the different insect and crustacean MTs thought to be responsible for methyl farnesoate biosynthesis: JH acid MT in insects (e.g. Shinoda and Itoyama 2003; Bellés et al. 2005) and farnesoic acid O-MT in crustaceans (e.g. Silva Gunawardene et al. 2002; Ruddell et al. 2003; Bellés et al. 2005). Despite their description by Bellés et al. (2005, page 186) as "orthologs," these insect and crustacean MTs show no significant similarity (Figure 3).

Furthermore, my preliminary BLAST searches of gene and protein databases (not shown) revealed a curious pattern for the JH acid MT: the only sequences that showed significant similarity to the *Drosophila melanogaster* sequence (see Figure 3) were other insect JH MTs and microbe sequences; no other animal sequences showed substantial similarity. Most of these microbe sequences were uncharacterized, but several were bacterial ubiquinone/methyltransferases!

Farnesoic acid O-methyltransferases (FA O-MTs) have been identified in several crustaceans (*e.g.* Figure 3), as well as in insects, including the honeybee *Apis mellifera* and the mosquito *Aedes aegypti*. These FA O-MT sequences are quite highly conserved among the arthropods (e.g. 44.8% amino acid similarity between *Scylla* mud crabs and *Aedes* mosquitos over the entire sequence; data not shown), but show comparatively low similarity with any insect JH Acid methyltransferases (JH Acid MTs; see above and Figure 3). Such levels of similarity are indicative of the fact that both are S-adenosyl-L-methionine (SAM)-dependent methyltransferases, a diverse family of enzymes found throughout prokaryotes and eukaryotes. Equally low levels of similarity are seen in comparisons within insect species between their JH Acid MT and FA O-MT genes. For example, *A. aegypti* JH Acid MT is 35% similar to *A. aegypti* FA O-MT over a 100 amino acid N-terminal stretch (data not shown), about the same similarity in the crustacean-insect comparison shown in Figure 3. A recent study (Burtenshaw et al. 2008) has demonstrated that the *D. melanogaster* ortholog of crustacean farnesoic acid O-methyltransferase is expressed in the ring gland, but that *in vitro* assays and genetic analyses show no evidence that

```
S.serrata FA O-MT        6  HGKTLRFQVKAAHDCHVAFTTGAEETDPMVEVFIG---------------   40
                            ||..||::              .|..|.::::..|
A.aegypti JH Acid MT    25  HGHLLRWK--------------EENEDSLLDIGCGSGDVLIDFVIPMVPP   60

S.serrata FA O-MT       41  ------GWEGAASAIRF--KKADDLVKV--DTPDIVTEAEYREF---WIA   77
                                  |.:.:...:||  |...|:..:  :|.||  |.:...|   |..
A.aegypti JH Acid MT    61  KRARVLGTDVSEQMVRFARKVHSDVENLFFETLDI--EGDISSFLNKWGC  108

S.serrata FA O-MT       78  VDH
                            .||
A.aegypti JH Acid MT   109  FDH
```

Fig. 3 Crustacean farnesoic acid O-methyltransferase (FA O-MT) is not orthologous with insect JH Acid methyltransferase (JH Acid MT). Alignment (using EMBOSS Align 2006-7 at the European Bioinformatics Institute; http://www.ebi.ac.uk/emboss/align/) of a mosquito, *Aedes aegypti*, JH acid methyltransferase (accession number DQ409061), and a mud crab, *Scylla serrata*, farnesoic acid O-methyltransferase (accession number DQ187991). Numbers (6-80; 25-111) refer to the amino acid positions out of 278 *(A. aegypti)* and 235 (*S. serrata*), respectively. Vertical lines define identical amino acid positions; pairs of dots identify chemically similar amino acids; single dots denote weak similarity; dashes indicate gaps inserted in the two sequences to preserve optimal alignment. These N-terminal portions of the two sequences are the only parts showing substantial similarity (32% with 19.4% identity). The remaining C-terminal regions (not shown) are only 12.2% similar (6.8% identical). This and all alignments referred to in the text were performed using the default settings in the "water" pairwise alignment method in EMBOSS.

this gene actually functions in JH biosynthesis in *D. melanogaster*. Although not discussed in this context by Burtenshaw and colleagues, their data seems more consistent with a scenario of independent evolution of crustacean MF and insect JH functions, as we hypothesize in this chapter.

These comparative genomic and functional data point away from the ancient hormone hypothesis, which would predict orthology between the enzymes catalyzing the farnesoic acid to MF conversion in crustaceans and insects. Instead, the crustacean enzyme known to catalyze the conversion of farnesoic acid to MF (FA O-MT) and the insect enzyme (JH Acid MT) are clearly not orthologous. While these results are at odds with the ancient hormone hypothesis[15], this pattern of non-orthology is exactly what the Baldwinian Assimilation and the HGT hypotheses would predict. The close similarity of JH acid MTs to microbe rather than other animal sequences

[15] It is formally possible, though, that insects and crustaceans shared a common (ancient) JH/MF biosynthetic pathway, but that one or the other taxa replaced their methyl-transferase after the insect/crustacean divergence. This would, therefore, be an example of "non-orthologous gene displacement," an evolutionary phenomenon not uncommon among bacteria (Koonin et al. 1996). Comparative analyses of the wide variety of genomes being currently studied should help us determine how common this process is in eukaryotes as well.

provides an intriguing indication that this enzyme may have been acquired in insects by horizontal gene transfer from a plant-associated microbe.

It may be objected that the overall architectures of the insect JH and crustacean MF biosynthetic pathways (see Figure 2) are too similar to have evolved independently in these two related arthropod groups. How can we evaluate this argument? Perhaps we can look at the production of JH-III (the active form in most insects) in sedge plants in the genus *Cyperus*. Is *Cyperus* JH-III produced by a similar biosynthetic pathway as JH in insects? Yes! First, some of the same exact intermediates (farnesol, MF) are found in sedges, and farnesol (e.g. in Rutaceae; Brophy and Goldsack 2005), farnesal [e.g. *Arabidopsis* (Brassicaceae); Crowell et al. 2007), farnesoic acid [e.g. *Xanthostemon* (Myrtaceae); Brophy et al. 2006], and MF [e.g. *Polyalthia viridis* (Annonaceae); Kijjoa et al. 1990] are found in taxonomically disparate plants. In fact, *Arabidopsis*, which produces both farnesol and farnesal (Crowell et al. 2007) is now known to have a *bona fide*, functional farnesoic acid methyltransferase, though it is non-orthologous to the two arthropod genes discussed here (Yang et al. 2006). Second, some of the aforementioned JH intermediates in plants are known to have negative impacts on insects that feed on them (reviewed in Hick et al. 1999), and are in some cases induced by insect feeding (e.g. Schnee et al. 2002). Furthermore, the final step in JH synthesis in sedges is catalyzed by a cytochrome p450 epoxidase (Bede et al. 2001), the same class of enzyme used in insect JH biosynthesis. Given this remarkable parallel evolution of JH biosynthesis in sedges and insects, and the presence of JH/MF intermediates in disparate plants, perhaps it is not so unlikely that similar enzymatic pathways could have evolved independently in crustaceans and insects as well.

Still, to fully evaluate the issues surrounding the origin of insect hormonal signaling, we await further detailed comparative biochemistry, endocrinology and genomics, as well as additional paleontological data. The results of such studies might substantially influence our thinking regarding the mechanisms of evolution, from the evolution of phenotypic plasticity and life histories to macroevolutionary questions concerning the origins of novelty.

ACKNOWLEDGEMENTS

Thanks to Douglas Emlen, Thomas Flatt, Lynn Riddiford, Till Luckenbach, David Epel, Amro Hamdoun, Klaus Hartfelder, Graeme Wistow, Douglas Whitman, David Borst, John Byers and Joerg Bohlmann for their comments on earlier versions of this chapter, and Conrad Labandeira for his helpful comments and for allowing me to cite his unpublished observations. Lynn Riddiford, Jim Truman, Hans Kelstrup and the Friday Harbor 2007 Arthropod Evolution Apprentices provided excellent comments late in the process. Thanks also to Bryan Cole for aid with alignments and BLAST searches, Chris Patton for scanning expertise, and Till Luckenbach for his help navigating a German article. Special thanks to the Seattle University Lemieux Library reference staff, and to Joseph Wible and the staff of the Miller Library at Hopkins Marine Station for invaluable help obtaining resources. Also, I am most grateful to Hopkins Marine Station (and in particular David Epel) for providing facilities and funding (NSF ESI-0454770) for the completion of this chapter.

References

Aguinaldo, A. M. A., J. M. Turbeville, L. S. Linford, M. C. Rivera, J. R. Garey, R. A. Raff and J. A. Lake. 1997. Evidence for a clade of nematodes, arthropods and other moulting animals. Nature, 387: 489–493.

Anke, H. and O. Sterner. 1991. Comparison of the antimicrobial and cytotoxic activities of twenty unsaturated sesquiterpene dialdehydes from plants and mushrooms. Planta Medica, 57: 344–346.

Asakawa, Y., M. Toyota, Y. Oiso and J. E. Braggins. 2001. Occurrence of polygodial and 1-(2,4,6-trimethoxyphenyl)-but-2-en-1-one from some ferns and liverworts: role of pungent components in bryophytes and pteridophytes evolution. Chemical and Pharmaceutical Bulletin (Tokyo). 49: 1380–1381.

Baker, M. E. 2005. Xenobiotics and the evolution of multicellular animals: Emergence and diversification of ligand-activated transcription factors. Integrative and Comparative Biology, 45: 172–178.

Baldwin, J. M. 1896. A new factor in evolution. American Naturalist, 30: 441–451.

Baldwin, J. M. 1902. Development and Evolution. Macmillan, New York.

Bede, J. C., P. E. A. Teal, W. G. Goodman and S. S. Tobe. 2001. Biosynthetic pathway of insect juvenile hormone III in cell suspension cultures of the sedge *Cyperus iria*. Plant Physiology, 127: 584–593.

Behmer, S. T. and W. D. Nes. 2003. Insect sterol nutrition and physiology: a global overview. Advances in Insect Physiology, 31: 1–72.

Bellés, X., D. Martín, and M-D. Piulachs. 2005. The mevalonate pathway and the synthesis of juvenile hormone in insects. Annual Review of Entomology, 50: 181–199.

Biggers, W. J. and H. Laufer. 1992. Chemical induction of settlement and metamorphosis of *Capitella capitata* Sp I (Polychaeta) larvae by juvenile hormone-active compounds. Invertebrate Reproduction & Development, 22: 39–46.

Biggers, W. J. and H. Laufer. 1999. Settlement and metamorphosis of *Capitella* larvae induced by juvenile hormone-active compounds is mediated by protein kinase C and ion channels. Biological Bulletin, 196: 187–198.

Bird, D. M. and H. Koltai. 2000. Plant parasitic nematodes: Habitats, hormones, and horizontally-acquired genes. Journal of Plant Growth Regulation, 19: 183–194.

Bohlmann J., J. Gershenzon and S. Aubourg. 2000. Biochemical, molecular genetic and evolutionary aspects of defense-related terpenoid metabolism in conifers. Recent Advances in Phytochemistry, 34: 109–149.

Bohlmann, J., G. Meyer-Gauen and R. Croteau. 1998. Plant terpenoid synthases: molecular biology and phylogenetic analysis. Proceedings of the National Academy of Sciences of the United States of America, 95: 4126–4133.

Bowers, W. S., H. M. Fales, M. J. Thompson and E. C. Uebel. 1966. Juvenile hormone: identification of an active compound from balsam fir. Science, 154: 1020–1021.

Bowers, W. S. 1997. Phytochemical defenses targeting the insect endocrine system. Acta Botanica Gallica, 144: 391–400.

Brand, J. M., J. W. Bracke, A. J. Markovetz, D. L. Wood and L. E. Browne. 1975. Production of verbenol pheromone by a bacterium isolated from bark beetles. Nature, 254: 136–137.

Brophy, J. J. and R. J. Goldsack. 2005. The leaf oils of *Coatesia* and *Geijera* (Rutaceae) from Australia. Journal of Essential Oil Research, 17: 169–174.

Brophy, J. J., R. J. Goldsack and P. I. Forster. 2006. A preliminary examination of the leaf oils of the genus *Xanthostemon* (Myrtaceae) in Australia. Journal of Essential Oil Research, 18: 222–230.

Bur, M. 1985. [Influence of plant growth hormones on development and reproduction of aphids (Homoptera: Aphidinea: Aphididae)]. Entomologia Generalis, 10: 183–200 [in German].

Burtenshaw, S. M., P. P. Su, J. R. Zhang, S. S. Tobe, L. Dayton and W.G. Bendena. 2008. A putative farnesoic acid O-methyltransferase (FAMeT) orthologue in *Drosophila melanogaster* (CG10527): Relationship to juvenile hormone biosynthesis. Peptides, 29: 242–251.

Byard, E., R. R. Shivers and D. E. Aiken. 1975. The mandibular organ of the lobster, *Homarus americanus*. Cell and Tissue Research, 162: 13–22.

Byers, J. A. 1981. Pheromone biosynthesis in the bark beetle, *Ips paraconfusus*, during feeding or exposure to vapours of host plant precursors. Insect Biochemistry, 11: 563–569.

Byers, J. A. 1983. Influence of sex, maturity and host substances on pheromones in the guts of the bark beetles, *Ips paraconfusus* and *Dendroctonus brevicomis*. Journal of Insect Physiology, 29: 5–13.

Byers, J. A. 1989. Chemical ecology of bark beetles. Experientia, 45: 271–283.

Byers, J. A. and G. Birgersson. 1990. Pheromone production in a bark beetle independent of myrcene precursor in host pine species. Naturwissenschaften, 77: 385–387.

Byers, J. A. and D. L. Wood. 1981. Antibiotic-induced inhibition of pheromone synthesis in a bark beetle. Science, 213: 763–764.

Cassagnau, P. and C. Juberthie. 1983. Neurohemal organs in Apterygota. pp. 281–318. *In* A. P. Gupta [ed.]. Neurohemal Organs of Arthropoda. Charles C. Thomas, Springfield, IL, USA.

Crowell, D. N., D. H. Huizinga, A. K. Deem, C. Trobaugh, R. Denton, S. E. Sen. 2007. *Arabidopsis thaliana* plants possess a specific farnesylcysteine lyase that is involved in detoxification and recycling of farnesylcysteine. The Plant Journal, 50: 839–947.

Davies, W. J. and H. G. Jones [eds]. 1991. Abscisic Acid Physiology and Biochemistry. Bios Scientific Publishers, Oxford, UK.

De Loof, A. 2008. Ecdysteroids, juvenile hormone and insect neuropeptides: recent successes and remaining major challenges. General and Comparative Endocrinology, 155: 3–13.

Dinan, L. 2004. Compilation of the literature reports for the screening of vascular plants, algae, fungi and non-arthropod invertebrates for the presence of ecdysteroids. In: R. Lafont, J. Harmatha, F. Marion-Poll, L. Dinan and I. D. Wilson [eds]. The Ecdysone Handbook, 3rd edition, on-line, http://ecdybase.org

Eales, J. G. 1997. Iodine metabolism and thyroid-related functions in organisms lacking thyroid follicles: are thyroid hormones also vitamins? Proceedings of the Society for Experimental Biology and Medicine, 214: 302–317.

Edwards, D., P. A. Selden, J. B. Richardson and L. Axe. 1995. Coprolites as evidence for plant-animal interaction in Siluro-Devonian terrestrial ecosystems. Nature, 427: 329–331.

Eidt, D. C. and C. H. A. Little. 1970. Insect control through host-insect asynchrony: a progress report. Journal of Economic Entomology, 63: 1966–1968.

Farrell, B. D. 1998. "Inordinate fondness" explained: Why are there so many beetles? Science, 281: 555–559.

Flatt, T., M-P. Tu and M. Tatar. 2005. Hormonal pleiotropy and the juvenile hormone regulation of *Drosophila* development and life history. BioEssays, in press.

Fraenkel, G. S. 1959. The raison d'etre of secondary plant substances. Science, 129: 1466–1470.

Gould, S. J. 1984. Challenges to neo-Darwinism and their meaning for a revised view of human consciousness. Tanner Lectures on Human Values. Clare Hall, Cambridge University, Cambridge, UK.

Habgood, K., H. Haas and H. Kerp. 2004. Evidence for an early terrestrial food web: coprolites from the Early Devonian Rhynie Chert. Transactions of the Royal Society of Edinburgh: Earth Sciences, 94: 371–389.

Hall, G. M., C. Tittiger, G. L. Andrews, G. S. Mastick, M. Kuenzli, X. Luo, S. J. Seybold and G. J. Blomquist. 2002. Midgut tissue of male pine engraver, *Ips pini*, synthesizes monoterpenoid pheromone component ipsdienol *de novo*. Naturwissenschaften 89: 79–83.

Harborne, J. B. 1991. Recent advances in the ecological chemistry of plant terpenoids, pp. 399–426. *In:* J. B. Harborne and F. A. Tomas-Barberan [eds]. Ecological Chemistry and Biochemistry of Plant Terpenoids. Clarendon Press, Oxford, UK. 1991.

Hatle, J. D. 2003. Physiology Underlying Phenotypic Plasticity and Polyphenisms: Introduction to the Symposium. Integrative and Comparative Biology, 43: 605–606.

Herout, V. 1970. Some relations between plants, insects and their isoprenoids, pp. 143–202. *In:* L. Reinhold and Y. Liwschitz [eds]. Progress in Phytochemistry, Vol. 2. Interscience Publishers, New York.

Heyland A. and J. Hodin. 2004. Heterochronic developmental shift caused by thyroid hormone in larval sand dollars and its implications for phenotypic plasticity and the evolution of non-feeding development. Evolution, 58: 524–538.

Heyland, A., J. Hodin and A. M. Reitzel. 2005. Hormone signaling in evolution and development: a non-model system approach. BioEssays, 27: 64–75.

Hick, A. J., M. C. Luszniak and J. A. Pickett. 1999. Volatile isoprenoids that control insect behaviour and development. Natural Products Reports, 16: 39–54.

Hinsch, G. W., and H. A. Hajj. 1975. The ecdysial gland of the spider crab, *Libinia emarginata* (L.), I. Ultrastructure of the gland in the male. Journal of Morphology, 145: 179–187.

Hodin, J. 2000. Plasticity and constraints in development and evolution. Journal of Experimental Zoology Part B, Molecular Development Evolution, 288: 1–20.

Jolivet, E., I. Delanglaisjeannin and J. F. Morotgaudry. 1991. Seaweed extracts—phytoactives properties and role in agriculture. Annee Biologique, 30: 109–126.

Kanazawa, A. 2001. Sterols in marine invertebrates. Fisheries Science, 67: 997–1007.

Kijjoa, A., M. M. M. Pinto, P. M. M. Pinho, B. Tantisewie, W. Herz. 1990. Clerodane derivatives from *Polyalthia viridis*. Phytochemistry, 29: 653–655.

Kobayashi, Y. and H. Ando. 1983. Embryonic development of the alimentary canal and ectodermal derivatives in the primitive moth, *Neomicropteryx nipponensis* issiki (Lepidoptera, Micropterygidae). Journal of Morphology, 176: 289–314.

Koonin, E. V., A. R. Mushegian and P. Bork. 1996. Non-orthologous gene displacement. Trends in Genetics, 12: 334–336.

Kroemer, K., R. Reski and W. Frank. 2004. Abiotic stress response in the moss *Physcomitrella patens*: evidence for an evolutionary alteration in signaling pathways in land plants. Plant Cell Reports, 22: 864–70.

Labandeira, C. C. 1998. Early history of arthropod and vascular plant associations. Annual Review of Earth and Planetary Sciences, 26: 329–377.

Labandeira, C. C. 2002a. Paleobiology of predators, parasitoids, and parasites: accommodation and death in the fossil record of terrestrial invertebrates, pp. 211–249. *In* M. Kowalewski and P.H. Kelley [eds]. The Fossil Record of Predation. Paleontological Society Papers, volume 8. The Paleontological Society: New Haven, CT.

Labandeira, C. C. 2002b. The history of associations between plants and animals, pp. 26–74, 248–261. *In* C. Herrera and P. Pellymr [eds]. Plant-Animal Interactions: An Evolutionary Approach. Blackwell Science, Oxford, UK.

Laufer, H. and W. J. Biggers. 2001. Unifying concepts learned from methyl farnesoate for invertebrate reproduction and post-embryonic development. American Zoologist, 41: 442–457.

Li, L., C. Li, G. I. Lee and G. A. Howe. 2002. Distinct roles for jasmonate synthesis and action in the systemic wound response of tomato. Proceedings of the National Academy of Sciences USA. 99: 6416–6421.

Martin, D., J. Bohlmann, J. Gershenzon, W. Francke and S. J. Seybold. 2003. A novel sex-specific and inducible monoterpene synthase activity associated with a pine bark beetle, the pine engraver, *Ips pini*. Naturwissenschaften, 90: 173–179.

Martin, D. M., J. Faldt and J. Bohlmann. 2004. Functional characterization of nine Norway Spruce TPS genes and evolution of gymnosperm terpene synthases of the TPS-d subfamily. Plant Physiology, 135: 1908–1927.

Nijhout HF. 1994. Insect Hormones. Princeton University Press, Princeton.

Niklas, K. J. 1997. The Evolutionary Biology of Plants. University of Chicago Press, Chicago.

Oritani, T. and H. Kiyota. 2003, Biosynthesis and metabolism of abscisic acid and related compounds. Natural Product Reports, 20: 414–425.

Palmer, J. D., D. E. Soltis and M. W. Chase. 2004. The plant tree of life: an overview and some points of view. American Journal of Botany, 91: 1437–1445.

Pearson, W. R. and D. J. Lipman. 1988. Improved tools for biological sequence analysis. Proceedings of the National Academy of Sciences USA, 85: 2444–2448.

Peña-Cortés, H. and L. Willmitzer. 1995. The role of hormones in gene activation in response to wounding, pp. 395–414. *In* P. J. Davies [ed.]. Plant Hormones: Physiology, Biochemistry and Molecular Biology. 2nd Edition. Kluwer Academic Publishers, Norwell, Massachusetts.

Pennock, J. F. 1977. Terpenoids in marine invertebrates, pp. 153–213. *In* T. W. Goodwin [ed]. International Review of Biochemistry, Vol. 14, Biochemistry of Lipids II. University Park Press, Baltimore, MD.

Pfennig, D. W. 1992. Proximate and functional causes of polyphenisms in an anuran tadpole. Functional Ecology, 6: 167–174.

Poulter, C. D. and H. C. Rilling. 1981. Prenyl transferases and isomerase, pp. 161–224. *In* J. W. Porter and S. L. Spurgeon [eds]. Biosynthesis of Isoprenoid Compounds, Vol. 2. Wiley & Sons, New York.

Robbins, W. E., J. N. Kaplanis, J. A. Svoboda and M. J. Thompson. 1971. Steroid Metabolism in Insects. Annual Review of Entomology, 16: 53–72.

Rothstein, M. 1968. Nematode Biochemistry IX. Lack of Sterol Biosynthesis in Free-Living Nematodes. Comparative Biochemistry and Physiology, 27: 309–317.

Ruddell, C. J., G. Wainwright, A. Geffen, M. R. H. White, S. G. Webster and H. H. Rees. 2003. Cloning, characterization, and developmental expression of a putative farnesoic acid O-methyl transferase in the female edible crab *Cancer pagurus*. Biological Bulletin, 205: 308–318.

Ryan, M. F. 2002. Insect Chemoreception. Kluwer, Dordrecht.

Sangster, T. A., S. Lindquist and C. Queitsch. 2004. Under cover: causes, effects and implications of Hsp90-mediated genetic capacitance. Bioessays, 26: 348–62.

Schnee, C., T. G. Köllner, J. Gershenzon and J. Degenhardt. 2002. The maize gene terpene synthase 1 encodes a sesquiterpene synthase catalyzing the formation of (E)-b-farnesene, (E)-nerolidol, and (E, E)-farnesol after herbivore damage. Plant Physiology, 130: 2049–2060.

Schultz, J. C. 2002. Shared signals and the potential for phylogenetic espionage between plants and animals. Integrative and Comparative Biology, 42: 454–462.

Schultz, J. C. and H. M. Appel. 2004. Cross-kingdom cross-talk: hormones shared by plants and their insect herbivores. Ecology, 85: 70–77.

Seybold, S. J., D. R. Quilici, J. A. Tillman, D. Vanderwel, D. L. Wood and G. J. Blomquist. 1995. *De novo* biosynthesis of the aggregation pheromone components ipsenol and ipsdienol by the pine bark beetles *Ips paraconfusus* Lanier and *Ips pini* (Say) (Coleoptera: Scolytidae). Proceedings of the National Academy of Sciences USA, 92: 8393–8397.

Shinoda, T. and K. Itoyama. 2003. Juvenile hormone acid methyltransferase: A key regulatory enzyme for insect metamorphosis. Proceedings of the Natural Academy Sciences USA, 100: 11986–11991.

Silva Gunawardene, Y. I. N., S. S. Tobe, W. G. Bendena, B. K. C. Chow, K. J. Yagi and S.-M. Chan. 2002. Function and cellular localization of farnesoic acid *O*-methyltransferase (FAMeT) in the shrimp, *Metapenaeus ensis*. European Journal of Biochemistry, 269: 3587–3595.

Sláma, K. and C. M. Williams. 1966. The juvenile hormone. V. The sensitivity of the bug, *Pyrrhocoris apterus*, to a hormonally active factor in American paper-pulp. Biological Bulletin, 130: 235–246.

Sláma, K., M. Romaòuk and F. Šorm. 1974. Insect Hormones and Bioanalogues. Springer-Verlag, New York.

Smyrniotopoulos, V., D. Abatis, L-A. Tziveleka, C. Tsitsimpikou and V. Roussis. 2003. Acetylene sesquiterpenoid esters from the green alga *Caulerpa prolifera*. Journal of Natural Products, 66: 21–24.

Stay, B. and S. S. Tobe. 2007. The role of allatostatins in juvenile hormone synthesis in insects and crustaceans. Annual Review of Entomology, 52: 277–299.

Svoboda, J. A., J. N. Kaplanis, W. E. Robbins and M. J. Thompson. 1975. Recent developments in insect steroid metabolism. Annual Review of Entomology, 20: 205–220.

Tallamy, D. W., C. A. Mullin and J. L. Frazier. 1999. An alternate route to insect pharmacophagy: The loose receptor hypothesis. Journal of Chemical Ecology, 25: 1987–1997.

Taylor, J. E., P. E. Hatcher and N. D. Paul. 2004. Crosstalk between plant responses to pathogens and herbivores: a view from the outside in. Journal of Experimental Botany, 55: 159–168.

Thaler, J. S. and R. M. Bostock. 2004. Interactions between abscisic-acid-mediated responses and plant resistance to pathogens and insects. Ecology, 85: 48–58.

Truman, J. W., W. S. Talbot, S. E. Fahrbach and D. S. Hogness. 1994. Ecdysone receptor expression in the CNS correlates with stage-specific responses to ecdysteroids during *Drosophila* and *Manduca* development. Development, 120: 219–234.

Tobe, S. S. and W. G. Bendena. 1999. The regulation of juvenile hormone production in arthropods, functional and evolutionary perspectives. Annals of the New York Academy of Sciences, 897: 300–310.

Toong, Y. C., D. A. Schooley and F. C. Baker. 1988. Isolation of insect juvenile hormone III from a plant. Nature, 333: 170–171.

Visscher, S. N. 1983. Effects of abscisic acid in animal growth and reproduction, pp. 553–579. *In:* F. T. Addicott [ed.]. Abscisic Acid. Praeger Scientific, New York.

Waddington, C. H. 1960. Evolutionary adaptation. *In* Sol Tax [ed.]. Evolution after Darwin, volume 1, pages 381–402. University of Chicago Press: Chicago.

Waddington, C. H. 1961. Genetic assimilation. Advances in Genetics, 10: 257–290.

Wheeler, D. E. and H. F. Nijhout. 2003. A perspective for understanding the modes of juvenile hormone action as a lipid signaling system. Bioessays, 25: 994–1001.

Wildon, D. C., J. F. Thain, P. E. H. Minchin, I. R. Gubb, A. J. Reilly, Y. D. Skipper, H. M. Doherty, P. J. O'Donnell and D. J. Bowles. 1992. Electric signaling and systemic proteinase inhibitor induction in the wounded plant. Nature, 360: 62–64.

Wyatt, G. R. 1997. Juvenile hormone in insect reproduction—a paradox? European Journal of Entomology, 94: 323–333.

Yang, Y., J. S. Yuan, J. Ross, J. P. Noel, E. Pichersky and F. Chen. 2006. *An Arabidopsis thaliana* methyltransferase capable of methylating farnesoic acid. Archives of Biochemistry and Biophysics, 448: 123–132.

Yesilada, E. and A. N. Bozcuk. 1995. [The effects of ABA and kinetin on the fecundity of *Drosophila melanogaster*]. Turkish Journal of Biology, 19: 37–44 [in Turkish].

Yesilada, E. and A. N. Bozcuk. 1996. [The effects of ABA and Kinetin on the developmental period of *Drosophila melanogaster*]. Turkish Journal of Biology, 20: 29–35 [in Turkish].

21

Phenotypic Plasticity and Evolvability: An Empirical Test with Experimental Evolution

Henrique Teotónio[1], Michael R. Rose[2], and Stephen R. Proulx[3]
[1] *Centro de Biologia do Desenvolvimento, Instituto Gulbenkian de Ciência, Oeiras Portugal; E-mail: teotonio@igc.gulbenkian.pt*
[2] *Department Ecology and Evolutionary Biology, University California Irvine USA; E-mail: mrrose@uci.edu*
[3] *Center for Ecology and Evolutionary Biology, University of Oregon Eugene USA; E-mail: proulx@proulxresearch.org*

Abstract

The relationship between phenotypic plasticity and the evolvability of a population, that is the population's capacity for evolutionary change, is critically dependent on the fitness costs of expressing particular phenotypes in particular environments and on evolutionary history. This relationship is the focus of the present chapter. We start by reviewing a model suggesting that phenotypic plasticity can either increase or decrease evolvability when there is a change in the environment, and then address some of the predictions made by the model using empirical data from experimental reverse evolution in *Drosophila melanogaster*. We find that phenotypic plasticity for fecundity facilitates adaptive evolution in our populations. We finish by discussing the stringent conditions that must be experimentally met to study the relationship between phenotypic plasticity and evolvability.

Introduction

Variation is a fundamental feature of natural systems, expressed at both individual and population levels, and both genetically and phenotypically. Phenotypic plasticity (PP) is one of the mechanisms generating phenotypic variation. PP can be defined as the expressed phenotype of individual genotypes, as a function of the environment (Scheiner 1993): plastic genotypes will express more phenotypic variation for an environmental range when compared to less plastic genotypes. As a consequence, populations composed of plastic genotypes are expected to show more phenotypic variation than populations composed of less plastic genotypes for the same environmental range (Sarkar and Fuller 2003). Environmental canalization on the other hand can be viewed as the converse of PP, with more canalized genotypes expected to express less phenotypic variation than less canalized genotypes (Gavrilets and Hastings 1994, Wagner et al. 1997).

Natural populations exhibit heritable variation for PP (West Eberhard 1989, Scheiner 1993, Bell 1997, Lynch and Walsh 1998, Schlichting and Pigliucci 1998). Selection for integrating environmental information into the developmental process can create a plastic response that results in more or less variable phenotypes. Much research has focused on determining the evolutionary conditions that lead to the origin and maintenance of variability for PP (Levins 1968, Via and Lande 1985, Scheiner 1993). It is thought that the expression of PP results from evolution in spatially heterogeneous and/or temporally fluctuating environments, the amount of heritable PP depending on interactions between environmental variation and life-cycle (Bull 1987, Meyers and Bull 2002). It is unclear, however, how commonly such theoretical expectations are fulfilled, because the available data (e.g., Via et al. 1995, de Jong and Crozier 2003) are usually insufficient to test the assumptions or predictions of the available models.

The consequences of heritable variation in PP are diverse, because it can act both as a constraint and a facilitator of evolution (cf. Whitlock 1995, Ancel 1999, 2000, Meyers and Bull 2002, Price et al. 2003). The relationship between PP and the evolvability of a population, that is the population's capacity for evolutionary change, is critically dependent on the fitness costs of expressing particular phenotypes in particular environments and on evolutionary history, including population structure. This relationship is the focus of the present chapter. We start by reviewing a model that suggests that PP both increases and decreases evolvability when there is a change in the environment, and then address some of the predictions made by the model using empirical data from experimental reverse evolution in *Drosophila melanogaster*.

Plasticity and Evolvability

Does phenotypic plasticity influence rates of evolutionary change? Theoretical research has found several conditions under which this appears to be true. For example, PP will be a constraint on evolution if populations have plastic genotypes that produce optimal phenotypes, even under environmental change. Under these conditions, evolutionary diversification in response to environmental change is less likely to occur (Hensen 2003, Price et al. 2003). A fit between phenotype and environment is maintained by plasticity, not genetic adaptation.

PP can also act as a facilitator of evolution. This idea dates back to the 19th century (e.g. Baldwin 1896). Recently the problem has been approached from a quantitative genetic standpoint (Ancel 1999, 2000, Price et al. 2003). For populations that are far away from the optimal phenotype due to an environmental change, PP can place a few genotypes close to these selective optima simply because it increases phenotypic variance (Baldwin 1896, Whitlock 1995, Ancel 1999, 2000, Price et al. 1993, Price et al. 2003). This allows the population to survive in the new environment long enough to adapt to it, increasing the likelihood of successful adaptation to the novel environment (Gomulkiewicz and Holt 1995).

Related to this last effect, the increase in phenotypic variance in more plastic populations will often smooth the adaptive landscape, the function describing the relationship between a population's mean fitness and mean phenotype (Simpson 1953, Lande 1976, 1979, Barton and Turelli 1989, Arnold et al. 2001). However, even for relatively simple fitness functions, the effect of smoothing the adaptive landscape on evolution will depend on where a particular population is in the fitness function. We illustrate this point in Figures 1–4. By smoothing the adaptive landscape, PP can reduce or even eliminate fitness valleys and thus allow faster evolution as shown in Figure 3. PP can increase the rate of adaptation to novel environments when it contributes to the genetic variance for fitness itself. [This effect depends on the way that phenotypic variance is translated into variance in fitness (Fisher 1958, Bull 1987, Price et al.1993, Whitlock 1995).] Once populations evolve phenotypes that move the population into the basin of attraction of a novel selective optimum, then the relationship between PP and evolvability is reversed. PP will then act as a constraint on further evolution because of two distinct factors. The first is that the strength of selection increases with distance from the selective optimum (Lande 1976, 1979, Arnold et al. 2001). The second factor is that stabilizing selection will erode high values of plasticity and thereby reduce the range of phenotypes being expressed (Via and Lande 1985, Price et al. 2003).

Considering all of the above arguments, it is no surprise that the relationship between PP and evolvability is highly non-linear. This relationship is greatly dependent on the developmental and physiological mechanisms that determine how phenotypes are related to fitness (Figure 2). Predictions about the relation between PP and evolvability are necessarily made on a local level, and with previous knowledge of the evolutionary history of the populations under study (Figures 1-3). In extreme cases, if the fitness landscapes are highly rugged and more than one character is considered, PP may have very little influence on evolvability on a global level (cf. Fisher 1958, Price et al. 1993, Coyne et al. 1997, Hensen 2003). Figure 4 shows that the non-linearity of the relationship between PP and evolvability is inherent to any natural system which has a fitness function with at least two selective optima of different heights. This is the model that we empirically address below with data from experimental evolution.

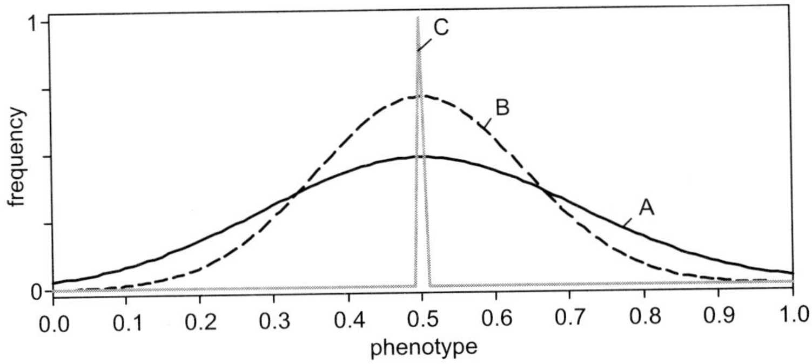

Fig. 1 *Frequency of phenotypes in populations.* This figure shows the complete ensemble of phenotypes generated by all the genotypes in three separate populations, according to their actual frequency. Here we assume that there is no genotype by environment interaction. The difference between the distributions indicated by the solid (A) and dashed (B) lines therefore reflect greater environmental variance (V_e) in the former case. The spike (C) in the middle of the figure shows the distribution that arises when there is no genetic variance (V_g) or V_e in a population. Note that V_e includes both phenotypic variation due to exogenous environmental factors and endogenous developmental variation that is not genetic in origin. Our interpretation is that distributions with higher V_e values reflect greater phenotypic plasticity, assuming no differences in exogenous conditions or genetic variation.

Finally, for populations to evolve towards novel selection optima, novel genotypic variants arising through mutation or recombination must appear if they are not already present when the environment changes. If these genotypic variants are more likely to appear in those genotypes which are also more phenotypically plastic, then PP will increase evolvability of a population by genetic assimilation (sensu Wadington 1953), and in this way counteract the effects just discussed above: more plastic population

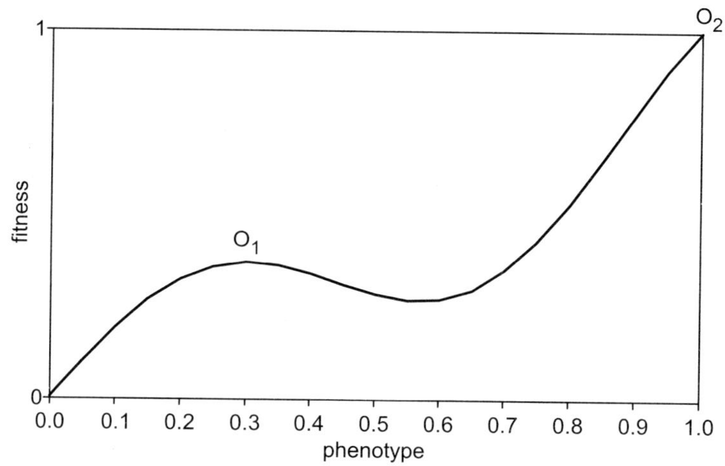

Fig. 2 *Fitness function.* This shows a fitness associated with specific phenotypic values. We make no assumption concerning the genotype, or genotypes, which may generate any particular phenotypic value. Specifically the same genotype may generate more than one phenotype and two genotypes may generate the same phenotype. In the case shown there are two local maxima for fitness at phenotypic value O_1 and O_2.

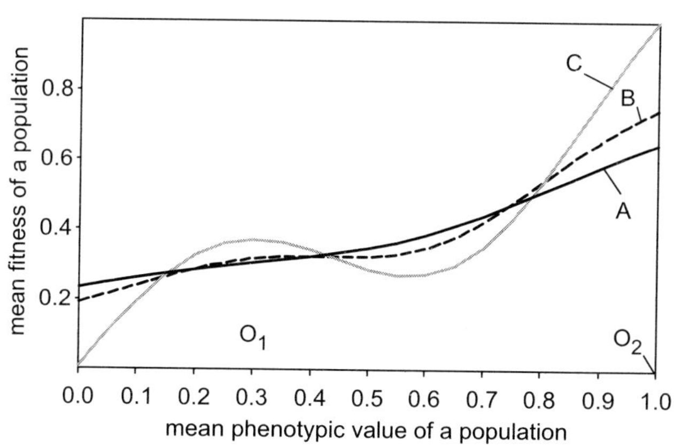

Fig. 3 *Adaptive landscapes.* This figure shows the relationship between the mean phenotypic value of a population and its mean fitness, which results from integrating the functions shown in Figures 1 and 2 (cf. Lande 1976, Arnold et al. 2001), across phenotypes and assuming no change in phenotypic variance. The three cases A, B, and C correspond to the phenotypic distributions A, B, and C shown in Figure 1, always assuming the fitness function of Figure 2. Note that the adaptive landscapes A and B do not have a fitness valley separating the two fitness optima, O_1 and O_2.

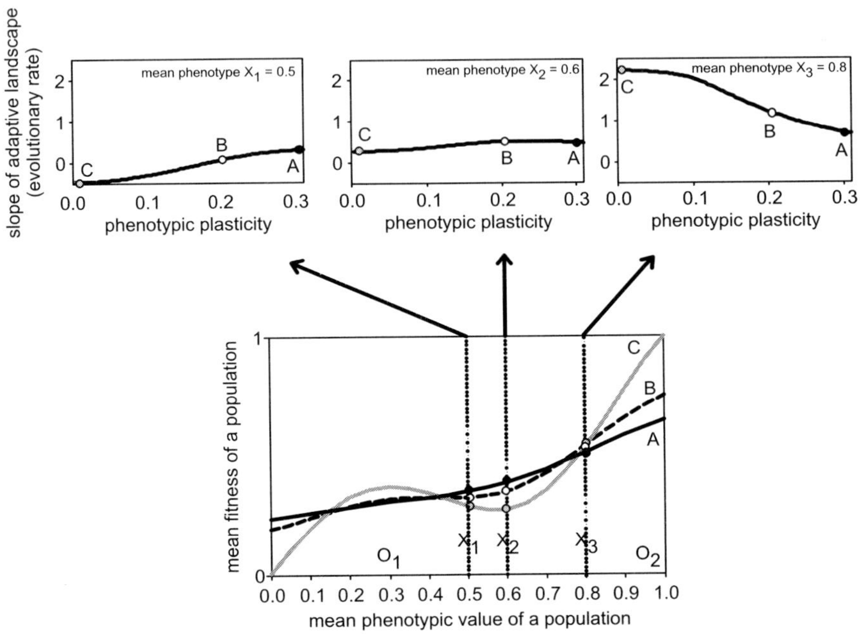

Fig. 4 *Evolvability as a function of phenotypic plasticity AND position on the adaptive landscape.* Here we assume that evolvability is determined solely by the first derivative of the mean fitness as a function of the mean phenotype, which we call the slope of the adaptive landscape. The top three panels show the slopes of the adaptive landscape at positions x_1, x_2, and x_3 proceeding from left to right. Note that the impact phenotypic plasticity has on the slope of the adaptive landscape is reversed for the cases of mean phenotype values x_1 and x_3. This shows that the impact phenotypic plasticity on evolvability can reverse in sign depending on position in adaptive landscape.

will then evolve faster. This matching between phenotypic variation generated by PP and novel adaptive genetic variation has been much discussed, but little evidence for its existence has been found (cf. Wagner 1996, Ancel and Fontana 2000, Rutherford and Lindquist 1998, Proulx and Phillips 2004).

In summary, PP can act as a facilitator of evolution in one of three non-exclusive ways. First, it may allow the persistence of a population following environmental changes. Second, it may reduce or eliminate fitness valleys which are otherwise un-crossable. Third, it may facilitate genetic assimilation when there is a correlation between PP and magnitude of genetic variation. By contrast, PP will act as a constraint if populations are already in the local vicinity of the optimal phenotype or by decreasing the amount of expressed genetic variance.

Because evolution is local, the conditions that favor an increase or a decrease in evolutionary rates with increased PP are quite restrictive (see also Ancel 2000). The first assumption made by the model developed above is that once the environment changes it will remain constant in time; at least until evolution to the novel optimum occurs. A second assumption is that PP will smooth the adaptive landscape of plastic populations. This is true for symmetric phenotypic distributions, such as those of Figure 1, but can be violated by highly skewed phenotypic distributions. Also, there must be heritable variation for PP itself if genetic assimilation is to take place, otherwise natural selection for the novel phenotypes could not be effective. Lastly, it is important to bear in mind that plasticity must come at a fitness cost in local environments, otherwise just a few plastic genotypes would always evolve to fixation.

Below we address some of the predictions made by the model presented in Figures 1–4 using data from a reverse evolution experiment.

Reverse Experimental Evolution in *Drosophila melanogaster*

Direct empirical assessment of the relationship between PP and evolvability requires populations that differ in heritable PP, all derived from a common recent ancestor, and facing the same novel environment. In such studies, evolutionary rates need to be obtained and direct or indirect measurements of additive genetic variance for fitness should be available. Comparative data from populations in nature are unlikely to supply enough appropriate, unconfounded, reproducible data (Leroi et al. 1994). Experimental evolution is more likely to provide the best means for hypothesis testing about PP, whether conducted in the field or in the laboratory.

Experimental evolution refers to an empirical approach where the experimenter controls the conditions under which evolution occurs, in a reproducible manner, and then observes its course in real time (see Wright 1977, Rose et al. 1996, Bell 1997). The experiments are usually done in the laboratory with replicated populations subjected to a particular environment and followed for many generations. This allows the experimenter to record the variables of interest as a time series for both experimental and control populations. Insects, in particular, have been a favorite group in evolution experiments (Wright 1977, Scheiner 1993, Rose et al. 1996).

The populations of *D. melanogaster* that we worked with are all derived from the same original ancestral population. In 1975, a sample from a natural population from Massachusetts USA was isolated and adapted at large population sizes to a laboratory life-cycle of two weeks (approximately 21h for embryogenesis, 5 days for larval development, 4.5 days for metamorphosis, 6 hours for sexual maturity and 3 days of adulthood before oviposition). From this original *Ancestral* population we have produced a selective radiation in the laboratory that has been ongoing for more than 20 years (reviewed in Rose et al. 2004). In 1980, five populations were established with breeding taking place at progressively later ages until the original life-cycle of two weeks was increased to 70 days (Rose 1984). These are here called the *Old* populations. Contemporaneously, five *Control* populations were maintained in the ancestral environment. Then in 1989, another set of five populations was derived from the Old populations and selected for increased starvation resistance at two weeks of age in the same 3–4-week life-cycle (the *Starvation* populations), while five other populations (the *Intermediate* populations) were selected over the same 3 to 4-week life-cycle as the Starvation populations but using a high food level (Rose et al. 1992). Finally in 1992, five other populations (the *Accelerated* populations) were derived from the Intermediate populations and selected for very early breeding and accelerated developmental time (Chippindale et al. 1997a). All 25 populations were maintained at high population sizes (N>500), without systematic inbreeding or hybridization, throughout their history (Teotónio and Rose 2000).

In 1997, 20 new populations were derived from the 20 established Old, Starvation, Intermediate, and Accelerated populations and returned to the environmental conditions of the Ancestral populations. We thus initiated an evolution experiment with a well-defined control: evolutionarily stable ultimate ancestors, the Control populations. This was the point of reference for testing hypotheses about reverse evolution. The Control populations had already been maintained for close to 450 generations in the ancestral environment, thus being the best representative of the ancestral phenotypic character states in the absence of cryonic samples.

Several life-history and morphological characters were followed in the reverse-evolved populations for 50 generations. Reverse evolution in these populations, particularly the degree of convergence, depends on previous evolutionary history (see Figure 5). Yet, for most studied characters, incomplete convergence to the ancestral states could not be explained by lack of genetic variation, accumulation of deleterious mutations, or the presence of genetic interactions (Teotónio and Rose 2000). Furthermore, adaptation occurred in all populations to a similar extent, as measured by competitive fitness assays such as male competitive performance against

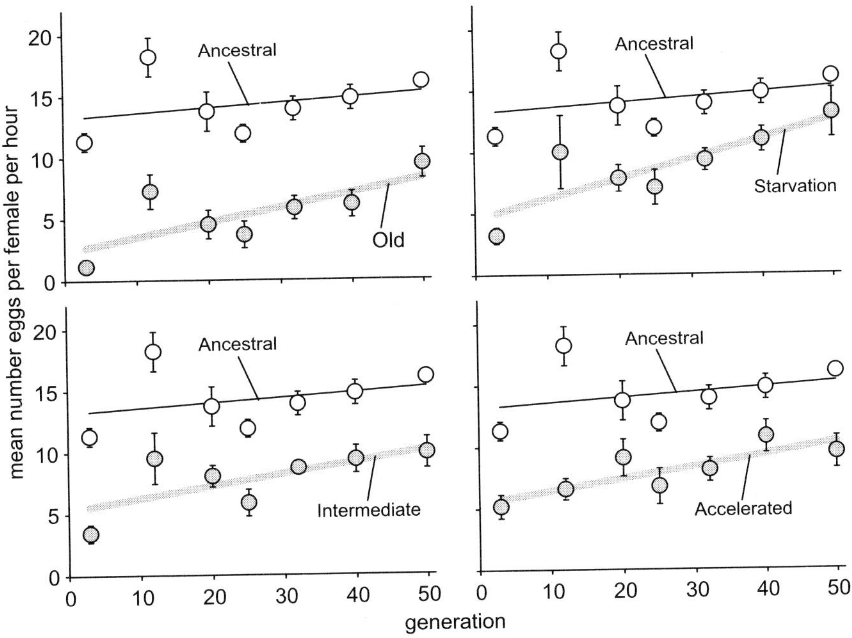

Fig. 5 *Fecundity reverse evolution in laboratory populations of Drosophila melanogaster.* Linear response of fecundity during one hour of egg laying, during 50 generations of reverse evolution. The mean of five replicate populations in each evolutionary group, Control/Ancestral, Old, Starvation, Intermediate, or Accelerated, is shown with associated standard error of the mean (Teotónio and Rose 2000). Only the Starvation populations converged to Control levels, all others remained differentiated after 50 generations of evolution. Regression analysis is shown for linear models fitted to the mean replicate population's response. All regressions significantly explain most of the variation except for the one describing the trajectory of the accelerated populations (see Teotónio and Rose 2000). In the present chapter we calculated evolutionary rates as the slope of a least-squares linear regression for *each* single population for a total of 25.

tester strains, as described in Teotónio et al. (2002). These results suggested that the fitness function in the ancestral environment is highly rugged, with several ways to combine phenotypic states to achieve similar levels of fitness.

In Teotónio et al. (2002), data on phenotypic plasticity for a fitness component, early fecundity, were reported. The ancestral environment is characterized by a two week generation in vials, with egg-laying to start the next generation taking place in vials under crowded conditions for approximately two hours (starting at exactly two weeks from the start of egg laying in the previous generation, less two hours), after which between

50 and 100 eggs are allowed to develop with abundant food in each rearing vial. In this ecological setting, natural selection seems to have favoured rapid egg-laying when flies are placed in the fresh food vials (Figure 6). Phenotypic plasticity for fecundity was measured as the number of eggs laid by each female as a function of the time allowed for egg laying (one, two or six hours). Observed PP was different for the differentiated populations not only in terms of scaling but also, and more importantly, in terms of genetic correlations among environments (Teotónio et al. 2002, Figure 6). Each of these reaction norms for egg-laying was inferred from the mean of five evolutionarily independent populations, so that all consistently observed

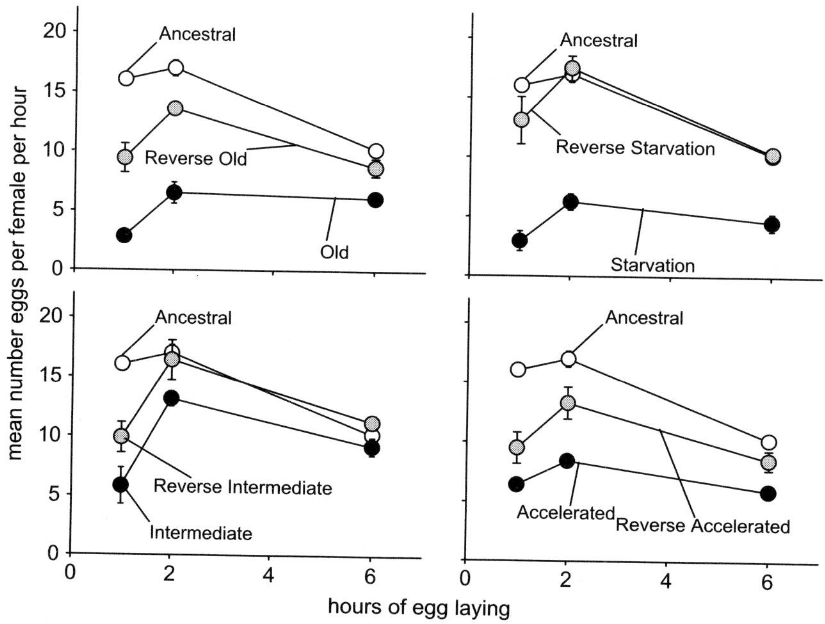

Fig. 6 *Reaction norms of fecundity at three different environments (1, 2, and 6 hours of allowed oviposition) for various populations.* The environmental conditions closely follow those used during reverse evolution in the ancestral environment (Teotónio et al. 2002). As in Figure 5, each point depicts the mean observation of five populations for each evolutionary group, with associated standard error. All assays were performed after 50 generations of reverse evolution. An ANCOVA model testing the hypothesis that the linear reaction norms have evolved homogeneously among different evolutionary groups was rejected indicating that PP for fecundity is eminently evolvable and context dependent (see Teotónio et al. 2002). Fecundity varies with environment (time allowed to oviposit), and thus is a plastic trait. In the present chapter phenotypic plasticity is measured as the range in the number of eggs laid across *all* three environments for each population.

changes had to be the result of natural selection. Thus, the Old, Starvation, Intermediate, and Accelerated populations varied in the amount of PP for fecundity. During reverse evolution of derivatives of these 20 populations, both the intercepts and slopes of the fecundity reaction norms evolved as a function of previous PP evolution (Teotónio et al. 2002).

In themselves, these results are not exceptional because other studies have reported changes in phenotypic plasticity for fitness components in relatively short evolutionary time spans, both for intercepts and slopes of linear reaction norms. For example, Chippindale et al. (1997b) described how the Old populations have high fecundity only when mated and given high protein content food, while the baseline Control populations have high fecundity as long as populations are maintained on such high protein food. Many examples like these are known, mostly from morphological phenotypes of uncertain relationship with fitness (cf. Scharloo 1991, Stearns et al. 1991. Scheiner 1993).

Evolvability and Plasticity during Experimental Reverse Evolution

The conditions for testing the model presented above and in Figures 1–4 are partially fulfilled by our laboratory evolution experiments on *Drosophila melanogaster*. In what follows, we determine if PP influenced evolvability when the differentiated populations, described above, were returned to, and subsequently evolved in the ancestral environment. Predictions that can be derived from other models, such as the prediction that plastic populations are more likely to persist under drastic environmental changes, or the prediction that the approach to the optimum will be slower for more plastic populations, are only briefly discussed (see above, Ancel 1999, 2000, Price et al. 2003).

The first condition to be fulfilled by our laboratory experiment is that the evolutionary trajectories of several populations undergoing reverse evolution for a character (early fecundity) showing PP are available. This allows measurement of evolutionary rates, as shown in Figure 5. We also show that PP for a character that is related to fitness (early fecundity) is heritable, precedes adaptation to the ancestral environment, and is eminently evolvable (Figure 6). It is important that no single group of populations is found to outperform all the other populations. Otherwise it is possible that a single genotype would be always favored. For example, Intermediate populations might have been predicted, a priori, to converge further toward the ancestral PP levels because they show character values more similar to the control ones before reverse evolution, but it was the Starvation populations that converged most closely to control fecundity values. The results suggest that there is a rugged fitness function, whose shape changes from the diversifying to the ancestral environments. In Figure 7 a measure of fitness (male competitive performance) of each population before and after reverse evolution, together with the control means, is shown against fecundity. The least squares quadratic fit of fitness to fecundity for all 45 populations is also shown, and this can be taken as the approximate fitness function in the ancestral environment. We use this function primarily

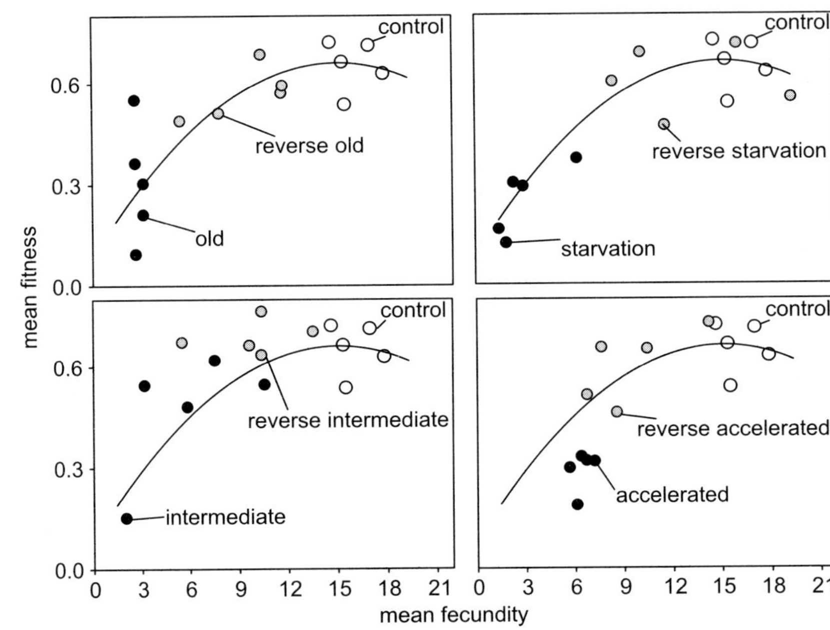

Fig. 7 *Fitness function in the ancestral environment.* Fitness is measured as the male competitive performance between each individual population and a standard genetically marked strain, by the number of fertilized genetically marked females (details in Teotónio et al. 2002). All assays were performed at generation 50 of reverse evolution, with differentiated populations being shown in black, reverse evolved populations by gray circles and control populations by white circles. ANOVA reveals that all differentiated populations converged to Control/Ancestral levels (Teotónio et al. 2002). The *x*-axis shows mean initial observed fecundity for each population as the one hour values of Figure 6. The least-squares quadratic fit is shown as a solid, curved line. The model used all populations and was highly significant ($F_{2,42}$=34, p<0.001, R^2=0.62).

to give an idea of the ruggedness of the fitness function. It appears that a single selective optimum is present along the phenotypic range observed and that all differentiated populations are within the basin of attraction of the novel selection optimum.

Given this fortunate situation, we can now test predictions relating PP to evolvability. With a single optimum fitness landscape, which appears to be the case, the standard theoretical prediction is that PP should be inversely proportional to evolvability, if PP smoothes the adaptive landscape. That is, more plastic populations should evolve *slower* than less plastic populations as shown in Figure 4, and below.

To test this prediction, initial fitness functions and phenotype distributions, as in Figures 1 and 2, must be accounted for, before the relationship between PP and evolutionary rates is tested. For example, according to quantitative genetics theory, the rate of evolutionary response is directly proportional to the additive genetic variance for fitness (Fisher 1958). In our populations, one therefore expects that some of the differences in observed evolutionary rates are due to initial fitness differences. This is exactly the pattern shown by populations undergoing reverse evolution. As shown in Figure 8, there is a negative relationship between evolutionary rates for fecundity and fitness. We used stepwise regression to statistically control for these fitness effects. However, we first tested whether different evolutionary groups had consistently different evolutionary rates. An ANOVA model with evolutionary group as a factor (with Control, Old, Starvation, Intermediate, and Accelerated selection regimes as distinct treatments) indicated that this was not the case ($F_{4,20}=2.13$, $p=0.11$): the selection history did not significantly affect evolutionary speed during reverse evolution. This justifies our next step, where all populations are grouped regardless of their differentiating evolutionary history. We measured PP as the range of fecundity values among all environments. This was based on the argument that this variable captures both the adaptive and non-adaptive components of PP (see above), while a measurement such as the slope of a linear regression of fecundity with egg-laying time is of more difficult interpretation (cf. Scheiner 1993).

In the stepwise regression, the evolutionary rate of response is first modeled as the dependent variable and both fitness and initial fecundity as the independent variables (see Figure 8). This model significantly explained variation in evolutionary rates ($F_{2,22}=9.38$, $p=0.001$). The residual evolutionary rates, from the estimated model, were then taken as the novel dependent variable and fitted to the variation in PP. This model was again

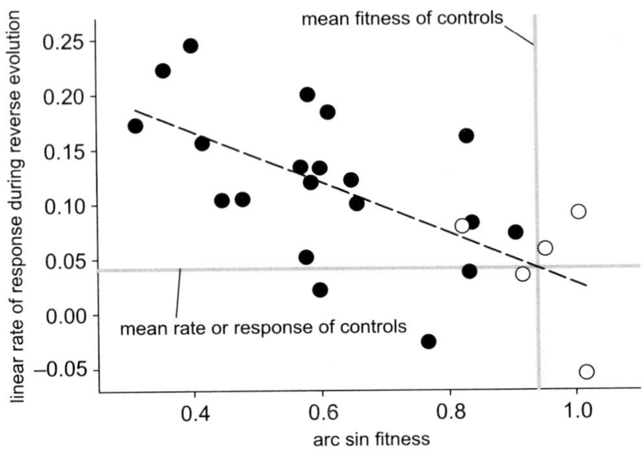

Fig. 8 *Fitness and evolvability.* Experimental results for relationship between rates of response to reverse evolution in fecundity, plotted against level of fitness for the differentiated populations, for each replicate separately. Fitness values were angular transformed for all analysis since they were taken as a proportion (Zar 1996). Twenty-five data points were used in the regression analysis, with control populations shown as white circles. The model significantly explains part of the variation in rates of response, shown by a dashed line ($F_{1,23}$=17.7; p=0.0003; R^2=0.44). Populations that were closer to the new selective optimum, that is, that had fitness levels closer to the Control/Ancestral populations, respond relatively slower than those that were farther away from it.

significant as shown in Figure 9 ($F_{1,23}$=4.4, p=0.05). However, rates of response are positively related to PP, which is the opposite of what was expected a priori. In summary, during experimental reverse evolution populations that showed more plasticity in fecundity evolved faster than less plastic populations, all else being equal. This is contrary to conventional expectations.

Comparisons with Simulated Data

Why is a positive relationship between PP and evolvability observed during reverse evolution? To better understand which conditions produce results similar to the experimental evolution, we conducted a simulation study based on the fitness function of Figure 2. We first generated an adaptive landscape for populations with various levels of PP, and then determined the rate of evolutionary response for a given initial phenotype (Figure 1 shows three such populations for three initial phenotypes). Random groups

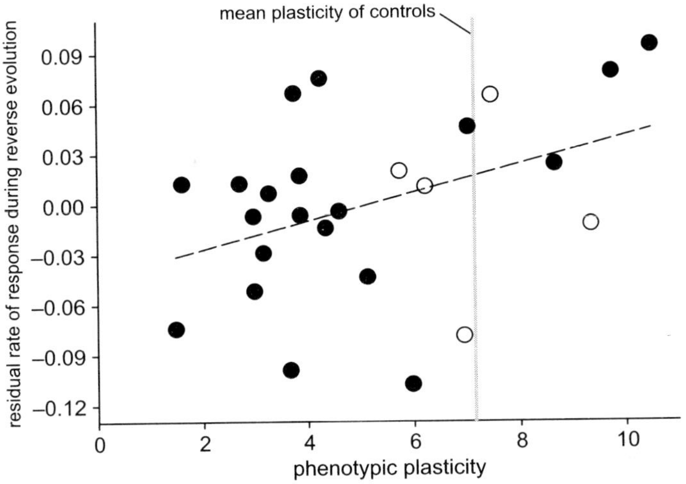

Fig. 9 *PP and evolvability during reverse evolution.* Residual rates of response during reverse evolution in fecundity, taken as the residuals of a regression model with fitness and fecundity as the independent variables, as a function of phenotypic plasticity (see main text for details). Differentiated populations are shown by black circles, while control populations by white circles. Regression analysis shown by a dashed line reveals that PP explains variation in residual rates of response, with more plastic populations responding faster than less plastic populations ($F_{1,23}$=4.4; p=0.048; R^2=0.16).

of populations that vary in both initial phenotype and PP were generated. These random populations were meant to represent the experimental populations at the beginning of reverse evolution. Since rates of response to selection can be calculated from adaptive landscapes using standard quantitative genetics theory, all the data is at hand to perform the same statistical analysis on simulated data as done on the experimental data.

The simulated fitness function has two local selective optima (O_1 and O_2 in Figure 2). This particular shape was chosen because it contains features that make it heuristic from an experimental evolution point of view: there is a global optimum, at phenotype value one (O_2), and some ruggedness to the rest of the landscape as represented by the second optimum, at around phenotype 0.3 (O_1). If populations start near the global optimum, only the shape of the landscape near the maximum has an effect on the evolutionary trajectories, while populations starting near the second peak can become trapped.

To examine the assertion made before that the relation between PP and evolvability is non-linear we estimated the statistical outcome of the stepwise regression in simulated data in particular phenotypic positions along the fitness landscape. To this end, 10,000 random groups of initial populations were generated by drawing the mean phenotype from a uniform distribution between zero and one, and the variance from a uniform distribution between 0 and 0.3. The amount of PP is equivalent in our model to amount of environmental variance (V_e). This simulated data was then subdivided into 10 subsets containing initial phenotypes in a range of 0.1. The simulation strategy explores how the initial state of the populations in phenotypic space determines the observed relationship between plasticity and rate of evolution. The overall pattern resulting from first regressing out both fitness and phenotype, as done before for the experimental data, and then fitting the residuals with PP, is shown in Figure 10. It is quite clear that there is a non-linear effect of PP on the rate of adaptation. Plasticity retards the rate of evolution both near and far from the highest fitness optimum (O_2). Only in the valley between the two peaks does plasticity have a positive effect on the rate of evolution.

A power analysis for each subset of the simulated data was then used to determine how often this effect would be detected using sample sizes similar to those available through experimental evolution. From each subset, 25 populations were selected at random and the stepwise regression analysis performed. This was done 40 times and the proportion of samples that showed a significant result was taken as the surrogate for statistical power. As shown (Figure 10), power was quite high for regions of phenotypic space that had a strong relationship between PP and rate of evolution. However, in an experimental system the starting populations are not likely to fall so neatly into a fixed range of starting phenotypes as in the simulated data. This will have the consequence of reducing the power to detect a relationship between PP and the rate of evolution (examples are shown in Figure 11a and 11b).

Could our populations be in a fitness valley at the start of reverse evolution and thus show a positive relationship between PP and evolvability? Given the data on the experimental fitness function, shown in Figure 7, this seems unlikely, but our simulations results suggest that this is possibly the best interpretation, if the experimental results are truly biologically significant. Another line of evidence in favor of this interpretation would come from the observation of the smoothing effect that PP is supposed to have on the adaptive landscapes, a crucial assumption for the model to work. The smoothing effect of the adaptive landscape in populations composed of more plastic genotypes can be indirectly assessed

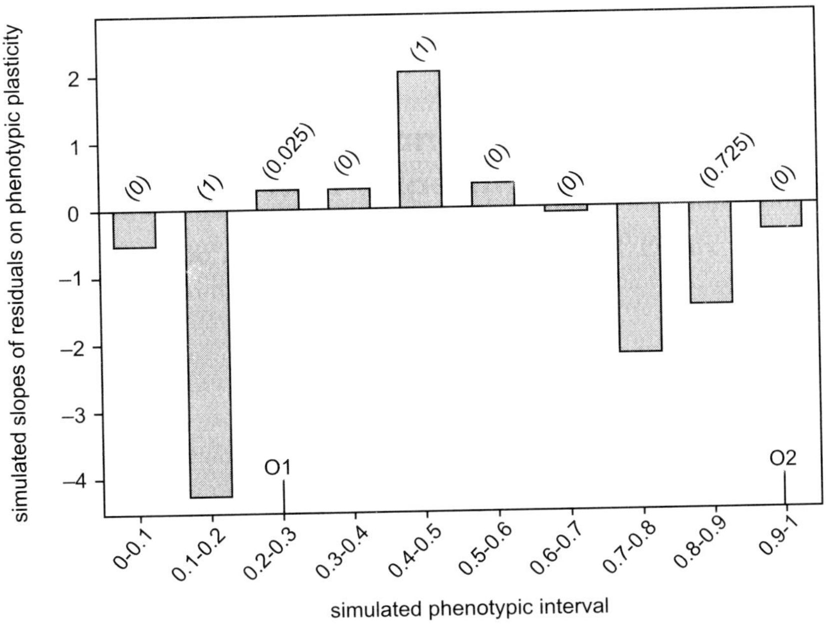

Fig. 10 *Simulated relation between PP and evolvability.* Bars represent the linear regression coefficient of a model with residual evolutionary rates (slopes) modeled against the predictor variable PP, after regressing out both fitness and phenotype. Each bar encompasses a phenotypic range of 0.1, using the phenotypic variance distributions of Figures 1 and 4 (0–0.3), and the fitness function of Figure 2 {W(p) = E^[–(1–p)] + [0.3 sin (2πp)]; with *W* being fitness and *p* phenotype, with local phenotype optima at O_1 and O_2} for approximately 975 simulated populations within each range. Numbers above each bar in parentheses show the results of a power analysis as explained in the text. Specific examples of this analysis for larger phenotypic ranges are given in Figure 11a and 11b.

by determining the correlation between the difference for mean fitness of each group of populations from the ancestral, and extent of phenotypic plasticity. This correlation should be positive in that more plastic genotypes should have a smaller fitness difference from the ancestral. In the experimental evolution data, this correlation is marginally significant in the predicted direction ($r_{Pearson}$=0.38, p=0.06). There is thus weak evidence that the adaptive landscape of these populations is smoothed by phenotypic plasticity.

Notwithstanding these considerations, other interpretations of the results can be given. The most obvious one is that it is possible that the experimental populations happen to show a positive relationship just by chance even if they are close to the novel selective optimum. Indeed, the power our model has to detect a positive (or for that matter negative)

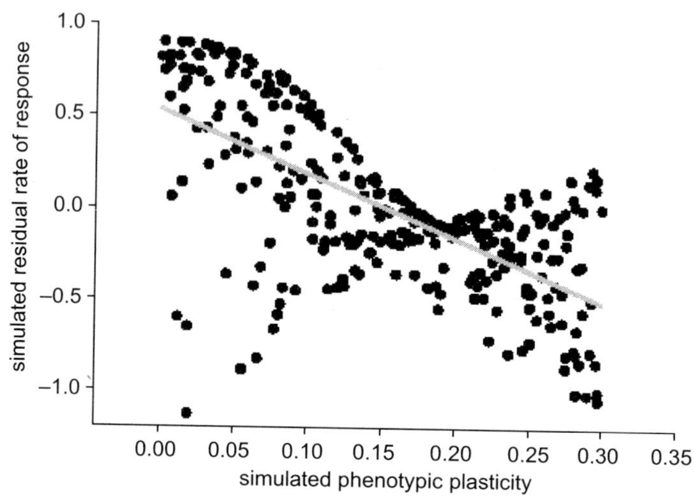

Fig. 11a *Simulated evolution near a selective optimum.* Simulated data from the model shown in Figures 1 to 4 was analyzed as the data from experimental reverse evolution. The simulated data were generated by selecting a value for genotype and plasticity at random from uniform distributions. The genotypes were drawn from a phenotypic range of 0.7–1.0 to place the populations near the higher peak (O_2), for a total of 300 observations. The variance parameter for plasticity was drawn from a range of 0.0–0.3. First step regression analysis with fitness and phenotype estimated a significant model for the dependent variable evolutionary rates, as measured by the slopes of the adaptive landscapes ($F_{2,297}$=56.6, p<0.001, R^2=0.27). Second step regression analysis of the residual evolutionary rates on phenotypic plasticity produced a significant model as well but with a poor fit, demonstrating the highly non-linear relationship between PP and evolvability, as shown by the gray line ($F_{1,298}$=226, p<0.001, R^2=0.43).

relationship is quite low, with an R^2 of about 0.16. Our analysis can only determine that the null hypothesis of no relationship is rejected.

Other interpretations for our findings can be given, not directly related to the model addressed in Figures 1–4. For example, PP can accelerate evolution if the propensity to be plastic somehow increases the propensity to acquire necessary genetic variation to evolve towards the novel peak; in other words, plastic genotypes more easily assimilate novel genetic variation [see Ancel and Fontana, 2000, for a relevant simulation study using RNA evolution of secondary structures]. This reasoning leads immediately to the following prediction: in phenotypically plastic genotypes, even if they are actively selected for, genetic variance for fitness is expected to be low initially, but increasing with the approach to the new selection optimum as genetic assimilation builds up. With regard to this, our

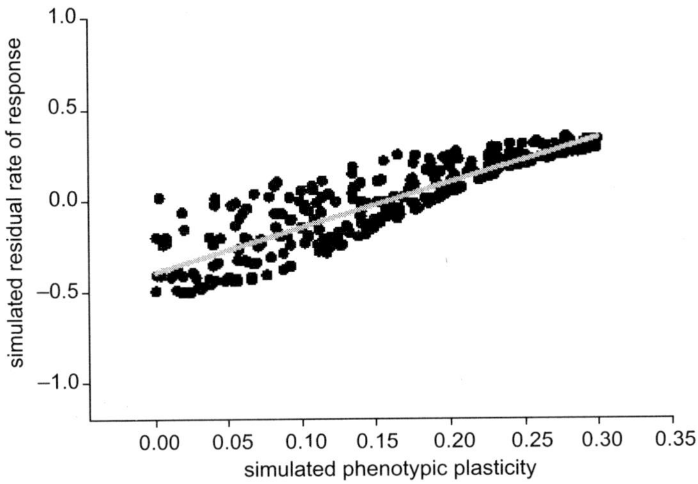

Fig. 11b. *Simulated evolution in the vicinity of a fitness valley.* Results are shown for populations that have genotypes situated on the fitness function valley, the genotypes being drawn from a uniform distribution with a range from 0.3–0.55, for a total of 300 observations. The variance parameter for plasticity was drawn from a range of 0.0–0.3 as in Figure 11a. First step regression analysis with fitness and phenotype as dependent variables estimated a significant but very poor fit model ($F_{2,297}$=3.3, p=0.04, R^2=0.02). Further analysis estimated that phenotypic plasticity does indeed reduce or eliminate the adaptive valleys by significantly explaining variation in residual evolutionary rates ($F_{1,298}$=1070, p<0.001, R^2=0.78).

data are silent because a trajectory of additive genetic variances for fitness would need to be estimated to test the hypothesis of genetic assimilation, which we did not do.

Lastly, less plastic populations could have accumulated deleterious mutations during diversifying evolution that would have prevented them from evolving as fast as phenotypically plastic populations. Conversely, more plastic populations could have maintained higher levels of additive genetic variance. Neither of these patterns appears to be relevant in these populations. Previous work demonstrated that response to selection for fecundity is unaffected by hybridization status (Teotónio and Rose 2000), and no fecundity enhancement seems to occur in hybrids between some of the differentiated and the control populations (Teotónio et al. 2004).

To summarize, there are indications that PP acts as a facilitator of adaptation during *Drosophila* reverse evolution, perhaps because PP smoothes the fitness function thereby reducing or eliminating a valley that may exist between the diversifying and the ancestral environment.

Testing Hypotheses about PP and Evolvability with Experimental Evolution

Theoretical models predict that PP expedites evolution only under fairly restrictive conditions, to the point that Ancel (2000) suggested that PP only expedites the initial evolutionary process until a few individuals reach the novel selective optima and then only when the phenotypic distributions are fairly distant from those same optima. This explanation does not require the smoothing of the adaptive landscape, and therefore PP does not have to be adaptive. It is difficult to test predictions such as these using experimental evolution since it would require having the same populations adapting in a highly rugged fitness function or to several different environments, with several control populations defining the selective optima.

Even so, we calculated the predicted time to convergence back to the ancestral phenotypic values, from Figure 5, as the generation at which each population converged or is predicted to converge to the confidence interval of the control population's linear response. We then analyzed time to convergence as we did for evolutionary rates, by first regressing fitness and phenotype and then taking the residuals to test for an association with phenotypic plasticity, but no significant results were obtained. Thus, no evidence exists for PP accelerating initial phenotypic evolution perhaps because these populations are fairly close to the ancestral optimum.

Two other predictions from the model were not addressed. One is the prediction that the approach to the selective optimum should be slower in more plastic populations than in less plastic populations, because plastic populations will have lower effective levels of genetic variance for fitness (Figure 1). Note that this is contrary to what occurs if genetic assimilation takes place. Also, an intermediate level of PP is expected to result in the highest levels of evolvability (Price et al. 2003). This occurs because of the balance between the initial effects of PP increasing the evolutionary rate in a novel environment and the retarding effects of PP as the population nears the new phenotypic optimum.

Our results suggest that future experiments should be designed to avoid some of the difficulties exposed by this study, difficulties that we will now enumerate. Our analysis took each population as the observational unit, but each population is composed of many different genotypes and thus the PP that we estimate should be seen as population PP (Sarkar and Fuller 2003). Variation in phenotypic plasticity is better estimated from variation among individual genotypes. The individual PP is usually obtained from clones of individual genotypes if possible or a family-based design when clonality is not possible (cf. Chippindale et al. 2001). The extent to which population PP can be explained by the same evolutionary mechanisms as individual PP is not known. However, the power of the experimental evolution approach is, in part, that it incorporates within population variance. Any natural system will have intra-population variance and so the population level approach discussed here may give the most empirical insight into evolution in nature.

Choosing the appropriate environmental range to estimate the reaction norm could potentially limit our ability to generalize findings (see Stearns et al. 1991). The experimental evolution approach once again allows the experimenter to determine the range of conditions that the populations are exposed to, but if the relationship between PP and evolvability is inherently non-linear, any experimental test of PP hypotheses must be local, using a restricted part of phenotypic space. Our discussion illustrated the point that it is fundamental to know where in phenotypic space the experimental populations are before adaptation, since the initial distribution of genotypes in a the population will be crucial for its future evolution.

Phenotypic plasticity is a direct product of developmental dynamics, being observable in insects possibly more than in any other group of organisms. PP will be selectively favored only to the degree that developmental processes can be modified to integrate environmental inputs and respond with an appropriate developmental trajectory. Because insects interact with their external environment during several developmental stages, the effects of plasticity at one stage may be amplified or dampened at another. Adaptive larval plasticity will evolve when environmental inputs at larval stages appropriately alter the developmental trajectory, changing life history features such as the rate of growth, timing of metamorphosis, and adult morphometrics so that fitness increases. At the adult stage, plasticity may be manifest in body size, relative size of reproductive tissue, timing of reproduction, and the rate of senescence. Because of the convolution of effects arising at different stages, PP at the larval stage may result in canalization at the adult stage (Proulx and Phillips 2004). Thus, the life-history stage at which traits are measured could affect our interpretation of the selective regime and the fitness function.

Conclusion

Phenotypic plasticity is the special property of genotypes that translates environmental variation into phenotypic variation. Therefore, a theory explaining and predicting patterns of biological diversity will need to incorporate into the same framework the evolution of phenotypic distributions generated by phenotypic plasticity together with other mechanisms that generate phenotypic variation (Lewontin 1974, Schlichting and Pigliucci 1998). Experiments that connect PP to evolvability may help achieve this goal. It appears, at least in the example given here, that PP can accelerate evolution by smoothing the adaptive landscape. This empirical result is a good starting point for a more satisfactory evolutionary biology of phenotypes.

Acknowledgments

The authors thank Patrick C. Phillips and Doug Whitman for comments on draft versions of the manuscript, and Doug Whitman for the invitation to participate in this volume. H. Teotónio is supported by FCT/FEDER (EU) and start-up funds from the Gulbenkian Foundation (Portugal). S. Proulx is supported by an NRSA award from NIH (USA).

References

Ancel, L. W. 1999. A quantitative model of the Simpson-Baldwin effect. J. Theor. Biol. 196: 197–209.

Ancel, L. W. 2000. Undermining the Baldwin expediting effect: does phenotypic plasticity accelerate evolution. Theo. Pop. Biol. 58: 307–319.

Ancel, L. W. and W. Fontana. 2000. Plasticity, evolvability, and modularity in RNA. J. Exp. Zool. (Mol. Dev. Evol.) 288: 242–283.

Arnold, S. J., M. E. Pfrender, and A. G. Jones. 2001. The adaptive landscape as a conceptual bridge between micro- and macroevolution. Genetica 112–113: 9–32.

Baldwin, J. M. 1896. A new factor in evolution. Am. Nat. 30: 441–451 Barton, N. H., and M. Turelli. 1989. Adaptive landscapes, genetic distance and the evolution of quantitative characters. Genet. Res. Camb. 49: 157–173.

Barton, N. H. and Turelli, M. 1987. Adaptive landscapes, genetic distance and the evolution of quantitative characters. Genet. Res. Camb. 49: 157–173.

Bell, G. 1997. Selection: The Mechanism of Evolution. Chapman and Hall, New York.

Bull, J. J. 1987. The evolution of phenotypic variance. Evolution 41: 303–315.

Chippindale, A. K., Alipaz, J. A., Chen, H.-W. and Rose M. R. 1997a. Experimental evolution of accelerated development in *Drosophila.* 1. Development speed and larval survival. Evolution 51: 1536–1551.

Chippindale, A. K., A. M. Leroi, H. Saing, D. J. Borash, and M. R. Rose. 1997b. Phenotypic plasticity and selection in *Drosophila* life history evolution. 2. Diet, mates and the cost of reproduction. J. Evol. Biol. 10: 269–293.

Chippindale, A. K., J. R. Gibson, and W. R. Rice. 2001. Negative genetic correlation for adult fitness between sexes reveals ontogenetic conflict in *Drosophila*. PNAS 98: 1671–1675.

Coyne, J. A., N. H. Barton, and M. Turelli. 1997. Perspective: a critique of Sewall Wright's shifting balance theory of evolution. Evolution 51: 643–671.

de Jong, G., and R. H. Crozier. 2003. A flexible theory of evolution. Nature 424: 16–17.

Fisher, R. A. 1958. The Genetical Theory of Natural Selection. Dover Publ. NY.

Gavrilets, S. and Hastings, A. 1994. A quantitative-genetic model for selection on developmental noise. Evolution 48: 1478–1486.

Gomulkiewicz, R., and R. D. Holt. 1995. When does natural selection prevent extinction? Evolution 49: 201–207.

Hensen, T. F. 2003. Is modularity necessary for evolvability? Remarks on the relationship between pleiotropy and evolvability. Biosystems 69: 83–94.

Lande, R. 1976. Natural selection and random genetic drift in phenotypic evolution. Evolution 30: 314–334.

Lande, R. 1979. Quantitative genetic analysis of multivariate evolution, applied to brain:body allometry. Evolution 33: 402–416.

Leroi, A. M., M. R. Rose, and G. V. Lauder. 1994. What does the comparative method reveal about adaptation? Am. Nat. 143: 381–402.

Levins, R. 1968. Evolution in Changing Environments. Princeton University Press, Princeton.

Lewontin, R. 1974. The Genetic Basis of Evolutionary Change. Columbia University Press, NY.

Lynch, M., and B. Walsh. 1998. Genetics and Analysis of Quantitative Traits. Sinauer, Sunderland.

Meyers, L. A., and J. J. Bull. 2002. Fighting change with change: adaptive variation in an uncertain world. TREE 12: 551–557.

Price T. D., A. Qvarnstrom, and D. E. Irwin. 2003. The role of phenotypic plasticity in driving genetic evolution. Proc. Royal Soc. Lond. B 270: 1433–1440.

Price, T., M. Turelli, and M. Slatkin. 1993. Peak shifts produced by correlated response to selection. Evolution 47: 280–290.

Proulx, S., and P. C. Phillips. 2004. The opportunity for canalization and the evolution of genetic networks. Am. Nat. 165: 147–162.

Rose, M. R. 1984. Laboratory evolution of postponed senescence in *Drosophila melanogaster*. Evolution 38: 1004–1010.

Rose, M. R., H. B. Passananti, and M. Matos [eds]. 2004. Methuselah Flies: A Case Study in the Evolution of Aging. World Scientific Publishing, Singapore.

Rose, M. R., T. J. Nusbaum, and A. K. Chippindale. 1996. Laboratory evolution: the experimental wonderland and the Cheshire Cat syndrome. *In* Adaptation Pp. 221–241 M. R. Rose, and G.V. Lauder [eds]. Academic Press, San Diego.

Rose, M. R., Vu, L. N., Park, S. V. and Graves, J. L. Jr. 1992. Selection of stress resistance increases longevity in *Drosophila melanogaster*. Exp Geront. 27: 241–250.

Rutherford, S. L. and S. Lindquist. 1998. Hsp90 as a capacitor for morphological evolution. Nature 396: 336–342.

Sarkar, S. and T. Fuller. 2003. Generalized norms of reaction for ecological developmental biology. Evo Dev 5: 106–115.

Scharloo, W. 1991. Canalization: genetic and developmental aspects. Annu. Rev. Ecol. Syst. 22: 65–93.

Scheiner, S. M. 1993. Genetics and evolution of phenotypic plasticity. Annu. Rev. Ecol. Syst. 24: 35–68.

Schlichting, C. D., and M. Pigliucci. 1998. Phenotypic Evolution, a Reaction Norm Perspective. Sinauer Assoc. Inc. Sunderland.

Simpson, G. G. 1953. The Major Features of Evolution. Columbia University Press, New York.

Stearns, S., G. de Jong, and B. Newman. 1991. The effects of phenotypic plasticity on genetic correlations. TREE 6: 122–126.

Teotónio, H. and Rose, M. R. 2000. Variation in the reversibility of evolution. Nature 408: 463–466.

Teotónio, H., Matos, M. and Rose, M. R. 2002. Reverse evolution of fitness in *Drosophila melanogaster*. J. Evol. Biol. 15: 608–617.

Teotónio H., Matos M. and Rose M. R. 2004 Quantitative genetics of functional characters in *Drosophila melanogaster* populations subjected to laboratory selection. J. Genet. 83, 265–277.

Via, S., and R. Lande. 1985. Genotype–environment interaction and the evolution of phenotypic. Evolution 39: 505–522.

Via, S., Gomulkiewicz, R., de Jong, G., Scheiner, S. M., Schlichting, C. D. and van Tienderen, P. H. 1995. Adaptive phenotypic plasticity: consensus and controversy. TREE 10: 212–217.

Wadington, C. H. 1953. The genetic assimilation of an acquired character. Evolution 7: 118–126.

Wagner, G. P., Booth, G., and Bagheri–Chaichian, H. 1997. A population genetic theory of canalization. Evolution 51: 329–347.

Wagner, A. 1996. Does evolutionary plasticity evolve? Evolution 50: 1008–1023.

West-Eberhard, M. J. 1989. Phenotypic plasticity and the origins of diversity. Annu. Rev. Ecol. Syst. 20: 249–278.

Whitlock, M. C. 1995. Variance induced peak shifts. Evolution 49: 252–259.

Wright, S. 1977. Evolution and the genetics of populations, vol 3, experimental results and evolutionary deductions. University of Chicago Press, Chicago.

Zar, J. H. 1996. Biostatistical Analysis, 3rd edition. Prentice Hall, Upper Saddle River.

Genus Index

Insect Family Index

Subject Index

D

H

M